FIFTH EDITION

CLAUDE A. VILLEE
Harvard University

WARREN F. WALKER, JR.
Oberlin College

ROBERT D. BARNES
Gettysburg College

GENERAL ZOOLOGY

W. B. SAUNDERS COMPANY
Philadelphia

Saunders College Publishing
210 West Washington Square
Philadelphia, PA   19105

Library of Congress Cataloging in Publication Data

Villee, Claude Alvin, 1917–

General zoology.

Includes bibliographies.

1. Zoology.   I. Walker, Warren Franklin, joint author.
   II. Barnes, Robert D., joint author.   III. Title.

QL47.2.V5 1978   591   77–80753

ISBN 0–7216–9039–4

Front cover illustration is a photomicrograph of a preovulatory oocyte of the rat, taken with Nomarsky optics, courtesy of Doctors Clara Szego and Robert Ezzell, University of California, Los Angeles.

General Zoology   ISBN   0-7216-9039-4

8901   147   9876

PREFACE

This edition represents a major revision of the text, the most extensive one yet undertaken. A survey of the courses in zoology and animal biology taught in this country revealed a remarkable disparity in their nature and in their relationship to other biological courses in the college or university. To make this book as useful as possible, and to provide the instructor with the greatest possible flexibility in planning his course, we have reorganized the sequence of subjects discussed. The book can be adapted to the teaching of any of several types of courses in general zoology. Although we expect that most instructors will use the text for a course in general zoology, some may elect to use Part I on cellular biology and Part II on animal development and heredity for a first-term course in cell biology and genetics. This could be followed by a second-term course on the biology of animal organisms using Part III, Animal Form and Function, and Part IV, the Animal Groups. Alternatively, Parts III and IV, with as much of the remainder of the book as desirable, could be used for a one-term course in animal biology when a previous introductory course in biology has dealt with the cellular and molecular aspects of the subject. As another alternative, Part IV, The Survey of the Animal Groups, together with Part V on ecology could provide the basis for a brief course in general zoology at the freshman or sophomore level. Clearly it is not necessary to use the entire book for any given course in animal biology.

In preparing the text we have endeavored to summarize the theories and facts that constitute modern zoological science. An appreciation of any science requires both a grasp of the product of the science – the facts that have been gained in that field – and an insight into the processes – the observations and experiments – by which such knowledge is acquired. The methods of science are introduced in Chapter 1, and throughout the text examples of experimental work are presented to illustrate current methods in zoological research.

The instructor may have neither the time nor the inclination to consider some of the material presented in the text. Each instructor, of course, emphasizes those topics he considers most important and most relevant. The text provides the interested student with an opportunity to read about subjects that may be omitted or considered only cursorily in the lectures and laboratory exercises. In discussing the many subjects that compose modern zoology, we have tried to distinguish between fact and theory and to cite some of the problems that remain for future zoologists to solve.

The introductory chapter describes the subsciences of zoology, the scientific method and the sources of scientific information and gives a brief history of the science of zoology. Chapter 2 outlines the chemical and physical concepts needed for an understanding of the dynamic and molecular aspects of cells and tissues. Chapter 3 discusses the structure and functions of the cell and its subcellular organelles, the exchange of materials between the cell and its environment via the plasma membrane and the types of tissues present in animal bodies. Chapter 3 has been extensively rewritten and includes a discussion of the cell cycle to introduce the subject of mitosis. The nature of enzymes, their role in cellular physiology and the principles of bioenergetics are discussed in Chapter 4.

Part II, a new part entitled Animal Development and Heredity, begins with a completely rewritten chapter on development. The chapters on genetics, 6 and 7, and on evolution, 8 and 9, are included in this section because they deal with general biological principles. Chapter 8, which discusses the concept of evolution, has been completely rewritten.

Part III, entitled "Animal Form and Function," is new, and the chapters composing it have been completely rewritten. This part attempts to point out the general problems of animal existence and the several common solutions to these problems that have been made by the various invertebrate and vertebrate animals. Its aim is to provide the student with an overview and with important generalizations that he can apply to new or less familiar animals encountered later. The comparative aspects of each system are discussed with ample coverage of the various invertebrate solutions to the problem, together with a thorough exploration of the anatomy and physiology of that system in the vertebrates. The section opens with Chapter 10, entitled "Symmetry, Form and Life Style," which introduces the comparative approach to form and function. It is followed by chapters on support and movement (11), nutrition and digestion (12), respiration and gas exchange (13), internal transport and circulatory systems (14), homeostasis and excretion (15), receptors and sense organs (16), nervous integration (17), hormonal integration (18), behavior (19) and a new chapter on reproduction (20). In these chapters, the structure and physiology of the mammal is emphasized but sufficient information is provided regarding lower vertebrates so that the mammals can be understood in evolutionary perspective. Many revisions have been made throughout this section to improve clarity of presentation and to keep the material current with recent advances in the subject. Notable changes have been made in the discussions of immunity, endocrines, hemostasis, vitamins, homeostasis, sense organs and reproduction.

The survey of animal phyla included in Part IV has two goals. It attempts to provide the student with an appreciation of the great diversity of animal life and to provide an understanding of the adaptive changes that have occurred in the evolutionary history of animals. The chapters begin with a concise summary of the main features of each phylum and end with the taxonomic classification of the members of the phylum.

The chapters describing the classes of vertebrates (Chapters 33 through 37) emphasize the adaptations of each class to its mode of life and the evolutionary relationships of the several groups. The chapter on primates and human evolution has been completely rewritten to reflect the impact of the numerous recent discoveries on human evolution stemming from East Africa. We are grateful to Professor Alan Walker of Harvard University, who kindly reviewed this chapter.

The final section of the book, Part V, is devoted to a discussion of animals and their environment. It includes a discussion of ecologic principles (Chapter 38) and a discussion of community and population ecology (Chapter 39) and ends with a chapter on human ecology (40).

In response to requests from students and teachers using the book, we have added a Glossary at the end of the book that gives the derivations and meanings of the more important zoological terms.

The preparation of this edition represented an unusual degree of cooperation among the three authors. Several meetings were held in Philadelphia and Woods Hole to discuss plans for the changes to be made and to review the manuscripts as they were written. Although each chapter was written originally by one of the three authors, the manuscript was read carefully by each of the other two, who made detailed suggestions for additions, deletions and revisions. Finally, all of the chapters were gone over by one of us to ensure uniformity of style and level of presentation.

Many new illustrations have been prepared for this edition and we are deeply grateful for the care and artistry with which Susan Heller, Ellen Cole, William Osborne and Mary Ann Nelson prepared the line drawings. We are indebted to Aldine Publishing Company, Vernon G. Applegate, Robert S. Bailey, Betty M. Barnes, Russell J. Barrnett, Angus Bellairs, Kurt Benirschke, C. M. Bogert, S. H. Camp and Company, Austin H. Clark, Allan D. Cruikshank, H. R. Duncker, Earl R. Edmiston, Alfred Eisenstadt, Frank Essapian, Don Fawcett, D. Fraser, W. H. Freeman Company, Golden Book Encyclopedia of Natural Science, J. N. Hamlet, Harcourt Brace Jovanovich, Harper & Row, Fritz Goro, C. Lynn Haywood, A. A. Knopf, Herbert Lang, Daniel Mazia, Jacques Millot, James W. Moffett, Peter Morrison, W. W. Norton and Company, Jean Luc Perret, J. D. Pye, Col. N. Rankin, E. S. Ross, P. R. Russell, Hugh Spencer, E. P. Walker, L. W. Walker, Time-Life Books, and Weidenfeld and Nicholson, who have kindly permitted us to use certain of their drawings and photographs.

Our special thanks are due to Richard Lampert, Biology Editor, who assisted greatly with the initial planning for this edition and to Robert Lakemacher, his successor, who assisted in many ways in the completion of the manuscript and its preparation for publication. We are indebted to the many members of the staff of the W. B. Saunders Company who have given so liberally of their time and care in preparing this revised edition. We want especially to thank Miss Debbie Patterson and Miss Patrice Lamb for their part in the preparation of this edition. Finally, we want to express our thanks to Kathleen Callinan for her assistance in preparing the Index.

CLAUDE A. VILLEE
WARREN F. WALKER
ROBERT D. BARNES

CONTENTS

PART TWO.   ANIMAL DEVELOPMENT AND HEREDITY

Chapter 5

Chapter 6

Chapter 19

Chapter 20

PART FOUR. ANIMAL GROUPS

Chapter 21

PART FIVE.   ANIMALS AND THEIR ENVIRONMENT

Chapter 38

ECOLOGY 872

Chapter 39

SYNECOLOGY: COMMUNITIES, BIOMES AND LIFE ZONES 890

Chapter 40

HUMAN ECOLOGY 922

GENERAL ZOOLOGY

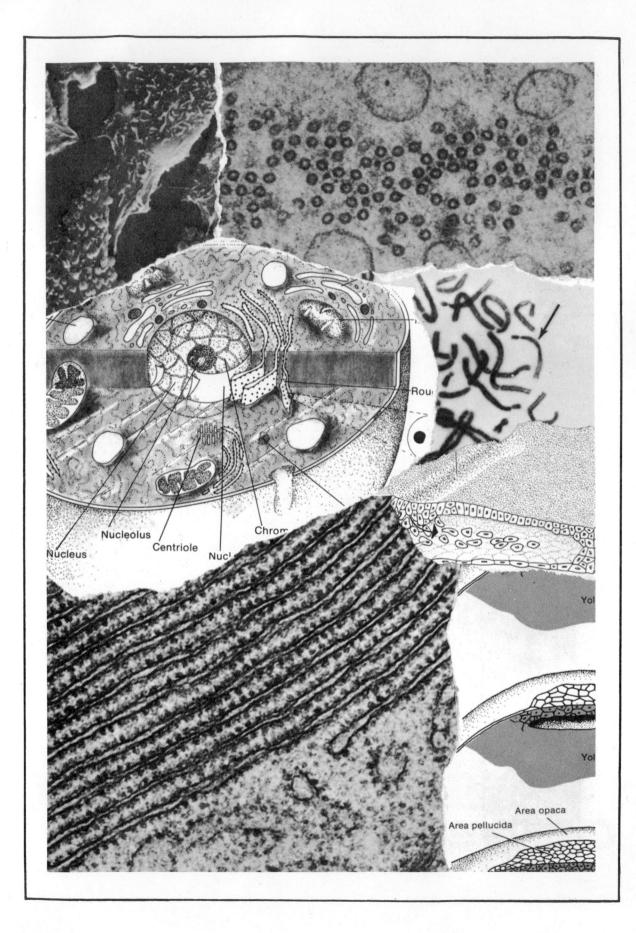

Nucleus

Nucleolus

Centriole

Nucl

Chrom

Rou

Area opaca

Area pellucida

Part One

THE CELLULAR BIOLOGY OF ANIMALS

INTRODUCTION

1.1 ZOOLOGY AND ITS SUBSCIENCES

Zoology, one of the biological sciences, deals with animals and the many different aspects of animal life. A "zoo" is a collection of animals, but a visit to a zoo, interesting though it is, can barely begin to suggest the enormous variety of animals living today. More than one million different kinds of animals have been described and classified—nearly 800,000 of these are insects! Each of these kinds has become adapted, through a long evolutionary process, to the particular set of environmental circumstances under which it survives best. In addition to the ones living at present, a host of other kinds of animals have lived in past ages but are now extinct.

Modern zoology concerns itself with much more than the simple recognition and classification of the many kinds of animals. It includes studies of the structure, function and embryonic development of each part of an animal's body; of the nutrition, health and behavior of animals; of their heredity and evolution; and of their relations to the physical environment and to the plants and other animals of that region.

Enough facts about animals and their ways are known to fill a whole library of books, and more information appears every year from the intensive researches of zoologists in the field and in the laboratory. No zoologist today can know more than a small fraction of this enormous body of knowledge. Zoology is now much too broad a subject to be encompassed by a single scientist or to be treated thoroughly in a single textbook. Most zoologists are specialists in some limited phase of the subject—in one of the subdivisions of zoology. The sciences of **anatomy, physiology** and **embryology** deal with the structure, function and development, respectively, of an animal. Each of these may be further subdivided according to the kind of animal investigated, e.g., invertebrate physiology, arthropod physiology, insect physiology or comparative physiology. **Parasitology** deals with those forms of life that live in or on and at the expense of other organisms. **Cytology** is concerned with the structure, composition and function of cells and their parts, and **histology** is the science of the structure, function and composition of tissues. The science of **genetics** investigates the mode of transmission of characteristics from one generation to the next and is closely related to the science of **evolution,** which studies the way in which new species of animals arise and how the present kinds of animals are related by descent to previous animals. The study of the classification of organisms, both animals and plants, is called **taxonomy.** The science of **ecology** is concerned with the relations of a group of organisms to their environment, including both the physical factors and the other forms of life which provide food or shelter for them, compete with them in some way, or prey upon them.

Some zoologists specialize in the study of one group of animals. There are **mammalogists, ornithologists, herpetologists** and **ichthyologists** who study mammals, birds, reptiles and amphibians, and fishes, respectively; **entomologists,** who investigate insects; **protozoologists,** who study the single-celled animals, and so on.

Advances in chemistry and physics have made possible quantitative studies of the molecular structures and events underlying biological processes. The term **molecular biology** has been applied to analyses of gene structure and function and genic control of the synthesis of enzymes and other proteins, studies of subcellular structures and their roles in regulatory processes within the cell, investigations of the mechanisms underlying cellular differentiation, and analyses of the molecular basis of evolution by comparative studies of the molecular structure of specific proteins—enzymes, hormones, cytochromes, hemoglobins—in different species.

The science of zoology thus includes both a tremendous body of facts and theories about animals and the means for learning more. The ultimate source of each fact is in some carefully controlled observation or experiment made by a zoologist. In earlier times, some scientists kept their discoveries to themselves, but there is now a strong tradition that scientific discoveries are public property and should be freely published. In a scientific publication the investigator must do more than simply say that he has made some particular discovery; he must give all of the relevant details of the means by which the discovery was made so that others can repeat the observation. It is this criterion of *repeatability* that makes us accept a certain observation or experiment as representing a true fact; observations that cannot be repeated by competent investigators are discarded.

When a scientist has made some new observation, or carried out a series of experiments that add to our knowledge in a field, he writes a report, called a "paper," in which he describes his methods in sufficient detail so that another worker can repeat them, gives the results of his observations, discusses the conclusions to be drawn from them, perhaps formulates a theory to explain them or discusses how they are explained by a previous theory, and finally indicates the place of these new facts in their particular field of science. The knowledge that his discovery will be subjected to the keen scrutiny of his colleagues is a strong stimulus for repeating the observations or experiments carefully before publishing them. He then submits his paper for publication in one of the professional journals in the particular field of his discovery. Several thousand zoological journals are published currently all over the world. Some of the more important American ones are the *Journal of Experimental Zoology, Journal of Cellular and Comparative Physiology, Biological Bulletin, Physiological Zoology, American Journal of Physiology, Anatomical Record, Ecology* and the journals devoted to research on a particular group of animals, such as the *Journal of Mammalogy.* The paper is read by one or more of the board of editors of the journal, all of whom are experts in the field. If it is approved, it is published and becomes part of "the literature" of the subject.

At one time, when there were far fewer journals, it might have been possible to read them all each month as they appeared, but this is obviously impossible now. Journals such as *Biological Abstracts* assist the hard-pressed zoologist by publishing, classified by fields, short summaries or abstracts of each paper published, giving the facts found, the conclusion reached, and an exact reference to the journal in which the full report appears. The ultimate in brevity is *Current Contents,* which simply lists the authors and titles of the papers published in several hundred journals, together with the name, volume and pages of the journal where the paper was published. A considerable number of journals devoted solely to reviewing the newer developments in particular fields of science have sprung up in the past 50 years; some of these are *Physiological Reviews, Quarterly Review of Biology, Nutrition Reviews, Annual Review of Physiology* and *Vitamins and Hormones.* The new fact or theory thus becomes widely known through publication in the appropriate professional journal and by reference in abstract and review journals and eventually may become a sentence or two in a textbook.

The professional societies of zoologists and the various special branches of zoology have annual meetings at which new discoveries may be reported. Two of the largest annual meetings are those of the American Institute of Biological Sciences and the Federation of American Societies for Experimental Biology. There are, in addition, national and international gatherings, called **symposia,** of specialists in a given field to discuss the newer findings and the present status of the knowledge in that field. For example, the discussions of the Cold Spring Harbor Symposia in Quantitative Biology, held each June at the Long Island Biological Laboratory in Cold Spring Harbor, are published and provide an excellent review of some particular field. A different subject is discussed each year.

1.2 THE SCIENTIFIC METHOD

The ultimate aim of each science is to reduce the apparent complexity of natural phenomena to simple, fundamental ideas and relations, to discover all the facts, and the relationships among them. The essence of the scientific method is the posing of questions and the search for answers, but they must be "scientific" questions, arising from observations and experiments, and "scientific" answers, ones that are testable by further observation and experiment. The Danish physicist Niels Bohr put it this way, "the task of science is both to extend the range of our experience and to reduce it to order."

There is, however, no single "scientific method," no regular, infallible sequence of events which will reveal scientific truths. Different scientists go about their work in different ways. George Sarton, in the *Study of the History of Science*, points out that "Even as all kinds of men are needed to build up a community, even so we need all kinds of scientists to develop science in every possible direction. Some are very sharp and narrow-minded, others broad-minded and superficial. Many scientists, like Hannibal, know how to conquer, but not how to use their victories. Others are colonizers rather than explorers. Others are pedagogues. Others want to measure everything more accurately than it was measured before. This may lead them to the making of fundamental discoveries, or they may fail, and by looked upon as insufferable pedants."

The ultimate source of all the facts of science is careful, close observation and experiment, free of bias and done as quantitatively as possible. The observations or experiments may then be analyzed, or simplified into their constituent parts, so that some sort of order can be brought into the observed phenomena. Then the parts can be reassembled and their interactions made clear. On the basis of these observations, the scientist constructs a **hypothesis,** a trial idea about the nature of the observation, or about the connections between a chain of events, or even about cause and effect relationships between different events. It is in this ability to see through a mass of data and construct a reasonable hypothesis to explain their relationships that scientists differ most and that true genius shows itself.

The role of a hypothesis is to penetrate beyond the immediate data and place it into a new, larger context, so that we can interpret the unknown in terms of the known. There is no sharp distinction between the usage of the words "hypothesis" and "theory," but the latter has, in general, the connotation of greater certainty than a hypothesis. A **theory** is a conceptual scheme which tries to explain the observed phenomena and the relationships between them, so as to bring into one structure the observations and hypotheses of several different fields. The theory of evolution, for example, provides a conceptual scheme into which fit a host of observations and hypotheses from paleontology, anatomy, physiology, biochemistry, genetics and other allied sciences.

A good theory correlates many previously separate facts into a logical, easily understood framework. The theory, by arranging the facts properly, suggests new relationships between the individual facts and suggests further experiments or observations which might be made to test these relationships. It may predict new phenomena that will be observed under certain circumstances and finally may provide the solution for practical problems. A good theory should be simple and should not require a separate proviso to explain each fact; it should be flexible, able to grow and to undergo modifications in the light of new data. A theory is not discarded because of the existence of some isolated fact which contradicts it, but only because some other theory is better able to explain all of the known data.

Once a hypothesis has been established, the rules of formal logic can be applied to deduce certain consequences. In physics, and to a lesser extent in the biological sciences, the hypotheses and deductions can be stated in mathematical terms, and far-reaching conclusions may be deduced. From these inferences, one can predict the results of other observations and experiments. Each hypothesis is ultimately kept, amended or discarded on the basis of its ability to make valid predictions. A hypothesis must be subject to some sort of experimental test—i.e., it must make a prediction that can be verified in some way—or it is mere speculation. Conversely, unless a prediction follows as the logical outgrowth of some theory it is no more than a guess.

The finding of results contrary to those predicted by the hypothesis causes the investigator, after he has assured himself of the validity of his observation, either to discard the hypothesis or to change it to account for both the original data and the new data. Hypotheses are constantly being refined and elaborated. There are few scientists who would regard any hypothesis, no matter how many times it may have been tested, as a statement of absolute and universal truth. It is rather regarded as the best available approximation to the truth for some finite range of circumstances.

The history of science shows that, although many scientists have made their discoveries by following the precepts of the ideal scientific method, there have been occasions on which important and far-reaching theories have resulted from making incorrect conclusions from erroneous postulates, or from the misinterpretation of an improperly controlled experiment! There are instances in which, in retrospect, it seems clear that all the evidence for the formulation of the correct theory was known, yet no scientist put the proper two and two together. And there are other instances in which scientists have been able to establish the correct theory despite an abundance of seemingly contradictory evidence.

In most scientific studies one of the ultimate goals is to explain the cause of some phenomenon, but the hard-and-fast proof that a cause and effect relationship exists between two events is really very difficult to obtain. If the circumstances leading to a certain event always have a certain factor in common in a variety of cases, that factor may be the cause of the event. The difficulty, of course, lies in making sure that the factor under consideration is the *only* one common to all the cases. This method of discovering the common factor in a series of cases that may be the cause of the event is known as the **method of agreement.** It can seldom be used as a valid proof because of this difficulty of being sure that it is indeed the only common factor. The simple observation that all people suffering from beriberi have diets which are low in thiamine is not proof that a deficiency of this vitamin causes the disease, for there may be many other factors in common.

Experiments based on the **method of difference** provide another way of elucidating cause and effect relations. If two sets of circumstances differ in only one factor, and the one containing the factor leads to an event and the other does not, the factor may be considered the cause of the event. For example, if two groups of rats are fed diets which are identical except that one contains all the vitamins and the second contains all but thiamine, and if the first group grows normally but the second fails to grow and ultimately develops polyneuritis, this would be a strong suggestion (but would not be acceptable as absolute proof) that polyneuritis, or beriberi in rats, is caused by a deficiency of thiamine. By using an inbred strain of rats that are as alike as possible in inherited traits, and by using littermates (brothers and sisters) of this strain, one could make certain that there were no hereditary differences between the controls (the ones getting the complete diet) and the experimentals (the ones getting the thiamine-deficient diet).

One might postulate that the thiamine-free diet does not have as attractive a taste as the one with thiamine, and the experimental animals simply eat less food, fail to grow, and develop the deficiency symptoms because they are partially starved. This source of error can be avoided by "pair-feeding," by pairing in some arbitrary way each control and experimental animal, then weighing the food eaten each day by each experimental animal and giving only that much food to the corresponding control member of the pair.

One of the more useful methods of detecting cause and effect relationships is the **method of concomitant variation.** If a variation in the amount of one given factor produces a parallel variation in the effect, the factor may be the cause. Thus, if several groups of rats were given diets with varying amounts of thiamine, and if the amount of protection against beriberi varied directly with the amount of thiamine in the diet, one could be reasonably sure that thiamine deficiency is the cause of beriberi.

It must be emphasized that it is seldom that we can be more than "reasonably sure" that X is the cause of Y. As more experiments and observations lead to the same result, the probability increases that X is the cause of Y. When experiments or observations can be made quantitative, when their results can be counted or measured in some way, the methods of statistical analysis provide a means for calculating the probability that Y follows X simply as a matter of chance. Scientists are usually satisfied that there is some sort of cause and effect relationship between X and Y if they can show that there is less than one chance in a hundred that the observed X-Y relationship could be due to chance alone. A statistical analysis of a set of data can never give a flat yes or no to a question; it can state only that something is very probable or very improbable. It can also tell an investigator approximately how many more times he must repeat the experiment to show with a given probability that Y is caused by X.

The proper design of experiments is a science in itself, and one for which only general rules can be made. In all experiments, the scientist must ever be on his guard against bias in himself, bias in the subject, bias in his instruments and bias in the way the experiment is designed.

Each experiment must include the proper **control group** (indeed some experiments require several kinds of control groups). The control group is one treated exactly like the experimental group in all respects but one, the factor whose effect is being tested. The use of controls in medical experiments raises the difficult question of the ethical justification of withholding treatment from a patient who might be benefited by it. If there is sufficient evidence that one treatment is indeed better than another, a physician would hardly be justified in further experimentation. However, the medical literature is full of treatments now known to be useless or even detrimental, which were used for many years, only to be abandoned finally as experience showed that they were ineffective and that the evidence which had originally suggested their use was improperly controlled. There is a time in the development of any new treatment when the medical profession is not only morally justified, but really morally required, to do carefully controlled tests on human beings to be sure that the new treatment is better than the former one.

In medical testing it is not sufficient simply to give a treatment to one group of patients and not to give it to another, for it is widely known that there is a strong psychological effect in simply giving a treatment of any sort. For example, a group of students in a large western university served as subjects for a test of the hypothesis that daily doses of extra amounts of vitamin C might help to prevent colds. This grew out of the observation that people who drink lots of fruit juices seem to have fewer colds. The group receiving the vitamin C showed a 65 per cent reduction in the number of colds contracted during the winter in which they received treatment as compared to the previous winter when they had no treatment. There were enough students in the group (208) to make this result statistically significant. In the absence of controls, one would have been led to the conclusion that vitamin C does help to prevent colds. A second group of students were given "placebos," pills identical in size, shape, color and taste to the vitamin C pills but without any vitamin C. The students were not told who was getting vitamin C and who was not; they only knew they were getting pills that might help to prevent colds. The group getting the placebos reported that they had a 63 per cent reduction in the number of colds! This controlled experiment thus showed that vitamin C had nothing to do with the decrease in the number of colds and that the reductions reported in both groups were either psychological effects or simply the result of a lesser amount of cold virus on the campus that year. In a "double blind" study the physician does not know which subject is receiving the experimental compound and which the placebo. Each patient is given pills with a different code number, and only after the experiment has been completed is the code revealed.

1.3 HISTORY OF ZOOLOGY

Our interest in animals is probably somewhat older than the human race, for the ape-men and man-apes that preceded us in evolution undoubtedly learned at an early time which animals were dangerous, which could be hunted for food, clothing or shelter, where these were to be found, and so on. Some of our prehistoric ancestors' impressions of the contemporary animals have survived in the cave paintings of France and Spain (Fig. 1.1). Some animals were regarded as good or evil spirits. Later man decorated pottery, tools, cloth and other objects with animal figures.

The early Egyptians had a wealth of knowledge about animals and had domesticated cattle, sheep, pigs, cats, geese and ducks. The Greek philosophers of the fifth and sixth centuries B.C., Anaximander, Xenophanes, Empedocles and others, speculated on the origin of the animals of the earth. One of the earliest classifications of animals is found in a Greek medical book of this time which classifies animals primarily as to whether or not they are edible. Aristotle (384–322 B.C.) was one of the greatest Greek philosophers and wrote on many topics. His *Historia animalium* contains a great deal of information about the animals of Greece and the nearby regions of Asia Minor. Aristotle's descriptions are quite good and are recognizable as those of particular animals living today. The breadth and depth of his zoological interests are impressive—he made a careful study of the development of the chick and of the breeding of sharks and bees, and he had notions about the functions of the human organs, some of which, not too surprisingly, were quite wrong. He presented an elaborate theory that animals have gradually evolved, based on a metaphysical belief that nature strives to change from the simple and imperfect to the more complex and perfect. His contributions to logic, such as the development of the system of inductive reasoning from specific observations to a generalization which explains them all, have been of inestimable value to all branches of science.

The Greek physician Galen (131–201 A.D.) was one of the first to do experiments and dissections of animals to determine structure and

Figure 1.1 Paintings by Upper Paleolithic man from the wall of the cavern at Lascaux, Dordogne, France. (Photo by Windels Montignac.) (From Villee, C. A.: Biology. 7th ed. Philadelphia, W. B. Saunders Co., 1977.)

functions (Fig. 1.2). The first experimental physiologist, he made some notable discoveries on the functions of the brain and nerves and demonstrated that arteries carry blood and not air. His descriptions of the human body were the unquestioned authority for some 1300 years, even though they contained some remarkable errors, being based on dissections of pigs and monkeys rather than of human bodies. Pliny (23–79 A.D.) and others in succeeding centuries compiled encyclopedias (Pliny's *Natural History* was a 37 volume work) regarding the kinds of animals and where they lived, which are remarkable mixtures of fact and fiction. Some of the ones written in the Middle Ages were called "bestiaries." The zoological books written in the Middle Ages are, almost without exception, copied from Aristotle, Galen and Pliny; no original observations were made to corroborate or refute the accuracy of these authorities.

The Renaissance in science began slowly with scholars such as Roger Bacon (1214–1294) and Albertus Magnus (1206–1280) who were interested in all branches of natural

Figure 1.2 Galen (131–201 A.D.). The founder of experimental physiology. (Courtesy of J.A.M.A., from the series Medical Greats.)

science and philosophy. The genius Leonardo da Vinci (1452–1519) was an anatomist and physiologist as well as a painter, engineer and inventor. He made many original observations in zoology, some of which came to light only much later when his notebooks were deciphered.

One of the first to question the authority of Galen's descriptions of human anatomy was the Belgian Andreas Vesalius (1514–1564), who was professor at the University of Padua in Italy (Fig. 1.3). By dissecting human bodies and making detailed, clear drawings of what he saw, Vesalius revealed many of the inaccuracies in Galen's descriptions of the human body. He published his observations and illustrations in *De Humani corporis fabrica* (On the Structure of the Human Body) in 1543. Since Vesalius dared to reject the authority of Galen, he was the object of much adverse criticism and was finally forced to leave his professorial post.

Just as Vesalius had emphasized the importance of relying on original observation rather than on authority in anatomy, so did William Harvey (1578–1657) in physiology (Fig. 1.4). Harvey, an English physician, received his medical training at the University of Padua, where Vesalius had taught. He returned to England and investigated the circulation of the blood. In 1628 he published *Exercitatio anatomica de motu cordis et*

Figure 1.4 William Harvey (1578–1657). He proved the circulation of the blood by the first quantitative physiologic demonstration. (From Garrison: History of Medicine. 4th ed. Philadelphia, W. B. Saunders Co., 1929.)

sanguinis in animalibus (Anatomical Studies on the Motion of the Heart and Blood in Animals). At that time blood was believed to be generated in the liver from food and to pass just once to the organs of the body where it was used up. The heart was believed to be nonmuscular and to be expanded passively by the inflowing blood. Harvey described, from direct observations on animals, how first the atria (auricles) and then the ventricles fill and empty by muscular contraction. He showed by experiment that when an artery is cut blood spurts from it in rhythm with the beating of the heart, and that when a vein is clamped it becomes full of blood on the side away from the heart and empty on the side toward the heart. He showed that valves in the veins permit blood to flow toward the heart but not in the reverse direction (Fig. 1.5). From these experiments he concluded that blood is carried away from the heart in arteries and back to the heart in veins. Furthermore, by measuring how much blood is delivered by each beat of the heart, and by measuring the number of heartbeats per minute, he could calculate the total flow of blood through the heart per minute or hour. This he found to be so great that it could not be generated anew in the liver but must be recirculated, used over and over again. This was the first quantitative physiological argument. He inferred that there

Figure 1.3 Andreas Vesalius (1514–1564). In his *De Humani corporis fabrica*, he established the basis for modern anatomy. (From Garrison: History of Medicine. 4th ed. Philadelphia, W. B. Saunders Co., 1929.)

Figure 1.5 Harvey's illustrations to demonstrate the direction of blood flow. *1*, The formation of "knots" at valves. *2*, Stripping a portion of a vessel to show that there is no back flow. *3* and *4*, Demonstration of valvular blocking of blood flow. (From Harvey: Exercitatio Anatomica de Motu Cordis et Sanguinis in Animalibus. Translation by C. D. Leake: Anatomical Studies on the Motion of the Heart and Blood, Springfield, Ill., Charles C Thomas, 1931.)

must be small vessels connecting arteries and veins to complete the circular path of the blood but, lacking a microscope, he was unable to see them. In later years he made a careful study of the development of the chick, published in 1651 as *Exercitationes de genera-*

tione animalium. In this he postulated that mammals, like the chick, develop from an egg.

The development of the compound microscope by the Janssens in 1590 and by Galileo in 1610 provided the means for attacking many problems in zoology and botany. Robert

Hooke (1635–1703), Marcello Malpighi (1628–1694), Antony van Leeuwenhoek (1632–1723) and Jan Swammerdam (1637–1680) were some of the first microscopists. They studied the fine structure of plant and animal tissues. Hooke was the first to describe the presence of "cells" in plant tissue, Leeuwenhoek was the first to describe bacteria, protozoa and sperm, and Malpighi was the first to describe the capillaries connecting arteries with veins. The light microscope has been modified and improved greatly in the past century, and our ability to see the fine structure of cells has been greatly extended by the invention of the phase microscope and of the transmission and scanning electron microscopes. These, with good resolution at magnifications as great as 100,000 to 500,000 diameters, have revealed a whole new level of complexity in the structure of all kinds of cells.

John Ray (1627–1705) and Linnaeus (Karl von Linné) (1707–1778) brought order into the classification of animals and plants and devised the binomial system (two names, genus and species) for the scientific naming of the kinds of animals and plants. Linnaeus first used this binomial system consistently in the tenth edition of his *Systema naturae* (1758).

Contributions to our understanding of the embryonic development of animals were made by Fabricius, the professor of Anatomy at Padua who taught William Harvey, and by Harvey, Malpighi and Kaspar Wolff (1759). Wolff proposed the theory of epigenesis, the concept that there is a gradual differentiation of structure during development from a relatively structureless egg. Karl Ernst von Baer (1792–1876) established the theory of germ layers and emphasized the need for comparative studies of development in different animals.

Following William Harvey, physiology was advanced by René Descartes (1596–1650), who was a philosopher rather than an experimenter. He believed that "animal spirits" are generated in the heart, stored in the brain, and pass through the nerves to the muscles, causing contraction or relaxation, according to their quantity. Charles Bell (1774–1842) and François Magendie (1783–1855) made notable contributions to our understanding of the function of the brain and spinal nerves. Johannes Muller (1801–1858) studied the properties of nerves and capillaries; his textbook of physiology stimulated a great deal of interest and research in the field. Claude Bernard (1813–1878) was one of the great advocates

of experimental physiology and contributed significantly to our understanding of the role of the liver, heart, brain and placenta. Henry Bowditch (1840–1911) discovered the "all-or-none" principle of the contraction of heart muscle and established the first laboratory for teaching physiology in the United States. Ernest Starling (1866–1927) made many contributions to the physiology of circulation and the nature of lymph, and with William Bayliss (1866–1924) elucidated the hormonal control of the secretion of pancreatic juice.

The Scottish anatomist John Hunter (1728–1793) and the French anatomist Georges Cuvier (1769–1832) were pioneers in the field of comparative anatomy, studying the same structure in different animals. Richard Owen (1804–1892) developed the concepts of homology and analogy. Cuvier was one of the first to study the structure of fossils as well as of living animals and is credited with founding the science of paleontology. Cuvier believed strongly in the unchanging nature of species and carried on bitter debates with Lamarck, who in 1809 proposed a theory of evolution based on the idea of the inheritance of acquired characters.

One of the most important and fruitful concepts in biology is the **cell theory,** which has gradually grown since Robert Hooke first saw, with the newly invented microscope, the dead cell walls in a piece of cork. The French biologist René Dutrochet clearly stated in 1824 that "all organic tissues are actually globular cells of exceeding smallness, which appear to be united only by simple adhesive forces; thus all tissues, all animal organs are actually only a cellular tissue variously modified." Dutrochet recognized that growth is the result of the increase in the volume of individual cells and of the addition of new cells. Two Germans, botanist M. J. Schleiden and zoologist Theodor Schwann, studied many different plant and animal tissues and are generally credited with formulating the cell theory, for they showed that cells are the units of structure in plants and animals, and that organisms are aggregates of cells arranged according to definite laws. The presence of a nucleus within the cell, now recognized as an almost universal feature of cells, was first described by Robert Brown in 1831.

Zoology, along with the other biological sciences, has expanded at a tremendous rate in the twentieth century, with the establishment of the subsciences of cytology, embryology, genetics, evolution, biochemistry, biophysics, endocrinology and ecology. The discoveries and new techniques of chemistry and physics have made possible new approaches to the biological sciences that have attracted the attention of many biologists. So many men have contributed to the growth of zoology that only a few in each field can be mentioned: Mendel, deVries, Morgan, Bridges and Müller in genetics, Darwin, Dobzhansky, Wright and Goldschmidt in evolution, Harrison and Spemann in embryology, and Crick, Watson, Kornberg and Nirenberg in molecular biology. Many others will be mentioned as these subjects are discussed in detail in the text.

The establishment and growth of the marine biological laboratories, such as the ones at Naples (Italy), Woods Hole (Mass.), Pacific Grove (Calif.), Friday Harbor (Wash.) and elsewhere, have played an important role in fostering research in zoological sciences. There are comparable stations for the study of fresh-water biology, such as the one at Douglas Lake, Michigan.

1.4 APPLICATIONS OF ZOOLOGY

Some of the practical uses of a knowledge of zoology will become apparent as the student proceeds through this text. Zoology is basic in many ways to the fields of human and veterinary medicine and public health, agriculture, conservation and to certain of the social sciences. There are esthetic values in the study of zoology, for a knowledge of the structure and functions of the major types of animals will greatly increase the pleasure of a stroll in the woods or an excursion along the seashore. Trips to zoos, aquariums and museums are also rewarding in the glimpses they give of the host of different kinds of animals. Many of these are beautifully colored and shaped, graceful or amusing to watch, but all will mean more to a person equipped with the basic knowledge of zoology which enables him to recognize them and to understand the ways in which they are adapted to survive in their native habitat.

ANNOTATED REFERENCES

Baker, J. J. W., and G. E. Allen: Hypothesis, Prediction, and Implication in Biology. Reading, Massachusetts, Addison-Wesley Publishing Co. Inc., 1968. Interesting examples of the application of the scientific method in biologic research.

Beveridge, W. I. B.: The Art of Scientific Investigation. New York, W. W. Norton & Co., 1957. One of the best over-all surveys of the scientific method.

Bronowski, J.: The Ascent of Man. Boston, Little, Brown and Company, 1974. A well-written and well-illustrated history of the relation of science to human culture.

Cannon, W. B.: The Way of an Investigator. New York, W. W. Norton & Co., 1945. An autobiography with many interesting anecdotes illustrating the application of the scientific method to medical research.

Conant, J. B.: Science and Common Sense. New Haven, Yale University Press, 1951. A general presentation of the methods of science; one of the classics in this field.

Feibleman, J. K.: Testing Hypotheses by Experiment. Perspect. Biol. Med., 4:91, 1960. An excellent brief discussion of scientific methods.

Gabriel, M. L., and S. Fogel: Great Experiments in Biology. New York, Prentice-Hall, Inc., 1955. By extensive quotations from the original papers, traces the development of some of the basic concepts of zoology.

Holton, G.: The Making of Modern Science—Biographical Sketches. Boston, Houghton Mifflin Co., 1971. A fascinating collection of biographical articles about scientists. The article by Robert Olby on Francis Crick is recommended reading in connection with "The Double Helix."

Kuhn, T.: The Structure of Scientific Revolutions. 2nd ed. Chicago, University of Chicago Press, 1970. An analysis of the methods by which science progresses.

Singer, C.: A History of Biology. Revised ed. New York, Abelard-Schumann, 1959. A general history of the biological sciences.

Watson, J. D.: The Double Helix. New York, Atheneum Publishers, 1968. An interesting insight into the motivations and thought processes of one of the classic molecular biologists; "How to become a Nobel Laureate."

Wilson, E. B.: An Introduction to Scientific Research. New York, McGraw-Hill Book Co., 1952. An excellent discussion in nontechnical terms of the methods of science and some of the factors to be considered in planning and carrying out scientific investigation.

THE PHYSICAL AND CHEMICAL BASIS OF LIFE

Most biologists are agreed that all the varied phenomena of life are ultimately explainable in terms of the same physical and chemical principles which define nonliving systems. A corollary of this is that when enough is known of the chemistry and physics of vital phenomena it may be possible to synthesize living matter. An opposite view, widely held by biologists until the present century, stated that some unique force, not explainable in terms of physics and chemistry, is associated with and controls life. Many of the phenomena that appeared to be so mysterious when first discovered have subsequently proved to be understandable without invoking a unique life force, and it is reasonable to suppose that future research will show that other aspects of life can also be explained by physical and chemical principles.

To differentiate the living from the nonliving and then to separate the living into plants and animals are difficult to do sharply and clearly. Organisms such as cats, clams and cicadas are clearly recognizable as animals, but sponges, for example, were considered to be plants until well into the nineteenth century, and there are single-celled organisms which, even today, are called animals by zoologists, plants by botanists, and protists by others. Even the line between living and nonliving is rather tenuous, for the viruses, too small to be seen with an ordinary light microscope, can be considered either the simplest living things or very complex, but nonliving, organic chemicals.

2.1 CHARACTERISTICS OF LIVING THINGS

All living things have, to a greater or lesser degree, the properties of specific organization, irritability, movement, metabolism, growth, reproduction and adaptation.

Organization. Each kind of living organism is recognized by its characteristic form and appearance; the adult organism usually has a characteristic size. Nonliving things generally have much more variable shapes and sizes. Living things are not homogeneous, but are made of different parts, each with special functions; thus the bodies of animals and plants are characterized by a specific, complex organization.

The structural and functional unit of both animals and plants is the **cell.** It is the simplest bit of living matter that can exist independently and exhibit all the characteristics of life. The processes of the entire organism are the sum of the coordinated functions of its constituent cells. These cellular units vary considerably in size, shape and function. Some of the smallest animals have bodies made of a single cell; the human body, in contrast, is made of countless billions of cells fitted together.

The cell itself has a specific organization and each kind of cell has a characteristic size and shape by means of which it can be recognized. A typical cell, such as a liver cell (Fig. 2.1), is polygonal in shape, with a **plasma**

Figure 2.1 Schematic drawing of a generalized animal cell.

membrane separating the living substance from the surroundings. Almost without exception, each cell has a **nucleus,** typically spherical or ovoid in shape, which is separated from the rest of the cell by a **nuclear membrane.** The nucleus, as we shall see later, has a major role in controlling and regulating the cell's activities. It contains the hereditary units or **genes.** A cell experimentally deprived of its nucleus usually dies in a short time; even if it survives for several days it is unable to reproduce.

The bodies of higher animals are organized in a hierarchy of increasingly complex levels: cells are organized into **tissues,** tissues into **organs,** and organs into **organ systems.**

Irritability.   Living things are irritable; they respond to stimuli, to physical or chemical changes in their immediate surroundings. Stimuli which are effective in evoking a response in most animals and plants are changes in light (either in its color, intensity or direction), temperature, pressure, sound, and in the chemical composition of the earth, water, or air surrounding the animal. In humans and other complex animals, certain cells of the body are highly specialized to respond to certain types of stimuli: the rods and cones in the retina of the eye respond to light, certain cells

in the nose and in the taste buds of the tongue respond to chemical stimuli, and special groups of cells in the skin respond to changes in temperature or pressure. In lower animals such specialized cells may be absent, but the whole organism responds to any one of a variety of stimuli. Single-celled amebas will respond by moving toward or away from heat or cold, certain chemical substances, or the touch of a microneedle. Indeed, many of the cells of higher animals have a similar generalized sensitivity.

Movement.   Living things are characterized further by their ability to move. The movement of most animals is quite obvious—they wiggle, swim, run or fly. The movements of plants are much slower and less obvious, but are present nonetheless. A few animals—sponges, corals, hydroids, oysters, certain parasites—do not move from place to place, but most of these have microscopic, hairlike projections from the cells, called **cilia** or **flagella,** to move their surroundings past their bodies and thus bring food and other necessities of life to themselves. The movement of an animal body may be the result of muscular contraction, of the beating of cilia or flagella, or of the slow oozing of a mass of cell substance (known as ameboid motion).

Metabolism. All living things carry on a wide variety of chemical reactions, the sum of which we call **metabolism.** There is no way of observing the occurrence of most of these chemical reactions without the aid of special apparatus. Elaborate physical and chemical equipment and substances labeled with radioactive or stable isotopes are used to trace in detail the paths of metabolism and their respective quantitative importance to the animal or plant under investigation. Such studies have shown that all cells are constantly taking in new substances, altering them chemically in a multitude of ways, building new cell components, and transforming the potential energy contained in large molecules of carbohydrates, fats and proteins into kinetic energy and heat as these substances are converted into simpler ones. The never-ending flow of energy within a cell, from one cell to another and from one organism to the next, is the very essence of life, one of the unique and characteristic attributes of living things. The rate of metabolism is affected by temperature, age, sex, general health and nutrition, by hormones, and by many other factors, even the time of day. The study of energy transformations in living organisms is termed **bioenergetics.**

Those metabolic processes in which simpler substances are combined to form more complex substances and which result in the storage of energy and the production of new cellular materials are termed **anabolic.** The opposite processes, in which complex substances are broken down to release energy, are called **catabolic.** Both types of metabolism occur continuously and are intricately interdependent so that they become, in practice, difficult to distinguish. Complex compounds of one sort may be broken down and their parts recombined in new ways to yield new compounds. The interconversions of carbohydrates, fats and proteins that occur continuously in most animal cells are examples of combined anabolic and catabolic processes. Furthermore, the synthesis of most molecules requires energy, so that some catabolic processes must occur to supply the energy to drive the anabolic reactions of these syntheses.

Growth. Both plants and animals grow; nonliving things do not. The increase in mass may be brought about by an increase in the size of the individual cells, by an increase in the *number* of cells or both. An increase in cell size may occur by the simple uptake of water, but this swelling is not generally considered to be growth. The term **growth** is restricted to those processes which increase the amount of living substance of the body, measured by the amount of nitrogen, protein or nucleic acid present. Growth may be uniform in the several parts of an organism, or, perhaps more commonly, growth is differential, greater in some parts than in others, so that the body proportions change as growth occurs.

Growth may occur throughout the life span of an organism or may be restricted to a part of it. One of the truly remarkable aspects of the process is that each organ continues to function while undergoing growth.

Reproduction. If there is any one characteristic that can be said to be the *sine qua non* of life, it is the ability to reproduce. Since individual animals grow old and die, the survival of the species depends upon the replacing of these individuals by new ones. Although at one time worms were believed to arise from horse hairs in a trough of water, maggots from decaying meat and frogs from the mud of the Nile, we now know that each can come only from previously existing ones. One of the fundamental tenets of biology is that "all life comes only from living things."

The classic experiment disproving the **spontaneous generation of life** was performed by an Italian, Francesco Redi, about 1680. Redi proved that maggots do not come from decaying meat by this simple experiment: He placed a piece of meat in each of three jars, leaving one uncovered, covering the second with a piece of fine gauze, and covering the third with parchment. All three pieces of meat decayed but maggots appeared on only the meat in the uncovered jar. A few maggots appeared on the gauze of the second jar, but not on the meat, and no maggots were found on the meat covered by parchment. Redi thus demonstrated that the maggots did not come from the decaying meat, but hatched from eggs laid by blowflies attracted by the smell of the decaying meat. Further observations showed that the maggots develop into flies which in turn lay more eggs. Louis Pasteur, about two hundred years later, showed that bacteria do not arise by spontaneous generation but only from previously existing bacteria. The submicroscopic filtrable viruses do not arise from nonviral material by spontaneous generation; the multiplication of viruses requires the presence of previously existing viruses.

The problem of the original source of life will be discussed later (p. 202), but it is likely that billions of years ago, when physical and chemical conditions on the earth's surface were quite different from those at present, the first living things *did* actually arise from nonliving material.

The process of reproduction may be as simple as the splitting of one individual into two. In most animals, however, it involves the production of specialized eggs and sperm which unite to form the zygote or fertilized egg, from which the new organism develops, In some animals, the liver flukes for example, reproduction involves several quite different forms, each of which gives rise to the next in succession until the cycle is completed and the adult reappears.

Adaptation. The ability of an animal or plant to adapt to its environment is the characteristic which enables it to survive the exigencies of a changing world. Each particular species can achieve adaptation either by seeking out a suitable environment or by undergoing modifications to make it better fitted to its present surroundings. Adaptation may involve immediate changes which depend upon the irritability of cells or the responses of the enzyme systems to inducers or repressors, or it may be the result of a long-term process of mutation and selection. It is obvious that no single kind of organism can adapt to all the conceivable kinds of environment; hence there will be certain areas where it cannot survive. The list of factors that may limit the distribution of a species is almost endless: water, light, temperature, food, predators, competitors, parasites, and so on.

2.2 THE ORGANIZATION OF MATTER: ATOMS AND MOLECULES

To get a more complete idea of what living matter is and what it can do, we must consider not only the easily visible macroscopic aspects of life and those that can be seen under the light or electron microscopes, but also those molecular patterns that lie far beyond the range of the microscope. To appreciate the latter requires some understanding of certain basic principles of physics and chemistry.

Matter and Energy. The universe consists of **matter** and **energy,** which are related by the famous Einstein equation, $E = mc^2$, where $E =$ energy, $m =$ mass and $c =$ the velocity of light, which is a constant. This equation provides the theoretical basis for the conversion of matter to energy that occurs within the sun or in an atomic bomb or nuclear reactor. In the familiar, everyday world, however, matter and energy are separate and distinguishable. Matter occupies space and has mass, and energy is the ability to produce a change or motion in matter—the ability to do work. Energy may take the form of heat, light, electricity, motion or chemical energy. We shall return (p. 63) to the kinds of energy transformations of importance in biological phenomena.

Atoms. Regardless of the form—gaseous, liquid or solid—that matter may assume, it is always composed of units called **atoms.** In nature there are 92 different kinds of atoms, ranging from hydrogen, the smallest, to uranium, the largest. In addition to these naturally occurring atoms, there are thirteen larger than uranium that have been man-made in a cyclotron or nuclear reactor. All of them, natural and synthetic, are much smaller than the tiniest particle visible under the miscroscope. By special scanning electron microscopy with magnifications of 5,500,000 times, Albert Crewe has been able to photograph certain larger atoms such as uranium and thorium. The structure and properties of atoms have been inferred from experiments made with many types of elaborate apparatus.

Once the atom was believed to be the ultimate, smallest unit of matter. Now physicists know that atoms are divisible into even smaller particles organized around a central core, as our solar system of planets is organized around the sun. The exact number and kind of these particles and their arrangement in the atom are the subject of continuing research. For our purposes we need consider only three types: **electrons,** which have a negative electric charge and an extremely small mass or weight; **protons,** which have a positive electric charge and are about 1800 times as heavy as electrons; and **neutrons,** which have no electric charge, but have essentially the same mass as protons.

The center of the atom, corresponding in position to the sun in our solar system, is the **nucleus,** containing protons and neutrons; it composes almost the total mass of the atom. Just as the solar system is mostly empty space, so is the atom, with electrons moving in the empty space around the nucleus. The electrons do not circle the nucleus in specific orbits, as the planets circle the sun; instead they whirl around the nucleus, now close to it, then further away. Each type of atom has a characteristic number of electrons whirling around the nucleus and characteristic numbers of protons and neutrons in the nucleus (Fig. 2.2). In all atoms, the number of protons in the nucleus equals the number of electrons circling around it, so that the atom as a whole is in a state of electrical neutrality. At least a dozen

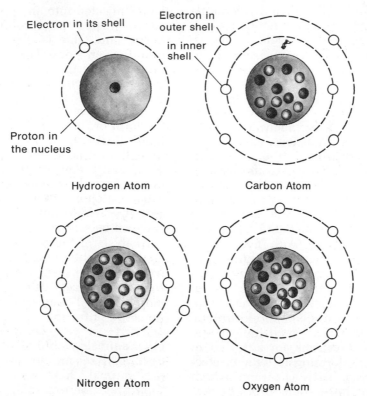

Figure 2.2 Diagrams of the structure of the atoms of the four chief elements of living matter; hydrogen, carbon, nitrogen and oxygen. The symbols used are ○, neutron; +, proton; ⊖, electron.

kinds of subatomic particles other than protons, neutrons and electrons are now known, but we need not consider them. The different kinds of matter reflect differences in the number and arrangement of these basic particles. Living systems are composed of exactly the same kinds of atoms, with the same kind of atomic structure, as nonliving systems.

Elements. An **element** is a substance composed of atoms, all of which have the same number of protons in the atomic nucleus and therefore the same number of electrons circling in orbitals. Most elements have a tendency to unite with other elements to form compounds, but a few, such as the "noble" gases, helium, neon, argon, krypton and xenon, occur in nature uncombined with other elements.

The unique aliveness of living things does not depend upon the presence of some rare or unique element. Four elements, carbon, oxygen, hydrogen and nitrogen, make up about 96 per cent of the material in the human body. Potassium, sulfur, calcium and phophorus are four other elements usually present to the extent of 1 per cent or more each. Since bone is largely composed of calcium and phosphorus, the amount of these elements is much greater in a bony animal than in a completely soft-bodied one. Smaller amounts of sodium, chlorine, iron, iodine, magnesium, copper, manganese, cobalt, zinc and a few others complete the list. These same elements are abundant in the atmosphere, in the sea and in the earth's crust. The phenomenon of life reflects the complexity of the interrelationships of these common, abundant elements.

For convenience in writing chemical formulas and reactions, chemists have assigned to each of the elements a symbol, usually the first letter of the name of the element: O, oxygen; H, hydrogen; C, carbon; N, nitrogen. A second letter is added to the symbol of those elements with the same initial letter: Ca, calcium; Co, cobalt; Cl, chlorine; Cu, copper, Na, sodium (Latin *natrium*).

Ions. The chemical properties of an element are determined primarily by the number and arrangement of electrons revolving in the outermost energy shell around the atomic nucleus and to a lesser extent by the number of electrons in the inner orbitals. These, in turn, depend upon the number and kind of particles, protons and neutrons, in the nucleus.

The number of electrons in the outermost shell varies from zero to eight in different kinds of atoms (Fig. 2.2). If there are zero or eight electrons in the outer shell the element is chemically inert and will not readily combine with other elements (e.g., the "noble" gases). When there are fewer than eight electrons, the atom tends to lose or gain electrons in order to achieve an outer shell of eight electrons. Since the number of protons in the atomic nucleus is not changed, this loss or gain of electrons produces an atom with a net positive or negative charge. Such electrically charged atoms are known as **ions.** Atoms with one, two or three electrons in the outer shell tend to lose them to other atoms and become positively charged ions because of the excess protons in the nucleus (e.g., Na^+, sodium ion; Ca^{++}, calcium ion). These are termed **cations** because they migrate to the cathode of an electrolytic cell. Atoms with five, six or seven electrons in the outer shell tend to gain electrons from other atoms and become negatively charged ions or **anions** (e.g., Cl^-, chloride ion). Anions migrate to the anode or positively charged electrode of an electrolytic cell. Because they bear opposite electric charges, anions and cations are attracted to each other and unite forming an ionic or electrostatic bond. Atoms such as carbon, which have four electrons in the outer orbit, neither lose nor gain electrons, but share them with adjacent atoms.

Isotopes. Physical research has shown that most elements·are composed of two or more kinds of atoms, which differ in the number of neutrons in the atomic nucleus. The different kinds of atoms of an element are called **isotopes** (iso = equal, tope = place), because they occupy the same place in the periodic table of the elements. All the isotopes of a given element have the same number of electrons circling the atomic nucleus.

Although the isotopes of a given element have the same chemical properties, they can be differentiated physically. Some are radioactive and can be detected and measured by the kind and amount of radiation they emit. Others can be differentiated by the slight difference in the mass of the atoms caused by an extra neutron in the nucleus.

Substances containing ^{15}N (heavy nitrogen) instead of ^{14}N, the usual isotope, or ^2H (heavy hydrogen, deuterium) instead of ^1H, will have a greater mass detectable with a **mass spectrometer.**

A tremendous insight has been gained into the details of the metabolic activities of cells by preparing some substance—sugar, for example—labeled with radioactive carbon (^{11}C or ^{14}C) or heavy carbon (^{13}C) in place of ordinary carbon (^{12}C). The labeled substances are fed or injected into an animal or plant, or cells are incubated in a solution containing the tracer, and the labeled products resulting from the cell's or organism's normal metabolic processes are then isolated and identified. By such experiments it has been possible to trace step by step the sequence of reactions undergone by a given compound and to determine the form or forms in which the labeled atoms finally leave the cell or organism. The rate of formation of bone, for example, and the effect of vitamin D and parathyroid hormone on this process, can be studied with the aid of radioactive calcium, ^{45}Ca. Many biological problems that could not be attacked in any other way can be solved by this method.

Analysis of the human body reveals that it contains about 50 per cent carbon, 20 per cent oxygen, 10 per cent hydrogen, 9 per cent nitrogen, 4 per cent calcium, 2.5 per cent phosphorus (P), 1 per cent potassium (K), 0.8 per cent sulfur (S), 0.4 per cent sodium (Na), and 0.4 per cent chlorine (Cl). Analyses of other animals would yield comparable results. Such analyses are not very informative unless the animal has some unusual element. Tunicates, for example, are unusual in that they contain a large amount of the element vanadium (V).

2.3 CHEMICAL COMPOUNDS

Most elements are present in living material as chemical compounds, rather than as free elements. A **chemical compound** is a substance composed of two or more different kinds of atoms or ions joined together; hence it can be decomposed or split apart into two or more simpler substances. The amounts of the elements in a given compound are always present in a definite proportion by weight. This reflects the fact that the atoms are joined together by chemical bonds in a precise way to form the compound. The assemblage of atoms held together by chemical bonds is called a **molecule.** A molecule is the smallest particle of a compound that has the composition and properties of a larger portion of it. A molecule is made of two or more atoms—which may be the same, as in a molecule of oxygen, O_2, or nitrogen, N_2—or they may be atoms of different elements. The properties of a chemical compound are usually quite different from the properties of its constituent elements. Each water molecule, for example, contains two atoms of hydrogen and one atom of oxygen, and the chemical properties of water are quite different from those of either hydrogen or oxygen. The chemical formula of water, H_2O, states the kinds of atoms that are present in the molecule and their relative proportions.

A molecule of the simple sugar, glucose, is composed of three kinds of atoms: six carbon atoms, twelve hydrogen atoms and six oxygen atoms; its formula in chemical shorthand is $C_6H_{12}O_6$. Any larger portion of glucose—a gram or kilogram—will also contain carbon, hydrogen and oxygen in these same proportions. If it were possible to divide a gram of glucose in half, then divide the half in half, then the quarters in half, and so on, eventually the glucose would be subdivided into its constituent molecules. Each of these molecules would have the same composition and properties as the original gram. However, if a glucose molecule were divided, the parts no longer would be sugar, but would be carbon, hydrogen and oxygen atoms, and these would have quite different properties.

The weight of any single atom or molecule is exceedingly small, much too small to be expressed in terms of grams or micrograms. These weights are expressed in terms of the atomic weight unit, the **dalton,** approximately the weight of a proton or neutron. An atom of the lightest element, hydrogen, weighs one dalton, a carbon atom weighs 12 daltons and oxygen has an atomic weight of 16 daltons. The **molecular weight** of a compound is the sum of the atomic weights of its constituent atoms. Thus the molecular weight of water, H_2O, is $(2 \times 1) + 16$, or 18 daltons. The molecular weight of glucose, $C_6H_{12}O_6$, is $(6 \times 12) + (12 \times 1) + (6 \times 16)$, or 180 daltons. The amount of a compound whose mass in grams is equivalent to its molecular weight is termed one **mole** (one mole of glucose is 180 g.). A one molar (1 M) solution contains one mole of a substance in one liter of water.

A large part of each cell is simply **water.** The percentage of water in human tissues ranges from about 20 per cent in bone to 85 per cent in brain cells. The water content is greater in embryonic and young cells and decreases as aging occurs. About 70 per cent of our total body weight is water; as much as 95 per cent of a jellyfish is water. Water has a number of important functions in living systems. Most of the other chemicals present are dissolved in it; they must be dissolved in water in order to react. Water also dissolves the waste products of metabolism and assists in their removal. Water has a great capacity for absorbing heat with a minimal change in its own temperature; thus, it protects the living

material against sudden thermal changes. Since water absorbs a large amount of heat as it changes from a liquid to a gas, the mammalian body can dissipate excess heat by the evaporation of sweat. Water's high heat conductivity makes possible the even distribution of heat throughout the body. Finally, water has an important function as a lubricant. It is present in body fluids wherever one organ rubs against another and in joints where one bone moves on another.

A **mixture** contains two or more kinds of atoms or molecules which may be present in varying proportions. Air is a mixture of oxygen, nitrogen, carbon dioxide and water vapor, plus certain rare gases such as argon. The proportions of these constituents may vary widely. Thus, in contrast to a pure compound, which has a fixed ratio of its constituents and definite chemical and physical properties, a mixture has properties which vary with the relative abundance of its constituents.

To learn more about the constituents of cells, biochemists have used very sensitive analytical techniques and have taken great pains to preserve the extremely labile substances present in these enormously complicated systems. To prevent the disappearance of certain substances it is necessary to quick-freeze a bit of excised tissue, or even a whole small animal, by dropping it directly into liquid nitrogen. Biochemical research has made it abundantly clear that the composition of every cell is constantly changing, that the cell constituents are in a "dynamic state." There is a continuous synthesis of large, energy-rich molecules and continual decomposition of these into smaller, energy-poor ones. Some of the most important compounds are present in cells only in extremely minute amounts at any given time, although the total amount formed and used in a 24 hour period may be quite large. An appreciation of this may be gained from the following consideration: when substances undergo chemical reactions in sequence (and almost all the reactions of importance biologically are sequences or "cycles"), such as $A \rightarrow B \rightarrow C \rightarrow D$, the rate of the whole process is controlled by the rate of the slowest reaction in the chain. For example, if reaction $A \rightarrow B$ is 10 times as fast as $B \rightarrow C$, and if $C \rightarrow D$ is 100 times as fast as $B \rightarrow C$, then the least reactive substance, B, will tend to accumulate and the most reactive one, C, will be present in the smallest amount. For this reason many of the most active and important substances are present in cells in extremely minute amounts. This, coupled with their chemical instability, has made their detection and isolation difficult.

The compounds present in cells are of two main types: **inorganic** and **organic.** The latter include all the compounds (other than carbonates) that contain the element carbon. The element carbon is able to form a much wider variety of compounds than any other element because the outer shell of the carbon atom contains four electrons, which can be shared in a number of different ways with adjacent atoms. At one time it was believed that organic compounds were uniquely different from other chemical substances and could be produced only by living matter. This hypothesis was disproved when the German chemist Wöhler succeeded in 1828 in synthesizing urea (one of the waste products found in human urine) from the inorganic compounds ammonium sulfate and potassium cyanate. Since that time thousands of organic compounds have been synthesized, some of which are quite complex molecules of great biological importance such as vitamins, hormones, antibiotics and drugs.

2.4 ELECTROLYTES: ACIDS, BASES AND SALTS

Among the inorganic compounds present in living systems are water, carbon dioxide, acids, bases and salts. An **acid** is a compound which releases hydrogen ions (H^+) when dissolved in water.* Acids turn blue litmus paper to red and have a sour taste. Hydrochloric (HCl) and sulfuric (H_2SO_4) are inorganic acids; lactic ($C_3H_6O_3$) from sour milk and acetic (CH_3COOH) from vinegar are two common organic acids. A **base** is a compound which releases hydroxyl ions (OH^-) when dissolved in water. Bases turn red litmus paper blue. Common inorganic bases include sodium hydroxide (NaOH) and ammonium hydroxide (NH_4OH).

For convenience in stating the degree of acidity or alkalinity of a fluid, the hydrogen ion concentration may be expressed in terms of pH, the logarithm of the reciprocal of the hydrogen ion concentration, $\log 1/[H^+]$. On this scale, a neutral solution has a pH of 7 (its hydrogen ion concentration is 0.0000001 or 10^{-7} molar), alkaline solutions have pH's ranging from 7 to 14 (the pH of 1 M NaOH), and acids have pH's from 7 to 0 (the pH of 1 M HCl). Most animal cells are neither strongly acid nor alkaline but contain a mixture of acidic and basic substances; their pH is about 7.0. Any considerable change in the pH of a cell is inconsistent with life. Since the scale is a logarithmic one, a solution with a pH of 6 has a hydrogen ion concentration 10 times as great as that of one with a pH of 7.

When an acid and a base are mixed, the hydrogen ion of the acid unites with the hydroxyl ion of the base to form a molecule of water (H_2O). The remainder of the acid (anion) combines with the rest of the base (cation) to form a **salt.** For example, hydrochloric acid (HCl) reacts with sodium hydroxide (NaOH) to form water and sodium chloride (NaCl) or common table salt:

$$H^+Cl^- + Na^+OH^- \rightarrow H_2O + Na^+Cl^-$$

A salt may be defined as a compound in which the hydrogen atom of an acid is replaced by some metal.

When a salt, an acid or a base is dissolved in water it separates into its constituent ions. These charged particles can conduct an electric current; hence these substances are known as **electrolytes.** Sugars, alcohols and the many other substances which do not separate into charged particles when dissolved, and therefore do not conduct an electric current, are called **nonelectrolytes.**

Cells and extracellular fluids contain a variety of mineral salts, of which sodium, potassium, calcium and magnesium are the chief cations (positively charged ions) and chloride, bicarbonate, phosphate and sulfate are the important anions (negatively charged ions). The body fluids of land vertebrates resemble sea water in the kinds of salts present and in their relative proportions, but the total concentration of salts is only about one-fifth as great as in sea water. Most biologists now believe that life originated in the sea. The cells of those early organisms became adapted to function optimally in the presence of this pattern of salts. As larger animals evolved and developed body fluids, this pattern of salts was maintained, even as some of the descendants migrated into fresh water or onto the land, for any marked change in the kinds of salts present would inhibit certain enzymes and place that kind of animal at a marked disadvantage in the competition for survival.

---

*Other definitions of acid and base, such as those by Brønsted or Lewis, may be useful in understanding certain more complex reactions. Brønsted defines an acid as a proton donor and a base as a proton acceptor.

Some animals have evolved kidneys and other excretory organs that selectively retain or secrete certain ions, thus leading to body fluids with somewhat different relative concentrations of salts. The concentration of each ion is determined by the relative rates of its uptake and excretion by the organism.

Although the concentration of salts in cells and in the body fluids is small, this amount is of great importance for normal cell functioning. The concentrations of the respective cations and anions are kept remarkably constant under normal conditions; any marked change results in impaired function and finally in death. A great many of the enzymes which mediate the chemical reactions occurring in the body require one or another of these ions—for example, magnesium, manganese, cobalt, potassium—as cofactors. These enzymes are unable to function in the absence of the ion. Normal nerve function requires a certain concentration of calcium in the body fluids; a decrease in this results in convulsions and death. Normal muscle contraction requires certain amounts of calcium, potassium and sodium. If a frog heart, for example, is removed from the body and placed in a solution of sodium chloride, it soon stops beating and remains in a relaxed state. If placed in a solution of potassium chloride, or in a mixture of sodium and calcium chloride, it ceases beating in the contracted condition. But if it is placed in a solution of the three salts in proper proportion it will continue to beat for hours. Under the proper conditions, the strength of the heartbeat is proportional to the concentration of calcium ions in the fluid bathing the heart; this method is sensitive enough to be used to measure the concentration of calcium ions. In addition to these several specific effects of particular cations, mineral salts serve an important function in maintaining the osmotic relationships between each cell and its environment.

2.5 ORGANIC COMPOUNDS OF BIOLOGICAL IMPORTANCE

The major types of organic substances present in cells are the carbohydrates, proteins, fats, nucleic acids and steroids. Some of these are required for the structural integrity of the cell, others to supply energy for its functioning, and still others are of prime importance in regulating metabolism within the cell. **Carbohydrates** and **lipids** are the chief sources of chemical energy in almost every form of life; **proteins** are structural elements but are of even greater importance as catalysts (enzymes) and regulators of cellular processes. **Nucleic acids** are of prime importance in the storage and transfer of information used in the synthesis of specific proteins and other molecules.

The basic pattern of the types of substances, and even their relative proportions, is remarkably similar for cells from the various parts of the body and for cells from different animals. A bit of human liver and the substance of an ameba each contain about 80 per cent water, 12 per cent protein, 2 per cent nucleic acid, 5 per cent fat, 1 per cent carbohydrate and a fraction of 1 per cent of steroids and other substances. Certain specialized cells, of course, have unique patterns of chemical constituents; the mammalian brain, for example, is rich in certain kinds of fats.

Carbohydrates. The simplest of the organic substances are the carbohydrates—the sugars, starches and celluloses—which contain carbon, hydrogen and oxygen in a ratio of approximately 1 C:2 H:1 O. Carbohydrates are found in all living cells, usually in relatively small amounts, and are important as readily available sources of energy. Both **glucose** (also known as dextrose) and **fructose** (also called levulose) are single sugars with the formula $C_6H_{12}O_6$. However, the arrangement of the atoms within the two molecules is different and the two sugars have somewhat different chemical properties and quite different physiological roles. Such differences in the molecular structures of substances with the same chemical formula are frequently found in organic chemistry. Chemists indicate the arrangement of atoms in a molecule by a **structural formula** in which the atoms are represented by their symbols—C, H, O, etc.—and the chemical bonds or forces which hold the atoms together are indicated by connecting lines. Hydrogen has one such bond; oxygen, two; nitrogen, three; and carbon, four. The structural formulas of glucose and fructose are compared in Figure 2.3. Note that the lower four carbon atoms in the two sugars have identical groups; only the upper two show differences.

Carbon atoms can unite with each other as well as with many other kinds of atoms and form an almost infinite variety of compounds. Carbon atoms linked together may form long chains (as in fatty acids), branched chains (certain amino acids), rings (purines and pyrimidines) and complex rings (steroids). Molecules are in fact three-dimensional struc-

```
        H                        H
        |                        |
        C=O                 H—C—O—H
        |                        |
   H—C—O—H                    C=O
        |                        |
  HO—C—H                   HO—C—H
        |                        |
   H—C—O—H                  H—C—O—H
        |                        |
   H—C—O—H                  H—C—O—H
        |                        |
   H—C—O—H                  H—C—O—H
        |                        |
        H                        H
     Glucose                  Fructose
```

Figure 2.3 Structural formulas of two simple sugars.

tures, not simple two-dimensional ones as these formulas suggest. There are more complex ways of representing the third dimension of the molecule. Since the properties of the compound depend in part on the exact nature of its three-dimensional structure, its **conformation,** such three-dimensional formulas are helpful in understanding the intimate relations between molecular structure and function.

Glucose is the only hexose which occurs in any quantity in the cells and body fluids of both vertebrates and invertebrates. The other carbohydrates eaten by vertebrates are converted to glucose in the liver. Glucose is an indispensable component of mammalian blood and is normally present in a concentration of about 0.1 per cent. A prolonged increase in the concentration of glucose in the blood, as in untreated or poorly controlled diabetes mellitus, can cause metabolic alterations and extensive damage to certain tissues such as the eye and the kidney. A reduced concentration of glucose in the blood leads to an increased irritability of certain brain cells, so that they respond to slight stimuli. They discharge nerve impulses which result in muscular twitches, convulsions, and finally unconsciousness and death. Brain cells use glucose as their prime metabolic fuel, and a certain minimum concentration of glucose in the blood is required to supply this. A complex physiological control mechanism, which operates like the "feedback" controls of electronic devices and which involves the liver, pancreas, pituitary and adrenal glands, maintains the proper concentration of glucose in the blood.

The double sugars, with the formula $C_{12}H_{22}O_{11}$, consist of two molecules of single sugar joined by the removal of a molecule of water. **Sucrose,** or table sugar, is a combination of glucose and fructose. Other common double sugars are **maltose,** composed of two molecules of glucose and **lactose,** composed of glucose and galactose. Lactose, found in the milk of all mammals, is an important item in the diet of the young of these forms. Fructose, the sweetest of the common sugars, is more than 10 times sweeter than lactose; sucrose is intermediate.

Most animal cells contain some **glycogen** or animal starch, the molecules of which are made of a very large number—thousands—of molecules of glucose joined together by the removal of an H from one and an OH from the next. Glycogen is the form in which animal cells typically store carbohydrate for use as an energy source in cell metabolism. The glycogen molecules within a living cell are constantly being built up and broken down. Glucose and other single sugars are not a suitable storage form of carbohydrate for, being soluble, they readily pass out of the cells. The molecules of glycogen, which are much larger and less soluble, cannot pass through the plasma membrane. Glycogen is typically stored intracellularly as microscopic granules, which can be made visible by special stains. Glycogen is readily converted into small molecules such as glucose-phosphate (p. 75) to be metabolized within the cell.

Cellulose, also composed of hundreds of molecules of glucose, is an insoluble carbohydrate which is a major constituent of the tough outer wall of plant cells. The bonds joining the glucose molecules in cellulose are different from those joining the glucoses in glycogen and are not split by the amylases that digest glycogen and starch.

Glucosamine and galactosamine, nitrogen-containing derivatives of the sugars glucose and galactose, are important constituents of supporting substances such as connective tissue fibers, cartilage and chitin, a constituent of the hard outer shell of insects, spiders and crabs.

Carbohydrates serve as a readily available fuel to supply energy for metabolic processes. Glucose is metabolized to carbon dioxide and water with the release of energy. A few carbohydrates combine with proteins (**glycoproteins**) or lipids (**glycolipids**) to serve as structural components of certain cells. **Ribose** and **deoxyribose** are five-carbon sugars that are components of ribonucleic acid (RNA) and deoxyribonucleic acid (DNA)

Fats. The term **fat** or lipid, refers to a heterogeneous group of compounds which share the property of being soluble in chloroform, ether or benzene, but are only very sparingly soluble in water. True fats are composed of carbon, hydrogen and oxygen, but have much less oxygen in proportion to the carbon and hydrogen than carbohydrates have. Fats have a greasy or oily consistency; some, such as beef tallow fat or bacon fat, are solid at ordinary temperatures, others, such as olive oil or cod liver oil, are liquid. Each molecule of fat is composed of one molecule of **glycerol** and three molecules of **fatty acid.** All such neutral fats, termed triacylglycerols, contain glycerol but may differ in the kinds of fatty acids present. Fatty acids are long chains of carbon atoms with a carboxyl group (−COOH) at one end. All fatty acids in nature have an even number of carbon atoms — palmitic has 16 and stearic, 18. Fatty acids with one or more double bonds are called **unsaturated.** Oleic acid has 18 carbons and one double bond (and hence has two less hydrogen atoms than stearic). A fat common in beef tallow, tristearin, $C_{57}H_{110}O_6$, has three molecules of stearic acid and one of glycerol (Fig. 2.4). Fats containing unsaturated fatty acids are usually liquid at room temperature, whereas **saturated** fats, such as tristearin, are solids.

Fats are important as biological fuels and as structural components of cells, especially cell membranes. Glycogen or starch is readily converted to glucose and metabolized to release energy quickly; the carbohydrates serve as short-term sources of energy. Fats yield more than twice as much energy per gram as do carbohydrates and thus are a more economical form for the storage of food reserves. Carbohydrates can be transformed by the body into fats and stored in this form — a restatement of the generally known fact that starches and sugars are "fattening." The reverse may also occur to some slight extent: Parts of the fat molecule may be converted into glucose and other carbohydrates. This has been shown by preparing fatty acids or glycerol labeled with radioactive or heavy carbon, feeding or injecting these into a rat or dog, and then isolating glucose from the blood or glycogen from the liver and demonstrating that these molecules now contain some of the labeled carbon atoms.

Certain kinds of fats, phospholipids and cholesterol, are important structural elements of the body. The plasma membrane around each cell and the nuclear membrane contain phospholipids as important constituents, and the myelin sheath around the nerve fibers (p. 61) has a high cholesterol content. Animals store fat as globules within the cells of adipose tissue. The layer of adipose tissue just under the skin serves as an insulator against the loss of body heat. Women tend to have a thicker layer of adipose tissue than do men and thus should be more tolerant to cold. Whales, which live in cold water and have no insulating hair, have an especially thick layer of fat (blubber) just under the skin for this purpose. The subcutaneous fat in humans keeps the skin firm, in addition to restricting the loss of body heat.

The fat deposits are not simply long-term stores of foodstuff used only in starvation, but are constantly being used and re-formed. Studies with labeled fatty acids showed that mice replace half of their stored fats in seven days.

Besides the true fats, composed of glycerol and fatty acids, lipids include several related substances that contain components such as phosphorus, choline and sugars in addition to fatty acids. The **phospholipids** are important constituents of the membranes of plant and animal cells in general and of nerve cells in particular. The fatty acid portion of the phospholipid molecule is **hydrophobic,** not soluble in water. The other portion, composed of glycerol, phosphate and a nitrogenous base such as choline, is ionized and readily water-soluble. For this reason, phospholipid molecules in a film tend to be oriented with the polar, water soluble portion pointing one way and the nonpolar, fatty acid portion pointing the other.

Figure 2.4   Structural formula of tristearin, a fat composed of glycerol (a) and three molecules of stearic acid (b). In the formula, $(CH_2)_{16}$ represents a chain of sixteen carbon atoms joined in a line, $-\overset{\displaystyle H}{\underset{\displaystyle H}{C}}-\overset{\displaystyle H}{\underset{\displaystyle H}{C}}- \ldots$, to each of which are attached two hydrogen atoms. (From Villee, C. A.: Biology. 7th ed. Philadelphia, W. B. Saunders Co., 1977.)

Figure 2.5   Structural formula of a sterol, cholesterol.

Beeswax, lanolin and other **waxes** contain a fatty acid plus an alcohol other than glycerol. **Cerebrosides,** as their name indicates, are fatty substances found especially in nerve tissue. They contain galactose, long chain fatty acids, and a long chain amino alcohol, sphingosine. The metabolic roles of these special fats are not clear at present.

Steroids.   Steroids are complex molecules containing carbon atoms arranged in four interlocking rings, three of which contain six carbon atoms each and the fourth of which contains five. Vitamin D, male and female sex hormones, the adrenal cortical hormones, bile salts and cholesterol are examples of steroids. **Cholesterol** (Fig. 2.5) is an important structural component of nervous tissue and other tissues, and the steroid hormones are of great importance in regulating certain aspects of metabolism.

Proteins.   Proteins differ from carbohydrates and true fats in that they contain nitrogen in addition to carbon, hydrogen and oxygen. Proteins typically contain sulfur and phosphorus also. Proteins are among the largest molecules present in cells and share with nucleic acids the distinction of great complexity and variety. Hemoglobin, the red pigment found in the blood of all vertebrates and many invertebrates, has the formula $C_{3032}H_{4816}O_{872}N_{780}S_8Fe_4$ (Fe is the symbol for iron). Although the hemoglobin molecule is enormous compared to a glucose or triolein molecule, it is only a small- to medium-sized protein. Many, indeed most, of the proteins within a cell are **enzymes,** biological catalysts which control the rates of the many chemical processes of the cell.

Protein molecules are made of simpler components, the **amino acids,** some 30 or more of which are known. Since each protein contains hundreds of amino acids, present in a certain proportion and in a particular order, an almost infinite variety of protein molecules is possible. Analytical tools such as the amino acid

analyzer permit one to determine the sequence of the amino acids in a given protein molecule. **Insulin,** the hormone secreted by the pancreas and used in the treatment of diabetes, was the first protein whose structure was elucidated. Work culminating in 1957 revealed the sequence of the 124 amino acids that compose the molecules of the enzyme ribonuclease, secreted by the pancreas.

It is possible to distinguish several different levels of organization in the protein molecule. The first level is the so-called **primary structure** which depends upon the sequence of amino acids in the **polypeptide chain.** This sequence, as we shall see, is determined in turn by the sequence of nucleotides in the RNA and DNA of the nucleus of the cell. A second level of organization of protein molecules involves the coiling of the polypeptide chain into a helix or into some other regular configuration. The polypeptide chains ordinarily do not lie out flat in a protein molecule, but they undergo coiling to yield a three-dimensional structure. One of the common secondary structures in protein molecules is the α-**helix,** which involves a spiral formation of the basic polypeptide chain. The α-helix is a very uniform geometric structure with 3.6 amino acids occupying each turn of the helix. The helical structure is determined and maintained by the formation of hydrogen bonds between amino acid residues in successive turns of the spiral.

A third level of structure of protein molecules involves the folding of the peptide chain upon itself to form globular proteins. Again, weak bonds such as hydrogen, ionic and hydrophobic bonds form between one part of the peptide chain and another part, so that the chain is folded in a specific fashion to give a specific overall structure of the protein molecule. Covalent bonds such as disulfide bonds (—S—S—) are important in the tertiary structure of many proteins. The biological activity of a protein depends in large part on the specific tertiary structure which is held together by these bonds. When a protein is heated or treated with any of a variety of chemicals, the tertiary structure is lost. The coiled peptide chains unfold to give a random configuration accompanied by a loss of the biological activity of the protein. This change is termed **denaturation.**

Those proteins comprising two or more subunits have a quaternary structure. This refers to the combination of two or more like or unlike peptide chain subunits, each with its own primary, secondary and tertiary structures, to form the biologically active protein molecule.

Each cell contains hundreds of different proteins and each kind of cell contains some proteins which are unique to it. There is evidence that each species of animal and plant has certain proteins which are different from those of all other species. The degree of similarity of the proteins of two species is a measure of their evolutionary relationship. Each species has a characteristic pattern of constituent proteins. This pattern differs at least slightly from that of related species and more markedly from those of more distantly related species. Because of the interactions of unlike proteins, grafts of tissue removed from one animal will usually not grow when implanted on a host of a different species, but degenerate and are sloughed off by the host. Indeed, even grafts made between members of the same species will usually not grow; only grafts between genetically identical donors and hosts—identical twins or members of a closely inbred strain—will grow.

Amino acids differ in the number and arrangement of their constituent atoms, but all contain an **amino group** (NH_2) and a **carboxyl group** (COOH). The amino group enables the amino acid to act as a base and combine with acids; the carboxyl group enables it to combine with bases. For this reason, amino acids and proteins are important biological "buffers" and resist changes in acidity or alkalinity. Amino acids are linked together to form proteins by **peptide bonds** between the amino group of one and the carboxyl group of the adjacent one (Fig. 2.6). Pure amino acids have a rather sweet taste. The proteins eaten by an animal are not incorporated directly into its cells but are first digested to the constituent

amino acids to enter the cell. Subsequently each cell combines the amino acids into the proteins which are characteristic of that cell. Thus, a human eats beef proteins in a steak, but breaks them down to amino acids in the process of digestion, then rebuilds them as human proteins—human liver proteins, human muscle proteins, and so on.

Proteins and amino acids may serve as energy sources in addition to their structural and enzymatic roles. The amino group is removed by an enzymatic reaction, **deamination,** and then the remaining carbon skeleton enters the same metabolic paths as glucose and fatty acids and eventually is converted to carbon dioxide and water by the Krebs tricarboxylic acid cycle (p. 72) and associated paths. The amino group is excreted as ammonia, urea, uric acid or some other nitrogenous compound, depending on the kind of animal. In prolonged fasting, after the supply of carbohydrates and fats has been exhausted, the cellular proteins may be used as a source of energy.

Animal cells can synthesize some, but not all, of the different kinds of amino acids; different species differ in their synthetic abilities. Humans, for example, are apparently unable to synthesize eight of these; they must either be supplied in the food eaten or perhaps synthesized by the bacteria present in the intestine. Plant cells apparently can synthesize all the amino acids. The ones which an animal cannot synthesize, but must obtain in its diet, are called **essential amino acids.** It must be kept in mind that these are no more essential for protein synthesis than any other amino acid, but are simply essential constituents of *the diet*, without which the animal fails to grow and eventually dies. Animals differ in their biosynthetic capabilities; what is an essential

Figure 2.6 Structural formulas of the amino acids glycine and alanine, showing, (a) the amino group and, (b) the acid (carboxyl) group. These are joined in a peptide linkage to form glycylalanine by the removal of water.

amino acid for one species may not be essential in the diet of another.

Nucleic Acids. Nucleic acids are complex molecules, larger than most proteins, and contain carbon, oxygen, hydrogen, nitrogen and phosphorus. They were first isolated by Miescher in 1870 from the nuclei of pus cells and gained their name from the fact that they are acidic and were first identified in nuclei. There are two classes of nucleic acid — one containing ribose and called **ribose nucleic acid** or **RNA,** and one containing deoxyribose and called **deoxyribonucleic acid** or **DNA.** There are many different kinds of RNA and DNA which differ in their structural details and in their metabolic functions. DNA occurs in the chromosomes in the nucleus of the cell and, in much smaller amounts, in mitochondria and chloroplasts. It is the primary repository for biological information. RNA is present in the nucleus, especially in the nucleolus, in the ribosomes, and in lesser amounts in other parts of the cell.

Nucleic acids are composed of units, called **nucleotides,** each of which contains a nitrogenous base, a five-carbon sugar and phosphoric acid. Two types of nitrogenous bases, purines and pyrimidines, are present in nucleic acids (Fig. 2.7). RNA contains the purines **adenine** and **guanine** and the pyrimidines **cytosine** and **uracil,** together with the pentose, ribose, and phosphoric acid. DNA contains adenine and guanine, cytosine and the pyrimidine **thymine,** together with deoxyribose and phosphoric acid. The molecules of nucleic acids are made of linear chains of nucleotides, each of which is attached to the next by bonds between the sugar part of one and the phosphoric acid of the next. The specificity of the nucleic acid resides in the specific sequence of the four kinds of nucleotides present in the chain. For example, CCGATTA might represent a segment of a DNA molecule, with C = cytosine, G = guanine, A = adenine and T = thymine.

An enormous mass of evidence now indicates that DNA is responsible for the specificity and chemical properties of the **genes,** the units of heredity. There are several kinds of RNA, each of which plays a specific role in the biosynthesis of specific proteins by the cell (p. 151).

Nucleotides and Coenzymes. Related structurally to nucleic acids but with quite different roles in cellular function are several mono- and dinucleotides. Each is composed of phosphoric acid, ribose and a purine or pyrimidine base, like the units composing the nucleic acids. Each of the bases may form a nucleoside triphosphate, with base, sugar and

Figure 2.7 Structural formulas of a purine, adenine; a pyrimidine, cytosine; and a nucleotide, adenylic acid. (From Villee, C. A.: Biology. 7th ed. Philadelphia, W. B. Saunders Co., 1977.)

Ribose — Phosphate — Phosphate — Ribose

Figure 2.8 Nicotinamide adenine dinucleotide (NAD). (From Villee, C. A.: Biology. 7th ed. Philadelphia, W. B. Saunders Co., 1977.)

three phosphate groups linked in a row. **Adenosine triphosphate,** abbreviated **ATP,** composed of adenine, ribose and three phosphates, is of major importance as the "energy currency" of all cells. The two terminal phosphate groups are joined to the nucleotide by energy-rich bonds, indicated by the symbol ~P. The biologically useful energy of these bonds can be transferred to other molecules; most of the chemical energy of the cell is stored in these ~P bonds of ATP, ready to be released when the phosphate group is transferred to another molecule. **Guanosine triphosphate, GTP,** is specifically required in certain steps in the synthesis of proteins; **uridine triphosphate, UTP,** is specifically required in certain steps in carbohydrate metabolism, e.g., glycogen synthesis; and **cytidine triphosphate, CTP,** is specifically required for certain steps in the synthesis of fats and phospholipids. All four nucleoside triphosphates are necessary for the synthesis of RNA, and the four deoxyribose nucleoside triphosphates—dATP, dGTP, dCTP and dTTP—are required for the synthesis of DNA.

Completing the list of nucleotides important in metabolic processes are the dinucleotides, NAD, NADP and FAD. **Nicotinamide adenine dinucleotide,** abbreviated **NAD** (and also called diphosphopyridine nucleotide, DPN), consists of nicotinamide, ribose and phosphate attached to an adenine nucleotide, phosphate, ribose and adenine (Fig. 2.8). NAD is of major importance as a primary electron and hydrogen acceptor in cellular oxidation reactions. Enzymes called **dehydrogenases** remove electrons and hydrogen from molecules such as lactic acid and transfer them to NAD, which in turn passes them on to other electron acceptors. **Nicotinamide adenine dinucleotide phosphate,** abbreviated **NADP** (and also called triphosphopyridine nucleotide, TPN), serves

as electron and hydrogen acceptor in certain other reactions. NADP is exactly like NAD except that it has a third phosphate group attached to the ribose of the adenine nucleotide. **Flavin adenine dinucleotide, FAD,** consists of riboflavin-ribitol-phosphate-phosphate-ribose-adenine and serves as a hydrogen and electron acceptor for certain other dehydrogenases. Notice that these dinucleotides have vitamins—nicotinamide or riboflavin—as component parts. These molecules, NAD, NADP and FAD, are termed **coenzymes;** they are cofactors required for the functioning of certain enzyme systems but are only loosely bound to the enzyme molecule and are readily removed. When they have accepted electrons and hydrogens, they are changed from their oxidized form, e.g., NAD, to their reduced form, NADH. They are converted back to the oxidized form when they transfer their electrons to the next acceptor in the chain of respiratory enzymes (p. 76).

In summary, the chemical components of the cells of animals and humans include water, inorganic salts and organic substances—nucleic acids, proteins, carbohydrates and fats, as well as many others. Carbohydrates and fats play some role in the structure of cells but are especially important as fuels; carbohydrates are readily available fuel, fats serve as long-term stores of energy. Nucleic acids have a primary role in storing and transmitting information. Proteins are structural and functional constituents of cells but may serve as fuel after deamination. The body can convert each of these substances into the others to some extent. The cell's contents compose a very complex, multicompartmental system. Many of the properties of cells—muscle contraction, ameboid motion and so on—involve changes in the biophysical state of this complex system.

ANNOTATED REFERENCES

Baker, J. J. W., and G. E. Allen: Matter, Energy and Life. 3rd ed. Reading, Mass., Addison-Wesley, .1974. A presentation of thermodynamic principles and their applications to studies of living systems.

Grunwald, E., and R. H. Johnsen: Atoms, Molecules and Chemical Change. Englewood Cliffs, N. J., Prentice-Hall, Inc., 1960. An introductory treatment of basic physics and chemistry.

White, E. H.: Chemical Background for the Biological Sciences. 2nd ed. Englewood Cliffs, N. J., Prentice-Hall, Inc., 1970. Interesting presentations of the structure of atoms and molecules and the nature of chemical reactions plus a variety of topics related to the subjects of this chapter.

For more detailed discussions of acids, bases, salts, and organic compounds and of the physical and chemical principles relating to life processes, consult one of the standard texts of college chemistry such as:

Jones, M. M., J. T. Netterville, D. O. Johnston and J. L. Wood: Chemistry, Man and Society. Philadelphia, W. B. Saunders Co., 1972.

Lee, G. L., H. O. Van Orden and R. O. Ragsdale: General and Organic Chemistry. Philadelphia, W. B. Saunders Co., 1971.

Watt, G. W., L. F. Hatch and J. J. Lagowski: Chemistry. New York, W. W. Norton & Co., 1964.

The Scientific American publishes excellent discussions of certain topics related to biology. Several hundred of these have been reprinted as "offprints" by Wm. Freeman Co., San Francisco. They are too numerous to cite individually but compose a rich source of collateral reading.

CELLS AND TISSUES

The living substance of all animals is organized into units called cells. Each cell contains a nucleus and is surrounded by a plasma membrane. Mammalian red blood cells lose their nucleus in the process of maturation, and a few types of cells such as those of skeletal muscles have several nuclei per cell, but these are rare exceptions to the general rule of one nucleus per cell. In the **protozoa** all of the living material is found within a single plasma membrane. These creatures may be considered to be unicellular, i.e., single-celled, or acellular, with bodies not divided into cells. Many protozoa have a high degree of specialization of form and function within this single cell (Fig. 3.1), and the single cell may be quite large, larger than certain multicellular organisms. Thus, it would be wrong to infer that a single-celled organism is necessarily smaller or less complex than a many-celled one.

3.1 THE CELL AND ITS CONTENTS

The term "cell" was applied by Robert Hooke, some 300 years ago, to the small, box-like cavities he saw when he examined cork and other plant material under the newly invented compound microscope. The important part of the cell, we now realize, is not the cellulose wall seen by Hooke, but the cell contents. In 1839 the Bohemian physiologist Purkinje introduced the term "protoplasm" for the living material of the cell. As our knowledge of cell structure and function has increased, it has become clear that the living contents of the cell compose an incredibly complex system of heterogeneous parts. Purkinje's term "protoplasm" has no clear meaning in a chemical or physical sense, but it may be used to refer to all the organized constituents of a cell.

Figure 3.1 Diagrams of an ameba (left) and *Epidinium* (right) to illustrate the range in complexity of the single-celled animals.

In this same year, 1839, a German botanist, Matthias Schleiden, and Theodor Schwann, his fellow countryman and a zoologist, formulated the generalization which has since developed into the **cell theory:** The bodies of all plants and animals are composed of cells, the fundamental units of life. The cell is both the structural and functional unit in all organisms, the fundamental unit possessing all the characteristics of living things. A further generalization, first clearly stated by Virchow in 1855, is that new cells can come into existence only by the division of previously existing cells. The corollary of this, that all cells living today can trace their ancestry back to the earliest living things, was stated by August Weismann about 1880.

The bodies of higher animals are made of many cells, differing in size, shape and functions. A group of cells which are similar in form and specialized to perform one or more particular functions is called a **tissue.** A tissue may contain nonliving cell products in addition to the cells themselves. A group of tissues may be associated into an **organ,** and organs into **organ systems.** For example, the vertebrate digestive system is composed of a number of organs: esophagus, stomach, intestine, liver, pancreas, and so on. Each organ, such as the stomach, contains several kinds of tissue—epithelium, muscle, connective tissue, nerves—and each tissue is made of many, perhaps millions, of cells.

A single-celled animal placed in the proper environment will survive, grow and eventually divide. For most single-celled animals, a drop of sea water or pond water will provide the environment required. It is more difficult to culture cells removed from a multicellular animal—a man, chick or frog. This was first accomplished in 1907 by Ross Harrison of Yale, who was able to grow cells from a salamander in a drop of nutrient medium containing blood plasma. Since then, many different kinds of cells from animals and plants have been cultured in vitro,* and many important facts about cell physiology have been revealed in this way.

3.2 THE PLASMA MEMBRANE

The outer surface of each cell is bounded by a delicate, elastic covering, an integral functional part of the cell, termed the **plasma membrane.** This is of prime importance in regulating the contents of the cell, for all nutrients entering the cell and all waste products or secretions leaving it must pass through this membrane. It hinders the entrance of certain substances and facilitates the entrance of others. Cells are almost invariably surrounded by a watery medium. This might be the fresh or salt water in which a small organism lives, the tissue sap of a higher plant, or the plasma or extracellular fluid of one of the higher animals.

The plasma membrane behaves as though it has ultramicroscopic pores through which certain substances pass. The size of these pores determines the maximal size of the molecules that can pass through the membrane. Factors other than molecular size, such as the electrical charge, if any, carried by the particles, the number of water molecules, if any, bound to the surface of the particle, and the solubility of the particle in lipids, may also be important in determining whether or not the substance will pass through the membrane. Infoldings of the plasma membrane may be continuous with channels that extend deep into the interior of the cell, providing pathways for the entrance of some materials and for the removal of secretory and excretory products.

All biological membranes appear to have a basically similar **lipid bilayer** structure. The central region of the membrane consists of two layers of phospholipid, each just one molecule thick, with their hydrophobic (water-repelling) tails aligned toward each other and with their polar head groups on the outside. The lipid bilayer is a mixture of phospholipid molecules in the fluid state. Special **membrane proteins** are associated with the lipid bilayer, some present only on the outer surface and others only on the inner surface. Some are found only *within* the membrane and still others extend completely *through* the lipid bilayer. Some of these membrane proteins are enzymes; others are receptors for hormones or other specific compounds. Certain of the proteins can move laterally within the membrane but are unable to rotate (Fig. 3.2).

The plasma membrane is much more than a simple cell envelope that prevents the ready movement of dissolved materials in either direction. It is an active, functional structure with enzymatic mechanisms that move specific molecules in or out of the cell against a concentration gradient. Multiple small evaginations of the plasma membrane, termed **microvilli,** are present in certain cells, such as those lining the kidney tubules (p. 353).

*In vitro, Latin *in glass.* The cells are removed from the animal body and incubated in glass vessels.

Figure 3.2 Diagram of the molecular architecture of biological membranes such as the plasma membrane. (From Villee, C. A.: Biology. 7th ed. Philadelphia, W. B. Saunders Co., 1977.)

High resolution electron micrographs of the plasma membrane (Fig. 3.3) show a dense-light-dense three-layered structure. The plasma membranes of animal, plant and bacterial cells and the membranes of many subcellular organelles all appear to have a similar three-layered structure.

Nearly all plant cells (but not most animal cells) have a thick **cell wall** made of cellulose that lies outside the plasma membrane. This cell wall is nonliving, secreted by the cell substance. It is pierced in many places with tiny holes through which the contents of one cell connect with the contents of the adjacent cell and through which materials can pass from one cell to the next. These tough, firm cell walls provide support for the plant body.

3.3 THE NUCLEUS AND ITS FUNCTIONS

The **nucleus** of the cell is usually spherical or ovoid. It may have a fixed position in the center of the cell or at one side, or it may be moved around as the cell moves and changes shape. The nucleus is separated from the cytoplasm by a nuclear membrane which controls the movement of materials into and out of the nucleus (Fig. 3.4). The electron microscope reveals that the nuclear membrane is double-layered and that there are extremely fine channels through the nuclear membrane through which the nuclear contents and cytoplasm are continuous (Fig. 3.5) and through which even large molecules of RNA may pass.

Figure 3.3 Electron micrograph of the cell membranes of intestinal cells showing the three-layered structure. Membrane, m; intercellular space, *is*. Magnification × 240,000. (From Villee, C. A.: Biology. 7th ed. Philadelphia, W. B. Saunders Co., 1977.)

Figure 3.4 *A*, Electron micrograph of the nucleus and surrounding cytoplasm of a frog liver cell. The spaghetti-like strands of the endoplasmic reticulum are visible in the lower right corner. Magnified 16,500 times. *B*, High power electron micrograph of mitochondria and endoplasmic reticulum within a rat liver cell. Granules of ribonucleoprotein (ribosomes) are seen on the strands of endoplasmic reticulum and structures with double membranes are evident within the mitochondria in the upper left corner and on the right. Magnified 65,000 times. (Electron micrographs courtesy of Dr. Don Fawcett.) (From Villee, C. A.: Biology. 7th ed. Philadelphia, W. B. Saunders Co., 1977.)

The nucleus is an important center for the control of cellular processes and is required for growth and for cell division. An ameba can survive for some days after the nucleus has been removed by a microsurgical operation, but it cannot grow and eventually dies. To demonstrate that it is the absence of the nucleus, not the operation itself, that causes the

Figure 3.5 Electron micrographs showing pores in the nuclear membrane. The endoplasmic reticulum is evident in both pictures; *B*, shows the ribosomes on the endoplasmic reticulum. *A*, Magnified 20,000 times; *B*, magnified 50,000 times. (Electron micrographs courtesy of Drs. Don W. Fawcett and Keith R. Porter.)

ensuing death, one can perform a **sham operation.** A microneedle is inserted into an ameba and moved around inside the cell to simulate the operation of removing the nucleus, but the needle is withdrawn without actually removing the nucleus. An ameba subjected to this sham operation will recover, grow and divide. A controlled experiment such as this, in which two amebas are subjected to the same operative trauma and the one with the nucleus lives whereas the one without the nucleus

dies, provides strong evidence of the vital role of the nucleus in regulating the metabolic processes that underlie growth and cell division.

A classic series of experiments demonstrating the role of the nucleus in the control of cell growth was performed by Hämmerling with the single-celled plant *Acetabularia mediterranea.* This marine alga, 4 to 5 cm. long, is mushroom-shaped, with "roots" and a stalk surmounted by a flattened, disc-shaped um-

Figure 3.6   Hämmerling's experimental demonstration of the production of an umbrella-regenerating substance by the nucleus of the alga, *Acetabularia.* Lower line, when a stalk from *A. crenulata* is grafted onto the base (white) of an *A. mediterranea* plant, the stalk regenerates an umbrella whose shape is that characteristic of *A. mediterranea* plants. (From Villee, C. A.: Biology. 7th ed. Philadelphia, W. B. Saunders Co., 1977.)

brella. The single nucleus is located near the base of the stalk. Hämmerling cut across the stalk (Fig. 3.6) and found that the lower part containing the nucleus could live and regenerate an umbrella. The upper part eventually died without regenerating a stalk and roots. In further experiments, Hämmerling first severed the stalk just above the nucleus, then made a second cut just below the umbrella. The section of stalk thus isolated, when replaced in sea water, was able to grow a partial or complete umbrella. This might seem to show that a nucleus is not necessary for regeneration. However, when Hämmerling cut off this second umbrella the stalk was unable to form a new one. From experiments such as these, Hämmerling concluded that the nucleus supplies some substance necessary for umbrella formation. This substance passes up the stalk and instigates umbrella growth. In the experiments described here, some of this substance remained in the stalk after the initial cuts, enough to produce one new umbrella. After that amount of "umbrella substance" was exhausted by the regeneration of an umbrella, no second regeneration was possible in the absence of a nucleus. What hypothesis would you make about the nature of "umbrella substance"? How would you go about testing your hypothesis?

A second species, *Acetabularia crenulata*, has a branched instead of a disc-shaped umbrella. When a piece of *crenulata* stalk (without a nucleus) was grafted onto the base of a *mediterranea* plant (containing a *mediterranea* nucleus) a new umbrella developed at the top of the stalk. The shape of the umbrella was determined not by the species of the stalk but by the species of the base (Fig. 3.6, lower part). The nucleus, through the action of its genes, provided the specific information that controlled the type of umbrella that was regenerated and could override the tendency of the stalk to form an umbrella characteristic of its own species.

When a cell has been killed by fixation with the proper chemicals and then stained with the appropriate dyes, several structures are visible within the nucleus (Fig. 3.7). Within the semifluid ground substance are suspended a fixed number of extended, linear, threadlike bodies called **chromosomes,** composed of DNA and proteins and containing the units of heredity, the **genes.** In a stained section of a nondividing cell the chromosomes typically appear as an irregular network of dark-staining threads and granules termed **chromatin.** Just prior to nuclear division these strands condense into compact, rod-shaped chromosomes which are subsequently distributed to the two daughter cells in exactly equal numbers. Each type of organism has a characteristic

Figure 3.7   Tissue sections of human adrenal gland stained to show cellular details; left, magnified 600 times; right, magnified 1500 times. (Courtesy of Dr. Kurt Benirschke.)

number of chromosomes present in each of its constituent cells. The fruit fly has 8 chromosomes, sorghum has 10, the garden pea, 14, corn, 20, the toad, 22, the tomato, 24, the cherry, 32, the rat, 42, humans, 46, the potato, 48, the goat, 60, and the duck, 80. The somatic cells of higher plants and animals each contain two of each kind of chromosome.

The 46 chromosomes in each human cell include two of each of 23 different kinds. They differ in their length, shape and the presence of knobs or constrictions along their length. In most species the morphological features of the different chromosomes are distinctive and permit the cytologist to distinguish the different pairs.

A cell with two complete sets of chromosomes is said to be **diploid.** Sperm and egg cells, which have only one of each kind of chromosome, one full set of chromosomes, are said to be **haploid.** They have just half as many chromosomes as the somatic cells of that same species. When the egg is fertilized by the sperm the two haploid sets of chromosomes are joined and the diploid number is restored.

The nucleus also contains one or more small spherical bodies, **nucleoli,** which are difficult to see in a living cell with an ordinary light microscope but are evident by phase microscopy. The cells of any particular animal have the same number of nucleoli. The nucleolus disappears when a cell is about to divide and reappears after division is complete. If the nucleolus is destroyed by carefully localized ultraviolet or x-irradiation, cell division is inhibited. This does not occur in control experiments in which regions of the nucleus other than the nucleolus are irradiated. The nucleolus plays a key role in the synthesis of the ribonucleic acid constituents of ribosomes.

3.4 CYTOPLASMIC ORGANELLES

Two small, dark-staining cylindrical bodies, called **centrioles,** are found in the cytoplasm near the nucleus of animal cells. With the electron microscope each centriole is revealed to be a hollow cylinder with a wall in which are embedded nine parallel, longitudinally oriented groups of microtubules, with three tubules in each group. The cylinders of the two centrioles are typically oriented with their long axes perpendicular to each other.

When cell division begins, the centrioles move to opposite sides of the cell. From each centriole there extends a cluster of raylike filaments called an **aster,** and between the separating centrioles a **spindle** forms, composed of threads of protein with properties similar to those of the contractile proteins in muscle, actin and myosin.

Those cells that bear cilia on their exposed surfaces have a structure at the base of each cilium, the **basal body,** that resembles the centriole in the presence of nine parallel tubules. Each cilium contains nine peripherally located longitudinal filaments and two centrally located ones. Like centrioles, basal bodies can duplicate themselves.

The cytoplasm may contain droplets of fat and crystals or granules of protein or glycogen which are simply stored for future use. In addition, it contains the metabolically active cell organelles, **mitochondria, endoplasmic reticulum** and **Golgi bodies.**

Mitochondria range in size from 0.2 to 5 micrometers and in shape from spheres to rods and threads (Fig. 3.4). Their number may range from just a few to more than 1000 per cell. When living cells are examined, their mitochondria appear to move, change shape and size, fuse with other mitochondria to form longer structures, or cleave to form smaller ones. Each mitochondrion is bounded by a double membrane, an outer smooth membrane and an inner one folded into parallel plates that extend into the central cavity and may fuse with folds from the opposite side. Each of these inner and outer membranes is composed of a bilayer of phospholipid molecules with associated membrane proteins. The shelf-like inner folds, termed **cristae,** contain the enzymes of the electron transmitter system, of prime importance in converting the potential energy of foodstuffs into biologically useful energy for cellular activities. The semifluid material within the inner compartment, the matrix, contains certain enzymes of the Krebs tricarboxylic acid cycle (p. 72). The mitochondria, whose prime function is the release of biologically useful energy, have been dubbed the "powerhouses" of the cell.

Cells, such as those of the pancreas that are especially active in protein synthesis, are crowded with the membranous labyrinth of the endoplasmic reticulum (Fig. 3.8); other cells may have only a scanty supply of such membranes. Two types are found, **granular** or **rough endoplasmic reticulum** to which are bound many **ribosomes,** and **agranular** or **smooth endoplasmic reticulum,** consisting of membranes alone. Bound to the granular reticulum or free in the cell matrix are many **ribosomes,** ribonucleoprotein particles on which proteins are synthesized. Both smooth and rough endoplasmic reticulum may be

Figure 3.8 Electron micrograph of the granular endoplasmic reticulum of a pancreatic acinar cell. This form of the reticulum consists of parallel arrays of broad flat sacs, or cisternae. The outer surface of their limiting membranes is studded with particles (15 nm.) of ribonucleoprotein (ribosomes). Osmium tetroxide fixation, × 85,000. (From Villee, C. A.: Biology. 7th ed. Philadelphia, W. B. Saunders Co., 1977.)

found in the same cell. The agranular endoplasmic reticulum may play some role in the process of cellular secretion. The tightly packed sheets of endoplasmic reticulum may form tubules some 50 to 100 nanometers in diameter. In other regions of the cell, the cavities of the endoplasmic reticulum may be expanded, forming flattened sacs called **cisternae.** The membranes of the endoplasmic reticulum divide the cytoplasm into a multitude of compartments in which different groups of enzymatic reactions may occur. The endoplasmic reticulum serves a further function as a system for the transport of substrates and products through the cytoplasm to the exterior of the cell and to the nucleus.

After mitochondria have been sedimented from homogenized cells by centrifugation, a heterogeneous group of smaller particles, termed **microsomes,** can be sedimented by centrifuging at about 100,000 times the force of gravity. Ribosomes can be separated from the rest of the microsomal fraction by treatment with appropriate detergents. Isolated ribosomes can then synthesize proteins in vitro

if supplied with the appropriate instructions in the form of messenger RNA and with an assortment of amino acids, an energy source and the other enzymes and transfer RNA's required (see Chapter 7). Ribosomes are ubiquitous, occurring in all kinds of cells from bacteria to higher plants and animals. Ribosomes contain RNA and protein and are composed of two nearly spherical subunits, which are combined to form the active protein synthesizing unit (Fig. 3.9). Ribosomes are synthesized in the nucleus and pass to the cytoplasm where they are active in synthesizing proteins. Ribosomes may be bound to the membranes of the endoplasmic reticulum, or they may be free in the matrix of the cytoplasm. In many cells, clusters of five or six ribosomes, termed **polysomes,** appear to be the functional unit that is effective in protein synthesis. It is estimated that a bacterial cell, such as *Escherichia coli,* contains some 6000 ribosomes and that the rabbit reticulocyte (precursor of the red blood cell) contains some 100,000 ribosomes. Ribosomes are remarkably uniform in their size, structure and composition, whether

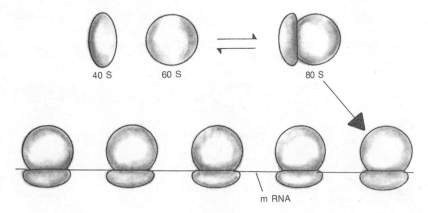

40 S 60 S 80 S

m RNA

Polysome

Figure 3.9 Diagram illustrating the assembly of the two subunits 40 S and 60 S to form a ribosome, 80 S, the subcellular organelle on which proteins are synthesized. The 80 S ribosomes attach to a messenger RNA strand to form a polysome. (From Villee, C. A.: Biology. 7th ed. Philadelphia, W. B. Saunders Co., 1977.)

isolated from bacterial or mammalian sources. They contain protein and RNA in nearly equal proportions, with little or no lipid. The protein has a fairly high content of basic amino acids.

One of the subunits of the ribosomes has a molecular weight of about 1,300,000 and the other has a molecular weight of about 600,000. The protein components of ribosomes from different cells are remarkably similar in their amino acid composition; however, the nucleotide composition of the RNA of ribosomes from different species varies considerably.

The microsomal fraction contains, in addition to the ribosomes involved in the synthesis of polypeptide chains, several less well characterized particles that contain enzymes involved in the metabolism of other types of compounds.

Golgi bodies, found in all cells except nature sperm and red blood cells, consist of an irregular network of canals. In the electron microscope (Fig. 3.10) Golgi bodies appear as parallel arrays of membranes without granules. The canals may be distended in certain regions to form small vesicles or vacuoles filled with material. The Golgi bodies are usually concentrated in the part of the cytoplasm near the centrioles and appear to serve as temporary storage places for proteins and other compounds synthesized in the endoplasmic reticulum. There is evidence to suggest that proteins made in the cisternae of the endoplasmic reticulum are sealed off in little packets of endoplasmic reticulum and travel to the Golgi bodies. There they are repackaged

in large sacks made of membranes from the Golgi bodies. In these packets they move to the plasma membrane, which fuses with the membrane of the vesicle, opening the vesicle and releasing its contents to the exterior of the cell.

Many, if not most, cells have hollow cylindrical cytoplasmic subunits termed **microtubules** (Fig. 3.11), which appear to be important in maintaining or controlling the shape of the cell. Microtubules play a role in such cellular movements as that of chromosomes by the mitotic spindle and serve as channels for the oriented flow of cytoplasmic constituents within the cell. Microtubules are also the major structural components of cilia and flagella.

Microtubules are composed of several types of proteins; **tubulin** is present in the largest amount. Tubulin is made of unlike α and β subunits that differ in their amino acid composition. Microtubules are long, hollow cylinders with a wall thickness of 4.5 to 7 nm. and diameters ranging from 20 to 30 nm. The thickness of the wall corresponds to the dimensions of the subunits and suggests that the walls are one molecule thick. Microtubules can grow by the addition of more α and β subunits or can shorten by the disassembly of subunits. The microtubules of nerve axons have a part in the rapid transport of proteins and other molecules such as the hormones of the posterior pituitary and the hypothalamic releasing factors down the axon to the tip.

Solid cytoplasmic **filaments** are present in some cells in addition to the hollow micro-

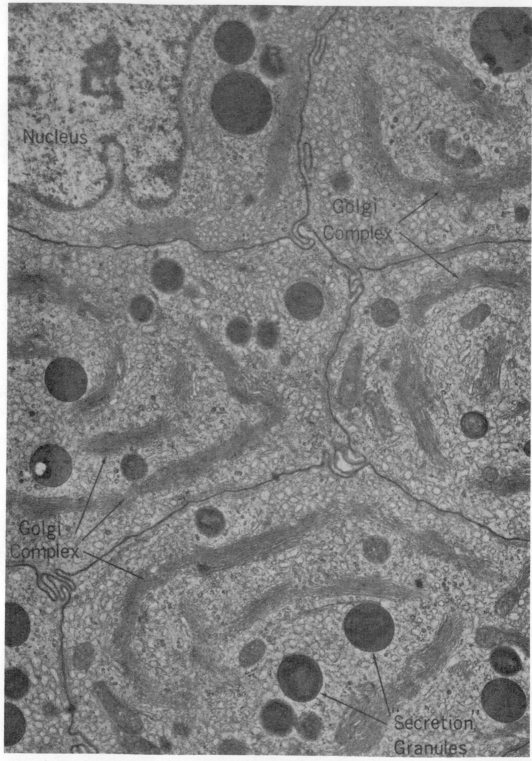

Figure 3.10 Electron micrograph of rabbit epididymis showing the extensive Golgi complex, evident in the parallel arrays of the membranes. Magnification × 9500. (From Fawcett, D. W.: The Cell. Philadelphia, W. B. Saunders Company, 1966.)

Figure 3.11 Electron micrographs of cytoplasmic microtubules. *A*, Microtubules of the mitotic spindle seen here in longitudinal section. The chromosomes are at the lower right. × 70,000. (Courtesy E. Roth.) *B*, The microtubules in transverse section present circular profiles. Those shown here are from the manchette of a mammalian spermatid. × 140,000. (From Villee, C. A., and V. G. Dethier: Biological Principles and Processes. Philadelphia, W. B. Saunders, 1972.)

tubules. These protein filaments play additional roles in cell structure and motion. The cytoplasm of the skeletal muscle fibers contains many long myofibrils, protein filaments that participate in muscle contraction.

Lysosomes are intracellular organelles about the size of mitochondria but less dense. They are membrane-bounded structures that contain a variety of enzymes capable of hydrolyzing the macromolecular constituents of the cell. In the intact cell these enzymes are segregated within the lysosome, presumably to prevent their digesting the contents of the cell. Rupture of the lysosome membrane releases the enzymes and accounts, at least in part, for the lysis of dead cells and the resorption of cells such as those in the tail of a tadpole during metamorphosis. Since they contain enzymes that can hydrolyze the major cellular constituents when they rupture and release them, the lysosomes have been termed "suicide bags" by the Belgian biochemist Christian De Duve.

The cytoplasm of certain cells, chiefly those of lower animals, contains **vacuoles,** bubble-like cavities filled with fluid and separated from the rest of the cytoplasm by a vacuolar membrane. Most protozoa, and the endoderm cells of cnidaria and flatworms, have **food vacuoles** in which food is digested. Digestive enzymes are secreted from the cytoplasm into the cavity of the vacuole, the food is digested, and the products of digestion are absorbed through the vacuolar membrane into the cytoplasm. The protozoa living in fresh water have the problem of eliminating the water which enters the cell constantly by osmosis (p. 55). These forms have evolved **contractile vacuoles,** which alternately fill with water from the adjacent cytoplasm and then eject the water to the surrounding environment.

Most animal cells are quite small, too small to be seen with the naked eye. The diameter of the human red blood cell is about 7.5 micrometers, but most animal cells have diameters ranging from 10 to 50 micrometers. There are a few species of giant amebas with cells about 1 mm. in diameter. The largest cells are the yolk-filled eggs of birds and sharks. The egg cell of a large bird such as a turkey or goose may be several centimeters across. Only the yolk of a bird's egg is the true egg cell; the egg white and shell are noncellular material secreted by the bird's oviduct as the egg passes through it.

The limit of the size of a cell is set by the physical fact that, as a sphere gets larger, its surface increases as the square of the radius but its volume increases as the cube of the radius. The metabolic activities of the cell are roughly proportional to cell volume. These activities require nutrients and oxygen, and release carbon dioxide and other wastes which must enter and leave the cell through its surface. The upper limit of cell size is reached when the surface area can no longer provide for the entrance of enough raw materials and the exit of enough waste products for cell metabolism to proceed normally. The limiting size of the cell will depend on its shape and its rate of metabolism. When this limit is reached the cell must either stop growing or divide.

Much has been learned in recent years of the role each of these particles plays in the economy of the cell. Cells are homogenized in special glass grinding tubes to break the cell membrane and release the intracellular structures. Then, by subjecting the homogenate to increasing amounts of centrifugal force in an ultracentrifuge, first the nuclei, then the mitochondria, and finally the microsomes can be sedimented separately. When these sedimented particles are examined in the electron microscope, they are found to have the same structure exhibited by comparable structures in the intact cell. The separated particles can then be suspended in suitable incubation media and their metabolism can be studied. Such separated mitochondria and microsomes will carry out many biochemical reactions, and much is now known about the functions of each of these particles. The liquid left after the homogenate has been subjected to high centrifugal force to sediment the microsomes contains many other enzymes which apparently exist in the cell more or less free in the ground substance.

Where, you may ask, is life localized—in the mitochondria? in the microsomes? in the ground substance? The answer, of course, is that life is not a function of any single one of these parts, but of the whole integrated system of many component parts, organized in the proper spatial relationship and interdependent on one another in a great variety of ways.

3.5 THE CELL CYCLE

The doubling of all the constituents of the cell followed by its division into two daughter cells is generally termed the **cell cycle.** A cell is "born" when its parent cell divides; it undergoes a cycle of growth and division and gives rise to two daughter cells. The doubling and halving processes in a population of cells are approximate rather than exact. For example, the size of connective tissue cells that are about to undergo division varies by about 12 per cent around the mean size. The size of the cell is limited in some way and a nondividing cell stops growing. It had been believed that growth was limited by the relative masses of nucleus and cytoplasm. This theory, proposed by Von Hertwig in 1908, suggested that when the ratio of the masses of nucleus and cytoplasm reaches a certain value the cell becomes unstable and cell division is triggered. It does seem to be true that the size of the cell is limited by the capacity of the single cell nucleus to support growth. If a cell is set up artificially containing two nuclei or two sets of chromosomes in a single nucleus, the cell can grow to twice the normal adult size. Cells exposed to an appropriate dose of x-rays can go through the cell cycle, but are unable to undergo division. The amount of genetic material in the nucleus increases and so does the size of the entire cell.

The cell cycle in a typical plant or animal cell requires about 20 hours for completion. Only an hour or so is devoted to mitosis; the rest of the time is required for interphase growth. Under optimal conditions of nutrition and temperature, the length of the cell cycle for any given kind of cell is constant. Under less favorable conditions it may be slowed, but it has not been possible to speed up the cell cycle and make cells grow faster. From this we infer that the duration of the cell cycle is the time required for carrying out some precise program that has been built into each cell. This program appears to include two parts: one having to do with the replication of the genetic material in the chromosomes and the other involving the doubling of all of the other constituents of the cell involved in growth.

The fertilized eggs of animals are unusual in that they rapidly undergo repeated divisions that may be completed in an hour or less. An egg cell is generously endowed with all of the kinds of molecules produced by growing cells and it divides repeatedly, yielding smaller and smaller cells. Its cell cycle does not involve growth, but simply preparation for division and the division process itself.

It was realized about 1950 that chromosomes undergo replication during the interphase and simply separate during mitotic division. Experiments using tritiated thymidine showed that DNA is synthesized during the interphase. The period of DNA replication during the interphase is termed the **S phase.** The time between mitotic division and the beginning of DNA replication is termed the **G_1 phase,** or "gap 1" phase. Following the completion of the S phase, the cell is usually not ready to divide immediately, but undergoes a second gap phase, the **G_2 phase.** The completion of G_2 is marked by the beginning of mitotic division, the **M phase.** In a typical cell cycle, the G_1 phase occupies about eight hours, the S phase six hours, the G_2 phase four and one-half hours, and the M phase one hour. Although the duration of each phase may vary somewhat, the greatest variation is found in the G_1 phase. Cells with very long cell cycles show most of the prolongation in the G_1 phase. Cells with very short cell cycles, such as eggs, have no measurable G_1 phase. In essentially all mammalian cells the S phase lasts six to eight hours and the G_2 phase three to five hours. The G_1 phase may be as short as a few hours or minutes or as long as several days or weeks. During the G_1 phase certain key processes occur that make it possible for the cell to enter the S phase and become committed to a future cell division (Fig. 3.12). In the S phase there is not simply a doubling of the amount of DNA in the nucleus, but an exact replication of each chromosome. The human cell contains 46 threads of DNA with a total length of 2 meters or more, all of which are stuffed into a nucleus about 5 micrometers in diameter. In the com-

plex process of replication an exact copy is made of each of these 46 threads. Replication does not simply begin at one end of each thread and travel along to the other end; rather, each thread undergoes replication in many segments according to a definite program. The segments do not replicate in tandem in any one chromosome, nor does any one chromosome complete its replication before the next chromosome begins. When the last segments have been replicated, DNA synthesis shuts down and does not resume until the next cycle. The G_2 phase includes the final steps in the cell's preparation for division, during which there is increased protein synthesis.

Some of the enzymes of the cell are synthesized at a constant rate throughout the cell cycle. A second group, the "step" enzymes, undergo abrupt increases at certain stages in the cell cycle, with different enzymes increasing at different stages. A third group, "peak" enzymes, increase at certain points in the cycle and thereafter decrease as though they were meant to carry out a particular job at a particular time and subsequently disappear. The enzymes involved in DNA synthesis are peak enzymes that appear only during the S phase. These observations imply that within the cell there are different programs for producing enzymes at certain times in the cell cycle. One of the basic dogmas of molecular genetics is that the production of an enzyme is the end result of a read-out of the gene for that enzyme. That there is a program for the production of enzymes during the cell cycle suggests that there is a program for the read-out of the genes for those enzymes. The reproducibility of the time table of the phases of the cell cycle suggests that certain genes determine the progression of the cell cycle from one phase to the next.

Scanning electron microscopy reveals that cells grown in culture have quite different external appearances during the different phases (Fig. 3.13). During the M phase the cells are spherical and not strongly attached to the substrate. As they enter the G_1 phase they begin to flatten, the cell surface shows bubble-like blisters and microvilli appear. Later in the G_1 phase the margins of the cells become thin and active and appear ruffled. During the S phase the cells flatten further and their surface becomes smooth. As the cell enters the G_2 phase it once again thickens, the surface shows ruffles and microvilli but not as many surface blisters as during the G_1 phase. Two important transition points occur in the cell cycle: one between the G_1 phase and the S phase when

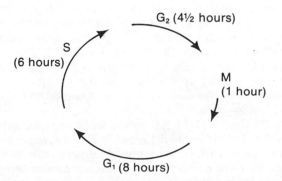

Figure 3.12 Diagram of the cell cycle showing durations of the four phases in a typical cell. (From Villee, C. A.: Biology. 7th ed. Philadelphia, W. B. Saunders Co., 1977.)

A

Figure 3.13 Hamster ovary cells in mitosis. A, The shape of the cell is changing from flat to round. A number of long, thin projections covering the cell surface help to secure the cell to the substrate. B, Flattened cells in late G_1 phase; the surfaces are covered with microvilli and blisterlike spheres (blebs). Several cells are establishing contact with each other. C, Increased flattening of cells having entered the S phase; no blebs now remain and microvilli are less abundant. "Ruffles" (light areas) appear on the edges of the cells. (Scanning electron micrographs by Keith R. Porter, David M. Prescott and Jearl F. Frye. From Mazia, D.: The cell cycle. Scientific American 230:54, 1974. © 1974 by Scientific American, Inc. All rights reserved.)

B

C

the replication of the chromosomes starts, and the other between the G_2 phase and the M phase when the chromosomes condense and mitosis begins. Certain animal viruses, the **Sendai viruses,** produce a change in the cell membrane, allowing two or more adjacent cells to fuse into a single cell. The new cell contains within a single membrane the cytoplasm and the nuclei of all the cells undergoing fusion. Quite different kinds of cells — even cells from different species of animals — can be fused by the action of the Sendai viruses, forming hybrids that will not only carry out normal processes but also will reproduce as hybrids.

By producing hybrids of cells in different phases of the cell cycle, we can ask whether there are certain signals that move a cell from one phase of the cycle to the next. When cells in the G_1 phase are fused with cells in the S phase, nuclei of the G_1 cells begin to make DNA long before they normally would. Thus a cell in the S phase contains something that triggers DNA synthesis, something that is absent from the cell that has not completed the G_1 phase.

Further insight has come from experiments with hybrids of different cells having cell cycles of different lengths. Hamster cells normally have a shorter G_1 phase than mouse cells, but in hybrids of mouse and hamster cells the mouse nuclei have a short G_1 phase like the hamster nucleus, and both nuclei enter the

S phase simultaneously. This permits the inference that whatever turns on DNA synthesis in the hamster nucleus also turns it on in the mouse nucleus. Once the nuclei enter the S phase the nucleus of each species follows its own timetable and its own replication program. Those parts of the chromosomes that replicate first in the parent cell also replicate first in the hybrids, and those that normally replicate last in the parent are last in the hybrid. This suggests that the signal for starting the S phase does not exert any control over what happens once the phase has started. Rather, the replication program characteristic of each species must be contained in the nucleus of that species' cells. Thus there is something that pervades the cell and is responsible for initiating the replication of chromosomes. Moreover, it can induce any nucleus to enter the S phase whether or not it is "ready." This signal for beginning chromosome replication might be some specific molecule, or it might be some specific alteration in the internal environment of the cell.

Entrance into the M phase is characterized by the condensation of the chromosomes; they become packed into threads that are visible under the microscope. When cells in the M phase are fused with cells in the G_2 phase, the chromosomes in the G_2 nucleus undergo condensation (Fig. 3.14). The chromosomes are double structures; they had undergone replication in the preceding S phase. However, when an M phase cell is fused with a cell in the G_1 phase, thereby forcing the chromosomes in the G_1 nucleus to condense before replication, the chromosomes are seen to be single rather than double structures. When an M phase cell is fused with a cell in the S phase, the chromosomes in the S phase nucleus condense, but they condense in small fragments. Thus something in the M phase cell forces chromosomes to condense even when they are not ready to

Figure 3.14 Prematurely condensed chromosomes (PCC) of Chinese hamster ovary cells. Cells synchronized in mitosis were fused with interphase cells with the help of U.V. inactivated Sendai virus. Metaphase chromosomes are darkly stained and well condensed; PCC are lightly stained with variable morphology. A, G_1 cell was fused with a mitotic cell. G_1-PCC (arrow) have a single chromatid and are greatly spiralized. B, Heterophasic binucleate cell formed by fusion of a cell in S phase with one in mitosis. S-PCC appear to be fragmented because of uneven condensation of S phase chromatin, dependent upon the state of DNA replication. Some chromosome regions have replicated; others have not yet started replication or are in the active process of replication. Replicated elements consist of double chromatids (arrow); unreplicated regions are seen as single chromatids. Actual site of replication (replicon) in the chromatin does not undergo condensation; hence it appears as a gap between the condensed regions. C, Product of fusion between G_2 and mitotic cells. G_2-PCC (arrow) are long and slender, with the two chromatids still attached to each other. (Courtesy of Dr. Potu N. Rao, The University of Texas System Cancer Center, M. D. Anderson Hospital and Tumor Institute, Houston, Tex.)

do so. Whatever it is, it appears to work in any kind of cell, even when fused with the cell of an animal only remotely related. If one cell contains condensed chromosomes in the M phase it can force premature chromosome condensation in the other cell.

An organism such as a higher animal, which can be viewed as a society of cells, includes some cells that will reproduce and others that will not. The cells of tissues that perform certain special services—those of the nervous system and the muscular system—do not in general reproduce at all. In tissues such as the skin, the blood-forming system and epithelial linings, new cells are produced at a rate which just compensates for the continued loss of old cells. The cells involved in immune reactions and the healing of wounds multiply at a rapid rate as a bodily response to some external provocation. Finally, in malignant cancer the cells no longer abide by the rules of governance of the organism and go through their cycles in an anarchic fashion. There is nothing special about the cell cycle in cancer cells compared with the cell cycle in normal cells. The tempo, the cycle and its phases are the same, but the difference is that cancer cells repeat the cycle without restraint.

A very simple rule emerges from these studies. Cells that do not divide never enter the S phase, and conversely, cells that enter the S phase almost always complete that phase and go on to divide. Thus replication of the chromosomes is a strong commitment ultimately to undergo division, and the major control of cell division lies in whether or not a cell enters into replication. However, cells that are not going to divide are not cells that have become stuck somehow in the transition between the G_1 and the S phase. Nondividing cells are probably best considered as ones that have not entered the cycle at all. The many kinds of specialized cells in the body are noncycling cells, making no progress at all toward cell division. Many of the cells that have left the cell cycle can be made to reenter it by placing them under specific culture conditions in the laboratory. Presumably agents that cause cancer must in some way cause noncycling cells to enter the cycle. The conversion of lymphocytes from noncycling to cycling cells is of importance in the immune response. The cells begin to grow and divide, producing cells that contribute to the formation of antibodies. Noncycling lymphocytes can be transformed into cycling ones in the laboratory by exposing them to **lectins,** specific kinds of plant proteins. When noncycling lymphocytes are artifi-

cially stimulated by exposure to lectins, nearly 24 hours pass before they enter the S phase and begin to replicate the chromosomes. The delay apparently is due to the need of these stimulated lymphocytes to produce the enzymes required for the replication process as a preliminary to entering the S phase. Thus they are not stuck in the G_1 phase, ready to go into the S phase when they are signaled. When stimulated, a noncycling cell must first do the things that a cycling cell does in the G_1 phase before it can enter the S phase and replicate its chromosomes.

3.6 MITOSIS

Because of the limitation on the size of individual cells, growth is accomplished largely by an increase in the *number* of cells. When a single-celled protozoan divides, the resulting two cells are separate individuals, members of a new generation. In multicellular animals, cell division results in an increase in the number of cells per individual. The process of cell division, called **mitosis,** is extremely regular and ensures the qualitatively and quantitatively equal distribution of the hereditary factors between the two resulting daughter cells. Mitotic divisions occur during embryonic development and growth; in the replacement of cells that wear out, such as blood cells, skin, the intestinal lining, and so on; and in the repair of injuries.

When a dividing cell is stained and examined under the microscope, dark-staining **chromosomes** are visible within the nucleus. Each consists of a central thread, the **chromonema** along which lie the **chromomeres**—small, beadlike, dark-staining swellings. It has been suggested that the chromomeres are, or contain, the genes, for breeding experiments have shown clearly that these hereditary units lie within the chromosome in a linear order. However, the correlation between chromomeres and genes is not regular; some chromomeres contain several genes and some genes have been located between chromomeres. Each chromosome has, at a fixed point along its length, a small clear circular zone called a **centromere** that controls the movement of the chromosome during cell division. As the chromosome becomes shorter and thicker just before cell division occurs, the centromere region becomes accentuated and appears as a constriction.

Cell division must be an extremely exact process to ensure that each daughter cell receives exactly the right number and kind of chromosomes. If we tamper experimentally with the mechanism of cell division, and the resulting cells receive more or less than the proper number of chromosomes, marked abnormalities of growth, and perhaps the death of these cells, will follow. Mitosis may be defined as the regular process of cell division by which each of the two daughter cells receives exactly the same number and the same kind of chromosomes that the parent cell contained. Although each chromosome appears to split longitudinally into two halves, each original chromosome has brought about the synthesis of an exact replica of itself immediately adjacent to it during the preceding S phase. The old and new chromosomes are identical in structure and function and at first lie so close to one another that they appear to be one. As mitosis proceeds and the chromosomes contract, the line of cleavage between them becomes visible. When the process is complete, the original and the new chromo-

somes separate and become incorporated into different daughter cells. The role of the complicated mitotic machinery is to separate the "original" and "replica" chromosomes and deliver them to opposite ends of the dividing cell so they will become incorporated into different daughter cells.

The term mitosis in a strict sense refers to the division of the nucleus into two daughter nuclei, and the term **cytokinesis** is applied to the division of the cytoplasm to form two daughter cells, each containing a daughter nucleus. Nuclear division and cytoplasmic division, although almost invariably well synchronized and coordinated, are separate and distinct processes.

Each mitotic division is a continuous process, with each stage merging imperceptibly into the next one. For descriptive purposes mitosis may be divided into four stages: **prophase, metaphase, anaphase** and **telophase** (Fig. 3.15). Between mitoses a cell is said to be in the resting stage. The nucleus

Figure 3.15 Mitosis in a cell of a hypothetical animal with a diploid number of six (haploid number = 3); one pair of chromosomes is short, one pair is long and hooked, and one pair is long and knobbed. A, Resting stage. B, Early prophase: centriole divided and chromosomes appearing. C, Later prophase: centrioles at poles, chromosomes shortened and visibly double. D, Metaphase: chromosomes arranged on the equator of the spindle. E, Anaphase: chromosomes migrating toward the poles. F, Telophase: nuclear membranes formed, chromosomes elongating, cytoplasmic division beginning. G, Daughter cells: resting phase.

is "resting" only with respect to division, however, for during this time it may be very active metabolically. It is difficult to visualize from a description or diagram of mitosis, or from examining a fixed and stained slide of cells, just how active a process cell division is. Motion pictures made by phase microscopy reveal that a cell undergoing division bulges and changes shape like a gunny sack filled with a dozen unfriendly cats.

Prophase. The chromatin threads condense and the chromosomes appear as a tangled mass of coiled threads within the nucleus. Early in prophase the threads are stretched maximally so that the individual chromomeres are visible. Later in prophase the chromosomes shorten and thicken and the chromomeres lie so close together that individual ones cannot be distinguished. Each part of the doubled chromosome is called a chromatid; the two chromatids are held together at the centromere, which remains single until the metaphase.

Early in prophase the centrioles migrate to opposite sides of the cell. Between the separating centrioles a **spindle** forms. The protein threads of the spindle are arranged like two cones base to base, broad at the center or equator of the cell and narrowing to a point at either end or pole. With a microneedle attached to a micromanipulator the spindle can be moved as a unit from one part of the cell to another. By appropriate techniques spindles may be isolated from dividing cells (Fig. 3.16).

At the end of prophase, the centrioles have gone to the opposite poles of the cell, the spindle has formed between them and the chromosomes have become short and thick.

Metaphase. When the chromosomes are fully contracted and appear as short, dark-staining rods, the nuclear membrane disappears and the chromosomes line up in the equatorial plane of the spindle. The short period during which the chromosomes are in this equatorial plane is known as the metaphase. This is much shorter than the prophase; although times for different cells vary considerably, the prophase lasts from 30 to 60 minutes or more, and the metaphase lasts only two to six minutes.

During the metaphase the centromere of each chromosome divides and the two chromatids become completely separate daughter chromosomes. The division of the centromeres occurs simultaneously in all the chromosomes, under the control of some as yet unknown mechanism. The daughter centromeres begin to move apart, marking the beginning of anaphase.

Anaphase. The chromosomes separate (Fig. 3.17), and one of the daughter chromosomes goes to each pole. The period during which the separating chromosomes move from the equatorial plate to the poles, known as the anaphase, lasts some three to 15 minutes. The spindle fibers apparently act as guide rails along which the chromosomes move toward the poles. Without such guide rails the chromosomes would merely be pushed randomly

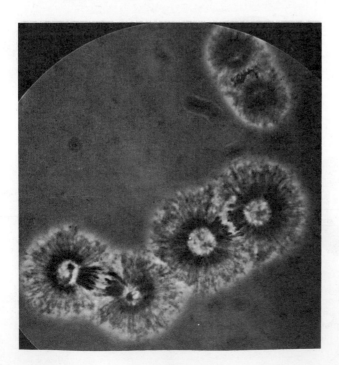

Figure 3.16 Photomicrograph of the mitotic apparatus isolated from dividing cells of a sea urchin embryo. Each mitotic apparatus includes spindle fibers, asters and chromosomes. A metaphase figure appears in the upper right and two anaphase figures below. (Courtesy of Daniel Mazia.) (From Villee, C. A.: Biology. 7th ed. Philadelphia, W. B. Saunders Co., 1977.)

Figure 3.17 Photographs of stages in mitosis in the white fish blastula. *A*, early prophase; *B*, later prophase; *C*, metaphase; *D*, two cells in early and late anaphase respectively; *E*, early telophase; and *F*, late telophase. The dark spot connecting the two cells in *E* is the remainder of the spindle. (Photographs courtesy of Dr. Susumu Ito.)

apart and many would fail to be incorporated into the proper daughter nucleus. The mechanism moving the chromosomes apart is not clear. Experiments indicate that some of the spindle fibers are contractile and can pull the chromosomes toward the poles. Spindles isolated from cells about to divide can be in-

duced to contract when ATP is added. The chromosomes moving toward the poles usually assume a V shape with the centromere at the apex pointing toward the pole. It appears that whatever force moves the chromosome to the pole is applied at the centromere. A chromosome that lacks a centromere, perhaps as a result of exposure to x-radiation, does not move at all in mitosis.

Telophase. When the chromosomes have reached the poles of the cell, the last phase of mitosis, telophase, begins. Several processes occur simultaneously in this period: a nuclear membrane forms around the group of chromosomes at each pole, the chromosomes elongate, stain less darkly, and return to the resting condition in which only irregular chromatin threads are visible, and the cytoplasm of the cell begins to divide. Division of the cytoplasm **(cytokinesis)** is accomplished in animal cells by the formation of a furrow which circles the cell at the equatorial plate and gradually deepens until the two halves of the cell are separated as independent daughter cells. The events of telophase require some 30 to 60 minutes for their completion.

The mitotic process results in the formation of two daughter cells from a single parent cell. Each daughter cell receives exactly the same number and kind of chromosomes, and of the units of heredity (genes) contained in these chromosomes, that were present in the parent cell. Since all the cells of the body are formed by mitosis from a single fertilized egg, each cell has the same number and kind of chromosomes, and the same number and kind of genes, as every other cell.

The speed and frequency of cell division vary greatly from tissue to tissue and from one animal to another. In the early stages of embryonic development, there may be only 30 minutes or so between successive cell divisions. In certain adult tissues, notably the nervous system, mitoses are extremely rare. In other adult tissues, such as the red bone marrow, where red blood cells are produced, mitotic divisions must occur frequently to supply the 10,000,000 red blood cells each human being produces every second of the day and night.

3.7 REGULATION OF MITOSIS

The factors which initiate and control cell division are not known exactly. The possible role of the ratio of cell surface to cell volume was discussed previously (p. 42). The ratio of *nuclear* surface to *nuclear* volume may also be important. Since normal cell function requires the transport of substances back and forth through the nuclear membrane, growth will eventually result in a state in which the area of the nuclear membrane is insufficient to meet the demands of the volume of cytoplasm. Cell division, by splitting the volume of cytoplasm into two parts and increasing the area of nuclear membrane, will restore optimal conditions.

The nucleus-cytoplasmic mass theory, proposed by von Hertwig in 1908, requires some changes, for factors other than cell mass are involved in controlling mitosis. Cells can be made to divide before they have doubled in size, and the eggs of many organisms grow to very large size before dividing. There is some evidence that the two cyclic nucleotides, cyclic AMP and cyclic GMP, play roles in regulating many cellular functions, including cell division. The two have opposite, antagonistic effects on the growth and division of certain cells grown in culture, cGMP stimulating growth and cAMP inhibiting it.

3.8 THE STUDY OF CELLULAR ACTIVITIES

Despite great differences in size, shape and location in the body, all cells have many metabolic activities in common. The myriad enzymes in each cell enable it to release energy by converting sugars, fats and proteins to carbon dioxide and water. Each cell synthesizes its own structural proteins and enzymes. Superimposed on this basic pattern of metabolism common to all cells may be other activities peculiar to each type of cell. For example, muscle cells have special contractile proteins, **myosin** and **actin;** particular digestive enzymes are produced by the cells lining the stomach and intestine; and the cells of the pituitary, adrenal and thyroid glands manufacture characteristic hormones.

Each of the many ways of studying cellular activity provides useful information about cell morphology and physiology. Living cells suspended in a drop of fluid can be examined under an ordinary microscope or with one equipped with **phase contrast lenses.** In this way one can study the movement of an ameba or a white blood cell, or the beating of the cilia on a paramecium. Cells from a many-celled animal—a frog, chick or man—can be grown by **"tissue culture"** for observation over a long period of time. A complex nutritive medium, made of blood plasma, an extract of embryonic tissues, and a mixture of salts, glucose, amino acids and vitamins, is prepared and sterilized. A drop of this is placed in a cavity on a special microslide, the cells to be cultured are added aseptically, and the cavity is sealed with a glass cover slip. After a few days the cells have exhausted one or more of the nutritive materials and must be transferred again to a fresh drop of medium. Connective tissue cells (fibroblasts) will survive in culture and undergo a relatively fixed number of mitoses (cell generations)—about 50—after which they can no longer divide. In contrast, cancer cells can survive in culture and undergo divisions indefinitely. Muscle cells in culture divide a few times and then fuse to form a muscle fiber, becoming contractile.

Cell morphology may be studied by using a bit of tissue that has been killed quickly with a special "fixative," then sliced with a machine called a microtome, and stained with special dyes. The stained thin slices, mounted on a glass slide and covered with a glass cover slip, are then ready for examination under the microscope. The nucleus, mitochondria and other specialized parts of the cell differ chemically and will combine with different dyes and be stained characteristic colors (Fig. 3.7). For observation in the electron microscope a bit of tissue is fixed with osmic acid, mounted in acrylic plastic for cutting in extremely thin sections, and then placed on a fine grid to be inserted into the path of the electron beam.

Both light microscopy and electron microscopy have revealed many details about cell structure (Fig. 3.18).

Some clue as to the location and functioning of enzymes within cells can be obtained by **histochemical** studies, in which a tissue is fixed and sliced thin by methods which do not destroy enzymic activity. Then the proper chemical substrate for the enzyme is provided and, after a specified period of incubation, some substance is added which will form a colored compound with one of the products of the reaction mediated by the enzyme. The regions of the cell which have the greatest enzymic activity will have the largest amount of the colored substance (Fig. 3.19). Methods have been devised which permit the demonstration and localization of a wide variety of enzymes. Such studies have given interesting insights into the details of cell function.

Another method of investigating cell function is to measure, by special microchemical analyses, the amounts of chemical used up or produced as a bit of tissue is incubated in an appropriate medium. In such experiments much has been learned of the roles in cell metabolism of vitamins, hormones and other

Figure 3.18 Electron micrograph of a typical mitochondrion from the pancreas of a bat, showing the cristae, matrix and matrix granules. Endoplasmic reticulum is seen at the upper left and some lysosomes at the lower right. ×79,000. (Courtesy of K. R. Porter.)

Figure 3.19 Histochemical demonstration of the location of the enzyme alkaline phosphatase within the cells of the rat's kidney. The tissue is carefully fixed and sectioned by methods which do not destroy the enzyme's activity. The tissue section is incubated at the proper pH with a naphthyl phosphate. Some hydrolysis of the naphthyl phosphate occurs wherever the phosphatase enzyme is located. The naphthol released by the action of the enzyme couples with a diazonium salt to form an intensely blue, insoluble azo dye which remains at the site of the enzymatic activity. The photomicrograph thus reveals the sites of phosphatase activity; i.e., the sites at which the azo dye is deposited. The cells of the proximal convoluted tubules (left) have a lot of enzyme; those of the loop of Henle (right) have little or no activity. (Courtesy of R. J. Barrnett.) (From Villee, C. A.: Biology. 7th ed. Philadelphia, W. B. Saunders Co., 1977.)

chemicals by adding these substances one by one and observing the resulting effects.

Every living cell, whether it is an individual unicellular animal or a single component of a multicellular one, must be supplied constantly with nutrients and oxygen. These materials are constantly being metabolized—used up—as the cell goes about its business of releasing energy from the nutrients to provide for its myriad activities. Some of the substances required by the cell are brought to it and taken in by complex active processes which require the expenditure of energy by the cell. Other substances are brought to the cell by the simpler, more easily understood physical process of **diffusion.** To understand this process, so important in many biological phenomena, we must first consider some of the basic physical concepts of energy and molecular motion.

3.9 ENERGY

Energy may be defined as the ability to do work, to produce a change in matter. It may take the form of heat, light, electricity, motion or chemical energy. Physicists recognize two kinds of energy: **potential energy,** the capacity to do work owing to the position or state of a body, and **kinetic energy,** the capacity to do work possessed by a body because of its motion. A rock at the top of a hill has potential energy; as it rolls downhill the potential energy is converted to kinetic energy.

Energy derived ultimately from solar radiation, trapped by photosynthetic processes in plants, is stored in the molecules of foodstuffs as the chemical energy of the bonds connecting their constituent atoms. This chemical energy is a kind of potential energy. When these food molecules are taken within a cell, chemical reactions occur which change this potential energy into heat, light, motion or some other kind of kinetic energy. Light is a kind of kinetic energy that may be thought of as the movement of photons or light quanta. All forms of energy are at least partially interconvertible, and living cells constantly transform potential energy into kinetic energy or the reverse (Table 3.1). If the conditions are suitably controlled, the amount

TABLE 3.1 ENERGY TRANSFORMATIONS IN CELLS

Transformation	Type of Cell
Chemical energy to electrical energy	Nerve, brain
Sound to electrical energy	Inner ear
Light to chemical energy	Chloroplast
Light to electrical energy	Retina of eye
Chemical energy to osmotic energy	Kidney
Chemical energy to mechanical energy	Muscle cell, ciliated epithelium
Chemical energy to radiant energy	Luminescent organ of firefly
Chemical energy to electrical energy	Sense organs of taste and smell

of energy entering and leaving any given system can be measured and compared. Such experiments have shown that energy is neither created nor destroyed, but simply transformed from one form to another. This is an expression of one of the fundamental laws of physics, the Law of the Conservation of Energy. Living things as well as nonliving systems obey this law.

3.10 MOLECULAR MOTION

The constituent molecules of all substances are constantly in motion. Despite the fact that wood, stone and steel seem very solid, their component molecules vibrate continuously within a very restricted space. The prime difference between solids, liquids and gases is the freedom of movement of the molecules present. The molecules of a solid are very closely packed and the forces of attraction between the molecules permit them to vibrate but not to move around. In the liquid state the molecules are somewhat farther apart and the intermolecular forces are weaker, so that the molecules can move about with considerable freedom. The molecules in the gaseous state are so far apart that the intermolecular forces are negligible and molecular movement is restricted only by external barriers. Molecular movement in all three states of matter is the result of the inherent heat energy of the molecules, the kinetic energy determined by the temperature of the system. By increasing this **molecular kinetic energy,** one can change matter from one state to another. When ice is heated it becomes water, and when water is heated it is converted to water vapor.

If a drop of water is examined under the microscope, the motion of its molecules is not evident. If a drop of India ink (which contains fine carbon particles) is added, the carbon particles move continually in aimless zigzag paths, for they are constantly being bumped by water molecules and the recoil from this bump imparts the motion to the carbon particle. The motion of such small particles is called **brownian movement,** after Robert Brown, an English botanist, who first observed the motion of pollen grains in a drop of water.

3.11 DIFFUSION

Molecules in a liquid or gaseous state will diffuse, that is, move in all directions until they are spread evenly throughout the space available. **Diffusion** may be defined as the movement of molecules from a region of high concentration to one of lower concentration brought about by their kinetic energy. The rate of diffusion is a function of the size of the molecule and the temperature. If a bit of sugar is placed in a beaker of water, the sugar will dissolve and the individual sugar molecules will diffuse and come to be distributed evenly throughout the liquid (Fig. 3.20). Each molecule tends to move in a straight line until it collides with another molecule or the side of the container; then it rebounds and moves in another direction. By this random movement of molecules, the sugar eventually becomes evenly distributed throughout the water in the beaker. This could be demonstrated by tasting drops of liquid taken from different parts of the beaker. If a colored dye is used in place of sugar, the process of diffusion can be observed directly. The molecules of sugar or dye continue to move after they have become evenly distributed throughout the liquid in the container; however, as fast as some molecules move from left to right, others move from right to left, so that an equilibrium is maintained.

Any number of substances will diffuse independently of each other. If a lump of salt is placed in one part of a beaker of water and a lump of sugar in another, the molecules of each will diffuse independently of the other and each drop of water in the beaker will eventually have some salt and some sugar molecules.

The rate of movement of a single molecule is several hundred meters per second, but each molecule can go only a fraction of a nanometer before it bumps into another molecule and rebounds. Thus the progress of a molecule in a straight line is quite slow. Diffusion is quite rapid over short distances but it takes a long time—days and even weeks—for a substance to diffuse a distance measured in centimeters. This fact has important biological implications,.for it places a sharp limit on the number of molecules of oxygen and nutrients that can reach an organism by diffusion alone. Only a very small organism that requires relatively few molecules per second can survive if it remains in one place and allows molecules to come to it by diffusion. A larger organism must have some means of moving to a new region or some means of stirring its environment to bring molecules to it. As a third alternative it may live in some spot where the environment is constantly moving past it—in a river, for example, or in the intertidal region at

Figure 3.20 Diffusion. When a cube of sugar is placed in water (A) it dissolves and its molecules become uniformly distributed throughout the water as a result of the molecular motion of both sugar and water molecules (B). When lumps of sugar and salt are placed in water (C), each type of molecule diffuses independently of the other, and both salt and sugar become uniformly distributed in the water (D).

the seashore. The larger land plants have solved this problem by developing an extensively branched system of roots which can tap a large area of the surrounding environment for the needed raw materials.

3.12 EXCHANGES OF MATERIAL BETWEEN CELL AND ENVIRONMENT

All nutrients and waste products must pass through the plasma membrane to enter or leave the cell. Cells are almost invariably surrounded by a watery medium—the fresh or salt water in which an organism lives, the tissue sap of a higher plant, or the plasma or extracellular fluid of a higher animal. In general, only dissolved substances can pass through the plasma membrane, but not all dissolved substances penetrate the plasma membrane with equal facility. The membrane behaves as though it had ultramicroscopic pores through which substances pass, and these pores, like

the holes in a sieve, determine the maximal size of molecule that can pass. Factors other than simple molecular size, such as the electric charge, if any, of the diffusing particle, the number of water molecules bound to the diffusing particle and its solubility in fatty substances, may also be important in determining whether or not the substance can pass through the plasma membrane.

A membrane is said to be **permeable** if it will permit any substance to pass through, **impermeable** if it will allow no substance to pass, and **differentially permeable** if it will allow some but not all substances to diffuse through. The nuclear and plasma membranes of all cells and the membranes surrounding food and contractile vacuoles are differentially permeable membranes. Permeability is a property of the *membrane*, not of the diffusing substance.

The diffusion of a dissolved substance through a differentially permeable membrane is known as **dialysis.** If a pouch made of collodion, cellophane or parchment is filled with a sugar solution and placed in a beaker of water, the sugar molecules will dialyze through the membrane (if the pores are large enough) and eventually the concentration of sugar molecules in the water outside the pouch will equal that within the pouch. The molecules then continue to diffuse but there is no net change in concentration, for the rates in the two directions are equal.

A different type of diffusion is observed if a membrane is prepared with smaller pores, so that it is permeable to the small water molecules but not to the larger sugar molecules. A pouch may be prepared of a membrane with these properties and filled with a sugar solution. The pouch is then fitted with a cork and glass tube and placed in a beaker of water so that the levels of fluid inside and outside the pouch are the same. The sugar molecules cannot pass through the membrane and so must remain inside the pouch. The water molecules diffuse through the membrane and mix with the sugar solution, so that the level of fluid within the pouch rises. The liquid within the pouch is 5 per cent sugar, and therefore only 95 per cent water; the liquid outside the membrane is 100 per cent water. The water molecules are moving in both directions through the membrane but there is a greater movement from the region of higher concentration (100 per cent, outside the pouch) to the region of lower concentration (95 per cent, within the pouch). This diffusion of water or solvent molecules through a membrane is called **osmosis,** and is illustrated diagrammatically in Figure 3.21.

If an amount of water equal to that originally present in the pouch enters, the solution in the pouch will be diluted to 2.5 per cent sugar and 97.5 per cent water, but the concentration of water outside the pouch will still exceed that inside and osmosis will continue. An equilibrium is reached when the water in the glass tube rises to a height such that the weight of the water in the tube exerts a pressure just equal to the tendency of the water to enter the pouch. Osmosis then occurs with equal speed in both directions through the differentially permeable membrane and there will be no net change in the amount of water in the pouch. The pressure of the column of water is called the **osmotic pressure** of the sugar solution. The osmotic pressure results from the tendency of the water molecules to pass through the differentially permeable membrane and equalize the concentration of water molecules on its two sides. A more concentrated sugar solution would have a greater osmotic pressure and would "draw" water to a higher level in the tube. A 10 per cent sugar solution would cause water to rise approximately twice as high in the tube as a 5 per cent solution.

It is evident from this discussion that dialysis and osmosis are simply two special forms of diffusion. Diffusion is the general term for the movement of molecules from a region of high concentration to a region of lower concentration, brought about by their kinetic energy. Dialysis is the diffusion of dissolved molecules (solutes) through a differentially permeable membrane, and osmosis is the diffusion of solvent molecules through a differentially permeable membrane. In biological systems the solvent molecules are almost universally water.

The salts, sugars and other substances dissolved in the fluid within each cell give the intracellular fluid a certain osmotic pressure. When the cell is placed in a fluid with the same osmotic pressure as that of its intracellular fluid, there is no net entrance or exit of water, and the cell neither swells nor shrinks. Such a fluid is said to be **isotonic** or isosmotic with the intracellular fluid of the cell. Normally, the blood plasma and body fluids are isosmotic with the intracellular fluids of the body cells. If the surrounding fluid contains more dissolved substances than the fluid within the cell, water will tend to pass out of the cell and the cell shrinks. Such a fluid is said to be **hypertonic** to the cell. If the surrounding fluid has a lower concentration of dissolved substances than the fluid in the cell, water tends to pass into the cell and the cell swells. This fluid is said to be **hypotonic** to the

Membrane

Sugar ● Water ○

Figure 3.21 Diagram illustrating osmosis. When a solution of sugar in water is separated from pure water by a semipermeable membrane which allows water but not the larger sugar molecules to pass through, there is a net movement of water molecules through the membrane to the sugar solution. The water molecules are diffusing from a region of higher concentration (pure water) to a region of lower concentration (the sugar solution).

cell. A solution of 0.9 per cent sodium chloride, 0.9 gm. per 100 ml. of water, sometimes loosely called "physiological saline," is isotonic to human cells.

A cell placed in a solution that is not isotonic with it may adjust to the changed environment by undergoing a change in its water content (by swelling or shrinking), so that it eventually achieves the same concentration of solutes as in the environment. Many cells have the ability to pump water or certain solute molecules into or out of the cell and in this way can maintain an osmotic pressure that differs from that of the surrounding medium. Amebas, paramecia and other protozoa that live in pond water, which is very hypotonic to their intracellular fluid, have evolved **contractile vacuoles** (see Fig. 3.1) which collect water from the interior of the cell and pump it to the outside. Without such a mechanism the cells would quickly burst from the water entering the cell.

The power of certain cells to accumulate selectively certain kinds of molecules from the environmental fluid is truly phenomenal. Human cells (and those of vertebrates in general) can accumulate amino acids so that the concentration within the cell is two to 50 times that in the extracellular fluid. Cells also have a much higher concentration of potassium and magnesium, and a lower concentration of sodium, than the environmental fluids. Certain primitive chordates, the tunicates (Chap. 32), can accumulate vanadium so that the concentration inside the cell is some 2,000,000 times that in the surrounding sea water. Seaweeds have a comparable ability to accumulate iodine. The transfer of water or of solutes in or out of the cell against a concentration gradient is physical work and requires the expenditure of energy. Some active physiological process is required to perform these transfers; hence a cell can move molecules against a gradient only as long as it is alive. If a cell is treated with some metabolic poison, such as cyanide, it quickly loses its ability to maintain concentration differences on the two sides of its plasma membrane.

3.13   TISSUES

In the evolution of both plants and animals, one of the major trends has been toward the structural and functional specialization of cells. The cells composing the body of one of the higher animals are not all alike, but are differentiated and specialized to perform certain functions more efficiently than an unspecialized animal cell could. This specialization has also had the effect of making the several parts of the body interdependent, so that an injury to, or the destruction of, cells in one part of the body may result in the death of the whole organism. The advantages of specialization are so great that they more than outweigh the disadvantages. The cells of the body which are similarly specialized are known as a tissue. A **tissue** may be defined as a group or layer of similarly specialized cells which together perform certain special functions. The study of the structure and arrangement of tissues is known as **histology**. Each tissue is composed of cells which have a characteristic shape, size and arrangement; the different types of tissue of the vertebrate body are readily recognized when examined microscopically. Certain tissues are composed of nonliving cell products in addition to the cells; connective tissue contains many fibers in addition to the fibroblasts or connective tissue cells, and bone and cartilage are made largely of proteins and salts secreted by the bone or cartilage cells.

The tissues of a multicellular animal may be classified in six major groups, each of which has several subgroups. These are epithelial, connective, muscular, blood, nervous and reproductive tissues.

Epithelial Tissues. Epithelial tissues are composed of cells which form a continuous layer or sheet covering the surface of the body or lining cavities within the body. There is usually a noncellular **basement membrane** underlying the sheet of epithelial cells. The epithelial cells in the skin of vertebrates are usually connected by small cytoplasmic processes or bridges. The epithelia of the body protect the underlying cells from mechanical injury, from harmful chemicals and bacteria, and from desiccation. The epithelial lining of the digestive tract absorbs water and nutrients for use in the body. The lining of the digestive tract and a variety of other epithelia produce and give off a wide spectrum of substances. Some of these are used elsewhere in the body and others are waste products which must be eliminated. Since the entire body is covered by an epithelium, all of the sensory stimuli pass through some epithelium to reach the specific receptors for those stimuli. The functions of epithelia are thus protection, absorption, secretion and sensation. The lining of the digestive tract, windpipe, lungs, kidney tubules and urinary bladder, and the outer layer of the skin are some familiar examples of epithelial tissues.

The cells in epithelial tissues may be flat, cuboidal or columnar in shape, they may be arranged in a single layer or in many layers, and they may have fine hairs or cilia on the free surface. On the basis of these structural characteristics epithelia are subdivided into the following groups.

Squamous epithelium is made of thin flattened cells the shape of flagstones or tiles (Fig. 3.22). It is found on the surface of the skin and the lining of the mouth, esophagus and vagina. The endothelium lining the cavity of blood vessels and the mesothelium lining the coelom are squamous epithelia. In the lower animals the skin is usually covered with a single layer of squamous epithelium, but in the higher animals the outer layer of the skin consists of stratified squamous epithelium, made of several layers of these flat cells.

The kidney tubules are lined with **cuboidal epithelium,** made of cells that are cube-shaped and look like dice (Fig. 3.22). Many other parts of the body, such as the stomach and intestines, are lined by cells that are taller than they are wide. An epithelium composed of such elongated, pillar-like cells is known as **columnar epithelium** (Fig. 3.22). Columnar epithelium may be simple, consisting of a single layer of cells, or stratified, composed of several layers of cells.

Either cuboidal or columnar epithelial cells may have cilia on their free surface. Ciliated cuboidal epithelium is found in the sperm ducts of earthworms and other animals, and ciliated columnar epithelium lines the ducts of the respiratory system of man and other air-breathing vertebrates. The rhythmic, concerted beating of the cilia moves solid particles in one direction through the ducts. Epithelial cells, usually columnar ones, may be specialized to receive stimuli. The groups of cells in the taste buds of the tongue or the olfactory epithelium in the nose are examples of sensory epithelium. Columnar or cuboidal epithelia may also be specialized for secreting certain products such as milk, wax, saliva, perspiration or mucus. The outer epithelium of most worms secretes a thin, continuous, noncellular protective layer, called the **cuticle,** which covers the entire body. Insects, spiders, crabs and other arthropods secrete a cuticle which may be quite thick and strengthened with deposits of chitin and salts. The hard protective shells of oysters and snails, composed of calcium carbonate, are secreted by epithelial cells in the mantles of these animals.

Connective Tissues. The connective tissues—bone, cartilage, tendons, ligaments, fibrous connective tissue and adipose tissue—support and bind together the other tissues and organs. Connective tissue cells characteristically secrete a nonliving material called the **matrix,** and the nature and function of each connective tissue is determined primarily by the nature of this intercellular matrix. The actual connective tissue cells may form only a small and inconspicuous part of

Figure 3.22 Diagram of the types of epithelial tissue and their location in the body.

the tissue. The matrix, rather than the connective tissue cells themselves, does the actual connecting and supporting.

Fibrous connective tissue consists of a thick, interlacing, matted network of fibers in which are distributed the cells that secreted the fibers (Fig. 3.23). There are three types of fibrous connective tissue, widely distributed throughout the body, which bind skin to muscle, muscle to bone, and so on. These include very delicate **reticular fibers;** thick, tough, unbranched, flexible, but relatively inelastic **collagen fibers;** and long, branched **elastic fibers. Adipose tissue** is rich in fat cells, specialized connective tissue cells which store large quantities of fat in a vacuole in the cytoplasm. Ligaments and tendons are specialized fibrous connective tissues. **Tendons** are composed of thick, closely packed bundles of collagen fibers, which form flexible cables that connect a muscle to a bone or to another muscle. A **ligament** is fundamentally similar in constitution to a tendon and connects one bone to another. An especially thick mat of

fibrous connective tissue is located in the lower layer of the skin of most vertebrates; when this is chemically treated—"tanned"—it becomes leather.

Connective tissue fibers contain a protein called **collagen,** which is rich in the amino acids glycine, proline and hydroxyproline. Treating the fibers with hot water converts some of the collagen into the soluble protein **gelatin.** The amino acid compositions of gelatin and collagen are nearly identical. Because so much connective tissue is present, nearly one-third of all the protein in the human body is collagen. The collagen units composing the fibers consist of a helix made of three peptide chains wound around each other in a supercable and joined by hydrogen bonds.

The supporting skeleton of vertebrates is composed of cartilage or bone. **Cartilage** appears as the supporting skeleton in the embryonic stages of all vertebrates, but in the adult is largely replaced by bone in all but sharks and rays. Cartilage can be felt in our bodies as the supporting framework of the auricle of the

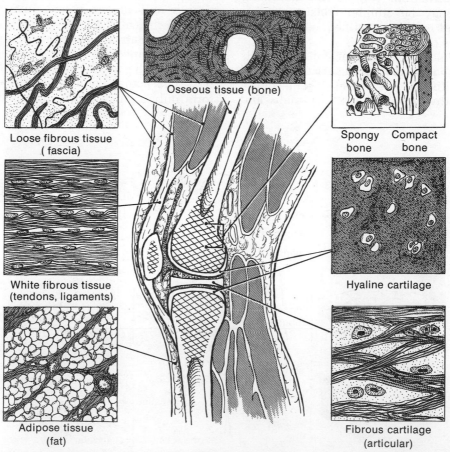

Figure 3.23 Diagram of the types of connective tissue and their location in the knee joint.

ear (the external ear flap) or the tip of the nose. It is made of a firm but elastic matrix secreted by cartilage cells which become embedded in the matrix (Fig. 3.23). These cartilage cells are alive; they may secrete collagenous fibers, or elastic fibers to strengthen the cartilage matrix.

Bone consists of a dense matrix composed of proteins, principally collagen, and calcium salts identical with the mineral hydroxyapatite, $Ca_3(PO_4)_2 \cdot CaCO_3$. About 65 per cent of the bone is made of this mineral. Bone cells (osteoblasts) remain alive and secrete a bony matrix, both protein and calcium salts, throughout life. The osteoblasts become surrounded and trapped by their own secretion and remain in microscopic cavities (lacunae) in the bone as living osteocytes (Fig. 3.23). The protein is laid down as minute fibers which contribute strength and resiliency, and the mineral salts contribute hardness to bone.

At the surface of each bone is a thin fibrous layer called the **periosteum** (peri, around; osteum, bone) to which the muscles are attached by tendons. The periosteum contains cells, some of which differentiate into osteoblasts and secrete protein and salts to bring about growth and repair. Most bones are not solid but have a marrow cavity in the center. The apparently solid matrix of the bone is pierced by many microscopic channels (haversian canals) in which lie blood vessels and nerves to supply the bone cells. The bony matrix is deposited, usually in concentric rings or lamellae, around these haversian canals. Each bone cell is connected to the adjacent bone cells and to the haversian canals by cellular extensions occupying minute canals (canaliculi) in the matrix. The bone cells obtain oxygen and raw materials and eliminate wastes by way of these canaliculi. The details of the architecture of a bone can be observed by grinding a slice of bone extremely thin and mounting it on a slide for inspection under a microscope. Bone contains not only bone-secreting cells but also bone-destroying cells. By the action of these two types of cells, the shape of a bone may be altered to resist changing stresses and strains. Bone formation and destruction is regulated by the availability of calcium and phosphate, by the presence of vitamin D, and by the hormone calcitonin and the parathyroid hormone, secreted by the thyroid and parathyroid glands. The marrow cavity of the bone may contain yellow marrow (largely a fat depot) or red marrow, the tissue in which red and certain white blood cells are formed.

Muscular Tissues. The movements of most animals result from the contraction of elongated, cylindrical or spindle-shaped cells, each of which contains many microscopic, elongate, parallel, contractile fibers called **myofibrils,** composed of the proteins myosin and actin. Muscle cells perform mechanical work by contracting, by getting shorter and thicker; they are unable to do work by pushing. Three types of muscle tissue are found in vertebrates: skeletal, cardiac and smooth (Fig. 3.24). **Cardiac muscle** is found only in the walls of the heart; **smooth muscle** in the walls of the digestive tract, the urinary and genital tracts, and the walls of arteries and veins; and **skeletal muscle** makes up the muscle masses which are attached to and move the bones of the body. Cardiac and skeletal muscle cells are among the exceptions to the rule that cells have but one nucleus; each of these cells has many nuclei. The nuclei of skeletal muscle cells have an unusual position, at the periphery of the cell, just below the plasma membrane; presumably, this is an adaptation to increase the efficiency of contraction. Skeletal muscle cells are extremely long, two or more centimeters in length; indeed, some investigators believe that the muscle cells extend from one end of the muscle to the other, so that their length is equal to that of the muscle. Muscle fibers range in thickness from 10 to 100 micrometers; continued strenuous muscle

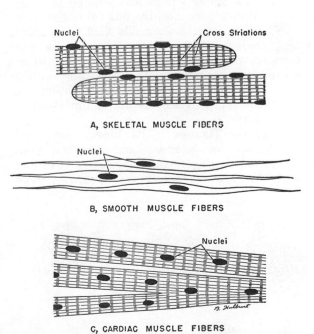

A, SKELETAL MUSCLE FIBERS

B, SMOOTH MUSCLE FIBERS

C, CARDIAC MUSCLE FIBERS

Figure 3.24 Types of muscle tissue. (From Villee, C. A.: Biology. 6th ed. Philadelphia, W. B. Saunders Co., 1972.)

TABLE 3.2 COMPARISON OF VERTEBRATE MUSCLE TISSUES

	Skeletal	Smooth	Cardiac
Location	Attached to skeleton	Walls of viscera: stomach, intestines, etc.	Wall of heart
Shape of fiber	Elongate, cylindrical, blunt ends	Elongate, spindle-shaped, pointed ends	Elongate, cylindrical; fibers branch and fuse
Number of nuclei per cell	Many	One	Many
Position of nuclei	Peripheral	Central	Central
Cross striations	Present	Absent	Present
Speed of contraction	Most rapid	Slowest	Intermediate
Ability to remain contracted	Least	Greatest	Intermediate
Type of control	Voluntary	Involuntary	Involuntary

activity increases the thickness of the fiber. The myofibrils of skeletal and cardiac muscle have alternate dark and light cross bands or **striations**. These striations appear to have some fundamental role in contraction, during which the dark band remains constant but the light band shortens. The contraction of skeletal muscles is generally voluntary, under the control of the will; that of cardiac and smooth muscles is involuntary. Cardiac muscle cells are striated but have centrally located nuclei. Smooth muscle cells are not striated, have pointed ends, and have centrally located nuclei. Smooth muscle contracts slowly but can remain contracted for long periods of time. The voluntary muscles of some invertebrates, such as the ones which close the shell of an oyster, are smooth muscles. Striated muscles can contract very rapidly but cannot remain contracted; a striated muscle fiber must relax and rest before it is able to contract again. The muscles of insects, spiders, crabs and other arthropods have cross striations and contract very rapidly. The distinguishing features of the three types of muscle are summarized in Table 3.2.

Vascular Tissues. The **blood,** composed of a liquid part—**plasma**—and of several types of **formed elements**—red cells, white cells and platelets—may be classified as a separate type of tissue or as one kind of connective tissue. The latter classification is based on the fact that blood cells and connective tissue cells originate from similar cells; however, the adult cells are quite different in structure and function. The **red cells** (erythrocytes) of vertebrates contain the red pigment hemoglobin, which has the property of combining easily and reversibly with oxygen. Oxygen, combined as oxyhemoglobin, is transported to the cells of the body in the red cells. Mammalian red cells are flattened, biconcave discs without a nucleus; those of other vertebrates are more typical cells with an oval shape and a nucleus.

The five different kinds of **white blood cells**—lymphocytes, monocytes, neutrophils, eosinophils and basophils (Fig. 3.25)—lack hemoglobin but move around and engulf bacteria. They can slip through the walls of blood vessels and enter the tissues of the body to engulf bacteria there. The fluid plasma transports a great variety of substances from one part of the body to another. Some of the substances transported are in solution, others are bound to one or another of the plasma proteins. The plasma of vertebrates is a light yellow color; in certain invertebrates the oxygen-carrying pigment is not localized in cells, but is dissolved in the plasma and colors it red or blue. **Platelets** are small fragments bro-

Figure 3.25 Types of white blood cells. *A,* basophil; *B,* eosinophil; *C,* neutrophil; *E–H,* a variety of lymphocytes; *I* and *J,* monocytes; *D,* a red blood cell drawn to the same scale. (From Villee, C. A.: Biology. 7th ed. Philadelphia, W. B. Saunders Co., 1977.)

ken off from cells in the bone marrow; they play a role in the clotting of blood (Chap. 14).

Nervous Tissues. Cells specialized for the reception of stimuli and the transmission of impulses are called **neurons.** A neuron typically has an enlarged cell body, containing the nucleus, and two or more cytoplasmic processes, the nerve fibers, along which the nerve impulse travels to the next neuron (Fig. 3.26). Nerve fibers vary in width from a few micrometers to 30 or 40 and in length from a millimeter or two to a meter or more. Two types of nerve fibers are distinguished: **axons,** which transmit impulses away from the cell body, and **dendrites,** which transmit them to the cell body. The junction between the axon of one neuron and the dendrite of the next neuron in the chain is called a **synapse.** At the synapse the axon and dendrite do not actually touch; there is a small gap between the two. Transmission of an impulse across the synapse involves a different mechanism from that moving an impulse along the nerve fiber. An impulse can travel across the synapse only from an axon to a dendrite; thus, the synapse serves as a valve to prevent the backflow of impulses. Neurons show widely diverse patterns in shape of the cell body and in number and length of dendrites and axons.

The cell bodies of neurons commonly occur in groups; there are columns of cell bodies in the spinal cord, sheets of cell bodies over the surface of parts of the brain, nodules of cell bodies ("nuclei") within the brain, and the ganglia of the cranial and spinal nerves. A **ganglion** is a group of nerve cell bodies located outside the central nervous system. A nerve consists of a group of axons and dendrites bound together by connective tissue. Each nerve fiber—axon or dendrite—is surrounded by a **neurilemma** or a **myelin sheath,** or both. The neurilemma is a delicate, transparent, tube-like membrane made of cells which envelop the fiber. The myelin sheath is made of noncellular, fatty material which forms a glistening white coat between the fiber and neurilemma. The myelin sheath is interrupted at fairly regular intervals along the nerve by constrictions called the nodes of Ranvier. Nerve fibers are either "medullated" and have a thick myelin sheath, or "nonmedullated" and have an extremely thin myelin sheath. Nerve fibers in the brain and spinal cord have a myelin sheath but no neurilemma; those in the autonomic nerves to the viscera, and the nerves of many invertebrates, are nonmyelinated and have a very thin or no myelin sheath but a neurilemma. The nerves to the skin and skeletal muscles of vertebrates have both a myelin sheath and a neurilemma surrounding them.

Figure 3.26 Diagrams of (*A*) an afferent neuron and (*B*) an efferent neuron. The arrows indicate the direction of the normal nerve impulse. (From King, B. G., and M. J. Showers: Human Anatomy and Physiology. 6th ed. Philadelphia, W. B. Saunders Co., 1969.)

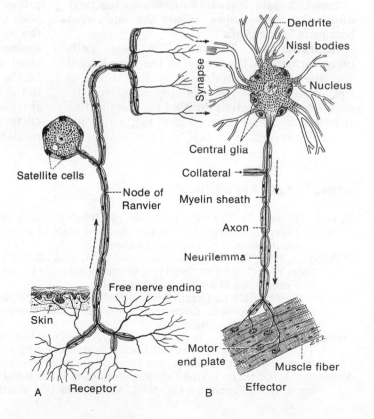

Dendrite
Nissl bodies
Nucleus
Central glia
Collateral →
Myelin sheath
Axon
Neurilemma
Synapse
Satellite cells
Node of Ranvier
Free nerve ending
Skin
Motor end plate
Muscle fiber
A Receptor
B Effector

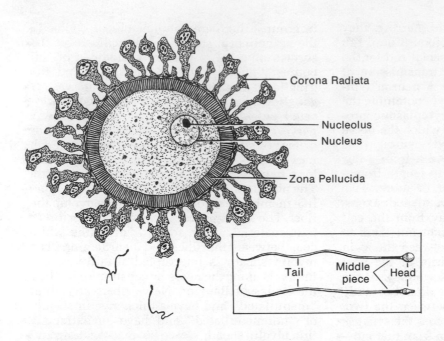

Corona Radiata

Nucleolus

Nucleus

Zona Pellucida

Tail   Middle   Head
        piece

Figure 3.27 Human egg and sperm, magnified 400 times. *Inset,* side and top views of a sperm, magnified about 2000 times. The egg is surrounded by other cells which form the corona radiata.

Nervous tissue contains, in addition to neurons, several different kinds of supporting cells called **neuroglia.** These have many cytoplasmic processes, and the cells and their processes form an extremely dense supporting framework in which the neurons are suspended. The neuroglia are believed to separate and insulate adjacent neurons, so that nerve impulses can pass from one neuron to the next only over the synapse, where the neuroglial barrier is incomplete.

Reproductive Tissues. The **egg cells** (ova) formed in the ovary of the female and the **sperm cells** produced by the testes of the male constitute the reproductive tissues—cells specially modified for the production of offspring (Fig. 3.27). Egg cells are generally spherical or oval and are nonmotile.

A typical egg has a large nucleus, called the germinal vesicle, and a variable amount of yolk in the cytoplasm. Shark and bird eggs have enormous amounts of yolk which provides nourishment for the developing embryo until it hatches from the shell. Sperm cells are small and modified for motility. A typical sperm has a long **tail,** the beating of which propels the sperm to its meeting and union with the egg. The **head** of the sperm contains the nucleus surrounded by a thin film of cytoplasm. The tail is connected to the head by a short **middle piece.** An **axial filament,** formed by the centriole in the middle piece, extends to the tip of the tail. Most of the cytoplasm is sloughed off as the sperm matures; this decreases the weight of the sperm and perhaps renders it more motile.

ANNOTATED REFERENCES

Bloom, W., and D. Fawcett: Textbook of Histology. 10th ed. Philadelphia, W. B. Saunders Co., 1975. One of the standard texts of histology with many superb illustrations of the structure of tissues.

DeRobertis, E. D. P., W. W. Nowinski and F. A. Saez: Cell Biology. 6th ed. Philadelphia, W. B. Saunders Co., 1976. Contains a detailed presentation of the structure and properties of a wide variety of cells.

Fawcett, D.: The Cell: Its Organelles and Inclusions. Philadelphia, W. B. Saunders Co., 1966. An excellent discussion of cell structure illustrated by electron micrographs of outstanding quality.

Hall, T. S.: A Source Book in Animal Biology. New York, McGraw-Hill Book Co., 1951. The development of the cell theory is presented in an interesting fashion by means of long quotations from some of the original scientific papers.

Kennedy, D. (Ed.): The Living Cell: Readings from the Scientific American. San Francisco, W. H. Freeman & Co., 1965. A rich collection of articles about the cell.

CELL METABOLISM

All cells have complex, efficient systems for transforming one type of energy into another by appropriate chemical reactions. The never-ending flow of energy within a cell, from one cell to another, and from one organism to another organism is the essence of life itself. Three major types of energy transformations can be distinguished in the biological world (Fig. 4.1). In the first, the radiant energy of sunlight is captured by the green pigment **chlorophyll,** present in green plants, and is transformed by the process of **photosynthesis** into chemical energy. This is used to synthesize carbohydrates and other complex molecules from carbon dioxide and water. The radiant energy of sunlight, a form of **kinetic energy,** is transformed into a type of **potential energy.** The chemical energy is stored in the molecules of carbohydrates and other food-stuffs as the energy of the bonds which connect their constituent atoms.

In a second type of energy transformation, the chemical energy of carbohydrates and other molecules is transformed by the process termed **cellular respiration** into the biologically useful energy of energy-rich phosphate bonds. This kind of energy transformation occurs in the mitochondrion. A third type of energy transformation occurs when the chemical energy of these energy-rich phosphate bonds is utilized by the cell to do work—the mechanical work of muscular contraction, the electrical work of conducting a nerve impulse, the osmotic work of moving molecules against a gradient, or the chemical work of synthesizing molecules for growth. As these transformations occur, energy finally flows to the environment and is dissipated as heat. Plants and animals have evolved some remarkably effective **energy transducers,** such as chloroplasts and mitochondria, to carry out these processes, together with control mecha-

Figure 4.1 Energy transformations in the biological world. (1) The radiant energy of sunlight is transformed in photosynthesis into chemical energy in the bonds of organic compounds. (2) The chemical energy of organic compounds is transformed during cellular respiration into biologically useful energy, the energy-rich phosphate bonds of ATP and other compounds. (3) The chemical energy of energy-rich phosphate bonds is utilized in cells to do mechanical, electrical, osmotic or chemical work. Finally energy flows to the environment as heat in the "entropy sink." (From Villee, C. A.: Biology. 7th ed. Philadelphia, W. B. Saunders Co., 1977.)

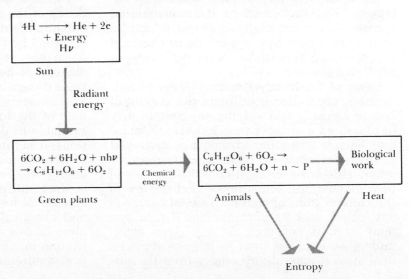

TABLE 4.1 ENERGY TRANSFORMATIONS
IN CELLS

Transformation	Type of Cell
Chemical energy to electrical energy	Nerve, brain
Sound to electrical energy	Inner ear
Light to chemical energy	Chloroplast
Light to electrical energy	Retina of eye
Chemical energy to osmotic energy	Kidney
Chemical energy to mechanical energy	Muscle cell, ciliated epithelium
Chemical energy to radiant energy	Luminescent organ of firefly
Chemical energy to electrical energy	Sense organs of taste and smell

nisms to regulate the transducers and enable the cells to adjust to variations in environmental conditions (Table 4.1).

The chemical reactions of cells, which provide for their growth, irritability, movement, maintenance and repair, and their reproduction, are collectively termed **metabolism**. The metabolic activities of animal, plant and bacterial cells are remarkably similar, despite the apparent differences of the organisms themselves. In all cells, sugars and related substances are continually being metabolized, via a large number of intermediate compounds, to water and carbon dioxide. During these conversions the energy of the sugar molecule is made available to drive other processes in the cell.

4.1 ENTROPY AND ENERGY

The branch of physics that deals with energy and its transformations, **thermodynamics,** is based on certain relatively simple principles which are universally applicable to chemical processes whether these occur in living or nonliving systems.

Laws of Thermodynamics. Under experimentally controlled conditions, the amount of energy entering and leaving any system may be measured and compared. It is always found that energy is neither created nor destroyed but is only transformed from one form to another. This is an expression of the first law of thermodynamics, sometimes called the **law of the conservation of energy:** the total energy of any object and its surroundings (i.e., a system) remains constant. As any given object undergoes a change from its initial state to its final state it may absorb energy from the sur-

roundings or deliver energy to the surroundings. The difference in the energy content of the object in its initial and final state must be just equalled by a corresponding change in the energy content of the surroundings. Heat is a convenient form in which energy may be measured which is why the study of energy has been called thermodynamics, i.e., heat dynamics. Heat, however, is not a useful way of transferring energy in biological systems, for living organisms are basically **isothermal** (equal temperature). There is no significant temperature difference between the different parts of the cell or between the different cells in a tissue. Stated in another way, cells do not act as heat engines. They have no means of allowing heat to flow from a warmer to a cooler body.

The **second law of thermodynamics** may be stated briefly as "the entropy of the universe increases." **Entropy** may be defined as a randomized state of energy that is unavailable to do work. The second law may be phrased that "physical and chemical processes proceed in such a way that the entropy of the system becomes maximal." Entropy then is a measure of randomness or disorder. In almost all energy transformations there is a loss of some heat to the surroundings, and since heat involves the random motion of molecules, such heat losses increase the entropy of the surroundings. Living organisms and their component cells are highly organized and thus have little entropy. They preserve this low entropy state by increasing the entropy of their surroundings. You increase the entropy of your surroundings when you eat a candy bar and convert its glucose to carbon dioxide and water and return them to the surroundings.

The force that drives all processes is the tendency of the system to reach the condition of maximum entropy. Heat is either given up or absorbed by the object to allow the system to reach the state of maximum entropy. The changes of heat and entropy are related by a third dimension of energy termed **free energy**. Free energy may be visualized as that component of the total energy of a system which is available to do work under isothermal conditions; it is thus the thermodynamic parameter of greatest interest in biology. Entropy and free energy are related inversely; as entropy increases during an irreversible process, the amount of free energy decreases. All physical and chemical processes proceed with a decline in free energy until they reach an equilibrium in which the free energy of the system is at a minimum and the entropy is at a maximum. Free energy is useful energy; entropy is degraded, useless energy.

4.2 CHEMICAL REACTIONS

A chemical reaction is a change involving the molecular structure of one or more substances; matter is changed from one substance, with its characteristic properties, to another, with new properties, and energy is released or absorbed. Hydrochloric acid, for example, reacts with the base, sodium hydroxide, to yield water and the salt, sodium chloride; in the process energy is released as heat. The chemical properties of HCl and NaOH are very different from those of NaCl and H_2O. In chemical shorthand a plus sign connects the symbols of the reacting substances, HCl and NaOH, and the products, NaCl and H_2O. An arrow indicates the direction of the reaction:

$$HCl + NaOH \rightarrow NaCl + H_2O + energy \text{ (heat)}$$

Most chemical reactions are reversible and this reversiblity is indicated by a double arrow: $\rightleftharpoons$.

Atoms are neither destroyed nor created in the course of a chemical reaction; thus the sum of each kind of atom on one side of the arrow must equal the sum of that kind of atom on the other side. This is an expression of one of the basic laws of physics, the **law of the conservation of matter.** The direction of a reversible reaction is determined by the energy relations of the several chemicals involved, their relative concentrations, and their solubility.

For each reaction the thermodynamic equilibrium constant, K, expresses the chemical equilibrium reached by the system. Thus for the reaction $A + B \rightleftharpoons C + D$,

$$K = \frac{[C] \times [D]}{[A] \times [B]}$$

The equilibrium constant for any given reaction is determined by the tendency of the reaction components to reach maximum entropy, or minimum free energy, for the system. The equilibrium constant, K, is related mathematically to the change in free energy of the components of the reaction: $\Delta G = -RT \ln K$. R is the gas constant, T the absolute temperature, and ln K is the natural logarithm of the equilibrium constant. The symbol ΔG represents the standard free energy change in kilocalories per mole (i.e., the gain or loss of free energy in calories as one mole of reactant is converted to one mole of product).

Examination of this equation reveals that when the equilibrium constant, K, is high, the standard free energy change, ΔG, is negative. Such a reaction will proceed with a decrease in free energy. However, when the equilibrium constant is less than one (e.g., 0.3) the reaction does not go far in the direction of completion, and the free energy change is positive. It is necessary to put energy into the system to transform one mole of reactant into one mole of product. When the equilibrium constant is 1, then the change in free energy is zero and the reaction is freely reversible.

The rate at which a chemical reaction occurs is determined by a number of factors, one of which is temperature. Each increase of 10° C. approximately doubles the rate of most reactions. This is true of biological processes as well as reactions in a chemist's test tube. This is further evidence that the chemical reactions of living things are fundamentally similar to those of nonliving ones. Whether or not an enzyme or any other catalyst is present in the system has no effect on either the equilibrium constant, K, or the free energy change, ΔG, undergone. Enzymes and other catalysts simply speed up the rate at which the system approaches equilibrium but do not change the equilibrium point itself.

The unit of energy most widely used in biological systems is the **Calorie,** which is the amount of heat required to raise one kilogram of water one degree centigrade (strictly speaking, from 14.5° to 15.5° C.). This unit, more correctly called a kilogram calorie or kilocalorie, is the unit commonly used in calculating the energy content of foods. Other forms of energy—radiant, chemical, electrical, the energy of motion or position—can be converted to heat and measured by their effect in raising the temperature of water.

Reactions which have a high equilibrium constant, K, and a negative standard free energy change, ΔG, are said to be **exergonic.** A reaction with a very low equilibrium constant and therefore a positive standard free energy change will not go to completion under standard conditions. Such processes are called **endergonic** processes. In biological systems, the endergonic processes must be coupled with some exergonic process so that the exergonic process delivers the required energy to the endergonic process. In such coupled systems, the endergonic process can occur only if the decrease in free energy of the exergonic process to which it is coupled is larger than the gain in free energy of the endergonic process.

Catalysis. Many of the substances that are rapidly metabolized by living cells are remarkably inert outside the body. A glucose solution, for example, will keep indefinitely in a bottle if it is kept free of bacteria and molds. It must be subjected to high temperature or to the action of strong acids or bases before it will decompose. Living cells cannot utilize conditions as extreme as these, for the cell itself would be destroyed long before the glucose, yet glucose is rapidly decomposed within cytoplasm at ordinary temperatures and pressures and in a solution which is neither acidic nor basic. The reactions within the cell are brought about by special agents known as **enzymes,** which belong to the class of substances known as catalysts.

A **catalyst** is an agent which affects the velocity of a chemical reaction without altering its end point and without being used up in the course of the reaction. The list of substances which may serve as a catalyst in one or more reactions is long indeed. Water is an excellent catalyst for many reactions. Pure, dry hydrogen gas and dry chlorine gas do not react when mixed, but if a slight trace of water is present they react with explosive violence to form hydrogen chloride. Metals such as iron, nickel, platinum and palladium, when ground into a fine powder, are widely used as catalysts in industrial processes such as the hydrogenation of cottonseed and other vegetable oils to make margarine or the cracking of petroleum to make gasoline. A minute amount of catalyst will speed up the reaction of vast quantities of reactants, for the molecules of catalyst are not exhausted in the reaction but are used again and again.

4.3 ENZYMES

Enzymes are protein catalysts, produced by living cells. They regulate the speed and specificity of the thousands of chemical reactions that occur within cells. Although enzymes are synthesized within cells, they do not have to be inside a cell to act as a catalyst. Many enzymes have been extracted from cells with their activity unimpaired. They can then be purified and crystallized and their catalytic abilities can be studied. Enzyme-controlled reactions are basic to all the phenomena of life: respiration, growth, muscle contraction, nerve conduction, photosynthesis, nitrogen fixation, deamination, digestion and so on.

Properties of Enzymes. Enzymes are usually named by adding the suffix -*ase* to the name of the substance acted upon; e.g., sucrose is split by the enzyme **sucrase** to give glucose and fructose. Enzymes are usually colorless, but they may be yellow, green, blue, brown or red. Most enzymes are soluble in water or dilute salt solution, but some, for example the enzymes present in the mitochondria, are bound together by lipoprotein (a phospholipid-protein complex) and are insoluble in water.

The catalytic ability of some enzymes is truly phenomenal. For example, one molecule of the iron-containing enzyme **catalase,** extracted from beef liver, will bring about the decomposition of 5,000,000 molecules of hydrogen peroxide (H_2O_2) per minute at $0°$ C. The substance acted upon by an enzyme is known as its **substrate;** thus, hydrogen peroxide is the substrate of the enzyme catalase.

The number of molecules of substrate acted upon by a molecule of enzyme per minute is called the **turnover number** of the enzyme. The turnover number of catalase is thus 5,000,000. Most enzymes have high turnover numbers, thus they can be very effective although present in the cell in relatively minute amounts. Hydrogen peroxide is a poisonous substance produced as a byproduct in a number of enzyme reactions. Catalase protects the cell by destroying the peroxide.

Hydrogen peroxide can be split by iron atoms alone, but it would take 300 *years* for an iron atom to split the same number of molecules of H_2O_2 that a molecule of catalase, which contains one iron atom, splits in one *second*. This example of the evolution of a catalyst emphasizes one of the prime characteristics of enzymes—they are very efficient catalysts.

Although enzymes in general catalyze specific reactions, they do differ in the number of kinds of substrates they will attack. **Urease** is an example of an enzyme which is absolutely specific. Urease decomposes urea to ammonia and carbon dioxide and will attack no substance other than urea. Most enzymes are not quite so specific, and will attack several closely related substances. **Peroxidase,** for example, will decompose several different peroxides in addition to hydrogen peroxide. A few enzymes are specific only in requiring that the substrate have a certain kind of chemical bond. The **lipase** secreted by the pancreas will split the ester bonds connecting the glycerol and fatty acids of a wide variety of fats.

In theory, enzyme-controlled reactions are reversible; the enzyme does not determine the *direction* of the reaction but simply accelerates the *rate* at which the reaction reaches equilibrium. The classic example of this is the action of the enzyme lipase on the splitting of fat, or union of glycerol and fatty acids. If one begins with a fat, the enzyme catalyzes the splitting of this to give some glycerol and fatty acids. If one begins with a mixture of fatty acids and glycerol, the enzyme catalyzes the synthesis of some fat. When either system has operated long enough, the same equilibrium mixture of fat, glycerol and fatty acid is reached:

$$\text{fat} \rightleftharpoons \text{glycerol} + 3 \text{ fatty acids}$$

The equilibrium point is determined by complex thermodynamic principles, which will not be discussed. Since reactions give off energy when going in one direction, it is obvious that an equivalent amount of energy in the proper form must be supplied to make the reaction go in the opposite direction.

To drive an energy-requiring reaction, some energy-yielding reaction must occur at about the same time. In most biological systems, energy-yielding reactions result in the synthesis of "energy-rich" phosphate esters, such as the terminal bonds of **adenosine triphosphate** (abbreviated as ATP). The energy of these energy-rich bonds is then available for the conduction of an impulse, the contraction of a muscle, the synthesis of complex molecules, and so on, much as the energy of a storage battery made by a generator is available for light, heat or running a motor. Biochemists use the term "coupled reactions" for two reactions which must occur together so that one can furnish the energy, or one of the reactants, needed by the other.

Enzymes generally work in teams in the cell, with the product of one enzyme-controlled reaction serving as the substrate for the next. We can picture the inside of a cell as a factory with many different assembly lines (and disassembly lines) operating simultaneously. Each of these assembly lines is composed of a number of enzymes, each of which catalyzes the reaction by which one substance is converted into a second. This second substance is passed along to the next enzyme, which converts it into a third, and so on along the line. From germinating barley seeds one can extract two enzymes that convert starch to glucose. The first, amylase, splits starch to maltose and the second, maltase, splits the double sugar maltose to two molecules of the single sugar glucose. Eleven different enzymes, working consecutively, are required to convert glucose to lactic acid. The same series of 11 enzymes is found in human cells, in green leaves and in bacteria.

Some enzymes, such as pepsin and urease, have been found to consist solely of protein. Many others, however, consist of two parts: one is protein (called the **apoenzyme**) and the other (called a **coenzyme**) is some smaller organic molecule, usually containing phosphate. Coenzymes can usually be separated from their enzymes and, when analyzed, have proved to contain some vitamin—thiamine, niacin, riboflavin, pyridoxine, etc.—as part of the molecule. This finding has led to the generalization that *all vitamins function as parts of coenzymes in the cell.*

Neither the apoenzyme nor the coenzyme alone has catalytic properties; only when the two are combined is activity evident. Certain enzymes require for activity, in addition to a coenzyme, the presence of one or more ions. Magnesium (Mg^{++}) is required for the activity of several of the enzymes in the chain which converts glucose to lactic acid. **Salivary amylase,** the starch-splitting enzyme of saliva, requires chloride ion as an activator. Most, if not all, of the elements required by plants and animals in very small amounts—the so-called **trace elements,** manganese, copper, cobalt, zinc, iron, and others—serve as enzyme activators.

Enzymes may be present in the cell either dissolved in the liquid phase or bound to, and presumably an integral part of, one of the subcellular organelles. A water extract of ground liver contains all the 11 kinds of enzymes necessary to convert glucose to lactic acid. The respiratory enzymes, which catalyze the metabolism of lactic acid and the carbon chains of fatty acids and amino acids to carbon dioxide and water, are integral parts of the mitochondria.

The Mechanism of Enzyme Catalysis. Many years ago Emil Fischer, the German organic chemist, suggested that the specificity of the relationship of an enzyme to its substrate indicated that the two must fit together like a lock and key. The idea that an enzyme combines with its substrate to form a reactive intermediate enzyme-substrate complex, which subsequently decomposes to release the free enzyme and the reaction products, was formulated mathematically by Leonor Michaelis more than 50 years ago. By brilliant inductive reasoning, he assumed that such a complex does form, and then calculated what the relationships between enzyme concentration, substrate concentration, and the velocity of the reaction should be. Exactly these relationships are observed experimentally, which is strong evidence of the validity of Michaelis' assumption that an enzyme-substrate complex forms as an intermediate.

Direct evidence of the existence of enzyme-substrate complexes was obtained by David Keilin of Cambridge University and Britton Chance of the University of Pennsylvania. Chance isolated a brown-colored peroxidase from horseradish and found that when this was mixed with the substrate, hydrogen peroxide, a green-colored enzyme-substrate complex formed. This, in turn, changed to a second, pale red complex which finally split to give the original brown enzyme and the products of the reaction. By observing the changes of color, Chance was able to calculate the rates of formation and of dissociation of this complex.

It is clear that when it is part of an enzyme-substrate complex, the substrate is much more reactive than it is when free. It is not clear, however, *why* this should be true. One explanation postulates that the enzyme unites with the substrate at two or more places, and the substrate molecule is held in a position which strains its molecular bonds and renders them more likely to break.

One approach to the study of enzymic action is to investigate the detailed structure of the enzyme molecule itself. It is now possible to determine the kinds of amino acids present in an enzyme molecule, their relative numbers and the sequence of amino acids in the peptide chain or chains composing the enzyme protein. Ribonuclease, which consists of 124 amino acids in a single chain that is looped on itself like a pretzel, was the first enzyme for which it was possible to give the exact sequence of amino acids in the entire molecule. Such analyses of the structure of the enzyme molecule may provide clues as to the mechanism of enzymic catalysis. It is probable that only a relatively small part of the enzyme molecule (termed the "active site") is involved in combining with the substrate. Studies of the sequence of amino acids and the three-dimensional structure of this part of the enzyme have been fruitful in throwing light on the mechanism of enzyme action. Studies of the constituents and topography of certain active sites have revealed the presence of certain amino acids (e.g., serine) or metal ions which actually participate in the enzyme-catalyzed reaction as proton donors or acceptors or as nucleophilic (electron-donating) or electrophilic (electron-attracting) agents. A metal ion present at the active site may bind with one or more groups on the substrate molecule and, by forming a chelated intermediate, participate in the strain-producing process.

4.4 FACTORS AFFECTING ENZYMIC ACTIVITY

Temperature. The velocity of most chemical reactions is approximately doubled by each 10 degree increase in temperature, and, over a moderate range of temperature, this is true of enzyme-catalyzed reactions as well. Enzymes, and proteins in general, are inactivated by high temperatures; the higher the temperature, the more rapidly the enzymic activity is lost. Native protein molecules exist at least in part as **spiral coils,** or helices, and the denaturation process appears to involve the unwinding of the helix. Most organisms are killed by exposure to heat because their cellular enzymes are inactivated. The processes of protein denaturation and enzyme inactivation show a striking parallelism and this is one bit of substantiating evidence that enzymes are proteins. The enzymes of humans and other warm-blooded animals operate most efficiently at a temperature of about 37° C. — body temperature — whereas those of plants and cold-blooded animals work optimally at about 25° C. Enzymes are generally not inactivated by freezing; their reactions continue slowly, or perhaps cease altogether at low temperatures, but their catalytic activity reappears when the temperature is again raised to normal.

Acidity. All enzymes are sensitive to changes in the acidity and alkalinity — the pH — of their environment and will be inactivated if subjected to strong acids or bases. Most enzymes exert their greatest catalytic effect only when the pH of their environment is within a certain rather narrow range. On either side of this optimum pH, as the pH is raised or lowered, enzymic activity rapidly decreases. The protein-digesting enzyme secreted by the stomach, pepsin, is remarkable in that its pH optimum is 2.0; it will work only in an extremely acid medium. The protein-digesting enzyme secreted by the pancreas, trypsin, in contrast, has a pH optimum of 8.5, well on the alkaline side of neutrality. Most intracellular enzymes have pH optima near neutrality, pH 7.0. This marked influence of pH on the activity of an enzyme is what would be predicted from the fact that enzymes are proteins. The number of positive and negative charges associated with a protein molecule, and the shape of the molecular surface, are determined in part by the pH. Probably only one particular state of the enzyme molecule, with a particular number of negative and positive charges, is active as a catalyst.

Concentration of Enzyme, Substrate and Cofactors. If the pH and temperature of an enzyme system are kept constant and if an excess of substrate is present, the rate of the reaction is directly proportional to the amount of *enzyme* present. This method is used, indeed, to measure the amount of some particular enzyme present in a tissue extract. If the pH, temperature and enzyme concentration of a reaction system are held constant, the initial reaction rate is proportional to the amount of *substrate* present, up to a limiting value. If the enzyme system requires a coenzyme or specific activator ion, the concentration of this substance may, under certain circumstances, determine the over-all rate of the enzyme system.

Enzyme Inhibitors. Enzymes can be inhibited by a variety of chemicals, some of which inhibit reversibly, others irreversibly. **Cytochrome oxidase,** one of the "respiratory enzymes," is inhibited by cyanide, which forms a complex with the atom of iron present in the enzyme molecule and prevents it from participating in the catalytic process. Cyanide is poisonous to humans and other animals because of its action on the cytochrome enzymes. One of the enzymic steps in the conversion of glucose to lactic acid is inhibited by fluoride ion and another by iodoacetate. These substances, and others, have been used as tools by biochemists to investigate the properties and sequences of enzyme systems.

Enzymes themselves may act as poisons if they get into the wrong place. As little as 1 mg. of crystalline trypsin injected intravenously will kill a rat. Certain snake, bee and scorpion venoms contain enzymes that destroy blood cells or other tissues.

4.5 RESPIRATION AND CELLULAR ENERGY

All the phenomena of life—growth, movement, irritability, reproduction, and others— require the expenditure of energy by the cell. Living cells are not heat engines; they cannot use heat energy to drive these reactions but must use chemical energy, chiefly in the form of energy-rich phosphate bonds, abbreviated ~P. These bonds have a relatively high **free energy of hydrolysis,** ΔG (i.e., the difference in the energy content of the reactants and the products after the bond is split is relatively high). The free energy of hydrolysis is *not* localized in the covalent bond joining the phosphorus atom to the oxygen or nitrogen atom. Thus the term "energy-rich phosphate bond" is actually a misnomer, but it is so deeply ingrained by long usage that it is not likely to be changed.

The term **cellular respiration** refers to the enzymic processes within each cell by which molecules of carbohydrates, fatty acids and amino acids are metabolized ultimately to carbon dioxide and water with the conservation of biologically useful energy. Many of the enzymes catalyzing these reactions are located in the cristae and walls of the mitochondria (see Fig. 3.18).

All living cells obtain biologically useful energy by enzymic reactions in which electrons flow from one energy level to another. For most organisms, oxygen is the ultimate electron acceptor; oxygen reacts with the electrons and with hydrogen ions to form a molecule of water. Electrons are transferred to oxygen by a system of enzymes, localized within the mitochondria, called the **electron transmitter system.**

Electrons are removed from a molecule of some foodstuff and transferred by the action of a specific enzyme to some primary electron acceptor. Other enzymes transfer the electrons from the primary acceptor through the several components of the electron transmitter system and eventually combine them with oxygen (Fig. 4.2). The chief source of energy-rich phosphate bonds, ~P, in the cell is from the flow of electrons through the acceptors and the electron transmitter system. This flow of electrons has been termed the "electron cascade," and we might picture a series of waterfalls over which electrons flow, each fall driving a water wheel, an enzymic reaction by which the energy of the electron is captured in a biologically useful form, that of the energy-rich phosphate bonds of ATP.

ATP is the "energy currency" of the cell. All the energy-requiring reactions of cellular metabolism utilize ATP to drive the reaction. Energy-rich molecules do not pass freely from one cell to another but are made at the site in which they are to be utilized. The energy-rich bonds of ATP that will drive the reactions of muscle contraction, for example, are produced in the muscle cells.

Processes in which electrons (e^-) are removed from an atom or molecule are termed **oxidations;** the reverse process, the addition of electrons to an atom or molecule, is termed **reduction.** An example of oxidation and reduction is the reversible reaction

$$Fe^{++} \rightleftharpoons Fe^{+++} + e^-$$

Figure 4.2 Diagram of the reactions of the "electron cascade," the succession of metabolic steps by which electrons are transferred from substrate to oxygen and the energy is trapped in a biologically useful form as energy-rich phosphate bonds, ~P. (From Villee, C. A.: Biology. 7th ed. Philadelphia, W. B. Saunders Co., 1977.)

The reaction toward the right is an oxidation (the removal of an electron) and the reaction toward the left is a reduction (the addition of an electron). Each oxidation reaction, in which an electron is given off, must be accompanied by a reduction, a reaction in which the electron is accepted by another molecule, for electrons do not exist in the free state.

The passage of electrons in the electron transmitter system is a series of oxidation and reduction reactions termed **biological oxidation.** When the energy of this flow of electrons is captured in the form of ~P, the overall process is termed **oxidative phosphorylation.** In most biological systems, two electrons and two protons (that is, two hydrogen atoms) are removed together and the process is known as **dehydrogenation.**

The specific compounds of the electron transmitter system that are alternately oxidized and reduced are proteins known as **cytochromes.** Each cytochrome contains a heme group similar to the one present in hemoglobin. In the center of the heme group is an iron atom which is alternately oxidized and re-

duced—converted from Fe^{++} to Fe^{+++} and back—as it gives off and takes up an electron.

Another component of the electron transmitter system, **ubiquinone** (also called coenzyme Q), consists of a head, a six-membered carbon ring, which can take up and release electrons, and a long tail composed of 10 repeating units. These repeating units, called isoprenoid groups, each consisting of five carbon atoms, are the basic structural units of molecules of rubber, sterols and steroids.

Lactic acid, the acid of sour milk, is an important intermediate in metabolism. It undergoes a dehydrogenation to yield pyruvic acid. Its molecule contains the structure

$$H-\underset{|}{\overset{|}{C}}-OH$$ from which two hydrogens can be

removed enzymically by **lactic dehydrogenase** (Fig. 4.3). In this reaction, and in all dehydrogenations, the electrons removed are transferred to a primary electron acceptor. The primary electron acceptor in this reaction is **nicotinamide adenine dinucleotide, NAD.** The product of the dehydrogenation reaction,

Figure 4.3 The enzymic conversion of lactic acid to acetyl coenzyme A. (From Villee, C. A.: Biology. 7th ed. Philadelphia, W. B. Saunders Co., 1977.)

pyruvic acid, cannot undergo dehydrogenation directly, for it does not have a structure suitable for attack by a dehydrogenase and must undergo further enzymic action to attain such a molecular structure.

The next reaction of pyruvic acid is a **decarboxylation,** the loss of carbon dioxide. Carbon dioxide is derived in biological systems only from carboxylic groups (COOH) by the process of decarboxylation. The product of the decarboxylation reaction undergoes a "make-ready" reaction with a large complex organic molecule, **coenzyme A,** which results in a substance with an H—C—OH group that can be dehydrogenated to give acetyl coenzyme A.

Another type of dehydrogenation, which involves a different molecular structure, is exemplified by the oxidation of succinic acid (Fig. 4.4). The dehydrogenation of molecules having a —CH_2—CH_2— group involves a **flavin** as the hydrogen and electron acceptor. Succinic acid is oxidized to fumaric acid in the course of which two of its hydrogens are transferred to a flavin. The product, fumaric acid, cannot be dehydrogenated directly but undergoes a make-ready reaction in which a molecule of water is added to yield malic acid.

Malic acid has an H—C—OH group which is suitable for dehydrogenation. The reaction catalyzed by malic dehydrogenase requires NAD

Figure 4.4 The oxidation of succinic acid. (From Villee, C. A.: Biology. 7th ed. Philadelphia, W. B. Saunders Co., 1977.)

as primary acceptor and yields oxaloacetic acid. Oxaloacetic acid may undergo several reactions, one of which is decarboxylation to yield pyruvic acid. The further metabolism of pyruvic acid to acetyl coenzyme A could occur by the reactions just discussed.

In the reactions by which carbohydrates, fats and proteins are oxidized, the cell utilizes these three simple types of reactions: dehydrogenation, decarboxylation and make-ready reactions. They may occur in different orders in different chains of reactions. All the dehydrogenation reactions are, by definition, oxidations — reactions in which electrons are removed from a molecule. The electrons cannot exist in the free state for any finite period of time and must be taken up immediately by other compounds, electron acceptors. Two of the primary electron acceptors of the cell are pyridine nucleotides — nicotinamide adenine dinucleotide (abbreviated NAD) and nicotinamide adenine dinucleotide phosphate (abbreviated NADP). The functional end of both pyridine nucleotides is the vitamin **nicotinamide.** The nicotinamide ring accepts one hydrogen ion and two electrons from a molecule undergoing dehydrogenation (e.g., lactic acid) and becomes reduced nicotinamide adenine dinucleotide, NADH, releasing one proton.

NAD and NADP serve as primary electron and hydrogen acceptors in dehydrogenation reactions involving substrates with the

$$H-\overset{\displaystyle |}{\underset{\displaystyle |}{C}}-OH$$ arrangement, such as the dehydrogenation of lactic or malic acid. The two pyridine nucleotides differ from one another in that NAD has two phosphate groups and NADP has three phosphate groups in the tail attached to the niacin ring. Most dehydrogenases specifically require either NAD or NADP as the hydrogen acceptor and will not work with the other; some enzymes are less specific and will work with either one, though they usually work more rapidly with one than the other.

Another primary hydrogen acceptor, **flavin adenine dinucleotide** (abbreviated **FAD**), serves in reactions involving the

$$-\overset{\displaystyle H}{\underset{\displaystyle H}{C}}-\overset{\displaystyle H}{\underset{\displaystyle H}{C}}-$$

structure, as in the dehydrogenation of succinic acid. In some reactions, **flavin mononucleotide (FMN),** which consists of part of the FAD molecule, may serve in its place. The FAD of succinic dehydrogenase is bound very

tightly to the protein part of the enzyme and cannot be removed easily. Such tightly bound cofactors are termed **prosthetic groups** of the enzyme. The pyridine nucleotide effective in the lactic dehydrogenase system, in contrast, is very loosely bound and is readily removed. Such loosely bound cofactors are termed **coenzymes.**

The reduced pyridine nucleotides, NADH or NADPH, cannot react with oxygen. Their electrons must be passed through the intermediate acceptors of the electron transmitter system before they can react with oxygen. The flavin primary acceptors usually pass their electrons to the electron transmitter system but some flavoproteins can react directly with oxygen. When this occurs, hydrogen peroxide, H_2O_2, is produced and no $\sim P$ is found. An enzyme that can mediate the transfer of electrons directly to oxygen is termed an **oxidase;** one that mediates the removal of electrons from a substrate to a primary or intermediate acceptor is termed a **dehydrogenase.**

4.6 THE TRICARBOXYLIC ACID CYCLE

The acetyl coenzyme A formed by the oxidation of lactic acid, or formed by the oxidation of fatty acids, undergoes a series of reactions involving dehydrogenation, decarboxylation and make-ready reactions which have been termed the **Krebs tricarboxylic acid cycle.** Acetyl coenzyme A undergoes a make-ready reaction by combining with oxaloacetic acid, which contains four carbons, to yield citric acid, which contains six carbons (Fig. 4.5).

Citric acid has neither an $H-\overset{\displaystyle |}{\underset{\displaystyle |}{C}}-OH$ nor a

$-CH_2-CH_2-$ group and cannot undergo dehydrogenation. Two additional make-ready reactions involving the removal and addition of a molecule of water yield isocitric acid which can undergo dehydrogenation at its

$H-\overset{\displaystyle |}{\underset{\displaystyle |}{C}}-OH$ group. The hydrogen acceptor is a

pyridine nucleotide, usually NAD, and the product is oxalosuccinic acid, which undergoes decarboxylation to yield α-ketoglutaric acid.

Just as pyruvic acid is converted to the coenzyme A derivative of an acid with one less carbon atom, α-ketoglutaric acid (five carbons) is converted to succinyl coenzyme A (four car-

Figure 4.5 The tricarboxylic acid cycle (Krebs citric acid cycle). (From Villee, C. A.: Biology. 7th ed. Philadelphia, W. B. Saunders Co., 1977.)

bons) by reactions requiring NAD, coenzyme A, thiamine pyrophosphate and lipoic acid as coenzymes. α-Ketoglutaric acid is metabolized by a dehydrogenation, a decarboxylation and a make-ready reaction using coenzyme A to yield succinyl coenzyme A.

The bond joining coenzyme A to succinic acid is an energy-rich one, $\sim$S, like the bond joining coenzyme A to acetic acid in acetyl coenzyme A. The energy of the bond of acetyl coenzyme A is utilized in bringing about the addition of the acetyl group to oxaloacetic acid. The energy in the $\sim$S bond in succinyl coenzyme A can be converted to an energy-rich phosphate bond, $\sim$P, in the form of ATP. The reaction of succinyl coenzyme A with inorganic phosphate yields succinyl phosphate and coenzyme A. The phosphate group is then transferred to ADP to form ATP and free succinic acid. This is an example of an energy-rich bond synthesized at the **substrate level** by reactions not involving the electron transmitter system. Only a small fraction of the total energy-rich bonds made in metabolism are formed by reactions such as this. Normal cells metabolizing in a medium containing oxygen synthesize most of their ATP by oxidative phosphorylation in the electron transmitter system.

The further oxidation of succinic acid to fumaric, malic and oxaloacetic acid completes the cycle, for the oxaloacetic acid is then ready to combine with another molecule of acetyl coenzyme A to form a molecule of citric acid. In the course of this cycle (Fig. 4.6), two molecules of CO_2 and eight hydrogen atoms are removed, and one molecule of $\sim$P is synthesized at the substrate level. The Krebs tricarboxylic acid (TCA) cycle, also called the citric acid cycle, is the final common pathway by which the carbon chains of carbohydrates, fatty acids and amino acids are metabolized.

Fatty Acid Oxidation. The carbon chain of a fatty acid is metabolized by a series of reactions which begins by a reaction with ATP and then with coenzyme A to yield the fatty acyl coenzyme A (Fig. 4.7). Palmitic acid, which contains 16 carbons, reacts to form palmityl coenzyme A. Palmityl coenzyme A undergoes a dehydrogenation between the second and third carbons of the chain. The hydrogen acceptor for this dehydrogenation is a flavin, for the group undergoing dehydrogenation is $-CH_2-CH_2-$. The product of this dehydrogenation undergoes a make-ready reaction, the addition of a molecule of water, to yield a molecule which has an H—C̶—OH group. This, in turn, undergoes a dehydrogenation utilizing NAD as the hydrogen acceptor (Fig. 4.7). The resulting molecule can undergo a make-ready

Figure 4.6 Diagram of the cyclic series of reactions, termed the Krebs citric acid cycle, by which the carbon chains of sugars, fatty acids and amino acids are metabolized to yield carbon dioxide. The reactions are designated as DH, dehydrogenations; DC, decarboxylations; and MR, make-ready. The over-all reaction effected by one "turn" of the cycle is summarized in the center. (From Villee, C. A.: Biology. 7th ed. Philadelphia, W. B. Saunders Co., 1977.)

Figure 4.7 Reactions in the oxidation of fatty acids. (From Villee, C. A.: Biology. 7th ed. Philadelphia, W. B. Saunders Co., 1977.)

reaction with coenzyme A to clip off a two-carbon unit, acetyl coenzyme A, and leave a carbon chain which is two carbons shorter. This molecule is in the activated state—it contains coenzyme A on its carboxyl group—and is ready to be dehydrogenated by the enzyme which uses flavin as a hydrogen acceptor. This repeating series of reactions, which includes dehydrogenations and make-ready reactions but not decarboxylations, cleaves a fatty acid chain two carbons at a time (Fig. 4.7). Seven such series of reactions will split palmitic acid to eight molecules of acetyl coenzyme A.

Glycolysis. Glucose and other sugars are also converted ultimately to acetyl coenzyme A. The series of glycolytic reactions begins, as with fatty acids, by one in which glucose is "activated." Glucose reacts with ATP to yield glucose-6-phosphate and ADP, a reaction catalyzed by the enzyme **hexokinase.** After this make-ready reaction, other make-ready reactions establish a configuration which can undergo dehydrogenation (Fig. 4.8). A rearrangement yields fructose-6-phosphate and the transfer of a second phosphate from ATP forms fructose-1, 6-diphosphate. Fructose-1,6-diphosphate is split by an enzyme, **aldolase,**

Figure 4.8 Diagram of the series of reactions by which glucose and other sugars are metabolized to pyruvic acid. Note that most of the steps are reversible, indicated by double arrows. Other steps are essentially irreversible; glucose-6-phosphate is converted to glucose by a separate enzyme, a glucose-6-phosphatase, and not by the hexokinase that catalyzes the conversion of glucose to glucose-6-phosphate. (From Villee, C. A.: Biology. 7th ed. Philadelphia, W. B. Saunders Co., 1977.)

into two three-carbon sugars, glyceraldehyde-3-phosphate and dihydroxyacetone phosphate. These two are interconverted by a separate enzyme.

Glyceraldehyde-3-phosphate reacts with a compound containing an —SH group which is part of the enzyme molecule to yield an H—C—OH group that can undergo dehydrogenation with NAD as the hydrogen acceptor. The product of the reaction, phosphoglyceric acid bound to the SH group of the enzyme, then reacts with inorganic phosphate to yield 1,3-

diphosphoglyceric acid and free enzyme —SH. The phosphate in carbon 1 is an energy-rich group which can react with ADP to form ATP. This, like the energy-rich phosphate of succinyl phosphate, is one made at the substrate level. The resulting 3-phosphoglyceric acid undergoes a make-ready rearrangement to yield 2-phosphoglyceric acid. Then, in an unusual reaction, an energy-rich phosphate is generated by the removal of water by a **dehydration,** rather than by the removal of two hydrogens, a **dehydrogenation.** The product, phosphopyruvic acid, can transfer its phosphate group to ADP to yield ATP and free pyruvic acid. This is the second energy-rich phosphate group generated at the substrate level in the metabolism of glucose to pyruvic acid.

Each glucose molecule yields two molecules of glyceraldehyde-3-phosphate and hence a total of four energy-rich phosphates is produced as glucose is metabolized to pyruvic acid. However, two energy-rich phosphates are utilized in the process, one to convert glucose to glucose-6-phosphate and the second to convert fructose-6-phosphate to fructose-1,6-diphosphate. The net yield in the process is $2 \sim P$ ($4 \sim P$ produced minus $2 \sim P$ used up in the reactions). Pyruvic acid is then metabolized to acetyl coenzyme A by the reactions described previously.

Amino Acid Oxidation. Amino acids are oxidized by reactions in which the amino group is first removed, a process called **deamination,** then the carbon chain is metabolized and eventually enters the tricarboxylic acid cycle. The amino acid alanine, for example, yields pyruvic acid when deaminated, glutamic acid yields α-ketoglutaric acid, and aspartic acid yields oxaloacetic acid. These three amino acids can enter the TCA cycle directly. Other amino acids may require several reactions in addition to deamination to yield a substance which is a member of the TCA cycle but ultimately the carbon chains of all the amino acids are metabolized in just this way.

The Electron Transmitter System. The major reactions by which biologically useful energy is produced occur when the electrons are transferred from the primary acceptors, such as the pyridine nucleotides, through the electron transmitter system and provide the energy to drive the reactions of oxidative phosphorylation. The electrons entering the electron transmitter system from NADH have a relatively high energy content. As they pass along the chain of enzymes they lose much of their energy, some of which is conserved in the form of ATP.

The enzymes of the electron transmitter system are located within the substance of the mitochondrion in the mitochondrial membranes. They are very closely related spatially and it seems likely that the electrons flow through a solid phase rather than between enzymes that are in solution. The components of the electron transmitter system are given (Fig. 4.2) in the order of their oxidation-reduction potentials, which range from −0.32 volt for the pyridine nucleotides to +0.81 volt for oxygen. However, it is not known whether a particular electron in passing from pyridine nucleotide to oxygen must go through each and every one of the intermediates or whether it might skip some steps. It probably has to pass through at least three different steps to account for the $3 \sim P$ which are made for each pair of electrons that pass from pyridine nucleotide to oxygen.

Oxidative phosphorylation is measured by the rate at which inorganic phosphate, P_i, is converted to ATP as NADH or some other substance undergoes oxidation. It is possible to prepare mitochondria by homogenizing the cells and separating the subcellular particles by centrifugation. Such mitochondria when removed carefully from the cell will carry out oxidative phosphorylation. It is even possible to disrupt mitochondria by ultrasonic vibration and obtain submitochondrial particles which will carry out oxidative phosphorylation. In these purified systems NADH will be oxidized and oxygen will be utilized only if ADP is present to accept the energy-rich phosphate produced by the flow of electrons. The flow of electrons is tightly coupled to the phosphorylation process and will not occur unless phosphorylation can occur also. This, in a sense, prevents waste, for electrons will not flow unless energy-rich phosphate can be formed.

The flow of electrons from pyridine nucleotides to oxygen, a total drop of 1.13 volts (from −0.32 to +0.81 volt), would yield 52 kilocalories per pair of electrons if the process were 100 per cent efficient. This may be calculated from the formula $\Delta G' = -nF \, \Delta E$, where $\Delta G'$ is the change in free energy, n is the number of electrons involved (2), F is the Faraday (23.04 kilocalories) and ΔE is the difference in the oxidation-reduction potentials of the reactants (1.13 volts). Under experimental conditions most cells will produce at most $3 \sim P$ per pair of electrons as the electrons pass from pyridine nucleotide to oxygen. Each $\sim P$ is equivalent to about 10 kilocalories. Hence, the $3 \sim P$ amount to about 30 kilocalories. The efficiency of the electron transmitter system may then be calculated as 30/52 or about 57 per cent.

TABLE 4.2 THE REACTIONS BY WHICH BIOLOGICALLY USEFUL ENERGY IS RELEASED

(1)		$C_6H_{12}O_6 + 2 \sim P \longrightarrow 2$ pyruvate $+ 2$ NADH $+ 4 \sim P$
(2)		2 pyruvate $\longrightarrow 2 CO_2 + 2$ acetyl CoA $+ 2$ NADH
(3)		2 acetyl CoA $\longrightarrow 4 CO_2 + 6$ NADH $+ 2 H_2FP + 2 \sim P$
(4)	Sum	$C_6H_{12}O_6 \longrightarrow 6 CO_2 + 10$ NADH $+ 2 H_2FP + 4 \sim P$
(5)		$C_6H_{12}O_6 + 6 O_2 \longrightarrow 6 CO_2 + 6 H_2O + 30 \sim P + 4 \sim P + 4 \sim P$

The amount of ATP present in any cell is usually rather small. In muscle cells, in which large amounts of energy may be expended in a short time during the contraction process, an additional substance, creatine phosphate, serves as a reservoir of $\sim$P. The terminal phosphate of ATP is transferred by the enzyme creatine kinase to creatine to yield creatine phosphate and ADP. The phosphate bond of creatine phosphate is an energy-rich bond. The $\sim$P of creatine phosphate must be transferred back to ADP, converting it to ATP, to be utilized in muscle contraction and other energy-requiring reactions.

The fact that phosphorylation is tightly coupled to oxidation or electron flow provides a system of control which can regulate the rate of energy production and adjust it to the momentary rate of energy utilization. In a resting muscle cell, for example, oxidative phosphorylation will occur until all of the ADP has been converted to ATP. Then, since there are no more acceptors of $\sim$P, phosphorylation must stop. Since oxidation is tightly coupled to phosphorylation, the flow of electrons will cease.

When the muscle contracts, the energy required is obtained from the splitting of the energy-rich terminal phosphate of ATP to yield ADP and inorganic phosphate plus energy. The ADP formed can then serve as an acceptor of $\sim$P, phosphorylation begins, and the flow of electrons to oxygen occurs. Oxidative phosphorylation continues until all of the ADP has been converted to ATP. Electric generating systems have an analogous control device which adjusts the rate of production of electricity to the rate of utilization of electricity.

Some interesting calculations of the over-all energy changes involved in metabolism in the human body have been made by E. G. Ball of Harvard University. Since the conversion of oxygen to water involves the participation of hydrogen atoms and electrons, the total flow of electrons in the human body can be calculated and expressed in amperes. From the average oxygen consumption of an adult 70 kg. man at rest — 264 ml. per minute — and the fact that each oxygen atom requires two hydrogen atoms and two electrons to form a molecule of water, Dr. Ball calculated that 2.86×10^{22} electrons flow from foodstuff, via dehydrogenases, primary acceptors and the cytochromes, to oxygen each minute in all the cells of the body. Since an ampere equals 3.76×10^{20} electrons per minute, this flow of electrons amounts to 76 amperes. This is quite a bit of current, for an ordinary 100 watt light bulb uses just a little less than 1 ampere.

The flow of electrons from substrate to oxygen involves a potential difference of 1.13 volts (from -0.32 to $+0.81$ volt). Since volts $\times$ amperes = watts, $1.13 \times 76 = 85.9$ watts. The human body, then, uses energy at about the same rate as a 100-watt light bulb, but differs from it in having a much larger flow of electrons passing through a much smaller voltage change.

The conversion of glucose to carbon dioxide and water yields about 4 Kcal./gm. Let us now dissect the over-all equation

$$C_6H_{12}O_6 + 6 O_2 \rightarrow 6 CO_2 + 6 H_2O + \text{energy}$$

and review where the energy is released (Table 4.2).

In glycolysis (reaction 1, Table 4.2), glucose is activated by the addition of 2 $\sim$P and converted to 2 pyruvate $+ 2$ NADH $+ 4 \sim$P. Then the two pyruvates are metabolized (reaction 2) to 2 CO_2 $+ 2$ acetyl CoA $+ 2$ NADH. Finally in the TCA cycle (reaction 3), the two acetyl CoA are metabolized to 4 CO_2 $+ 6$ NADH $+ 2 H_2FP + 2 \sim$P.

These reactions can be added together (reaction 4) by eliminating items that are present on both sides of the arrows. Then, since the oxidation of NADH in the electron transmitter system yields 3 $\sim$P per mole, the 10 NADH = 30 $\sim$P. The oxidation of H_2FP yields 2 $\sim$P per mole and 2 $H_2FP = 4 \sim$P. Summing these we see that the complete aerobic metabolism of 1 mole (180 gm.) of glucose yields 38 $\sim$P (reaction 5). Each $\sim$P is equivalent to about 10 kilocalories and the 38 $\sim$P = 380 kilocalories.

When a mole of glucose is burned in a calorimeter some 690 kilocalories are released as heat. The metabolism of glucose in an animal cell releases 380/690, or about 55 per cent, of the total energy as biologically useful energy, $\sim$P. The remainder of the energy is dissipated as heat.

Figure 4.9   Electron micrograph of a mitochondrion from a pancreatic centroacinar cell. The double-layered unit membrane is evident in the smooth outer membrane (om) and in the inner membrane (im) which folds to form the cristae (mc). Magnification × 207,000. (Courtesy of G. E. Palade; from De Robertis, E. D., W. W. Nowinski, and F. A. Saez: Cell Biology. Philadelphia, W. B. Saunders Company, 6th ed. 1975.)

4.7 THE MOLECULAR ORGANIZATION OF MITOCHONDRIA

Mitochondrial shapes range from spherical to elongated, sausage-like structures, but an average mitochondrion is an ellipsoid about 3 μm long and a little less than 1 μm in diameter. Some very generously endowed large cells such as the giant ameba, *Chaos chaos*, have several hundred thousand mitochondria, and an average mammalian liver cell has about one thousand. The mitochondrial protein accounts for about 20 per cent of the total protein of the liver cell.

The inner membrane of the mitochondrion is folded repeatedly (Fig. 4.9) to form the shelflike cristae extending into, or all the way across, the central cavity. Both the inner and outer mitochondrial membranes are lipid bilayers containing protein molecules.

The enzymes of the TCA cycle have been found in the soluble matrix within the mitochondrion. The enzymes of the electron transmitter system are tightly bound to the inner membrane. Each group of these electron transmitter enzymes, termed a **respiratory assembly,** is one of the fundamental units of cellular activity. It has been estimated that the mitochondrion of the liver cell contains some 15,000 respiratory assemblies and that they make up about one quarter of the mass of the mitochondrial membranes. Thus the mitochondrial membrane is not just a protective skin but an important functional part of the mitochondrion.

High resolution electron micrographs of mitochondria, made by Humberto Fernández-Moran at the University of Chicago, have demonstrated the presence of small particles on the outer surface of the outer membrane and the inner surface of the inner membrane (Fig. 4.10). The particles on the inner membrane typically have a spherical knoblike head, a cylindrical stalk and a base plate. These particles may be sites of the enzyme reactions carried out by mitochondria. The particles of the inner membrane are not the respiratory assemblies, as was suggested when they were first found, but they do contain one or more "coupling factors" necessary for oxidative phosphorylation.

It seems clear that much of the cell's biologically useful energy, ATP, is generated by enzyme systems located in the *inner* membrane of the mitochondrion, yet most of the energy utilized by the cell is required for processes

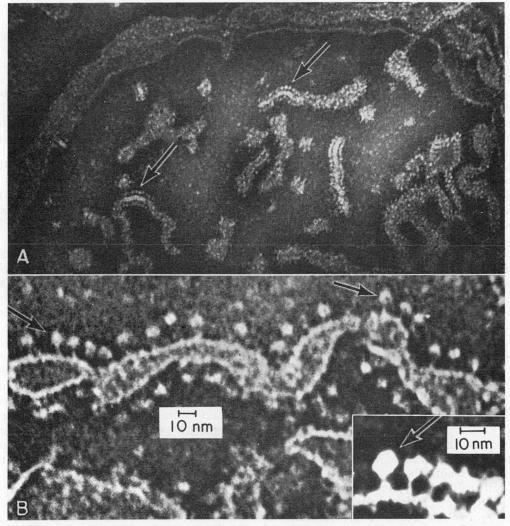

Figure 4.10 Electron micrograph of a mitochondrion swollen in a hypotonic solution and negatively stained with phosphotungstate, *A,* Isolated cristae, × 85,000. *B,* At higher magnification, × 500,000, the elementary particles attached by a stalk to the surface of the cristae are evident. *Inset:* Magnified × 650,000, the polygonal shape of the elementary particle and its slender stalk are clearly visible. (From De Robertis, E. D., F. A. Saez, and E. M. De Robertis, Jr.: Cell Biology. 6th ed. Philadelphia, W. B. Saunders Co., 1975. Courtesy of H. Fernández-Moran.)

that take place *outside of* the mitochondrion. ATP is used in the synthesis of proteins, fats, carbohydrates, nucleic acids and other complex molecules, in the transport of substances across the plasma membrane, in the conduction of nerve impulses, and in the contraction of muscle fibers, all of which are reactions occurring largely or completely outside of the mitochondrion in other parts of the cell. Several mechanisms, too complex to be considered here, have evolved to mediate the transfer of ~P across the mitochondrial membrane.

A single cell may have a thousand or more mitochondria and the function of each one must be controlled appropriately to generate the amount of energy required by the cell at any given moment. The rate at which ~P is utilized by a cell may vary over a remarkably wide range as the cell becomes active or quiescent. This has been measured in the muscle cells of a frog. One hundred grams of frog muscle utilizes 1.6μ moles of ~P per minute when quiescent and 3300μ moles of ~P per minute in a state of tetanus (continuous contraction). The rate of production of ATP in the cell is controlled in large part by the rate of utilization of ATP by the cell. The flow of electrons in the electron transmitter system is tightly coupled to phosphorylation, and oxidative phosphorylation can occur only when there is ADP to be converted to ATP. These facts provide the basis for the system by which a cell's utilization of ATP, which produces ADP, is used to regulate the rate at which ATP is produced. In addition, the structure and biological activity of some of the enzymes involved in glucose oxidation are affected by the concentration of ADP present. In this way an increased concentration of ADP can lead to increased activity of these enzymes and an increased production of ATP. The problem of the nature and interrelations of these various biological control systems is very much in the forefront in biology today.

4.8 THE DYNAMIC STATE OF CELLULAR CONSTITUENTS

The body of an animal or human appears to be unchanging as days and weeks go by and it would seem reasonable to infer that the component cells of the body, and even the component molecules of the cells, are equally unchanging. In the absence of any evidence to the contrary, it was generally held, until about 1937, that the constituent molecules of animal and plant cells were relatively static and that,

once formed, they remained intact for a long period of time. A corollary of this concept is that the molecules of food which are not used to increase the total cellular mass are rapidly metabolized to provide a source of energy. It followed from this that one could distinguish two kinds of molecules: relatively static ones that made up the cellular "machinery," and ones that were rapidly metabolized and thus correspond to cellular "fuel."

However, in 1937 Rudolf Schoenheimer and his colleagues at Columbia University began a series of experiments in which amino acids, fats, carbohydrates and water, each suitably labeled with some "heavy" or radioactive isotope, were fed to rats. Schoenheimer's experiments, which have been confirmed many times since, showed that the labeled amino acids fed to the rats were rapidly incorporated into body proteins. Similarly, labeled fatty acids were rapidly incorporated into the fat deposits of the body, even though in each case there was no increase in the *total* amount of protein or fat. Such experiments have demonstrated that the fats and proteins of the body cells—and even the substance of the bones— are constantly and rapidly being synthesized and broken down. In the adult the rates of synthesis and degradation are essentially equal so that there is little or no change in the total mass of the animal body. The distinction between "machinery" molecules and "fuel" molecules becomes much less sharp. The machinery is constantly being overhauled and some of the machinery molecules are broken down and used as fuel.

The one exception to the rule of molecular flux is provided by the molecules of DNA that constitute the genes within the nucleus of the cell. Experiments with labeled atoms have shown that the molecules of DNA are remarkably stable and are broken down and resynthesized only very slowly, if at all. The amount of DNA per nucleus is constant, and new molecules of DNA must be synthesized before a cell can divide. The stability of the DNA molecules may be important in ensuring that hereditary characters are transmitted to succeeding generations with as few chemical errors as possible. In contrast, the molecules of RNA undergo constant synthesis and degradation; indeed, the amount of RNA per cell may vary within wide limits.

From the rate at which labeled atoms are incorporated it has been calculated that one-half of all the tissue proteins of the human body are broken down and rebuilt every 80 days. The proteins of the liver and blood serum are replaced very rapidly, one-half of them being synthesized every 10 days. Some enzymes in the liver have half-lives as short as two to four hours. The muscle proteins, in contrast, are replaced much more slowly, one-half of the total number of molecules being replaced every 180 days. The celebrated aphorism of Sir Frederick Gowland Hopkins, the late English biochemist, sums up this concept very succinctly: "Life is a dynamic equilibrium in a polyphasic system."

4.9 BIOSYNTHETIC PROCESSES

Our discussion thus far has dealt with processes that break down molecules of foodstuffs and conserve their energy in the biologically useful form of ~P. Cells have the ability to carry out an extensive array of biosynthetic processes utilizing the energy of ATP and, as raw materials, some of the five-, four-, three-, two- and one-carbon compounds that are intermediates in the metabolism of glucose, fatty acids, amino acids and other compounds.

This whole subject of intermediary metabolism, by which an enormous variety of compounds is synthesized, is much too complicated to be discussed in detail here. The enzyme controlling each reaction is genetically determined and the over-all maze of enzyme reactions by which any compound is synthesized includes a variety of self-adjusting control mechanisms to regulate and integrate the reactions. Several basic principles of cellular biosynthesis can be distinguished:

1. Each cell, in general, synthesizes its own proteins, nucleic acids, lipids, polysaccharides and other complex molecules and does not receive them preformed from other cells. Muscle glycogen, for example, is synthesized within the muscle cell and is not derived from liver glycogen.

2. Each step in the biosynthetic process is catalyzed by a separate enzyme.

3. Although certain steps in a biosynthetic sequence will proceed without the use of energy-rich phosphate, the over-all synthesis of these complex molecules requires chemical energy. Why should this be true? Can you relate this to your understanding of the concept of entropy?

4. The synthetic processes utilize as raw materials relatively few substances, among which are acetyl coenzyme A, glycine, suc-

cinyl coenzyme A, ribose, pyruvate and glycerol.

5. These synthetic processes are in general not simply the reverse of the processes by which the molecule is degraded but include one or more separate steps which differ from any step in the degradative process. These steps are controlled by different enzymes and this permits separate control mechanisms to govern the synthesis and the degradation of the complex molecule.

6. The biosynthetic process includes not only the formation of the macromolecular components from simple precursors but their assembly into the several kinds of membranes that compose the outer boundary of the cell and the intracellular organelles. Each cell's constituent molecules are in a dynamic state and are constantly being degraded and synthesized. Thus, even a cell that is not growing, not increasing in mass, uses a considerable portion of its total energy for the chemical work of biosynthesis. A cell that is growing rapidly must allocate a correspondingly larger fraction of its total energy output to biosynthetic processes, especially the biosynthesis of protein. A rapidly growing bacterial cell may use as much as 90 per cent of its total biosynthetic energy for the synthesis of proteins.

Many of the steps in biosynthetic processes involve the formation of peptide bonds, glycosidic bonds and ester bonds, but these bonds are *not* formed by reactions in which water is removed. The biosynthesis of sucrose in the cane sugar plant, for example, does not proceed via

$$glucose + fructose \rightleftharpoons sucrose + H_2O$$

This reaction would require energy, some 5.5 kilocalories per mole, to go to the right if all reactants were present in the concentration of 1 mole per liter. However, the concentration of glucose and fructose in the plant cell is probably less than 0.01 mole per liter, whereas the concentration of water is very high, about 55 moles per liter. Thus, the equilibrium point of the reaction under these conditions would be very far to the left.

Instead, one or more of the reactants is activated by a reaction with ATP in which the terminal phosphate is enzymatically transferred to glucose with the conservation of some of the energy of ATP. The glucose phosphate, with a higher energy content than free glucose, can react with fructose via another enzyme-catalyzed reaction to yield sucrose and inorganic phosphate.

$$\text{ATP} + \text{glucose} \rightarrow$$
$$\text{ADP} + \text{glucose-1-phosphate}$$
$$\text{glucose-1-phosphate} + \text{fructose} \rightarrow$$
$$\text{sucrose} + \text{phosphate}$$
Sum: $\text{ATP} + \text{glucose} + \text{fructose} \rightarrow$
$$\text{sucrose} + \text{ADP} + \text{P}_i$$

This reaction proceeds to the right because there is a net decrease in free energy. The 10 kilocalories of the ~P bond are used to supply the 5.5 kilocalories needed to assemble the glucose and fructose into sucrose; the over-all decrease in free energy is 4.5 kilocalories per mole. Since water is not a product of this reaction, the high concentration of water in the cell does not inhibit it.

The same principle applies to the synthesis of the peptide bonds of proteins, the ester bonds of lipids and nucleic acids, and the glycosidic bonds of polysaccharides, but in many of these not one but *two* terminal phosphates of ATP are transferred as a unit in the reaction by which one of the molecules is activated. The activated molecule may be X ~PP or X ~AMP, depending on which part of the ATP is transferred to the substrate molecule. In either case the reaction ultimately involves the hydrolysis of 2 ~P with an energy utilization of 20 kilocalories per mole.

$$X + \text{ATP} \rightarrow X \sim\text{PP} + \text{AMP}$$
$$X \sim\text{PP} + Y \rightarrow XY + \text{PP}_i$$
$$\text{PP}_i + \text{H}_2\text{O} \rightarrow \text{P}_i + \text{P}_i$$
Sum: $X + Y + \text{ATP} \rightarrow XY + \text{AMP} + 2\text{P}_i$

4.10   BIOLUMINESCENCE

A number of animals, and some molds and bacteria as well, have an enzymic mechanism for the production of light. Luminescent animals are found among the protozoa, sponges, cnidaria, ctenophores, nemerteans, annelids, crustaceans, centipedes, millipedes, beetles, flies, echinoderms, mollusks, hemichordates, tunicates and fishes. From this wide and irregular distribution of the light-emitting ability it is clear that the enzymes for luminescence have appeared independently in a number of different evolutionary lines. It is sometimes difficult to establish that a given organism is itself luminescent; in a number of instances animals once believed to be luminescent have been shown instead to contain luminescent bacteria. When the bacteria are removed the animal is no longer able to emit light. Several different exotic East Indian fish have **light organs** under their eyes in which live luminous bacteria (Fig. 4.11). The light organs contain long cylindrical cells which are well provided with blood vessels to supply an adequate amount of oxygen to the bacteria. The bacteria emit light continuously and the fish have a black membrane, somewhat similar to an eyelid, that can be drawn up over the light organ to turn off the light. How the bacteria come to collect in the fish's light organ, as they must in each newly hatched fish, is a complete mystery.

Some animals have accessory lenses, reflectors and color filters with the light-producing organ, and the whole complex assembly is like a lantern. Certain shrimp have such complicated light-emitting organs.

The production of light is an enzyme-controlled reaction, the details of which differ in different organisms. Bacteria and fungi produce light continuously if oxygen is available. Most luminescent animals, in contrast, give out flashes of light only when their luminescent organs are stimulated. The name **luciferin** has been given to the material which is oxidized to produce light and **luciferase** to the enzyme which catalyzes the reaction. The luciferins from the crustacean *Cypridina* and from the firefly *Photinus* have been isolated and crystallized and shown to be quite different chemically.

The oxidation of luciferin by luciferase can occur only in the presence of oxygen. It is pos-

Figure 4.11 *Anomalops katoptron*, a luminescent fish from the waters of the Malay Archipelago. The crescent-shaped luminescent organs below the eyes are equipped with reflectors. (After Steche.) (From Villee, C. A.: Biology. 7th ed. Philadelphia, W. B. Saunders Co., 1977.)

sible to extract luciferin and luciferase from a firefly, mix the two in a test tube with added Mg^{2+} and adenosine triphosphate, and demonstrate the emission of light in the test tube. The energy for the reaction is supplied by the ATP, and under certain conditions the amount of light emitted is proportional to the amount of ATP present. This system can be used to measure the amount of ATP in a tissue extract.

The amount of light produced by certain luminescent animals is amazing. Many fireflies produce as much light, in terms of lumens per square centimeter, as do modern fluorescent lamps. Different kinds of animals may emit lights of different colors, red, green, yellow or blue. One of the more spectacular luminescent creatures is the "railroad worm" of Uruguay, the larva of a beetle, which has a row of green lights along each side of its body and a pair of red lights on its head. The light produced by luminescent organisms is entirely in the visible part of the spectrum; no ultraviolet or infrared light is produced. Since very little heat is given off in the process, bioluminescence has been called "cold light."

What advantage an animal derives from the emission of light can only be guessed at. For deep-sea animals, which live in perpetual darkness, light organs might be useful to enable members of a species to recognize one another, to serve as a lure for prey or as a warning for would-be predators. Experiments have shown that the light emitted by fireflies serves as a signal to bring the two sexes together for mating. The light emitted by bacteria and fungi probably serves no useful purpose to the organisms, but is simply a by-product of oxidative metabolism, just as heat is a by-product of metabolism in other plants and animals.

ANNOTATED REFERENCES

Baldwin, E. B.: Dynamic Aspects of Biochemistry. 4th ed. New York, Cambridge University Press, 1964. A detailed and technical but very well-written and interesting account of cellular metabolism.

Dixon, M., and E. C. Webb: The Enzymes. 2nd ed. New York, Academic Press, 1964. Presents, in considerable depth and detail, the structure and kinetic properties of enzymes together with a classification of enzymes by function.

Harvey, E. N.: Bioluminescence. New York, Academic Press, 1952. A classic presentation of the interesting phenomenon of "living light."

Henderson, L. J.: The Fitness of the Environment. New York, Macmillan, 1913. Advanced the thesis that the environment had to have certain physical and chemical characteristics for life to develop.

Lehninger, A. L.: Bioenergetics. 2nd ed. New York, Benjamin, 1971. An excellent discussion of the principles of thermodynamics and their application to biological systems.

Schoenheimer, R.: The Dynamic State of the Body Constituents. Cambridge, Harvard University Press, 1949. Describes the classic experiments that demonstrated the rapid renewal of the chemical constituents of tissues.

The standard college textbooks of biochemistry may be consulted for more detailed discussions of the topics presented in this chapter. These include:

McGilvery, R. W.: Biochemical Concepts, Philadelphia, W. B. Saunders Co., 1975.

Lehninger, A.: Biochemistry. 2nd ed. New York, Worth Publishing Co., 1975.

Metzler, D. E.: Biochemistry. New York, Academic Press, 1977.

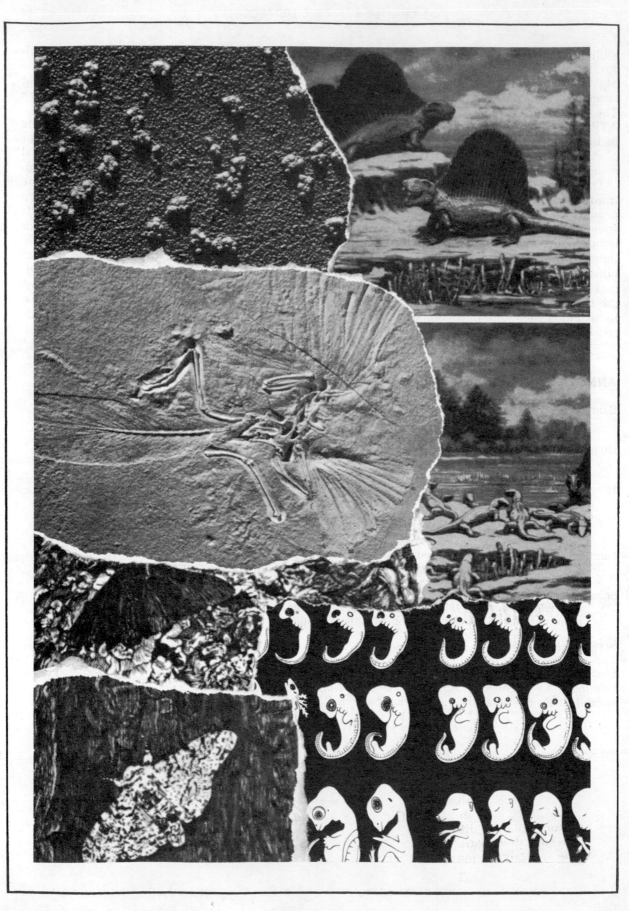

Part Two

ANIMAL DEVELOPMENT AND HEREDITY

DEVELOPMENT

The division, growth and differentiation of a fertilized egg into the remarkably complex and interdependent system of organs that is the adult animal is certainly one of the most fascinating of all biological phenomena. Not only are the organs complicated and reproduced in each new individual with extreme fidelity of pattern but also many of these organs begin to function while they are still developing. The human fetal heart begins to beat during the fourth week of gestation, long before its development is complete.

The pattern of cleavage, blastula formation and gastrulation is seen with various modifications in all multicellular animals. The details of later development are quite different in animals of different phyla but are similar in more closely related forms. The main outlines of human development can be discerned by studying the embryos of rats, pigs, chickens and even frogs. The development of humans and other primates is characterized by the early development of the embryonic membranes, the amnion, chorion and allantois.

5.1 DEVELOPMENTAL STAGES

The development of a new animal begins with the formation of eggs and sperm in members of the parental generation. During gametogenesis (p. 88) meiosis reduces the number of chromosomes from the diploid to the haploid condition, and there is a random selection of the specific genes included in the gametes that will be united in the new individual and guide its development. The egg cell grows and accumulates in its cytoplasm a variety of substances that will be used in metabolic processes in the early stages of development of the fertilized egg. The sperm matures and changes from a spherical cell to one adapted to move actively to the egg and fertilize it.

The fertilization of the egg, the second stage of development, begins with the processes that bring the parents together and stimulate them to release their gametes. The penetration of the egg by the sperm is followed by the activation of the fertilized egg so that it begins to develop. During the activation process the positions of the several kinds of organ-forming material within the egg's cytoplasm may undergo marked changes.

The fertilized egg then undergoes a rapid succession of mitotic divisions termed **cleavage;** these result in a hollow ball of cells, a **blastula,** with a single layer of cells, the **blastoderm,** surrounding the **blastocoele cavity.** During the cleavage divisions the size of the embryo does not increase and the cleavage cells, or blastomeres, become successively smaller with each division. Since glycogen and other nutrients stored in the cytoplasm of the egg are metabolized to provide energy for the cleavage divisions, the mass of the blastula (its dry weight) may be less than the mass of the unfertilized egg.

The next stage of development, **gastrulation,** results in the appearance of two or more layers of cells, the **germinal layers,** from the single-layered blastoderm. The cavity of the double-walled gastrula formed by invagination is called the **archenteron** (primitive gut), and the opening from the archenteron to the exterior is the **blastopore.** The bodies of higher animals consist of several layers of tissues and organs, all of which can be traced back to the three germinal layers, **ectoderm, mesoderm** and **endoderm.** The outer ectoderm forms the epidermis of the skin and the nervous system;

the innermost layer, the endoderm, forms the lining of the gut and the digestive glands; and the middle layer, the mesoderm, is the source of the muscles, the vascular system, the lining of the coelomic cavity, the reproductive organs, the excretory organs and the bones and cartilages of the internal skeleton.

The archenteron, or a portion of it, eventually gives rise to the cavity of the digestive tract. The blastopore develops into the oral opening in cnidarians and becomes the anal opening in echinoderms and chordates. In mollusks, annelids and arthropods, the blastopore becomes divided into two openings, one of which becomes the mouth and the other the anus. The central portion of the digestive tract is formed from endoderm, but typically the ectoderm invaginates at both oral and anal ends and fuses with the endoderm. Thus the most anterior portion of the digestive tract, the **stomodeum,** or mouth cavity, and the most posterior portion of the digestive tract, the proctodeum adjoining the anus, are lined with ectoderm rather than endoderm.

Development proceeds with **organogenesis,** during which the three germ layers divide into smaller masses of cells, each destined to form a specific organ or portion of the adult. Each organ is first discernible as a group of cells, a rudiment, which becomes segregated from the other cells of the embryo. Organogenesis is followed by growth and differentiation of the cells of the organ rudiments. Eventually the cells acquire the specific structures and chemical and physiological properties that enable them to perform their functions so that the developing animal can take up an independent existence.

The young organism emerging from the egg may be a miniature copy of the adult, or it may lack certain organs and differ to some extent from the adult in form as well as in size. In some species the form emerging from the egg has special organs, not present in the adult, that are needed for survival of the young in its particular mode of existence. Such a young animal, termed a **larva,** may have a life style quite different from the adult. Eventually the larva undergoes a relatively rapid and marked change of form, **metamorphosis,** and achieves the adult form. During metamorphosis the special larval characteristics are lost and the adult characteristics appear. The young and adult animals continue to undergo changes with time which we call **aging** and which must be considered part of the developmental processes.

5.2 MEIOSIS

The process of mitosis, which ensures that each daughter cell will receive exactly the same number and kind of chromosomes that the parent cell had, was discussed earlier (p. 46) together with the features of the cell cycle. The constancy of the chromosome number in the cells of successive generations of organisms is ensured by the process of **meiosis,** which occurs during the formation of eggs or sperm in animals. The process of meiosis involves a pair of cell divisions during which the chromosome number is reduced to one-half, so that each gamete receives only half as many chromosomes as are contained in the somatic cells of the parent. When two gametes unite in fertilization the fusion of their nuclei reconstitutes the 2N number of chromosomes. In meiosis the members of each pair of chromosomes separate and pass to different daughter cells. As a result of this, each gamete contains one and only one of each kind of chromosome; in other words, it contains one complete set of chromosomes. This is accomplished by the pairing, or **synapsis,** of the like chromosomes and a separation of the members of the pair, with one going to each pole of the dividing cell. The like chromosomes that pair during meiosis are called **homologous chromosomes.** They are identical in size and shape, have identical chromomeres along their length and contain similar hereditary factors, or genes. A set of one of each kind of chromosome is called the 1N, or **haploid, number.** A set of two of each kind is called the 2N, or **diploid, number.** The haploid number for the human is 23 and the diploid number is 46. Gametes—eggs and sperm—have the haploid number. Fertilized eggs (zygotes) and all the cells of the body developing from the zygote have the diploid number. A fertilized egg gets exactly half of its chromosomes and half of its genes from its mother and the other half from its father. Only the last two cell divisions, which result in mature functional eggs or sperm, are meiotic; all other cell divisions are mitotic.

The process of meiosis consists of two cell divisions, the first and second meiotic divisions, which occur in succession (Fig. 5.1). Each of these includes prophase, metaphase, anaphase and telophase stages, but there are important differences between mitosis and meiosis, especially in the prophase of the first meiotic division. In this the chromosomes first appear as long, thin threads, becoming shorter and thicker, as in mitosis. The

X-over occurs here

Figure 5.1 Meiosis in a hypothetical animal with a diploid chromosome number of six. It has three pairs of chromosomes, of which one is short, one is long with a hook at the end, and one is long and knobbed. *A*, Early prophase of the first meiotic division: chromosomes begin to appear. *B*, Synapsis: the pairing of the homologous chromosomes. *C*, Apparent doubling of the synapsed chromosomes to form groups of four identical chromatids, tetrads. *D*, Metaphase of the first meiotic division: the tetrads lined up across the equatorial plate. *E*, Anaphase: the chromatids migrating toward the poles. *F*, Telophase of the first meiotic division. *G*, Prophase of the second meiotic division. *II*, Metaphase of the second meiotic division. *I*, Anaphase of the second meiotic division. *J*, Mature gametes, each of which contains only one of each kind of chromosome.

homologous chromosomes undergo synapsis while they are still elongate and thin. The homologous chromosomes pair longitudinally; they come to lie close together side by side along their entire length and are twisted around each other. After synapsis, or pairing, has occurred, the chromosomes continue to shorten and thicken. Each one becomes visibly double, consisting of two chromatids, as in mitosis. This doubling has occurred in the S phase before meiosis begins.

At the end of the first meiotic prophase the chromosomes have doubled and undergone synapsis to yield a bundle of four homologous chromatids called a **tetrad.** Each pair of chromosomes gives rise to a bundle of four so the number of tetrads equals the haploid number of chromosomes. In human cells there are 23 tetrads (and a total of 92 chromatids) at this stage. The centromeres have not divided and there are only two centromeres for the four chromatids of each tetrad.

While these events are occurring, the two centrioles pass to opposite poles, a spindle forms between the centrioles, and the nuclear membrane dissolves. The tetrads line up around the equator of the spindle and the cell is said to be in metaphase. In the anaphase of the first meiotic division the daughter chromatids formed from each chromosome, still united by the centromere, separate and move toward opposite poles. Thus the homologous chromosomes of each pair, but not the daughter chromatids of each chromosome, are separated in anaphase 1. This differs from mitotic anaphase in which the centromeres do divide and the daughter chromatids pass to opposite poles. At telophase the haploid number of double chromosomes is present at each pole.

In most animals there is no clear interphase between the two meiotic divisions. The chromosomes do not divide into daughter chromatids and there is no synthesis of DNA as there is in the S phase between mitotic divisions.

The chromosomes do not form chromatin threads; instead the centriole divides, a new spindle forms in each cell at right angles to the spindle of the first division, and the haploid number of double chromosomes lines up on the equator of each spindle. The telophase of the first meiotic division and the prophase of the second meiotic division are usually of rather short duration. The lining up of the double chromosomes on the equator of the spindle constitutes the metaphase of the second meiotic division. The metaphases of the first and second meiotic divisions can be distinguished because in the first the chromosomes are arranged in bundles of four and in the second the chromosomes are arranged in bundles of two. The centromeres divide and the daughter chromatids, now chromosomes, separate and move to opposite poles. Thus in the telophase of the second meiotic division in humans, 23 chromosomes, one of each kind, arrive at each pole. The cytoplasm then di-

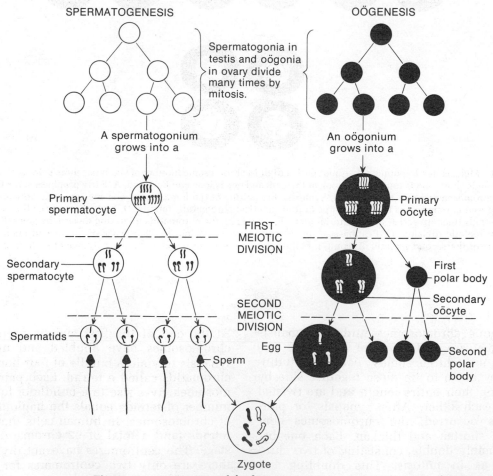

Figure 5.2 Comparison of the formation of sperm and eggs.

vides, nuclear membranes form, and the chromosomes gradually elongate and become chromatin threads.

The two successive meiotic divisions yield four nuclei, each of which has one and only one of each kind of chromosome, a haploid set. The members of the homologous pairs of chromosomes are segregated into separate daughter cells. The four cells resulting from the two meiotic divisions are now mature gametes and do not undergo any further mitotic or meiotic divisions.

Fundamentally the same process occurs in the meiotic divisions in the testis, which result in sperm, and in the meiotic divisions in the ovary, which result in eggs, but there are some important differences in detail.

5.3 SPERMATOGENESIS

A typical testis consists of thousands of cylindrical sperm tubules in each of which develop billions of sperm. The walls of the sperm tubules are lined with primitive, unspecialized germ cells called **spermatogonia.** Throughout embryonic development and during early postnatal development the spermatogonia divide mitotically, giving rise to additional spermatogonia to provide for the growth of the testis. After sexual maturity some spermatogonia begin to undergo spermatogenesis, the formation of mature sperm, while others continue to divide mitotically and produce more spermatogonia for later spermatogenesis. In most wild animals there is a definite breed-

Figure 5.3 Diagram of part of a section of a human seminiferous tubule to show the stages in spermatogenesis and in the transformation of a spermatid into a mature sperm. (From Villee, C. A.: Biology. 7th ed. Philadelphia, W. B. Saunders Co., 1977.)

ing season, either in spring or fall, during which the testis increases in size and spermatogenesis occurs. Between breeding seasons the testis is small and contains only spermatogonia. In man and most domestic animals spermatogenesis occurs throughout the year once sexual maturity is reached.

Spermatogenesis begins with the growth of the spermatogonia into larger cells known as **primary spermatocytes** (Fig. 5.2). These divide (first meiotic division) into two **secondary spermatocytes** of equal size which in turn undergo the second meiotic division to form four **spermatids** of equal size. The spermatid, a spherical cell with a generous amount of cytoplasm, is a mature gamete with the haploid number of chromosomes. A complicated process of growth and differentiation, though not cell division, converts the spermatid into a functional sperm. The nucleus shrinks in size and becomes the head of the sperm. (Fig. 5.3), while the sperm sheds most of its cyto-

plasm. Some of the Golgi bodies congregate at the front end of the sperm and form a cap, the **acrosome.** This contains enzymes which may play a role in penetrating the cell membrane of the egg.

The two centrioles of the spermatid move to a position just in back of the nucleus. A small depression appears on the surface of the nucleus, and one of the centrioles, the proximal centriole, takes up a position in the depression at right angles to the axis of the sperm. The second, or distal, centriole, just behind the proximal centriole, gives rise to the **axial filament** of the sperm tail (Fig. 5.4). Like the axial filament of flagella, it consists of two longitudinal fibers in the middle and a ring of nine pairs, or doublets, of longitudinal fibers surrounding the two.

The mitochondria move to the point at which head and tail meet and form a small **middle piece** that provides energy for the beating of the tail. Most of the cytoplasm of the spermatid is discarded as **residual bodies** that are taken up by phagocytosis by the **Sertoli cells** in the seminiferous tubules. The mature sperm retains only a thin sheath of cytoplasm surrounding the mitochondria in the middle piece and the axial filament of the tail.

The spermatozoa of various animal species

Figure 5.4 A, Diagram of the head and middle piece of a mammalian sperm, greatly enlarged, as seen in the electron microscope. B and C, Top and side views of a sperm seen by light microscopy. (From Villee, C. A.: Biology. 7th ed. Philadelphia, W. B. Saunders Co., 1977.)

Figure 5.5 Spermatozoa from different species of vertebrates, illustrating the differences in size and shape. 1, Gastropod. 2, Ascaris. 3, Hermit crab. 4, Salamander. 5, Frog. 6, Chicken. 7, Rat. 8, Sheep. 9, Man. (From Villee, C. A.: Biology. 7th ed. Philadelphia, W. B. Saunders Co., 1977.)

may be quite different. There are great variations in the size and shape of the tail and in the characteristics of the head and middle piece (Fig. 5.5). The sperm of a few animals such as the parasitic round worm *Ascaris* have no tail and move instead by ameboid motion. Crabs and lobsters have a curious tailless sperm with three pointed projections on the head which stick to the surface of the egg, holding the sperm securely in place. The middle piece uncoils like a spring and pushes the nucleus of the sperm into the egg cytoplasm, thus accomplishing fertilization.

5.4 OÖGENESIS

The ova, or eggs, develop in the ovary from immature sex cells, **oögonia.** Early in development the oögonia undergo many successive mitotic divisions to form additional oögonia, all of which have the diploid number of chromosomes. In many animals, notably the vertebrates, the oögonia and oöcytes are surrounded by a layer of follicle cells. In the human this occurs early in fetal development, and by the third month the oögonia begin to develop into **primary oöcytes.** When a human female is born, her two ovaries contain some 400,000 primary oöcytes that have attained the prophase of the first meiotic division. These primary oöcytes remain in prophase until the woman reaches sexual maturity; then as each follicle matures the first meiotic division resumes and is completed at about the time of ovulation (15 to 45 years after meiosis began!).

The events occurring in the nucleus—synapsis, the formation of tetrads, and the separation of the homologous chromosomes—are similar to those occurring in spermatogenesis, but the division of the cytoplasm is unequal, resulting in one large cell, the **secondary oöcyte,** which contains the yolk and nearly all the cytoplasm and one small cell, the first **polar body,** which consists of practically nothing but a nucleus (Fig. 5.2). It was named a polar body before its significance was understood because it appeared as a small speck at the animal pole of the egg.

In the second meiotic division the secondary oöcyte divides unequally to yield a large **oötid** with essentially all the yolk and cytoplasm, and a small second polar body, both of which have the haploid number of chromosomes. The first polar body may divide at about the same time into two additional polar bodies. The oötid undergoes further changes, but no further cell division, to become a mature ovum. The three small polar bodies soon disintegrate so that each primary oöcyte gives rise to just one ovum in contrast to the four sperm formed from each primary spermatocyte. The unequal cytoplasmic division ensures that the mature egg will have enough cytoplasm and stored yolk to survive if it is fertilized. The primary oöcyte in a sense puts all of its yolk in one ovum; the egg has neatly solved the problem of reducing its chromosome number without losing the cytoplasm and yolk needed for development after fertilization.

The composition of the stored nutrients in the yolk varies from one species to another but usually includes proteins, phospholipids and neutral fats organized in granules. The amount and distribution of yolk in an egg plays an important role in determining the pattern of development that ensues following fertilization. Eggs of animals such as sea urchins and mammals, with small amounts of yolk evenly distributed within the cytoplasm, are called **isolecithal** or **homolecithal.**

Many animals, including flatworms, snails, clams and most vertebrates, have eggs with the yolk concentrated at the lower, or **vegetal, pole.** Such eggs are termed **telolecithal.** The upper, or **animal, pole** contains the nucleus and less yolky cytoplasm. Frogs have moderately telolecithal eggs, but those of reptiles and birds are extreme, with as much as 90 per cent of the egg composed of yolk. A cap of nonyolky cytoplasm containing the nucleus rests on the animal pole. Two proteins found in the yolks of both amphibian and avian eggs are **phosvitin** and **lipovitellin.** Phosvitin is a curious protein; about half of its amino acid residues are phosphorylated serine. Lipovitellin is a larger protein that contains about 17 per cent lipid. In amphibians and birds, phosvitin is synthesized in the maternal liver and transported in the blood to the ovary. Once taken up by the egg, it may be phosphorylated further, which makes it less soluble and causes it to precipitate in the oöcyte.

The eggs of arthropods, especially insects, have a different pattern of yolk distribution and are termed **centrolecithal.** The yolk is concentrated in the center of the egg, and the cytoplasm is present as a thin layer on the entire surface of the egg. In addition, an island of cytoplasm in the center of the egg contains the nucleus.

5.5 FERTILIZATION

Developing eggs eventually reach a state of maturation at which they are capable of fusion with a sperm, a process known as *fertilization.* By various means, described in more detail in Chapter 20, sperm and eggs of the same species are brought into proximity with each other. The swimming movements of the sperm bring it into contact with the egg. Each gamete contributes to a complex series of cellular changes that result in the passage of the sperm through the egg membrane and outer envelopes. The acrosome releases enzymes that aid in penetration of the envelopes; in some species, the acrosome forms a projecting filament that is the first part of the sperm to contact the egg membrane. The cell membrane of the egg fuses with that of the sperm on either side of the head, in effect incorporating the sperm within the cytoplasm of the egg. The tail may remain outside.

Within the egg cytoplasm, the compact nuclear material of the sperm head swells to form a **sperm,** or **male pronucleus,** that moves toward the **egg,** or **female pronucleus.** The two pronuclei either fuse together to form a zygote nucleus or each contributes its chromosomes to the first cleavage spindle.

Sperm penetration activates the egg, throwing its metabolic machinery, so to speak, from neutral into forward gear. The egg may now complete its meiotic divisions, for in many animals, egg meiosis halts at various points until fertilization occurs. The egg then prepares for the cleavage divisions that lead to the formation of the embryo.

The union of one haploid set of chromosomes from the sperm with another haploid set from the egg, which occurs in fertilization, reestablishes the diploid chromosome number. Thus the fertilized egg, or zygote, and all of the body cells developing from it by mitosis have the diploid number of chromosomes. In each individual exactly half of the chromosomes and half of the genes come from the mother and the other half from the father. All of the phenomena of mendelian genetics depend upon these simple facts. Because of the nature of gene interaction, the offspring may resemble one parent more than the other, but the two parents make essentially equal contributions to its inheritance.

5.6 CLEAVAGE AND GASTRULATION

The entrance of the sperm into the egg (Fig. 5.6) initiates a rapid series of changes—the completion of the meiotic divisions, the fusion of male and female pronuclei, the complex movements of the egg's cytoplasmic constituents and the beginning of the cleavage divisions. After fertilization, the rates of oxygen consumption and protein synthesis in the eggs of certain species rise sharply. Some evidence supports the hypothesis that the RNA present in the fertilized egg, which directs protein synthesis during the early stages of cleavage, was synthesized in the oöcyte before fertilization and hence is a product of the maternal genotype rather than of the embryo's genotype. The RNA is masked in the unfertilized egg, perhaps by combination with protein, and is unavailable for translation into a peptide chain. The ribosomes of the unfertilized egg are generally single, and polyribosomes appear only after fertilization has occurred. One of the effects of fertilization appears to be an unmasking of the maternal messenger RNA so that it can be translated in protein synthesis. Later, at about the time of gastrulation, the genome of the embryo is transcribed to form embryonic messenger RNA, which codes for proteins that play a role in the gastrulation process and in later development.

After fertilization the zygote divides repeatedly by a series of mitotic cell divisions termed **cleavage.** The pattern of cell division is determined largely by the amount of yolk present in the egg. The plane of the first cleavage division of an isolecithal egg passes through the animal and vegetal poles of the egg, yielding two equal cells, or **blastomeres** (Fig. 5.7). The second cleavage division passes through animal and vegetal poles, at right angles to the first, and divides the two cells into four. The third cleavage division is horizontal—its plane is at right angles to the planes of the first two cell divisions and the

Figure 5.6 Diagram of the stages in the process of fertilization, the union of the egg and sperm.

AMPHIOXUS   AMPHIBIAN   BIRD   MAMMAL

unsegmented

polar bodies — zona pellucida

2-cell

4-cell

8-cell

blastula-exterior

blastula-vertical section

blastocoel   subgerminal space   outer inner cell mass

early gastrula longitudinal section

archenteron

archenteron   yolk

Figure 5.7 Stages in cleavage and early gastrulation in eggs of chordates. *A, Amphioxus* (holoblastic cleavage, isolecithal egg with little yolk). *B,* Frog (holoblastic cleavage, moderately telolecithal egg with much yolk). *C,* Bird (meroblastic discoidal cleavage, telolecithal egg with much yolk). *D,* Mammal (holoblastic cleavage, isolecithal egg with essentially no yolk). (From Storer and Usinger: General Zoology, 3rd ed. Copyright, 1957, by McGraw-Hill Book Co., Inc.)

embryo is split into four cells above and four cells below this line of cleavage. Further divisions result in embryos containing 16, 32, 64, 128 cells and so on, until a hollow ball of cells, the **blastula,** is formed. The wall of the blastula consists of a single layer of cells, and the fluid-filled cavity in the center of the sphere is termed the blastocoele.

The single-layered blastula is converted into a double-layered sphere, a **gastrula,** by the invagination of a section of one wall of the blastula (Fig. 5.8). This eventually meets the opposite wall and obliterates the original blastocoele. The new cavity of the gastrula is the **archenteron,** and the opening of the archenteron to the exterior, the **blastopore,** marks the site of the invagination that produced the

gastrula. The outer of the two walls of the gastrula is the ectoderm, which eventually forms the skin and nervous system. The inner layer of cells lining the archenteron is mainly the presumptive endoderm, which will form the digestive tract and its outgrowths, such as the liver, pancreas and lung, plus the presumptive notochord and presumptive mesoderm, which will form the remaining organs of the body. As the gastrula elongates in the anterior-posterior axis the presumptive notochord is stretched into a longitudinal band of cells occupying the mid-dorsal part of the inner layer. The presumptive mesoderm forms two longitudinal bands of cells, one on each side of the presumptive notochord. The remainder of the lateral, ventral and anterior

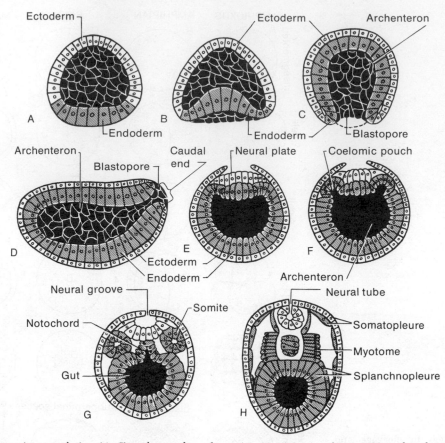

Figure 5.8 Stages in gastrulation (A–C) and mesoderm formation (F–H) in *Amphioxus*. Note that the mesoderm forms by the budding of pouches from the archenteron.

parts of the inner layer are presumptive endoderm cells.

Cleavage and gastrulation are markedly modified in telolecithal eggs by the presence of the large amount of yolk. The cleavage divisions of the cells originating from the lower part of the frog's egg are slowed by the inert yolk so that the blastula consists of many small cells at the animal pole and a few large cells at the vegetal pole (Fig. 5.7). The lower wall of the blastula is much thicker than the upper one, and the blastocoele is flattened and displaced upward. In the bird's egg only the small disc of cytoplasm at the animal pole undergoes cleavage divisions (Fig. 5.10). The lower, yolk-filled part of the egg never cleaves. The shallow cavity under the dividing cells is called the subgerminal space since it is not homologous to the blastocoele of the frog egg. Gastrulation occurs in both frog and chick eggs and an archenteron is formed, but the process is greatly modified by the presence of the yolk. Gastrulation in the frog involves the invagination of the yolk-filled cells of the vegetal pole,

the turning in of cells at the dorsal lip of the blastopore **(involution)** and the growth of ectoderm down and over the cells of the vegetal pole **(epiboly)** (Fig. 5.9).

The continued mitotic divisions during gastrulation lead to more cells and more nuclear material but little or no change in the total volume or mass of the embryo. The rate of metabolism, as measured by the rate of oxygen consumption, increases two- or threefold over the rate during cleavage, presumably to supply the biologically useful energy for the morphogenetic movements and for the sharply increased rate of synthesis of messenger RNA and of proteins.

Cleavage in centrolecithal eggs, such as those of insects, begins with the division of the nucleus in the central island of cytoplasm (Fig. 5.11). After several nuclear divisions without cytoplasmic division the nuclei migrate out from the center of the egg. Each nucleus is surrounded by a bit of the original central cytoplasm. When the nuclei reach the surface of the egg the cytoplasm surrounding

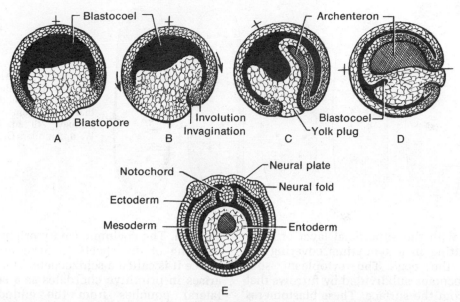

Figure 5.9 *A–D*, Successive stages in gastrulation and mesoderm formation in Amphibia. *E*, Transverse section of an early neurula stage.

Figure 5.10 Successive stages in the cleavage of a hen's egg. *A*, Cleavage is restricted to a small disc of cytoplasm on the upper surface of the egg yolk, called the blastodermic disc. *B*, A subgerminal space appears beneath the blastodermic disc, separating it from the unsegmented yolk. *C*, The blastodermic disc cleaves into an upper epiblast and a lower hypoblast separated by the blastocoele. (From Villee, C. A.: Biology. 7th ed. Philadelphia, W. B. Saunders Co., 1977.)

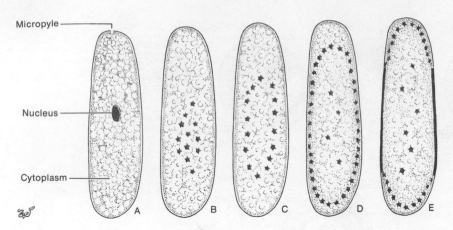

Micropyle

Nucleus

Cytoplasm

Figure 5.11 A, The centrolecithal egg of a beetle. B–E, Stages in the cleavage of the centrolecithal egg. (From Villee, C. A.: Biology. 7th ed. Philadelphia, W. B. Saunders Co., 1977.)

them fuses with the superficial layer of cytoplasm, resulting in a syncytium covering the surface of the egg. The cytoplasm subsequently becomes subdivided by furrows that extend in from the surface. These blastomeres are connected to the yolk mass for a time but eventually become separated from it; the constituents of the yolk are gradually used up as nutrients for the developing embryo. This stage can be compared to the formation of the blastula even though there is no cavity comparable to the blastocoele. The blastoderm surrounds a mass of uncleaved yolk rather than a cavity.

In all animals except sponges and cnidaria, which never develop beyond the gastrula stage, a third layer of cells, the mesoderm, develops between ectoderm and endoderm. In annelids, mollusks and certain other invertebrates the mesoderm develops from special cells that are differentiated early in cleavage (p. 99). These migrate to the interior and come to lie between the ectoderm and endoderm. They then multiply to form two longitudinal cords of cells that develop into sheets of mesoderm between the ectoderm and en-

doderm. The coelomic cavity originates by the splitting of the sheets to form pockets, and hence it is called a schizocoele. The mesoderm arises in primitive chordates as a series of bilateral pouches from the endoderm (Fig. 5.8). These lose their connection with the gut and fuse one with another to form a connected layer. The cavity of the pouch is retained as the coelom, called an **enterocoele** because it is derived indirectly from the archenteron. The mesoderm in amphibia is formed in part from the roof of the archenteron and in part from cells that roll in at the lips of the blastopore and move anteriorly as a sheet between ectoderm and endoderm (Fig. 5.9). In birds and mammals a thickened band of ectoderm and endoderm cells, the primitive streak, develops on the surface of the developing embryo, marking the longitudinal axis of the embryo (Fig. 5.12). This is homologous to the dorsal lip of the blastopore of lower forms. At the primitive streak cells migrate in from the surface, proliferate and spread as a sheet of mesoderm between the ectoderm and endoderm. The primitive streak is a dynamic structure and persists even though the cells com-

Edge of area pellucida — Primitive streak
— Hensen's node
Epiblast

Figure 5.12 Gastrulation in the bird. The anterior half of the area pellucida of a chick embryo is cut transversely to show the migration of mesodermal and endodermal cells from the primitive streak. (From Villee, C. A.: Biology. 7th ed. Philadelphia, W. B. Saunders Co., 1977.)

Hypoblast
Migrating cells

TABLE 5.1 DERIVATIVES OF GERM LAYERS AND MAMMALIAN DEVELOPMENT

Ectoderm	Endoderm	Mesoderm
Epidermis of the skin	Lining of gut	Muscles—smooth, skeletal and cardiac
Hair and nails	Lining of trachea, bronchi and lungs	Dermis of skin
Sweat glands	Liver	Connective tissue, bone and cartilage
Brain, spinal cord, ganglia, nerves	Pancreas	Dentin of teeth
Receptor cells of sense organs	Lining of gallbladder	Blood and blood vessels
Lens of the eye	Thyroid, parathyroid and thymus glands	Mesenteries
Lining of mouth, nostrils and anus	Urinary bladder	Kidneys
Enamel of teeth	Lining of urethra	Testes and ovaries

posing it are constantly changing as they migrate.

However the mesoderm may originate, it typically forms two sheets which grow laterally and anteriorly between the ectoderm and endoderm; one sheet becomes attached to the inner endoderm and the other to the outer ectoderm. The cavity between the two becomes the coelom, or body cavity. This splitting of the mesoderm permits the development of two independent sets of muscles—one in the body wall adapted for locomotion and the other in the gut wall adapted for churning and moving the contents of the gut.

The sheets of mesoderm grow ventrally, and the ones from either side meet in the ventral midline; the coelomic cavities on the two sides then fuse into one. The mesoderm grows dorsally along each side of the notochord and neural tube and becomes differentiated into segmental blocks of tissue, the **somites,** from which the main muscles of the trunk develop. Other mesodermal cells become detached from the inner border of the somites, migrate inward, surround the notochord and neural tube, and develop into the vertebrae. The kidneys and their ducts and the gonads and their ducts are derived from the mesoderm originally located between the somites and the coelom.

The primitive skeleton of the chordate, the **notochord,** is a flexible, unsegmented, longitudinal rod found in the dorsal midline of all chordate embryos. It is formed at the same time and in a similar way as the mesoderm, as an outgrowth of the roof of the archenteron, from the dorsal lip of the blastopore or from the primitive streak. Later in the development of vertebrates the notochord is replaced by the vertebral column derived from part of the mesoderm.

The contributions of each germ layer to the development of a typical mammal are summarized in Table 5.1.

5.7 SPIRAL CLEAVAGE AND ITS EVOLUTIONARY IMPORTANCE

The acoelomate Platyhelminthes and Nemertea, and several eucoelomate phyla, including the Mollusca and Annelida, share a pattern of early embryonic development called **spiral cleavage.** This pattern combines a precise system of cell division with a predetermined fate of the subsequent cells. Within any one species, the pattern is constant, but the pattern may vary a bit from one species to another. In a typical pattern of spiral cleavage the first and second cell divisions are meridional and at right angles to each other, forming four cells of nearly equal size (Fig. 5.13).

The third division is **oblique.** The mitotic spindles are neither vertical nor horizontal but inclined to one side (one of them is indicated by the solid line in the 8–cell stage, Fig. 5.13). As a result, the upper four cells are displaced circularly so that each upper cell touches *two* lower cells. The third division is also *unequal,* separating four upper small **micromeres** from four lower large **macromeres.** The four micromeres are called the **first quartette.** The fourth division is also oblique but always in the opposite direction from the third (Fig. 5.13). The first quartette forms eight cells; the macromeres divide unequally, producing again an upper tier of small cells, the **second quartette,** and a lower tier of four macromeres.

The fifth division continues the pattern. The axes of division are oblique in the direction of those of the third division. Each of the original first quartette cells (indicated by "1" in Figure 5.13) now divides to produce four, for

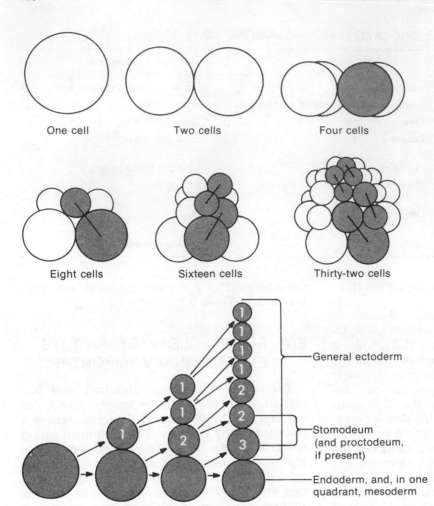

Figure 5.13 Spiral cleavage. One quadrant (the progeny of one cell of the four-cell stage) is shaded. Lines indicate the axes of the preceding mitoses. The lower diagram shows the fates of the cells of one quadrant. Numbers indicate the three quartettes or their progeny.

a total of 16 cells. Each of the original second quartette cells (indicated by "2" in Figure 5.13) now divides to produce two, for a total of eight cells. The macromeres divide unequally to form a **third quartette** of micromeres and a lower tier of four macromeres. Divisions continue to be oblique, alternating to one side and then the other. If the 8, 16, and 32 cell stages are studied carefully, it should be evident that the embryo is twisted by this process, first one way and then the other. This is particularly evident to an observer looking down on the top of the embryo when the macromeres are resting firmly on the bottom of the dish. It is this phenomenon that gives the pattern its name, spiral cleavage. It should be evident that the progeny of each cell of the four-cell stage remain together in one quadrant of the embryo.

Gastrulation usually begins after the sixth or seventh cleavage. In all the phyla showing spiral cleavage the macromeres of the 32-cell stage and all their progeny pass into the interior. Usually all of the mesoderm develops from one of the macromeres, which may be slightly larger than the others, while the other three become endoderm. The three quartettes form all of the ectoderm and its derivatives.

The occurrence of spiral cleavage in so many phyla suggests that it is a basic pattern of development in three-layered animals. The concept of a "main line" of evolution emerges, proceeding from the flatworms and nemerteans to the mollusks and annelids. The members of some phyla, such as the rotifers, gastrotrichs and nematodes, show further specialization and modification toward rigid embryonic fates and cell constancy, while others, including the chordates and echinoderms, show simplification and a loss of rigidity in early development.

5.8 MORPHOGENETIC MOVEMENTS

Gastrulation in all of the types of embryos involves the movement, or migration, of cells, which occurs in specific ways and leads to specific arrangements of cells. These morphogenetic movements involve considerable parts of the embryo, which stretch, fold, con-

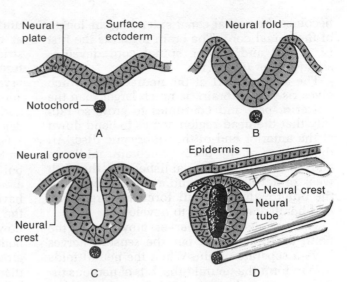

Figure 5.14 A series of cross sectional diagrams through the surface ectoderm to show the formation of the neural tube and neural crest. (After Arey.)

tract or expand. The movements do not appear to be analogous to the contraction of muscles or to some sort of ameboid movement of the whole embryo. The forces driving these movements and the factors controlling their direction are unknown. If removed from the embryo and grown in organ culture, a section of presumptive ectoderm from the early gastrula of the frog will expand actively just as it does *in situ* in contributing to the movements of gastrulation. A bit of the blastopore lip transplanted from one embryo into another will invaginate and form an archenteric cavity independent of the archenteron of the host embryo. Invagination involves not only the movement of cells but also changes in their shape—contraction at certain ends of the cells and expansions at others.

The movements of the cells and the positions they take up are guided, at least in part, by **selective affinities** of certain cells that can be demonstrated if the cells of an early embryo are disaggregated and then incubated in various combinations. Epidermal cells become concentrated on the exterior of the cell mass and mesodermal cells take up a position between the epidermis and endoderm. Neural plate cells form hollow vesicles resembling a neural tube or brain vesicle, and mesodermal cells tend to arrange themselves around coelomic cavities. This sorting out of mixtures of cells is believed to result from their selective affinities. When cells touch as a result of random movement, they may remain in contact if held together or they may move apart if not bound strongly. The specific affinities of different cells are related to the specific kinds of glycoproteins present in the cell membrane. These substances enable the cells to recognize "like" and "unlike" cells. The cells stick to like cells but not unlike ones.

5.9 DEVELOPMENT OF THE NERVOUS SYSTEM

Although the two-week-old human embryo is a simple, flat disc, the two-month-old embryo has nearly all of its structures, at least in rudimentary form. At about the third week the cells in the ectoderm just over the notochord in front of the primitive streak become a thickened **neural plate.** The center of this becomes depressed and forms the **neural groove** (Fig. 5.14), and the outer edges of the plate rise in two longitudinal **neural folds** that meet at the anterior end and appear, when viewed from above, like a horseshoe. These folds gradually come together at the top, forming a hollow **neural tube.** The cavity at the anterior part of this neural tube becomes the ventricles of the brain, and the cavity in the posterior part

Figure 5.15 A dorsal view of a human embryo, showing the closure of the neural tube between somites 4 and 7 and the open neural folds anterior and posterior to this. (From Villee, C. A.: Biology. 7th ed. Philadelphia, W. B. Saunders Co., 1977.)

becomes the neural canal extending the length of the spinal cord. The brain region is the first to appear and a long spinal cord develops slightly later (Fig. 5.15).

The anterior part of the neural tube, which gives rise to the brain, is much larger than the posterior part and continues to grow so rapidly that the head region comes to bend down at the anterior end of the embryonic disc. All the regions of the brain—forebrain, midbrain and hindbrain—are established by the fifth week of development, and a week or two later the outgrowths that will form the large cerebral hemispheres begin to develop.

The various motor nerves grow out of the brain or spinal cord, but the sensory nerves have a separate origin. When the neural folds fuse to form the neural tube, bits of nervous tissue, the **neural crest,** are left over on either side of the tube (Fig. 5.14). These migrate downward from their original position and form the dorsal root ganglia of the spinal nerves and the postganglionic sympathetic neurons. From sensory cells in the dorsal root ganglia, dendrites grow out to the sense organs and axons grow in the spinal cord. Other neural crest cells migrate and become the cells of the adrenal medulla and of the neurilemma sheath of peripheral neurons.

A pair of saclike protrusions, the **optic vesicles,** appear on the lateral walls of the forebrain and grow laterally. The base of the vesicle becomes constricted as the optic nerve. The optic vesicle comes in contact with the inner surface of the overlying epidermis, then flattens out and invaginates to form a double-walled **optic cup.** The inner, thicker layer of the cup becomes the sensory **retina** of the eye, and the outer, thinner layer of the cup becomes the pigment layer of the retina. When the optic cup touches the overlying epidermis it stimulates the latter to develop into a **lens** rudiment. In birds and mammals the epidermis thickens and folds in to produce a pocket. This pinches off and forms a lens vesicle that lies in the opening of the optic cup and is surrounded by the iris, formed from the rim of the optic cup. The cells on the inner side of the lens vesicle become columnar and then are transformed into long fibers. Their nuclei degenerate and the cytoplasm becomes hard, transparent and able to refract light.

5.10 DEVELOPMENT OF BODY FORM

The conversion of the two-week-old flat disc into a roughly cylindrical embryo is accomplished by three processes: (1) the growth of the embryonic disc, which is more rapid than the growth of the surrounding tissue; (2) the underfolding of the embryonic disc, especially at the front and rear ends; and (3) the constriction of the ventral body wall to form the future umbilical cord and to separate the embryo proper from the extra-embryonic parts. In addition, the body begins to separate into head and trunk, and the pectoral and pelvic appendages appear.

Growth is rapid at the anterior end of the embryonic disc, and soon the head region bulges forward from the original embryonic area. The tail, which even human embryos have at this stage, bulges to a lesser extent over the posterior end. The sides of the disc grow downward, eventually to form the sides of the body. The embryo becomes elongated because growth is more rapid at the head and tail ends than laterally. The enlarging of the embryo has been compared to the increase in size of a soap bubble blown from a pipe, which, as it grows, swells out in all directions above the mouth of the pipe (the yolk sac). What is to become the mouth and heart originally lies in front of the embryonic disc, and as the disc grows and bulges over the tissues in front, the mouth and heart swing underneath to the ventral side. A similar underfolding occurs at the posterior end. By such growth and underfolding the lateral and eventually the ventral walls of the body are formed, and the embryo becomes more or less cylindrical in shape.

When the embryo is still a simple disc its entire undersurface is open to the yolk cavity. As the body walls fold, a foregut and a hindgut (which form, respectively, the anterior and posterior parts of the digestive tract) are cut off from the yolk sac but remain connected to it by the yolk stalk. As the embryo grows and folds, the amnion also grows to enclose it, finally constricting the yolk stalk and the body stalk (with its allantois and blood vessels) together into a single cylindrical tube, the **umbilical cord.** This takes place about four weeks after development has commenced and allows the embryo to float free in the liquid-filled amniotic cavity, connected to the chorion and placenta only by the umbilical cord (Fig. 5.16).

The month-old embryo, which is about 5 mm. long, is now recognizable as a vertebrate of some kind. It has become cylindrical, with a relatively large head region and with prominent gills and a tail. Meanwhile, blocks of muscle known as somites are forming rapidly in the mesoderm on either side of the notochord and the beating heart is present as a large bulge on the ventral surface behind the gills. The arms and legs are still mere buds on the sides of the body.

By the end of six weeks the embryo is about 12 mm. long, the head begins to be differen-

Figure 5.16 *A–D,* Successive stages in the development of the umbilical cord and body form in the human embryo. The solid lines represent layers of ectoderm; the dashed lines, mesoderm; and the dotted lines, endoderm. (From Villee, C. A.: Biology. 7th ed. Philadelphia, W. B. Saunders Co., 1977.)

tiated, the arms and legs have grown out, but the tail and gills are still present.

At the end of two months of growth, when it is 25 mm. long, the embryo begins to look definitely human. The face has begun to develop, showing the rudiments of eyes, ears and nose. The arms and legs have developed, at first resembling tiny paddles, but by this stage the beginnings of fingers and toes are evident (Fig. 5.17). Some of the bones are beginning to ossify and the organ systems have differentiated. The tail, which was prominent during the fifth week of development, has begun to shorten and be concealed by the growing buttocks. As the heart moves posteriorly on the ventral side and the gill pouches become less conspicuous, a neck region appears. Now most of the internal organs are well laid out so that development in the remaining seven months consists mostly of an increase in size and the completion of some of the mi-

nor details of organ formation (Fig. 5.18).

The embryo is about 75 mm. long after three months of development, 250 mm. long after five months and 50 cm. long after nine months. During the third month the nails begin to form, and the sex of the fetus can be distinguished; by four months the face looks quite human; and by five months, hair appears on the body and head. During the sixth month, eyebrows and eyelashes appear. After seven months, the fetus resembles an old person with red and wrinkled skin. During the eighth and ninth months, fat is deposited under the skin, causing the wrinkles partially to smooth out; the limbs become rounded, the nails project at the finger tips, the original coat of hair is shed, and the fetus is "at full term," ready to be born (Fig. 5.19). The total gestation period, or time of development, for human beings is about 280 days from the beginning of the last menstrual period before conception until the time of birth.

Figure 5.17 Stages in the development of the human arm (upper row) and leg (lower row) between the fifth and eighth weeks. (From Villee, C. A.: Biology. 7th ed. Philadelphia, W. B. Saunders Co., 1977.)

Figure 5.18   A graded series of human embryos. Note the characteristic position of the arms and legs in the four month fetus. (From Arey, L. B.: Developmental Anatomy. 7th ed. Philadelphia, W. B. Saunders Company, 1974.)

Figure 5.19   A diagrammatic section through the uterus showing the placenta and the fetus shortly before birth. (From Villee, C. A.: Biology. 7th ed. Philadelphia, W. B. Saunders Co., 1977.)

Figure 5.20 Ventral views of successive stages in the development of the heart. See text for discussion. (From Villee, C. A.: Biology. 7th ed. Philadelphia, W. B. Saunders Co., 1977.)

5.11 FORMATION OF THE HEART

Many organs develop in the embryo without having to function at the same time, but the heart and circulatory system must function while undergoing development. The heart forms first as a simple tube from the fusion of two thin-walled tubes beneath the developing head. In this early condition it is essentially like a fish heart, consisting of four chambers arranged in a series: the **sinus venosus**, which receives blood from the veins; the single **atrium**; the single **ventricle**; and the **arterial cone**, which leads to the aortic arches.

Initially the heart is a fairly straight tube, with the atrium lying posterior to the ventricle; but since the tube grows faster than the points to which its front and rear ends are attached, it

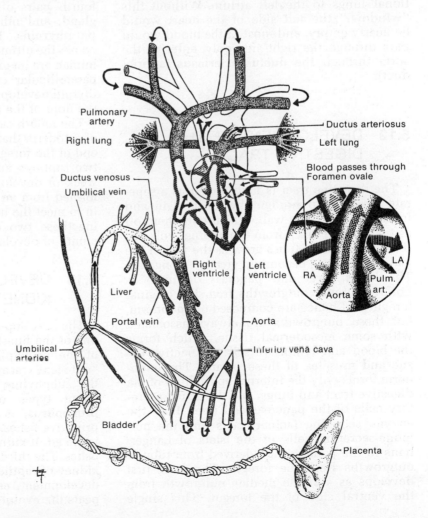

Figure 5.21 The human circulatory system before birth. The structures peculiar to the fetal circulatory system are the umbilical arteries and veins to the placenta; the arterial duct (ductus arteriosus), which connects the pulmonary artery and aorta; and the venous duct (ductus venosus), which connects the umbilical vein and the inferior vena cava. The oval window (foramen ovale) connects the right and left atria. (From Villee, C. A.: Biology. 7th ed. Philadelphia, W. B. Saunders Co., 1977.)

bulges out to one side (Fig. 5.20). The ventricle then twists in an S-shaped curve down and in front of the atrium, coming to lie posterior and ventral to it as it does in the adult. The sinus venosus gradually becomes incorporated into the atrium as the latter grows around it, and most of the arterial cone is merged with the wall of the ventricle.

When it first appears, the embryonic heart is a single structure with only one of each chamber, whereas the adult heart is a double pump, with separate right and left atria and ventricles. This separation prevents the mixing of aerated blood from the lungs with non-aerated blood from the rest of the body. The heart begins separating into four chambers at an early stage. The two ventricles are completely separated by the end of the second month. The atria are partly separated, but complete separation does not occur until after birth, when the **oval window** between them finally closes (Fig. 5.21). Before birth this must be kept open to permit blood to enter the left side of the heart, for in the fetus only a small amount of blood passes through the nonfunctional lungs to the left atrium. Without this "window," the left side of the heart would be nearly empty, and most of the blood would pass through the right side only, entering the aorta through the **ductus arteriosus** (arterial duct).

5.12 DEVELOPMENT OF THE DIGESTIVE TRACT

The digestive tract is first formed as a separate foregut and hindgut by the growth and folding of the body wall, which cuts them off as two simple tubes from the original yolk sac (Fig. 5.16). These tubes grow as the rest of the embryo grows, becoming greatly elongated. The lungs, liver and pancreas originate as hollow, tubular outgrowths from the original foregut and hence are composed of endoderm, but these outgrowths are always associated with some mesodermal tissue, which forms the blood and lymph vessels, connective tissue and muscles of these organs. The endoderm forms only the internal epithelium of the digestive tract and lungs, and the actual secretory cells of the pancreas and liver. Both the enzyme-secreting (acinar) cells and the hormone-secreting cells of the islets of Langerhans in the pancreas are derived from tubular outgrowths from the foregut. The lung first develops as a single median outgrowth from the ventral side of the foregut. This single tube, the forerunner of the trachea, soon branches into two tubes, the rudiments of the bronchi, which in turn divide repeatedly and eventually give rise to the complex structure that is the adult lung.

The most anterior part of the foregut flattens out to become, in cross section, a flattened oval, rather than a circle, and develops into the **pharynx.** In the pharyngeal region a series of five paired pouches, the **gill pouches,** bud out laterally from the endoderm and meet a corresponding set of inpocketings from the overlying ectoderm. In lower vertebrates, such as fishes, the two sets of pockets fuse to make a continuous passage from the pharynx to the outside—the **gill slits,** which function as respiratory organs. In the human and other higher vertebrates this normally does not occur; the pouches exist but are nonfunctional vestiges that give rise to other structures or disappear. For example, the first pair of pouches becomes the cavities of the middle ear and their connection with the pharynx, the **eustachian tubes.** The second pair of pouches becomes a pair of tonsils, while parts of the third and fourth pairs of pouches become the **thymus gland,** and other parts of them become the **parathyroids.** The fifth pair of pouches becomes the ultimobranchial bodies, which in the human are incorporated into the thyroid as the parafollicular cells secreting calcitonin. The thyroid develops from a separate outgrowth on the floor of the pharynx.

The mouth cavity arises as a shallow pocket of ectoderm that grows in to meet the anterior end of the foregut; the membrane between the two ruptures and disappears during the fifth week of development. Similarly, the anus is formed from an ectodermal pocket that grows in to meet the hindgut; the membrane separating these two disappears early in the third month of development.

5.13 DEVELOPMENT OF THE KIDNEY

The development of the kidney provides one of the finest and most clear-cut examples of the principle of recapitulation, i.e., that "ontogeny recapitulates phylogeny." Within the subphylum of vertebrates are three different types of kidney. The earliest, or **pronephros,** is the adult kidney of certain primitive fishes. The second, or **mesonephros,** is the adult kidney of amphibia and the higher fishes. The third, the **metanephros,** is the adult kidney of reptiles, birds and mammals. But in development, each of the higher animals repeats the evolutionary sequence of this organ.

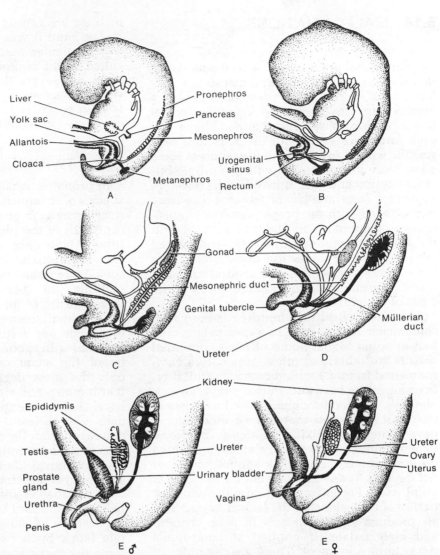

Figure 5.22 Diagrams showing stages in the development of the urinary and reproductive systems in the human. *A*, Early in the fifth week of development; *B*, early in the sixth week; *C*, seventh week; *D*, eighth week. *E* ♂, three months, male; *E* ♀, three months, female. (From Villee, C. A.: Biology. 7th ed. Philadelphia, W. B. Saunders Co., 1977.)

Thus frog embryos first develop a pronephros, which functions during early embryonic life, before the permanent mesonephros develops; and the human develops first a nonfunctional pronephros, then a mesonephros, which may be functional during fetal life, and finally the permanent metanephros (Fig. 5.22). The three kidneys develop one after another in both time and space, each new kidney lying posterior to the previous one.

The pronephros, which in the human embryo consists of about seven pairs of rudimentary kidney tubules, develops in the mesoderm and degenerates during the fourth week of embryonic life. From the tubules a pair of wolffian ducts grows back to the hindgut and connects with it.

The tubules of the mesonephros originate during the fourth week, reach their height at the end of the seventh week, and degenerate by the sixteenth week. These tubules connect with the ducts left by the degenerated pronephros and empty into them. In the female the mesonephros and its ducts degenerate completely except for a few nonfunctional remnants, but in the male some of the tubules remain and are converted into the epididymides, while the ducts become the vasa deferentia.

The metanephros of reptiles, birds and mammals develops as a pair of buds from the ducts of the mesonephros. The **ureter** and collecting tubules of the kidney develop from these buds, while the Bowman's capsules and convoluted tubules develop from the same sort of mesoderm that formed the tubules of the pronephros and mesonephros at a more anterior point. Later, these two portions unite to form the kidney tubules of the adult. The metanephros begins forming during the fifth week and is practically complete by the sixteenth week.

5.14 MALFORMATIONS

In view of the extreme complexity of the developmental process it is indeed remarkable that it occurs so regularly and that so few malformations occur. The development of a human arm involves the formation of 29 bones, each of a specific size and shape and each forming a joint with the next in a very specific way. It involves the formation of some 40 or more muscles, each with the proper origin, insertion and size. It involves the development of a large number of motor and sensory nerves, each with the proper synaptic connections on the motor end plates of muscles or in the sensory receptors in the skin, tendons and joints. And it involves the development of a large number of arteries and veins arranged in a fairly specific pattern to supply each of the parts of the arm.

About one child in a hundred is born with some major defect, such as a cleft palate, club foot or spina bifida. Some of these congenital defects are inherited; others result from environmental factors. Experiments with fruit flies, frogs and mice have shown that x-rays, ultraviolet rays, temperature changes and a variety of chemical substances may induce alterations in development. The kind of defect produced depends on the time in development at which the environmental agent is applied and does not depend to any great extent on the kind of agent used. For example, x-rays, the administration of cortisone and the lack of oxygen will all produce similar defects in mice—harelip and cleft palate—if applied at comparable times in development. There are certain critical periods in development, during which particular organs are differentiating and growing most rapidly and are most susceptible to interference.

According to old superstitions, such abnormalities are caused by fright or injury to the mother, and it was even supposed that injury to a particular part of the mother's body resulted in the malformation of that part of the fetus. However, the injuries commonly believed to cause malformations usually occur later in pregnancy, long after the organs have completed differentiation.

5.15 TWINNING

In primates, whales, horses and many other species of mammals, offspring are usually produced singly. In other animals, more than one (up to 25 in the pig) are produced in a single litter. About once in every 88 human births, two individuals are delivered at the same time. More rarely, three, four, five and even six children are born simultaneously. About three-fourths of multiple births are the result of the simultaneous release of two eggs, both of which are fertilized and develop. Such fraternal (**dizygotic**, "two egg") twins may be of the same or different sex and have only the same degree of family resemblance that brothers and sisters born at different times have. They are entirely independent individuals with different hereditary characteristics and result from the fertilization of two or more eggs ovulated simultaneously.

In contrast, identical (**monozygotic**, "one egg") twins are formed from a single fertilized egg that at some early stage of development divides into two (or more) independent parts, each of which develops into a separate fetus. Such twins, of course, are of the same sex, have identical hereditary traits and are so similar that it is difficult to tell them apart. Monozygotic twinning may occur in any of several ways. The two blastomeres produced by the first cleavage may separate and each become an embryo, the inner cell mass

Figure 5.23   Two methods of monozygotic twinning: A, by division of inner cell mass; B, by the formation of two primitive streaks. In the latter case, the twins would be connected to a single yolk sac, as in C. (After Arey.)

may divide, or two primitive streaks may form on a single embryonic disc (Fig. 5.23). Twins have been produced experimentally by the first method in lower vertebrates. This method is less likely to occur in mammals than in others, for the mammalian egg and cleavage stages are surrounded by a strong membrane, the zona pellucida, which should prevent the blastomeres from separating.

A particularly interesting example of twinning occurs in the armadillo. This animal always has quadruplets and the four individuals are always of the same sex. The fact that only one corpus luteum is found in the ovary, which means that only one follicle and egg matured and ovulated, indicates that all are monozygotic twins. When the blastocyst is examined we find that the inner cell mass has subdivided into four parts.

Occasionally, identical twins develop without separating completely and are born joined together (Siamese twins). All grades of union have been known to occur, from almost complete separation to fusion throughout most of the body, so that only the head or the legs are double. Sometimes the two twins are of different sizes and degrees of development, one being quite normal, while the second is only a partially formed parasite on the first. Such errors of development usually die during or shortly after birth.

5.16 CHANGES AT BIRTH

Great changes take place within a short time in the newborn mammal. Hitherto it has received both food and oxygen from the placenta; now its own digestive and respiratory systems must function. Correlated with these changes are major changes in the circulatory system; the umbilical arteries and veins are cut off, and the blood flow through the heart and lungs is altered.

It is believed that the first breath of the newborn mammal is initiated by the accumulation of carbon dioxide in the blood after the umbilical cord is cut; this stimulates the respiratory center in the medulla. The resulting expansion of the lungs enlarges its blood vessels, which previously were partially collapsed, and blood from the right ventricle flows in ever-increasing amounts through the vessels of the lung, instead of through the arterial duct that connected the pulmonary artery and aorta during fetal life. The resulting increase in blood returning from the lungs to the left atrium results in the closing of the valvelike opening of the oval window (Fig. 5.24) These changes take place within a short period after birth, and eventually the flap of the oval window grows into place and the arterial duct degenerates, so that the adult pattern of circulation

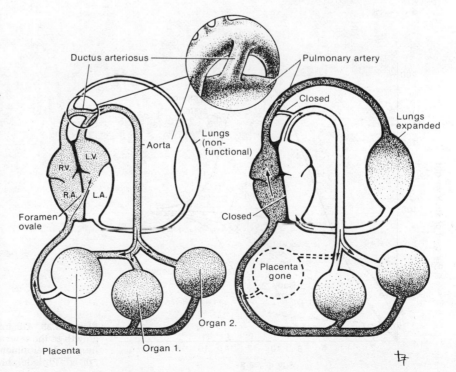

Figure 5.24 Changes in the human circulatory system at birth. *Left*, The circulatory system of the fetus; compare with Figure 5.21. *Right*, The circulatory system of the newborn child. Aerated blood is shown in white, nonaerated blood in dark stippling, and a mixture of the two in lighter stippling. In the embryo most of the blood entering the right atrium reaches the aorta either via the oval window (foramen ovale) or via the arterial duct (ductus arteriosus) *(inset)*. The changes at birth are: (1) the loss of the placenta; (2) the expansion of the lungs; (3) the closing of the foramen ovale and (4) the closing and degeneration of the arterial duct. (From Villee, C. A.: Biology. 7th ed. Philadelphia, W. B. Saunders Co., 1977.)

is established. Occasionally the oval window fails to close or the arterial duct fails to degenerate and there is a mixing of oxygenated and nonoxygenated blood. In humans this results in a "blue baby," a child whose skin has a purplish hue because of inadequate oxygenation of its blood. Delicate surgical procedures, developed after years of practice operations on dogs, now make it possible to operate on the heart itself and cure this condition.

5.17 POSTNATAL DEVELOPMENT

Development does not, of course, cease at birth. At birth the teeth and genital organs of the human infant are only partly formed and the body proportions are quite different from those of the adult. The head is proportionately much larger early in development than in the adult. The head composes about half of the length of the two-month-old human fetus, but its growth terminates early in childhood so that the head of the adult is proportionately smaller than that of the newborn. The arms attain their proportionate size shortly after birth but the legs attain theirs only after some 10 years of growth. The last organs to mature in humans are the genitals, which do not begin to grow rapidly until 12 to 14 years after the infant is born.

The degree of maturity and self-sufficiency of the newly hatched bird or newly born mammal varies widely from one species to another. Baby chicks and ducks can run around and eat solid food just after hatching, but baby robins are blind, have very few feathers and cannot stand. The newborn guinea pig has fur and teeth and can eat solid food. Newborn rats, mice and humans are quite helpless and require a lot of parental care to survive. The developmental processes that occur after birth involve some multiplication and differentiation of cells but in large part they involve the growth of cells formed earlier. The weight or size of an animal or plant as a function of time usually yields an S-shaped growth curve, one remarkably like the growth curve of a population of individuals (p. 891). Although biologists can identify some of the factors ("environmental resistance") that stabilize the size of a population at a certain level (p. 893), very little is known of the factors that lead to the cessation of cell multiplication and growth. Some plants and animals do not, in fact, stop growing but continue to grow, though perhaps at a slower pace, all through their lives.

The human egg is about 100 micrometers in diameter, just barely visible to the naked eye. A baby at birth is about 50 centimeters long, roughly 5000 times as long as the egg. In developing from an infant to an adult the height increases only an additional three and one-half times, to about 175 centimeters. The maximal rate of linear growth occurs before birth, in the fourth month of fetal life. There is a

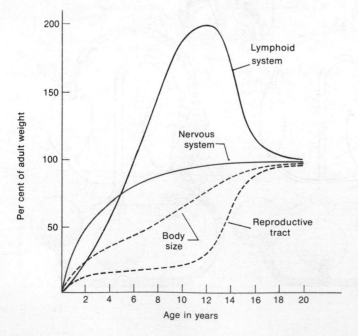

Figure 5.25 Diagram showing the relative rates of growth of the several different organ systems during human development. (From Villee, C. A.: Biology. 7th ed. Philadelphia, W. B. Saunders Co., 1977.)

final growth spurt at the time of adolescence which reaches its peak at about age 12 years in girls and about age 14 in boys.

Each structure and organ has its characteristic rate of growth. The growth rates of the various organs can be assigned to one of four types (Fig. 5.25). The growth curve of the skeleton follows that of the body as a whole. The brain and spinal cord grow relatively rapidly early in childhood and nearly reach their adult size by the age of nine. Lymphoid tissue, including the thymus, has a third type of growth curve, reaching a maximum at age 12 that exceeds the adult value. It subsequently undergoes involution until about age 20, when it attains adult values. The fourth pattern of growth is shown by the reproductive system, which grows very slowly until age 12 or thereabouts and then undergoes rapid growth at puberty.

ANNOTATED REFERENCES

Arey, L. B.: Developmental Anatomy. 7th ed. Philadelphia, W. B. Saunders Co., 1974. A standard text on human development, emphasizing morphology.

Balinsky, B. I.: An Introduction to Embryology. 4th ed. Philadelphia, W. B. Saunders Co., 1975. A well-written and well-illustrated account of animal development, with excellent discussions of gametogenesis, fertilization and early development.

Bell, E.: Molecular and Cellular Aspects of Development. New York, Harper and Row, Publishers, 1965.

Bodemer, C. W.: Modern Embryology. New York, Holt, Rinehart and Winston, 1968. A good general text of embryology including the details of organogenesis and a synthesis of the experimental analysis of cellular differentiation.

DeHaan, R. L., and H. Ursprung: Organogenesis. New York, Holt, Rinehart and Winston, 1965. An excellent collection of papers by many investigators in the field.

Metz, C. B., and A. Monroy (eds.): Fertilization. New York, Academic Press, 1967. Fine discussions of sperm motility and metabolism and of the process of fertilization.

Milne, L. J., and M. J. Milne: The Mating Instinct. Boston, Little Brown and Co., 1954. An interesting essay on the mating process in certain animals.

Page, E. W., C. A. Villee, and D. B. Villee: Human Reproduction. 2nd ed. Philadelphia, W. B. Saunders Co., 1976. A general account of human reproduction ranging from the behavioral to the endocrine and biochemical aspects of the subject.

Rugh, R., and L. B. Shettles: From Conception to Birth: The Drama of Life's Beginnings. New York, Harper and Row, 1971. A popularly written book, with superb illustrations.

CHAPTER 6

PRINCIPLES OF HEREDITY

6.1 HISTORY OF GENETICS

It must have been thousands of years ago when man first made one of the fundamental observations of heredity—that "like tends to beget like." But his curiosity as to why this is true and how it is brought about remained unsatisfied until the beginning of the present century. A number of breeders, such as Kölreuter, who worked with tobacco plants about 1770, crossed different varieties of plants and produced hybrids. Kölreuter recognized that parental characters were transmitted by both the pollen and the ovule. Mendel's careful work with peas revealed the fundamental principles of heredity, but the report of his work, published in 1866, was far ahead of its time. It is clear that his work was known to a number of the leading contemporary biologists, such as the botanist Nägeli, but in the absence of our present knowledge of chromosomes and their behavior, its significance was unappreciated.

The chromosomal details of mitosis were described by Eduard Strasburger in 1876. Eduard van Beneden (1887) discovered the process of meiosis and understood its significance. Earlier that same year Weismann had pointed out, simply from theoretical considerations, that the chromosomal number in gametes must be half of that in somatic cells. It is conceivable that some brilliant theoretical biologist with these facts at hand might have postulated that, if hereditary factors were units located in the chromosomes, the mating of different parental types would yield offspring in predictable ratios. However, no such synthesis was made, and the existence of these definite ratios of the types of offspring resulting from a given mating remained to be demonstrated experimentally.

In 1900, three different biologists, working independently—de Vries in Holland, Correns in Germany and von Tschermak in Austria—rediscovered the phenomenon of regular, predictable ratios of the types of offspring produced by mating pure-bred parents. They then found Mendel's published report and, realizing his priority in these discoveries, gave him credit for his work by naming two of the fundamental principles of heredity **Mendel's laws.**

With the genetic and cytologic facts at hand, W. S. Sutton and C. E. McClung independently came to the conclusion (1902) that the hereditary factors are located in the chromosomes. They also pointed out that, since there are many more hereditary factors than chromosomes, there must be more than one hereditary factor per chromosome. By 1911, T. H. Morgan was able to postulate, from the regularity with which certain characters tended to be inherited together, that the hereditary factors (which he named "genes") were located in the chromosomes in linear order, "like the beads on a string."

6.2 MENDEL'S DISCOVERIES

Gregor Johann Mendel (1822–1884) was an Austrian abbot who spent some eight years breeding peas in the garden of his monastery at Brünn, now part of Czechoslovakia. He succeeded in reaching an understanding of the basic principles of heredity because (1) he studied the inheritance of single contrasting characters (such as green versus yellow seed color, wrinkled versus smooth seed coat), instead of attempting to study the complete inheritance of each organism; (2) his studies were quantitative; he counted the number of each type of offspring and kept accurate records of his crosses and results; and (3) by design or by good fortune, he chose a plant, and particular characters of that plant, that gave him clear ratios. If he had worked with other plants or with certain other traits of peas, he would have been unable to get these ratios. Now that the principles of heredity have been established, the explanation for these more complicated types of inheritance is clear.

Mendel established pure-breeding strains of peas with contrasting characters—yellow seed coat versus green seed coat, round seeds versus wrinkled ones—and then made crosses of the contrasting varieties. He found that the offspring of a cross of yellow and green all had yellow seed coats; the result was the same whether the male or the female parent had been the yellow one. Thus, the character of one parent can "dominate" over that of the other, but which of the contrasting characters is dominant depends upon the specific trait involved, not upon which parent contributes it. This observation, repeated for several different strains of peas, led Mendel to the generalization—the "law of dominance"—that when two factors for the alternative expression of a character are brought together in one individual, one may be expressed completely and the other not at all. The character which appears in the first generation is said to be **dominant;** the contrasting character is said to be **recessive.**

Mendel then took the seeds produced by this first generation of the cross (called the **first filial generation,** abbreviated F_1), planted them and had the resulting plants fertilize themselves to produce the second filial generation, the F_2. He found that both the dominant and the recessive characters appeared in this

generation, and upon counting the number of each type (Table 6.1) he found that, whatever set of characters he used, the ratio of plants with the dominant character to those with the recessive character was very close to 3:1. From such experiments Mendel concluded that (1) there must be discrete unit factors which determine the inherited characters; (2) these unit factors must exist in pairs; and (3) in the formation of gametes the members of these pairs separate from each other, with the result that each gamete receives only one member of the pair. The unit factor for green seed color is not affected by existing for a generation within a yellow seeded plant (e.g., the F_1 individuals). The two separate during gamete formation and, if a gamete bearing this factor for green seed coat fertilizes another gamete with this factor, the resulting seed has a green color. The generalization known as Mendel's first law, the **law of segregation,** may now be stated as : Genes exist in pairs in individuals, and in the formation of gametes each gene separates or segregates from the other member of the pair and passes into a different gamete, so that each gamete has one, and only one, of each kind of gene.

In other experiments Mendel observed the inheritance of two pairs of contrasting characters in a single cross. He mated a pure-breeding strain with round yellow seeds and one with wrinkled green seeds. The first filial generation all had round yellow seeds, but when these were self-fertilized he found in the F_2 generation all four possible combinations of seed color and shape. When he counted these he found 315 round yellow seeds, 108 round green seeds, 101 wrinkled yellow seeds, and 32 wrinkled green seeds. There is a close approximation of a 3:1 ratio for seed color (416 yellow to 140 green) and for seed shape (423 round to 133 wrinkled). Thus, the inheritance of seed color is independent of the inheritance of seed shape; neither one affects the other. When the two types of traits are considered

TABLE 6.1 AN ABSTRACT OF THE DATA OBTAINED BY MENDEL FROM HIS BREEDING EXPERIMENTS WITH GARDEN PEAS

Parental Characters	First Generation	Second Generation	Ratios
Yellow seeds × green seeds	All yellow	6022 yellow : 2001 green	3.01 : 1
Round seeds × wrinkled seeds	All round	5474 round : 1850 wrinkled	2.96 : 1
Green pods × yellow pods	All green	428 green : 152 yellow	2.82 : 1
Long stems × short stems	All long	787 long : 277 short	2.84 : 1
Axial flowers × terminal flowers	All axial	651 axial : 207 terminal	3.14 : 1
Inflated pods × constricted pods	All inflated	882 inflated : 299 constricted	2.95 : 1
Red flowers × white flowers	All red	705 red : 224 white	3.15 : 1

together, it is clear that there is a ratio of 9 with two dominant traits (yellow and round): 3 with one dominant and one recessive (round and green): 3 with the other dominant and recessive (yellow and wrinkled): 1 with the two recessive traits (green and wrinkled). Mendel's second law, the **law of independent assortment,** may now be given as: The distribution of each pair of genes into gametes is independent of the distribution of any other pair.

6.3   CHROMOSOMAL BASIS OF THE LAWS OF HEREDITY

Each cell of every member of a given species of animal or plant contains a definite number of chromosomes; the constancy of the chromosome number is assured by the precise and regular events of mitotic division (p. 46). Many widely different species of animals and plants have the same number of chromosomes. It is not the number of chromosomes, but the nature of the hereditary factors within them, that differentiates species.

The constancy of the chromosome number in successive generations of the same species is assured by the precise separation of the members of the pairs of homologous chromosomes in the meiotic divisions leading to the formation of gametes. The normal number of chromosomes for somatic cells is reconstituted in fertilization, when the egg and sperm nuclei fuse.

The laws of heredity follow directly from the behavior of the chromosomes in mitosis, meiosis and fertilization. Within each chromosome are numerous hereditary factors, the **genes,** each of which controls the inheritance of one or more characteristics. Each gene is located at a particular point, called a **locus** (plural, loci), along the chromosome. Since the genes are located in the chromosomes, and each cell has two of each kind of chromosome, it follows that each cell has two of each kind of gene. The chromosomes separate in meiosis and recombine in fertilization and so, of course, do the genes within them. We currently believe that the genes are arranged in a linear order within the chromosomes; the **homologous chromosomes** have similar genes arranged in a similar order. When the chromosomes undergo synapsis during meiosis (p. 88) the homologous chromosomes become attached point by point and, presumably, gene by gene.

6.4   GENES AND ALLELES

Studies of inheritance are possible only when there are two alternate, contrasting conditions, such as Mendel's yellow and green peas or round and wrinkled ones. These contrasting conditions, inherited in such a way that an individual may have one or the other but not both, were originally termed allelomorphic traits. At present the terms allele and gene are used more or less interchangeably; both refer to the hereditary factor responsible for a given trait. The term allele emphasizes that there are two or more alternative kinds of genes at a specific locus in a specific chromosome.

Brown and black coat color are allelomorphic traits in guinea pigs. Each body cell of the guinea pig has a pair of chromosomes which contain genes for coat color; since there are two chromosomes, there are two genes per cell. A "pure" black guinea pig (one of a pedigreed strain of black guinea pigs) has two genes for black coat, one in each chromosome, and a "pure" brown guinea pig has two genes for brown coat. The genes themselves, of course, have no color; they are neither brown nor black. The brown gene controls the formation of an enzyme involved in one of the steps in the production of a brown pigment in the hair cells, whereas the black gene produces a different enzyme which alters the chemical reactions in the synthesis of pigment and results in the production of black pigment. In working genetic problems, letters are conventionally used as symbols for the genes. A pair of genes for black pigment is represented as **BB,** and a pair of genes for brown pigment by **bb.** A capital letter is used for one gene and the corresponding lower case letter is used to represent the gene for the contrasting trait, the allele.

6.5   A MONOHYBRID CROSS

The usage of genetic terms and some of the basic principles of genetics can be illustrated by considering a simple monohybrid cross, that is, a cross between two individuals that differ in a single pair of genes. The mating of a "pure" brown male guinea pig with a "pure" black female guinea pig is illustrated in Figure 6.1. During meiosis in the testes of the male the two **bb** genes separate so that each sperm has only one **b** gene. In the formation of ova in the female the **BB** genes separate so that each ovum has only one **B** gene. The fertilization of

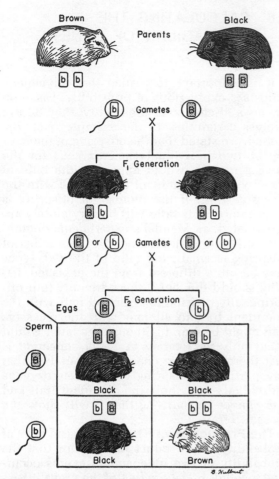

Figure 6.1 An example of a monohybrid cross: the mating of a brown with a black guinea pig. (From Villee, C. A.: Biology. 7th ed. Philadelphia, W. B. Saunders Co., 1977.)

this egg by a **b** sperm results in an animal with the genetic formula **Bb.** These guinea pigs contain one gene for brown coat and one for black coat. What color would you expect them to be—dark brown, gray or perhaps spotted? In this instance they are just as black as the mother. The gene for black coat color is said to be **dominant** to the gene for brown coat color. It will produce black body color even when only one dose of the black gene is present. The brown gene is said to be **recessive** to the black one; it will produce brown coat color only when present in double dose (**bb**). We may define recessive genes as ones which will produce their effects only when an individual has two of them that are identical, and dominant genes as those which will produce their effects even when only one is present in an individual. The phenomenon of dominance supplies part of the explanation as to why an individual may resemble one of his parents

more than the other despite the fact that both make equal contributions to his genetic constitution. In one species of animal, black coat color may be dominant to brown while in another species brown might be dominant to black. The particular genetic relations that obtain in any given species must be determined by experiment.

An animal or plant with two genes exactly alike, two blacks (**BB**) or two browns (**bb**), is said to be **homozygous** or "pure" for the character. An organism with one dominant and one recessive gene (**Bb**) is said to be **heterozygous** or "hybrid." Thus, in the mating under consideration the black and brown parents were homozygous, **BB** and **bb**, respectively, and the offspring in the F_1 were all heterozygous, **Bb. Recessive genes** are those which will produce their effect only when homozygous; a **dominant gene** is one which will produce its effect whether it is homozygous or heterozygous.

In the process of gamete formation in these heterozygous black F_1 guinea pigs, the chromosome containing the **B** gene undergoes synapsis with, and then separates from, the homologous chromosome containing the **b** gene, so that each sperm or egg has a **B** gene or a **b** gene. No sperm or egg is without one or the other and none has both. Since there are two kinds of eggs and two kinds of sperm, the mating of two of these heterozygous black guinea pigs permits four different combinations of eggs and sperm. Gametes containing **B** genes and ones containing **b** genes are formed in equal numbers. There is no special attraction or repulsion between an egg and a sperm containing similar genes; an egg containing a **B** gene is just as likely to be fertilized by a **B** sperm as by a **b** sperm. The four possible combinations occur with equal frequency, which can be determined by algebraic multiplication:

$$(\tfrac{1}{2}B + \tfrac{1}{2}b) \text{ eggs} \times (\tfrac{1}{2}B + \tfrac{1}{2}b) \text{ sperm}$$
$$= \tfrac{1}{4}BB + \tfrac{1}{2}Bb + \tfrac{1}{4}bb.$$

The possible combinations of eggs and sperm may be represented in a "Punnett square" (Fig. 6.1), devised by the English geneticist R. C. Punnett. The possible types of eggs are written across the top of the Punnett square and the possible types of sperm are arranged down its left side, then the squares are filled in with the resulting zygote combinations. Three-fourths of the offspring are either **BB** or **Bb,** and consequently have a black coat color, and one-fourth are **bb,** with a brown coat color.

This 3:1 ratio is characteristically obtained in the second generation of a **monohybrid cross,** i.e., a mating of two individuals which differ in a single trait governed by a single pair of genes. The genetic mechanism responsible for the 3:1 ratios obtained by Mendel in his pea breeding experiments is now evident.

The appearance of an individual with respect to a certain inherited trait is known as its **phenotype.** The organism's genetic constitution, usually expressed in symbols, is called its **genotype.** In the cross we have been considering, the phenotypic ratio of the F_2 generation is 3 black coated guinea pigs: 1 brown coated guinea pig and the genotypic ratio is 1 **BB**:2 **Bb**:1 **bb.** The phenotype may be a morphological characteristic — shape, size, color — or a physiological characteristic such as the presence or absence of a specific enzyme required for the metabolism of a specific substrate. Guinea pigs with the genotypes **BB** and **Bb** are alike phenotypically; they both have black coats. One-third of the black guinea pigs in the F_2 generation of the mating of black × brown are homozygous, **BB**, and the other two-thirds are heterozygous, **Bb.** The two types can be differentiated by making a **test cross,** by mating them with homozygous brown (**bb**) guinea pigs. If all of the offspring are black, what inference would you make about the genotype of the black parent? If any of the offspring are brown, what conclusion would you draw regarding the genotype of the black parent? Can you be more certain about one of these two inferences?

This sort of testing is of great importance in the commercial breeding of animals or plants where the breeder is trying to establish a strain that will breed true for a certain characteristic. When the individuals in a breeding stock are selected on the basis of their own phenotypes, the breeding program is not maximally effective because it does not differentiate between homozygous and heterozygous individuals. The method of **progeny selection** is one in which a breeder tests the genotypes of his breeding stock by making test matings and observing the offspring. If the offspring are superior with respect to the desired trait, the parents are thereafter used regularly for breeding. Two bulls, for example, may look equally healthy and vigorous, yet one will have daughters with qualities of milk production that are distinctly superior to those of the daughters of the other bull. By a method of progeny selection one breeding program raised the average annual egg production of a flock of hens from 114 to 200 eggs over a period of eight years.

6.6 CALCULATING THE PROBABILITY OF GENETIC EVENTS

It is important to realize that all genetic ratios are expressions of probability, based on the laws of chance or probability; they do not express certainties. In the examples just discussed, we stated that the offspring of the mating of two individuals heterozygous for the same gene pair would appear in the ratio of three with the dominant trait to one with the recessive trait. If the number of offspring is large enough this ratio will be very closely approximated, as Mendel's experiments demonstrated (Table 6.1). However, if the number of offspring is small, the ratio of the two types may be quite different from the expected 3:1. Why should this be? If there are only four offspring, any distribution from all four with the dominant trait to all four with the recessive trait might be found, although the latter would occur only very rarely. A better statement is that there are three chances in four ($3/4$) that any particular offspring of two heterozygous individuals will show the dominant trait and one chance in four ($1/4$) that it will show the recessive trait.

The Product Law.   One of the basic laws of probability, the **Product Law,** states that the probability of two independent events occurring together is the product of the probabilities of each occurring separately. Using this principle we can calculate how frequently in a large series of matings, if each mating results in exactly four offspring, all four would show the recessive trait. How often in the mating of two heterozygous black guinea pigs, **Bb,** would one expect to get a litter of four brown guinea pigs? The probability that the first one will be brown is $1/4$. The probability that the second one will be brown is also $1/4$. The fertilization of each egg by a sperm is an independent event and we use the Product Law to calculate the combined probability of two (or more) events occurring together. Thus the probability that all four of the offspring of any given mating will be brown is $1/4 \times 1/4 \times 1/4 \times 1/4$, or $1/256$.* In other words, there is one chance in 256 that all four guinea pigs will have brown coat color!

The probability that any given offspring will show the dominant trait, black coat color, is $3/4$. We can, in a similar fashion, calculate the

---

*For ease in multiplication, the fractions may be converted to decimals, but the fraction $1/256$ emphasizes that there is one chance in 256 of obtaining the desired result.

probability that all four guinea pigs in the litter will be black. This is ¾ × ¾ × ¾ × ¾, or 81 chances in 256 that all four guinea pigs will have black coat color. You must not assume that if you actually mated 256 pairs of heterozygous guinea pigs and each had four offspring you would be guaranteed one litter of four brown-coated offspring. A great many people have lost vast sums of money in gambling by making a similar mistake! If you have made 255 such matings without getting a set of four brown offspring, what is the probability of getting four brown guinea pigs on the 256th try? You might be misled into thinking it is bound to happen, but in fact there is still only one chance in 256 that it will occur, since each of these matings is an independent event.

How many different ways are there of getting three black and one brown guinea pig in a given mating? A look at Figure 6.2 shows that there are four ways: The first to be born could be brown and the next three would be black. Or, the second one born could be brown and the first, third and fourth black. The other two possibilities are that the brown one would be the third or the fourth to be born. To calculate the probability that three of the four offspring will be black and one brown we must multiply the number of possible combinations by the probability of each type. There are $4 \times \frac{3}{4} \times \frac{3}{4} \times \frac{3}{4} \times \frac{1}{4}$, or $\frac{108}{256}$ chances that there will be three black guinea pigs and one brown guinea pig in a litter of four. It may be surprising at first that the 3:1 ratio is actually obtained in less than half of the total number of litters of four. However, when we add up the total numbers of black and brown offspring from a large number of matings, the 3:1 ratio is more and more closely approximated.

How many different ways are there of getting two black and two brown guinea pigs in a litter of four? There are six ways, as shown in

Figure 6.2. From this you can calculate that the probability that two of the four offspring will be black and two brown is $6 \times \frac{3}{4} \times \frac{3}{4} \times \frac{1}{4} \times \frac{1}{4}$, or $\frac{54}{256}$. You can see that there are four different combinations that give one black and three brown. The first born may be black and the rest brown, or the second will be black and all others brown, or the third or the fourth will be black. The probability of getting one black guinea pig and three brown guinea pigs is $4 \times \frac{3}{4} \times \frac{1}{4} \times \frac{1}{4} \times \frac{1}{4}$, or $\frac{12}{256}$.

All probabilities are expressed as fractions ranging from zero, expressing an impossibility, to one, expressing a certainty. Probabilities can be multiplied or added just like any other fractions. In this example there are no other possibilities; the four guinea pigs must be either four black, or three black and one brown, or two black and two brown, or one black and three brown, or four brown. The sum of the five probabilities will add up to one — $\frac{81}{256} + \frac{108}{256} + \frac{54}{256} + \frac{12}{256} + \frac{1}{256} = \frac{256}{256} = 1$.

The color of the iris of the human eye is inherited by several pairs of genes, but one pair is the primary factor differentiating brown eye color from blue. The gene for brown eye color, **B**, is dominant to the gene for blue, **b.** If two heterozygous brown-eyed people marry, what is the probability that they will have a blue-eyed child? Clearly, there is one chance in four that any child of theirs will have blue eyes. Each mating is a separate, independent event and its result is not affected by the results of any previous matings. If these two brown-eyed parents have had three brown-eyed children and are expecting their fourth child, what is the probability that it will have blue eyes? Again, the unwary might guess that this one must have blue eyes, but in fact there is still only one chance in four of its having blue eyes and three chances in four that it will have brown·eyes.

All four black	Three black and one brown	Two black and two brown	One black and three brown	All four brown
3/4 × 3/4 × 3/4 × 3/4	4 × 3/4 × 3/4 × 3/4 × 1/4	6 × 3/4 × 3/4 × 1/4 × 1/4	4 × 3/4 × 1/4 × 1/4 × 1/4	1/4 × 1/4 × 1/4 × 1/4
= 81/256	= 108/256	= 54/256	= 12/256	= 1/256

Figure 6.2 The various ways in which one might obtain exactly three black guinea pigs and one brown guinea pig in a litter of four. (From Villee, C. A.: Biology. 7th ed. Philadelphia, W. B. Saunders Co., 1977.)

All genetic events are governed by the laws of probability. The prediction of any single event—e.g., predicting the characteristics of any single offspring—is highly uncertain; however, if the number of events is large enough, the laws of probability provide a reasonable prediction of the fraction of these events that will be of one type or the other.

6.7 INCOMPLETE DOMINANCE

In many different species and for a variety of traits it has been found that one gene is not completely dominant to the other. Heterozygous individuals have a phenotype which can be distinguished from that of the homozygous dominant; it may be intermediate between the phenotypes of the two parental strains. Such pairs of genes are said to be **codominant** or to show incomplete dominance. The mating of red short-horn cattle with white ones yields offspring which have an intermediate, roan-colored coat. The mating of two roan-colored cattle yields offspring in the ratio of 1 red:2 roan:1 white; thus the genotypic and phenotypic ratios are the same; each genotype has a recognizably different phenotype. If you saw a white cow nursing a roancolored calf, what would you guess was the coat color of the calf's father? Is there more than one possible answer?

This phenomenon of **incomplete dominance** is found with a number of traits in different animals and with some human characteristics. Studies of a number of human diseases inherited by recessive genes—sickle cell anemia, Mediterranean anemia, gout, epilepsy and many others—have shown that the individuals who are heterozygous for the trait have slight but detectable differences from the homozygous normal individual.

6.8 CARRIERS OF GENETIC DISEASES

Careful investigation of the phenotypes of many gene pairs has revealed slight differences between the homozygous dominant and the heterozygous individual. In the human many inherited diseases are transmitted by recessive genes, and it is important to be able to distinguish the homozygous normal individual from the heterozygous individual who may superficially appear to be normal but is a **carrier** for the trait. The mating of two car-riers, two heterozygous individuals, would provide one chance in four for the appearance of a homozygous recessive individual showing the inherited disease. The human trait **sickle cell anemia** is inherited by a single pair of genes, but an individual must have two such genes to show the anemia. Thus the gene for sickle cell anemia may be said to be recessive to the normal gene. The red cells of a person with sickle cell anemia are shaped like a sickle or a half moon, whereas normal red cells are biconcave discs. The sickle cells contain hemoglobin molecules with a slightly different molecular structure from that found in normal red cells. The hemoglobin molecules in an individual with sickle cell anemia have the amino acid valine instead of glutamic acid at position 6, the sixth amino acid from the amino terminal end, in the β chain. The substitution of an amino acid with an uncharged side chain (valine) for one with a charged side chain (glutamate) makes the hemoglobin less soluble, and it tends to form crystals that break the red cell. The red cells of sickle cell anemics are more fragile than those of normal individuals. Individuals heterozygous for the sickling gene have a mixture of normal and abnormal hemoglobins in their red cells; about 45 per cent of the total hemoglobin has the abnormal chemical constitution. Their red cells do not usually undergo sickling but can be made to do so when the amount of oxygen in the blood is reduced. They are said to have sickle cell "trait." This provides a very simple test, using only a drop of blood, by which the heterozygote can be distinguished from the homozygous normal.

The human disease **phenylketonuria** is also inherited by a single pair of genes. The homozygous recessive individual lacks the enzyme needed to convert one amino acid, phenylalanine, to another, tyrosine. The phenylalanine accumulates in the tissues and causes retarded mental development. The heterozygote appears perfectly normal, but when given a standard amount of phenylalanine in his diet, may accumulate more in his blood and excrete more in his urine. Unfortunately the results of such tests are not always clear-cut in distinguishing between homozygous normal and heterozygous carrier individuals; the values for the two groups overlap considerably. However, a very simple test, adding ferric chloride to the urine in the diapers of the newborn, distinguishes the homozygous phenylketonuric, who can then be kept on a diet containing a minimal amount of phenylalanine. This is effective in minimizing the amount of mental retardation as the infant develops.

6.9   A DIHYBRID CROSS

Frequently a geneticist must analyze the inheritance of two or more traits in the same group of individuals. A mating that involves individuals differing in two traits is called a **dihybrid cross.** The principles involved and the procedure of solving problems are exactly the same in monohybrid and in dihybrid or trihybrid crosses. In the latter the number of types of gametes is greater and the number of types of zygotes is correspondingly larger.

When two pairs of genes are located in different (nonhomologous) chromosomes, each pair is inherited independently of the other; that is, each pair separates during meiosis independently of the other. When a black, short-haired guinea pig (**BBSS,** short hair is dominant to long hair) and a brown, long-haired guinea pig (**bbss**) are mated, the **BBSS** individual produces gametes all of which are **BS.** The

bbss guinea pig produces only **bs** gametes. Each gamete contains one and only one of each kind of gene. The union of **BS** gametes and **bs** gametes yields only individuals with the genotype **BbSs.** All of the offspring are heterozygous for hair color and for hair length and all are phenotypically black and short-haired.

When two of the F₁ individuals are mated, each produces four kinds of gametes in equal numbers, **BS, Bs, bS, bs;** thus 16 combinations are possible among the zygotes (Fig. 6.3). There are 9 chances in 16 of obtaining a black, short-haired individual, 3 chances in 16 of obtaining a black, long-haired individual, 3 chances in 16 of obtaining a brown, short-haired individual and 1 chance in 16 of obtaining a brown long-haired individual.

In a similar fashion problems involving three pairs of genes may be solved. An individual heterozygous for three pairs of genes will yield eight types of gametes in equal numbers. The

Figure 6.3 An example of a dihybrid cross: the mating of a black, short-haired guinea pig and a brown, long-haired one, illustrating independent assortment. (From Villee, C. A.: Biology. 7th ed. Philadelphia, W. B. Saunders Co., 1977.)

union of these eight types of eggs and eight types of sperm will yield 64 possible zygotes in the F_2 generation. In peas, as Mendel demonstrated, yellow seed color (**Y**) is dominant to green (**y**), smooth seeds (**S**) are dominant to wrinkled (**s**), and tall plants (**T**) are dominant to dwarf (**t**). The mating of a homozygous yellow, smooth, tall plant (**YYSSTT**) with a homozygous green, wrinkled, dwarf plant (**yysstt**) will produce offspring all of which are yellow, smooth and tall (**YySsTt**). When two of these F_1 plants are mated, F_2 offspring are produced in the ratio of 27 yellow, smooth, tall : 9 yellow, smooth, dwarf : 9 yellow, wrinkled, tall : 9 green, smooth, tall : 3 yellow, wrinkled, dwarf : 3 green, wrinkled, tall : 3 green, smooth, dwarf : 1 green, wrinkled, dwarf. Draw a Punnett square to verify these numbers.

6.10 DEDUCING GENOTYPES

The science of genetics resembles mathematics in that when the student has mastered the few basic principles involved he can solve a wide variety of problems. These basic principles include: (1) Inheritance is biparental; both parents contribute to the genetic constitution of the offspring. (2) Genes are not altered by existing together in a heterozygote. (3) Each individual has two of each kind of gene, but each gamete has only one of each kind. (4) Two pairs of genes located in different chromosomes are inherited independently. (5) Gametes unite at random; there is neither attraction nor repulsion between an egg and a sperm containing identical genes.

In working genetics problems, it is helpful to use the following procedure:

1. Write down the symbols used for each pair of genes.

2. Determine the genotypes of the parents, deducing them from the phenotypes of the parents and, if necessary, from the phenotypes of the offspring.

3. Derive all of the possible types of gametes each parent would produce.

4. Prepare the appropriate Punnett square and write the possible types of eggs across its top and the types of sperm along its side.

5. Fill in the squares with the appropriate

Cat A, short-haired   Short-haired kitten

Short-haired male cat   Cat B, long-haired   Long-haired kitten

Cat C, short-haired   Long-haired kitten

Figure 6.4 An example of problem-solving in genetics: deducing parental genotypes from the phenotypes of the offspring. See text for discussion.

genotypes and read off the genotypic and phenotypic ratios of the offspring.

As an example of the method of solving a problem in genetics, let us consider the following: The length of fur in cats is an inherited trait; the gene for long hair (l), as in Persian cats, is recessive to the gene for short hair (L) of the common tabby cat. Let us suppose that a short-haired male is bred to three different females, two of which, A and C, are short-haired and one, B, is long-haired (Fig. 6.4). Cat A gives birth to a short-haired kitten, but cats B and C each produce a long-haired kitten. What offspring could be expected from further mating of this male with these three females?

Since the long-haired trait is recessive we know that all the long-haired cats must be homozygous. We can deduce, then, that cat B and the kittens produced by cats B and C have the genotype ll. All the short-haired cats have at least one L gene. The fact that any of the offspring of the male cat has long hair proves that he is heterozygous, with the genotype Ll. The kitten produced by cat B received one l gene from its mother but must have received the other from its father. The fact that cat C gave birth to a long-haired kitten proves that she, too, is heterozygous and has the genotype Ll. It is impossible to decide, from the data at hand, whether the short-haired cat A is homozygous LL or heterozygous Ll. A test cross with a long-haired male would be helpful in deciding this. Further mating of the short-haired male with cat B would give half long-haired and half short-haired kittens, whereas further mating of the short-haired male with cat C would give three times as many short-haired kittens as long-haired ones.

6.11 THE GENETIC DETERMINATION OF SEX

The sex of an organism is a genetically determined trait. There is an exception to the general rule that all homologous pairs of chromosomes are identical in size and shape: the so-called **sex chromosomes.** In one sex of each species of animals there is either an unpaired chromosome or an odd pair of chromosomes, the two members of which differ in size and shape. In most species the females have two identical chromosomes, called X chromosomes, and males have either a single X chromosome or one X plus a generally somewhat smaller one called the Y chromosome. The existence of these unpaired chromo-

somes was discovered by C. E. McClung, in 1902, when he was studying the process of meiosis in the testes of grasshoppers. He made the shrewd guess that these might play some role in sex determination. In a few animals, the butterflies and birds, the system is reversed and the male has two X chromosomes and the female one X and one Y. The Y chromosome usually contains few genes and in most species the X and Y chromosomes are distinguished by their different size and shape. Yet in meiosis the X and Y chromosomes act like homologous chromosomes; they undergo synapsis, separate, pass to opposite poles, and become incorporated into different gametes (Fig. 6.5). Human males have 22 pairs of ordinary chromosomes, called **autosomes,** one X and one Y chromosome, whereas females have 22 pairs of autosomes and two X chromosomes.

The experiments of C. B. Bridges revealed that the sex of fruit flies, *Drosophila*, is determined by the ratio of the number of X chromosomes to the number of haploid sets of autosomes. Males have one X and two haploid sets of autosomes, a ratio of 1:2, or 0.5. Females have two X and two haploid sets of autosomes, a ratio of 2:2, or 1.0. By genetic techniques possible in fruit flies, Bridges established abnormal flies with one X and three sets of autosomes. These flies, with a ratio of 0.33, had all their male characteristics exaggerated; Bridges called them "supermales." Other abnormal individuals, with three X and two sets of autosomes, were "superfemales," with all the female characteristics exaggerated. Individuals with two X chromosomes and three sets of autosomes, a ratio of 0.67, were intersexes, with characters intermediate between those of normal males and normal females. All of these unusual flies, supermales, superfemales and intersexes, were sterile.

In man, and perhaps in other mammals, maleness is determined in large part by the presence of the Y chromosome. An individual with an XXY constitution is a nearly normal male in external appearance, though with underdeveloped gonads (Klinefelter's syndrome). An individual with one X but no Y chromosome has the appearance of an immature female (Turner's syndrome). It is possible to determine the "nuclear sex" of an individual by careful microscopic examination of some of his cells. Individuals with two X chromosomes have a "chromatin spot" at the edge of the nucleus which is evident in cells from the skin or from the mucosal lining of the mouth (Fig. 6.6). Cells from male individuals with only one X chromosome do not show a chromatin spot.

Diploid cells
of parents

Female   Male

X   X   X   Y

Synapsis

Anaphase of
meiotic
division

Egg:
one type

Sperm:
two types

Sperm

Eggs

Offspring
equal
numbers
of males
and
females

Female   Male

Figure 6.5 Diagram illustrating the transmission of the sex chromosomes of the fruit fly.

The chromatin spot represents one of the two X chromosomes, which becomes condensed and dark-staining. The other X chromosome, like the autosomes, is a fully extended thread not evident by light microscopy. Mary Lyon has suggested that only one of the two X chromosomes in the female is active; the condensed chromatin spot represents the inactive X chromosome. Which of the two becomes inactive in any given cell is a matter of chance; thus the cells in the body of a female are of two types, according to which X chromosome is inactive. Since the two X chromosomes may have different genetic complements, the cells

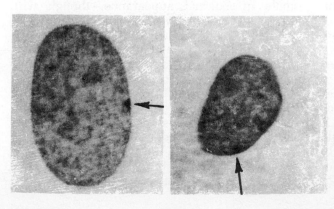

Figure 6.6 Sex chromatin in human fibroblasts cultured from the skin of a female. The chromatin spot at the periphery of each nucleus is indicated by the arrow. (Feulgen, ×2200. Courtesy Dr. Ursula Mittwoch, Galton Laboratory, University College, London.) (From Villee, C. A.: Biology. 7th ed. Philadelphia, W. B. Saunders Co., 1977.)

may differ in the effective genes present. Mice have several genes for coat color in the X chromosome, and females heterozygous for two such genes may show patches of one coat color in the midst of areas of the other, a phenomenon termed variegation. The inactivation of one X chromosome apparently occurs early in embryonic development and thereafter all the progeny of that cell have the same inactive X chromosome. Although one X chromosome appears to be inactive, there are marked abnormalities of development when one X chromosome is completely missing from the chromosomal complement of the cell, as in the XO condition of Turner's syndrome.

All the eggs produced by XX females have one X chromosome. Half of the sperm produced by XY males contain an X chromosome and half contain a Y chromosome. The fertilization of an X-bearing egg by an X-bearing sperm results in an XX, or female, zygote, and the fertilization of an X-bearing egg by a Y-bearing sperm results in an XY, or male, zygote. Since there are equal numbers of X- and Y-bearing sperm, there are equal numbers of male and female offspring. In human beings, there are approximately 107 males born for every 100 females, and the ratio at conception is said to be even higher, about 114 males to 100 females. One possible explanation of the numerical discrepancy is that the Y chromosome is smaller than the X chromosome, and a sperm containing a Y chromosome, being a little lighter and perhaps able to swim a little faster than a sperm containing an X chromosome, would win the race to the egg slightly more than half of the time. Both during the period of intrauterine development and after birth, the death rate among males is slightly greater than that among females, so that by the age of 10 or 12 there are equal numbers of males and females. In later life there are more females than males in each age group.

6.12 SEX-LINKED CHARACTERISTICS

The X chromosome contains many genes, and the traits controlled by these genes are said to be **sex-linked,** because their inheritance is

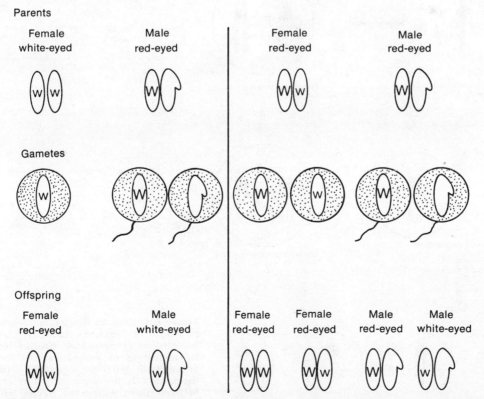

Figure 6.7 Diagram illustrating sex-linked inheritance, the inheritance of red versus white eye color in fruit flies. See text for discussion.

linked with the inheritance of sex. The Y chromosome contains very few genes, so that the somatic cells of an XY male contain only one of each kind of gene in the X chromosome instead of two of each kind as in XX females. A male receives his single X chromosome, and thus all of his genes for sex-linked traits, from his mother. Females receive one X from the mother and one from the father. In writing the genotype of a sex-linked trait it is customary to write that of the male with the letter for the gene in the X chromosome plus the letter Y for the Y chromosome. Thus AY would represent the genotype of a male with a dominant gene for trait "A" in his X chromosome.

The phenomenon of sex-linked traits was discovered by T. H. Morgan and C. B. Bridges in the fruit fly, *Drosophila*. These flies normally have eyes with a dark red color, but Morgan and Bridges discovered a strain with white eyes. The gene for white eye, **w**, proved to be recessive to the gene for red eye, **W**, but in certain types of crosses the male offspring had eyes of one color and the female offspring had eyes of the other color. Morgan reasoned that

the peculiarities of inheritance could be explained if the genes for eye color were located in the X chromosome; later work proved the correctness of this guess. Crossing a homozygous red-eyed female with a white-eyed male (**WW** × **wY**) produces offspring all of which have red eyes (**Ww** females and **WY** males). But crossing a homozygous white-eyed female with a red-eyed male (**ww** × **WY**) yields red-eyed females and white-eyed males (**Ww** and **wY**) (Fig. 6.7).

In man, **hemophilia** (bleeder's disease) and **color blindness** are sex-linked traits. About four men in every hundred are color-blind, but somewhat less than one per cent of all women are color-blind. Only one gene for color-blindness produces the trait in males, but two such genes (the trait is recessive) are necessary to produce a color-blind female.

Not all the characters which differ in the two sexes are sex-linked. Some, the **sex-influenced traits,** are inherited by genes located in autosomes rather than X chromosomes, but the expression of the trait, the action of the gene which produces the phenotype, is altered by

Two homologous chromosomes undergo synapsis in meiosis.

Centromere

Each chromosome duplicates to form two chromatids.

Centromeres duplicate, second meiotic division follows.

First meiotic division

Crossing over between a pair of chromatids.

Four haploid gametes produced; here two crossover and two noncrossover gametes.

Figure 6.8 Diagram illustrating crossing over, the exchange of segments between chromatids of homologous chromosomes. Crossing over permits recombination of genes (e.g., vB and Vb); the farther apart genes are located on a chromosome, the greater is the probability that crossing over between them will occur. (From Villee, C. A.: Biology. 7th ed. Philadelphia, W. B. Saunders Co., 1977.)

the sex of the animal, presumably by the action of one of the sex hormones. The presence or absence of horns in sheep, mahogany-and-white spotted coat versus red-and-white spotted coat in Ayrshire cattle, and pattern baldness in man are examples of such sex-influenced traits.

6.13 LINKAGE AND CROSSING OVER

Each species of animal has many more pairs of genes than it has pairs of chromosomes. Obviously there must be many genes per

chromosome. Humans have 23 pairs of chromosomes, some large and some smaller, but thousands of pairs of genes. The chromosomes are inherited as units—they pair and separate during meiosis as units; thus, all the genes in any given chromosome tend to be inherited together and are said to be linked. If the chromosomal units never changed, the traits would always be inherited together and linkage would be absolute. However, during meiosis when the chromosomes are pairing and undergoing synapsis, homologous chromosomes may exchange entire segments of chromosomal material, a process called **crossing over** (Fig. 6.8). This exchanging of segments

Figure 6.9 Diagram of a cross involving linkage and crossing over. The genes for vestigial versus normal wings and black versus gray body in fruit flies are linked; they are located in the same chromosome. (From Villee, C. A.: Biology. 7th ed. Philadelphia, W. B. Saunders Co., 1977.)

Figure 6.10 Diagram illustrating the principle of inferring the order of genes on a chromosome from their crossover values.

occurs at random along the length of the chromosome. Several exchanges may occur at different points along the same chromosome at a single meiotic division. It follows that the greater the distance is between any two genes in the chromosome, the greater will be the likelihood that an exchange of segments between them will occur.

In fruit flies the pair of genes **V** for normal wings and **v** for vestigial wings and the pair of genes **B** for gray body color and **b** for black body color are located in the same pair of chromosomes. They tend to be inherited together and are said to be linked. What would you predict the offspring would be like from a cross of a homozygous **VVBB** fly with a homozygous **vvbb** fly? They are all flies with gray bodies and normal wings and have the genotype **VvBb**. When one of these F_1 heterozygotes is crossed with a homozygous **vvbb** fly (Fig. 6.9), the offspring appear in a ratio which differs from that of the ordinary test cross for a dihybrid. If the two pairs of genes were not linked but were in different chromosomes, the offspring would appear in the ratio of ¼ gray-bodied, normal-winged : ¼ black-bodied, normal-winged : ¼ gray-bodied, vestigial-winged : ¼ black-bodied, vestigial-winged flies. If the genes were completely linked and no exchange of chromosomal segments occurred, then only the parental types—flies with gray bodies and normal wings and flies with black bodies and vestigial wings—would appear among the offspring, and these would be present in equal numbers (Fig. 6.9). However, there is an exchange of segments between the locus of gene **V** and the locus of gene **B**. Because of this crossing over of part of the chromosomes, some gray-bodied, vestigial-winged flies and some black-bodied, normal-winged flies (the cross-over types) appear among the offspring (Fig. 6.9). Most of the offspring, of course, resemble the parents and are gray, normal or black, vestigial. In this particular instance, crossing over occurs between these two points in this chromosome in about one cell in every five or in 20 per cent of the total undergoing meiosis. In such crosses, about 40 per cent of the offspring are gray flies with normal wings. Another 40 per cent are black flies with vestigial wings. Ten per cent

are gray flies with vestigial wings, and 10 per cent are black flies with normal wings. The distance between two genes in a chromosome is measured in "cross-over units" or "centimorgans" (named in honor of T. H. Morgan,

Figure 6.11 Diagram of the X chromosome of a fruit fly as seen in a cell of the salivary gland together with a map of the loci of the genes located in the X chromosome, with the distances between them as determined by frequency of crossing over. (Hunter and Hunter: College Zoology.)

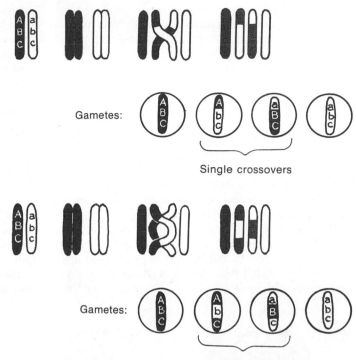

Gametes:

Single crossovers

Gametes:

Double crossovers

Figure 6.12 Diagram illustrating the results of single and double crossing over in a pair of homologous chromosomes. Note that the simultaneous occurrence of two crossovers in a given chromosome leads to a reconstitution of the original combination of genes, e.g., A and C.

the American geneticist who studied crossing-over in the fruit fly) which represent the percentage of crossing over that occurs between them. Thus, **V** and **B** are said to be 20 centimorgans apart.

In a number of species the frequency of crossing over between specific genes has been measured. All of the experimental results are consistent with the hypothesis that genes are present in a linear order in the chromosomes. Thus, if the three genes A, B and C occur in a single chromosome, the amount of crossing over between A and C is either the sum of, or the difference between, the amounts of crossing over between A and B and B and C. For example, if the crossing over between A and B is five units and between B and C is three units, the crossing over between A and C will be found to be either eight units (if C lies to the right of B) or two units (if C lies between A and B) (Fig. 6.10). By putting together the results of a great many such crosses, detailed maps of the location of specific genes on specific chromosomes have been made (Fig. 6.11). The mapping of human chromosomes is very difficult and has been proceeding slowly; several loci in the X chromosome have been mapped.

Since crossing over occurs at random, more than one cross-over may occur in a single chromosome at a given time. One can observe among the offspring only the frequency of

recombination of characters, not the actual frequency of cross-overs. The frequency of crossing over will be somewhat larger than the observed frequency of recombination because the simultaneous occurrence of two crossovers between two particular genes leads to the reconstitution of the original combination of genes (Fig. 6.12).

6.14 THE INTERACTIONS OF GENES

The relationship between the genes and their phenotypes discussed so far is simple and clear: each gene produces a single trait. Genetic research with many different kinds of animals and plants has revealed that the relationship between gene and trait may be quite complex. Several pairs of genes may interact to affect the production of a single trait; one pair of genes may inhibit or reverse the effect of another pair; or a given gene may produce different effects when the environment is altered in some way. The genes are inherited as units but may interact with one another in some complex fashion to produce the trait.

Complementary Genes. Two independent pairs of genes which interact to produce a trait in such a way that neither dominant can produce its effect unless the other is present too are called **complementary genes.** The presence of at least one dominant gene from each pair produces one character; the alternative condition results from the absence of either dominant or of both dominants.

Several different varieties of sweet peas with white flowers are known, and the mating of most white-flowered plants produces only white-flowered offspring. However, when plants from two particular white-flowered varieties were crossed, all the offspring had purple flowers! When two of these purple F_1 plants were crossed, or when they were self-fertilized, an F_2 generation was produced in the ratio of 9 purple to 7 white (Fig. 6.13). Subsequent analysis has shown that two pairs of genes located in different chromosomes are involved; one (**C**) regulates some essential step in the production of a raw material, and the

other (**E**) controls the formation of an enzyme which converts the raw material into purple pigment. The homozygous recessive **cc** is unable to synthesize the raw material, and the homozygous recessive **ee** lacks the enzyme to convert the raw material into purple pigment. One of the white-flowered varieties was genotypically **ccEE**—lacked the gene for the synthesis of raw material—and the other was **CCee**, without the gene for the enzyme required for pigment synthesis. Crossing **CCee** and **ccEE** produces an F_1 generation all of which are **CcEe** and have purple flowers because they have both raw material and enzyme for the synthesis of the pigment. A pure-breeding variety of purple-flowered sweet peas could be established by self-fertilization of a plant with the genotype **CCEE**.

Supplementary Genes. The term **supplementary genes** is applied to two independent

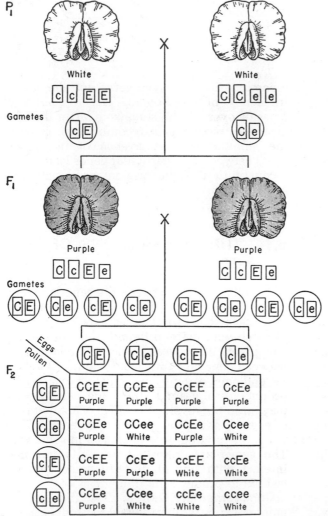

Phenotypes: 9 purple : 7 white

Figure 6.13 Diagram of a cross, illustrating the action of complementary genes, the two pairs of genes which regulate flower color in sweet peas. At least one C gene and one E gene must be present to produce a colored flower. The absence of either one or both results in a white flower. (From Villee, C. A.: Biology. 7th ed. Philadelphia, W. B. Saunders Co., 1977.)

pairs of genes which interact in the production of a trait in such a way that one dominant will produce its effect whether or not the second is present, but the second gene can produce its effect only in the presence of the first. In guinea pigs, in addition to the pair of genes for black versus brown coat color (**B** and **b**), the gene **C** controls the production of an enzyme which converts a colorless precursor into the pigment melanin and, hence, is required for the production of any pigment at all in the coat. The homozygous recessive, **cc**, lacks the enzyme, no melanin is produced, and the animal is a white-coated, pink-eyed **albino,** no matter

what combination of **B** and **b** genes may be present. The eyes have no pigment in the iris and the pink color results from the color of the blood in the tissues of the eye. The mating of an albino, **ccBB,** with a brown guinea pig, **CCbb,** produces offspring all of which are genotypically **CcBb** and have black-colored coats! When two of these F_1 black guinea pigs are mated, offspring appear in the F_2 in the ratio of 9 black: 3 brown : 4 albino. Make a Punnett square to prove this.

Some combination of complementary and supplementary genes may be involved in the inheritance of a single trait. The dominant

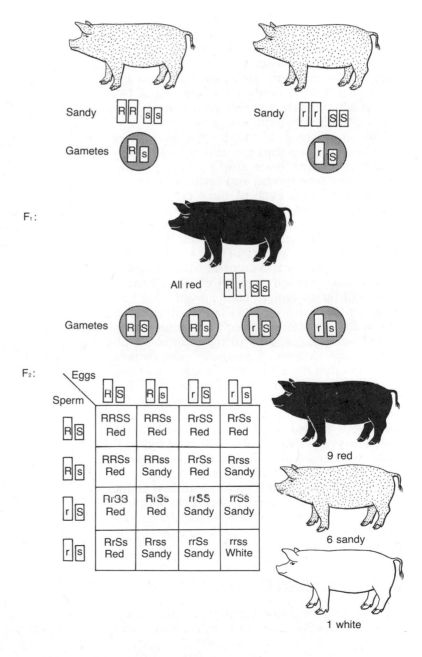

Figure 6.14 Diagram of the mode of inheritance of coat color in Duroc-Jersey pigs, illustrating inheritance by "mutually supplementary" genes.

genes **C** and **R** are both necessary for the production of red kernels in maize, and the absence of either dominant results in white-colored kernels. There is, in addition, a **P** gene which produces purple-colored kernels if both **C** and **R** genes are present. The **P** gene is supplementary to the other two pairs of genes and **C** and **R** are complementary.

The coat color of Duroc-Jersey pigs represents a slightly different type of gene interaction. Two independent pairs of genes (**R-r** and **S-s**) regulate coat color; at least one dominant of each pair must be present to give the full red-colored coat. Partial color, sandy, results when only one type of dominant is present, and an animal which is homozygous for both recessives (**rrss**) has a white-colored coat. The mating of two different strains of sandy-colored pigs, **RRss** × **rrSS**, yields offspring all of which are red, and the mating of two of these red F_1 individuals produces an F_2 generation in the ratio of 9 red : 6 sandy : 1 white (Fig. 6.14). Genes which interact in this fashion have been termed "mutually supplementary."

The inheritance of comb type in poultry provides an interesting example of genic interaction. Leghorns have single combs, Wyandottes have rose combs, and Brahmas have pea combs (Fig. 6.15) Each of these types is true breeding. Suitable crosses demonstrate that the gene for rose comb (**R**) is dominant to single (**r**) and that the gene for pea comb (**P**) is also dominant to its allele (**p**) for single comb. However, when a pea-combed fowl is mated with a rose-combed one, all of the offspring have a different type of comb, resembling half of a shelled walnut and called walnut. When two of these walnut-combed F_1 individuals are mated, offspring appear in the ratio of 9 walnut : 3 pea : 3 rose : 1 single. We can deduce from this that the genotype of a single-combed fowl must be **rrpp;** a pea-combed fowl is either **PPrr** or **Pprr;** a rose-combed fowl is either **ppRR** or **ppRr,** and a walnut comb develops in animals with at least one **P** and one **R** gene. Thus, the genotypes **PPRR, PpRR, PPRr** and **PpRr** all yield walnut combs. Certain Malay varieties of chicken have walnut combs.

It is clear that there is nothing unusual about the method of inheritance of any of these genes; the phenotypic ratios observed are simply the result of some variation in the interaction of the genes in the production of the phenotype.

6.15 POLYGENIC INHERITANCE

Many human characteristics, height, body form, intelligence and skin color, and many

Figure 6.15 Diagram of the inheritance of comb types in chickens. See text for discussion.

commercially important characters of animals and plants, such as milk production in cows, egg production in hens, the size of fruits, and the like, are not separable into distinct alternate classes and are not inherited by single pairs of genes. Nonetheless these traits are governed by genetic factors; there are several, perhaps many, different pairs of genes which affect the same characteristic. The term **polygenic inheritance** is applied to two or more independent pairs of genes which affect the same character in the same way and in an additive fashion. When two varieties which differ in some trait controlled by polygenes are crossed, the F_1 are very similar to one another and are

TABLE 6.2 POLYGENIC INHERITANCE OF SKIN COLOR IN MAN

		AaBb	AaBb
Parents..		(Mulatto)	(Mulatto)
Gametes...		AB Ab aB ab	AB Ab aB ab

Offspring:
    1 with 4 dominants — AABB — phenotypically Negro
    4 with 3 dominants — 2 AaBB and 2 AABb — phenotypically "dark"
    6 with 2 dominants — 4 AaBb, 1 AAbb, 1 aaBB — phenotypically mulatto
    4 with 1 dominant — 2 Aabb, 2 aaBb — phenotypically "light"
    1 with no dominants — aabb — phenotypically white

usually intermediate in the expression of this character between the two parental types. Crossing two F_1 individuals yields a widely variable F_2 generation, with a few members resembling one grandparent, a few resembling the other grandparent, and the rest showing a range of conditions intermediate between the two.

The inheritance of human skin color was carefully investigated by C. B. Davenport in Jamaica. He concluded that the inheritance of skin color is controlled by two pairs of genes, **A-a** and **B-b**, inherited independently. The genes for dark pigmentation, **A** and **B**, are incompletely dominant, and the darkness of the skin color is proportional to the sum of the dominant genes present. Thus, a full Negro has four dominant genes, **AABB**, and a white person has four recessive genes, **aabb.** The F_1 offspring of a mating of white and Negro are all **AaBb**, with two dominant genes and a skin color (mulatto) intermediate between white and Negro. The mating of two such mulattoes produces offspring with skin colors ranging from full Negro to white (Table 6.2). A mulatto with the genotype **AaBb** produces four kinds of eggs or sperm with respect to the genes for skin color: **AB, aB, Ab** and **ab.** From a Punnett square for the mating of two doubly heterozygous mulattoes (**AaBb**) it will be evident that there are 16 possible zygote combinations: one with four dominants (black), four with three dominants (dark brown skin), six with two dominants (mulatto), four with one dominant (light brown skin) and one with no dominants (white skin). The genes **A** and **B** produce about the same amount of pigmentation and the genotypes **AaBb, AAbb** and **aaBB** produce the same phenotype, mulatto skin color.

This example of polygenic inheritance is fairly simple, for only two pairs of genes appear to be involved. With a larger number of pairs of genes, perhaps 10 or more, there are so many classes and the differences between them are so slight that the classes are not distinguishable; a continuous series is obtained. The inheritance

of human stature is governed by a large number of pairs of genes, with shortness dominant to tallness. Since height is affected not only by these genes but also by a variety of environmental agents, there are adults of every height from perhaps 140 cm. (55 inches) up to 203 cm. (84 inches). If we measure the height of 1083 adult men selected at random and draw a graph

Figure 6.16 An example of a "normal curve," or curve of normal distribution: the heights of 1083 adult white males. The blocks indicate the actual number of men whose heights were within the unit range. For example, there were 163 men between 67 and 68 inches in height. The smooth curve is a normal curve based on the mean and standard deviation of the data. (From Villee, C. A.: Biology. 7th ed. Philadelphia, W. B. Saunders Co., 1977.)

of the number having each height, we will obtain a bell-shaped normal curve, or **curve of normal distribution** (Fig. 6.16). It is evident that there are few extremely tall or extremely short men, but many of intermediate height. If we measured the heights of 1000 adult American women and plotted the measurements in the same way, the data would generate a curve with a similar shape, but the mean would be less; women, on the average, are shorter than men. If we measured the heights of 1000 men and women, what sort of curve would be generated by the data?

All living things show comparable variations in certain of their characteristics. If one were to measure the length of 1000 shells from the same species of clam, or the weight of 1000 hens' eggs, or the amount of milk produced per year by 1000 dairy cows, or the intelligence quotient (I.Q.) of 1000 grade school children, and make graphs of the number of individuals in each subclass, one would obtain a normal curve of distribution in each instance. The variation is due in part to the action of polygenes and in part to the effects of a variety of environmental agents.

When a geneticist attempts to establish a new strain of hens that will lay more eggs per year, or a strain of turkeys with more breast meat, or a strain of sheep with longer, finer wool, he selects individuals which show the desired trait in greatest amount for further breeding. There is a limit, of course, to the effectiveness of selective breeding in increasing some desirable trait or in decreasing some undesirable one. When the strain becomes homozygous for all the genetic factors involved, further selective breeding will be ineffective.

The inheritance of certain traits depends not only on a single pair of genes which determines the presence or absence of the trait but also on a number of polygenes which determine the extent of the trait. For example, the presence or absence of spots in the coat of most mammals is determined by a single pair of genes. The size and distribution of the spots, however, are determined by a series of polygenes. The term **modifying factors** has been suggested for polygenes which affect the degree of expression of another gene.

6.16   MULTIPLE ALLELES

In all the types of inheritance discussed so far, there have been only two possible alleles, one dominant and one recessive gene, which could be represented by capital and lower case letters respectively. In addition to a dominant and a recessive gene, there may be one or more additional kinds of gene found at the same location in the chromosome that affect the same trait in an alternative fashion. The term **multiple alleles** is applied to three or more different kinds of gene, three or more alternative conditions at a single locus in the chromosome, each of which produces a distinctive phenotype. In the population as a whole, there are distributed three or more different alleles, but each individual has any two, and no more than two, of the possible types of alleles, and any gamete has only one. The members of an allelic series are indicated by the same letter, with suitable distinguishing superscripts.

One series of multiple alleles which affects coat color in rabbits includes the dominant gene **C** for normal coat color, the recessive gene **c** which produces albino coat color when homozygous, and two other alleles, c^h and c^{ch}. The gene c^h, when homozygous, produces the "Himalayan" pattern of white coat over the body but with a dark color on the tips of the ears, nose, tail and legs. The gene c^{ch}, when homozygous, produces the "chinchilla" pattern of light gray fur all over the body. These alleles may be arranged in the series **C**, c^{ch}, c^h and **c,** in which each gene is dominant to the succeeding genes but recessive to the preceding ones. In other series of multiple alleles the genes may be incompletely dominant so that the heterozygote has a phenotype intermediate between those of its two parents, or one which is some combination of the two parental phenotypes.

Multiple alleles govern the inheritance of the human blood groups O, A, B and AB (p. 324). The three alleles of the series, a^A, a^B and **a,** regulate the kind of agglutinogen in the red blood cells (Table 6.3). Gene a^A produces agglutinogen A, gene a^B produces agglutinogen B, and gene **a** produces no agglutinogens. Gene **a** is recessive to the other two, but neither a^A nor a^B is dominant to the other; each produces its characteristic agglutinogen independently of the other. Transfusions of blood from one person to another are successful only when the two bloods are compatible, when the agglutinins in the plasma of the recipient do not react with the agglutinogens in the red cells of the donor to cause agglutination, clumping of the red cells. People with type O blood (no agglutinogens in their red cells) are known as "universal donors"; their blood can be transfused into the veins of persons with any of these blood groups. People with type AB blood

TABLE 6.3 THE INHERITANCE OF THE HUMAN BLOOD GROUPS

Blood Group	Genotypes	Agglutinogen in Red Cells	Agglutinin in Plasma	Can Give Blood to Groups	Can Receive Blood from Groups
O	aa	none	a and b	O, A, B, AB	O
A	$a^A a^A$, $a^A a$	A	b	A, AB	O, A
B	$a^B a^B$, $a^B a$	B	a	B, AB	O, B
AB	$a^A a^B$	A and B	none	AB	O, A, B and AB

are called "universal recipients"; they have no agglutinins in the plasma and hence their plasma will not cause agglutination of the red cells from any person.

Since blood types are inherited, and do not change in a person's lifetime, they are useful indicators of parentage. In cases of disputed parentage, genetic evidence can show only that a certain man or woman *could be* the parent of a particular child, and never that he *is* the parent. In certain circumstances, however, the genetic evidence can definitely exclude a particular man or woman as the parent of a given child. Thus, if a child of blood group A is born to a type O woman, no man with type O or type B blood could be its father (Table 6.4). Could an AB man be the father of an O child? Could an O man be the father of an AB child? Could a type B child with a type A mother have a type A father? a type O father?

We now know of 11 different sets of blood groups, inherited by different pairs of genes, all of which are helpful in establishing paternity. The most important of these are the Rh alleles, which determine the presence or absence of a different agglutinogen, the **Rh factor,** first found in the blood of rhesus monkeys. There are actually several alleles at the **Rh** locus, but to simplify matters we shall consider just two: **Rh,** which produces the Rh positive antigen, and the recessive **rh,** which does not produce

the antigen. Genotypes **RhRh** and **Rhrh** are phenotypically Rh positive and genotype **rhrh** is phenotypically Rh negative. An Rh negative woman married to an Rh positive man may have an Rh positive child. If some blood manages to pass across the placenta from the fetus to the mother it will stimulate the formation, in her blood, of antibodies to the Rh factor. Then, in a subsequent pregnancy, some of these Rh antibodies may pass through the placenta to the child's blood, and react with the Rh antigen in the child's red cells. The red cells are agglutinated and destroyed and a serious, often fatal, anemia, called **erythroblastosis fetalis,** ensues. The introduction of fetal red cells into the maternal blood stream occurs during parturition, or, to a lesser degree, during spontaneous or induced abortion. The production of maternal anti-Rh antibodies to these fetal antigens can be blocked by injecting small amounts of anti-D gamma globulin into Rh negative women just after childbirth or abortion.

Extensive surveys have shown that approximately 41 per cent of native white Americans are type O, 45 per cent are type A, 10 per cent are type B, and 4 per cent are type AB. The frequency of the blood groups in other races may be quite different; American Indians, for example, have a low frequency of group A and a high frequency of group B. No one blood type is characteristic of a single race; the racial differences lie in the *relative frequency* of the several blood types. Studies of the relative frequencies of the blood groups found in different races living today and in mummies and prehistoric skeletons have provided valuable evidence as to the relationships of the present races of man.

6.17 INBREEDING AND OUTBREEDING

It is commonly believed that the mating of two closely related individuals—brother and sister or father and daughter—is harmful and leads to the production of monstrosities. The marriage of first cousins is forbidden by law in

TABLE 6.4 EXCLUSION OF PATERNITY BASED ON BLOOD TYPES

Child	Mother	Father Must Be of Type	Father Cannot Be of Type
O	O	O, A, or B	AB
O	A	O, A, or B	AB
O	B	O, A, or B	AB
A	O	A or AB	O or B
A	A	A, B, AB, or O	——
B	B	A, B, AB, or O	——
A	B	A or AB	O or B
B	A	B or AB	O or A
B	O	B or AB	O or A
AB	A	B or AB	O or A
AB	B	A or AB	O or B
AB	AB	A, B, or AB	O

some states. Carefully controlled experiments with many different kinds of animals and plants have shown that there is nothing harmful in the process of inbreeding itself. It is, in fact, one of the standard procedures used by commercial breeders to improve strains of cattle, corn, cats and cantaloupes. It is not necessarily a bad practice in the human species. In all animals or plants it simply tends to make the strain homozygous.

All natural populations of individuals are heterozygous for many traits; some of the hidden recessive genes are for desirable traits, others are for undesirable ones. **Inbreeding** will simply permit these genes to become homozygous and lead to the unmasking of the good or bad traits. If a stock is good, inbreeding will improve it; but if a stock has many undesirable recessive traits, inbreeding will lead to their phenotypic expression.

The crossing of two completely unrelated strains, called **outbreeding,** is another widely used genetic maneuver. It is frequently found that the offspring of such a mating are much larger, stronger and healthier than either parent. Much of the corn grown in the United States is a special hybrid variety developed by the United States Department of Agriculture from a mating of four different inbred strains. Each year, the seed to grow this uniformly fine hybrid corn is obtained by mating the original inbred lines. If the hybrid corn were used in mating, it would give rise to many different kinds of corn since it is heterozygous for many different traits. The mule, the hybrid offspring of the mating of a horse and donkey, is a strong, sturdy animal, better adapted for many kinds of work than either of its parents. This phenomenon of **hybrid vigor,** or **heterosis,** does not result from the act of outbreeding itself, but from the heterozygous nature of the F_1 organisms which result from outbreeding. Each of the parental strains is homozygous for certain undesirable recessive traits, but the two strains are homozygous for *different* traits, and each one has dominant genes to mask the undesirable recessive genes of the other. As a concrete example, let us suppose that there are four pairs of genes, **A, B, C** and **D;** the capital letters represent the dominant gene for some desirable trait, and the lower case letters represent the recessive gene for its undesirable allele. If one parental strain is then **AAbbCCdd** and the other **aaBBccDD,** the offspring will all be **AaBbCcDd** and have all of the desirable and none of the undesirable traits. The actual situation in any given cross is undoubtedly much more complex and involves many pairs of genes.

6.18 PROBLEMS IN GENETICS

Since many students find that solving problems is helpful in learning the principles of genetics we have provided the following:

1. In peas, the gene for smooth seed coat is dominant to the one for wrinkled seeds. What would be the result of the following matings: heterozygous smooth × heterozygous smooth? Heterozygous smooth × wrinkled? Heterozygous smooth × homozygous smooth? Wrinkled × wrinkled?

2. In peas, the gene for red flowers is dominant to the one for white flowers. What would be the result of mating heterozygous red-flowered, smooth-seeded plants with white-flowered, wrinkled-seeded plants?

3. The mating of two black short-haired guinea pigs produced a litter which included some black long-haired and some white short-haired offspring. What are the genotypes of the parents and what is the probability of their having black short-haired offspring in subsequent matings?

4. Human color blindness is a sex-linked, recessive trait. What is the probability that a woman with normal vision whose husband is color-blind will have a color-blind son? a color-blind daughter? What is the probability that a woman with normal vision whose father was color-blind but whose husband has normal vision will have a color-blind son? a color-blind daughter?

5. The gene for white eye color (w) in fruit flies is sex-linked and recessive to normal red eye color (W). Give the results of mating (a) a heterozygous red-eyed female with a red-eyed male, (b) a white-eyed female with a red-eyed male and (c) a heterozygous red-eyed female with a white-eyed male.

6. A blue-eyed man, both of whose parents were brown-eyed, marries a brown-eyed woman whose father was blue-eyed and whose mother was brown-eyed. Their first child has blue eyes. Give the genotypes of all the individuals mentioned and give the probability that the second child will also have blue eyes.

7. Outline a breeding procedure whereby a true-breeding strain of red cattle could be established from a roan bull and a white cow.

8. Suppose you learned that shmoos may have long, oval or round bodies and that matings of shmoos gave the following results:

long × oval gave 52 long and 48 oval
long × round gave 99 oval
oval × round gave 51 oval and 50 round

oval × oval gave 24 long, 53 oval and 27 round

What hypothesis about the inheritance of shmoo shape would be consistent with these results?

9. A mating of an albino guinea pig and a black one gave six white (albino), three black and three brown offspring. What are the genotypes of the parents? What kinds of offspring, and in what proportions, would result from the mating of the black parent with another animal that has exactly the same genotype as it has?

10. Mating a red Duroc-Jersey hog to sow A (white) gave pigs in the ratio of 1 red : 2 sandy : 1 white. Mating this same hog to sow B (sandy) gave 3 red : 4 sandy : 1 white. When this hog was mated to sow C (sandy) the litter had equal numbers of red and sandy piglets. Give the genotypes of the hog and the three sows.

11. A walnut-combed rooster is mated to three hens. Hen A (walnut-combed) has offspring in the ratio of 3 walnut : 1 rose. Hen B (pea-combed) has offspring in the ratio of 3 walnut : 3 pea : 1 rose : 1 single. Hen C (walnut-combed) has only walnut-combed offspring. What are the genotypes of the rooster and the three hens?

12. The size of the egg laid by one variety of hens is determined by three pairs of genes; hens with the genotype **AABBCC** lay eggs weighing 90 gm. and hens with the genotype **aabbcc** lay eggs weighing 30 gm. Each dominant gene adds 10 gm. to the weight of the egg. When a hen from the 90 gm. strain is mated with a rooster from the 30 gm. strain, the hens in the F_1 generation lay eggs weighing 60 gm. If a hen and rooster from this F_1 generation are mated, what will be the weights of the eggs laid by the hens of the F_2 generation?

13. Mrs. Doe and Mrs. Roe had babies at the same hospital and at the same time. Mrs. Doe took home a girl and named her Nancy. Mrs. Roe received a boy and named him Harry. However, she was sure that she had had a girl and brought suit against the hospital. Blood tests showed that Mr. Roe was Type O, Mrs. Roe was type AB, Mr. and Mrs. Doe were both type B, Nancy was type A and Harry was type O. Had an exchange occurred?

14. A woman who is type O and Rh negative is married to a man who is type AB and Rh positive. The man's father was type AB and Rh negative. What are the genotypes of the man and woman and what blood types may occur among their offspring? Is there any danger that any of their offspring may have erythroblastosis fetalis?

ANNOTATED REFERENCES

Carlson, E.: The Gene: A Critical History. Philadelphia, W. B. Saunders Co., 1966. Traces the development of the concept of the gene.

Herskowitz, I. H.: Genetics. 4th ed. Boston, Little, Brown & Co., 1976. A standard, college-level text.

Strickburger, M. W.: Genetics. 3rd ed. New York, Macmillan, 1975. A standard, college-level text.

Sturtevant, A. H.: History of Genetics. New York, Harper & Row, 1965. An authentic account of the rise of genetics by a man who played a prominent role in the development of this important science.

Whitehouse, H. L.: Towards an Understanding of the Mechanism of Heredity. 2nd ed. New York, St. Martin's Press, 1965. A discussion of genetics in terms of its hypotheses and how they were tested.

MOLECULAR AND MATHEMATICAL ASPECTS OF GENETICS

Our modern concepts of genetics originated with the rediscovery of Mendel's Laws in 1900. Since that time, geneticists have been attempting to determine the physical structure and chemical composition of the hereditary units—the genes—and to discover the mechanisms by which they transfer biological information from one cell to another and control the development and maintenance of the organism. During the first half of this century, much was learned about the complexity of the molecular structure of proteins. As a result, nearly all biochemists assumed that any complex biological unit with such marked specificity as the gene must also be a protein. There was great difficulty, however, in explaining how protein molecules could be duplicated precisely, as genes must be with each cell division. The realization that genetic information is transferred by nucleic acid molecules rather than by protein molecules arose gradually. Despite the recent rapid advances in our understanding of many facets of biochemical genetics, our knowledge of the molecular organization of chromosomes, especially the complex chromosomes of higher plants and animals, is still incomplete.

Our present belief that DNA and RNA are the primary agents for the transfer of biological information came about gradually, culminating in 1953 in the proposal by James Watson and Francis Crick of a model of the DNA molecule that explained how it could transfer informa-

tion and undergo replication. This proposal stimulated an enormous flood of research and has led to the present **"central dogma"** of biology: Genes are composed of DNA and are located within the chromosomes. Each gene contains information coded in the form of a specific sequence of purine and pyrimidine nucleotides within its DNA molecule. The unit of genetic information, called a **codon,** is a group of three adjacent nucleotides that specify a single amino acid in a polypeptide chain. Thus the genetic code is a **"triplet"** code. The DNA molecule consists of two complementary chains of polynucleotides twisted about each other in a regular helix and joined by hydrogen bonds between specific pairs of purine and pyrimidine bases. The DNA molecule is replicated when the two strands of the helix separate, and each acts as a template for the formation of a new complementary strand. Each pair of strands, one old and one new, then twist together to form two daughter helices.

7.1 THE CHEMISTRY OF CHROMOSOMES

Chromosomes contain DNA, RNA, histones and other proteins. There are about 6×10^{-9} milligram of DNA per nucleus in somatic cells and 3×10^{-9} milligram of DNA per nucleus in egg or sperm cells (Table 7.1). In tissues known

TABLE 7.1 AMOUNT OF DEOXYRIBONUCLEIC ACID (DNA) PER NUCLEUS IN ANIMAL TISSUES, EXPRESSED AS Mg. $\times 10^{-9}$

Species	Sperm	Red Cell	Liver	Heart	Kidney	Pancreas	Spleen
Shad	0.91	1.97	2.01				
Carp	1.64	3.49	3.33				
Brown trout	2.67	5.79					
Toad	3.70	7.33					
Frog		15.0	15.7				
Chicken	1.26	2.49	2.66	2.45	2.20	2.61	2.55
Dog			5.5		5.3		
Rat			9.47	6.50	6.74	7.33	6.55
Ox	3.42		7.05		6.63	7.15	7.26
Human	3.25	7.30	10.36		8.6		

to be polyploid (having more than two sets of chromosomes per nucleus) the amount of DNA per nucleus corresponds to the multiple of chromosomes known to be present. Tetraploid cells, with four sets of chromosomes, have 12×10^{-9} milligram of DNA per nucleus. The amounts of DNA and histone are essentially the same in all the cells of a given organism, whereas the amounts of RNA and of several other proteins vary considerably from cell to cell. That the amount of DNA, like the number of genes, is constant in all the cells of the body, and the amount of DNA in germ cells is only half the amount in somatic cells, is evidence that DNA is an essential part of the gene. From the amount of DNA per cell, one can estimate the number of nucleotide pairs per cell and thus the amount of genetic information present (Table 7.2).

TABLE 7.2 THE AMOUNT OF DNA PER CELL IN ANIMAL AND PLANT CELLS AND IN VIRUS PARTICLES

	DNA (Mg. $\times 10^{-9}$ per Cell)	Nucleotide Pairs per Cell
Mammals	6	5.5×10^9
Birds	2	2×10^9
Reptiles	5	4.5×10^9
Amphibia	7	6.5×10^9
Fish	2	2×10^9
Insects	0.17–12	0.16×10^9
Crustacea	3	2.8×10^9
Mollusks	1.2	1.1×10^9
Echinoderms	1.8	1.7×10^9
Sponges	0.1	0.1×10^9
Higher plants	2.5–40	2.3×10^9
Fungi	0.02–0.17	0.02×10^9
Algae	3	2.8×10^9
Bacteria	0.002–0.06	2×10^6
T_2 bacteriophage	0.00024	2.2×10^5
λ bacteriophage	0.00008	7×10^4
Papilloma virus	—	6×10^3

The chromosomes of higher animals and plants (but not bacteria) contain five different types of histones present in approximately equimolar amounts, with the exception of histone I, which is present in one-half the amount of the others. The phosphate groups of the DNA molecules bind to the lysine and arginine residues of the histones by salt linkages; the DNA-histone complex is called **chromatin.** It now appears that the basic unit of a chromosome, termed a **nucleosome,** consists of a globular set of eight histones, two of each of the four types, to which is attached a segment of DNA about 200 nucleotides long. This globule is a solid structure with no holes, and the DNA is believed to be wound around the outside of the globule. The location and the role of histone I is still uncertain. Evidence indicates that all of the DNA of any given chromosome is present in a single long polynucleotide molecule. In between successive nucleosomes a portion of the DNA is free and connects one nucleosome with the adjacent one. In addition to histones, the chromosome contains a variety of acidic proteins which are believed to function in controlling gene activity by determining which genes will be expressed at any given time in any given cell.

The chromosomes of prokaryotic organisms appear to be enormous single molecules of DNA, with molecular weights of 1.2×10^8 (bacteriophage) or 2×10^9 (*Escherichia coli*). In these organisms the molecule of DNA appears to be circular, without any free ends. It has been suggested that this circularity prevents replication of the DNA and that the DNA molecule is converted from the circular to a linear form for only a brief moment during the replication process. The individual DNA molecules in the chromosomes of higher organisms are not circular, for electron micrographs published in 1966 by Dr. Margit Nass showed that the DNA in the nuclei of mouse fibroblasts is in linear fibrous strands 10 to 15 micrometers long. The mitochondria of many, perhaps all, cells contain a small amount of DNA (about 1 or 2 per cent of the total), and Dr. Nass's electron micrographs revealed that the mitochondrial DNA consists of circular strands some 5 micrometers long.

7.2 THE ROLE OF DNA IN HEREDITY

The first direct evidence that DNA can transmit genetic information came from the experiments of Avery and his coworkers with the "transforming agents" isolated from pneumococci and certain other bacteria. A transforming agent isolated from strain III of pneumococcus (the bacteria causing pneumonia) will, when added to culture medium, transform strain II organisms into strain III pneumococci. This is a "permanent" transformation; the strain III organisms so produced multiply to give only strain III offspring. Similar transforming agents have been found in other types of bacteria, and in each case, when they have been isolated and characterized, they have been found to be pure DNA.

DNA has been shown to be the carrier of genetic information in bacterial viruses that consist of a "head," made up of a protein membrane around a DNA "core," and a tail composed of protein. The virus becomes attached to the bacterial cell by the tip of its tail. The DNA present in the head of the virus is transferred inside the bacterial cell, but most of the tail and the head membrane remain outside. The viral protein on the outside of the bacterial cell can be sheared off by stirring the infected bacteria in a Waring Blendor. The bacteria containing the viral DNA will then produce a large number of virus particles identical with the one used for infection. Thus, most of the viral protein can be eliminated without interfering with the replication process, indicating that the protein is not important in the formation of new virus particles.

Experiments using labeled atoms have shown further that the atoms of the protein in the parent virus do not appear in the progeny, whereas more than 50 per cent of the atoms of the parental DNA do appear in the progeny. Within the bacterial cell the nucleic acid leads to the production both of additional nucleic acid cores and of protein coats, and many virus particles are formed and released when the infected bacterial cell bursts. In other experiments pure nucleic acid has been prepared from plant viruses and shown to have the viral activity characteristic of that species of virus.

Further evidence that DNA is the carrier of genetic information comes from experiments in which genetic recombination has been demonstrated between different strains of bacteria. On rare occasions two bacteria come together, a cytoplasmic bridge forms between them, and DNA passes across this bridge from the donor to the recipient cell. The amount of DNA transferred is proportional to the amount of genetic information transferred as measured on a genetic map. If the bacteria are left undisturbed, as much as one-third of the genetic material is transferred from donor to recipient. However, if bacteria are placed in a Waring Blendor at various times after the cytoplasmic bridge has formed and the cells are separated, varying amounts of DNA and of genetic material will be transferred. The proportionality between the amount of DNA and the length of the genetic map transferred indicates that genetic information is contained in DNA.

Bacterial genes may be transferred passively from one bacterium to another by a bacteriophage, a process termed **transduction.** As a virus particle forms within the host bacterium, it may enclose and come to contain a small segment of the bacterial genetic material along with the 'phage DNA. When the 'phage is subsequently released, it becomes attached to a new bacterium and injects the segment of bacterial chromosome from the previous host into the new host along with its own 'phage DNA. The segment of bacterial DNA may undergo crossing over with the new host's chromosome and thus incorporate genes from the previous host strain. Since only DNA is transferred in this way, this is further evidence confirming the hypothesis that genes are DNA.

Nucleic acids absorb ultraviolet light very strongly with a maximum at 260 nanometers. The production of mutations by ultraviolet light is also maximal at 260 nanometers. When one compares the number of mutations produced per unit of energy delivered and the wave length at which the energy is delivered, an **action spectrum** for mutations is obtained. There is a close correlation between the action spectrum for the production of mutations and the absorption spectrum for nucleic acids. The simplest explanation of this phenomenon is that genes are composed of nucleic acids, that mutations are produced as nucleic acids absorb energy, and that the absorbed energy is effective in changing the nucleic acid molecule to yield a mutant gene.

7.3 THE WATSON-CRICK MODEL OF DNA

Highly purified DNA has been prepared from a wide variety of animals, plants and bacteria and found in each case to consist of a sugar—**deoxyribose**—phosphoric acid and nitrogenous bases. Four major kinds of bases are found in DNA: two purines, **adenine** and **guanine,** and two pyrimidines, **cytosine** and **thymine.** The ratios of these purine and pyrimidine bases are the same in the DNA from all the cells of a given species but may differ in samples of DNA taken from different species (Table 7.3). In all samples of DNA the total amount of purines equals the total amount of pyrimidines $(A + G = T + C)$, the amount of adenine equals the amount of thymine $(A = T)$, and the amount of guanine equals the amount of cytosine $(G = C)$. DNA from mammalian tissues is generally rich in adenine and thymine and relatively poor in guanine and cytosine, whereas nucleic acid from bacterial sources is, in general, rich in guanine and cytosine and relatively poor in adenine and thymine.

Further clues about the structure of the DNA

TABLE 7.3 RELATIVE AMOUNTS OF PURINES AND PYRIMIDINES IN SAMPLES OF DNA

Source	Adenine	Guanine	Cytosine	Thymine
Beef thymus	29.0	21.2	21.2	28.5
Beef liver	28.8	21.0	21.1	29.0
Beef sperm	28.7	22.2	22.0	27.2
Human thymus	30.9	19.9	19.8	29.4
Human liver	30.3	19.5	19.9	30.3
Human sperm	30.9	19.1	18.4	31.6
Hen red cells	28.8	20.5	21.5	29.2
Herring sperm	27.8	22.2	22.6	27.5
Wheat germ	26.5	23.5	23.0	27.0
Yeast	31.7	18.3	17.4	32.6
Vaccinia virus	29.5	20.6	20.0	29.9
Bacteriophage T_2	32.5	18.2	16.7	32.6

molecule came from x-ray diffraction studies carried out by Rosalind Franklin in the laboratory of M. H. F. Wilkins. When a pure crystal of DNA or some other substance is bombarded with x-rays, the x-rays are diffracted, or bent, in specific directions as they pass through the substance. The amount and nature of the bending of the x-rays depends on the structure of the molecule itself. The pattern of x-ray diffraction (Fig. 7.1), although incomprehensi-

A   B

Figure 7.1 X-ray diffraction photographs of suitably hydrated fibers of DNA, showing the so-called B configuration. A, Pattern obtained using the sodium salt of DNA. B, Pattern obtained using the lithium salt of DNA. This pattern permits a most thorough analysis of DNA. The diagonal pattern of spots (reflections) stretching from 11 o'clock to 5 o'clock and from 1 o'clock to 7 o'clock provides evidence for the helical structure of DNA. The elongated horizontal reflections at the top and bottom of the photographs provide evidence that the purine and pyrimidine bases are stacked 0.34 nm. apart and are perpendicular to the axis of the DNA molecule. (Courtesy of Biophysics Research Unit, Medical Research Council, King's College, London; from Villee, C. A.: Biology. 7th ed. Philadelphia, W. B. Saunders Co., 1977.)

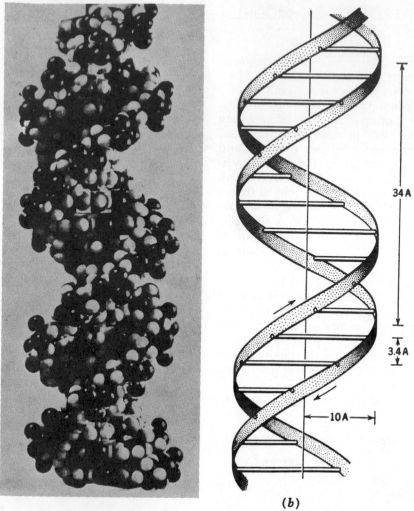

(b)

Figure 7.2 Photograph of a molecular model of deoxyribonucleic acid (through the courtesy of Dr. M. H. F. Wilkins). A schematic drawing of this two-stranded structure is shown on the right, together with certain of its dimensions in Angstrom units. (Anfinsen: Molecular Basis of Evolution, John Wiley & Sons.)

ble to the novice, provides to the experienced eye a number of clues about the structure of the molecule. From such x-ray diffraction pictures, Franklin and Wilkins inferred that the nucleotide bases, which are flat molecules, are stacked one on top of the other like a group of saucers. These x-ray diffraction patterns showed three major periodicities in crystalline DNA, one of 0.34 nm., one of 2.0 nm. and one of 3.4 nm.

On the basis of the analytical results and the x-ray diffraction patterns, Watson and Crick proposed in 1953 a model of the DNA molecule (Fig. 7.2). The great usefulness of this model in providing a chemical explanation for the biological properties of DNA led to the awarding of Nobel prizes to Watson and Crick in 1962.

The studies of Pauling and other chemists provided a great deal of information about the

exact distances between atoms that are bonded together in a molecule, the angles between the bonds of a given atom and the sizes of the atoms. Using this information Watson and Crick began to build scale models of the component parts of DNA and then fit them together to agree with the various experimental data. It had been known that the adjacent nucleotides in DNA are joined in a chain by phosphodiester bridges which link the 5′ carbon of the deoxyribose of one nucleotide with the 3′ carbon of the deoxyribose of the next nucleotide. It seemed clear to Watson and Crick that the 0.34 nm. periodicity found by Franklin and Wilkins corresponded to the distance between successive nucleotides in the chain. It was a reasonable guess that the 2.0 nm. periodicity corresponded to the width of the chain. To explain the 3.4 nm. periodicity, they postulated, as Pauling had, that the chain was coiled in a helix. A **helix** is formed by winding a chain around a cylinder; in contrast, a **spiral** is formed by winding a chain around a cone. This 3.4 nm. periodicity corresponded to the distance between successive turns of the helix. A chain can be wound around a cylinder either loosely or tightly; this corresponds to the steepness of the pitch of the screw in the helix. Since 3.4 nm. is just 10 times the 0.34 nm. distance between the successive nucleotides, it was clear that each full turn of the helix contained 10 nucleotides. From these data Watson and Crick could calculate the density of a chain of nucleotides coiled in a helix 2 nm. wide, with turns that were 3.4 nm. long. Such a chain would have a density only half as great as the known density of DNA. Consequently, they postulated that there were two chains—a **double helix** of nucleotides—that made up the DNA molecule.

The next problem was to determine the spatial relationships between the two chains that make up the double helix. Having tried a number of arrangements with their scale models, they found that the best fit with all the data was with one in which the two nucleotide helices were wound in opposite directions (Fig. 7.3), with the sugar-phosphate chains on the outside of the helix and the purines and pyrimidines on the inside, held together by hydrogen bonds between bases on the opposite chains. These hydrogen bonds hold the chains together and maintain the helix. A double helix can be visualized by imagining the form that would result from taking a ladder and twisting it into a helical shape, keeping the rungs of the ladder perpendicular. The sugar and phosphate molecules of the nucleotide chains make up the railings of the ladder, and the rungs are formed by the nitrogenous bases held together by hydrogen bonds.

Further studies of the possible models made it clear to Watson and Crick that each crossrung must contain one purine and one pyrimidine. The space available with the 2.0 periodicity would accommodate one purine and one pyrimidine, but not two purines, which would be too large, and not two pyrimidines, which would not come close enough together to form proper hydrogen bonds. Further examination of the detailed molecule showed that although a combination of adenine and cytosine was the proper size to fit as a rung on the ladder, they could not be arranged in such a way that they would form proper hydrogen bonds. A similar consideration ruled out the pairing of guanine and thymine; however, adenine and thymine

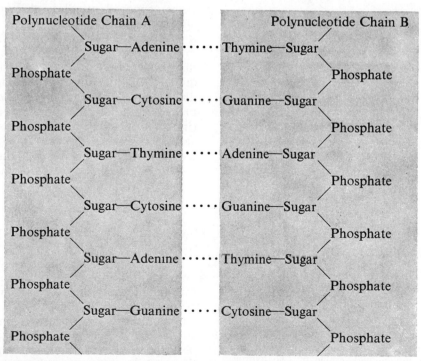

Figure 7.3 Schematic diagram of a portion of a DNA molecule, showing the two polynucleotide chains joined by hydrogen bonds (......). The chains are not flat, as represented here, but are coiled around each other in helices (see Fig. 7.2). (From Villee, C. A.: Biology. 7th ed. Philadelphia, W. B. Saunders Co., 1977.)

would form hydrogen bonds, and guanine and cytosine could form hydrogen bonds. The nature of the hydrogen bonds requires that adenine pair with thymine and that guanine pair with cytosine. This concept of **specific base pairing** provided the basis for Chargaff's rules that the amounts of adenine and thymine in any DNA molecule are always equal and the amounts of guanine and cytosine are always equal. Two hydrogen bonds can form between adenine and thymine, and three hydrogen bonds between guanine and cytosine (Fig. 7.4). The specificity of the kind of hydrogen bond that can be formed ensures that for every adenine in one chain there will be a thymine in the other chain. Similarly for every guanine in the first chain there will be a cytosine in the second chain. Thus the two chains are complementary

to each other; i.e., the sequence of nucleotides in one chain dictates the sequence of nucleotides in the other. The two strands are **antiparallel**; they extend in opposite directions and have their terminal phosphate groups at opposite ends of the double helix.

The most distinctive properties of the genetic material are that it carries information and it undergoes replication. The Watson-Crick model explains how DNA molecules may carry out these two functions. When a DNA molecule undergoes replication, the two chains separate and each one brings about the formation of a new chain, which is complementary to it. Thus two new chains are established (Fig. 7.5). The nucleotides in the new chain are assembled in a specific order, because each purine or pyrimidine in the original chain forms hydrogen bonds with the complementary pyrimidine or purine nucleotide triphosphate from the surrounding medium and lines them up in a complementary order. Phosphate ester bonds are formed by the reaction catalyzed by **DNA polymerase** to join the adjacent nucleotides in the chain, and a new polynucleotide chain results (Fig. 7.6). The new and the original chains then wind around each other, and two new DNA molecules are formed. Each chain, in other words, serves as a template or a mold against which a new partner chain is synthesized. The end result is two complete double chain molecules, each identical to the original double-chain molecule.

A second prime function of DNA, in addition to its role in replication, is the transcription of the information contained in its specific sequence of nucleotides sometime between cell divisions. The product of the transcription process, messenger RNA, then combines with ribosomes to carry out the synthesis of enzymes and other specific proteins. Thus we can visualize how each gene can lead to the production of a specific enzyme. We shall return to this subject in another section.

7.4 THE GENETIC CODE

The Watson-Crick model of the DNA molecule implied that genetic information is transmitted by some specific sequence of its constituent nucleotides. In 1954, George Gamow, an imaginative physicist, was one of the first to suggest that the minimum coding relation between nucleotides and amino acids would be three nucleotides per amino acid. Four nucleotides taken two at a time provide for only 16 combinations ($4^2 = 16$), whereas four nucleo-

Figure 7.4 Diagram showing the specific pairing of adenine and thymine (above), joined by two hydrogen bonds (······), and of guanine and cytosine (below), linked by three hydrogen bonds. (From Villee, C. A.: Biology. 7th ed. Philadelphia, W. B. Saunders Co., 1977.)

Figure 7.5 Diagrammatic scheme of how DNA molecules may undergo replication in the reaction catalyzed by the enzyme DNA polymerase. (From Villee, C. A.: Biology. 7th ed. Philadelphia, W. B. Saunders Co., 1977.)

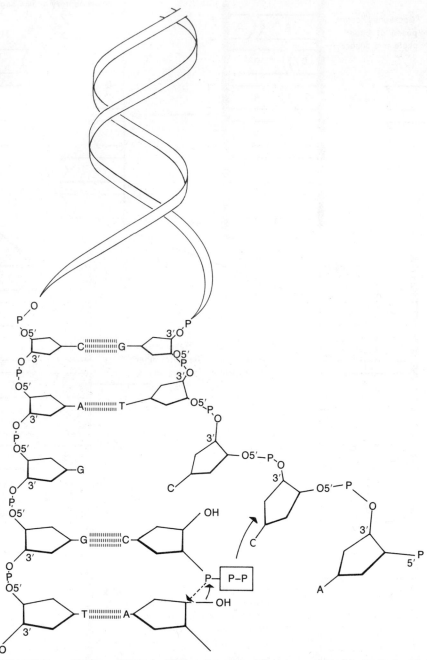

Figure 7.6   Mechanism of the replication of DNA by DNA polymerase. The two strands of the DNA double helix are shown separating. The left one, which runs from the 5′ phosphate to 3′ OH, is being copied starting from the bottom. The newly synthesized chain begins with a 5′ phosphate. In the new chain, as in the old, adenine forms base pairs with thymine and cytidine forms base pairs with guanine. The new phosphodiester bond is made between adjacent bases in the forming chain by an attack by the 3′ OH group of the deoxyribose on the bond between the inner phosphate and the outer two phosphates of the adjacent trinucleotide. The two outer phosphates are split off as inorganic pyrophosphate. (From Villee, C. A.: Biology. 7th ed. Philadelphia, W. B. Saunders Co., 1977.)

tides taken three at a time provide for 64 combinations ($4^3 = 64$). At first glance, this would seem to provide many more code symbols than are needed, since there are only 20 different amino acids. It was believed at one time that some of these 64 combinations were simply "nonsense" codes that did not specify any amino acid. However, there is now strong evidence that all but three of the 64 combinations do, in fact, code for one or another amino acid, and that as many as six different nucleotide triplets may specify the same amino acid.

The fundamental characteristics of the genetic code are now well established; it is a triplet code with three adjacent nucleotide bases, termed a **codon,** specifying each amino acid (Table 7.4). For some years the question of whether the code was overlapping or not remained unsolved. For example, is the sequence CAG, AUC, GAC read only as CAG AUC, GAC, or can it also be read CAG, AGA, GAU, AUC, UCG, CGA, GAC? Is each nucleotide part of one codon or three? The amino acid sequences of each of the several mutant forms of the hemoglobin molecule have been analyzed. In each, only a single amino acid in the peptide chain is substituted. In contrast, if the code were overlapping and a given nucleotide were part of three adjacent codons, we would expect three adjacent amino acids to be changed. An overlapping code would restrict the possible orders of amino acids in a peptide. Thus the amino acid specified by CAG could be followed only by AG(X), by one of two amino

acids. It is clear from analyses of amino acid sequences in peptides that there are no such restrictions on the sequences possible. Finally, experiments with synthetic polynucleotides having known base sequences have shown conclusively that the code is not overlapping.

The code is commaless; no "punctuation" is necessary, since the code is read out beginning at a fixed point and the entire strand is read, three nucleotides at a time, until the read-out mechanism comes to a specific "termination" code, which signals the end of the message. The nature of the signal, "begin reading here," at least in bacteria, also appears to be specified by a specific sequence of bases.

Experimental Evidence for a Triplet Code. From a mathematical analysis of the coding problem, Crick concluded early in 1961 that three consecutive nucleotides in a strand of messenger RNA provide the code that determines the position of a single amino acid in a polypeptide chain. Experimental evidence to support this was quickly forthcoming from the laboratory of Nirenberg and Matthaei. Using purified enzyme systems, they studied the incorporation of specific labeled amino acids into protein under the direction of artificial messenger RNAs of known composition. Nirenberg prepared a synthetic polyuridylic acid (UUUUU), using the enzyme polynucleotide phosphorylase. When this artificial messenger RNA was added to a system of purified enzymes for the synthesis of proteins, phenylalanine, and no other amino acid, was incorporated into protein; the polypeptide that resulted contained only phenylalanine. The inference that UUU is the code for phenylalanine was inescapable. Further similar experiments by Nirenberg and by Severo Ochoa showed that polyadenylic acid provided the code for lysine and that polycytidylic acid coded for proline. Making mixed nucleotide polymers (such as poly AC) and using them as artificial messengers made possible the assignment of many other nucleotide combinations to specific amino acids.

These experiments did not reveal the order of the nucleotides within the triplets, but this has been inferred from other kinds of experiments. Nirenberg and Leder discovered that even when no protein synthesis is occurring, specific amino acyl transfer RNA molecules will be attached to ribosomes when messenger RNA is present. Fortunately this effect does not require a long molecule of mRNA, which would be difficult to synthesize. Indeed, synthetic messenger RNA molecules as short as trinucleotides will suffice to promote the binding of

TABLE 7.4 THE GENETIC CODE: THE SEQUENCE OF NUCLEOTIDES IN THE TRIPLET CODONS OF MESSENGER RNA WHICH SPECIFY A GIVEN AMINO ACID

First Position (5' end)	Second Position	Third Position (3' end)			
		U	C	A	G
U	U	Phe	Phe	Leu	Leu
	C	Ser	Ser	Ser	Ser
	A	Tyr	Tyr	Terminator	Terminator
	G	Cys	Cys	Terminator	Try
C	U	Leu	Leu	Leu	Leu
	C	Pro	Pro	Pro	Pro
	A	His	His	Glu·NH$_2$	Glu·NH$_2$
	G	Arg	Arg	Arg	Arg
A	U	Ileu	Ileu	Ileu	Met
	C	Thr	Thr	Thr	Thr
	A	Asp·NH$_2$	Asp·NH$_2$	Lys	Lys
	G	Ser	Ser	Arg	Arg
G	U	Val	Val	Val	Val
	C	Ala	Ala	Ala	Ala
	A	Asp	Asp	Glu	Glu
	G	Gly	Gly	Gly	Gly

specific amino acyl transfer RNAs to the ribosomes. It is possible to synthesize trinucleotides of known sequence; using these, the coding assignment of all 64 possible triplets has been determined. For example, GUU, but not UGU or UUG, induces the binding of valine transfer RNA to ribosomes. UUG induces the binding of leucine transfer RNA. Since GUU and UUG code for different amino acids, it follows that the reading of the code in the messenger RNA strand makes sense only in one direction.

Codon-Amino Acid Specificity. Careful examination of the coding relationships (Table 7.4) shows that there is a pattern to the "degeneracy" in the code. Bernfield and Nirenberg found in 1965 that the binding of phenylalanine transfer RNA is induced by both UUU and UUC. The binding of serine transfer RNA is induced by both UCG and UCU, and proline transfer RNA is bound by CCC or CCU. In all these the two alternative triplets are identical except for the substitution of one pyrimidine for the other (C or U) at the 3′ end. In other instances, adenine and guanine can be interchanged in the 3′ position of the triplet. For example, the code for lysine may be either AAG or AAA. For a number of amino acids the first two nucleotides of the codon are specific, but any of the four nucleotides may be present in the 3′ position. All of the four possible combinations will code for the same amino acid. In these, although the code may be read three nucleotides at a time, only the first two nucleotides appear to contain specific information. Only methionine and tryptophan have single triplet codes; all the other amino acids are specified by from two to as many as six different nucleotide triplets.

That messenger RNA is read three nucleotides at a time was firmly established by experiments carried out by Khorana and his colleagues. A poly UC messenger containing the regularly alternating base sequence UCU-CUCUC was synthesized and used in a protein-synthesizing system. The resulting polypeptide contained a regular alternation of serine and leucine. Mathematical analysis of this result shows that the coding unit must contain an odd number of bases. Khorana then synthesized the nucleotide sequence AAGAA-GAAGAAG. When this nucleotide polymer was used as template in a protein-synthesizing system, the result was either polylysine (AAG), polyglutamate (GAA) or polyarginine (AGA). The type of peptide synthesized depends on which nucleotide in the polynucleotide chain happens to be read first (Fig. 7.7). This "frame shift" effect (as in a movie camera) can be accounted for only if the chain is read in sequence, three nucleotides at a time, beginning from a fixed point. There are a few instances in which a specific nucleotide at the 5′ end of the codon can be changed and still provide a code for the same amino acid. For example, both UUG and CUG provide a code for leucine. Arginine is coded by six nucleotide triplets, two of which differ in containing A instead of C at the 5′ end. It appears that the middle nucleotide is the most informative one

Figure 7.7 Diagram illustrating the "frame-shift" effect in the translation of the synthetic polynucleotide AAGAAGAAGAAG. This led to the synthesis of polylysine, polyglutamate or polyarginine, depending on which nucleotide was read first. When read as AAG, polylysine was formed; when read as AGA, polyarginine was formed; and when read as GAA, polyglutamate was formed. The diagram emphasizes the three different codon-anticodon relationships possible. (From Villee, C. A.: Biology. 7th ed. Philadelphia, W. B. Saunders Co., 1977.)

in the triplet, the 5′ nucleotide is the next most informative, and the 3′ nucleotide is least specific.

Coding Degeneracy. The biological significance of the finding that the code is "degenerate," that more than one codon may specify a given amino acid and lead to its insertion into a growing peptide chain, is not fully understood. We must keep in mind that what has been termed a code is in fact a complex physical chemical system that depends for its operation on the specificity of several reactions catalyzed by enzymes. The sequences of nucleotides in DNA will have no meaning except in relation to a specific set of amino acyl transfer RNAs, each of which is synthesized by a specific enzyme. The protein is then synthesized on the ribosome by yet another enzyme system operating on the enzyme–messenger RNA–ribosome–transfer RNA complex. The specificity of the system as a whole depends on the ability of the enzyme to recognize specific amino acyl transfer RNAs when these are present at the ribosomal binding site.

Although it is known that there may be more than one transfer RNA for a given amino acid, it is not yet clear whether the number of amino acyl transfer RNA molecules equals the number of codons for that specific amino acid. Do the six different codons for serine correspond to six different serine-transfer RNA molecules, or is there only a single serine tRNA that responds to any of the six mRNA codons? Crick's "wobble" hypothesis (Fig. 7.8) states that the first two bases of a transfer RNA anticodon undergo hydrogen bonding specifically with the first two bases of the messenger RNA codon in antiparallel. However, the third base can undergo unusual base pairing—i.e., it can "wobble." His theory predicts that the inosine residue of alanine transfer RNA may pair either

with U, C or A in a messenger codon. This could account for three of the four codons for alanine and would require only one additional transfer RNA to recognize the GCG code. The organisms that have been studied so far appear to have a set of at least 40 different transfer RNAs. Messenger RNA is read from the 5′ end toward the 3′ end; the codon at the 5′ end corresponds to the N-terminal amino acid of the polypeptide. Three of the 64 codons, UAA, UAG and UGA, do not specify any amino acid. Brenner suggested that these are "terminator" triplets, which signal the end of the polypeptide chain and cause the protein chain to become detached from the ribosome.

Universality of the Code. There is good reason to believe that the genetic code is universal; that is, a given codon specifies the same amino acid in the protein-synthesizing systems of all organisms from viruses to the human. Synthetic poly U, for example, causes synthesis of polyphenylalanine in cell-free systems derived from bacteria, wheat germ or rabbit reticulocytes. Perhaps even more convincing is the experimental finding that RNA isolated from tobacco mosaic virus and used as a template in the protein-synthesizing system derived from *Escherichia coli* synthesizes proteins that are similar to native tobacco mosaic virus protein. RNA from the chick oviduct used as the template in a protein-synthesizing system of ribosomes and transfer RNA from *E. coli* resulted in the synthesis of **ovalbumin,** the characteristic protein of egg white. More indirect but quite persuasive evidence of universality can be obtained by looking carefully at the replacements of one amino acid for another found in the more than 60 mutant forms of human hemoglobin that have been analyzed. All of these replacements are ones that would be expected if there was a single base transfor-

Figure 7.8 Diagram illustrating Crick's "wobble" hypothesis. The first two bases of the transfer RNA anticodon (CG) form the expected specific hydrogen bonds with the first two bases of the messenger RNA codon in antiparallel; however, the third base of the anticodon (I) can base pair with U, C or A in a messenger codon. This theory accounts for three of the four known codons for alanine. (From Villee, C. A.: Biology. 7th ed. Philadelphia, W. B. Saunders Co., 1977.)

mation between the triplets assigned to these amino acids on the basis of their behavior in the *E. coli* cell-free, protein-synthesizing system. The inference is that the code is the same in human beings and in bacteria.

One apparent exception to this generalization proved ultimately to be a striking confirmation of it. The amino acid sequence of a mutant form of hemoglobin, hemoglobin I, was initially reported as involving a substitution of aspartate for lysine. This would require a change in two of the three nucleotides in the codon, a "double mutation." This would be possible theoretically but quite improbable statistically. However, a reinvestigation of the amino acid composition showed that the changed amino acid was glutamate, and a substitution of glutamate for lysine requires a change in only one of the three nucleotides in the codon.

Cytochrome c, a protein constituent of the electron transmitter system, has been isolated from several organisms, and the amino acid sequence of each peptide has been determined. The differences from one species to another are very small, and the amino acid substitutions are those that would be expected if a single nucleotide were substituted for another in that codon. Similar analyses of a number of other proteins have given similar results, and this indirect evidence for the universality of the genetic code is very strong.

Colinearity. DNA is a linear polynucleotide chain, and a protein is a linear polypeptide chain. The sequence of amino acids in the peptide chain is dictated by the order of the corresponding nucleotide bases in codons in the messenger RNA, and this in turn is determined by the sequence of nucleotides in one of the two polynucleotide chains of the DNA molecule. Changing the sequence of nucleotides in the DNA molecule produces a corresponding change in the sequence of amino acids in peptides. The DNA molecule and the resulting polypeptide chain are said to be **colinear.**

This concept of colinearity was implicit in the original Watson-Crick model of the DNA molecule. Direct evidence of colinearity came from analyses by Charles Yanofsky of the genetic control of the enzyme tryptophan synthetase in bacteria.

Tryptophan synthetase is composed of four subunits, two A chains and two B chains. The A chain, studied by Yanofsky, is a polypeptide containing 267 amino acids. Yanofsky carefully mapped the genetic location of each of a large number of mutants with altered A chains. He then collected A-chain polypeptides from each of the mutant strains and analyzed them to determine which amino acid had been changed. His analyses showed that the relative position of the changed nucleotide within a gene, as determined by genetic analysis, corresponds to the relative position of the altered amino acid in the peptide chain of the enzyme molecule, as determined by direct chemical analysis of the peptide. Each amino acid substitution could be accounted for by a change in a single nucleotide in a codon. Glycine (GGA), for example, was replaced by glutamate (GAA) in the tryptophan synthetase of one mutant and by arginine (AGA) in another. Other examples of colinearity have been derived from analyses of the genetic control of hemoglobin synthesis in man.

It is now clear that DNA does not control the production of a polypeptide by any direct interaction with the amino acid. Instead, it forms an intermediate template, an RNA molecule, which in turn directs the synthesis of the peptide chain. The DNA has been compared to a master model that is carefully preserved in the nucleus and used only to synthesize secondary working models which pass out to the ribosomal mechanism in the cytoplasm and are utilized for the actual synthesis of proteins.

The coding relationships between DNA, RNA and protein involve (1) the **replication** of DNA to form new DNA; (2) the **transcription** of DNA to form a messenger RNA template; and (3) the **translation** of the code of the messenger RNA template into the specific sequence of amino acids in a protein. The arrows in the formula indicate the direction of transfer of genetic information:

$$\text{DNA} \xrightarrow{\text{transcription}} \text{RNA} \xrightarrow{\text{translation}} \text{Protein}$$
$$\big\downarrow \text{replication}$$
$$\text{DNA}$$

7.5　THE SYNTHESIS OF DNA: REPLICATION

The synthesis of DNA is catalyzed by an enzyme system, **DNA polymerase,** first isolated from the cells of *Escherichia coli* by Kornberg and colleagues in 1957. DNA polymerase requires the triphosphates of all four deoxyribonucleosides (abbreviated dATP, dGTP, dCTP and dTTP) as substrates, magnesium ions, and a small amount of high molecular weight DNA polymer to serve as a primer and template for the reaction. The product of the reaction is more DNA polymer and a molecule of pyrophosphate for each molecule of deoxyribonucleotide incorporated.

$$\left.\begin{array}{l}\text{dATP}\\\text{dGTP}\\\text{dCTP}\\\text{dTTP}\end{array}\right\}_n \xrightarrow[\text{DNA polymerase}]{\substack{\text{DNA}\\\text{Mg}^{++}}} \text{DNA} + n\text{PPi}$$

The nucleoside triphosphate attacks the free 3′ hydroxyl of the last deoxyribose in the chain and forms an ester bond, freeing a molecule of pyrophosphate (Fig. 7.6). By appropriate experiments with ^{14}C-labeled nucleotides, Kornberg could show that the ratios of A:T and of G:C in the DNA synthesized were the same as the corresponding ratios in the DNA used as primer. This suggested that the DNA produced is a copy of the primer DNA, as predicted by the Watson-Crick model.

The DNA polymerase from *Escherichia coli* can use template DNA prepared from a wide variety of sources—bacteria, viruses, mammals and plants—and will produce DNA with a nucleotide ratio comparable to that of the template used. Thus the sequence of nucleotides in the product is dictated by the sequence in the primer and not by the properties of the polymerase or the ratio of the substrate molecules present in the reaction mixture. Using a more highly purified enzyme, Kornberg was able in 1968 to synthesize biologically active viral DNA using viral DNA as primer. The DNA produced would infect bacteria just like "live" viruses.

Khorana and his colleagues prepared synthetic deoxyribonucleotide polymers containing adenylic and cytidylic acids and other polymers containing alternating thymidylic and guanylic acids. Neither one alone could serve as a template for DNA polymerase, but a mixture of the two, which forms a synthetic double-stranded helix with conventional pairing of the bases, can serve as a template.

The DNA template appears to have two functions in the DNA polymerase system: first, to provide 3′-OH groups which are free to serve as the growing end or primer of the polymer and second, to provide coded information. A double-stranded molecule is required because each strand of the pair serves as a template for the extension of the complementary strand and as a primer for its own extension. The DNA-like polymer that is produced by the action of the DNA polymerase in the presence of double-stranded template is also double-stranded and has the same base composition as the template DNA. The ratios of the bases are those predicted by the Watson-Crick model.

In the cells of higher organisms, the synthesis of DNA occurs only during the interphase, when chromosomes are in their extended form and are not readily visible. Thus, if an enzyme similar to the Kornberg enzyme catalyzes the synthesis of DNA in vivo, there must be some sort of biological signal which initiates DNA synthesis at this time and turns it off at other times. It appears that both the enzyme DNA polymerase and the substrates, dATP, dGTP, dCTP and dTTP, are present all the time, and hence there must be some change in the DNA template which initiates DNA synthesis and then turns it off.

Semi-conservative Replication. During DNA replication two new strands are formed, each complementary to one of the existing DNA strands in the double-stranded helix. The double-stranded helix unwinds; one strand provides a template for one new one and the other original strand also provides a template for a second new strand. This is called a **semi-conservative** mechanism: the two original strands of DNA are retained in the product, one in each of the two daughter helices.

Strong experimental evidence supports the concept that each eukaryotic chromosome is a single DNA molecule. The DNA molecules of the fruit fly *Drosophila* are as long as 2.1 cm., just the length of the longest chromosome. A question that immediately comes to mind is how can a eukaryotic cell manage to replicate such an enormous molecule in the allotted time during the cell cycle (the S phase). The answer is that replication does not simply begin at one end of the molecule and proceed to the other. Rather, replication begins at many sites along the chromosome (some two micrometers apart on the average in the rapidly dividing fertilized egg) and proceeds in *both* directions at about the same rate, one micrometer per minute.

The experiment of Meselson and Stahl (Fig. 7.9) provided evidence that DNA replication is carried out by a semi-conservative mechanism, at least in bacteria. Bacteria were grown for several generations in a medium containing heavy nitrogen, ^{15}N. Thus, all of the nitrogen in the DNA of those bacteria was ^{15}N. When a sample of the DNA was isolated and centrifuged in a cesium chloride density gradient, the DNA collected at a level which reflected the presence of the heavy nitrogen in the DNA molecules.

The bacteria were then removed from the ^{15}N

Bacteria growing in ¹⁵N. All its DNA is heavy.

Transfer to ¹⁴N medium.

Continued growth in ¹⁴N medium.

DNA isolated from the cells mixed with CsCl solution (6M; density~1.7) and placed in ultracentrifuge cell. Solution centrifuged at very high speed for ~48 hours.

DNA molecules move to positions where their density equals that of the CsCl solution.

$\sigma = 1:65$ $\sigma = 1:80$ Centrifuge cell

Greater concentration of CsCl at the outside is due to its sedimention under the centrifugal force.

Location of heavy DNA
¹⁴N-¹⁵N hybrid DNA
Light DNA

The location of DNA molecules within the centrifuge cell can be determined by ultraviolet optics. DNA solutions absorb strongly at 2600Å.

Before transfer to ¹⁴N.

One cell generation after transfer to ¹⁴N.

Two cell generations after transfer to ¹⁴N.

Figure 7.9 Diagram of the experiment of Meselson and Stahl which indicated that DNA is replicated by a semi-conservative mechanism: the two original strands of DNA are retained in the product, one in each daughter helix. (From Villee, C. A.: Biology. 7th ed. Philadelphia, W. B. Saunders Co., 1977.)

medium, placed in a medium containing ordinary nitrogen, ¹⁴N, and allowed to divide once in this medium. When some of the DNA from this generation was isolated and centrifuged, all the DNA was lighter and had a density corresponding to its being half labeled with ¹⁵N and half labeled with ¹⁴N. If the Watson-Crick theory is correct and replication is semi-conservative, this result would be expected because one strand of the double-stranded DNA in each organism is ¹⁵N-labeled and the other is ¹⁴N-labeled.

When the organisms were allowed to divide again in ¹⁴N medium, each molecule of progeny DNA again received one parental strand and made one new strand containing ¹⁴N. As a result, some double-stranded DNA containing only ¹⁴N was formed and appeared as a light DNA on centrifugation. The parental ¹⁵N-containing strands made complementary strands containing ¹⁴N and appeared on centrifugation with a density characteristic of the half ¹⁵N, half ¹⁴N double-stranded state. Thus the original parental strands of DNA are not dispersed or split apart during the replication process but are conserved and passed on to the next generation of cells. Each strand of the parental double helix is conserved in a different daughter cell, hence the process is termed semi-conservative.

If the replication of the strands begins as the strands begin to untwist, Y-shaped molecules of DNA should be evident during the replication process. Such Y-shaped regions have been found by autoradiography of chromosomes of *Escherichia coli* labeled with ³H-thymidine.

7.6 TRANSCRIPTION OF THE CODE: THE SYNTHESIS OF RNA

Unlike DNA, molecules of RNA do not usually have complementary base ratios, and the inference is that RNA is not a double helix like DNA but is single-stranded. It is a long, unbranched molecule containing four kinds of ribonucleotides linked by 3'-5' phosphodiester bonds. RNA contains ribose instead of deoxyribose and uracil rather than thymine.

Three kinds of RNA molecules are required for protein synthesis: **messenger RNA,** which transmits genetic information from the DNA molecule in the nucleus to the cytoplasm; **ribosomal RNA,** which makes up a large portion of the cytoplasmic particles called **ribosomes** on which protein synthesis occurs; and **transfer RNA,** which acts as an adaptor to bring amino acids into line in the growing polypeptide chain in the appropriate place.

Messenger RNA is synthesized by a **DNA-dependent RNA polymerase** found first in the nuclei of rat liver and subsequently in other animals, bacteria and plants. It requires DNA as a template and uses as substrate the four triphosphates of the ribonucleosides commonly found in RNA. The products are RNA and inorganic pyrophosphate. The reaction system can utilize single-stranded or native DNA or synthetic deoxyribopolynucleotides as template.

Transfer RNA and ribosomal RNA are also produced in the nucleus by DNA-dependent RNA-synthesizing systems and are transcribed from complementary deoxynucleotide sequences in DNA.

The RNA that is produced by the use of these carefully defined templates is exactly that predicted by the kinds of base pairing permitted in the Watson-Crick model. Although the DNA template is double-stranded and contains two different but complementary template sequences (which would have quite different genetic information), it appears that only one DNA strand is selected for transcription and only one kind of mRNA is produced. The molecular basis for this distinction between the two strands is unknown.

Electron microscopy has established that most cells contain an extensive system of tubules with thin membranes termed the endoplasmic reticulum. Associated with the endoplasmic reticulum or floating freely in the cytoplasm are small particles composed of RNA and protein termed ribosomes. The RNA

of ribosomes after extraction with phenol to remove protein appears to consist of two components with molecular weights of about 600,000 and 1,300,000.

Molecules of transfer RNA are considerably smaller than the molecules of messenger or ribosomal RNA. Each functions as a specific adaptor in protein synthesis, binding to and identifying one specific amino acid; i.e., each of the 20 amino acids is attached to one or more specific kinds of transfer RNA. One portion of the nucleotide sequence in transfer RNA represents an **anticodon,** a nucleotide triplet complementary to the codon in messenger RNA that specifies that amino acid. Transfer RNA is unusual in that it contains not only the four usual ribonucleotides, adenylic, guanylic, cytidylic and uridylic, but also small amounts of unusual nucleotides such as **6-methylamino adenylic acid, dimethylguanylic acid** and **thymine ribotide.**

Transfer RNAs are polynucleotide chains of some 70 nucleotides. Each of the several kinds of transfer RNA has an identical sequence of nucleotides, CCA, at the 3' end to which the amino acid is attached. Each in addition has guanylic acid at the opposite, 5', end of the nucleotide chain. The chain is doubled back on itself, forming three or more loops of unpaired nucleotides; the folding is stabilized by hydrogen bonds between complementary bases in the intervening portions of the chain to form a

Figure 7.10 A diagram of the three dimensional clover leaf structure of transfer RNA. One loop contains the triplet anticodon which forms specific base pairs with the mRNA codon. The amino acid is attached to the terminal ribose at the 3' OH end, which has the sequence CCA of nucleotides. Each transfer RNA also has guanylic acid, G, at the 5' end (P). The pattern of folding permits a constant distance between anticodon and amino acid in all transfer RNAs examined.

TABLE 7.5 AGREEMENT BETWEEN THE ANTICODONS FOUND IN ISOLATED tRNAs AND THE ANTICODONS PREDICTED FROM THE GENETIC CODE

Amino Acid	mRNA Codons →	Complements ←	Observed Anticodons* ←
Alanine	GpCpA	UpGpC	
	GpCpG	CpGpC	
	GpCpC	GpGpC	IpGpC
	GpCpU	ApGpC	
Phenylalanine	UpUpC	GpApA	OMGpApA
	UpUpU	ApApA	
Tyrosine	UpApU	ApUpA	
	UpApC	GpUpA	Gpψpా A
Serine	ApGpC	GpCpU	
	ApGpU	ApCpU	
	UpCpC	GpGpA	IpGpA
	UpCpU	ApGpA	
	UpCpA	UpGpA	
	UpCpG	CpGpA	
Valine	GpUpU	ApApC	
	GpUpC	GpApC	IpApC
	GpUpA	UpApC	
	GpUpG	CpApC	

*The anticodons contain some "unusual" bases, including I (inosine), OMG (2′-O-methylguanosine) and ψ (pseudouridine).

double helix (Fig. 7.10). The loop nearest the amino acid acceptor (CCA) has seven nucleotides, with cytidine, pseudouridine and thymidine at positions 21, 22 and 23 from the CCA end. The triplet that is complementary to the codon is located in a seven-membered middle loop, and is preceded by uridine and followed by adenosine or a modified adenosine in the loop. Another unusual base, dimethylguanosine, is found eight positions before the anticodon, at the base of the larger (eight to 12 nucleotides) loop near the 5′ end. The folding results in a constant distance between the anticodon and the amino acid in all of the tRNAs examined so far. The triplet anticodons found in these tRNAs agree very well (Table 7.5) with the complements predicted to the known mRNA codons for those amino acids.

The first transfer RNA to be completely analyzed was the transfer RNA for alanine derived from yeast cells. This dramatic achievement of Robert Holley and his colleagues was recognized with the awarding of a Nobel Prize to Holley in 1968. The alanine tRNA has 77 nucleotides arranged in a unique sequence. Nine of these are unusual bases, with one or more methyl groups which are added enzymatically after the nucleotides have been linked by phosphodiester bonds. Certain of these unusual bases cannot form conventional base pairs and may serve to disrupt the base pairing in other parts of the transfer RNA. This could expose specific chemical groups on the tRNA which form secondary bonds to messenger RNA, or to the ribosome or perhaps to the enzyme needed to attach the specific amino acid to its specific transfer RNA molecule. The exact sequence of nucleotides in several other transfer RNAs is now known. This implies that the sequence of the nucleotides in the genes that specify each of these particular kinds of transfer RNA is also known. The sequence in the genic DNA can be inferred from the Watson-Crick rules of specific base pairing. The gene specifying the transfer RNA for alanine would also be 77 nucleotides long. A complete, double-stranded DNA molecule with this sequence of 77 nucleotides was synthesized in 1970 by Khorana and was found to be transcribed to give alanine tRNA. This was the first gene to be synthesized.

7.7 REVERSE TRANSCRIPTION

The central dogma states that biological information flows from DNA to RNA to protein. An important exception to this rule was discovered by Howard Temin in 1964 when he found that infection with certain RNA tumor viruses, such as the Rous sarcoma virus, is blocked by inhibitors of DNA synthesis and by inhibitors of DNA transcription. This suggested that DNA synthesis and transcription are required for the multiplication of RNA tumor viruses and that information flows in the reverse direction; that is, from RNA to DNA. Temin proposed that a DNA provirus is formed as an intermediate in the replication of these RNA tumor viruses and their cancer-producing effect. Temin's hypothesis required a new kind of enzyme—one that would synthesize DNA using RNA as a template. Just such an enzyme was discovered by Temin and by David Baltimore in 1970, discoveries for which they received the Nobel Prize in 1975. This RNA-directed DNA polymerase (also called reverse transcriptase) has been found to be present in all RNA tumor viruses.

An infecting RNA virus binds to and enters the host cell. The viral RNA is uncoated and enters the cell nucleus. There the viral reverse transcriptase forms a (−) DNA strand complementary to the (+) RNA, and subsequently a (+)

DNA strand is synthesized from the (−) DNA strand. The complementary (±) DNA strands, the provirus, are integrated into the host cell's genome. After cell division the virus is transcribed and the viral proteins are formed in the cytoplasm as this RNA is translated. Viral (+) RNA molecules are formed and incorporated into mature virus particles with a protein coat.

The question of whether viruses cause human cancer is being investigated intensively in many laboratories at present. Suggestive evidence that viruses probably do play a role in certain human cancers is now at hand. First, human leukemias, sarcomas, lymphomas and breast adenocarcinomas have been shown to contain large RNA molecules similar to those of tumor viruses that cause cancer in mice. Second, these human cancer cells contain particles with reverse transcriptase activity. Third, some human cancer cells have virus-like sequences of nucleotides in their DNA, sequences not found in the DNA of comparable normal cells.

7.8 THE SYNTHESIS OF A SPECIFIC POLYPEPTIDE CHAIN

Activation of Amino Acids. The first step in the synthesis of a peptide requires the activation of the amino acid by an enzyme system in the cytoplasm. There is a separate specific amino acid–activating enzyme for each amino acid. The enzyme first catalyzes the reaction of the amino acid (aa) and ATP to form the amino acid adenylic acid compound (aa-AMP) and release pyrophosphate. The same enzyme then catalyzes the transfer of the amino acid to the specific transfer RNA for the amino acid, yielding the transfer RNA-amino acid compound and free adenylic acid. The amino acid is attached at the end of the transfer RNA which contains cytidylic, cytidylic and adenylic acid; the amino acid is attached to the ribose of the terminal adenylic acid. If these three nucleotides are removed the transfer RNA is unable to function.

The Role of Transfer RNA. Francis Crick had predicted on theoretical grounds that some sort of nucleic acid molecule must serve as an adaptor in the course of protein synthesis. He argued that since there is no simple steric correspondence between a polynucleotide chain and a polypeptide chain which would enable the nucleotide sequence to specify the amino acid sequence directly, the amino acids

might be lined up in appropriate register by means of small RNA adaptor molecules. These adaptor molecules could assemble at specific places on the nucleic acid template by their complementary sequences of nucleotides. This hypothesis has now been validated by a wide variety of experiments. For example, Chapeville prepared a complex of cysteine with its specific transfer RNA (cysteinyl-tRNA^{cys}) and then converted the cysteine to alanine while it was still bound to the transfer RNA (alanyl-tRNA^{cys}). When this was added to a protein-synthesizing system, a polypeptide was made which contained *alanine* at the sites in the peptide chain where cysteine should have been. Thus these experiments provided direct proof that the ordering of specific amino acids into their appropriate place is dictated by the specific transfer RNAs and not by the amino acid that is bound to it.

Ribosomal Functions. The amino acid bound to its specific transfer RNA is transferred to the ribosomes. The role of the ribosome is to provide the proper orientation of the amino acid-transfer RNA precursor, the messenger or template RNA, and the growing polypeptide chain, so that the genetic code on the template or messenger RNA can be read accurately. There are some 15,000 ribosomes in a rapidly growing cell of *E. coli*, each with a molecular weight of nearly 3,000,000. These ribosomes account for nearly one-third of the total mass of the cell. Only one polypeptide chain can be formed at a time on any given ribosome.

The template for the synthesis of a specific protein is supplied by the messenger RNA formed on one strand of the double helix of DNA. The messenger RNA undergoes processing in the nucleus, during which a long tail of polyadenylic acid containing about 100 molecules of adenylic acid is added. This reaction is an enzymic reaction utilizing ATP as the donor of the adenylate. The poly A-rich RNA then passes out of the nucleus and becomes associated with the ribosomes. The poly A tail may play some role in transport through the nuclear membrane or in protecting the messenger RNA from destruction by ribonuclease.

Ribosomes from different kinds of cells may differ somewhat in their mass, in the composition of their RNA and in the ratio of RNA to protein, but there is a general similarity in their structures. Ribosomes can be separated into two subunits if placed in a solution with a low concentration of magnesium ion. The subunit structure is apparent in electron micrographs; the smaller subunit seems to sit like a cap on the flat surface of the larger subunit. There are

21 proteins and one RNA in the smaller subunit of bacterial ribosomes and 35 proteins and two RNA molecules in the larger subunit. Each subunit can be carefully separated into its constituent RNA and proteins. The 21 proteins and the RNA that compose the smaller subunit of bacterial ribosomes will undergo spontaneous reassembly and form a fully functional subunit. Similarly, the 35 proteins and the two RNA molecules of the larger subunit will reassemble spontaneously to form a fully functional subunit. These findings demonstrate that all the information needed for the correct assembly of

the ribosome from its parts is contained in the structure of those parts; no factor from outside the ribosome is required. This indicates that the formation of the ribosome within the living cells is also a self-assembly process. The ribosomes of higher organisms tend to be somewhat larger than the bacterial ones and are composed of two or four subunits. Each ribosome contains several dozen kinds of proteins bound to the RNA. Ribosomes have GTPase activity, which appears to play an important but undetermined role in the transfer of amino acids from transfer RNA to the forming peptide

Figure 7.11 Electron micrographs of polyribosomes isolated from the reticulocytes of a rabbit. The upper preparation was shadowed with gold, and the lower one was stained with uranyl nitrate. The electron micrographs show that polyribosomes tend to occur in clusters of four, five or six, and that the clusters are connected by a thin strand of mRNA. (Photographs courtesy of Dr. Alexander Rich.)

chain. Protein-synthesizing particles somewhat similar to ribosomes are also found in the nucleus, in the chloroplasts of plant cells and in mitochondria.

Protein synthesis has been studied intensively in preparations of rabbit reticulocytes, which are engaged primarily in making just one protein—hemoglobin. Alexander Rich and his coworkers showed that the ribosomes most active in protein synthesis are those that interact in clusters of five or more (Fig. 7.11). These clusters, termed **polysomes,** are held together by the strand of messenger RNA. Peptide chains are formed by the sequential addition of amino acids beginning at the N-terminal end, the end having a free amino group. Electron micrographs suggest that individual ribosomes become attached to one end of a polysome cluster, and gradually move along the messenger RNA strand as the polypeptide chain attached to it increases in length by the sequential addition of amino acids (Fig. 7.12). Thus each ribosome appears to ride along the extended messenger RNA molecule, "reading the message" as it goes and in some way bringing the transfer RNA molecule charged with its specific amino acid into line at the right position. After it completes the reading of one molecule of messenger RNA and releases the polypeptide that it has synthesized, the ribosome appears to jump off the end of one messenger RNA chain, move to a new messenger RNA and begin reading it (Fig. 7.13). The growing peptide chain always remains attached to its original ribosome; there is no

transfer of a peptide chain from one ribosome to another. Several ribosomes may be working simultaneously on a single strand of messenger RNA, each reading a different part of the message.

The incorporation of the first amino acid of a new polypeptide chain requires a special mechanism, since there is no preexisting polypeptide-transfer RNA complex to accept it at the combining site. The N-terminal amino acids of proteins are often substituted at their alpha amino end with an acetyl group or a formyl group. There is, in fact, a special enzyme that adds a formyl group to methionine after methionine has been linked to its transfer RNA. Such N-formyl methionine molecules are incorporated preferentially at the N-terminal end of peptide chains, whereas methionines without the formyl group are incorporated in intermediate positions in the chain. Two different types of methionyl-transfer RNA may be involved, only one of which can receive a formyl group.

Three stages—initiation, elongation and termination—can be distinguished in the process of protein synthesis. The overall process requires the coordinated action of more than 100 different macromolecules, including mRNA, all the specific tRNAs, activation enzymes, initiator factors, elongation factors and termination factors, in addition to the ribosomes themselves, each made of subunits and many kinds of protein and RNA. In *Escherichia coli,* in which the process has been studied in greatest detail, messenger RNA, formylme-

Figure 7.12 The postulated mechanism by which a polypeptide chain is synthesized. Each ribosome "rides" along the messenger RNA, reading and translating the genetic message. Amino acids are added as dictated by the specific base-pairing of the mRNA codon and the tRNA anticodon. (Afer Watson, J. D.: The Molecular Biology of the Gene. New York, W. A. Benjamin, 1965.)

Figure 7.13 Diagram of the postulated mechanism of protein synthesis on the ribosome, illustrating the relationship between the triplet code of the DNA helix, the complementary triplet code of messenger RNA and the complementary triplet code (anticodon) of transfer RNA. Molecules of tRNA charged with specific amino acids are depicted coming from the right, assuming their proper place on mRNA at the ribosome, transferring the amino acid to the growing peptide chain, and then (*left*) leaving the ribosome to be recharged with amino acids for further reactions. The growing polypeptide chain remains attached to its original ribosome. (From Villee, C. A.: Biology. 7th ed. Philadelphia, W. B. Saunders Co., 1977.)

thionyl tRNA and the smaller ribosomal subunit come together to form an "initiation complex." The formylmethionyl tRNA recognizes the special initiation sequence on the messenger RNA (which is not necessarily at the 5′ end of the molecule). The larger ribosomal subunit then joins the complex to form a complete ribosome, and this is ready for the next phase. The elongation process begins with the binding to the complex of the next succeeding amino acyl tRNA, which is recognized by the specific codon next in line. The elongation cycle continues with the formation of a peptide bond, the elimination of the uncharged tRNA, and the transfer of the tRNA containing the peptide from one site within the ribosome to the other. This leaves the second site ready to accept the next amino acyl tRNA molecule. Both the binding of the amino acyl tRNA and the translocation from one site to the other require GTP. A specific protein elongation factor is required for the binding of the amino acyl tRNA to the ribosome. The synthesis of the peptide chain is terminated by "release factors" that recognize

the terminator codons UAA, UGA and UAG. This leads to the hydrolysis of the bond between the polypeptide and the transfer RNA. When the peptide chain is completed and released from the ribosome, the ribosome dissociates into its two subunits, the larger and smaller subunits.

The several steps involved in protein synthesis can be specifically inhibited by certain toxins and antibiotics. Streptomycin, for example, inhibits the initiation process, tetracyclines inhibit the binding of amino acyl tRNAs to the smaller subunit, and erythromycin binds to the larger subunit and inhibits the transfer of the growing peptide chain and its tRNA from one site within the ribosome to the next.

All the processes of gene replication, gene transcription and protein synthesis depend upon the formation of specific, though relatively weak, hydrogen bonds between specific pairs of purine and pyrimidine bases. The specificity of these bonds ensures the remarkable accuracy of the process; mistakes in base pairing occur less than one time in a thousand.

To determine in which direction synthesis of the peptide chain proceeds, tritium-labeled leucine was added to the hemoglobin-synthesizing system of rabbit reticulocytes in pulses of different duration. With a very long pulse, the labeled leucine was evenly distributed in all parts of the polypeptide; however, with short pulses, the labeled leucine was found only in those parts of the molecule synthesized last, and these are at the C-terminal end. Thus the chain is synthesized beginning at the amino-terminal end and proceeding in sequence to the carboxyl-terminal end. In the cell-free system the synthesis of the complete α chain of hemoglobin, which is 141 amino acids long, requires 1.5 minutes; i.e., about two amino acids are added each second.

Two different enzymes are required to carry out the transfer of the amino acids from transfer RNA to the peptide linkage in the peptide chain. One enzyme binds the amino acid transfer RNA and the other is a synthetase that forms the peptide bond. The reactions that require GTP are the binding of the tRNA to the ribosome and the translocation of the amino acid-tRNA complex from one site in the ribosome to the other, rather than the synthesis of the peptide bond. GTP is used and the products are GDP and inorganic orthophosphate. In cell-free systems some 50 molecules of GTP are hydrolyzed for each amino acid transferred to a growing peptide. Presumably the protein-synthesizing system within the intact cell does not make such profligate use of its $\sim$ P!

An overview of the several steps involved in the synthesis of a specific polypeptide chain is provided in Figure 7.14. Clearly much is yet to be learned about the biosynthesis of proteins. Even the best cell-free protein-synthesizing system operates at a rate only 0.01 as rapid as that within an intact, living cell.

7.9 CHANGES IN GENES: MUTATIONS

Although genes are remarkably stable and are transmitted to succeeding generations with great fidelity, they do from time to time un-

Figure 7.14 Overview of the process by which biological information is transferred from DNA via RNA to specific polypeptides. The peptide subunits are then assembled into multichain proteins. (From Villee, C. A.: Biology. 7th ed. Philadelphia, W. B. Saunders Co., 1977.)

dergo changes called **mutations.** After a gene has mutated to a new form this new form is stable and usually has no greater tendency to mutate again than the original gene. A mutation has been defined as any inherited change not due to segregation or to the normal recombination of unchanged genetic material. Mutations provide the diversity of genetic material which makes possible a study of the process of inheritance, and investigations of the nature of the mutation process have provided important clues as to the nature of the genetic material itself.

Some mutations, termed **chromosomal mutations,** are accompanied by a visible change in the structure of the chromosome. A small segment of the chromosome may be missing (a **deletion**) or be represented twice in the chromosome (a **duplication**) (Fig. 7.15). A segment of one chromosome may be transferred to a new position on a new chromosome (a **translocation**), or a segment may be turned end for end and attached to its usual chromosome (an **inversion**). Point mutations or gene mutations involve small changes in molecular structure that are not evident under the microscope. These gene mutations involve some change in the sequence of nucleotides within a particular section of the DNA molecule, usually the substitution of one nucleotide for another in a given codon.

From your knowledge of the DNA molecule, you might predict that replacing one of the purine or pyrimidine nucleotides by an analogue such as azaguanine or bromouracil would result in mutation. In several experiments in which such analogues were incorporated into bacteriophage DNA, no mutations were evident. Because of the degeneracy of the genetic code (Table 7.4), a number of changes in base pairs could occur without changing the amino acid specified. In other experiments, the incorporation of bromouracil into DNA did lead to an increased rate of mutation. Other chemicals known to be mutagenic include nitrogen mustards, epoxides, nitrous acid and alkylating agents. These are all chemicals which can react with specific nucleotide bases in the DNA and change their nature. When an analogue is incorporated into DNA it may lead to mistakes in the pairing of nucleotides during subsequent replication processes. For example, when bromouracil is incorporated into DNA in place of thymine it will pair with guanine rather than with adenine, the normal pairing partner of thymine. This would lead to the substitution of a G-C pair of nucleotides at the point in the double helix previously occupied by an A-T pair of nucleotides (Fig. 7.16).

The presence of mutagenic materials may increase the frequency of mistakes in nucleotide base pairing which would result in the production of DNA molecules containing only natural bases, but with an altered base order. The change in normal base order is significant, because during subsequent replications, the altered base sequence will be reproduced by the normal process of DNA synthesis. The definition of mutation includes the restriction that the change that has been introduced into the DNA molecule must be propagated subsequently for an indefinite number of times.

"Spontaneous" gene mutations may result from errors in base pairing, during the replication process; thus the A-T normally present at a given site may be replaced by G-C, C-G or T-A. The altered DNA will be transcribed to give an altered mRNA, and this will be translated into a

Figure 7.15   Diagram illustrating the types of mutations which involve changes in the structure of the chromosome.

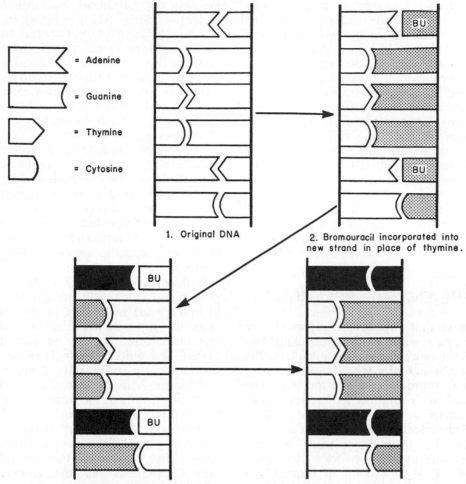

1. Original DNA

2. Bromouracil incorporated into new strand in place of thymine.

3. Strand with bromouracil leads to production of new strand with guanine paired to the bromouracil.

4. New, mutant DNA which contains no analogue bases, but has nucleotide sequence different from original, with GC pairs in place of AT.

Figure 7.16 Diagrammatic scheme of how an analogue of a purine or pyrimidine might interfere with the replication process and cause a mutation, an altered sequence of nucleotides in the DNA, indicated in black. The nucleotides of the new chain at each replication are indicated by the dotted blocks. In this instance, two new GC pairs are indicated. Probably a single substitution of a GC pair for an AT pair would be sufficient to cause a mutation if it occurred in the triplet code at a point that changed the amino acid specified, i.e., in the first or second base of the triplet, or in the third member of certain triplets (see Table 7.4). (From Villee, C. A.: Biology. 7th ed. Philadelphia, W. B. Saunders Co., 1977.)

peptide chain with one amino acid different from the normal sequence. If the altered amino acid is located at or near the active site of the enzyme, the altered protein may have markedly decreased or altered catalytic properties. If the altered amino acid is elsewhere in the protein, it may have little or no effect on the properties of the enzyme and may thus go undetected. The true number of gene mutations may be much greater than the number observed.

If a single nucleotide pair were inserted into or deleted from the DNA molecule, it would shift the reading of the genetic message, alter all the codons lying beyond that point, and

change completely the nature of the resulting peptide chain and its biological activity. Thus if the normal sequence is CAG TTC ATG, read (CAG) (TTC) (ATG), the insertion of a G between the two T's results in CAG, TGT CAT G, read (CAG) (TGT) (CAT) (G . . .).

Gene mutations can be induced not only by exposing the cell to certain chemicals but by a variety of types of radiation – x-rays, gamma rays, cosmic rays, ultraviolet rays, and the several types of radiation that are by-products of atomic power. How radiation may lead to changes in base pairs is not clear, but the radiant energy may react with water molecules to release short-lived, highly reactive free radicals that attack and react with specific bases. Mutations occur spontaneously at low, but measurable, rates which are characteristic of the species and of the gene; some genes are much more prone to undergo mutation than others. Natural radiations, such as cosmic rays, probably play some role in causing "spontaneous mutations," but other factors, such as errors in base pairing, play a role. The rates of spontaneous mutation of different human genes range from 10^{-3} to 10^{-5} mutations per gene per generation. Since we have a total of some 2.3×10^4 genes, this means that the total mutation rate is on the order of one mutation per person per generation. Each one of us, in other words, has some mutant gene that was not present in either of our parents.

7.10 GENE-ENZYME RELATIONS

If we assume that a specific gene leads to the production of a specific enzyme by the method outlined above, we must next inquire how the presence or absence of a specific enzyme may affect the development of a specific trait. The expression of any structural or functional trait is the result of a number, perhaps a large number, of chemical reactions which occur in series, with the product of each reaction serving as the substrate for the next: $A \rightarrow B \rightarrow C \rightarrow D$. The dark color of most mammalian skin or hair is due to the pigment melanin (D), produced from dihydroxyphenylalanine (dopa) (C), produced in turn from tyrosine (B) and phenylalanine (A). Each of these reactions is controlled by a particular enzyme; the conversion of dopa to melanin is mediated by tyrosinase. Albinism, characterized by the absence of melanin, results from the absence of tyrosinase. The gene for albinism (a) does not produce the enzyme tyrosinase but its normal allele (A) does.

The earliest attempts to connect the action of a particular gene with a particular reaction were studies of the inheritance of flower colors, in which the specific flower pigments could be extracted and analyzed. Other studies of the inheritance of coat color in mammals and eye color in insects were also able to relate specific genes with specific enzymic reactions in the synthesis of these pigments.

A major advance was made when George Beadle and Edward Tatum conceived the idea of looking for mutations in the mold *Neurospora* that interfere with the reactions by which chemicals essential for its growth are produced. The wild type *Neurospora* requires as nutrients only sugar, salts, inorganic nitrogen and biotin. A mixture of these comprises the so-called "minimal medium" for the growth of the wild type *Neurospora* (Fig. 7.17). Exposure of the conidia (haploid asexual spores) to x-rays or ultraviolet rays produces many mutations. After irradiation the mold is supplied with a "complete" medium, an extract of yeast which contains all the known amino acids,

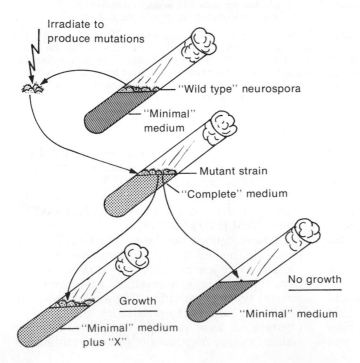

Irradiate to produce mutations

"Wild type" neurospora

"Minimal" medium

Mutant strain

"Complete" medium

Growth

No growth

"Minimal" medium plus "X"

"Minimal" medium

Figure 7.17 The method of producing and testing biochemical mutants in *Neurospora*. See text for discussion.

Figure 7.18 Pathway by which phenylalanine and tyrosine are metabolized. Mutants that interfere with the production of the enzymes that catalyze each of these steps may result in "inborn errors of metabolism," such as phenylketonuria, albinism and alkaptonuria. (From Villee, C. A.: Biology. 7th ed. Philadelphia, W. B. Saunders Co., 1977.)

vitamins, purines, pyrimidines, and so on. Any nutritional mutant produced by the irradiation is thus able to survive and reproduce to be tested subsequently.

Some of the cells from the irradiated mold are then placed on minimal medium. If these are unable to grow we know that a mutant has been produced which is unable to produce some compound essential for growth. By trial and error, by adding substances to the minimal medium in groups or singly, the required substance is identified. In each instance genetic tests show that the mutant strain produced by irradiation differs from the normal wild type by a single gene, and chemical tests show that the addition of a single chemical substance to the minimal medium will enable the mutant strain to grow normally. The obvious inference is that each normal gene produces a single enzyme which regulates one step in the biosynthesis of this particular chemical; the mutant gene does not produce this enzyme. It has been possible in certain instances to show that the particular enzyme can be extracted from the cells of normal *Neurospora* but not from cells

of the mutant strain. The synthesis of each of these nutritional substances involves a number of different steps, each mediated by a separate, gene-controlled enzyme. An estimate of the minimal number of steps involved can be obtained from the number of different mutants that will interfere with the production of that substance.

Similar one-to-one relationships of gene, enzyme and biochemical reaction in humans were first described by the English physician A. E. Garrod in 1908. **Alkaptonuria** is a trait, inherited by a recessive gene, in which the patient's urine turns black on exposure to air. The urine contains homogentisic acid, a normal intermediate in the metabolism of phenylalanine and tyrosine. The tissues of normal people have an enzyme which oxidizes homogentisic acid so that it is eventually excreted as carbon dioxide and water (Fig. 7.18). Alkaptonuric patients lack this enzyme because they lack the gene which controls its production. As a result, homogentisic acid accumulates in the tissues and blood and spills over into the urine. Garrod used the term "inborn errors of metabolism" to describe alkaptonuria and comparable conditions such as phenylketonuria and albinism.

The question of whether the genes normally operate so as to lead to the production of the *maximum* number of enzymes all the time has been given consideration in recent years. From a variety of experimental evidence it would appear that this is not the case. Each gene is probably "repressed" to a greater or lesser extent under normal conditions and then, in response to some sort of environmental demand for that particular enzyme, the gene becomes "derepressed" and leads to an increased production of the enzyme. When a single gene is fully derepressed it can lead to the synthesis of fantastically large amounts of enzymes — one enzyme may compose 5 to 8 per cent of the total protein of the cell! If all enzymes were produced at this same fantastically high rate, metabolic chaos would result. Thus, the phenomena of gene repression and derepression would appear to be necessary to prevent this chaos and to provide a means for increasing or decreasing the rate of synthesis of one particular enzyme in response to variations in environmental requirements.

The control mechanism postulated by Jacob and Monod from their studies of β-galactosidase in E. coli states that in addition to the structural genes that provide the code for the synthesis of specific proteins there are "regulator genes" that code for the synthesis of "repressors"; that the repressors may be active or inactive depending on whether they are bound to small molecules, the inducers or corepressors; and that active repressors bind to operator sites in the DNA and turn off the transcription of the adjacent structural genes. The structural genes under the control of one repressor are termed an operon.

There is clear evidence that the synthesis of many proteins in E. coli and in other organisms is not influenced by substances such as corepressors and inducers in the external environment; either their operator sites are turned on all the time or their genes are not associated with an operator site.

7.11 GENES AND DIFFERENTIATION

One of the important unsolved problems of modern biology is the nature of the mechanisms that regulate developmental processes so that each organ appears at the proper time and in the proper spatial relation to the other organs. How can a single cell, a single fertilized egg, give rise to the many different types of cells that make up the adult organism, cells that differ so widely in their structure, functions and chemical properties? The advances in biochemical genetics in the past 25 years have permitted new intellectual and experimental approaches to this problem.

We now have a detailed working hypothesis as to how biological information is transferred from one generation of cells to the next and how this information may be transcribed and translated in each cell so that specific enzymes and other proteins are synthesized. The operation of this system would produce a multicellular organism in which each cell would have the same assortment of enzymes as every other cell. Additional hypotheses are needed to account for (1) the means by which the *amount* of any given enzyme produced in a cell is regulated; (2) the control of the *time* in the course of development when each kind of enzyme appears; and (3) the mechanism by which *unique patterns* of proteins are established in each of the several kinds of cells in a multicellular organism despite the fact that they all contain identical quotas of genetic information.

The Preformation Theory and Epigenesis. A theory widely held by early embryologists was that the egg or sperm contained a completely formed but minute germ which simply grew and expanded to form the adult. This **preformation theory** was gradually displaced by the contrasting theory of **epigenesis,** which stated that the unfertilized egg is structureless, not organized, and that development proceeds by the progressive differentiation of parts. However, development is not simply epigenetic. Certain potentialities, though not structures, may be localized in certain regions of the egg and early embryo; this restricts the developmental possibilities of that part. When the embryos of echinoderms or chordates are separated experimentally at the two- or four-cell stage, each of the separated cells will form an embryo complete in all details, although smaller than normal. However, when embryos of annelids or mollusks are separated at the two-cell stage, neither cell can develop into a whole embryo. Each cell develops only into those structures it would have formed normally — half an embryo, perhaps, or some part of one. This localization of potentialities eventually occurs in the development of all eggs; it simply occurs earlier in some than in others.

Some sort of chemical or physiological dif-

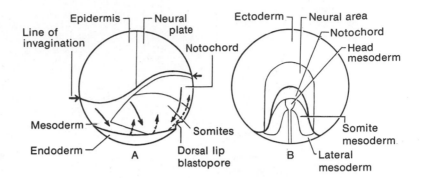

Figure 7.19 Embryo maps. A, Lateral view of a frog gastrula showing the presumptive fates of its several regions. B, Top view of chick embryo showing location in the primitive streak stage of the cells which will form particular structures of the adult.

ferentiation must be present before any structural differentiation is visible, but the basic problem of how the chemical differentiation arises remains unsolved. By appropriate experiments it has been possible to map out the location of these potentialities in frog, chick and other embryos (Fig. 7.19).

Differential Nuclear Division. Cellular differentiation might be explained if genetic material were parceled out differentially at cell division, and the daughter cells received different kinds of genetic information. Although there are a very few clear instances of differential nuclear division in animals such as *Ascaris* and *Sciara*, this does not appear to be a general mechanism of differentiation. By and large, genes are neither lost nor gained during developmental processes. The generalization that the mitotic process ensures the exact distribution of genes to each cell of the organism is valid. Thus the differences in the kinds of enzymes and other proteins found in different cells of the same organism must arise by differences in the activity of the same set of genes in different cells. The experiments of Briggs and King showed that even the nucleus from a differentiated cell taken from an advanced stage of embryonic development can, when placed in an enucleated egg, lead to the development of a normal embryo. Thus it clearly has retained a full set of genetic information.

Differential Gene Activity. Some striking evidence regarding the differential activity of genes comes from cytologic studies of insect tissues. In certain insect tissues the chromosomes undergo repeated duplication, and the daughter strands line up exactly in register, locus by locus, so that characteristic bands appear along the length of the **giant chromosomes.** When these bands are examined carefully, either in the same tissue at different times or in different tissues at the same time, certain differences in appearance become evident. A particular section of a chromosome may have the appearance of a diffuse puff (Fig. 7.20). Histochemical tests and autoradiographic evidence from experiments using tritium-labeled precursors have shown that the puff consists of RNA. It has been inferred that genes show this puffing phenomenon when they become active, and that the puff represents the messenger RNA produced by the active gene in that band. The appearance of puffs at certain regions in the chromosome can be correlated with specific cellular events, such as the initiation of molting and pupation.

The turning on and off of the synthesis of specific proteins—differentiation at the molecular level—could occur by some process involving the genic DNA, the transcription of the DNA to form messenger RNA, the processing of the messenger RNA and its transport to the cytoplasm, the combination of the messenger RNA with the ribosome during protein synthesis, or even some transformation of the ultimate protein product. In view of the many kinds of DNA that are present in the genic complement of a cell in a multicellular plant or animal, we might ask what prevents that cell from continuously producing all the tremendous variety of messenger RNAs and their corresponding protein products that are possible. What determines which molecules of DNA are to be transcribed at any given moment in a given cell?

A mechanism that controls the transcription of DNA to regulate the production of messenger RNA would probably be the most economical one biologically, for it would clearly be to the cell's advantage not to have its ribosomes encumbered with nonfunctional molecules of messenger RNA. A cell optimally should produce only those kinds of messenger RNA that will code for the specific proteins required at that moment.

The continued synthesis of any protein requires the continued synthesis of its corre-

Figure 7.20 Diagrams illustrating changing appearance of a polytene chromosome of a salivary gland of *Chironomus tentans* as a chromosome puff gradually appears. The material comprising the puff has been shown by histochemical tests and by autoradiography with tritium labeled uridine to be largely ribonucleic acid. (From Beermann: Chromosoma, 5:139–198, 1952.)

sponding messenger RNA. Each kind of messenger RNA has a half-life ranging from a few minutes in certain microorganisms to 12 or 24 hours or longer in mammals. Although each molecule of RNA template can serve to direct the synthesis of many molecules of its protein, the RNA is eventually degraded and must be replaced. This provides a mechanism by which a cell can alter the kind of protein being synthesized as new types of messenger RNA replace the previous ones. Thus the cell can respond to exogenous stimuli with the production of new types of enzymes.

Enzyme Induction. The induction of enzymes by environmental stimuli has been cited as a model for embryonic differentiation. Bacteria, and to some extent animal cells, respond to the presence of certain substrate molecules by forming enzymes to metabolize them. Jacques Monod has suggested that extracellular or intracellular influences may initiate or suppress the synthesis of specific enzymes, thus affecting the chemical constitution of the cell and leading to differentiation. As the embryo develops, the gradients established as a result of growth and cell multiplication could result in quantitative and even qualitative differences

in enzymes. The induction or inhibition of one enzyme could lead to the accumulation of a chemical product that would induce the synthesis of a new enzyme and confer a new functional activity on these cells.

Morphogenesis appears to be too complex a phenomenon to be explained in terms of any single process, such as enzyme induction. Enzymes can be induced in an embryo by the injection of an appropriate substance. Adenosine deaminase, for example, has been induced in the chick embryo by the injection of adenosine; however, no enzyme has been induced that is not normally present to some extent in the embryo. Adult tissues such as the liver may show marked differences in their enzymic activities, differences that might be the result of adaptations comparable to those seen in bacteria. Adaptive changes in enzymes, however, are temporary and reversible, whereas differentiation is a permanent, essentially irreversible process.

The DNA of the genes not being transcribed at any given moment may be bound to a histone or to an acidic nuclear protein that makes the DNA unavailable for the transcription system. The hypothesis that in the nucleus some genes are free and can be transcribed whereas others are bound and not transcribable is supported by a variety of experimental evidence.

Other biologists have postulated that DNA may be transcribed to form messenger RNA but that the mRNA is "masked" and inactive as a template for protein synthesis until it is subsequently unmasked by a separate process. This form of control may operate especially during early embryonic development. The mRNA synthesized in the oöcyte from the maternal genome may remain masked and inactive until activated by the fertilization process and freed to undergo translation.

Studies of the synthesis of RNA in the nucleus have shown that a large fraction, perhaps 80 per cent, of the total RNA synthesized is destroyed without leaving the nucleus; only about 20 per cent of the nuclear RNA is identical with RNA present in the cytoplasm. Some have suggested that this "heterogeneous" nuclear RNA serves in some way to regulate genic action. There appears to be far more DNA in the cells of multicellular organisms than is necessary to serve as template for the mRNA used in directing the synthesis of protein. Is the rest of the DNA part of some enormous, complex regulatory system that controls the activity of the DNA that does produce mRNA? Others interpret the rapid turnover of nuclear RNA as indicating that all genic DNA is transcribed into RNA all the time, but that much of the RNA produced is rapidly destroyed before leaving the nucleus. Only the RNA that is processed by the addition of a polyadenylic acid tail survives and passes to the cytoplasm. This second interpretation implies that gene action is regulated not during transcription but subsequently, between transcription and translation, by some process which selectively stabilizes certain kinds of RNA.

Isozymes. The kinds of enzymes present in different cells and tissues of the same organism may show marked qualitative and quantitative differences. Mammalian liver cells, for example, have a glucose-6-phosphatase and can convert glycogen and other precursors to free glucose, whereas skeletal muscle cells lack this enzyme. Enzymes that catalyze the same reaction in different tissues may differ in molecular size, in their amino acid sequences, in their immunologic properties and in their responses to hormones and other control mechanisms.

Even within a single tissue or a single cell, multiple molecular forms of an enzyme, termed **isozymes,** may be found. All these proteins catalyze the same general reaction but have distinct chemical and physical properties. The different molecular forms may bear a different net charge and thus can be separated by electrophoresis.

The lactic dehydrogenase isozymes are composed of subunits, four of which combine to form the active enzyme. There are two major kinds of subunits, A and B, each of which is a polypeptide chain with a specific, gene-determined sequence of amino acids. The entire molecule is analogous to a hemoglobin molecule, which is composed of two α and two β polypeptides; however, in the lactic dehydrogenase molecule *any* combination of the two types of subunits is permissible. The combination of two kinds of subunits taken four at a time (A_4, A_3B, A_2B_2, AB_3, B_4) accounts for the five isozymes of lactic dehydrogenase that are typically observed when the enzyme is extracted from a tissue and subjected to electrophoresis.

It appears that there are two genes, one for each of the subunits. Different types of tissues have characteristic ratios of the different chains in their tetramers, presumably reflecting differences in the relative activities of the two genes and differences in the rates of synthesis of the two subunits. What controls the relative activities of the two genes remains unknown. A very curious observation, and one that requires explanation, is that the lactic dehydrogenase in the breast muscle of the chick changes during embryonic development from a pure B_4 isozyme, through a series of intermediates, to a pure A_4 type in the adult.

Organizers. When a piece of the dorsal lip of the blastopore of a frog gastrula is excised and implanted beneath the ectoderm of a second gastrula, the tissue heals in place and causes a second brain, spinal cord and other parts to develop at that site, so that a double embryo, or closely joined Siamese twins, results (Fig. 7.21). Many tissues show similar abilities to organize the development of an adjoining structure. The eye cup will initiate the formation of a lens from overlying ectoderm even if it is transplanted to the belly region where the cells would normally form belly epidermis. Such experiments indicate that development involves a coordinated series of stimuli and responses, each step determining the succeeding one. The term "organizer" is applied to the region of the embryo having this property and also to the chemical substance released by that region that passes to the adjoining tissue and directs its development.

It had been widely accepted that organizers could transmit their inductive stimuli only

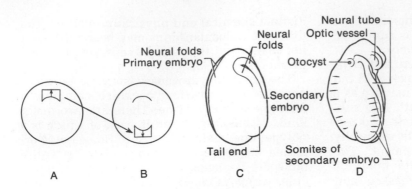

Figure 7.21 The induction of a second frog embryo by the implantation of the dorsal lip of the blastopore from embryo *A* onto the belly region of embryo *B*. Embryo *B* then develops through stage *C* to a double embryo, *D*.

when in direct contact with the reactive cells, but Niu and Twitty showed that induction may be mediated by diffusible substances that can operate without direct physical contact of the two tissues. Niu and Twitty grew small clusters of frog ectoderm, mesoderm and endoderm cells in tissue culture and found that ectoderm alone would never differentiate into nerve tissue. Ectoderm cells placed in a medium in which mesoderm cells had been grown for the previous week did differentiate into chromatophores and nerve fibers. No comparable differentiation occurred when ectoderm cells were placed in cultures that had contained endoderm. It appears that inductive tissues such as notochord and mesoderm cells contain and release diffusible substances, nucleoproteins, which can operate at a distance and induce the differentiation of ectoderm.

Morphogenetic Substances. Further evidence of diffusible morphogenetic substances is provided by Grobstein's experimental analysis of the differentiation of kidney tubules. Nephrogenic mesenchyme is normally induced to form renal tubules by the tip of the ureter. If the two rudiments are separated by treatment with trypsin and the dissociated cells are grown in tissue culture, they will reaggregate and form tubules. Certain other tissues, the ventral portion of the spinal cord, for example, also proved to be efficient inducers of renal tubules. Grobstein then separated the inducing and reacting tissues by cellophane membranes of varying thicknesses and porosities. The inducing substance can pass through membranes up to 60 micrometers thick and with pores 0.4 micrometer in diameter. Electron micrographs revealed that induction occurred even when there was no cellular contact through the fine pores; the inducing principle is diffusible. It appears to be a large molecule and is at least in part protein for it is inactivated by trypsin.

Other experiments by Grobstein have reem-

phasized that extrinsic factors as well as nuclear factors may play a role in the process of differentiation. For example, embryonic pancreatic epithelium will continue to differentiate in organ culture only in the presence of mesenchyme cells. This requirement can be met not only by pancreatic mesenchyme from the mouse but also by mesenchyme from a variety of other sources, even embryonic chick mesenchyme. The mesenchyme can be replaced by a chick embryo juice, and the active principle of the embryo juice appears to be a protein, for it is inactivated by trypsin but not by ribonuclease or deoxyribonuclease. It can be sedimented by high speed centrifugation and appears to be a large protein. This factor is only weakly effective in causing the differentiation of salivary gland epithelium in culture and is ineffective in inducing the formation of kidney tubules or cartilage from their respective mesenchymes. Grobstein's experiments indicate that there is a spectrum of protein factors, each of which is more or less specific for the differentiation of one kind of cell.

The morphogenetic role of steroids in development has been explored by Dorothy Price. When the entire reproductive tract of a fetal male rat is dissected and explanted, it grows normally in culture if the testes are present and in their normal spatial relationship to the tract. If both testes are removed from the explanted reproductive tract, its development is inhibited and the accessory organs, the vas deferens, seminal vesicles and prostate, do not differentiate. If the removed testes are replaced by pellets of testosterone implanted in their place, development of the explanted tract proceeds normally (Fig. 7.22). Thus testosterone can diffuse, at least over short distances, and induce the development of the male reproductive tract.

Suspensions of dissociated, individual healthy cells can be prepared by treating a tissue briefly with a dilute solution of trypsin.

Figure 7.22 Schematic representation of fetal male reproductive tracts explanted after about 17 days of development. A, The tract at 17 days, at the time of explantation. B–F, Tracts cultured for 4 days, and hence at 21 days of development. B, Tract cultured with both testes present; the seminal vesicles and prostate glands developed. C, Tract cultured without testes showed regression of the wolffian ducts; no seminal vesicles and only a few prostate buds developed. D, Tract cultured with one testis in place developed normally. E, Tract cultured with one testis but spread apart showed some regression of the wolffian duct of the side without the testis, and the seminal vesicle on that side was smaller or absent. F, Tracts cultured with no testes but with testosterone micropellets present developed with normal wolffian ducts, seminal vesicles and prostatic buds. Müllerian duct (M); wolffian duct (W); seminal vesicles (S); prostate gland (P). (Reproduced by permission from Price and Pannabecker. *In* Wolstenholme, G. E. W., and Millar, S. C. P. (eds.): *Ciba Foundation Colloquia on Aging in Transient Tissues.* London, J. and A. Churchill, 1956.) (From Villee, C. A.: Biology. 7th ed. Philadelphia, W. B. Saunders Co., 1977.)

When the suspension of cells is placed in tissue culture medium the cells may reaggregate and continue to differentiate in conformity with their previous pattern. The cells in tissue culture reaggregate not in a chaotic mass but in an ordered fashion, forming recognizable morphogenetic units. The cells appear to have specific affinities, for epidermal cells join with each other to form a sheet, disaggregated kidney cells join to form kidney tubules, and so on (Fig. 7.23). The question of how one cell recognizes another cell and joins with it to form a tubule, sheet or other structural unit is indeed a fascinating one.

Disaggregated cells growing in culture can be subjected to a variety of experimental procedures to analyze the factors controlling their differentiation. Dissociated epidermal cells grown in the usual culture medium will reaggregate and differentiate as a stratified

Figure 7.23 A mass of neural plate cells and epidermal cells from the ventral surface were removed from an amphibian embryo and disaggregated. The two types of cells subsequently reaggregated in tissue culture to form a heterogeneous mass. Finally each type of cell migrated in an appropriate fashion so that the medullary plate cells rounded up and formed a neural tubelike structure in the center, whereas the epidermal cells migrated to the periphery of the mass and joined to form a new epidermis.

squamous epithelium. But if vitamin A is added to the culture medium, or if the dissociated cells are exposed to vitamin A for as short a time as 15 minutes, the cells reaggregate and differentiate into a columnar epithelium with mucus-secreting goblet cells. Vitamin A under these circumstances has a strong morphogenetic effect.

Embryonic tissues growing in vivo have differential sensitivities to changes in nutrients, to the presence of inhibitors and antimetabolites and to various environmental agents. Any of these factors, applied during the appropriate critical period in development, may change the course of development and differentiation and mimic the phenotype of a mutant gene, producing what is termed a **phenocopy.**

A striking demonstration that the same set of genes operating in dissimilar environments may have different morphological effects was provided by experiments with three races of frogs found in Florida, Pennsylvania and Vermont. Each of these races normally develops at a speed that is adapted to the length of the usual spring and summer season in its locale. Southern frogs develop slowly and northern frogs develop more rapidly. When northern frogs are raised under southern environmental conditions their development is overaccelerated, whereas when southern frogs are raised under northern conditions their development is overretarded.

When an egg is fertilized with a sperm from a different race and the original egg nucleus is removed before the sperm nucleus can unite with it, it is possible to establish a cell with "northern" genes operating in "southern" cytoplasm or the reverse. Northern genes in southern cytoplasm resulted in poorly regulated development; the animal's head grew more rapidly than the posterior region and became disproportionately large. Southern genes introduced into northern cytoplasm led again to poorly regulated development but the head rather than the posterior region was retarded and disproportionately small. Genes from the Pennsylvania race of frog acted as "northern" with Florida cytoplasm but as "southern" with Vermont cytoplasm. The same set of genes had diverse morphological effects when they operated in different cytoplasmic environments.

Cellular differentiation may involve the differential activation of specific genes in different tissues; it may involve mechanisms affecting protein synthesis but operating at the ribosomal level or at the cell surface where the transport of substances into or out of the cells is regulated.

Differentiation may be controlled at least in part by influences originating outside the cell, by "organizers" from neighboring cells in early differentiation, by the mesenchymal proteins studied by Grobstein or by hormones from distant cells. Such systemic influences participate in the integration of the differentiation of individual cells into the larger pattern of differentiation of the tissues of the whole organism. Eventually it should be possible for developmental biologists to bridge the gap between studies of development at the level of the whole organism and studies at the molecular level and trace in detail the sequence of events from the initial action of the gene to the final expression of the phenotype. Our descriptive knowledge of the phenomena of fertilization, cleavage, gastrulation, morphogenesis and organogenesis is quite extensive, but our understanding of the fundamental molecular mechanisms involved in each of these processes is indeed rudimentary. This is one of the most exciting areas of biological research at present.

7.12 LETHAL GENES

Certain genes produce such a tremendous deviation from the normal development of an organism that it is unable to survive. The presence of these **lethal genes** can be detected by certain upsets in the expected genetic ratios. For example, some mice in a certain strain had yellow coat color, but experimenters found it impossible to establish a true-breeding strain with yellow coat. Instead, when two yellow mice were bred, offspring were produced in the ratio of 2 yellow: 1 nonyellow. A yellow mouse bred to a black mouse gave half yellow mice and half black mice among the offspring. Then investigators noticed that the litters of yellow × yellow matings were somewhat smaller than other litters of mice, being only about three-quarters as large. They reasoned that one-quarter of the embryos, those homozygous for yellow, did not develop. When the uterus of the mother was opened early in pregnancy the abnormal embryos, those homozygous for the yellow trait, were found. Embryos homozygous for yellow color begin development, then cease developing, die and are resorbed.

"Creeper" fowl have short legs and short wings and when two "creeper" fowl are bred the offspring are in the ratio of two "creepers" to one normal. One-quarter of the embryos—those homozygous for "creeper"—have marked abnormalities of the vertebrae and spinal cord and die without hatching.

These lethal genes, yellow and "creeper," produce a phenotypic effect when heterozygous and hence are said to be dominant. Many, perhaps most, of the lethal genes appear to have no effect when heterozygous but cause death when homozygous and are called recessive lethal genes. These can be detected only by special genetic techniques. When wild populations of fruit flies and other organisms are analyzed, the presence of many recessive lethals is revealed. In the light of our present theory about the relations between genes and development we can suppose that a lethal gene is a mutant which causes the absence of some enzyme of primary importance in intermediary metabolism. The absence of this enzyme prevents the proper development of the organism.

Each recessive gene described so far produces its trait when it is homozygous, and each dominant gene produces its effect when it is homozygous or heterozygous, but other genes are known which do not always produce their expected phenotypes. Genes which always produce the expected phenotype are said to have complete penetrance. If only 70 per cent of the individuals of a stock homozygous for a certain recessive gene show the character phenotypically, the gene is said to have 70 per cent penetrance. The term **penetrance** refers to the statistical regularity with which a gene produces its effect when present in the requisite homozygous (or heterozygous) state. The percentage of penetrance of a given gene may be altered by changing the conditions of temperature, moisture, nutrition, and so forth under which the organism develops.

Some stocks which are homozygous for a recessive gene may show wide variations in the appearance of the character. Fruit flies homozygous for a recessive gene which produces shortening and scalloping of the wings exhibit wide variations in the degree of shortening and scalloping. Such differences are known as variations in the **expressivity** or expression of the gene. The expressivity of the gene may also be altered by changing the environmental conditions during the organism's development. In view of the long and sometimes tenuous connection between the gene in the nucleus of the cell and the final production of the trait, it is easy to understand why the expression of the trait might vary or why the mutant trait might be completely absent.

7.13 THE MATHEMATICAL BASIS OF GENETICS: THE LAWS OF PROBABILITY

The discussions of heredity in Chapter 6 were concerned with inheritance in individuals, with the appearance of the offspring of two individuals with specific traits. Geneticists may also be concerned with the genetic characteristics of a population as a whole. It is possible, with the aid of simple mathematical methods, to make inferences about the mode of inheritance of a trait from its distribution in a population. All genetic events are governed by the laws of probability and, although the outcome of any single event is highly uncertain, in a large number of events the laws of probability provide a reasonable prediction of the fraction of those events which will be of one type or the other. In tossing a coin, where the probability, p, of obtaining a "heads" is one chance in two, or ½, one cannot predict the outcome of any *single* toss of the coin. But in 100 tosses about 50 will come up "heads" and 50 will come up "tails." Probabilities are usually expressed as the fraction obtained by dividing the number of "favorable" events by the total number of possible events.

When expressed in this fashion, the limits of probability are from 0 to 1. A probability of 0 indicates that the event is impossible; there is no favorable possibility. A probability of 1 represents a certainty; that is, all the possible events are favorable ones. If you are engaged in a game of chance in which the favorable event involves turning up a "three" on a die, the probability of this favorable event is one in six. If a bag contains 10 red, 40 black and 50 white marbles the chance of picking a single red marble out of the bag is ten out of one hundred or ¹⁄₁₀.

Three types of probabilities can be distinguished. **A priori probabilities** are those which can be specified in advance from the nature of the event. For example, in flipping a coin the probability of obtaining a head is one in two and in casting a die the probability of obtaining a six is one in six; these probabilities are independent of whether the event actually occurs. In contrast, **empiric probabilities** are obtained by counting the number of times a given event occurs in a certain number of trials. For example, if a surgeon performs a certain type of operation on 500 people and 40 of them succumb, then the probability of death in this type of operation is 40 in 500 or 0.08. Such empiric probabilities have to be used in many fields of research where there is no theoretical basis, no *a priori* basis, for predicting the outcome. This type of probability is used in setting up the "risk tables" widely used by insurance firms. If a scientist collects data about the number of individuals with a certain trait in a given population and wants to know the probability that these numbers agree with the ratio expected on the basis of some genetic theory (1:1 or 3:1, etc.), he uses the methods of **sampling probability.**

If two events are independent the probability of their coinciding is the *product* of their individual probabilities. For example, the probability of obtaining a head on the first toss of a coin is $\frac{1}{2}$ and the probability of obtaining a head on the second toss of a coin (an independent event) is $\frac{1}{2}$. The probability of obtaining two heads on successive tosses of the coin is the product of their probabilities, $\frac{1}{2} \times \frac{1}{2}$, or $\frac{1}{4}$. There is one chance in four of obtaining two heads on two successive tosses of a coin. This "product rule" of probability also holds for three or more independent events. For example, the probability of choosing at random an individual who is male, has blood group A and was born in June is $0.5 \times 0.4 \times 0.084 = 0.0168$.

The probability that one or another of two mutually exclusive events will occur is the sum of their separate probabilities. For example, in rolling a die the probability that the die will come up *either* two or five is $\frac{1}{6} + \frac{1}{6} = \frac{1}{3}$.

The application of these considerations to genetics is illustrated in Figure 7.24. Let us consider the probability that a child will inherit from his father the particular allele that the father inherited in turn from his father, the child's grandfather. The father has two alleles, A^P and A^M, obtained from the grandfather and grandmother respectively. He will pass onto his son one of these two. The "favorable"

event is the transmission of the A^P gene and the total number of events is two; thus the probability that the son will receive from his father the same allele that the father received from his grandfather is one in two. The probability that the child will receive from his father the allele that the father obtained from the grandfather (A^P) and also will obtain from his mother the allele that she received from the grandmother (A^M) is the product of their independent occurrences, $\frac{1}{2} \times \frac{1}{2} = \frac{1}{4}$. The probability that the child will obtain either the two alleles from the two grandfathers, $A^P A^{P'}$, or the two alleles from the two grandmothers, $A^M A^{M'}$, is the sum of their independent probabilities, $\frac{1}{4} + \frac{1}{4} = \frac{1}{2}$.

7.14 POPULATION GENETICS

The question that sometimes puzzles beginning geneticists is: Why, if brown eye genes are dominant to blue eye genes, haven't all the blue eye genes disappeared? The answer lies partly in the fact that a recessive gene, such as the one for blue eyes, is not changed by having existed for a generation next to a brown eye gene in a heterozygous individual, **Bb.** The remainder of the explanation lies in the fact that as long as there is no selection for either eye color, that is, as long as people with blue eyes are just as likely to marry and have as many children as people with brown eyes, successive generations will have the same proportion of blue- and brown-eyed people as the initial one.

A brief excursion in mathematics will show why this is true. If we consider the distribution of a single pair of genes, **A** and **a,** any member of the population will have the genotype **AA, Aa** or **aa.** No other possibilities exist. Now let us suppose that these genotypes are present in the population in the ratio of $\frac{1}{4}$**AA**:$\frac{1}{2}$**Aa**:$\frac{1}{4}$**aa.** If all the members of the population select their

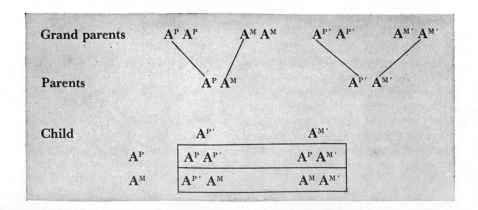

Figure 7.24 An example of the application of the laws of probability to genetics, illustrating both the "product law" of independent events and the "sum law" of mutually exclusive events. See text for discussion. (From Villee, C. A.: Biology. 7th ed. Philadelphia, W. B. Saunders Co., 1977.)

TABLE 7.6 THE OFFSPRING OF THE RANDOM MATING OF A POPULATION COMPOSED OF ¼**AA**, ½**Aa** AND ¼**aa** INDIVIDUALS (From Villee, C. A.: Biology. 7th ed. Philadelphia, W. B. Saunders Co., 1977.)

Mating Male Female	Frequency	Offspring
AA × AA	1/4 × 1/4	1/16 AA
AA × Aa	1/4 × 1/2	1/16 AA + 1/16 Aa
AA × aa	1/4 × 1/4	1/16 Aa
Aa × AA	1/2 × 1/4	1/16 AA + 1/16 Aa
Aa × Aa	1/2 × 1/2	1/16 AA + 1/8 Aa + 1/16 aa
Aa × aa	1/2 × 1/4	1/16 Aa + 1/16 aa
aa × AA	1/4 × 1/4	1/16 Aa
aa × Aa	1/4 × 1/2	1/16 Aa + 1/16 aa
aa × aa	1/4 × 1/4	1/16 aa
	Sum:	4/16 AA + 8/16 Aa + 4/16 aa

mates at random without regard to whether they are **AA, Aa** or **aa,** and if all of the types of pairs produce, on the average, comparable numbers of offspring, the succeeding generations will also have genotypes in the ratio ¼**AA**:½**Aa**:¼**aa**. This can be demonstrated by putting down all the possible types of matings, the frequency of their occurrence at random, and the kinds and proportions of offspring produced by each type of mating. When all the types of offspring are summed, it will be found that the next generation will also have genotypes in the ratio ¼**AA**:½**Aa**:¼**aa** (Table 7.6).

G. H. Hardy, an English mathematician, and G. Weinberg, a German physician, independently observed in 1908 that the frequencies of the members of a pair of allelic genes in a population are described by the expansion of a binomial equation. If we let p be the proportion of **A** genes in the population and q be the proportion of **a** genes in the population, since a gene must be either **A** or **a**, $p + q = 1$. Thus, if we know the value of either p or q, we can calculate the value of the other.

When we consider all of the matings in any given generation, a p number of **A**-containing eggs and a q number of **a**-containing eggs are fertilized by a p number of **A**-containing sperm and a q number of **a**-containing sperm: $(pA +$

$qa) \times (pA + qa)$. The proportion of the types of offspring of all these matings is described by the algebraic product: $p^2AA + 2pqAa + q^2aa$. If p, the frequency of gene **A**, equals ½, then q, the frequency of gene **a**, equals $1 - p$ or $1 - ½$ or ½. From the formula, the frequency of genotype **AA**, p^2, equals $(½)^2$ or ¼; the frequency of **Aa**, $2pq$, equals $2 \times ½ \times ½$, or ½, and the frequency of **aa**, q^2, equals $(½)^2$ or ¼. Any population in which the distribution of alleles **A** and **a** conforms to the relation $p^2AA + 2pqAa + q^2aa$ is in **genetic equilibrium.** The proportions of these alleles in successive generations will be the same (unless altered by selection or mutation).

7.15 GENE POOLS AND GENOTYPES

The genetic constitution of a population of a given organism is termed the **gene pool.** Stated differently, all the genes of all the individuals in a population make up the gene pool. This may be contrasted to the **genotype,** which is the genetic constitution of a single *individual.* Any individual may have only two alleles of any given gene. In contrast, the gene pool of the population may contain any number of different alleles of a specific gene. The A, B, and O blood groups (p. 132) are inherited by three alleles, a^A, a^B, and **a**. In the population there are three alleles, but any given individual can have no more than two of the three.

The frequencies of the O, A, B and AB blood groups have been measured in several different populations (Table 7.7), and from these we can calculate the gene frequencies for a^A, a^B and a. Calculating the gene frequencies for a three-allele system is a little more complex mathematically than for a two-allele system, and we shall not go into it here. Expansion of the equation $(p a^A + q a^B + r a)^2$, in which p, q and r represent the frequencies of genes a^A, a^B and a, respectively, gives $p^2 a^A a^A + 2 pq a^A a^B + 2 pr a^A a + q^2 a^B a^B + 2 qr a^B a + r^2 aa$. The frequency of the a^B allele varies most strikingly (Table

TABLE 7.7 THE FREQUENCY OF THE ABO BLOOD GROUPS IN DIFFERENT POPULATIONS (From Villee, C. A.: Biology. 7th ed. Philadelphia, W. B. Saunders Co., 1977.)

Population	Frequency of Blood Groups				Gene Frequencies		
	O	A	B	AB	I^A	I^B	i
Northern Europeans	.40	.45	.10	.05	.29	.08	.63
African blacks	.42	.24	.28	.06	.13	.23	.64
American Indians	.67	.29	.03	.01	.17	.01	.82

7.7), with a twentyfold difference between the frequency of the allele in the African Blacks and the American Indians.

The gene pools of different populations may differ in the ratios or proportions of the specific alleles. One population may have the alleles **A** and **a** in a ratio of 0.5 to 0.5. Another population of the same species may have the two alleles in the ratio 0.7**A**:0.3**a**. If all the individuals in this second population have equal chances of surviving to adulthood and equal chances of producing gametes, then 70 per cent of the sperm produced by the entire population of males would have gene **A** and 30 per cent would have gene **a**. Similarly, 70 per cent of the eggs produced by the entire population of females would contain gene A and 30 per cent would have gene **a**. The random union of the eggs and sperm would result in offspring in the ratio of 0.49**AA** + 0.42**Aa** + 0.09**aa**. Notice that the gene pool in the offspring is identical to the gene pool of the parents!

	sperm	
eggs	.7A	.3a
.7A	.49AA	.21Aa
.3a	.21Aa	.09aa

By similar calculations you can show that the next generation, and each succeeding generation, will contain an identical gene pool, 0.7**A** and 0.3**a**. The three kinds of genotypes will be present in the ratio 0.49**AA**:0.42**Aa**:0.09**aa** in succeeding generations provided that (1) there are no mutations for **A** or **a**; (2) the three kinds of genotypic individuals have equal probabilities of surviving, mating and producing offspring, and there is no selection of mates according to these genotypes; and (3) the population of individuals is large enough.

This principle that a population is genetically stable in succeeding generations is termed the **Hardy-Weinberg Principle.** Since the principle was established, J. B. S. Haldane, R. A. Fisher and Sewall Wright have developed mathematical methods for analyzing the inheritance of a given trait in a population. Subsequently it has become clear that the process of evolution, stated in the simplest terms, represents departure from the Hardy-Weinberg principle of genetic stability. Evolution involves changes in the gene pool of a population that result from mutations and selection. Thus an understanding of the Hardy-Weinberg principle is of prime importance in understanding the mechanism of evolutionary change.

The values of p and q, which are **gene frequencies**, cannot be measured directly. However, since the recessive phenotype can be distinguished you can determine q^2, the frequency of genotype **aa**. From this you can calculate the gene frequencies q (which is the square root of q^2) and p (which is $1 - q$). Finally you can calculate the frequencies of the other genotypes, p^2**AA** and $2pq$**Aa**. To calculate the number of individuals in a population who are genetic carriers for a given trait (i.e., are heterozygotes, **Aa**) you need to know only that it is inherited by a single pair of genes and the frequency with which the homozygous recessive individuals appear in the population.

Albinos are individuals with no pigment at all in their skin or hair. **Albinism** is an inherited trait in which the individual lacks a specific enzyme, **tyrosinase.** Tyrosinase catalyzes one of the reactions involved in the production of the dark pigment **melanin.** Albinism is inherited by a single pair of genes, and albinos, the homozygous recessive individuals, occur about once in 20,000 births. From this fact you can calculate that the frequency of **aa** individuals (q^2) is $1/20{,}000$. From the value of q^2 you can determine q by taking its square root. The square root of $1/20{,}000$ is about $1/141$. Since $p = 1 - q$ or $1 - 1/141$, $p = 140/141$. You now have values for both p and q, and can calculate the value of $2pq$, which represents the frequency of the genetic "carrier" **Aa** individuals: $2 \times 140/141 \times 1/141 = 1/70$. Surprising as it may seem, one person in 70 is a **carrier** of albinism, although only one person in 20,000 is homozygous and displays the trait. At first glance it may seem odd that there are so many carriers in a population that contains so few homozygous recessives, yet reflecting on the mathematical relations involved should lead you to realize that this must be true. When q is small (such as $1/141$), then q^2 will be very small, but $2pq$ will be much larger.

The ability or inability to taste **phenylthiocarbamide** (PTC) and related compounds with the N—C=S group is an inherited trait. Some
|
people find that this substance has a bitter taste; others report it to be completely tasteless. L. H. Snyder examined 3643 people and found that 70.2 per cent were "tasters" and 29.8 per cent were "nontasters" for PTC. If this trait is inherited by a single pair of genes, with "tasting" dominant to "nontasting,"

TABLE 7.8 THE INHERITANCE OF THE ABILITY TO TASTE PHENYLTHIOCARBAMIDE

Marriage	No. of Families	Offspring		Proportion of Nontasters	
		Tasters	Nontasters	Observed	Calculated*
taster × taster	425	929	130	0.123	0.124
taster × nontaster	289	483	278	0.336	0.354
nontaster × nontaster	86	5	218	0.979	1.0

*From the Hardy-Weinberg Law, it follows that among marriages of parents with unlike traits (i.e., taster × nontaster), the fraction of the offspring with the recessive trait is $q/1 + q$. Among marriages of parents with like traits (i.e., taster × taster), the fraction of offspring with the recessive trait, nontaster, is $(q/1 + q)^2$. Since 29.8 per cent of the population were nontasters, $q^2 = 0.298$, $q = 0.545$, $q/1 + q = 0.354$ and $(q/1 + q)^2 = 0.124$. (From Villee, C. A.: Biology. 7th ed. Philadelphia, W. B. Saunders Co., 1977.)

then the mathematics underlying population genetics would predict that in a population of marriages of tasters with tasters, 12.4 per cent of the children will be nontasters (Table 7.8). It similarly predicts that in a population of marriages of tasters with nontasters, 35.4 per cent of the children will be nontasters. In Snyder's study the percentages actually found were 12.3 per cent and 33.6 per cent, respectively. From this close agreement, we may conclude that the original assumption, that the tasting-nontasting trait is inherited by a single pair of genes, is correct. Nontasters, although constituting some 30 per cent of the white population, are very rare in populations of Negroes, Eskimos and American Indians. Thus the **T-t** gene pools in these populations are quite different.

As another example of the use of the Hardy-Weinberg Principle, let us consider the inheritance of the blood groups M, MN and N. In the United States, among the white population 29.16 per cent have blood type M, 49.58 per cent have blood type MN, and 21.26 per cent have blood type N. Applying the Hardy-Weinberg Principle to these data, $q^2 = 0.2126$, from which we calculate that $q = 0.46$. To estimate the value for p, we subtract 0.46 from 1 and arrive at 0.54. The square of this estimate of p, $(0.54)^2$, is equal to 0.29 and $2pq = 2 \times 0.54 \times 0.46$ or 0.49. The excellent agreement between the values observed and the theoretical values is evidence that M and N blood types are inherited by a single pair of genes, with neither gene being dominant to the other (they may be termed **codominants**) so that the heterozygote shows the two kinds of blood antigens.

7.16 HUMAN CYTOGENETICS

Many of the basic principles of genetics were discovered by experiments in lower organisms

in which it was possible to relate genetic data with cytologic events. This involved making smears of cells and examining them under the microscope to see the number and the structure of the chromosomes present. Some of the organisms used in genetics, such as the fruit fly *Drosophila*, have few chromosomes (four pairs) and, at least in certain cells of the body (for example, the salivary glands), the chromosomes are quite large and the details of their structure are readily evident.

Although the normal human chromosome number is 46, some rare instances of abnormal chromosome numbers have been reported. These usually are associated with some change in the phenotype of the individual. Changes in chromosome number are termed *ploidy*. An individual may be polyploid, having one or more complete extra sets of chromosomes, or he may have one or two extra chromosomes with the total chromosome number of 47 or 48. A severely defective male child was found to be a *triploid* individual with a total of 69 chromosomes. He had 66 autosomes, two X chromosomes and one Y chromosome. It seems likely that this zygote was formed from a normal haploid egg which was fertilized by an unusual diploid sperm or from an exceptional diploid egg fertilized by a normal haploid sperm.

Nondisjunction refers to the failure of a pair of homologous chromosomes to separate normally during the reduction division. Two X chromosomes, for example, might fail to separate and both might enter the egg nucleus, leaving the polar bodies with no X chromosome. Alternatively, the two joined X chromosomes might go into the polar body, leaving the female pronucleus with no X chromosome. Nondisjunction of the XY chromosomes in the male might lead to the formation of sperm which have both an X and a Y chromosome or to sperm with neither an X nor a Y chromosome. Chromosomal nondisjunction may occur during either the first or second meiotic division; it may also occur during mitotic divisions and lead to the establishing of a group of abnormal cells in an otherwise normal individual.

Cytogenetic studies have clarified the origin of one of the more distressing abnormal human conditions, that of **Down's syndrome,** or "mongolism." Individuals suffering from this have abnormalities of the face, eyelids, tongue and other parts of the body, and are greatly retarded in both their physical and mental development. The term "mongolism" was originally applied to this condition because affected individuals often show a fold of the eyelid similar to that typical of members of the Mongolian race. Down's syndrome is a relatively common congenital malformation, occurring in 0.15 per cent of all births. It had been known for some time that the appearance of Down's syndrome is related to the age of the mother and that it increases greatly with maternal age. For example, Down's syndrome is a hundredfold more likely in the offspring of women 45 years or older than in the offspring of mothers under 19. The occurrence of Down's syndrome, however, is independent of the age of the father, and it is also independent of the number of preceding pregnancies in the woman. Cytogenetic studies revealed that individuals with Down's syndrome have one extra small chromosome 21, a total of 47. The presence of this extra small chromosome is believed to arise by nondisjunction in the maternal oöcyte.

Down's syndrome should be inherited as though it were a dominant gene, since an afflicted individual would form gametes half of which have the normal complement of 23 chromosomes and half of which have 24 chromosomes. In the rare cases where persons with Down's syndrome have had offspring they have produced normal and afflicted children in about equal proportions.

Another condition caused by an upset in chromosome number is that of individuals who are outwardly nearly normal males but have small testes. They produce few or no sperm, they have seminiferous tubules which are very aberrant in appearance, and they usually have gynecomastia (a tendency for formation of female-like breasts). This condition, called **Klinefelter's syndrome,** usually becomes apparent only after puberty, when the small testes and gynecomastia may bring the individual to the attention of his physician. The cells of these individuals show a chromatin spot, and at one time they were thought to be XX individuals, i.e., genetic females. However, when their chromosomes were examined cytologically and counted it was found that they have 47 chromosomes; their cells have *two* X and one Y chromosome. The fact that they are nearly normal males in their external appearance emphasizes the male-determining effect of the Y chromosome in man.

Another condition resulting from changes in chromosome number is **Turner's syndrome,** in which the external genitalia, though feminine, are those of an immature female. The

TABLE 7.9 SOME HUMAN CHROMOSOMAL ABNORMALITIES
(From Page, E. W., Villee, C. A. and Villee, D. B.: Human Reproduction. 2nd ed. Philadelphia, W. B. Saunders Co., 1976)

Abnormality	Genetic Features	Clinical Aspects
Turner's syndrome (gonadal dysgenesis)	XO	Short stature, streak ovary, juvenile female genitalia, poorly developed breasts
Klinefelter's syndrome	XXY	Gynecomastia, small testes
Triple X females	XXX	Two "Barr bodies" present, fairly normal females but secondary sex characteristics may be poorly developed
Down's syndrome	Trisomy 21	Epicanthal folds, protruding tongue, hypotonia, mental retardation
Trisomy 18	Trisomy 18	Mental retardation, multiple congenital malformations
D trisomy	Trisomy 15	Mental retardation, severe multiple anomalies, cleft palate, polydactyly, central nervous system defects, eye defects
Translocation mongolism	15/21, 21/22, or 21/21 translocation	Mongolism, clinically similar to trisomy 21
Philadelphia chromosome	Deletion of one arm of chromosome 21	Chronic granulocytic leukemia
Orofaciodigital syndrome	Translocation of part of chromosome 6 to 1	Defects of upper lip, palate, and mouth, stubby toes with short nails
Cri du chat syndrome	Deletion of short arm of chromosome 5	Mental retardation, facial anomalies

internal reproductive tract is present and resembles that of an immature but perfectly formed female. The uterus is present but small, and the gonads may be absent. The cells of these individuals are "chromatin negative," which suggests that they are males. However, they have only one X chromosome but no Y chromosome. This type of disorder again emphasizes the importance of the Y chromosome in determining the male characteristics.

An individual with an extra chromosome, with three of one kind, is said to be **trisomic,** and an individual lacking one of a pair is said to be **monosomic.** Thus individuals with Down's syndrome are trisomic for chromosome 21, and individuals with Turner's syndrome are monosomic for the X chromosome. The features of certain human chromosomal abnormalities are summarized in Table 7.9.

In summary, the primary genetic material in all living systems is nucleic acid. The genetic material is DNA in all organisms except a few of the viruses in which only RNA is present. Genetic information is transferred from one generation to the next in the form of specific sequences of nucleotides in the nucleic acid chain. The number of nucleotides that form a unit in this informational code, a codon, is three. A gene is a localized sequence of nucleotides in a nucleic acid which carries the information for the synthesis of a peptide chain with a specific sequence of amino acids. The function of the gene is to transfer this specificity to some nongenic macromolecule, presumably messenger RNA. The model of DNA proposed by Watson and Crick provides for gene specificity, gene replication and gene mutation. A gene mutation is believed to be an alteration of the specific sequence of nucleotides in the nucleic acid molecule which leads to the formation of a different type of protein or perhaps prevents the formation of that protein. Each gene is believed to control the production of a single specific protein.

ANNOTATED REFERENCES

Frisch, L. (Ed.): The Genetic Code. Cold Spring Harbor Symposium on Quantitative Biology. Cold Spring Harbor, N.Y. Vol. 31, 1966. A rich source of facts and theories detailing the state of our knowledge of biochemical genetics as of June 1966.

Hartman, P. E., and S. R. Suskind: Gene Action. Englewood Cliffs, N.J. Prentice-Hall, Inc., 1964. A brief paperback presentation of molecular genetics.

Ingram, V.: The Biosynthesis of Macromolecules. New York, W. A. Benjamin, 1965. A clear, logically presented discussion of the genetic control of protein synthesis.

Markert, C. L.: Developmental Genetics. Englewood Cliffs, N.J., Prentice-Hall, Inc., 1964. A summary of the application of current genetic theories to the problem of differentiation.

Turpin, R., and J. Lejeune: Human Afflictions and Chromosomal Aberrations. London, Pergamon Press, 1969. An excellent source of information on human genetics.

Watson, J. D.: Molecular Biology of the Gene. 3rd ed. New York, W. A. Benjamin, 1976. A masterful summary of biochemical genetics based on experiments with bacteria and viruses.

Wolstenholme, G. E. W., and C. M. O'Connor: The Biochemistry of Human Genetics. Boston, Little, Brown and Company, 1959. The proceedings of a Ciba symposium on the chemical aspects of human inheritance.

CHAPTER 8

THE CONCEPT OF EVOLUTION

An immense variety of forms of life inhabit every conceivable place on land and in the water. They exhibit tremendous variations in size, shape and degree of complexity and in methods of obtaining food, of evading predators and of reproducing their kind. How all these species came into existence, how they came to have the particular adaptations which make them peculiarly fitted for survival in a particular environment, and why there are orderly degrees of resemblance between forms which permit their classification in genera, orders, classes and phyla are fundamental problems of zoology. From detailed comparisons of the structures of living and fossil forms, from the sequence of the appearance and extinction of species in times past, from the physiological and biochemical similarities and differences between species, and from the analyses of heredity and variation in many different animals and plants has come one of the great unifying concepts of biology, that of **evolution.**

The concept of evolution can emerge logically and naturally from our understanding of the principles of genetics. This is not, however, how it arose historically. It came as a result of countless observations of similarities and differences in structures and functions of the various kinds of animals and plants in different parts of the world. It came from Charles Darwin's profound insight into the arrangement of the pieces of the puzzle of how these similarities and differences may have arisen.

8.1 WHAT IS EVOLUTION?

The term evolution means an unfolding, or unrolling, a gradual, orderly change from one state to the next. The planets and stars, the topography of the earth, and the chemical compounds of the universe have undergone gradual, orderly changes sometimes called **inorganic evolution.** The principle of **organic evolution,** now universally accepted by biologists, simply applies this concept to living things: all the various plants and animals living today have descended from simpler organisms by gradual modifications which have accumulated in successive generations.

Evolution is continuing to occur. In the last few hundred thousand years, hundreds of species of animals and plants have become extinct and other hundreds have arisen. The process is usually too gradual to be observed, but there are some remarkable examples of evolutionary changes which have taken place within historic times. For example, some rabbits were released early in the fifteenth century on a small island near Madeira called Porto Santo. There were no other rabbits and no carnivorous enemies on the island and the rabbits multiplied at an amazing rate. In 400 years they became quite different from the ancestral European stock; they were only half as large, had a different color pattern, and were more nocturnal animals. Most important, they could not produce offspring when bred with members of the European species. They were, in fact, a new species of rabbit.

8.2 DEVELOPMENT OF IDEAS ABOUT EVOLUTION

The idea that the present forms of life have arisen from earlier, simpler ones was far from new when Charles Darwin published *The*

Origin of Species in 1859. The oldest speculations about evolution are found in the writings of certain Greek philosophers: Thales (624–548 B.C.), Anaximander (588–524 B.C.), Empedocles (495–435 B.C.) and Epicurus (341–270 B.C.). The spirit of this age of Greek philosophy was somewhat similar to that of our own age, for simple, natural causes were sought to explain all phenomena. Since they knew very little biology, however, their ideas about evolution were extremely vague and can scarcely be said to foreshadow our present theory of organic evolution. Aristotle (384–322 B.C.), who was a great biologist as well as a philosopher, knew a great deal about animals and plants and wrote detailed, accurate descriptions of many of them. He observed that organisms could be arranged in graded series from lower to higher and drew the correct inference that one evolved from the other. However, he had the metaphysical belief that the gradual evolution of living things occurred because nature strives to change from the simple and imperfect to the more complex and perfect. Aristotle's evolutionary notion did not become generally accepted, partly because the ancients had not observed that species change and partly because Western culture had already accepted the Judeo-Christian ideas of creation and the fixity of the earth and its creatures.

With the Renaissance, interest in the natural sciences quickened and the increasing knowledge of the many kinds of animals led more and more scientists to consider the concept of evolution more favorably. Among these were Hooke (1635–1703), Ray (1627–1705), Buffon (1707–1788), Erasmus Darwin (1731–1802) and Lamarck (1744–1829). Even before the Renaissance, men had discovered shells, teeth, bones and other parts of animals buried in the ground. Some of these corresponded to parts of familiar living animals, but others were strangely unlike any known form. Many of the objects found in rocks high in the mountains, far from the sea, resembled parts of marine animals. In the fifteenth century, the versatile artist and scientist, Leonardo da Vinci, gave the correct explanation of these curious finds, and gradually his conclusion that they were the remains of animals that had existed at one time but had become extinct was accepted. This evidence of former life suggested to some people the theory of **catastrophism**—the idea that a succession of catastrophes, fires and floods, have periodically destroyed all living things, followed each time by the origin of new and higher types by acts of special creation.

Three Englishmen in the eighteenth and early nineteenth centuries laid the foundations of modern geology, and by their careful, cogent arguments advanced the theory of **uniformitarianism** to replace the concept of catastrophism. In 1785 James Hutton developed the concept that the geologic forces at work in the past were the same as those operating now. He arrived at this conclusion after a careful study of the erosion of valleys by rivers and the formation of sedimentary deposits at the mouths of rivers. He demonstrated that the processes of erosion, sedimentation, disruption and uplift, carried on over long periods of time, could account for the formation of fossil-bearing rock strata. The publication of John Playfair's *Illustrations of the Huttonian Theory of the Earth* in 1802 gave further explanation and examples of the idea of uniformitarianism in geologic processes. Sir Charles Lyell, one of the most influential geologists of his time, finally converted most of the contemporary geologists to the theory of uniformitarianism by the publication of his *Principles of Geology* (1832). A necessary corollary of the idea that slowly acting geologic forces have worn away mountains and filled up seas is that geologic time has been immensely long. This idea, completely revolutionary at the time, paved the way for the acceptance of the theory of organic evolution, for the process of evolution requires an extremely long time.

The earliest theory of organic evolution to be logically developed was that of Jean Baptiste de Lamarck, the great French zoologist whose *Philosophie Zoologique* was published in 1809. Lamarck, like most biologists of his time, believed that all living things are endowed with a vital force that controls the development and functioning of their parts and enables them to overcome handicaps in the environment. He believed that any trait acquired by an organism during its lifetime was passed on to succeeding generations—that acquired characters are inherited. Developing the notion that new organs arise in response to the demands of the environment, he postulated that the size of the organ is proportional to its use or disuse. The changes produced by the use or disuse of an organ are transmitted to the offspring and this process, repeated for many generations, would result in marked alterations of form and function. Lamarck explained the evolution of the giraffe's long neck by suggesting that some short-necked ancestor of the giraffe took to browsing on the leaves of trees, instead of on grass, and that, in reaching up, it stretched and elongated its neck. The offspring, inheriting the longer neck, stretched still farther, and the process was repeated until the present long neck was achieved.

Both Buffon and Erasmus Darwin had similar ideas about the role in evolution of the direct response of the organism to its environment but had not expressed them so clearly. This theory, called **Lamarckism,** provides an explanation for the remarkable adaptation of many plants and animals to their environment but is completely unacceptable because of the overwhelming genetic evidence that acquired characteristics cannot be inherited. The theoretical distinction between somatoplasm and germ plasm made by Weismann (1887) refuted all theories of evolution based on the inheritance of acquired characters. Acquired characters are present only in the body cells (somatoplasm) and not in the germ cells (germ plasm), and only traits present in the germ plasm are transmitted to the next generation.

8.3 BACKGROUND FOR "THE ORIGIN OF SPECIES"

Charles Darwin made two great contributions to the body of scientific knowledge: he presented a wealth of detailed evidence and cogent arguments to show that organic evolution had occurred, and he formulated a theory, that of **natural selection,** to explain the mechanism of evolution.

Darwin was born in 1809 and was sent at the age of 15 to study medicine at the University of Edinburgh. Finding the lectures intolerably dull, he transferred, after two years, to Christ's College, Cambridge University, to study theology. Many of Darwin's friends at Edinburgh were interested in geology and zoology, and at Cambridge he joined a circle of friends interested in collecting beetles. Through them he came to know Professor Henslow, the naturalist. Shortly after leaving college, and upon the recommendation of Professor Henslow, Darwin was appointed naturalist on the ship *Beagle,* which was to make a five-year cruise around the world preparing navigation charts for the British Navy. The *Beagle* left Plymouth in 1831 and cruised slowly down the east coast and up the west coast of South America. While the rest of the company mapped the coasts and harbors, Darwin studied the animals, plants and geologic formations of both coastal and inland regions. He made extensive collections of specimens and copious notes of his observations. The *Beagle* then

spent some time at the Galapagos Islands, west of Ecuador, where Darwin continued his observations of the flora and fauna, comparing them to those on the South American mainland. These observations convinced Darwin that the theory of special creation was inadequate and set him to thinking about alternative explanations.

Upon his return to England in 1836, Darwin spent his time assembling the notes of his observations for publication and searching for some reasonable explanation for the diversity of organisms and the peculiarities of their distribution. As Darwin wrote in his notebook:

On my return home in the autumn of 1836 I immediately began to prepare my journal for publication, and then saw how many facts indicated the common descent of species. . . . In July (1837) I opened my first notebook for facts in relation to the origin of species, about which I had long reflected, and never ceased working for the next twenty years. . . . Had been greatly struck from the month of March on character of South American fossils, and species on Galapagos Archipelago. These facts (especially latter) origin of all my views. . . .

In October (1838), that is fifteen months after I had begun my systematic inquiry, I happened to read for amusement *Malthus on Population,* and being well prepared to appreciate the struggle for existence which everywhere goes on, from long-continued observation of the habits of animals and plants, it at once struck me that under these circumstances favorable variations would tend to be preserved, and unfavorable ones to be destroyed. The result of this would be the origin of new species. Here then I had at last got a theory by which to work.

Darwin spent the next 20 years accumulating data from many fields of biology, examining it critically, building up a tremendous body of facts that demonstrated that evolution had occurred and formulating his arguments for natural selection. As Darwin was pondering his ideas, Alfred Russel Wallace, a naturalist and explorer who was studying the flora and fauna of Malaya and the East Indies, was similarly struck by the diversity of living things and the peculiarities of their distribution. Like Darwin, he happened to read Malthus' treatise and came independently to the same conclusion, that evolution occurred by natural selection. In 1858 Wallace sent a manuscript to Darwin, and asked him, if he thought it of sufficient interest, to present it for publication. Darwin's friends persuaded him to present Wallace's paper along with an abstract of his own views, which he had prepared and circulated to a few friends several years earlier, to a meeting of the Linnaean Society in July, 1858. This group felt that Darwin's long study of the question entitled him to priority, and Wallace accepted this. Darwin's monumental *On the Origin of Species by Means of Natural Selection* was published in November, 1859.

The time was ripe for the formulation and acceptance of the theory of organic evolution. The publication of Lyell's *Principles of Geology* and the subsequent acceptance of the idea of geologic evolution, the publication of Malthus' ideas on population growth and pressure and the struggle for existence, together with the vast accumulation of information about the distribution of living and fossil forms of life, and studies of comparative anatomy and embryology, all showed the inadequacy of the theory of special creation. Because the time was ripe, Darwin's theory rapidly gained acceptance.

8.4 THE THEORY OF NATURAL SELECTION

The theory of **natural selection,** as developed by Darwin and Wallace, is based on a combination of observations well known at the time and certain inferences drawn from them.

1. It could be observed that organisms have a very high reproductive potential. More organisms of each kind are born than can possibly obtain food and survive. If all of the offspring of any species remained alive and reproduced, they would soon crowd all others from the earth.

2. It is also a fact that this high reproductive potential is not achieved, since the population size of each species remains fairly constant or fluctuates within limits.

3. It follows from these observations that, despite some chance elimination of offspring, those that are better fitted to the particular conditions would survive in greater number than those that are less fit. Consider a hungry bass swimming into a school of minnows. Those in the path of the bass are more likely to be eaten than those on the periphery. This is a random event dependent on the path of the bass and having nothing to do with fitness. There will be some elimination and survival that will depend purely on chance. However, of those minnows in the path of the bass, some may have more efficient sense organs or superior powers of locomotion that enable them to escape. The fitter minnows in the bass's path survive, and the less fit are eaten.

4. Darwin was also aware of the existence of numerous inherited variations, although he and his contemporaries did not understand the basis for variation or inheritance.

5. Given variation and "survival of the fittest" (a phrase coined by Herbert Spencer, a contemporary philosopher, and not used by Darwin), Darwin and Wallace further deduced that the surviving individuals would transmit their superior qualities to the next generation. Superior features would accumulate generation after generation, lead to adaptive change and result in the "origin of species."

Natural selection operates in many ways. Nineteenth century Darwinians emphasized the direct combat between individuals of the same or different species, which they saw as leading to the survival of the fittest. Herbert Spencer viewed nature as "red in tooth and claw." Although this type of selection certainly occurs, we now recognize that the essence of natural selection is not so much survival of the fittest as it is greater reproductive success of the fitter. The basis of natural selection is differential reproduction because those individuals best adapted to existing conditions produce more viable offspring. Many factors can be involved in this. Some individuals may grow faster than others and monopolize favorable living space; some may be more tolerant of desiccation (in an environment where water is scarce) and reproduce in greater numbers than those who need more water.

Darwin's theory of natural selection appeared to be so reasonable that most biologists soon adopted it. Yet the notion came under criticism in the late nineteenth and early twentieth centuries. It depends on a continuing source of inherited variation. Experiments on beans made by Johannsen in 1909 showed that natural selection will not affect an inbreeding population after the first few generations, for inherited variation quickly is exhausted. As mutations and the sources of variation became better understood, it appeared to many geneticists that more knowledge of the causes of mutation would itself explain evolution. Natural selection appeared to be unnecessary. T. H. Morgan, one of the pioneers in explaining inheritance, espoused this view.

Beginning in the 1930's a synthesis of evolutionary views from different biological disciplines began to emerge. Among the leaders in promoting an integration were the late Theodosius Dobzhansky, a pioneer student of population genetics, George Gaylord Simpson, a noted vertebrate paleontologist, and Ernst Mayr, a systematist and field naturalist who contributed an understanding of species as they exist in nature. The studies of these and others have gradually led to our contemporary theory of evolution, which is often known as **Neo-Darwinism** because natural selection is a key but no means the only ingredient.

8.5 SPECIES, POPULATIONS AND GENE POOLS

Some evolutionary scholars have pointed out that Darwin's book deals not so much with the origin of species as with the origin of adaptations, partly because the species concept was not fully understood when he wrote. Many definitions of a species have been proposed but the most productive, from the perspective of evolutionary theory, is the concept of a **biological species**. In sexually reproducing organisms, a species is a group of individuals that have the ability to reproduce with one another and are reproductively isolated from other such groups in nature. Corollaries of this concept are that members of the same species share morphological, physiological and ecological features, have a common geographic range and have common isolating mechanisms that prevent them from crossing with other species under natural conditions. Exceptions occur. Occasionally, different species cross, or individuals of the same species from widely separated parts of the range will not cross successfully even when brought together. If species are evolving groups, some exceptions are to be expected.

Species are realities of nature, not abstractions. Living matter is not a continuum but is broken up into distinct units (lions, tigers, house cats) that are reproductively isolated from each other under normal conditions. There probably are limits to the amount of variability that can exist within a species. Reproductive isolation ensures that these limits are not exceeded and avoids the waste that might result from disharmonious recombinations of incompatible genetic material.

Members of a species are not distributed uniformly throughout their range, but rather occur in clusters spatially separated to some extent from other clusters. The bull frogs of one pond form a cluster separated from those in an adjacent pond. Some exchanges occur by migrations between ponds, but the frogs in one pond are more likely to reproduce with those in the same pond. Members of a species tend to be distributed in local interbreeding populations, often called **demes** (after one of the townships of ancient Attica). The territorial limits of any particular deme may be vague and difficult to define, yet the concept is useful for purposes of theoretical analysis.

Members of a deme share a common **gene pool** (Fig. 8.1) that is the aggregation of all of the genes and their alleles of all of the individ-

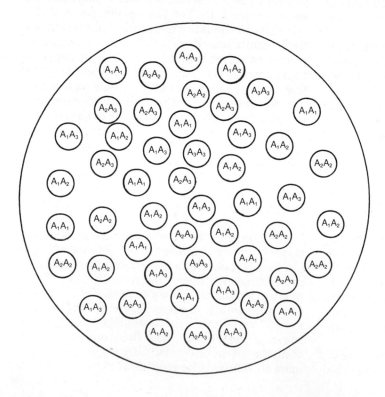

Figure 8.1 A diagram illustrating the concept of the gene pool. Diploid individuals are represented by small circles. Each has only two members of any set of alleles, but the pool may contain many (A_1, A_2, A_3, . . .). The allele frequencies in this case are $A_1 = 0.40$, $A_2 = 0.35$, $A_3 = 0.25$. (From Villee, C. A., and V. G. Dethier: Biological Principles and Processes. 2nd ed. Philadelphia, W. B. Saunders Co., 1976.)

uals of the population (see Section 7.15). Each individual is genetically unique and has a specific genotype. In simple terms, evolution is a change in the types and frequencies of specific alleles in the gene pools of populations and species. The gene pool of a population is the smallest evolutionary unit. You and I as individuals are not evolving, but the gene pool from which we obtained our genes and to which we contribute genes may be changing because of differential reproduction, among other reasons.

8.6 VARIATION IN LOCAL POPULATIONS

Changes in the types and frequency of genes in gene pools, whether by natural selection or other means, is not possible unless there is a source of inherited variation: a large reservoir of variability in the population, the introduction of new variations, or both. It is axiomatic that variation is the raw material for evolutionary change. Three sources of variability are available to most populations: (1) recombination, (2) migration and (3) mutation. **Recombination** is simply the intermixing of genetic traits already in the population through sexual reproduction. New genetic material is not introduced, but new combinations of genes may interact differently and result in new phenotypes. The effect of recombination can be surprisingly great. Nine different genotypes are generated in a dihydrid cross (AaBb × AaBb) involving only two genes on different chromosomes and each with only two alleles. If we were dealing with five genes on different chromosomes and six alleles per gene (and these are not exceptionally high numbers), the number of different genotypes possible would be 4,084,101. Some of the combinations that are generated may be adaptively superior, and natural selection could favor changes in linkage groups or other mechanisms that would perpetuate them.

Migration of individuals between demes of the same species also does not introduce new variation into the species, but the frequency of alleles in one deme can be changed by the influx of individuals from a different deme with a different gene pool. The United States has been a melting pot for racial groups, each of which originally had many attributes of a deme. American Blacks have a frequency of the Rh° allele (one of the alleles of the Rh blood group, see Section 6.16) of 0.45, in contrast to African Blacks in whom the frequency is 0.63. The reduction of the frequency in American Blacks has been attributed to an influx over many generations of alleles from the white population in which the frequency of Rh° is very low, 0.03. Frequencies have undoubtedly changed in the white population too, for migration has been in both directions, but the effect would not be so evident in the larger white population, and it is also more difficult to estimate an ancestral base point for comparison.

New variations are introduced into a species only through **mutation,** the ultimate source of variation on which evolutionary changes depend. Mutations result from a change in the nucleotide base pairs of the genetic code, from a rearrangement of genes within chromosomes so their interactions produce different effects (Section 7.9), or from a change in the number of chromosomes. Most mutations occur spontaneously at measurable rates, and a given gene may mutate in different directions, thereby giving rise to a series of multiple alleles ($A \rightarrow a_1 \rightarrow a_2 \rightarrow a_3 \ldots$). A gene also may mutate back to its original form ($a_1 \rightarrow A$). Typically, an equilibrium is established between mutations in different directions.

A mutation is a random change with respect to the direction of evolution. If a population is adapting to a dry environment, mutations appropriate for dry conditions are no more apt to occur than ones for wet conditions or ones that have no relationship to the environmental problem. Since mutation is a random change, its effect upon the members of a population well adapted to its current environment is more likely to be deleterious than beneficial. If a watch is keeping good time, a random change in its mechanism is more likely to upset its timekeeping ability than improve it. If mutations are random and frequently deleterious, how can they be an important source of variation for adaptive evolutionary change? The answer is complex.

First of all, not all mutations are deleterious; some are neutral and a few are beneficial. Studies have shown that mutations that bring about significant departures from the phenotypic norm and are most likely to be deleterious are rather rare (Fig. 8.2). Most mutations produce small changes in the phenotype that often are detectable only by biochemical techniques. This is the case with isozymes, which are proteins with essentially the same enzymatic activity but with slightly different molecular configurations. By acting against seriously abnormal phenotypes, natural selection quickly eliminates, or reduces to low frequencies,

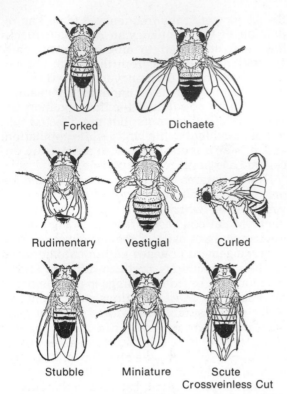

Forked   Dichaete

Rudimentary   Vestigial   Curled

Stubble   Miniature   Scute
                        Crossveinless Cut

Figure 8.2 Some wing and bristle mutants in the fruit fly, *Drosophila melanogaster*. (Drawn by E. M. Wallace: Sturtevant and Beadle: An Introduction to Genetics.)

large deleterious mutations. Small mutations, even ones with slightly deleterious phenotypic effects, have a better chance of being incorporated into the gene pool where at some later time they may be helpful to the population.

Whether a mutant is deleterious or beneficial must be evaluated in the context of all of its phenotypic effects in the particular environment in which it acts. Many genes are **pleiotropic**, affecting more than one trait. The mutant allele for sickle cell anemia produces an altered hemoglobin, HbS, that is less soluble than normal hemoglobin, HbA, especially at low oxygen tensions. Most individuals homozygous for the allele die. Individuals heterozygous for the sickle cell allele are more resistant to a particularly virulent type of malaria caused by the protozoan *Plasmodium falciparum* than are individuals homozygous for the allele for normal hemoglobin. In certain parts of Africa, India and Southern Asia, where this disease is prevalent, heterozygous individuals survive in greater numbers. In a heterozygous individual each allele produces its specific kind of hemoglobin, and the red blood cells contain the two kinds in roughly

equivalent amounts. Such cells do not undergo sickling except at very low oxygen tensions, but the red cells containing the HbS are resistant to infection with the malarial organism; this renders carriers partially immune to this type of malaria. Each of the two types of homozygous individuals is at a disadvantage. Those homozygous for the sickling allele are very likely to die of anemia; those homozygous for the normal allele may die of malaria. The frequency of the sickle cell allele reaches 40 per cent in the gene pools of black populations in parts of Africa. Resistance to falciparum malaria is of no advantage in North America, and the frequency of the allele in the gene pools of American Blacks has been reduced to 4 or 5 per cent, partly because of dilution of the pool through migration from other populations (as with the Rh° allele) and partly by selection against it.

Pleiotropic genes and those whose effects vary with the environmental context are very common. This suggests that a mutant that is deleterious in a well adapted organism living in a stable environment may become beneficial if the environment changes. In the perspective of hundreds or millions of years, most environments change greatly.

Identical mutations recur time and time again in evolving populations. Beneficial ones have been incorporated into the gene pool; very deleterious ones have been eliminated; neutral and mildly deleterious ones may be carried. There are genetic mechanisms, involving incompletely understood interactions between different genes, that permit certain attributes of a mutant to be expressed, whereas its more deleterious effects may be suppressed. Some dominant-recessive interactions of a set of alleles are not dependent so much on the alleles themselves as on modifying genes at other loci with which they interact. Mutants whose deleterious effects have been suppressed can be carried in gene pools. When environmental conditions change, their expression may be changed. Mechanisms of this type enable gene pools to carry tremendous reservoirs of variability, much of it at low frequency and much of it even hidden. This has been demonstrated in experiments with the fruit fly *Drosophila*. A population is established from offspring of a single female fly fertilized in the wild (Fig. 8.3). Her progeny has a certain variability and by appropriate selection procedures it is possible to either increase or decrease certain traits, such as the mean number of bristles on a certain part of the thorax, and to extend the range of variability beyond that in the initial popula-

Figure 8.3 Mean bristle number on a part of the thorax in lines of *Drosophila melanogaster* derived from the progeny of a single female with 36 bristles. Generations are plotted on the abscissa; bristle number on the ordinate. Solid lines represent generations during which selection for high or low bristle number were made. Dashed lines represent subcultures that were continued without selection. See text sections 8.6 and 8.9 for further explanation. (From Mayr, E.: Populations, Species and Evolution: An Abridgment of Animal Species and Evolution. Cambridge, Harvard University Press, 1970. Based on experiments by Mather and Harrison.)

tion. Since the change occurs in relatively few generations and can go in either direction, stored genetic variability rather than new mutations must be the source of variation. Genetic variability stored within a population gives it the capacity to respond to different selective forces, but eventually the supply of variability must be replenished by new mutation.

8.7 SHAPING VARIATION

Populations have a large reservoir of variation that, as the Hardy-Weinberg principle shows (see Section 7.14), remains in equilibrium provided the population is large, mating is random, natural selection is not operating and no new variation is introduced. Populations have a tendency to remain in equilibrium for many generations. This is true especially for a large population well adapted to a stable environment.

Evolution represents change and this requires shifting the Hardy-Weinberg equilibrium toward a new equilibrium value. Two major factors that may alter the equilibrium are natural selection and sampling errors in small populations. Natural selection, which is defined as nonrandom reproduction and transmission of traits to succeeding generations, may occur at any of a number of different times in the life cycle of an organism. There may be nonrandom mating (which Darwin called sexual selection), nonrandom fecundity (differences in the number of offspring produced)

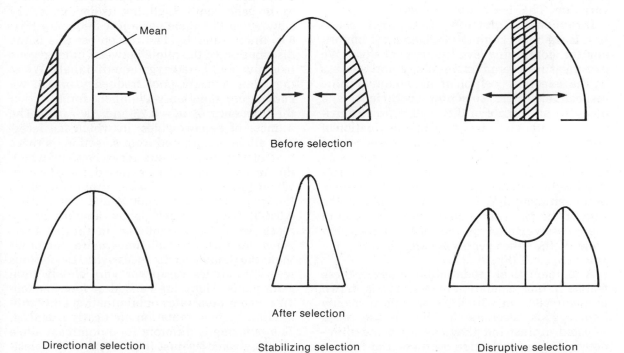

Figure 8.4 A diagram showing the effects of different types of natural selection on the distribution of variation within a deme. Abscissa represents variation; ordinate and number of individuals. Hatched areas represent variants selected against; arrows, the direction of selection.

or nonrandom survival of offspring to the critical reproductive age. The last is the most common and it frequently involves subtle interactions between organisms and the environment in which they live. The fate of organisms after their offspring have been reared is of little evolutionary consequence.

Selection favors certain individuals and reduces the reproductive effectiveness of others. In addition to operating on the level of individuals, selection may affect groups of related individuals as units. This has been called **group selection,** and the size of the group may range from a few kin to demes or even species.

Selection can have several different outcomes dependent upon the interaction between a deme and its environment (Fig. 8.4). (1) If an environment is changing in one direction, selection will shift the mean and range of variation of the population toward adaptation to the new conditions. This is **directional selection.** (2) In a stable environment, natural selection favors those individuals best adapted to existing conditions. Extreme variants in the population are eliminated, the range of variability is reduced, and there is a tighter clustering of variation around the mean. This is **stabilizing selection.** (3) Occasionally the environment favors two or more different morphological types within the same deme. Selection promotes a bimodal or polymodal distribution of variation. This is **disruptive selection.**

Directional Selection. Directional selection is the most common and the most important for continuing evolutionary change. We can measure reproductive failure and success by the selection coefficient and adaptive value respectively. The **selection coefficient** is a measure on a scale of 0 to 1 (0 = no elimination, 1 = total elimination) of the elimination of one genotype relative to the more successful ones. **Adaptive value** measures relative success and is the complement (1–S) of the selection coefficient. Each genotype has both a selection coefficient and an adaptive value, and the sum of the two must equal 1. If 40 per cent of the recessive genotype (**aa**) is eliminated, but none of the other genotypes are, then S = 0.4 and 1–S = 0.6 (Fig. 8.5).

A mathematical model of the effect of these factors in one generation of selection is shown in Figure 8.5, in which we assume that only homozygous recessive individuals are being selected against but they are not being eliminated (**aa** has a selection coefficient of 0.4 and an adaptive value of 0.6). If the initial allele frequency of **A** $(p) = 0.700$ and **a** $(q) = 0.300$, we can compute the initial genotype frequencies

by applying the Hardy-Weinberg equation (p. 171) as shown on the top line. After one generation of selection on the assumptions made, the frequency of **A** (p) increased slightly to 0.726 and that of **a** (q) dropped to 0.274, and this is reflected in a change in the frequencies of the genotypes.

Any selection coefficient and adaptive value can be assigned to the genotypes (provided the two add up to 1), and the allele frequency in the next generation can be computed by using the scheme shown in Figure 8.5. Algebraic formulae also can be determined for different situations. In the case discussed (homozygous recessives are selected against but are not lethal) the value for q in generation $n + 1 =$

$$\frac{q_n (1 - Sq_n)}{1 - Sq_n^2} = \frac{0.3 \times 0.88}{0.964} = 0.274$$

The change in q (Δq) between generations n and $n + 1 =$

$$q_{n+1} - q_n = 0.274 - 0.300 = -0.026$$

The frequency of allele **a** decreased 2.6 per cent in one generation of selection.

By analyzing the rate of change over several generations, the outcome of directional selection over many generations can be predicted. The frequency of a recessive allele being selected against drops rapidly when its frequency in the gene pool is high, but its decline is very slow when the allele has a low frequency (Fig. 8.6, lines a and b). The reason for this is the distribution of the allele between homozygous recessive and heterozygous individuals. When the allele **a** has a pool frequency of 0.5, we know from the Hardy-Weinberg formula that the frequency of aa = 0.25 and Aa = 0.50. The number of heterozygotes, in which the recessive allele is sheltered from selection, is twice that of the homozygous recessives, in which the recessive allele is exposed to selection. When the frequency of **a** has dropped to 0.01, the frequency of **aa** = 0.0001 and that of **Aa** = 0.0198. Now a much larger amount of the recessive allele is sheltered in the heterozygotes, for they are 198 times more numerous than the homozygous recessives. Deleterious recessive alleles disappear very slowly from gene pools. Many are carried at low frequencies where their rates of elimination and introduction by new mutation are nearly balanced. The outcome is different for a dominant allele being selected against (Fig. 8.6, line c). Selecting against **AA** or **Aa** individuals is the same as selecting for **aa** individuals. When the frequency of **A** is very high, there are few **aa** indi-

	AA	Aa	aa	Total
Genotype frequencies in generation n A(p) = 0.700; a(q) = 0.300	$p^2 = 0.490$	$2pq = 0.420$	$q^2 = 0.090$	$(p+q)^2 = 1$
Selection coefficient (S)	0	0	0.4	—
Adaptive value (1 − S)	1	1	0.6	—
Surviving to reproduce (Initial freq. × Ad. Val.)	$p^2(1-S) = 0.490$	$2pq(1-S) = 0.420$	$q^2(1-S) = 0.054$	$1 - Sq^2 = 1 - 0.036 = 0.964$
Frequencies of reproducing genotypes	$\dfrac{p^2(1-S)}{1-Sq^2} = \dfrac{.49}{.964} = .508$	$\dfrac{2pq(1-S)}{1-Sq^2} = \dfrac{.42}{.964} = .436$	$\dfrac{q^2(1-S)}{1-Sq^2} = \dfrac{.054}{.964} = .056$	$= 1$
Value of p in reproducing individuals $= p^2 + pq = 0.508 + 0.218 = 0.726$ Value of q in reproducing individuals $= q^2 + pq = 0.056 + 0.218 = 0.274$ $\Big\} = 1$				
Genotype frequencies in generation n + 1	$p^2 = 0.527$	$2pq = 0.398$	$q^2 = 0.075$	$= 1$

Figure 8.5 A mathematical model in algebraic and numerical terms showing the effect of one generation of selection against some homozygous recessive individuals. Starting genotype frequencies (generation n) are shown on the first line, assuming the initial allele frequencies are A (p) = 0.700 and a (q) = 0.300. Genotypes that survive to reproduce (fourth line) are computed by multiplying their initial frequency by their adaptive value. Since some individuals have been eliminated, the population is only 0.964 of its original size. The surviving genotypes are expressed as a frequency of the surviving population in the fifth line. Since the original Hardy-Weinberg equilibrium has been upset and a new one not yet established, the frequency of the alleles in the survivors must be computed as shown on the sixth line rather than by $\sqrt{q^2}$. Given the new allele frequencies, new equilibrium genotype frequencies can be computed for generation n + 1 (last line).

Figure 8.6 The outcome of directional selection: a = selection against a recessive lethal allele; b = selection against a recessive allele with a selection coefficient of 0.5; c = selection against a dominant allele with a selection coefficient of 0.5. (From Levine, L.: Biology of the Gene. St. Louis, C. V. Mosby Company, 1973.)

A

Figure 8.7 A, Drawing from a photograph of a melanic and peppered form of the moth, *Biston betularia*, resting on a lichen-covered tree trunk in a rural area. B, Similar individuals on a soot-covered tree trunk in an industrial area. (From Ehrlich, P. R., R. W. Holm, and D. R. Parnell: The Process of Evolution. New York, McGraw-Hill Book Company, 1974, after H. B. D. Kettlewell.)

B

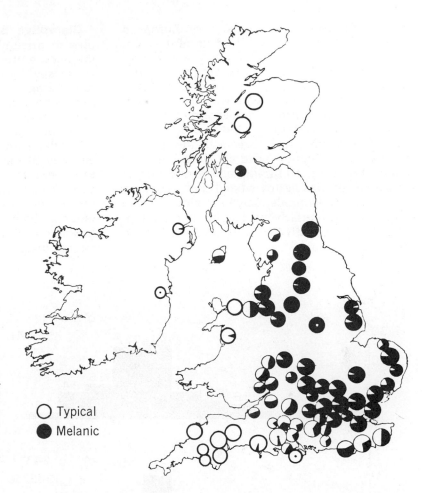

Figure 8.8 The distribution of peppered (typical) and melanic forms of *Biston betularia* in Great Britain and Ireland. Frequencies of the melanic form are very high in the London area and in the industrial Midlands. (From Ehrlich, P. R., R. W. Holm and D. R. Parnell: The Process of Evolution. New York, McGraw-Hill Book Co., 1974.)

○ Typical
● Melanic

viduals to be favored. The frequency of **A** decreases slowly at first, then more rapidly as the frequency of **a**, and hence **aa**, individuals increases.

Directional selection has been studied in many laboratory and field situations. Most of the peppered moths (*Biston betularia*) in rural parts of England have a black and white peppered wing color, and only a few are all black or melanic (Figs. 8.7 and 8.8). The situation is reversed in industrial regions. Why? Moths rest by day on tree trunks where some are eaten by birds. Prof. H. B. D. Kettlewell of Oxford University postulated that the peppered pattern is less conspicuous on the lichen-covered trees of rural areas, whereas the melanic form is less conspicuous on the soot-covered trees of industrial regions. To test this hypothesis, hundreds of male moths of each type were raised, marked with a spot of paint under their wings, and released in both rural and industrial areas; the survivors were recaptured after a certain period of time by attracting them with light or virgin females. The percentage of each type recaptured in one set of experi-

ments is shown in Table 8.1. Significantly more melanic forms survived in the industrial areas and peppered forms in the rural areas. Natural selection shapes the gene pool of each deme to the local conditions. In some localities, selection operates in one direction; in other localities, in the opposite direction.

Stabilizing Selection. An instance of stabilizing selection has been described by G. H. W.

TABLE 8.1 PER CENT RECAPTURE OF PEPPERED AND MELANIC FORMS OF *BISTON BETULARIA* AFTER EXPOSURE TO SELECTION IN INDUSTRIAL AND RURAL AREAS*

Environment	Per cent Recapture	
	Peppered Form	Melanic Form
Industrial area	16	34
Rural area	12	6

*(From an experiment performed by H. B. D. Kettlewell.)

Sein, who examined the size of European moles (as reflected by skull length of museum specimens) before and after a very severe winter of 1946 to 1947, during which the ground froze very deeply and stayed frozen for a longer time than normal. Frequency distributions of the skull size of moles captured before and after that winter (Fig. 8.9) showed that the modal class (the size class with the greatest number of individuals) did not shift, but, as is characteristic of stabilizing selection, the range of variability was narrowed and clustered more tightly about the mode. Large individuals, who would be particularly susceptible to food deprivation, were affected more severely than small individuals.

Disruptive Selection. The clearest examples of disruptive selection involve **Batesian mimicry,** a type of mimicry in which an edible insect species (the **mimic**) has a color pattern resembling that of a distasteful **model** species. Many experiments have demonstrated that bird predators learn to recognize, and tend to avoid, a conspicuously colored insect that is distasteful or dangerous. Selection favors conspicuous patterns in distasteful species because predators learn to avoid them quickly. An edible species that mimics a distasteful model shares in the immunity of the model, provided the models significantly outnumber the mimics. In some localities in Africa there are three different distasteful species of butterfly (Fig.

Figure 8.9 Frequency distributions of skull length of European moles before and after the severe winter of 1946 to 1947. Stabilizing selection has occurred. (From Mayr, E.: Animal Species and Evolution. Cambridge, Harvard University Press, 1963, after G. H. W. Sein.)

Figure 8.10 Disruptive selection and Batesian mimicry in the African swallowtail butterfly, *Papilio dardanus*. Top figure, male of *P. dardanus*; right-hand column, three distasteful models from one locality (*Danais chrysippus, Amauris niavius, A. echeria*); left-hand column, female mimics of *P. dardanus* from the same locality. (From Ehrlich, P. R., R. W. Holm, and D. R. Parnell: The Process of Evolution. New York, McGraw-Hill Book Co. 1974. after R. C. Punnett.)

8.10). Certain females of the edible swallowtail butterfly (*Papilio dardanus*) mimic one model; other females mimic other models. Selection has favored varieties of the swallowtail that resemble one or the other model species and has operated against intermediates. The single population has been disrupted into three different types. The mimicry of several model species is an advantage because it reduces the chance for mimics to outnumber models. If edible mimics were to become very numerous, the conditioning and occasional positive reinforcement of the predators would decrease and they would recognize that much of the time individuals with the conspicuous color pattern are acceptable.

Male swallowtails do not mimic any species, probably because they must have a constant color pattern to be accepted as mates by females. The loss of some males is of less bio-logical consequence than the loss of egg-laying females.

Negative and Positive Selection.   Natural selection has two facets, for it both eliminates unfit individuals and favors fit ones. Critics have emphasized its negative aspects and claim that without selection we would have today all of the existing varieties together with those that have been eliminated. In doing so critics overlook the creative aspects of evolution that are derived from a selective elimination and survival. Selection is the only force known that can bring genetic variation into harmony with the environment and lead to adaptation. By eliminating less favorable genes and alleles, selection changes the composition of the gene pool in a favorable direction and increases the probability that the favorable genes and alleles responsible for an adaptation will come together in the same individuals. Sir Garvin de

Beer, the noted British biologist, pointed out on the occasion of the centenary of *The Origin of Species* that "the effects of the natural selection are the reverse of chance when considered *ex post facto*; . . . they channel random variation into adaptive directions and thereby simulate the appearance of purposive change. . . . Natural selection has been paradoxically defined as a mechanism for generating an exceedingly high degree of improbability."*

Genetic Drift. The variability in a gene pool also can be shaped, but in a random way, by sampling errors if the population is small. The Hardy-Weinberg equilibrium is a statistical law and therefore is valid only when the number of individuals is large. Assume you have a gene pool in which the alleles **A** and **a** each have a frequency of 0.5. This can be simulated by using a bowl of 1000 beads, one half red and one half blue. In a random sample of 100 beads, it is unlikely that you will have exactly 50 of one and 50 of the other, but if the sample is as large as 100 it is also unlikely that the deviation from 50–50 will be great. In a small sample the probability of a significant departure from the true values is increased. If the sample size is 2, the probability that both will be red is $0.5 \times 0.5 = 0.25$. The effect is the same when a population is very small; there is a reasonable probability that its variability will deviate significantly from that in the population from which it was derived. If the population remains small for many generations, sampling errors accumulate and the population's variability drifts in a random way. One allele may be eliminated from the population and another may become fixed (frequency = 1). Sewall Wright termed this phenomenon **genetic drift.**

Drift can be demonstrated in mathematical and computer models and also can be seen in some small, isolated populations. In a small Pennsylvania Dunkard community studied by Glass, the frequency of many alleles departed significantly from those in the surrounding population, from which the Dunkards isolated themselves, as well as from the European community from which the Dunkards were descended. The allele **a** for the blood group O has a frequency of 0.59 in the Dunkards, 0.70 in the surrounding population and 0.64 in the ancestral German community. Glass attributed the departure to genetic drift.

There is some dispute as to how often genetic drift is a major factor in evolutionary change, but it is important when new demes are established. Usually only a few individuals cross a species border and establish a new population on the other side of a barrier. They carry with them only a small and random sample of the gene pool of the deme from which they came. This is known as the **founder effect.** We believe that it is significant for the formation of new species around the border of a parental species, as occurs, for example, on oceanic islands.

8.8 MAINTENANCE OF VARIABILITY

New variation is introduced continually into populations by mutation, but gene pools are not overloaded with variation because it also is reduced and shaped by natural selection and genetic drift. Variation is not eliminated by these forces, however, because other forces tend to maintain a reservoir of variability.

Heterozygotic Superiority. Individuals heterozygous for certain alleles may be better equipped for survival than either homozygote. The reasons for this phenomenon, known as **heterosis,** are not always clear. Sometimes each allele promotes the synthesis of a protein, and the slight differences between the proteins may enable them to complement each other and work as a biochemical team. In some cases an unfavorable effect of each allele in the homozygote is not expressed in the presence of another allele in the heterozygote (see Section 6.17). A special case is seen in the case of balanced lethals. Of the many alleles at the **T locus** in mice (so named because it has an obvious effect on the tail) the **T** and **t** mutants are lethal when homozygous (**TT** or **tt**), but a heterozygous mouse (**Tt**) is viable, although without a tail. It requires the presence of one wild allele at this locus (symbolized +) for a viable tailed mouse. The other allele may be **T, t** or one of the other alleles in this series, and the mouse would be +**T,** +**t,** or +−.

Since the reproduction of the favored heterozygotes produces homozygotes of each type as well as more heterozygotes, it is impossible for any allele to be eliminated from the gene pool. An equilibrium between the alleles, which is dependent on the relative fitness of the two homozygotes, eventually is established. Equilibrium is reached when the frequency of **A** (p) times the selection coefficient of individuals homozygous for this allele (S_1) equals the frequency of **a** (q) times the selection coefficient as **aa** individuals (S_2): $pS_1 = qS_2$. By substituting $1 - q$ for p, the formula becomes $(1-q)S_1 = qS_2$. By solving for q we find its equilibrium value ($\hat{q}$):

$$\hat{q} = \frac{S_1}{S_1 + S_2}.$$

*Sir Garvin de Beer, *The Darwin-Wallace Centenary,* Endeavour, April, 1958.

All genotypes will be retained in the population in a certain balance. The less favorable homozygotes constitute the **genetic load** of the population.

The term **balanced polymorphism** describes populations in which various types of individuals are maintained in the same frequency in successive generations. Heterosis is a very common cause for balanced polymorphism. It maintains variability and gives the population a great deal of evolutionary plasticity. If conditions become different, so that the selective coefficients of the homozygotes change, a new equilibrium can be established quickly. Dobzhansky and his students have studied chromosomal polymorphism in a Western fruit fly, *Drosophila pseudoobscura*. The many gene arrangements on the third chromosome have been named after localities where they were discovered: Standard, Arrowhead, Chiricahua, Santa Cruz, Tree Line. They can be treated as though they were different alleles. An individual diploid fly may be homozygous for any one of these chromosomes (Standard/Standard) or heterozygous (Standard/Chiricahua, Tree Line/Santa Cruz, etc.) A particular population contains many, or sometimes all, of the gene arrangements at frequencies that best fit the population to the local conditions. As conditions change over the geographic range of the species, the frequencies in the local populations change (Fig. 8.11).

Ecological Factors. A diversity of ecological factors over a species range also can promote variability, as we have seen in the case of the British peppered moth.

Figure 8.11 Chromosomal polymorphism in the third chromosome pair of *Drosophila pseudoobscura*. Frequencies for five gene arrangements (inversions) are shown for nine populations that lie along the Cascade and Sierra Nevada ranges. (From Dobzhansky, T., and C. Epling: Carnegie Institute of Washington, publication number 554, 1944.)

8.9 INTEGRITY OF THE GENE POOL

The genes and alleles in a gene pool have come together in different combinations in individuals through generations of sexual reproduction and natural selection. Their interactions to produce well adapted individuals and a well adapted population have been tested many times. Any genes or alleles that interact very unfavorably with others have been eliminated or their effects modified. The remaining genes and alleles interact in harmony and are said to be **coadapted.**

The existence of coadaptive interactions can be demonstrated when something upsets them. In selecting strongly for high or low bristle number in *Drosophila* (Fig. 8.3), Mather and Harrison found that the line with low bristle numbers died after 35 generations. Sterility began to appear in the line with high bristle numbers after 20 generations. Selection was stopped for several generations, and bristle number decreased. When selection was resumed, bristle number increased to the low 50's and fluctuated around this level despite continued selection. Subcultures taken from both high and low lines before they reach maximal or minimal numbers of bristles maintained their new values without further selection. In interpreting these data, Mather and Harrison pointed out that they were not working with one set of alleles for high or low bristle number, but rather they were selecting for interacting gene complexes. If selection moved a complex too far too fast, disharmonious interactions leading to death or sterility began to appear. If selection did not go too far (as in the subcultures) or was stopped for a while (the high line), new and viable interactions could be established. Plant and animal breeders have discovered similar phenomena. When one character (high yield, for example) is selected for strongly, undesirable "correlated responses," such as a reduction in disease resistance, frequently occur.

Genes and alleles in a pool do interact with one another in complex ways that we do not understand fully, and their interactions give the pool of the population or species a certain cohesiveness and integrity. During the formation of a new species a population separates from the parental pool, loses its cohesiveness with it and establishes a new and different pool of coadapted genes.

8.10   SPECIATION

Central to the question of how species are formed is an understanding of the isolating mechanisms that prevent closely related species living in the same geographic area from mating with each other. We call these **sympatric species** to distinguish them from **allopatric species,** which are closely related species living in widely separated geographical areas. Sympatric species are very common. Bull frogs, green frogs, wood frogs, leopard frogs and pickerel frogs, all belonging to the genus *Rana,* may be found in or near the same pond. They and other sympatric species are prevented from mating by many reproductive isolating mechanisms.

Isolating Mechanisms.   Isolating mechan-

isms that prevent potential mates of different species from mating (termed "premating") are of several types. (1) Different sympatric species commonly live or breed in different habitats. Wood frogs breed in temporary woodland ponds; bull frogs, in larger, permanent bodies of water. They are separated by **habitat isolation.** (2) Sympatric species frequently breed at different times of the year or times of day, so are kept apart by **temporal isolation.** The fruit flies *Drosophila pseudoobscura* and *D. persimilis* are sympatric over much of their range, but *pseudoobscura* is more active in the afternoon and *persimilis* in the morning. (3) Most species of animals have a courtship behavior, so mating between species normally is prevented by **ethological isolation. Courtship** is an exchange of signals between a male and a female. A male approaches a female and gives a sign of some sort, which may be visual, auditory or chemical. If the female belongs to the same species, she recognizes the signal and returns one. The encouraged male gives another signal and further exchanges of signals eventually result in both partners being stimulated sufficiently to copulate (Fig. 8.12). If courtship starts between members of two different species, one partner may not recognize one of the signals, fail to respond, and courtship will stop.

When premating isolating mechanisms fail, as they occasionally do, **postmating isolating mechanisms** may prevent effective crossing. (1) The sperm of one species may die in the reproductive tract of the other or be unable to fertilize the egg. Isolation is achieved by **gamete mortality** or **incompatability.** (2) Eggs may be fertilized by foreign sperm but the hybrids may die during development because of disharmonic genic interactions. Isolation is achieved by **hybrid mortality.** Nearly all of the

Figure 8.12 Courtship differences between two species of fruit fly: *A, Drosophila melanogaster, B, Drosophila simulans.* Type of activity is shown on the vertical axis; duration in seconds, on the horizontal axis. (From Mayr, E.: Populations, Species and Evolution. Cambridge, Harvard University Press, 1970, after A. Manning, 1959.)

hybrids die in the embryonic stage when eggs of a bull frog are fertilized artificially with sperm of a leopard frog. (3) Hybrids of an interspecific cross may develop to maturity but may be adaptively inferior to either parental species because their mixture of characters is not well suited to either parental habitat. Isolation is achieved by **hybrid inferiority.** (4) Occasionally, interspecific hybrids not only survive but also are unusually vigorous and exhibit heterosis. The mule, a hybrid of a male ass and a female horse, is a familiar example. The hybrids are typically **sterile** because of synaptic failure during meiosis (p. 88) and are prevented from either perpetuating themselves or back crossing with either parental species.

There are many isolating mechanisms. Seldom is any one infallible, but most sympatric species are isolated very effectively in their native habitats by the effect of some combination of mechanisms. If the environment of sympatric species is altered by bringing them into the laboratory or a zoo, interspecific crosses take place more easily. Allopatric species normally are separated geographically, but their reproductive isolating mechanisms can be tested by bringing them together. Frequently the postmating mechanisms are more effective than the premating ones.

Geographic Speciation. A new species arises, of course, from one or more preexisting species. We can imagine several ways by which this can occur. One of the most important is **geographic,** or **allopatric, speciation.** This is such a slow process that none of us can see it from beginning to end in our lifetime, but we can find examples of the steps we believe take place. An essential element in geographic speciation is **geographic variation.** Most species do vary geographically in their structure, physiology, ecology and behavior because selection works in different directions in different parts of the geographic range. The northern populations of a warm-blooded species tend to have a larger body size in the Northern Hemisphere than southern populations because of a more favorable surface-volume relationship. When a sphere increases in size, mass increases as the cube of the linear dimension, whereas surface area increases as the square of the linear dimension. Animals are not spheres, but the same principle applies. A large animal has more heat-producing mass relative to its heat-losing surface than a small one. This particular phenomenon is widespread and is known as **Bergmann's rule.** A different sort of geographic variation is seen in the white-crowned sparrow and some other birds. North-

ern populations are migratory, whereas southern ones are not. Such characteristics clearly are adaptive, but geographic variation in a feature such as the number of scales along the belly of a species of snake does not appear to be adaptive in itself. Many such characters may be pleiotropic manifestations of a gene complex that is adaptive in other ways. Others may be the result of genetic drift or other random changes. Although geographically different parts of a species tend to diverge, gene flow by migration between adjacent demes prevents the species from breaking into two reproductively isolated parts (Fig. 8.13A). The species as a whole has a unity and cohesiveness maintained by gene flow and the sharing of many interacting gene complexes.

Frequently, geographic variation is gradual over the range of the species, but the change may be particularly rapid in a relatively narrow geographic band, which usually correlates

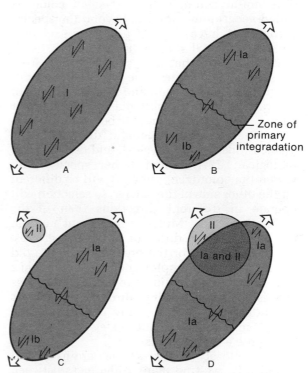

Figure 8.13 Steps in geographic speciation. The ovals and circles represent ranges of species I and II; large arrows, the direction of natural selection; half arrows, gene flow. *A,* A wide ranging species showing geographic variation resulting from different selection pressures. *B,* The species has differentiated into subspecies. *C,* A geographic isolate has separated from one subspecies and differentiated into a different, allopatric species. *D,* The range of the new species expands so that this species becomes partly sympatric with the ancestral species.

with an abrupt environmental change or with a semibarrier that restricts but does not stop gene flow between parts of a species. Regions of this type are used to subdivide a species into **subspecies** (Fig. 8.13B). At subspecies borders we find a zone of **primary integration** where hybridization between the two subspecies takes place.

Species spread out geographically until they reach a region where environmental conditions exceed the range of tolerance of members of the species. Such a region constitutes a **barrier.** Depending on the species, this might be a mountain range, a desert, a forest, a large body of water, and so on. Barriers are seldom absolute, and a few members of a species somehow cross them and establish a colony on the other side. A few individuals may stray high into a mountain pass and reach a valley on the other side of the range. Oceanic islands have been colonized by individuals blown out from the mainland in a storm or carried out on mats of vegetation drifting in ocean currents. When the borders of a species are closely examined, it is not uncommon to find small colonies, called **geographic isolates,** beyond the species border that have little if any gene exchange with the parental species (Fig. 8.13C).

Many believe that new species evolve from the geographic isolates. The few colonizers will carry with them a small and random sample of the variability of the parental species (founder effect). The population will be small for several generations and genetic drift will take place. Probably the physical, and certainly the biological, environment (other species with whom the colonizers interact) will be different on the other side of the barrier, so selection will operate in different directions than in the parental species. Mutations occur at random and these will be different in the two groups. Conditions exist for the breakdown of the old coadaptive gene complex and the evolution of a new interacting genetic system. Mayr describes the isolates as passing through a **"genetic revolution."** Doubtless many isolates become extinct in the process, but the few that succeed constitute a new species.

When closely related species are examined genetically, we find many differences between them in the genes they carry, in allele frequencies, and often in chromosome arrangements (inversions, translocations). Richard Goldschmidt argued in the 1940's that the chromosome differences were the essence of speciation and that speciation could not take place unless macromutations of this type occurred. He felt that macromutations created a significant departure, a "hopeful monster," that would at once be reproductively isolated from the parental species. Biologists today recognize that chromosome rearrangements often are important in species differentiation but do not ascribe more importance to them than to other genetic changes. Frequently chromosome changes do not produce the large morphological departures visualized by Goldschmidt.

The development of reproductive isolation is believed to be a by-product of the genetic revolution that the geographic isolates pass through. The newly formed species is at first allopatric, living in a different geographic area than the parental species. It may remain so, and there are many examples of closely related species that replace one another geographically. But sometimes the new species invades the range of the parental species and becomes sympatric with it (Fig. 8.13D). If the new species mates with the parental species at all, the hybrids probably will be inferior in some way because of unfavorable interactions between different gene complexes. If this is the case, natural selection will intensify any slight reproductive differences between them. Those individuals whose behavior, ecology or other attributes favor their mating with their own kind (**homogamy**) will have a greater reproductive success than those less discriminating individuals who waste their reproductive effort and gametes in producing inferior hybrids.

When two closely related species meet and compete, natural selection also intensifies incipient ecological and morphological differences between them. It is unlikely, for example, that two different bird species will be equally effective in feeding on the same range of seed size. If the bill size of one is slightly larger, this species will be more effective in feeding upon the larger seeds. Individuals of the species with the larger bills will be favored by selection because they are more likely to avoid competing directly with birds feeding on the smaller seeds. Over many generations, one species tends to exclude the other from feeding on the particular seed size to which it is becoming specialized. We call this **competitive exclusion.** The accompanying intensification of morphological differences, in this case bill size, is called **character displacement.**

An excellent example of the results of geographic speciation and subsequent overlap of ranges is seen in a group of birds known as **Darwin's finches** that inhibit the Galapagos Islands, located on the equator 600 miles west of Ecuador (Fig. 8.14). There are now 14

Figure 8.14 Map of the Galapagos Islands. Numbers indicate the number of species of Darwin's finches that occur on the different islands. Some species are also found on many different islands. (From Bowman, R. I.: University of California Publications in Zoology, vol. 58, 1961.)

species of finches in the archipelago, differing primarily in feeding habits and in bill size and shape. Several species occur on the same islands, and they are reproductively isolated by courtship differences in which bill display is an important signal. The Galapagos are volcanic islands that have never been connected to the mainland. They were colonized originally by chance by one or a few ancestral species of finch. Those on each island were geographically isolated and diverged as they adapted to the food and other resources of their island. Occasionally, different species, or incipient species, came together on the same island. Competitive exclusion and character displacement intensified differences between them, but the exact form that character displacement took on different islands varied according to the particular species that were in competition with each other. Four species of ground finch of the genus *Geospiza* occur on the large central islands (Fig. 8.15). Three differ in bill size and in the size of seeds they usually eat. Of the two small finches with similar bill sizes, one lives in the

arid lowlands and the other lives in the humid uplands. Not all of these species occur in the smaller, arid, outlying islands. The species present have divided up the available niches between them, and their bills are modified accordingly (Fig. 8.15).

Competitive exclusion and character displacement that accompany geographic speciation promote the divergence of an ancestral species into many different ones specialized for different foods and modes of life. We call this **adaptive radiation.** The species of *Geospiza* are an example of adaptive radiation on a small scale. On a larger scale, placental mammals have radiated from primitive, semiarboreal, insect-eating types (Fig. 8.16). Most bats are flying insect-eaters, wolves are terrestrial carnivores, seals are aquatic carnivores, horses are grazing herbivores, rats are gnawing herbivores, and so on.

Sympatric Speciation. Sympatric species can arise by geographic speciation, and many biologists believe that all sympatric species evolved in this way. Others raise the question

Ecological niche	Central Islands	Outlying Islands		
		Tower	Culpepper	Hood
Large Ground Finch	G. magnirostris	G. magnirostris	G. conirostris darwini	
Cactus Ground Finch	G. scandens	G. conirostris propinqua		G. conirostris conirostris
Small Ground Finch (in arid low lands)	G. fuliginosa	G. difficilis difficilis	G. difficilis spetentrionalis	G. fuliginosa
Small Ground Finch (in humid uplands)	G. difficilis debilirostris	No humid uplands present		

Figure 8.15 Competitive exclusion and character displacement of bill size in ground finches of the genus *Geospiza* on representative Galapagos Islands. (Adapted from Lack, D.: Darwin's Finches. Cambridge University Press, 1947.)

as to whether or not a group of populations living in one geographic region but in different ecological situations can be pulled apart by strong, divergent selection. Some populations of one species may be adapting to forest conditions, others to the conditions in adjacent meadows. Since there is gene flow between them, can divergence proceed so far that hybrids between forest and meadow species are inferior? If so, reproductive isolation would result and we would have two species. The formation of two species within one geographic region by divergent selection is called **sympatric speciation.** The question is unresolved. We sometimes find swarms of closely related sympatric species, each adapted to a different food or habitat, living together in an area where it is difficult to imagine the amount of past geographic isolation needed to explain their evolution by geographic speciation. The numerous species of cichlid fishes in Lake Victoria is a case in point. A compelling argument against sympatric speciation is that we cannot

point to examples of the postulated intermediate stages. On the other hand we do find illustrations of all stages in geographic speciation leading to the formation of sympatric species. Perhaps if we knew more about the geologic history of Lake Victoria and changes in its water level, we would find the basis for the physical separation needed for geographic speciation.

Speciation by Polyploidy. Geographic and sympatric speciation, if the latter occurs, are gradual processes requiring hundreds or thousands of years. One very rapid mode of speciation is by an abrupt change in the number of chromosome sets going, for example, from the diploid (2N) number to the tetraploid (4N) condition. The most common way this occurs, known as **allopolyploidy,** involves the hybridization of two different species and the continuation of the hybrids as a third species. Let **AA** symbolize the entire chromosome complement of one diploid species; **BB**, the chromosomes of another diploid species (Fig. 8.17). A

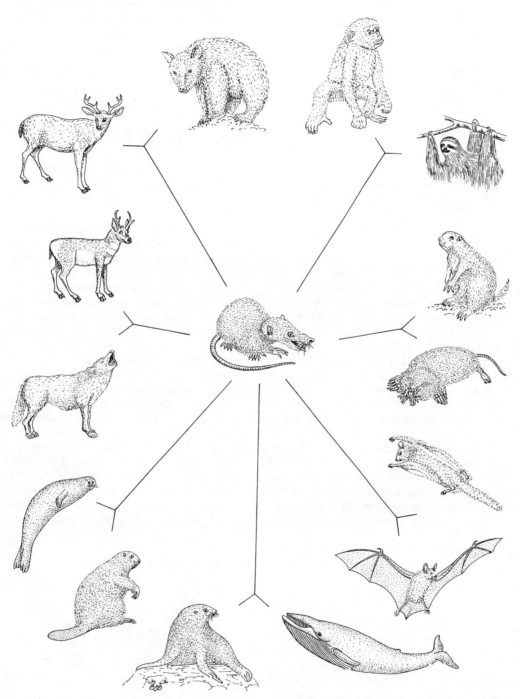

Figure 8.16 Adaptive radiation in placental mammals. The various species shown have evolved from the common insectivorous ancestor depicted in the center. Each of the descendants has become adapted to a different mode of life.

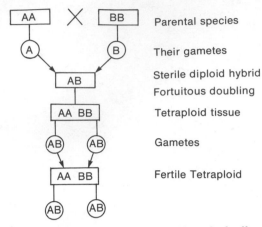

Figure 8.17 Diagram of one method by which allopolyploidy can occur. AA and BB symbolize the diploid chromosome complement of two different species.

hybrid between the two species has the chromosome composition **AB**. It is sterile because the **A** and **B** chromosomes do not synapse normally in the meiotic divisions preceding gamete formation. However, in an occasional mitotic division, the replicated chromosomes fail to separate and a tetraploid cell (**AABB**) is formed. This cell gives rise by mitosis to a clone of tetraploid cells. When certain of these cells begin to differentiate into reproductive cells, normal meiosis occurs because one set of **A** chromosomes can synapse with the other set of **A** chromosomes, and likewise **B** with **B**. Diploid gametes (**AB**) are produced, and their union forms a tetraploid zygote that develops into a tetraploid individual. Tetraploid plants can cross normally and perpetuate themselves, but they cannot cross easily with either diploid parental species.

Botanists estimate that approximately one-half of the species of flowering plants have evolved through allopolyploidy. As one example, there are species of violet with every multiple of 6 chromosomes (=n) from 12 (=2n) to 54 (=9n). Allopolyploidy is less common in animals, although an analysis of chromosome numbers in related species sometimes suggests allopolyploidy or the related phenomenon of self-polyploidy (**autopolyploidy: AA → AAAA**). Plant species are isolated from each other by habitat to a greater extent than animal species. Floods, landslides, forest fires and other natural disasters disturb the habitat. Different plant species recolonizing the area meet and hybridize, for they are not separated by behavioral differences. Because of the way they grow, plants also have more opportunities than ani-

mals for the fortuitous doubling of chromosome number in tissue that will give rise to spores and gametes. Most plants produce both male and female reproductive organs, and the sex of a tissue is determined by genic interactions. Sex in higher animals is based on chromosome balances (see Section 6.11). Polyploidy can upset these balances easily and lead to the formation of sterile intersexes and supersexes.

8.11 TRANSPECIFIC EVOLUTION

Evolution is a continuing process that over millions of years has resulted in the formation not only of new species, but also of new genera, orders, classes and even new phyla of organisms. Long-term processes leading to the formation of the higher taxonomic categories are referred to as transpecific evolution. Fundamentally, transpecific evolution is a continuation over millions of years of the processes of speciation. Geographic speciation can lead to the adaptive radiation of an ancestral group. The resulting multiplication of species adapted to different modes of life is called **divergent evolution,** or **cladogenesis** (Fig. 8.18). In the adaptive radiation of different groups, unrelated species adapt to similar environmental conditions and come to resemble each other, at least superficially. This is **convergent evolution.** Adaptation to an aerial life requires wings. Wings have evolved independently in insects, birds and bats (Fig. 8.19) and in an extinct group of flying reptiles (the pterosaurs, see Fig. 34.22). As another example, adaptation to active swimming requires a streamlined body shape and fins. Fish, dolphins and an ex-

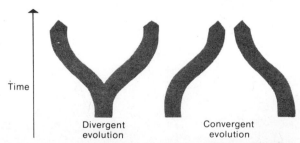

Figure 8.18 Diagram illustrating the difference between divergent and convergent evolution. A single stock may branch to give two diverging stocks that become more and more different as evolution proceeds. In convergent evolution two stocks originally quite different come to resemble each other more and more as time passes because they occupy a comparable habitat and become adapted to similar conditions. (From Villee, C. A.: Biology. 6th ed. Philadelphia, W. B. Saunders Co., 1972.)

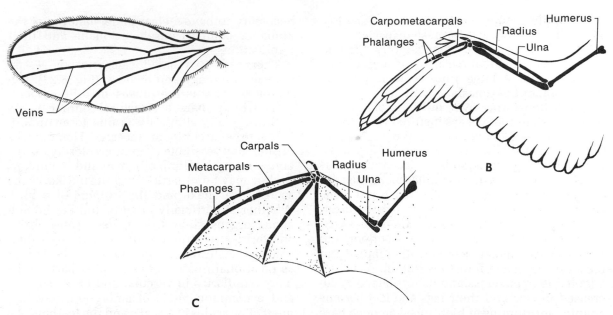

Figure 8.19 Convergent evolution in the wings of (A) a fly, (B) bird, and (C) bat. All are airfoils and superficially alike, but underlying differences reflect an independent evolution.

Figure 8.20 Stages in the evolution of the horse, illustrating (top) the changes in size and shape, (second row) the bones of the fore and hind feet, (third row) the skull, and (bottom) the grinding surfaces of the second upper molar tooth. *Eohippus* is a synonym of *Hyracotherium*. There were many side lines of horse evolution in addition to the line leading to the contemporary *Equus*.

tinct reptile group (the ichthyosaurs, see Fig. 34.15) have converged in these respects.

Phyletic Speciation. One slightly different type of speciation that we also encounter when we deal with a large time scale is **phyletic speciation,** or **anagenesis,** in which a species as a whole transforms over millions of years to such an extent that we are justified in calling it a new species. During much of the past 70 million years, the climate of North America became dryer and grasslands replaced many forests. Selection pressures gradually changed the adaptations of horses. The ancestral one, *Hyracotherium,* was a terrier-sized, forest-dwelling animal with three functional toes well suited to support it on the soft forest floor, and low-crowned molar teeth well adapted for browsing upon soft forest vegetation. As horses adapted to open, grassland conditions, they increased in size and their legs and feet became adapted for running at high speed on open hard ground (Fig. 8.20). The distal portion of their legs elongated, and the number of toes was reduced to one. Their molar teeth enlarged faster than other parts of the body, and cement spread from the roots over the surfaces of the teeth and into the valleys between ridges on the teeth surfaces. The teeth became high-crowned and well adapted for grinding hard grasses. Their premolar teeth came to resemble the molar teeth. These and other morphological changes were very gradual, but certainly they were caused by genetic changes. If we could bring generations separated by millions of years together, genetic incompatibility probably would preclude successful crossing even if mating occurred. The process, however, has been so gradual that if we could have observed it, we would have found it difficult to draw species lines. Pragmatically we use the incompleteness of the fossil record and draw our lines where we find significant morphological discontinuities.

Phylogeny. Determining the history of long-term evolutionary changes and the relationships between organisms is termed **phylogeny.** We use discontinuities of increasing magnitude first to group similar species into one genus, then similar genera into one family, and so on up the hierarchy of taxonomic categories to phyla and kingdoms (Chapter 21). Morphological similarities and differences in living species and fossils have been the primary tool of phylogeny, but biochemical, genetic and other criteria now are being used as well.

In order to determine the relationship between animals, and the changes in organ systems that have occurred during evolution, it is necessary to have a clear understanding of the kinds of comparisons being made and their implications.

There are two broad categories of resemblance: homology and analogy (Fig. 8.21). **Homologous organs** are organs in different animals which share a structural similarity. By definition this similarity is due to an inheritance from a common ancestor. Homologous organs may resemble each other closely, as the bones of the forelimb of a man and a monkey, or they may be superficially quite unlike, as the scales of a snake and the feathers of a bird. Whether superficially alike or unlike, homologous organs share certain deepseated, basic similarities in structure that derive from their common ancestry. This includes such features as positional relationships to other parts of the body, similarity in vascular and nerve supply, and a common mode of embryonic development. The scales of a snake and the feathers of a bird resemble each other closely in their early stages of embryonic development. It should also be clear from these examples that homologous organs may, or may not, have similar functions.

Analogous organs are organs in different animals with a functional similarity that is in no way related to a common ancestry; that is to say, analogous organs are nonhomologous organs with a similar function. Often analogous organs have little gross structural resemblance to each other for example, the gills of a crayfish and the lungs of a human being. Other analog-

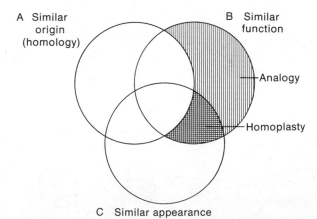

Figure 8.21 A Venn diagram of the relationship between different types of resemblances. *A,* A set of organs having the same evolutionary origin (homologous organs); *B,* a set with the same function; *C,* a set resembling each other superficially. Analogy is defined as that portion of set *B* (vertical hatches) that does not intersect with set *A;* homoplasty (horizontal hatches) is the intersection of analogous organs with ones in set *C.*

ous organs, such as the wings of insects and vertebrates, may have a superficial structural resemblance (Fig. 8.19). Their common function, flight, has only one structural solution, the evolution of some sort of an airfoil. Analogous organs with a structural resemblance are said to be **homoplastic.** Homoplastic resemblances usually result from convergent evolution.

Evolutionary relationships must be traced through a careful identification of homologous organs, for, by definition, they imply descent from a common ancestor. A potential source of error, however, is to confuse homoplastic organs, which have evolved independently, with homologous ones. Usually, homoplastic resemblances are superficial and not basic. In the wing example it is clear that the bones supporting the wings of the bird and the bat have a basic resemblance, i.e., are homologous, whereas the veins supporting the insect wing are an entirely different sort of structure. The implication is that the bird and the bat had a common ancestor in which the pattern for the bones in the pectoral appendage had already been established; the insect evolved a wing quite independently and from different materials. Another source of confusion is that a pair of organs may be homologous at one level of comparison and homoplastic at another. Although supported by homologous bones, the flying surface of the bird wing is composed of feathers, whereas the bat's wing is composed of a skin membrane. At this level of comparison the two wings are homoplastic. All of this suggests that birds and bats had a remote common ancestor with a certain arrangement of bones in the pectoral appendage, but that the wings of birds and bats evolved independently in the two groups; bats did not evolve from birds.

Adaptive Shifts. Although transpecific evolution appears to be explicable in terms of the various types of speciation we have been considering, there are aspects that need closer examination. One is the major adaptive shifts that have occurred many times in the course of evolution. Terrestrial vertebrates, as one example, at some point evolved from fish, and the adaptive shift from water to land would seem to be at least on an order of magnitude more complex than the adaptive shifts we see in geographic speciation. When closely analyzed, large adaptive shifts can be resolved into a series of smaller ones, yet it is generally agreed that several factors must coincide.

One factor is a pressure upon some populations of an ancestral species to shift to a somewhat different habitat or mode of life, or both.

Given a choice, most organisms tend to stay in the habitat and ecological niche to which they are best adapted, but an increase in population size or a constriction of the favorable environment occasionally forces some populations into environments that at first may be marginal to them.

As a second factor, the habitat or mode of life into which some populations shift must be relatively vacant; that is, not already exploited fully by some other species that certainly would be better adapted. New and at first relatively vacant habitats and modes of life arise from time to time by the extinction of former occupants and by the continual expansion of life and diversification of living organisms. The evolution of grasses created an opportunity for the evolution of grazing herbivores. The evolution of antelope and other large herbivorous mammals created the opportunity for the evolution of lions and other large carnivores.

Given the existence of a pressure to shift and the availability of an incompletely exploited habitat or mode of life, a major adaptive shift does not occur unless the shifting organisms are viable in the new situation. At first they will not be well adapted to their new conditions, but they must be able to survive. The new arrivals must have some degree of **preadaptation;** that is, while in their original habitat, they must have evolved certain features that enable them to exploit an available new situation. Preadaptive features are not evolved in anticipation of a shift. Usually they are adaptive in some way to the ancestral environment, yet at the same time open up new evolutionary possibilities. The fish ancestors of terrestrial vertebrates probably lived in bodies of fresh water subject to stagnation or seasonal drought and drying up. Lungs evolved in this context as accessory respiratory organs. At the same time lungs enabled these fish to survive on land. The land was available at this time, for there were no other vertebrates upon it and only a few invertebrate groups. A constriction of the favorable fresh water environment probably forced some fish to spend longer and longer periods on shore, feeding or escaping aquatic predators.

Directed Evolution, Extinction. Once an adaptive shift has been made, organisms improve their adaptation to the new conditions, and this may continue for millions of years. As paleontologists of the first half of this century examined the fossil record, they believed that they could detect an evolutionary momentum. An evolutionary trend once started appeared to continue unchanged, despite environmental fluctuations, for millions of years. The gradual

Figure 8.22 Lateral view of the skull of the Pleistocene saber-toothed tiger, *Smilodon*; length about 12 inches. (From Romer, A. S.: Vertebrate Paleontology. Chicago, University of Chicago Press, 1966, after Matthew.)

increase in horse size and the gradual reduction in the size of their lateral toes appeared to be examples. Some trends seemed to go beyond the optimal condition, and an organ reached an overdeveloped size that was detrimental. Saber-toothed tigers of the last Ice Age had huge canines (Fig. 8.22), and some paleontologists suggested that they interfered with jaw closing and eating and were responsible for the extinction of the group. The name **orthogenesis** (straight line development) has been given to trends of these types. The implication is that extraordinary forces beyond the ones we have been considering are necessary to explain them.

George Gaylord Simpson and others have questioned an orthogenetic reading of the fossil record and have pointed out that such a reading was an early conclusion based on a fossil record far less complete than we now have. Modern data indicate that the evolution of the horse was not a straight line but was more like a bush with many branching lines. We can assemble certain species and make a straight line, but in doing so we ignore many others, including some in which there was a decrease in size. The jaws of saber-toothed tigers did not become nonfunctional. The huge canines were an adaptation to kill and feed upon large herbivores, such as mammoths and mastodons (relatives of the elephant). These cats became extinct when their food supply became extinct.

Orthogenesis is sometimes simulated because the direction of environmental change and natural selection may be the same for millions of years. Also, as organisms begin to specialize for a particular mode of life, horses for grazing and running on hard ground, for example, they frequently evolve certain attributes that cannot be reversed easily. Lost toes cannot be regained. A commitment to a particular mode of life is made, and the more specialized the adaptation, the greater the commitment. Adaptation toward running and grazing continues until it reaches a high level of perfection, and it becomes increasingly difficult for a major change in evolutionary direction to occur.

Specialized organisms survive so long as the environment and mode of life to which they are adapted continue. Environmental changes frequently lead to the extinction of organisms highly adapted to a limited mode of life. Those less specialized or that have evolved features (such as lungs in fishes) that open up new evolutionary horizons are the ones that survive environmental crises and continue.

8.12 THE ORIGIN OF LIFE

The modern theories of mutation, natural selection and population dynamics provide us with a satisfactory explanation of how the present-day animals and plants evolved from previous forms by descent with modification. The question of the ultimate origin of life on this planet has been given serious consideration by many different biologists. Some have postulated that some kind of spores or germs may have been carried through space from another planet to this one. This is unsatisfactory, not only because it begs the question of the ultimate source of these spores, but also because it is extremely unlikely that any sort of living thing could survive the extreme cold and intense irradiation of interplanetary travel. Evidence for life in other parts of the cosmos has come from the discovery in 1961 of what appear to be fossils of microscopic organisms, somewhat like algae, in meteorites. This does not provide evidence, however, that living organisms could be transported through space in a meteorite.

The spontaneous origin of living things at the present time is believed to be extremely improbable. Francesco Redi's experiments showed, about 1680, that maggots do not arise de novo from decaying meat and laid to rest the old superstitition that animals could appear by spontaneous generation. Some 200 years later Louis Pasteur showed conclusively that microorganisms such as bacteria do not arise by spontaneous generation but come only from previously existing bacteria. Other investigators have shown since then that even the smallest organisms, filterable viruses, do not come from nonviral material by spontaneous generation. The multiplication of viruses requires the presence of previously existing viruses. Although the spontaneous generation of life at present is unlikely, it is most probable that billions of years ago, when chemical and physical conditions on the earth's surface were quite different from those at present, the first living things did arise from nonliving material.

This concept, that the first things did evolve from nonliving things, and suggestions as to what the sequence of events may have been were put forward by Pflüger, J. B. S. Haldane, R. Beutner, and particularly by the Russian biochemist A. I. Oparin, in his book, *The Origin of Life* (1938). According to the currently accepted Kant-Laplace hypothesis, our solar system originated about four and one-half billion years ago through the rotation and condensation of a vast cloud of interstellar dust and gases. The early atmosphere, which was formed by volcanic gases, was strongly reducing and contained large quantities of methane, ammonia, nitrogen and water vapor. Probably lesser amounts of hydrogen and carbon dioxide were present. Most carbon was in metallic carbides, and most oxygen was in various oxides in the earth's crust. Conditions existed for the formation of many organic compounds. Twenty-two different amino acids have been isolated from South African Precambrian rocks that are at least 3.1 billion years old, and traces of living organisms have been found in rocks about three billion years old.

A number of reactions are known by which organic substances can be synthesized from inorganic ones. Metallic carbides could react with water to form acetylene, which would subsequently polymerize to form compounds with long chains of carbon atoms. High-energy radiation such as cosmic rays can catalyze the synthesis of organic compounds. This was shown by Melvin Calvin's experiments in which solutions of carbon dioxide and water were irradiated in a cyclotron and formic, oxalic and succinic acids, which contain 1, 2 and 4 carbons respectively, were obtained. These, as you know, are intermediates in certain metabolic pathways of living organisms. Irradiating solutions of inorganic compounds with ultraviolet light or passing electric charges through the solution to simulate lightning also produce organic compounds. Harold Urey and Stanley Miller in 1953 exposed a mixture of water vapor, methane, ammonia and hydrogen gases

to electric discharges for a week and demonstrated the production of amino acids such as glycine and alanine, together with other complex organic compounds. Similar reactions doubtless occurred in the earth's prebiotic atmosphere. Amino acids and other compounds could be produced in nature at the present time by lightning discharges or ultraviolet radiations; however, any organic compound produced in this way might undergo spontaneous oxidation or it would be taken up and degraded by molds, bacteria and other organisms.

The details of the chemical reactions that could give rise, without the intervention of living things, to carbohydrates, fats and amino acids have been worked out by Oparin and extended by Calvin and others. Most, if not all, of the reactions by which the more complex organic substances were formed probably occurred in the sea in which the inorganic precursors and the organic products of the reaction were dissolved and mixed. The sea became a sort of prebiotic, organic soup. This must have thickened in certain limited regions of the earth by evaporation, freezing or adsorption on the surface of mineral particles. Molecules collided with increasing frequency, reacted and aggregated to form new molecules of increasing size and complexity. As more has been learned of the role of specific hydrogen bonding and other weak intermolecular forces in the pairing of specific nucleotide bases and the effectiveness of these processes in the transfer of biological information, it has become clear that similar forces could have operated early in evolution before "living" organisms appeared.

The known forces of intermolecular attraction and the tendency for certain molecules to form liquid crystals provide us with means by which large, complex, specific molecules can form spontaneously. Oparin suggested that natural selection can operate at the level of these complex molecules before anything recognizable as life is present. As the molecules came together to form colloidal aggregates, these aggregates began to compete with one another for raw materials. Some of the aggregates which had some particularly favorable internal arrangement would acquire new molecules more rapidly than others and would eventually become the dominant types.

Once some protein molecules had formed and had achieved the ability to catalyze reactions, the rate of formation of additional molecules would be greatly stepped up. Next, in combination with molecules of nucleic acid, these complex protein molecules probably acquired the ability to catalyze the synthesis of molecules like themselves. These hypothetic, autocatalytic particles composed of nucleic acids and proteins would have some of the properties of a virus, or perhaps of a freeliving gene. The next step in the development of a living thing was the addition of the ability of the autocatalytic particle to undergo inherited changes — to mutate. Then, if a number of these free genes had joined to form a single larger unit, the resulting organism would have been similar to certain present-day viruses. A major step in early evolution was the development of a protein-lipid membrane around this aggregate which permitted the accumulation of some and the exclusion of other molecules. All the known viruses are parasites that can live only within the cells of higher animals and plants. However, a little reflection will suggest that free-living viruses, ones which do not produce a disease, would be very difficult to detect; such organisms may indeed exist.

The first living organisms, having arisen in a sea of organic molecules and in an atmosphere devoid of oxygen, presumably obtained energy by the fermentation of certain of these organic substances. These heterotrophs could survive only as long as the supply of organic molecules in the organic soup, accumulated from the past, lasted. Before the supply was exhausted, however, the heterotrophs evolved further and became autotrophs, able to make their own organic molecules by chemosynthesis or photosynthesis. One of the byproducts of photosynthesis is gaseous oxygen, and it is likely that all the oxygen in the atmosphere was produced and is still produced in this way. It is estimated that all the oxygen of our atmosphere is renewed by photosynthesis every 2000 years and all the carbon dioxide molecules pass through the photosynthetic process every 300 years. All the oxygen and carbon dioxide in the earth's atmosphere are the products of living organisms and have passed through living organisms over and over again during the course of evolution.

An explanation of how an autotroph may have evolved from one of these primitive, fermenting heterotrophs was presented by N. H. Horowitz in 1945. According to Horowitz' hypothesis, an organism might acquire, by successive mutations, the enzymes needed to synthesize complex from simple substances, in the reverse order to the sequence in which they are used in normal metabolism. Let us suppose that our first primitive heterotroph required

organic compound Z for its growth. Substance Z, and a variety of other organic compounds, Y, X, W, V, U, etc., were present in the organic sea broth which was the environment of this heterotroph. They had been synthesized previously by the action of nonliving factors of the environment. The heterotroph would survive nicely as long as the supply of compound Z lasted. If a mutation occurred which enabled the heterotroph to synthesize substance Z from substance Y, the strain of heterotroph with this mutation would be able to survive when the supply of substance Z was exhausted. A second mutation, which established an enzyme catalyzing a reaction by which substance Y could be made from the simpler substance X, would again have great survival value when the supply of Y was exhausted. Similar mutations, setting up enzymes enabling the organism to use successively simpler substances, W, V, U, . . . and finally some inorganic substance, A, would eventually result in an organism able to make substance Z, which it needs for growth, out of substance A by way of all the intermediate compounds. When, by other series of mutations, the organism was able to synthesize all of its requirements from simple inorganic compounds, as the green plants can, it would have become an autotroph. Once the first simple autotrophs had evolved, the way was clear for the further evolution of the vast variety of green plants, bacteria, molds and animals that inhabit the world today.

These considerations lead us to the conclusion that the origin of life, as an orderly natural event on this planet, was not only possible but it also was almost inevitable. Furthermore, with the vast number of planets in all the known galaxies of the universe, many of them must have conditions which permit the origin of life. It is probable, then, that there are many other planets in other solar systems on which life as we know it exists. Wherever life is possible, it should, if given enough time, appear and ramify into a wide variety of types. Some of these may be quite dissimilar from the ones on this planet, but others may be quite like those found here; some may, perhaps, be like ourselves. Living things on other planets might have a completely different kind of genetic code, or might be composed of elements other than carbon, hydrogen, oxygen and nitrogen.

It seems unlikely that we will ever know how life originated, whether it happened only once or many times, or whether it might happen again. The theory (1) that organic substances were formed from inorganic substances by the action of physical factors in the environment; (2) that they interacted to form more and more complex substances, finally enzymes, and then self-reproducing systems ("free genes"); (3) that these "free genes" diversified and united to form primitive, perhaps virus-like hetero-trophs; and (4) that autotrophs then evolved from these heterotrophs, has the virtue of being quite plausible. Many of the parts of this theory have been subjected to experimental verification.

ANNOTATED REFERENCES

Blum, H. F.: Time's Arrow and Evolution. Revised ed. Gloucester, Mass., Peter Smith, 1969. A discussion of theories of the origin of life.

Darwin, C.: The Origin of Species. 1859. Available in a number of recent reprint editions. This classic is well worth sampling for its clear, logical arguments and for its wealth of examples.

Dobzhansky, T.: Genetics of the Evolutionary Process. New York, Columbia University Press, 1970. A rewriting and extension of his classic book on Genetics and the Origin of Species. The original edition (1937) was instrumental in the development of NeoDarwinism.

Ehrlich, P. R., R. W. Holm, D. R. Parnell: The Process of Evolution. 2nd ed. New York, McGraw-Hill Book Company, 1974. A standard textbook and useful reference on the mechanisms of evolution.

Goldschmidt, R. B.: The Material Basis of Evolution. New Haven, Yale University Press, 1940. Presents in detail Goldschmidt's views on the importance of large mutations in the evolution of species.

Irvine, W.: Apes, Angels and Victorians. New York, McGraw-Hill Book Co., 1955. Presents a clear picture of the impact of the theory of evolution on Victorian England and a vivid portrayal of Thomas Huxley's championing of Darwin's theory.

Keosian, J.: The Origin of Life. 2nd ed. New York, Van Nostrand-Reinhold Publishing Co., 1968. An interesting updating of Oparin's and Calvin's theories.

Lack, D.: Darwin's Finches. Cambridge, Cambridge University Press, 1947. A classic case study of speciation.

Mayr, E.: Populations, Species, and Evolution. Cambridge, Harvard University Press, 1970. An abridgement and revision of his earlier book on Animal Species and Evolution. A scholarly account of evolutionary theory with an emphasis upon systematic and field aspects.

Oparin, A. I.: The Origin of Life. New York, Macmillan, 1938. A translation of Oparin's classic arguments as to how life may have evolved from nonliving systems.

Orgel, L. E.: The Origins of Life. New York, John Wiley and Sons, 1973. An account for the general reader of the origin of our solar system, the origin of life on earth, and the possibilities of extraterrestrial life.

Osborn, H. F.: From the Greeks to Darwin. New York, Macmillan, 1913. An interesting and classic account of the history of ideas on evolution.

Savage, J. M.: Evolution. New York, Holt, Rinehart and Winston, 1963. A readable, relatively brief presentation of the evidence for and theories of evolution.

Simpson, G. G.: The Meaning of Evolution, Revised Edition. New Haven, Yale University Press, 1967. An excellent, nontechnical presentation of evolutionary concepts with an emphasis upon the paleontological aspects.

Stebbins, G. L., Jr.: Processes of Organic Evolution. Englewood Cliffs, N.J., Prentice-Hall, 1966. A clear, literate paper-back summary of evolution and natural selection.

Williams-Ellis, A.: Darwin's Moon. London and Edinburgh, Blackie & Son, 1966. An excellent biography of Alfred Russel Wallace with a vivid portrayal of Wallace's life and times which reemphasizes his solid contributions to the advancement of the theory of evolution.

CHAPTER 9

THE EVIDENCE FOR EVOLUTION

The evidence that organic evolution has occurred is so overwhelming that no one who is acquainted with it has any doubt that new species are derived from previously existing ones by descent with modification. The extensive fossil record provides direct evidence of organic evolution and gives the details of the evolutionary relationships of many lines of descent. Most of the facts underlying the several subdivisions of biology acquire significance and make sense only when viewed against the background of evolution.

9.1 THE FOSSIL EVIDENCE

The evidence of life in former times is now both abundant and diverse. The science of **paleontology,** which deals with the finding, cataloguing and interpretation of **fossils,** has aided immensely in our understanding of the lines of descent of many vertebrate and invertebrate stocks. The term "fossil" (Latin *fossilium,* something dug up) refers not only to the bones, shells, teeth and other hard parts of an animal's body which may survive but also to any impression or trace left by previous organisms. In view of the large number of fossils that have been found, it is sobering to realize that only a small fraction of all the organisms that ever lived have been preserved as fossils, and only a small fraction of these fossils have been dug up and studied to date.

Footprints or trails made in soft mud, which subsequently hardened, are a common type of fossil. From such remains one can infer something of the structure and body proportions of the animals that made them. For example, the tracks of an amphibian from the Pennsylvanian period that were discovered in 1948 near Pittsburgh revealed that the animal moved by hopping rather than by walking, for the footprints lay opposite each other in pairs.

The commonest vertebrate fossils are skeletal parts. From the shape of bones and the position of the bone scars which indicate points of muscle attachment, paleontologists can make inferences about an animal's posture and style of walking, the position and size of its muscles, and hence the contours of its body. Careful study of fossil remains has enabled paleontologists to make reconstructions of what the animal must have looked like in life (Figs. 8.20 and 9.1).

In one interesting and striking type of fossil, the original hard parts, or more rarely the soft tissues of the body, have been replaced by minerals, a process called **petrifaction.** Iron pyrites, silica and calcium carbonate are some of the common petrifying minerals. The petrified muscle of a shark more than 300,000,000 years old was so well preserved by petrifaction that not only the individual muscle fibers but also even their cross striations could be observed in thin sections under the microscope. A famous example of the process of petrifaction is the Petrified Forest in Arizona.

Molds and casts are superficially similar to petrified fossils but are produced in a different way. **Molds** are formed by the hardening of the material surrounding a buried organism, followed by the decay and removal of the body of the organism. The mold may subsequently be filled by minerals which harden to form **casts** which are exact replicas of the original struc-

Figure 9.1 An example of a fossil, the remains of *Archaeopteryx*, a tailed, toothed bird from the Jurassic period. (Courtesy of the American Museum of Natural History.)

tures. Some animal remains have been exceptionally well preserved by being embedded in tar, amber, ice or volcanic ash. The remains of woolly mammoths, deep frozen in Siberian ice for more than 25,000 years, were so well preserved that the meat was edible!

The formation and preservation of a fossil require that some structure be buried. This may take place at the bottom of a body of water or on land by the accumulation of wind-blown sand, soil or volcanic ash. Many of the people and animals living in Pompeii were preserved almost perfectly by the volcanic ash from the eruption of Vesuvius. Sometimes animals were trapped and entombed in a bog, quicksand or asphalt pit, such as the famous La Brea tar pits in Los Angeles which have provided superb fossils of Pleistocene animals.

9.2 THE GEOLOGIC TIME TABLE

The earth's crust consists of five major rock strata, each subdivided into minor strata lying one on top of the other. These layers were generally formed by the accumulation of sediment — sand or mud — at the bottom of oceans, seas or lakes. Each rock stratum contains certain characteristic fossils which can now be used to identify deposits made at the same time in different parts of the world. Geologic time has been divided, according to the succession of these rock strata, into eras, periods and epochs

(Table 9.1). The duration of each period or epoch can be estimated from the thickness of the sedimentary deposits, although, of course, the rate of deposition was not exactly the same in different places and at different times.

The layers of sedimentary rock should occur in the sequence of their deposition, with the newer later strata on top of the older earlier ones, but subsequent geologic events may have changed the relationship of the layers. Not all of the expected strata may occur in some particular region, for that land may have been exposed rather than submerged during one or more geologic ages. In some regions the strata formed previously have subsequently emerged, been washed away, and then relatively recent strata have been deposited directly on very ancient ones. Certain sections of the earth's crust, in addition, have undergone massive foldings and splittings, so that early layers have come to lie on top of later ones. The age of a rock stratum may be determined by a study of its fossils, for some kinds of fossils were deposited in only one era or period.

Rock deposits are now dated largely by taking advantage of the fact that certain radioactive elements are transformed into other elements at rates which are slow and essentially unaffected by the pressures and temperatures to which the rock has been subjected. Half of a given sample of uranium will be converted to a special isotope of lead in 4.5 billion years. Hence, by measuring the proportion of uranium and lead in a bit of crystalline rock, its age can

TABLE 9.1 GEOLOGIC TIME TABLE

Era	Period	Epoch	Duration in Millions of Years	Time from Beginning of Period to Present (Millions of Years)	Geologic Conditions	Plant Life	Animal Life
Cenozoic (Age of Mammals)	Quaternary	Recent	0.011	0.011	End of last ice age; climate warmer	Decline of woody plants; rise of herbaceous ones	Age of man
		Pleistocene	1	1	Repeated glaciation; four ice ages	Great extinction of species	Extinction of great mammals; first human social life
	Tertiary	Pliocene	12	13	Continued rise of mountains of western North America; volcanic activity	Decline of forests; spread of grasslands; flowering plants, monocotyledons developed	Man evolved from manlike apes; elephants, horses, camels almost like modern species
		Miocene	12	25	Sierra and Cascade mountains formed; volcanic activity in northwest U.S.; climate cooler		Mammals at height of evolution; first manlike apes
		Oligocene	11	36	Lands lower; climate warmer	Maximum spread of forests; rise of monocotyledons, flowering plants	Archaic mammals extinct; rise of anthropoids; forerunners of most living genera of mammals
		Eocene	22	58	Mountains eroded; no continental seas; climate warmer		Placental mammals diversified and specialized; hoofed mammals and carnivores established
		Paleocene	5	63			Spread of archaic mammals
Rocky Mountain Revolution (Little Destruction of Fossils)							
Mesozoic (Age of Reptiles)	Cretaceous		72	135	Andes, Alps, Himalayas, Rockies formed late; earlier, inland seas and swamps; chalk, shale deposited	First monocotyledons; first oak and maple forests; gymnosperms declined	Dinosaurs reached peak, became extinct; toothed birds became extinct; first modern birds; archaic mammals common
	Jurassic		46	181	Continents fairly high; shallow seas over some of Europe and Western U.S.	Increase of dicotyledons; cycads and conifers common	First toothed birds; dinosaurs larger and specialized; primitive mammals
	Triassic		49	230	Continents exposed; widespread desert conditions; many land deposits	Gymnosperms dominant, declining toward end; extinction of seed ferns	First dinosaurs, pterosaurs and egg-laying mammals; extinction of primitive amphibians

Table 9.1 continued on opposite page.

be measured. In this way, the oldest rocks of the earliest geologic period are calculated to be about 3,500,000,000 years old and the latest Cambrian rocks to be 500,000,000 years old. These dates have been confirmed by newer methods in which the radioactive decay of rubidium 87 (half-life 47 billion years!) and potassium 40 has been utilized to measure the ages of micas and feldspars. Events in more recent times can be dated quite accurately by the

TABLE 9.1 GEOLOGIC TIME TABLE (*Continued*)

Era	Period	Epoch	Duration in Millions of Years	Time from Beginning of Period to Present (Millions of Years)	Geologic Conditions	Plant Life	Animal Life
	Appalachian Revolution (Some Loss of Fossils)						
Paleozoic (Age of Ancient Life)	Permian		50	280	Continents rose; Appalachians formed; increasing glaciation and aridity	Decline of lycopods and horsetails	Many ancient animals died out; mammal-like reptiles, modern insects arose
	Pennsylvanian		40	320	Lands at first low; great coal swamps	Great forests of seed ferns and gymnosperms	First reptiles; insects common; spread of ancient amphibians
	Mississippian		25	345	Climate warm and humid at first, cooler later as land rose	Lycopods and horsetails dominant; gymnosperms increasingly widespread	Sea lilies at height; spread of ancient sharks
	Devonian		60	405	Smaller inland seas; land higher, more arid; glaciation	First forests; land plants well established; first gymnosperms	First amphibians; lung-fishes, sharks abundant
	Silurian		20	425	Extensive continental seas; lowlands increasingly arid as land rose	First definite evidence of land plants; algae dominant	Marine arachnids dominant; first (wingless) insects; rise of fishes
	Ordovician		75	500	Great submergence of land; warm climates even in Arctic	Land plants probably first appeared; marine algae abundant	First fishes, probably fresh-water; corals, trilobites abundant; diversified mollusks
	Cambrian		100	600	Lands low, climate mild; earliest rocks with abundant fossils	Marine algae	Trilobites, brachiopods dominant; most modern phyla established
	Second Great Revolution (Considerable Loss of Fossils)						
Proterozoic			1000	1600	Great sedimentation; volcanic activity later; extensive erosion, repeated glaciations	Primitive aquatic plants—algae, fungi	Various marine protozoa; towards end, mollusks, worms, other marine invertebrates
	First Great Revolution (Considerable Loss of Fossils)						
Archeozoic			2000	3600	Great volcanic activity; some sedimentary deposition; extensive erosion	No recognizable fossils; indirect evidence of living things from deposits of organic material in rock	

decay of carbon 14, which has a half-life of 5568 years.* Relatively short periods of geologic time are estimated by measuring the rate at which waterfalls recede upstream as they wear away the rocks over which they tumble or by counting the annual deposits of clay on the bottoms of ponds and lakes.

Between the major eras there were widespread geologic disturbances, called **revolutions,** which raised or lowered vast regions of the earth's surface and created or eliminated shallow inland seas. These revolutions produced great changes in the distribution of sea

---

*Organic carbon is derived by CO_2 fixation from atmospheric CO_2 and the ratio of ^{12}C to ^{14}C in living organisms is the same as that in the atmosphere. No exchange of carbon atoms with the atmosphere occurs after death and the ^{14}C in the body is slowly transformed into ^{14}N. The age of organic remains can be estimated from their $^{12}C/^{14}C$ ratio and the half-life of ^{14}C.

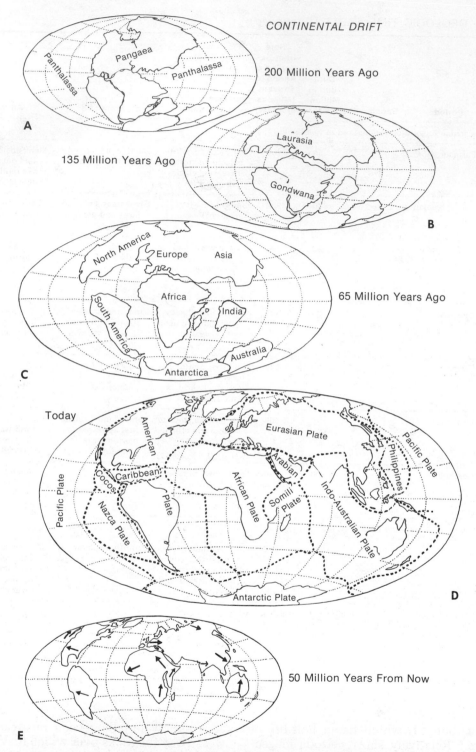

Figure 9.2 Continental drift. *A*, The supercontinent Pangaea of the Triassic period, about 200 million years ago. *B*, Break-up of Pangaea into Laurasia (Northern Hemisphere) and Gondwana (Southern Hemisphere) 135 million years ago in the Cretaceous period. *C*, Further separation of land masses, which occurred in the Tertiary period, 65 million years ago. Note that Europe and North America are still joined and that India is a separate land mass. *D*, The continents today. *E*, Projected positions of the continents in 50 million years. (From Norstog, K., and R. W. Long: Plant Biology. Philadelphia, W. B. Saunders Co., 1976.)

and land organisms and wiped out many of the previous forms of life. The Paleozoic era ended with the revolution that raised the Appalachian mountains and, it is believed, killed all but 3 per cent of the forms of life existing then. The Rocky Mountain revolution (which raised the Andes, Alps and Himalayas as well as the Rockies) annihilated most reptiles of the Mesozoic.

The raising and lowering of portions of the earth's crust result from the slow movements of the enormous tectonic plates that compose the crust and float on the underlying molten core. The movements of these plates have produced tremendous changes in the earth's geography (Fig. 9.2). These movements are continuing and build up tensions in the earth's crust that, when released, result in earthquakes.

9.3 THE GEOLOGIC ERAS

Archeozoic Era. The rocks of the oldest geologic era are very deeply buried in most parts of the world but are exposed at the bottom of the Grand Canyon and along the shores of Lake Superior. The oldest geologic era, the Archeozoic, begins not with the origin of the earth, but with the formation of the earth's crust, when rocks and mountains were in existence and the processes of erosion and sedimentation had begun. The Archeozoic era lasted about two billion years, about as long as all the succeeding eras combined. It was characterized by widespread volcanic activity and large upheavals which resulted in the raising of mountains. The heat, pressure and churning associated with the movements of the earth's crust probably destroyed most of whatever fossil remains there may have been, but a few traces of life remain. Scattered through the Archeozoic rocks are flakes of graphite, pure carbon, which are probably the transformed remains of plants and animal bodies. Although graphite can originate inorganically, its distribution in the rocks suggests that it was formed organically. If the amount of graphite in these rocks can be taken as a measure of the amount of living things in the Archeozoic, and there are reasons for believing that this is justified, then life must have been abundant in the Archeozoic seas, for there is more carbon in these rocks than in the coal beds of the Appalachians.

Proterozoic Era. The second geologic era, which lasted about one billion years, was characterized by the deposition of large quantities of sediment and by at least one great period of glaciation during which ice sheets stretched to within 20 degrees of the equator. There was less volcanic activity in this than in the preceding era, and the rocks are better preserved. Only a few fossils have been found in Proterozoic rocks but they show not only that life was present but also that evolution had proceeded quite far before the end of the era. Plants and animals were differentiated, multicellular forms had evolved from unicellular ones and some of the major groups of plants and animals had appeared. Sponge spicules, jellyfish, and the remains of fungi, algae, brachiopods and annelid worm tubes have been found in Proterozoic rocks. Several rich deposits of Pre-Cambrian fossils have been found in South Australia. These include jellyfish, corals, segmented worms and two animals with no resemblance to any known fossil or living form. The bodies of these animals were soft and were strengthened only by spicules of calcium carbonate.

Paleozoic Era. A second great revolution ended the Proterozoic era. During the ensuing 370,000,000 years of the Paleozoic every phylum and class of animals except birds and mammals appeared. Some of these animals appeared and became extinct in a short time (geologically speaking), and their fossils provide convenient markers by which rocks of the same era in different localities can be correlated.

The fossil deposits of the first three periods of the Paleozoic era, the Cambrian, Ordovician and Silurian, were mostly laid down in the seas. Large shallow seas covered most of the continents during these three periods and they teemed with life. Many of these forms had hard skeletons or armor coverings which left a good fossil record. The organisms living in the Cambrian were so varied and complex that they must have evolved from ancestors dating back to the Proterozoic era. Apparently, both plants and animals lived in the sea, and the land was a curious lifeless waste until the Ordovician, when plants became established on land. The Cambrian seas contained small floating plants and animals that were eaten by primitive shrimplike crustaceans and swimming annelid worms. The sea floor was covered with simple sponges, corals, echinoderms growing on stalks, snails, pelecypods and primitive cephalopods. An exceptionally well-preserved collection of Cambrian fossils was found in the mountains of British Columbia; it included annelids, crustaceans and a connecting link similar to peripatus. The most numerous animals were brachiopods and trilobites. Brachiopods, sessile, bivalved plankton feeders, flourished in the Cambrian and the rest of the Paleozoic. One of the present-day brachiopods, *Lingula*, is the oldest known genus of animals and is almost identical with its Cambrian ancestors. The trilobites (see Fig. 29.7) were primitive arthropods with flattened, elongated bodies covered dorsally by a hard shell. The shell had two longitudinal grooves that divided the body into three lobes. On the ventral side of the body was a pair of legs on each somite but the last, and each leg was biramous, had an outer gill branch and an inner walking or swimming branch. Most trilobites were only 5 to 8 cm. long but the largest was about 60 cm. They reached their peak of importance in the late Cambrian and then dwindled and became extinct in the Permian.

Evolution since the Cambrian has been characterized by the elaboration and ramification of the lines already present rather than by the establishment of entirely new forms. The original, primitive members of most lines were replaced by more complex, better adapted ones. The Ordovician seas contained, among other forms, giant cephalopods, squidlike animals with straight shells 5 to 7 meters long and 30 cm. in diameter. The Ordovician seas were apparently warm, for certain corals which grew only in warm waters lived as far north as Ontario and Greenland. The first vertebrates, the jawless, limbless, armored, bottom-dwelling fishes called **ostracoderms** (p. 768), appeared in the Ordovician. These lived in fresh water and their bony armor may have served as a defense against their chief predator, the carnivorous giant arachnids called **eurypterids** (p. 676). Two important events of the Silurian were the evolution of land plants and of the first air-breathing animals, primitive scorpions.

The evolution of the vertebrates from ostracoderms to placoderms, cartilaginous and bony fishes, amphibians, reptiles, birds and mammals will be traced in Chapters 32 to 37. The Devonian seas contained corals, sea lilies and brachiopods in addition to a great variety of fishes. Trilobites were still present but were declining in numbers and importance. The first land vertebrates, the amphibians called **labyrinthodonts,** appeared in the latter part of the Devonian; this period also saw the first true forests of ferns, "seed ferns," club mosses and horsetails and the first wingless insects and millipedes.

The Mississippian and Pennsylvanian periods are frequently grouped together as the Carboniferous, for during this time there flourished the great swamp forests whose remains gave rise to the major coal deposits of the world. The earliest stem reptiles appeared in the Pennsylvanian and from these there evolved in the succeeding Permian period a group of early, mammal-like reptiles, the **pelycosaurs,** from which the mammals eventually evolved (Fig. 9.3). Two important groups of winged insects, the ancestors of the cockroaches and the ancestors of the dragonflies, evolved during the Carboniferous.

The Permian period was characterized by widespread changes in topography and climate. The land began to rise early in the period, so that the swamps and shallow seas were drained, and the Appalachian revolution that ended the period, together with widespread glaciation, killed off a great many kinds of animals. The trilobites finally disappeared and the brachiopods, stalked echinoderms, cephalopods and many other kinds of invertebrates were reduced to small, unimportant, relict groups.

Mesozoic Era. The Mesozoic era began some 230,000,000 years ago and lasted some 167,000,000 years. It is subdivided into the Triassic, Jurassic and Cretaceous periods. During the Triassic and Jurassic most of the continental area was above water, warm and fairly dry. During the Cretaceous the Gulf of Mexico expanded into Texas and New Mexico, and the sea once again overspread large parts of the continents. There were great swamps from Colorado to British Columbia (Fig. 9.4). In the latter part of the Cretaceous the interior of the North American continent was further submerged and cut in two by the union of a bay from the Gulf of Mexico and one from the Arctic Sea. The Rocky Mountain revolution ended the Cretaceous with the upheaval of the Rockies, Alps, Himalayas and Andes mountains. The Mesozoic was characterized by the tremendous evolution, diversification and specialization of the reptiles and is commonly called the Age of Reptiles. Mammals originated in the Triassic and birds in the Jurassic. Most of the modern orders of insects appeared in the Triassic, and snails, bivalve mollusks and sea urchins underwent important evolutionary changes.

At the end of the Cretaceous a great many reptiles became extinct; they were apparently unable to adapt to the marked changes brought about by the Rocky Mountain revolution. As the climate became colder and drier, many of the plants which served as food for the herbivorous reptiles disappeared. Some of the herbivorous reptiles were too large to walk about on land when the swamps dried up. The

Figure 9.3 Texas in the Permian period, about 280,000,000 years ago. Various pelycosaurs are shown. Some had large fins, others were essentially like lizards. In the lower illustration is a salamander-like amphibian with a flat, triangular skull. (Copyright, Chicago Natural History Museum, from the painting by Charles R. Knight.)

smaller, warmblooded mammals which appeared were better able to compete for food, and many of these ate reptilian eggs. The demise of the many kinds of reptiles was probably the result of a combination of a whole host of factors rather than any single one.

Cenozoic Era. The Cenozoic era, extending from the Rocky Mountain revolution to the present, is subdivided into the earlier Tertiary period, which lasted some 62,000,000 years, and the present Quaternary period, which includes the last million or million and one-half years.

The Tertiary is subdivided into five epochs, the Paleocene, Eocene, Oligocene, Miocene and Pliocene. The Rockies, formed at the beginning of the Tertiary, were considerably eroded by the Oligocene, and the North American continent had a gently rolling topography. Another series of uplifts in the Miocene raised the Sierra Nevadas and a new set of Rockies, and resulted in the formation of the western deserts. The climate of the Oligocene was rather mild, and palm trees grew as far north as Wyoming. The uplifts of the Miocene and Pliocene and the successive ice ages of the Pleistocene killed off many of the mammals that had evolved.

The last elevation of the Colorado Plateau, which initiated the cutting of the Grand Canyon, occurred almost entirely in the short

Figure 9.4   Western Canada in the Cretaceous period, about 110,000,000 years ago. The land was low, well watered and covered with numerous swamps. Most of the dinosaurs were harmless, plant-eating Ornithischians, reptiles with birdlike pelvic bones. Two types of duck-billed dinosaurs can be seen—three large, uncrested ones in the upper portion, and two kinds of crested ones in the lower portion. In the upper right foreground is a heavily armored, four-footed dinosaur covered with bony plates and spines. In the upper right and lower left background are ostrich dinosaurs—tall slender animals, with the general proportions of an ostrich, but with short forelegs and a long, slender tail. (Copyright, Chicago Natural History Museum, from the painting by Charles R. Knight.)

Pleistocene and Recent epochs, the two subdivisions of the Quarternary period. Four periods of glaciation occurred in the Pleistocene, between which the sheets of ice retreated. At their greatest extent, these ice sheets extended as far south as the Missouri and Ohio rivers and covered 4,000,000 square miles of North America. The Great Lakes, which were carved out by the advancing glaciers, changed their outlines and connections several times. It is estimated that at one time, when the Mississippi river drained lakes as far west as Duluth and as far east as Buffalo, its volume was more than 60 times as great as at present. During the Pleistocene glaciations enough water was removed from the oceans and locked in the vast sheets of ice to lower the level of the oceans as much as 100 meters. This created land connections, highways for the dispersal of many land forms, between Siberia and Alaska at Bering Strait and between England and the continent of Europe. Many mammals, including the sabertoothed tiger, the mammoth and the giant ground sloth, became extinct in the Pleistocene after primitive man had appeared.

The fossil record available today makes it impossible to doubt that the present species arose from previously existing, different ones. For many lines of evolution the individual steps are well known; other lines have some gaps which remain to be filled by future paleontologists.

Even if the remarkably detailed fossil record did not exist, studies of the anatomy, physiology and biochemistry of modern animals and plants, their development and cytogenetics, and the manner in which they are distributed over the earth's surface have provided incontrovertible proof that organic evolution has occurred.

9.4 THE EVIDENCE FROM TAXONOMY

The science of taxonomy began long before the doctrine of evolution was accepted; indeed the founders of scientific taxonomy, Ray and Linnaeus, were firm believers in the fixity, the unchangingness, of species. Present-day taxonomists are concerned with the naming and describing of species primarily as a means of discovering or clarifying evolutionary relationships. This is based upon the assumption that the degree of resemblance in homologous structures is a measure of the degree of relationship. The characteristics of living things are such that they can be fitted into a hierarchical scheme of categories, each more inclusive than the previous one—species, genera, families, orders, classes and phyla. This can best be interpreted as proof of evolutionary relationship. If the kinds of animals and plants were not related by evolutionary descent, their characteristics would be present in a confused, random pattern and no such hierarchy of forms could be established.

The classification of living organisms into well-defined groups is possible because most of the intermediate forms have become extinct. If representatives of every type of animal and plant that ever lived were still living today, there would be many series of intergrading forms, and the division of these into neat taxonomic categories would be difficult indeed. The present-day species have been compared to the terminal twigs of a tree whose main branches and trunk have disappeared. The fascinating puzzle for the taxonomist is to reconstruct the missing branches and put each twig on the proper branch.

9.5 THE EVIDENCE FROM MORPHOLOGY

A study of the details of the structure of any particular organ system in the diverse members of a given phylum reveals a basic similarity of form which is varied to some extent from one class to another. The skeletal, muscular, circulatory and excretory systems of the vertebrates provide especially clear illustrations of this principle, but this is generally true of all systems in all phyla. Only those similarities based on homologous organs (p. 200) can be used in classification. A seal's front flipper, a bat's wing, a cat's paw, a horse's front leg and a human hand, though superficially dissimilar and adapted for quite different functions, nevertheless are homologous organs. Each consists of almost the same number of bones, muscles, nerves and blood vessels arranged in the same pattern, and their mode of development is very similar. The existence of such homologous organs implies a common evolutionary origin.

Many species of animals have organs or parts of organs which are small or lacking some essential part and are either functionless or used in a way quite different from their well-developed homologue in another species. There are more than 100 such **vestigial organs** in the human body, including the appendix, the coccyx (fused tail vertebrae), the wisdom teeth, the nictitating membrane of the eye, body hair and the muscles that move the ears (Fig. 9.5). Such organs are the remnants of ones which were functional in the ancestral forms, but when some change in the environment rendered the organ no longer necessary for survival it gradually became reduced to a vestige. This appears at first glance to be an application of Lamarck's idea of the role of "use and disuse" of an organ in evolution, but the underlying mechanism is quite different. Mutations for the decrease in the size and functional importance of an organ are occurring constantly; as long as the organ is necessary for survival, such mutations are lethal and eliminate their possessors. But if the

Figure 9.5   Diagrams of some of the vestigial organs of the human body.

organ is no longer needed for survival, such mutations will not be lethal, and they may accumulate and lead to the reduction of the organ.

9.6   THE EVIDENCE FROM COMPARATIVE PHYSIOLOGY AND BIOCHEMISTRY

The study of the physiological and bio-chemical traits of organisms generally requires complex apparatus and is more difficult than the direct observation of morphological charac-teristics. Yet, as such studies have been made using a wide variety of animal types, it has be-come clear that there are functional similarities and differences which parallel closely the mor-phological ones. Indeed, if one were to estab-lish taxonomic relationships based on physio-logical and biochemical characters instead of on the usual structural ones, the end result would be much the same.

The fundamental similarity of the chemical constituents and patterns of enzymes present in cells of different animals was presented in Chapter 4. There are, however, certain chemi-cal constituents, certain enzymes and certain hormones that are found in some animals and and not in others. The distribution of these bio-chemical characters strongly parallels the evo-lutionary relationships inferred from other evidence.

The blood serum of each species of animal contains certain specific proteins. The de-gree of similarity of these serum proteins can be determined by **antigen-antibody** reactions. To perform the test, an experimental animal, usu-ally a rabbit, is injected with a small amount of the serum—for example, a sample of human serum. The proteins of the injected serum are foreign to the rabbit's blood and hence act as antigens. The plasma cells of the rabbit respond by producing antibodies specific for human blood-protein antigens. These antibodies are then obtained by withdrawing blood from the rabbit and allowing it to clot; the antibodies are in the serum. When a dilute sample of this serum is mixed with a drop of human serum, the antibody for human serum reacts with the human serum antigen and produces a visible precipitation. The strength of the reaction can be measured by making successive dilutions of the human serum, mixing each dilution with a fresh sample of the antibody solution (the rabbit serum), and observing at what point the precipitation no longer occurs. When serum from an animal other than man is mixed with rabbit serum containing antibody for human serum proteins, there is either no precipitation at all, or else a precipitation occurs only with concentrated antigen solutions. By testing, in turn, the sera of a variety of animals with rab-bit serum containing antibodies for human serum proteins, the degree of similarity be-tween the proteins can be determined. If the serum of another animal contains proteins which are similar to those of man, a precipita-tion will occur. In this way, man's closest "blood relations" have been found to be the great apes, and then, in order, the Old World monkeys, the New World monkeys, and finally the tarsioids. The serum of the lemur gives the smallest amount of precipitation when mixed with antibodies specific for human serum.

The biochemical relationships of a variety of forms tested in this way correlate with and complement the relationships determined by other means. Cats, dogs and bears are closely related, as determined by this test; cows, sheep, goats, deer and antelopes constitute another closely related group. This test reveals that there is a closer relationship among the modern birds than among the mammals, for all of the several hundred species of birds tested give strong and immediate reactions with serum containing antibodies for chicken serum. From other tests it was concluded that birds are more closely related to the crocodile line of reptiles than to the snake-lizard line, which corroborates the paleontologic evidence. Similar tests of the sera of crustaceans, insects and mollusks have shown that those forms regarded as being closely related from morphological or paleontological evidence also show similarities in their serum proteins.

Investigations of the sequence of amino acids in the α and β chains of hemoglobins from different species have revealed great similarities, of course, and specific differences, the pattern of which demonstrates the order in which the underlying mutations, the changes in nucleotide base pairs, must have occurred in evolution. The evolutionary relationships inferred

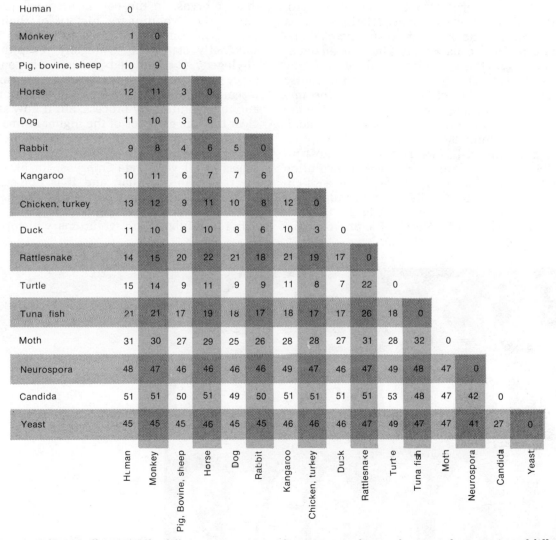

Figure 9.6 A diagram illustrating the differences in amino acid sequences in the cytochrome c's from a variety of different species of animals and plants. The numbers refer to the number of different amino acids in the cytochrome c's of the species compared. (From Dayoff, M. O., and R. V. Eck: Atlas of Protein Sequence and Structure. Silver Springs, Maryland, National Biomedical Research Foundation, 1968.)

from these studies agree completely with those based on morphological studies. Analyses of the amino acid sequence in the protein portion of the cytochromes provide further concurring evidence of evolutionary relationships (Fig. 9.6). Thus evidence of evolutionary relationships can be adduced from similarities and differences in *molecular* structure as well as from gross morphology.

The properties of specific enzymes, which reflect their molecular structure, also provide evidence of evolutionary relationships. The pattern and rates of reaction of lactate dehydrogenase and certain other enzymes with the normal pyridine nucleotide coenzyme (NAD) and with analogues of NAD can be used to demonstrate evolutionary relationships.

It might seem unlikely that an analysis of the urinary wastes of different species would provide evidence of evolutionary relationship, yet this is true. The kind of waste excreted depends upon the particular kinds of enzymes present, and the enzymes are determined by genes, which have been selected in the course of evolution. The waste products of the metabolism of purines, adenine and guanine are excreted by man and other primates as uric acid, by other mammals as allantoin, by amphibians and most fishes as urea, and by most invertebrates as ammonia. Vertebrate evolution has been marked by the successive loss of enzymes required for the stepwise degradation of uric acid. Joseph Needham made the interesting observation that the chick embryo in the early stages of development excretes ammonia, later it excretes urea, and finally it excretes uric acid. The enzyme uricase, which catalyzes the first step in the degradation of uric acid, is present in the early chick embryo but disappears in the later stages of development. The adult frog excretes urea, but the tadpole excretes ammonia. These are biochemical examples of the principle of recapitulation.

9.7 THE EVIDENCE FROM EMBRYOLOGY

The importance of the embryologic evidence for evolution was emphasized by Darwin and brought into even greater prominence by Ernst Haeckel in 1866 when he developed his theory that embryos, in the course of development, repeat the evolutionary history of their ancestors in some abbreviated form. This idea, succinctly stated as "Ontogeny recapitulates phylogeny," stimulated research in embryology and focused attention on the general resemblance between embryonic development and the evolutionary process, but it now seems clear that the embryos of the higher animals resemble the *embryos* of lower forms, not the adults as Haeckel had believed. The early stages of all vertebrate embryos, for example, are remarkably similar, and it is not easy to differentiate a human embryo from the embryo of a fish, frog, chick or pig (Fig. 9.7).

In recapitulating its evolutionary history in a

Fish Salamander Turtle   Chicken   Pig   Cow   Rabbit   Man

Figure 9.7 Comparison of early and later stages in the development of vertebrate embryos. Note the similarity of the earliest stages of each.

few days, weeks or months the embryo must eliminate some steps and alter and distort others. In addition, some new characters have evolved which are adaptive and enable the embryo to survive to later stages. For example, mammalian embryos, which have many early characteristics in common with those of fish, amphibians and reptiles, have other structures which enable them to survive and to develop within the mother's uterus rather than within an egg shell. Such secondary traits may alter the original characters common to all vertebrates so that the basic resemblances are blurred. The concept of recapitulation must be used with due caution, rather than rigorously, but it does provide an explanation for many otherwise inexplicable events in development.

Studies of the embryonic forms may provide the only means for identifying the relationships of certain organisms. *Sacculina*, for example, is an extremely aberrant barnacle which parasitizes crabs. The adult form is a saclike structure which sends processes into the tissues of the host to absorb nourishment. It resembles no other organism and its relationship became clear only when it was found that its larva is like that of other barnacles until it becomes attached to the abdomen of the host. Then it loses it appendages and other structures and becomes the adult, saclike creature.

The concept of recapitulation is very helpful in understanding the curious and complex development of the vertebrate circulatory and excretory systems. It is also useful, when not taken too literally, in getting a broad picture of the whole of development. Thus, the fertilized egg can be compared to the putative single-celled flagellate ancestor of all animals, and the blastula can be compared to a colonial protozoan or to some hypothetic blastula-like animal which has been postulated to be the ancestor of all Metazoa. Haeckel believed that the ancestor of coelenterates and all the higher animals was a gastrula-like organism with two layers of cells and a central cavity connected by a blastopore to the outside. After gastrulation, development follows one of two main lines. In the echinoderms and chordates the blastopore becomes the anus or comes to lie near the anus. In the annelid-mollusk-arthropod line the blastopore becomes the mouth or comes to lie near the mouth. In both lines the mesoderm develops between the ectoderm and endoderm. In the chordate-echinoderm line the mesoderm develops, at least in part, as pouches from the primitive digestive tract, whereas in the annelid-mollusk line the mesoderm usually originates from special cells differentiated early in development.

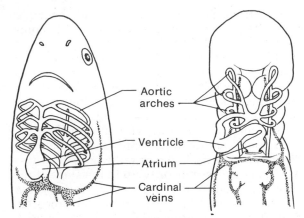

Figure 9.8 Ventral views of the heart and aortic arches of a human embryo (right) and an adult shark (left). Both have a single atrium and single ventricle, several aortic arches, and anterior and posterior cardinal veins emptying into the heart.

All chordate embryos develop, shortly after the mesoderm begins to appear, a dorsal hollow nerve cord, a notochord and pharyngeal pouches. The early human embryo at this stage resembles a fish embryo, with gill pouches, pairs of aortic arches, a fishlike heart with a single atrium and ventricle, a primitive fish kidney and a well-differentiated tail complete with muscles for wagging it (Fig. 9.8). At a slightly later stage the human embryo resembles a reptilian embryo. Its gill pouches regress; a new kidney, the mesonephros, forms and the pronephros disappears or becomes incorporated into other structures; and the atrium becomes divided into right and left chambers. Still later in development, the human embryo develops a mammalian, four-chambered heart and a third kidney, the metanephros. During the seventh month of intrauterine development the human embryo, with its coat of embryonic hair and in the relative size of body and limbs, resembles a baby ape more than it resembles an adult human.

Our increasing understanding of physiologic genetics provides us with an explanation of the phenomenon of recapitulation. All chordates have in common a certain number of genes which regulate the processes of early development. As our ancestors evolved from fish, through amphibian and reptilian stages, they accumulated mutations for new characteristics but kept some of the original "fish" genes, which still control early development. Later in development the genes which the human shares with amphibians influence the course of development so that the embryo resembles a frog embryo. Subsequently, some of the genes which we have in common with reptiles come into control. Only after this do most of the peculiarly mammalian genes exert their influence, and these are followed by the action of genes we have in common with other primates. The anthropoid apes, which have the most immediate ancestors in common with us, have the most genes in common with us and their development is identical with ours except for some fine details. A pig or rat, whose ancestors are the same as ours only up to the stage of the primitive placental mammals, has fewer genes in common and has developmental processes that diverge at an earlier time. In general, during development the general characteristics that distinguish phyla and classes appear before the special characteristics that distinguish genera and, finally, species. Within each phylum, the higher forms pass through a sequence of developmental stages which are similar to those of lower forms but achieve a different final form by adding changes at the end of the original sequence and by altering certain of the earlier embryonic stages they share with the lower forms.

9.8 THE EVIDENCE FROM GENETICS AND CYTOLOGY

For the past several thousand years man has been selecting and breeding animals and plants for his own uses, and a great many varieties, adapted for different purposes, have been established. These results of artificial selection provide striking models of what may be accomplished by natural selection. All of our breeds of dogs have descended from one, or perhaps a very few, species of wild dog or wolf, yet they vary so much in color, size and body proportions that if they occurred in the wild they would undoubtedly be considered separate species. They are all interfertile and are known to come from common ancestors, so they are regarded as varieties of a single species. A comparable range of varieties has been produced by artificial selection in cats, chickens, sheep, cattle and horses. Plant breeders have established by selective breeding a tremendous variety of plants. From the cliff cabbage, which still grows wild in Europe, have come cultivated cabbage, cauliflower, kohlrabi, Brussels sprouts, broccoli and kale.

Geneticists have been able to trace the ancestry of certain modern plants by a combination of cytological techniques in which the morphology of the chromosomes is compared and by breeding techniques which compare the kinds of genes and their order in particular chromosomes in a series of plants. In this way, the present cultivated tobacco plant, *Nicotiana tabacum*, was shown to have arisen from two species of wild tobacco, and corn was traced to teosinte, a grasslike plant which grows wild in the Andes and Mexico. The cytological details of the structure of the giant chromosomes of the salivary glands of fruit flies have been of prime importance in unraveling the evolutionary history of the many species of *Drosophila*.

9.9 THE EVIDENCE FROM THE GEOGRAPHIC DISTRIBUTION OF ORGANISMS

In the course of the voyage of the *Beagle*, Darwin was greatly impressed by his observations that plants and animals were *not* found everywhere that they could exist if climate and topography were the only factors determining their distribution. Central Africa, for example, has elephants, gorillas, chimpanzees, lions and antelopes, while Brazil, with a similar climate and other environmental conditions, has none of these but does have prehensile-tailed monkeys, sloths and tapirs, animals not found in Africa. The facts of biogeography, the geographic distribution of plants and animals, were of prime importance in leading both Darwin and Alfred Russel Wallace to the conclusion that organic evolution had occurred by natural selection. The present distribution of organisms, and the sites at which their fossil remains are found, are understandable only on the basis of the evolutionary history of each species and the movements of the tectonic plates of the earth's crust.

The **range** of each species is that particular portion of the earth in which it is found. The range of a species may be restricted to a few square miles or less or, as with humans, may include almost the entire earth. In general, the ranges of closely related species or subspecies are not identical, nor are they widely separated, but are adjacent and separated by a barrier of some sort. This generalization was stated by David Starr Jordan and is known as **Jordan's rule.** The explanation for this should be clear from the discussion of the role of isolation in species formation. A single species cannot be subdivided as long as interbreeding can occur throughout the whole population. But when some barrier is interposed between two parts of the population so that interbreeding is prevented, the two populations will, in the subsequent course of time, accumulate different gene complexes.

One of the fundamental assumptions of biogeography is that each species of animal or plant originated only once. The place where this occurred is known as its **center of origin.** The center of origin is not a single point but the range of the population when the new species was formed. From this center of origin each species spreads out, under the pressure of an increasing population, until it is halted by a barrier of some kind: a physical one such as an ocean, mountain or desert, an environmental one such as unfavorable climate, or a biological barrier such as the absence of food or the presence of other species which prey upon it or compete with it for food or shelter.

As one might expect, regions which have been separated from the rest of the world for a long time, such as South America and Australia, have a unique assemblage of animals and plants. Australia has a population of monotremes and marsupials that is found nowhere else. Australia became separated from Malaya during the Mesozoic, before placental mammals evolved, and its primitive mammals were not eliminated, as were the monotremes and most of the marsupials in the other parts of the world, by the competition of the better adapted placental mammals. The Australian marsupials evolved into a wide variety of forms, each adapted to some particular combination of environmental factors.

The kinds of animals and plants found on oceanic islands resemble, in general, those of the nearest mainland, yet include species found nowhere else. Darwin studied the flora and fauna of the Cape Verde Islands, some 400 miles west of Dakar in Africa, and of the Galapagos Islands, a comparable distance west of Ecuador. On each archipelago the plants and the nonflying animals were indigenous, but those of Cape Verde resembled African species and those of the Galapagos resembled South American ones. Organisms from the neighboring continent migrated or were carried to the island and subsequently evolved into new species. The animals and plants found on oceanic islands are only those that could survive the trip there. There are, for example, no frogs or toads on the Galapagos and no native terrestrial mammals, even though conditions would favor their survival.

There are many facts of the present-day distribution of animals and plants which can be explained only by knowledge of their history. Alligators, for example, are found only in the rivers of southeastern United States and in the Yangtse River in China. Sassafras, tulip trees and magnolias are found only in the eastern United States, Japan, and eastern China. The explanation for these curious patterns of distribution lies in the fact that early in the Cenozoic era the northern hemisphere was much flatter than at present and the North American continent was connected with eastern Asia by a land bridge at what is now Bering Strait. The climate of the whole region was much warmer than at present, and fossil evidence shows that alligators, magnolia trees and sassafras were distributed over the entire region. Later in the Cenozoic, as the Rockies increased in height, the western part of North America became much colder and drier. During the Pleistocene the ice sheets moving down from the north met the desert and mountain regions of western North America, and the animals and plants that had lived in that region either became extinct or migrated. In southeastern United States and in eastern China were regions untouched by the glaciations and here the alligators and magnolia trees survived. Because the alligators and magnolias of the two regions have been separated for many million years, they have had the opportunity to accumulate different random mutations. They are, thus, slightly different but closely related species of the same genera.

9.10 THE BIOGEOGRAPHIC REALMS

Careful studies of the distribution of plants and animals over the earth have revealed the existence of six major biogeographic realms, each characterized by the presence of certain unique organisms (Fig. 9.9). These realms were originally defined on the basis of the distribution of mammals, but they have proved to be valid for many other kinds of animals and plants as well. The various parts of each realm may be widely separated and have quite different conditions of climate and topography, but it has been possible, during most geologic eras, for organisms to pass more or less freely from one part to another. In contrast, the six realms are separated from each other by major barriers of sea, desert or mountains.

The **Palearctic** realm includes Europe, Africa north of the Sahara desert, and Asia north of the Himalaya and Nan-Ling mountains, plus Japan, Iceland and the Azores and Cape

Figure 9.9 The biogeographic realms of the world with some of their characteristic animals. (From Villee, C. A.: Biology. 7th ed. Philadelphia, W. B. Saunders Co., 1977.)

(Continued on opposite page)

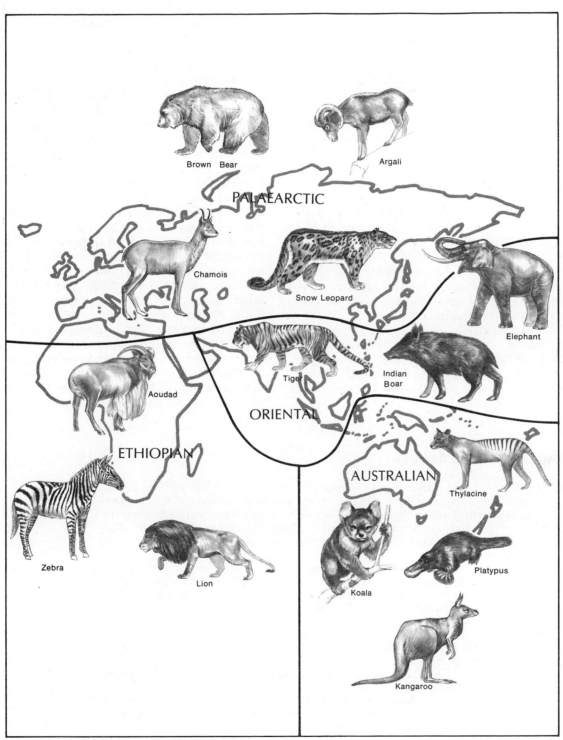

Figure 9.9 (continued)

Verde Islands. The animals indigenous to the Palearctic are moles, deer, oxen, sheep, goats, robins and magpies.

The **Nearctic** realm includes Greenland and North America north of the northern plateau of Mexico. This contains many of the same animals as the Palearctic, plus species of mountain goats, prairie dogs, opossums, skunks, raccoons, bluejays, turkey buzzards and wrentits found nowhere else. The land bridge connecting North America and Asia at Bering Strait in former geologic times permitted the migration back and forth of many kinds of animals and plants. The flora and fauna of the Palearctic and Nearctic realms are similar in many respects and the two are sometimes combined as the **Holarctic** region.

The **Neotropical** realm consists of South America, Central America, southern Mexico and the islands of the West Indies. Its distinctive fauna includes alpacas, llamas, prehensile-tailed monkeys, blood-sucking bats, sloths, tapirs, anteaters and a host of bird species— toucans, puff birds, tinamous and others— found nowhere else in the world.

The part of Africa south of the Sahara, plus the island of Madagascar, composes the **Ethiopian** realm. The gorilla, chimpanzee, zebra, rhinoceros, hippopotamus, giraffe, aardvark, and many birds, reptiles and fishes live only in this realm.

The **Oriental** realm includes India, Ceylon, southeast Asia, southern China, the Malay peninsula and some of the islands of the East Indies— the Philippines, Borneo, Java and Bali. Some of the animals peculiar to it are the orangutan, black panther, Indian elephant, gibbon and tarsier.

Australia, New Zealand, New Guinea, and the remaining islands of the East Indies, those east of Celebes and Lombok, make up the **Australian** realm. The line separating the Oriental and Australian realms, known as Wallace's Line, separates Bali and Lombok, goes through the straits of Macassar between Borneo and Celebes, and passes east of the Philippines. Although the island of Bali and Lombok are separated by a channel only 30 kilometers wide, their respective animals and plants are more unlike than are those of England and Japan, almost on the opposite sides of the world from each other. Native to the Australian realm are the duck-billed platypus, echidna, kangaroo, wombat, koala bear, and other marsupials. Its assortment of curious birds includes the cassowary and emu, the lyrebird, cockatoo and bird-of-paradise.

Why certain animals appear in one region yet are excluded from another in which they are well adapted to survive (and in which they flourish when introduced by man) can be explained only by their evolutionary history.

ANNOTATED REFERENCES

Baldwin, E. B.: An Introduction to Comparative Biochemistry. 4th ed. Cambridge, Cambridge University Press, 1964. A classic exposition of some of the biochemical similarities among animals that aid in defining evolutionary relationships.

Berrill, N. J.: The Origin of Vertebrates. London, Oxford University Press, 1955. Contains excellent descriptions of fossil vertebrates and presents the theory that vertebrates evolved from ascidians.

Bryson, V., and H. J. Vogel (Eds.): Evolving Genes and Proteins. New York, Academic Press, 1965. The proceedings of a symposium held at Rutgers University, containing a series of articles which compare the structures of specific proteins and nucleic acids in different species and show how this evidence may be used to deduce evolutionary relationships.

Colbert, E. H.: Evolution of the Vertebrates. New York, John Wiley & Sons, Inc., 1955. A text recommended for its presentation of information regarding vertebrate fossils.

Dodson, E. O., and P. Dodson: Evolution: Process and Product. 2nd ed. New York, Reinhold Publishing Corp., 1976. A good introductory text describing the general field of evolutionary phenomena.

Jukes, T. H.: Molecules and Evolution. New York, Columbia University Press, 1966. A readable introductory text that relates modern genetics and the evolution of specific molecular structures.

Moore, R. C., C. G. Lalicker and A. G. Fischer: Invertebrate Fossils. New York, McGraw-Hill Book Co., 1952. An excellent source of information regarding extinct invertebrate animals.

Raymond, P. E.: Prehistoric Life. Cambridge, Harvard University Press, 1939. A very readable description of the important vertebrate fossils.

Romer, A. S.: Vertebrate Paleontology. 3rd ed. Chicago, University of Chicago Press, 1966. A comprehensive, well-written text, highly recommended as a source book for further reading.

Shrock, R. R., and W. H. Twenhofel: Principles of Invertebrate Paleontology. New York, McGraw-Hill Book Co., 1953. An excellent text describing the evolution of the invertebrates.

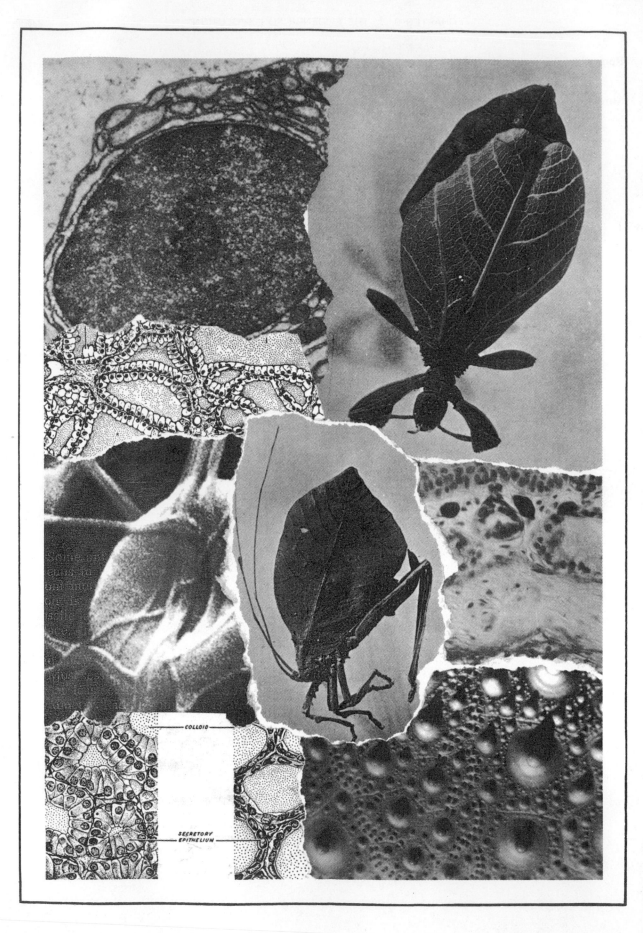

COLLOID

SECRETORY
EPITHELIUM

Part Three

ANIMAL FORM AND FUNCTION

SYMMETRY, FORM AND LIFE STYLE

There is a tremendous diversity of animal life with an almost endless variety of structural and functional modifications. The million or more species described so far are distributed almost everywhere over the earth's surface and are adapted for many different habitats and life styles. Yet through all of this diversity a number of common ground plans can be distinguished. This is true because most animals ultimately share a common evolutionary origin; different groups have diverged at different points along evolutionary lines and thus share many features that developed in earlier ancestral forms. In addition, many unrelated groups have invaded and occupied similar environments and habitats and have thus been subjected to similar selection pressures. Not surprisingly, similar solutions to common problems (convergent evolution) appear frequently in animal evolution.

The following chapters will explore some of the problems of animal existence and some of the solutions that animals have evolved during the more than two billion years they have occupied this planet. We will largely be concerned with multicellular animals, but from time to time reference will be made to the unicellular animal-like protozoans. This chapter will serve to introduce the general architecture of animals and its relationship to animal life styles.

10.1 MOTILITY

Certainly the most distinguishing feature of animals is their ability to move. Although there are many forms that live attached to the substratum, such sessile species have evolved from motile ancestors and can still move parts of the body.

Most animals are bilaterally symmetrical, a type of symmetry in which the body can be divided through only one plane to produce two mirror image halves. **Bilateral symmetry** is a natural adaptation to animal motility. The front, or **anterior,** end of the body is directed forward and meets the environment first. It is thus different from the rear, or **posterior,** end. The **ventral** body surface facing the ground, or bottom **(substratum),** is different from the opposite, or **dorsal,** surface directed upward. The two **lateral** body surfaces contact the environment in much the same way and are similar.

The plane of bilateral symmetry dividing the body into two identical halves passes from anterior to posterior and from dorsal to ventral (Fig. 10.1). This plane is called the **midsagittal** plane (or section). Any plane lateral to the midline is said to be **parasagittal.** A **cross,** or **transverse,** plane passes between lateral surfaces dorsal to ventral. The **frontal** plane passes from anterior to posterior but between lateral surfaces and is perpendicular to the other two planes.

Associated with motility and bilateral symmetry is the development of a head, or **cephalization.** Sense organs and associated nervous tissues tend to be more highly concentrated at the anterior end than elsewhere on the body. The mouth is also most frequently located at the anterior end and associated with the head. However, there are animals in which the mouth is located more posteriorly along the body axis.

Figure 10.1 Diagram to illustrate transverse, sagittal and frontal planes in a bilaterally symmetrical animal.

Some animals are **sessile,** attached by some means to the substratum. They have evolved from motile ancestors, and the sessile condition is clearly a secondary one. Although sessile animals cannot move from place to place, most possess free-swimming larval stages, and almost all can move some part of the body.

Most sessile animals are radially symmetrical or at least exhibit a tendency toward **radial symmetry.** Radial symmetry is an arrangement of similar parts around a central body axis (Fig. 10.2). Such a symmetry is adaptive for a sessile life style, for the organism can meet its environment equally from all directions. Plants typically exhibit a radial body plan.

Many zoologists have concluded that the first multicellular animals probably were radial even though they were motile. They probably moved through the water with one pole forward (Fig. 10.3A) and would thus have possessed anterior and posterior ends but no dorsal and ventral surfaces. From such ancestors the sessile and radial hydras, sea anemones and corals may have evolved. If so, their radial symmetry would be *primary*; i.e., would be like that of their immediate ancestors. The sea stars and sea urchins living today are radial but motile (Fig. 10.4). Moreover, their radial symmetry appears to be *secondary*; i.e., the ancestors of these animals were bilateral and free-swimming, for the free-swimming development stages, or larvae, of sea stars and sea urchins are bilateral (Fig. 10.4). In the evolutionary history of the group, some ancestral form perhaps became attached and the more adaptive radial symmetry evolved. Such a stage is suggested by the living sea lilies (Fig.

Figure 10.2   Radial symmetry. A, Sea anemone, an animal with primary radial symmetry. B, Sessile barnacles, bilateral animals with some secondary radial features. (By Betty M. Barnes.)

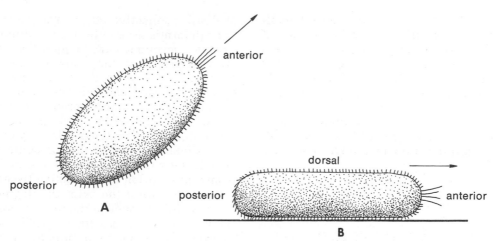

Figure 10.3 Hypothetical ancestral metazoans. *A,* Free-swimming, radially symmetrical ancestor. *B,* Creeping, bilaterally symmetrical ancestor.

10.4). The sea stars and sea urchins perhaps evolved from sessile forms that lost the attached condition and became secondarily motile.

Aside from echinoderms (sea stars, sea urchins and relatives), most attached animals with a bilateral ancestry show only a slight tendency toward radial symmetry. Barnacles, for example, are basically bilateral, but the outer protective plates are arranged in a ring around the body (Fig. 10.2 *B*).

If the first animals were radial even though

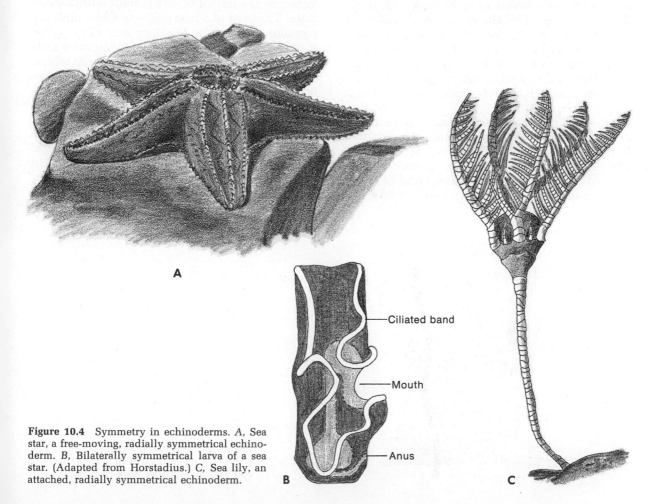

Figure 10.4 Symmetry in echinoderms. *A,* Sea star, a free-moving, radially symmetrical echinoderm. *B,* Bilaterally symmetrical larva of a sea star. (Adapted from Horstadius.) *C,* Sea lily, an attached, radially symmetrical echinoderm.

Ciliated band

Mouth

Anus

motile, what led to the evolution of bilaterality? Perhaps some early population of the ancestral radial form took up an existence close to the bottom, creeping or even swimming over the surface (Fig. 10.3B). They would already have been elongate with an anterior and posterior end. With the assumption of a bottom habit, one surface became directed toward the bottom. Ventral and dorsal sides could then have differentiated, and the animal would have become bilateral.

10.2 ANIMAL ARCHITECTURE

The vast majority of animals have bilateral symmetry, with a relatively uniform basic design. The mouth is typically located at the anterior end and the gut tube parallels the anterior-posterior axis of the body.

The animal's outer casing is provided by the body wall, covered on the exterior by the **integument,** or skin. The integument is composed of one or more layers of epithelial cells (Figs. 10.5 and 10.6). Certain of the epithelial cells may be specialized for the formation of such structures as hairs, spines, feathers and nails. In a number of animal groups, such as crabs and clams, all or part of the integumentary epithelium secretes an outer nonliving shell or a cuticular covering.

In some animals, including vertebrates, a layer of connective tissue underlies the epithelial portion of the integument (Figs. 10.6 and 10.7). Thus the vertebrate integument consists of an outer **epidermis** composed of stratified squamous epithelium and an inner **dermis**

of dense connective tissue. The basal layer of the epidermis is a columnar epithelium, the **stratum germinativum,** which by continual mitosis produces the cells to the outside. These cells change shape as they shift outward, becoming first cuboidal and then flattened (squamous). During this process they also accumulate the horny insoluble protein **keratin.** The outermost cells are continually or periodically sloughed off. The dermis contains blood vessels, nerve endings, glands and often fat deposits, in addition to connective tissue.

Pigment within the skin, the degree of vascularity and, in some animals, the presence of refractive granules or surface striations interact to produce brilliant coloration. Some color patterns conceal the animal from predators or enable an animal to lie unseen as it awaits its prey, but others are used for species recognition or as warning signals to frighten another animal or establish territory. Squids, shrimp, crabs and lower vertebrates, such as the frog (Fig. 10.6), may have several kinds of pigment contained in stellate cells known as **chromatophores.** Changes in general color tone are effected by the migration of pigment within these cells. The skin of a frog darkens when melanin pigment streams into the processes of the cells; it lightens when the pigment concentrates near the center of the cell (Fig. 10.6). Pigment migration is controlled by nerves in some vertebrates and by hormones in others. Chromatophores are rare in mammals, but melanin is present both within and between basal cells of the epidermis. Some melanin is present in the skin of all human beings (except albinos, p. 160), but it is especially abundant in the skin of Blacks.

The integument is the barrier between the ex-

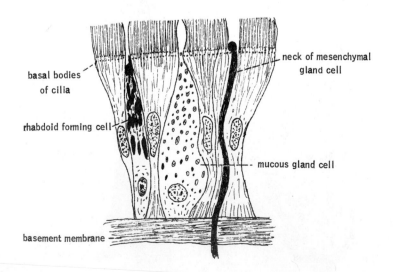

basal bodies of cilia

neck of mesenchymal gland cell

rhabdoid forming cell

mucous gland cell

basement membrane

Figure 10.5 Section through the integument of a flatworm composed of a single layer of epithelial cells. (After Bock from Hyman.)

Stratum corneum
Stratum germinativum
Chromatophore
Mucous gland
Blood vessel
Epidermis
Dermis
Nerve

Figure 10.6 A photomicrograph of a vertical section through the skin of a frog.

Capillary loop in papilla
Sebaceous gland
Arrector pili m.
Hair follicle
Hair matrix
Sweat gland
EPIDERMIS
DERMIS
SUB-CUTANEOUS TISSUE

Figure 10.7 Diagrammatic vertical section through human skin. (Modified from Pillsbury, Shelley and Kligman; Manual of Cutaneous Medicine, 1961.)

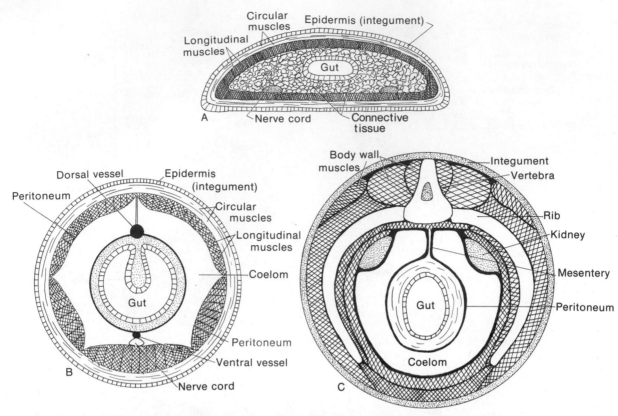

Figure 10.8 Diagrammatic cross sections through the middle of the body of three bilateral metazoans. *A*, Flatworm; *B*, earthworm; *C*, vertebrate.

ternal environment and the inner environment of the animal. In addition to an obviously important role in protection, it may also be involved in various regulatory functions that will be discussed in later chapters.

Beneath the integument the body wall is composed of one to several muscle layers; the number and arrangement of the layers vary from group to group. Earthworms, for example, possess an outer circular and an inner longitudinal layer of muscle (Fig. 10.8*B*). The vertebrate body wall is composed of distinct muscles that are oriented in various ways in different regions of the body (Fig. 10.8*C*).

The space between the body wall and the more central gut tube may be filled with tissue or fluid. In more primitive animals, such as flatworms, the space is filled with connective tissue (Fig. 10.8*A*). The fluid-filled condition is of much wider occurrence. The fluid lies within a body cavity, or **coelom,** which is lined by a layer of epithelial cells, called **peritoneum** (Fig. 10.8*B*). Some internal organs bulge into

the coelom but from behind the peritoneum; i.e., they are **retroperitoneal;** other organs, such as the gut of vertebrates, push so far into the coelom that they are suspended by a fold of peritoneum. The suspension is called a **mesentery** (Fig. 10.8*C*).

The structural design of many animals involves **metamerism,** the division of the body into a linear series of similar parts, or *segments* (Fig. 10.9). The segmental repetition of parts includes both internal and external structures. The anterior part of the body bearing the sense organs and brain is not a segment, nor is the terminal section, which bears the anus. These nonsegmental parts are called the **acron** and **pygidium** respectively. The acron makes up a major part of a metameric animal's head. New segments are formed at the rear, in front of the pygidium. Metamerism is believed to have evolved twice in the Animal Kingdom, once in the ancestors of the segmented worms (annelids) and arthropods and once in the ancestors of the vertebrates. In both it is believed to have

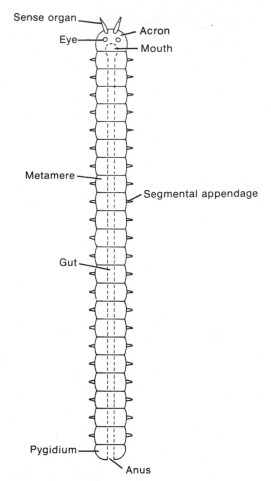

Figure 10.9 Diagram of a metameric animal, such as an annelid worm.

represented an adaptation for locomotion and will be discussed in the next chapter.

There has been a tendency in all groups of metameric animals for a varying number of anterior trunk segments to join the acron, forming a more complex "head." Indeed, in vertebrates, the acron is vestigial at most, and the "head" is actually composed of a large number of anterior segments.

10.3 SIZE

The size of multicellular animals varies over a wide range. At one extreme are microscopic forms; at the other are animals of vast bulk, such as elephants and whales. Many structural and functional features of animals are corre-

lated with their size. As size increases, volume (or weight*) increases by the cube, whereas surface area increases by the square of the linear dimension (Fig. 10.10). Thus in large animals there is proportionately less surface area to volume than in small animals, and various modifications have evolved that provide for the inadequacy. Very small animals with a large ratio of surface to volume may be subject to greater loss of heat and water than larger species.

There is a common misconception that small animals are always primitive and large animals are specialized. Small size, or miniaturization, has also been an important part of the adaptive evolution of a number of animal groups, such as mites and bryozoans.

10.4 COLONIAL ORGANIZATION

The majority of animals are solitary; i.e., individuals of a given species live independently of each other except for sexual pairing. Other kinds of animals cannot exist except in close association; such forms are said to be **colonial.** Colonial organization has evolved numerous times in the Animal Kingdom but is one of two types. In many colonial animals, including the familiar corals, the individuals within the colony are connected anatomically (Fig. 10.11). All individuals arise by budding except the first, which develops from the fertilized egg. In the colonies of other animals such as ants, bees, termites, baboons and humans, the individuals are not anatomically connected but live together and are functionally or behaviorally interdependent. Such groups of animals are usually said to be *social* colonies.

Polymorphism, or the specialization of forms for the division of labor, characterizes many colonial species. Individuals have become specialized for functions such as feeding, defense or reproduction. The worker and soldier castes of ants are more familiar examples of polymorphism in social insects (Fig 29.58), but structurally united colonies may also exhibit relatively complex polymorphism. Not all colonial animals are polymorphic, but polymorphism occurs only in species with colonial organization.

*Changes in weight and volume are usually proportional.

Cube Side	Surf. Area	Vol.	Surf. Area:Vol. Ratio
1 cm.	1 sq. cm.	1 c. c.	1:1
10 cm.	100 sq. cm.	1000 c. c.	1:10
100 cm.	10,000 sq. cm.	1,000,000 c. c.	1:100
1000 cm.	1,000,000 sq. cm.	1,000,000,000 c. c.	1:1000

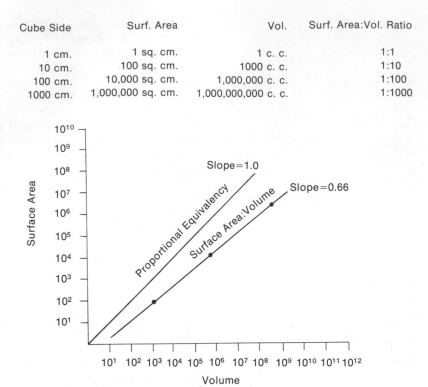

Figure 10.10 Relationship of surface area to volume in four cubes of increasing size. When plotted logarithmically the slope is 0.66, as opposed to a slope of 1.0 if the relationship were proportional.

Figure 10.11 Part of a coral colony. The individuals, some of which are contracted and some of which are expanded, are connected by a lateral sheet of tissue.

10.5 PREDICTABILITY IN ANIMAL DESIGN

The sea is the ancestral environment of animals. From the sea there have been numerous invasions to fresh water and land. Of these three major environments the sea is in many ways the most uniform and least stressful. Oxygen is usually adequate, there is no danger of desiccation (drying up), and the salt content of the sea is somewhat similar to that of extracellular fluids.

The salt content of fresh water, on the other hand, is far below that of tissue fluids. Pools, ponds and even streams may periodically dry up, streams fluctuate in velocity and turbidity, and oxygen may be absent or at very low levels, as at the bottom of some lakes. On land oxygen is plentiful, but desiccation is an ever present danger. Each of these three major environments poses different problems of existence, and a variety of different structural and functional adaptations have evolved in the animals inhabiting them.

Many structural and functional adaptive themes appear over and over again in animal design, for they are often related to size, life style or environment. A knowledge of these relationships can have great predictive value. Thus if the size of an animal, its environment and something of its life style are known, it is possible to predict, or make educated guesses, about many aspects of its structure and physiology. The following chapters will provide a basis of understanding for making many such predictions.

ANNOTATED REFERENCES

The following general references contain a great deal of information on the anatomy and physiology of organ systems.

Alexander, R. McN.: Animal Mechanics. London, Sidgewick and Jackson, 1968. Numerous physical factors that apply to the structure and movement of animals are considered.

Alexander, R. McN.: The Chordates. Cambridge, Cambridge University Press, 1975. An excellent analysis in functional terms of the structure of selected chordates.

Barnes, R. D.: Invertebrate Zoology. 3rd ed. Philadelphia, W. B. Saunders Company, 1974. A very useful textbook and reference on the anatomy, physiology, ecology and classification of invertebrates.

Beck, W. S.: Human Design. New York, Harcourt Brace Jovanovich, 1971. A carefully integrated account of form and function in man from the molecular to the organ level.

Bloom, W., and D. W. Fawcett: A Textbook of Histology. 10th ed. Philadelphia, W. B. Saunders Co., 1975. The microscopic and ultrastructure of the cells and tissues of mammalian organ systems are thoroughly described.

Fretter, V. and Graham, A.: A Functional Anatomy of Invertebrates. New York, Academic Press, 1976. A review of major invertebrate groups with an emphasis upon the function of their major anatomical features.

Gordon, M. S., et al.: Animal Function: Principles and Adaptations. New York, Macmillan, 1968. A useful textbook of comparative physiology.

Guyton, A. C.: Textbook of Medical Physiology. 5th ed. Philadelphia, W. B. Saunders Co., 1976. A detailed consideration of mammalian physiology is presented in this standard textbook.

Prosser, C. L. (Ed.): Comparative Animal Physiology. 3rd ed. Philadelphia, W. B. Saunders Co., 1973. A very valuable source book on the physiology of both invertebrates and vertebrates.

Romer, A. S.: The Shorter Version of the Vertebrate Body. 4th ed. Philadelphia, W. B. Saunders Co., 1971. The morphological aspects of the evolution of vertebrates are thoroughly considered in this widely used textbook of comparative anatomy.

Ruch, T. C., and H. D. Patton (Eds.): Physiology and Biophysics. 20th ed. Philadelphia, W. B. Saunders Co., 1974. An advanced text and source book on mammalian physiology.

Schmidt-Nielsen, K.: Animal Physiology, Adaptation and Environment. Cambridge, Cambridge University Press, 1975. A physiological account of the interrelationships between animals and their environments with emphasis upon oxygen, food, temperature, water, movement and integration.

Wessels, N. K. (Ed.): Vertebrate Adaptations. San Francisco, W. H. Freeman and Company, 1969. A collection of articles from the Scientific American, including many cited in subsequent chapters. Each group of articles on an adaptive topic is preceded by a general introduction by the editor.

Young, J. Z., and M. J. Hobbs: The Life of Mammals. 2nd ed. New York, Oxford University Press, 1975. Structure and function are carefully considered in this thorough account of the gross and microscopic anatomy and the physiology of mammals.

Young, J. Z.: The Life of Vertebrates. 2nd ed. Oxford, Clarendon Press, 1962. One or more chapters are devoted to the anatomy, physiology and evolution of each of the classes of vertebrates in this excellent text.

SUPPORT AND MOVEMENT

Each organism requires a suitable place in which to live, obtain food and other essential compounds, eliminate wastes and reproduce. These activities are facilitated by the animal's ability either to move through its environment or, if sedentary, to help move the environment across it. There are many kinds of movement; a common one involves skeletal elements that support the body and transfer muscle forces.

11.1 SKELETONS

Functions of Skeletons. Most animals have some sort of skeleton, within or on the outside of their bodies, a major function of which is to maintain body shape and provide support. The support may be derived from hard tissues or from the turgidity of a liquid. Support is particularly important in large organisms and those that live out of water. Aquatic animals, especially small ones, can survive without a skeleton because water provides a great deal of buoyancy. Only a few terrestrial animals, certain worms and slugs, lack a skeleton; little support is afforded by air, which is much less dense than water.

Support, however, is not the only skeletal function, for a skeleton often is present in small animals in which support is not needed. Fish only a few millimeters long and insects less than one millimeter in length have skeletons. Protection of the organism or certain of its internal parts frequently is an important skeletal function. Sessile animals in particular need protection, for they cannot retreat to shelter if the need arises. Depending on the requirements of the particular species, skeletons may also store calcium, phosphorus and

other important minerals, provide a site for the manufacture of blood cells, and transmit forces from muscles to the point of force application. Skeletal parts are used in many movements, including biting and locomotion.

Skeletal Materials and Types. Most skeletons are composed of hard parts, hence they are relatively rigid except at joints. Calcium phosphates and carbonates are widely used and are found in groups ranging from protozoans to vertebrates. Calcium is present as calcium carbonate in invertebrate skeletons but as calcium phosphate in vertebrate skeletons. Salts of magnesium, strontium and other substances are sometimes used. Siliceous compounds form skeletons in some protozoans and sponges. Skeletons may be composed of specialized organic compounds or complexes of organic and inorganic materials. The latter are particularly strong, for the inorganic crystals resist compression, and the organic component resists tension and provides elasticity. **Spongin** fibers contribute to the skeletons of many sponges; **conchiolin** is a part of the molluscan shell; **chitin** makes up most of the arthropod skeleton; and **collagen** is an important constituent of vertebrate cartilage and bone (p. 58).

In most invertebrates the hard materials are secreted on the outside of the body, hence they form **exoskeletons.** In contrast, vertebrates (together with some invertebrates) have skeletons located within the body wall or deeper body tissues. These are **endoskeletons.** Not all skeletons are composed of hard materials. In the earthworm, for example, a turgid column of liquid, called a **hydroskeleton,** in the gastrovascular cavity, coelom, blood vessels or other body spaces can provide support or induce movement.

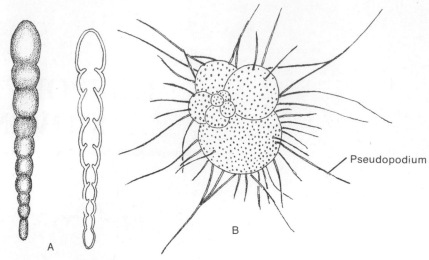

Figure 11.1   Foraminiferans. *A*, Shell of *Rheophax nodulosa* (entire and in section); *B*, living *Globigerina*. (After Cushman from Calvez.)

A few representative skeletons are considered here to illustrate their variety, advantages and limitations; others will be seen in the survey of the Animal Kingdom.

Some Exoskeletons.   Foraminifera, marine protozoans closely related to amebas, secrete a chambered, calcareous shell that is perforated by many small openings through which cytoplasmic processes, **pseudopodia,** extend (Fig. 11.1). The shell is composed at first of a single chamber with an opening at one end. When the animal outgrows the first chamber, it secretes a new one. This process continues, producing a multichambered shell occupied by

Figure 11.2   Skeletons of madreporarian, or stony, corals. The cups in which the polyps are located are widely separated in *A* and close together in *C. D,* A brain coral, in which the polyps are arranged in rows. (By Betty M. Barnes.)

one individual. The shape and arrangement of the shell varies according to the species. Accumulations of foraminifera shells are important components of many marine sediments and limestones.

Stony corals are made up of thousands of interconnected individuals, **polyps,** that secrete a calcium carbonate skeleton beneath themselves and around their sides, so that the lower part of the polyp typically is located in a skeletal cup (Fig. 11.2). Radiating calcareous septa in the bottom of the cups project up into folds at the base of the polyps. As polyps grow and multiply, new material is secreted and added to material already present. Different species of coral display various growth forms. Some are low and encrusting, others are upright and branching. Coral reefs are composed largely of the exoskeletons of innumerable generations of stony corals. Living individuals are found only on the surface of the upper part of the reef, but the reef may extend down hundreds or even thousands of feet.

Chitons, snails, clams and nearly all other mollusks are covered by a shell secreted over most of the animal's surface by a modified body wall known as the **mantle.** Molluscan shells have evolved into many shapes: a single spiral in snails, bivalved in clams and chambered in the nautilus. The first formed part of the shell is a horny protein called **conchiolin,** but as the shell grows it is strengthened by secretions of calcium carbonate. This occurs most actively around the edge of the mantle, although some new material is added to older parts of the shell. Thus the shell increases in diameter and thickness at the same time. The surface of the shell, the **periostracum,** is composed of conchiolin, and beneath this there are several layers of calcium carbonate deposited within an organic framework. The calcium carbonate may be in the form of prisms or lathlike layers of aragonite and calcite. The innermost layer is sometimes very smooth and pearl-like, or nacreous.

Arthropods, including crabs, insects and spiders, have an exoskeleton, or **cuticle,** composed of the polysaccharide **chitin** and protein

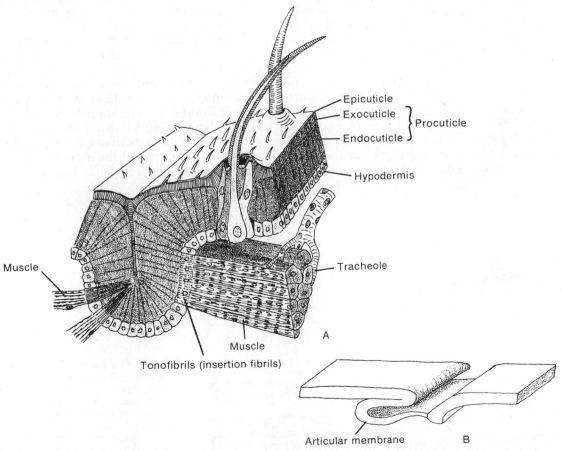

Figure 11.3 *A,* Three dimensional section of an arthropod integument (insect). *B,* An intersegmental joint. Articular membrane is folded beneath segmental plate. (*A* after Weber from Kaestner; *B* after Weber from Vandel.)

bound together to form a complex glycoprotein. Calcareous materials are incorporated into the exoskeletons of crabs and other crustaceans. The exoskeleton covers the entire body surface and often extends inward to line much of the digestive tract, as well as the air-transporting tubes (tracheae) of insects.

The exoskeleton is secreted by an underlying layer of integumentary epithelial cells known as the **hypodermis** (Fig. 11.3A). It is composed of a thin, surface **epicuticle** and a much thicker **procuticle,** which is subdivided into an **exocuticle** and **endocuticle.** The epicuticle, especially of terrestrial species, contains waxes; when these are absent, the exoskeleton is relatively permeable to water and gases. The exocuticle is stronger than other parts, for it has been tanned; its molecular structure has been stabilized by reactions with phenols and by the formation of additional cross linkages in the protein molecules. Sensory processes are borne on the exoskeleton, and it frequently is perforated by a few pores through which secretions are discharged.

The exoskeleton of the arthropod is divided into a series of strong, tubelike segments connected to each other by thinner, folded **articular membranes** where movement occurs (Fig. 11.3B). The skeleton protects the body and provides attachment for the muscles that lie inside of it. So long as the animal remains small, the skeleton is very strong in relation to its weight, for a cylinder can support more weight without bending than can a solid rod composed of the same amount of material. This advantage is quickly lost as the animal gets larger and the radius of the cylinder increases, for a wide cylinder must have dispro-

portionately thicker walls to resist collapse. Most arthropods are small, and only a few aquatic crabs and lobsters attain moderate size.

A major disadvantage of the arthropod exoskeleton is its effect upon growth. Periodically, the hypodermis detaches from the exoskeleton and secretes enzymes—chitinase and proteinases—into a **molting fluid** that accumulates beneath the skeleton (Fig. 11.4). The untanned endocuticle is gradually digested. As this is occurring, a new skeleton is secreted beneath the old one. Waxes and other lipids in its epicuticle prevent it from being digested. Eventually, the old exoskeleton ruptures along molting lines, which represent lines of weakness where the endocuticle of which they were composed has been digested, and the animal pulls out of its old encasement. The new skeleton is soft and pliable for a few hours or days and is stretched to accommodate the increased size of the animal. Animals that have recently molted are very vulnerable to predators, for they are soft and their skeletons, being weak, do not transmit muscle forces well. Insects avoid this problem, in part, by molting and growing only as larvae; adult insects do not grow.

Some Endoskeletons. An endoskeleton does not protect the entire organism as effectively as an exoskeleton, but it transmits muscle forces well, and the animal need not undergo molting in order to grow. Although they are not limited to large animals, endoskeletons are particularly effective for support in large animals such as vertebrates. For example, one set of cylinders around the outside of our body would have to be very thick, heavy and cum-

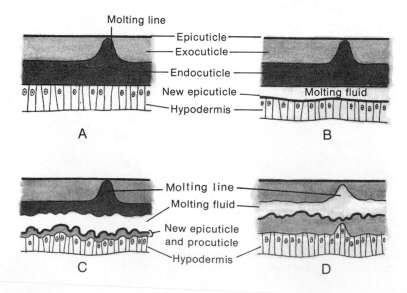

Figure 11.4 Molting in an arthropod. A, The fully formed exoskeleton and underlying epidermis between molts. B, Separation of the epidermis and secretion of molting fluid and the new epicuticle. C, Digestion of the old endocuticle and secretion of the new procuticle. D, The animal just before molting, encased within both new and old skeleton.

Figure 11.5 The internal, siliceous skeleton of a radiolarian. (Courtesy of E. Giltsch, Jena.)

bersome, but a series of smaller internal cylinders and rods can provide the support needed with a minimum of weight.

Endoskeletons are found in many different phyla. Most radiolarians, marine protozoa closely allied to amebas and foraminifera, have very delicate internal skeletons composed of elaborate silica lattices (Fig. 11.5). Radiolarians are relatively large planktonic organisms; some reach several millimeters in diameter. Their skeletons provide a large area with minimal weight and help to keep these organisms afloat. These skeletons are an important component of ocean deposits, especially at great depths and pressures where the calcareous shells of foraminifera dissolve.

Sponges are supported and their shape maintained by a network of internal mineral spicules or organic spongin fibers, or both, that

Figure 11.6 Endoskeleton of echinoderms. *A*, Surface view of the lattice-like arrangement of the ossicles of a starfish body wall after removal of overlying tissues. *B*, Similar view of a portion of a sea urchin test, showing the tubercles to which spines articulate, paired perforations for tube feet and junction lines between ossicles. *C*, Lateral view of the common Atlantic sea urchin, *Arbacia punctulata*, showing long spines. (*A*, after Fisher from Hyman; *B* and *C*, courtesy of Betty M. Barnes.)

are secreted by wandering cells called **amebocytes.** Depending upon the class of sponge, the spicules are composed of calcium carbonate or of silicon dioxide. It is the spongin skeleton that we use in natural bath sponges.

The body walls of starfish, sea urchins and most other echinoderms are supported by in-

ternal calcareous pieces, or **ossicles,** that are interconnected in a lattice-like array in the wall of starfish but form a nearly solid internal shell, or **test,** in sea urchins. The test is perforated by many small openings for various organs (Fig. 11.6A and B). Spines that develop beneath the epidermis attached to many ossicles are long and mobile in sea urchins for

Figure 11.7 A and B, Diagrammatic frontal section through two annelid segments. External arrows indicate direction of force exerted by body wall muscles. Internal arrows indicate direction of force exerted by coelomic fluid pressure: (A) during circular muscle contraction and (B) during longitudinal muscle contraction. C, Diagram showing mode of locomotion of an earthworm. Segments undergoing longitudinal muscle contraction are marked with larger dot and drawn twice as wide as those undergoing circular muscle contraction. The forward progression of a segment during the course of several waves of circular muscle contraction is indicated by the horizontal lines connecting the same segments. (After Gray and Lissmann.)

which they serve as organs of locomotion (Fig. 11.6C).

The skeleton of fishes, frogs, cats and other vertebrates is an endoskeleton composed of many individual bones and cartilages joined to each other in such a way as to provide a strong but moveable framework. Some parts—for example, the bony scales of fishes and the plates of a skull—may lie in a very superficial position, yet they develop within or just beneath the skin. The vertebrate skeleton is discussed in detail in Section 11.2.

Hydroskeletons. Hydroskeletons may occur in conjunction with other skeletons, or they may provide the primary skeletal support. The body cavity of the earthworm, its **coelom,** is filled with a **coelomic fluid.** Contraction of circular muscles in the body wall increases the pressure of the fluid, stiffens the body and forces it to elongate, stretching longitudinal muscles in the body wall in the process (Fig. 11.7A). Contraction of longitudinal muscles also increases coelomic pressure but causes the body to shorten and widen as circular muscles are stretched (Fig. 11.7B). In an unsegmented, wormlike, coelomate animal the coelomic fluid forms a continuous hydroskeleton, so that alternate contraction of circular and longitudinal muscles results in lengthening and shortening of the entire animal. The partitioning of the coelom and muscles into discrete segments in the earthworm and most other annelids enables body lengthening and shortening to be confined to a few segments at a time. Locomotor movements can be more precise. Alternate waves of elongation and contraction move posteriorly along the length of the animal (Fig. 11.7C). First the front of the body is elongated and pushed forward. Then this part shortens and widens, bristles in the body wall anchor it to the substratum, and more posterior parts of the body are pulled forward. The segmentation, or metamerism, of annelids probably evolved as a burrowing mechanism. It is basically a modification of the coelom and body wall muscles, but the metanephridia, ganglia and blood vessels are also segmentally arranged (p. 325).

The segmented exoskeleton of arthropods normally is moved by antagonistic sets of muscles within it. The segments in spider legs are an exception. They are bent, or flexed, upon each other by flexor muscles, but there are no antagonistic muscles that can straighten, or extend, the limb. Extension results from the high liquid pressure within the open circulatory spaces of the limb. Here we see an interesting interaction between an exoskeleton and

hydraulic pressure that extends the limbs and holds them in this position.

Amphioxus, one of the lower chordates, has a **notochord,** a longitudinal cord of cells filled with liquid; its turgidity is maintained by a firm, surrounding connective tissue sheath. The notochord is a hydroskeleton and is analogous to a rubber tube filled with water under pressure. It can bend yet is firm enough to provide support and resist shortening of the body. Contraction of the segmented, longitudinal muscles on either side of the body wall results, therefore, in side to side undulatory movements that sweep posteriorly along the body. All chordates, including vertebrates, have a notochord as larvae or embryos. It is replaced in most vertebrate adults by the vertebral column. Metamerism of the muscles of chordates evolved independently of annelid-arthropod segmentation. It too is related to locomotion but, in this case, to lateral undulations, rather than to burrowing.

Another hydroskeletal mechanism is seen in the copulatory organ, or **penis,** of many male animals. The penis of a mammal, for example, is flaccid most of the time, but during copulation it enlarges and becomes turgid as a result of an enlargement of arteries and an influx of blood into numerous vascular spaces. Distension of vascular spaces presses on veins leaving the penis and temporarily slows down venous return. Other hydroskeletons occur occasionally in different groups, e.g., the foot of a clam and the tube feet of echinoderms.

11.2 THE VERTEBRATE SKELETON

Parts of the Skeleton. During embryonic development most of the vertebrate skeleton is composed of cartilage, but the cartilage is replaced by bone in most adults. This bone is called **cartilage replacement bone** to distinguish it from the **dermal bone** that develops within or just beneath the skin without any cartilaginous precursor. These two types of bone differ only in their mode of development; they are similar histologically. Dermal bone forms the bony scales of fishes, which will be considered with the groups they characterize, and plates that contribute to the skull and pectoral girdle.

For purposes of description, the skeleton can be divided into a **somatic skeleton,** located in the body wall and appendages, and a deeper **visceral skeleton,** associated primitively with the pharynx wall and gills. The somatic skeleton contains both dermal and cartilage replacement bones, but only cartilage replacement bones are in the visceral skeleton. The skeleton can be further subdivided as follows:

Somatic skeleton (skeleton of the body wall)
  Axial skeleton (vertebral column, ribs, sternum and most of the skull)
  Appendicular skeleton (girdles and limb bones)
Visceral skeleton (skeleton of the pharyngeal wall, primitively associated with the gills)

The Fish Skeleton. The parts of the skeleton can be seen more clearly in a dogfish (a small shark) than in terrestrial vertebrates (Fig. 11.8). The dogfish skeleton is typical of the skeleton of primitive vertebrates, except that the skeleton is entirely cartilaginous. The failure of the dogfish's skeleton to ossify is believed to represent the retention of an embryonic condition rather than a primitive adult condition. The vertebral column is composed of vertebrae, each of which has a biconcave **centrum** that develops around and largely replaces the embryonic notochord. Dorsal to each centrum is a **vertebral arch** surrounding the **vertebral canal** in which the spinal cord lies, short **ribs** attach to the vertebrae, a sternum is absent, and the individual vertebrae

are rather loosely held together. A strong vertebral support is not required in the aquatic environment.

Most of the skull of the dogfish is an odd-shaped box of cartilage encasing the brain and major sense organs. This belongs to the axial skeleton and is known as the **chondrocranium.** It forms the core of the skull of all vertebrates. Other basic components of a vertebrate skull include the anterior arches of the visceral skeleton and dermal bones that encase the chondrocranium and anterior visceral arches. These dermal bones have been lost during the evolution of cartilaginous fishes, but they were present in the fishes ancestral to tetrapods.

The visceral skeleton consists of seven pairs of >-shaped visceral arches. The arches are hinged at the apex of the >; they are interconnected ventrally but are free dorsally. Each arch lies in the wall of the pharynx and supports gills in very primitive vertebrates. In jawed vertebrates the first or **mandibular arch** becomes enlarged and, together with associated dermal bones, forms the upper and lower jaws. The entire jaw in the dogfish is derived from the mandibular arch; there are no surrounding dermal bones. The second or **hyoid arch** of the dogfish helps to support the jaws. Its dorsal portion, known as the **hyoman-**

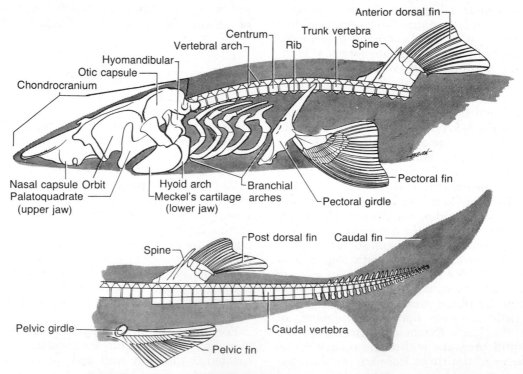

Figure 11.8 A lateral view of the skeleton of a dogfish.

dibular, extends as a prop from the otic capsule (the part of the chondrocranium housing the inner ear) to the angle of the jaw. The third to seventh visceral arches are known as branchial arches; they support the gills, and gill slits lie between them.

The appendicular skeleton is very simple in the dogfish. A U-shaped bar of cartilage, the pectoral girdle, lies in the body wall posterior to the gill region and supports the pectoral fins. The pelvic girdle is a transverse bar of cartilage in the ventral body wall anterior to the termination of the digestive tract. It supports the pelvic fins but is not connected with the vertebral column.

The Tetrapod Skeleton. The numerous changes that have occurred in the evolution of the skeleton from primitive fishes are correlated with the shift from water to land, and with the greater activity of the tetrapod. They can be understood by a study of the frog and human skeletons (Figs. 11.9 and 11.10). The vertebral column must support the weight of the body in all tetrapods, and it has become much stronger. It is thoroughly ossified, and the individual vertebrae are strongly united by overlapping articular processes borne on the vertebral arches. A well-developed vertebral spine extending dorsally from the arch, and lateral transverse processes serve for the attachment of ligaments and muscles. Ribs articulate on the centrum and transverse processes.

Correlated with changes in the method of locomotion and the independent movement of various parts of the body, there is more regional differentiation of the vertebral column. All tetrapods have at least one neck or cervical vertebra, the atlas, that articulates with the skull, a series of trunk vertebrae, one or more sacral vertebrae that articulate with

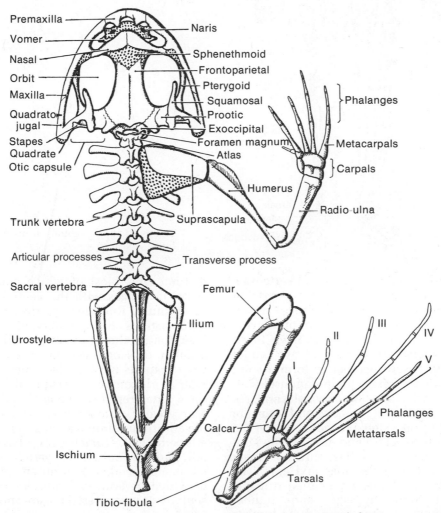

Figure 11.9 A dorsal view of the frog's skeleton. Major cartilaginous areas are stippled. Roman numerals refer to digit numbers. (After Parker and Haswell.)

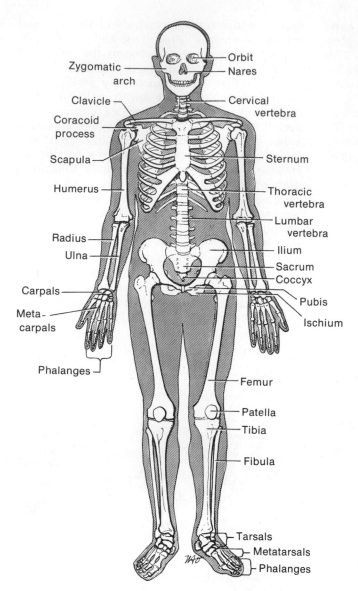

Zygomatic arch
Orbit
Nares
Clavicle
Cervical vertebra
Coracoid process
Scapula
Sternum
Humerus
Thoracic vertebra
Lumbar vertebra
Radius
Ilium
Ulna
Sacrum
Coccyx
Carpals
Pubis
Meta- carpals
Ischium
Phalanges
Femur
Patella
Tibia
Fibula
Tarsals
Metatarsals
Phalanges

Figure 11.10 A ventral, or anterior, view of the human skeleton.

the pelvic girdle, and tail, or **caudal vertebrae.** Amphibians have a single cervical vertebra and only one sacral vertebra. Frogs, which have a jumping method of locomotion, have a short trunk in which short ribs have fused onto the transverse processes, and two caudal vertebrae have fused together to form a solid **urostyle** on which certain jumping muscles attach. Mammals have seven cervical vertebrae of which the first two, the atlas and **axis,** are modified to permit extensive movement of the head. The skull rocks up and down at the joint between the skull and atlas; turning movements occur between the atlas and axis. The vertebral column of the mammalian trunk is differentiated into thoracic and lumbar regions. We have 12 **thoracic** and six **lumbar vertebrae.** Only the thoracic vertebrae bear

distinct **ribs,** most of which connect via the costal cartilages with the ventral breast bone, or **sternum.** Rudimentary ribs, present in the other regions during embryonic development, fuse onto the transverse processes. The number of sacral vertebrae has increased as the tetrapods have evolved more effective terrestrial locomotion. Typically, amphibians have one, reptiles two and mammals three. Our greater number (five) is correlated with the additional problems of support inherent in a bipedal gait. Human beings have no external tail, and the four caudal vertebrae that remain are fused together to form an internal **coccyx** to which certain anal muscles attach. Representative human vertebrae are shown in Figure 11.11.

The skull of a frog or human (Figs. 11.9

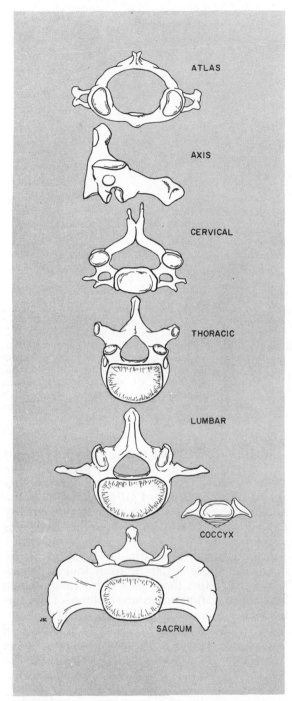

Figure 11.11 Types of mammalian vertebrae as represented by those of a human being. The atlas is seen in lateral view; the others are viewed from the anterior end. (From Villee, C. A.: Biology. 7th ed. Philadelphia, W. B. Saunders Co., 1977.)

and 11.10) can be divided into a portion housing the brain and encasing the inner ear, the **cranium** and the bones forming the jaws and encasing the eyes and nose, the **facial skeleton.** The mammalian brain and cranium

are very large. The eyes are lodged in sockets known as **orbits.** A **temporal fossa** lies posterior to each orbit in mammals and contains certain of the powerful jaw muscles. It is bounded laterally and ventrally by a handle-like bar of bone, the **zygomatic arch,** from which other muscles extend to the lower jaw. External nostrils, or **nares,** lead to the paired nasal cavities and internal nostrils, or **choanae,** lead from these cavities. The choanae of frogs open into the roof of the front part of the mouth cavity, but in mammals a bony **hard palate** extends the nasal cavities more posteriorly so the choanae open into the pharynx. This permits a mammal to chew and manipulate food in the mouth while continuing to breathe; breathing need be stopped only momentarily when the food is swallowed. An endotherm needs more food and oxygen than an ectotherm, since it must be able to sustain a high rate of metabolism and generate enough heat to maintain a constant body temperature. An **external acoustic meatus** leads to the eardrum, which is sunk beneath the body surface in mammals. There are many foramina for blood vessels and nerves; most conspicuous is the large **foramen magnum** through which the spinal cord enters the skull.

Many of the individual bones forming the skull of the frog and human are shown in Figures 11.9 and 11.12. As the brain grew larger during the course of mammalian evolution, the cartilage replacement bones of the chondrocranium no longer completely encased it. They form a ring of bone around the foramen magnum (the **occipital bone**), encase the inner ear (part of the **temporal bone**), and form the floor of the cranium. The sides and roof of the cranium are completed by dermal bones such as the **frontal** and **parietals** and by a portion of the mandibular arch known as the **alisphenoid.** The last is a cartilage replacement bone.

Of particular importance are changes in the jaws and in the small bones of the inner ear that transmit pressure waves from the eardrum to the inner ear. The jaws of frogs and most vertebrates are formed of dermal bones that encase the cartilaginous mandibular arch, but the jaw joint lies between the posterior ends of the rods forming the arch, that is between the dorsal palatoquadrate and ventral Meckel's cartilage. Usually a **quadrate bone** ossifies from the posterior end of the palatoquadrate and an **articular bone** from the end of Meckel's cartilage. Recall that the dorsal part of the hyoid arch (the hyomandibular) helps to suspend the palatoquadrate in the dogfish

Figure 11.12 Components of the human skull, hyoid and larynx. Dermal bones have been left plain, chondrocranial derivatives are hatched, those parts of the embryonic visceral skeleton that disappear are stippled, parts of the visceral arches that persist are shown in black. Roman numerals refer to visceral arches and their derivatives. (Modified after Neal and Rand.)

(Fig. 11.8). In most tetrapods this element has been transformed into the **stapes,** a slender rod of bone extending from the eardrum to the part of the skull housing the inner ear (Fig. 11.9), but it still retains a connection with the quadrate. During the evolution of mammals the jaw mechanism became stronger and the bite more powerful. The dentary, a dermal bone of the lower jaw, enlarged and contacted the squamosal bone (part of the mammalian temporal bone) of the upper part of the skull (Fig. 11.12). A new joint evolved between these bones, and the original jaw joint, located just posterior to the new one, became redundant. The articular and quadrate, which had become quite small, were overspread by the eardrum and incorporated in the middle ear as the outer two auditory ossicles, **malleus** and **incus** respectively. The incus of mammals, like the quadrate from which it evolved, connects with the stapes. Mammals have three auditory ossicles which form a pressure amplifying leverage system (Chap. 16). It is of interest to observe that the auditory ossicles have the same relationship to each other as their homologues in fish.

The ventral part of the hyoid arch, together with the remains of the third visceral arch,

forms the **hyoid bone** (a sling for the support of the tongue) and the **styloid process** of the skull, to which the hyoid is connected by a ligament. With the loss of gills in tetrapods, the remaining visceral arches have become greatly reduced, but parts of them form the cartilages of the larynx.

Although the appendicular skeleton of the dogfish is quite different from that of terrestrial vertebrates, there is a close resemblance between the appendicular skeleton of the terrestrial vertebrates and their ancestral crossopterygian fishes (Fig. 11.13). The humerus of the frog or human arm and the femur of the leg (Figs. 11.9 and 11.10) represent the single proximal bone of the crossopterygian fin; the **radius** and **ulna,** or **tibia** and **fibula,** the next two bones. These bones are fused in frogs as an adaptation for jumping. The **carpals** or **tarsals, metatarsals** or **metacarpals,** and **phalanges** of the hand or foot are homologous with the more peripheral elements of the crossopterygian fin. We tetrapods have a single bone in the proximal part of the appendage followed by two bones in the second part because this pattern was established by our piscine ancestors.

The girdles of tetrapods are necessarily

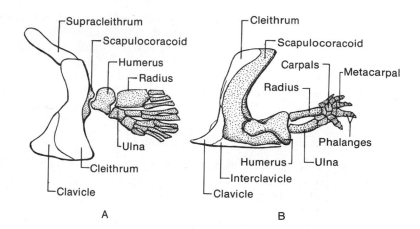

Figure 11.13 Lateral views of the appendicular skeleton of a crossopterygian (A) and ancestral amphibian (B) to show the changes that occurred in the transition from water to land. Dermal bones have been left plain; cartilage replacement bones are stippled. (A modified after Gregory; B after Romer.)

stronger than those of fish. The pectoral girdle is bound onto the body by muscles, but the pelvic girdle extends dorsally and is firmly attached to the vertebral column. A **pubis, ischium** and **ilium** are present on each side of the pelvic girdle, though all have fused together in the adult. Our pectoral girdle includes a **scapula,** a **coracoid process,** which is a distinct bone in most lower tetrapods, and a **clavicle.** The clavicle is the only remnant of a series of dermal bones that are primitively associated with the girdle. All other girdle bones are cartilage replacement bones.

11.3 JOINTS

Where bones come together, they do not simply abut one upon the other but are united by **joints** of varying degrees of complexity. The structure of joints is correlated with the functions the bones perform and the degree of movement between them. Three broad categories can be recognized: immovable joints, slightly movable joints and freely movable joints.

Bones of the cranium (e.g., parietal and frontal, Fig. 11.12) provide a solid protective housing for the brain and are united by immovable joints known as **sutures.** The connective tissue, **periosteum,** which covers the surfaces of all bones, forms a septum that dips between the bones. The bones can grow at their periphery, but they remain firmly articulated with each other at all times.

The vertebrae must form a firm supporting beam, yet a beam that is capable of limited bending. Most vertebrae are united by slightly movable **symphyses.** An elastic pad of fibrocartilage separates them and they are held together by **ligaments**—tough bands of connective tissue that extend across the articulation.

Since the appendages are used as flexible struts in locomotion, a great deal of motion occurs between most of their bones. We find freely movable articulations of several kinds: ball and socket joints between the appendages and the girdles, a hinge joint at the knee, a combination of a hinge and pivot joint at the elbow, and so on. All are often called **synovial joints,** for they contain a synovial cavity that permits free movement. Structurally they are quite complex (Fig. 11.14). The adjacent bones are usually shaped so that a protuberance of one fits into a depression in the other. This, of course, makes the bones less likely to disarticulate during movement. **Articular cartilages** cover the ends of the bones and provide a smooth, somewhat elastic surface. A **synovial membrane** lines the synovial cavity and produces, probably by filtration from blood vessels within it, a lymphlike synovial fluid that lubricates the articulation. Stiffness

Figure 11.14 A diagrammatic section through a synovial joint. Ligaments, not shown, cross the joint peripheral to the articular capsule and provide additional strength to the joint in certain areas.

of joints in old age results in part from a reduction in the amount of this fluid. The periosteum covering the bones extends across the joint as an **articular capsule,** ligaments strengthen the joint, and it is also supported by the tonus of surrounding muscles.

11.4 THE BASIS FOR MOVEMENT

The movement of an animal, or of materials within an animal, is dependent upon changes in cell shape (ameboid movement), the beating of cytoplasmic processes (cilia and flagella) or the contraction of muscle fibers. Although superficially different, all of these types of motion are energy-consuming processes in which chemical energy stored in the energy-rich phosphate bonds of adenosine triphosphate is converted into the mechanical energy of contractile proteins.

Ameboid Movement. In many cells one or more processes, known as **pseudopodia,** develop on the cell surface, and the rest of the

cell flows in their direction. This is known as ameboid movement because it occurs in all amebas and other members of the Sarcodina, but the phenomenon is widespread and is observed in the amebocytes of sponges, many of the coelomic and blood corpuscles of other invertebrates and in certain white blood cells of vertebrates. It also occurs in embryonic tissues such as mesenchyme, in wound healing and in many cell types in tissue culture. The pseudopodia may be used to surround and engulf a food particle (Fig. 11.15A to D), or they may attach to the substratum and be used in locomotion.

During ameboid movement, several pseudopodia may start to form on different parts of the cell, but usually one becomes dominant, and the cell advances in that direction. It should be emphasized that there is no permanent front end, for the dominant pseudopod may be on any surface. Light microscope studies have shown that the cytoplasm of an ameba can be divided into a semirigid **ectoplasm** beneath the cell membrane and a

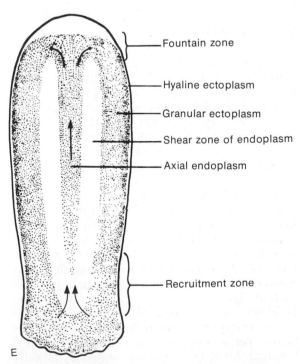

Figure 11.15 Ameboid movement. A to D, an ameba engulfing a flagellate by extending pseudopodia; stages are at about one second intervals. E, Endoplasm flowing into an advancing pseudopod. (E, modified from Allen, R. D. "Ameboid Movement," in Brachet, J., and A. E. Mirsky (Eds.): "The Cell," Academic Press, New York, 1961.)

deeper, more fluid **endoplasm** (Fig. 11.15E). During pseudopod formation, the deepest, axial endoplasm flows forward, separated from the ectoplasm by a sheer zone of endoplasm. It then spreads laterally in the "fountain zone" near the front of the pseudopod, becomes semirigid and adds to the advancing sleeve of ectoplasm. Concurrently, at the posterior end of the cell, in the "recruitment zone," ectoplasm is converted to the forward-moving stream of axial endoplasm.

Electron and fluorescent microscopy, together with biochemical studies, are beginning to reveal the nature of the force involved in the cytoplasmic flow and its point of application. It appears that the basis for ameboid movement, as well as for other types of cell movement, is essentially the same as that in muscle contraction, for it involves contractile proteins such as **actin** and **myosin.** Actin molecules are in a diffuse state in the endoplasm, but they polymerize and become more filamentous in the fountain zone. As they become filamentous, they interact with myosin molecules, and this leads to a contraction. The contraction probably involves a sliding of actin and myosin filaments past each other, but this phenomenon is best understood in muscle (see below). Endoplasm is pulled forward and added to the ectoplasm, which represents contracted or still contracting actin and myosin. This contraction exerts additional force on the endoplasm. It is not known how long actin and myosin remain in the contracted state, but they become dissociated in the recruitment zone. Actin is converted into a less structured state and moves forward again in the endoplasm. Local changes in the availability of calcium and magnesium ions appear to be responsible for the changes in the state of actin. As in muscle cells, the concentration of calcium must attain a certain threshold level for actin and myosin to interact and contract. When the concentration is less than this critical threshold, actin and myosin dissociate.

Ciliary and Flagellar Movement. Another type of motion involves the beating of slender, hairlike, cytoplasmic processes projecting from the cell surface. These are called **cilia** when each cell has one to many short processes and **flagella** when each has only one or a few longer processes, but it is now recognized that the fundamental structure of cilia and flagella is the same. Cilia and flagella are nearly ubiquitous in the Animal Kingdom, where they are used to move a cell through liquid or liquid along the surface of cells. They occur in the ciliates and flagellates among the protozoa, in the collar cells of sponges, on many epithelial surfaces ranging from the gastrovascular cavity of coelenterates to the lining of vertebrate reproductive ducts and in the tails of spermatozoa of most animals. Only arthropods lack motile cilia or flagella, but ultrastructural comparisons reveal that certain of their sensory cells are modified cilia.

Ciliary and flagellar movement has been analyzed by means of very high speed cinephotography. It turns out to be very complex and also variable among different species. Sometimes undulatory waves move along the flagellum, but successive waves are oriented 90 degrees to each other. This imparts a helical twist or rotary action to the flagellum, so that it acts as a propellor (Fig. 11.16A). A flagellum may pull a cell forward by a tip-to-base undulation or push it by a base-to-tip undulation. The beat of a cilium is more planar; the action is oarlike, with a rather "stiff-armed" effective stroke and a "curling" or feathered recovery stroke in a plane different from that of the effective stroke (Fig. 11.16B).

If there are many cilia upon a cell (it is estimated that paramecium has several thousand), their action must be integrated if movement is to be effective. Often traveling, or **metachronal, waves** pass along the surface (Fig. 11.16C). If a paramecium encounters an obstacle, ciliary action is temporarily reversed; the animal backs away from the obstacle, turns and again moves forward.

Electron microscopy reveals that each cilium contains a sheaf of nine double microtubules, situated just beneath the cell membrane, and a core of two single microtubules (Fig. 11.17). Radial spokes extend between the outer doublets and the two central microtubules. The mechanism of bending is not certain, but it is known that the walls of the microtubules are made up of protein filaments composed of **tubulin.** Arms composed of the protein **dynein** extend between the outer doublets and form temporary cross-bridges between them. The arrangement is similar in many ways to that in muscle filaments (see below), although the contractile proteins are different. According to the sliding microtubule hypothesis, the release of energy from ATP and the successive forming of dynein bridges cause one or more microtubules to slide past others. In a sense, the dynein bridges of one double microtubule "walk along" an adjacent doublet. The resulting shearing forces are resisted by the radial spokes, so that some of the sliding action is converted to a bending action.

Cilia and flagella are associated with basal

Figure 11.16 Ciliary and flagellar movement. *A*, Three successive stages in the undulatory movement of a flagellum. *B*, Successive stages in the oarlike action of a cilium; effective stroke shown in white and recovery stroke in black. In both *A* and *B*, water is moved in the direction of the large arrow and the cell moves in the opposite direction. *C*, Scanning electron microscope photograph of fixed metachronal waves in a ciliated protozoan. (*A* and *B* from Satir, P.: How cilia move. Scientific American *231*(4):44–52, (Oct.) 1974; *C*, from Prosser after Tamm, S. L., and G. A. Horridge, Proc. R. Soc. Lond. [Biol.], *175*:219–333, 1970.)

bodies **(kinetosomes),** which are essentially similar to the centrioles associated with the mitotic spindles (p. 48). Basal bodies are related to the development and growth of the cilia and may be involved in the integration of the activity of different cilia. Basal bodies are interconnected in a complex way in paramecium.

Muscle Cells and Their Contraction. Ameboid and ciliary action are efficient for individual cells and small animals, and ciliary

motion is also important in special situations in larger animals, such as moving a sheet of mucus on mammalian respiratory epithelium. Most movements of larger animals, however, as well as the movement of food and gases in and out of their bodies, the circulation of blood and other internal transport processes, result from muscular activity.

Muscle is a biocontractile system in which cells, or parts of them, are elongated and specialized to develop tension along their axis of

Figure 11.17 Structure of a cilium. *A,* Electron microscope photograph of cross sections through the gill cilia of the mussel, *Anodonta cataracta,* magnified 110,000 ×. *B,* Diagram of a cross section. (*A,* from Gibbons, I. R.: The relationship between the fine structure and direction of beat in gill cilia of a lamellibranch mollusc. J. Biophys. Biochem. Cytol. *11:*179–205, 1961; *B,* from Afzelius, B.: A human syndrome caused by immotile cilia. Science *193:*317, 1976. Copyright 1976 by the American Association for the Advancement of Science.)

Figure 11.18 Diagrams of the sliding filament hypothesis of muscle contraction. *A,* Portion of a muscle cell or fiber. *B,* Enlargement of a part of a single myofibril to show the cross banding. *C,* Enlargement of one sarcomere of a myofibril to show the myofilaments in their relaxed configuration. *D,* Comparable view of a contracted sarcomere. *E,* Transverse section through *D* at the level of the arrow to show that each thick filament is surrounded by six thin ones. (Adapted from H. E. Huxley.)

elongation. Contractile stalks are found in some protozoans, and relatively unspecialized contractile cells occur in sponges around the opening through which water is discharged from the central cavity. As animals and their movements become more complex, simple contractile cells are replaced by well differentiated muscular tissue. Yet simple contractile cells persist in certain higher animals in which a few isolated contractile elements perform the needed function. The myoepithelial cells that aid in the discharge of secretions from mammalian sweat glands are an example.

As explained earlier (p. 59), vertebrates have smooth, cardiac and skeletal muscular tissues. The cells of these tissues contain contractile fibrils, and the contractile elements of the last two types are so arranged that the muscle fibers have a cross-striated appearance when seen under a light microscope. Smooth and striated muscle fibers are also found among invertebrates. As in vertebrates, invertebrate striated fibers occur where brief, rapid actions are required. Rapidly swimming jellyfish have bands of striated fibers, and nearly all of the musculature of arthropods is striated.

The mechanism of muscle contraction is best understood from studies of vertebrate striated muscle. The cytoplasm of a striated muscle fiber is packed with many longitudinal **myofibrils,** which can be seen with a light microscope (Fig. 11.18 A and B). Electron microscopy and biochemical analysis show that each myofibril is composed, in turn, of many longitudinal **myofilaments** of two types: **thick filaments** composed of the protein **myosin** and **thin filaments** composed primarily of **actin** but containing small amounts of **tropomyosin** and **troponin** (Fig. 11.18C). The denser, thick filaments are in register with each other and account for the darker, anisotropic (in polarized light) **A bands.** The thin filaments, also in register, account for the lighter, isotropic **I bands.** The two types of filament partially overlap, which explains the somewhat denser and less dense (Hensen's disk, or **H zone**) portions of the A bands. **Z lines,** to which the thin filaments attach, cross the entire myofibril in the center of the I bands. The portion of a myofibril between Z lines constitutes a **sarcomere.** Very high resolution electron microscopy shows that the thick filaments bear processes that extend as **cross-bridges** to the thin filaments.

When striated muscle contracts, the I bands narrow, and the H zones may disappear as the Z lines move closer together (Fig. 11.18D and E). The degree of narrowing of the I bands is a function of the extent of the contraction. Such observations in the late 1950's led H. E. Huxley and Jean Hanson of Cambridge University and King's College, London to propose the sliding filament hypothesis of muscle contraction, a theory that has since been elaborated by many investigators. During contraction, connections appear to be repeatedly formed between the cross-bridges of the thick filaments and the active sites on the thin filaments. It is as though the bridges were hands that grasp the thin filaments and pull them deeper into the array of thick filaments. This would explain the narrowing of the I bands. Recent work suggests that the active sites on the actin molecules in relaxed muscle are covered and blocked by the troponin-tropomyosin molecular complexes. When a nerve impulse reaches a muscle fiber, calcium ions, which have been bound to intracellular membranes within the muscle cell, are relased. These ions now bind to the troponin-tropomyosin complexes, change their molecular configuration, expose the active actin sites and permit contraction to occur. During relaxation, calcium ions are released and sequestered in other parts of the cell.

The portion of the myosin molecules in the cross-bridges acts as an ATPase, and energy for the reaction is provided by the breakdown of ATP when the thick and thin filaments interact. Despite its importance, muscles do not store much ATP. Energy for the resynthesis of ATP from ADP is provided first by energy released from other organic phosphates that are stored in greater quantity: **creatine phosphate** in vertebrate and some invertebrate muscles and **arginine phosphate** in most invertebrates. High energy phosphates are restored by energy derived from the breakdown of muscle **glycogen.** In the absence of sufficient oxygen for the complete oxidation of glycogen, as during strenuous exercise, about 7 per cent of the energy stored in glycogen can be released by the anaerobic breakdown of glycogen to **lactic acid.** But, lactic acid accumulates and eventually must be oxidized to CO_2 and water (Section 4.6). The amount of O_2 required to fully oxidize the accumulated lactic acid represents an **oxygen debt.** One continues to breathe deeply after vigorous exercise as this debt is repaid. Energy released from the oxidation of lactic acid restores the energy-rich phosphates and resynthesizes some of the lactic acid back to glycogen. The existence of this mechanism, which permits a great deal of energy to be expended in the temporary absence of sufficient oxygen, is of great importance to animals, for it permits them to contract muscles forcibly and repeatedly in emergencies without having the contractions limited by the rate at which oxygen can be delivered to the tissues.

The contraction of smooth muscles utilizes the same basic mechanism as that of striated muscle. Smooth muscle contains thin filaments of actin. Myosin is present but apparently is not organized into well defined thick filaments.

11.5   MUSCLES

The muscles of most animals consist of large groups of muscle fibers bound together and invested by connective tissue through which nerves and blood vessels run. However, some of the very small muscles of invertebrates consist of only a few muscle fibers, with little connective tissue between them. Collagen, a major constituent of vertebrate connective tissue, is absent in arthropods.

Muscles usually attach to skeletal elements, which are moved as the muscles shorten and lengthen. In arthropods, muscles attach directly to the exoskeleton (Fig. 11.3*A*), or inward extensions of the skeleton, but in vertebrates the attachments to bones are by the in-

vesting connective tissue fibers, some of which become continuous with the connective tissue **periosteum** covering bones and some of which penetrate the bones. Often the connective tissue between muscle and bone forms a cordlike **tendon** (Fig. 11.19*A*), sometimes a sheetlike **aponeurosis**, and sometimes muscle fibers come so close to the bone that the attachment is referred to as **fleshy.**

When a muscle contracts, one end, its **origin,** remains relatively fixed in position, whereas the other, its **insertion,** pulls upon a bone that is drawn toward the muscle's origin. Origin and insertion are convenient descriptive terms, but it should be emphasized that the force developed at each end of the muscle is the same. Sometimes movement may take place at each end, or movement may occur at either end, according to the circumstances.

Since muscles perform work only by contracting, they must work against antagonistic forces. The muscle that holds the shell of a clam closed works against the force of an elastic spring ligament in the shell hinge. When the muscle relaxes, the ligament opens the valves of the shell and stretches the muscle to its original length. Most muscles are arranged in antagonistic sets, such as the biceps and triceps of the arm (Fig. 11.19*A*). Contraction

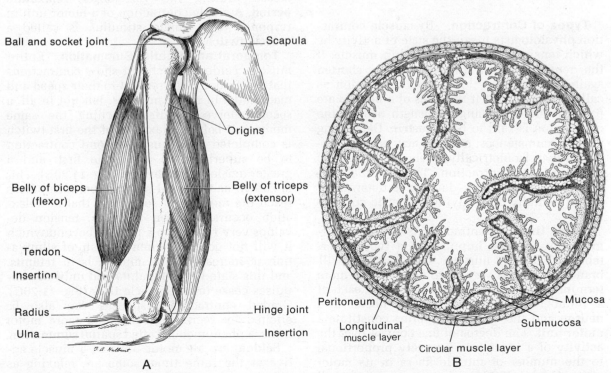

Figure 11.19   Arrangement of muscles. *A*, Antagonistic skeletal muscles attached to bones in the human arm. *B*, Antagonistic layers of smooth muscle forming part of the wall of the mammalian intestine.

of the biceps causes a decrease of the angle between upper arm and forearm, an action called **flexion,** and contraction of the triceps increases this angle, or **extends** the forearm. Other common sets of antagonistic actions are **protraction** and **retraction,** which describe respectively the forward or backward movement of the entire limb of a quadruped at the shoulder or hip, and **adduction** and **abduction,** which describe the pulling of a part toward or away from a point of reference. For limb movements, the reference is the midventral line of the body.

Not all muscles attach to bones. Many form a mass of interlacing muscle fibers, as in the foot of a snail, or form layers in the walls of hollow organs (Fig. 11.19B). These fibers do not have definable origins and insertions but pull upon each other and the surrounding soft tissues so as to change the size and shape of the organ of which they are a part. Restoration of the original length of the fibers is achieved by the pull of fibers having a different direction, as the two layers of muscle in the intestine wall, or from the filling of a hollow organ, as in a urinary bladder.

11.6　PHYSIOLOGY OF MUSCLE CONTRACTION

Types of Contraction. By muscle contraction physiologists mean the state of activity by which tension is developed within muscle. If the tension causes the muscle to shorten against a constant load, the contraction is called **isotonic,** but if the ends of a muscle are fixed so that no change in length occurs, the contraction is said to be **isometric.** Depending upon circumstances, muscles may contract isotonically or isometrically, or show attributes of both types of contraction. The maximal tension that muscles can develop is determined experimentally by studying isometric contraction.

Motor Units. Contraction is initiated normally by a nerve impulse. In a typical vertebrate skeletal muscle, one **neuron** will branch in such a way as to have one or more terminals, called **motor end plates,** on each of a number of different muscle fibers. The neuron and the fibers it supplies constitute a **motor unit.** The degree of fine control over the activity of a muscle is inversely proportional to the number of muscle fibers in its motor units. Ocular muscles that move our eyeball may have as few as two or three fibers to a unit, whereas some of the leg muscles that hold our body erect contain several hundred per unit.

Muscle Twitch. When a nerve impulse reaches a motor end plate, a **transmitter substance,** usually **acetylcholine,** is released by the neuronal ending and initiates changes leading to the activation of the muscle. The drug **curare,** the chief ingredient of the arrow poison used by South American Indians, blocks the myoneural junction and paralyzes the muscle. However, a curare-paralyzed muscle can still be caused to contract by direct electrical stimulation, a demonstration that muscle is independently irritable. Activation of muscle fibers involves a depolarization of their cell membranes, which is reflected by an **action potential** and certain chemical changes. Muscle fibers may be activated by a single nerve impulse, but two or more in rapid succession may be needed. The first impulse initiates changes that are built upon by subsequent ones, if they arrive before the effects of the first have worn off. The early impulse is said to **facilitate** activation.

Activation requires a few milliseconds, so there is a delay, or **latent period,** between the application of a stimulus and the onset of contraction (Fig. 11.20A). Tension develops during the longer **contraction period** and decreases during the yet longer **relaxation period.** A single contraction of a motor unit in response to a single stimulus is called a **muscle twitch.**

Temporal and Spatial Summation. Entire muscles seldom twitch but show contractions that are graded with respect to their speed and magnitude. In many muscles, but not in all, a second nerve impulse reaching the same motor unit before relaxation of the first twitch is completed will cause a second contraction to be superimposed upon the first, and a greater tension will develop (Fig. 11.20B). This is known as **temporal summation.** If impulses reach the motor unit so rapidly that no relaxation occurs between twitches, tension develops very rapidly to a plateau beyond which it will not go even when the rate of stimulation is increased. The muscle is in **tetanus,** and this state will be maintained until the impulses cease or the muscle tires (Fig. 11.20C). Graded contractile responses can also be elicited by recruiting more and more motor units, a phenomenon called **spatial summation.**

Seldom are all motor units in a muscle active at the same time; some are relaxing as others contract. In this way a muscle can sustain a contraction for a long time without fa-

Figure 11.20 Isometric contraction in a motor unit. *A,* A twitch; *B,* temporal summation of two twitches; *C,* tetanus. Vertical arrows indicate points of stimulation.

tigue. Most muscles in our body are under a state of partial contraction, known as **tonus,** all of the time. This maintains organs in proper spatial relationship to each other. A muscle under slight tension can also react more rapidly and contract more strongly than one that is completely relaxed.

Muscle Electricity and Heat. A weak electric current, the action potential, is produced during muscle contraction, and heat is also generated, for only about 20 per cent of the energy released is used in mechanical work. Some animals make special use of these accompanying phenomena. The electric organs of electric eels and certain other fishes are modified muscle tissue in which the cells are specialized to maximize production of an electric current. In different species, the electric organs may be used in electrolocation, defense or the stunning of prey.

Heat is simply dissipated in most animals, but in the warm-blooded, or endothermic, birds and mammals, heat derived from muscle contraction is the main source of the body heat needed to sustain high rates of metabolism. Special mechanisms have evolved to conserve or dissipate heat as the situation warrants and to maintain body temperature at high and constant levels (p. 836).

11.7 VERTEBRATE PHASIC AND TONIC MUSCLES

Vertebrate skeletal muscles are not all alike. Some are known as phasic muscles, adapted for rapid responses of short duration, whereas others, the tonic muscles, are better suited for slow, sustained contractions. **Phasic,** or **twitch, muscles** are composed of muscle fibers that are supplied by large, rapidly conducting neurons, and there is a single motor end plate per muscle fiber (Fig. 11.21*C*). Phasic fibers respond in an **all-or-none** fashion. If the stimulus attains a certain threshold value, the fibers in the motor unit will twitch. The fibers respond maximally or not at all, for a wave of depolarization spreads throughout them. They cannot be activated again until after a brief recovery period. Temporal summation, therefore, cannot occur, but spatial summation can. The frog's sartorius muscle, used in jumping, is a phasic muscle.

Tonic muscles are composed of fibers that are darker in color, for they contain a great deal of oxygen-binding **myoglobin,** a hemoglobin-like compound. Tonic fibers are supplied by smaller, slower conducting neurons that have multiple motor end plates along each muscle fiber. These fibers do not give an

A. VERTEBRATE

B. CRUSTACEAN

Muscle fiber

Phasic   Tonic

C. VERTEBRATE

Slow

Fast

Inhibitory

D. CRUSTACEAN

Figure 11.21 Patterns of nerve supply in vertebrate and crustacean muscles. A, Vertebrate muscles are supplied by large nerves composed of many neurons; B, crustacean muscles are supplied by small nerves that may contain as few as two neurons. C, Pattern of neuron endings on individual muscle fibers in vertebrate phasic and tonic muscle and D, in crustacean muscle. (Modified from Schmidt-Neilsen, K.: Animal Physiology. Cambridge, Cambridge University Press, 1975.)

all-or-none response to a threshold stimulus, for activation may be limited to a few sarcomeres. Repeated stimulation leads to temporal summation until all of the fibers in a motor unit are activated. Spatial summation also occurs. Relaxation of tonic fibers is 50 or more times slower than that of phasic fibers. Postural muscles are examples of tonic skeletal muscles. Some vertebrate skeletal muscles are composed entirely of one or the other type of fiber, but often both are present. Smooth muscles are also tonic, but they have an entirely different pattern of innervation. Neurons do not terminate on every muscle cell; rather, activation spreads slowly from innervated cells to others.

11.8 SOME SPECIALIZATIONS OF INVERTEBRATE MUSCLES

Molluscan Catch Muscle. The adductor muscle that closes the shell of a clam is divided into two parts in many species. One is striated, acts as a phasic muscle and quickly closes an open shell. The other, which is smooth, is tonic and holds the shell closed. An unusual attribute of the tonic muscle is its ability to hold the shell closed for many hours, or even days, and sometimes against a strong external force, such as a starfish might exert in trying to pull the shell open. It is as though there were some sort of "catch" mechanism that maintains the contracted state, but how it works is not fully understood. It is known that the catch muscle contains relatively large filaments of paramyosin. Some investigators have postulated that paramyosin stabilizes the actomyosin cross-bridges so that they are not released. Others hypothesize that continued nerve impulses are necessary to maintain contraction. Discovery that oxygen consumption

continues during catch, albeit at very low levels, adds support to this theory.

Polyneuronal Innervation of Crustacean Muscle. Although arthropods resemble vertebrates in having phasic and tonic skeletal muscle fibers, the muscles differ greatly in their innervation (Fig. 11.21). Arthropod muscles have been studied primarily in the claws of crabs and other crustaceans. Entire muscles often are supplied by only two (or a few) neurons, and one neuron may extend to two or more muscles. As in vertebrate tonic muscle, each muscle fiber receives multiple terminations from a neuron, but in crustaceans two or more different neurons terminate on each fiber. In some cases as many as five may innervate a single fiber. At least one is inhibitory; the others induce varying degrees of fast or slow contraction. The final response of a muscle depends on the rate of stimulation of the different neurons and the interaction between them. These complexities enable a few neurons and muscle fibers to generate as wide a range of speed, force and duration of muscle contraction as is brought about by the much larger nervous and muscular systems of vertebrates.

Insect Indirect Flight and Tymbal Muscles. The indirect flight and tymbal (sound producing) muscles of insects are of interest, for they contract at frequencies far faster than can be induced by nerve impulses. A midge may beat its wings 2000 times per second, and some tymbal membranes oscillate as fast as 7000 times per second! Muscle activation, recovery and reactivation by neuronal stimulation require more time than this. Indirect flight muscles, which are found among the various kinds of flies, wasps and beetles, do not attach directly to the wings, as muscles do in slower flying insects, but rather to the walls of the thorax (Fig. 11.22). Vertical fibers depress the thorax roof, or **tergum,** against a fulcrum

Figure 11.22 Diagrams of the indirect flight muscles of an insect as seen in cross sections of the thorax. *A,* Contraction of vertical muscles lowers the tergum and raises the wings. *B,* Contraction of longitudinal muscles raises the tergum and lowers the wings. (After Ross.)

formed by the lateral thoracic walls and hence raise the wings. Longitudinal fibers shorten the thorax in an anteroposterior direction, which increases its height and lowers the wings. These two sets of muscles are antagonistic to each other and to the elasticity of the thoracic wall. A nearly isometric contraction of one set develops a tension that causes a sudden change in thorax shape, and this stretches the other set of muscles. The stretching of this set, rather than another nerve impulse, induces its contraction. Much of the activity is **myogenic;** that is, it is inherent within the muscles themselves. Nerve impulses are necessary to keep the muscles in an active state, but there may be 20 or more wing beats for each nerve impulse.

11.9   THE VERTEBRATE MUSCULAR SYSTEM

In describing the vertebrate muscular system and tracing its evolution, it is convenient to group the muscles into **somatic muscles,** associated with the body wall and appendages, and **visceral muscles,** associated with the pharynx and other parts of the gut tube. This grouping parallels the major divisions of the skeletal system. Somatic muscles are striated and under voluntary control. Most of the visceral muscles are smooth and involuntary; however, the visceral muscles associated with the visceral arches, called branchial muscles, are striated and some are under voluntary control.

Evolution of Somatic Muscles. Most of the somatic musculature of fishes consists of segmental **myomeres** (Fig. 11.23). This is an effective arrangement for bringing about the lateral undulations of the trunk and tail that are responsible for locomotion. The muscles of the paired fins are very simple and in many fishes consist of little more than a single dorsal **extensor,** or **abductor,** that pulls the fin up and caudally and a ventral **flexor,** or **adductor,** that pulls the fin down and anteriorly.

The transition from water to land entailed major changes in the somatic muscles. The appendages became increasingly important in locomotion, and movements of the trunk and tail less important. The segmental nature of the trunk muscles is largely lost, although traces can be seen in the **rectus abdominis** that extends longitudinally on each side of the midventral line (Figs. 11.24 and 11.25). Back

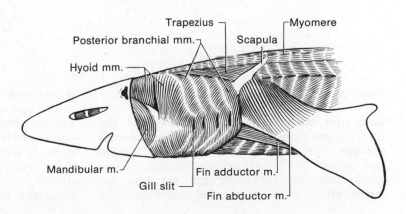

Figure 11.23 A lateral view of the anterior muscles of a dogfish. (Modified after Howell.)

Figure 11.24 Superficial skeletal muscles of the frog in á dorsal (left side of figure) view and a ventral (right side) view.

Figure 11.25 An anterior view of certain of the superficial muscles of a human being.

muscles remain powerful, for they play an important role in supporting the vertebral column and body, but trunk muscles on the flanks form thin sheets, such as the **external oblique,** that support the abdominal viscera and assist in breathing movements. Some trunk muscles, including the **serratus anterior,** attach onto the pectoral girdle and in a

quadruped help transfer body weight to the girdle and appendage. The primitive single fin extensor and flexor became divided into many components, and these became larger and more powerful. Despite the complexity of tetrapod appendicular muscles, it is possible to divide them into a dorsal group that evolved from the fish extensor and a ventral group derived from the flexor. Our **latissimus dorsi** and **triceps** (Fig. 11.25), for example, are dorsal appendicular muscles, whereas the **pectoralis** and **biceps** are ventral appendicular muscles.

Evolution of Branchial Muscles. Branchial muscles are well developed in fishes and are grouped according to the visceral arches with which they are associated (Fig. 11.23): **mandibular muscles** and certain of the **hyoid muscles** are concerned with jaw movements; most of the rest, with respiratory movements of the gill apparatus. Bran-

chial muscles obviously become less important in tetrapods, for the gills are lost and the visceral arches are reduced. Nevertheless, certain ones are retained. Those of the mandibular arch remain as the **temporalis, masseter** and other jaw muscles (Figs. 11.24 and 11.25). Most of those of the hyoid arch move to a superficial position and become the **facial muscles** that are responsible for facial expressions. Those of the remaining arches are associated with the pharynx and larynx and some, e.g., the **sternocleidomastoid** and

TABLE 11.1 MAJOR ATTACHMENTS AND ACTIONS OF IMPORTANT FROG MUSCLES

Muscle	Origin	Insertion	Action
Jaw Muscles			
Temporalis	Dorsal surface of skull	Mandible anterior to jaw joint	Closes jaws
Depressor mandibulae	Dorsal surface of skull, connective tissue of back	Mandible posterior to jaw joint	Open jaws
Submaxillary	Medial surface of mandible	Longitudinal connective tissue septum between submaxillaries of opposite sides	Raises floor of mouth and pharynx during breathing and swallowing
Trunk Muscles			
Obliquus externus	Connective tissue of back	Longitudinal midventral connective tissue septum	Supports abdomen, helps compress lungs during breathing
Rectus abdominis	Sternum	Anteroventral surface of pelvic girdle	Supports abdomen, flexes back
Longissimus dorsi	Posterior end of skull, dorsal surface of vertebrae	Urostyle and ilium	Extends back
Coccygeoiliacus	Urostyle	Ilium	Helps extend back
Foreleg Muscles			
Dorsalis scapulae	Surface of scapula and suprascapula	Proximal end of humerus	Abducts arm, helps protract arm
Latissimus dorsi	Dorsolateral surface of trunk	Proximal end of humerus	Abducts arm, helps retract arm
Triceps	Base of scapula, shaft of humerus	Proximal end of radioulna	Extends forearm
Deltoideus	Base of scapula, clavicle	Proximal end of humerus	Adducts arm, protracts arm
Pectoralis	Ventral surface of trunk and sternum	Proximal end of humerus	Adducts arm, retracts arm
Coracoradialis	Ventral surface of coracoid	Proximal end of radioulna	Flexes forearm
Hindleg Muscles: Thigh			
Triceps femoris (rectus anticus, vastus internus and externus)	Ilium and capsule of hip joint	Proximal end of tibiofibula	Extends shank
Iliofibularis Semimembranosus Gracilis major and minor	Posterodorsal part of pelvic girdle	Proximal end of tibiofibula	Flexes shank, during shank extension helps retract and abduct thigh
Sartorius	Anteroventral part of pelvic girdle	Proximal end of tibiofibula	Flexes shank, helps adduct and protract thigh
Adductor magnus Adductor longus	Ventral part of pelvic girdle	Distal portion of femur	Adducts thigh
Hindleg Muscles: Shank			
Gastrocnemius	Distal end of femur	Plantar aponeurosis and terminal segment of each toe	Extends foot
Tibialis posticus	Shaft of tibiofibula	Proximal end of tarsals	Helps extend foot
Peroneus Tibialis anticus longus	Distal end of femur	Proximal end of tarsals	Flexes foot, helps extend shank

Figure 11.26 A comparison of the vertebrate endo-skeleton and joint with the arthropod exoskeleton and joint.

trapezius, become important muscles associated with the pectoral girdle.

Origins, insertions and actions of the major muscles of the frog are presented in Table 11.1 as an example of the complexity of the muscular system of a tetrapod. Most of these muscles are illustrated in Figure 11.24.

11.10 BIOMECHANICS OF THE MUSCULOSKELETAL SYSTEM

Muscles frequently attach to skeletal elements that serve as lever arms to transmit the force to a point of application. The force is applied across a moveable joint (Fig. 11.26). Sometimes the muscle itself crosses the joint, but often the joint is crossed by a tendon or, in arthropods, by a tendon-like **apodeme** composed of cuticle.

The point of rotation, or **fulcrum,** of a lever system is at or near the proximal end of the bone. In most retractor and flexor systems

(Fig. 11.27), the point of application of the **in-force** by muscles and the point of delivery of **out-force** by the lever system lie on the same side of the fulcrum, but the in-force is applied closer. The forces are **vector quantities** because they have both magnitude and direction. The **in-lever arm** (the perpendicular distance between the fulcrum and the line of muscle action) is shorter than the **out-lever arm** (the perpendicular distance from the fulcrum to the distal end of the lever system). This affects the magnitude of the forces, for when the lever system is in equilibrium the product:

$$\text{in-force} \times \text{in-lever arm} = \text{out-force} \times \text{out-lever arm.}$$

Since the in-lever arm is the shorter, it is evident that the in-force developed by the muscles must exceed the out-force delivered. The system is mechanically inefficient from the point of view of in-force and out-force relationships, but it provides for compactness and relatively high velocity. **Velocity,** also a vector quantity since it has both magnitude and di-

Figure 11.27 Forelimbs of a horse and an armadillo drawn to the same size to show how changes in the in-lever and out-lever arm relationships adapt the limb lever system for velocity or force. (After J. Maynard Smith, from Young, J. Z.: Life of Mammals. 2nd ed. New York, Oxford University Press, 1975.)

rection, is related to the length of the lever arms too, because a point on a lever farther from the fulcrum moves faster than one near the fulcrum. This relationship is expressed in the formula

$$\text{in-velocity} \times \text{out-lever arm} =$$
$$\text{out-velocity} \times \text{in-lever arm}.$$

It follows from these force and velocity relationships that force and velocity can be altered by changing the length of the in- and out-lever arms relative to each other, but force and velocity are inversely related. In a horse, the ratio of in-lever to out-lever is much lower than in an armadillo (Fig. 11.27). Consequently, the distal end of a horse's limb, which is adapted for running, can move faster but with relatively less force than an armadillo's limb; whereas the armadillo's limb, adapted for burrowing, can deliver relatively more force but at the expense of velocity.

In most protractor and extensor lever systems, for example the extension of the human arm by the triceps muscle (Fig. 11.19), the in-lever (perpendicular distance from the elbow joint to the line of action of the triceps) and out-lever (perpendicular distance from the elbow joint to the hand) are on opposite sides of the fulcrum rather than on the same side, but the principles of force and velocity discussed above still apply.

11.11 LOCOMOTION

Animals exhibit many kinds of locomotion, but certain principles apply to all. To move, an animal must support itself and develop a thrust against the surrounding medium (water or air) or the substratum (ocean bottom or ground).

Most terrestrial animals support themselves with a skeleton. Much of the support for aquatic animals, even for those with a skeleton, comes from the water. If the **density** (mass per unit volume) of the animal is the same as that of the surrounding water, the animal has **neutral buoyancy** and neither rises nor sinks. Many aquatic organisms have a composition sufficiently similar to water so that their density is nearly the same and little energy need be expended to keep afloat. Others have tissues denser than water so tend to sink, but this tendency can be mitigated by including within their mass some material less dense than water. Examples are the gas-filled "sail" of the Portuguese man-of-war, a colonial

cnidarian related to jellyfish, the numerous small chambers of gas in the cuttlebone of the cuttlefish, a molluscan relative of the squid, and the **swim bladder** of most bony fish. Gas can be secreted into or reabsorbed from the fish's swim bladder by means of special glands associated with the circulatory system as the animal respectively sinks or rises in the water, so the fish maintains a neutral buoyancy over a wide range of depth. Other aquatic organisms may have a great deal of oil in their tissues. This is true for many small planktonic organisms. Sharks, which do not have swim bladders, have very large livers with a high oil content. The oil, of course, serves as an important energy reserve as well as reducing the density of the animal.

The motion of an animal is derived from pushing against the surrounding medium or substratum, which, in accordance with Newton's third law of motion, generates an equal and opposite thrust against the animal. It is part of this latter force that moves the animal forward. For example, when a fish pushes its tail back against the water with a particular force (Fig. 11.28), the water pushes against

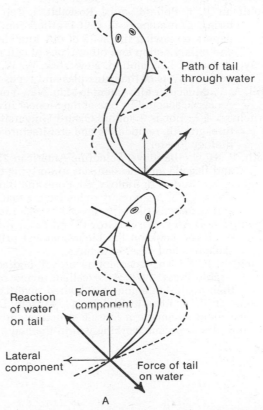

Figure 11.28 Diagram of a dogfish swimming to illustrate the thrust of the tail against the water and the reaction of the water on the tail. (Modified from Marshall and Hughes.)

the tail with an equal but opposite force that can be resolved into lateral and forward components. As the tail moves from side to side, the lateral components cancel each other, and the forward components propel the animal. Newton's third law applies in all locomotion; it is the nature of the thrust that varies. Frequently, the tail, trunk or one or more appendages move back against the medium or substratum. The cilia of paramecium or the flagellum of a sperm tail push against the water. The up and down undulations of a leech's body or the lateral undulations of a snake's trunk push against the water or ground. The thrusts of a bird's or insect's wings generate both an upward force that supports the animal and a forward one that advances it. The legs of arthropods or terrestrial vertebrates push against the ground. Sometimes the thrust comes from a change in body shape. An ameba forms pseudopodia by changing shape. Thrusts are developed by the creeping foot of a snail or the burrowing foot of a clam, by changes in foot shape and, as with an earthworm, by changing its body shape. Some animals use jet propulsion to thrust against the medium. A squid, for example, can move very rapidly by emitting a powerful jet of water from its funnel.

Although the basic principles of locomotion are the same for all animals, there is much variation that relates to the animal's mode of life and the environment in which it lives. In addition to providing support and propulsion, the locomotor mechanisms of aquatic and flying animals must provide for stability against roll, pitch and yaw and for reduced friction. The number and combination of feet placed on the ground at one time by a terrestrial animal represents an interaction between the amount of support and thrust optimum for a particular body form and velocity. Some examples of locomotion are considered in further detail in the survey of the Animal Kingdom.

ANNOTATED REFERENCES

Additional information on the skeleton and movement can be found in the general references cited at the end of Chapter 10.

Goldman, R. T. Pollard, and J. Rosenblum: Cell Motility. Cold Spring Harbor Laboratories, Conferences in Cell Proliferation, 1976. A three volume collection of papers reviewing the state of our knowledge on muscle contraction, ameboid and ciliary action and other types of cell motility.

Gray, J.: Animal Locomotion. New York, W. W. Norton and Company, 1968. A thorough analysis of the principles and types of animal locomotion.

Hill, A. V.: Muscular Movement in Man. New York, McGraw-Hill Book Company, 1927. A classic account by one of the pioneer investigators of muscle physiology.

Pritchard, J. J.: Bones. London, Oxford University Press, 1974. The microscopic structure, growth, remodeling and architecture of bone are summarized in this Oxford Biology Reader.

Satir, P.: How cilia move. Scientific American 231:44, (Oct.) 1974. The beating of cilia and flagella and mechanisms underlying their action are described.

Simkiss, K.: Studies in Biology. 53. Bone and Biomineralization. London, The Camelot Press, 1975. A summary of the factors that control mineralization of invertebrate and vertebrate tissues prepared for the Institute of Biology.

Thompson, D'Arcy, W.: On Growth and Form. Revised ed. New York, Macmillan, 1942. A classic book on the biomechanical principles that determine the shapes of skeletons and other structures.

Trueman, E. R.: The Locomotion of Soft-bodied Animals. Bristol, England, Edward Arnold Press, 1975. An excellent review of crawling, swimming, burrowing and their evolution among soft-bodied invertebrates.

Wilkie, D. R.: Studies in Biology. 11. Muscle. London, The Camelot Press, 1968. An excellent summary of muscle structure and action from the biochemical level to muscles as organs working within the body.

ANIMAL NUTRITION

Foods are organic compounds used in the synthesis of new biomolecules and as fuels in the production of cellular energy. Most foods are fatty acids, glycerols, sugars and amino acids. Foods, or organic nutrients, are the same for all organisms, whether they are bats, oak trees or fungi.

Organisms differ not in the nature of their foods but in the way they acquire them. **Autotrophs,** which include all green plants, algae, some protozoans and some bacteria, synthesize their foods from inorganic compounds. Photosynthesis, the production of carbon compounds from CO_2 and H_2O using light as an energy source, is the most common and most important mode of autotrophic nutrition.

Heterotrophs acquire their foods from the cells or the organic products of other organisms. All animals are heterotrophs and, since they consume the cells or cellular products of other organisms, are often said to be **holozoic.** Indeed, the mode of nutrition is an important basis for defining animals. Animals can be defined as motile, multicellular heterotrophs; this definition would limit them to the Metazoa. If Protozoa are included, the concept of "animals" becomes much more difficult to define, for protozoans are neither multicellular nor are they all heterotrophic. Some animal parasites, as well as many bacteria and fungi, although heterotrophic, are not holozoic. They absorb food from the body of their host or digest organic materials externally and absorb the products (e.g., mold on bread). They are said to be **saprozoic** or **saprophytic.**

12.1 DIGESTION AND ABSORPTION

Digestion. Digestion is a necessary part of heterotrophic nutrition. The large molecules of carbohydrates, fats and proteins consumed as parts of cells and tissues must be broken down into smaller constituent units, such as sugars and amino acids, in order to be transported across cell membranes in the process of absorption. Even if membrane transport of large molecules were not a problem, the organic compounds synthesized by an autotroph are often not the same as those consumed as food. Digestion is therefore necessary before reassembly can occur.

Digestion involves the addition of water to the molecule being fragmented, a reaction termed **hydrolysis.** Although hydrolysis is an exothermic reaction, only a small amount of energy is released, and enzymes must catalyze the reaction in order for it to take place rapidly.

Enzymes are more or less specific in cleaving certain types of bonds, and thus different enzymes attack different types of food molecules or different parts of the molecule. However, many digestive enzymes are not as substrate-specific as are intracellular enzymes. This is an advantage in digestion, for food will generally include a wide variety of similar but not identical compounds.

Most animals from sponges to humans possess the same general types of enzymes. The major classes of enzymes are as follows:

$$COOH-CH_2-CH_2-\underset{\underset{CO}{|}}{\overset{\overset{NH}{|}}{CH}}$$

$$+ (H)(OH)$$

$$\underset{\underset{COOH}{|}}{\overset{\overset{NH}{|}}{HO-\langle\ \rangle-CH_2-CH}}$$

$$\xrightarrow{\quad\quad}$$
$$\xleftarrow{\text{enzyme, peptidase}}$$

$$COOH-CH_2-CH_2-\underset{\underset{COOH}{|}}{\overset{\overset{NH}{|}}{CH}}$$

Glutamic acid

$$HO-\langle\ \rangle-CH_2-\underset{\underset{COOH}{|}}{\overset{\overset{NH_2}{|}}{CH}}$$

Two terminal amino acids   Tyrosine
of a protein

Proteases cleave the peptide bonds of proteins. **Exopeptidases** separate the terminal amino acid from the rest of the protein molecule. **Endopeptidases** attack the center of the molecule, splitting it into two smaller fractions. The point of cleavage depends upon the enzyme. Trypsin, for example, can cleave only those peptide bonds adjacent to the amino acids arginine and lysine.

Carbohydrases digest carbohydrates. **Polysaccharidases** digest high molecular weight carbohydrates. **Amylase,** which breaks down starch, is the principal animal polysaccharidase, but some animals also possess **cellulase,** which breaks down cellulose. **Oligosaccharidases** split low molecular weight trisaccharides and disaccharides into simple sugars. Each requires a different enzyme; the disaccharides sucrose and maltose are hydrolysed by sucrase and maltase.

Lipases split the ester bonds of fats, separating the fatty acids from glycerol.

Digestive enzymes are commonly released in an inactive form, **zymogen,** and conditions at the site of digestion bring about activation of the zymogen. For example, pepsin, an endopeptidase, is secreted in the stomach of humans and other vertebrates as inactive pepsinogen. The acidity of the stomach, produced by secretions of hydrochloric acid, converts pepsinogen to pepsin.

The secretion of protein-splitting enzymes in an inactive form and the presence of a mucous lining of the digestive tract are important adaptations for preventing these enzymes from digesting the tissues that secrete them.

Absorption. After digestion, the smaller food compounds are absorbed by cells of the digestive tract by diffusion or active transport. The mechanisms for the active transport of food products into the intestinal mucosal cells of vertebrates are not completely known. Absorption continues after most of the food in the intestinal lumen has been taken up and any concentration gradient that would have favored diffusion has been reversed. There are some surprising variations in the absorption rates of similarly sized molecules; if glucose, for example, is given an arbitrary absorption rate of 100, galactose has a rate of 115 and fructose has a rate of 44. Poisons that interfere with the metabolism of mucosal cells greatly reduce the rate and amount of absorption.

12.2 EVOLUTION OF THE ANIMAL GUT

We do not know what the first multicellular animals, or metazoans, were like. They evolved long before the existing fossil record, and their small size and soft bodies would have made preservation very unlikely. Many zoologists speculate that they were small, radially symmetrical, slightly elongate animals (Fig. 10.3 A). A layer of flagellated cells covered the outside of the body, and a solid mass of cells filled the interior. It is quite probable that there was neither mouth nor gut. The exterior flagellated cells provided for locomotion and also were involved in feeding. Small food particles, such as bacteria, algae, protozoans and cell fragments, were engulfed by pseudopodia. The food particle became enclosed within a food vacuole located in the cell, and digestive enzymes passed from the cytoplasm into the vacuole, where digestion occurred **intracellularly.** The products of digestion were absorbed from the vacuole, which became progressively smaller. Any undigested wastes were discharged to the exterior. Absorbed food material reached interior cells by diffusion.

Although we can only speculate about the feeding mechanisms of the first multicellular animals, there are organisms living today that feed in this manner. Many of the animal-like unicellular protozoans digest their food within food vacuoles, and food capture may be by engulfment, as in the amebas. Among multicellular animals, sponges have neither mouth nor gut, and the small particles on which they feed are digested intracellularly. Although the organization of the sponge body is very different from that of our hypothetical metazoan, sponges probably diverged early from the main line of animal evolution, before the development of the gut.

All other animals possess a mouth and gut as organs of heterotrophic nutrition. If a gut is not present, the gutless condition is secondary; i.e., the gut has been lost. There are no living or fossil forms that tell us anything about how a gut cavity was first formed, but the early gut was probably a simple sac with only one opening, the mouth. A digestive system of this type is found in many of the more primitive animals, such as the little flatworms *Stenostomum* and *Macrostomum* (Fig. 12.1).

These little freshwater animals, less than one mm. in length, swim or crawl about over debris in pools and ponds. The ventral mouth leads into a simple ciliated tube, the pharynx, which in turn opens into a large simple sac, the **intestine** or **enteron.** These flatworms feed on other small animals, which they swallow whole (ingest). When the prey reaches the intestinal sac, certain cells produce enzymes that pass into the cavity and digest proteins **extracellularly.** The prey soon fragments, and the resulting particles are phagocytosed by other intestinal cells, where digestion is com-

pleted intracellularly. Indigestible wastes are **egested** back out of the mouth.

Note the difference in the size of the food of these flatworms and that of sponges and our hypothetical metazoan. The evolution of a gut cavity makes possible the utilization of larger food masses. This in turn demands that digestion be extracellular, at least in part.

The more familiar planarians are larger flatworms, in which the gut sac is highly branched, facilitating the diffusion of absorbed food material to various parts of the body (Fig. 25.2). Planarians also have a muscular tubular pharynx that can be extended out of the mouth into the body of the prey or into dead animal remains. Special pharyngeal enzymes aid in penetration by the pharynx, and the contents of the prey are sucked up into the intestine.

Hydras, sea anemones and corals also have a digestive cavity with only one opening (Fig. 13.3). They feed on other small animals that, after being swallowed whole, are first digested extracellularly and then intracellularly.

Most animals possess a digestive tract with separate intake and exit openings, mouth and anus. Such an arrangement permits simultaneous and continuous processing of food as it passes along the gut tube. One process need not be halted to permit another to take place. Parts of the gut tube may be restricted to specific functions, and this has led to regional specialization. The following are the most commonly encountered specializations of the animal digestive tract:

Buccal, or Mouth, Cavity. The anterior region of the gut into which the mouth opens. The buccal cavity commonly contains teeth, jaws, salivary glands and other structures con-

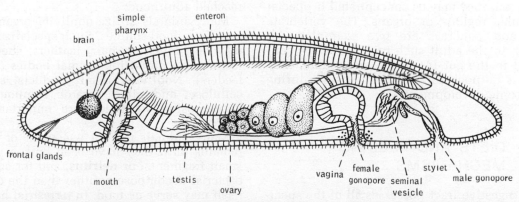

Figure 12.1 Semidiagrammatic sagittal section of the freshwater flatworm *Macrostomum*. (After Ax, from Barnes, R. D.: Invertebrate Zoology. 3rd ed. Philadelphia, W. B. Saunders Company, 1974.)

cerned with feeding and food intake, or **inges-tion.**

Pharynx. An anterior region of the gut that is usually muscular and frequently specialized for the ingestion of food.

Esophagus. A tubular section that transports food to more posterior regions of the gut.

Crop. An area specialized for temporary food storage.

Gizzard. A region of the gut specialized for the mechanical breakdown, or **trituration,** of food into smaller particles prior to digestion. The walls are muscular, and the surface facing the lumen often bears hard plates, teeth or ridges.

Stomach. A dilated, sometimes muscular, region of the gut, within which storage and digestion take place. In some animals the stomach may also be a site of absorption.

Intestine. A tubular region involved in digestion and absorption. Feces are usually formed in the posterior part of the intestine.

Rectum. A terminal part of the gut in which feces are formed and stored prior to elimination.

Unfortunately, terminology based on function has not always been used in the naming of gut regions. Early comparative anatomists often named regions of other animals after their supposed similarities to regions of the human digestive tract. Thus the crayfish has a cardiac and a pyloric stomach (the upper and lower regions of the human stomach are termed cardiac and pyloric) that are in fact a combination crop, gizzard and stomach.

Two other kinds of specialization of the gut deserve mention. Parts of the gut tube, especially the intestine, commonly contain structural modifications that increase the surface area. The walls may be infolded or bear fingerlike projections; or, there may be outpocketings called ceca present.

Secretory cells are usually a part of the gut lining, but they may be concentrated in special glandular regions or organs. The vertebrate liver and pancreas are two examples. Although in the adult such organs may be connected to the gut by only a small duct, they arise as outpocketings of the gut tube during embryonic development.

12.3 DIETS AND FEEDING MECHANISMS

No digestive tract possesses all of the specializations described above. Whatever specializations are present depend largely upon (1) the diet and (2) the feeding habits of the animals. The feeding habits and diet are related in turn to the animal's life style; i.e., whether it is free moving or attached, lives in burrows or in tubes and so forth.

Kinds of Diets. Animals have exploited virtually all types of organic food sources. **Herbivorous** animals utilize plants as food. Most herbivores feed on only a few, or in some cases on only one, plant species. The plants that serve as food depend upon the size, habitat and geographical distribution of the animal. Moreover, the diet may be restricted to certain parts of the plant, such as roots, leaves, sap, nectar and seeds. We generally associate herbivorous species with either vascular or flowering plants, but there are many species of animals that feed on algae, fungi or bacteria. These restrictions account for the great diversity of herbivorous diets.

Carnivorous animals feed on the bodies of other animals. Like herbivores, carnivores are usually restricted to the kinds of food found in their habitat and geographical range. Although most carnivores usually consume all or most of the bodies of their prey, some eat only certain parts, such as blood, skin and so on. Animal parasites are actually carnivores, but the injury they inflict on the prey, or **host,** is usually not sufficient to cause death. Some parasites may utilize the prey over a long period of time before eventually bringing about death. Certain wasps, for example, deposit their eggs on the bodies of caterpillars, which do not die until after the wasps have reached adulthood.

Many animals utilize nonliving organic substances as food sources. Such specialized diets may be limited to hair, feathers, shed skin, feces or decomposing animal bodies (carrion feeders). Nonliving plant products, such as cellulose, and other organic compounds of decomposing vegetation are important food sources.

Many animals do not use plant or animal remains until they have been broken up into small fragments, or **detritus,** and for some the bacterial decomposers rather than the detritus itself may serve as food. In terrestrial habitats, such as a forest floor, detritus is called **humus** and consists mostly of plant remains. In the sea, detritus derived from various sources is at first suspended in the sea water but gradually sinks to the bottom and becomes mixed with sand grains. The deposited material is an important food source for many animals.

The diet of some animals is augmented by or dependent upon other organisms, or **symbionts,** that live within them. For example, many sea anemones and reef corals contain large numbers of unicellular golden-brown algae within their body tissues. Part of the organic carbon produced by the photosynthetic activity of the algae is taken up by the host as glycerol. Unicellular green algal symbionts in certain hydras and a few other animals provide their hosts with glucose. The organic compounds obtained from these algae only supplement the diets of their hosts, which also feed in more conventional ways.

Although large numbers of animals are herbivorous or feed on plant products, such as wood, relatively few animals possess the enzymes to digest cellulose. Cellulose is most commonly digested by bacterial symbionts. In cattle and other ruminants, for example, grass or hay undergoes bacterial fermentation in the rumen, one of four chambers of the stomach (Fig. 12.2). Food is regurgitated from time to time as the animal ruminates, or chews its cud. The rumen, which has a capacity of up to approximately 200 liters, contains a large colony of bacteria and other microorganisms that play a twofold role in the animal's nutrition. They produce **cellulases,** which split the β-glucosidic bonds of cellulose, and other enzymes that convert the glucose to smaller units, chiefly acetic acid, that are absorbed directly from the rumen. As the microorganisms multiply, they synthesize amino acids and proteins. The bacteria are "harvested" and digested by the cow. Eventually, material from the rumen passes through the reticulum and omasum, where much of the water is pressed out of the food mass, and into the abomasum. The abomasum alone contains gastric glands.

Termites and wood roaches feed upon wood but depend upon symbiotic protozoan flagellates rather than bacteria for cellulose digestion.

Feeding Mechanisms. The food of animals is obtained through a variety of feeding mechanisms. Although similar diets and feeding habits have evolved many times in different groups, the feeding processes are not identical. Each mechanism utilizes the special features of the body plan of the particular group of animals involved. Most animal feeding mechanisms, however, fall within one of the following categories:

Cropping and Sucking Herbivores. The buccal cavity of many herbivores is equipped with a variety of structures, such as jaws, teeth, blades and scrapers, that crop and ingest fragments of plant tissue. The ingestive organs of other herbivores are designed for piercing plant tissue and sucking out sap or cell contents. The herbivore gut, which may contain colonies of cellulose-digesting organisms, is often longer than that of carnivores. The difference in gut length associated with diet is especially striking in the development of frogs. Algae-feeding herbivorous tadpoles have a long, greatly coiled gut, but during the metamorphosis of the tadpole to the insect-feeding adult form, the gut becomes much shorter.

Raptorial Feeding. Raptorial feeding (capture of prey) is generally associated with a carnivorous diet. Buccal structures are often involved in seizing and holding prey, but many animals utilize other parts of the body. The prey is either swallowed whole or is ingested in fragments after being bitten and torn. The digestive secretions contain a high percentage of proteases and digestion is usually rapid.

Parasitism. Parasites are specialized carnivores. Those that attach to the outside of the host **(ectoparasites)** are adapted for clinging and generally have ingestive organs for sucking blood or tissue fluids that are obtained by piercing or biting the integument. Those that live within the body of the host **(endopara-**

Figure 12.2 Course of food through the stomach of a cow.

sites) may also obtain food by sucking, or they may absorb simple food compounds directly from the host.

Suspension Feeding. Many aquatic animals feed on minute plants, animals and detritus suspended in water. The suspended plants and animals are collectively known as **plankton.** Suspension feeding is a rather indefinate term that refers more to the source of food than to the method of feeding. The most common type of suspension feeding is **filter feeding,** in which suspended food is filtered or screened from the water. Filter feeding typically involves three processes: (1) creation of a water current; (2) filtering by some specialized part of the body; and (3) transport of the collected food from the filter to the mouth. Fanworms, a group of marine tube-dwelling worms, are good examples of filter feeders (Fig. 12.3). In these worms, a circle of feather-like structures, **radioles,** projects from the head in the form of a funnel. Cilia on the radioles produce a water current that passes down into the funnel, and as water passes between the radioles, food particles are trapped in mucus on the radiole surface (the filter). Tracts of cilia, different from those producing the water current, transport food particles down the radioles to the mouth.

The term ciliary feeding should be distinguished from the term filter feeding: not all filter-feeders employ cilia; and not all of the animals using cilia are filter feeders.

Deposit Feeding. The organic detritus de-

Figure 12.3 Filter feeding in the annelid fanworm *Sabella. A,* Water current passing through radioles projecting from opening of tube. *B,* Water current (*large arrows*) and ciliary tracts (*small arrows*) over a section of one radiole. The spheres A, B and C indicate the different sizes of particles sorted. (After Nicol, from Newell, R. C.: Biology of Intertidal Animals. American Elsevier Publishing Co., Inc., 1970.)

radiole

ventral sac

sand tube

A

FRONTAL CILIARY STREAM

LATEROFRONTAL CILIARY STREAM

LONGITUDINAL FOOD GROOVE

ABFRONTAL CILIARY STREAM

B

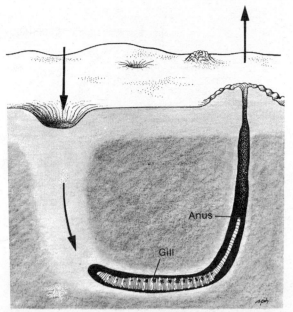

Figure 12.4 The lugworm, *Arenicola,* in burrow. Arrows indicate the direction of water flow produced by the worm. The worm ingests the column of sand on the left, through which water is filtered. Pile of sand at burrow opening is defecated castings. (From Barnes, R. D.: Invertebrate Zoology. 3rd ed. Philadelphia, W. B. Saunders Company, 1974.)

posited on the bottom of aquatic habitats and mixed with sand grains is utilized as a food source by many animals, called deposit feeders. Some, such as the marine lugworms, are **nonselective deposit feeders.** Lugworms live upside down in L-shaped burrows and ingest the substratum, both sand grains and deposited

detritus, at the end of the burrow (Fig. 12.4). During passage through the gut, the organic material is digested away from the mineral component. Like many other nonselective deposit feeders, lugworms periodically back up to the surface and defecate the mineral matter in conspicuous piles, called **castings.** Although terrestrial, earthworms are in part nonselective deposit feeders.

Some deposit-feeding animals ingest only organic material. The marine terebellid worms are such **selective deposit feeders** (Fig. 12.5). The terebellids possess a large number of long, delicate head tentacles that project out over the substratum from the tubes or burrows in which worms live. Light organic particles adhere to mucus on the tentacles and are transported back to the mouth by cilia, as well as by contraction of the tentacles.

Feeding habits and diets greatly affect the gut modifications discussed earlier. The close interrelationship between all of the factors can be seen by comparing two related species with different feeding habits. A finch and an owl are related closely enough so that gut modifications correlate more with differences in diet than with differences in phylogenetic position. A seed-eating finch possesses a heavy beak especially adapted for cracking seeds (Fig. 12.6). The edges are sharp and the interior surface is ridged. During feeding, seeds first pass down to the stomach, but as the stomach fills, the crop, which in birds is a side pocket of the esophagus, opens and receives the seeds. Later, contrac-

Figure 12.5 *A,* The annelid worm *Amphitrite* at aperture of burrow with tentacles outstretched over substratum. (From Barnes, R. D.: Invertebrate Zoology. 3rd ed. Philadelphia, W. B. Saunders Company, 1974.) *B,* Cross section through tentacle of *Terebella lapidaria* creeping over substratum. *C,* Section of tentacle of *Terebella lapidaria* rolled up to form a ciliary gutter and transporting deposit material. *D,* A tentacle being wiped by one of the lips. (After Dales.)

Figure 12.6 *A–F*, Removal of hull from seed in finch beak. Lower mandible can move forward and backward (*A* and *B*) and from side to side (*D* and *E*). (Redrawn from Ziswiler, from Welty, J. C.: The Life of Birds. 2nd ed. Philadelphia, W. B. Saunders Company, 1975.) *G*, Digestive system of a pigeon; that of a finch would be similar. (After Schimkewitsch and Stresemann, from Welty, J. C.: The Life of Birds. 2nd ed. Philadelphia, W. B. Saunders Company, 1975.)

tions of the crop send seeds to the stomach. The upper part of the stomach is glandular and produces proteases in an acid medium, as in most other vertebrates. The lower portion of the stomach is modified as a gizzard with heavy muscular walls. The interior keratinaceous surface is ridged. Ingested mineral particles, or grit, lodge between the ridges and aid in crushing and grinding seeds.

In contrast to finches, owls are raptorial and feed on mice and other small rodents, which are swallowed entire. The prey is caught and killed with the powerful talons and short curved beak (Fig. 12.7). There is no crop, and the glandular part of the stomach is well developed. The gizzard is reduced to a valve, which prevents hair and bones from entering the intestine. These indigestible items are regurgitated. The intestine is shorter than that of the finch.

The insects provide many similar illustrations. Grasshoppers, for example, have a pair of jawlike mandibles on either side of the mouth. In grass-eating species, the mandibles have opposing chisel-like cutting surfaces and more basally flattened grinding molar surfaces (Fig. 12.8). The grass is manipulated and aligned by a pair of maxillae and a labium located behind the mandibles.

Pieces of food removed by the mandibles are moistened with salivary secretions. They pass through a pharynx and an esophagus into a crop, where they are temporarily stored (Fig. 12.8C). The contents of the crop slowly pass into a proventriculus, a short gizzard-like region with projecting cuticular teeth. Digestion and absorption take place in the ventriculus, or midgut. Wastes are carried posteriorly by a tubular intestine to the terminal rectum, which opens to the exterior through the anus.

Leafhoppers and squash bugs have the same feeding appendages as grasshoppers, but they are highly modified for piercing and sucking (Fig. 12.9A and B). A needle-like stylet formed by the mandibles and maxillae contains two channels, one for conveying salivary secretions outward and one for the passage of plant juices into the gut. The elongated labium functions as a guide for the stylet.

A

B

C

D

Figure 12.7 *A–C,* Owl eating a mouse. Only owls and parrots can lift food to the beak with the foot. *D,* Pellets of undigestible material regurgitated by an owl. (Photos by C. R. Austing, from Welty, J. C.: The Life of Birds. 2nd ed. Philadelphia, W. D. Saunders Company, 1975.)

The pharynx is modified as a pump for drawing in fluid, which then passes through an esophagus to the midgut. The proventriculus is reduced to a valve.

Fluid feeders take in large amounts of water.

In leafhoppers, the intestine loops forward over the anterior part of the midgut, and much of the ingested water passes directly across the gut walls into the intestinal lumen, bypassing the remainder of the midgut (Fig. 12.9C).

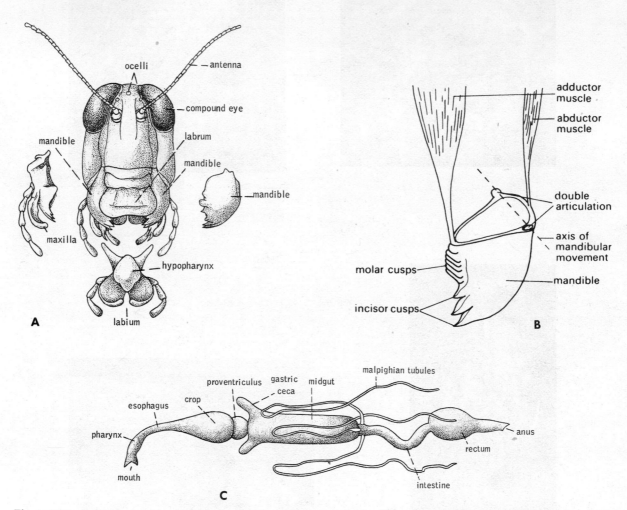

Figure 12.8   *A*, Front view of a grasshopper, showing mouthparts. *B*, Mandible of a grasshopper that feeds on grass, show-
ing cutting and grinding surfaces. (After Snodgrass from Chapman, R. F.: The Insects: Structure and Function. American
Elsevier Publishing Co., Inc., 1969.) *C*, Lateral view of digestive tract of a chewing insect, such as a grasshopper. (After
Snodgrass.)

Figure 12.9 *A*, Head of a hemipteran insect penetrating a plant surface with its beak. *B*, Section through beak. *C*, Digestive tract of a spittlebug. (*A* after Kullenberg; *B* after Poisson; *C* after Snodgrass, from Chapman, R. F.; The Insects: Structure and Function. American Elsevier Publishing Co., Inc., 1969.)

12.4 THE VERTEBRATE PATTERN

Mouth. The basic pattern of the digestive system is similar in all vertebrates. In very primitive vertebrates the mouth is unsupported by jaws, but most vertebrates have jaws and a good complement of teeth that aid in obtaining food.

Teeth are similar in structure to the scales of sharks, which are composed of enamel- and dentin-like materials, and are believed to have evolved from bony scales. A representative mammalian tooth (Fig. 12.10) consists of a **crown** projecting above the gum, a **neck** surrounded by the gum and one or more **roots** embedded in sockets in the jaws. The crown is covered by a layer of **enamel.** Enamel, which is the hardest substance in the body, consists almost entirely of crystals of calcium salts. Calcium, phosphate and fluoride are important constituents of enamel, and all must be present in the diet in suitable amounts for proper tooth development and maintenance. The rest of the tooth is composed of **dentin,** a substance similar in composition to bone. In the center of the tooth is a **pulp cavity,** containing blood vessels and nerves. A layer of **cement** covers much of the root and holds the tooth firmly in place in the jaw.

In mammals, the teeth are differentiated into several types that are used for the seizing and mechanical breakdown of food (p. 838). Mammalian teeth, unlike those of lower vertebrates, are not continuously replaced by new sets. Humans first develop a set of deciduous, or **milk, teeth**—two incisors, one canine and two premolars on each side of each jaw. These are later replaced by **permanent teeth;** in addition, three molars develop on each side of each jaw behind the premolars. The molars last throughout life and are not replaced (Fig. 12.10).

A fish can easily manipulate and swallow food, for the flow of water aids in carrying it back into the pharynx. Oral glands and a muscular tongue are poorly developed in fishes. The evolution of these structures accompanied the transition from water to land, and they became more elaborate in the higher terrestrial vertebrates. The **tongue** of the frog and the anteater is a specialized food gathering device (Fig. 12.11), but in most mammals its chief function is to manipulate food in the mouth and to aid in swallowing. In many mammals, the tongue pushes the food between the teeth so that it is thoroughly masticated and mixed with saliva. The food is then shaped into a ball, a **bolus,** and moved into the pharynx by raising the tongue (Fig. 12.13). The tongue also has numerous microscopic taste buds; the human tongue is of great importance in speech.

In addition to a liberal sprinkling of simple glands in the lining of the mouth cavity, mammals have evolved several pairs of conspicuous **salivary glands** that are connected to the mouth by ducts. The location of the human **parotid, mandibular** and **sublingual glands** is

Figure 12.10 The teeth of humans. *Left,* a vertical section through a molar tooth; *right,* types of teeth. (From Jacob and Francone: Structure and Function in Man. 3rd ed. Philadelphia, W. B. Saunders Co., 1974.)

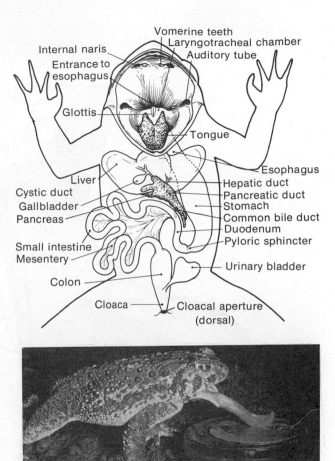

Figure 12.11   A ventral view of the frog's digestive system. The liver lobes have been turned forward to show the gallbladder. Tongue action is shown in the inset. Notice that the tongue is stretched to over half the length of the body. (Inset from Van Riper in Natural History, Vol. LXVI, No. 6.)

shown in Figure 12.12. In primitive terrestrial vertebrates, oral glands simply secrete a mucous and watery fluid that lubricates the food, and this is still the major function of saliva. The saliva of most mammals and of a few other terrestrial vertebrates contains **salivary amylase,** which splits starch and glycogen into the disaccharide maltose. Chloride ions present in the saliva are necessary to activate amylase. Its pH optimum is close to neutrality, so its action is eventually stopped by the acidic gastric juice of the stomach. But, since it takes one-half hour or longer for the food and gastric juice to become thoroughly mixed, 40 per cent or more of the starches are split before the amylase is inactivated.

Pharynx and Esophagus. Part of the human pharynx lies above the **soft palate** (Fig. 12.13) and receives the internal nostrils, or **choanae,** and the openings of the pair of eustachian, or **auditory, tubes** from the middle ear cavities. Another part lies beneath the soft palate just posterior to these parts and leads to the esophagus and larynx. Passage of food into the pharynx initiates a series of reflexes: The muscular soft palate rises and

prevents food from entering the nasal cavities; breathing momentarily stops; the larynx is elevated and the epiglottis swings over the glottis, preventing food from entering the larynx; the tongue prevents food from returning to the mouth; and muscular contractions of the pharynx move the bolus into the esophagus.

The pharynx of the terrestrial vertebrate is a rather short region in which the food and air passages cross, but in fishes it is a more extensive area associated with the gill slits. Successive waves of contraction and relaxation of the muscles, known as **peristalsis,** propel the bolus down the esophagus to the stomach. The muscles relax in front of the food and contract behind it. When the food reaches the end of the esophagus, the cardiac sphincter, which closes off the entrance to the stomach, relaxes and allows it to enter. The esophagus is generally a simple conducting tube, but in some animals its structure has been modified for storage. The crop of the pigeon, for example, is a modified part of the esophagus.

Stomach. The stomach and intestines of vertebrates are suspended by mesenteries

Parotid duct

Parotid gl.

Sublingual gl.

Submandibular gl.

Site of 1st constriction

Esophagus

Left bronchus crossing
in front of esophagus

Site of 2nd constriction

Site of 3rd constriction

Diaphragm

Spleen

Stomach

Liver

Splenic flexure

Transverse colon

Gallbladder

Hepatic flexure

Pancreas

Descending colon

Ascending colon

Small intestine

Cecum

Appendix

Sigmoid colon

Rectum

Anus

Figure 12.12   The human digestive system. (From Gardner, W. D., and W. A. Osburn: Structure of the Human Body. 2nd ed. Philadelphia, W. B. Saunders Company, 1973.)

Figure 12.13 Diagrams showing the shifts in the position of the tongue, soft palate and epiglottis when a bolus of food is swallowed. (Slightly modified after Villee, C. A.: Biology, 7th ed. Philadelphia, W. B. Saunders Company, 1977.)

within the large abdominal coelom. Blood vessels and nerves reach the viscera through the mesenteries. The presence of a coelom, together with a small amount of watery coelomic fluid, facilitates the expansion, contraction and other functional movements of the viscera.

The **stomach** is usually a J-shaped pouch (Figs. 12.11 and 12.12) whose chief functions are the storage and mechanical churning of food and the initiation of the enzymic hydrolysis of proteins. Lampreys, lungfishes and some other primitive fishes do not have a stomach, and the absence of this organ is thought to have been a characteristic of the ancestral vertebrates. The early vertebrates were probably filter feeders that fed more or less continuously on minute food particles that could be digested by the intestine alone. Presumably, the evolution of jaws and the habit of feeding less frequently and on larger pieces of food required that there be an organ for storage and initial conversion of this food into a state in which it could be digested further in the intestine. In most vertebrates both mechanical and chemical breakdown of food begins in the stomach.

After food enters the stomach, the **cardiac sphincter** at its anterior end and the **pyloric sphincter** at the posterior end close. Muscular contractions of the stomach churn the food, breaking it up mechanically and mixing it with the gastric juice secreted by tube-shaped **gastric glands.** The gastric juice contains hydrochloric acid and the proteolytic enzyme **pepsin.** In addition, **rennin** is particularly abundant in the stomach of young mammals and causes the milk protein casein to coagu-

late and remain in the stomach long enough to be acted on by pepsin. For centuries, rennin has been extracted from the stomachs of calves and used to curdle milk; this is an important step in the manufacture of cheese.

When the food has been reduced to a creamy material known as **chyme,** and most of the microorganisms that entered the stomach with it have been killed by the action of the gastric juices, the pyloric sphincter opens and the food passes into the small intestine. The most fluid food passes first. Indeed, upon entering the stomach, water passes almost immediately into the intestine. The acidic food enters the intestine in spurts and is quickly neutralized by the alkaline secretions flowing into the intestine from the liver and pancreas.

Liver and Pancreas. The liver and pancreas are large, glandular outgrowths from the anterior part of the intestine (Figs. 12.11 and 12.12). The **liver,** in fact, is the largest organ of the body. Its cells continually secrete **bile,** which passes through hepatic ducts into the **common bile duct** and then up the cystic duct into the **gallbladder.** Bile does not enter the intestine immediately, for a sphincter at the intestinal end of the bile duct is closed until food enters the intestine. Contraction of the wall of the gall bladder, stimulated by the hormone pancreozymin (p. 285), forces bile out. The bile that is finally poured into the intestine is concentrated, for a considerable amount of water and some salts are absorbed from the bile in the gall bladder.

Although bile contains no digestive enzymes, it nevertheless has a twofold digestive role. Its alkalinity, along with that of the pancreatic secretions, neutralizes the acid food entering the intestine and creates a pH favorable for the action of pancreatic and intestinal enzymes. Its bile salts emulsify fats, breaking them up into smaller globules and thereby providing more surfaces on which fat-splitting enzymes can act. These salts are also essential for the absorption of fats and fat-soluble vitamins (A, D, E, K). Most of the bile salts are not eliminated with the feces but are absorbed from the intestine with the fats and return to the liver by the blood stream to be used again.

The color of bile (green, yellow, orange or red in different species) results from the presence of **bile pigments** derived from the breakdown of hemoglobin in the liver. The bile pigments are converted by enzymes of the intestinal bacteria to the brown pigments responsible for the color of the feces. If their excretion is prevented by a gallstone or some other obstruction of the bile duct, they are reabsorbed by the liver and gall bladder, the feces are pale, and the pigments accumulate in the skin, giving it the yellowish tinge characteristic of jaundice.

In addition to its involvement with digestion, the liver performs many other functions, some of which will be discussed in later sections.

The **pancreas** is an important digestive gland, producing quantities of enzymes that act upon carbohydrates, proteins, fats and nucleic acids. These enzymes enter the intestine by way of a pancreatic duct that joins the common bile duct. In some vertebrates an accessory pancreatic duct empties directly into the intestine. The pancreas contains patches of hormone-producing tissue, the **islets of Langerhans,** which will be considered in a later section.

Intestine. Most of the digestive processes, and virtually all of the absorption of the end products of digestion, occur in the intestine. Enzymes acting in the intestine are produced by the pancreas, by many of the epithelial cells lining the intestine, and, in birds and mammals, by small, tube-shaped **intestinal glands.** Adequate surface area for absorption is made available by the length of the intestine and by outgrowths and foldings of the lining.

The structural details of the intestine vary considerably among vertebrates. Primitive fishes have a short, straight intestine extending posteriorly from the stomach. Since its internal surface is increased by a helical fold known as the spiral valve, it is called a **valvular intestine.** Terrestrial vertebrates have lost the spiral valve and make up for this by an increase in the length of the intestine, which becomes more or less coiled. The tetrapod in-

testine has become further differentiated into an anterior **small intestine** and a posterior **large intestine** (Figs. 12.11, and 12.12). The first part of the small intestine is the **duodenum** and, in mammals, the two succeeding parts are the **jejunum** and **ileum.** The large intestine, or **colon,** of the frog and most vertebrates leads to a posterior chamber known as the **cloaca.** The cloaca, which also receives the products of the urinary and reproductive systems, opens on the body surface by the **cloacal aperture.** In mammals, the cloaca has become divided into a ventral part, which receives the urogenital products, and a dorsal *rectum,* which opens on the body surface at the **anus.** A blind pouch called the **cecum** is present at the junction of small and large intestines of mammals. This is very long in herbivores such as the rabbit and horse and contains colonies of bacteria that digest cellulose. Humans have a small cecum with a vestigial **vermiform appendix** on its end. An **ileocecal** valve located between the small and large intestine prevents bacteria in the colon from moving back up into the small intestine.

A transverse section of the small intestine of a mammal illustrates the microscopic structure of the digestive tract (Figs. 12.14 and 12.15). There is an outer covering of **visceral peritoneum** (the serous coat), a layer of **smooth muscle,** a layer of vascular connective tissue, the **submucosa** and, finally, the innermost layer, the **mucosa** (or mucous membrane). The outer fibers of the muscular coat are usually described as longitudinal; the inner, as circular. Actually, both layers are spiral; the outer is an open spiral and the inner, a tight spiral. The relaxations and contractions of these layers are responsible for the peristaltic and churning movements that mix the food with digestive enzymes and move it along the digestive tract. The mucosa consists of a layer of smooth muscle, connective tissue and, finally, the simple columnar epithelium next to the lumen. Many mucus-producing **goblet cells** are present in the lining epithelium, and their secretion helps to lubricate the food and protect the lining of the intestine. In the small intestine of mammals and birds, the mucosa bears numerous minute, finger-shaped **villi** containing blood capillaries and small lymphatic vessels. Circular folds in the intestinal mucosa, villi and microvilli form an enormous surface area for absorption; the surface area of the human small intestine may be as much as 550 square meters. The villi are moved about by the muscle layer in the mucosa, the **muscularis mucosae,** and by strands of smooth muscle that extend into them. Tube-

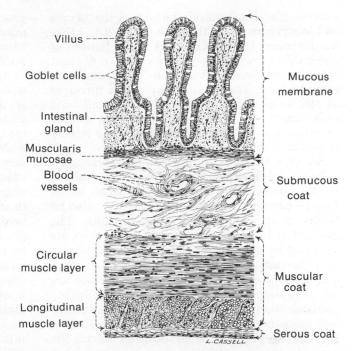

Villus

Goblet cells

Intestinal
gland

Muscularis
mucosae

Blood
vessels

Mucous
membrane

Submucous
coat

Circular
muscle layer

Muscular
coat

Longitudinal
muscle layer

Serous coat

L. CASSELL

Figure 12.14 Diagrammatic cross section of a portion of the small intestine. Numerous capillaries and a lymphatic vessel enter each villus. (From King, B. G., and M. J. Showers: Human Anatomy and Physiology. 6th ed. Philadelphia, W. B. Saunders Company, 1969.)

Striated
Border

Terminal Web

Terminal
Bar

Figure 12.15 Electron micrograph of part of the surface of two epithelial cells from a villus to show the striated border composed of numerous microvilli. The terminal bar is a specialization of the plasma membrane of adjacent cells that unites them tightly and closes the intercellular space. Original photograph, ×30,000. (From Bloom, W., and D. W. Fawcett: A Textbook of Histology. 10th ed. Philadelphia, W. B. Saunders Company, 1975.)

shaped intestinal glands known as the **crypts of Lieberkühn** lie at the base of the villi. Mitotic division of epithelial cells in the bottom of the crypts continually produces new cells that migrate outward over the villi and are sloughed off. There is a rather rapid turnover of epithelial cells, for an individual cell lasts only about 48 hours. The intestinal glands produce many digestive enzymes and a watery intestinal fluid.

Starches not broken down in the stomach by salivary amylase are soon converted to maltose in the neutral environment of the small intestine by **pancreatic amylase.** There may also be traces of amylase in the intestinal juice. The disaccharides maltose, sucrose and lactose are acted on by specific enzymes: **maltase, sucrase** and **lactase** (Table 12.1). These and other enzymes produced by the intestinal cells are not completely free in the main part of the intestinal lumen, for cell-free extracts of intestinal juice contain very little enzyme. Rather, the enzymes appear to be localized among the numerous microvilli that compose the striated border of the intestinal epithelium (Fig. 12.8). The disaccharides are hydrolyzed as they enter this border; all yield glucose; sucrose and lactose also yield fructose and galactose, respectively. It is in the form of these monosaccharides that most carbohydrates are absorbed.

The endopeptidases **trypsin** and **chymotrypsin** are secreted by the pancreas as inactive precursors, **trypsinogen** and **chymotrypsinogen.** Their activation is initiated by **enterokinase,** an intestinal enzyme that splits off a portion of the trypsinogen molecule and converts it to trypsin. Once formed, trypsin helps to continue the activation of trypsinogen, and it alone activates chymotrypsinogen. The various endopeptidases—pepsin, trypsin, chymotrypsin—split bonds adjacent to specific amino acids; their concerted action is necessary for the fragmentation of protein molecules. Pepsin only splits bonds adjacent to phenylalanine or tyrosine; trypsin, bonds next to arginine or lysine; and chymotrypsin, bonds next to leucine, methionine, phenylalanine, tyrosine or tryptophan. Exopeptidases secreted by the pancreas and small intestine attack the terminal peptide bonds, stripping off one amino acid after another. Exopeptidases known as **carboxypeptidases** split off the terminal amino acid with the free carboxyl group (p. 27); **aminopeptidases** separate the terminal amino acid with the free amino group (NH_2). The free amino acids resulting from the digestion of proteins are then absorbed into the blood stream.

Fats emulsified by bile salts are attacked by esterases that cleave the ester bonds joining fatty acids to glycerol. **Lipase,** most of which is secreted by the pancreas, although some comes from the intestinal glands, is the principal esterase of vertebrates. The mixture of bile salts, fatty acids and partly digested fats

TABLE 12.1 MAJOR DIGESTIVE ENZYMES OF VERTEBRATES

Produced by	Enzyme	Substrate Acted upon	Product
Salivary glands	Amylase	Starch	Maltose (double sugar)
Gastric glands	Pepsinogen converted to pepsin	Proteins	Peptides
	Rennin	Casein	Coagulated casein
Pancreas	Amylase	Starch	Maltose
	Trypsinogen converted to trypsin	Proteins	Peptides
	Chymotrypsinogen converted to chymotrypsin	Chymotrypsinogen	Chymotrypsin
		Proteins	Peptides
	Exopeptidases	Peptides	Amino acids
	Lipase	Emulsified fat	Fatty acids and glycerol
	Nucleases	Nucleic acids	Nucleotides
Intestinal glands	Amylase	Starch	Maltose
	Maltase	Maltose	Glucose (single sugar)
	Sucrase	Sucrose (double sugar)	Glucose and fructose
	Lactase	Lactose (double sugar)	Glucose and galactose
	Enterokinase	Trypsinogen	Trypsin
	Exopeptidases	Peptides	Amino acids
	Lipase	Emulsified fats	Fatty acids and glycerol
	Nucleases	Nucleic acids	Nucleotides

collectively emulsifies fats further into particles, many of which are small enough to be absorbed directly.

The absorption of fats and fatty acids presents a special problem, for unlike the other end products of digestion, they are not water soluble. Apparently, their uptake is facilitated by their combination with bile salts, for this makes a soluble complex. Bile salts are freed within the mucosal cells, and the fatty acids and glycerol are combined with phosphate to form **phospholipids.** Unhydrolyzed fat and most of the phospholipids enter the lymphatic capillaries, but the other absorbed materials enter the capillaries of the blood vessels.

Although most of the water ingested and existing in the gastrointestinal secretions is absorbed in the small intestine, the material that enters the large intestine is still quite fluid. Most of the remaining water and many salts are absorbed as the residue passes through the colon. If the residue passes through very slowly and too much water is absorbed, the feces become very dry and hard and **constipation** may result; if it goes through very rapidly and little water is absorbed, **diarrhea** results. Many bacteria reside in the colon and synthesize a variety of vitamins, which are absorbed from the colon. The bacteria reproduce very rapidly, and many are eliminated. As much as 25 per cent of the feces may consist of bacteria.

The Control of Digestive Secretions. Each is secreted at an appropriate time: We salivate when we eat, and gastric juice is produced when food reaches the stomach. The control of these digestive secretions is partly nervous and partly endocrine. The smell of food or its presence in the mouth stimulates sensory nerves that carry impulses to a salivation center in the medulla of the brain. From there the impulses are relayed along motor nerves to the salivary glands, which then secrete.

The control of gastric secretion is more complex. Years ago the famous Russian physiologist Pavlov performed an experiment in which he brought the esophagus of a dog to the surface of the neck and severed it. When the dog ate, the food did not reach the stomach, yet some gastric juice was secreted provided that the **vagus nerve,** which carries motor fibers to the stomach and other internal organs, was left intact. If the vagus nerve was cut, this secretion did not occur. This experiment proved that the control of gastric secretion was at least partly nervous. Subsequently, it was discovered that if the vagus was cut but food was

permitted to reach the stomach, a considerable flow of gastric juice was produced. Obviously, the vagus nerve is not the only means of stimulating the gastric glands. Further investigation revealed that when partly digested food reaches the pyloric region of the stomach, certain of the mucosal cells produce the hormone **gastrin,** which is absorbed into the blood through the stomach wall and ultimately reaches the gastric glands, stimulating them to secrete.

Other factors are also involved. Certain foods, particularly partly digested meats, release substances of unknown composition that stimulate gastric secretion. The absorption of certain products from the small intestine, among them alcohol, further stimulates gastric secretion. These are among the reasons for restricting meats and alcohol from the diets of ulcer patients.

Gastric secretion is reduced and finally stopped as the stomach empties and food enters the duodenum. When food, especially fats, enters the duodenum, the duodenal mucosa produces the hormone **enterogastrone** which, on reaching the stomach, inhibits the secretion of the gastric glands and slows down the churning action of the stomach. This not only helps to prevent the stomach from digesting its own lining but also enables fatty foods to stay for a longer period in the duodenum where they can be acted on by bile salts and lipase.

One of the first hormones to be discovered was **secretin,** which initiates pancreatic secretion. In 1902 Bayliss and Starling were investigating the current belief that the secretion of pancreatic juice was under nervous control. They found that the pancreas secreted its juice when acid food entered the small intestine even though the nerves to and from the intestine were cut. A stimulant of some sort apparently traveled in the blood. The injection of acids into the blood stream had no effect, so they reasoned that some stimulating principle must be produced by the intestinal mucosa upon exposure to acid foods. When they injected extracts of such a mucosa into the circulatory system, the pancreas secreted. It is now recognized that secretin itself only stimulates the pancreas to produce a copious flow of a bicarbonate-rich alkaline liquid that neutralizes the acid material entering from the stomach. A second hormone produced by the intestinal mucosa, **pancreozymin,** is necessary to stimulate the release of pancreatic enzymes. Pancreozymin also causes the gall bladder to contract and release the bile. Vagal stimulation also plays a role in the release of bile.

12.5 REGULATION OF FOOD SUPPLY

Autotrophic organisms can regulate the acquisition of different food compounds by regulating their rates of synthesis. Regulation by controlling diet is much more difficult for heterotrophic organisms. Most animals feed discontinuously, which results in more food being acquired than can be utilized at one time, and the food ingested at a particular feeding may contain just adequate amounts of some substances and excess amounts of others.

These problems are overcome by storage and interconversion of the several types of food. Glycogen, the typical animal carbohydrate, functions as a short-term storage product, one that is called up rapidly when available sugars in the blood or elsewhere are depleted. Glycogen is stored in most vertebrate cells but is especially abundant in liver and muscle. Significantly, all of the blood draining the intestine passes through the liver by way of the **hepatic portal** venous system. In the course of passage through the liver, some of the excess sugars absorbed from the intestine are taken up by the liver cells and converted to glycogen. The hormone **insulin,** produced by clusters of pancreatic endocrine cells, the **islets of Langerhans,** controls glycogen synthesis. A rise in blood sugar stimulates the pancreatic cells to produce insulin. Insulin is transported via the blood stream throughout the body where it induces glycogen synthesis in muscle cells and in the liver. The reverse reaction, the conversion of glycogen to glucose, is regulated by another pancreatic hormone, **glucagon,** and by **epinephrine.** However, muscle cells lack the enzyme to convert glucose-6-phosphate to glucose, so that muscle glycogen can only be used as an energy store for muscle cells.

Food taken in as carbohydrate or protein can be converted to fats for storage, and then utilized subsequently as a source of energy when the animal is not feeding. Fats stored in the liver or adipose tissue undergo constant breakdown and resynthesis, even though the total amount stored may change very little over long periods of time. The cells of adipose tissue are modified connective tissue cells; most of each is occupied by a large fat globule. Fat cells are located in various places, depending upon the group of animals. In vertebrates, fat is accumulated beneath the dermis of the skin, around the liver and heart and in the mesenteries of the gut. Most fatty acids and glycerol are resynthesized into fat during absorption, and the fat is transported as small globules, the **chylomicrons,** by the lymphatic system to the blood vessels. Within two or three hours after absorbing a fatty meal, the chylomicrons disappear from the blood; some are taken up by liver cells, others are digested within the blood stream by **lipoprotein lipase.** Lipoprotein lipase is produced in great quantities by the fat depots of the body, and it is believed that most of the hydrolyzed fat is quickly absorbed and resynthesized by these tissues.

Proteins are not usually stored in animals, but in cases of extreme starvation many animals will utilize their tissue proteins, sacrificing muscles, gonads and other organs.

The ability of animals to convert food compounds from one form to another is of great importance in balancing the uncertainties of diet with internal food requirements. Excess sugars can be converted and stored as glycogen or can be converted to fatty acids and stored as fats. However, most animals cannot synthesize the polyunsaturated fatty acids they require; these must be obtained in the diet.

Like fatty acids, some but not all of the 20 amino acids used in protein synthesis can be synthesized by animals. The laboratory rat, for example, can synthesize 10 of the 20. The large amount of excess amino acids that may be absorbed from the diet, especially in carnivorous animals, cannot be stored but can be used as fuels or converted to other compounds, such as sugar or fatty acids, via pathways shown in Figure 12.16. Protein metabolism will be discussed further in Chapter 15.

12.6 NON-FOODS IN THE DIET

An animal's diet not only provides food but also is the principal source of other substances essential to cellular constituents. The diet is a major source of water, both free water and water from metabolism; indeed, there are many animals that never drink water, obtaining it entirely from food. The diet is also the primary source of inorganic ions, such as calcium, phosphate, potassium, sodium, chloride, iodide and others involved in various cell functions and in the formation of skeletons and shells.

Traces of copper are needed as a component of certain enzyme systems and for the proper utilization of iron in hemoglobin. Traces of manganese, molybdenum, cobalt and zinc are required as components of certain enzyme sys-

Figure 12.16 Diagram of metabolic pathways by which food compounds are interconverted, stored or used as fuels. Keto acids in unshaded boxes; amino acids in shaded boxes. (Modified in part from Lehninger.)

tems; zinc, for example, is a component of carbonic anhydrase, alcohol dehydrogenase and a number of other enzymes. The role of fluoride ion in the prevention of dental caries is well known.

Still another provision of the animal diet is vitamins. Vitamins are relatively simple organic compounds required in small amounts in the diet. They differ widely in their chemical structure but have in common the fact that they cannot be synthesized in adequate amounts by the animal and hence must be present in the diet. All plants and animals require these same substances to carry out specific metabolic functions, but organisms differ in their ability to synthesize them; thus, what is a "vitamin" for one animal or plant is not necessarily one for another. Only humans, monkeys and guinea pigs require vitamin C, **ascorbic acid,** in their diets; other animals can synthesize ascorbic acid from glucose. The mold *Neurospora* requires the vitamin biotin.

Insects cannot synthesize cholesterol, and it might be argued that for insects cholesterol is a vitamin.

The vitamins whose role in metabolism is known—niacin, thiamine, riboflavin, pyridoxine, pantothenic acid, biotin, folic acid and cobalamin—have been found to be constituent parts of one or more coenzymes. Two major groups of vitamins can be distinguished: those that are soluble in lipid solvents, the **fat-soluble vitamins** A, D, E and K; and those that are readily soluble in water, the **water-soluble vitamins** C and the B complex. The lack of any one of these vitamins produces a deficiency disease with characteristic symptoms, e.g., scurvy (lack of ascorbic acid), beriberi (lack of thiamine), pellagra (lack of niacin) and rickets (lack of vitamin D).

Thiamine pyrophosphate is the coenzyme for the oxidative decarboxylation of α-keto acids, such as pyruvate and α-ketoglutarate. It is also the coenzyme for transketolase. **Riboflavin,** as riboflavin monophosphate and flavin adenine dinucleotide, is required for the reactions of electron transport in the mitochondria and for certain oxidations in the endoplasmic reticulum. Riboflavin occurs in most foods and is synthesized by the intestinal bacteria, so that deficiencies of riboflavin are quite rare. **Pyridoxine,** as pyridoxal phosphate, is the coenzyme for many different reactions involving amino acids—transamination, decarboxylation to amines and so forth—and for glycogen synthetase. **Biotin** is a coenzyme for the reactions in which carbon dioxide is added to an organic molecule, such as the conversion of pyruvate to oxaloacetate, and the carboxylation of acetyl coenzyme A to form malonyl coenzyme A (the first step in the biosynthesis of fatty acids). It is widely distributed in foods, and only individuals who eat raw eggs in large quantities are likely to become deficient in biotin. Egg white contains a protein, **avidin,** that forms a tight complex with biotin and prevents its functioning.

Niacin is part of the coenzymes nicotinamide adenine dinucleotide (NAD) and nicotinamide adenine dinucleotide phosphate (NADP), which are cofactors for many dehydrogenases and serve as hydrogen acceptors or donors. The vitamin **folic acid** appears in the coenzyme tetrahydrofolic acid, a coenzyme for many reactions involving one-carbon transfers, and in biopterin, the coenzyme for the conversion of phenylalanine to tyrosine. **Cobalamin** (vitamin B_{12}) serves as the coenzyme in certain reactions involving one-carbon transfers such as the transfer of methyl groups, CH_3.

Cobalamin is synthesized by bacteria but not by higher plants or animals—thus this is a "vitamin" for the green plants as well as for animals. **Ascorbic acid** is known to play a role in the hydroxylation of proline to form hydroxyproline during collagen formation, but the overall function of this vitamin is still unknown.

Vitamin A, or **retinol**, is converted in the retina to retinal, a component of the light-sensitive pigment, visual purple. A defiency of retinol may lead to "night-blindness" and a severe deficiency will lead to xerophthalmia, a blindness due to the abnormal deposition of keratin as a film over the cornea. Keratin is also deposited in other mucous membranes of individuals suffering from xerophthalmia. Vitamin A undoubtedly has metabolic functions other than its conversion to retinal in the visual process but these are unknown. Plants synthesize β-carotene, an orange-yellow pigment, which can be split to give two molecules of retinol. Large amounts of retinol are stored in the liver, enough to supply a person for several years.

Vitamin D, **cholecalciferol,** plays a role in the movement of calcium ions through membranes, perhaps by stimulating the synthesis of a specific protein required in the transport process. Cholecalciferol can be formed in the skin from a precursor, cholesta-5,7-dienol, by the action of ultraviolet light, which cleaves the B ring of the precursor molecule. Thus cholecalciferol is a "vitamin" only if the person isn't exposed to an adequate amount of sunlight. Vitamin D is converted to 25-hydroxycholecalciferol by an enzyme in the liver and then to 1,25-dihydroxycholecalciferol by an enzyme in the kidney. The biologically active, hormone-like molecule 1,25-dihydroxycholecalciferol stimulates the synthesis of a calcium transport protein in the intestinal mucosa, and this substance is responsible for the increased uptake of calcium from the intestinal contents. Excessive doses of vitamin D are toxic, causing hypercalcemia and the deposition of calcium in soft tissues. An excess of vitamin A is also toxic. Acute poisoning with vitamin A can occur by eating the livers of polar bears, a pound of which contains a 20-year supply of retinol for a human. Chronic toxicity is more common and results from prolonged overdosage with vitamin preparations.

A number of similar substances, referred to as **vitamin K,** play a role in the normal coagulation of blood by promoting the synthesis of prothrombin in the liver. These compounds are found in many kinds of food and are synthesized by intestinal bacteria; thus a deficiency of vitamin K is more often associated with some abnormality of its absorption than with a lack in the diet. Vitamin E, or α-**tocopherol,** plays some unknown role in metabolism. Male animals made deficient in tocopherol undergo degenerative changes in the testes and become sterile, and eggs from vitamin E-deficient hens fail to hatch. The skeletal muscles of vitamin E-deficient animals eventually undergo degeneration. Tocopherol may serve as an antioxidant, protecting certain labile cellular components from being oxidized by removing intermediate free radicals. Some investigators believe that the effects of tocopherol deficiency can be traced to the accumulation of fatty acid peroxides, which react with and destroy other cellular components. There is some evidence that vitamin E plays a role in the mitochondrial electron transport system, but the nature of that role is unclear.

A summary of the common vitamins needed by humans, their characteristics, and their associated deficiency disease is presented in Table 12.2.

12.7 FUEL UTILIZATION: METABOLIC RATES AND ENERGY

Foods that are not stored or utilized in the synthesis of new cell components are utilized as fuel in the production of cellular energy, a process termed respiration. As described in Chapter 4, cell respiration is a process of energy transfer. Substrates such as glucose are metabolized to carbon dioxide and water, and

TABLE 12.2 COMMON VITAMINS

Vitamins	Common Sources	Function	Disease and Symptom if Deficient in diet
A, Retinol	Butter, eggs, fish liver oils. Carotene in plants can be converted to A	Maintenance of epithelial cells, component of visual pigments	Scaly skin, easy infection, night blindness, xerophthalmia
B Complex* B₁, Thiamine	Yeast, meat, whole grain, eggs, milk, green vegetables	Coenzyme in decarboxylation, carbohydrate metabolism	Beriberi: nerve and muscle degeneration
B₂, Riboflavin	Same as B₁, colon bacteria	Coenzyme in cellular oxidations, electron transport	Stunted growth, cracked skin
B₆, Pyridoxine	Same as B₁	Coenzyme in amino acid metabolism	Dermatitis
G, Niacin	Same as B₁	Coenzyme in cellular oxidations	Pellagra: inflammation of skin, nervous disorders
B₁₂, Cobalamin, and Folic Acid	Same as B₁	Coenzymes in DNA synthesis, cell division	Anemia (Red cell precursors are most rapidly dividing cells.)
C, Ascorbic Acid	Citrus fruits, fresh vegetables. Destroyed on cooking	Collagen formation, maintenance of connective tissues and capillary walls	Scurvy: bleeding gums, swollen joints, general weakness
D, Cholecalciferol	Eggs, meat, liver oil. Cholesta-5,7-dienol in skin converted to D on exposure to ultraviolet light	Absorption and utilization of calcium and phosphorus	Rickets. weak bones, defective teeth
E, Alpha-tocopherol	Green vegetables, wheat germ, vegetable oils	Maintenance of reproductive cells	Sterility in poultry, rats, and possibly man
K, Several Naphthoquinone Compounds	Green vegetables, colon bacteria	Synthesis of prothrombin in liver, hence normal blood clotting	Bleeding

*Other B complex vitamins (pantothenic acid, biotin) are seldom deficient in human diets.

some of the energy is conserved in the energy-rich phosphates of ATP. ATP in turn is the energy source for most of the endothermic reactions of the cell. These energy transfers are only about 40 per cent efficient and most of the energy loss is in the form of heat. Since the heat conversion factor is constant and since energy production is an aerobic process, the amount of heat produced and the amount of oxygen utilized are relatively precise reflections of the amount of fuel consumed, or the animal's **metabolic rate.**

These relationships are further influenced by activity. Muscular activity obviously demands more energy than that required in the resting state or for what is termed the **standard metabolic rate.** But the standard metabolic rates of animals are not the same. More active animals tend to have higher standard metabolic rates than do less active species.

The heat produced in cell metabolism is eventually dissipated across the body surface, but the rate of loss varies in different animals. Most animals are **ectothermic;** their rate of heat loss is so high and their rate of heat production so low that the body temperature is set by that of the external environment rather than by internal metabolism. The body temperatures of such animals fluctuate with the temperature of the environment and are said to be **poikilothermic.** Ectothermic and poikilothermic are actually different ways of describing the same condition and can be used synonymously.

Birds and mammals are **endothermic.** Their body temperature is derived from internal heat production. Heat loss is less than in ectotherms, and the rate of internal heat production is greater. Body temperature does not fluctuate and is independent of environmental temperatures. Animals with constant body temperature are said to be **homeothermic.**

Birds and mammals are the only living continuously endothermic animals, but it is now believed that many dinosaurs may have been warm-blooded endotherms (see p. 811). Many other animals, such as some lizards, certain active fish and some insects, are intermittently endothermic.

Homeothermism is maintained in birds and mammals by regulating heat production and by controlling heat loss across the body surface. Fur and feathers act as insulating barriers, and in aquatic mammals, such as seals, porpoises and whales, there is a thick layer of fat associated with the dermis of the skin. The sweat glands of mammals serve to cool the skin. Sweat is poured out onto the surface of the skin, and some of the heat required for evaporation is removed from the body. Panting in dogs functions in the same way. It is significant that sweat glands are largely restricted to certain large animals, such as primates, horses, and camels. Very small animals, such as rodents, moles and shrews, lack sweat glands, but possess a very great surface area for heat loss for their size.

Various controls come into operation at temperature extremes. On a hot day, loss of excess internal heat must be facilitated. Blood flow into the skin is increased through the opening of arterioles. Sweating may occur if sweat glands are present. Activity may be reduced, and the animal may remain in its burrow or den.

At low temperatures heat loss must be curtailed. Blood flow to the skin is reduced. Shivering may occur, the muscle contraction thereby increasing internal heat production. The insulation of many temperate and arctic animals is very effective against heat loss.

Behavioral modifications are also important at lower temperatures. Many animals avoid exposure by remaining in burrows or retreats. Bears and some other mammals will sleep for long periods during the winter, but this is not hibernation. True hibernators are small mammals with high metabolic rates, such as some rodents, hamsters, bats and hedgehogs, that undergo a marked state of torpor at certain lower critical temperatures. The metabolic rates decline greatly, heart beat slows and the body temperature drops to a low level. The energy saving is very great, and the animal slowly utilizes a store of fat acquired before hibernation. Some birds also hibernate, and there are a few hummingbirds and bats (mammals) that exhibit a daily state of torpor. Adaptations for thermoregulation will be described in more detail in Chapters 34, 35 and 36.

If the oxygen consumptions per gram of body weight (a measure of fuel utilization and metabolic rate) of different species within the same group of animals are compared, striking differences are evident. Note that by using oxygen consumption per gram of body weight, it is possible to compare animals that are greatly different in size, such as rats and horses. When the rates of different mammals are plotted against the mass of the animal, as in Figure 12.17, it can immediately be seen that metabolic rate and size are inversely related. For example, the metabolic rate of a shrew is many times greater than that of an elephant. This difference is also reflected in the fact that a shrew consumes its entire body weight in food every 24 hours (an obvious impossibility for an elephant).

Figure 12.17 Rate of oxygen consumption of various mammals plotted against body weight. Ordinate has an arithmetic scale, but abscissa is logarithmic. (From Schmidt-Nielsen, K.: Animal Physiology: Adaptation and Environment. Cambridge, Cambridge University Press, 1975.)

Why is the metabolic rate of small animals greater than that of large ones? Among mammals, homeothermism is certainly a factor. More internal heat must be produced to compensate for the greater heat loss from the larger surface area relative to volume in small species. But this is not the complete explanation. When metabolic rates are plotted against body size using a logarithmic scale, a proportional relationship will have a slope of 1.0; i.e., an increase in one will result in an equivalent increase in the other (Fig. 12.18). If the increase in metabolic rate in small mammals were simply related to heat loss through increase in surface area, then the slope should parallel the slope for the relationship of surface area to volume, a slope of 0.67 (volume increases by the cube and surface area by the square). Instead, the actual slope is about 0.75. Moreover, the slope is the same for ani-

Figure 12.18 Oxygen consumption rates per unit of body weight of various organisms plotted against body weight, using log coordinates. The regression lines (slope) are about 0.75. (After Hemmingsen, from Schmidt-Nielsen, K.: Animal Physiology: Adaptation and Environment. Cambridge, Cambridge University Press, 1975.)

mals that do not maintain a constant body temperature, poikilotherms, as it is for homeotherms. Thus, we can make the generalization that metabolic rate increases with decrease in body size. But the generalization applies to *all* animals, indeed to all organisms, and not just to mammals. The reasons for this, however, are not yet fully understood.

ANNOTATED REFERENCES

In addition to the titles listed below, many of the general references listed at the end of Chapter 10 cover the topic of nutrition and related subjects.

Jennings, J. B.: Feeding, Digestion, and Assimilation in Animals. 2nd ed. New York, Pergamon Press, 1973. A brief account of all aspects of animal nutrition.

Jørgensen, C. B.: The Biology of Suspension Feeding. New York, Pergamon Press, 1966. A detailed review of filter feeding and other types of suspension feeding.

GAS EXCHANGE

The energy required for the myriad activities of cells is derived primarily from the reactions of biological oxidation (p. 70). The essential feature of these reactions is the transfer of hydrogen atoms or their electrons from donors to acceptors, accompanied by the transfer of energy to phosphate ester bonds. The ultimate electron acceptor in the metabolism of most animal cells is oxygen, which is converted to water. For these processes to continue, the supply of oxygen must be renewed constantly, because only small amounts of oxygen can be stored in blood or in tissues, and the carbon dioxide produced must be removed. Thus, the survival of each cell requires the continuous exchange of gases with its environment.

13.1 GASES

The Need for Oxygen. As discussed in Chapter 12, animals differ greatly in their need for energy and oxygen. The basal metabolism is higher for endothermic birds and mammals than it is for ectothermic vertebrates and invertebrates. Our basal metabolism, expressed as oxygen consumption per gram of body weight per hour, is 0.25 ml. O_2 per gram per hour, a toad's at 15°C is 0.05 ml. O_2 per gram per hour and a lobster's at 15°C is 0.04 ml. O_2 per gram per hour. Basal metabolism is influenced by size (Fig. 12.18) and also by sex. Men have a higher basal metabolism than women of the same age and weight. Metabolism is greatly elevated in all animals during physical activity. Our oxygen consumption during exercise is 15 to 20 times its basal value, and that of an insect is 50 to 200 times greater when flying than at rest. These and other factors affecting metabolism determine an animal's need for oxygen and affect many aspects of its mode of life, including sensitivity to oxygen deprivation, mechanisms of gas exchange and the range of environments in which it can live.

Availability of Oxygen. The availability of oxygen varies in different habitats. Dry air contains about 21 per cent oxygen and 0.03 per cent carbon dioxide; the remainder is nitrogen with traces of other gases. The availability of a gas is expressed as a **partial pressure,** which is the pressure of that gas in a mixture of gases. This pressure is calculated by multiplying the total pressure of the mixture of gases by the percentage of the particular gas in the mixture. Air at sea level has a pressure of 760 mm. Hg and is 21 per cent oxygen; hence the partial pressure of oxygen at sea level is $760 \times 0.21 = 159.60$ mm. Hg.

The partial pressure of oxygen is decreased when water vapor is a gaseous component of air, as within the lungs of many animals. It is also lower at high altitudes, where the total atmospheric pressure is reduced. At 6000 meters (19,685 feet), the partial pressure of oxygen is reduced from its sea level value of 159 mm. Hg to 80 mm. Hg. Although the air at 6000 meters is still 21 per cent oxygen, the low partial pressure of oxygen reduces its availability because it affects the amount of oxygen that will go into solution. Oxygen must be in solution before it can be used by an organism or its cells. Aquatic organisms derive their oxygen from oxygen dissolved in water. In terrestrial animals, oxygen first dissolves in the moist surface of the body or lung and then enters the blood or body fluids. When a gas and liquid are in contact, an equilibrium is reached when the rate at which molecules pass from the gas to the liquid equals the rate at which they pass from the liquid to the gas. The higher the partial pressure of oxygen in air, the more oxygen that will go into solution and be available to the cells.

The tendency of a gas to escape from solution, its **tension,** is numerically equal to the partial pressure of the gas in the air with which it is in equilibrium. The tension of a gas is an important force in determining its movement between liquids and between liquid and air; but it does not measure the quantity of the gas in solution. Quantity is affected not only by tension, but also by the solubility of the particular gas and by factors that affect solubility, such as temperature, salinity and pH. Fresh water at 20°C contains 6.57 ml. of oxygen per liter. It is important to note that aquatic animals have less oxygen available to them than do terrestrial ones who are living in an environment that is 21 per cent oxygen (i.e., 210 ml. of oxygen per liter).

Oxygen Conformity and Regulation. Some animals, known as **oxygen conformers,** have the ability to adjust their level of metabolism and oxygen consumption according to the availability of oxygen. The annelid worm, *Glycera,* for example, lowers its metabolism and oxygen consumption gradually as the water in its burrow becomes stagnant during low tide. As the tide comes in, oxygen consumption goes up. Many coelenterates and most parasitic worms are also conformers. Oxygen conformity is an adaptation that enables some animals to adjust to a variable supply of oxygen to the tissues, either because of great variations in environmental oxygen or because of limitations in the animal's ability to extract oxygen from the environment or distribute it within the body.

Most animals are **oxygen regulators.** Their oxygen consumption, although varying with such factors as temperature, body size and level of activity, is independent of the availability of environmental oxygen unless the oxygen level drops below a certain partial pressure known as the **critical level.** Below this, the animal's oxygen consumption necessarily declines rapidly, and death may follow. Oxygen regulators normally live in environments where the oxygen supply is well above their critical level. They cannot live long or be active in environments where the supply of oxygen falls below this level. The critical level is adaptive for each animal. It is higher in terrestrial animals than in aquatic ones; it is higher in species adapted to life in swift-moving streams than in ones adapted to life in stagnant ponds.

13.2 STEPS IN GAS EXCHANGE

To obtain oxygen and eliminate carbon dioxide, an animal must have a **respiratory membrane,** a moist, thin and permeable surface exposed to the environment, through which gases can diffuse. An animal must also have some mechanism to move gases between the respiratory membrane and its cells. The way these two conditions are met varies greatly among animals, depending upon their size, oxygen needs, the environment in which they live and their evolutionary history.

In some animals the general body surface has the properties of a respiratory membrane, especially when the animal is small, but often the respiratory membrane is limited to parts of a **respiratory organ,** such as a gill lamella, the end of a tracheal tubule or a lung alveolus (Fig. 13.1). **Gills** are defined as respiratory organs whose surface area is increased by evaginations of a surface, usually the external. The evaginations may be platelike lamellae or threadlike filaments. **Lungs** are relatively large, often saclike, organs that extend into the body from a limited part of the body surface or from the digestive tract. Their respiratory surface is often increased by many pocket-like folds, such as the **alveoli** of mammalian lungs. The **tracheae** of insects are also inward extensions but differ from lungs in that there are a great many separate invaginations, so that no body cell is very far from the respiratory membrane in the **tracheoles** at the ends of the trachea. All of these respiratory organs are found in both aquatic and terrestrial animals, although gills are more common in aquatic animals and lungs and tracheae in terrestrial ones.

The respiratory organ contains the respiratory membrane, may protect it in various ways, and provides a means for **ventilation,** that is, for moving the external environment, water or air, across it. Breathing is the movement of air through a respiratory organ. If the gases are not already in water, they will go into solution on the moist surface of the respiratory membrane and cross it by following diffusion gradients; i.e., by moving from an area of high tension to one of lower tension (see Section 3.11). Since oxygen is being consumed within the body, it has a lower tension on the inner surface of the respiratory membrane than on the outer surface and will diffuse into the body. Carbon dioxide, which has a higher tension in the body than in the external environment, will diffuse out.

In a small animal, diffusion may be the only force needed to move gases between the respi-

Figure 13.1 A diagram of some of the types of respiratory organs present in animals.

ratory membrane and the rest of the body. Protozoans and the cells of sponges, cnidarians and many small worms are either in contact with the external environment or sufficiently close to it so that diffusion is adequate for moving gases through their bodies. However, diffusion is relatively slow (its rate is inversely proportional to distance), and as a practical matter, cells must lie within 0.05 mm. of an oxygen source unless their metabolic rate is very low indeed. Large, active animals have some system of **bulk flow** whereby gases are carried passively in a moving liquid (blood, for example) that circulates between the cells

and the respiratory membrane (Fig. 13.2). Gases may be in solution in the blood or be chemically bound to carriers, such as hemoglobin. In tracheal systems, ventilation of the respiratory membrane and distribution of gases by diffusion or bulk flow to the cells are combined in an extensive system of tubules. Although gases are moved frequently by bulk flow to and from the body cells or their vicinity, the final exchange is always by diffusion. Because cells consume oxygen and produce carbon dioxide, the gradient between them and gases in the blood or tracheoles is maintained.

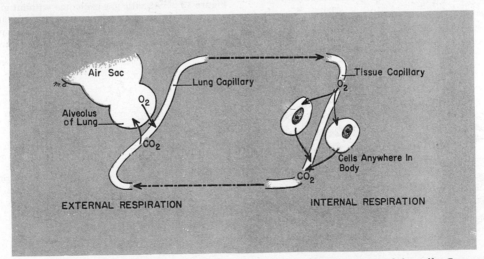

Figure 13.2 Diagram illustrating the transportation of gases between a respiratory organ and the cells. Gas exchange at the respiratory organ is frequently called external respiration; gas exchange with the cells is called internal respiration.

13.3 ENVIRONMENTAL GAS EXCHANGE IN AQUATIC ANIMALS

Most small aquatic organisms have no special respiratory organs. Their body surface, in some cases supplemented by the lining of the digestive tract or other body spaces, is the only respiratory membrane that they need, since they possess a very favorable surface area to volume ratio. Flagellated or ciliated cells frequently move water across the organism's surface or through its cavities so that there is a continual supply of oxygen. Currents through a digestive tract are used, of course, in feeding (Chapter 12), but gas exchange with the lining cells also occurs. Flagellated collar cells move water through the numerous cavities of a sponge. Flagella or cilia move water in and out of the extensive, and often branching, gastrovascular cavity of cnidarians and primitive worms (Fig. 13.3A). Sea squirts, *Amphioxus*

and other lower chordates exchange gases at the body surface and with the current of water carried by cilia through their pharynx and many gill slits. This is primarily a feeding current and gills are not present in the gill slits until the vertebrate stage of evolution.

Larger and more active aquatic animals have respiratory organs to which the respiratory membranes are confined or that supplement other surfaces. The organs are usually gills; however, the sea cucumber, a soft-bodied echinoderm, has a pair of extensively branching **respiratory trees** that have evaginated from the posterior end of its digestive tract to extend into the coelom (Fig. 13.3B). Since they extend into the body, they can be regarded as lunglike organs. A pumping action of the cloaca moves a ventilating current of sea water into and out of the respiratory trees.

Characteristics of Gills. Gills are organs composed of many delicate lamellae, or filaments; their existence is possible in an aquatic environment because the density of water pro-

A

B

Figure 13.3 Aquatic gas exchange without gills. *A*, Gases are exchanged in the coelenterate *Hydra* through the body surface and the lining of the gastrovascular cavity. Arrows show water movement. *B*, The respiratory trees of a sea cucumber.

Figure 13.4 Theoretical diagrams to illustrate the effects of counter current or parallel flow on the exchange of oxygen between water and blood. The equilibrium attained in parallel flow would be somewhere above 50 per cent saturation of the blood because of the presence of hemoglobin, yet the blood becomes less fully saturated than it would during counter current flow. (Modified after Hughes.)

vides adequate support. Gills would tend to dry out, collapse and clump together in air. Their large surface area is necessary to provide an adequate supply of oxygen, for water contains relatively little dissolved oxygen. A large volume of water must also move across the gills, because the oxygen in the water adjacent to the gill surface is depleted rapidly by diffusion into the body. The mudpuppy, an aquatic salamander, and certain other animals slowly wave their gills, but more often water is moved across them. Moving the gills in water requires a considerable expenditure of energy, because water is a rather dense medium. Moving water across the gills in one direction, rather than back and forth in the manner of air moving into and out of a lung, can reduce energy expenditure, and sometimes energy can be saved by exposing the gills to an environmental current or by utilizing other body currents and movements. Clams and mussels combine a feeding and respiratory current. Rapidly swimming fishes, such as tuna, move water across their gills primarily by their forward locomotion.

Gases diffuse rapidly between the water crossing the gills and body fluids circulating through them. The cuticle, if one is present, and the epithelial layers separating the water from the body fluids are very thin. The effectiveness of the gills is increased in some mollusks and crustaceans and in most fishes by the evolution of a **counter current exchange** mechanism in which water flows across the gills in a direction opposite to the flow of blood through them. A difference in oxygen tension

between the water and blood is maintained throughout the system (Fig. 13.4). Blood leaving a gill has become nearly fully saturated with oxygen. It is exposed to water just starting across it and flowing in the opposite direction. This water is fully saturated with oxygen, hence oxygen diffuses into the blood. At the other end of a gill, the water has given up most of its oxygen, but it is exposed to blood just entering the gill that contains even less oxygen. Virtually all of the oxygen in the water can be extracted by the blood

Gases diffuse freely across gills and so do many other small, soluble molecules, including water, salts and nitrogenous wastes, such as ammonia and urea. The elimination of nitrogenous wastes is advantageous, but the loss or gain of water and salts may pose osmotic problems that must be met (Chapter 15).

A special problem of aquatic life is that the water to which the gills and body surface are exposed has a high capacity to absorb and conduct heat. Water continually drains away the heat produced by metabolic activity. No aquatic invertebrates, and only a few very active fishes, can maintain a body temperature higher than that of their environment.

Invertebrate Gills. Many sorts of gills and gill-like structures, some simple, others quite elaborate, are found among invertebrates. Polychaete annelid worms have paired lateral projections, the **parapodia,** that are used in locomotion, but parts may also be modified as gills.

Molluscan gills are composed of many filaments, arranged in different ways in the

Figure 13.5 *A*, Diagrammatic lateral view of a primitive mollusk to show the location of the gills. *B*, Section in the plane A—B of *A* to show the structure of one gill plate; dashed arrows indicate the direction of blood flow.

various groups. In primitive mollusks, one or more pairs of gills lie in the mantle cavity formed by the overhanging shell and body wall (Fig. 13.5 *A* and *B*). Each gill consists of an array of triangular plates attached to either side of the central axis. Cilia on the plates draw a current of water into the mantle cavity, dorsally between the plates, and finally back out of the cavity. Vascular channels in the gill axis carry a bloodlike liquid into and out of the thin plates where gas exchange with the water occurs.

Most aquatic arthropods have gills that are modified appendages or outgrowths closely associated with appendages. In crayfishes and crabs, the gills arise from the base of the appendages and adjacent body wall and extend up into a **branchial chamber** that lies on each side of the animal between the body wall and a part of the carapace that has grown laterally and ventrally over the gills (Fig. 13.6*A* and *B*). Each gill consists of a central axis to which are attached either lamellae or filaments, depending on the species (Fig. 13.6*C* and *D*) The ventilating current is produced by the rapid sculling action of the **gill bailer**, a semilunar-shaped process of one of the oral appendages (the second maxilla). The gill bailers drive two exhalant water streams out of the branchial chambers to each side of the mouth. Water enters the branchial chambers of the crayfish at the back of the carapace and

between the legs. In crabs, the carapace is tightly sealed along its ventral margin, and the inhalant apertures are at the base of the great claws (chelipeds). Channels in each gill axis carry blood to and from spaces in the lamellae; gas exchange occurs across the thin walls of the lamellae.

Vertebrate Gills. Fishes and many amphibians exchange gases with the environment by means of gills. Some fish larvae and larval amphibians have **external gills,** filamentous processes that extend outward from the side of the head near the openings of the gill slits. Adult fishes have **internal gills** situated within the slits. In most species the slits are covered by a fold of the body, the **operculum,** which forms an **opercular chamber** lateral to them. The gills themselves consist of lamellae attached to a visceral arch at the base of each gill (Fig. 13.7*A*). Secondary folds perpendicular to the lamellae contain capillary beds, and it is here that gas exchange with the water occurs. Exchange is particularly effective because of a countercurrent flow between blood and water (Fig. 13.7*B*).

A pumping action of the mouth and pharynx causes a ventilating current of water to flow across the gills (Fig. 13.7 *C* and *D*). During inspiration, both the pharynx and opercular chambers expand, causing a decrease in pressure within them relative to the surrounding water. These chambers are acting

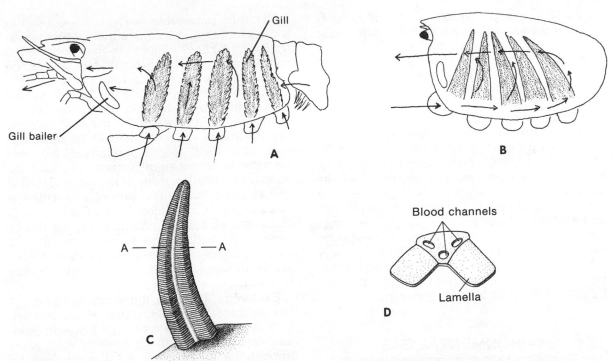

Figure 13.6 Gills of decapod crustacea. Lateral view of the cephalothorax of a crayfish (A) and crab (B) to show the location of their gills and the ventilating current of water through the branchial chamber. The carapace is drawn as though it were transparent. C, One gill of a crab consists of a stack of lamellae attached to a central axis. D, Cross section of a crab gill. (C and D modified from Calman.)

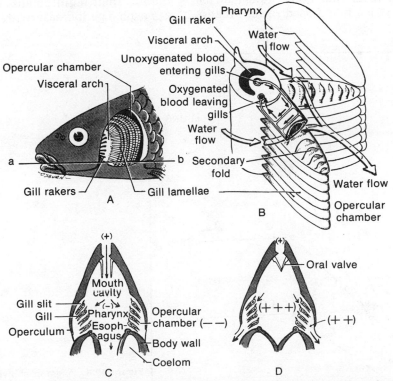

Figure 13.7 Gills of fishes. A, The operculum has been cut away to show the gills in the opercular chamber. B, An enlargement of a portion of one gill. C and D, Frontal sections through the mouth and pharynx in the plane of line a–b in A. Inspiration occurs in C and expiration in D. Relative water pressures in the various parts of the system are shown by + and −.

as suction pumps. Water is drawn in only through the mouth, for a thin membrane on the free edge of the operculum acts as a valve, preventing entry in this direction. Once in the pharynx, the water passes across the gills into the opercular chamber as a result of the lower pressure there. Expiration begins with the closure of the mouth, or of oral valves, and a contraction of the pharynx and opercular chamber, so that the pressure in them exceeds that of the surrounding water. These chambers are now acting as force pumps. The membrane at the edge of the operculum is pushed open and water is discharged. During both inspiration and expiration there is a pressure gradient moving water across the gills. Water does not enter the esophagus, for this is collapsed except when swallowing occurs. Food and other particles are prevented from clogging the gills by gill rakers, which act as strainers.

13.4 ENVIRONMENTAL GAS EXCHANGE IN TERRESTRIAL ANIMALS

A particular advantage of terrestrial life is that oxygen is much more abundant than in water. A problem is that body water can be lost easily through any exposed surface that is moist, thin, permeable and vascular enough to serve as a respiratory membrane. Most terrestrial animals have adapted to these conflicting conditions by evolving respiratory organs, either trachea or lungs, in which the respiratory membranes lie deep within the body in spaces with a high humidity, so that water loss is minimized. Since air is 21 per cent oxygen, the volume of air that needs to be moved in and out of these spaces is relatively low, and no more is moved than is needed to meet the animal's metabolic needs at the particular time and circumstances. The rate of ventilation is more carefully regulated than in animals utilizing gill ventilation. Air is light and can be moved with little energy expenditure. In some animals diffusion is adequate. Oxygen, for example, diffuses 300,000 times faster in air than in water.

Exchange Through the Body Surface. A few terrestrial animals, those whose metabolism is not high and who live in damp habitats, can exchange all or some of their gases through the body surface. Earthworms have no special respiratory organs, and gases diffuse through their vascular body surface. Although frogs and toads have lungs, they exchange a substantial fraction of gas through their thin, moist and vascular skin. Since they are ectotherms, their requirements for metabolism and gas exchange increase with temperature. More

Figure 13.8 A graph showing the amount of gas exchanged through the lungs and skin of a toad, *Bufo americanus*, at different temperatures. (From Hutchison, Whitford and Kohl, 1968, Physiological Zoology, 41:65–85, University of Chicago Press.)

carbon dioxide is eliminated through the skin than through the lungs at all temperatures, but the lungs are more important than the skin for obtaining oxygen at all but very low temperatures (Fig. 13.8). Certain woodland salamanders (the plethodonts), commonly found under rocks and logs, have lost their lungs, and all of their gas exchange occurs through the skin and the lining of the mouth and pharynx.

Gills. Gills can be used for gas exchange in air, provided they lie in chambers in which a high humidity can be maintained. They are usually confined to animals that venture onto the land only temporarily or whose evolutionary history has recently brought them to the land. Some species of crayfish make brief foraging excursions onto the land at night. Coconut crabs, which live on the land as adults and even climb trees, retain reduced gills, but the moist branchial chambers have been converted into lungs. In pill bugs, small terrestrial crustaceans frequently found under stones and rotting logs, part of the abdominal appendages form gills and part an operculum, or gill covering, that helps reduce loss of body water by evaporation. *Synbranchus*, a South American swamp fish, frequently leaves the water; every 10 to 15 minutes it inflates its opercular chambers with air and then keeps them closed as gases are exchanged with the gills and vascular lining of the chambers.

Tracheal Systems. Tracheal systems are the commonest gas exchange organs of terrestrial arthropods. Most body segments of insects have paired lateral apertures, the **spiracles,** that lead into a system of **tracheal tubules** (Fig. 13.9A and B). Filtering devices prevent small particles from clogging the system. Valves usually are present and can open and close as the animal's needs vary. The tracheae have a more or less ladder-like pattern with interconnecting transverse and longitudinal trunks. They terminate in minute tubules, the **tracheoles,** which are generally less than one micrometer in diameter (Figure 13.9C). The tracheoles are formed by special tracheole cells. They contain the respiratory membrane, and they permeate all of the tissues of the body. Some even penetrate muscle cells whose oxygen needs are very high. A cuticle lines the entire system, but only that in the tracheae is shed during molting.

Cuticular rings, **taenidia,** strengthen and hold the tracheae open. In small insects, diffusion is the only force needed to exchange gases because it is so rapid through the air-filled tubules. Only the terminal portions of the tracheoles contain liquid. Larger and more active insects have a ventilating system that includes **air sacs** that can be compressed and expanded by the action of body muscles (Fig. 13.9D). A synchronized pattern of spiracle opening and closing keeps a unidirectional flow of air moving through the larger tracheae.

A few insects are aquatic and either come to the surface to get air or have evolved devices to obtain oxygen from the water. Some diving beetles have a dense mat of hairlike cuticular processes that are nonwettable and entrap a permanent bubble of air. The spiracles open only into this bubble, and gases diffuse between it and the water. The air bubble acts as a gill! Some aquatic insect larvae have more conventional gills into which the tracheae extend.

A tracheal system usually combines ventilation of the respiratory membrane with distribution of gases to and from the tissues. Blood is not used for gas transport, but does, of course, provide the cells with food and removes nitrogenous wastes. Because of the large role that diffusion plays, tracheal systems are suitable only for small animals: insects, centipedes, millipedes and small spiders. Large spiders have lungs, or a combination of lungs and tracheae, and gases are transported by blood.

Invertebrate Lungs. Lungs are the respiratory organs of the larger terrestrial invertebrates and of nearly all terrestrial vertebrates, the lungless plethodont salamanders being an exception. Invertebrate lungs are known as **diffusion lungs,** for gases move in and out of them primarily by diffusion, although some body movements often help in ventilation. One group of mollusks, the pulmonate snails, have evolved lungs. The gills normally contained within the mantle cavity have been lost, and the wall of the cavity itself has become very vascular and serves as a single lung. It opens to the surface by a small pore, the **pneumostome,** that can be opened or closed according to the gas exchange needs of the animal. The lung may have evolved in snails living in environments where they became exposed, at least temporarily, to air. Land snails, of course, have retained the ancestral lung, but it has also been kept by many snails that secondarily became more aquatic. Our common fresh water snails are pulmonate and come to the surface for air.

Scorpions and the larger spiders have **book lungs.** One or more pairs of lungs have invaginated from the underside of the abdomen and each opens to the surface by a spiracle. Each book lung contains many leaflike lamel-

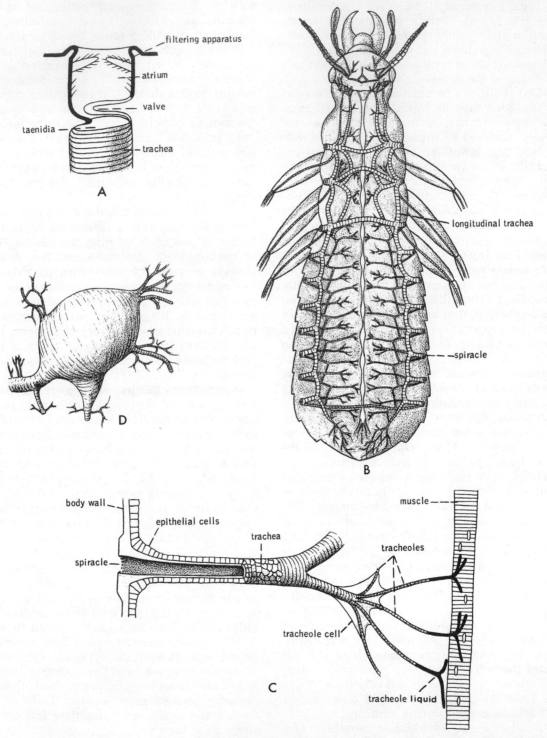

Figure 13.9 *A*, A spiracle with atrium, filtering apparatus and valve. (After Snodgrass.) *B*, A tracheal system of an insect. (After Ross.) *C*, Diagram showing relationship of spiracle and tracheoles with tracheae. (Modified from Ross.) *D*, An air sac. (After Snodgrass.)

Figure 13.10 Section through a book lung of a spider.

lae held apart by bars that enable air to circulate freely (Fig. 13.10). Gases are exchanged between the air and liquid circulating within the lamellae.

Vertebrate Lungs. The lungs of vertebrates are called **ventilation lungs** because air is moved in and out of them by a pumping action of adjacent body parts. They usually develop as a bilobed outgrowth from the floor of the pharynx, extend posteriorly, grow around the digestive tract and come to lie in the dorsal part of the pleuroperitoneal cavity. This cavity is the part of the vertebrate coelom that contains the lungs and digestive organs; the pericardial cavity is the part that surrounds the heart.

Lungs probably evolved as an accessory respiratory organ in swamp-dwelling fish that lived in ponds where the oxygen tension was low or that even dried up occasionally. Con-

temporary lung fish live in tropical environments similar to this. The lungs have been converted to a **swim bladder** in most fishes descended from ancestral lunged fish. The swim bladder is primarily a hydrostatic organ (see Section 11.11); however, the gas within it is mostly oxygen, so it is also a reserve that can be used when the fish cannot obtain enough oxygen through its gills.

Amphibians evolved from primitive fishes with lungs. Their aquatic larvae exchange gases through gills, but lungs appear and begin to function in late larvae that surface to gulp air. Lungs are the primary adult respiratory organs. They are saclike, and their internal surface area is increased somewhat by pocketlike folds (Fig. 13.11), but the surface area is far smaller than in higher terrestrial vertebrates. Amphibians also have a relatively ineffective method of ventilating the lungs. A frog,

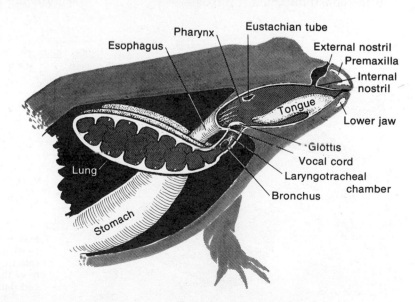

Figure 13.11 A diagrammatic longitudinal section of the respiratory system of the frog.

for example, moves air in and out of its lungs by a three phase **buccopharyngeal pump.** During the first phase, the floor of the pharynx is lowered and fresh air entering through the nostrils goes into the distended posteroventral portion of the buccopharyngeal cavity (Fig. 13.12). The **glottis,** which is a slitlike opening from the pharynx to the laryngotracheal cavity and lungs, remains closed during this phase but opens in the second one. A slight contraction of the flanks then drives stale air from the lungs, across the dorsal portion of the buccopharyngeal cavity and out the nostrils. There is little mixing of stale and fresh air. Finally, in phase three the nostrils are closed by the lower jaw pushing against the moveable premaxillary bones (Fig. 13.12), the glottis remains open, and a contraction of the pharynx drives the fresh air into the lungs. This type of pump, which utilizes fishlike pharyngeal movements, provides the animal with most of its oxygen, but cannot eliminate sufficient carbon dioxide. As we have seen, contemporary amphibians supplement their pulmonary ventilation by the diffusion of gases through the skin. The need to supplement pulmonary respiration by cutaneous respiration, and the loss of body water that this entails, is a factor that has prevented amphibians from fully exploiting the terrestrial environment. Cutaneous respiration also places a limit on the maximum size of amphibians; small organisms have more body surface relative to mass than large ones. Higher terrestrial vertebrates have evolved lungs with greater internal surface areas and more efficient ways of ventilating the lungs. They have thus been able to dispense with cutaneous respiration and the constraints that it imposes.

In mammals (Fig. 13.13), air is drawn into the paired **nasal cavities** through the **external nostrils,** or **nares.** These cavities are separated from the mouth cavity by a hard palate, and the animal can breathe while food is in its mouth. The surface area of the cavities is increased by a series of ridges known as **conchae,** and the nasal mucosa (in addition to having receptors for smell) is vascular and ciliated and contains many mucous glands. In the nasal cavities the air is warmed and moistened and minute foreign particles are entrapped in a sheet of mucus, which is carried by ciliary action into the pharynx where it is swallowed or expectorated. Inspired air is moistened in primitive terrestrial vertebrates, such as the frog, but cold-blooded tetrapods do not need so much conditioning of the air as birds and mammals.

Air continues through the **internal nostrils,** or **choanae,** passes through the **pharynx** and enters the **larynx,** which is open except when food is swallowed. The raising of the larynx during swallowing can be demonstrated by placing your hand on the Adam's apple, the external protrusion of the larynx. The **epiglottis** flips back over the entrance of the larynx when it is raised.

The larynx is composed of cartilages derived from certain of the visceral arches and serves both to guard the entrance to the windpipe, or **trachea,** and to house the **vocal cords** (Fig. 13.14). The vocal cords are a pair of folds in the lateral walls of the larynx. They can be brought close together, or be moved apart, by the pivoting of laryngeal cartilages connected to their dorsal ends. The **glottis** of mammals is the opening between the vocal cords. When the cords are close together, air expelled from

Figure 13.12 Diagram of the three stages in the ventilation of the frog's lungs. Black arrows indicate the course of fresh and stale air; white arrows the movements of the floor of the pharynx and flanks.

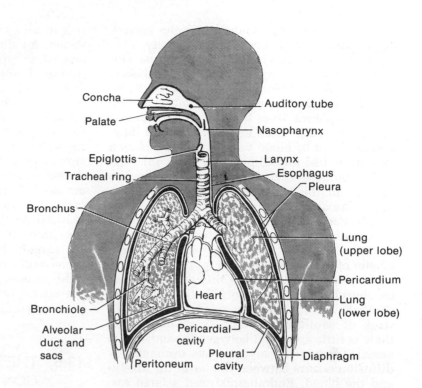

Figure 13.13 The human respiratory system. Details of the alveolar sacs, drawn enlarged here, are shown in Figure 13.15.

the lungs vibrates them, and they in turn vibrate the column of air in the upper respiratory passages just as the reed in an organ pipe vibrates the air in the pipe. This phenomenon is called **phonation,** or the production of a vocal sound. **Speech** is the shaping of the vocal sounds into patterns that have meaning for us; it is accomplished by the pharynx, mouth, tongue and lips. Muscle fibers in the vocal cords and larynx control the tension of the cords and the pitch of our voice. We nor-

mally speak when air is expired. Ventriloquists have trained themselves to speak during inspiration.

The **trachea** extends down the neck and finally divides into **bronchi** that lead to the pair of **lungs.** Unlike the esophagus, which is collapsed except when a ball of food is passing through, the trachea is held open by **C**-shaped cartilaginous rings, and air can move freely through it. Its mucosa continues to condition the air.

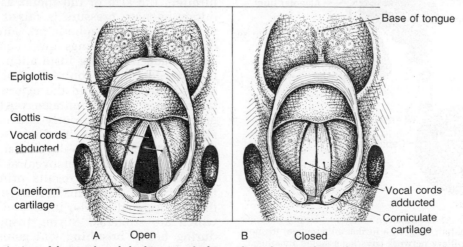

Figure 13.14 A view of the vocal cords looking into the larynx from above. *A,* Normal position of vocal cords; *B,* position of cords during speech. (From Jacob, S. W. and C. A. Francone: Structure and Function in Man. 2nd ed. W. B. Saunders Co., 1970.)

The lungs of amphibians lie in the antero-dorsal part of the pleuroperitoneal cavity. This cavity has become divided in mammals. The pleural cavities lie within the chest, or **thorax,** and are separated from the peritoneal cavity by a muscular **diaphragm.** A coelomic epithelium, the **pleura,** lines the pleural cavities and covers the lungs. Each bronchus enters a lung, accompanied by blood vessels and nerves, in a mesentery-like fold of pleura (Fig. 13.13).

The bronchi branch profusely within the lungs, and the walls of the respiratory passages become progressively thinner (Fig. 13.13). Each passage eventually terminates in an **alveolar sac** whose walls are so puckered by pocket-shaped **alveoli** that it resembles a cluster of miniature grapes (Fig. 13.15). The alveolar walls are extremely thin and could not be detected with certainty until they were studied with the electron microscope. A network of capillaries, which is so dense that there is little space left between the individual vessels, covers the alveoli. Gas exchange by diffusion occurs between the air in the alveoli and the blood. Endotherms need a large surface for gas exchange to sustain their high level of metabolic activity. The alveolar surface of human lungs is estimated to be 55 square meters.

A thin film of lipoprotein termed "surfactant," secreted by type II pneumocytes in the alveoli, acts as a detergent to reduce surface tension within the alveoli, and this is impor-tant in several ways. It helps in water conservation, for the lower the surface tension, the less water will be drawn from the capillaries into the lumina of the alveoli. A reduced surface tension also reduces the amount of muscular effort needed to expand the alveoli as air is inspired. Finally, the lipoprotein helps to equalize forces acting in the alveoli during the different phases of respiration. Surface tension is great in a sphere of small radius and decreases as the radius increases. The alveoli are small at the end of expiration and surface tension is therefore high, but the tension is reduced by the concomitant concentration of the lipoprotein in the reduced surface area of the alveoli. Surface tension decreases when the alveoli expand as air is inspired and their radii increase, but this is compensated for by the attenuation of the lipoprotein in the expanding surface area of the alveoli.

13.5 THE MECHANICS AND CONTROL OF BREATHING

Mammalian lungs are ventilated by changing the size of the thoracic cavity and, consequently, the pressure within the lungs. The lungs follow the movements of the chest wall, for they are at once separated from it and united to it by the adhesive force of a thin layer of fluid that lies within the pleural cavities. During normal, quiet **inspiration,** the size of the thorax is increased slightly, intrapulmonary pressure falls to about 3 mm. Hg below atmospheric pressure, and air passes into the lungs until intrapulmonary and atmospheric pressures are equal. During normal **expiration,** the size of the thorax is decreased, intrapulmonary pressure is raised to about 3 mm. Hg above atmospheric pressure, and air is driven out of the lungs until equilibrium is again reached. During inspiration, the thorax is enlarged by the contraction of the dome-shaped **diaphragm** and the **external intercostal muscles.** The diaphragm pushes the abdominal viscera posteriorly and increases the length of the chest cavity (Fig. 13.16); the external intercostals raise the sternal ends of the ribs and expand the dorsoventral diameter of the chest. Expiration results primarily from the relaxation of the inspiratory muscles and the elastic recoil of the lungs and chest wall, which are stretched during inspiration. But during heavy breathing, antagonistic expiratory muscles can decrease the size of the thoracic cavity. Contraction of **abdominal**

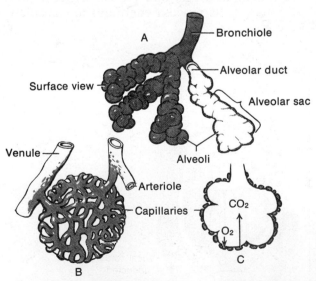

Figure 13.15 A, Termination of the respiratory passages in the mammalian lung; B, a further enlargement to show the dense capillary network covering a single alveolus; C, an alveolus in section. Alveoli have a diameter of 0.2 to 0.3 mm.

Figure 13.16 Mechanics and control of breathing. *A,* The elevation of the ribs and depression of the diaphragm during inspiration increases the size of the chest cavity, indicated by the black area. *B,* A diagram of the nervous mechanism for controlling the rhythm of breathing. See text for explanation.

muscles forces the abdominal viscera against the diaphragm and pushes it anteriorly; **internal intercostals** pull the sternal ends of the ribs posteriorly.

The lungs of an adult man can hold about 6 liters of air, but in quiet breathing they contain only about half this amount, of which 0.5 liter is exchanged in any one cycle of inspiration and expiration. This half liter of **tidal air** is mixed with the 2.5 liters of air already in the lungs. Vigorous respiratory movements can lower and raise the intrapulmonary pressure 60 mm. Hg below and above atmospheric pressure, and under these conditions 4 to 5 liters of air can be exchanged. This maximum is known as the **vital capacity.** There is always, however, at least a liter of **residual air** left in the lungs to mix with the tidal air, for the strongest respiratory movements cannot collapse all the alveoli and respiratory passages. Since the inspired air always mixes with a certain amount of stale air already in the lungs, alveolar air has a lower partial pressure of oxygen and a higher partial pressure of carbon dioxide than atmospheric air. The high water vapor content of alveolar air also lowers its partial pressure of oxygen.

Respiratory movements are cyclic and are controlled by inspiratory and expiratory centers (collectively called the **respiratory center**) in the medulla of the brain. The respi-

ratory center has a rhythmic activity alternately sending out impulses to the inspiratory and expiratory muscles. The basis for this rhythmicity is not completely understood, but a neuronal feed-back mechanism does play an important role. The inspiratory center sends out impulses along the nerves to the inspiratory muscles (neuron #1, Fig. 13.16*B*), and we breathe in. The alveoli fill with air, become stretched, and the resultant sensory impulses traveling to the respiratory center inhibit inspiration (#2). At the same time, impulses that were initiated in the inspiratory center and took a rather circuitous route within the brain reach the expiratory center (#3) and stimulate it to send impulses out to the expiratory muscles (#4). As we exhale, inhibiting impulses stop coming from the stretch receptors in the lungs, and the inspiratory center becomes active again.

The respiratory center receives many other signals indicating important changes in the body and modifies the depth and rate of breathing accordingly. Increased metabolism during exercise, for example, results in an increased carbon dioxide content of the blood. This in turn increases the excitability of the respiratory center, and we breathe more rapidly and deeply. The same thing happens when we voluntarily hold our breath. Since the lungs are not being ventilated, carbon dioxide accumulates in the alveolar air and blood and eventually reaches a level that activates the respiratory center, and we breathe again involuntarily. You cannot suffocate by holding your breath. The accumulation of carbon dioxide in the blood of a newborn baby normally initiates its breathing.

The oxygen content of the blood is less important than the carbon dioxide content in stimulating the respiratory center. However, chemoreceptive cells in the carotid bodies, associated with the carotid arteries supplying the head, are activated by any marked drop in oxygen content. Impulses initiated by them stimulate the inspiratory center.

Receptors in the larynx and trachea can also affect respiration. If food inadvertently enters these passages, the receptors are stimulated and a very vigorous expiration, i.e., a cough, results. The cough reflex is one of many safeguards in the body that are activated if something goes wrong with the primary control mechanism, in this case the swallowing reflex.

13.6 GAS TRANSPORT

The tracheal system of arthropods described above brings oxygen from the body surface directly to the tissues and removes carbon dioxide. Other animals that are too large to exchange gases by diffusion alone have a transport system that carries gases between the respiratory membrane and the tissues. Coelomic fluid performs this function in a few invertebrates; for example, some polychaete worms and echinoderms. Coelomic fluid is moved through the body cavity by contractions of the body wall in annelids and by ciliary action in echinoderms. Most animals utilize the circulatory system and its fluids (blood or hemolymph) for gas transport. A small amount of oxygen and carbon dioxide can be carried in physical solution in coelomic fluid, blood or hemolymph. Most of these gases, however, enter into chemical reactions with materials in the fluids and are carried as other compounds or are loosely bound to other compounds. The compounds that bind oxygen and carbon dioxide are usually colored **respiratory pigments.** Hemoglobin, which gives the red color to our blood, is the most familiar one. If human blood were simply water, it could carry only about 0.2 ml. of oxygen and 0.3 ml. of carbon dioxide in each 100 ml. The properties of hemoglobin enable blood to carry 100 or more times as much gas: 20 ml. of oxygen and 30 to 60 ml. of carbon dioxide per 100 ml.

Figure 13.17 The structure of a heme. (From Prosser, C. L.: Comparative Animal Physiology. 3rd ed. Philadelphia, W. B. Saunders Co., 1973.)

Respiratory Pigments. All respiratory pigments are metal-containing proteins. The protein portion of a **hemoglobin** molecule is a **globin** consisting in most vertebrates of four peptide chains, typically two α chains and two β chains. **Heme,** a porphyrin composed of four pyrroles with an iron atom at the center, is attached to each of the four chains (Fig. 13.17). The heme is the same in all hemoglobins and is also very similar to the heme in the cytochromes of the electron transport system (p. 70). The globins differ greatly in the number of peptide chains and their amino acid composition, and these differences impart characteristic properties to the numerous varieties of hemoglobin. Hemoglobin, the most widespread of the respiratory pigments, is found in nearly all vertebrates, the sea cucumber, the smaller crustacea, some insect larvae, many annelids, a few mollusks and some of the primitive worms. It is contained within the red blood cells, or **erythrocytes,** of vertebrates but is in solution in the circulating fluids in most of the invertebrates. The hemoglobin of a few polychaetes is carried in coelomic corpuscles. Having hemoglobin within cells prevents the undesirable osmotic effects that it would have in free solution, and the chemical environment with which hemoglobin molecules react can be more carefully regulated. When hemoglobin is in solution in body fluids, osmotic problems are ameliorated by combining many small units into fewer large ones. The hemoglobin molecule of Arenicola, a marine polychaete, is a polymer containing 96 heme units, whereas vertebrate hemoglobin, which is contained within cells, has no more than four.

Hemocyanin is next in abundance to hemoglobin. It is found in mollusks, apart from clams and their allies, in the larger crustaceans and in a few arachnids, including the horseshoe crab, *Limulus.* It is a large molecule carried in solution and gives the blood a slightly bluish tinge. Unlike hemoglobin, hemocyanin does not contain a porphyrin, and oxygen is carried between two copper atoms, which are a part of the protein chain. Other respiratory pigments are **hemerythrin,** a pink iron protein very similar to hemocyanin, which is always carried in cells, and **chlorocruorin,** a green-colored iron porphyrin protein, almost identical to hemoglobin, which is carried in solution. These last two are limited to a few invertebrate groups.

The variety of respiratory pigments and their unusual distribution among animals indicate that they have evolved independently in a number of different groups. Polychaete worms are particularly variable, and a few have both hemoglobin and chlorocruorin in solution in their blood plasma. The pigments meet the precise gas transport needs of the particular animal in its natural environment.

Gas Transport by Hemoglobin. As oxygen diffuses from the respiratory organ into the blood, it combines with hemoglobin (Hb) to form **oxyhemoglobin** (HbO_2):

$$Hb + O_2 \rightleftharpoons HbO_2.$$

During the formation of oxyhemoglobin, one molecule of oxygen (O_2) forms a loose bond with one of the four iron atoms in the molecule of hemoglobin. Subsequently, additional molecules of oxygen bind to the other three iron atoms. Since a single mammalian erythrocyte contains as many as 265,000,000 molecules of hemoglobin, a tremendous amount of oxygen can be carried. The reaction is reversible, and hemoglobin releases oxygen when blood reaches the tissues where the oxygen tension is low.

The combination of oxygen with hemoglobin and its release from oxyhemoglobin are controlled by the oxygen tension, but carbon dioxide has an important influence. Up to 20 per cent of the carbon dioxide entering the blood in the tissues reacts with oxyhemoglobin (Fig. 13.18*A*, top line), helps release oxygen, and combines with hemoglobin to form **carbaminohemoglobin** ($HbCO_2$). This reaction does not have any marked effect upon the pH of the blood.

The remainder of the carbon dioxide reacts with water to form **carbonic acid** (Fig. 13.18*A*, second line, H_2CO_3). This reaction occurs much more rapidly in the erythrocytes than in the plasma because they contain the enzyme

Figure 13.18 Diagrams of the major chemical reactions in a red blood cell that are related to the transport of carbon dioxide and oxygen. *A,* Carbon dioxide enters a red cell in a tissue capillary and oxygen is released. *B,* the opposite reactions occur within a cell in a lung capillary.

carbonic anhydrase, which speeds up the reaction some 1500 times. Most of the carbonic acid, in turn, dissociates into its constituent hydrogen and bicarbonate ions. The hydrogen ions produced combine with oxyhemoglobin, and this facilitates the dissociation of oxyhemoglobin and the release of oxygen to the tissues (third line of diagram). Potassium ions (K+), which are loosely associated with oxyhemoglobin, are also released as oxyhemoglobin takes up hydrogen ions to become reduced hemoglobin. Some of the bicarbonate ions combine with potassium within the erythrocyte (fourth line of diagram), but most of them diffuse out of the cell into the plasma, where they combine with sodium. As the cell loses these negative ions, chloride ions (Cl⁻) diffuse in, so the electrochemical balance of the cell is maintained. Thus most of the carbon dioxide is carried as carbonic acid and bicarbonate ions.

These reactions associated with the transport of carbon dioxide also maintain the pH of the cell close to neutrality. Oxyhemoglobin is a stronger acid than hemoglobin, and the conversion of oxyhemoglobin (HbO_2) to hemoglobin (Hb) would tend to raise the pH within the red cell (make it more alkaline). The formation and dissociation of carbonic acid would tend to lower the pH within the red cell (make it more acid). These two opposing phenomena tend to balance each other and the pH of the erythrocyte is maintained essentially unchanged at about pH 7.4.

The reactions in the blood with oxygen are reversible, and their direction is determined by the relative carbon dioxide and oxygen tensions. In the tissues, carbon dioxide is continually produced, oxygen continually consumed. There is a relatively high concentration of carbon dioxide, so the reactions move in the direction of binding carbon dioxide and releasing oxygen. As blood passes through the respiratory organ, oxygen diffuses in and carbon dioxide diffuses out. There is a relatively high oxygen tension so the reactions move in the direction of binding oxygen and releasing carbon dioxide as shown in Figure 13.18B.

These reactions enable the blood to carry a great deal more oxygen and carbon dioxide than it could in simple physical solution; they prevent the pH of the blood from changing greatly; and they facilitate the release of oxygen in the tissues of the body and the release of carbon dioxide in the lungs.

Oxygen Dissociation Curves. The properties of hemoglobin are such that the amount of oxygen that is bound is not directly proportional to the oxygen tension; a graph of the relationship gives an S-shaped **oxygen dissociation curve** (Fig. 13.19). Hemoglobin becomes more fully saturated with oxygen in the lung, for example, where its tension (pO_2) is about 100 mm. Hg, than it would if the relationship between saturation and tension were linear. Hemoglobin also gives up its oxygen more readily as oxygen tension drops in the tissues of the body. Our body tissues normally have an oxygen tension of about 40 mm. of Hg, but this falls to 20 mm. or less during strenuous exercise.

Hemoglobin has the capacity to deliver a great deal of oxygen to the tissues, and this capacity is increased by the effect of carbon dioxide. Carbon dioxide lowers the capacity of hemoglobin to bind with oxygen; i.e., it shifts the oxygen dissociation curve to the right (Fig. 13.20). This shift is known as the **Bohr effect.** The differences between the oxygen dissocia-

Figure 13.19 The combination of oxygen with hemoglobin as a function of oxygen tension. When blood is fully saturated, all of the oxygen binding sites are occupied. The dashed line shows the relation that would exist if hemoglobin bound oxygen as a linear function of oxygen tension. Because of the interactions between the four heme groups in a single hemoglobin molecule, the binding of oxygen to hemoglobin follows the **S**-shaped curve.

Figure 13.20 The effect of carbon dioxide tension ($_pCO_2$) on the delivery of oxygen to the tissues. The arrow A indicates the amount of oxygen released from hemoglobin as the $_pO_2$ falls from that of arterial blood (100 mm. Hg) to that of venous blood (40 mm. Hg). The arrow B indicates the additional amount of oxygen released because of the greater $_pCO_2$ in venous blood.

tion curves for arterial blood with low carbon dioxide tension and venous blood with high carbon dioxide tension illustrate how much more oxygen can be delivered to the tissues by a given amount of blood as it takes up carbon dioxide in the tissues and gives it off in the lungs.

Changes in the structure of the respiratory pigments adapt them to different conditions because they affect their oxygen-binding and -releasing capacities. Mammalian **fetal hemoglobin,** for example, has an oxygen dissociation curve that lies to the left of that of adult hemoglobin. It has a higher affinity for oxygen at any given tension, and this facilitates the

transfer of oxygen from maternal to fetal blood. Muscles frequently contain a special **myoglobin** whose curve is also shifted to the left, and this facilitates the transfer of oxygen from blood to the muscles.

The properties of the hemoglobins of different animals vary according to the requirements of the particular species and the environment in which it lives (Fig. 13.21). Comparisons are made by determining the oxygen tension needed for a 50 per cent saturation of the blood (P_{50}) under stated conditions of temperature and pH (or amount of carbon dioxide in the blood). The higher the P_{50} value, the more the oxygen dissociation curve is

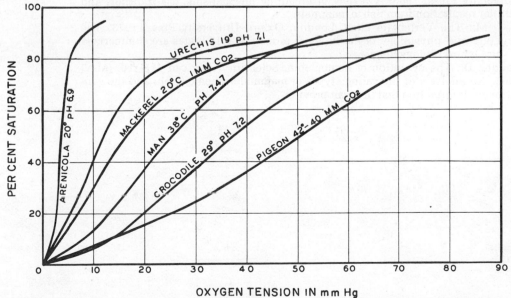

Figure 13.21 Oxygen dissociation curves for representative animals. These are equilibrium curves that lie between the venous and arterial curves. *Arenicola* and *Urechis* are marine polychaete worms. (From Prosser, C. L.: *Comparative Animal Physiology.* 3rd ed. Philadelphia, W. B. Saunders Co., 1973.)

shifted to the right. As a curve shifts to the right, the hemoglobin's affinity for oxygen is lowered. A higher oxygen tension is required to saturate the blood, but more oxygen can be delivered to the tissues, provided the respiratory organ and environment are capable of nearly fully saturating the blood. For example, if human and pigeon blood become only 95 per cent saturated in the lungs and unload oxygen in the tissues at a tension of 30 mm. of Hg (Fig. 13.21), the bird can deliver more oxygen (saturation drops from 95 per cent to about 25 per cent) than the human being (satu-

ration drops from 95 per cent to 60 per cent). In general, very active animals, such as birds, have curves shifted to the right. Small endothermals have curves shifted to the right too, for their relative surface area, heat loss and level of metabolism are higher than for a large endotherm. Aquatic animals that must obtain their oxygen in an environment where the amount is relatively low have hemoglobin with a high oxygen affinity; i.e., their curves are shifted to the left. *Arenicola*, a burrowing marine polychaete worm, needs hemoglobin with a very high affinity for oxygen.

ANNOTATED REFERENCES

Additional information on gas exchange can be found in the general references on vertebrate organ systems cited at the end of Chapter 10.

Andersen, H. T.: Physiological adaptations in diving vertebrates. Physiological Review 46:212, 1966. An excellent summary of the respiratory, vascular and other adaptations made by terrestrial vertebrates that have readapted to an aquatic mode of life.

Comroe, J. H., Jr.: The lung. Scientific American 214:57, (Feb.) 1966. A fascinating summary of pulmonary anatomy and physiology.

Conference for National Cooperation in Aquatics: The New Science of Skin and Scuba Diving. New York, Association Press, 1962. This standard and widely used manual includes excellent discussions on the physiology of diving.

Fulton, J. F.: Selected Readings in the History of Physiology. Springfield, Ill., Charles C Thomas, 1930. Interesting accounts of the discovery of respiratory processes are included in this very readable book.

Harrison, R. J., and G. L. Kooyman: Diving in Marine Mammals. London, Oxford University Press, 1971. An Oxford Biology Reader summarizing anatomical and physiological adaptations for diving.

Hughes, G. M.: Comparative Physiology of Vertebrate Respiration. Cambridge, Harvard University Press, 1963. A review of breathing mechanisms, gas transport and cellular respiration from fish to mammals.

Hughes, G. M.: The Vertebrate Lung. London, Oxford University Press, 1973. The structure and physiology of the lungs of terrestrial vertebrates are summarized in this Oxford Biology Reader.

Zuckerkandle, D.: The evolution of hemoglobin. Scientific American 212:110, (May) 1965. An analysis of the various types of mammalian hemoglobin and the evolutionary information that they can provide.

INTERNAL TRANSPORT

14.1 FUNCTIONS OF TRANSPORT SYSTEMS

Materials are transported within cells by diffusion, perhaps assisted by cytoplasmic streaming. In this chapter we will consider the ways in which materials are brought to and removed from the environment in which cells live so that the cells are maintained in a steady state, or in dynamic equilibrium, as materials are exchanged. The composition of the fluid surrounding cells varies only within narrow limits. For protozoa and small aquatic animals the surrounding fluid is simply the pond water or puddle in which they live; in larger animals it is the **interstitial fluid** between the cells. Interstitial fluid contains the largest component of body water outside of the cells (Fig. 14.1). Other extracellular water lies within blood vessels **(plasma water)** or has been secreted by cells into various epithelial spaces, such as the coelom or lumen of the digestive tract **(transcellular water).** All of these fluids are involved in the transport process. As cells metabolize they obtain oxygen, glucose, minerals and other needed substances from the interstitial fluid and discharge carbon dioxide, nitrogenous wastes and other by-products of metabolism into it.

The problem of internal transport is one of moving materials between the interstitial fluid bathing the cells and the sites where needed materials enter the body and wastes are removed. Diffusion alone is adequate in very small animals but is supplemented in most larger animals by the active transport of materials through body cavities of one kind or another or through a circulatory system. The de-

gree of development and complexity of the transport system is very much related to the size of the animal and its rate of metabolism. An animal cannot be large or sustain a high rate of metabolism without a very efficient circulatory system that can quickly supply the

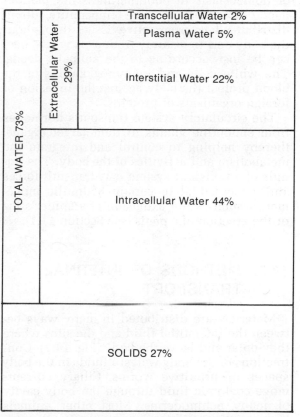

Figure 14.1 The distribution of solids and water in the adult mammalian body, expressed as a percentage of the fat-free total body weight.

interstitial fluid with the multitude of materials the cell needs and rapidly remove waste products.

Cells can tolerate only slight changes in the chemical and physical properties of their immediate fluid environment. By carrying materials to and from the interstitial fluid, the transport system plays an important role in **homeostasis;** that is, in maintaining the constancy of the environment around the cells. This is not as great a problem in small marine organisms as it is in others, for the composition of interstitial fluid and sea water are very similar. However, the external environment of terrestrial and fresh water species is very different from their internal environment. The constancy of the internal environment must be maintained despite changes in the external environment and despite pressures for change in the internal environment resulting from metabolism. Hemoglobin plays an important role in buffering the blood (see Section 13.6); that is, in maintaining a pH close to neutrality, despite perturbations as gases are exchanged. Blood performs similar functions with respect to the content of water, salts and other components of interstitial fluid. An additional aspect of homeostasis in endotherms is the maintenance of a constant body temperature. Blood distributes heat from active tissues throughout the body and to or away from sites where heat can be lost, according to the animal's needs. The white cells and immunoglobulins of the blood protect the body against the invasion of foreign organisms or proteins.

The circulatory system transports hormones from endocrine glands to their target organs, thereby helping to control and integrate the metabolism and activities of the body. The liquids of a transport system may transmit forces and are essential in certain hydraulic movements, such as the extension of a spider's leg or the erection of a penis (see Section 11.1).

14.2 METHODS OF INTERNAL TRANSPORT

Materials are distributed in many ways between the interstitial fluid and the sites where they enter and leave the body (Fig. 14.2). Contractions of the body wall stir fluids in the body spaces of primitive worms. Ciliary currents move coelomic fluid through the body cavity of many echinoderms. Most other animals have a distinct circulatory system through which blood or hemolymph is moved by the contractions of certain blood vessels, specialized hearts or adjacent body muscles. In the octopus and other cephalopod mollusks, and in annelid worms, vertebrates, and other animals with **closed circulatory systems,** blood moves from one or more hearts through arteries to extensive networks of capillaries in the body tissues and returns through veins. The capillaries are very small, thin-walled vessels, and small molecules diffuse easily between the blood and the interstitial fluid. Most mollusks and arthropods, and some of the lower chordates (tunicates) have **open circulatory systems.** Liquid moves from the heart through arteries directly to spaces among the cells. Collectively these spaces form a **hemocoel,** and the liquid moving through them is most appropriately called **hemolymph,** for it combines properties of blood and lymphlike interstitial fluid. Many of the tissue spaces are quite diffuse, whereas others are organized into special channels, but the channels do not have the lining of endothelial cells that characterizes blood vessels. Hemolymph in them is in direct contact with body cells.

A closed circulatory system is not necessarily more advanced than an open system, but it permits a higher pressure and more rapid flow of blood that distributes materials more quickly between body organs. The octopus and squid can be very active mollusks partly because they have closed systems, whereas clams, snails and other mollusks with open systems are sluggish, slow-moving creatures. Animals with closed systems are able to regulate the flow of materials to the tissues more precisely than animals with open systems. By dilating and constricting appropriate vessels, blood can be shunted to particularly active tissues, for example, to muscles during rapid locomotion, whereas less active tissues will have a reduced blood flow through them. Circulating fluids and materials are distributed more evenly to all of the tissues in animals with open systems.

14.3 PLASMA AND INTERSTITIAL FLUID

Adult human beings have about 5.7 liters of **blood.** Our blood, as well as that of other animals, is composed of a liquid component, the **plasma,** and various formed elements that are carried in the plasma, among them red blood cells **(erythrocytes),** white blood cells **(leukocytes)** and **platelets** (Fig. 14.3). Inverte-

Figure 14.2 Methods of internal transport. Water circulation through the cavities of a sponge (A) and cnidarian (B). C, Movement of body fluids within the body cavity of a primitive worm. Closed circulatory system of an annelid worm (D) and vertebrate (F). E, Open circulatory system of a crustacean. (From Prosser, C. L.: *Comparative Animal Physiology*. 3rd ed. Philadelphia, W. B. Saunders Co., 1973.)

brates have other types of corpuscles and amebocytes. The plasma itself is about 90 per cent water, 7 to 8 per cent soluble plasma proteins, 1 per cent electrolytes and the remaining 1 to 2 per cent is composed of a variety of substances: (1) nutrients such as glucose, amino acids, lipids and vitamins; (2) metabolic intermediates such as pyruvate and lactate; (3) nitrogenous wastes, including urea and uric acid; (4) dissolved gases; and (5) hormones. Many of these molecules are free in solution; others, including some of the trace metals, are bound to specific transport proteins. The primary plasma proteins are albumins, globulins and fibrinogen. The albumins, present in high concentrations, have an important osmotic effect and function in drawing water into the blood from the inter-

stitial fluid (see Section 14.13). Some globulins are transport proteins, some are important factors in blood clotting (Section 14.5), and many of those in the gamma globulin fraction of the plasma are antibodies that provide immunity from infectious diseases (Section 14.7). Fibrinogen plays a key role in blood clotting. Most of the plasma proteins are synthesized in the liver, but antibodies are made by lymphoid tissues and tissue plasma cells.

The plasma is a complex liquid in dynamic equilibrium with other body fluids. Many substances move in and out of it, but its composition and properties remain remarkably constant. We have considered the important role of the liver in storing and releasing glucose so that its concentration in the plasma varies

Figure 14.3 Photomicrograph of human blood cells. Many erythrocytes and one large leukocyte (a neutrophil) can be seen. The single dark dot and the clump of several dots are platelets. (From Zucker, Scientific American, Vol. 204, No. 2.)

only within narrow limits (see Section 12.5). Plasma proteins and salts help hemoglobin buffer the blood. Plasma is an important component of the system that maintains a careful balance between the intake of materials from the environment, their storage and removal from storage, their utilization by the cells and the elimination of wastes or products that may be present in excess of the body's needs.

Plasma and interstitial fluid are separated by the endothelial wall of the capillaries, and they differ in composition only in the materials that do not pass through this wall or that do not pass through readily. Most of the blood cells remain in the plasma, but certain leukocytes that are capable of ameboid movement squeeze between the cells of capillary walls and enter the tissue spaces. Most of the large protein molecules remain in the plasma, but some enter the interstitial fluid. Other components of the plasma and interstitial fluid diffuse readily through the capillary wall, so their concentration in the two compartments tends toward equilibrium. Animals with open circulatory systems have no capillaries; no distinction can be made between the fluids in the blood vessels and interstitial spaces. We speak of it all as hemolymph.

14.4 RED BLOOD CELLS

The respiratory pigments of invertebrates are carried free in solution or within corpuscles of various kinds that circulate in the blood, hemolymph or coelomic fluid (see Section 13.6). The vertebrate respiratory pigment, hemoglobin, is always located within red blood cells, or **erythrocytes.** Most vertebrates have oval-shaped, nucleated erythrocytes, but those of mammals lose their nuclei as they develop, and become biconcave discs. Such a shape provides more surface area than a sphere of equal volume, and the increased surface area, in turn, facilitates the passage of gases and other materials through their plasma membrane. Red cells are the most numerous of the cellular elements of the blood, there being about 5,000,000 per cubic millimeter in adult human blood.

Mature mammalian erythrocytes do not survive indefinitely. Experiments which involve tagging them with radioactive iron show that they have an average life span of 120 days. Cells lining the blood spaces of the spleen and liver eventually engulf or **phagocytize** the red cells and digest them. The iron of the heme is salvaged by the liver and is reused, but the rest of the heme is metabolized to bile pigments and excreted. Under normal circumstances, the erythrocyte population is in a steady state with new ones being synthesized as rapidly as the old are destroyed. But if delivery of oxygen to the tissues is reduced, as during hemorrhage or when ascending to a high altitude, more cells are made available so that delivery of oxygen to the tissues is restored to normal levels. Additional erythrocytes can be released immediately into the circulating blood from reserve stores in the spleen, but the reduced oxygen tension also triggers a set of reactions that leads to an increased rate of synthesis of erythrocytes. Cells in the kidney in particular respond to a lower oxygen tension by synthesizing a hormone known as **erythropoietin.** This passes in the blood to the red bone marrow where it stimulates the first stage in red cell production, the formation of hemocytoblasts from primordial stem cells. In lower vertebrates, red cells are produced in vascularized connective tissues of the kidney, liver and spleen. These sites are important during the embryonic development of mammals, but the red bone marrow is the primary source of erythrocytes in the adult.

Erythrocyte destruction and production are surprisingly rapid. From the total number of red cells in the body and their average life span, one can calculate that about 2,500,000 are made and destroyed each second of the day and night. If the rate of production of cells or of hemoglobin decreases, some type of **anemia** results. Anemia is characterized by a decrease in the number of red cells per cubic millimeter of blood, by a decrease in the amount of hemoglobin per red cell, or both. In pernicious anemia the number of erythrocytes steadily decreases. Eating large quantities of liver increases the rate of red cell formation, for liver is rich in vitamin B_{12}, which is necessary for normal erythrocyte development. A person with pernicious anemia cannot absorb enough B_{12}, even though the requisite amount may be present in the diet, because the lining of his stomach does not secrete enough "intrinsic factor," necessary for the absorption of B_{12}. If an excess is made available by giving foods especially rich in B_{12}, enough can be absorbed. It is now known that vitamin B_{12} is a necessary coenzyme in DNA synthesis, which precedes cell division. Since erythrocyte precursors are among the most rapidly dividing cells in the body, a deficiency in vitamin B_{12} is first manifested by anemia.

14.5 HEMOSTASIS

The loss of blood or hemolymph through an injury can severely impair the distribution of materials to and from the body cells, and animals with circulatory systems have evolved mechanisms for **hemostasis**; that is, for stopping the flow of blood after an injury. Because of its great medical importance, hemostasis has been studied most intensively in vertebrates, and especially in mammals, but the phenomenon occurs in invertebrates as well.

Hemostasis in all animals involves a short-term closure of the blood vessels, followed by the formation of a clot, which by its contraction holds the injured tissues together until healing occurs. In animals with closed circulatory systems, nerve reflexes cause small broken vessels to constrict, thereby reducing blood flow. Many crustaceans with open systems can reflexly discard a limb that may be caught by a predator. The limb is severed at a distinct **autotomy plane,** and a specialized membrane here constricts around the perforation so there is very little loss of hemolymph.

Temporary closure of vessels in most animals is also achieved by a clumping, or agglutination, of blood cells in the injured vessel or hemolymph space. In lower vertebrates the cells that clump are **thrombocytes,** but in mammals **platelets** play this role. Platelets are non-nucleated fragments of cytoplasm that bud off from giant cells in the bone marrow. They number about 300,000 per cubic millimeter in our blood and survive for eight to 10 days. Serotonin released by platelets at the injury site prolongs vasoconstriction, and released adenosine diphosphate stimulates the platelets to swell and agglutinate.

Clot formation in vertebrates is initiated by **thromboplastin** and other substances released by injured tissue and by thrombocytes or platelets. Thromboplastin triggers a series of proenzyme-enzyme transformations that has been compared to an "enzyme cascade." The conversion of one proenzyme to an active enzyme triggers the conversion of a second one and so on through at least five steps, several of which require the presence of calcium ions. When whole blood is removed from the body, clotting can be prevented by adding substances such as sodium citrate that bind calcium ions. One advantage of this complexity is the amplification that results. The few molecules involved in the first step affect more and more at each successive step until the reaction finally involves countless billions. The final enzyme in the initial phase of the clotting reaction, **prothrombinase**, in the presence of phospholipids released by the platelets and calcium ions, then converts a plasma globulin, **prothrombin**, into **thrombin** by splitting off certain peptide fragments. Thrombin, in turn, acts as a proteolytic enzyme and catalyzes the conversion of the soluble protein **fibrinogen** into insoluble **fibrin.** This involves the removal of two peptides from the fibrinogen molecule to form activated **fibrin monomer.** Fibrin monomer polymerizes rapidly into long fibrin threads which form a delicate network within

Figure 14.4 Scanning electron micrograph of a portion of a human blood clot showing an erythrocyte enmeshed in a network of fibrin. (From Science, Vol. 173, No. 3993, 1971. Cover Picture. Photograph by Emil Bernstein, the Gillette Company Research Institute. Copyright 1971 by the American Association for the Advancement of Science.)

which the formed elements of the blood are trapped (Fig. 14.4). A **clot** has formed. The clot subsequently contracts and squeezes out the liquid phase, called **serum**. Serum differs from plasma in that it lacks fibrinogen and cannot clot. Vitamin K, necessary for clotting, does not enter into this series of reactions directly but is essential for the production of prothrombin in the liver.

The clotting reactions can be summarized in the following scheme:

Clotting of hemolymph has been described for many arthropods. It, too, appears to involve enzymic changes leading to the precipitation of soluble proteins, but it is not yet well understood.

Many safeguards reduce the chance of clots occurring accidentally in the blood vessels. Specific inhibitors in the blood prevent activation of each of the numerous enzymes unless a critical concentration of the activating substances is achieved. A chance accumulation

Thromboplastin from platelets and injured tissues

$$
\begin{aligned}
&\left.\begin{array}{l}
\longrightarrow \\
\longrightarrow \\
\longrightarrow \\
\longrightarrow
\end{array}\right\} \text{Activation of various} \\
&\qquad\qquad\quad\ \text{clotting factors} \\[4pt]
&\longrightarrow \text{Prothrombinase} \\[6pt]
&\qquad\quad + Ca^{++} \text{ and platelet phospholipid} \\[6pt]
&\text{Prothrombin} \longrightarrow \text{Thrombin + peptides} \\[4pt]
&\text{Fibrinogen} \xrightarrow{\ \text{Thrombin}\ } \text{Fibrin monomer + peptides} \\[4pt]
&\text{Fibrin monomer} \xrightarrow{\ \text{Polymerization}\ } \text{Fibrin threads}
\end{aligned}
$$

of activating substances is unlikely because these are removed by the liver. Finally, a fibrinolysin system, which dissolves fibrin in clots that may start to form, can be activated. However, despite these mechanisms, a clot, known as a **thrombus,** may develop in a vessel and can be very serious if it plugs a vessel that supplies a vital area.

In the hereditary disease **hemophilia** there is a deficiency of one of the factors necessary for the formation of prothrombinase, so that clots do not form and the slightest scratch may lead to fatal bleeding. This disease attracted special attention because it appeared in several different European royal families and was apparently inherited from Queen Victoria of England.

14.6 WHITE BLOOD CELLS

In addition to cells containing respiratory pigments and ones involved with hemostasis, the circulating fluids of animals contain a variety of colorless cells, or corpuscles, that perform many different functions. The circulatory system transports them to the particular part of the body where they may be needed. Many are important in body defense or wound healing; a few among invertebrates have an excretory role. In vertebrates, these cells are known as white blood cells, or **leukocytes.** Five distinct types can be recognized: **lymphocytes, monocytes, neutrophils, eosinophils** and **basophils** (Fig. 3.25). They differ in the size and shape of the nucleus and in the size and nature of the granules present in the cytoplasm. Collectively, there are only about 7000 leukocytes per cubic millimeter in human blood. They are produced in the lymph nodes, the spleen and red bone marrow, and survive from one to four days. Although they are passively carried by the blood, most leukocytes can also creep about by ameboid motion. This enables them to squeeze between the cells of the capillary walls and enter the tissues where their physiological functions are performed. Many are lost from the body through the lungs, digestive tract and kidneys.

These cells are important in protecting the body against disease organisms. Bacteria penetrating through a wound produce toxins which, together with factors in injured tissue, cause blood vessels in the infected area to dilate and become more permeable to liquids and to white blood cells. The area becomes inflamed and swollen; vast numbers of leukocytes migrate into the region and begin to engulf and digest bacteria and destroyed cells. Neutrophils are among the first cells to become active, but if the infection is severe they break down and contribute, along with the bacteria and tissue debris, to the **pus** that forms. As time goes on, monocytes and lymphocytes transform into large phagocytic cells known as **macrophages**, and they clean up the area. Other lymphocytes transform into fibroblasts which, along with fibroblasts in the connective tissue, repair the injury by producing fibers and forming **scar tissue.**

14.7 IMMUNITY

The phagocytosis of invading bacteria and viruses is a primary body defense that is developed in many invertebrates and all vertebrates. Birds, mammals and, to a lesser extent, lower vertebrates have evolved a second line of defense, for they have the capacity to synthesize special proteins, known as **antibodies,** which combat foreign substances **(antigens)** that enter the body. Antigens are usually proteins associated with invading bacteria or viruses, but any foreign protein and certain foreign polysaccharides and DNA may induce antibody formation. Usually only a specific part of the antigen molecule, known as the **immunological specific determinant group,** causes a specific antibody to be produced. Some antigens have several different determinant groups and hence induce synthesis of different antibodies, each specific to its determinant group. Antibodies are immunoglobulins that combine with the antigens and neutralize them; they may cause the invading microorganisms to clump, or **agglutinate,** thereby effectively preventing a further penetration of the body; they may cause the invading microorganisms to break up and dissolve (a phenomenon known as **lysis**); or they may make the invaders more susceptible to phagocytosis. The antigen-antibody reaction is generally very specific. Antibodies that have developed in response to mumps viruses, for example, will not combine with other antigens. The configuration of the antigen and antibody molecules is such that only antibodies that have developed in response to a given antigen can fit on the surface of the antigen and react with it.

Once the body has produced antibodies in response to an antigen, it retains the capacity to produce similar antibodies for many years. If a subsequent invasion of the same type of antigen occurs during this period, antibodies specific for it may be present and more can be synthesized very quickly by sensitized cells. The body has developed an **immunity** to this particular antigen. The immunity that is acquired as a result of having once had mumps, smallpox and certain other infectious diseases lasts a very long time, generally for life. The immunity to certain other diseases lasts for a much shorter time and, after it is lost, one can get the disease again.

However, it is not necessary to contract a disease to develop an immunity to it. During the late eighteenth century, Edward Jenner observed that milkmaids and others who handled the udders of cows infected with cowpox never got smallpox. In 1796, he took a bit of the material from the pustules of an infected cow and scratched it into the skin of a person. Individuals so treated acquired a mild disease but thereafter were immune to smallpox. Cowpox is caused by the **vaccinia** virus; smallpox, by a different but related **variola** virus. Vaccinia does not cause serious disease in humans, but is similar enough to variola so that antibodies that develop in response to it are effective in combating variola. Jenner's experiments were the beginning of the **vaccination technique.** Since then, many kinds of vaccines have been developed. Usually a related and less virulent microorganism, which could serve as the basis of a vaccine, is not available, but vaccines can be produced by rendering the disease organisms harmless by appropriate treatment and injecting them. Although the organisms are incapable of causing the disease, they are still capable of inducing antibody formation.

The production of antibodies in response to an antigen that enters the body, either through disease or vaccination, is known as an **active immunity.** Active immunities are typically long-lasting. Temporary protection can be given a person who has been exposed to a new and dangerous antigen by injecting an antiserum containing immunoglobulins from an organism that has already produced the appropriate antibodies. A person bitten by a poisonous snake, for example, can be given antiserum obtained from a horse that has developed antibodies specific to this snake venom. The horse has developed an active immunity in response to successive injections of small amounts of the snake venom, but the immunity conferred on the person receiving the antiserum is a **passive immunity.** A newborn infant receives a passive immunity to certain diseases by the passage of maternal antibodies through the placenta, or in the early milk. Passive immunities last only a few days or weeks.

Intensive research in recent years has greatly increased our understanding of the structure and action of antibodies and of the cells that make them. Lymphocytes are the key cells in recognizing antigens and initiating a response to them. Two populations of lymphocytes exist in the body. Although both are derived from common stem cells in bone marrow, the two undergo maturation processes in different tissues before entering lymph nodes and the general circulation. **T-lymphocytes** are processed in the **thymus,** an organ of pharyngeal pouch origin located at the base of the neck and in the adjacent parts of the thoracic cavity in young mammals. The thymus atrophies after maturity. **B-lymphocytes** were first identified in chickens where they undergo their processing in the **bursa of Fabricius,** a pouchlike diverticulum of the cloaca. The site of processing of mammalian B-lymphocytes is not entirely clear, for mammals have no bursa, but it appears to include aggregations of lymphocytes associated with the digestive tract: tonsils, appendix and islands of lymphoid tissue in the intestinal wall (Fig. 14.5). The two types of lymphocytes can be distinguished in scanning electron micrographs, for B-lymphocytes have a rough, fimbriated surface, whereas T-lymphocytes have a relatively smooth surface (Fig. 14.6).

B- or T-lymphocytes that are competent to respond to a particular antigen stimulus first multiply rapidly. Some of the progeny cells remain as lymphocytes with the capacity to react again to the particular stimulating antigen. They give the body an immunological memory. It can respond more rapidly to a second encounter with the same antigen. In other respects the reactions of the two cells differ. Some of the progeny of the stimulated B-lymphocytes develop an elaborate granular endoplasmic reticulum capable of protein synthesis (Fig. 14.7), lodge in the tissues and, now known as **plasma cells,** synthesize appropriate antibodies. The antibodies are released into the blood and carried as part of the gamma-globulin fraction. This **humoral antibody response** combats bacterial and some viral infections.

T-lymphocytes, on exposure to appropriate antigens, are transformed into **lymphoblasts** that produce antibody-like substances that are not released into the circulation. Rather, the lymphoblasts themselves react directly with the antigen, usually fungi, some viruses and foreign tissues. This **cell-mediated response** is

Figure 14.5 Diagram of the development of the two types of mammalian immune systems from stem cells in the bone marrow. (From Frobisher, M., R. Hinsdill, K. Crabtree, and C. Goodheart: *Fundamentals of Microbiology*. 9th ed. Philadelphia, W. B. Saunders Co., 1974.)

responsible for the rejection of skin and other tissue grafts. Most adult vertebrates will not accept a homograft from another individual that is genetically very different but will accept autografts from other parts of the body or isografts from an identical twin. When performing an organ transplant, the surgeon tries to obtain an organ from an individual as genetically similar as possible to the recipient, and even then steps must be taken to suppress the activity of the T-lymphocytes.

The existence of two populations of lymphocytes is supported by many lines of evidence. There are certain rare human diseases in which one or the other type of immune system has failed to develop properly. Some individuals who are very susceptible to bacterial infections lack plasma cells and the ability to produce circulating antibodies. Other individuals, particularly prone to fungal and certain viral infections, lack T-lymphocytes but have B-lymphocytes, plasma cells and circulating antibodies. The experimental removal of the thymus of rats and mice early in their life also leads to a deficiency in T-lymphocyte differen-

tiation. Such individuals will accept skin grafts from genetically different members of the species.

Although there are two distinct types of immune response, humoral and cell-mediated, there is evidence of cooperation between them. Under certain circumstances T-lymphocytes appear to release a factor that facilitates the B-lymphocytes' response.

Our understanding of the mechanism whereby the lymphocytes that respond to the antigenic stimulus produce the appropriate antibody is far from complete. An early **instructive theory** postulated that antigen can enter any lymphocyte not previously exposed to antigen, act as a template, and instruct the cell's protein-synthesizing machinery to make antibodies that will combine properly with the antigen. This theory, however, runs counter to what we now know about the essential role of genes and messenger RNA in protein synthesis. Protein (antigen) cannot act as a template for the synthesis of more protein. More recent evidence supports the **clonal selection theory** proposed by Sir Macfarlane Burnet in the late

Figure 14.6 Scanning electron micrographs of a B-lymphocyte (A) and a T-lymphocyte (B) showing surface differences. (From Polliack, A., N. Lampden, B. D. Clarkson, E. DeHarven, Z. Bentwich, F. P. Siegal, and H. G. Kunkel: Identification of human B and T lymphocytes by scanning electron microscopy. J. Exp. Med., *138*:607, 1973.)

Figure 14.7   Electron micrograph of a human plasma cell showing the hypertrophied granular endoplasmic reticulum that is associated with protein synthesis. (From Bloom and Fawcett after Van Breeman.)

1950's. According to this view, random somatic mutations occur in the lymphocytes as they multiply during embryonic development. Hundreds of thousands of different lines of lymphocytes are generated, each with the capacity to make a specific antibody in response to a specific antigen. When an antigen enters the body certain lymphocytes already exist that have the competence to respond to it, for there is a veritable dictionary of different lymphocytes within the body. The appropriate lymphocytes are stimulated (or selected) by the antigen, multiply rapidly and form a clone of identical cells that synthesize the antibodies needed to combat the antigen. The time required for this accounts for the latent period between an infection and the body's response.

The clonal selection theory helps us to understand how the body distinguishes between "self" and "nonself." Antibodies are synthesized in response to foreign proteins (**"nonself"**) but normally not in response to the body's own proteins ("**self**"). During embryonic development, when the immune system is not completely formed, lymphocytes encounter the various body tissues. If a mutation has given a lymphocyte the capacity to react with one of the body's own proteins, a mild re-

action will occur, and the lymphocyte will be eliminated before it has had a chance to multiply and produce other cells of its type. By birth the individual will have a wide assortment of lymphocytes capable of reacting only with foreign antigens.

Considerable evidence supports this notion. If foreign antigens are injected into an individual during fetal or neonatal life before the immune system has matured, the body assumes they are "self," and lymphocytes that start to react against them are eliminated. It is possible, for example, to inject a newborn mouse of strain X with cells from strain Y. The Y cells are incorporated as "self," and at a later time the X strain mouse will accept a skin graft from a Y mouse, something it would not normally do. The X mouse has developed an **acquired immunological tolerance.** A converse situation sometimes develops in which certain of the body's own tissues are no longer recognized as "self." Certain types of arthritis and multiple sclerosis are **autoimmune diseases** in which antibodies are synthesized that destroy some of the body's own proteins.

Much is now known about the structure of antibodies and the ways they react with antigens. Most are Y-shaped molecules with the two upper arms containing the active sites. When antigens attach to them, the stem of the molecule is changed in such a way that it becomes reactive. In some cases inactive **complement enzymes** in the blood attach to the antigen-antibody complex, become active and lyse the foreign protein. In **allergic reactions** certain antibodies that have previously developed in response to an antigen (in this case called an **allergen**) become attached to **mast cells** in the tissues. Re-exposure to the specific allergen activates the antibody-mast cell complex, and the mast cells produce **histamine** and other substances that are responsible for the allergic symptoms. Antihistamines can alleviate the symptoms but do not prevent the antigen-antibody reaction that causes histamine production.

14.8 THE ABO BLOOD GROUPS

When the practice of transfusing blood from one person to another was begun, it was found that the transfusions were sometimes successful, but more often they were not and erythro-cytes in the blood of the recipient would clump (agglutinate), with fatal results. Careful analysis by Landsteiner at the beginning of this century showed that specific antigenic proteins, called A and B, might be present in the plasma membranes of the erythrocytes. These antigens are called **agglutinogens,** since they may cause agglutination of the red cells. Some individuals have protein A, some B, some both A and B and some neither. Naturally occurring antibodies **(agglutinins)** specific for these agglutinogens, and designated a and b, may be present in the plasma. An individual with red cells containing a certain agglutinogen does not, of course, have the agglutinin specific for it in his plasma. If he did, his own red cells would be agglutinated.

Four groups of persons, O, A, B and AB, can be recognized according to the presence or absence of these agglutinogens and agglutinins (Fig. 14.8). Before a transfusion is made, both donor's and recipient's cells are typed by adding them to serum containing an a agglutinin (anti-A serum) and to serum containing a b agglutinin (anti-B serum), and observing which, if either, causes the cells to clump. Transfusions between members of the same

Blood group	Agglutinogen in erythrocyte	Agglutinins in plasma	Reaction with	
			Anti-A serum	Anti-B serum
O Universal donor	none	a and b		
A	A	b	agglutination	
B	B	a		agglutination
AB Universal recipient	A and B	none	agglutination	agglutination

Figure 14.8   The ABO blood groups. The two right-hand columns show the appearance of the red cells of each group when exposed to anti-A and anti-B serum. The blood group to which an individual belongs is identified by the antiserum in which its red cells agglutinate; if there is no agglutination with either antiserum, the individual belongs to group O.

group are perfectly safe, and transfusions between different groups are also safe provided that the donor's erythrocytes do not contain an agglutinogen that will react with the recipient's agglutinins. The agglutinins in the donor's plasma become so diluted in the recipient that they have no effect and they may be disregarded unless an unusually large transfusion is given. Members of Group O, who have neither of the agglutinogens, can give blood to members of any group and are "universal donors." But since their plasma contains both of the agglutinins, they can receive blood only from members of their own group. Members of Group AB, in contrast, have neither agglutinin, and can receive blood from members of any group. Since they have both agglutinogens, they can give blood only to members of their own group. They are "uni-

versal recipients." Members of Groups A and B can give blood to members of Group AB and receive from members of Group O. The inheritance of these blood groups was considered in Section 6.16.

14.9   INVERTEBRATE CIRCULATORY PATTERNS

Most primitive worms lack a circulatory system, but there is one present in many of the higher invertebrate phyla: mollusks, annelids, arthropods, echinoderms. The circulatory system of most echinoderms is rather rudimentary, and a system of coelomic canals is responsible for internal transport.

Annelids.   Annelid worms are the major in-

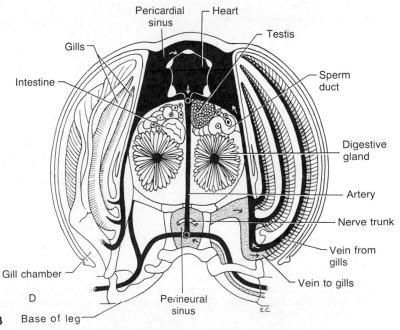

Figure 14.9 Representative invertebrate circulatory systems. Arrows indicate the direction of blood or hemolymph flow. *A*, The closed system of an annelid as represented by a portion of the polychaete *Nereis*. Anterior is toward the right. The paired circumesophageal vessels that carry blood from the dorsal to the ventral vessel are not shown. (Modified from Nicoll.) *B*, The open system of an arthropod as seen in a cross section through the thorax of a crayfish. (After Howes.)

vertebrate group with a closed circulatory system. A pulsating **dorsal vessel** conveys blood anteriorly (Fig. 14.9 A). Near the front of the body one or more pairs of **circumesophageal vessels** carry the blood around the gut to a **ventral vessel.** In some annelids, the circumesophageal vessels pulsate and act as accessory hearts. Blood flows posteriorly in the ventral vessel and is distributed by branches to capillary networks in the tissues, thence back into the dorsal vessel. Typically there is a single branch from the ventral vessel to the gut and a pair of branches to the nephridia, body wall and parapodia in each body segment.

Arthropods. Arthropods, although related to annelids, have evolved an open circulatory system. The **heart,** which may be long and tubular or compact and limited to the thorax, is situated dorsally within an expanded portion of the hemocoel known as the **pericardial sinus** (Fig. 14.9B). Hemolymph returning from tissue spaces enters the pericardial sinus rather than going directly into the heart. Hemolymph enters the heart through paired lateral openings, the **ostia,** which contain valves that prevent it from leaving. **Anterior** and **posterior aortae** leave the front and back of the heart and lead into branches that carry hemolymph to tissue spaces in the various organs. The tissue spaces drain into sinuses, often including a large **perineural sinus** around the ventral nerve cord, whence the hemolymph passes through channels in the gills and back to the pericardial sinus. Insects have a well developed circulatory system, but its functions do not include the transport of gases, which are carried instead by the tracheal system (see Section 13.4).

14.10 VERTEBRATE CIRCULATORY PATTERNS

Primitive Fishes. The circulatory system is closed in all vertebrates. It has undergone many changes during the evolution of vertebrates, most of which are correlated not only with the shift from gills to lungs as the site of external gas exchange that occurred during the transition from water to land, but also with the development of the efficient, high pressure circulatory system necessary for an active terrestrial vertebrate.

In a primitive lungless fish (Fig. 14.10), all of the blood entering the heart from the veins has a low oxygen and a high carbon dioxide content; i.e., it is venous blood. The heart consists of a **sinus venosus,** a single **atrium,** a single **ventricle** and a **conus arteriosus** arranged in linear sequence. The muscular contraction of the heart increases the blood pressure, which is very low in the veins, and sends the blood out through an artery, the **ventral aorta,** to five or six pairs of **aortic arches** that extend dorsally through capillaries in the gills to the **dorsal aorta.** Carbon dioxide is removed and oxygen is added as the blood flows through the gills; i.e., it changes to arterial blood (see Section 13.3). The dorsal aorta distributes this through its various branches to all parts of the body.

Blood pressure decreases as blood flows along because of the friction between the blood and the lining of the vessels. Pressure is reduced considerably as the blood passes through the capillaries of the gills, for friction is greatest in vessels of small diameter. Mean blood pressure in the ventral aorta of a dogfish

Figure 14.10   The major parts of the circulatory system of a primitive fish. *1,* Sinus venosus; *2,* atrium; *3,* ventricle; *4,* conus arteriosus of the heart. The aortic arches are numbered with Roman numerals. Only traces of the first aortic arch remain in the adults of most fishes.

during heart contraction, for example, is about 30 mm. Hg; that in the dorsal aorta is about 20 mm. Hg. This relatively low blood pressure in the aorta becomes even lower when the blood reaches the capillaries in the tissues. Circulation in primitive fishes is rather sluggish and not conducive to great activity.

Veins drain the capillaries of the body (where blood pressure is further reduced) and lead to the heart, but not all veins go directly to the heart. In primitive fish, blood returning from the tail first passes through capillaries in the kidneys before entering veins leading to the heart. Veins that drain one capillary bed and lead to another are called portal veins, and these particular veins are known as the **renal portal system.** Other veins, known as the **hepatic portal system,** drain the digestive tract and lead to capillary-like passages in the liver. Since much of the blood returning to the heart has passed through one or the other of these portal systems in addition to the capillaries in the gills and tissues, its pressure is near 0 mm. Hg.

It is not difficult to appreciate the significance of a hepatic portal system, since the liver plays such an important role in the metabolism of foods, but the adaptive significance of a renal portal system in primitive vertebrates is less clear.

Amphibians. Correlated with the shift from gills to lungs, many changes occurred in the heart and aortic arches (Fig. 14.11). The aortic arches were reduced in number, the first two and the fifth being lost. Those that remain are no longer interrupted by gill capillaries. In a primitive tetrapod, such as the frog, the third pair of aortic arches forms part of the **internal carotid arteries** supplying the head; the fourth, the **systemic arches** leading to the. dorsal aorta; and the sixth, the **pulmocutaneous arches** leading to the lungs and skin. New veins, the **pulmonary veins,** return aerated blood from the lungs to the heart. The heart now receives blood from both the body and lungs. Though blood streams from the body and lungs are separated in the frog by a divided atrium, the two converge in a single ventricle (Fig. 14.12). Recent experiments using indicators to identify the streams have shown that, although there is some mixing, a high degree of separation is achieved. The slight differences in the times the streams enter the ventricles, a spongy ventricular muscle that helps hold them apart, the deflective action of a spiral valve within the conus arteriosus, and the internal partitioning of the arterial trunks that lead from the conus to the vessels

supplying the body and lungs, all play a role in separating the two streams. The pulmocutaneous arch receives primarily right atrial blood relatively low in oxygen content, the systemic arch receives a rather complete mixture of blood, and the carotid arch receives primarily left atrial blood with a high oxygen content. The mixture of left and right atrial blood going to the body is not as detrimental as may at first appear, because some of the right atrial blood comes from the skin where aeration also occurs. When the frog is under water the lungs are not used, and the lack of a divided ventricle is probably an advantage because it allows aerated blood returning to the right atrium from the skin to be distributed directly to the body. The lack of a divided ventricle in modern amphibians may be an adaptation related to the evolution of cutaneous respiration. Living lungfish have a partial interventricular septum; this provides a more complete separation of the blood coming from the body and that from the lungs.

The loss of gills and the evolution of a double circulation through the heart result in a much higher blood pressure in the arteries of a primitive tetrapod than in a fish. Blood in the dorsal aorta of a frog has a mean pressure of about 30 mm. Hg, 50 per cent more than that of the dogfish. Although this increases the circulatory rate, the benefit is somewhat offset by the fact that the blood delivered to the tissues is mixed to some extent and does not contain so much oxygen, relatively, as it did in a fish. The pattern of the major arteries and veins of a frog is shown in Figure 14.12.

Reptiles and Birds. Higher tetrapods depend upon their lungs for external gas exchange. Since no gas exchange occurs in the skin, there is no mixing of aerated blood from the skin with blood from the body. The mixing in the heart of arterial blood from the lungs with venous blood from the body is lessened in reptiles by a partial division of the ventricle and by a complex, tripartite division of the conus (Fig. 14.11). In a lizard, for example, the venous blood from the body enters the right side of the heart, and most of it is sent to the lungs via the pulmonary arteries. After returning to the left side of the heart, this blood, now rich in oxygen content, is distributed to the head and also to the trunk via the right fourth aortic arch. Venous blood on the right side of the heart that did not go to the lungs mixes with oxygen-rich blood and is sent to the trunk via the left fourth aortic arch. It has been suggested that this is a mechanism that permits some venous blood to by-pass the lungs

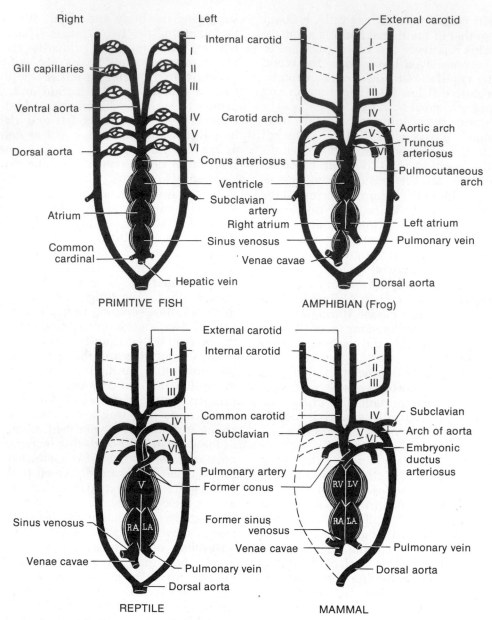

Figure 14.11 Diagrams of the heart and aortic arches to show the changes that occurred in the evolution from primitive fishes to mammals. All are ventral views. The heart tube has been straightened so that the atrium lies posterior to the ventricle.

at an evolutionary stage when the lungs may not be able to handle the volume of blood returning from the body to the heart. The head and brain receive oxygen-rich blood; the trunk, whose tissues are less sensitive to oxygen deprivation, receives a mixed blood. Birds have no mixing of blood at all, for their ventricle is completely divided and they have lost the left fourth aortic arch of their reptilian ancestors.

Mammals. Mammals too have a completely divided ventricle, so there is no mixing

of oxygen-rich and oxygen-depleted bloods (Fig. 14.11). Venous blood from the body enters the **right atrium,** into which the primitive sinus venosus has become incorporated. Arterial blood from the lungs enters the **left atrium.** The atria pass the blood on to the **right** and **left ventricles,** respectively. The ventricular walls of mammals are more muscular than those of lower vertebrates and can increase the blood pressure considerably. The primitive conus arteriosus has become completely divided, part contributing to the pul-

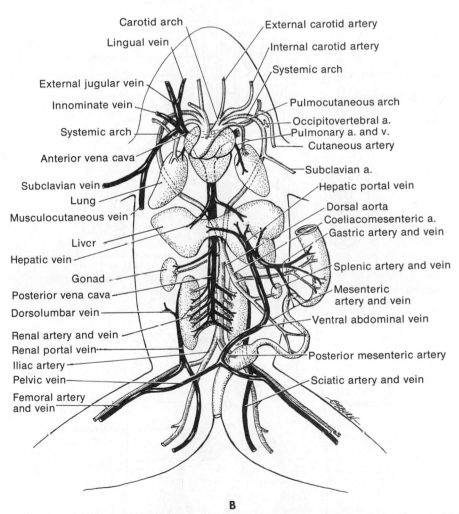

Figure 14.12 *A*, Ventral view of a dissection of the heart of a frog showing the partial separation that occurs in the ventricle and conus arteriosus between blood high in oxygen content entering from the lungs and left atrium and less highly oxygenated blood entering from the skin, body and left atrium. *B*, Ventral view of the major arteries and veins of the frog. Veins are shown in black; arteries are white. Certain of the anterior veins have been omitted from the right side of the drawing and certain of the anterior arteries from the left side.

monary artery leading from the right ventricle to the lungs and the rest to the arch of the aorta leading from the left ventricle to the body.

The sixth pair of aortic arches form the major part of the mammalian **pulmonary arteries,** and the third pair contribute to the **internal carotid arteries.** But it will be observed in Figure 14.11 that only the left side of the fourth arch, known as the **arch of the aorta,** leads to the dorsal aorta. The right fourth arch contributes to the right **subclavian artery** to

the shoulder and arm and does not connect posteriorly with the aorta.

The major change in the veins is the complete loss of a renal portal system. Blood from the tail and posterior appendages enters a **posterior vena cava,** which continues forward to the heart. It receives blood from the kidneys but does not carry blood to them. An **anterior vena cava** drains the head and arms. The hepatic portal system is still present. The pattern of the major human arteries and veins is shown in Figure 14.13.

Figure 14.13   The major arteries and veins of a human being as seen in an anterior view.

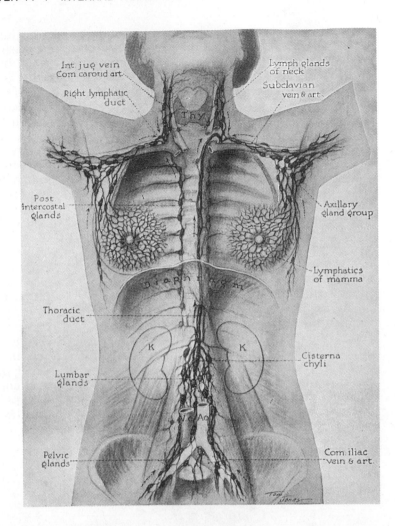

Figure 14.14 Major lymphatic vessels and nodes in the trunk of a human female. (Courtesy of S. H. Camp Co.)

These evolutionary changes have resulted in a very efficient circulatory system. Mammals have relatively more blood than lower vertebrates; it is distributed under greater pressure, and there is no mixing of arterial and venous blood. A human being, for example, has 7.6 ml. of blood per 100 gm. of body weight compared with 2 ml. per 100 gm. in a fish. The mean pressure in our dorsal aorta is about 100 mm. Hg.

As blood pressures have increased during the evolution of the circulatory system, more liquids and plasma proteins have escaped from the capillaries into the interstitial fluid than have been returned by the veins. A separate **lymphatic system** evolved; this plays a role in returning fluid and plasma proteins from the tissues to the main part of the circulatory system. Lymphatic vessels arise as outgrowths from the veins and, in general, tend to parallel the veins and ultimately empty into them.

The system reaches its greatest development in mammals (Fig. 14.14). **Lymphatic capil-** laries occur in most of the tissues of the body. They are more permeable than the capillaries between the arteries and veins, and pressures within them are exceedingly low. Their high permeability makes them the most likely route for the spread of microorganisms or cancer cells within the body. **Lymph nodes** lie at many points where small lymphatic vessels converge. They are major sites for the production of lymphocytes, and cells within them can phagocytize invading bacteria or respond to them by initiating antibody production. Lymphatic vessels draining the hind limbs and the pelvic and abdominal regions finally converge upon a lymph sac, the **cisterna chyli,** located posterior to the diaphragm and dorsal to the aorta. The largest lymphatic vessel of the body, the **thoracic duct,** extends anteriorly from this point, receives the drainage from the left side of the anterior parts of the body, and finally enters the left brachiocephalic vein. A smaller **right lymphatic duct** drains the right anterior parts of the body and enters the right brachiocephalic vein. When a lymphatic system is well developed, heart, arteries and veins are sometimes called the **cardiovascular system,** to distinguish this part of the circulatory system from the lymphatic system.

14.11 FETAL AND NEONATAL CIRCULATIONS

The placenta of the mammalian fetus, rather than the digestive tract, lungs and kidneys, is the site for exchange of materials. This, together with the fact that the vessels in the lungs of the fetus are not developed enough to handle the total volume of blood that is circulating through the body, requires certain differences in the fetal circulation (Fig. 14.15). Blood rich in oxygen returns from the placenta in an **umbilical vein,** passes rather directly through the liver via the **ductus venosus,** and enters the posterior vena cava, where it is mixed with blood returning from the posterior half of the fetus. The posterior vena cava empties into the right atrium, which also receives venous blood from the head by way of the anterior vena cava.

The lungs cannot accommodate all of this blood early in development, yet a large volume of blood must pass through all chambers of the heart to ensure their normal development. The entrance of the posterior vena cava is directed toward an opening, the **foramen ovale,** in the partition separating the two atria. Most of the blood from the posterior vena cava tends to go through this into the left atrium, thence to the left ventricle and out to the body through the arch of the aorta. This blood bypasses the lungs yet permits the left side of the heart, which otherwise would receive little blood from the collapsed lungs, to function and develop normally. The rest of the blood from the posterior vena cava enters the right ventricle along with the blood from the anterior vena cava and starts out the pulmonary artery toward the lungs. However, only a fraction of this blood passes through the lungs to return to the left atrium and mix with blood from the posterior vena cava. Most of the blood in the pulmonary artery goes through another by-pass, the **ductus arteriosus,** to the dorsal aorta. The ductus arteriosus represents the dorsal part of the left sixth aortic arch (Fig. 14.11). Since the ductus arteriosus enters the

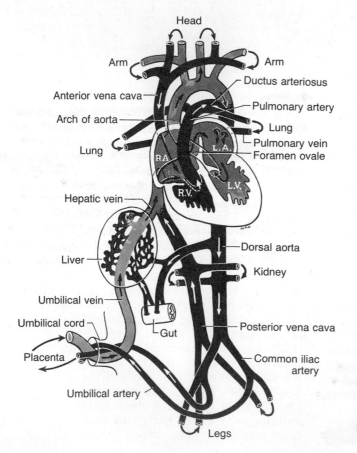

Figure 14.15 Circulation in a fetal mammal. The shading gives some indication of the mixing of the blood, though there is more mixing than can be indicated diagrammatically. The lightest shading represents blood with the highest oxygen content; the darkest shading, blood with the lowest oxygen content. (Modified after Patten.)

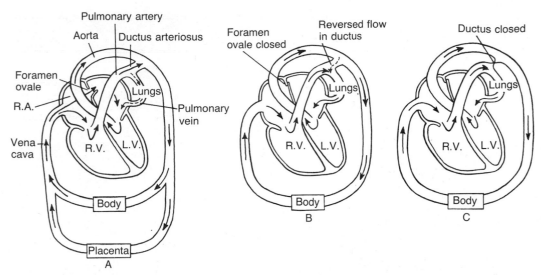

Figure 14.16 Circulatory changes at birth: A, fetal; B, neonatal; C, adult circulation. Solid arrows represent major blood circuits; broken arrows, lesser circuits.

aorta after the arteries to the head have been given off, the head receives the blood with the highest oxygen content. After the entrance of the ductus arteriosus, the blood in the aorta is highly mixed. This is the blood that is distributed to the rest of the body and, by way of **umbilical arteries,** to the placenta.

Throughout fetal life the lungs and most of the vessels within them are collapsed. The resistance to blood flow through the lungs (pulmonary resistance) is greater than the resistance to flow through the body and placenta (systemic resistance). As a consequence there is a large flow of blood through the placenta and a small flow through the lungs. The volume of blood returning to the left atrium is less than the volume entering the right atrium. This facilitates the opening of the valve in the foramen ovale and the continued by-passing of the lungs (Fig. 14.16A).

These conditions are immediately reversed at birth. The placenta is expelled and blood volume in the systemic circuit is reduced. Carbon dioxide accumulates in the fetal blood, activating the respiratory system; the lungs fill with air, and the pulmonary vessels that were collapsed open up. Resistance to blood flow in the pulmonary circulation is now less than that in the systemic circulation. More blood flows through the lungs and returns to the left atrium than during fetal life. The valve in the foramen ovale is pushed against the interatrial septum and soon adheres to it by the growth of tissue. This by-pass of the lungs is cut off.

The circulatory pattern is now very close to the adult condition except that the ductus arteriosus remains open (Fig. 14.16B). Since pul-

monary resistance is less than systemic resistance, the direction of flow in the ductus arteriosus is reversed. Some of the blood that has already been through the lungs and is leaving the heart in the arch of the aorta flows back to the lungs through the ductus arteriosus. This pattern of circulation, which lasts from several hours to a day or two in the human infant, is known as the **neonatal circulation.** Experiments on newborn lambs show that this reversal of flow, and the consequent double aeration of some of the blood at a time when the lungs are not functioning at maximal efficiency, is of great significance. If the ductus arteriosus is experimentally tied off during this period, the hemoglobin is 10 to 20 per cent less saturated with oxygen than it normally is.

Muscles in the wall of the ductus arteriosus eventually contract and stop all blood flow through it, and the adult pattern is established (Fig. 14.16C). Eventually, the duct becomes permanently occluded by the growth of fibrous tissue into its lumen. The stimulus for these changes is unknown, but if the duct remains open, an undue strain is placed upon the heart, for it must pump this extra amount of blood that is recirculating through the lungs in addition to the normal amount of blood to the tissues.

14.12 THE PROPULSION OF BLOOD AND HEMOLYMPH

Hearts. All animals with circulatory systems have a muscular pump of some type that

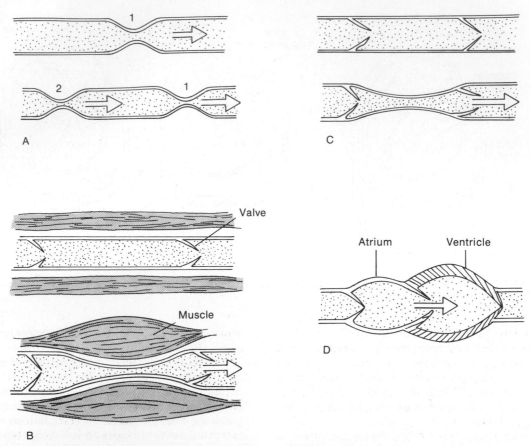

Figure 14.17 Circulatory pumping mechanisms. *A*, A wave of peristaltic contraction (1) travels along the peristaltic tubular heart and is soon followed by another (2). *B*, Contraction and bulging of surrounding muscles squeeze upon the valved vein of a vertebrate. *C*, A portion of the valved tubular heart of an insect. *D*, The chambered heart of a mollusk or vertebrate. The atrium is contracting, and the ventricle is filling. Arrows indicate the direction of blood flow.

propels the blood or hemolymph. In its simplest form this is a pulsating vessel in which blood is pushed along by waves of peristaltic contraction (Fig. 14.17*A*). Valves are not present. **Peristaltic tubular hearts** of this type are found in some primitive worms. Valves are present within other pumping mechanisms and insure that blood moves only in one direction. The motive force comes either from the contraction of surrounding body muscles that bulge and squeeze the vessel (Fig. 14.17*B*) or from muscles within the wall of the vessel itself (Fig. 14.17*C* and *D*). The former mechanism occurs in the veins of vertebrates and is important in returning low pressure venous blood to the heart. The latter mechanism is exemplified by the long, **valved tubular hearts** of insects and the **chambered hearts** of invertebrates and vertebrates. Some chambered hearts have only a single enlarged chamber, as in crustaceans, but more often there are two or more in series with valves between them, as in mollusks and vertebrates. The heart may lie

within a cavity that facilitates its contraction and expansion. The **pericardial sinus** of arthropods is a part of the hemocoel, but the **pericardial cavity** of mollusks and vertebrates is a part of the coelom. The various types of pumping mechanism are not mutually exclusive, and it is not unusual to find more than one in the same animal. In addition to the primary heart, accessory hearts frequently develop where an increase in blood or hemolymph pressure is necessary: the base of the gills in some mollusks or the base of the legs in some arthropods.

Usually, the residual pressure in fluid returning to a chambered heart is sufficient for the blood to enter. The first chamber of the heart is thin-walled, so that it easily expands as fluid enters. It contracts with force sufficient only to drive the fluid into succeeding chambers, where enough pressure is developed to deliver the blood or hemolymph to the tissues and back to the heart. Sometimes, however, the pressure is so low in vessels or spaces just before the heart that the heart, or its first chamber, must actively expand and literally suck fluid in. Arthropods have a **suction heart.** Thin muscle strands extend from the outside of the heart to the wall of the hemolymph-filled pericardial sinus in which it lies. Their contraction enlarges the single-chambered heart, and blood is sucked in through the ostia. Valves at the bases of the aortae leaving the heart close, so that hemolymph is not drawn back into the heart from them. When the heart contracts, they open and valves in the ostia close. Cartilaginous fishes, such as sharks, have evolved a different type of suction heart. The heart lies within a pericardial cavity surrounded by cartilages and large muscle masses. The cavity does not collapse when the ventricle of the heart contracts as it pumps blood to the body; therefore, as the ventricle gets smaller, pressure is decreased within the pericardial cavity. The thin-walled sinus venosus and atrium expand into the area of reduced pressure and suck blood in from the veins.

The Mammalian Heart. The heart of mammals and most other vertebrates is not a suction heart, for it receives blood under low pressure from the veins. The heart lies within the pericardial cavity and is covered with a smooth coelomic epithelium, the **visceral pericardium.** It is lined by the simple squamous epithelium, the **endothelium,** which lines all parts of the circulatory system. The rest of its wall is composed of **cardiac muscle,** which is unique in that its fibers branch and anastomose profusely (see Fig. 3.24). Prominent cross bands occur periodically. For a long time their nature was a puzzle, but electron micrographs (Fig. 14.18) reveal that cells uniting end-to-end interdigitate complexly with each other in this region. This specialization joins the cellular units more tightly and enables them to withstand better the tensions that are continually developed as the heart beats. The musculature of the atria is separated from that of the ventricles, but the individual muscle cells or fibers in each are more intimately united than in other types of muscle. Atria or ventricles respond as a unit. Any stimulus that is strong enough to elicit a response will elicit a total response. Thus, the atria and ventricles follow the "all-or-none" law that applies to individual motor units of phasic skeletal muscles (see Section 11.7).

During a heart cycle, the atria and ventricles contract and relax in succession. Contraction of these chambers is known as **systole;** relaxation, as **diastole.** Ventricular systole is

Figure 14.18 Electron micrograph of a portion of the end-to-end union (intercalated disc) of two cardiac muscle cells or fibers to show the interdigitation of their plasma membranes. Parts of three myofibrils can be seen in each muscle fiber. (From Bloom and Fawcett.)

very powerful and drives the blood out into the pulmonary artery and arch of the aorta under high pressure. Since the muscle fibers of the ventricles are arranged in a spiral, the blood is not just pushed out but is virtually wrung out of them. When the ventricles relax, their elastic recoil reduces the pressure within them, and blood enters from the atria. Atrial contraction does not occur until the ventricles are nearly filled with blood. The atria are primarily antechambers that accumulate blood during ventricular systole.

Blood being pumped by the heart is prevented from moving backward by the closure of a system of valves (Fig. 14.19). One with three cusps, known as the **tricuspid valve,** lies between the right atrium and ventricle; one with two cusps, the **bicuspid valve,** between the left chambers. These valves operate automatically as pressures change, opening when atrial pressure is greater than ventricular, closing when ventricular pressure is greater. **Tendinous cords** extend from the free margins of the cusps to the ventricular wall and prevent them from turning into the atria during

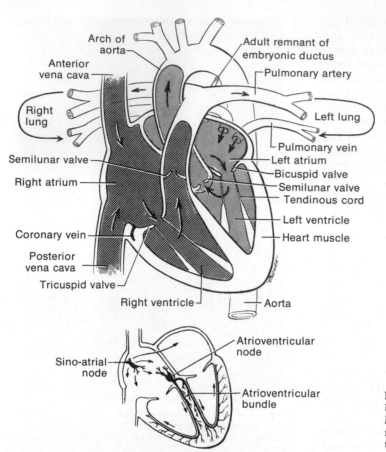

Figure 14.19 The adult mammalian heart. *Upper*, course of blood through the heart; *lower*, distribution of the specialized cardiac muscle that forms the conducting system of the heart.

the powerful ventricular contractions. When the ventricles relax, blood in the pulmonary artery and aorta, which is under pressure, tends to back up into them. This closes the pocket-shaped **semilunar valves** at the base of these vessels that prevent blood from them returning to the ventricles. Abnormalities in the structure of the valves occurring congenitally or produced by disease may prevent their closing properly. Blood then leaks back during diastole; the leaking blood is heard as a "heart murmur."

Heart Beat and Its Integration. Heart musculature has an inherent capacity for beating, and a heart, if properly cultured, will continue to beat rhythmically when excised from the body. An integrating system within the heart initiates the beat and stimulates successive chambers in turn. In mammals, each contraction is initiated in the **sinoatrial node,** or "pacemaker"—a node of specialized cardiac muscle **(Purkinje fibers)** located in that part of the wall of the right atrium into which the primitive sinus venosus is incorporated (Fig. 14.19). The impulse spreads to all parts of the atria and to another node of Purkinje fibers, the **atrioventricular node,** from which it con-

tinues along pathways of Purkinje fibers to all parts of the ventricles. Since there is no muscular connection between atria and ventricles, an impulse can reach the ventricular muscles only through the Purkinje fibers. It does so very rapidly, so that ventricular contraction begins at the apex of the heart and spreads quickly toward the origin of the great arteries leaving the heart.

The rhythmicity of the sino-atrial node results from the leakage of positively charged sodium ions, which are abundant outside all cells, through the plasma membranes of the cells of the node. The resting electrical potential of the membrane is reversed and the cells become active. A similar phenomenon occurs when nerve cells and other muscle fibers become active. As sodium ions are pumped out of the cell, the original resting potential of the membrane is restored. The unique feature of the cells of the sino-atrial node is that sodium ions begin to leak in again after the resting potential has been re-established without the specific stimulus that is needed to activate nerve or other muscle cells. The cycle of inward leakage and outward pumping of sodium ions accounts for the rhythmicity of the sino-atrial node.

Cardiac Output and Control. Cardiac output, defined as the volume of blood or hemolymph pumped to the body in a unit of time, is adapted to body size and the changing metabolic needs of a particular animal. Theoretically, output can be varied by changes in heart size, and hence volume, or by changes in the rate of heart beat (the pulse). One might expect small animals to have relatively larger hearts than large animals because their metabolism is relatively higher (surface-volume relationships again), but heart size appears to vary little among closely related animals of different size. The heart of a shrew and the heart of an elephant each constitute about 0.6 per cent of body weight. Volume of output, however, can be increased slightly during exercise, for the increased pressure and more rapid return of venous blood stretches the heart musculature. This causes the heart to contract with greater force and to send out the greater volume of blood received during each period of atrial diastole. Within physiological limits, the greater the tension on cardiac (or any other) muscle, the more powerful will be its contraction. This capacity of the heart to adjust its output per stroke to the volume of blood delivered to it is known as *Starling's law of the heart.* Pulse rate is the more important variable in controlling output. Small animals meet their greater metabolic needs in part by a very rapid heart beat. The pulse rate of a shrew at rest is 600 beats per minute, whereas that of an elephant is only 25 per minute. Rate may increase several times during exercise, but the rate of increase is not directly proportional to the increased oxygen needs. The blood itself delivers more oxygen because the oxygen tension in active tissues drops, and a greater proportion of the oxygen carried is unloaded by the blood (see Section 13.6).

The hearts of annelids and mollusks are primarily **myogenic.** That is, heart beat initiates in the heart musculature itself, and rate of beat is influenced by extrinsic factors that affect the musculature directly. Increases in temperature or of the pressure of the fluid entering the heart cause an acceleration. Inhibiting and accelerating neurons terminate on the molluskan heart but are of less importance in regulating beat.

The hearts of arthropods and vertebrates, although having an underlying myogenic rhythm, are **neurogenic** in their control. Nerve impulses terminating on the heart normally initiate the beat, and complex reflexes control acceleration and deceleration. In mammals, motor neurons that increase or decrease the heart rate extend from a **vasomotor center** in the medulla of the brain to the sino-atrial node. Sensory impulses from many parts of the body influence the vasomotor center. For

Figure 14.20 Diagram of major nerve reflexes controlling heart output and blood pressure. Motor nerve fibers have been drawn more heavily than sensory fibers. The motor nerves to the heart are parts of the autonomic nervous system, which is discussed more fully in Chapter 17.

example, receptors in the right atrium are stimulated by the increase in the pressure of the venous blood returning to the heart which occurs during exercise. They travel to the vasomotor center (1, Fig. 14.20) and, by inhibiting the depressor nerve fibers and activating accelerator fibers, cause an increased number of nerve impulses to go out on the accelerator nerve (2). The rate of heart beat and strength of contraction increase to accommodate the greater return of venous blood. Blood pressure also rises and, if it exceeds a certain threshold, pressure receptors in the aortic arch and other vessels near the heart are activated. Impulses travel to the vasomotor center (3) and, again by an appropriate combination of inhibition and excitation, the rate of impulses traveling down the accelerator nerve is decreased and an increased number go out on the depressor nerve (4). Heart rate and force of contraction are reduced to appropriate levels. Other reflexes are initiated by chemoreceptors. A reduction in the level of oxygen in the arterial blood is detected by receptors in the carotid body (5), and this can lead to an increase in heart rate (6). Yet other extrinsic factors, such as temperature, carbon dioxide content of the blood and one's emotional state, can influence heart rate by acting on the vasomotor center.

The heart of a normal adult human being who is not exercising sends about 70 ml. of blood per beat out into the aorta. At the normal rate of 72 beats per minute, this is a total output of 5 liters per minute, which is approximately equivalent to the total amount of blood in the body. A similar observation made in 1628 by William Harvey led him to conclude that the blood recirculates. Until that time it was believed that blood was continually produced in the liver, pumped to the tissues and consumed. Harvey's calculations showed that the amount of blood pumped by the heart each hour was much more than could possibly be produced and consumed. He made the correct inference that the blood must recirculate, even though he could not see the microscopic capillaries that connect arteries and veins.

Although a large volume of blood flows through the cavities of the heart, this blood may not provide for the metabolic needs of the heart musculature. In mammals, a pair of **coronary arteries** arise from the base of the arch of the aorta and supply capillaries in the heart wall. This capillary bed is drained ultimately by a **coronary vein** that empties into the right atrium. Obviously, any damage to the coronary vessels, the plugging of one of the larger arteries by a thrombus, for example, could have serious consequences, for the heart muscles cannot function without a continuing supply of oxygen and food.

14.13   THE PERIPHERAL FLOW OF BLOOD AND HEMOLYMPH

Arteries.   Arteries are vessels leading from the heart to capillaries or to hemocoel spaces. Usually the fluid within them is rich in oxygen and low in carbon dioxide, but the gas content of vessels depends on the site of external gas exchange. The pulmonary artery of an adult mammal and the ventral aorta of a fish contain blood low in oxygen, for the blood is on the way to the respiratory membrane. The arteries of vertebrates are lined with endothelium and have a relatively thick wall containing elastic connective tissue and smooth muscle. Invertebrate arteries are similar but lack muscle unless they are pulsating vessels. The walls of the large arteries leaving the heart of vertebrates are richly supplied with elastic tissue. The force of each ventricular systole forces blood into the arteries and stretches them to accommodate it. During diastole, the elastic recoil of the first part of the artery to expand helps to push the blood into the adjacent part of the artery, which in turn expands. If the arteries were rigid pipes, they would deliver blood to the tissues in spurts that coincided with ventricular systole. The blood would pound like steam rushing into empty radiator pipes. The elasticity of the larger arteries transforms what would otherwise be an intermittent flow into a steady flow. The alternate stretching and contracting of the arteries travels peripherally, very rapidly (7.5 meters per second) and can be detected as the **pulse,** but the blood itself does not move as fast.

The smaller arteries (Fig. 14.21), and especially the **arterioles** preceding the capillaries, contain a relatively large amount of smooth muscle, and they regulate the supply of blood to the various organs. **Vasodilator** and **vasoconstrictor nerves** supply these muscles, causing them to relax or contract. If a region of the body becomes very active, its small arteries enlarge, and the blood flow through them is increased. If an area is not particularly active, its small arteries constrict, and blood flow is reduced.

As the arteries extend to the tissues, they branch and rebranch. At each branch, the lumen becomes smaller, but the total cross

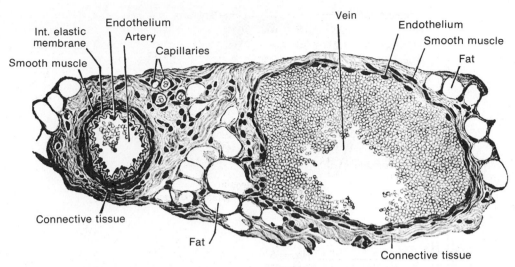

Figure 14.21 A drawing of a cross section through a small artery and an accompanying vein. Several capillaries lie between them. (Modified from Bloom and Fawcett.)

sectional area of all these branches increases greatly. The velocity of blood flow, therefore, decreases, for the blood, like a river widening out and flowing into a lake, is moving into an area that grows larger and larger. The mean blood pressure is also decreased continually because of the friction of the blood moving in the vessels (Fig. 14.22). Blood pressure continues to decrease as the blood flows through the capillaries and veins. The rate of flow, however, increases as the blood passes from the capillaries to the venules, and as these smaller veins lead into fewer larger ones. The blood is now moving into a smaller and smaller area and, like water flowing out of a lake into a narrowing river, moves faster and faster.

Capillary Exchange. In animals with

Figure 14.22 Variation in blood velocity and pressure in different parts of the cardiovascular system. The velocity does not return to its original value, for the cross sectional area of the veins is greater than the cross sectional area of the arteries. The blood pressure in the veins near the heart is less than atmospheric pressure because of the negative pressure within the thorax.

Figure 14.23 Capillary exchange. *A*, Diagram of the exchange of materials between the blood and interstitial fluid in a capillary bed. *B*, Graph of the forces within a capillary responsible for the movement of water in and out. The forces graphed are net forces, i.e., the difference between the true hydrostatic (Hp) and colloidal osmotic pressures (Cop) in the plasma and comparable forces (which are rather low) in the interstitial fluid.

open circulatory systems, the hemolymph flows from arteries into the hemocoel and directly bathes all of the tissues and cells. Cells are separated from the hemolymph only by their plasma membranes. In animals with closed systems, blood flows from arteries into capillaries, and exchanges must occur between the blood and interstitial fluid through the capillary walls (Fig. 14.23A). Capillaries are small, numerous and exceedingly thin-walled vessels. Their diameter is about that of blood cells, and they frequently are spaced very close together. Studies of sections of skeletal muscle suggest that there are some 240,000 capillaries per square centimeter. Blood does not flow evenly through all parts of a capillary bed. Within a capillary bed there are certain **thoroughfare channels** through which some blood flows from arterioles to venules all of the time. Tiny, muscular **precapillary sphincters** surround the beginning of the capillaries that branch off of these channels, so parts of

the bed may be open or closed according to the needs of the tissues.

Capillaries are somewhat more permeable than the plasma membrane of cells and behave as though they had pores about 8 nm. in diameter (in contrast to the 0.8 nm. diameter pores in the plasma membrane). Oxygen, glucose, amino acids, various ions and other needed substances easily diffuse through the capillary walls into the interstitial fluid following their concentration gradients. Conversely, carbon dioxide, nitrogenous wastes and other by-products of metabolism easily diffuse into the blood.

Water moves each way, and maintaining the proper amount in the interstitial fluid involves a combination of forces. As we have seen, most of the large plasma proteins remain in the blood, where they exert an osmotic force called the **colloidal osmotic pressure** that draws water into the capillaries. An opposite force, the blood or **hydrostatic pressure,** forces water out. At the arterial end of the capillary bed, hydrostatic pressure exceeds the colloidal osmotic pressure, so that there is a net force (called the **filtration pressure**) that moves water out of the capillaries (Fig. 14.23B). As blood moves through the capillary bed, friction causes a great reduction of the hydrostatic pressure. Some reduction in the colloidal osmotic pressure also occurs because of the loss of some plasma proteins, but the reduction is not as great. At the venous end of a capillary bed, colloidal osmotic pressure exceeds hydrostatic pressure so that there is a net force (the **absorption pressure)** that draws water into the capillaries. The absorption pressure is frequently a bit lower than the filtration pressure, but the tissues do not flood. Capillaries have somewhat more absorption surface at their venous end through which more water can enter. Terrestrial vertebrates, whose blood pressure is somewhat higher than that of fishes, also have a system of **lymphatic capillaries** (p. 331).

Venous and Lymphatic Return. Veins are vessels that return blood to the heart. In animals with open systems the venous channels do not have an endothelial lining, although they may be lined with connective tissue. Their structure in vertebrates is fundamentally similar to that of arteries, though a vein is larger and has a much thinner and more flaccid wall than its companion artery (Fig. 14.21). Since they are larger, the veins hold more blood than the arteries and are an important reservoir for blood. Lymphatic vessels have even thinner walls. Valves present in both veins and lymphatics permit the blood and lymph to flow only toward the heart.

Though blood pressure is low in the veins (Fig. 14.22) and lowest in the large veins near the heart, it is still the major factor in the return of blood to the heart in a mammal. Two other factors assist it. One is the fact that the elastic lungs are always stretched to some extent and tend to contract and pull away from the walls of the pleural cavities. This creates a slight subatmospheric or negative pressure within the thoracic cavity that is greatest during inspiration. The larger veins, of course, pass through the thorax, and the reduction of pressure around them decreases the pressure within them and increases the pressure gradient. The other factor is that the contraction and relaxation of body muscles exert a "milking" action on the veins and force the blood toward the heart. All these factors increase during exercise, which makes for a more rapid return of blood and an increased cardiac output.

The return of lymph is dependent upon similar forces. The interstitial fluid itself has a certain pressure derived from the flow of liquid out of the capillaries. This establishes a pressure gradient in the lymphatics that is made steeper by the negative intrathoracic pressure. The "milking" action of surrounding muscles and, for lymphatics returning from the intestine, the movement of the villi, help considerably. Some lower vertebrates have lymph "hearts" — specialized pulsating segments of lymphatic vessels.

ANNOTATED REFERENCES

Attention is again directed to the general references on vertebrate organ systems cited at the end of Chapter 10.

Cooper, M. D., and A. R. Lawton, III: The development of the immune system. Scientific American 234:58, (Nov.) 1974. The differentiation of T- and B-lymphocytes and their vital roles in immune responses are discussed.

Harvey, W.: Anatomical Studies on the Motion of the Heart and Blood. Springfield, Ill., Charles C Thomas, 1931. This is one of a number of available translations of Harvey's important studies in 1628 which established the circulation of the blood and introduced the experimental method into biology.

Krogh, A.: The Anatomy and Physiology of Capillaries. New Haven, Yale University Press, 1922. The classic study on this subject written by one of the pioneer investigators of capillary circulation.

Mayerson, H. S.: The lymphatic system. Scientific American 208:80, (June) 1963. The importance of this second drainage system of the tissues is thoroughly discussed.

Muir, A. R.: The Mammalian Heart. London, Oxford University Press, 1971. The structure and arrangement of cardiac muscle and the control of its activity are emphasized in this Oxford Biology Reader.

Weiner, A. S.: Blood Groups and Blood Transfusions. 3rd ed. Springfield, Ill., Charles C Thomas, 1943. An old but classic treatise which traces the history of the discovery of blood groups and their applications to problems of transfusion, anthropology, disputed paternity and forensic medicine.

Wiggers, C. J.: The heart. Scientific American 196:87, (May) 1957. A fine discussion of the activities of the heart and the safety factors that enable it to continue operating even though partially impaired by coronary disease.

Zweifach, B. W.: The microcirculation of the blood. Scientific American 200:54, (Jan.) 1959. A discussion of the factors that control circulation in capillary beds.

CHAPTER 15

REGULATION OF INTERNAL BODY FLUIDS

Living organisms are composed mostly of water. A 70 kilogram adult human contains about 49 liters of water; about 35 liters are within his cells, and the rest is extracellular body fluids. Some of these extracellular fluids lie in the minute spaces between cells; others may be located in special cavities, such as the coelom; and a large amount may be present in the blood vascular system (p. 313). These extracellular fluids are a dynamic part of the internal environment of the animal. They contain salts and proteins (Fig. 15.1); they are in osmotic balance with the intracellular environment; and they are routes for the passage of food, gases and wastes. The composition of body fluids, both intracellular and extracellular, must be maintained within relatively narrow limits, regardless of the external environment.

The regulation of internal body fluids poses a number of physiological problems, as the following story illustrates. Two sailors adrift in a life boat are survivors of a ship that had sunk in the South Pacific. They are dying of thirst,

	EXTRACELLULAR FLUID	INTRACELLULAR FLUID
Na⁺	137 mEq/l.	10 mEq/l.
K⁺	5 mEq/l.	141 mEq/l.
Ca⁺⁺	5 mEq/l.	0 mEq/l.
Mg⁺⁺	3 mEq/l.	62 mEq/l.
Cl⁻	103 mEq/l.	4 mEq/l.
HCO₃⁻	28 mEq/l.	10 mEq/l.
Phosphates	4 mEq/l.	75 mEq/l.
SO₄⁻⁻	1 mEq/l.	2 mEq/l.
Glucose	90 mgm. %	0 to 20 mgm. %
Amino acids	30 mgm. %	200 mgm. % ?
Cholesterol Phospholipids Neutral fat	0.5 gm. %	2 to 95 gm. %
Po₂	35 mm. Hg	20 mm. Hg ?
Pco₂	46 mm. Hg	50 mm. Hg ?
pH	7.4	7.1 ?

Figure 15.1 Diagram illustrating the chemical composition of extracellular fluids and intracellular fluid. (Redrawn after Guyton, A. C.: Function of the Human Body, 4th ed. Philadelphia, W. B. Saunders Company, 1974.)

surrounded by water that they cannot drink. The salt content of sea water is about 3.5 per cent, well above the maximal salt content of urine that the human kidney can excrete. Under the hot tropical sun, they sweat and lose more water. A line from the boat enables them to catch fish for food. But the fish protein, when used as an energy source, releases nitrogenous waste that requires more water in order to be excreted. Both sailors die. If they had cut up the fish and drunk the fluids squeezed out through a piece of shirt, they would have survived considerably longer.

This story serves to illustrate four factors that complicate regulation of internal body fluids: (1) the salt content of the environment in which an animal lives: (2) the water demands of special physiological processes; (3) the need to maintain internal salt balance; and (4) the elimination of nitrogenous wastes. These complications are interrelated, but since excretory organs are involved with other aspects of regulation, in addition to the elimination of wastes, we will begin with excretion.

15.1 NITROGENOUS WASTES

Protein Metabolism. The continual breakdown of organic compounds within living systems results in *metabolic wastes*. A major waste is carbon dioxide, which is derived from the decarboxylation of metabolites at various points in cell respiration. The degradation of nucleic acids and proteins produces nitrogenous wastes, most of which are formed in the utilization of amino acids as fuels. Amino acids are deaminated in a reaction that replaces the amino group with an oxygen atom, converting the amino acid to a keto acid.

$$
\begin{array}{ccc}
CH_3 & & CH_3 \\
| & & | \\
HC\,NH_2 & \longrightarrow & C{=}O \quad +NH_3 \\
| & & | \\
COOH & & COOH \\
\text{Alanine} & & \text{Pyruvic acid \quad Ammonia}
\end{array}
$$

The keto acid is then metabolized in the TCA cycle as a source of biologically useful energy or is converted to other compounds. Pyruvic, α-ketoglutaric and oxaloacetic acids are keto acids and principal points of entry of the carbon skeletons of amino acids into the respiratory pathways (Fig. 12.16). Each amino acid has a corresponding α-keto acid, and the processes of deamination and transamination convert the carbon chain of the amino acid into the corresponding α-keto acid. Some of the α-keto acids, such as pyruvate, α-ketoglutarate and oxaloacetate, feed directly into the TCA cycle. Other α-keto acids may require a series of reactions with many intermediate steps before resulting in a compound that enters the respiratory cycle. Note in Figure 12.16 that not all compounds enter as keto acids.

Most of the pathways for the oxidation of amino acids are different from those involved in the synthesis of amino acids. There are, however, some shared reversible reactions. The addition of an amino group to α-ketoglutaric acid yields the amino acid glutamic acid.

$$
\begin{array}{ccc}
COOH & & COOH \\
| & & | \\
CH_2 & & CH_2 \\
| & & | \\
CH_2 \quad +NH_3 \longrightarrow & & CH_2 + H_2O + NAD^+ \\
| & & | \\
C{=}O \quad + & & H{-}C\,NH_2 \\
| & NADH+H^+ & | \\
COOH & & COOH \\
\alpha\text{-Ketoglutaric acid} & & \text{Glutamic acid}
\end{array}
$$

The liver and kidney are major sites of amino acid metabolism in vertebrates. (Remember that the liver first receives the products of digestion from the intestine.)

Forms of Nitrogenous Wastes. The amino group removed in deamination appears as a molecule of ammonia, a toxic compound that cannot be permitted to accumulate and must be eliminated. Many animals, including most fish, eliminate the ammonia directly (ammonotelic). Other animals eliminate more complex but less toxic nitrogenous wastes synthesized from ammonia, a process that requires energy, and thus has an ATP cost. Urea, a simple compound synthesized from carbon dioxide, two amino groups and three ATPs, is the waste product of mammals, amphibians and some fish. Urea synthesis takes place within the liver and involves a cyclic series of reactions and intermediate compounds, including ornithine, citrulline and arginine (Fig. 15.3).

Uric acid, the principal waste product of birds, terrestrial reptiles and land snails, is produced in humans and other mammals from the degradation of nucleic acids. It is a slightly soluble purine, produced by the oxidation of adenine and guanine (Fig. 15.2). Ammonia, urea and uric acid are the most common forms of nitrogenous wastes eliminated by animals; spiders, however, excrete guanine.

$$NH_3 \qquad CH_4ON_2 \qquad C_5H_4O_3N_4 \qquad C_5H_6ON_5$$

AMMONIA	UREA	URIC ACID	GUANINE
Most aquatic animals	Sharks and rays Some fish Amphibians Mammals	Land snails Insects Turtles Lizards Snakes Birds	Spiders

Figure 15.2 The principal nitrogenous waste products of various animal groups.

15.2 EXCRETORY ORGANS

Very small aquatic animals have sufficient surface area in relation to volume to permit the elimination of ammonia by simple diffusion to the exterior. Even large aquatic animals may lose some ammonia by way of the gill surface (p. 297); however, animals with a volume greater than a few cubic centimeters require excretory organs for the removal of their nitrogenous wastes.

Excretory organs share a number of structural and functional features. They are typically tubular or saccular structures adapted for concentrating wastes. In one type, the inner end of the tubule opens to the coelom (Fig. 15.4A).

Wastes are filtered from the blood into the coelomic fluid, which passes in turn into the excretory tubule. As the coelomic fluid passes down the tubule, it is subjected to varying degrees of **selective reabsorption.** Some organic substances, such as sugars and amino acids, may be reabsorbed, and, depending on the environment of the animal, there may be some reabsorption of salts and water. In addition to selective reabsorption, secretion of wastes by the tubule wall may occur. The secretion and reabsorption of materials from the lumen of the tubule are facilitated by the blood or blood vessels surrounding the tubule. As a result of selective reabsorption and secretion, the wastes within the tubular fluid be-

Figure 15.3 Cycle of reactions involved in the synthesis of urea. Two amino groups are required. One is provided by a molecule of ammonia, which combines with CO_2 and ATP in the mitochondria to form carbamoyl phosphate. The other amino group is provided by aspartate, which received it by transamination from glutamate. When urea is split off from arginine, the ornithine that remains is free to combine with carbamoyl phosphate to form more citrulline.

Figure 15.4 Models of the two principal types of excretory organs in animals, based on their function. In type *A*, the excretory tubule opens into the coelom, and wastes are concentrated from coelomic fluid, which receives filtrate from the blood. In type *B*, the excretory tubule is closed, and wastes are concentrated from blood filtrate.

come more concentrated, and the final mixture of water, wastes and salts is expelled to the exterior as **urine.**

In a second type of excretory organ, the tubule does not open into the coelom (Fig. 15.4*B*). Rather, the blind end of the tubule or saccule receives filtrate directly from the blood. The blood filtrate is modified and concentrated by selective reabsorption and secretion during its passage through the tubule. In neither type of excretory organ are wastes filtered from the blood. Remember that blood filtrate is not waste; it is simply the fluid portion of blood (minus large molecules and cells) that can pass through the capillary wall. *Excretory organs do not filter wastes; they concentrate them.*

The excretory organs of earthworms belong

to the first type. Each of the two tubules, or **metanephridia,** per segment arises in a ciliated funnel, or **nephrostome,** which opens into the coelom of the preceding segment (Fig. 15.5). The tubule perforates the septum and extends back into the following segment, where it may loop or coil before passing through the body wall. The external opening is called a **nephridiopore.**

Snails and clams possess very similar metanephridia, but since these animals are not metameric, only one or two are present.

Some small animals, including flatworms and rotifers, possess blind tubules called protonephridia. The blind end bears a tuft of cilia or a flagellum and is called a flame bulb or flame cell, depending upon its structure (Fig. 15.6). The wall of this terminal part of the tubule is partially perforated by slits. A cell membrane covers each slit, much like a pane of glass over a window. The protonephridial tubule opens to the exterior by way of a nephridiopore. Although often considered to be excretory organs, these tubules are usually not involved in the excretion of nitrogenous wastes but may function as pumps in the removal of excess water.

The excretory organs of crayfish, called green glands, belong to the second type. Each of these paired organs consists of a saccule located in the head and bathed in blood (Fig. 15.7). Filtrates pass from the blood through the wall of the sac and then down a tubule. Selective reabsorption occurs in certain parts of the tubule and bladder, and the resulting urine is expelled to the outside through an opening at the base of each antenna.

Malpighian tubules of insects and spiders are also blind-ending tubules bathed in blood (Fig. 12.8). Unlike crustacean green glands,

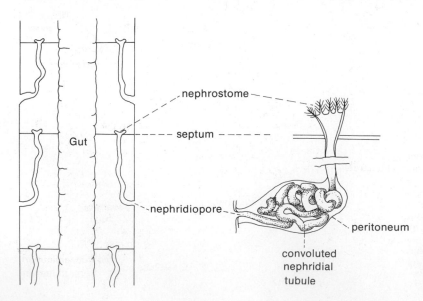

Figure 15.5 Annelid metanephridia. *A*, Diagrammatic dorsal view of two segments showing position of nephridia. *B*, One nephridium of *Nereis vexillosa*, a marine annelid in which the nephridial tubule is greatly coiled. (Modified from Jones.)

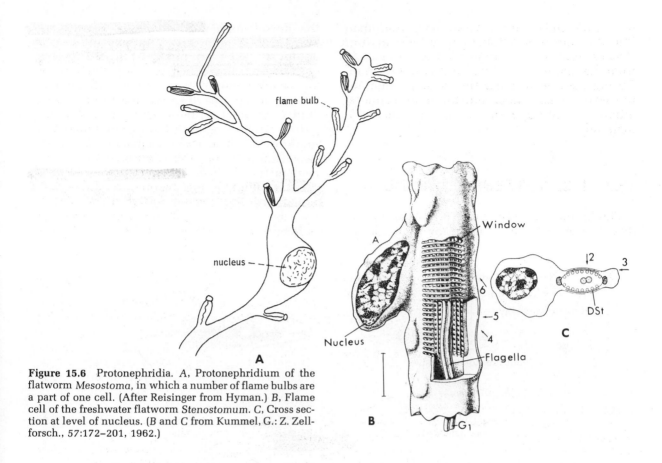

Figure 15.6 Protonephridia. *A*, Protonephridium of the flatworm *Mesostoma*, in which a number of flame bulbs are a part of one cell. (After Reisinger from Hyman.) *B*, Flame cell of the freshwater flatworm *Stenostomum*. *C*, Cross section at level of nucleus. (*B* and *C* from Kummel, G.: Z. Zellforsch., *57*:172–201, 1962.)

Figure 15.7 Antennal gland (excretory organ) of a decapod crustacean. *A*. Entire excretory organ. Nephridiopore opens onto base of second antenna. *B*. Diagrammatic section through end sac wall, showing site of filtration. (Redrawn from Kummel by A. P. M. Lockwood, 1967: Physiology of Crustacea. W. H. Freeman and Co., San Francisco.)

they can number from two to many and open into the intestine rather than to the exterior. Wastes concentrated from the blood thus pass from the tubules into the intestine and rectum. During passage through the rectum, much of the water is absorbed, and the pasty urine is eliminated through the anus with the fecal material.

15.3   THE VERTEBRATE KIDNEY

The kidneys of vertebrates are paired organs that lie dorsal to the coelom on each side of

the dorsal aorta. All vertebrate kidneys are composed of units called kidney tubules, or nephrons, which are blind-ending and receive a filtrate from the blood. However, the number and arrangement of the nephrons differ among the various groups of vertebrates. Comparative studies have led to the conclusion that in ancestral vertebrates each kidney contained one nephron for each body segment that lay between the anterior and posterior ends of the coelom (Fig. 15.8A). These nephrons drained into a **wolffian,** or **archinephric, duct** that continued posteriorly to the cloaca. Such a kidney may be regarded as a complete kidney, or **holonephros,** for it extends the entire length of

Figure 15.8   A comparison of the evolution and embryonic development of the kidney and its ducts. *A, B* and *C*: The evolutionary sequence of kidneys. *A*, Hypothetical ancestral vertebrate with a holonephros; *B*, a fish with an opisthonephros; *C*, a reptile with a metanephros. *D* and *E*, The developmental sequence of kidneys in a reptile. *F*, A mammalian embryo in which the cloaca is becoming divided by the growth of the fold indicated by the arrow. The ventral part of the cloaca contributes to the urethra in the male. It becomes further subdivided in the female and contributes to both urethra and vagina. In both sexes, the dorsal part of the cloaca forms the rectum.

the coelom. A holonephros is found today in the larvae of certain cyclostomes but not in any adult vertebrate.

In the kidney of the adult fish and amphibian (Fig. 15.8*B*), the most anterior tubules have been lost, some of the middle tubules are associated with the testis, and there is a concentration and multiplication of tubules posteriorly. Such a kidney is known as a posterior kidney, or **opisthonephros.**

Reptiles, birds and mammals (Fig. 15.8*C*) have lost all the middle tubules not associated with the testis and have an even greater multiplication and posterior concentration of tubules. The number of nephrons is particularly large in birds and mammals; their high rate of metabolism yields a large amount of wastes. It is estimated that humans have about 1,000,000 nephrons per kidney, whereas certain salamanders have less than 100. The tubules producing urine drain into a **ureter,** which evolved as an outgrowth from the wolffian duct. The wolffian duct itself has been taken over completely by the male genital system. The kidney of the higher vertebrates is known as a **metanephros.**

The evolutionary sequence of kidneys is holonephros, opisthonephros and metanephros. In the embryonic development of vertebrates, we find a slightly different sequence, but one that also involves a posterior concentration of kidney functions (Fig. 15.8*D* and *E*). In an early embryo of a reptile, for example, segmentally arranged tubules appear dorsal to the anterior end of the coelom, form the wolffian duct and disappear. These transitory tubules constitute a **pronephros.** Then, a middle group of tubules, known as the **mesonephros,** appear and connect with the wolffian duct. These function during much of embryonic life, but when the metanephric tubules develop, all of the mesonephric tubules are lost, except for those associated with the testes. The embryonic sequence of kidneys in the development of a higher vertebrate is pronephros, mesonephros and metanephros.

Most tetrapods have a **urinary bladder** that develops as a ventral outgrowth from the cloaca. Generally, the excretory ducts from the kidneys lead to the dorsal part of the cloaca, and urine must flow across it to enter the bladder; however, in mammals (Fig. 15.8*F* and

Renal arteries & veins

Adrenal gland

Cortical substance

Pyramid
Medullary substance
Minor calyx
Major calyx

Pelvis

Left kidney

Aorta

Ureter

Inferior mesenteric artery

Inferior vena cava

Common iliac artery & vein

Bladder

Urethra

Figure 15.9 The human excretory system as seen in a ventral view. The right kidney has been sectioned to show the internal structures. (From Villee, C. A.: Biology. 7th ed. Philadelphia, W. B. Saunders, 1977.)

15.9) the ureters lead directly to the bladder, and the bladder opens to the body surface through a short tube, the **urethra.** The cloaca becomes divided and disappears as such in all but the most primitive mammals. The dorsal part of the cloaca forms the rectum, and the ventral part contributes to the urethra of higher mammals (Fig. 15.8F).

Urine is produced continually by the kidneys and is carried down the ureters by peristaltic contractions. It accumulates in the bladder, for a smooth muscle sphincter at the entrance of the urethra and a striated muscle sphincter located more distally along the urethra are closed. Urine is prevented from backing up into the ureters by valvelike folds of mucous membrane within the bladder. When the bladder becomes filled, stretch receptors are stimulated, and a reflex is initiated that leads to the contraction of the smooth muscles in the bladder wall and the relaxation of the smooth muscle sphincter. Relaxation of the striated muscle sphincter is a voluntary act.

Nephron Structure. The excretory ducts and the urinary bladder are important adjuncts to the kidneys, but the essential work of the system, the selective removal of materials from the blood, is performed by the individual kidney tubules. The mammalian nephron may be taken as an example. The proximal end of each nephron (Fig. 15.10) is known as **Bowman's capsule**. It is a hollow ball of squamous epithelial cells, one end of which has been pushed in by a knot of capillaries called a **glomerulus.** Bowman's capsule and the glomerulus constitute a **renal corpuscle.** The rest of the nephron is a tubule, largely composed of cuboidal epithelial cells and subdivided in mammals into a **proximal convoluted tubule,** a **loop of Henle** and a **distal convoluted tubule.** A **collecting tubule** receives the drainage of several nephrons and leads to the **renal pelvis,** an expansion within the kidney of the proximal end of the ureter (Fig. 15.9). The location of the different parts of a nephron within the kidney and their relationship to blood vessels have important functional consequences. As shown in Figure 15.10, the renal corpuscles and convoluted tubules lie in the outer **cortex** of the kidney, and a dense capillary network surrounds the convoluted tubules; the loops of Henle extend toward the center, or **medulla,** of the kidney. Most of the human nephrons extend only a short distance into the medulla, but about one-fifth of them (the **juxtamedullary nephrons**) have long loops of Henle that

Figure 15.10 A diagram of mammalian nephrons and associated blood vessels. A juxtamedullary nephron is shown on the left; a cortical nephron on the right. (After Smith.)

extend, along with the collecting tubules and capillary loops, far into the medulla. It is the convergence of these structures into subdivisions of the renal pelvis (the **calyces**) that forms the renal **pyramids** (Fig. 15.9).

Urine Formation. The kidney tubules produce **urine,** a watery solution containing waste products of metabolism removed from the blood but not plasma substances needed by the organism. The most abundant nitrogenous waste in humans and other mammals is urea, but lesser amounts of ammonia, uric acid and creatinine are present. The yellowish color of urine is due to **urochrome**, a pigment derived from the breakdown of hemoglobin and, hence, related to the bile pigments.

That the first step in urine formation is **glomerular filtration** was demonstrated in the 1920's by Dr. A. N. Richards who developed a micropipette technique for removing and analyzing minute samples of fluid from the lumen of Bowman's capsule. Water, various ions and small organic molecules, including simple sugars and amino acids as well as nitrogenous wastes, pass easily through the semipermeable membranes of the renal corpuscle. Filtration permits all small molecules that are free in the plasma to pass through, and they appear in the filtrate in the same concentration as in the plasma. Blood cells and large molecules, including fats and plasma proteins, remain in the blood. The only small molecules to be held in the blood are those bound to plasma proteins; among these are iron and other trace minerals and certain vitamins.

Estimates of the filtration rate are obtained by the use of **inulin,** a polysaccharide from artichokes that is filtered easily and is not removed or added to the filtrate as the fluid continues down the kidney tubules. For example, when inulin is injected intravenously until the plasma concentration reaches 1 mg. per ml., 130 mg. of inulin appear in the urine per minute. For this to occur, 130 ml. of plasma must be filtered each minute. This amounts to 188 liters of filtrate per day! Such a large volume is not surprising when one realizes that the kidneys have an unusually rich blood supply. They receive about one-quarter of the cardiac output in a mammal, which enables them to process the total blood volume every four or five minutes.

Materials are filtered from the glomeruli, as they are in other capillary beds, but the arrangement of the blood vessels is such that more material passes out of the glomerulus than returns to it. A glomerulus lies between an **afferent arteriole,** a branch of a renal arte-ry, and an **efferent arteriole**, which leads to capillaries on other parts of the tubule (Fig. 15.10). The efferent arteriole is the smaller of the two, and this ensures a high **filtration pressure** in the glomerular capillaries to drive water and small solute molecules from the blood. The filtration pressure exceeds the pressures within the kidney tubule and the osmotic pressure of the blood (both of which promote the return of fluid to the blood), by 20 mm. of mercury. As a consequence, there is a substantial net production of glomerular filtrate.

The volume of urine in humans is only about 1 per cent of the glomerular filtrate, and nearly all of the glucose, amino acids and ions present in the glomerular filtrate are taken back into the blood by **tubular reabsorption** as the filtrate passes down the tubules. Other substances may be added to the filtrate by **tubular secretion.** Two important techniques have helped to elucidate these processes. One is a study of **clearance rates.** For example, each minute human kidneys excrete an amount of urea equal to that present in 75 ml. of plasma. It can be said that 75 ml. of plasma has been cleared of urea, or that urea has a clearance rate of 75 ml. per minute. Other substances have different clearance rates. From inulin studies we know that the filtration rate is 130 ml. per minute. If the clearance rate of a freely filterable substance (such as urea) is less than this, some of this substance must be reabsorbed in the tubules; if the clearance rate exceeds 130 ml. per minute, then some of the substance must be added by tubular secretion.

Sites of reabsorption and secretion have been determined by the **stop-flow technique.** A catheter is inserted into the renal pelvis and clamped. As liquid in the tubules backs up, pressure increases and filtration stops, but the different parts of the tubules continue to act on the now stationary column of liquid. After a suitable interval, the catheter is opened and a series of samples are taken and analyzed. The first samples come from the collecting tubules, and successively later ones come from more and more proximal parts of the tubules. From such studies it is concluded that in mammals virtually all of the glucose and amino acids that pass into the filtrate, and some urea, are reabsorbed in the proximal convoluted tubule. Sodium, chloride and bicarbonate ions are reabsorbed in both the proximal and distal convoluted tubules. More than 99 per cent of the water in the glomerular filtrate is reabsorbed; this occurs at many levels in the tubules. Creatinine, ammonia, hydrogen and potassium ions and various drugs (penicillin) are among the few substances added to the filtrate by tubular secretion in mammals, and most of this occurs in distal parts of the tubule. Certain teleosts have lost their renal corpuscles, and tubular secretion is their only means of eliminating waste products.

Reabsorption, which plays such an important role in urine formation, involves both the passive diffusion of materials back into the capillaries surrounding the tubules and the active uptake of materials by the tubular cells and their secretion into the blood against a concentration gradient. This, of course, requires the expenditure of energy by the tubular cells. Apart from urea, most of the reabsorption of solutes that takes place in the proximal convoluted tubule is an active process because the concentration of these substances in the filtrate in this region is the same as their concentration in the blood. Those materials that can be actively reabsorbed are taken back in varying amounts, depending upon their concentration in the blood. If the concentration of one of these materials in the blood and glomerular filtrate rises above a certain level, known as the **renal threshold,** not all of it will be reabsorbed into the blood from the tubule, and the amount present in excess of the renal threshold will be excreted. The quantitative value of the renal threshold differs for different substances. In **diabetes mellitus** the impaired cellular utilization of glucose leads to a high concentration of glucose in the blood; the renal threshold for glucose (about 150 mg. of glucose per 100 ml. blood) is exceeded; and sugar appears in the urine in large amounts.

The osmotic pressure of the body fluids is controlled by the amount of salts that are returned to the blood from the glomerular filtrate; the pH of the fluids is regulated by the elimination or retention of basic and acidic substances.

Water is reabsorbed passively. As solutes are actively taken out of the filtrate in the proximal tubule, the water in the filtrate tends to become more concentrated than it is in the blood. However, water molecules passively diffuse out as fast as solutes are pumped out; hence the concentration of the filtrate remains the same as that of the blood. About 80 per cent of all the water taken back is reabsorbed in this way in the proximal tubule, but the tubular filtrate cannot be made more concentrated (i.e., to contain less water) than the blood by this mechanism.

In most of the lower vertebrates the urine is not more concentrated than the blood; however, birds and mammals do produce a hyperosmotic urine. The unique feature of the mammalian nephron is the loop of Henle, and for a long time it has been implicated in the ability of these vertebrates to produce a very concentrated urine, but the mechanism of its operation has only recently been discovered. One important factor is that the descending and ascending limbs of the loop of Henle lie parallel to each other, so that the direction of flow of fluid in one is opposite to that in the other (Fig. 15.11). Another factor is the active transport of sodium ions out of the ascending limb into the interstitial fluid and the passive diffusion of these ions back into the descending limb. Together these factors result in a countercurrent multiplying mechanism. Sodium

Figure 15.11 Diagram of the countercurrent multiplying mechanism of the mammalian kidney. The general direction of fluid movement is shown by broken arrows; active sodium transport, by heavy black arrows; passive sodium transport, by small black arrows; water movements, by white arrows. Numerals show the relative concentration of osmotically active solutes. In hundreds (add 00), they express the concentrations in milliosmols per liter.

pumped out of the ascending limb goes right back into the descending limb. The recycling of sodium in the loop of Henle, together with additional sodium being continually brought to the loop in the glomerular filtrate, results in an accumulation of sodium in the loop of Henle and the surrounding interstitial fluid of the medulla. A concentration gradient from cortex to medulla is established as shown in Figure 15.11. The degree of accumulation depends upon the length of the loop; there is more opportunity to pump out sodium in a long loop. The importance of the capillary loops associated with the juxtamedullary nephrons is that here too we have a countercurrent mechanism that permits most of the sodium that starts out of the medulla in the blood to diffuse back into the blood entering the medulla. This, combined with the rather sluggish rate of blood flow through these vascular loops, means that little sodium is carried away from the medulla by this route.

The osmotic gradient established by these mechanisms makes it possible for additional water to be reabsorbed passively and for a hypertonic urine to be produced. Water simply follows the osmotic gradient, moving from an area of low osmotic pressure (high concentration of water) to one of high osmotic pressure (low concentration of water). The glomerular filtrate, which was isotonic to the

blood is the proximal tubule, loses water as it passes down into the loop of Henle. It does not regain this water as it ascends the loop of Henle, for the cells of the ascending limb of the loop have a low permeability to water. However, the filtrate becomes more dilute because of the large amount of sodium pumped out. By the time the filtrate reaches the distal tubule it is again isotonic, or in some cases hypotonic, to the blood. The filtrate now descends through the medulla again, this time in the collecting tubule, loses additional water, and becomes very hypertonic.

Kidney Regulation. The amount of material reabsorbed or excreted by the kidney depends to some extent on the rate of glomerular filtration. If this is too high, certain essential substances are flushed through the tubules before reabsorption is completed; if the filtration rate is too low, some products normally excreted may be reabsorbed. Filtration pressure, and hence rate, can be varied within limits by the constriction or relaxation of the afferent arterioles leading to the glomeruli. The **juxtaglomerular apparatus** enables the kidney to monitor the nature of its output and modify filtration pressure accordingly. Shortly before it empties into the collecting tubule, a distal convoluted tubule passes between the afferent and efferent arterioles (Fig. 15.12) and comes into an intimate relationship with mod-

Figure 15.12 A diagram of the juxtaglomerular apparatus. (From Ham: Histology. J. B. Lippincott Co.)

ified muscular cells (juxtaglomerular cells) around the afferent arteriole. These cells contain granules identified as a vasoconstrictor substance. The degree of constriction of the juxtaglomerular cells appears to be influenced by the concentration of materials in the distal tubule.

The amount of water excreted and hence the volume of body fluids are also very much affected by the **antidiuretic hormone**, released by the posterior lobe of the pituitary (p. 440). This increases the permeability of the cells of the collecting tubule to water, so that more water is reabsorbed into the blood. If an excess of water is present in the body fluids, the blood volume and pressure increase. This raises the glomerular filtration pressure, and more filtrate is produced. An increase in the amount of water in the tissue fluid inhibits the release of the antidiuretic hormone, the permeability of collecting tubule cells is lowered, and less water is reabsorbed. Increased production of filtrate and decreased reabsorption of water rapidly bring the volume of body fluids down to normal. If the volume of body fluids falls below normal, as in a severe hemorrhage, these factors work in the opposite direction: less glomerular filtrate is produced, more water is reabsorbed, and the volume of body fluid is soon raised to normal.

The osmotic pressure of the tubular contents also affects the amount of water removed. If a large amount of salts or sugars is being eliminated, the osmotic pressure of the tubular contents is increased and less water can be reabsorbed. The urine volume is greater when there is a large amount of osmotically active substances in the urine, as after a large intake of salt or in diabetes mellitus.

15.4 OSMOTIC REGULATION IN MARINE ANIMALS

The amount of salt in the open ocean is about 35 parts per thousand (35%o), or about 3.5 per cent; little change in salt content is believed to have occurred since animals first evolved in the ancient seas. In estuaries and bays the salinity may be considerably lower, depending upon the inflow of fresh water from rivers and streams. The term **brackish** refers to estuarine waters that are more saline than fresh but less saline than the open ocean.

With the exception of such vertebrates as fishes, porpoises and whales, the internal body fluids of most marine animals are isosmotic with sea water. The salt content of their blood,

coelomic fluid and intercellular fluids is about the same as that of their marine environment, although the content of specific ions may be somewhat different. This does not apply to intracellular fluids. They are isosmotic with extracellular fluids, but their osmotic pressure is influenced by the large amount of organic compounds present. Intracellular osmotic pressures are commonly adjusted with free amino acids.

If the salinity of the external environment changes slightly, the concentration of salts in the body fluids of the animal also changes. Such marine animals are said to be **osmoconformers;** i.e., the salt concentration of their internal environment conforms to that of the surrounding external environment.

Since a minimal internal salt content is necessary for life, osmoconformers require environments of relatively high salinity. They are **stenohaline**, restricted to a narrow range of salinities, usually near that of the open ocean.

Osmoconformers, like aquatic animals in general, eliminate their nitrogenous wastes as ammonia. Ammonia is highly soluble and, since it is toxic, a large amount of water must be available so that it can be kept at low levels in the urine. Water supply is not a problem for most aquatic animals. As might be expected, the urine of osmoconformers is isosmotic with both their internal body fluids and the surrounding sea water.

Not all marine animals are osmoconformers and stenohaline. If a spider crab and a blue crab are placed in an aquarium in which the salinity is slowly lowered, their blood salts gradually fall; i.e., both act as osmoconformers (Fig. 15.13). With further decline in environmental salt concentrations, the spider crab dies but the blue crab does not. If the blue crab's blood salts are analyzed, they are found to be at higher levels than the salts in the environment; it holds onto its blood salts; it does not conform to the level of environmental salts. It is **osmoregulating.**

Marine osmoregulators can invade or inhabit estuarine waters, where salinities are low. They can usually tolerate a wide range of salinities and are said to be **euryhaline.**

The distribution of marine invertebrate animals within the Chesapeake Bay reflects their osmoregulating abilities and euryhaline character. Sea stars and sea urchins, which are stenohaline and osmoconformers, are only found near the mouth of the Chesapeake, where salinities are high. At Annapolis, about halfway up the Chesapeake, where the salinity is about 1.5 per cent, the marine fauna is limited to osmoregulators — various species of worms,

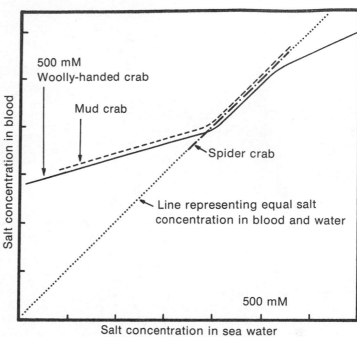

Figure 15.13 The osmoregulatory ability of three crabs. The osmotic concentration of blood changes with the concentration of salts in the surrounding sea water. The spider crab cannot osmoregulate, and the osmotic concentration of its blood conforms to the salt concentration of the sea water. Below and above certain concentrations, the crab dies. The blood of both the mud crab and the woolly-handed crab conforms to an upper range of external salt concentration, but at the lower concentrations the crabs maintain higher osmotic concentrations in the blood by absorbing salts through the gills. These two crabs can live in brackish estuarine environments where the spider crab can not. The woolly-handed crab can also excrete salt at high blood concentrations. The osmoregulatory ability of the blue crab would be similar to that of the mud crab. (Adapted from Schmidt-Nielsen.)

snails, clams and stinging nettles. At the upper end of the Chesapeake, where the salinity is very low, only a few marine animals, such as blue crabs and oysters, can exist.

How and where does osmoregulation take place? The excretory organs would be a logical site for osmoregulation if the excretory tubules were able to reabsorb salts and produce a dilute urine; i.e., one hypo-osmotic to the blood. This would solve two problems for an animal living in a dilute environment: the retention of salts and the removal of excess water that would tend to diffuse inward. Some animals can use their excretory organs to osmoregulate in this way, but many cannot. When osmoregulating crabs are placed in dilute sea water, the urine flow may greatly increase as the animal eliminates the excess of water diffusing inward. But in the process, salts are lost, for most crabs can only produce a urine isosmotic or only slightly hypo-osmotic with their blood. For example, in normal sea water the European crab *Carcinus* excretes a volume of urine equivalent to 3.6 per cent of its body weight each day; in sea water diluted to 1.4 per cent salinity it eliminates a urine equivalent to 33.0 per cent. Replacement of salt lost in the urine is provided by the gills, which actively pick up salt from the ventilating current. Thus, in osmoregulating crabs the green glands are the water pumps and the gills are centers of salt replacement.

The Vertebrate Problem. The salt content of the internal body fluids of most vertebrates is about 1 per cent and may reflect their freshwater origin or an invasion of fresh water early in their evolutionary history. Marine vertebrates are thus believed to be immigrants to the sea. In any event, their internal salt content is only about one-third that of sea water, and water tends to diffuse outward, especially across surfaces such as the gills. They are therefore living in a physiological desert, and water conservation is a major problem.

Marine bony fish cannot produce urine more concentrated than their blood, for the kidney tubules lack loops of Henle. Thus the kidney cannot function in osmoregulation. Marine fish obtain their water with food and by drinking sea water. The excess salt (Na^+ and Cl^-) taken into the blood is excreted by active transport across the gill epithelium. The kidneys remove magnesium and sulfate ions, in addition to ammonium ion (NH_4^+), which is the nitrogenous waste of many fish (Fig. 15.14).

Other marine vertebrates solve the problem of living in a hyperosmotic medium differently. Sharks produce urea as their nitrogenous waste and retain it in their internal body fluids in such high concentration that the osmotic pull of water inward is in equilibrium with the outward diffusion resulting from the higher external salt concentration. The gills of sharks are not permeable to urea as are those in most fishes. The concentration of urea in the body fluids of sharks may reach 100 times that in mammals, a level far exceeding the tolerance of other vertebrates. The salts that ac-

Figure 15.14 Osmoregulation in bony fish. The body fluids of fish have a salt content of about 1 per cent. Fish are thus hyposmotic to sea water and hyperosmotic to freshwater.

cumulate through both food intake and inward diffusion across surfaces such as the gills are removed by the kidneys and by a special salt-excreting gland attached to the caudal end of the shark intestine.

Sea birds and sea turtles excrete excess salt from a pair of salt glands in their heads. In sea birds, these glands open into the nasal passageways; in sea turtles, they open into the eyes. Turtles literally weep salty tears. The glands excrete only in response to an internal salt load, and the concentration of salt in the excreted fluid can reach a level almost twice that of sea water. Only birds and mammals can excrete a urine saltier than the blood, and bird urine is only twice as concentrated as the blood, which is still less than the concentration of salt in sea water.

Seals, porpoises and whales obtain much of their water from their food; some food (such as mollusks and crustaceans) is isosmotic with sea water, and some (such as fish) is less concentrated than sea water. However, marine mammals can utilize the kidney for osmoregulation. They have very long loops of Henle, which make it possible to produce a urine that has a salt concentration greater than that of sea water.

Humans cannot drink sea water, because they cannot produce a urine with a salt content greater than approximately 2.2 per cent. If we were to drink a liter of sea water, it would require one and one-third liters of urine water to eliminate the excess salt. Moreover, the magnesium ions in sea water cause diarrhea, and additional water is lost. The sailors in our story could have prolonged their lives by drinking fish juices because the internal body fluids of fish, like those of humans and other vertebrates, have a salt content of about 1 per cent.

15.5 OSMOREGULATION IN FRESHWATER ORGANISMS

The salt concentration in fresh water varies with the water source but is always very low. The external environment is thus very hypo-osmotic to the internal body fluids of freshwater animals, and they must cope with the tendency of water to diffuse into the body, of salts to diffuse outward, and of the internal body fluids to lose salts through excretion.

Given these conditions, the following generalizations are not unexpected: (1) All freshwater animals are osmoregulators. (2) Excretory organs function as water pumps, and the urine is usually hypo-osmotic to the internal body fluids as a result of selective reabsorption of salts. (3) Salt concentration of internal body fluids is maintained at the lowest level compatible with the animal's metabolic needs. Thus, less energy is required to maintain internal salt balance. (4) Nitrogenous wastes are usually eliminated as ammonia, since plenty of water is available to remove it in dilute concentrations. However, freshwater fish and amphibians more commonly excrete nitrogenous wastes as urea.

Freshwater fish excrete a hypo-osmotic urine but still suffer some salt loss. Replacement of salt is provided by gills that, in contrast to those of marine fish, pick up salts from the ventilating current (Fig. 15.14). Crayfish osmoregulate similarly. The green glands produce a hypo-osmotic urine, and gills are also the site of salt replacement.

In crayfish, as well as in some other crustaceans living in fresh or brackish water, the exoskeleton is less permeable than that of marine forms. This reduces, but does not eliminate, the inward diffusion of water.

Freshwater clams and snails have extremely low internal salt concentrations, only about 0.33 per cent, in comparison with about 1 per cent in freshwater fish. Their nephridia produce a copious hypo-osmotic urine amounting to approximately 50 per cent of the body weight per day, and the concentration of salts in the urine is about one-half of that of the blood. Salts are actively absorbed across the body surface.

Most very small freshwater animals have no specialized excretory organs. The surface area is adequate to permit elimination of ammonium ions by outward diffusion. Nevertheless, they must osmoregulate. Freshwater sponges and protozoans such as amebas and *Paramecium* possess water-pumping organelles called contractile vacuoles or water expulsion vesicles. The vacuole slowly accumulates water and, when full, collapses to the outside through a temporary opening. *Paramecium* possesses a contractile vacuole at each end of the body. The vacuoles fill through tiny radiating canals and empty through a canal to the outside. The rate of pulsation is governed by the rate at which water is taken in by diffusion and feeding.

Rotifers are microscopic multicellular animals that are very common in freshwater (Fig. 26.2). The water pumps of these animals are the protonephridial tubules described earlier. Excess water from the internal body fluids passes through the fenestrations of the flame cells and is driven down the tubule to the outside by the beating of the flagella.

15.6 OSMOREGULATION IN TERRESTRIAL ANIMALS

The danger of desiccation is a major threat to animals living on land. Water is lost in urine and in feces, but evaporation is the principal route of water loss. At any given temperature, a given volume of air is capable of holding a certain amount of water vapor. The difference between the amount of water vapor actually present and the amount that would be if the air were saturated is known as the **saturation deficit.** The saturation deficit determines the steepness of the evaporation gradient between the animal and the external environment and varies with season, climate, habitat and time of day, all of which affect the distribution and behavior of land animals.

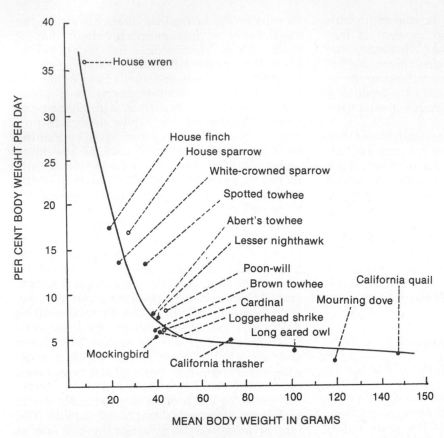

Figure 15.15 Daily water loss in different species of birds at 25°C. plotted against mean total body weight. The water loss is stated as a percentage of body weight. (Modified from Bartholomew, G. A., and T. C. Cade, Auk *80*: 504–539, 1963.)

The general body surface and the gas exchange surface are the two principal sites of evaporation. In a snake, for example, 64 per cent of the water loss by evaporation may occur across the skin and 36 per cent through gas exchange in the lungs.

Water loss by evaporation poses an even greater problem for small animals than for large, because the ratio of surface area to volume increases with decreasing size. This relationship is illustrated in birds in Figure 15.15. At 25° C. a 20 gram sparrow is subject to several times the water loss of a 100 gram owl.

Water lost must be replaced with water gained by drinking and eating and, in some animals, by direct uptake of water vapor in the surrounding air.

Adaptive Strategies. The many different groups of land animals, such as vertebrates, spiders, insects and snails, are representatives of separate invasions of the terrestrial environment. Within each of these groups various adaptations have evolved that increase the likelihood of maintaining a balance between water lost and gained. The principal adaptive strategies are described as follows:

1. Reduction of evaporation by
   A. Internal gas exchange organs. The gas exchange organs of most land animals are located internally or at least within a protective covering where the thin, moist exchange surface is less likely to dry up and there is less water lost by evaporation.
   B. Modification of the integumental barrier. Loss of water by evaporation across the general body surface is reduced by various modifications of the skin that make the integumental barrier more impervious to the outward passage of water.
   C. Occupation of humid habitats. Subterranean burrows and spaces beneath stones and logs and in leaf mold, where saturation deficits are lower than in exposed locations, are less stressful environments for maintaining water balance.
   D. Nocturnal activity. Many species move out of protective retreats only at night, when the danger of desiccation is greatly reduced.
2. Reduction of water loss from excretion by
   A. Production of insoluble nitrogenous wastes. Some land animals excrete their nitrogenous wastes in the form of uric acid or guanine, which have very low solubility and thus require little water for removal.

B. Production of a hyperosmotic urine. Mammals, birds and insects are able to excrete urine with a salt concentration greater than that of the blood; such a concentrated urine conserves water.

C. Low urine output. Some terrestrial animals conserve excretory water by greatly reducing urine flow and utilizing other avenues for the elimination of nitrogenous wastes.

3. Reduction of water loss from egestion. Some animals, especially those living in deserts, conserve water by defecating a relatively dry feces.

4. Toleration of internal water loss. Most animals can tolerate relatively limited fluctuations in the levels of internal water. For example, if a human loses more than 12 per cent of his body water, death may result. But there are a few animals that as an adaptation to water stress can tolerate the loss of more than half of their body water, which can be replaced later when water is again available.

5. Dependency upon water of oxidation. Significant amounts of water are liberated in various reactions of cell respiration. The oxidation of 100 grams of glucose, for example, yields 60 grams of water. A few desert animals survive entirely on this water source.

Most terrestrial animals employ only some of the adaptations just described, but the combination enables the species to balance water loss with gain in its particular environment.

A few groups of animals commonly considered terrestrial actually live in highly specialized aquatic microhabitats. Soil nematodes, or roundworms, for example, live in the water films around soil particles. There are also a number of groups of minute animals, including nematodes, that live in the water films of moss. When the moss is dry, they can survive for long periods of time in a highly desiccated state.

Earthworms are poorly adapted land animals and are only active within their burrows in moist soil or at the surface at night. When the soil becomes dry, they move to deeper levels, where they may lose water and become dormant. Earthworms excrete ammonia and urea, and the nephridia can, in the event of excess water, produce a hypo-osmotic, but not a hyperosmotic, urine.

Land snails and slugs are also subject to great evaporative water loss across the skin. Moreover, additional water is lost in the secretion of the mucous trail over which they crawl. However, these animals are able to tolerate considerable desiccation. The European land snail *Helix* can survive a water loss equivalent to 50 per cent of its body weight, and the slug *Limax* can survive an 80 per cent loss. They excrete uric acid, which conserves some water. Nevertheless, snails and slugs are confined to humid habitats or exposure only at night, when saturation deficits are low.

Spiders, insects and pill bugs are all small terrestrial arthropods, each representative of a different past invasion of land. Spiders and insects are highly successful terrestrial animals, as their great numbers attest. Contributing to their success has been the evolution of a waxy layer on the surface of the exoskeleton. When exposed to the same saturation deficit, their evaporation per square centimeter is comparable to that of mammals and is only a fraction of that of snails and slugs. The internal tracheal tubules, which are the gas exchange organs of insects as well as of many spiders, reduce respiratory evaporation. Insects also conserve water by excreting uric acid and producing a hyperosmotic urine via the malpighian tubules and rectum. In fact, insects are the only animals other than birds and mammals that produce a hyperosmotic urine. Spiders excrete guanine, which has relatively low solubility in water. Nevertheless, very small spiders, pos-

TABLE 15.1 COMPARISON OF WATER LOSS BY EVAPORATION FROM THE BODY SURFACE OF DIFFERENT ANIMALS

	Evaporation* (μg)
Earthworm	400
Garden snail, active	870
Garden snail, inactive	39
Salamander	600
Frog	300
Human (not sweating)	48
Rat	46
Water snake	41
Pond turtle	24
Box turtle	11
Iguana lizard	10
Gopher snake	9
Desert tortoise	3
Cockroach	49
Flour mite	2
Tick	0.8

*Micrograms per cm.² body surface per hour per mm. Hg saturation deficit. (After Schmidt-Nielsen, K.: The neglected interface: the biology of water as a liquid gas system. Q. Rev. Biophys. 2:283, 1969.)

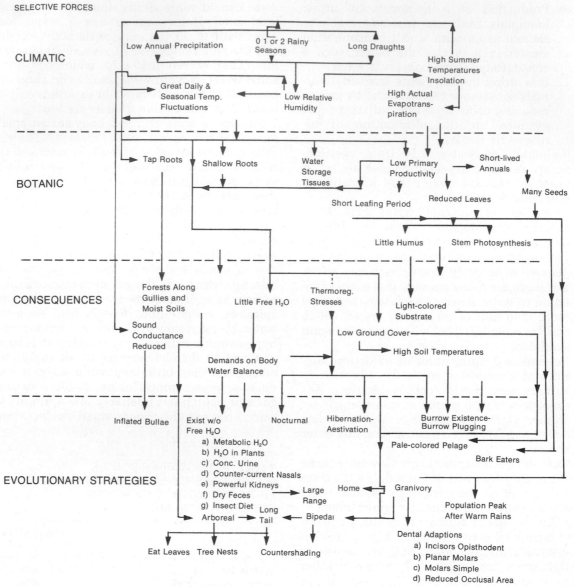

Figure 15.16 Adaptations of desert rodents and their relationship with various environmental factors. (From Mares, M. A.: Paleobiology, 2:39–63, 1976.)

sessing a large surface area in relation to volume, are confined to leaf mold and other more humid habitats.

Pill bugs are not insects but crustaceans, relatives of crabs and shrimps (Fig. 29.31). They are mostly nocturnal animals, living beneath stones and wood and in leaf mold. In contrast to spiders and insects, they lack a waxy outer covering and are subject to much greater water loss. They are ammonotelic, like most aquatic animals, but are remarkable in that they excrete gaseous ammonia rather than ammonium ions. The gills are the principal sites of excre-

tion, and urine output from the green glands is greatly reduced.

Frogs and toads suffer the highest rate of evaporative water loss of all terrestrial vertebrates, a rate far higher than that of insects and spiders (Table 15.1). In spite of the warty skin of the toad, it loses water about as rapidly as the frog. Nocturnal habits and protective environments are the principal defenses of these animals against desiccation. Both excrete urea, which is less toxic than ammonia but very soluble.

Most reptiles are much better adapted for

life on land than amphibians. The skin of the snake, for example, is highly resistant to water loss from evaporation, and its nitrogenous wastes are excreted as uric acid. Uricotelism probably first evolved in reptiles as an adaptation for reproduction on land, since it permits accumulation within the egg of nitrogenous wastes produced by the developing embryo.

Birds excrete uric acid. As in reptiles, uric acid is important in conserving water. The integument of the bird, with its covering of feathers, reduces water loss, and the avian kidneys can produce a hyperosmotic urine.

Mammals are also well adapted for life on land. The skin is an effective barrier against evaporation, about as effective as that of insects and spiders. Mammals excrete the soluble waste urea, but most mammals live where sources of drinking water and water in food are adequate. Moreover, the mammalian kidney can produce a hyperosmotic urine. Human beings can produce a urine that contains as much as 6 per cent urea and 2.2 per cent salts, providing considerable conservation of water. On the other hand, mammals can also produce a dilute (hypo-osmotic) urine when it is beneficial.

The greatest number of adaptations for water balance are found in desert animals. The extremely sparse water supplies and great saturation deficits of the desert environment require that the animals living there exploit all of the available water resources and preserve their internal body fluids. Figure 15.16 outlines the adaptive strategies of desert plants and animals, relating them to each other and to various environmental stresses.

Kangaroo rats, mammals abundant in and common to southwestern America, have been studied more extensively than any other desert animal. These small rodents live in subterranean burrows and come to the surface only at night. They rarely drink, since standing water is almost never available. They eat only dry food, mostly seeds, and rely almost entirely on the water produced in the oxidation of food. There are no sweat glands. The feces contain little moisture, and their very long loops of Henle can produce a highly concentrated urine, containing as much as 23 per cent urea and 7 per cent salt (the maximum in human urine is 6 per cent urea and 2.2 per cent salt). Note that kangaroo rats could drink sea water!

Camels exhibit a very different adaptive strategy. They are unable to store water (the hump contains stored fat) but can tolerate great water loss. Most mammals can tolerate a loss of only about 20 per cent of the body weight, humans only 12 per cent. Camels can survive a 40 per cent depletion. They can drink an enormous amount of water at one time, as much as one-third the body weight in 10 minutes. There then may follow a gradual loss that can take the animal through a prolonged period without water. Camels have sweat glands, but sweating commences at a higher internal body temperature than in other mammals. Moreover, camels begin the day with a lower early morning temperature than do other mammals.

ANNOTATED REFERENCES

Many of the references listed at the end of Chapter 10 provide excellent coverage of the topic of regulation of internal body fluids and related subjects.

CHAPTER 16

RECEPTORS AND SENSE ORGANS

In order to survive each organism has evolved some means for making appropriate, meaningful, adaptive responses to specific changes in the environment. This requires that the organism have receptors or sense organs to detect the changes in the environment, systems of nerves and endocrine organs to integrate and coordinate the information received and to trigger the response, and effectors, such as muscles, glands, melanocytes, nematocysts, electric organs or luminescent structures, to carry out the responses. Protozoa are sensitive to many kinds of stimuli, as evidenced by their negative responses to bright lights, certain chemicals and electric currents. For survival on a more complex level of existence the metazoa have evolved a variety of specialized receptor cells, each sensitive to one type of stimulus. Sense organs enable their possessors to obtain the information needed as they search for food, find and attract a mate and escape enemies; they are of great importance in the survival of the individual and of the species.

The receptor cells within sense organs can be remarkably sensitive to the appropriate stimulus. The human eye can be stimulated by an extremely weak beam of light, the human ear can detect very faint sounds, and the human taste buds can detect very dilute solutions of compounds such as vinegar. The sense organs of some animals are sensitive to stimuli that are completely ineffective in humans. Dogs and cats can hear high pitched whistles inaudible to us. Bats emit very high pitched noises and are guided in flight by the echoes that rebound from objects in their path. They can catch insects, guided by the echoes from

their small prey; some insects can also hear these high pitched noises and take appropriate evasive action.

Simply put, the function of any receptor cell is to detect a very small amount of energy in a specific form and to transmit the information to the nervous system. The ability of an organism, such as a human being, to perceive different kinds of stimuli (light, sound, touch and so on) is a function of the specificity of the receptors and of their specific connections within the nervous system. The nerve impulse initiated in a sensory neuron by a specific receptor is not specific and is the same no matter where it originates. Differences in the intensity of the stimulus are coded in the number of nerve fibers activated and in the number of impulses passing along a given nerve fiber. Awareness of the sensation depends on the precise part of the brain to which the impulse passes.

The various types of receptors can be classified in several ways: whether they are dispersed (the receptors for touch in the skin) or concentrated (the photoreceptors in the retina of the eye), on the basis of their structure or on the basis of the kind of energy change they monitor (mechanoreceptors, chemoreceptors, photoreceptors, and so on).

16.1 RECEPTOR MECHANISMS

Although humans are traditionally believed to have five senses—touch, smell, taste, sight and hearing—some of these can be divided into several distinct senses. The general sense of touch, for example, includes touch, pain, pressure, cold and heat. The ear contains an organ for sensing equilibrium and rotation, in addition to the one for sensing sound waves. We also can sense the tension of muscles, the movement of joints and internal conditions, such as thirst, hunger, nausea, pain and orgasm. The receptors for such senses are located in the muscles, joints, viscera, throat and other places. Some of the sensory hairs of arthropods and the receptor parts of the rods and cones in the vertebrate eye appear to be modified cilia, for they have structural features that are similar.

Receptor cells have the dual function of detecting change and transmitting information concerning the nature of the change to the central nervous system. Very little useful information would be available to an organism if it responded indiscriminately to all kinds of environmental changes. In the course of evolution, sense organs have become highly specific for one kind of environmental stimulus: one organ detects light, another detects mechanical pressure, a third detects certain chemicals and so on. No sense organ would be very useful if it responded only to gross changes in the environment. However, if it were so sensitive that it responded to every moving molecule or electron, it would transmit only noise. Each sense organ has thus evolved specificity and an optimal, not necessarily maximal, sensitivity, so that it maintains an optimal ratio of signal to noise. Sense organs should have the capacity for recording not only "on" and "off" but also rate, magnitude and direction of change.

Each sense organ has a specialized structure that may consist of not only one or more receptor cells but also a variety of accessory tissues. The receptor cells of the human eye are the rods and cones; the accessory structures are the cornea, lens, iris and ciliary muscles (Fig. 16.17). Some of the capacities of a sense organ are intrinsic to the receptor cells; others are conferred upon the sense organ by the accessory structures. Without a lens an eye would be incapable of detecting many events. Accessory structures may also act as filters that limit the performance of receptors. For example, ultraviolet light is not seen by the human eye because it is filtered out by the cornea and lens; the retina itself is not insensitive to ultraviolet light. In contrast, the eyes of insects have no filters for ultraviolet light, and the detection of ultraviolet is important in their lives.

The importance of the accessory structures is especially evident in the mechanical senses, in which they modify sensitivity and determine the nature of the effective stimulus. Among invertebrates, for example, one type of mechanoreceptor may be sensitive only to gross touch because it is associated with a stout, rigid spine, whereas another type may be an auditory receptor because it is connected to a long filamentous hair that is moved by sound waves. The hairs on the paired posterior appendages of crickets (the cerci) are examples of this type of auditory organ.

Receptors are usually parts of nerve cells, the axons of which extend directly into the central nervous system or connect synaptically with one or more interneurons connected with the central nervous system. Other receptors, such as the human taste buds, are modified epithelial cells connected to one or more sensory nerve cells (Fig. 16.13). In those sense organs in which the receptor is a primary neuron, this one cell both detects and transmits the information to the central nervous system. In organs in which the receptor is an epithelial cell, it detects, but the information is transmitted by the associated neuron. Each receptor receives a small amount of energy from the environment and is specialized to receive one particular form of energy more efficiently than others. Rods and cones absorb the energy of photons of certain specific energies and wave lengths. Temperature receptors respond to radiant energy transferred by radiation, conduction or convection. Taste and smells are detected by the potential energy in the mutual attraction and repulsion of molecules.

The various kinds of environmental energy cause the receptors to perform biological work. This work transforms metabolic energy into electrical energy. These relationships are exemplified by a very simple sense organ, the tactile hair of an insect. This hair, plus its associated cells, is a complete sense organ (Fig. 16.1). The **bipolar neuron** at its base is the receptor; the dendrite is attached to the base of the hair, near the socket; the axon passes directly to the central nervous system without synapsing. In its unstimulated state this neuron maintains a steady resting potential; i.e., there is a potential difference between the inside and outside of the neuron. This potential difference exists because the ionic compositions of the fluids on each side of the semipermeable membrane are different. The difference is maintained by the sodium pump and by metabolic work performed by the cell. When the hair is touched its shaft moves in the socket and mechanically deforms the den-

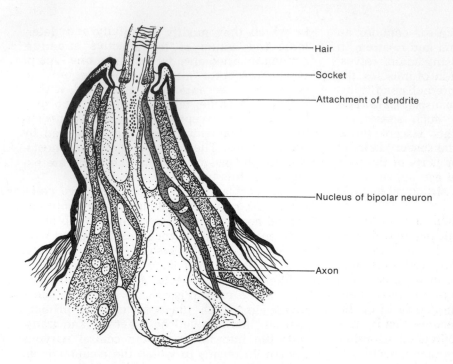

—Hair

—Socket

—Attachment of dendrite

—Nucleus of bipolar neuron

—Axon

Figure 16.1 A tactile hair from a caterpillar, showing the attachment of the dendrite of the bipolar neuron (the mechanoreceptor) at the point where the shaft of the hair enters the socket. (From Hsu, F.: Étude cytologique et comparée sur les sensilla des insectes. La Cellule 47:1–60, 1938.)

drite. This stimulus, the deformation of the cell, increases the permeability of the neuronal membrane to ions, with the result that the potential difference between the two sides of the membrane decreases, disappears or increases. If it decreases or disappears the cell is said to be **depolarized.** If it increases, the cell is said to be **hyperpolarized.** The state of depolarization caused by the stimulus is termed the **receptor potential.** The receptor potential spreads relatively slowly down the dendrite, decaying exponentially. When a special area of the cell near the axon, the **axon hillock,** becomes depolarized, **action potentials** are generated. The action potentials then travel along the axon to the central nervous system.

The primary receptor detects an event in the environment (movement of the hair), generates electrical energy at the expense of its metabolic energy (the receptor potential) and transmits information (action potentials) to the central nervous system.

The relations between the stimulus, the receptor potential and the action potentials are summarized in Figure 16.2. The amplitude and duration of the receptor potential are related to the strength and duration of the stimulus; a strong stimulus causes a greater depolarization of the receptor membrane than does a weak one. The action potentials are repetitive and the frequency at which they are generated is related to the magnitude of the recep-

Stimulus
(change in
environment)

Excitation process
in receptor

Nerve impulse
(in afferent fiber)

Sensation

Figure 16.2 A diagram showing the relations among the stimulus, the receptor potential, the action potential and sensation. (From Adrian, E. D.: The Basis of Sensation. London, Chatto & Windus Ltd., 1949.)

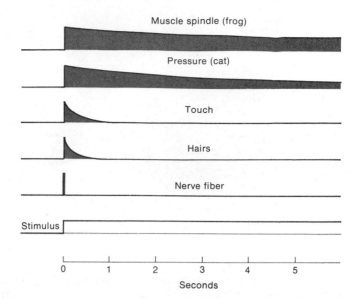

Figure 16.3 A diagram showing the relation between the stimulus and the different rates of adaptation for different receptors and a nerve fiber. The heights of the curves indicate the rates of discharge of action potentials. (From Adrian, E. D.: The Basis of Sensation. London, Chatto & Windus Ltd., 1949.)

tor potential. The strength of the stimulus is reflected in the frequency of the action potentials. The amplitude of each action potential bears no relation to the stimulus. It is characteristic of the particular neuron under the usual recording conditions.

The action potential (p. 396) is an "all-or-none" phenomenon. The receptor potential, in contrast, is a **graded response.** Once a stimulus has triggered a receptor to generate action potentials, the stimulus has no further control over them. Even though the stimulus may continue unabated, neither the receptor potential nor the action potentials will continue unchanged (Fig. 16.3). The receptor potential gradually falls, and the frequency of the action potentials decreases. This phenomenon is termed **adaptation.** Some receptors adapt very rapidly and completely; others do so more slowly.

16.2 SENSORY CODING AND SENSATION

The stimulation of any sense organ or sensory receptor initiates a type of coded message composed of action potentials transmitted by the nerve fibers and decoded in the central nervous system. Impulses from the sense organs may differ in (1) the total number of fibers transmitting; (2) the specific fibers carrying action potentials; (3) the total number of action potentials passing over a given fiber; (4) the frequency of the action potentials passing over a given fiber or (5) the time relations between action potentials in different specific

fibers. How the sense organs initiate different codes and how the brain analyzes and interprets them to produce the various sensations are not yet understood. It is important to remember that all action potentials are qualitatively the same. Light of wave length 400 nanometers (blue), sugar molecules (sweet) and sound waves of 440 hertz (A above middle C) all cause action potentials to the brain via the appropriate nerves; these action potentials are identical. How can the organism then assess its environment accurately? The qualitative differentiation of stimuli must depend on the sense organ itself or the brain, or both. In fact it depends upon both. Our ability to discriminate red from green, hot from cold or red from cold is due to the fact that particular sense organs and their individual receptor cells are connected to specific cells in particular parts of the brain.

The frequency of the repetitive action potential along sensory nerve fibers codes the intensity of the stimulus. Since each receptor normally responds to but one category of stimuli, a message arriving in the central nervous system along this nerve is interpreted as meaning that a particular stimulus has occurred. Interpretation of the message and, in the case of human beings, the quality of sensation depend on which central interneurons receive the message. Sensation, when it occurs, occurs in the brain. Rods and cones do not see—only the combination of rods, cones and centers in the brain can see. Furthermore, many sensory messages never give rise to sensations. For example, when chemoreceptors in the carotid sinus and the hypothalamus sense internal changes in the body, appropriate physiological adjust-

Inner optic chiasma

Outer optic chiasma

Retina

Inner synaptic region

Outer synaptic region

Layer of ganglion cells

Figure 16.4 An example of neural circuitry in the optic lobes of insects. This is a diagrammatic representation of the optic lobe of the blowfly *Calliphora vomitoria*. (From Cajal, S. R., and D. Sanchez: Contribucion al conocimiento de los centros nerviosos de los insectos. Trabajas del Lat. de Investig. Biologicas (University of Madrid) 13:1–167, 1915.)

ments are made, but our consciousness is not stirred.

Since only those nerve impulses that reach the brain can result in sensations, any blocking of the passage of the impulse along the nerve fiber by an anaesthetic has the same effect as removing the original stimulus entirely. The sense organs, of course, will continue to initiate impulses that can be detected by the proper electrical apparatus, but the anaesthetic prevents them from reaching their destination.

Another method of coding information, termed **cross-fiber patterning,** is used in olfactory organs. It is unlikely that the olfactory organ contains a specific receptor for each of the thousands of individual odors that can be recognized. There is evidence that only a limited number of categories of receptors exist, each of which responds to a spectrum of odors. There is no rigid specificity because the spectra overlap. Perception of different characteristic odors probably depends upon the pattern of response of all of the fibers responding together.

The temporal pattern of action potentials generated in a single neuron may serve as a code for different stimuli. Single taste receptors in flies, for example, generate action potentials at an even, regular frequency when the stimulus is salt but generate irregular frequencies when the stimulus is acid.

The axon of a sensory neuron in invertebrates typically extends to the central nervous system without synapsing. In these circumstances the message generated at the periphery arrives unaltered. In the compound eye, as in most vertebrate sense organs, many interneurons are interposed between the receptor and the central nervous system (Fig. 16.4). The vertebrate retina or olfactory bulb has an exceptionally complicated neural circuitry. As a consequence of all of these synaptic connections, the original message is altered and may lose or gain some of its information. The message that finally arrives at the brain has been well censored by interneurons and bears even less resemblance to the original stimulus than did the action potentials from the receptor.

16.3 MECHANORECEPTORS

Receptors are customarily classified according to the nature of their effective stimuli. Thus there are mechanoreceptors, chemoreceptors, photoreceptors, thermoreceptors and electroreceptors. Each of these receptors consists of one or more specialized cells or free nerve endings that respond as miniature transducers, for, like the piezoelectric crystal in a phonograph pick-up, they convert one form of energy into another. Each receptor is affected by the type of energy to which it is attuned and converts it into an electric current, the **receptor potential,** which in turn initiates the action potentials of the nerve impulse. Unlike nerve or muscle, the activity of a receptor is not an "all-or-none" phenomenon but varies with the strength of the stimulus. The receptor potential must attain a certain threshold to initiate an action potential. If the intensity of the receptor potential exceeds the threshold, it will initiate additional action potentials, additional nerve impulses.

Mechanoreceptors are sensitive to stretch, compression or torque imparted to tissues by the weight of the body, the relative movement of parts, the gyroscopic effects of moving parts and the impact of the substratum or the surrounding medium (air or water). Mechanoreceptors are concerned with enabling an organism to maintain its primary body attitude with respect to gravity and with maintaining postural relations (information that is essential for all forms of locomotion and for all coordinated and skilled movements). Mechanoreceptors provide information about the shape, texture, weight and topographical relations of objects in the external environment; they are necessary for the operation of certain internal organs; and they supply information about the presence of food in the stomach, feces in the rectum, urine in the bladder and a fetus in the uterus. Mechanoreceptors may be categorized as tactile, proprioceptive and auditory.

The Tactile Sense

Among the simplest tactile receptors are the **tactile hairs** of invertebrates. The tactile hair of an insect is a **phasic receptor** (i.e., it responds only when the hair is moving). When the hair is displaced, a receptor potential develops and a few action potentials are generated, but all activity ceases when motion ceases, even though the hair is maintained in the displaced position.

The remarkable tactile sensitivity of human beings, especially in the fingertips and lips, is due to a large and diverse number of sense organs in the skin (Fig. 16.5). By comparing the distribution of the different types of sense organs and the types of sensations produced, it has been found that free nerve endings are responsible for pain perception, that basket nerve endings around hair bulbs (Meissner's corpuscles and Merkel's discs) are responsible

for touch, that the end bulbs of Krause and Ruffini's endings are responsible for sensations of cold and warmth and that pacinian corpuscles mediate the sensation of deep pressure.

The pacinian corpuscle has been studied extensively. The bare axon is surrounded by lamellae interspersed with fluid. Compression causes displacement of the lamellae, which provides the deformation stimulating the axon. Even though the displacement is maintained under steady compression, the receptor potential rapidly falls to zero and action potentials cease. This is a phasic receptor responding to velocity.

Proprioception

Among invertebrates the sense organs most commonly concerned with relaying postural information are hairs, plates (campaniform organs) and other modified cuticular structures. These are **tonic** (static) sense organs. Unlike phasic receptors, the receptor potential is maintained (though not at a constant magnitude) as long as the stimulus is present and action potentials continue to be generated. Thus there is continued information about the position of the organ concerned.

Each human muscle, tendon and joint is equipped with proprioceptors sensitive to muscle tension and stretch. By means of these sense organs we can perform manual acts such as dressing or tying knots even when our eyes are closed. Impulses from the proprioceptors are extremely important in ensuring the har-

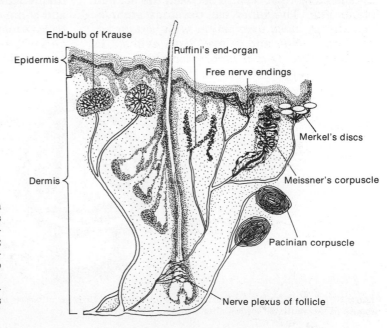

Figure 16.5 Diagrammatic section through the skin showing the types of sense organs present. The sense organs respond to the following stimuli: cold–end-bulbs of Krause; warmth–Ruffini's end-organs; touch–Meissner's corpuscles and Merkel's discs; deep pressure–pacinian corpuscles; and pain–free nerve endings. (From Villee, C. A.: Biology. 7th ed. Philadelphia, W. B. Saunders Company, 1977.)

monious contraction of the different muscles involved in a single movement. Without them, complicated skillful acts would be impossible. Impulses from these organs are also important in the maintenance of balance. Proprioceptors are probably more numerous and more continuously active than any of the other sense organs, although we are less aware of them than we are of the others. We can get some idea of what life without proprioceptors would be like when an arm or leg "goes to sleep"—a feeling of numbness results from the lack of proprioceptive impulses.

The Lateral Line System in Fishes

Fishes have a lateral line system that is sensitive to currents, changes in pressure and vibrations of low frequency. It consists of groups of hair cells similar to those found in the ear (p. 371). Most of these are arranged in a longitudinal canal extending the length of the trunk and tail and in a series of canals that ramify over the head, but often other groups of hair cells are scattered over the body surface. The canals are imbedded in the skin but connect with the surface through a series of pores (Fig. 16.6). The system is sometimes described as one of "distant touch," because the fish can detect an object in its approach or an object's approach toward itself by the resulting deflection of certain other groups of hair cells. Neurons from these receptors enter an acustico-lateralis area of the brain along with neurons from the ear, which suggests that there is close relationship between the ear and lateral line. The inner ear develops embryologically in close association with certain lateral line canals, and it may have evolved in the same way. Larval amphibians have a lat-

Figure 16.6   Lateral line organs of a fish. *A*, Surface view of the anterior end of the catfish, *Nemachilus*. Circles indicate the position of the lateral line; small dots, isolated groups of hair cells. *B*, Diagrammatic vertical section through a lateral line canal and several isolated groups of hair cells. Water currents (small arrows) and the deflection of groups of hair cells caused by an approaching object are shown. (From Dijkgraaf, S.: Bau und Funktionen der Seitenorgane und des Ohrlabyrinths bei Fischen. Experientia *8*: 205–216, 1952.)

eral line system, but it is lost during metamorphosis. Terrestrial vertebrates never have this system at all.

Phonoreception

Phonoreception, or hearing, refers to the detection of pressure waves resulting from a mechanical disturbance some distance away. The human ear can detect sound waves with frequencies of from 20 to about 20,000 hertz, and some animals—bats, for example—can detect frequencies of over 100,000 hertz. Frequencies of less than 20 hertz are usually perceived as vibrations and not as sound. Mammals, birds

Figure 16.7   The left ear of a fish seen in a lateral view. Only an inner ear is present, embedded within spaces in the otic capsule of the skull. (Modified after Kingsley.)

and some reptiles have a **cochlear duct,** an elongated cul-de-sac extending from the sacculus concerned with phonoreception. Fishes have a homologous, but very much smaller, diverticulum known as the **lagena** (Fig. 16.7). The rudimentary nature of this structure, together with early experiments in which fishes were shown to be unresponsive to sounds made in the air, led to the conclusion that they could not hear. Later it was realized that most airborne sounds are reflected by the air-water surface. However, sounds that are generated in the water travel rapidly and far. With the development of underwater listening and sound-generating devices investigators have discovered that aquatic organisms produce and respond to a great many sounds. Since the tissues of fishes are mostly water, they are essentially transparent to a sound generated in the water; the sound waves pass right through without interruption. For a fish to detect the waves, there must be some tissue that will respond to the sound waves differently than the whole body. In most fishes this need is served by a large otolith in the sacculus. Catfish and certain other fish utilize an air-filled swim bladder as a hydrophone. The swim bladder, which evolved as a modification of a lunglike structure, is located in the dorsal part of the body cavity. Pressure waves passing through the fish's body change the volume of the swim bladder, and these movements are transmitted to the sacculus by way of a chain of small bones derived from the vertebrae. This mechanism is very effective, for the fishes that have this type of mechanism can respond to a wider range of frequencies (from 60 to 6000 hertz) than other fishes, and they can discriminate better between the different kinds of sounds.

Phonoreception in Tetrapods

In the tetrapods, a part of the membranous labryinth of the ear, usually the lagena or cochlear duct, is specialized for phonoreception. A variety of devices have evolved to transmit either ground or airborne vibrations to the phonoreceptor. Frogs have an external **tympanic membrane** (Fig. 16.8) that responds to vibrations in the air and a bone, the stapes, that transmits the vibrations across the middle ear cavity from the tympanic membrane to the oval window (fenestra ovalis) in the otic capsule. The oval window communicates with the inner ear.

The hearing apparatus of mammals is similar but more elaborate (Fig. 16.9). The outer ear consists of two parts: the skin-covered cartilaginous pinna, or auricle, and the auditory canal (external auditory meatus) leading from it to the middle ear. In humans, the pinna, or visible ear, is of some slight use for directing sound waves into the canal, but in other animals, such as the cat, the larger, movable pinna is very important. It acts as an ear trumpet that concentrates and increases slightly the pressure of the sound waves that vibrate the tympanic membrane, or eardrum, which is situated in a protected site at the internal end of the auditory canal. The middle ear is a small chamber containing three tiny bones: the hammer-shaped **malleus,** the anvil-shaped **incus** and the stirrup-shaped **stapes,** arranged in sequence. These transmit sound waves across the middle ear cavity to the oval window. The stapes evolved from a part of the hyoid arch of fishes, and the malleus and incus were derived from the posterior part of the mandibular arch when a new jaw joint

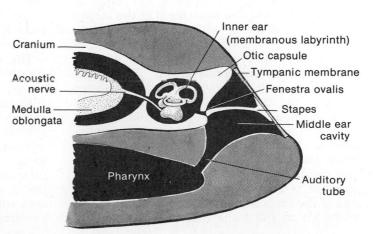

Figure 16.8 A diagrammatic cross section through the head of a frog to show the ear and its relation to surrounding parts.

Figure 16.9 The mammalian ear. *A*, Schematic drawing of the outer, middle and inner ear of a human being. *B*, Diagram of the cochlea as though it were uncoiled. *C*, An enlarged cross section through the cochlea. The cochlear duct and other parts of the membranous labyrinth are filled with endolymph and are shown in white; perilymph is black with white stipples.

evolved anterior to the previous one. The three auditory ossicles, or earbones, form a lever system that reduces the displacement amplitude but increases the pressure amplitude of the sound waves. The movement of the stapes against the membrane within the oval window is only about one-half as extensive as the movement of the tympanic membrane, but the force of the movement is greater. Two small muscles derived from visceral arch muscles are attached to the ossicles and reflexly dampen their oscillation to sounds of very high intensity. They form a protective mechanism for the ear somewhat analogous to the iris of the eye. The major pressure amplification results from the fact that the area of the eardrum is nearly 20 times greater than the area of the membrane in the oval window. Virtually all of the force that impinges on the tympanic membrane reaches the membrane in the oval window, and since this membrane is smaller, the force per unit area is increased. Pressure amplification is essential for an efficient transfer of energy from the light compressible external air to the dense incompressible liquid of the inner ear.

The narrow **auditory,** or **eustachian, tube** connects the middle ear to the pharynx and serves to equalize the pressure on the two sides of the eardrum. If the middle ear were completely closed, any variation in atmospheric pressure would cause a pronounced and painful bulging or caving in of the eardrum. The pharyngeal end of the eustachian tube normally is collapsed and closed so that we do not become unpleasantly aware of our own voice. The auditory tube is opened during yawning or swallowing, and during an abrupt ascent or descent in an elevator or airplane, such actions help to prevent a cracking sensation of the eardrums produced by the changes in the atmospheric pressure.

The inner ear consists of a complicated group of interconnected canals and sacs often referred to, most appropriately, as the membranous **labyrinth.** The part of the labyrinth concerned with hearing is a spirally coiled tube of two and one-half turns, resembling a snail's shell, called the **cochlea.** The cochlea consists of three canals separated from each other by thin membranes and coming almost to a point at the apex. The oval window is connected to the base of one of these tubes, the **vestibular canal.** At the base of the tympanic canal is another opening, the round window, covered by a membrane that also leads to the middle ear. These two canals are interconnected at the apex of the cochlea and are filled with a fluid known as **perilymph.** Between the two lies a third canal, the **cochlear duct,** filled with **endolymph** and containing the actual organ of hearing, the **organ of Corti.** This structure consists of five rows of cells extending the entire length of the coiled cochlea; each cell is equipped with hairlike projections extending into the cochlear canal. Each organ of Corti contains about 24,000 hair cells. The hair cells rest upon the **basilar membrane** separating the cochlear duct from the tympanic canal (Fig. 16.10). Overhanging these cells is the **tectorial membrane,** attached along one edge to the membrane on which the hair cells rest and with its other edge free. The hair cells initiate impulses in the fibers of the auditory nerve.

Pressure waves produced by the stapes at the oval window pass through the vestibular canal, cross the cochlear duct and escape through the tympanic canal and the round window. The basilar membrane, which supports the organ of Corti, is set in vibration. The basilar and the tectorial membranes are hinged at different places; hence, a pressure wave causes a slight differential movement between them and develops a shearing force that stimulates the intervening hair cells, which are mechanoreceptors. The ear is extremely sensitive: at certain frequencies it can detect any displacement of these membranes that is less than the diameter of a hydrogen atom. Sensory neurons of the acoustic nerve extend from the hair cells into the brain.

How sounds are analyzed by the cochlea is a complex process not completely understood. It is well established that pitches or tones of different frequencies are detected in different regions of the cochlea. According to the current hypothesis, a movement of the stapes against the liquid in the vestibular canal causes the proximal part of the basilar membrane to move toward the tympanic canal. This sets up a tension in the basilar membrane and initiates a wave that travels along the membrane. The mechanism is analogous to a wave sent along a whip by a jerking movement at the proximal end. The wave travels along the entire basilar membrane, but the physical properties of the membrane, including its width and elasticity, change progressively, so that the length and amplitude of the wave varies. Waves of different frequency reach maximal amplitude at different points along the cochlea. Long wave lengths, or low notes, cause a maximal displacement near the apex of the cochlea, whereas shorter wave lengths, or high notes, cause a maximal displacement at the proximal end of the basilar membrane. Von Békésy demonstrated the essential truth of this hypothesis by direct observations of the ears of fresh cadavers. More intense sounds, in addition to stimulating hair cells to move vigorously, cause a longer length of the basilar membrane to vibrate, but the displacement peak remains the same for any given frequency. Two musical instruments playing the same note have dissimilar qualities because of differences in the number and kinds of overtones, or harmonics, that are present. Both instruments stimulate the same part of the basilar membrane, but they vary with respect to the other parts of the membranes that are stimulated simultaneously. Loud sounds cause resonance waves of greater amplitude and lead to a more intense stimulation of the hair cells and the initiation of a greater

Position on Basilar Membrane

Figure 16.10 Displacement of the basilar membrane resulting from a traveling wave of a specific frequency. The shape of the wave at two instances in time is shown. Dashes outline the entire envelope of the wave. For any one frequency, maximum displacement of the membrane occurs only at one point. (From Case after Von Békésy.) (From Von Békésy, G.: J. Acoust. Soc. Amer. 19:452, 1947.)

number of impulses per second passing over the auditory nerve to the brain.

Careful histologic work has shown that the nerve fibers from each particular part of the cochlea are connected to particular parts of the auditory area of the brain, so that certain brain cells are responsible for the perception of high tones whereas other cells are responsible for the perception of low tones.

The normal human ear is just about as efficient a hearing device as anything could possibly be, for, like the eye, it has evolved to the point where any further increase in sensitivity would be useless. If the ear were more sensitive, it would pick up the random movement of the air molecules, which would result in a constant hiss or buzzing. If the eye were more sensitive, a steady light would appear to flicker, because the eye would be sensitive to the individual photons impinging on it.

There is little fatigue connected with hearing. Even though it is constantly assailed by noises, the ear retains its acuity, and fatigue disappears after a few minutes. When one ear is stimulated for some time by a loud noise, the other ear also shows fatigue; that is, it loses acuity, indicating, not unexpectedly, that some of the fatigue is in the brain rather than in the ear.

Impaired hearing or deafness may be caused by injuries or malformations of either the sound-transmitting mechanisms of the outer, middle or inner ears or of the sound-perceiving mechanisms of the latter. The external ear may become obstructed by wax secreted by the glands in its wall, the middle ear bones may become fused after an infection, or the inner ear or auditory nerve may be injured by a local inflammation or by the fever accompanying some disease. Conduction deafness can be corrected by a hearing aid that amplifies vibrations enough to be transmitted directly through the skull bones to the cochlea. If the auditory nerve or the cochlea is damaged, the resulting deafness usually cannot be corrected. Continued exposure to loud, high-pitched noises, as with a boilermaker or a member of a rock band, may destroy certain hair cells in the cochlea, leading to partial deafness.

The hair cells lining the cochlear duct are mechanoreceptors that respond with the generation of a receptor potential when they are stimulated by the movements of the basilar membrane.

Relatively few animals have a sense of hearing. The vertebrate ear began as an organ of equilibrium, the cochlea being a later evolutionary outgrowth of the saccule that reached full development only in mammals. The human ear is indeed a curious evolutionary hodgepodge: the cells sensitive to sound are apparently adaptations of cells sensitive to the motion of liquids; the middle ear and eustachian tube were originally part of the respiratory apparatus of fish; the stapes was originally a structure that attached the jaws of primitive fishes to the cranium; and the malleus and incus are the remnants of the lower and upper jaws, respectively, of our ancestral fish. In the jawless fish, ancestors of our ancestral fish, the structures were part of the support for the gills. Thus respiratory organs became eating organs first and then organs for hearing. This is an example of one of the fundamental patterns of evolution—the reshaping of old organs to perform new functions rather than the creation of completely new structures.

16.4 MUSCLE SPINDLES

The mammalian muscle spindle is a versatile stretch receptor and illustrates how sensory performance may be modified by the consequences of its own action. In the muscles of higher vertebrates there are, in addition to the usual striated muscle fibers termed **extrafusal fibers,** special **intrafusal fibers** associated with sensory nerve endings. A bundle of intrafusal fibers, together with their sensory endings, is termed a **muscle spindle** (Fig. 16.11). Intrafusal fibers are striated except in the region of the nucleus. There are two kinds of nerve endings: flower-spray endings belonging to a thin afferent nerve and annulospiral endings belonging to a thick afferent nerve. In the region of attachment of the extrafusal muscle fibers to the tendon there is another sense organ, the **tendon** or Golgi **organ.** There are two sets of motor neurons to the muscle: the alpha efferents innervate the ordinary, or extrafusal, muscle fibers, and the gamma efferents innervate the intrafusal fibers. For violent muscular contraction, impulses from the central nervous system come mostly by way of the alpha efferents. If, as a consequence, the muscle is stretched excessively, the Golgi organ and muscle spindle are stimulated. Messages from the Golgi organ pass up the sensory nerves to a point within the spinal cord where they synapse with the alpha efferents. They inhibit the alpha efferents and the muscle stops contracting; thus tension is kept within bearable limits, and the muscle is kept at a constant length under a specific load. In the production

Figure 16.11   A diagram of the muscle receptor system showing the relations between the intrafusal and extrafusal fibers, the tendon organ, the spindle receptors and the motor innervation. (From Villee, C. A.: Biology. 7th ed. Philadelphia, W. B. Saunders Company, 1977.)

Figure 16.12 Golgi spindle receptors have a series relationship to the muscle fibers. Spindles are in parallel. Pull on the muscle (A) increases the rate of firing of both receptors. Active contraction of the muscles (B) will cause an increase in discharge of the Golgi tendon organ and a decrease in rate of discharge from the spindle. (From Ochs, S.: Elements of Neurophysiology. New York, John Wiley & Sons, Inc., 1965.)

of slow voluntary movements, impulses from the central nervous system pass down the gamma efferents to the intrafusal fibers, which begin a slow, graded contraction. As a consequence, the muscle spindle is stimulated and sends impulses to the synapse with the alpha efferent, which is excited to cause the extrafusal fibers to contract. Since the intrafusal fibers are connected in parallel with the other fibers, rather than in series, as are the Golgi organs, they become slack (Fig. 16.12). The spindle then stops exciting the alpha efferents and the muscle ceases contracting. The net result is to cause the muscle to come to a new state of tension; thus muscle tone is maintained and precise voluntary movements can be made.

While the muscle has been contracting, it has been stretching its antagonist. Naturally, the spindle of the antagonist muscle excites it to contract. If it were to continue to do so in the face of the pull being exerted upon it, the excessive strain would stimulate its Golgi organ. This inhibits its contraction. Thus antagonistic muscles are prevented from fighting each other.

16.5 CHEMORECEPTORS

Throughout the Animal Kingdom, many sexual, reproductive, social and feeding activities are regulated or influenced by specific chemical aspects of the environment. Insects, for example, use a great many chemicals in communication, for defense from predators and for the recognition of specific foods. Many vertebrates employ chemical secretions to mark territory, to attract their sexual partners or to defend themselves. Chemoreception is

also involved in the tracking of prey by carnivores and in the detection of carnivores by their intended prey.

A common chemical sense is widely distributed among aquatic animals. This sense responds rather nonspecifically to the presence of dilute acids, bases or salts. Free nerve endings in the skin are the chemoreceptors involved, and their stimulation usually results in a sensation of pain. Among mammals the sense is restricted to the moist areas of the body. In contrast, specific and highly sensitive chemoreceptive systems compose the senses of taste and smell. These are easily distinguished in ourselves and other terrestrial orga-

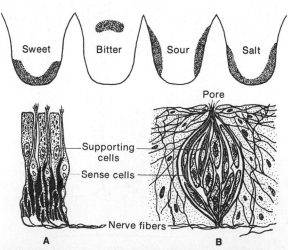

Figure 16.13 *Above,* The distribution on the surface of the tongue of taste buds sensitive to sweet, bitter, sour and salt. *Below*: A, Cells of the olfactory epithelium of the human nose. B, Cells of a taste bud in the epithelium of the tongue. (From Villee, C. A.: Biology. 7th ed. Philadelphia, W. B. Saunders Company, 1977.)

Opening
at tip

Thin-walled
cavity

Thick-walled
cavity

Dendrites of
chemoreceptors

Socket of
hair

Cuticula

Hypodermis

Cuticular sheath
enclosing dendrites

Axons

Vacuole

Trichogen
(hair-forming cell)

Tormogen
(socket-forming cell)

Bipolar neurons

Figure 16.14 Diagram of a chemo-receptive hair of the blowfly. Three of the five neurons are shown. (From Dethier, V. G.: Insects and the concept of motivation. *In* Levine, D. (ed.): Nebraska Symposium on Motivation. Lincoln, University of Nebraska Press, 1966.)

nisms, but as one examines aquatic organisms, especially those lower in the phylogenetic scale, it becomes increasingly difficult to decide what is taste and what is olfaction. Human beings depend primarily on visual or auditory cues. We use our chemical senses much less than other mammals do and tend to minimize their importance.

The organs of taste are budlike structures located in mammals on the tongue and soft palate, but in lower vertebrates they are found in many parts of the mouth and pharynx, and even in some tissues on the skin of the head. There is very rapid replacement of taste cells in the human, each cell being replaced within 10 to 30 hours. Each taste cell has at its free surface a border of microvilli, many of which project into a tiny pore connecting with the fluids bathing the surface of the tongue. There are no taste hairs as was once believed. Each taste cell is an epithelial cell and is the receptor. The connections with the nerve cells are complex, for each taste cell is innervated by more than one neuron. Some neurons may connect with one taste cell and others with many. The complexity of connections renders

difficult an interpretation of taste sensory physiology.

Traditionally there are four basic tastes: sweet, sour, salt and bitter. To this must now be added water. It is true that the greatest sensitivity to each of these tastes is restricted to a certain area of the human tongue—not all of the papillae are restricted in their sensitivity to a single taste modality (Fig. 16.13). Some are specific to salt, acid or sugar, but most of the taste buds respond to two or more categories of taste solutions. Thus the detection and processing of information in the taste organs of the tongue are very complicated. Taste discrimination probably depends on a code that consists of cross-fiber patterning; that is, each receptor responds to more than one kind of chemical but no two respond exactly alike, so that the total pattern of messages going to the brain is different for different solutions.

Flavor does not depend on the perception of taste alone. It is compounded of taste, smell, texture and temperature. Smell affects flavor because odors pass from the mouth to the nasal chamber by way of the choanae.

16.6 THE SENSE OF TASTE IN INSECTS

One of the most thoroughly studied organs of taste is the taste hair of the fly (Fig. 16.14). The terminal segments of the legs and the mouth parts of flies, moths, butterflies and certain other insects are equipped with very sensitive hairs and pegs. In the fly each one of these contains four taste receptors and a tactile receptor. All are primary sensory neurons. One taste receptor is more or less specific to sugars, one to water and two to salts. If water is placed on one hair of a thirsty fly, action potentials generated by the water cell pass directly to the central nervous system and cause the fly to respond by extending its retractible proboscis and drinking. Similarly, sugar on one hair stimulates the sugar receptor and causes the fly to feed. Salts cause the fly to reject the solution.

The Sense of Smell (Olfaction)

The sense of smell of terrestrial vertebrates is served by primary neurons located in the nasal epithelium in the upper part of the nasal cavity (Fig. 16.15). Each of these neurons has a short axon that passes through the cribriform plate and immediately synapses with other neurons. The complexity of this neural circuitry is illustrated in Figure 16.16. In the rabbit, for example, there are some 10^8 receptors. Twenty-six thousand of these synapse in the olfactory bulb of the brain with each **glomerulus,** and each glomerulus connects with 24 **mitral cells** and 68 **tufted cells.** The possibilities for processing olfactory data generated by the receptors even before they reach the cerebrum are enormous. In contrast to the sensations of taste, the various odors cannot be assigned to specific classes. Each substance has its own distinctive smell. The olfactory organs respond to remarkably small amounts of a substance. The synthetic substitute for the odor of violets, ionone, can be detected by most people when it is present to the extent of one part in more than 30 billion parts of air. The sense of smell is rapidly fatigued, and air originally having a powerful stimulus may seem odorless after a few minutes. This fatigue is specific for the particular substance producing it. Thus, receptors that have become insensitive to one substance will react quite normally to another. This suggests that there are many different kinds of sense cells, each specific for a particular chemical. Some people either completely lack a sense of smell or are able to smell some substances but not others.

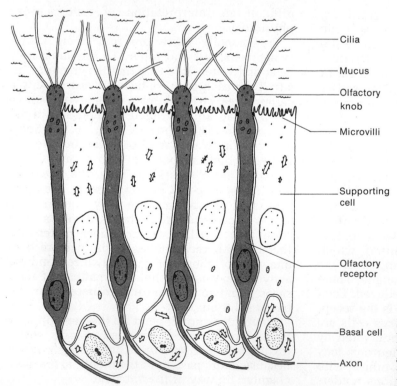

Cilia

Mucus

Olfactory knob

Microvilli

Supporting cell

Olfactory receptor

Basal cell

Axon

Figure 16.15 Simplified diagram of olfactory epithelium indicating the various cellular components. (From Moulton, D. G., and L. Beidler: Structure and function in the peripheral olfactory system. Physiol. Rev. 47:4, 1967.)

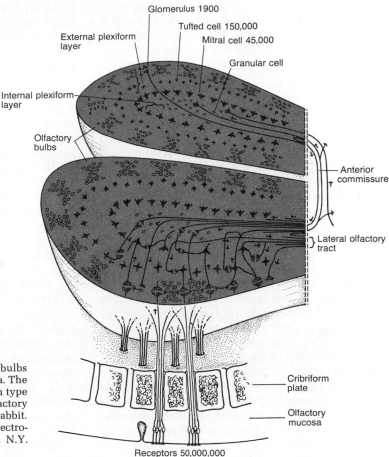

External plexiform
layer

Glomerulus 1900

Tufted cell 150,000

Mitral cell 45,000

Granular cell

Internal plexiform
layer

Olfactory
bulbs

Anterior
commissure

Lateral olfactory
tract

Cribriform
plate

Olfactory
mucosa

Receptors 50,000,000

Figure 16.16 Structure of the olfactory bulbs and their relations to the nerves and mucosa. The figures are estimates of the numbers of each type of cell in an olfactory bulb and in the olfactory mucosa lining one nasal cavity of the rabbit. (From Moulton, D. G., and D. Tucker: Electrophysiology of the olfactory system. Ann. N.Y. Acad. Sci. 116:380–428, 1964.)

16.7 PHOTORECEPTORS

Light-sensitive cells exist in almost all organisms. Even protozoa respond to changes in light intensity, usually moving away from the source of light. Some protozoa have eye spots that are more sensitive to light than the rest of the cell. The circumference of the bell of the jellyfish contains clusters of photoreceptor cells that form simple eyes, or **ocelli.** These can detect general light intensity. Flatworms, such as planaria, have photoreceptors, bowl-shaped structures containing black pigment at the bottom of which are clusters of light-sensitive cells. These are shaded by the pigment from light coming from all directions except above and slightly to the front. This arrangement enables the planaria to detect the direction of the source of light, but the planarian eye cannot form images. Planaria have other light-sensitive cells over the entire body surface and will continue to react to light after the photoreceptors have been removed, although more slowly and less accurately.

A necessary first step in the evolution from

photoreceptor to true eye was the development of a lens to concentrate light on a group of photoreceptors. As better lens systems evolved, the photoreceptor became able to form images. The most highly developed eyes are found in insects, crabs, lobsters and other arthropods, in squids, octopuses and other cephalopods and in the vertebrates. Two fundamentally different types of eyes have evolved: the **camera eye** of the vertebrates and cephalopods and the **compound,** or **mosaic, eye** of the arthropods. The eyes of invertebrates are usually specializations of the skin and only in vertebrates is part of the eye, the retina, derived partly from the embryonic brain and partly from the skin. The function of the eye of most animals, and undoubtedly its original function, was to provide information about the general light intensity of the environment. Such information may have important implications for an animal. A low light intensity might mean that the animal is safely under cover or that it is dusk or night, at which time activity is increased or decreased. A point of high light intensity such as that produced by the sun or at the opening of a burrow or at the water surface might provide a cue for orientation. A sudden change in light intensity might indicate the passing of a predator and initiate a withdrawal or escape reflex.

16.8 THE VERTEBRATE EYE

The ancestral vertebrates had eyes of two types: a median eye on top of the head, which probably distinguished only between light and dark, and a pair of image-forming eyes on the sides of the head. Cyclostomes and a few reptiles retain functional median eyes, but in most groups it has become a small **pineal body** attached to the top of the brain. The mammalian pineal gland is a small organ with an endocrine role. Although the lateral image-forming eyes of different groups of vertebrates vary in their adaptation for seeing beneath water, in the air and under varying light intensities, all are similar in their major features. The eyes of all mammals, for example, are oval-shaped organs constructed on the principles of a simple camera.

The analogy between a vertebrate eye and a camera is complete: the eye (Fig. 16.17) has a **lens** that can be focused for different distances, a *diaphragm* (the **iris**) that regulates the size of the light opening (the **pupil**) and a light-sensitive **retina** that is located at the rear of the eye corresponding to the film of the camera. Next to the retina is a sheet of cells filled with black pigment that absorbs extraneous light and prevents internally reflected

light from blurring the image (cameras are also black on the inside). This sheet, the **choroid coat,** contains blood vessels that nourish the retina. The outer coat of the eyeball, the **sclera,** is a tough, opaque, curved sheet of connective tissue that helps protect the inner structures and maintain the rigidity of the eyeball. On the front surface of the eye, this sheet becomes the thinner, transparent **cornea** through which light enters. The surface of the cornea is covered with a layer of stratified epithelium, the **conjunctiva,** which is continuous with the epidermis.

A transparent elastic lens located just behind the iris bends the light rays coming in, bringing them to a focus on the retina. The lens is aided by the curved surface of the cornea and by the refractile properties of the liquids inside the eyeball. The cavity between the cornea and the lens is filled with a watery substance, the **aqueous humor;** the larger chamber between the lens and the retina is filled with a more viscous, gelatinous fluid, the **vitreous humor.** Both fluids are important in maintaining the shape of the eyeball. The aqueous humor is secreted by the ciliary body, a doughnut-shaped structure that attaches the ligament holding the lens to the eyeball. It is drained through the canal of Schlemm at the base of the cornea. The aqueous humor helps to nourish the cornea and lens, which lack blood vessels, and by maintaining the intraocular pressure, it helps to maintain the shape of the eyeball. An imbalance between the rates of production and drainage of the aqueous humor can lead to increased intraocular pres-

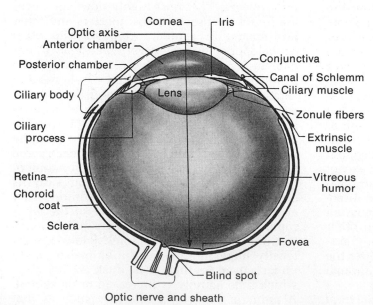

Figure 16.17 Diagram of a section through a mammalian eye.

sure and the disease **glaucoma,** in which the pressure flattens and eventually injures the retina. A number of zonule fibers extend from the ciliary body to the lens and help to hold the lens in place. Muscles within the ciliary body affect the lens' shape through the zonule fibers and focus the image sharply on the retina.

The amount of light entering the eye is regulated by the iris, a ring of muscle that appears in the human as blue, green or brown, depending on the amount and nature of the pigment present. The iris is composed of two sets of muscle fibers; the one arranged circularly contracts to decrease the size of the pupil and the one arranged radially contracts to increase the size of the pupil. The response of these muscles to changes in light intensity is not instantaneous but requires some 10 to 30 seconds to adapt either to a very dimly lighted area or to the high light intensity of full sunlight.

Each eye has six muscles stretching from the surface of the eyeball to various points in the bony socket of the skull. These enable the entire eye to be moved and oriented in a given direction. In the human and other animals with binocular vision these muscles are innervated in such a way that the eyes normally move together and focus on the same area. In terrestrial vertebrates a pair of movable eyelids cover the eyeballs, and the cornea is kept moist, cleansed and perhaps nourished by the secretions from tear glands. Tears are drained from the median corner of the eye by a lacrimal duct that leads into the nasal cavity. Many

mammals, such as cats, have a **nictitating membrane,** a third eyelid in the median corner of the eye. It is moved passively over the cornea when the eyeball is retracted slightly and aids in cleaning and protecting the eye. This membrane is reduced to a vestigial **semilunar fold** in the human.

The light-sensitive part of the vertebrate eye is the **retina,** a hemisphere made up of an abundance of receptor cells called, according to their shape, **rods** and **cones.** In the human eye there are some 125,000,000 rods and 6,500,000 cones. In addition, the retina contains many sensory and connector neurons and their axons. Curiously, the light-sensitive cells are located at the *back* of the retina and directed away from the source of light. To reach them, light must pass through several layers of neurons. This seemingly inappropriate arrangement is explained by the fact that the eye develops as an outgrowth of the brain and folds in such a way that the sensitive cells eventually lie on the farthermost side of the retina (Fig. 16.18). The polarity of the light-sensitive cells is retained during their various developmental maneuvers. The fact that the retina and optic nerves are developmentally parts of the brain also explains why at least two types of afferent neurons, bipolar and ganglion cells, are involved in transmitting impulses from the rods and cones. Chains of neurons are common in brain tracts, but in most peripheral nerves only one neuron extends from a receptor cell to the brain or spinal cord.

In the center of the retina of the human eye, directly in line with the center of the cornea

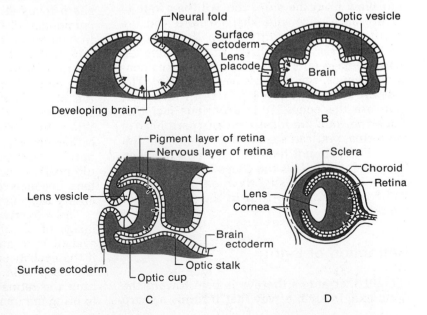

Figure 16.18 The development of the eye. *A,* Cross section through an embryo in which the anterior portions of the neural folds are closing to form the brain; *B,* the optic vesicles evaginate from the sides of the forebrain; *C,* an optic cup develops from each optic vesicle and the lens forms from adjacent surface ectoderm; *D,* the choroid, sclera and part of the cornea develop from surrounding mesoderm. Arrows indicate the original polarity of the ectoderm cells. (*D* from Romer.)

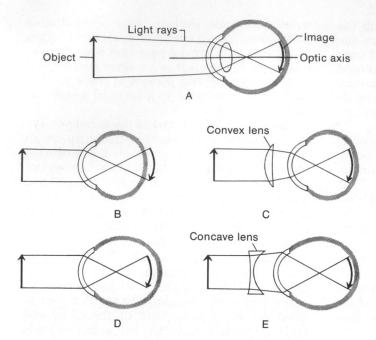

Figure 16.19   Image formation by the eye. *A,* Normal eye; *B,* far-sighted eye; *C,* far-sighted eye corrected by a convex lens; *D,* near-sighted eye; *E,* near-sighted eye corrected by a concave lens.

and lens (the optical axis), is the region of keenest vision, a small depressed area called the **fovea.** Here are concentrated the light-sensitive cones responsible for vision in bright light, for the perception of detail and for color vision. The other light-sensitive cells, the rods, are more numerous in the periphery of the retina, away from the fovea. These function in twilight or dim light and are insensitive to colors. You ordinarily are not aware that only those objects more or less directly in front of the eye can be perceived in color, but this can be demonstrated by a simple experiment. Close one eye and focus the other on some point straight ahead. As a colored object is gradually brought into view from the side, you will be aware of its presence, size and shape before you are aware of its color. Only when the object is brought closer to the direct line of vision, so that its image falls on a part of the retina containing cones, can its color be determined. The rods are actually more sensitive in dim light than are the cones. It is a curious fact that since the rods are located in the periphery of the retina, you can see an object better in dim light if you do not look at it directly (for then its image will fall on the cones in the center of the retina) but look slightly to one side, so that its image falls on the rods in the periphery of the retina.

Refraction of Light

Light that enters the eye is bent toward the optic axis in such a way that it forms a sharp inverted image upon the retina (Fig. 16.19). The lens is important in bending light rays, but the cornea and aqueous and vitreous humors are also involved. The cornea is the major refractive agent in terrestrial vertebrates, for the difference between the refractive index of air and the cornea is greater than that between any of the other refractive media. The action of the cornea places the image approximately on the retina; the lens brings it into sharp focus.

When the eye is at rest, distant objects are in focus. The refractive power of the eye must be increased in viewing a near object or its image will be blurred, for the image will come into sharp focus at a point behind the retina. Accommodation for near vision is accomplished in mammals by the contraction of muscles within the ciliary body. This brings the point of origin of the zonule fibers a bit closer to the lens and releases their tension. The front of the elastic lens bulges out slightly and its refractive powers are increased accordingly. When the ciliary muscles are relaxed, intraocular pressure pushes the wall of the eyeball outward and increases the tension of the zonule fibers, and the lens is flattened a bit. The lens becomes less elastic with age, and our ability to focus on near objects decreases.

The refractive parts of the eye form a sharp image of an object on the retina only if the eyeball is of an appropriate length. In theory, if the eyeball is shorter than normal, as in far-sighted people, the image of an object falls behind the retina. Accommodation is necessary to bring the image into focus, and the power of accommodation may not be great enough to focus on a near object. This can be corrected by placing a convex lens in front of the eye (Fig. 16.19C). Nearsighted people have eyeballs that are longer than normal and the image falls short of the retina. This can be corrected by placing a concave lens in front of the eye (Fig. 16.19E).

The Chemistry of Vision

The light strikes the rods and cones and activates them; they, in turn, initiate nerve impulses. The outer segment of each rod contains an elaboration of the membrane system of the cell, and a great deal of the pigment **rhodopsin** is associated with these membranes. All visual pigments have a common plan, a chromophore (retinal) bound to a protein (opsin). The combination is known as rhodopsin, or **visual purple.** Rhodopsin is the visual pigment in the rods of most vertebrates and in the retinal cells of insects. The cones contain iodopsin, a visual pigment consisting of the same chromophore (retinal) but a different protein. Iodopsin is found in humans, chickens, cats, snakes, frogs and crayfish. Fresh water fish have different visual pigments. Their rod pigment, **porphyropsin,** consists of retinal$_2$ plus opsin. The cone pigment, **cyanopsin,** consists of retinal$_2$ plus a different protein. Cyanopsin is found in the eyes of fresh water fish, tortoises and tadpoles. It is interesting that euryhaline eels, salmon and trout have both rhodopsin and porphyropsin, but the pigment commonly associated with spawning (in fresh water) predominates. The sea lamprey has mostly rhodopsin on its downstream migration to the sea and mostly porphyropsin when swimming upstream as a sexually mature adult. Amphibians change from porphyropsin to rhodopsin when they develop from tadpole to adult.

Retinal is the aldehyde of Vitamin A (retinol) and is formed by an oxidation catalyzed by alcohol dehydrogenase. Quanta of light striking rods or cones trigger a chemical change in retinal that results in the emission of a nerve impulse by the receptor cell. These structures are ready to discharge, having been charged with the requisite energy by internal chemical reactions.

The eye is called upon to respond to an enormous range of light intensities, and the rhodopsin system is peculiarly adapted to provide for a wide range of responses. The eyes of mollusks, arthropods and vertebrates, which arose quite independently in the course of evolution, all utilize the same basic chemical reaction—the conversion by light of the *cis* form of retinal to the *trans* form. This involves a rearrangement of the molecular structure, which can occur very rapidly. Light energy converts rhodopsin (with the cis form of retinal) into **lumirhodopsin,** an unstable compound containing the trans form of retinal. It decays first into **metarhodopsin** and then into free retinal and opsin (Fig. 16.20). The light energy is needed only for the cis-trans isomerization; the other changes do not require light. Some of the changes affect, in a way not yet clear, the permeability of the membrane with which rhodopsin is associated and lead to the development of the receptor potential. Many believe that photoisomerization itself is the activating process, partly because of its great speed and the speed with which the light affects the rods. It can be shown that a single quantum of light can be absorbed by a single molecule of rhodopsin and lead to the excitation of a single rod. When the eye is exposed to a flash of light lasting only one-millionth of a second, the eye sees an image of light that persists for nearly one-tenth of a second. This is the length of time that the retina remains stimulated following a flash and presumably reflects the length of time that lumirhodopsin persists in the rods. This persistence of images in the retina enables your eye to fuse the successive flickering images on a cinema or a television screen, and you have the impression of seeing a continuous moving picture.

A cycle of breakdown and resynthesis of rhodopsin goes on continuously if the eyes are exposed to any light, but the equilibrium point shifts to adapt for seeing in very dim or very bright light. Low levels of rhodopsin are present in the rods of an eye adapted to bright light, for much has been broken down into trans retinal, some of which has been converted to vitamin A. Most of the retinal involved in these reactions is contained in the pigment cells of the retina. In an eye adapted for seeing in dim light a maximal amount of rhodopsin is present in the rods, and some of the reserves of retinal have been used in its synthesis. Shifts between fully light- and dark-adapted retinas are gradual processes requiring some minutes. When the eye is suitably shielded from light, the breakdown of rhodopsin is prevented and its concentration gradually builds up until essentially all of the opsin has been converted to rhodopsin. The sensitivity of the eye to light, a function of the amount of rhodopsin present, can increase a thousandfold if the eye is dark-adapted for a few min-

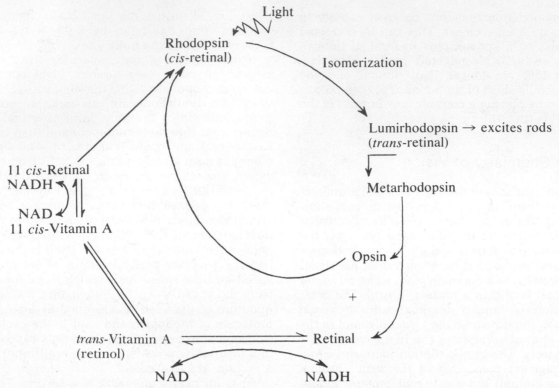

Figure 16.20 The rhodopsin-retinal system underlying the sensitivity of rods to light. (From Villee, C. A.: Biology. 7th ed. Philadelphia, W. B. Saunders Company, 1977.)

utes and about a hundred thousand-fold if the eye is dark-adapted for as long as an hour.

The chemistry of the cones and of color vision itself is less understood, but it is known that the cones contain a light-sensitive pigment, iodopsin, composed of retinal and a different opsin. The cones are considerably less sensitive to light than are rods and cannot provide vision in dim light. The prime function of the cones is to perceive colors. The evidence from psychological tests is consistent with the hypothesis that there are three different types of cones that respond respectively to blue, green and red light. This has been substantiated by the demonstration that three kinds of color receptors—red, green and blue—can be isolated from human and monkey retinas. Each can respond to light with a considerable range of wave lengths; the green cones, for example, can respond to light of any wave length from 450 to 675 nanometers (this includes blue, green, yellow, orange and red light), but they respond to green light more strongly than to any of the others. Intermediate colors, other than blue, green and red, are perceived by the simultaneous stimulation of two or more types of cones. According to this

theory, yellow light (i.e., light with a wave length of 550 nanometers) stimulates green and red cones to an approximately equal extent, and this is interpreted by the brain as yellow color. Colorblindness results when one or more of the three types of cones is absent because of the lack of the gene necessary for the formation of that kind of cone or cone pigment.

Organization of the Retina

The rods and cones of the retina synapse with bipolar cells (Fig. 16.21), which in turn synapse with ganglion cells. The axons of the ganglion cells cross the retina, come together at the blind spot and pass to the brain as the **optic nerve.**

The synaptic relationships of rods and cones differ in an important way. Many rods—as many as several hundred in the peripheral part of the retina—converge on a single bipolar cell, and a number of bipolar cells may converge on a single ganglion cell. Far fewer cones converge on a single bipolar cell, and in the fovea many of the cones synapse with a single bipolar and ganglion cell. The dif-

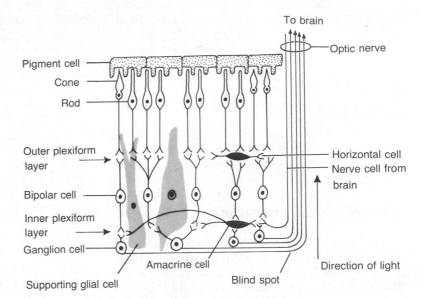

Figure 16.21 A diagrammatic vertical section through the retina to illustrate major types of interconnections among the component cells.

ference in neuronal pathways, together with a difference in the threshold of illumination, explains why rods are more effective than cones in receiving light of low intensity. Stimuli from a number of rods, each of which is below the threshold of a bipolar cell, converge upon a single cell, have an additive effect upon it and may activate it. Much less summation, or none at all, occurs in the more "private line" system of the cones. Since the greatest density of rods and the greatest degree of convergence occur in the periphery of the retina, one can see best in dim light by looking out of the side of the eye, so that the image falls to the side of the fovea. Rods are particularly abundant in the eyes of nocturnal vertebrates.

Although rod vision is more sensitive to dim light than cone vision, it also follows from their different neuronal pathways that rod vision is less acute. The eye cannot distinguish which rods are receiving light among a number of rods that converge on a single bipolar cell. In contrast, cone vision in the fovea is extremely acute, because most of the cones have a direct line to the brain. Distinct pathways to the brain from cones sensitive to different wave lengths of light are also neces-

sary for an animal to distinguish and interpret colors. Rod vision may be compared to a high speed, coarse grain, black-and-white photographic film and cone vision to a slower, fine grained, color film.

There are some further neuronal complexities within the retina (Fig. 16.21). A network of **horizontal cells** interconnects rods and cones and the bipolar cells at the level of the synapses between the receptors and bipolar cells. A second network of **amacrine cells** interconnects the bipolar cells at the level of their synapses with the ganglion cells. Some of these interconnections appear to be inhibitory and others facilitatory. Ganglion cells have a certain background rate at which they discharge impulses, even in the dark. If a beam of light impinges on the retina, ganglion cells receiving impulses from the activated rods and cones first increase their rate of discharge and then taper it off nearly to the background level (Fig. 16.22). When the light is turned off, the rate of discharge may stop momentarily before the background rate resumes. Ganglion cells receiving from the area of the retina immediately adjacent to the illuminated area respond in the opposite way. They are inhibited and cease their background discharge during the

Figure 16.22 Differences in the rate of discharge of impulses of a ganglion cell receiving from an illuminated part of the retina (Excitation), and one receiving from an adjacent nonilluminated part (Lateral Inhibition). Impulses are represented by the vertical lines. (Modified from Guyton, A. C.: Textbook of Medical Physiology. 5th ed., W. B. Saunders Co., 1976.)

period of illumination of their neighbor's field. When the light goes off they briefly increase their rate of discharge. These mechanisms would appear to increase the contrast between the areas of the retina receiving different intensities of illumination. Although most of the cells in the retina carry impulses to the brain, a few neurons carry impulses from the brain to the retina. These fibers probably have a role in enhancing or suppressing retinal activity, but we know very little about them.

Eyes of Other Vertebrates

The eyes of all vertebrates are essentially alike, but those of primitive vertebrates differ from mammalian eyes in several respects, for the problems associated with sight beneath water are not identical with those in the air. The water itself cleans and moistens the eye, and fishes have not evolved movable eyelids or tear glands. Secondly, the refractive index of water is nearly the same as that of the cornea, so the cornea of a fish's eye does not bend light rays very much. Most refraction is accomplished by the lens, which is nearly spherical and hence has a greater refractive power than the oval lens of tetrapods. It is interesting in this connection that the lens of a frog's eye flattens a bit during metamorphosis when a change in environment occurs. Finally, the method of accommodating for near and far vision is different, for in the fishes and amphibians the lens is moved back and forth in a camera fashion and does not change shape.

The Cephalopod Eye. The eye of a cephalopod, such as a squid, octopus or nautilus, is highly developed and parallels the eye of vertebrates to a striking degree (Fig. 16.23). The cephalopod eye has a retina, a movable lens, an iris diaphragm and a cornea, but the photoreceptors in the retina are directed toward the source of light instead of away, as in vertebrates. The cephalopod eye has a large number of photoreceptor cells, which suggests that object discrimination is possible. To discriminate an object requires that there be enough photoreceptors present so that the neuron light points can form an object when put together. Behavioral studies suggest that cephalopods are very nearsighted animals.

16.9   THE COMPOUND EYE OF ARTHROPODS

Most arthropods have eyes. Some are simple, having only a few photoreceptors; others are large, with thousands of retinal cells. A transparent lens-cornea is the usual skeletal contribution to the eye, and the focus is usually fixed since the immovable lens is continuous with the surrounding exoskeleton. Insects and many crustaceans, such as crabs and shrimp, have a **compound eye,** so called because it is composed of many long cylindrical units, each of which possesses all of the elements for light reception. Each unit, called an **ommatidium,** contains an outer cornea, a middle crystalline cone and an inner **retinula** (Fig. 16.24). On the external surface, the cornea of each ommatidium is distinct and forms one facet of the compound eye. The cornea functions as the lens and the crystalline cone funnels light

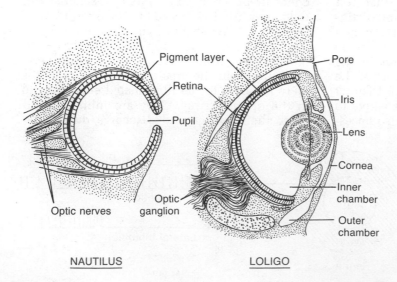

NAUTILUS   LOLIGO

Figure 16.23 Cephalopod eyes. *Left,* The pinhole camera type in *Nautilus.* (After Borradaile et al.) *Right,* The lens type complete with shutter (iris) in *Loligo.* (After Williams.)

Figure 16.24 A, Insect ommatidia, showing a diurnal type (*left*) and a nocturnal type (*right*). In the diurnal type, the pigment is shown in two positions, adapted for very dark conditions on the left side and for relatively bright conditions on the right. B, Nocturnal type of eye adapted for dark conditions, showing how light can be concentrated upon one rhabdome from several lenses. If the pigment moved downward, light from peripheral lenses would be screened out. (From Villee, C. A.: Biology. 7th ed. Philadelphia, W. B. Saunders Company, 1977.)

down to the retinula, which is equivalent to the retina and is composed of a rosette of seven monopolar neurons. Their inner photosensitive surfaces (rhabdomeres) together form a central **rhabdome.** Movable screening pigment is commonly present between adjacent ommatidia.

The term "compound eye" implies that each ommatidium forms a separate image, but this is untrue and misleading. The seven retinula cells function as a single photoreceptor unit and together transmit a single signal. The image formed by the entire eye, although sometimes called mosaic, depends upon all of the signals transmitted by the ommatidia, and thus is really no different from that of other types of eyes. Compound eyes may function differently in bright light than they do in weak light. In bright light the screening pigment is extended between the ommatidium and each retinula responds only to the light received by its facet and crystalline cone (Fig. 16.24 A). This type of image, called an **apposition image,** is believed to be especially effective in detecting movement, for slight changes in the position of the moving object will stimulate different ommatidia.

In weak light the screening pigment is contracted, and the light received by an ommatidium can cross over to adjacent units; thus, a retinula may be fired by the light received through several ommatidia (Fig. 16.24 B). Under these conditions the eye is said to form a **superposition image,** but it is unlikely that there is any object discrimination at all. Rather, the eye is detecting changes in general light intensity or the position of a bright light source or shadow. Although the eyes of some crustaceans can adapt to bright or dim light, the compound eyes of any given insect are usually adapted for functioning in either bright or weak light, but not in both. Most species are either

diurnal or nocturnal or live in habitats in which there is reduced light; hence, the eyes usually function under only limited light conditions. Eyes especially adapted for functioning in bright light may have a large amount of screening pigment, and the long crystalline cone is very close to the retinula. Eyes adapted for functioning in dim light may have little screening pigment, and the short crystalline cone is some distance from the retinula, permitting light rays to cross over from one ommatidium to another (Figure 16.24 B).

Visual acuity varies greatly and depends upon the number of photoreceptors present, regardless of the type of eye. The eyes of some hunting spiders and the compound eyes of many insects and some crabs appear to be capable of detecting objects. In all of these species the number of photoreceptors is very great. The eye of the dragon fly, for example, has 10,000 ommatidia, and the eye of a wolf spider has 4500 photoreceptors. However, when one compares these numbers with the 130 million photoreceptors in the human eye, it is evident that the images formed in even the most highly developed arthropod eye must be very crude. Color discrimination has been demonstrated in several arthropods and is an important adaptation of bees and other insects that depend on flowers as a food source.

Although the compound eye of the arthropod forms only coarse images, it compensates for this by being able to follow flicker to high frequencies. Flies are able to detect flickers of up to approximately 265 per second, whereas the human eye can detect flickers of only 45 to 53 per second. Because flickering lights fuse above these values, we see motion pictures as smooth movement and the ordinary 60 cycle light in a room as a steady light. To an insect, both motion pictures and light must flicker horribly. Because the insect has such a high critical flicker fusion rate, any movement of prey or enemy is immediately detected by one of the eye units. Hence, the compound eye is peculiarly well suited to the arthropod's way of life.

Compound eyes are superior to our eyes in two other respects. They are sensitive to different wave lengths of light from the red into the ultraviolet, and they are able to analyze the plane of polarization of light. An insect can see well into the ultraviolet, and its world of color is much different from ours. Since different flowers reflect ultraviolet to different degrees, two flowers that appear identically white to us may appear strikingly different to insects. How the world might appear to an insect with ultraviolet vision can be appreciated by viewing a landscape through a television camera with an ultraviolet transmitting lens. The sky that appears equally blue to us in all quadrants reveals quite different patterns to an insect because the plane of polarization of the light is not the same in all parts of the sky, and hence the insect's eye can detect the difference. Honeybees and some other arthropods employ this ability as a navigational aid.

16.10 THERMORECEPTORS

Cells sensitive to variations in temperature are found in a wide variety of animals. Paramecia will avoid warm or cold water and will collect in a region where the temperature is moderate. Some insects have thermoreceptors either in the antennae or all over the body. Insects that suck blood from warm-blooded animals are attracted to their prey by the temperature gradients nearby. Blood sucking bugs, for example, are much less able to find their prey after their antennae, which contain their thermoreceptors, have been removed. Fish have fairly sensitive thermoreceptors, for a change of only 0.5 degree C. will change the behavior of sharks and bony fish. Certain snakes have sensitive thermoreceptors, usually located in pits on the sides of their heads, by which they can detect the presence and location of warm-blooded prey.

16.11 ELECTRORECEPTORS

Certain skeletal muscles of fish, such as the electric ray or electric eel, have in the course

Figure 16.25 Electrolocation in the fish *Gymnarchas*. The fish generates an electric field, its tail negative to its head (*A*). Objects that conduct less (*B*) or more (*C*) electricity than the surrounding water distort the electric field and this is detected by special sense organs along the sides of the fish. (From Villee, C. A., and Dethier, V. G.: Biological Principles and Processes. 2nd ed. Philadelphia, W. B. Saunders Co., 1976.)

of evolution developed into electric organs capable of discharging pulses of electric energy. The electric eel, *Electrophorus*, can general pulses in excess of 500 volts. In other fishes the pulses of electricity given off are very weak. Researchers have found that these weak pulses are part of an electric guidance system. Certain species of tropical fish live in water that is so turbid that their eyes are essentially useless. As a means of orienting themselves, they set up a pulsating electric field around their bodies by means of their special electric organs. The electric field generated has the fish's tail negative with respect to its head. Any object in the environment having an electrical conductivity different from that of the water will distort the field, and these changes can be detected by the fish by special sense organs in its lateral line system (Fig. 16.25).

16.12 THE IMPORTANCE OF KNOWING WHICH WAY IS UP

It is clearly important for every animal to know at all times which way is up, so that it can contract is muscles or beat its cilia to move appropriately. When the organ in our inner ear responsible for equilibrium is malfunctioning, we become dizzy and unable to stand upright. Organs of balance, called **otocysts** or **statocysts,** are found in most phyla of animals, even in the cnidaria (Fig. 16.26). These are usually hollow spheres of sense cells, in the middle of which is a **statolith,** a particle

of sand or calcium carbonate pressed by gravity against certain mechanoreceptors. As the animal's body changes position, the statolith is pressed against different sense cells, and the animal is then stimulated to regain its orientation with respect to gravity. The statocysts of jellyfish are associated at specific sites between the scallop-like folds of the margin of the bell to form a **rhopalium.** The jellyfish carry out swimming movements that are primarily vertical; if the bell becomes tilted, stimuli from the statocysts bring about asymmetrical contractions and the bell is righted. The horizontal movements of the jellyfish depend largely upon water currents. Decapods, such as crabs and lobsters, have a statocyst at the base of each first antenna that opens by a pore to the exterior. The animals use sand grains as statoliths and must replace them after each molt. A newly molted crayfish given iron filings instead of sand grains will place them within the statocyst chamber and then will orient itself to a magnetic field rather than to gravity.

Membranous Labyrinths. All vertebrates have the ability to perceive differences in the orientation of their bodies with respect to their surroundings and to maintain their equilibrium. Although vision and proprioceptive impulses from the muscles and joints play a part, this ability is primarily a function of the inner ear. The inner ear contains a complex of membranous walls, sacs and canals called the membranous labyrinth, filled with a liquid **endolymph** and surrounded by a protective liquid cushion, the **perilymph.** The membranous labyrinths of nearly all vertebrates include the saccule, utricle and three semicircular canals.

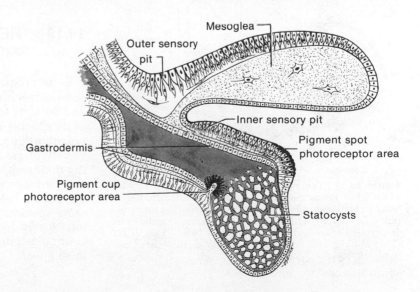

Figure 16.26 Section through a rhopalium showing the hood and the various sensory areas. (Modified from Hyman after Schewiakoff.)

The **utricle** and **saccule** are small hollow sacs containing patches of sensitive hair cells and small ear stones, or **otoliths,** made of calcium carbonate. Normally the pull of gravity causes the otoliths to press against particular hair cells, stimulating them to initiate impulses to the brain via sensory nerve fibers at their bases. When the head is tipped, the otoliths press upon the hairs of other cells and stimulate them.

Each of the three **semicircular canals** consists of a semicircular tube connected at both ends to the utricle (Fig. 16.27). The canals are so arranged that each is at right angles to the other two. At one of the openings of each canal into the utricle is a small bulblike enlargement, the **ampulla,** containing a clump of hair cells similar to those in the saccule and utricle but lacking otoliths. These cells are stimulated by movements of the endolymph within the canals. When the head is turned, there is a lag in the movement of the fluid within the canals; the hair cells move in relation to the fluid and are stimulated by its flow. This stimulation produces not only the awareness of rotation but also certain reflex movements in response to it, movements of the eyes and head in a direction opposite to the original rotation. The three canals are located in three different planes; hence movement in any direction will stimulate the movement of fluid in at least one of the canals.

Differences in the position of the head (**static equilibrium**) affect the way gravity pulls the otoliths on the underlying hair cells in the utricle and saccule (and we constantly know which way is up). Rapid forward movement (**linear acceleration**) causes the otoliths, which have more inertia than the fluid, to push back upon certain hair cells. During sudden turns of the head in various planes (**angular acceleration**), the endolymph of the semicircular canals, because of its inertia, does not move as fast as the hair cells in the ampulla, and their differential rate of movement stimulates the hair cells. We cannot distinguish between static equilibrium and linear acceleration on the basis of signals from the ear alone; inputs from other sensory organs are also necessary.

The Halteres of Flies.   Millions of years before humans invented the gyroscope, flies had evolved a balancing organ to stabilize flight. Any flying machine must maintain stability if it is to be controllable in the air. Flies must be able to control lift and to stabilize in all three planes of rotation; that is, they must correct for pitch, roll and yaw. Flies accomplish this with information derived from the **halteres,** a pair of marvelously modified hind wings. Each haltere is a heavy mass of tissue on a thin stalk (Fig. 16.28) and resembles an Indian club. The base is folded and articulated in a complicated fashion and is equipped with some 418 mechanoreceptors. These respond to strains produced in the cuticle by gyroscopic torque generated by the beating of the halteres. These oscillating masses generate forces at the base of the stalk as the whole fly rotates. They probably do not act as stabilizing gyroscopes of the sort placed in ships to offset the movement of the waves. Their action is indirect in that their mechanoreceptors signal the central nervous system to make the necessary corrections in flight.

16.13   THE RECOGNITON OF FRIEND AND FOE

Each animal must constantly analyze the signals reaching it by its various sense organs and be alert to the presence of other members of its species, as well as of other species in the neighborhood. Different kinds of animals use a variety of kinds of communications in analyzing their neighbors. Competition for food, mating partners or nesting space may result in aggressive behavior between members of the same species. When one animal is approached too closely by another, its first response is to assume a defensive posture. The members of most species of arthropods and vertebrates ex-

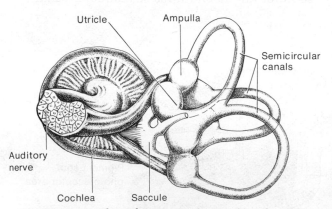

Figure 16.27   The right semicircular canals and cochlea of an adult human, shown dissected free of surrounding bone and enlarged about five times, seen from the inner and posterior side. Note that the plane of each semicircular canal is perpendicular to those of the other two. (From Villee, C. A.: Biology. 7th ed. Philadelphia, W. B. Saunders Company, 1977.)

A **B**

Figure 16.28 *A*, Ventral view and *B*, dorsal view of a model of the left haltere of the blowfly *Lucilia sericata*. Rows of sense organs are visible at the base. Magnification × 130. (Courtesy of J. W. S. Pringle.) (From Villee, C. A.: Biology. 7th ed. Philadelphia, W. B. Saunders Co., 1977.)

hibit an "individual distance," the minimal distance required to separate two animals without a fight ensuing. It can be pictured as the space around an animal into which only sexual partners may enter without eliciting aggressive behavior. Even the members of the opposite sex of the same species may elicit aggressive behavior and fighting at times other than the mating season. The photoreceptors and the sense of sight may be involved in recognizing neighbors. Organisms such as fireflies and certain deep sea fishes have luminescent organs that give off bursts of light by which members of the species can recognize each other and communicate. Other animals, such as crickets and frogs, emit sounds that are interpreted by other members of the species as warning or mating signals. Moths, butterflies and many mammals emit specific compounds that are carried in the air and subsequently picked up and interpreted by other members of the species as signals for mating or other activities. These chemical substances, termed **pheromones,** are particularly important in regulating the mating of certain moths (p. 458). Finally, certain insects and other arthropods, as well as certain vertebrates, recognize other members of their species and distinguish them from enemies by the sense of touch, using their antennae or mechanoreceptors in the skin. These are the evolutionary precursors of human encounter groups.

ANNOTATED REFERENCES

Alpern, M., M. Lawrence, and B. Wolsk: Sensory Processes. Belmont, California, Brooks-Cole Publishing Co., 1967. A general discussion of the perception of sensory stimuli.

Dethier, V. G.: The Hungry Fly. Cambridge, Harvard University Press, 1976. Describes the author's classic studies of the sensory systems of insects.

Goldsmith, T. H., and T. D. Bernard: The Visual System of Insects. *In* Rockstein, M. (Ed.): The Physiology of Insecta. Vol. 2. New York, Academic Press, 1965.

Gregory, R. L.: Eye and Brain: The Psychology of Vision. 2nd ed. New York, McGraw-Hill Book Company, 1973. A fine discussion of the role of the brain in the perception of sensory stimuli.

Katz, B.: Nerve, Muscle and Synapse. New York, McGraw-Hill Book Company, 1966. A recommended account of the functions of the nervous system.

Oatley, K.: Brain Mechanisms and Mind. New York, E. P. Dutton, 1972. Oatley, an experimental psychologist, has written a well-illustrated survey for the general reader.

Quarton, G. C., T. Melnechuk, and F. O. Schmitt (Eds.): The Neurosciences: A Study Program. New York, Rockefeller University Press, 1967. A fine series of chapters, each written by a different expert, covering a wide range of topics related to neurobiology. An excellent reference book.

Stevens, S. S., and F. Warshovsky: Sound and Hearing. New York, Time-Life Science Library, 1965. A well-illustrated account of the sense of hearing.

Thompson, R. F.: Introduction to Physiological Psychology. New York, Harper and Row Publishers, Inc., 1975. A text of the biological foundations of psychology, written for the undergraduate student.

NERVOUS SYSTEMS AND NEURAL INTEGRATION

17.1 IRRITABILITY AND RESPONSE

A fundamental property of all cells is response to stimuli. Cells are said to be irritable, and in response to a stimulus, waves of excitation are conducted along their surfaces. Such waves of excitation are even conducted, although very slowly, by eggs and plant cells. The nerve cell, or **neuron**, represents an evolutionary adaptation for the rapid transmission of a wave of excitation. They are found in all multicellular animals except sponges and collectively constitute the **nervous system**. This system makes possible the rapid integration of internal functions and adjustments to changes in the external environment. The evolution of the nervous system is closely connected with the evolution of animal locomotion. Indeed, heterotrophic nutrition, muscular movement and nervous integration are three distinct and interrelated features of animal design. (Although sponges have no nerve cells, waves of excitation can be conducted from cell to cell at a rate of about 1 cm. per minute). There are spindle-shaped contractile cells around the openings of the pores that have been termed "independent effectors" because they respond to touch by contracting and thus combine sensory and motor functions. The simplest special coordinating system is the nerve net of the cnidaria (p. 402). A diffuse network of nerve fibers is found all over the body; a few types of sea anemones and medusae have rudimentary nerve trunks composed of aggregations of nerve fibers. Conduction in the nerve net progresses in all directions; the fibers are not actually fused together, but impulses pass from one fiber to the next in either direction.

17.2 THE NEURON

The structural and functional unit of the nervous system of all multicellular animals is the neuron (Fig. 17.1). The average neuron is slightly less than 0.1 mm. in diameter, but it may be several meters long. Traditionally, it is pictured as having three parts—the axon, the cell body and the dendrite; a long **axon** emerges from one end of the cell body and bushy **dendrites** emerge from the other. However, there are so many exceptions to this generalization that the usefulness of this anatomical subdivision is rather limited. Functional distinctions are more accurate, for there are three functional parts of the neuron. The dendrites constitute the part of the neuron specialized for receiving excitation, whether from environmental stimuli or from another cell. The axon is specialized to distribute or conduct excitation away from the dendritic zone. It is generally long and smooth but may give off an occasional collateral. Within the central nervous system, neurons are surrounded and supported by cells termed **neuroglia.** Outside of the central nervous system, the neuron is wrapped in **Schwann cells.** Each axon ends in a distribution or emissive apparatus termed

Figure 17.1 Diagram of a variety of receptor and effector neurons, arranged to illustrate the idea that impulse origin, rather than cell-body position, is the most reasonable focal point for the analysis of neuron structure in functional terms. Thus, the impulse conductor or axon may arise from any response generator structure, whether transducing receptor terminals or synapse-bearing surfaces (dendrites, cell body surface or axon hillock). The interior of the cell body (chromidial neuroplasm, or perikaryon) is conceived of as related primarily to the outgrowth of axon and dendrites and to metabolic functions other than membrane activity. Thus, the position of the perikaryon in the neuron is not critical with respect to the "neural aspects" of neuron function—namely, response generation, conduction and synaptic transmission. Except for the stretch neuron of the crayfish, the neurons shown are those of vertebrates. (Redrawn from Bodian, D.: The generalized vertebrate neuron. Science 137:323–326, 1962.)

the **telodendria**. The **cell body**, which contains the nucleus, is concerned with metabolic maintenance and growth of the cell and may be situated anywhere with respect to the other parts. In the sensory nerves from the skin, for example, it is situated on an offshoot of the axon. Neurons exhibit an enormous variety in structure, and their different functional and anatomical types are characteristically found in various parts of the nervous system.

It is the axon that is usually responsible for the tremendous length of a neuron. The axon of a sensory cell, barely 0.1 mm. in diameter, located in the toe of a giraffe traverses a distance of several meters before ending in the spinal cord. The bundling together of many axons makes up the nerves and nerve trunks seen in an anatomical dissection. A common connective tissue sheath surrounds the nerves.

Neurons are classified as **sensory** neurons, **motor** neurons or **interneurons** on the basis of their function. Sensory **(afferent)** neurons are either receptors (olfactory receptors) or connectors of receptors (taste receptors) that conduct information from the receptor endings. Motor **(efferent)** neurons conduct in-

formation to the effectors, such as muscles, glands, electric organs and light organs. Interneurons connect two or more neurons and usually lie entirely within the central nervous system. In contrast, sensory and motor neurons typically have one of their endings in the central nervous system and the other close to the animal's external or internal environment.

The cell bodies of neurons are typically grouped together in masses called **ganglia.** In the simplest sense, a ganglion is any aggregation of neural cell bodies. The **dorsal root ganglia** of vertebrates (Fig. 17.2) are collections of cell bodies of sensory neurons, and the vertebrate **autonomic ganglia** are groups of motor neuron cell bodies. A ganglion is commonly an aggregation of neuronal cell bodies plus interneurons. It is a place where different neurons connect with one another and where much integration may occur. The aggregations of nerve cell bodies within the central nervous system of vertebrates are termed **nuclei.**

In the central nervous system of all animals the cellular part and the fibrous part of the neurons are separated into two zones. In vertebrates, the **gray matter** (usually on the inside but also on the outside in higher brain centers) contains cell bodies plus axons and dendrites. The **white matter** consists of axons plus their myelin sheaths. In invertebrate nerve cords, the outside consists solely of cell bodies and the inside consists of fibers.

A neuron consists of the usual cellular components: nucleus, cytoplasm that extends to the outermost branches and a cell or plasma membrane enclosing everything. Enveloping the axon outside the central nervous system is a cellular sheath, the **neurilemma**, composed of Schwann cells. These cells, migrating in from the mesenchyme, line up along axons and wrap around them. On some axons the Schwann cell lays down within its folds a spiral wrapping of insulating fatty material called **myelin** (Figs. 17.3 and 17.4). At the gaps, or **nodes of Ranvier,** between adjacent Schwann cells the axon is free of myelin. Axons lying within the brain and spinal cord have no neurilemma sheaths, and their myelin is provided by satellite cells (oligodendrocytes) rather than by Schwann cells. Nerves consisting of heavily myelinated fibers, such as those in the brain and spinal cord and those to skin and skeletal muscle, are white in appearance. Those with little or no myelin are gray.

The roles of the neurilemma and myelin sheaths are not completely understood, but there is clearly an interdependence between the neuron and its sheath cells. An axon separated from its cell body by a cut soon degenerates. A hollow tube of Schwann cells remains, but the myelin eventually disappears. As long as the cell body of the neuron has not been injured, it is capable of regenerating a new axon. Sprouting begins within a few days after the cut (Fig. 17.5). The growing axon enters the old sheath tube and proceeds along it to its final destination in the central nervous system or periphery. Axons can grow in the absence of a sheath if some conduit is provided for them. They can be made to grow within sections of blood vessels or extremely fine plastic tubes. The length of time required for regeneration depends on how far the nerve has to grow and may require as much as two years. Regeneration following a cut within the spinal cord or brain is very feeble and may be totally absent.

It is a remarkable fact that each regenerating axon of a cut nerve finds its way back to its former point of connection, whether this be a specific connection in the central nervous system or a specific muscle or sense organ in the periphery. If during the early stages of development of an amphibian one transplants an extra limb bud next to the normally developing limb, both will grow to maturity. The extra limb moves synchronously with the normal one. Anatomical examination reveals that the nerve that innervates the normal limb sends

Figure 17.2 Diagram of the primary types of sensory and motor neurons of the spinal nerves and their connections with the spinal cord. For convenience, the sensory neurons are shown on the left and the motor neurons on the right, though both kinds are found on each side of the body. (From Villee, C. A.: Biology. 7th ed. Philadelphia, W. B. Saunders Co., 1977.)

Figure 17.3 Sheath cells on neuron. *Upper:* Dissection of a myelinated nerve fiber. *Lower:* Envelopment of axis cylinder by a sheath cell. (After B. B. Geren, 1954; from Ballard, W. W.: Comparative Anatomy and Embryology. New York, The Ronald Press Company, 1967.)

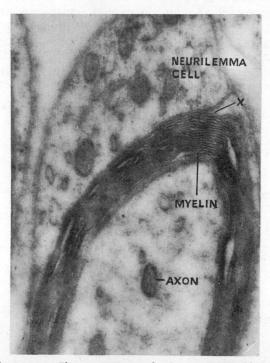

Figure 17.4 Electron micrograph of sciatic nerve fiber of a seven-day-old mouse showing the development of the myelin sheath by the folding of the cell walls of the neurilemma. The spiral infolding of the cell membrane of the neurilemma cell is visible as it is forming the thick, compact, many-layered myelin sheath. Note that at X the layers of myelin are continuous with the cell membrane of the neurilemma. Magnification ×83,000. (Courtesy of Dr. Betty G. Uzman.) (From Villee, C. A.: Biology. 7th ed. Philadelphia, W. B. Saunders Co., 1977.)

out branches to the extra one. The extra limb exerts some stimulating influence on the growing nerves to produce more branches and some directive influences as well.

After the optic nerve of the frog is cut, the axons of retinal cells regenerate and make proper connections within the central nervous system. This is confirmed by behavioral studies; when a frog with a regenerated optic nerve sees a fly it strikes at it accurately. This visual-motor performance indicates that there is a topological correspondence between points in the retina and visual areas of the central nervous system. If an eye is excised and rotated 180 degrees before reimplantation, the frog will strike 180 degrees off target. This aberration is explained by the fact that the retinal cells regenerate to their proper destinations in the central nervous system, but the retina itself is upside down with respect to the normal environment.

The principal function of the myelin sheath is to provide for saltatory conduction (p. 399), but it may have other functions as well. It is affected by the neuron it surrounds, for when the distant nerve cell body is cut off from its axon, the myelin surrounding the axon begins degenerating within a few minutes. The neuron thus has some trophic function that is necessary for the well-being of the Schwann cell. It is now clear that the nervous system has, in addition to its role of transmitting im-

Figure 17.5 Distal part of nerve fiber five minutes after cutting shown at the left. Retraction of the central end of the cut axon one hour later. Budding at third and fourth day. Slender axon grows into the distal degenerated stumps (upper part of figure) at days 5 and 7. (Courtesy of J. Z. Young; from Young, J. F.: Factors influencing the regeneration of nerves. Advances Surg. 1:165, 1949.)

pulses, important trophic relations with all the organs it innervates. The presence of nerves is essential for the regeneration of amputated amphibian limbs, for the normal maintenance of taste buds and for the continued functional integrity of muscles. Something other than impulses is transported along the long cellular extensions of the neurons, and it has been demonstrated that there is active flow of cytoplasm away from the cell body.

17.3 THE NERVE IMPULSE

The study of the nature of the nerve impulse has been fraught with special difficulties, because nothing visible occurs when an impulse passes along a nerve. Only with the develop-

ment of microchemical techniques was it possible to show that the nerve fiber expends more energy, consumes more oxygen and gives off more carbon dioxide and heat when an impulse is transmitted than it does in the resting state. This indicates that metabolic, energy-requiring processes are involved in the conduction of an impulse or in the recovery of a nerve after conduction, or both. The transmission of a nerve impulse obeys the "all-or-none" law: the conduction of the impulse is independent of the nature or strength of the stimulus starting it, provided that the stimulus is strong enough to start an impulse. The energy for the conduction of the impulse comes from the nerve, not from the stimulus, so that although the speed of the conducted impulse is independent of the strength of the stimulus, it is affected by the state of the nerve fiber. Drugs or low temperature can retard or prevent the transmission of an impulse. The impulses transmitted by all types of neurons are essentially the same. That one impulse results in a sensation of light, another in a sensation of pain and a third in the contraction of a muscle is a function of the way the nerve fibers are connected and not of any special property of the impulses.

Our present knowledge of the nature of the nerve impulse has been derived in large part from experiments using the large axons found in squid, crayfish and certain worms. The giant axon of the squid is large enough, nearly 1 mm. in diameter, that investigators can introduce microelectrodes and micropipets into the substance of the nerve fiber and measure the electrical potential across the nerve membrane. According to the present **membrane theory** of nerve conduction, the electrical events in the nerve fiber are governed by the **differential permeability** of the neuronal membrane to sodium and potassium ions, and these permeabilities are regulated by the **electric field** across the membrane. The interaction of these two factors, differential permeability and electric field, leads to a requirement for a critical threshold of change in order for excitation to occur. Excitation is a regenerative release of electrical energy from the nerve membrane, and the propagation of this change along the fiber is the **action potential,** the brief all-or-none electrochemical depolarization of the membrane.

In order to understand what happens when a nerve impulse passes along the fiber, we must first have a clear picture of the state of the resting nerve fiber. The resting neuron is a long, cylindrical tube whose plasma membrane separates two solutions of different chemical composition with the same total number of ions. In the external medium, sodium and chloride ions predominate; within the cell, potassium and various organic ions predominate. In the giant axon of the squid, the prototype for all axons, sodium is approximately 10 times more concentrated outside the membrane than inside. Chloride is five to 10 times more concentrated outside, and potassium is 30 to 40 times more concentrated inside. The neuronal membrane, which is about five nanometers thick, has high electrical resistance and capacity and low selective ionic permeability.

Potassium and chloride ions diffuse relatively freely across this membrane, but the permeability to sodium is low. Potassium tends to

leak out of the neuron and sodium tends to leak in, but because of the selective permeability of the membrane, potassium tends to leak out faster than sodium leaks in. This, plus the fact that the negatively charged organic ions within the cell cannot get out, causes an increasingly negative charge on the inside of the membrane. As the inside becomes more negative in relation to the outside, the exit of potassium ions is impeded. Ionic conditions would eventually change and come to a new equilibrium if something were not done to counteract this leakage of ions. The steady state is maintained by the **sodium pump**, which actively transports sodium ions from the inside to the outside against a concentration and electrochemical gradient. The sodium pump requires energy, utilizing ATP derived from metabolic processes within the nerve cell. The pump can be turned off experimentally by poisoning the nerve cell with a metabolic inhibitor, such as cyanide. The differential distribution of ions on the two sides of the membrane results in a potential difference of 60 to 90 millivolts across the membrane, the **resting membrane**

Figure 17.6 Diagram illustrating the membrane theory of nerve transmission. *A*, Resting nerve, showing the polarization of the membrane with positive charges on the outside and negative charges on the inside. *B–D*, Successive stages in the conduction of a nerve impulse, showing the wave of depolarization of the membrane and the accompanying action potential propagated along the nerve. (From Villee, C. A.: Biology. 7th ed. Philadelphia, W. B. Saunders Company, 1977.)

Figure 17.7 Generation of the nerve impulse. *A*, Increase in Na^+ permeability as membrane depolarizes. *B*, Na^+ and K^+ movement across membrane during action potential. *C*, Diagram illustrating flow of electrotonic current in vicinity of action potential (shaded), and examples of extracellular and intracellular recording of action potentials. With a penetrating microelectrode the entire action potential is recorded; whereas with an extracellular electrode only the positive overshoot is detected. (*A* and *B* adapted from Katz, B.: The Croonian Lecture: The transmission of impulses from nerve to muscle, and the subcellular unit of synaptic action. Proc. Roy. Soc. Biol. 155:455, 1962.)

potential, with the interior being negatively charged with respect to the exterior (Fig. 17.6).

The extrusion of sodium ions is accompanied by the entrance of potassium ions; there appears to be an exchange of cations at the cell surface, with a potassium ion entering for each sodium ion extruded. The relatively low ionic permeability of the membrane is such that even when the pump is poisoned with cyanide, many hours elapse before the concentration gradients of sodium and potassium across the membrane disappear.

Electrical studies of the **cable properties** of the nerve fiber show that the axon could hardly serve as a passive transmission line because its cable losses are enormous. When a weak signal is applied to the fiber, one too small to excite its usual relay mechanism, the signal fades out within a few millimeters of its origin. The nerve impulse could not be propagated over the long distances in the nerve unless there were some process to boost the signal repeatedly. The excitatory process regenerates and reamplifies the signal at each point along the nerve fiber. The cable proper-

ties of the nerve allow a change in electrical potential to spread along the nerve fiber for a short distance and to stimulate the excitatory process in the adjacent portion of the nerve.

The electrochemical potential energy of the resting potential forms the basis for the generation and transmission of the action potential (Fig. 17.7). The permeability of the membrane depends upon the magnitude of the transmembrane electrical potential. Although the permeability of the membrane to sodium is very low at the usual resting membrane potential, it increases as the membrane potential decreases. This permits the leakage of sodium ions down an electrochemical gradient into the interior of the nerve. This further decreases the membrane potential and further increases the permeability to sodium. The process thus is self-reinforcing and progressive, resulting in the upward deflection of the action potential. The entering sodium ions drop the transmembrane potential to zero and beyond, to about −40 or −50 millivolts. After one or two milliseconds, the permeability of sodium decreases and potassium begins to move out. This movement leads to the restoration of the resting potential; that is, to the repolarization in the membrane. When the membrane is completely repolarized, the permeability to potassium becomes normal and the excess sodium that entered during the process is slowly removed by the sodium pump. The actual quantities of ions that move in and out during the passage of an action potential are so small that there is no detectable change in the concentration of either ion in the fiber during an impulse.

Because of the changes in permeability that accompany the depolarization of the nerve membrane, the fiber cannot immediately transmit a second impulse. This period of inexcitability, termed the **absolute refractory period**, is brief and lasts until normal permeability relations have been restored. *A nerve impulse, then, is a wave of depolarization that passes along the nerve fiber. The change in membrane potential in one region renders the adjacent region more permeable, and the wave of depolarization is transmitted along the fiber.* The entire cycle of depolarization and repolarization requires only a few milliseconds.

Propagation of the action potential depends upon **electrotonic currents** that flow ahead of the nerve impulse. The process has some similarities to the charging and discharging of a condenser. At any point where an action potential has been generated, current flows ahead within the neuron, through the membrane and returns to the outer surface (Fig. 17.7C). The circuit is completed by current flowing in the solution that bathes the nerve. If the nerve is immersed in mineral oil, so that only a thin film of conductive fluid bathes the nerve, the external resistance is greatly increased and the rate of conduction of impulses is greatly decreased. The effectiveness of electrotonic currents in propagating the impulse depends upon the magnitude of the current and the resistance of the neuronal membranes, the cytoplasm and the surrounding medium. These factors determine at what distance from the active sites the membrane permeability will be sufficiently increased to start the regenerative sodium entry. The concept that the action potential is propagated by electrotonic currents moving ahead of it is termed the **local circuit theory of propagation.**

The rate of conduction of impulses increases as the diameter of the axon increases because the internal resistance declines. As a result, large nerve fibers conduct impulses more rapidly than do small ones. Giant nerve fibers have evolved in various members of the Animal Kingdom, although not in the higher vertebrates. The giant axons of the squid, crayfish and earthworm conduct impulses rapidly and have been studied intensively. They generally

Figure 17.8 Saltatory conduction in a myelinated axon. *A,* Recording at and between successive nodes of Ranvier demonstrates electrotonic transmission between nodes, as impulse appears instantaneously at all points within a node while diminishing in magnitude. At the node, a time delay occurs, and the impulse regains initial magnitude, showing the node to be the site of the active impulse-generating process. *B,* Support to the idea that electrotonic current flowing from a previous active site exits and excites at the next node comes from a demonstration that current (μA = microamperes) required to excite the axon is least at the nodes. (From Case, J.: Sensory Mechanisms. New York, The MacMillan Company, 1966.)

serve the purpose of conducting danger signals, for which speed rather than detailed information is critical.

In the vertebrates, a high rate of conduction is achieved by a different evolutionary development. The myelin sheath surrounding the neuron insulates and prevents the flow of current between the fluid external to the sheath and the fluid within the axon. At the **nodes of Ranvier** there are gaps in the sheath where there is no insulation. At these points free ionic communication between the inside and outside of the membrane is possible. Impulses are generated only at the nodes, and nerve impulses leap from one node to the next (Fig. 17.8). This type of transmission, termed **saltatory conduction**, enables a myelinated nerve that is only a few micrometers in diameter to conduct impulses at velocities of up to 100 meters per second, whereas the largest unmyelinated axons, nearly one millimeter in diameter, conduct with velocities of 20 to 50 meters per second.

Results from the experiments by the English physiologist Adrian, published in 1926, provided the explanation as to how the nervous system transmits differences in intensity. By applying graded stimuli to an isolated sense organ and amplifying and measuring the impulses in its nerve. Adrian showed that variations in the intensity of the stimulus lead to the variations in the frequency with which the impulses are transmitted: the stronger the stimulus, the more impulses per second. This principle is true for both vertebrate and invertebrate sensory nerves. In contrast, a single impulse in a vertebrate motor nerve elicits a single twitch of all phasic muscle fibers in the motor unit. The differences in the strength of contraction of the muscle as a whole are due to variations in the total number of motor units actively contracting at any given moment. In the motor nerves of many invertebrates, however, the frequency of the impulses does not affect the contraction strength of the innervated muscle. A single nerve impulse will, in general, not stimulate the muscle to contract. At least two successive impulses are required and the strength of contraction is inversely proportional to the interval between the two. In many arthropods all the muscle fibers in a given muscle are innervated by branches of a single axon. A single impulse in the axon will not produce contraction but repeated impulses will. The tension in the muscle increases with the frequency of the stimulation. It would appear that although the arrival of a single impulse at the nerve-muscle junction is unable to bring about muscle contraction, it does affect the junction in such a way as to make it possible for a second impulse to do so if it arrives soon enough after the first. This phenomenon is known as **facilitation.**

Neuronal Functions

The idea that a nerve fiber is simply a sort of cytoplasmic telephone wire has rapidly changed; we have come to appreciate that neurons are very dynamic and metabolically active cells. These cells carry out the intracellular transport of materials over long distances by the active cytoplasmic streaming and undulations of the axons. Neurons may be true secretory cells, producing and secreting neurohormones such as vasopressin, oxytocin and the hypothalamic-releasing factors. The regeneration of amputated limbs in salamanders requires specific neuronal influences and will not occur if the nerves are removed. When a mammalian axon is severed, a new cytoplasmic process grows out of its stump and advances by an ameboid tip at the rate of 3 or 4 mm. per day. This tip carries on pinocytosis and shows a remarkable ability to find and recognize its proper connection. After the tip has made contact with its proper distal portion, the mass of the axon is brought back to normal by a complex synthetic process accompanied by striking morphological changes in the cell body of the neuron. The cell body contains an abundant accumulation of endoplasmic reticulum and ribosomes, detected by light microscopy in the 19th century and termed **Nissl substance.** The Golgi apparatus of neurons is also abundant and well developed. The cytoplasm of the axon contains longitudinally oriented **neurofibrils** and **neurotubules** that are associated with the intracellular transport of molecules.

Changing the concentration of sodium ions on the outside of the membrane is accompanied by a decrease in the height of the action potential. For example, a decrease in the concentration of sodium ions on the outside of the membrane is accompanied by a decrease in the height of the action potential. To explain these observations, Hodgkin suggested that the change in membrane potential is due to the sudden penetration of sodium ions into the interior of the cell. The action potential is related to some physical change in the plasma membrane that results in an increased permeability to sodium ions. Direct measurements of Na^+ and K^+, using isotopically labeled ions, showed that the **sodium flux** inward increases twentyfold during stimulation and the **potassium flux** outward increases about ninefold. Thus the nerve impulse is accompanied by a sudden influx of sodium down an electrochemical gradient followed by an efflux of potassium down an electrochemical gradient that abolishes or reverses the resting potential. The actual number of ions that move through the membrane during a one millisecond pulse is very small and does not significantly change the concentration of Na^+ or K^+ ions on the two sides of the membrane. The change is detected primarily by its effect on the membrane potential.

17.4   TRANSMISSION AT THE SYNAPSE

The nervous systems of invertebrate and vertebrate animals are composed of individual discontinuous neurons. This requires some mechanism for the transfer of the neural message from the axon of one neuron to the den-

drite of the next or, at the neuromuscular junction, to the muscle. The junction between the axon of one neuron and the dendrite of the next is termed a **synapse**. At some specialized synapses, transmission is accomplished electrically. The arrival of the action potential at the end of the axon of the presynaptic neuron sets up electric currents in the external fluid of the synaptic gap. These currents, in turn, stimulate the dendrite of the postsynaptic cell to generate an action potential. Thus the transmission and the generation of the action potential take place in the postsynaptic cell in basically the same way as in a single nerve fiber. Electrical synapses of this sort are found in parts of the nervous system of the crayfish and the fish. Transmission across the giant synapse in an abdominal ganglion of the nerve cord of the crayfish is accomplished electrically. The membrane contact of this special synapse is able to act as a rectifier and allow current to pass easily from the axon of a connector neuron to the dendrite of a motor neuron but not in the reverse direction.

At most synapses, however, a gap of some

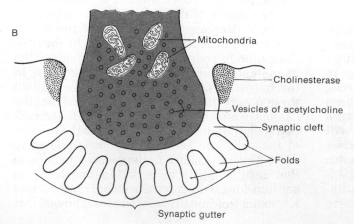

Figure 17.9 Diagram of a neuromuscular junction showing (*below*) the vesicles of acetylcholine in the sole foot of the neuron that penetrates into the membrane of the muscle fiber. This membrane becomes extensively folded to form the "synaptic gutter" in the plasma membrane of the muscle cell. The transmission of a nerve impulse from one neuron to a second neuron across a synapse is believed to involve a similar structural and functional junction. (From Villee, C. A.: Biology. 7th ed. Philadelphia, W. B. Saunders Co., 1977.)

Figure 17.10   Electron micrograph of a synaptic ending in the stimulated olfactory bulb of the rat. The synaptic ending contains three mitochondria (*mi*) and several synaptic vesicles (*sv*). The zone of contact between the two neurons is indicated by the two arrows. The nerve membranes appear to be thickened at "active points" (*ap*) in this zone of contact. (From De Robertis, E., and A. Pellegrino De Iraldi: Anat. Rec. *139*:299, 1961.)

20 nm. separates the two plasma membranes, and the impulse is transmitted across this gap by special chemical transmitters. A specific chemical is synthesized by the neuron and released from the tip of the axon (**neurosecretion**) when a nerve impulse reaches it. This diffuses across the synaptic gap and attaches to specific molecular sites in the dendrite (an example of **chemoreception**), producing a change in the properties of the membrane of the dendrite and leading to the initiation of a new nerve impulse. The chemical transmitted passes from axon to dendrite by simple diffusion. Over the short distance involved, diffusion is rapid enough to account for the speed of transmission observed at the synapse. Transmission at the neuromuscular junction and certain other synapses involves the secretion and chemoreception of acetylcholine. This potent stimulant causes a local depolarization of the membrane of the muscle cell, which sets up propagated impulses in the membrane and causes a contraction of the muscle fiber (Fig. 17.9). **Curare** prevents the transmission of impulses from nerve to muscle, specifically at this type of synapse, by combining with the receptors for acetylcholine and preventing their normal reaction with it.

The sympathetic postganglionic fibers accelerate the heart rate by releasing **norepinephrine.** Such fibers are termed **adrenergic,** whereas those that secrete acetylcholine are termed **cholinergic.** In the synaptic area are potent enzymes: acetylcholinesterase, which specifically hydrolyzes and inactivates acetylcholine, and monoamine oxidase, which oxidizes and inactivates norepinephrine. These enzymes prevent the continuous stimulation of a dendrite or muscle by the neurotransmitter material.

Acetylcholine is released by motor nerves in discrete tiny packets, each of which contains about 1000 molecules. The mechanism that releases acetylcholine requires calcium ions and is inhibited by magnesium ions. The transmitter substance is stored within the nerve endings in small intracellular structures that discharge their entire contents to the surface. Electron micrographs (Fig. 17.10) of the tips of the neurons at the synapse reveal masses of **synaptic vesicles** that appear to be the sites of storage of the neurotransmitters.

Thus the arrival of the nerve impulse leads to the liberation of the contents of one or more of these vesicles into the synaptic space. How an action potential induces the release of a packet of acetylcholine molecules from its vesicle is still unknown, but the existence of these vesicles provides a satisfactory explanation for the polarity of the synapse; that is, for the fact that a nerve impulse will travel in one direction but not in the reverse between axon and dendrite. Norepinephrine is concentrated in the synaptic vesicles of adrenergic fibers and is released by the arrival of an action potential. Other neurons within the central nervous system may have other neurochemical transmitters, such as **serotonin** or **dopamine.**

Synapses are important functionally because they are points at which the flow of impulses through the nervous system is regulated. Not every impulse reaching a synapse is transmitted to the next neuron. The synapses, by regulating the route of nerve impulses through the nervous system, determine the response of the organism to specific stimuli. The amount of synaptic resistance can be modified by nerve impulses. One impulse might cancel out the effect of another, a process known as **inhibition**. The opposite condition, whereby one impulse strengthens another, is called **facilitation**. These two processes are of prime importance in effecting the integration of body activities. Inhibition and facilitation can occur only in the synapse, since once an impulse starts along a neuron it can be neither stopped nor accelerated. Information transmitted along the axon as a spike potential is coded and decoded at the synapse by processes involving neurosecretion and chemoreception, graded thresholds and the facilitation and summation of inhibitory and excitatory impulses. These are the substrates of animal behavior and the basis of the complex human processes of learning, memory and intelligence.

17.5 EVOLUTION AND ORGANIZATION OF THE NERVOUS SYSTEM

The presence of a nervous system provides an animal with the means for receiving a variety of information from the outside environment and about its internal environment, for coordinating these bits of information and for responding appropriately. A well developed nervous system is associated with motility and large body size. An animal with no nervous system at all might be motile but very small (e.g., protozoa), and one with a poorly developed nervous system might be large but sessile (e.g., sponges and some cnidaria), but all animals that are motile and large have a nervous system capable of rapid transmission of the nerve impulse. The range of behavior of an animal is roughly proportional to the complexity of its nervous system.

Nervous systems function as transducers and amplifiers; they convert one form of energy into another, and they can take in a small signal and put out a much larger one. The central nervous system of the higher animal operates as an analytical computer.

Nerve Net Systems. In the primitive nerve net systems of the cnidarians and the acoels, the tips of the neurons are undifferentiated as to axon and dendrite, and excitation can be transmitted across the synapse in either direction. Thus each end of the neuron can both secrete a neurotransmitter and receive excitation from the adjacent neuron by its chemoreceptors.

The cnidarian system includes sensory neurons, motor neurons, and interneurons; there are some connections from sensory to motor neurons, but usually the connections are via interneurons. As a result of this nerve net arrangement, the conduction of impulses is generally diffuse, and excitation tends to spread in all directions from the initial point. However, some cnidarian nerve nets permit elaborate response patterns and can give rise to complex and coordinated behavior patterns.

The neurons of higher animals are specialized into three functionally distinct regions. The **generator region** is usually restricted to the dendrites and is characterized by specializations of the neuronal membrane that permit it to be excited by neurotransmitters from an adjacent neuron. The **conductile region** makes up most of the axon of the neuron. The **transmissional region** is composed of the tips of the axon and is specialized for the secretion of neurotransmitters, such as acetylcholine and norepinephrine. The cell body, or soma, usually does not participate in electrical activity. When a nerve is stimulated by an electrical stimulus, the resulting impulse is propagated along the axon (the conductile region) in both directions. The action potential propagating toward the transmissional region is termed **orthodromic**, and the impulse transmitted towards the dendrite is termed the **antidromic** action potential. Under normal conditions a nerve impulse always arises at the portion of the conductile region adjacent to the generator region (by a process that occurs in that region) and is conducted orthodromically to the transmissional region.

The generator regions of receptor neurons are activated by a variety of mechanical, chemical, thermal and other stimuli, and the generator regions of interneurons and motor neurons are stimulated through synaptic transmission by a neurotransmitter secreted by the adjacent neuron. In response to the neurotransmitter, the generator region may set up one or more propagated action potentials or nerve impulses in the conductile region.

The generator region of the neuron acts as an amplifier, generating an electric current greater than that acted on it. The generator region may also act as a transducer, converting one type of energy state or flow into a different type.

Most of the central synapses of invertebrates are **neuropiles** composed of interdigitating membranes from pre- and postsynaptic cells. A few examples are known of synapses across which transmission occurs electrically rather than by neurotransmitters; e.g., synapses in the giant fibers of annelids and crustaceans. For this to occur, the synaptic area must be large enough to minimize synaptic membrane resistance and to permit the current flowing through the postsynaptic membrane to exceed the minimum signal strength required to elicit a response.

The neurons of many types of animals have neurosecretory functions in addition to that in generating and conducting nerve impulses. The hormones produced have a variety of functions. Some nerves secrete what appears to be a carrier protein, termed **neurophysin**, to which the hormones are bound. Specific functions of neurohormones have been demonstrated in annelids, crustaceans, insects and vertebrates. The neurosecretory neurons typically have axons that terminate on or near a blood vessel or blood sinus. Neurohormones in certain annelids regulate the conversion of young worms into the sexually active reproductive forms. In crustaceans, neurohormones regulate carbohydrate metabolism, molting and metamorphosis, water regulation, sexual maturation and the movement of pigment in chromatophores.

A nervous system of the nerve net type with diffuse conduction is sufficient to meet the needs of sessile, radially symmetrical animals, such as the cnidarians, but not to meet the needs of motile, bilaterally symmetrical animals. Several trends in the evolution of the nervous system are evident: the appearance of synapses that permit transmission of impulses in one direction only to minimize or eliminate the diffuse conduction of the nerve net system; the aggregation of nerve processes to form one or more longitudinal bundles of nerve fibers; a concentration of cell bodies into groups (ganglia); and the further concentration of ganglia at the anterior end of the body (cephalization), correlated with the appearance of sense organs, to form a brain. The brain and longitudinal cords make up the central nervous system characteristic of the higher animals. Smaller bundles of nerves from the central system to the distal regions of the body constitute the peripheral nervous system. Thus nerves are composed of bundles of fibers of sensory or motor neurons, and ganglia are clusters of nerve cell bodies.

The most primitive flatworms, the acoels, have a nerve net system similar to that of the cnidarians and ctenophores that lies near the surface of the body at the base of the epidermis. Other acoels and flatworms have a nervous system that is sunk in beneath the muscle layer and organized in five or so pairs of cords arranged radially (Fig. 17.11). An aggregation of nervous tissue surrounds the statocyst at the anterior end of the body, the first hint perhaps of a "brain." In other flatworms and certain related groups of animals, there is a tendency for the number of pairs of nerve cords to be reduced. The Nemertea have two long lateral nerve cords and a smaller middorsal one. The flatworm nervous system lacks ganglia.

With further evolution, two types of nervous systems have developed: the dorsal, hollow, single nerve cord found in chordates and the ventral, solid, usually paired nerve cords with segmental ganglia characteristic of annelids and arthropods. The evolutionary development of the nervous system has included an enormous multiplication of units, diversification of form and specialization of function; an increasing complexity of interneuronal connections; a gathering together of neurons with common or associated functions into groups; and a centralization and cephalization of neural tissues.

The nervous systems of annelids, arthropods and mollusks typically include a brain and a longitudinal ventral nerve cord. The nervous system of the earthworm is composed of neurons, with axons and dendrites arranged in definite nerve cords and fibers (Fig. 28.21). Its system is differentiated into central and peripheral nervous systems with sensory neurons, motor neurons and interneurons

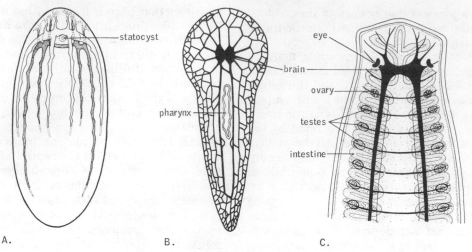

Figure 17.11 Turbellarian nervous systems. *A*, Radial arrangement of nerves in the acoel *Anaperus*. (After Westblad.) *B*, Ventral submuscular nerve net of polyclad, *Gnesioceros*. *C*, Anterior end of the nervous system of the planarian *Procerodes*. (Modified after Lang.) (From Barnes, R. D.: Invertebrate Zoology. 3rd ed. Philadelphia, W. B. Saunders Co., 1974.)

joined by synapses, so that nerve impulses travel in one direction only. This enables the central nervous system to act as an integrator, selecting certain incoming sensory impulses and passing them to effectors while inhibiting or suppressing others. The earthworm has a central nerve cord extending the entire length of the body that enables its separate segments to move in a coordinated fashion. Each segment has a pair of ganglia, a collection of nerve cell bodies. The most anterior part of the cord is an enlarged ganglion, a "brain" that sends impulses down the cord to coordinate movements. After the brain has been removed, the earthworm can move almost as well as before, but it persists in futile efforts to go ahead instead of turning aside when it comes to an obstacle. The brain, therefore, appears to be necessary for adaptive movements. The nervous systems of arthropods and mollusks are generally similar to that of the earthworm. In all of these, the nerve cord is ventral to the digestive system and is solid. Although the annelids and the more primitive arthropods have a pair of ganglia in each segment of the body, the ganglia of the higher arthropods tend to be fused and accumulate in the anterior part of the body.

Despite the small size of many invertebrates, their nervous systems are enormously complex. The crayfish system, for example, contains nearly 100,000 neurons, and the housefly has more than one million neurons. The sense organs of insects and decapods—the compound eye, taste organs and mechanical sense organs concerned with postural relationships and tactile discrimination—are highly devel-

oped. The synaptic connections in these animals are incredibly complex, and the brain and ganglia are jammed with integrating interneurons.

The nervous system and sense organs of cephalopods are among the most highly developed of all invertebrates. The ganglia, which in other mollusks are located at different points along the nerve cords, are concentrated at the anterior end to form a complex brain housed within a capsule. Functional centers of the brain have been identified, and the behavior of the octopus and other cephalopods has been studied intensively.

17.6   ORGANIZATION OF THE VERTEBRATE NERVOUS SYSTEM

The neurons of the vertebrate nervous system can be assigned either to the **central nervous system**, which includes the brain and spinal cord, or to the **peripheral nervous system**, which includes the nerves that extend between the central nervous system and the receptors or effectors. The functional organization of the nervous system can be introduced by considering a specific example. When you touch a hot object (Fig. 17.12), receptors in the skin are stimulated and they initiate impulses in afferent neurons. These neurons are part of a spinal nerve and extend into the spinal cord, where they synapse with interneurons. The interneurons in turn transmit impulses to efferent neurons that extend from the cord and carry impulses back through the spinal nerve to a group of extensor muscles in your hand. Their contraction withdraws your hand from the stove. For the movement to be effective, the antagonistic flexor muscles should relax, which involves the inhibition of impulses going to these muscles. Normally, some impulses pass to all of the muscles of the body continually and cause a partial contraction, called muscular **tonus**. Such a stimulus and response is a simple **spinal reflex**, and the neuronal pathway along which the impulse travels is called a **reflex arc**.

A reflex is an innate, stereotyped, automatic response to a given stimulus that depends only on the anatomical relationships of the neurons involved. The impulse need not pass through any of the higher centers of the brain in order for the response to occur. An impulse may be carried to the cerebral cortex by other interneurons, and you then become aware of the stimulus and may decide to do something about it—perhaps withdraw the entire arm or turn off the stove. Reflexes are the functional units of the nervous system, and many of our activities are the result of them. They are important in controlling heart rate, blood pressure, breathing, salivation, movements of the digestive tract and so on. When we step on something sharp or come in contact with something hot we do not wait until the pain is experienced by the brain and then, after delib-

eration, decide what to do. Our responses are immediate and automatic. The foot or hand is being withdrawn by reflex action *before* the pain is experienced. Many of the more complicated activities of our daily lives, such as walking, are regulated to a large extent by reflexes.

The reflexes present at birth and common to all human beings are called **inherited reflexes**. Others, acquired later as the result of experience, are called **conditioned reflexes**. The minimum anatomical requirements for reflex behavior are a sensory neuron with a receptor to detect the stimulus connected by a synapse to a motor neuron that is attached to a muscle or some other effector. This simplest type of reflex arc is termed **monosynaptic** because there is only one synapse between the sensory and motor neurons. More typically reflex arcs include one or more interneurons between the sensory and motor neurons. A simple reflex in which stimulation of a receptor produces contraction of a single muscle is typified by the **knee jerk**. When the tendon of the knee cap is tapped, and thereby stretched, receptors in the tendon are stimulated; an impulse travels over the reflex arc, up to the spinal cord and down again; and the muscle attached to the tendon contracts, resulting in a sudden straightening or extending of the leg.

A number of reflexes involve the connections of many interneurons in the spinal cord.

Figure 17.12 Diagram of the neurons involved in the passage of impulses to and from the brain and the relationship of these to the neurons in a reflex arc (shown simplified).

Figure 17.13 Diagrams of important types of neuronal interrelationships. *A*, A divergent pathway; *B*, a convergent pathway; *C*, a multiple chain circuit; *D*, a closed chain circuit. A given neuron may be involved in more than one of these pathways.

Not only does the spinal cord function in the conduction of impulses to and from the brain but it also plays an important role in the integration of reflex behavior. Its importance in this respect is demonstrated in a "spinal" animal, one whose brain has been destroyed or removed. If a piece of acid-soaked paper is applied to the back of a spinal frog, one leg will come up and flick the paper away. No matter how many times the piece of paper is replaced on the skin, the leg will come up and flick it away. This response, involving many muscles working in a coordinated fashion, is purely reflex and demonstrates one of the chief characteristics of a reflex: fidelity of repetition. A frog with a brain might make the response two or three times, but eventually it would do something else, perhaps hop away. Most reflexes have some survival value for the animal. The anatomical configuration responsible for the reflex was selected during evolution because of this survival value.

The degree of complexity of a reflex varies with the number of spinal segments involved. The entire reflex arc may be contained within a single spinal segment. Many reflexes, however, require interneuronal connections between two or more segments of the spinal cord. The scratch reflex, the righting reflex and the walking reflexes involve the input of sensory information from many segments of the cord. Complex reflexes may even involve parts of the brain, such as the midbrain and medulla.

The integration and correlation of the body's activities involve pathways with many neurons, not just two or three, as in the simpler reflexes. Some pathways are **divergent** (Fig. 17.13*A*); that is, the axon of one neuron may branch many times, each branch synapses with a different neuron, and these in turn may branch further. Such an arrangement permits a single impulse to exert an effect over a wide area. Indeed, a single impulse may ultimately activate a thousand or more neurons. In **convergent** pathways (Fig. 17.13*B*) neurons from many different areas converge upon a single neuron or group of neurons. The response of the last neuron in a convergent pathway is the result of the interaction of a variety of excitatory and inhibitory influences. Thus, convergent pathways are important as the basis for the integrative activity of the nervous system.

Many neuronal circuits involve the passage of impulses only as long as the first neuron continues to be stimulated. When that stimulation ceases, the passage of impulses stops. Other arrangements of neurons in the nervous system ensure the continuation of the impulse for a period of time after the stimulus has stopped. In a **multiple chain circuit** (Fig. 17.13*C*) the first neuron is stimulated momentarily, an impulse travels to a terminal neuron and also by way of a branch to a second neuron. The second neuron is stimulated and a moment later sends a second impulse to the same terminal neuron and also, via a branch, to a third neuron that is stimulated and sends yet a third impulse to the same terminal neuron. If a great many neurons are involved, the terminal neuron will receive a whole series of impulses, and continue to receive them for some time after the initial stimulus has stopped. One or more branches of the neurons in a **closed chain circuit** (Fig. 17.13*D*) transmit impulses back to a point near the beginning of the circuit. Once such a circuit is activated, impulses can continue indefinitely unless the neurons become fatigued or are inhibited. Such circuits may form the basis for the spontaneous activity in the inspiratory center and other similar centers in the brain.

Considerable complexity of behavior can be

developed when the number of synapses in the reflex arc is large. For example, in the flexion reflex different responses can be elicited as the strength of the stimulus is increased. In experiments with dogs in which the connection between the spinal cord and the brain has been cut, a weak noxious stimulus applied to the foot pad results in withdrawal of the foot. An increase in stimulus strength causes a strong flexion of the lower leg. Further increase in the strength of the stimulus elicits flexion of the upper leg as well. This spreading of response with increase in stimulus strength is termed **irradiation.**

Although one might, in theory, construct a responsive animal simply out of reflexes, in the higher animals reflexes account for only a small portion of total behavior. In some of the simpler animals, however, reflexes play an important role, and consequently the behavior of these animals is relatively simple, rigid and stereotyped. Much of the behavior of starfish, for example, can be explained in terms of reflex arcs. Starfish can extend and retract their tube feet and make postural movements associated with ambulation. The extension and retraction are unoriented reflex responses. The apparent coordination of all the tube feet retracting in unison when a wave washes over the animal is simply the sum of individual responses to a common stimulus. Many aspects of crustacean behavior are reflex responses—the withdrawal of the eye stalk, the opening and closing of the claws and the movements associated with escape, defense, feeding and copulation.

17.7 THE PERIPHERAL NERVOUS SYSTEM

Emerging from the brain and spinal cord and connecting them with every receptor and effector in the body are the paired cranial and spinal nerves that make up the **peripheral nervous system**. Cranial and spinal nerves are bundles of nerve fibers—axons and dendrites. The only nerve cell bodies present in the peripheral nervous system are those of the sensory neurons aggregated into ganglia near the brain or spinal cord and of certain motor neurons of the autonomic system, discussed further on.

Spinal Nerves

There is a pair of peripheral nerves for each body segment of the vertebrate body. Afferent and efferent neurons lie together in most of a spinal nerve, but near the cord the nerve splits into a dorsal and ventral root and the neurons are segregated (Fig. 17.14). The **dorsal root** contains the afferent neurons and bears an enlargement, the **dorsal root ganglion**, that contains their cell bodies. The cell bodies of afferent neurons are nearly always located in ganglia on both spinal and cranial nerves. The afferent neurons enter the spinal cord and terminate in synapses with the dendrites or cell bodies of interneurons. Most of these cell bodies are located in the dorsal portion of the gray matter of the cord. The **ventral root** con-

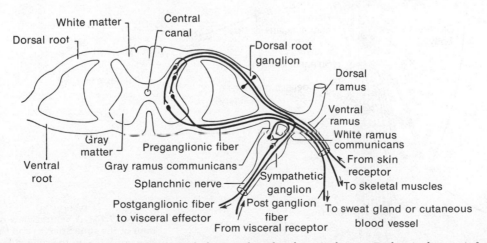

Figure 17.14 A diagrammatic cross section through the spinal cord and a spinal nerve. Each spinal nerve is formed by the union of dorsal and ventral roots and divides laterally into several branches (rami) going to different parts of the body. The dorsal ramus contains the same types of neurons as the ventral ramus.

tains the efferent neurons; their cell bodies nearly always lie in the ventral portion of the gray matter of the cord.

The spinal nerves of all vertebrates are essentially alike, although in the most primitive vertebrates the roots do not unite peripherally and some of the efferent neurons leave the cord in the dorsal root. In most vertebrates the roots unite to form a spinal nerve that subsequently divides into a **dorsal ramus**, which supplies the skin and muscles in the dorsal part of the body, a **ventral ramus**, which innervates the lateroventral parts of the body, and usually one or more **communicating rami** to the visceral organs. Afferent and efferent neurons are found in each ramus. Humans have 31 pairs of spinal nerves; those supplying the receptors and effectors of the limbs are larger than the others, and their ventral rami are interlaced to form a complex network, or **plexus**, from which nerves extend to the limbs (Fig. 17.15).

Cranial Nerves

The nerves from the nose, eye and ear have evolved with the organs of special sense. They are composed entirely of afferent fibers, except for a few efferent neurons in the optic and vestibulocochlear (auditory) nerves that extend to the sense organs and may modulate their activity. The remaining cranial nerves contain large numbers of both afferent and efferent fibers, and they are considered to be serially homologous with the separate roots of the spinal nerves of primitive vertebrates. Some of them are essentially the cephalic counterparts of dorsal roots, others are the counterparts of ventral roots. The location of the cell bodies of the neurons of cranial nerves and of the endings within the brain follows the pattern described for spinal neurons.

Reptiles, birds and mammals have 12 pairs of cranial nerves, omitting the minute and poorly understood nervus terminalis. Though

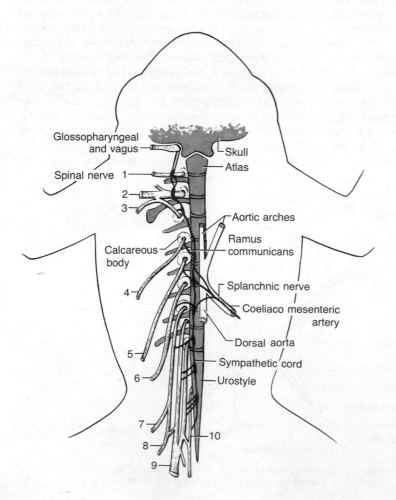

Figure 17.15 A ventral view of the ventral rami of the spinal nerves and the sympathetic cord lying on the right side of the vertebral column of a frog. (Modified after Gaupp.)

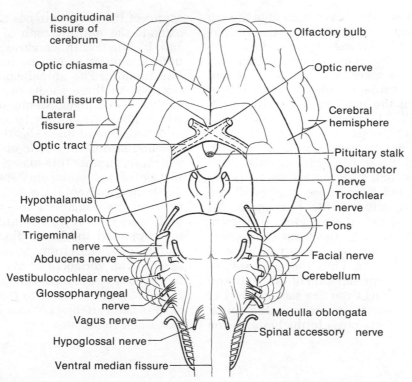

Figure 17.16 A ventral view of the brain of a sheep. The stumps of all but the first pair of cranial nerves are visible. The olfactory nerves consist of the processes of olfactory cells (cf. Fig. 16.16), which enter the olfactory bulbs in many small groups that cannot be seen with the unaided eye. The rhinal fissure separates the ventral olfactory portion of each cerebral hemisphere from the rest of the hemisphere. The paths of the optic fibers in the optic chiasma have been indicated by broken lines.

distributed to the nasal mucosa, this nerve is not olfactory. The other cranial nerves and their distribution are shown in Table 17.1 and their stumps can be seen in an illustration of a sheep brain (Fig. 17.16).

Fishes and amphibians have only the first 10 pairs of cranial nerves and lack the spinal accessory and hypoglossal nerves. The trigeminal, facial, glossopharyngeal and vagus nerves of fishes are primarily associated with the muscles of the visceral arches, and, as shown in Table 17.1, they supply the derivatives of this musculature in the higher vertebrates. Muscles may change shape and function tremendously during the course of evolution, but their innervation remains remarkably constant. One of the most important cranial nerves, the tenth, or **vagus**, forms part of the autonomic system and innervates the internal organs of the chest and upper abdomen.

17.8 THE CENTRAL NERVOUS SYSTEM: THE SPINAL CORD

The tubular spinal cord, surrounded and protected by the neural arches of the vertebrae,

has two important functions: to transmit impulses to and from the brain and to act as a reflex center. In cross section, two regions are evident: an inner, butterfly-shaped mass of **gray matter** made up of nerve cell bodies and an outer mass of **white matter** made up of bundles of axons and dendrites (Fig. 17.17). The whiteness of these bundles is due to the myelin sheaths of the axons and dendrites. The ends of the axons and dendrites, present in the central gray matter, have no myelin sheath. The "wings" of the gray matter are divided into two dorsal and two ventral horns. The latter contain the cell bodies of motor neurons whose axons pass out through the spinal nerves to the muscles. All of the other neurons in the spinal cord are interneurons.

The axons and dendrites of the white matter are segregated into bundles with similar functions: the **ascending tracts** transmit impulses to the brain, and the **descending tracts** carry impulses from the brain to the effectors. Neurologists have carefully noted the symptoms of patients with injured spinal cords and later correlated these with the particular tracts found to be damaged when the patient's nervous system was examined after death. From these observations they have been able to map out the location and functions of the various tracts (Fig. 17.17). For example, the dorsal columns of the white matter transmit impulses originating in the sense organs of muscles, tendons and joints through which we are aware of the position of the parts of the body. In advanced syphilis, these columns may be destroyed, so that the patient cannot tell where his arms and legs are unless he looks at them. He must watch his feet in order to walk.

One curious fact, still not satisfactorily explained, emerged from these studies of the location and function of the fiber tracts. All the fibers in the spinal cord *cross over* from one side of the body to the other somewhere along their path from sense organ to brain or from brain to muscle. Thus the right side of the brain controls the left side of the body and receives impressions from the sense organs of the left side. Some fibers cross in the spinal cord itself and others cross in the brain. In the center of the gray matter is a small canal that extends the entire length of the neural tube and is filled with **cerebrospinal fluid**. This fluid is similar to plasma but contains much less protein. The spinal cord and brain are wrapped in three sheets of connective tissue known as **meninges**. Meningitis is a disease in which these wrappings become infected and inflamed. One of these sheaths (**dura mater**) is fastened against the bone of the cranium and vertebrae; another (**pia mater**) is located on the surface of the brain and spinal cord; and a third, the **arachnoid**, lies between. The spaces between the arachnoid and the pia are filled with more cerebrospinal fluid, so that the spinal cord (and the brain) floats in this liquid and is protected from bouncing against the bone of the vertebrae or the skull with every movement. An **extradural space**, in which lie blood vessels, connective tissue and fat, may exist in the vertebral column between the dura mater and the bone.

TABLE 17.1 CRANIAL NERVES OF MAN

Nerve	Origin of Afferent Neurons	Distribution of Efferent Neurons
I, Olfactory	Olfactory portion of nasal mucosa (smell)	
II, Optic	Retina (sight)	A few to retina
III, Oculomotor	A few fibers from proprioceptors in extrinsic muscles of eyeball (muscle sense)	Most fibers to four of the six extrinsic muscles of eyeball, a few to muscles in ciliary body and pupil
IV, Trochlear	Proprioceptors in extrinsic muscles of eyeball	Another extrinsic muscle of eyeball
V, Trigeminal	Teeth, and skin receptors of the head (touch, pressure, temperature, pain); proprioceptors in jaw muscles	Muscles derived from musculature of first visceral arch, i.e., jaw muscles
VI, Abducens	Proprioceptors in an extrinsic muscle of eyeball	One other extrinsic muscle of eyeball
VII, Facial	Taste buds of anterior two-thirds of tongue (taste)	Muscles derived from musculature of second visceral arch, i.e., facial muscles; salivary glands; tear glands
VIII, Vestibulocochlear	Semicircular canals, utriculus, sacculus (sense of balance); cochlea (hearing)	A few to cochlea
IX, Glossopharyngeal	Taste buds of posterior third of tongue; lining of pharynx	Muscles derived from musculature of third visceral arch, i.e., pharyngeal muscles concerned in swallowing; salivary glands
X, Vagus	Receptors in many internal organs: larynx, lungs, heart, aorta, stomach	Muscles derived from musculature of remaining visceral arches (excepting those of pectoral girdle), i.e., muscles of pharynx (swallowing) and larynx (speech); muscles of gut, heart; gastric glands
XI, Spinal Accessory	Proprioceptors in certain shoulder muscles	Visceral arch muscles associated with pectoral girdle, i.e., sternocleidomastoid and trapezius
XII, Hypoglossal	Proprioceptors in tongue	Muscles of tongue

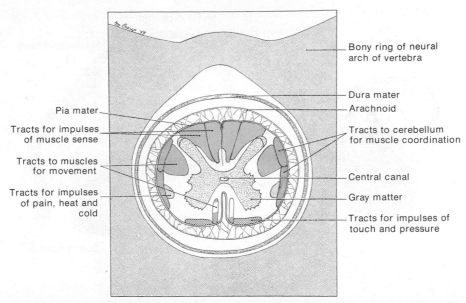

Figure 17.17   Cross section of the spinal cord surrounded by the bony vertebra, showing the meninges (*dura mater, pia mater* and *arachnoid*), the gray matter and some of the important nerve tracts of the white matter. (From Villee, C. A.: Biology. 7th ed. Philadelphia, W. B. Saunders Co., 1977.)

17.9   THE CENTRAL NERVOUS SYSTEM: THE BRAIN

The brain is the enlarged anterior end of the neural tube. In the human, the enlargement is so great that the resemblance to the spinal cord is obscured, but in lower vertebrates, the relationship of brain to spinal cord is clearer. The brain develops as a series of enlargements of the anterior end of the embryonic neural tube (Fig. 17.18). In an early embryo there are three swellings, the forebrain, midbrain and hindbrain, but both the forebrain and the hindbrain are later subdivided so that five regions are present in the adult. The forebrain divides into **telencephalon** and **diencephalon**. The telencephalon differentiates into a pair of **olfactory bulbs**, which receive the endings of olfactory cells, and a pair of **cerebral hemispheres**, which become the major brain center in the higher vertebrates. The lateral walls of the diencephalon become the **thalamus**, its roof the **epithalamus** and its floor the **hypothalamus**. Fibers in the optic nerves cross below the hypothalamus and form an **optic chiasma** (Fig. 17.16). All the optic nerves cross to the opposite side of the brain in most vertebrates, but only half of them cross in mammals. The pituitary gland is attached to the hypothalamus just posterior to the optic chiasma, and the pineal body is attached to the epithalamus. There is no further division in the midbrain, or **mesencephalon**, but its roof differentiates into a pair of **optic lobes** in all vertebrates. In addition to the optic lobes, or **superior colliculi**, the roof of the mesencephalon of mammals bears a pair of **inferior colliculi**. The hindbrain divides into a **metencephalon**, the dorsal portion of which forms the **cerebellum**, and a **myelencephalon**, which becomes the **medulla oblongata** (Table 17.2).

The Ventricular System

The central canal of the spinal cord extends into the brain and is continuous with several large interconnected chambers known as **ventricles**. A lateral ventricle lies in each cerebral hemisphere and is connected with the third

TABLE 17.2 BRAIN REGIONS AND DERIVATIVES

Brain Region	Major Derivatives
Forebrain (Prosencephalon)	
Telencephalon	Olfactory bulbs
	Cerebral hemispheres
Diencephalon	Epithalamus, Pineal body
	Thalamus
	Hypothalamus, Pituitary gland (part)
Midbrain (Mesencephalon)	Superior colliculi
	Inferior colliculi
Hindbrain (Rhombencephalon)	
Metencephalon	Cerebellum
	Pons
Myelencephalon	Medulla oblongata

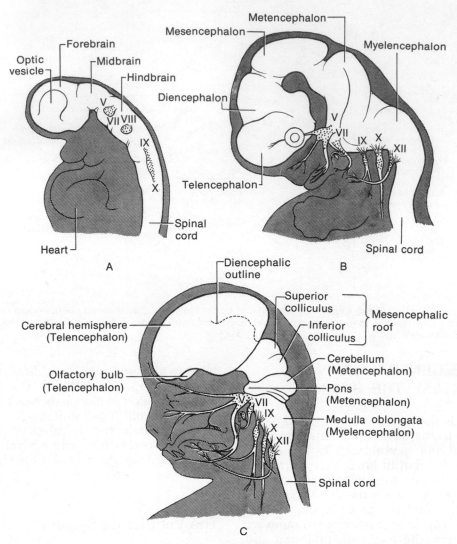

Figure 17.18 Three stages in the development of the human brain. *A,* The three primary brain regions can be recognized in an embryo that is about three and one-half weeks old. *B,* All five brain regions are evident in an embryo seven weeks old. *C,* The various structures found in a fully developed brain are beginning to differentiate in an embryo eleven weeks old. (After Patten.)

ventricle in the thalamus by an **interventricular foramen of Monro**. The **cerebral aqueduct of Sylvius** extends from the third ventricle through the mesencephalon to a fourth ventricle in the metencephalon and medulla oblongata. All of these cavities are filled with cerebrospinal fluid, much of which is produced by vascular **choroid plexuses** that develop in the thin roof of the medulla and diencephalon and in the lateral ventricles of mammals. Cerebrospinal fluid escapes from the brain through foramina in the roof of the medulla and slowly circulates in the spaces between the layers of connective tissue, the meninges, that encase the brain and spinal cord. The cerebrospinal fluid lies in the sub-

arachnoid space, between the arachnoid and pia mater, and extends far into the nervous tissue in **perivascular spaces** that surround the blood vessels penetrating the brain. It is produced continuously and re-enters the circulatory system by filtering into certain venous sinuses located in the dura mater covering the brain. The cerebrospinal fluid forms a protective liquid cushion for the brain and spinal cord and may help nourish the tissues of the central nervous system. The perivascular spaces serve some of the functions of a lymphatic system by returning plasma protein that may have escaped from capillaries in the brain. Nervous tissue does not have a lymphatic drainage.

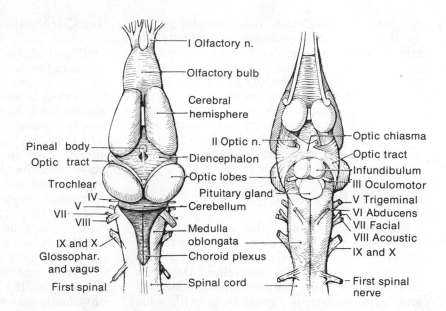

Figure 17.19 *Left,* A dorsal view and *right,* a ventral view of the brain of the frog. (Modified after Gaupp.)

Labels (dorsal view, left):
I Olfactory n.
Olfactory bulb
Cerebral hemisphere
Pineal body
Optic tract
Trochlear
IV
V
VII
VIII
IX and X
Glossophar. and vagus
First spinal

Labels (center):
II Optic n.
Diencephalon
Optic lobes
Pituitary gland
Cerebellum
Medulla oblongata
Choroid plexus
Spinal cord

Labels (ventral view, right):
Optic chiasma
Optic tract
Infundibulum
III Oculomotor
V Trigeminal
VI Abducens
VII Facial
VIII Acoustic
IX and X
First spinal nerve

Medulla Oblongata

The medulla oblongata (Figs. 17.19 and 17.20) lies between the spinal cord and the rest of the brain and is fundamentally the same in all vertebrates. The gray columns of the spinal cord extend into the medulla, but within the brain they become discontinuous, breaking up into discrete islands of cell bodies known as **nuclei.** The dorsal nuclei receive the afferent neurons from cranial nerves that are attached to this region and contain the cell bodies of afferent interneurons. These are sensory nuclei, just as the dorsal columns of the cord are sensory columns. The ventral nuclei contain the cell bodies of the efferent neurons of the cranial nerves and hence are motor nuclei. In mammals, the reflexes that regulate

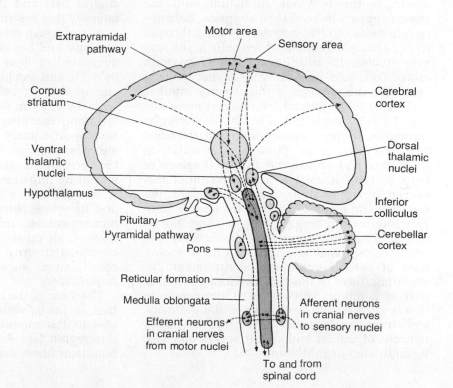

Figure 17.20 Diagrammatic lateral view of the human brain on which some of the major masses of gray matter, the reticular formation and certain important pathways have been projected.

Labels:
Extrapyramidal pathway
Corpus striatum
Ventral thalamic nuclei
Hypothalamus
Pituitary
Pyramidal pathway
Pons
Reticular formation
Medulla oblongata
Efferent neurons in cranial nerves from motor nuclei
Motor area
Sensory area
Cerebral cortex
Dorsal thalamic nuclei
Inferior colliculus
Cerebellar cortex
Afferent neurons in cranial nerves to sensory nuclei
To and from spinal cord

the rate of heart beat, the diameter of arterioles, respiratory movements, salivary secretion, swallowing and many other processes are mediated by these nuclei. Afferent impulses come into the sensory nuclei and are relayed by the interneurons to the motor nuclei, and efferent impulses go out to the effectors. Motor and sensory nuclei associated with other cranial nerves are also found in the metencephalon and the mesencephalon.

Reticular Formation

Between the motor and sensory nuclei throughout the brain stem is a network of thousands of cell bodies and their processes known as the **reticular formation**. Dendrites and axons of these neurons branch profusely and have numerous connections with each other and with many of the sensory and motor pathways passing to and from the brain (Fig. 17.20). Certain of these neurons receive input from as many as 4000 other neurons and feed out to 25,000 other neurons. The reticular formation is believed to be a very primitive part of the brain, from which the long ascending and descending fiber tracts and the various centers have evolved. Convergence and interaction of neurons from many of the sense organs, the cerebellum, the cerebrum, the thalamus, the hypothalamus and other areas occur in the reticular formation, and the region appears to be able to suppress, enhance or otherwise modify impulses passing through it. For example, branches from cells in the sensory nuclei, in addition to synapsing with motor cells and ascending to higher centers, feed into this system. If the sensory input is important to the animal (an unusual noise, the cry of a baby at night and so on), the reticular formation, in turn, sends nonspecific impulses to the cerebral cortex. These fan out widely in the cortex and arouse the animal if asleep or help to keep it alert if awake. Without this source of stimulation, the cortex does not function and cannot interpret the specific sensory information brought to it. In short, the reticular formation appears to be involved in what we call consciousness; that is, an awareness of oneself and one's environment. Descending fibers in the reticular formation form part of a motor pathway from the cerebrum that is known as the **extrapyramidal pathway**, and impulses passing down to the motor neurons of cranial and spinal nerves may also be facilitated or inhibited.

The Cerebellum and Pons

All vertebrates have a **cerebellum** that develops in the dorsal part of the mesencephalon and is a center for balance and motor coordination. It is small in many of the lower vertebrates, such as the frog (Fig. 17.19), in which muscle movements are not complex but is very large in birds and mammals. Impulses enter it from most of the sense organs, but those from the proprioceptive organs in the muscles and the parts of the ear concerned with equilibrium play a particularly prominent role (Fig. 17.20). The cerebellum keeps track of the orientation of the body in space and the degree of contraction of the skeletal muscle. In mammals this information is sent by way of part of the reticular formation and thalamus to the cerebrum, where voluntary movements are initiated. Copies, so to speak, of the motor directives sent out by the cerebrum to the muscles are sent to the cerebellum, which monitors the responses to the body and returns corrective signals to the cerebrum or, in some cases, directly to the muscle. Much of the gray matter of the mammalian cerebellum lies on the surface, where there is more room for the increased number of cell bodies. The surface is also complexly folded, which further increases the surface area available for cell bodies.

The mammalian cerebellum consists of a central part and two hemispheres extending laterally that resemble pine cones in shape. Its gray surface is made up of the cell bodies of neurons, and beneath is a mass of white tissue composed of fiber tracts connecting with the medulla and the higher parts of the brain. The size of the cerebellum is roughly correlated with the amount of muscular activity; it regulates and coordinates muscle contraction and is proportionately large in extremely active animals, such as birds. Removal or injury of the cerebellum is accompanied not by paralysis but by impairment of muscle coordination. A bird without a cerebellum is unable to fly and its wings thrash about jerkily. When the human cerebellum is injured by a blow or disease, all muscular movements become uncoordinated. Any activity requiring delicate coordination, such as threading a needle, is impossible.

The floor of the metencephalon is unspecialized in lower vertebrates and simply contributes to the medulla oblongata. In mammals, this region has evolved into a **pons**, a thick bundle of fibers extending crossways from one cerebellar hemisphere to the other and carrying impulses from one to the other, thus coordinating muscle movements on the two sides of the body. The pons also contains nuclei that relay impulses from the cerebrum to the cerebellum.

The Optic Lobes

The optic lobes of fishes and amphibians (Figs. 17.19 and 17.21) receive the termination of all of the fibers in the optic nerve and the projections from most other sense organs. This sensory information is integrated, and motor impulses are sent to the appropriate efferent neurons. The optic lobes are the master integrating center of the brain, insofar as these vertebrates have such a center. The cerebral hemispheres of the lower vertebrate are small and concerned almost exclusively with integrating olfactory impulses. In reptiles, some of the visual and other types of sensory data are sent to the cerebral hemispheres, and the cerebrum begins to assume certain functions of the optic lobes (Fig. 17.21). Still more sensory information is sent to the cerebral hemispheres of birds and mammals, and the hemispheres of mammals have taken over most of the functions of the optic lobes. In mammals the optic lobes (superior colliculi) remain relatively small centers that regulate the movements of the eyeballs and pupillary and accommodation reflexes. A pair of inferior colliculi are present posterior to the superior colliculi in mammals. They are a center for certain auditory reflexes. The reflex constriction of the pupil when light shines on the eye and the pricking up of a dog's ears in response to sound are controlled by reflex centers in the superior and inferior colliculi. The midbrain also contains a cluster of nerve cells regulating muscle tonus and posture.

Thalamus and Hypothalamus

In front of the midbrain the central canal widens and becomes the third ventricle, the roof of which contains a cluster of blood vessels (**choroid plexus**) secreting the cerebrospinal fluid. The thick walls of the third ven-

Figure 17.21 Diagrammatic lateral views of the brain of a fish, a reptile and a mammal to show the increasing importance of the cerebral hemispheres and the decreasing importance of the optic lobes as integration centers. The shift in optic pathways is shown; a similar shift occurs for other sensory modalities.

tricle, the **thalamus**, are relay centers for sensory impulses. Fibers from the spinal cord and lower parts of the brain synapse here with other neurons going to the various sensory areas of the cerebrum. The thalamus appears to regulate and coordinate the external manifestations of emotions. Thus, by stimulating the thalamus a sham rage can be produced in a cat—the hair stands on end, the claws protrude, the back becomes humped and other signs of anger are evinced. However, as soon as the stimulation stops, the appearance of rage ceases. The thalamus has become enlarged in the course of evolution as the cerebral hemispheres have assumed a dominant role in integrating the activities of the body. All the sensory impulses that go to the cerebrum, except those from the olfactory organ, are relayed in nuclei in the dorsal part of the thalamus. Many motor impulses descending from the cerebrum go directly to the motor nuclei and columns, but some are relayed in nuclei in the ventral part of the thalamus.

The thalamus is more than just a relay station, for considerable processing of the sensory input occurs here. The thalamic nuclei have many connections with each other, with the hypothalamus, with the reticular formation and with many parts of the cerebral hemispheres. Pain, temperature and certain other sensory modalities reach the level of conscious awareness in the thalamus. Other sensations are to some extent sorted out in this region, and a determination is made as to which of these we will concentrate on among the hundreds of stimuli continually impinging upon us. The thalamus influences the manner in which we view many stimuli; that is, the different degree of agreeableness or disagreeableness that we may place upon the same

type and intensity of stimuli at different times. An experimental rat with an electrode implanted in the "pleasure center" in the thalamus will push a lever that stimulates the center thousands of times per hour. Other areas appear to be "punishing centers," and an experimental animal will avoid activating an electrode implanted in one of them. The hypothalamus is an important center for the control of many autonomic functions, with centers regulating body temperature, appetite, water balance, carbohydrate and fat metabolism, blood pressure and sleep. It is curious that the anterior part of the hypothalamus prevents a rise in body temperature and the posterior part prevents a fall. The hypothalamus controls certain functions of the anterior lobe of the pituitary, such as the secretion of gonadotropins, by producing **releasing factors**. It also produces the hormones **oxytocin** and **vasopressin**, which are released in the posterior lobe of the pituitary. The parts of the hypothalamus that produce neurosecretions are neuroendocrine transducers, for they convert nerve signals into endocrine signals.

Cerebral Hemispheres

The parts of the brain considered so far have to do with unlearned automatic behavior determined by the fundamental structure of these parts, which is essentially the same from fish to human. The **cerebral hemispheres,** the largest and most anterior parts of the human brain, have a basically different function—controlling learned behavior. The complex psychological phenomena of consciousness, intelligence, memory, insight and the interpretation of sensations have their physio-

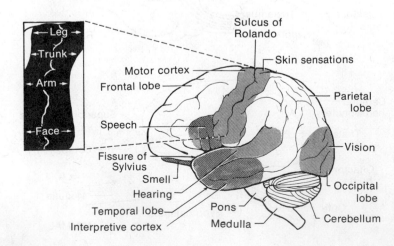

Figure 17.22 Important cortical areas of the human brain as seen in a lateral view. Most association areas of the cortex have not been hatched.

A   B

Figure 17.23 Vertical sections through (*A*) the sensory cortex receiving from the skin and (*B*) the motor cortex. The proportions of the cortex related to the various body regions have been emphasized by drawing body parts in similar proportions. (From Penfield and Rasmussen: The Cerebral Cortex, published by Macmillan.)

logical basis in the activities of the neurons of the cerebral hemispheres.

As the cerebral hemispheres assumed the dominant role in nervous integration during the course of evolution, they enlarged and grew posteriorly over the diencephalon and mesencephalon (Fig. 17.21). A layer of gray matter has developed on the surface of the cerebrum and forms a gray cortex that provides more area for the increased number of cell bodies. The cortex is also complexly folded, having numerous ridges (**gyri**) with furrows (**sulci**) between them, and this further increases the surface area. Over 12 billion neurons and even more neuroglial cells are present.

Parts of the cerebral hemispheres are still concerned with their primitive function of olfactory integration, but their great enlargement is correlated with the evolution of other integrative centers (Fig. 17.22). Impulses from the eyes, ears, skin and many other parts of the body are carried to the cerebral cortex by afferent interneurons after being relayed in the thalamus. The impulses terminate in specific parts of the cerebral cortex; their locations have been determined by correlating brain injuries with loss of sensation and also by electrical stimulation during brain operations. Many human brain operations are performed under local anaesthesia, and the patient can describe the sensations that are felt when particular regions of the cerebral cortex are stimu-

lated. Impulses from the skin terminate in the gyrus located just posterior to the central **fissure of Rolando**, a prominent sulcus extending down the side of each hemisphere and dividing the hemisphere into an anterior **frontal** and a posterior **parietal lobe**. The sensory areas of the skin are projected upside down. Impulses from the head are conducted to the lower part of the gyrus, whereas those from the feet reach the upper part. The extent of the area receiving impulses from any part of the body is proportional to the number of sense organs in that part of the body, not to the surface area of that part. Thus, the area receiving impulses from the hands is as extensive as the area receiving impulses from the trunk (Fig. 17.23*A*).

Impulses from the ear are carried to part of the **temporal lobe** that is separated from the frontal and parietal lobes by the **lateral fissure of Sylvius**. Impulses from the eye are received in the **occipital lobe**, which lies just posterior to the parietal lobe. The path of the optic fibers of mammals is an exception to the generalization that most afferent impulses cross at some point during their ascent to the brain. Half of the fibers in each optic nerve cross in the optic chiasma and end on the opposite side of the brain, but the other half do not. Thus, destruction of one occipital lobe results in inability to perceive images that fall on half of each retina rather than complete loss of vision in one eye.

Figure 17.24 A sagittal section of the human brain and its surrounding meninges. Cerebrospinal fluid is produced by the choroid plexuses, circulates as indicated by the arrows and finally enters a venous sinus in the dura mater. (Modified after Rasmussen.)

Appropriate motor impulses to the skeletal muscles are initiated in response to all the sensory data that enter the cerebrum. The cell bodies of the efferent interneurons are contained in the motor cortex, which lies just anterior to the sulcus of Rolando. The motor cortex is subdivided, in the manner of the adjacent sensory cortex, into areas associated with different parts of the body. Fibers to the hand occupy a large portion, for the muscles that control finger movements contain more motor units than most (Fig. 17.23*B*). This is correlated with the intricacy of our finger movements. Many efferent interneurons pass directly to the motor nuclei of the brain and to the motor columns of the spinal cord, crossing to the opposite side along the way (Fig. 17.20). This direct pathway to the lower motor neurons is called the **pyramidal pathway**. It is phylogenetically a new pathway, being present only in mammals, and reaching its greatest development in primates. Skilled and learned movements are, for the most part, mediated by this pathway. Many other motor impulses leave the cortex by way of the **extrapyramidal pathway**, which involves the relay of motor impulses in a mass of gray matter (the **corpus striatum**) situated deep within each cerebral hemisphere and additional

relays in the thalamus and reticular formation of the brain stem. This is phylogenetically an older system and is associated with grosser movements, automatic postural adjustments and stereotyped responses.

When all the areas of known function are plotted, they cover almost all of the rat's cerebral cortex, a large part of the dog's and a moderate amount of the monkey's, but only a small part of the total surface of the human cortex. The remainder, known as **association areas,** is made up of neurons that are not directly connected to sense organs or muscles but supply interconnections between the other areas. Billions of association neurons are arranged in most parts of the cortex in six vertical layers, with numerous interconnections and with the afferent neurons bringing impulses into the cortex and the efferent ones carrying them out. Such an organization makes possible the exceedingly complex functional interrelationships between different parts of the cortex and between the cortex and subcortical regions. Information in one hemisphere can also affect the other by impulses crossing on commissural fibers that extend between them. A particularly large commissure, known as the **corpus callosum** and found only in placental mammals, can be seen in a sagittal section of the brain (Fig. 17.24). All of these interconnections permit the integration of many different sensory modalities, the comparison of current sensory input with information stored in the brain, and the composition of a motor response appropriate to the present requirements of the organism. In some way, the association areas integrate into a meaningful unit all the diverse impulses constantly reaching the brain so that the proper response is made. They interpret and manipulate the symbols and words by which our thought processes are carried on. When disease or accident destroys the functioning of one or more association areas, the condition known as **aphasia** may result, in which the ability to recognize certain kinds of symbols is lost. For example, the names of objects may be forgotten, although their functions are remembered and understood.

Brain Waves

Metabolism is invariably accompanied by electrical changes, and the electrical activity of the brain can be recorded by an **electroencephalograph**. To obtain recordings, electrodes are taped to different parts of the scalp and the activity of the underlying parts of the cortex is measured. Electroencephalographic tracings show that the brain is continuously active; **alpha waves**, the most regular manifestations of activity, emanate from the visual areas at the back of the brain when the subject is resting quietly with his eyes shut. These waves occur rhythmically at a rate of nine or 10 per second and have a potential of about 45 microvolts. When the eyes are open, the waves disappear and are replaced by more rapid, irregular waves. That the latter are produced by objects seen can be demonstrated by presenting the eyes with some regular stimulus, such as light blinking at regular intervals, and observing that brain waves with a similar rhythm appear. During sleep, the brain waves are altered, become slower and larger (have a greater potential), as the subject falls into deeper and deeper unconsciousness. The dreams of a sleeping subject are mirrored in flurries of irregular waves.

The Autonomic Nervous System

The heart, lungs, digestive tract and other internal organs, together with the ciliary and iris muscles of the eye, the small muscles asso-

TABLE 17.3 SELECTED EFFECTS OF AUTONOMIC STIMULATION*

Organ	Sympathetic Stimulation	Parasympathetic Stimulation
Skin		
Hair muscles	Contraction	....
Sweat glands	Secretion	....
Eye		
Iris sphincter	....	Contraction
Iris dilator	Contraction	....
Circulatory system		
Heart (rate and force)	Increased	Decreased
Coronary arteries	Dilation	Constriction
Most other arteries	Constriction	Dilated
Lung bronchi	Dilation	Constriction
Digestive organs		
Muscles of stomach and		
intestine	Decreased peristalsis	Increased peristalsis
Salivary glands	Some mucus secretion	Secretion
Gastric and intestinal glands	....	Secretion
Pancreas	....	Secretion
Liver	Bile flow inhibited Glucose released	Bile flow stimulated
Urinary organs		
Urinary bladder muscles	Relaxation	Contraction
Bladder sphincter	Contraction	Relaxation
Adrenal medulla	Secretion	....

*Dotted line indicates the absence of innervation.

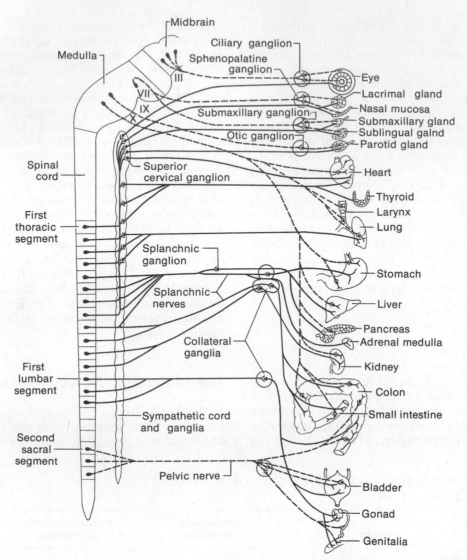

Figure 17.25 The human autonomic nervous system. Sympathetic fibers are drawn in solid lines; parasympathetic fibers in broken lines. The sympathetic fibers that go to the skin are not shown. (After Howell.)

ciated with hairs and many of the glands of the body, are innervated by a special set of peripheral nerves collectively called the **autonomic nervous system**. This system is composed of two parts: the **sympathetic** and **parasympathetic** nerves.

The organs innervated by the autonomic system function automatically, requiring no thought on our part; indeed, they cannot be controlled voluntarily. The autonomic nervous system is usually defined as a motor system, and the afferent fibers that return from internal organs are not part of this system, even though they may be in a nerve composed largely of autonomic fibers. The autonomic system is distinguished from the rest of the nervous system by several features. There is no voluntary

control by the cerebrum over these nerves. Most organs receive a *double* set of fibers—one set via the sympathetic nerves and one via the parasympathetic nerves. Impulses from the sympathetic and parasympathetic nerves typically have antagonistic effects on the organ innervated. Thus if one speeds up an activity, the other decreases it. These effects are summarized in Table 17.3.

Another peculiarity of the autonomic system is that motor impulses reach the effector organ from the brain or spinal cord not by a single neuron, as do those in all other parts of the body, but by a relay of two neurons. The cell body of the first neuron in the chain, the **preganglionic neuron**, is located in the brain or spinal cord; that of the second neuron, the **postganglionic neuron**, is located in a ganglion somewhere outside the central nervous system (Fig. 17.25). The cell bodies of the postganglionic neurons of the sympathetic nerves are close to the spinal cord; those of the parasympathetic nerves are close to or actually within the walls of the organs they innervate.

The Sympathetic System

The sympathetic system consists of nerve fibers whose preganglionic cell bodies are located in the lateral portions of the gray matter of the spinal cord. Their axons pass out through the ventral roots of thoracic and anterior lumbar nerves. These fibers may synapse with the postganglionic fibers in the sympathetic ganglia along the sympathetic cord, or they may continue from the sympathetic cord through the splanchnic nerves to **collateral ganglia** located at the base of the coeliac and mesenteric arteries. Postganglionic sympathetic fibers continue from the ganglia to the organs they supply. The sympathetic ganglia are paired, and a chain of 18 is present on each side of the spinal cord from the neck to the abdomen. The axons of some of the postganglionic neurons pass from the sympthetic ganglion back to the spinal nerve and through it to innervate sweat glands, the muscles that make the hair stand erect and the muscles in the walls of blood vessels. The axons of other postganglionic neurons pass from the sympathetic ganglia of the neck up to the salivary glands, the iris of the eye and the pineal gland.

The Parasympathetic System

The parasympathetic system consists of fibers originating both in the brain and emerging by way of the third, seventh, ninth and especially the tenth, or vagus, nerves and in the sacral nerves two, three and four and joining to form the pelvic nerve (Fig. 17.25). The vagus nerve emerges from the medulla and passes down the neck to the chest and abdomen, innervating the heart, respiratory system and digestive tract as far as the small intestine. The large intestine and the urinary and reproductive systems are innervated by parasympathetic fibers from the pelvic nerves. The iris of the eye, the sublingual and submaxillary glands and the parotid gland are innervated by the third, seventh and ninth cranial nerves, respectively. These nerves all contain the axons of preganglionic neurons in the chain; the ganglia of the parasympathetic system are located in or near the organs innervated, so that the axons of the postganglionic neurons are all relatively short.

Sympathetic and parasympathetic systems usually have opposite effects upon the organs innervated. Sympathetic stimulation speeds up the rate and increases the force of the heart beat, causes most peripheral arteries to constrict (thereby increasing blood pressure), increases the glucose content of the blood and, in general, has effects that enable the body to adjust to conditions of stress. It inhibits the activity of the digestive tract. Parasympathetic stimulation speeds peristalsis of the digestive tract and similar vegetative processes, but it slows down the heart rate and decreases blood pressure.

ANNOTATED REFERENCES

Aidley, D. J.: The Physiology of Excitable Cells. New York, Cambridge University Press, 1974. Presents discussions of the nature of the nerve impulse and of some of the experiments that have led to our present theories regarding the nerve impulse.

Bitterman, M. E.: The Evolution of Intelligence. Scientific American, *212*:92, (Jan.) 1965. An interesting report on the qualitatively different sorts of intelligence found in different classes of vertebrates.

Bullock, T. H., and G. A. Horridge: Structure and Function in the Nervous Systems of Invertebrates. San Francisco, W. H. Freeman Co., 1969. A two volume comprehensive treatise covering the nervous systems of a variety of invertebrate animals.

Calder, N.: The Mind of Man. New York, The Viking Press, Inc., 1970. An interesting report of recent research on the functions of the brain for the layman.

Freedman, R., and J. E. Morris: The Brains of Animals and Man. New York, Holiday House, Inc. 1972. A well-illustrated discussion of modern research on brain function.

Gardner, E.: Fundamentals of Neurology. 6th ed. Philadelphia, W. B. Saunders Co., 1975. A fine, concise account of the morphology and physiology of the human nervous system.

Hodgkin, A. L.: The Conduction of the Nerve Impulse. Springfield, Illinois, Charles C
    Thomas, 1964. An excellent summary of the experiments that led to our present
    concept of the mechanism of nerve conduction.
Noback, C. R.: The Human Nervous System. New York, McGraw-Hill Book Co. 1967. An
    excellent and superbly illustrated discussion of the nervous system and sense
    organs. Structural and functional aspects of the subject are closely integrated.
Sherrington, C. S.: Integrative Action of the Nervous System. New Haven, Yale Univer-
    sity Press, 1948. A true classic in neurobiology, written by an outstanding pioneer
    in the field.

ENDOCRINE SYSTEMS AND HORMONAL INTEGRATION

The activities of the various parts of the bodies of higher animals are integrated by the nervous system, discussed in the previous chapter, and by the hormones of the endocrine system. Nerves and sense organs enable an animal to adapt very rapidly—with responses measured in milliseconds—to changes in the environment. The swift responses of muscles and glands are typically under nervous control. The **hormones** secreted by the glands of the endocrine system diffuse or are transported by the blood stream to other parts of the body and coordinate their activities. The responses under endocrine control are in general somewhat slower—measured in minutes, hours or weeks—but longer lasting than those under nervous control. The long-range adjustments of metabolism, growth and reproduction are typically under endocrine control. Hormones play a key role in maintaining within narrow limits the concentrations of glucose, sodium, potassium, calcium, phosphate and water in the blood and extracellular fluids of human beings and other vertebrates. Hormonal regulation and integration of cellular processes occur in insects, crustaceans, annelids, mollusks and certain other invertebrates as well as in the vertebrates.

18.1 ENDOCRINE GLANDS

Endocrine glands secrete their products into the blood stream rather than into a duct leading to the exterior of the body or to one of the internal organs, as do exocrine glands. For this reason they are sometimes referred to as ductless glands or glands of internal secretion. The thyroid, parathyroids, pituitary and adrenals function only in the secretion of hormones and are truly ductless glands. The pancreas, ovaries and testes have both external secretions via ducts and internal secretions carried by the blood stream. The pancreas is really two functionally separate organs combined in one structure; it produces digestive enzymes in addition to hormones. In some of the lower vertebrates, the two parts of the pancreas are anatomically separate.

The term hormone was originated in 1904 by the British physiologists W. M. Bayliss and E. H. Starling, who were studying the control of the exocrine function of the pancreas by secretin, a substance produced in the duodenal mucosa. They defined a hormone as "any substance normally produced by the cells in one part of the body and carried by the blood stream to distant parts which it affects for the good of the body as a whole." As currently used, the term hormone refers to a special chemical messenger that is produced by some restricted region of an organism and diffuses or is transported by the blood stream to another region of the organism, where it is effective in very low concentrations in regulating and coordinating cellular activities. Most hormones are carried in the blood from the site of production to the site of action, but neurohormones may pass down an axon, and prostaglandins may be transferred in the seminal fluid. This definition of hormones in-

Figure 18.1 Diagram illustrating the differences between endocrine secretion (*A*), neurosecretion (*B*) and neurotransmission (*C*). In (*A*) and (*B*), H represents hormone and R represents receptor. (From Villee, C. A.: Biology. 7th ed. Philadelphia, W. B. Saunders Company, 1977.)

cludes some remarkably diverse chemical compounds: amino acids and amines, peptides, proteins, fatty acids, purines, steroids and gibberellins. It seems unlikely that all of these diverse compounds would affect cell function by the same or similar mechanisms. Indeed, it is becoming apparent that some, perhaps many, hormones have several independent mechanisms of action by which they regulate cellular activities. All of the hormones are required for normal body function, and each must be present in a certain optimal amount. Either a hyposecretion (deficiency) or hypersecretion (excess) of any one may result in a characteristic pathological condition.

The impulses traveling in sensory neurons, motor neurons and interneurons are all similar, simple action potential spikes. The type of information transmitted depends on where the impulse originates and ends. The kind of perception generated in the brain depends on where the sensory stimulus originated—in the eye, ear, nose, skin or internal organs. In the endocrine system, the information transferred depends not only on where it originates and ends but also on the kind of material being transported. In any endocrine system we can distinguish three parts—the secreting cell, the

transport mechanism and the target cell—each characterized by a greater or lesser degree of specificity (Fig. 18.1). In general, each type of hormone is synthesized and secreted by a specific kind of cell. Some hormones are transported in the blood in solution, but most are bound to some protein component of the serum. Some are bound nonspecifically to albumin; others are selectively bound to specific high affinity proteins. The hormones are taken up and bound by specific receptors in their target cells. Neurohormones are produced in neurons, pass down the axon and are secreted at the axonal tip (Fig. 18.1B), a process termed neurosecretion (p. 416).

18.2 HORMONES AND RECEPTORS

A few hormones, such as thyroxine and growth hormone, affect the metabolic condition of nearly every cell of the body; they respond to the presence of the hormone and show an altered metabolic state when deprived of the hormone. Most hormones, however, elicit a detectable response only in certain cells of the body. With some hormones only a few types of cells respond; with others a broader spectrum of response can be observed. An engrossing question at present is "how does the target cell recognize its appropriate hormone?" Each type of target cell contains a specific protein, a **receptor,** that forms a tight and specific combination with one type of hormone. The receptors for several hormones have been at least partially purified. The first interaction of the hormone with any component of the target cell is with the receptor protein, and it is the specificity of the latter that determines which hormone is effective in which cell. The blood and interstitial fluid bathing each cell in the body contain the entire spectrum of hormones, but the specific receptor enables the cell to pick out one specific hormone and ignore all the others. The

hormone-receptor interaction is characterized by high affinity, high specificity and low capacity; the receptors are readily saturated by low concentrations of hormone.

Hormones are remarkably effective substances; a very small quantity produces a marked effect in the structure and function of one or another part of the body. Only small amounts of a hormone are secreted by the endocrine gland at any one time, and the amount circulating in the blood is very small. Even the amount excreted each day in the urine is not very large. Because of this, the isolation of a pure hormone can be a difficult job indeed. To obtain a few milligrams of pure estradiol, one of the female sex hormones, more than two tons of pig ovaries were extracted. When the hormone secreted by a given gland has been extracted and its chemical structure has been determined, it can be prepared synthetically to be used in treating diseases resulting from a deficiency of that hormone.

Much has been learned about endocrine function by careful observation of the symptoms of human diseases resulting from the hypo- or hypersecretion of hormones. Further information has been derived from the careful study of strains of rats, mice and other animals with particular genetic endocrine abnormalities—dwarf mice, obese mice, diabetic mice and so on.

The location of the human endocrine glands is shown in Figure 18.2. Their relative position in the body is much the same in all of the vertebrates. The source and physiological effects of the principal mammalian hormones are listed in Table 18.1.

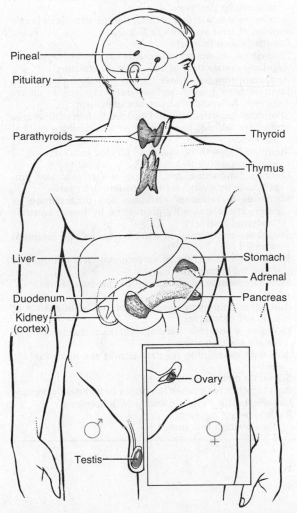

Figure 18.2 The approximate locations of the major endocrine glands in men and women. (From Villee, C. A.: Biology. 7th ed. Philadelphia, W. B. Saunders Company, 1977.)

18.3 MOLECULAR MECHANISMS OF HORMONE ACTION

Any theory of the molecular mechanisms by which a given hormone produces its effects in specific tissues must account for the high degree of specificity of many hormones and for the remarkable degree of biological amplification inherent in hormonal processes. Hormones circulate in the blood stream in very low concentrations—steroid hormones at concentrations of about 10^{-9} M and peptide and protein hormones at concentrations of about 10^{-12} M. The several current theories regarding the mechanism of hormone action suggest that the hormone first combines with a specific receptor protein. The receptors for steroid hormones are soluble proteins located in the cytoplasm that combine with the steroid and

TABLE 18.1 VERTEBRATE HORMONES AND THEIR PHYSIOLOGIC EFFECTS
(From Villee, C. A.: Biology. 7th ed. Philadelphia, W. B. Saunders, 1977.)

Hormone	Source	Physiologic Effect
Thyroxine	Thyroid gland	Increases basal metabolic rate
Parathyroid hormone (PTH)	Parathyroid glands	Regulates calcium and phosphorus metabolism
Calcitonin	Parafollicular cells of thyroid	Antagonist of PTH
Insulin	Beta cells of islets in pancreas	Increases glucose utilization by muscle and other cells, decreases blood sugar concentration, increases glycogen storage and metabolism of glucose
Glucagon	Alpha cells of islets in pancreas	Stimulates conversion of liver glycogen to blood glucose
Secretin	Duodenal mucosa	Stimulates secretion of pancreatic juice
Pancreozymin	Duodenal mucosa	Stimulates release of bile by gallbladder and release of enzymes by pancreas
Epinephrine	Adrenal medulla	Reinforces action of sympathetic nerves; stimulates breakdown of liver and muscle glycogen
Norepinephrine	Adrenal medulla	Constricts blood vessels
Cortisol	Adrenal cortex	Stimulates conversion of proteins to carbohydrates
Aldosterone	Adrenal cortex	Regulates metabolism of sodium and potassium
Dehydroepiandrosterone	Adrenal cortex	Androgen; stimulates development of male sex characters
Growth hormone	Anterior pituitary	Controls bone growth and general body growth; affects protein, fat and carbohydrate metabolism
Thyrotropin (TSH)	Anterior pituitary	Stimulates growth of thyroid and production of thyroxine
Adrenocorticotropin (ACTH)	Anterior pituitary	Stimulates adrenal cortex to grow and produce cortical hormones
Follicle-stimulating hormone (FSH)	Anterior pituitary	Stimulates growth of graafian follicles in ovary and of seminiferous tubules in testis
Luteinizing hormone (LH)	Anterior pituitary	Controls production and release of estrogens and progesterone by ovary and of testosterone by testis
Prolactin (LTH)	Anterior pituitary	Maintains secretion of estrogens and progesterone by ovary; stimulates milk production by breast; controls "maternal instinct"
Oxytocin	Hypothalamus, via posterior pituitary	Stimulates contraction of uterine muscles and secretion of milk
Vasopressin	Hypothalamus, via posterior pituitary	Stimulates contraction of smooth muscles; antidiuretic action on kidney tubules
Melanocyte-stimulating hormone (MSH)	Anterior lobe of pituitary	Stimulates dispersal of pigment in chromatophores
Testosterone	Interstitial cells of testis	Androgen; stimulates development and maintenance of male sex characters
Estradiol	Cells lining follicle of ovary	Estrogen; stimulates development and maintenance of female sex characters
Progesterone	Corpus luteum of ovary	Acts with estradiol to regulate estrous and menstrual cycles
Prostaglandins	Seminal vesicle and other tissues	Stimulate uterine contractions
Chorionic gonadotropin	Placenta	Acts together with other hormones to maintain pregnancy
Placental lactogen	Placenta	Has effects like prolactin and growth hormone
Relaxin	Ovary and placenta	Relaxes pelvic ligaments
Melatonin	Pineal gland	Inhibits ovarian function

Figure 18.3 A comparison of current concepts of the receptors for steroid hormones (H_s) and peptide hormones (H_p). Note that the steroid hormone enters the cell, binds with a soluble receptor, passes into the nucleus and is attached to the chromatin, stimulating RNA synthesis. In contrast, the peptide hormone remains outside of the cell, binding to a receptor on the plasma membrane. This activates a membrane-bound adenyl cyclase, producing cyclic AMP. The latter stimulates a cascade of protein kinases, which phosphorylate an enzyme, thereby activating (or inactivating) it. (From Villee, C. A.: Biology. 7th ed. Philadelphia, W. B. Saunders Company, 1977.)

transport it into the nucleus (Fig. 18.3). In contrast, the receptors for peptide and protein hormones are located on the plasma membranes of the target cell. The peptide hormone is bound to the receptor on the surface of the cell, and it is believed that this combination stimulates in some fashion the adenyl cyclase on the inner side of the plasma membrane, resulting in an increased production of cyclic AMP. An important intracellular function of cyclic AMP is to activate one or more of the protein kinases within the cell. The protein kinases in turn phosphorylate other proteins, which convert an inactive protein to an active enzyme or perhaps an active enzyme to an inactive phosphorylated form. For example, phosphorylase kinase is activated when phosphorylated, whereas glycogen synthetase becomes inactivated when phosphorylated.

The effects of hormones in facilitating the entrance of certain substrates into the cell, as for example in the uptake of glucose by muscle cells stimulated by insulin, have suggested that the hormone combines with some protein or other component of the cell membrane. This leads to a change in the molecular architecture of the membrane and hence in its permeability to specific substrates.

Another hypothesis states that the hormone increases the enzymic activity of a protein. The stimulation of adenyl cyclase and the increased production of cyclic 3′, 5′ adenylic acid from ATP induced by several protein hormones are examples of this. The cyclic AMP is regarded as an intracellular "second messenger" that mediates the effect of the hormone. Epinephrine, for example, stimulates the adenyl cyclase of liver cells (Fig. 18.4). The resulting cyclic AMP activates a protein kinase that transfers a phosphate group from ATP to a third enzyme, phosphorylase kinase, and activates it so that it can in turn convert an inactive fourth enzyme, phosphorylase b, to active phosphorylase a. The latter then catalyzes the production of glucose-1-phosphate from glycogen. At each of these successive steps in the enzymic cascade, there is an amplification of tenfold to a hundredfold, just as in the sequence of enzymes involved in blood clotting. The cascade effect permits a very small amount of epinephrine to lead to the production of a very large amount of glucose-1-phosphate.

A third general hypothesis suggests that the hormone enters the nucleus and activates certain previously repressed genes (Fig. 18.5). This leads to the production of additional or new kinds of messenger RNA that code for the synthesis of specific new proteins. The messenger RNA undergoes processing in the nucleus and a long tail of polyadenylic acid is added. The polyadenylic acid-messenger RNA passes out of the nucleus to the ribosomes and there serves as a template for the synthesis of proteins. This theory also accounts for the marked amplification of hormonal effects because a small amount of hormone, by turning on the transcription of a specific gene, could result in the production of many molecules of messenger RNA and many, many more molecules of protein. The generalization emerging from the many studies of hormonal action is that protein and peptide hormones bind to receptors on the cell membrane and stimulate adenyl cyclase. The product of adenyl cyclase (cyclic AMP) stimulates protein kinases that activate enzymes by phosphorylating them. All of these phenomena occur in the cytoplasm. In contrast, steroid hormones are bound to cytoplasmic receptors and transfer as a receptor-steroid complex into the nucleus, where they activate the transcription of certain genes. The resulting polyadenylate messenger RNA codes for the synthesis of specific proteins, specific enzymes that are responsible for the physiological changes resulting from hormonal action. Evidence

supporting the latter hypothesis is derived from studies of the effects of estradiol on the uterus, of estradiol and progesterone on the oviduct, and of dihydrotestosterone on the seminal vesicle and prostate.

It is not clear whether hormones are typically used up as they carry out their regulatory action in the target tissue. Estradiol, however, is not used up or changed chemically as it stimulates the growth of the uterus. Hormones bound to their receptors appear to be relatively stable, but hormones circulating in the blood have relatively short biological half-lives. They are inactivated and eliminated from the body and must be replaced by new hormone molecules synthesized in the endocrine glands.

It seems unlikely that all hormones have a common molecular mechanism by which their effects are produced. Indeed, there is evidence that certain hormones produce their effects not by a single mechanism but by several different mechanisms acting in parallel within a single cell. The theories current at any given moment tend to reflect our general knowledge of cellular and molecular biology. Those implicating effects of hormones on cell membranes have given way to those implicating an effect altering the activity of an enzyme, and these have

Figure 18.4 The sequence of enzymic events by which epinephrine or glucagon stimulates adenyl cyclase and brings about the synthesis of 3′,5′-adenosine monophosphate (cyclic AMP). This in turn activates a protein kinase that phosphorylates phosphorylase kinase, which in turn phosphorylates and activates phosphorylase. This finally brings about the cleavage of glycogen and the secretion of glucose. (From Villee, C. A.: Biology. 7th ed. Philadelphia, W. B. Saunders Company, 1977.)

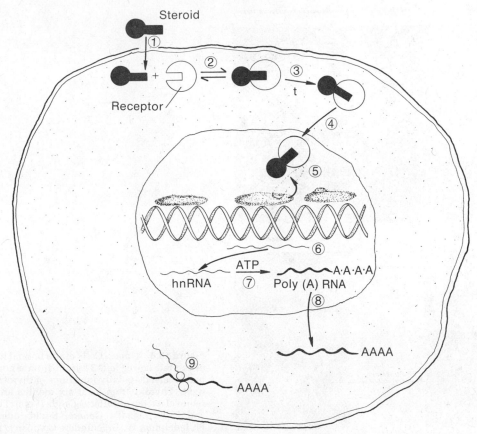

Figure 18.5 Successive steps in the postulated mechanism by which a steroid hormone stimulates the synthesis of specific proteins in its target cells. The steroid enters the cell (1) and binds to a specific receptor protein (2). The steroid-receptor complex undergoes a temperature-dependent transformation process (3), enters the nucleus (4) and binds to the chromatin (5). This in some way facilitates the transcription of the DNA in that section of the chromatin (6). The resulting "heterogeneous nuclear RNA" undergoes processing with ATP (7), and a long polyadenylic acid tail is added. The poly (A) rich mRNA leaves the nucleus (8) and serves as the template for the synthesis on the ribosome (9) of the specific protein. (From Villee, C. A.: Biology. 7th ed. Philadelphia, W. B. Saunders Company, 1977.)

been replaced in turn by theories involving the effects of the hormone on the genetic mechanism that result in the synthesis of specific kinds of RNA and proteins.

18.4 THE THYROID GLAND

All vertebrates have a bilobed **thyroid gland** located in the neck. In mammals the two lobes are located on either side of the larynx and are joined by a narrow isthmus of tissue extending across the ventral surface of the trachea near its junction with the larynx. The thyroid has an exceptionally rich blood supply, which reflects its function as an endocrine gland. It develops as a ventral outgrowth of the pharynx, but the connection is usually lost early in development. The gland is composed of cuboidal epithelial cells arranged in hollow spheres one cell thick. The cavity of each folli-

cle is filled with a gelatinous colloid secreted by the epithelial cells lining it (Fig. 18.6).

Follicle cells have the remarkable ability to accumulate iodide from the blood. This is used in the synthesis of **thyroglobulin,** a protein secreted into the colloid and stored. Proteolytic enzymes secreted in the lysosomes of the thyroid cells hydrolyze thyroglobulin to its constitutent amino acids, one of which is **thyroxine,** a derivative of the amino acid tyrosine containing 65 per cent iodine. Thyroxine passes into the blood stream, where it is transported loosely bound to certain plasma proteins. In tissues, thyroxine, which contains four atoms of iodine per molecule, may be converted to triiodothyronine, which contains one less atom of iodine and is several times more active than thyroxine. It is not yet clear whether the hormone active at the cellular level is only triiodothyronine or whether thyroxine may also be effective.

The first clue to thyroid function came from

SECRETORY EPITHELIUM   BLOOD VESSEL   INTERFOLLICULAR CONNECTIVE TISSUE

FAT CELL   COLLOID   VACUOLE

COLLOID

SECRETORY
EPITHELIUM

Figure 18.6 *Upper,* Cells of the normal thyroid gland of the rat. *Lower left,* Thyroid from a normal rat that had received 10 daily injections of thyrotropin. *Lower right,* Thyroid from a rat six months after complete removal of the pituitary gland. (From Turner, C. D., and J. T. Bagnara: General Endocrinology. 6th ed. Philadelphia, W. B. Saunders Company, 1976.)

observations made by a British physician, Sir William Gull, in 1874, when he noted the association of spontaneous decreased function of the thyroid with puffy dry skin, dry brittle hair and mental and physical lassitude. A Swiss surgeon, Kocher, removed the thyroid from a series of patients and noted that they developed the same symptoms as Gull's patients. In 1895, using a calorimeter to measure the rate of metabolism in patients by the amount of heat they produced, Magnus-Levy found that patients with **myxedema** (Gull's disease) had notably lower metabolic rates than normal. When these patients were fed thyroid tissue, their metabolic rate was raised toward normal. This led to the idea that the thyroid secretes a hormone that regulates the metabolic activity of all body cells.

The role of the thyroid hormone in all vertebrates is to increase the rate of oxidative, energy-releasing processes in all body cells. The amount of energy released by an organism at rest under standard conditions, measured in a calorimeter by the amount of heat given off or calculated from the oxygen consumed, is decreased in thyroid deficiency and increased when thyroxine is administered or when the

thyroid gland is overactive. Complete removal of the thyroid gland from the mammal reduces the animal's metabolic rate to half the normal value, and the body temperature decreases slightly. Individual tissue slices from an animal with thyroid deficiency show a reduced metabolic rate, a decreased oxygen consumption and decreased utilization of carbohydrates, fats and proteins when incubated in vitro. Since nutrients are metabolized at a lower rate, they tend to be stored and the hypothyroid animal tends to become obese.

By its action on metabolic processes, thyroxine has a marked influence on growth and differentiation. Removing the thyroid of a young animal causes decreased body growth, retarded mental development and delayed or decreased differentiation of gonads and external genitalia. All of these changes are reversed by the administration of thyroxine. The metamorphosis of frog and salamander tadpoles into adults is controlled by the thyroid. The removal of the larval thyroid completely prevents metamorphosis, and administering thyroxine to tadpoles causes them to metamorphose prematurely into miniscule adults (Fig. 18.7). The effect of thyroxine on amphibian metamorphosis appears to be not simply a secondary result of its effect on metabolism, for tadpole metabolism can be increased by dinitrophenol, but premature metamorphosis does not occur.

The production and discharge of thyroxine is regulated by **thyroid stimulating hormone** (TSH), secreted by the anterior lobe of the pituitary. In 1916, P. E. Smith found that the removal of the pituitary of frog tadpoles produces deterioration of the thyroid and prevents metamorphosis. Similar pituitary control of thyroid function has been demonstrated in rats, humans and other mammals. The secretion of TSH by the pituitary is regulated in part by the amount of thyroxine in the blood. Thus, a decreased production of thyroxine by the thyroid leads to lowered concentration of thyroxine in the blood stream. This stimulates the pituitary to release TSH, which passes to the thyroid and raises its output of thyroxine. When the concentration of thyroxine in the blood is returned to normal, the release of TSH is decreased. By this feedback mechanism, the output of thyroxine is kept relatively constant and the basal metabolic rate is kept within the normal range.

A deficiency in the amount of thyroxine secreted by the thyroids in an adult results in myxedema, characterized by a low metabolic rate and decreased heat production. The body temperature may drop to several degrees below normal so that the patient constantly feels cold. His pulse is slow and he is physically and mentally lethargic. His appetite usually remains normal, and since the food consumed is not used up at the normal rate, there is a tendency toward obesity. The skin becomes waxy and puffy, owing to the deposition of mucous fluid in the subcutaneous tissues, and the hair usually falls out. Myxedema responds well to the administration of thyroxine or dried thyroid gland. Since thyroxine is not digested appreciably by the digestive juices or the bacteria in the gut, it can be given by mouth.

Hypothyroidism results when the diet does not contain enough iodine for the synthesis of thyroxine. The gland compensates for the insufficiency by increasing in size under stimulation by TSH, and the resulting goitrous enlargement may become a large disfiguring mass weighing more than one kilogram. The symptoms accompanying the goiter resemble those of myxedema but are milder. This type of goiter occurs in areas where the soil lacks iodine or in regions remote from the sea where seafood, which is rich in iodine, is unobtaina-ble. The incidence of such simple goiters has been greatly reduced by modern packing and shipping methods, which permit the transport of seafood to all parts of the country, and by the addition of iodine as potassium iodide to table salt.

Thiouracil and related compounds are goitrogenic. They block the reactions by which iodide is oxidized and fixed onto the tyrosine molecule, thus decreasing the production of thyroxine and secondarily increasing the production and release of TSH from the pituitary.

Hypothyroidism present from birth is known as **cretinism.** Children suffering from the disease are dwarfs of low intelligence who never mature sexually (Fig. 18.8A). If treatment with thyroxine is begun early, normal physical growth and mental development can be effected.

The overproduction of thyroid hormone produces Graves' disease or **exophthalmic goiter** (Fig. 18.8C). The thyroid may be enlarged or may be of nearly normal size, but it produces an excessive amount of hormone. The pa-

Figure 18.7 The effect of thyroid feeding upon the tadpoles of *Rana catesbiana. A,* The untreated control, which was killed at the end of the experiment. The metamorphosed animal at the lower right (G) was killed two weeks after starting the feeding of thyroid gland. The remaining animals (B to F) were removed from the experiment at intervals during this period. Note the effect of thyroid substances on the metamorphosis of the mouth, tail and paired appendages. (From Turner, C. D.: General Endocrinology. 3rd ed. Philadelphia, W. B. Saunders Company, 1960.)

Figure 18.8 *A*, A cretin. *B*, Simple goiter. *C*, Exophthalmic goiter. (*A* and *B* from Selye: Textbook of Endocrinology, published by Acta Endocrinologia, Inc.; *C* from Houssay: Human Physiology, published by McGraw-Hill Book Company, Inc.)

tient's basal metabolic rate increases to as much as twice the normal amount. The excessively rapid heat production causes the hyperthyroid person to feel uncomfortably warm and to perspire profusely. Because the food he eats is used up quickly, he tends to lose weight even on a high caloric diet. High blood pressure, nervous tension, irritability, muscular weakness and tremors are symptomatic of the condition, but probably the most characteristic symptom is the protrusion of the eyeballs, called exophthalmos, which gives the patient a wild, staring expression (Fig. 18.8C). The swelling of the gland as the result of hyperthyroidism is known as exophthalmic goiter in order to distinguish it from the sim-

ple goiter caused by insufficient dietary iodine. Identical symptoms can be caused by feeding thyroid substance or thyroxine to normal people.

Hyperthyroidism can be treated by surgically removing some of the thyroid gland, by killing the cells with X-rays, by administering the drug thiouracil, which inhibits the synthesis of thyroxine or by injecting radioactive iodine, ^{131}I. The radioactive iodine is accumulated by the thyroid and its radiation destroys the cells.

In addition to the follicular cells that secrete thyroxine, the thyroid contains parafollicular cells that secrete **calcitonin,** a peptide containing 32 amino acids. This hormone acts with parathyroid hormone to regulate the concentration of calcium in the blood. Its effects oppose those of parathyroid hormone; it inhibits bone resorption and leads to a decrease in the concentration of calcium in the blood and body fluids.

18.5 THE PARATHYROID GLANDS

Imbedded in the thyroid glands of terrestrial vertebrates are small masses of tissue called **parathyroid glands.** There are usually two pairs of parathyroids, which develop as outgrowths of the third and fourth pairs of pharyngeal pouches. Each gland consists of solid masses and cords of epithelial cells, rather than of spherical follicles as in the thyroid. The hormone secreted by these cells, **parathyroid hormone** (PTH), regulates the concentrations of calcium and phosphorus in the blood and body fluids. It is essential for life; the complete removal of the parathyroid results in death in a few days. PTH is a single peptide chain containing 84 amino acids with a molecular weight of 8500. It is inactivated by proteolytic enzymes and cannot be administered orally. Parathyroid hormone promotes the absorption of calcium from the lumen of the intestine, the release of calcium from the bones and the reabsorption of calcium from the glomerular filtrate in the kidney tubules. Parathyroidectomy produces a decreased concentration of calcium in the serum, a decreased excretion of phosphorus and a resulting increased amount of phosphorus in the serum. The parathyroidectomized animal is subject to muscular tremors, cramps and convulsions—a condition known as **tetany,** which results from the decreased concentration of calcium in the body fluids. Injecting a solution of calcium stops the tetanic convulsions, and further convulsions can be prevented by repeated administration of calcium.

Parathyroid deficiencies are rare, occurring occasionally when the glands are removed inadvertently during an operation on the thyroid or when degeneration results from an infection. The administration of PTH cannot be used for the long-term treatment of parathyroid deficiencies, for the patient becomes refractory to repeated injections of the hormone. The deficiency can be treated successfully by a diet rich in calcium and vitamin D and low in phosphorus.

Hyperfunction of the parathyroid, induced by a tumor of the gland, is characterized by high calcium and low phosphorus content of the blood and by increased urinary excretion of both calcium and phosphorus. The calcium comes, at least in part, from the bones, and as a result, the bones become demineralized, soft and easily broken. The increased concentration of calcium in the body fluids results in deposits of calcium in abnormal places—the kidneys, intestinal wall, heart and lungs. The disease can be treated by removing the excess parathyroid tissue surgically or by destroying it with X-rays.

18.6 THE ISLET CELLS OF THE PANCREAS

Scattered among the pancreatic acinar cells that secrete the digestive enzymes are clusters of endocrine cells, the **islets of Langerhans,** quite different in appearance and staining properties. They have a richer supply of blood vessels than the acinar cells and no associated ducts. The islet cells can be differentiated into several types by the staining reactions of their cytoplasmic granules. The pancreas develops as two outgrowths from the duodenum that grow together and fuse in most vertebrates. The islet cells develop as buds from the pancreatic ducts and eventually lose all connection with the ducts. In some bony fishes, the acinar and islet tissues form spatially separate organs. The pancreas of the cyclostome is ductless and is located in the wall of the intestine or in the liver.

In 1886, two German investigators, Minkowski and Von Mering, were studying the role of the pancreas in digestion by removing the gland surgically from dogs and noting the subsequent effects of digestive functions. The caretaker for the animals noticed that their urine attracted swarms of flies to the cages. Upon analysis, large amounts of sugar were found in the dogs' urine, and the resemblance to human diabetes was recognized. Diabetes had been known since the first century A. D., but its cause was unknown and no treatment was effective. A short time before Minkowski and Von Mering performed their experiments, the cure for myxedema—the feeding of thyroid glands—had been discovered, and it was hoped that a similar feeding of the pancreas or the injection of pancreatic extracts would cure diabetes. After 1892, when the discoveries were published, many scientists tried to prepare effective extracts of pancreas, but none of the preparations was very effective and many were toxic. The digestive enzymes of the pancreas destroyed the hormone before it could be extracted and purified. Finally, in 1922, two Canadians, Banting and Best, obtained an active substance by tying the pancreatic ducts, waiting several weeks for the acinar cells to degenerate, and then making an extract from the remaining islet cells. Since the islets cells develop before the enzyme-producing cells in embryonic animals, they were also able to obtain active extracts from fetal pancreas. The first preparation of pure crystalline insulin was made in 1927 by Abel. The present commercial insulin is extracted from beef, sheep or hog pancreas by an acid alcohol method that rapidly inactivates the proteolytic enzymes.

Insulin, secreted by the β cells of the pancreas, consists of two peptide chains joined by disulfide bonds. The pioneering work of Fred Sanger and his colleagues at the University of Cambridge determined the sequence of the 21 amino acids in one chain and the 30 amino acids in the other. Insulin was the first protein whose amino acid sequence was determined, and these studies earned Sanger a Nobel prize, for they not only clarified the structure of insulin but also established methods by which the amino acid sequence of other proteins could be determined. Insulin is synthesized in the β cells of the pancreas as a single peptide composed of 84 amino acids. This compound, **proinsulin,** undergoes folding; three disulfide bonds are formed; and a C peptide containing 33 amino acids is removed from the center of the original chain by hydrolytic cleavage. This leaves two peptide chains joined by the disulfide bridges. Proinsulin is an example of a prohormone, and it has been discovered that several other peptide hormones are synthesized within cells as prohormones that undergo partial cleavage to yield smaller peptides with full hormonal activity. Steiner showed in 1976 that the peptide synthesized on the ribosomes of the β cells is an even longer chain, **preproinsulin,** that loses a portion of the amino terminal part of the chain to become proinsulin.

Most commercial preparations of insulin contain a second pancreatic hormone that increases blood sugar concentration instead of decreasing it as insulin does. This hormone, **glucagon,** has been separated from insulin, crystallized and found to be a single peptide chain composed of 27 amino acids. Glucagon is secreted by the α cells of the pancreatic islets, and insulin is secreted by the β cells.

Insulin and glucagon take part in the regulation of carbohydrate metabolism, along with certain hormones secreted by the pituitary, adrenal medulla and adrenal cortex. By way of cyclic AMP, glucagon activates the enzyme **phosphorylase,** which is involved in the conversion of liver glycogen to blood glucose and thus raises the concentration of glucose in the blood. Insulin increases the rate of conversion of blood glucose to intracellular glucose-6-phosphate, thereby decreasing the concentration of glucose in the blood, increasing the storage of glycogen in skeletal muscle and increasing the metabolism of glucose to carbon dioxide and water. A deficiency of insulin decreases the utilization of sugar, and the alterations in carbohydrate metabolism that result secondarily produce many other changes in the metabolism of proteins, fats and other substances.

The surgical removal of the pancreas or its hypofunction in **diabetes mellitus** impairs glucose utilization and results in elevated concentrations of glucose in the blood **(hyperglycemia).** The patient excretes large amounts of glucose in the urine **(glycosuria)** because the concentration of sugar in the blood exceeds the renal threshold (p. 352). Extra water is required to excrete this sugar, and the urine volume increases; as a result the patient tends to become dehydrated and thirsty. The tissues, unable to obtain enough glucose from the blood, convert protein into carbohydrate. Much of this is excreted and there is a steady loss of weight. Fat deposits are mobilized and the lipids are metabolized. The concentration of lipids in the blood may increase to the point where the blood has a milky appearance. The fatty acids are not metabolized completely but tend to accumulate as partially oxidized **ketone bodies,** such as acetoacetic acid. The ketone bodies are volatile and have a sweetish smell that gives a peculiar and characteristic odor to the breath of diabetics. Ketone bodies are acidic and must be excreted in the urine, causing an acidosis (decrease in the alkaline reserves of the body fluids).

Untreated diabetes is ultimately fatal because of the acidosis, the toxicity of the accumulated ketone bodies and the continuous loss of weight. The injection of insulin alleviates all of the diabetic symptoms. (The patient is enabled to utilize carbohydrates normally and the other symptoms disappear.) However, since the action of insulin persists for a short time only, repeated injections are necessary. Insulin does not "cure" diabetes, since the pancreas does not begin secreting its hormone again, but continued injections prevent the appearance of the symptoms and enable the diabetic to lead a normal life. The effect of an injection of insulin lasts for only a short time, a day at most, for the injected insulin is gradually destroyed in the tissues, as is the insulin secreted by the patient's pancreas. Long-lasting insulins, such as protamine zinc insulin and globin insulin, reduce the number of injections needed by most diabetics to one a day.

The administration of a large dose of insulin to a normal or diabetic person causes a marked decrease in the concentration of blood sugar. The nerve cells, which require a certain amount of glucose for normal function, become hyperirritable and then fail to respond as the glucose level decreases. The patient becomes bewildered, incoherent and comatose and may die unless glucose is administered. There are rare cases of pancreatic tumors that oversecrete insulin and cause recurring attacks of convulsions and unconsciousness by reducing the concentration of glucose in the blood. This condition is treated by surgically removing a part of the pancreas.

The secretion of insulin and glucagon is controlled by the concentration of glucose in the blood. When the concentration of glucose in the blood increases (after a meal, for example), the secretion of insulin is stimulated and it acts to restore the glucose level to normal. A rise in the concentration of insulin in the blood can be detected within two minutes after an experimentally induced rise in blood glucose. A major effect of insulin is a dramatic increase in the transport rate of glucose into skeletal muscle and adipose tissue. This leads to a decrease in the concentration of glucose in the blood and removes the stimulus for the secretion of insulin. The secretion of glucagon is stimulated by low concentrations of glucose. As the glucose content of the blood falls below the optimal range, the release of glucagon from the α cells of the pancreas is stimulated, the glycogen phosphorylase system of the liver is stimulated by the glucagon, and the glucose released by the liver returns the concentration of glucose in the blood to normal.

18.7 THE ADRENAL (SUPRARENAL) GLANDS

The small paired adrenal glands of mammals are located at the anterior end of each kidney. In humans, they weigh only about 12

grams but have a richer supply of blood vessels per mass of tissue than any other organ of the body. Each adrenal consists of an outer, pale, yellowish-pink **cortex** and a dark, reddish-brown inner **medulla**. In cyclostomes and fishes the two parts are spatially separate. In amphibians, reptiles and birds their anatomic relations are quite variable, and the two parts are interspersed. Cortical tissue develops from coelomic mesoderm near the mesonephric kidneys, whereas the medullary tissue is ectodermal, derived from the neural crest cells that also form the sympathetic ganglia.

The cells of the adrenal medulla are arranged in irregular cords and masses around the blood vessels (Fig. 18.9). The medulla secretes two closely related hormones: **epinephrine** (also called adrenalin) and

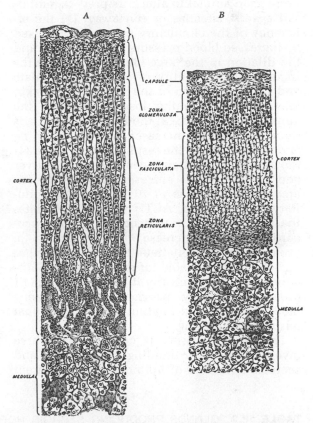

Figure 18.9 Sections through the adrenal cortex and medulla of (A) normal and (B) hypophysectomized rats. Since the functional capacity of the adrenal cortex is conditioned by the release of ACTH, hypophysectomy results in tremendous shrinkage of the cortex. The medulla is not influenced by hypophysectomy. Both sections are drawn to scale. (From Turner, C. D., and J. T. Bagnara: General Endocrinology. 6th ed. Philadelphia, W. B. Saunders Company, 1976.)

norepinephrine. These are comparatively simple amines derived from the amino acid tyrosine. Epinephrine produces an increase in heart rate, in blood pressure, in blood glucose, in the rate at which blood coagulates and a decrease in the glycogen content of liver and muscle. It causes goose flesh and dilation of the pupils of the eye and of most blood vessels but constriction of those of the skin with the result that the skin becomes pale. Norepinephrine has much weaker effects on blood sugar and heart rate but is a more powerful vasoconstrictor.

The secretions of the adrenal medulla function during emergencies to reinforce and prolong the action of the sympathetic nervous system. Epinephrine secretion is increased by stresses such as cold, pain, trauma, emotional states and certain drugs. The following physiologic changes resulting from the action of the sympathetic nerves and epinephrine could prepare an animal to attack its prey, defend itself against enemies or run away: (1) the efficiency of the circulatory system is increased by increased blood pressure and heart rate and the dilation of the large blood vessels; (2) the increase in the ability of blood to coagulate and the constriction of the vessels in the skin tend to minimize the loss of blood if the animal is wounded; (3) the intake of oxygen is increased by the increased rate of breathing and the dilation of the respiratory passages; (4) the mobilization of glycogen stores of the liver and muscle makes glucose available for energy; (5) the release of ACTH from the pituitary is stimulated. ACTH in turn stimulates the release from the adrenal cortex of glucocorticoids that increase the conversion of proteins to carbohydrate and make further carbohydrate available for quick energy. Epinephrine is used clinically in treating asthma (it dilates respiratory passages), in increasing blood pressure and in stimulating a heart that has stopped beating.

The adrenal cortex is composed of three layers of cells surrounding the medulla and secretes a number of hormones with different types of activity. The cortex is composed of the outer **glomerulosa,** the middle **fasciculata** and the inner **reticularis** zones (Fig. 18.9). The cells of the fasciculata are believed to be the most active in hormone production. Human embryos and those of other primates have adrenals as large as the kidneys, owing to the presence of a large mass of cells, the **fetal zone,** interposed between cortex and medulla. The fetal zone regresses and disappears shortly after birth. The adrenal cortex produces and secretes (1) **glucocorticoids,** which promote the conversion of proteins to carbohydrates; (2) **mineralocorticoids,** which regulate sodium and potassium metabolism; and (3) **androgens,** which have male sex hormone activity. The most potent glucocorticoid is **cortisol,** and the most potent mineralocorticoid is **aldosterone.** Deoxycorticosterone is also an effective regulator of salt and water metabolism and is widely used clinically.

Steroids are produced from cholesterol, which in turn is synthesized by the union of isoprenoid units. The cholesterol content of the adrenal cortex exceeds that of many other organs; as much as 5 per cent of the wet weight of the gland may be cholesterol. Steroids are synthesized from cholesterol not only in the adrenal cortex but also in the testis, ovary and placenta (Table 18.2). The steroids produced by the adrenal cortex are composed of 21 or 19 carbons arranged in four connected rings. Three of the rings have six carbons and the fourth has five (Fig. 18.10). A sequence of enzymic reactions converts cholesterol to pregnenolone, and other enzymes convert this to progesterone. Progesterone is not only one of the primary female sex hormones but also an important intermediate in the synthesis of both glucocorticoids and mineralocorticoids. In the synthesis of glucocorticoids a hydroxyl group is added at position 17 as a first step. This is followed by additional hydroxylations at Carbons 11 and 21. Hydroxylations of progesterone at positions 21, 11 and 18 yield aldosterone, the major mineralocorticoid. The removal of the short side chain of pregneno-

TABLE 18.2 GLANDS PRODUCING STEROID HORMONES

Adrenal Cortex	Ovary	Testis	Placenta
Cortisol	Estradiol	Testosterone	Progesterone
Deoxycorticosterone	Progesterone	Androstenedione	Estradiol
Aldosterone	Androgens	Estradiol	
Androsterone		Estrone	
Dehydroepiandrosterone		Corticoids	
Estradiol			

Figure 18.10 The sequence of reactions by which androgens, estrogens and corticoids may be synthesized from progesterone. (From Villee, C. A.: Biology. 7th ed. Philadelphia, W. B. Saunders Company, 1977.)

lone, Carbons 20 and 21, yields dehydroepiandrosterone, the major androgen of the adrenal.

The human disease resulting from decreased secretion of adrenocortical hormones was first described in 1855 by the English physician Thomas Addison. **Addison's disease** is usually caused by a tubercular or syphilitic infection of the cortex that destroys its cells. The hypofunction of the adrenal in Addison's disease or its complete surgical removal is characterized by low blood pressure, muscular weakness, digestive upsets, increased excretion of sodium and chloride in the urine, increased retention of potassium in body fluids and cells and a peculiar bronzing of the skin caused by the deposition of melanin. There is a marked decline in the concentration of blood sugar and in the glycogen content of liver, muscle and other tissues. Experimental evidence

shows that the animal's ability to produce carbohydrates from proteins is greatly impaired. The loss of sodium produces an acidosis, and the loss of body fluid leads to lower blood pressure and decreased rate of blood flow. The appetite for food and water decreases and there is loss of weight. The upsets in the digestive tract include diarrhea, vomiting and pain. Muscles are more readily fatigued and less able to do work. The basal metabolic rate decreases, and the animal is less able to withstand exposure to cold and other stresses. Death ensues within a few days after complete adrenalectomy. Patients with Addison's disease are treated by the oral or intramuscular administration of a natural adrenal steroid, such as cortisol, or of a synthetic steroid, such as dexamethasone or 2α-methyl, 9α-fluorocortisol acetate. The latter is extremely

Figure 18.11 *Left,* The normal control of adrenal function by ACTH. A releasing factor from the hypothalamus stimulates the pituitary to secrete ACTH, which increases the synthesis of cortisol in the adrenal cortex. The cortisol then inhibits the secretion of ACTH by the pituitary. *Right,* In one type of adrenal cortical virilism, the adrenals lack an enzyme and the conversion of progesterone to cortisol is greatly reduced. The production of cortisol cannot be increased by ACTH, and ACTH secretion is not inhibited. The secretion of ACTH remains high; this results in increased adrenal size and the secretion of androgens, such as testosterone, instead of cortisol. (From Villee, C. A.: Biology. 7th ed. Philadelphia, W. B. Saunders Company, 1977.)

potent in both mineralocorticoid and glucocorticoid activities.

The development and function of the adrenal cortex is regulated by **adrenocorticotropic hormone,** ACTH, secreted by the anterior lobe of the pituitary. ACTH increases the size of the adrenal cortex by stimulating the synthesis of RNA and protein. In addition, it stimulates the production of corticoids by increasing the activity of one or more of the enzymes involved in the conversion of cholesterol to cortisol. Stimulation of the adrenal cortex by ACTH leads to an increased production of cortisol and an increased concentration of that hormone in the blood. This increased concentration of cortisol inhibits the secretion of ACTH by the pituitary. Cortisol may act directly by inhibiting the synthesis of ACTH in the pituitary or indirectly by decreasing the production of corticotropin-releasing hormone, CRH, by the hypothalamus.

The primary stimulus that elicits the secretion of cortisol is some sort of physical stress—an injury, a painful disease, exposure to heat or cold—that generates impulses that are forwarded to the hypothalamus. The hypothalamus secretes CRH, which passes along the hypothalamo-hypophyseal portal system of blood vessels directly to the anterior lobe of the pituitary. The CRH stimulates the appropriate cells in the pituitary to secrete ACTH, and this is carried in the blood to the adrenal cortex where it stimulates the production and

release of cortisol. The mobilization of amino acids and lipids from peripheral tissues and the process of gluconeogenesis in the liver provide substrates for the repair of the damage and decrease the stimulus that led to the production of the releasing factor. The increased concentration of cortisol in the blood acts in the hypothalamus or pituitary, or both, to decrease the production and release of ACTH by negative feedback control.

An inherited defect of any one of the several enzymes involved in the synthesis of cortisol from pregnenolone and cholesterol may lead to enlargement of the adrenal cortex. The commonest defect is a deficiency of the enzyme catalyzing the insertion of a hydroxyl group at Carbon 21 of the steroid. Such a defect leads to an accumulation of intermediates, such as 17-hydroxyprogesterone, in the biosynthetic pathway. Since these cannot be converted to cortisol, some are converted to androgens such as androstenedione (Fig. 18.11). This may be converted either in the adrenal or elsewhere in the body into testosterone, the most potent androgen. The lack of cortisol to inhibit ACTH production in the pituitary results in the oversecretion of ACTH. The negative feedback control cannot operate, and the adrenal grows larger and secretes even more androgens. Eventually, the individual becomes virilized. Female fetuses lacking the enzyme have external genitalia that are masculinized to varying degrees, but male fetuses with the enzymic defect may show little or no abnormality at birth. After birth, early virilization in both males and females is manifested by enlargement of the phallus, early development of pubic and axillary hair, lowering of the voice and other effects of androgens. Patients with this condition, **adrenal cortical hyperplasia,** can be treated by injecting cortisol to turn off the production of ACTH by the pituitary.

18.8 THE PITUITARY GLAND

The pituitary gland, or **hypophysis cerebri,** is an endocrine gland lying in a small depression on the floor of the skull just below the hypothalamus, to which it is attached by a narrow stalk. Its only known function is the secretion of hormones. The pituitary has a double origin: a dorsal outgrowth **(Rathke's pouch)** from the roof of the mouth grows up and surrounds a ventral evagination (the **infundibulum)** from the hypothalamus (Fig. 18.12). Both parts are of ectodermal origin. Rathke's pouch soon loses its connection to the mouth, but the connection to the brain (the **infundibu-**

lar stalk) remains. The hypophysis has three lobes: the anterior and intermediate lobes derived from Rathke's pouch and a posterior lobe from the infundibulum. The pituitary, like the adrenal, is a double gland whose parts have quite different functions. The anterior lobe has no nerve fibers and is stimulated to release its hormones by hormonal factors reaching it through its blood vessels. The anterior lobe receives a double blood supply — arterial and portal. Some branches of the internal corotid artery pass directly to the pituitary; others serve a capillary bed around the infundibular stalk and the median eminence of the hypothalamus (Fig. 18.13). Portal veins from these capillaries pass down the infundibular stalk and empty into those surrounding the secretory cells of the anterior lobe. They provide a direct route by which releasing factors secreted by the hypothalamus pass directly to the anterior lobe and control the secretion of its hormones.

The anterior lobe contains at least five types of cells differing in shape, size, staining properties and the kind of granules present in the cytoplasm. It seems likely that each type pro-

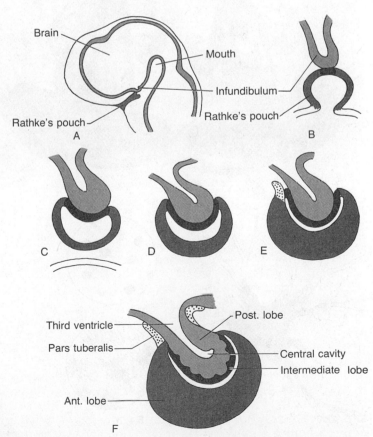

Figure 18.12 The development of the pituitary gland. *A,* Sagittal section through head of young embryo. *B–F,* Sagittal sections of successive stages of developing pituitary gland.

duces and secretes a different kind of hormone. The cell secreting growth hormone has been identified as a rounded cell whose cytoplasm is packed with dense, round acidophilic granules. These have been isolated and shown to have a high content of growth hormone. The cells secreting prolactin stain deeply with carmine stain and contain granules that are larger and more ovoid than the granules in the growth hormone–secreting cells. The intermediate lobe contains basophil cells smaller than those of the anterior lobe, some with and some without granules.

The posterior lobe is composed of nonmyelinated nerve fibers and branching cells that contain brownish cytoplasmic granules. Its two hormones, **oxytocin** and **vasopressin** (also known as **antidiuretic hormone,** ADH), are not produced in the posterior lobe but are secreted by neurosecretory cells in the supraoptic and paraventricular nuclei of the brain, pass down the axons of the hypothalamic-hypophyseal tract and are stored and released by the posterior lobe. Investigations have shown that there

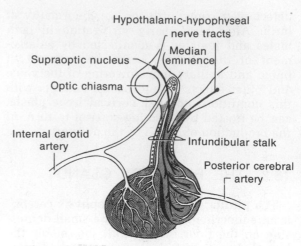

Figure 18.13   Blood supply of the pituitary gland.

are two types of neurons in these nuclei; one type produces only oxytocin and the other only vasopressin. This is the basis of the "one neuron–one hormone" hypothesis: a neuron produces oxytocin or vasopressin, but not both. Both the supraoptic and paraventricular

Figure 18.14 Integumentary adaptations in normal frogs (*Rana pipiens*). *A*, Light-adapted animal; *B*, dark-adapted animal. (Turner, C. D.: General Endocrinology. 2nd ed. Philadelphia, W. B. Saunders Co., 1955.) *C*, A chromatophore, greatly magnified, showing the pigment. *D*, A section of skin of frog adapted to a warm light environment. *E*, Skin adapted to a cool dark environment.

Figure 18.15 The effects of hypophysectomy in the rat. *A*, Normal littermate control; *B*, littermate hypophysectomized when 36 days of age. These photographs were made at 144 days of age, when the control animal weighed 264 gm. and the hypophysectomized rat weighed 80 gm. A^1, A^2 and A^3 are thyroid, adrenals and ovaries from normal animal; B^1, B^2 and B^3 are thyroids, adrenals and ovaries from hypophysectomized animal. Note marked differences in size. (From Turner, C. D.: General Endocrinology. 2nd ed. Philadelphia, W. B. Saunders Company, 1955.)

nuclei have about equal numbers of the two kinds of neurons.

The brilliant work of Vincent du Vigneaud, for which he was awarded the Nobel Prize in 1955, led to the isolation of these two hormones, the determination of their molecular structure, and their synthesis. Each is a peptide containing nine amino acids, seven of which are the same for both hormones. It is of considerable interest that these two substances, with quite different physiologic properties, differ in only two amino acids. Oxytocin stimulates the contraction of the uterine muscles and sometimes is injected during childbirth to contract the uterus. Vasopressin causes a contraction of smooth muscles; its contraction of the muscles in the walls of arterioles causes a general increase in blood pressure. It also regulates the reabsorption of the water by the kidney tubules (see Section 15.3).

An injury of the brain nuclei, the posterior lobe or the connecting nerve tracts may lead to a deficiency of antidiuretic hormone (vasopressin) and the condition known as **diabetes insipidus.** This disease, characterized by the failure of the kidney to concentrate urine, results in the patient's excreting as much as 30 or 40 liters of urine daily and hence suffering from excessive thirst. Injection of vasopressin does not cure the disease, just as injection of insulin does not cure diabetes mellitus, but it does relieve all the symptoms, and by repeated

Figure 18.16 The effect of growth hormone on the dachshund. Top, normal dog. Bottom, dog injected with growth extract for a period of six months. (From Evans, Simpson, Meyer and Reichert.)

injections of vasopressin the patient can live a normal life. In one form of diabetes insipidus the receptors in the kidney for ADH are missing, and hence the injection of antidiuretic hormone does not alleviate the disorder.

The intermediate lobe of the pituitary secretes a **melanocyte-stimulating hormone** (MSH) that darkens the skin of fishes, amphibians and reptiles by dispersing the pigment granules in the chromatophores. The skin of a frog becomes darkened in a cool dark environment and light-colored in a warm light place (Fig. 18.14). Hypophysectomy produces a permanent blanching of the skin and injection of MSH causes darkening. The location of the pigment in the chromatophores is controlled directly by the amount of MSH present, not by nerves. The pituitaries of birds and mammals are rich in MSH, but MSH has little or no effect on their pigmentation.

The anterior lobe of the pituitary secretes growth hormone **(somatotropin)**, thyroid-stimulating hormone **(TSH)**, adrenocorticotropic hormone **(ACTH)**, follicle-stimulating hormone **(FSH)**, luteinizing hormone **(LH)** and prolactin (luteotrophic hormone, **LTH**). The importance of these hormones is demonstrated by the marked abnormalities that follow hypophysectomy: the cessation of growth in young animals, regression of gonads and reproductive organs and atrophy of the thyroid and adrenal cortex (Fig. 18.15).

Growth hormone was the first pituitary hormone to be described. As early as 1860 it was recognized that gigantism was correlated with an enlargement of the pituitary. A growth-promoting extract of beef pituitary was made by Evans and Long in 1921, and pure growth hormone was isolated by C. H. Li in 1944. Human growth hormone is a single peptide chain of 191 amino acids. It controls general body and bone growth and leads to an increase in the amount of cellular protein (Fig. 18.16). Overactivity of the pituitary during the growth period leads to very tall but well proportioned persons, and underactivity leads to small persons of normal body proportions, called **midgets.** After normal growth has been completed, hypersecretion of growth hormone produces **acromegaly,** characterized by the thickening of the skin, tongue, lips, nose and ears and by growth of the bones of the hands, feet, jaw and skull. The other bones have apparently lost their ability to respond to growth hormone.

In a race of hereditary dwarf mice, the pituitary lacks the type of cell that secretes growth hormone. These animals can be induced to grow to normal size by implanting a pituitary from a normal mouse. Growth hormones from different species have somewhat different amino acid compositions and differ in effectiveness. Beef growth hormone will cause growth in rats but not in humans or monkeys. Growth hormone prepared from human or monkey pituitaries will stimulate growth in humans and monkeys. This hormone affects the rates of a number of metabolic processes leading to increased protein synthesis, the mobilization of lipid from tissues, an increased lipid concentration in the blood, an increased deposition of glycogen in liver and muscle and an increased concentration of glucose in the blood. It is curious that a man in his 80's has about as much growth hormone in his pituitary as does a rapidly growing child. Growth hormone is rapidly degraded after it has been secreted into the blood stream; its biological half-life is about 25 minutes.

Adrenocorticotropic hormone, ACTH, is a peptide containing 39 amino acids in a single chain. The first 13 of these amino acids have a sequence identical to the peptide chain in melanocyte-stimulating hormone, MSH. ACTH is synthesized rapidly, little is stored in the pituitary, and it is rapidly removed from the plasma. Its biological half-life is less than 20 minutes. This hormone stimulates the growth of the adrenal cortex and its production of adrenocorticosteroids. The ACTHs produced by different species have similar but not absolutely identical peptide chains. The substitutions of amino acids in the ACTHs of different species are restricted to the region from the 25th to the 32nd residue. A synthetic polypeptide chain of the first 23 amino acids is fully active in stimulating the secretion of adrenal corticoids, but it does not react with antibodies developed in the rabbit against the ACTH of the pig. Thus it appears that the usual immunological determinants of ACTH are located in the carboxyl tail of the peptide.

The atrophy of the thyroid that follows extirpation of the pituitary can be prevented by the administration of an extract containing **thyroid-stimulating hormone** (TSH). The injection of TSH in a normal animal causes growth of the thyroid and thickening of the follicle cells, with the result that they become columnar rather than cuboidal. TSH is a basic glycoprotein with a molecular weight of about 25,000. Its carbohydrate components include mannose, sucrose and N-acetyl galactosamine.

Follicle-stimulating hormone (FSH) and **luteinizing hormone** (LH) are both glycoproteins. FSH, with a molecular weight of about 32,000, contains about 8 per cent of carbohy-

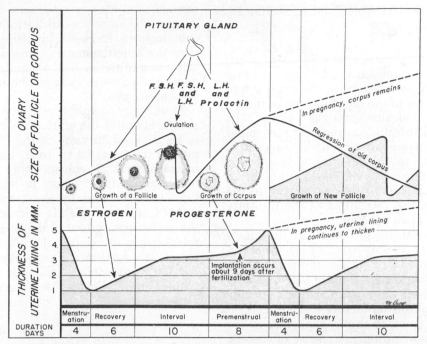

Figure 18.17 The menstrual cycle in the human female. The solid lines indicate the course of events if the egg is not fertilized; the dashed lines indicate the course of events when pregnancy occurs. The actions of the hormones of the pituitary and ovary in regulating the cycle are indicated by arrows. (From Villee, C. A.: Biology, 7th ed, Philadelphia, W. B. Saunders Company, 1977.)

drate, including some sialic acid, which apparently is required for biological activity, for FSH is no longer hormonally active after the sialic acid has been removed by treating the hormone with the enzyme neuraminidase. LH is a slightly smaller protein with a molecular weight of about 26,000. The LHs from different species have carbohydrate contents ranging from 4.5 per cent in humans to 11 per cent in sheep. A third gonadotropin, **prolactin** or luteotrophic hormone, is a protein with a single peptide chain of 198 amino acids but no carbohydrate components.

The three glycoprotein hormones of the pituitary—TSH, FSH and LH—are each composed of two subunits, termed α and β. The α subunits of all three are very similar peptides with 96 amino acids in the chain, whereas the β subunits are distinctive and appear to be responsible for the biological specificity of the hormone. The hormones can be separated into their subunits, which have little or no biological activity, and then recombined to give full activity. The combination of a TSH α subunit with an FSH β subunit results in a protein with FSH activity.

The ovaries or testes of a hypophysectomized young animal never become mature, and they neither produce gametes nor secrete enough sex hormones to develop the second-

ary sex characters. Hypophysectomy of an adult results in involution and atrophy of the gonads. Both FSH and LH are necessary for achieving sexual maturity and for the regulation of the estrous or menstrual cycles. The effect of FSH is primarily on the development of graafian follicles in the ovaries; it does not cause any significant production of estradiol. Luteinizing hormone is also necessary for the development of follicles, and it controls the release of ripe eggs from the follicle, the formation of corpora lutea and the production and release of estrogens and progesterone (Fig. 18.17). Prolactin, or LTH, maintains the secretion of estrogens and progesterone by the ovary and stimulates the secretion of milk by the breast after parturition. It is effective only after the breast has been appropriately stimulated by estrogen and progesterone. Prolactin induces behavior patterns leading to the care of the young (the "maternal instinct") in mammals and other vertebrates. Roosters treated with prolactin will take care of chicks, taking them to food and water, sheltering them under their wings and protecting them from predators.

FSH and LH control the development and functions of the testis. FSH increases the size of the seminiferous tubules, and LH stimulates the interstitial cells of the testis to produce the male sex hormone.

18.9 HYPOTHALAMIC RELEASING HORMONES

The control of pituitary function, which ensures that the proper amount of each of these hormones will be released at the proper moment in response to the requirements of the organism, is a complex process indeed. The release of each trophic hormone is controlled in part by the concentration of the respective target hormone in the circulating blood. For example, the release of ACTH is inhibited by cortisol. This mechanism ensures that in the normal animal the secretions of the pituitary and its target organ are kept in balance.

The hypothalamus provides an important control of pituitary function. Axons from certain centers in the hypothalamus end in the median eminence in the floor of the third ventricle. The tips of these axons secrete specific releasing or inhibiting factors that are carried by the portal veins to the hypophysis, where they stimulate or inhibit the release of specific pituitary hormones. The first of the hypothalamic substances to be identified as a distinct entity was CRH, or **corticotropin releasing hormone,** which evokes the release of ACTH. The release of CRH from the hypothalamus is inhibited by glucocorticoids and increased by adrenalectomy. The first releasing hormone to be isolated and synthesized was TRH, or **thyrotropin releasing hormone.** This is a tripeptide composed of pyroglutamate, histidine and proline. The synthetic compound is as active biologically as the natural tripeptide and causes the release of TSH from the anterior pituitary. The administration of TRH causes the release of prolactin, as well as TSH, and thus TRH may be a physiologically important regulator of the release of both.

Another substance isolated from the median eminence, **gonadotropin releasing hormone** (GnRH), induces the release of both FSH and LH from the pituitary tissue. GnRH has been isolated and identified as a decapeptide with the sequence pyroglu-his-trp-ser-tyr-gly-leu-arg-pro-glyNH$_2$. A synthetic decapeptide with this sequence has properties identical to the natural hormone's for regulating the synthesis and release of both FSH and LH. Some investigators believe that there are separate FSH-releasing and LH-releasing hormones, but others conclude that a single GnRH regulates the release of both.

The release of growth hormone in the pituitary is stimulated by a **growth hormone releasing hormone** or a GHRH. This has also been found to be a decapeptide, with the sequence val-his-leu-ser-ala-glu-glu-lys-glu-ala. A growth hormone–inhibiting hormone, or **somatostatin,** has been identified and purified from the hypothalamus. Somatostatin is a peptide containing 14 amino acids, with the sequence ala-gly-cys-lys-asn-phe-phe-trp-lys-thr-phe-thr-ser-cys.

Thus the concept that the pituitary is the master gland of the endocrine system has had to be changed in view of the finding that the master pituitary is, in turn, the slave of the hypothalamus. However, both the hypothalamus and pituitary may be regulated by the hormones secreted by the target glands—the thyroid, adrenal, gonads and so on. Thus, in the endocrine system there is no master-slave relationship to be discerned; instead, the system is an egalitarian society composed of a federated group of autonomous organs, each of which can provide feedback controls for certain other glands.

18.10 THE PINEAL

The **pineal gland,** a small round structure on the upper surface of the thalamus lying between the cerebral hemispheres, is derived embryologically as an outgrowth of the brain. This gland secretes **melatonin,** a methoxy indol synthesized from tryptophan converted to 5-hydroxy tryptamine (serotonin). Serotonin is acetylated at the amine nitrogen, and a methyl group is added to the 5-hydroxy group by hydroxy indol-O-methyl transferase, an enzyme found only in the pineal. The enzyme is stimulated by norepinephrine released by the tips of the sympathetic nerves that extend to the pineal from a cervical sympathetic ganglion. Light falling on the retina of the eye increases the synthesis of melatonin by the pineal. A small nerve, the **inferior accessory optic tract,** passes from the optic nerve through the medial forebrain and connects with the sympathetic nervous system. The pineal secretion inhibits ovarian functions either directly or by way of an effect on the pituitary. Girls blind from birth undergo puberty earlier than normal, apparently because they lack the inhibitory effect of melatonin.

18.11 THE TESTES

Although Berthold concluded in 1849 from his experiments with roosters that the testis produces a blood borne substance needed for the development of male sex characteristics, the major androgen, testosterone, was not identified until 1935. Testosterone has a general protein anabolic effect on metabolism, inducing growth by stimulating the formation of cell proteins. The administration of androgens leads to an increase in body weight, owing to the synthesis of protein in muscle and, to a lesser extent, in the liver and kidney.

Testosterone and other androgens stimulate the development and maintenance of the secondary male characters; the enlargement of the external genitals; the growth of the accessory glands, such as the prostate and seminal vesicles; the growth of the beard and body hair and the deepening of the voice. The secondary sex characters of other animals—the antlers of deer and the combs, wattles and plumage of birds—are also controlled by androgens. Male sex hormones are responsible, in part, for increased libido in both sexes and for the development of mating behavior.

The removal of the testis (castration) of an immature male prevents the development of the secondary sex characters. A castrated man, a **eunuch,** has a high-pitched voice, beardless face and small genitals and accessory glands. Castration was practiced in the past to provide guardians for harems and sopranos for choirs. Many domestic animals are castrated to make them more placid. The injection of testosterone into a castrated animal restores all the sex characters to normal. The anal fin of the male mosquito fish, *Gambusia,* is differentiated into a penis-like organ used to transfer sperm to the female. It fails to develop if the fish is castrated but appears if the castrate male or the female is treated with testosterone.

It should be clearly understood that males produce female sex hormones, estrogens, and that females produce androgens. One of the richest sources of female sex hormones is the urine of stallions. The normal differentiation of the sex characters is a function of a balance between the two.

The failure of the testes to descend normally from the body cavity to the scrotal sac, called **cryptorchidism,** produces sterility but has little or no effect on the production of testosterone. Microscopic examination of an undescended testis shows that the cells in the seminiferous tubules regress, but the interstitial cells are normal. The cells of the seminiferous tubules are particularly susceptible to heat, and the temperature of the body cavity, 3 or 4 degrees higher than that of the scrotal sac, destroys them. It is probable that the elevated temperature during a prolonged fever makes a man

sterile for some time. In rodents, rabbits and some other mammals the testes remain in the body cavity except during the breeding season when they descend into the scrotal sac.

The removal of the pituitary causes regression of both interstitial cells and seminiferous tubules of the testes. Androgen secretion is decreased and the secondary sex characters regress. Normal development of the seminiferous tubules in spermatogenesis apparently requires the combined action of FSH, LH and testosterone. The administration of excessive amounts of testosterone or estrogen may produce regression of the testes, presumably by inhibiting the release of FSH and LH from the pituitary.

The cyclic growth and regression of the testes in animals with periodic breeding seasons appear to be mediated by the pituitary. Such animals have very low amounts of circulating gonadotropin in the nonbreeding season. Changes in the ambient temperature or in the amount of daily illumination produce stimuli that are mediated by the brain and hypothalamus to induce gonadotropin secretion by the pituitary and the consequent growth and functional state of the testes.

18.12 THE OVARIES

The ovaries of vertebrates are endocrine organs as well as the source of eggs; they produce the steroid hormones **estradiol** and **progesterone.** Some mammalian ovaries produce a third hormone, the protein **relaxin,** which softens the ligaments of the pubic symphysis at the time of parturition.

In the human, the major sources of female sex hormones are the cells lining the ovarian follicles and those of the corpus luteum formed after ovulation has occurred (Fig. 18.18). The follicle cells secrete primarily estradiol, and the luteal cells secrete primarily progesterone. Estradiol regulates and controls the body changes in the female at the time of puberty or sexual maturity: the broadening of the pelvis, development of the breasts, growth of the uterus, vagina and external genitalia, change in voice quality and onset of the menstrual cycle. Progesterone is required for the completion of each menstrual cycle, for the implantation of the fertilized egg in the uterus and for the development of the breasts during pregnancy.

Histochemical evidence indicates that the thecal cells of the follicle are the source of estrogens and that they, plus the granulosa cells

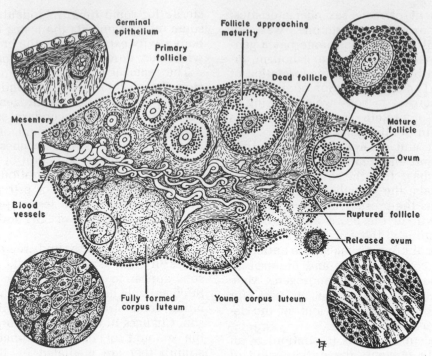

Figure 18.18   Stages in the development of an egg, follicle and corpus luteum in a mammalian ovary. Successive stages are depicted clockwise, beginning at the mesentery. Insets show the cellular structure of the successive stages. (From Villee, C. A.: Biology. 7th ed. Philadelphia, W. B. Saunders Company, 1977.)

of the corpus luteum, are the source of progesterone. The primary estrogenic hormone is 17β estradiol; other estrogens, such as estrone and estriol, may be metabolites of estradiol.

It is a curious fact that the primary female sex hormone, estradiol, with 18 carbons is synthesized from the 19-carbon male sex hormone, testosterone. This in turn is synthesized from the second female sex hormone, progesterone, which has 21 carbons. Progesterone is the precursor not only of androgens and estrogens but also of the glucocorticoids and mineralocorticoids in the adrenal cortex.

18.13 THE ESTROUS AND MENSTRUAL CYCLES

The females of most mammalian species show cyclic periods of the sex urge and will permit copulation only at certain times, known as periods of **estrus** or "heat," when conditions are optimal for the union of egg and sperm. Most wild animals have one estrous period per year; the dog and cat have two or three; and rats and mice have estrous periods every four or five days. Estrus is characterized by heightened sex urge, ovulation and changes in the lining of the uterus and vagina. After

estrus, the uterine lining thickens and its glands and blood vessels develop to provide an optimal environment for the implanting embryos. In contrast, the cycle of the primates is characterized by periods of vaginal bleeding called **menstruation** that result from the degeneration and sloughing of the endometrial lining of the uterus. Primates, unlike other mammals, show little or no cyclic change in the sex urge and permit copulation at any time in the menstrual cycle.

A key event in the estrous cycles of lower mammals and the menstrual cycles of primates is **ovulation,** the release of a mature egg from a follicle in the ovary (Fig. 18.17). This must be released when sperm are likely to be present in the oviduct and when the lining of the uterus, the **endometrium,** is in the proper condition to permit implantation of the fertilized egg. The coordination of these events involves some half dozen hormones in addition to neural mechanisms. Ovulation is triggered by a surge of LH, secreted by the pituitary in response to GnRH, secreted by the hypothalamus (Fig. 18.19). This in turn is triggered by a surge of estradiol, which operates by positive feedback control to induce the release of GnRH and LH. Prolonged treatment of the female with a constant amount of either estradiol or progesterone, as in the oral contraceptive pills, can block the LH surge and thus block ovulation. It thus appears that a sharp rise in estradiol must occur to affect the positive feedback center in the hypothalamus so that the pituitary is induced to secrete LH. This theory is supported by experiments in which a single dose of estrogen induced ovulation in animals whose ovaries had been primed with suitable doses of FSH and LH.

The pattern of gonadotropin release is cyclic in the female mammal but noncyclic in the male. In the rat, the male-female patterns of gonadotropin release are determined in early postnatal life. The hypothalami of both genetic male, XY, and female, XX, rats will develop the female cyclic pattern of release unless exposed to testosterone early in postnatal life. However, human females with congenital adrenal cortical hyperplasia who have been exposed to large amounts of androgen during fetal life subsequently develop normal ovulatory cycles if treated with cortisol. There is evidence to suggest that both men and women have a tonic center in the arcuate nucleus, the cells of which produce GnRH at a relatively constant

rate. Other neurons in the suprachiasmatic nucleus of women constitute the cyclic center; this responds to the estrogen surge and secretes the GnRH, which regulates the LH surge that causes ovulation. GnRH is released at the tips of the axons located in the median eminence. From there it passes into capillaries of the hypophyseal portal vessels and down the pituitary stalk to the anterior pituitary. The vessels break up into a second network of capillaries, and releasing hormone passes to the cells of the pituitary.

The lining of the uterus is almost completely sloughed at each menstrual period and thus is thinnest just after the menstrual flow. At that time, under the influence of FSH, one or more of the follicles in the ovary begin to grow rapidly (Fig. 18.17). The follicular cells produce estradiol, which stimulates the growth of the endometrium and some growth of the uterine glands and blood vessels. Ovulation occurs in response to the LH surge from the pituitary, and in the human this occurs about 14 days before the beginning of the next menstrual period. The corpus luteum develops and,

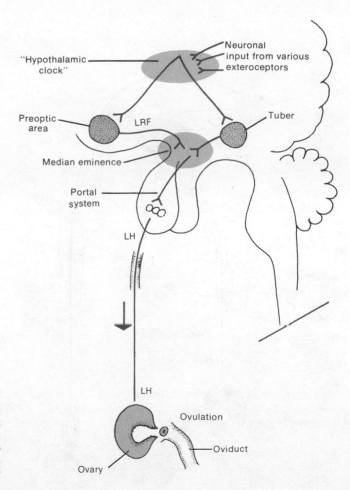

Figure 18.19 Spontaneous ovulation in humans and other primates. The secretion of releasing factors from the hypothalamus is stimulated not by nerve impulses coming from the vagina but by a surge of estradiol from the ovary, modulated by some sort of endogenous biological clock in the hypothalamus. (From Villee, C. A.: Biology. 7th ed. Philadelphia, W. B. Saunders Company, 1977.)

under the stimulus of LH and prolactin, secretes progesterone and estradiol, which promote the further growth of the endometrium. The endometrial glands become secretory and the blood vessels become long and coiled. Progesterone decreases the activity of the uterine muscles and brings the uterus into a condition such that the developing embryo formed from the fertilized egg can undergo implantation and development. Progesterone inhibits the development of other follicles. If fertilization and implantation do not occur, the corpus luteum begins to regress and secretes less progesterone; the endometrium, no longer provided with enough progesterone to be maintained, begins to slough off, resulting in menstruation.

In the rabbit, cat, ferret, mink and certain other mammals, ovulation occurs reflexly in response to the stimulus of the act of copulation (Fig. 18.20). Nerves in the lining of the vagina send impulses up the spinal cord to the brain and stimulate the hypothalamus to

secrete GnRH. The stimulus for ovulation may be a single copulation, as in the rabbit, or a minimum of 19 per day, as required by the short-tailed shrew. Direct electrical stimulation of the appropriate regions of the hypothalamus, the **tuber cinereum** and **preoptic region,** can cause ovulation in the rabbit, cat or monkey. In the rabbit the neural stimulation of coitus causes the hypothalamus to produce GnRH, which passes via the portal system to the anterior pituitary and causes it to release a surge of LH. This passes to the ovary and causes the follicle to rupture, thereby releasing the ovum. Such animals are termed **reflex ovulators.**

A second and larger group of mammals are **spontaneous ovulators;** their ovulation is stimulated not by coitus but by the series of feedback hormonal controls described in the human. In these ovulators, the timing and frequency of ovulation may be influenced by environmental factors. A female rat kept under normal conditions of day and night

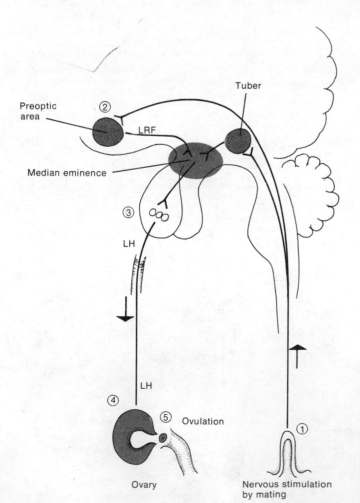

Figure 18.20 Diagram of the nervous pathways involved in the reflex ovulation in an animal such as the rabbit. The stimulation of receptors in the vagina by the mating act generates impulses that pass to the hypothalamus and bring about the secretion of releasing factors. These pass to the pituitary and cause the release of luteinizing hormone, which goes via the blood stream to the ovary and initiates ovulation, the release of the egg. (From Villee, C. A.: Biology. 7th ed. Philadelphia, W. B. Saunders Company, 1977.)

lighting will ovulate early in the morning, between 1:00 and 2:30 A.M. If rats are kept for two or more weeks under artificial conditions in which the periods of light and dark are reversed, the time of ovulation will be shifted 12 hours. Rats exposed to continuous light eventually stop ovulating and go into constant estrus, with persistent vaginal cornification. Primates are spontaneous ovulators, and there is ample evidence that the rhythm of the human menstrual cycle and ovulation can be influenced by environmental factors. Nurses on night duty and airline stewardesses who travel long distances to different time zones frequently note perturbations in their menstrual cycles.

Oral Contraceptives

Estrogen and progesterone block ovulation not by a direct effect on the ovary but by preventing the secretion of gonadotropin-releasing hormone by the hypothalamus. If a woman were to ovulate after she was pregnant and if this second egg were also fertilized, she would then have two fetuses in her uterus — one several months younger than the other. At parturition both would be ejected, and the younger fetus would be at a severe disadvantage. This has been prevented by the evolution of a hormonal system that prevents ovulation once conception has occurred. The high levels of progesterone and estrogen that are maintained throughout pregnancy prevent the release of an LH surge.

The oral contraceptives contain synthetic estrogens and progestins that prevent ovulation in a similar manner. Natural estradiol and progesterone are rapidly metabolized in the body, but synthetic hormones with slightly altered molecular structures are metabolized more slowly and hence persist for a longer time. The oral contraceptives, like the natural hormones, inhibit the secretion of releasing hormones and LH and thus prevent ovulation. A woman taking "the pill" has no midcycle surge of LH and FSH and does not ovulate.

The oral contraceptives in general use are combination pills, containing both an estrogen and a progestin. These are taken daily for a period of 21 days followed by a break of seven days, during which menstruation occurs. Gonadotropin secretion is suppressed, preventing ovulation. In addition, the oral contraceptive prevents the normal maturation of the endometrium and produces an altered cervical mucus hostile to penetration by sperm.

Another type of pill, containing only progestin, is taken every day throughout the cycle and, although ovulation may occur, fertilization is prevented by the changes in the nature of the cervical mucus and the endometrium.

18.14 THE HORMONES OF PREGNANCY

If the egg has been fertilized and implants in the endometrium, the cells of the trophoblast in the developing placenta secrete **chorionic gonadotropin.** Its strong luteinizing and luteotrophic activities maintain the maternal corpus luteum and stimulate its continued secretion of progesterone. By the 16th week or so of pregnancy in the human, the placenta itself produces enough progesterone so that the corpus luteum is no longer needed and undergoes involution. Thus the placenta is an endocrine organ that produces hormones similar to those of the ovary and pituitary, in addition to being an organ for the support and nourishment of the fetus. These placental hormones, together with those of the maternal and fetal endocrine glands, control the many adaptations necessary for the continuation and successful termination of pregnancy. The placenta does not merely accumulate a hormone made elsewhere, for bits of placenta grown in tissue culture produce both protein and steroid hormones

One of the earliest signs of pregnancy is the appearance of chorionic gonadotropin in the blood and urine. The peak of its production is reached in the second month of pregnancy, after which the amount in blood and urine decreases to low levels (Fig. 18.21). Several pregnancy tests involve the detection of this gonadotropin in a sample of the urine from a woman suspected of being pregnant. A very sensitive radioimmunoassay test for chorionic gonadotropin can diagnose pregnancy just a few days after implantation. The human placenta, and probably the placentas of other mammals, secretes another protein hormone, **placental lactogen,** with properties somewhat similar to those of pituitary growth hormone and prolactin. In addition, the placenta secretes both estrogens and progesterone, which reinforce the ovarian hormones in the maintenance of pregnancy.

In some animals, such as the rabbit, the placenta is a significant source of relaxin. This protein hormone, also produced by the ovary, functions to relax the ligaments of the pelvis in order to facilitate the birth of the young.

Figure 18.21 Hormone levels in blood and urine during pregnancy.

Relaxin is effective only after the connective tissue of the pubic symphysis has been sensitized by the action of estradiol. Relaxin also inhibits the motility of the uterine muscles.

The production of estrogens and progesterone increases gradually throughout pregnancy and reaches a peak just before or at the time of parturition. The factors that determine the onset of labor, the expulsion of the fetus from the uterus, remain a mystery. There are many hormonal changes that occur close to the time of parturition—decreases in estrogen and progesterone, increases in certain hormones secreted by the fetal adrenal and changes in prostaglandin production in the uterus—but whether these are causes, effects or unrelated phenomena remains to be determined.

18.15 HORMONAL CONTROL OF LACTATION

The growth and development of the mammary glands after puberty and the production and secretion of milk after parturition are controlled by a complex sequence of hormonal events. The initial development of the breast and the proliferation of glandular elements are stimulated by estradiol and require the presence of insulin and growth hormone. Progesterone stimulates the further development of the glands at and after puberty. During pregnancy the additional development of the mammary glands is stimulated by estradiol and by progesterone produced primarily in the placenta. The alveoli and ducts of the mammary glands continue to develop and become secretory; however, progesterone inhibits the production of milk, and it is only after parturition and the sudden decrease in progesterone in the blood that occurs with the loss of placenta that milk production can begin. Throughout pregnancy the production of pituitary gonadotropins is inhibited by the placental gonadotropins, but with the expulsion of the placenta at birth, the pituitary begins to secrete large quantities of prolactin that stimulate the breast to produce milk (Fig. 18.22). If the breast is regularly emptied of milk, the pituitary will continue to secrete prolactin, but if milk is not removed, secretion will stop. Nerve impulses from the nipple to the hypothalamus stimulate the secretion of prolactin-releasing hormone, which passes to the pituitary and increases the production of prolactin. ACTH, TSH and growth hormone may also play a role in controlling both the growth of the mammary glands and their production of milk.

The secretion of milk from the alveolar glands into the milk duct is under the control of prolactin. The transport of milk from the alveolus to the nipple, where it can be removed by the suckling infant, is triggered by the **milk ejection reflex,** a neurohormonal reflex involving oxytocin. Shortly after the infant is put to breast, the mammary gland seems suddenly to fill with milk, which comes under pressure and may spurt from the nipple. Milk flow can occur in anticipation of the suckling reflex, and, in contrast, stress or discomfort can inhibit this flow so that less milk is obtained by the suckling child. The rapid increase in pressure within the mammary gland is due to the sudden expulsion from the alveoli of milk that has been synthesized previously. The movement of milk results from the contraction of myoepithelial cells that squeeze the alveoli and expel their contents. The act of suckling, or the milking of a cow or mare, triggers nerve impulses from the receptors in the nipple that pass up the spinal cord to the hypothalamus and cause the release of oxytocin from the posterior lobe of the pituitary. In the mammary gland, oxytocin stimulates the myoepithelial cells to contract, increasing intramammary pressure and bringing about the ejection of milk. Thus the milk ejection response is a reflex, but unlike ordinary reflexes the afferent arc is nervous but the efferent arc is hormonal. Like ordinary neural reflexes, the response can be conditioned, which explains ejection before milking in response to stimuli associated with nursing.

During lactation the release of gonadotropins is suppressed with the result that ovulation occurs only occasionally and menstrual periods are usually irregular.

18.16 PROSTAGLANDINS

There is great interest at present in the prostaglandins, derivatives of 20-carbon polyunsaturated fatty acids. Prostaglandins have a characteristic five-carbon ring in the middle of the chain and hydroxyl or ketone groups on certain carbons (Fig. 18.23). Many different prostaglandins have been discovered or synthesized and their chemistry is quite complex. Although prostaglandins were first found in semen and are produced by the seminal vesicle, it is now clear that they are produced by a great many types of tissues and released into the blood stream. They are synthesized from such polyunsaturated fatty acids as arachidonic, a 20-carbon fatty acid chain with four double bonds. The production rate in the human, the amount synthesized per day, was measured by Samuelson in Sweden. In four normal adult males, the production rate

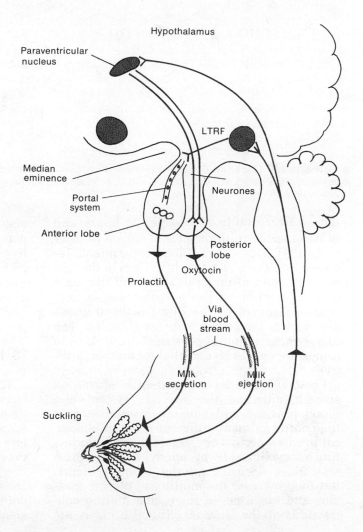

Figure 18.22 A diagram of the hormonal controls that stimulate the production and secretion of milk in the mammary gland. (From Villee, C. A.: Biology. 7th ed. Philadelphia, W. B. Saunders Company, 1977.)

Figure 18.23 The biosynthesis of prostaglandins from arachidonic acid, a polyunsaturated fatty acid with a 20-carbon chain. (From Page, E. W., C. A. Villee, and D. B. Villee: Human Reproduction. 2nd ed. Philadelphia, W. B. Saunders Company, 1976.)

ranged from 109 to 226 μg per 24 hours, and in two normal adult females, it ranged from 23 to 48 μg per 24 hours. These experiments indicate that there is a sex difference in the rate of production of prostaglandins and that only a small fraction of the polyunsaturated fatty acids consumed in the diet (about 10 grams per day) is converted to prostaglandins. The high concentration of prostaglandin in human semen appears to be essential for normal fertility.

Prostaglandins have a variety of effects on smooth muscles, the nervous system and blood pressure and have been implicated in the control of many different types of biological events — perhaps by regulating the production of cyclic AMP by adenyl cyclase. Prostaglandins decrease arterial blood pressure, but they increase the motility of uterine muscles and are used in increasing uterine contractions at the time of childbirth. They are also used to induce abortions earlier in pregnancy. Prostaglandins also increase the motility of intestinal muscles and may cause severe cramps, nausea, vomiting and diarrhea. In a number of species certain prostaglandins have a sedative, tranquilizing effect on the animal.

18.17 ARTHROPOD HORMONES

The hormones and neurosecretions of crustaceans and insects are probably better known than those of other invertebrates. In these forms there is experimental evidence for a battery of hormones that regulate chromatophore expansion, molting, growth, sexual reproduction and development. A number of the crustacean hormones, such as those controlling chromatophore expansion, are released from a gland in the eye stalk. These hormones have been studied extensively, since the eye stalk can be removed readily and extracts of the eye stalk can be prepared and tested.

Some insects, such as the grasshopper, pass through a series of successive molts during development. With each molt they look more like the adult, but there is no striking change in appearance with any one molt (Fig. 18.24). In contrast, moths, butterflies, flies and many

other insects pass through successive stages that are quite unlike one another. From the egg hatches a wormlike **larva**—called a caterpillar (moths) or maggot (flies)—that crawls around, eats voraciously and molts several times, each time becoming a larger larva. The last larval molt forms a **pupa,** which neither moves nor feeds. The moth and butterfly larva spins a cocoon and pupates (molts to form a pupa)

INCOMPLETE METAMORPHOSIS   COMPLETE METAMORPHOSIS

Figure 18.24 Comparison of gradual, or incomplete, metamorphosis of a grasshopper and complete metamorphosis of a cecropia moth. (From Villee, C. A.: Biology. 6th ed. Philadelphia, W. B. Saunders Company, 1972.)

within it. During pupation the structures of the larva are broken down and used as raw materials for the formation of parts of the adult animal. Each part—legs, wings, eyes and antennae—develops from a group of cells, called a **disc,** that develop directly from the egg. They have never been a functional part of the larva but remain quiescent during the larval period. During the pupal stage these discs grow and differentiate into the adult structures, but they remain collapsed and folded. When the adult hatches out of the pupal case, blood is pumped into these collapsed structures, they unfold and inflate, and the chitinous exoskeleton hardens. This striking change in appearance from larva to adult is termed **metamorphosis.** The grasshopper, which undergoes a gradual change in form from larva to adult, is said to exhibit incomplete metamorphosis.

The larva and adult insect not only have a different appearance but they also have quite different modes of life. The butterfly larva eats leaves; the adult drinks nectar from flowers. The mosquito larva lives in ponds and eats algae and protozoa; the adult sucks the blood of humans and other mammals. The adults of some species, such as the mayfly, live only

a few hours, just long enough to mate and lay eggs.

Because its exoskeleton is firm and rigid, an insect can grow and change its appearance only by molting; that is, by periodically shedding its chitinous exoskeleton and growing a new one. Before shedding the old exoskeleton, the insect develops a new one underneath (Fig. 18.25); some of the materials of the old exoskeleton are salvaged and used in the synthesis of the new one. A line appears along which the old skeleton will break, and the insect swells up by swallowing water or air, bursting the old exoskeleton and stretching the new one to its full size. At this time, while it lacks its hard exoskeleton, the insect is likely to dry out or be eaten by a predator.

The molting of insects and other arthropods is under hormonal control. The major organ initiating molting in insects is the **intercerebral gland** on the surface of the brain (Fig. 18.26). Axons from these neurosecretory cells pass posteriorly to the **corpus cardiacum,** composed of the expanded tips of these axons. The intercerebral gland secretes several hormones that regulate various aspects of body activity. One of these, **prothoracicotropic hormone,** is released from the corpora car-

Figure 18.25 Diagram of the process of molting in an arthropod. A, The fully formed exoskeleton and underlying epidermis between molts. B, Separation of the epidermis and the secretion of molting fluid and a new epicuticle. C, Digestion of the old secondary chitinous layer and secretion of a new primary chitinous layer. D, The situation just before the old exoskeleton is shed. (Villee, C. A. and V. G. Dethier: Biological Principles and Processes. 2nd ed. Philadelphia, W. B. Saunders Company, 1976.)

Figure 18.26   The endocrine glands of the cockroach. Those on the left lie in the head dorsal to the esophagus; the prothoracic gland, right, is in the ventral part of the prothorax, among the muscle cells. (After Bodenstein, D.: Recent Progress In Hormone Research. Vol. 10, New York, Academic Press, Inc., 1954.)

diaca and initiates the molting process by stimulating a second endocrine gland, the prothoracic, to secrete **ecdysone.** The conditions that stimulate the activity of the intercerebral glands have been studied in the blood-sucking bug *Rhodnius prolixus.* Stretch receptors in its abdomen are stimulated when the bug has a large meal of blood, and these stimulate the intercerebral glands to activity. *Rhodnius* feeds only infrequently, but one large meal is enough to support the metabolic activity associated with a molt.

The prothoracic gland, a diffuse set of strands of large ectodermal cells in the ventral part of the prothorax, secretes ecdysone, a steroid synthesized from cholesterol. Ecdy-

sone differs from the steroids secreted by vertebrates in that it retains the side chain of cholesterol (Fig. 18.27); it has five hydroxyl groups, two of which, at positions 14 and 25, are quite unusual. It has a ketone group adjacent to a double bond but in ring B instead of in ring A, as in the vertebrate steroids. More than 500 kilograms of silk worms must be extracted to obtain 25 milligrams of ecdysone. There is a striking parallelism between the stimulation of production of ecdysone by prothoracicotropic hormone and the stimulation of production of adrenal steroids by ACTH. Ecdysone stimulates the hypodermis to secrete molting fluid, thus leading to molt and metamorphosis. The same system of prothora-

Figure 18–27   The molting hormone of insects, ecdysone.

cicotropic hormone and ecdysone initiates the molt from larval to pupal stage.

In 1960 Clever and Karlson found that within 15 minutes of injecting minute amounts of ecdysone into the larva of the midge *Chironomus* the puffing or swelling of a specific region of a particular chromosome occurred. Dipteran insects may have in certain tissues giant chromosomes composed of many chromatids. In these giant chromosomes, individual bands, corresponding to specific genetic loci, are visible under the microscope and can be identified. The puffing of certain bands in the chromosomes occurs at specific times in the course of normal development and is known to represent the synthesis of RNA at those sites (Fig. 18.28). Injecting ecdysone causes puffing at band I-18C and the subsequent production of the enzyme **dopa decarboxylase** in epidermal cells. Dopa decarboxylase catalyzes the conversion of DOPA to N-acetyl dihydroxyphenylethylamine, an agent involved in the hardening of the cuticle.

This suggestion that hormones may control the transcription of specific genes is being investigated by many biologists.

The control of metamorphosis involves a third set of endocrine organs, the **corpora allata,** small glands located in the head, just behind the corpora cardiaca. If these glands are removed from a young larva, it will undergo metamorphosis at the next molt, even though this may be one or more molts too soon. Conversely, if the corpora allata from young larvae are transplanted into older larvae about to undergo metamorphosis at the next molt, the juvenile form is retained at the molt. The corpora allata secrete **juvenile hormone,** $C_{18}H_{30}O_3$, which has been characterized as a derivative of a fatty acid. It is probably an isoprenoid derivative of mevalonic acid and is unusual in having ethyl side chains. A substance extracted from certain American paper products (e.g., *The New York Times*) prepared from the woods of certain trees has marked juvenile hormone activity. It

10μ

Figure 18.28 Diagram illustrating the changing appearance of a many-stranded (polytene) chromosome in a salivary gland of *Chironomus tentans* as a chromosome puff gradually appears. The material that makes up the puff has been shown by histochemical tests and by autoradiography with tritium labeled uridine to consist largely of RNA. (From Beermann: Chromosoma, 5:139–198, 1952.)

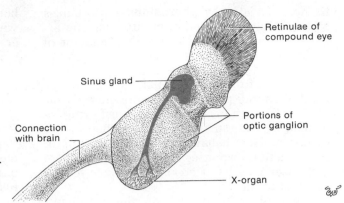

Figure 18.29 A diagram showing the location of the sinus gland and X organ in the eyestalk of the crab. (Villee, C. A., and V. G. Dethier: Biological Principles and Processes, 2nd ed. Philadelphia, W. B. Saunders Company, 1976.)

is conjectured that the trees protect themselves from some types of insects by secreting a juvenile hormone-like material that prevents the larvae from becoming adults and reproducing. It has also been suggested that we use juvenile hormone as an insecticide to prevent the multiplication of insects. Under test conditions, spraying caterpillars, or the foliage on which they are feeding, with solutions containing juvenile hormone has prevented their pupating normally. Instead of molting into giant larvae, they die.

Juvenile hormone inhibits metamorphosis but permits molting to occur, thus ensuring that the larva will molt several times and reach a large size before pupating. Juvenile hormone is not produced during the last larval stage, hence pupation can occur at the ensuing molt.

The corpora allata secrete hormones that affect other phenomena in certain groups of insects. The deposition of yolk in eggs and the secretions of the accessory sex glands are under the control of hormones from the corpora allata. In some cockroaches the male is attracted to the female by a pheromone (p. 458) that she produces and secretes under the control of a hormone from the corpora allata.

Some pupae, such as those of the giant silkworm *Platysamia cecropia*, remain in a state of arrested development or dormancy (called **diapause**) over the winter. If a newly formed pupa is kept at 24°C., it will remain inactive for five or six months but eventually will begin to develop. If a newly formed pupa is chilled to 4°C. for six weeks and then kept at 24°C. it begins at once to develop. Thus, chilling initiates further development by stimulating the release of prothoracicotropic hormone from the intercerebral gland. If intercerebral glands are removed and chilled and then transplanted into a diapausing pupa that has not been chilled, the pupa begins development.

The molting process in crustaceans is under the control of a hormone that is accumulated in the **sinus glands** in the eyestalk (Fig. 18.29). These glands secrete several hormones, one of which initiates molting. The sinus glands are composed of the expanded tips of axons surrounding a blood sinus. The cell bodies of these axons are located some distance away, in the eyestalk in the **X organ**. The hormone is produced in the cell bodies of the X organ and passes along the axons to the sinus gland, where it is stored and released. When the eyestalk of a crab or other crustacean is removed, a molt usually occurs, indicating that the hormone made in the X organ and secreted by the sinus gland tends to inhibit molting. The **Y organ**, composed of diffuse strands of ectodermal cells at the base of the large muscles of the mandibles, is an endocrine gland that produces the hormone that induces molting. Ecdysone prepared from insects will induce molting when injected into crustaceans, and it is likely that the molting hormone produced by the Y organ is ecdysone or a very similar steroid. The hormone of the X organ inhibits molting by preventing the secretion of molting hormone (ecdysone) by the Y organ. The X organ secretes, in addition to the antimolting hormone, hormones that affect the distribution of pigment in the compound eyes and the pigmentation of the body and several that influence metabolism and reproduction.

18.18 PHEROMONES

In recent years it has been appreciated that the behavior of animals may be influenced not only by hormones (chemicals released into the internal environment by endocrine glands that regulate and coordinate the activities of other tissues) but also by **pheromones**—substances released into the *external* environment by *exocrine* glands that influence the behavior of other members of the same species. We are accustomed to thinking that information can be transferred from one animal to another by sight or sound; pheromones represent a means of communication, a means of transferring information, by smell or taste. Pheromones evoke specific behavioral, developmental or reproductive responses in the recipient; these responses may be of great significance for the survival of the species.

Some pheromones act on the recipient's central nervous system and produce an immediate effect on its behavior. Among these **releaser** pheromones are the **sex attractants** of moths and the **trail pheromones** and **alarm substances** secreted by ants. Other pheromones, termed **primer** pheromones, act more slowly and trigger a chain of physiological events in the recipient that affect its growth and differentiation, including the regulation of the growth of locusts and the control of the number of reproductives and soldiers in termite colonies.

The sex attractants of moths provide some of the more spectacular examples of pheromones. Among those that have been isolated and identified are **bombykol,** a 16-carbon alcohol with two double bonds, secreted by female silkworms, and **gyplure,** 10-acetoxy-Δ^7— hexadecenol, secreted by female gypsy moths. The male silk moth has an extremely sensitive receptor in his antennae for sensing the attractant. When an investigator records the nerve impulses coming from the antennae, he finds that these electroantennagrams show specific responses to bombykol and not to other substances. The male silk moth cannot determine the direction of the source by flying up a concentration gradient, because the molecules are nearly uniformly dispersed except within a few meters of the source. Instead, he responds to the stimulus by flying *upwind* to the source. With a gentle wind, the bombykol given off by a single female moth covers an area several thousand meters long and as much as 200 meters wide. An average silkworm contains some 0.01 mg. of bombykol. It can be shown experimentally that a male responds appropriately when as little as 10,000 molecules of attractant are allowed to diffuse from a source 1 cm. away from him. He can have received only a few hundred of these molecules, perhaps less. Thus, the amount of attractant in one female could stimulate more than one billion males! The attractants, generally hydrocar-

bons, contain 10 to 17 carbons in the chain, which provides for the specificity of the several kinds of attractants.

The sex attractant of the American cockroach is not a long chain alcohol like bombykol or gyplure but has a central three-carbon ring to which methyl groups and a propanoxy group are attached. Sex attractants have been tested as possible specific insecticides. By putting sex attractant on stakes placed every 10 meters in a large field, investigators could blanket the air with sex attractant, thus confusing the males and greatly decreasing the probability of their finding females and mating with them.

The fire ants, when returning to the nest after finding food, secrete a trail pheromone that marks the trail so that other ants can find their way to the food. The trail pheromone is volatile and evaporates within two minutes, so that there is little danger of ants being misled by old trails. Ants also release alarm substances when disturbed, and this (rather like ringing the bell in a firehouse) in turn transmits the alarm to ants in the vicinity. These alarm substances have a lower molecular weight than the sex attractants and are less specific, so that members of several different species respond to the same alarm substance.

Worker bees, on finding food, secrete **geraniol,** a 10-carbon, branched chain alcohol, in order to attract other worker bees to the food. This supplements the information conveyed by their wagging dance (p. 475). Queen bees secrete 9-ketodecanoic acid, which, when ingested by worker bees, inhibits the development of their ovaries and their ability to make royal cells in which new queens might be reared. This substance also serves as a sex attractant to male bees during the queen's nuptial flight.

In colonial insects, such as ants, bees and termites, pheromones play an important role in regulating and coordinating the composition and activities of the population. A termite colony includes morphologically distinct queen, king, soldiers and nymphs or workers. All develop from fertilized eggs; however, queens, kings and soldiers each secrete inhibitory substances, pheromones, that act on the corpus allatum of the nymphs and prevent their developing into the more specialized types. If the queen dies, there is no longer any "antiqueen" pheromone released and one or more of the nymphs develop into queens. The members of each colony will permit only one queen to survive, devouring any excess ones. Similarly, the loss of the king termite or a reduction in the number of soldiers permits other nymphs to develop into the specialized castes to replace them. Males of migratory locusts secrete a substance from the surface of their skin that accelerates the growth of young locusts.

There are examples of primer pheromones in mammals, as well as in insects. When four or more female mice are placed in a cage, there is a greatly increased frequency of pseudopregnancy. If their olfactory bulbs are removed, this effect disappears. When more females are placed together in a cage, their estrous cycles become very erratic. However, if one male mouse is placed in the cage, his odor can initiate and synchronize the estrous cycles of all the females (the "Whitten effect") and reduce the frequency of reproductive abnormalities. Even more curious is the finding (the "Bruce effect") that the odor of a strange male will block pregnancy in a newly impregnated female mouse. The nerve impulses from the nose pass to the hypothalamus and block the output of prolactin-releasing factor. The subsequent lack of prolactin leads to regression of the corpora lutea and the failure of the fertilized ova to implant.

The question of whether or not there are human pheromones remains unanswered, but of interest in this respect is the observation of the French biologist J. LeMagnen that the odor of 15-hydroxypentadecanoic acid is perceived clearly only by sexually mature females, and that it is perceived most sharply at about the time of ovulation! Males and immature females are relatively insensitive to this substance, but male subjects became more sensitive to it after an injection of estrogen.

Analysis of the menstrual cycles of the students at an American women's college showed a statistically significant tendency for the increasing synchronization of menstrual cycles among roommates and close friends. The study ruled out several possible explanations for the phenomenon but suggested some pheromonal effect between girls who were together much of the time. The study also demonstrated a significant shortening of the menstrual cycles of women who were with male companions three or more times per week; those who dated twice a week or less had longer and more irregular cycles. This appears to be analogous to the Whitten effect, but whether it involves a pheromone is unknown.

ANNOTATED REFERENCES

Barrington, E. J. W.: Introduction to General and Comparative Endocrinology. Oxford, Clarendon Press, 1963. A comparative treatment of endocrine principles.

Drill, V. A.: The Oral Contraceptives. New York, McGraw-Hill Book Co. 1966. A clear discussion of how the oral contraceptives work and what their major side effects may be.

Gorbman, A., and H. A. Bern: A Textbook of Comparative Endocrinology. New York, John Wiley and Sons, Inc., 1962. An excellent presentation of the broad evolutionary aspects of endocrine systems in vertebrate and invertebrate animals.

Kistner, R.: The Pill. New York, Delacorte Press, 1969. A well-rounded discussion of the pros and cons of oral contraceptives.

Litwack, G. (Ed): Biochemical Actions of Hormones. Vols. 1–3. New York, Academic Press, Inc., 1971, 1972 and 1975. An advanced monograph concerned with the mechanisms by which hormones exert their effects.

McKerns, K. W. (Ed): The Gonads. New York, Appleton-Century-Crofts, 1969. A fine source book with chapters by 46 authorities. The papers were presented at a symposium, and the transcript of the discussion provides further insight into the hormonal aspects of reproduction.

Tepperman, J.: Metabolic and Endocrine Physiology. 2nd ed. Chicago, Year Book Medical Publishers, 1968. A clear presentation of the metabolic effects of the mammalian hormones.

Turner, C., and J. Bagnara: General Endocrinology. 6th ed. Philadelphia, W. B. Saunders Co., 1976. An excellent introductory text covering the basic biological aspects of endocrinology.

Villee, D. B.: Human Endocrinology: A Developmental Approach. Philadelphia, W. B. Saunders Company, 1975. A well-written account of human endocrinology, pointing up the changes in the endocrine system from prenatal life to senescence.

CHAPTER 19

BEHAVIOR

Much of what an animal does can be analyzed in terms of specific behavior patterns—specific coordinated sequences of neuromuscular activity. These typically occur in response to some change in the environment and may result in the movement of the animal's entire body or some part of it. A dog may wag its tail, a bird may sing and a butterfly may release a volatile sex attractant. Movements of the entire body are involved in more complex behavior patterns, such as the stalking movements of a cougar, the courtship movements of a stickleback fish or the crawling of a snail. Some behavior patterns result in an animal's not moving: a pointer freezes when it has detected a pheasant, a possum plays dead in the presence of an enemy and a young bird remains motionless on the ground in the presence of a predator. All animals exhibit behavior patterns, although in some, such as sponges, clams and bryozoa, the detection of behavior patterns may require extreme patience on the part of the observer. Behavior patterns commonly involve muscles, cilia, flagella or pseudopodia as effectors, but they may involve responses such as the emission of light, the release of the secretion of an exocrine or endocrine gland or the generation of electricity in an electric organ.

19.1 BEHAVIORAL ADAPTATIONS

Certain general types of behavior can be detected in many different groups of animals, and the behavior of diverse groups may be influenced by certain common factors. Behavior is one means by which an animal may adapt to changes in the environment. The physiologic basis of behavior resides in the manifold activities of the animal's nervous system. All movements may ultimately be analyzed in terms of the patterns of nervous activity and the patterns of neuromuscular connections that are involved in their production.

A behavior pattern is typically triggered in response to some variation in the environment—some change in the light, temperature, humidity, oxygen, carbon dioxide, pH or texture of its surroundings. Animals may make behavioral responses to changes in their biotic environment; specific behavior patterns may be triggered by the presence of certain other organisms. Predators seek prey; the prey attempts to evade the predator. Changes in an animal's internal environment may also initiate specific behavior patterns. When a cat has not eaten for some time, the concentration of glucose in its blood decreases and the stomach increases its motility. In response to these and other stimuli, the animal becomes restless, moves about and looks for food. When it has eaten, its internal conditions are changed, the restlessness ceases and the animal may either groom itself or sleep.

Behavioral responses are adaptive either for the survival of the individual or the survival of the species. Certain behavioral responses may lead to the death of the individual but increase the likelihood of survival of the species through the survival of the offspring. Each animal's behavior patterns must enable it to live long enough to reproduce. It must avoid deleterious environments, predators, parasites and competition from members of its own species. It must obtain an adequate supply of raw materials for its biosynthetic processes and an adequate supply of energy for its metabolic machinery. At the appropriate time it may need to locate and mate with another of its kind and perhaps subsequently guard and educate the young. No two species accomplish all of these ends by the same patterns of behavior. Behavior, indeed, is just as diverse as biological structure and is just as characteristic of a given species as its size, form, color or odor.

Behavior patterns are determined by the capabilities of the organism's **receptor, effector** and **nervous systems,** which have in turn been determined by evolution. What happens in response to a change in the environmental variable, the stimulus, depends on how the receptor, effector and nervous systems are interconnected and integrated. Their organization may be relatively direct, so that the stimulus triggers a simple set of muscle actions that proceed automatically once started. As muscular action proceeds, it may be constantly monitored by other stimuli that it has generated in the body, which therefore steer and guide it. Alternatively, the stimulus may trigger a complex response that is completely programmed genetically in the central nervous system. The response mechanism in other behavioral systems may involve both learned and genetically programmed components.

19.2 REFLEXES

In the simplest sort of response the stimulus triggers a muscle or a set of muscles that respond in a predictable, unvarying fashion. This behavior is determined only by the presence of the appropriate sense organs connected with a particular set of muscles. In the **knee-jerk,** or **stretch reflex,** a tap on the tendon below the knee cap stretches the attached muscle. The stretching stimulates the spindle organ (Fig. 19.1). Impulses from this receptor neuron are conducted along its axon to the spinal cord where they pass by a synapse to a

motor neuron that stimulates the muscle to contract, producing the knee jerk. The knee-jerk reflex is an example of behavior dependent simply upon the presence of a stretch receptor connected with a specific muscle by a **monosynaptic reflex arc.**

The nature of the reflex response may be altered by the characteristics of the synapses connecting the incoming and outgoing segments of the circuit. There is delay at the synapse, and the stronger the stimulus, the shorter the delay. This relates the onset of the behavior to the strength of the stimulus. In addition, the phenomenon of **after-discharge** may appear: the synapse may continue to be active after the stimulus has ceased. A cockroach stimulated to run in response to a touch on its anal cercus may continue to run for some seconds after the stimulus has ceased. The continued running is due in part to after-discharge at a synapse in the ventral nerve cord.

Synapses also exhibit the phenomenon of **temporal summation.** If an investigator applies to a fly's mouth parts a sugar solution that is concentrated enough to stimulate taste hairs but too dilute to evoke a behavioral response, the proboscis is not extended. However, if this stimulation is quickly followed by another weak stimulation (which would not itself evoke a response), proboscis extension does occur.

When two or more neurons synapse with the same postsynaptic neuron, there may be **spatial summation.** Two incoming impulses, one in each presynaptic neuron and each too weak to trigger the discharge of the motor neuron, may add together and be effective.

The reflex arc underlying the flexion reflex involves two synapses—one between the sen-

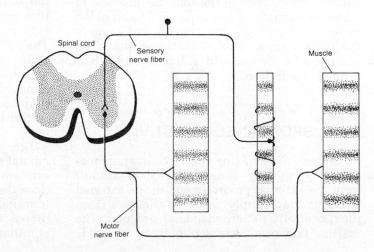

Figure 19.1 A diagram of the anatomy of the stretch reflex, a monosynaptic reflex arc. (After Van der Kloot, W. G.: Behavior. New York, Holt, Rinehart and Winston, Inc., 1968.)

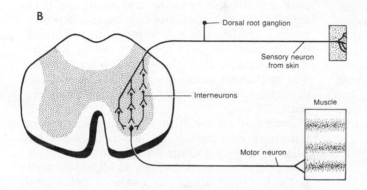

Figure 19.2 A, The most direct nerve pathway between a stretch receptor and a motor neuron innervating the antagonist muscle. B, A polysynaptic reflex arc between sensory neurons in the skin and motor neurons. (After Van der Kloot, W. G.: Behavior. New York, Holt, Rinehart and Winston, Inc., 1968.)

sory neuron and interneuron and one between interneuron and motor neuron (Fig. 19.2). The interneuron is inhibitory, and when stimulated by sensory input from one muscle, it inhibits activity in the antagonistic muscle. Considerable complexity of behavior can be developed when the number of synapses in the reflex arc is increased. Increasing the strength of the stimulus elicits different responses in the flexion reflex. In experiments with dogs in which the connection between the spinal cord and brain has been cut, a weak noxious stimulus applied to the foot pad results in withdrawal of the foot. An increase in stimulus strength causes a strong flexion of the lower leg. A stronger stimulus elicits flexion of the upper leg, as well. This spreading of response with increase in stimulus strength is termed **irradiation.**

19.3 SPONTANEOUS ACTIVITY

Patterns of behavior may occur spontaneously, completely independently of stimuli from the external environment, or the external stimulus may simply serve to release a complex, centrally patterned neural program. The marine lugworm, *Arenicola,* lives in an L-shaped burrow in the mud and keeps the burrow open by rhythmic movements (Fig. 12.4). These arise not in response to environmental stimuli but are initiated by a series of spontaneously active clocks in the various ganglia of the worm. The brain is not necessary for these activities to occur, and isolated fragments of the worm show the same kind of movements.

Other behavior programs are built into the central nervous systems of invertebrates and need only be triggered or released by specific stimuli. Wind blowing on special receptors on the head of the locust initiates impulses from the brain to the thoracic ganglion; these in turn send impulses to the motor nerves innervating the wing muscles and result in flight. The pattern of impulses from the thoracic ganglion determines the sequence of the flight muscles. Even when all the muscles have been removed, so that there is no feedback from proprioceptors in the wings, normal neural patterns are delivered by the thoracic ganglion.

The singing of crickets requires environmental stimuli only to initiate it. The complex series of muscle contractions needed to produce the various kinds of songs are largely determined in the brain. This inference was drawn from experiments in which electrical stimulation of one area of the cricket's brain

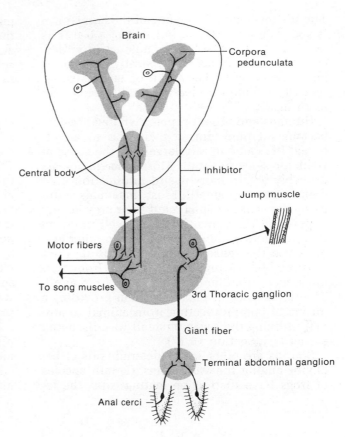

Figure 19.3 The main interactions between ganglia that control singing outbursts in grasshoppers and crickets. Electrical stimulation of a nucleus called the corpora pedunculata in the brain causes a burst of normal song and simultaneously inhibits the motor neurons involved in jumping. The central body in the brain organizes the patterns of the song, and stimulation here produces abnormal noises. Stimulation in the 3rd thoracic ganglion excites motor neurons in a meaningless pattern. The alerting mechanism of the anal cerci, which are sensitive to sounds and air movements, is inhibited during song, thus preventing the animal's alarming itself by its own noise. (After Huber, F.: Invertebrate Nervous Systems, Wiersma, C. A. G. (Ed.). Chicago, University of Chicago Press, 1967.)

triggered normal song rhythms, whereas stimulation of another area produced atypical songs (Fig. 19.3).

Many birds that nest on the ground, such as the graylag goose, retrieve any egg that happens to roll out of the nest by placing their bill on the far side of the egg and rolling the egg back into the nest with side-to-side movements of the bill that prevent the egg from slipping away (Fig. 19.4). When the egg is replaced with a cylinder that does not wobble, the side-to-side compensatory movements cease. Alternatively, if the egg or the cylinder is removed once the retrieving movement has started, the bill is still drawn back to the breast. This is a fixed action pattern that, once elicited by the stimulus (the sight of an egg outside the nest), continues independently.

The side-to-side movements are steered by feedback stimuli coming from the wobbling egg. They simply modify the orientation of the basic pattern.

19.4 STIMULI AND SIGNS

In the terms of electronics, receptors are "narrow-pass filters," transmitting a narrow range of environmental energies. The receptors of different species may pass different portions of the range. The honeybee sees a far different world than we do, for its eye is sensitive to ultraviolet but not to red light. In addition, bees can detect differences in the plane of polarization of light in the sky, so that what appears uniform in color to us is patterned to the bee.

Figure 19.4 A, Greylag goose retrieving an egg that is outside the nest. This movement is very stereotyped in form and used by many ground-nesting birds. B, The goose attempts to retrieve a giant egg in precisely the same fashion. (From Lorenz, K. Z. and N. Tinbergen: Z. Tierpsychol. 2, 1, 1938.)

The flicker-fusion frequency of the human eye is about 50 per second, but that of a bee is as much as 250 per second. Thus, an incandescent lamp that looks like a steady light to us, but is actually flickering 50 or 60 cycles per second, will appear clearly as a flickering light to an insect.

Humans and other primates whose ancestors became adapted during evolution to an arboreal life came to rely largely on the sense of sight for sensory input. Most other mammals probably obtain more useful information from their sense of smell. Most mammals hunt using the sense of smell, and their prey largely depend on their own sense of smell to detect the predators. Mammals and certain other animals use the sense of smell in defining their territory, in distinguishing friends from enemies, male from female and young from adult. The importance of pheromones in providing a means of communicating from animal to animal utilizing the sense of smell was discussed earlier (see Section 18.17).

Sounds are other major determinants of behavior among the vertebrates. Certain species of frogs have distinctive mating calls; the female of each species will respond only to those mating calls from the males of its species and will ignore those of others. Similarly, the songs of birds provide determinants for a wide range of behavior, mating calls and alarm calls.

There are marked differences in the sensory capacities of animals throughout the Animal Kingdom. The bat emits sounds with frequencies far beyond the range of the human ear. These sounds bounce off any insect flying in the vicinity, and by **echolocation** the bat is able to home in on the insect and catch it. Certain moths can hear these ultra high-frequency cries of the bat and take evasive action. Porpoises utilize sonar for underwater navigation by echolocation, and spiders can interpret the vibrations produced on their webs by ensnared insects. A rattlesnake can sense heat well enough to strike in the dark at the heat-producing object.

The central nervous sytems of many organisms are capable of filtering out environmental stimuli that may be of little value. If an animal were to process all the sensory information it could receive, its central nervous

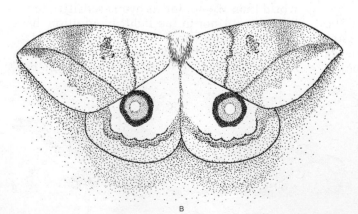

Figure 19.5 The moth *Automeris coresus* (A) at rest and (B) displaying the vivid eyespots on its hind wings in response to a light touch. (Modified from Blest: Behaviour 11:257, 1957.)

system would be swamped. Certain segments of the retina of the frog's eye are specialized to react to small, dark, convex objects. The principal item in the diet of frogs is flies, and frogs respond more readily to small, dark, circular moving objects near them than to larger, more diffuse or distant objects. Such **sign stimuli** often serve as triggers for fixed action patterns of behavior. When quick action is essential, as in escaping from a predator, a danger sign is more useful than a detailed description of the danger. Alarm signs, whether they are sights or sounds, are usually simple and contrast sharply to the environment. Small birds typically show an immediate flight reaction to animals with large eyes (understandably, considering that their major predators, owls and hawks, have large eyes). Small birds flee any staring eye, and many moths, the prey of birds, have evolved eye-shaped spots on their hind wings. These "eyes" are concealed by the forewings when the moth is at rest, but spring into view (Fig. 19.5) when the moth is disturbed and extends its forewings.

By varying systematically the characters of a stimulus, the behavioral scientist can discover the specific "sign." In the spring, male **sticklebacks** establish territories from which they drive other males. By presenting males with various models, it was found that the sign of the male is a red belly (Fig. 19.6); other aspects of morphology are unimportant. Courtship, expressed as a zigzag dance, is elicited by a swollen belly, the sign of the female.

The natural stimulus may not be the most effective one. An experimenter, by exaggerating the best features of a stimulus, may devise "superoptimal" stimuli that are more effective in electing a response than the normal stimulus.

19.5 METHODS OF STUDYING BEHAVIOR

The neurophysiologic approach to behavior involves analyses of neural function; the neurophysiologist removes, stimulates, and records the activity of neurons. By placing an electrode on the nerve leading from some sense organ and making a recording, the experimenter can determine what kinds of sensory input are available to the animal. By monitoring neuronal activity he can determine what stimuli will actually generate action potentials in a given sense organ (Fig. 19.7). In this way, he can distinguish between an animal that is incapable of detecting a particular

Figure 19.6 A series of models used in tests of aggression with male sticklebacks. The four crude imitations colored red below released more attacks than the accurate model, which lacked red. (After Tinbergen, N.: The Study of Instinct. London, Oxford University Press, 1951.)

stimulus and another animal that can detect the stimulus but simply is not responding under the given test conditions.

Complex patterns of behavior can be produced by electrical stimulation of certain parts of the central nervous system. Although the enormous complexity of the central nervous system makes it difficult to interpret the results of such experiments, the location of the neural pathways involved in the control of certain movements can be established. When an electrode is implanted in the brain of an anesthetized animal and the animal is allowed to wake and move around freely, it can be stimulated successively with impulses of different strengths. Observation of the relation between the area stimulated and the behavior or mood observed in the animal provides some insight into the operation of the brain (Fig. 19.8). Stimulation of a certain portion of the thalamus in a number of mammalian species can be used as a reward for behavior. Human subjects report that this part of the brain is a pleasure center; that electrical stimulation of the area does not produce sensations of sight, sound, light or odor, but simply of pleasure. Stimulation of this center has been used to a limited extent to relieve suicidal tendencies in extremely manic-depressive individuals. However, a rat with an electrode implanted in its thalamus that it can activate by pressing a bar at one end of the cage slowly dies of starvation and exhaustion because it neither sleeps nor eats but simply

0.1 sec

Figure 19.7 Nervous activity of the tympanic nerve of a noctuid moth. The upper trace of activity is of the left "ear" of the moth, the lower trace the right. The response seen is due to the approach of an echolocating (moth-eating) bat. In records (A) and (B), the difference in left and right tympanic activity indicates that the bat is approaching from the left. A burst of activity (C) is the response to a bat "buzz" rather close to the moth. (Courtesy of K. D. Roeder.)

Figure 19.8 Electrical stimulation of attack behavior. The rooster is stimulated by electrodes implanted in a particular part of its brain. It starts to move toward the keeper and, as stimulation is continued, it attacks. (Courtesy of E. von Holst.)

pleasures itself to death by repeatedly pressing the bar.

In a second method, the investigator makes systematic changes in the sensory input of an animal and observes the changes in output in some segment of its behavioral repertoire. He then makes a diagrammatic representation of the relationship between input and output, usually relying upon feedback loops and other cybernetic phenomena.

In addition to these two methods of studying behavior, which might be described as neurophysiological because they actually concentrate on the neuronal circuitry underlying behavior, there are two widely used methods of studying behavior itself. The psychological school of behaviorists, led by B. F. Skinner at Harvard, has attempted to establish general principles of behavior and especially of learning. Skinner's studies of training rats and pigeons to do certain tasks, using immediate reward or reinforcement when the task has been accomplished, has been the basis of programmed learning. It is the goal of these studies, carried out in the laboratory, to control all the experimental variables so that the effect of one can be measured.

One of the central concepts that emerged from studies by the comparative psychologists was the stimulus-response theory. This attempted to explain all behavior as a series of reactions to stimuli. Although the importance of stimuli and responses in determining behavior patterns is clearly apparent, animals are more than a collection of sets of stimulus-response reactions. The experimental psychologist confines his subject to carefully defined experimental conditions, such as a **maze** or a **shuttle box,** in which the subject learns to move from one side to another to avoid an electric shock, or a **Skinner box,** in which it learns to press a lever or peck at a key to receive a reward of food or drink. The aim of the experimental psychologist is to establish laws of behavior that will describe how an animal's behavior changes after specific kinds of practice, reward or punishment are experienced. In this way the efficacy of one or another experimental situation in producing learning can be measured.

Another system of behavioral studies, founded in the late 1930's by Konrad Lorenz and Niko Tinbergen, is termed **ethology.** The ethologists insisted that animal behavior be studied under conditions that are as nearly natural as possible. They have explored the rich variety of adaptive behavior patterns shown by birds, fish and insects and have been especially concerned with analyzing the factors controlling social behavior, including aggressive, reproductive and parental behavior.

Members of the ethologic school insist that the natural behavior of a species be carefully observed before one begins an experimental analysis of the factors involved. The ethologist is greatly concerned with the biological significance of behavior patterns and how these patterns have evolved. Many of the biologically important behavior patterns studied by ethologists appeared to be primarily the product of genetic expression and less the result of learning; initially, perhaps the ethologists overemphasized the innate qualities of behavior. Both ethologists and psychologists recognize that the behavior an animal shows at any moment is the result of past selective forces. These determine a particular morphologic, biochemical and neuronal structure in an animal. The functional characteristics of this structure are continually altered by experience, by hormones and by other physiologic changes. The sum of these factors results in the maintenance of the animal in its proper condition, and stimuli from the biotic and the abiotic environment can then elicit particular behavior patterns.

19.6 DETERMINANTS OF BEHAVIOR: MOTIVATION

What an animal does at any given time is determined by the stimuli impinging upon it and by its motivation. The nature of the stimuli to which the animal will respond and the latency of its response depends upon its motivation. Obviously, a hungry dog will respond differently from one that has just been fed when both are presented with a dish of food and some alternative. Physiologic research has revealed further types of interaction between external stimuli and motivation. Each stimulus may evoke a response within the brain by its specific sensory pathways. In addition, it may elicit many less specific responses. In the vertebrate, incoming sensory pathways have sidebranches, or collaterals, extending to a diffuse series of fiber tracts, the **reticular formation,** which is connected to the brain's higher centers (Fig. 19.9). Impulses from the reticular formation arouse the brain's higher centers into action. Thus, any stimulus may not only evoke a specific response pertaining to itself but also change the animal's state of arousal and responsiveness to other stimuli, both those that are related to the first and those that are unrelated to it.

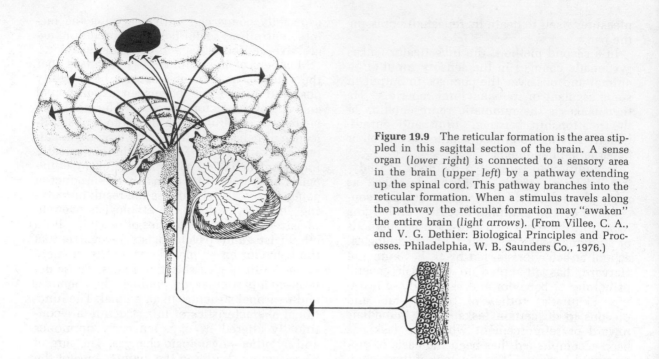

Figure 19.9   The reticular formation is the area stippled in this sagittal section of the brain. A sense organ (*lower right*) is connected to a sensory area in the brain (*upper left*) by a pathway extending up the spinal cord. This pathway branches into the reticular formation. When a stimulus travels along the pathway the reticular formation may "awaken" the entire brain (*light arrows*). (From Villee, C. A., and V. G. Dethier: Biological Principles and Processes. Philadelphia, W. B. Saunders Co., 1976.)

A given stimulus applied to the same animal at different times does not always evoke the same response. The relationship between stimulus and response may be altered by fatigue, maturation, learning, motivation and even by the season of the year. Some examples of motivated behavior are feeding, drinking, courting, nest-building, territorial defense and care of the young. Perhaps the most striking feature of motivational behavior is that it is highly specific to a particular state and to particular stimuli. In general, a hungry animal will not court and an animal in a highly receptive sexual state will not feed.

Each type of specific motivation causes the animal to orient its behavior towards some specific goal. The pattern begins with a searching phase, usually called **appetitive behavior.** This does not imply that the animal is consciously "searching" or that it necessarily "knows" what the goal is. It simply means that there is a phase of undirected behavior which, in the normal course of events, will lead to a particular source of further stimulation. Only rarely can a hungry fox simply get up and eat; ordinarily it must hunt, and the behavior patterns used in hunting may be many and diverse. It is frequently impossible to identify the nature of appetitive behavior until the actual goal can be observed. An animal searching for food, water or a mate may behave quite similarly, but with each the stimuli required to

bring the search to an end are specific. When a fox hunting for food encounters prey, it then switches to an oriented type of behavior; when the food is seized, eating ensues. At this time the various searching patterns give way to a series of responses directed at the goal. These responses, called **consummatory acts,** may be stereotyped, fixed action patterns. Eating is the consummatory act of feeding behavior, drinking relates to thirst, copulation to sexual behavior and so on. Consummatory acts are normally followed by a period of quiescence during which the animal is no longer responsive to stimuli from the goal; thus, a well fed animal shows no further appetitive behavior even though the appropriate stimuli may be present. Quiescence relates only to one type of behavior; the animal may be actively pursuing some other sort of goal. Frequently, responsiveness gradually builds up again so that there is a fairly direct relationship between the threshold level of responsiveness for a given set of goal stimuli and the time that has elapsed since the previous performance of the consummatory act.

The **hypothalamus** exerts a great influence on feeding and drinking behavior, sexual behavior, sleep, maternal behavior and behavioral responses to temperature change. It is generally considered to be the control center for much of what we call motivated behavior. Feeding behavior, for example, is controlled by feeding and satiety centers. Stimulating the **satiety center** in the ventral medial region of the hypothalamus decreases the amount of food eaten. Stimulating the **feeding centers** in the lateral hypothalamus causes the animal to eat (even if it has just had a full meal).

Experiments with goats have demonstrated the role of the hypothalamus in controlling motivated drinking behavior. When a goat (or any other mammal) needs water, the deficit of water in the blood stimulates special cells in the lateral area of the hypothalamus. Impulses from these cells stimulate the pituitary to release antidiuretic hormone. This hormone results in the resorption of more water by the kidneys and also stimulates the animal to look for water and to drink it. When a very small amount of hypertonic salt solution is injected with a fine needle directly into this area of the hypothalamus, the goat drinks enormous quantities of water, even though it had previously satisfied its thirst and needed no water. It will even drink water adulterated with high concentrations of salt or quinine.

Motivation can be measured only indirectly. For example, to determine the motivation towards feeding behavior one can measure the amount of food eaten, how bitter the food can be made before it will be refused by the hungry animal, how fast the animal will run toward food, how strongly it will pull toward food when fitted with a harness, the level of electric shock that will be tolerated by the animal to get to the food, or the rate at which the animal will press a bar to obtain a food reward. Motivational behavior is a response to a need to restore a condition of homeostasis. When homeostatic balance has been restored, there is generally some sort of sensory feedback that signals that the need has been satisfied.

Hormones may alter behavior by stimulating the development of organs employed in behavior, by affecting the early development of the nervous system, by producing changes in peripheral organs concerned with the generation of sensory input to the central nervous system, by influencing special centers in the central nervous system or by exerting general nonspecific effects on the animal as a whole.

Thyroidectomy, gonadectomy, adrenalectomy and hypophysectomy—each enhances the nest-building behavior of rats by leading to a lowered body temperature. One well studied example of hormonally induced changes in peripheral organs affecting behavior is the effect of **prolactin** on parental feeding by doves. "Pigeon milk," on which doves feed their young, is actually sloughed epithelium of the crop sac. Prolactin causes the crop to enlarge just before the young hatch. Sensory stimulation from the crop at this time is an important factor in initiating parental feeding, for the application of a local anesthetic to the crop reduces parental feeding.

The role of hormones in the development and expression of behavior has been demonstrated in many vertebrates by castrating young animals and observing the absence of certain adult or sexual behavior later on. Injecting sex hormones into sexually immature animals will induce sexual behavior prematurely. Sexual behavior in spayed animals is restored by injecting or implanting the appropriate sex hormones.

Sexually mature female cats usually have three estrous cycles a year. When in estrus, a female cat in the presence of a male will elevate the rump, deflect the tail to one side and make treading movements with the hind legs. A cat in *anestrus* will fight if a male approaches too closely.

19.7 BEHAVIOR PATTERNS

Feeding. All animals show characteristic behavior patterns in finding and consuming food. Even the oyster and other sessile filter feeders that do not hunt their food utilize sense organs, filtering mechanisms and coordinated muscle contractions to obtain particles of food of the correct size. Most animals search actively for specific kinds of food; when the animal's sense organs detect a certain constellation of physical characteristics, the object is treated as food and is eaten. Some insects feed on particular plants only when specific receptors, located on hairs on their mouth parts, are stimulated by a particular chemical compound secreted by the plant. Frogs are visually stimulated by small, dark, moving forms such as flies. The identification of such forms by the frog's nervous system appears to result from the properties of special interneurons in the visual pathway; these fire only when specific types of movement occur across the frog's retina. Young birds initially peck at any small round object that contrasts with the background. By trial and error, their pecking movements become more accurate and discriminating, and they learn not to peck at objects, such as pebbles, which have no nutritive value.

Predatory Behavior. The predator-prey relationship, in which one animal kills another for food, is of special interest because of the interactions between the attack patterns of the predator and the defense patterns of the prey. Lions and leopards show a complex set of movements as they move along downwind of a herd of antelope, creep toward one animal and run it down with a short burst of speed. If the antelope is not caught during this short burst, it usually escapes altogether.

Many birds are attracted to the visual patterns presented by insects, their normal prey. In an evolutionary response to this, a number of larval and adult insects have "eyespot" patterns that are suddenly exposed when the insect is disturbed, creating a "noninsect" pattern (Fig. 19.5). Other insects resemble their

Figure 19.10   Cryptic resemblance of several different types of insects to leaves. (Courtesy of H. B. Cott.)

Figure 19.11 Chemical defense systems in arthropods. The bombardier beetle produces a noxious chemical spray when attacked by ants (*D*) or a toad (*A*, *B* and *C*). When the toad tastes the beetle's spray, it rejects the insect (*C*). The beetle *Eleodes* does a headstand as it ejects its defensive spray (animal to the left in *F*); a mimic of *Eleodes*, *Megasica* also does the headstand, although it has no defensive chemical spray (animal to the right in *F*). The grasshopper mouse avoids the spray of *Eleodes* by sticking the beetle's abdomen toward the ground (*E*). (Courtesy of T. Eisner.)

environment closely and become immobile when disturbed (Fig. 19.10). Predators are usually attracted by movement, and simply being still is excellent defense for the prey. In a number of arthropods (whip scorpions, many insects and most millipedes) and some vertebrates (toads, skunks), behavior patterns involving chemical warning and defense devices have evolved as a means of protection against predators (Fig. 19.11).

A common type of predator-prey interaction, termed **protean behavior,** occurs when the prey appears to go through a sequence of possible escape patterns at random and the predator cannot predict the exact path of the prey's movements. A stickleback fish being chased by a duck follows an erratic path that is very difficult to follow and apparently impossible to predict.

Orientation. The behavioral activities of animals are in large measure directed by sensory information concerning the environment. If a sequence of muscle contractions is to bring about a biologically effective movement, the animal must have information about where it is and where its limbs are. Input from proprioceptors indicates the position of the limbs in relation to the body, and input from other sense organs indicates the body's position in space.

The octopus can be trained to attack only when stimulated by certain visual patterns. However, when its gravitational sensory input is cut off by removal of the statocysts, the octopus has great difficulty responding to the visual patterns in the correct way. The central nervous system must integrate its visual and its gravitational information before the proper responses can occur.

Most animals orient themselves in a particular way in the area in which they move about. Amphipods living on the beach use visual stimuli (the sun during the day, the moon at night) as indicators of compass direction to determine their way back to the ocean. This was demonstrated by experiments in which screens and mirrors were used to change the apparent position of the sun or moon. Certain wolf spiders live along river banks and escape from enemies by running across the water. These spiders also use the position of the sun to determine the direction of escape. It appears that the correct escape direction is learned, for when young spiders are moved to a new location, their responses are altered so that they move in the correct direction in relation to the water in the new locality.

Mammals frequently mark the limits of their **home range** (the area of their daily movements) or **territory** (the area an animal defends) by chemical marks. Some regularly use the secretions of special scent glands to mark the limits of their territory. The familiar sight of a dog at a fire hydrant or shrub is not so much the result of pressure on the dog's bladder as it is an attempt to provide a chemical marker for later orientation.

In addition to those movements involved in daily life, many vertebrates and some arthropods have special behavior patterns of migration, during which they cover relatively great distances, usually moving to a different type of environment. Certain environmental changes bring about specific physiologic changes in the animals; these are followed by the extended periods of locomotor activity that constitute the migration. In locusts, the increased tactile contacts due to crowding bring about physiologic changes that precede the formation of migratory swarms. Spiny lobsters may use tactile stimuli to become entrained in long migrating columns.

The migrations of birds have fascinated many investigators. Increasing day length in the spring stimulates the hypothalamus to secrete gonadotropin-releasing hormones (see Section 18.9); these stimulate the pituitary to secrete gonadotropins, which cause the gonads to enlarge and secrete sex hormones, initiating the northward migrations. The length of the migrations and the route taken vary from one species to another. How birds navigate is not completely known. Experiments suggest that some species use the position of the sun or the pattern of stars to navigate; others may use visual landmarks, and still others may be able to detect the magnetic fields of the earth and "fly by compass."

Aggression. Competition for food, mating partners and nesting space may result in aggressive behavior patterns, primarily between members of the same species. When an animal is approached too closely by another, its first response is often a defensive posture (Fig. 19.12). Most arthropods and vertebrates have an **individual distance,** the minimum distance required to separate two animals without a fight ensuing. It can be pictured as a space around an animal into which only sexual partners may enter without eliciting aggressive behavior.

This space is distinct from an animal's **terri-**

Figure 19.12 Defensive display of the ghost crab, *Ocypode.* The bright white mani of the chelipeds are ordinarily held under the body, out of sight. (Courtesy of K. Daumer.)

Figure 19.13 Fighting among rattlesnakes. The snakes do not bite one another but wrestle, pushing against each other with their heads (a and b) or ventral parts of their bodies (c and d) until one is pinned (e). (From The Fighting Behavior of Animals by I. Eibl-Eibesfeldt, Copyright © 1961 by Scientific American, Inc. All rights reserved.)

tory, the total environmental area it will defend. In its territory, an animal obtains the food, nest materials and space needed for the rearing of its young. Since only members of the same species are likely to utilize the same portions of the environment, territorial defense is usually shown only toward members of the same species, and aggressive interactions are especially frequent at territorial boundaries. Such "**boundary encounters**" are usually longer and more complex than fights occurring well within an individual's territory. One animal will quickly retreat from a second when the meeting occurs within the second's territory. At the boundary between territories, neither animal retreats and complex patterns of behavior may be exchanged. The animals may show ambivalent behavior patterns as they are simultaneously motivated to attack and retreat. Under such confusing conditions, the animals often do neither but instead do something totally unrelated to the situation.

Herring gulls, after a series of aggressive postures at a territorial boundary, may either pull grass from the ground, an activity usually associated with nest building, or go through motions similar to grass pulling. Such **displacement activities** may be quite ineffective under the circumstances. Grass pulling has secondarily become part of the gull's aggressive repertoire.

An important characteristic of intraspecific fighting is that it rarely results in any physical damage to the combatants. Through evolutionary selection the acts have become **ritualized**—the movements have a social communicatory function, and the interactions are mainly exchanges of signals or **displays.** The wrestling of rattlesnakes (Fig. 19.13) produces even less damage to the participants than the shows put on by "wrestlers" on televison. When one individual wins an encounter, the loser frequently assumes a submissive posture that inhibits further attack by the winner. In some species of hermit crabs, exchanges of aggressive movements stop when one or both crabs suddenly squat down on the sand. Fighting may, in some species, continue until severe damage is done (cat and cock fights are justly famous); equally paired mice will sometimes fight to the point of bloodshed; large male seals regularly lose blood in fights for the possession of a harem of females. The great majority of intraspecific fights, however, do not result in physical damage.

Dominance Hierarchies. In many vertebrates and a few arthropods, the social interactions of the members of a local group are well organized and help to decrease aggressive activity among the members of the group. In the simplest form of such a social hierarchy (the "**peck order**"), one individual, α, is dominant over all others, a second, β, is dominant over all except α and so on. Dominance is expressed by winning in aggressive interaction, by having first chance at the available food and mating partners and by having first choice of nesting locations. The other animals are subordinate and either assume a submissive posture or retreat from the more dominant individual.

An animal's position in a hierarchy may be a function not only of its size but also of its sex, individual temperament, immediate physiologic condition and behavioral interactions outside of its social group. A chicken can maintain distinct positions in a number of groups at the same time, but its positions in all groups can be altered by interactions with chickens outside these groups. Thus, when a chicken loses an encounter with another very aggressive chicken outside its usual social group, it may lose subsequent encounters within its social group.

Hierarchies have been observed in a number of invertebrate groups, but the organization within small groups of crayfish appears to be one of the clearest examples of individual

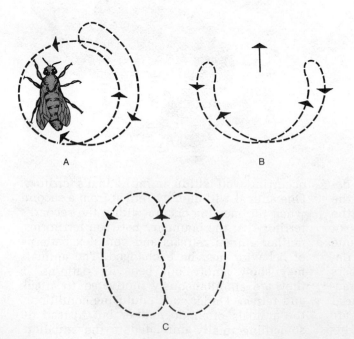

Figure 19.14 Three kinds of communication dances performed by honey bees. *A*, The round dance; *B*, the sickle dance; *C*, the waggle dance. (From von Frisch, K. and M. Lindauer: *In* McGill, T. (Ed.): Readings on Animal Behavior. New York, Holt, Rinehart and Winston, Inc., 1965.)

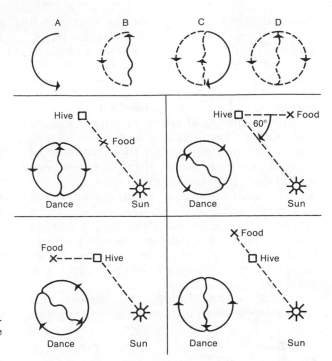

Figure 19.15 Communication among honey bees: Indication of the direction of food from the hive by the waggle dance.

recognition-dependent structure. Straight-line hierarchies develop after a small number of fights, and subsequent interactions are limited to exchanges of displays of mild threats.

Communication. Animals often communicate by behavior patterns which mutually affect one another's behavior; typically, the behavior concerns either mating or aggression. One of the classic examples of nonaggressive, nonsexual communication is the means by which honeybees direct other members of the colony to a source of food. When a worker bee has found a source of food, it collects a sample and flies back to the hive. It transmits information to other members of the colony by the kind of "dance" it does on a vertical surface of the hive. If the food is near the hive, the honeybee circles first in one direction and then in the other in a "round dance" (Fig. 19.14). Other bees follow the dancing forager and pick up from her the scent of the flower visited. The other worker bees then fly out and search in all directions near the hive.

If the food is located at a greater distance from the hive, the bee goes through a half circle, then moves in a straight line, waggling its abdomen from side to side and finally moves in a half circle in the other direction. During the straight portion of the dance, the bee produces a series of sounds. The angle that the straight, waggling portion of the dance makes with the vertical is the same as the

angle of flight to the food in relation to the sun (Fig. 19.15). Information on the distance to be traveled in that direction is also transmitted by the pattern of the "waggle dance." There is a correlation between the number of waggle dances executed in a given time and the distance of the food from the hive. An even closer correlation, however, exists between distance and the sound pulses the dancing bee produces. Bees can sense the vibrations of the substratum, and sound would seem to be a good way to communicate in the dim interior of the hive.

The distance to be traveled is "calculated" by the bee on the basis of the flight to the food source not the return trip. Moreover, the instructions apparently include corrections for wind and large obstacles. The dancing bee also indicates the richness of the food source by how long and vigorously it repeats the dance. Other bees gain further information about the food sources by the smell of the flowers the bee has recently visited. The more primitive social bees, the stingless bees, lay down an odor trail, a series of chemical marks, to the food source; in addition, the collector bee guides other individuals along the trail.

Fire ants communicate information about food sources by a pheromone (p. 458) produced in a special **Dufor's gland** located in the abdomen of the worker ant. When one ant has located a source of food, it gathers what it can and starts back toward the nest. As it moves along the ground, it periodically extrudes its stinger and places a small amount of pheromone on the substratum. When another fire ant comes across this trail of chemical dashes it becomes very active, follows the trail and is led to the source of the food.

Other examples of communication between members of a group can be seen in the schooling of fish and the group flight patterns of birds. The importance of visual cues has been established for schooling in some fish, but what controls the sudden group movements of a flock of birds is still a mystery. In both fishes and birds it appears that grouping is a defense against predators. Predators are often able to catch isolated individuals but are confused by the many movements of the individual animals making up a group.

Courtship and Mating. All bisexual animals possess mechanisms for getting egg and sperm together to continue the existence of the species. For some sessile marine animals, such as clams, urchins and tunicates, courtship and mating are simply the releasing of sperm or eggs into the surrounding waters (where external fertilization takes place) when the animals are chemically stimulated by the presence of a few gametes in the water.

Changes in the physical environment induce (via hormones) physiologic changes in many species which prepare the animals so that subsequent specific stimuli will elicit the appropriate mating patterns. In other species the presence and movements of individuals of the opposite sex initiate the physiologic changes. Mutual exchanges of early courtship movements initiate and synchronize physiologic changes so that the sexes are ready to mate at the same time.

Further precopulatory behavior patterns (or repetition of the ones effective in physiologic

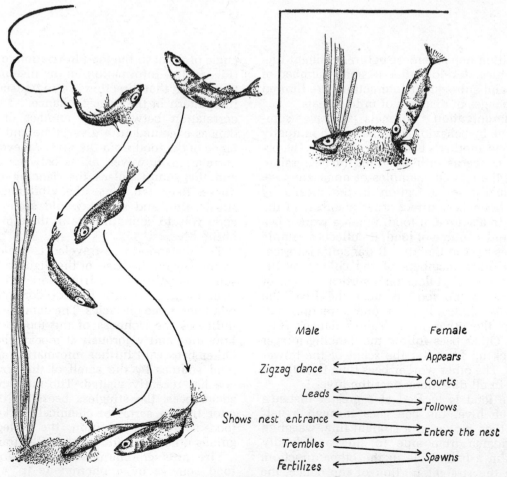

Male	Female
	Appears
Zigzag dance	
	Courts
Leads	
	Follows
Shows nest entrance	
	Enters the nest
Trembles	
	Spawns
Fertilizes	

Figure 19.16 Courtship in the three-spined stickleback. The appearance of the female in breeding condition elicits a zigzag dance by a territory-holding male. The female follows the male, who leads her to the entrance of the nest he has built. When the female is in the nest, the male's tactile prodding elicits egg laying (inset). (After Tinbergen, N.: The Study of Instinct. London, Oxford University Press, 1951.)

Figure 19.17 Courtship in one species of bower bird. The male builds a bower of sticks (*A*) and decorates it with pebbles and berries (*B*). The female bower bird enters the construction (*C*), and, after the male displays, she sits (*D*) and the birds mate (*E*). The female leaves, builds the nest (*F*) and rears the young birds by herself. (Courtesy of E. T. Gilliard.)

priming) serve two important functions: (1) they decrease aggressive tendencies and (2) they establish species and sexual identification. The first function is needed if two animals are to avoid fighting long enough to mate. Where this condition has not been completely fulfilled, special adaptations are necessary. For example, the male praying mantis is smaller than the female and subject to predatory attack by her. However, mating can continue even after the male's head has been bitten off by the female, since the nervous activities controlling copulatory movements are centered in an abdominal ganglion. In fact, removing the head enhances copulatory behavior by eliminating inhibitory impulses originating in the brain.

In some cockroaches and other primitive insects the female feeds on a special secretion on a part of the male's back, which diverts her attention during copulation. The males of certain predatory flies present the females with little packages of food wrapped in silken threads. The males of other species divert the female's attention by presenting her with an empty balloon of silk. The balloon both decreases the female's aggressive tendencies and identifies the sex and species of the male presenting it. A web of silk is used by some pseudoscorpions to form a "hallway" to guide

the female. In these species the intersex aggressiveness is very great, and these offerings by the male play an important role in neutralizing it.

The complex pattern of reproductive behavior of the **stickleback** has been studied intensively as a model (Fig. 19.16). After the male has migrated and staked out his territory, he builds a nest and courts any egg-laden female who enters his territory. In courting, he performs a special zigzag dance. The female follows the male and he leads her to the nest, at which point he directs his head toward the entrance. The pointing stimulates the female to enter the nest, and she in turn stimulates the male to tremble. His trembling stimulates the female to spawn and the male fertilizes the eggs. The presence of the eggs stimulates his sexual behavior. The tendency towards sex and aggression can be measured by presenting to a male on his home territory either another male or a female fish confined in a glass tube. If the fish presented to him is a receptive female with a swollen belly, the male performs the zigzag dance. A count of the number of bites or zigzags per unit time is a measure of the respective strengths of the male's sexual and aggressive tendencies.

The characteristic pattern of singing of a territorial male bird advertises his presence. The song patterns of each species are distinct and presumably function both to keep other males of that species away from a male's territory and to attract a female of the species to the singing male. Certain fishes, frogs and insects also have mating calls which provide effective cues for the discrimination of species. This ethologic isolating mechanism prevents the wasting of genetic material.

The bowerbirds of tropical jungles have a very complex courtship behavior (Fig. 19.17) involving a fixed sequence of sign stimuli flashed from partner to partner. In an arena, a special territory where mating takes place, the male sings and displays his bright colors in special short flight movements. In other species, the male bowerbird's plumage is less brilliantly marked, but he builds and decorates an elaborate display area, thus attracting a female not by his own color but by the form and color of pebbles and berries added to the display area. After mating, the female bowerbird builds the nest and cares for the brood.

Most species of birds nest and rear their young in the territory; the male typically shares the tasks of incubating the eggs and feeding the young birds.

Parental Care. Most animals give little or no care to their offspring. This lack of parental behavior is usually correlated with the production of great numbers of eggs and sperm. Some reptiles and insects place their eggs in special burrows or nests where development is completed, but the parents do not aid the young after they hatch. Female crayfish and other crustaceans attach their eggs to their abdominal pleopods and aerate them until the eggs hatch. The young of some scorpions remain with the mother, clinging to her back until after their first molt. Care-giving behavior between one generation and the next is well developed in the social insects and in birds and mammals. Many young birds respond to tactile, acoustic and visual stimuli given by the adult by gaping—stretching upward toward the parent with open mouth. The parent responds to this by feeding the young (some

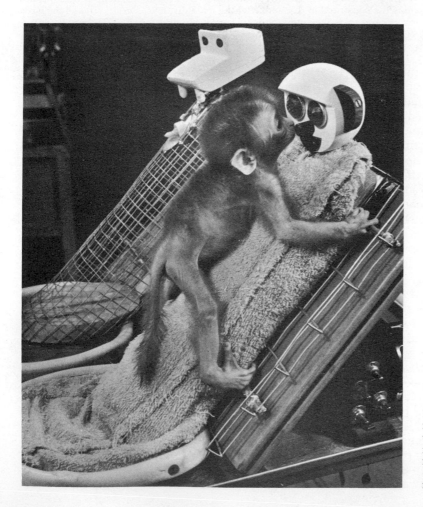

Figure 19.18 Young rhesus monkey, reared without a real mother, clinging to the surrogate used in raising it. (Photography by Gordon Coster. *In* Love in infant monkeys by H. F. Harlow. Copyright © 1959 by Scientific American, Inc. All rights reserved.)

baby birds have brightly colored oral cavities that serve as sign stimuli to the parents).

The extensive exchanges between mother and young in primates often have long-term effects on the behavioral development of the young. Rhesus monkeys deprived of their normal relationship with mothers or siblings never develop normal social and sexual relationships. Monkeys reared by artificial mothers do not interact normally with other members of the species nor are they able to mate when mature.

The elements of the maternal-infant relationship were studied by Dr. Harry Harlow using surrogate mothers of cloth and wire (Fig. 19.18). He found that both bodily contact and visual stimuli are important to this relationship. Young monkeys formed strong emotional attachments to cloth mothers, ran to them when frightened by a mechanical teddy bear beating a drum and would explore strange objects in the room only as long as the "mother" was present as a psychologic refuge. Monkeys without a real or appropriate surrogate mother fled to a corner of the room, clasped their heads and bodies and rocked convulsively back and forth. Their actions resembled the behavior of autistic children.

The young of some primates receive attention from several of the adult females of the troop, which is the social unit consisting of several adult males, three to four adult females and their offspring. Initially the mother will not allow others to hold the infant, but soon it is held and carried by other females, each of which nudges, touches, smells and licks it. If the infant squeals or begins to struggle, another female (usually the mother) comes over and takes it. The touch and color of the fur and the movements of the mother elicit a clinging response in the infant; the infant's color pattern elicits the interest and movements of the female, who then facilitates clinging. These mutual responses of mother and newborn are very important in the development of the social behavior and organization of primate groups.

19.8 CYCLIC BEHAVIOR AND BIOLOGICAL CLOCKS

Many animals exhibit behavior patterns that recur with a regular periodicity—each year, each lunar month, each day or with the tides. Some behavior patterns are regulated by a complex interaction of several types of periods. In recent years there has been much interest in these cycles and in their regulation by some sort of intrinsic "biological clocks." The organisms in the temperate zones of the world typically show marked seasonal cycles of activity. Most fishes, birds and mammals have a single breeding season in the spring; deer and sheep breed in the fall. Certain animals migrate at specific times to higher latitudes or altitudes. Some animals show a pronounced decrease in activity and metabolism during the winter, a phenomenon termed **hibernation.** In the tropics, although there is no sharp change in temperature from one month to the next, a period of heavy rainfall usually alternates with a period of little or no rain. The cycles of plants and animals in such regions are geared to these changes in the environment.

Other cycles of animal, and perhaps plant, activity reflect the phases of the moon. The most striking ones are those in marine organisms that are tuned to the changes in the tides due to the phases of the moon. The swarming of the Palolo worm at a particular time of the year is governed by a combination of annual, lunar and tidal rhythms (p. 498). The Atlantic fireworm swarms in the surface waters around Bermuda for 55 minutes after sunset on days of the full moon during the three summer months. The grunion, a small pelagic fish of the Pacific coast of the United States, swarms from April through June on those three or four nights when the spring tide occurs. At precisely the high point of the tide the fish squirm onto the beach, deposit eggs and sperm in the sand and return to the sea in the next wave. By the time the next tide reaches that portion of the beach 15 days later, the young fish have hatched and are ready to enter the sea. The menstrual cycles of humans and apes also follow lunar periods.

Organisms (even the various parts of an organism) usually do not function at a constant rate over the entire 24 hours of a day. Repeated sequences of events occur at about 24 hour intervals in the lives of many organisms. These have been termed "circadian rhythms" (*circa*=about; *dies*=day). It is important to note that when the organism is placed under constant conditions, these rhythms have a period of *about* 24 hours but almost never *exactly* 24 hours. In nature the activity cycle is lined up with the cycle of physical events that determine the length of the day. The daily changes in light are the signals used by most organisms as a "zeitgeber" (*zeit* = time; *geben* = to give).

Some animals are **diurnal,** showing their

greatest activity during the day, whereas others are **nocturnal** and most active during the hours of darkness. Still others are **crepuscular,** having their greatest activity during the twilight hours. Certain insects exhibit diurnal variations in pigmentation and continue to show these when placed in continuous darkness. Some spiders show a diurnal rhythm of the secretion of neurohormones; preparations of nervous tissues isolated from the animal and maintained in vitro continue to show this cyclic production of neurohormones. A diurnal cycle of the deposition of glycogen in the liver of the rabbit and mouse has been detected. Many of our physiologic processes exhibit circadian rhythms. Our body temperature is highest at 4:00 or 5:00 P.M. and lowest (as much as 1° C. lower) at 4:00 or 5:00 A.M. The secretions of a number of hormones, the heart rate and blood pressure and the rate of excretion of sodium and potassium all show circadian peaks and valleys. Even parasites within the bodies of their hosts have persistent rhythms. The nematode that causes elephantiasis in humans remains in the circulation of the deep tissues during the day and migrates to the peripheral circulation at night. This behavior is synchronized with that of the intermediate host, a nocturnal mosquito that transmits the parasite from one human host to the next.

The diurnal cycles of many organisms are firmly entrenched and are not easily changed even by prolonged exposure to constant light, to constant darkness or to abnormal cycles of light and darkness (a cycle having 8 or 10 hours instead of 24, for example)

Many different kinds of animals have evolved biological clocks by means of which their activities are adapted to the regularly recurring changes in the external physical environment, and perhaps to changes in their internal milieu as well. These clocks, together with other signals from the external environment, indicate to the organism that time of day most appropriate for some activity or physiologic process. Birds and bees have highly developed timing devices and can use them in navigation. Bees are not only able to find their way from hive to feeding ground using the direction of the sun as a reference, but they also can make suitable corrections for the sun's position as the day advances; that is, they can correct their celestial navigation for the time of day, just as man does.

At one time it was believed that the timing mechanisms of the biological clocks operated on a sort of "hourglass" principle, that some regularly recurring stimulus, such as sunrise, triggered a physiologic process requiring a certain length of time for its completion, at which time the signal was sent to the appropriate organ. More recent evidence favors the concept that the timing mechanisms involve some sort of cyclic oscillation. Biological clocks do require a small but definite amount of biologically useful energy for their operation; they cannot be primarily dependent on enzyme-catalyzed reactions because they show little or no change with temperature. The biological clock could involve some sort of biophysical mechanism, such as an alteration in the physical state of a complex molecule like a protein, or an alternation of the tension and relaxation of some physical structure.

Snails and other marine organisms whose activities vary with the tide continue to show the cyclic variations in activity when removed to an aquarium and protected from changes in light, temperature and other factors. This persistence of rhythmic changes in activity, coordinated with the cyclic changes in the environment from which the animal was removed, is strong evidence for the existence of biological clocks.

The keying of these endogenous rhythms to changes in the local environment is obviously of supreme adaptive importance. They enable certain adjustments in physiologic and biochemical processes to occur before the time of the change in the environment. Thus, the organism's activities are synchronized with the various periodicities in the environment. Although the cycles are set and persist under constant conditions, they can be reset if necessary. If animals that breed in the spring are transported from the northern to the southern hemisphere, their cycle eventually shifts over to coincide with the occurrence of spring in their new home. Members of the jet set, flying east or west to widely different time zones, eventually change their pattern of sleeping and wakefulness and their many other circadian rhythms. In short, all cycles are set by local time.

19.9 SOCIAL BEHAVIOR

The behavior of animals must be adjusted not only to factors in the physical environment but also to other individuals, both of the same species and of other species. In the broadest sense, any interaction between two or more individuals constitutes **social behavior.** Usually, however, social relationships imply interactions among members of the same species.

The mere presence of more than one individual does not mean that the behavior is social. Many factors of the physical environment bring animals together in **aggregations,** but interactions, if they occur here, are apt to be circumstantial. A light at night is a stimulus that causes large numbers of moths and other insects to collect around it. The high humidity under a log is a common stimulus causing aggregations of wood lice. A water hole in the dry savanna is a common focus for large numbers of birds and mammals. None of these groups is, strictly speaking, social.

Much of what we recognize as social behavior is based, directly or indirectly, on mating behavior. Except in parthenogenetic species, male and female must come together in a cooperative fashion. Although this might appear to be a simple matter to arrange, there are in fact some major obstacles that must be circumvented. Elaborate patterns of behavior have evolved to overcome these.

First of all, eggs and sperm must be produced at the same time. This means that male and female must find each other even if fertilization in that species is external, as in fish. It also means that both individuals must be in a reproductive state. The synchrony of time and place is accomplished by common stimuli from the external environment and by the stimulatory effect of one sex upon the other. Many fish, birds and mammals are brought into synchronous reproductive readiness by increasing day length, which acts on the pituitary to release gonadotropic hormones that in turn affect behavior. The increasing water temperature in spring induces the hormonal changes in the stickleback which predispose the fish to migrate into shallow water. The visual stimulation provided by the vegetation there releases reproductive behavior.

In many species of animals, family life—a long-term relationship between male, female and young—is a natural consequence of sexual relations. The basis of family relationships is the provision of shelter, food and defense for the young. Parents and young must react to one another in a variety of ways. Many birds will not feed the young unless they gape or beg. Young are stimulated by the sight of the parent's head over the nest, vibration of the nest by the alighting adult or, in the case of the sea gull, by a "mew-call." The gull chick is further stimulated by the parent's bill, a long thin structure with a red spot at the tip. Larger groups may consist of several to

many families or of individuals no longer tied to a family, as in flocks of birds, schools of fish and herds of animals. There are several adaptive advantages of grouping: A group is more alert to danger than a single individual. What one member may miss another sees, and the whole group takes flight. Often several unrelated species group together. In mixed groups of baboons and impala, the keen-sighted baboons spot visible danger and the keen-scented impala detect hidden danger. A group offers protection against attack. Male musk oxen make a ring around the young and female musk oxen when the group is threatened. A group is capable of communal attack as well as communal defense. The most frequently observed form of communal attack is the mobbing of crows, owls or hawks by groups of smaller birds. Communal hunting has been developed to a fine art by wolves and lions. The origin of a group and its continued cohesiveness depend upon a constant flow of information within the group.

Reproductive fighting and **dominance fighting** are aspects of group behavior that may also have adaptive value.

Reproductive fighting, directed toward reproductive rivals, serves to space out the breeding pairs. The end result is a more even distribution of food, more nesting and breeding space and the prevention of multiple copulations and their attendant waste. This kind of spacing provides each pair with its own **territory** from which it drives trespassers. The size of a territory may vary from the very small domain of a nesting gull to the extensive hunting territory of a large carnivore. The size of the territory of gulls and many other colonially nesting birds is the diameter of a circle from within which one bird can reach out to peck another without having to stir from its nest. This type of territory is nest-centered.

The territory of most other species is more than the breeding area proper; it is the area which supplies food, and its size is in part a function of the nature of the food. Territorial size may also be determined by topography (i.e., the actual amount of space available) but does not appear to change with variations in the abundance of individuals.

Animals with highly organized societies may have a **colony range** rather than a pair territory. Troops of baboons may claim a range of three to six square miles, but this may overlap the home range of an adjacent troop. The location and extent of ranges are set by group tradition. Territorial boundaries are learned by each generation. As long as food and water are available, troops may never meet. Arboreal species, prevented by the foliage from seeing their neighbors, are spaced out by vocal signals. Howler monkeys, for example, begin each day by howling for about 30 minutes to advertise their position to other troops. When there are shortages, as when only one water hole is available for baboons, the troops may come together, but they do not mix.

Other social animals, exemplified by the prairie dogs, inhabit territories that are subdivided into fixed and stable family units. The resemblance to human social life in this respect is striking. Whereas troops of baboons resemble human nomadic tribal life, prairie dog colonies resemble urban apartment living.

Another way of sharing the available environment may be called **temporal spacing.** It is achieved by the establishment of a dominance hierarchy in a group that occupies common territory. No individual calls an area his own or occupies it for more than a brief period; however, not all spots are equally available to all individuals. The best of everything comes first to dominant members of a group, and members lower on the social ladder must wait their turn. Social order is generally the rule in animals that live in groups. Among vertebrates it attains its highest development in troops of primates. A male achieves a dominant position in the group by his physical condition, age, aggressiveness and fighting ability and by the rank of his mother. He then commands the immediate surrounding space. Once achieved, his position is held by symbolic threats rather than by actual fighting. These threats are ritualized behavior patterns developed from preparatory movements for fighting, as when a man shakes a clenched fist at his enemy.

The hierarchy is further complicated in baboons by the fact that a number of males may join together in a governing clique. There may be an outside male who can outfight any individual member, but the group combines to stand against the intruder.

Each animal comes to know his rank in the social order and, if challenged by his superior, makes ritualized submission by raising his rump while the dominant asserts his dominance by mounting. Grooming, which is basically a hygienic measure, has acquired social significance in the more aggressive species of primates. The dominant member is more often groomed than grooming.

The cohesiveness of a primate group and the intensity of the dominance hierarchy are directly related to the level of potential danger to which the species is subjected. Gorillas, for ex-

ample, do not face the same dangers as baboons, are much less aggressive animals (despite their ferocious appearance) and lead a considerably more relaxed life. There are dominant males who lead, but there is little violence. Baboons and macaques, on the other hand, are subjected to many dangers in their environment; they are, accordingly, very aggressive.

19.10 SOCIAL INSECTS

Among the oldest and most highly developed societies are those of insects. Three hundred million years ago these societies were already in existence. In the course of evolution social habits arose at least 24 different times. The establishment of social life in these usually short-lived animals was made possible by the prolongation of parental life. First the progeny depended on the parents, then the parents on the progeny.

Various gradations between solitary and social life are evident in existing species. Some solitary insects, such as mosquitoes and dragonflies, drop their eggs in some body of water and go about their business. Butterflies and flesh-flies lay their eggs on food suitable for the young. The solitary digger wasps dig a hole, provision it with food, such as paralyzed spiders and caterpillars, lay an egg, seal the entrance and depart. The subsocial insects represent a further step toward social habit; they provide a mass of food for each egg but remain to guard the nest or young. In some species of dung beetles the female collects and rolls a ball of dung, excavates a burrow, drops the ball in, lays eggs and departs. In other species the male assists by guarding the dung balls while the female excavates. In still another species both sexes dig chambers, stock them with dung on which the female lays eggs and then guard them until the young hatch. At this time all disperse.

True social insects forage for food for the colony continuously, and the young cooperate in caring for the next generation. The most highly developed and complex of these are the social termites, ants and bees. All of these are matriarchies; all members of the colony are the offspring of a single female and hence all have very similar genotypes.

The greatest diversity is found among ants. The army ants have no permanent nest but travel in long marauding columns and rest in temporary bivouacs. There are ants that kill off other species, carry off their young and are thenceforth fully dependent on the "slaves" that develop from these larvae. Other ants collect nectar and feed it to special members of the colony that become living honey casks hanging from the roofs of underground chambers. The ants that feed on the honeydew produced by aphids carry these "ant cows" into overwintering quarters and protect them from predators. There are leaf-cutter, or fungus-growing, ants that cut leaves and carry them underground to serve as a substrate for growing pure strains of fungus, which is eaten by the ants.

The cohesiveness and organization of insect societies often prompts comparison with human societies, and indeed some notable similarities are readily discernible: (1) Insect societies arise from discrete families (i.e., one female and her offspring) and by swarming frequently send off new colonies. Human societies originally began as discrete families, but this is not repeated phylogenetically. (2) Humans have tradition and social heredity; insects, in this case ants, bequeath fungi and real estate (hunting grounds). (3) Humans use tools; insects do not, with the rare exception of the ant that uses its larvae as spinning shuttles to fasten leaves together with larval silk. (4) Humans control the environment of their dwellings; bees and termites also control the temperature of their nests. (5) Humans domesticate animals; ants "domesticate" aphids. (6) Human beings have a language; bees and ants communicate by dances and pheromones.

A basic difference between human and insect societies is in the manner in which the two adapt to different requirements. The human has evolved learning and abstract intelligence whereby he meets different situations. The adaptation provides him with a flexible society. Insects evolved different genotypes and different castes (soldiers, workers and so on), each of which meets a specific need. Their society is, therefore, a rigid one.

19.11 LEARNING

Learning is a relatively long-lasting adaptive change in behavior resulting from experience. It is usually defined by exclusion; that is, it is a modification of behavior that cannot be accounted for by sensory adaptation, central excitatory states, endogenous rhythms, motivational states or maturation. Learning exhibits many forms and is not a unitary phenomenon. Laboratory experiments have shown that the members of almost every animal phylum can undergo learning phenomena, and field observations have shown that learning is important in a variety of natural situations with a wide variety of animals. Learning in different animal species has different characteristics and may involve different mechanisms. There are several categories of learning: habituation, classic conditioning, operant conditioning, trial and error, insight learning and imprinting.

Habituation is perhaps the simplest form of learning, for it does not involve the acquisition of new responses but rather the loss of old ones. An animal gradually stops responding to stimuli that have no relevance to its life, ones that are neither rewarding nor punishing. Birds soon ignore the scarecrow that put them to flight when it was first placed in a field. The taming of animals represents a common form of habituation. The nature of the processes underlying habituation is obscure, but it must be a property of the central nervous system and not of the sense organ.

All other types of learning consist of strengthening responses that are significant to the animal. The simplest of these is **classic reflex conditioning,** or Pavlovian conditioning. Pavlov's classic experiments with conditioning in dogs frequently dealt with the salivation reflex. If food is placed in a dog's mouth, the dog salivates. By placing a tube in the salivary duct Pavlov could collect and measure all the saliva produced in response to a given stimulus. He then gave the dog standard stimuli by puffing known amounts of meat powder into its mouth through a tube. A standard amount of meat powder caused the secretion of a certain amount of saliva. Pavlov then associated the unconditioned stimulus, the meat powder, with another stimulus, the ringing of a bell or the ticking of a metronome. Initially the second stimulus caused no response, but after a number of pairings of the bell and meat powder stimuli, the saliva began to drip from the tube when the bell was rung and before the meat powder was administered. Pavlov called the second stimulus the conditioned stimulus and the response to it, salivation in response to a bell or metronome, the **conditioned reflex.** Pavlov's experiments showed that almost any stimulus could act as a conditioned stimulus provided it did not produce too marked a response itself.

Conditioned reflexes of this type have been observed in many kinds of animals from earthworms to chimpanzees. Birds become conditioned to avoid the black and orange caterpillars of the cinnibar moth, which have a very bad taste. The birds associate the bad taste with the orange and black color pattern and avoid not only the caterpillars but also certain wasps and other black and orange colored insects.

In **operant conditioning** a particular act is rewarded (or punished) when it occurs, thus increasing (or decreasing) the probability that the act will be repeated. A central feature of operant conditioning is the concept that the animal must play a role in bringing about the response which is rewarded or punished. In the simplest type, a rat presses a bar in his cage and is rewarded with food. Initially the pressing of the bar occurs by chance, but on successive trials the rat learns that pressing the bar provides him with food. The rat might at first press the bar with its nose or foot, but the experimenter can determine which of these will be rewarded. He can arrange matters so that only when the rat presses the bar with his left front foot will he receive the reward of food. The experimenter can indeed set up his experimental cage in such a way that the rat is punished, perhaps with an electric shock, if he presses the bar with anything other than his left front foot. Thus operant training may be either reward training (food) or escape or avoidance training (avoiding the electric shock). The **trial and error learning** that occurs in nature is probably more like operant conditioning than classic conditioning. In both types of training only the *probability* of the occurrence of a certain act under a given set of conditions is changed.

Some generalizations can be made about associative learning, as exemplified by these conditioning experiments. First, the time relations are critical. If the unconditioned stimulus precedes the conditioned stimulus, there is little or no conditioning. If the conditioned stimulus precedes the unconditioned stimulus by more than a second, little or no conditioning results. Thus, **contiguity** of conditioned and unconditioned stimuli is required for associative learning. Another relevant feature is **repetition.** The more often the conditioned stimulus and the unconditioned stimulus are paired, the stronger is the acquired conditioned response. The amount of saliva produced by Pavlov's dogs in response to the bell or metronome increased with each trial. A rat learning to run a maze makes fewer and fewer mistakes with each repetition. The rate at which an animal learns is described by a **learning curve,** generated by plotting the time taken to complete the task at each trial or the number of errors committed at each trial against the number of trials. Repetition finally produces a degree of learning beyond which

there is no improvement; however, training beyond this point (overtraining) does make the response more resistant to extinction. **Extinction** is the decay of learning in the absence of reinforcement. Without reinforcement the conditioned response may disappear completely.

Two other concepts associated with conditioning are generalization and discrimination. **Generalization** refers to the common observation that an animal conditioned to one stimulus will also be conditioned to closely related stimuli. The closer the two resemble each other, the better the response will be to the new stimulus. If a dog is conditioned to respond in some way to a 1000 cps tone, it will respond somewhat to a 500 cps tone or to a 1500 cps tone. Its response to a 100 cps or a 2000 cps tone would be poorer. The opposite process is termed **discrimination.**

A type of learning in which an animal learns without reinforcement and later puts the information to good use has been termed **latent learning.** An example of this is exploration. Most animals spend a lot of time exploring new environments, familiarizing themselves with their surroundings. Ants, bees and wasps make **orientation flights** around a nest they have built to learn its position. Some wasps have been shown to be able to learn the essential landmarks around their nest in an orientation flight lasting only nine seconds.

Insight learning, or insight reasoning, is considered to be the highest form of learning. It is defined as the ability to combine two or more isolated experiences to form a new experience tailored to a desired goal. If an animal is blocked from obtaining food he can see and selects an appropriate detour to it on his first try, he has probably used insight to solve the problem. Only monkeys and chimpanzees seem to be able to succeed on their first attempt in situations such as this; even dogs and raccoons usually fail on their first attempt. One of the classic examples of insight learning is that of the chimpanzee who piled up boxes or fitted two bamboo poles together to get some bananas hung out of reach. The chimpanzee figured out the solution without being taught.

There are several other types of problem-solving tests used in measuring an animal's ability to learn. These are called **learning set tests.** In one, **oddity principle learning,** the animal must pick out the odd member of a set of three, no matter what the objects are (Fig. 19.19). In another, **the delayed reaction set,** the animal is allowed to see which of two cups has food hidden under it. Later, after a delay,

Figure 19.19 A rhesus monkey performing an oddity-principle test. (From Stone, K. P. (Ed.): Comparative Psychology. Englewood Cliffs, N.J., Prentice-Hall, Inc., 1951.)

it has to lift the correct cup. In a third type of **learning set test,** an animal is presented with two dissimilar objects, one of which always has food under it. Repeated trials are necessary before the animal immediately goes to the object concealing the reward. This discrimination problem is repeated with many different pairs of objects. Although each problem is just as difficult as the first, the animal usually performs with increasing accuracy on repeated trials. Finally, when presented with two objects, the animal lifts one. If the food is there, it chooses that object on each sub-

sequent presentation. If the food is not there, it lifts the other object and then always chooses that one first on subsequent presentations. The animal finally gets the idea, or forms a learning set, that the food is always under a particular object. If it picks the object correctly the first time, there is no need to look further. The rates at which various mammals form discrimination learning sets vary tremendously (Fig. 19.20)

Imprinting is a form of learning first described in birds but now known to occur in sheep, goats, deer, buffalo and other animals

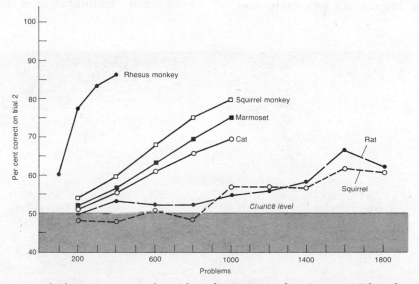

Figure 19.20 The rate at which various mammals can form discrimination learning sets. With each new problem the animal's choice on the first trial has to be random, but if it has learned the principle behind the problems, trial 2 should be correct. Note how long it is before the scores of rats or squirrels on trial 2 become better than chance or 50 per cent. Many monkeys reach almost 100 per cent within 400 problems. (From Warren: The Behavior of Nonhuman Primates. Vol. 1. New York, Academic Press, 1965.)

whose young are able to walk around at birth. It is a phenomenon whereby a young animal becomes "attached" to the first moving object it sees (or hears or smells) and reacts to it as it would toward its mother. Konrad Lorenz, who investigated this phenomenon extensively in the 1930's, described imprinting as a unique learning process whereby the young of precocial birds (those able to walk just after hatching) form an attachment with a mother figure. Normally this is their actual mother, but they will become attached to almost any moving object they see during a brief critical period shortly after hatching (Fig. 19.21). This following response is of considerable adaptive value in keeping the young bird close behind its parent and well within the protective range of the parent. Lorenz found that the young bird's choice of a mother figure during this imprinting period also affected its choice of a sexual partner when it matured. Young birds reared by a foster mother of another species cease to follow her when they develop and become independent. However, when they became sexually mature, they court and attempt to mate with birds of the foster mother's species.

Imprinting occurs in mammals, as well as in birds. Orphan lambs reared by humans follow them about and show little interest in other sheep. Harlow's studies with rhesus monkeys show very clearly that the development of normal sexual and social responses depends on their being reared with their mother or with siblings during infancy. There is little doubt that something similar to imprinting occurs in humans; human infants are extremely sensitive to maternal deprivation from about 18 months to three years of age.

The Physical Basis of Memory and Learning. Learning in some way involves the storage of information in the nervous system and its retrieval on demand. Many of the recent studies of learning have dealt with attempts to discover its physical basis. Somewhere in the nervous system there must be stored a more or less permanent record of what has been learned that can be recalled on future occasions. This record is termed the memory trace, or **engram.**

Electrical stimulation of the cerebral cortex of a patient undergoing brain surgery can cause vivid recollection of long-forgotten events. From this it was inferred that the items of memory, the engrams, were filed away somehow in specific places in the brain. Some years ago Karl Lashley investigated the retention of maze learning in rats by removing portions of the cortex after the rats had learned to solve various problems. The essence of Lashley's results was that the extent of the memory removed by the operation was a function of how much of the cortex was removed and not of what specific part of the cortex was removed. Lashley concluded that the cortex was equipotential and that engrams were not sorted out at specific cortical sites but were in some way present throughout its substance. He speculated that memory might be some system of impulses in reverberating circuits.

Other investigations of the role of the cerebral hemispheres in memory have used the **"split-brain" technique** of R. W. Sperry. The

Figure 19.21   A remote-controlled decoy duck was used by Hess and his colleagues in imprinting experiments. Here both decoy and duckling move about freely rather than on a runway. (From "Imprinting" in Animals by E. H. Hess. Copyright © 1958 by Scientific American, Inc. All rights reserved.)

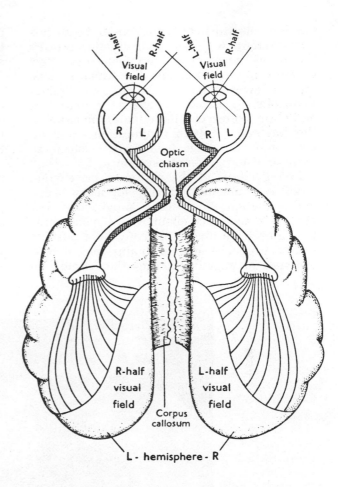

Figure 19.22 Split brain technique. The right and left halves of the brain have been completely separated by cutting the corpus callosum and the optic chiasm. (Redrawn from Sperry, R. W.: Science, *133*:1749–1757, 1961.)

cerebral hemispheres are separated by a cut down the midline that severs the **corpus callosum,** a large band of transverse fibers that passes ventrally and links the two hemispheres (Fig. 19.22). If a subject learns something using one eye (the other eye is covered), there is no problem when the eye coverings are reversed—the subject can make discriminations with its "untrained" eye. Fibers from the left and right halves of the retina cross over in the optic chiasma, and the visual cortex in each cerebral cortex receives information from both eyes. If the optic chiasma is cut, something learned by the right eye is still recognized by the left eye as long as the corpus callosum is intact. However, if the corpus callosum is cut also, half of the brain no longer knows anything learned visually by the other half of the brain after the operation. These split-brain animals are of great interest to psychologists, for they are in effect animals with two separate brains. It is possible to train the two halves of the brain to make diametrically opposite discriminations. The left half may learn that the red circle conceals the food, whereas the right half learns that the green

circle conceals the food. These experiments show that one function of the corpus callosum is to transfer engrams (or information used in forming engrams) from one hemisphere to the other and that storage is eventually bilateral. The two cerebral cortices may be linked by fiber tracts lower in the brain, but they apparently are not effective in transferring engrams.

Investigations of the memory system of mammals have shown that memories of recent events, **"short-term" memory,** have properties that differ somewhat from memories of events further in the past, **"long-term" memory.** People suffering from concussion are often unable to recall the events that immediately preceded their accident but have normal recall of events in the past. When their memory returns, events are remembered roughly in the order of their occurrence. This phenomenon has been termed **retrograde amnesia.** It appears that each learning trial sets up a process in the brain that relies on the continuous activity of certain neurons. This **consolidation process** results in the formation of an engram and its storage somewhere in the cortex. Concussion, electroshock and similar treatments can interfere with the consolidation process that forms the engrams, but they have no effect on fully formed engrams. The **hippocampus**, on the lower inner margin of each cerebral hemisphere, is connected to the hypothalamus and to the cortex via the reticular formation of the forebrain. If both hippocampi are damaged, human beings have severe defects in their short-term memory system.

Investigations of the nature of the engram have been pursued actively for many years, and although they have provided a fair amount of evidence about what the engram is *not*, they have given relatively little positive information. The areas of brains known to be the sites of engram storage—the mammalian

cerebral cortex and the optic lobes of the octopus—typically have many self-exciting or reverberating circuits. In these, neurons are arranged in a loop so that impulses conducted along the axon eventually reach the dendrite or cell body. One earlier theory suggested that engrams were represented by continuous activity within such reverberating circuits. These loops may well be involved in the consolidation process, but it is unlikely that they could be the basis of long-term storage. Nerve cells are efficient transmitters over brief periods, but the accurate repetition of a very specific pattern of impulses could hardly continue over many minutes, much less days or years. Clear evidence to the contrary was provided by experiments in which rats were

Figure 19.23 Photograph (A) and diagram (B) of a salmon fixed in place with an electrode in its olfactory lobe so that a recording of its action potentials can be made as water from different rivers is placed in its nostril. (Photograph and diagram courtesy of Dr. Kiyoshi Oshima, Kyoto University.)

cooled down to 0° for periods of one hour. At this temperature all electrical activity in the brain ceased, yet when the animals were re-warmed they retained their memory of events prior to the cooling as well as normal non-cooled rats.

Another theory suggests that memory is not the continuous activity of the special circuit, but that the electrical activity associated with a learning trial "wears a path" which will facilitate transmission along that specific path the next time the events occur.

Several features of the engram suggest that it is some sort of permanent growth process, which in turn suggests that protein synthesis may be involved in some way in the conversion of short-term memory into the long-term engram.

Experiments with both goldfish and mice show that **puromycin,** which inhibits the synthesis of proteins (p. 155) interferes with the consolidation of memory but not with short-term learning. Established long-term memories can be normal in puromycin-treated animals; hence, it is unlikely that the drug affects the mechanism by which memories are recalled. These experiments suggest that short-term memory, memory immediately following training, is stored by neural activity, but this is shortly followed by a storage phase involving protein synthesis. This storage phase, which may last for several days, converts the short-term memory into the stable, long-term memory. This conclusion can be only tentative, for puromycin is not specific and may affect processes that are necessary for the formation of memories but only indirectly connected with it.

The deciphering of the genetic code led some investigators to the theory that memory might have a basis in the synthesis of RNA—that engrams might consist of specific sequences of nucleotides in RNA molecules.

There have been reports of the transfer of specific conditioned reflexes in planaria and fish by the administration of RNA extracts. Studies by Hyden and others suggest that there are changes in the structure of RNA in specific neurons that are involved in specific kinds of learning—for example, in the neurons of the vestibular nuclei of rats that have undergone training in a task involving balance.

The homing in of salmon on the specific tributary in which they were born depends primarily on olfactory cues; the salmon smells some distinctive aroma of his home river, remembers it and subsequently homes in on it. Salmon have been prepared in the laboratory with electrodes in their olfactory lobes to record neuronal activity and with a catheter in the nasal orifice (Fig. 19.23). When water from the salmon's home river is perfused through the fish's nose, a clear burst of nerve impulses is detected in the olfactory lobe. When water from other rivers is perfused through the nose, little or no electrical activity is detected in the electrodes buried in the olfactory lobe. The experimenters prepared an RNA extract of the olfactory lobe of a salmon that homed in on river A and injected that into the olfactory lobe of a salmon that homed in on river B. Within 48 hours the olfactory lobe of this second salmon showed clear bursts of electrical activity, recorded by the indwelling electrode, when water from river A was perfused into the salmon's nostril! This suggests that some sort of memory was transferred from one salmon to another via the RNA extract. There have been similar reports of the transfer of training from one animal to another by the injection of RNA extracts, but negative reports in other instances. There still is no conceptual framework as to how memory might be coded in RNA, in protein or in any other kind of molecule, but the subject is being attacked experimentally with considerable vigor.

ANNOTATED REFERENCES

The literature relating to the field of animal behavior is exceptionally rich and the titles selected here are only a few of the many excellent sources available.

Bastock, M.: Courtship: A Zoological Study. London, Heinemann Educational, 1967. An interesting book describing courtship and reproductive behavior in a variety of animal types.

Beach, Frank: Hormones and Behavior. New York, Paul B. Hoeber, 1947. A classic study of reproductive behavior.

Carthy, J. D.: The Behavior of Arthropods. New York, Academic Press, 1968. A fascinating summary of studies of mechanisms controlling insect behavior.

Dethier, V. G., and E. Stellar: Animal Behavior. 3rd ed. Englewood Cliffs, N. J., Prentice-Hall, Inc., 1970. A concise survey of the subject.

Etkin, William: Social Behavior and Organization Among Vertebrates. Chicago, University of Chicago Press, 1964. A description of social behavior from the ecological and evolutionary points of view.

Gaito, J.: Macromolecules and Behavior. Amsterdam, North-Holland Publishing Co., 1966. Interesting reading about the chemical basis of behavioral phenomena.

Hinde, R. A.: Animal Behavior. 2nd ed. New York, McGraw-Hill Book Co., 1970. A good general text combining ethological and psychological viewpoints.

Hirsch, J. (Ed.): Behavior-Genetic Analysis. New York, McGraw-Hill Book Co., 1967. A multi-author account of the inheritance of behavior patterns.

Lorenz, K.: On Aggression. New York, Harcourt, Brace & World, Inc., 1966. A classic volume discussing aggressive behavior.

Manning, A.: An Introduction to Animal Behavior. 2nd ed. Reading, Mass., Addison-Wesley, 1972. A shorter introduction to the subject.

Marler, P., and W. J. Hamilton: Mechanisms of Animal Behavior. New York, John Wiley & Sons, Inc., 1966. An excellent general text with detailed accounts of animal behavior.

McGaugh, J. L., N. M. Weinberger, and R. E. Whalen: Psychobiology. San Francisco, W. H. Freeman & Co., 1967. A text by many authors that discusses a wide variety of investigations in animal behavior.

McGill, T. E.: Readings in Animal Behavior. New York, Holt, Rinehart and Winston, 1965. A multi-author text that provides samples of current thought and research on animal behavior.

Morgan, C. T.: Physiological Psychology. 3rd ed. New York, McGraw-Hill Book Co., 1960. An excellent text presenting the physiologic approach to an analysis of behavior.

Schrier, A. M., H. F. Harlow and F. Stollnitz (Eds.): The Behavior of Nonhuman Primates. New York, Academic Press, 1965. A well-written account of the experimental analysis of primate behavior.

Slucken, W.: Imprinting and Early Learning. London, Methuen & Co., 1964. An account of experiments and observations on early learning.

Thorpe, W. H.: Learning and Instinct in Animals. Cambridge, Harvard University Press, 1956. A classic exposition of theories of learning.

Tinbergen, N.: Social Behavior in Animals. New York, John Wiley & Sons, 1953. A classic book on ethology.

Wells, M. J.: Brain and Behavior in Cephalopods. London, Heinemann Educational, 1962. An interesting description of the experimental analysis of behavior in the octopus.

Wilson, E. O.: Sociobiology. Cambridge, Harvard University Press, 1975. The biological basis of social behavior with a wealth of information about animal studies.

REPRODUCTION

The survival of each species of animal requires that its individual members multiply, producing new individuals to replace those killed by predators, parasites or old age. One of the fundamental tenets of biology, *omne vivum ex vivo* (all life comes only from living things), is an expression of this basic characteristic of all living things—their ability to reproduce. Reproduction at the molecular level is a function of the unique capacity of the nucleic acids for self-replication, which depends on the specificity of the relatively weak hydrogen bonds between pairs of nucleotides. At the level of the whole organism, reproduction ranges from the simple fission of bacteria (a process that does not involve sex at all) to the incredibly complex structural, functional and behavioral processes in the higher animals. Human reproduction involves not only the genetic transfer of biological information from one generation to the next but also the endocrine regulation of the development of the genital tracts, oögenesis, ovulation and spermatogenesis, as well as the intricate behavior patterns that ensure that eggs and sperm are released at the same time and in the same place in order that they meet to form the fertilized egg, or zygote. This is followed by the complex processes of development and differentiation by which a zygote becomes an adult organism. The reproductive process is cyclic, and we could begin a study of reproduction at any point in it.

The details of the reproductive process vary tremendously from one kind of organism to another, but we can distinguish two basically different types—asexual and sexual. In **asexual reproduction** a single parent splits, buds or fragments to give rise to two or more offspring that have hereditary traits identical with those of the parent. Even higher animals may reproduce asexually; indeed, the production of identical human twins by the splitting of a single fertilized egg is a kind of asexual reproduction.

In contrast, **sexual reproduction** involves two parents; each contributes a specialized reproductive cell, a gamete, and these fuse to form the fertilized egg. The **egg** is typically large and nonmotile, with a store of nutrients to support the development of the embryo that results when the egg is fertilized. The **sperm** is typically small and motile, adapted to swimming actively to the egg by beating its long whiplike tail. Sexual reproduction is biologically advantageous, for it makes possible the combination of the best inherited characteristics of the parents and provides for the possibility that some of the offspring may be better adapted to survive than either parent was. Evolution can proceed much more rapidly and effectively with sexual, than with asexual, reproduction.

20.1 ASEXUAL REPRODUCTION

Perhaps the simplest form of asexual reproduction is the splitting of the body of the parent into two more or less equal parts, each of which becomes a new, independent, whole organism (Figs. 20.1 and 20.2). This form of reproduction, termed **fission,** occurs chiefly among the protists, the single-celled animals and plants. The cell division involved is mitotic. Hydras and certain other cnidaria reproduce by **budding;** a small part of the parent's body becomes differentiated and separates from the rest. This part develops into a complete new individual and may take up independent existence, or the buds from a single parent may remain attached as a colony of many individuals.

Figure 20.1 Asexual reproduction in euglena, paramecium and ameba. In each, one cell divides mitotically to give rise to two cells.

Salamanders, lizards, starfish and crabs can grow a new tail, leg or other organ if the original one is lost. When this ability to regenerate the whole from a part is extremely marked, it becomes a method of reproduction. The body of the parent may break into several pieces, and each piece then develops into a whole animal by regenerating the missing parts. Such reproduction by **fragmentation** is common among the flatworms. Starfish have the ability to regenerate an entire new starfish from a single arm.

Some animals reproduce asexually by spores, special cells with resistant coverings adapted to withstand unfavorable environmental conditions, such as excessive heat, cold or desiccation. The parasitic protozoa, the Sporozoa, typically live as parasites within the cells of the host and reproduce by spores. Sporozoa cause some serious diseases, such as malaria in humans and coccidiosis in poultry. The malaria organism, *Plasmodium,* has a complex life cycle involving mosquitos and either birds or mammals (Fig. 20.3). Through the bite of an infected mosquito the malaria organisms are introduced into the blood stream of the bird or mammal and enter the red blood

cells. Within the red cell, each *Plasmodium* grows and divides by multiple fission into some 12 to 24 spores, each of which enters the blood stream when the red cell, after swelling, disintegrates. The spores released infect new red cells and the process is repeated. The simultaneous breakdown of billions of red cells causes the malarial chill, followed by fever as the toxic substances released penetrate to other organs of the body. If an uninfected mosquito bites the infected bird or mammal, it becomes infected by sucking up some of the plasmodia with the blood. Inside the mosquito's stomach, a complicated process of sexual reproduction occurs that involves the formation of eggs and sperm, fertilization and multiple fission of the resulting zygote to yield new infective spores. Some of these then migrate into the mosquito's salivary glands, where they are ready to infect the next individual bitten.

Figure 20.2 An electron scanning micrograph of an early stage of binary fission in the protozoan, *Didinium nasutum.* This type of asexual reproduction is common among protozoa. (Courtesy of Dr. Eugene B. Small; from Villee, C. A.: Biology. 7th ed. Philadelphia, W. B. Saunders Co., 1977.)

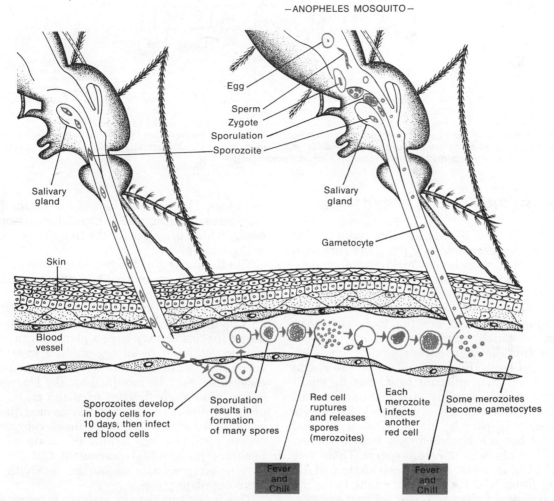

Figure 20.3 Diagram of the life cycle of the sporozoan *Plasmodium*, which causes malaria in humans and other mammals. An infected mosquito (*left*) bites a human and injects sporozoites of *Plasmodium* into the blood stream. These undergo sporulation and reproduce asexually within the red blood cells of the host. Periodically, the infected red cells rupture, and the new crop of merozoites released then infect other red cells. The bursting of the red cells releases toxic substances that cause the periodic fever and chill. Some merozoites develop into gametocytes, which can infect another mosquito when one bites the human. The gametocytes develop into eggs and sperm (*right*) within the mosquito and undergo sexual reproduction. The resulting zygote by sporulation produces a host of sporozoites that migrate to the salivary glands and are ready to be injected when the mosquito bites the next human. (From Villee, C. A.: Biology. 7th ed. Philadelphia, W. B. Saunders Co., 1977.)

20.2 SEXUAL REPRODUCTION

Sexual reproduction is characterized by the development of a new individual from a zygote produced by the fusion of two gametes—an egg and a sperm. Certain protozoa have a complicated process of sexual reproduction in which two individuals come together and fuse temporarily along their oral surfaces (Fig. 20.4). The nucleus of each individual divides several times before one of the resulting daughter nuclei migrates across to the other protozoan and fuses with one of its nuclei. The two protozoa then separate and each reproduces asexually by fission. Paramecia are not differentiated morphologically into sexes, but Sonneborn has shown that there are as many as eight distinct, genetically determined, mating types. A member of one group will mate only with a member of some other group.

Pairing   1st meiotic   2nd meiotic   Haploid   Mutual   Nuclear fusion
          division   division   mitosis   fertilization

Figure 20.4 Sexual reproduction in paramecium. Two individuals with diploid micronuclei undergo conjugation (*left*). After meiosis (*second and third figures*), three of the nuclei degenerate and the fourth nucleus divides by mitosis (*fourth figure*). The organisms undergo mutual fertilization; one nucleus passes from each organism to the other. This is followed by fusion of the haploid nuclei to form a new diploid nucleus in each of the two organisms (*last figure*). The original macro nuclei disappear. The two new diploid nuclei divide several times by mitosis and eventually establish both the new micronuclei and the new macronuclei. (See Figure 22.7 for additional details.)

20.3   REPRODUCTIVE SYSTEMS.

Although the details of the reproductive process vary tremendously from one animal to another, it is possible to make some generalizations about animal reproductive systems that may be helpful in understanding the variations. The basic components of the male reproductive system are the male gonad, or **testis,** in which sperm are produced and the **sperm duct** that transports sperm to the exterior of the body (Fig. 20.5). Parts of the sperm duct, or of areas adjacent to it, may be modified for specific reproductive functions. A part of the duct may be given over for sperm storage; this part is often called a **seminal vesicle,** but the human organ termed the seminal vesicle does not store sperm. There may be glandular areas for the production of seminal fluid, which serves as a vehicle for the sperm and may also activate, nourish and pro-

tect them. The terminal part of the sperm duct may open onto or into a copulatory organ, a **penis,** which provides for the transfer of sperm to the female.

The basic parts of the female system are the female gonad, or **ovary,** and the **oviduct,** a tube for the transport of eggs to the exterior (Fig. 20.5). Like the sperm duct, the oviduct may be modified in any of several ways in different species. It may have a glandular portion for the secretion of an egg shell, a case or a cocoon. A section of the oviduct, a **seminal receptacle,** may be modified for the storage of sperm after their transfer from the male. Another portion of the oviduct may be modified as a **uterus** for egg storage or for the development of the fertilized egg within the body of the female. The terminal portion of the oviduct may be adapted as a **vagina** for receiving the male copulatory organ.

The male and female systems of most ani-

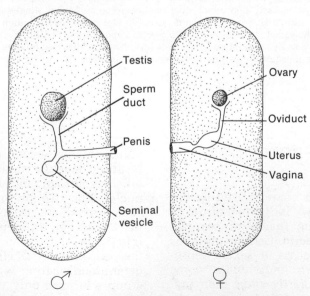

Testis

Sperm duct

Penis

Seminal vesicle

Ovary

Oviduct

Uterus

Vagina

♂   ♀

Figure 20.5 Diagrams of generalized male and female reproductive systems.

mals do not possess all of these modifications, and they cannot be considered as part of any kind of progressive sequence of evolutionary changes. Rather, the presence or absence of some specific adaptation of the reproductive system is correlated with the circumstances of reproduction: whether the animals live in the sea, in fresh water or on land; whether fertilization is external or internal; whether the eggs are liberated singly into the water to develop or are deposited onto the bottom in envelopes; and whether or not the developing individual passes through a larval stage. In view of the great range of reproductive conditions that exist within any large diverse group of animals, it is to be expected that the reproductive systems would display a corresponding range of modifications, some of which may have evolved independently a number of times.

Most species of fishes utilize a rather primitive form of reproduction in which the female sheds into the water enormous numbers of eggs, which may or may not be fertilized by sperm from some passing male of the same species. Thereafter, both male and female fishes ignore the eggs. In such fishes, the uterus is missing and the male does not have a copulatory organ. In other fishes, however, internal fertilization takes place before egg laying. In the sea catfish, for example, the male anal fin is used as an intromittent organ. The fertilized eggs are deposited in a clump that the male picks up and carries in his mouth during most of their incubation (presumably fasting and yielding not to temptation!).

Many of the lower animals are **hermaphroditic;** both ovaries and testes are present in the same individual, and it produces both eggs and sperm. Some hermaphroditic animals, such as the parasitic tapeworms, are capable of self-fertilization. (Does this violate the generalization that sexual reproduction involves two individuals?) Since a particular host animal may be infected with but one parasite, hermaphroditism is an important adaptation for the survival of the parasitic species. Most hermaphrodites, however, do not reproduce by self-fertilization; in the earthworm, for example, two animals copulate and each one inseminates the other. In other species, for example, the oyster, self-fertilization is impossible because the testes and ovaries produce gametes at different times in the animal's life cycle.

The reproductive systems of different species have a fundamentally similar plan, but many variations on the theme are evident. The gonads and their ducts may be single, paired or multiple, perhaps present in several segments of the body. Sperm produced in the coiled seminiferous tubules of the testis are transported by a series of ducts to the exterior and suspended in seminal fluid secreted by glands associated with the reproductive tract. Seminal fluid contains glucose and fructose, which serve as substrates for sperm metabolism, buffers that protect the sperm from the acids normally present in the urethra and female tract, and mucous materials that lubricate the passages through which the sperm travel.

In many vertebrates a number of accessory structures have developed to facilitate the transfer of sperm from the male to the female reproductive tract and to provide a place in the female tract for the development of the fertilized egg. These structures have evolved either from or in close association with the urinary system, and the two together are frequently termed the urogenital system. In the male mammal, for example, the **vasa deferentia** empty into the **urethra,** which also transports urine from the bladder to the outside. This urethra is surrounded by the external reproductive organ, the **penis,** which consists of three columns of **erectile tissue,** spongy venous spaces that become filled with blood during sexual excitement in order to produce an erection of the penis.

Eggs are produced in the ovaries of the female and are typically surrounded and nourished by follicle or nurse cells during their development. At the time of **ovulation,** the eggs are released from the ovary into the abdominal cavity, whence they pass into the funnel-shaped end of the oviduct. Eggs are moved along the oviduct by the peristaltic contractions of its muscular wall or by the beating of cilia lining the lumen of the duct. The yolk of the bird or reptile egg is formed while the egg is still in the ovary, but the egg white and shell are added by glands in the wall of the oviduct. The oviducts may open directly to the exterior, or they may expand into a terminal duct, the uterus, a thick-walled muscular pouch in which the young develop. In mammals, the uterus is connected with the exterior by the **vagina,** which is adapted to receive the penis of the male during copulation. Female mammals have a **clitoris,** the homologue of the male penis, just anterior to the opening of the vagina; it contains sense organs and erectile tissue that becomes engorged with blood during sexual excitement.

20.4   FERTILIZATION

Most aquatic animals deliver their eggs and sperm directly into the surrounding water, and **fertilization,** the union of egg and sperm, occurs there by chance. This primitive and rather uncertain method of uniting the gametes is called **external fertilization.** Such animals usually have no accessory sex structures.

In other animals, fertilization occurs within the body of the female, usually in the oviduct, after the sperm have been transferred from the male by copulation or some other means. This method of **internal fertilization** requires some cooperation between the two sexes, and many species have evolved elaborate patterns of mating behavior to ensure that the two sexes are brought together, mate at the most appropriate time and take care of the resulting offspring. A variety of secondary sex characteristics have evolved and serve as stimuli to attract the opposite sex and to elicit a mating response.

In the human, sperm deposited in the vagina during intercourse travel up the vagina and into the uterus largely by the force of the muscular contractions of the walls of these organs. Most of the sperm become lost in the journey, but a few hundred find their way to the openings of the fallopian tubes and swim up them. Although sperm motility may play little or no role in transporting sperm from the vagina to the oviduct, it probably is important in penetrating the egg. If fertilization of the egg takes place, it usually occurs in the upper third of the fallopian tube. Only one of the hundreds of millions of sperm deposited at each ejaculation fertilizes a single egg.

The eggs of some species of lower animals secrete **fertilizin,** an important constituent of the jelly coat surrounding the egg. Fertilizin

causes the sperm to clump together and stick to the surface of the egg. Other extracts of egg jelly stimulate sperm motility and respiration and prolong sperm viability. Each human egg is surrounded by a layer of cells, the **corona radiata,** derived from the follicle. By the time the ovum has reached the fallopian tube, it has completed the first meiotic division and extruded the first polar body. It is probable that hyaluronidase from the seminal plasma and the hydrolytic enzymes of the lysosomes in the head of the sperm play a role in the penetration of the sperm through the corona radiata and the vitelline membrane surrounding the egg. Once one sperm has entered the egg there is a change in the surface layer of the ovum that prevents the entrance of other sperm. The sperm either leaves its tail outside the egg or sheds it shortly after entering the egg cytoplasm; only the nuclear material of the head and the centriole remain (Fig. 20.6). The ovum completes the second meiotic division and extrudes the second polar body. The head of the sperm swells to form the **male pronucleus,** and the nucleus of the ovum becomes the **female pronucleus.** The fusion of the haploid male pronucleus with the haploid female pronucleus forms the nucleus of the zygote and restores the diploid number of chromosomes.

The eggs of many species can be stimulated to cleave and develop without fertilization. The development of an unfertilized egg into an adult is known as **parthenogenesis** (virgin birth). Some species of arthropods apparently consist solely of females that reproduce parthenogenetically. In other species, parthenogenesis occurs for several generations and then some males are produced that develop and mate with the females. The queen honey bee is fertilized by a male just once during her lifetime, in her "nuptial flight." The sperm are

Figure 20.6 Diagram of the stages in the process of fertilization, the union of the egg and sperm.

stored in a pouch connected with the genital tract and enclosed by a muscular valve. If sperm are released from the pouch as she lays eggs, fertilization occurs and the eggs develop into females—queens and workers. If the eggs are not fertilized, they develop into males—drones.

Changes in temperature, pH or the salt content of the surrounding water or chemical or mechanical stimulation of the egg itself will stimulate the eggs of many species to parthenogenetic development. A variety of marine invertebrates, frogs, salamanders and even rabbits have been produced parthenogenetically. The resulting adult animals are generally weaker and smaller than normal and are infertile.

The females of all birds, most insects and many aquatic invertebrates lay eggs from which the young eventually hatch; such animals are said to be **oviparous** (egg-bearing). In contrast, mammals produce small eggs that are kept in the uterus and provided with nutrients from the mother's blood until they can exist independently, at least to some extent. Such animals are said to be **viviparous** (live-bearing). In certain other species—some insects, sharks, lizards and certain snakes—the female is **ovoviviparous;** she produces large yolk-filled eggs that are retained with her reproductive tract for a considerable period of development. The developing embryo forms no close connection with the wall of the oviduct or uterus and receives no nourishment from the mother.

The number of eggs produced by each female of a given species and the chance that any particular egg will be fertilized and develop to maturity are inversely related. In the evolution of the vertebrates from fish to mammals, the trend has been toward the production of fewer eggs and the development of better parental care of the young. The cod and salmon produce millions of eggs each year, but only a small number of these ever become adult fish. In contrast, mammals have few offspring but take good care of them with the result that the majority attain maturity. Fish and amphibians generally do not take care of developing eggs, which are simply deposited in water and left to develop completely unaided. The eggs of reptiles are usually laid in earth or sand and develop there, warmed by the sun but without parental care. Birds, in contrast, have a complex behavior pattern for nest building, incubating the eggs by sitting on them and caring for the newly hatched youngsters. The mammalian egg develops within the mother's uterus, where it is safe from predators and harmful environmental factors. Most mammals have a strong "maternal instinct" to take care of the newborn until they can shift for themselves. This pattern of behavior is controlled by hormones secreted by the pituitary gland and ovary.

20.5 MATING BEHAVIOR

Most animals breed only during a relatively brief season of the year, and the production and release of eggs and sperm must be synchronized if fertilization is to occur. Typically the males and females of a species are triggered by some environmental cue, such as a change in the photoperiod or the ambient temperature, seasonal rainfall or specific relations of tidal and lunar cycles (cf. the Palolo worm, p. 498). In some species the males and females not only must be brought to full sexual activity at the same time but also must be induced to move or migrate to specific mating or breeding grounds. The migration of salmon upstream to breed, the migration of eels to a breeding ground in the Central Atlantic, the migration of the gray whale to Baja California and the migrations of turtles and birds are examples of such movements. Animals that live a solitary life during most of the year—seals, penguins and certain sea birds—come together for brief periods of mating at specific times and places. The members of many species fight at every encounter, and some signal is necessary to hold this aggressive behavior in abeyance in order that mating may occur.

The specific reproductive synchronization of a particular male for a particular female frequently involves some sort of **courtship behavior** (Figs. 19.16 and 19.17). The courtship, usually initiated by the male, may be a very brief ceremony, or in certain species of birds may last for many days. The courtship behavior serves two additional roles: it tends to decrease aggressive tendencies, and it establishes species and sexual identification; i.e., it identifies a member of the same species but of the opposite sex. Special structural and functional adaptations have evolved in some species for which aggression seems to be especially difficult to control (see Section 19.7).

The song patterns of birds and the mating calls of certain fishes, frogs and insects provide effective cues for the discrimination of the species. Female frogs are attracted to calling males of their own species but not to calling males of related species.

20.6 SOME ADAPTATIONS FOR REPRODUCTION

The most primitive reproductive systems are found in certain marine animals. The system is simple, and the gonoducts, the sperm ducts or oviducts, function only to transport the large number of gametes to the exterior. The gonads of polychaetes, for example, are not distinct organs but are masses of developing gametes that appear as projections or swellings of the peritoneum in certain segments. In the primitive worms, most of the segments produce gametes but in others, gametes appear only in certain posterior segments. The gametes are shed from the site of production into the coelomic cavity, and when the worm is sexually mature the coelom is packed with eggs or sperm. Although some polychaetes do have special gonoducts, in most species the gametes leave the body through the metanephridial pore or by the breakdown of the body wall. This escape of gametes by the ruptured body wall is found in some nereids and other polychaete worms that become pelagic when they are sexually mature. After the body wall has ruptured the adults die.

Epitoky

Copulatory phenomena occur in a few of the polychaetes, but in most of them the eggs are fertilized in the surrounding sea water after the eggs and sperm have been shed. In many species of polychaetes, specialized reproductive individuals appear that differ in various secondary sexual characteristics from the nonsexual form. This phenomenon is termed **epitoky,** and the epitokous individuals may differ in the shape of the head, the structure of the parapodia, the size of the segments, the musculature of the segments and other features. The part of the body that bears gametes may be strikingly modified so that the body of the worm appears to be divided into two quite different segments. The Samoan palolo worm, *Palola viridis,* has an unmodified anterior portion of the body, whereas the posterior epitokal region is a very long chain of narrow posterior segments with an eye spot on the ventral side of each. Typically, epitokous polychaetes swim to the surface when the eggs and sperm are shed; this type of behavior is known as **swarming.** In this way sexually mature individuals congregate in one area of the sea, which increases the likelihood that fertiliza-

tion will occur. The female produces a pheromone that attracts the male and stimulates him to shed his sperm. The presence of the sperm, in turn, stimulates the female to shed her eggs. The male of the polychaete *Autolytus* swims in circles around the female, touching her with his antennae and releasing sperm. This behavior triggers the female to shed her eggs.

Swarming

Swarming often coincides with certain lunar periods, and the most striking examples are displayed by the palolo worms. The Samoan palolo worm lives in rock and coral crevices below the low tide mark and releases its epitokes in October or November at the beginning of the last lunar quarter. The Atlantic palolo, *Eunice schemacephala,* lives in habitats in West Indian waters similar to those of the Samoan palolo; however, it swarms in July during the first and third quarters of the lunar cycle. At three or four in the morning during these periods the worm backs out of its burrow; the sexual epitokal region breaks free and swims to the surface where it undergoes spiral swimming movements. By dawn, the ocean surface is covered with sexual bodies, and when the sun rises the epitokes rupture, eggs and sperm are released and fertilization follows immediately. The fertilized egg develops into a ciliated larval stage within 24 hours, and the larvae sink to the bottom within three days. The physiological factors that control swarming and its periodicity are as yet poorly understood. Swarming appears to be under hormonal control, with neurosecretory cells in the brain playing a primary role in inhibiting the formation of an epitokal portion of the body. The relation of the lunar phases to swarming periods is different in different species, which makes any hypothesis based simply on light intensity difficult to establish.

In most of the lower marine animals, the eggs contain a relatively small amount of yolk sufficient to supply energy and raw materials for development to the larval stage, at which time feeding begins and the larva obtains an outside source of food material. The larvae are dispersed primarily by ocean currents; the locomotor activity of the minute larva is concerned with feeding or with maintaining position vertically in the water. At a certain time in development, the larva settles on the bottom in response to various substrate stimuli characteristic of the habitat in which they live as adults.

Although such a primitive reproductive system was satisfactory despite the high mortality rate of eggs, embryo and larvae, changing evolutionary conditions led to a number of modifications and adaptations in reproductive phenomena. Clearly, copulatory behavior and internal fertilization greatly increase the probability that eggs will be fertilized, but this requires a much more complex reproductive system, more specialized gonoducts and an intromittent organ in the male, together with elaborate behavioral phenomena to bring the two sexes together without their fighting or eating each other. At present we can only speculate as to how this was achieved.

Hermaphroditism

The adaptive significance of hermaphroditism in groups of animals that are sessile is clear, for when animals are widely dispersed, any individual that settles nearby is a potential mate. Thus, although most crustaceans are motile and dioecious (have separate sexes), the sessile barnacles are hermaphroditic. However, there are many exceptions to this rule, and there are motile hermaphroditic forms and sessile dioecious forms. In some species of hermaphroditic animals, each individual first develops male gonads and functions as a male, and later in time the male gonad atrophies and a female gonad appears. This phenomenon is termed **protandry.** Some limpets and slipper shells among the mollusks are protandric. Individual slipper shells tend to live in clusters stacked one upon another, permitting the penis of the upper individual to reach the female gonopore of the individual below. The young specimens are always males, but after a period of transition the male reproductive tract degenerates, and the animal then develops into a female. The sex of each individual appears to be influenced in part by the sex ratio of the entire association. An older male will remain a male if it is attached to a female, but if it is removed or isolated it will develop into a female. In other hermaphroditic mollusks, such as the sea hare *Aplysia*, there is copulation with mutual sperm transfer. In this group, chains of copulating animals form in which any given member of the chain receives sperm from the animal behind it and provides sperm for the one in front.

Even for the simplest form of reproduction with external fertilization it has been necessary to evolve adaptations so that eggs and sperm are released simultaneously. Thus, adaptations for the synchronization of reproductive function in the male and the female are important from the lowest to the highest form. An enormous variety of neural and hormonal mechanisms have evolved to achieve synchrony in different phyla of species. Almost all of these depend ultimately on some environmental cues, such as a change in the photoperiod, the ambient temperature, the supply of nutrients and lunar and diurnal cycles.

Egg Deposition

Most animals do not simply release their eggs to be dispersed into the environment. The eggs are enclosed within one or another type of envelope and deposited on the bottom or attached to some object. Egg deposition occurs in many marine animals and in almost all freshwater animals, for in fresh water the current of the stream would tend to sweep away the eggs. The inhabitants of many lakes arrive there as eggs or larvae from the streams that feed into the lakes. The egg coverings of terrestrial animals are important to reduce desiccation. An egg that has been enveloped with a membrane or shell must at some period in development undergo hatching. In contrast, the eggs that are simply released into the surrounding ocean and are fertilized there do not undergo hatching because they have no membrane or shell. The females of species that deposit eggs must have some modification of the gonoducts to produce the material for the egg envelopes or cases. The oviduct of a bird, for example, is greatly modified to produce the egg white and the hard calcareous shell around the egg. The eggs of many polychaetes and mollusks are laid in mucous masses that are attached to objects in the water.

Parental care of the eggs, termed "brooding," occurs in a wide variety of animals. Many polychaetes brood their eggs. The tube dwelling polychaetes brood them attached within the tube; other species use a respiratory water channel, the cavity of the operculum or a secreted sac attached to the ventral surface of the body. The freshwater *Nereis limnicola* broods its eggs internally, and its larvae develop within the coelom. One polychaete, *Ctenodrilus*, is reported to be truly viviparous, for the embryo's nutrition is supplied by maternal blood vessels. Viviparity evolved from the ovoviviparous condition and is found in some scorpions, some sharks and bony fishes, a few lizards and, of course, mammals.

An adaptation to terrestrial life during the

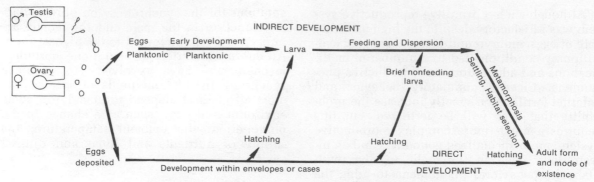

Figure 20.7   A comparison of direct and indirect development.

evolution of the reptiles was the development of eggs surrounded by a leathery or calcareous shell to protect the developing embryo from desiccation. The eggs of reptiles and birds are self-sufficient, completely enclosed systems termed **cleidoic** ("boxlike") **eggs.** The extraembryonic membranes and shells are essentially watertight so that the water of the amniotic fluid does not diffuse away. All of the nutrients required for the development of the embryo until it hatches as a miniature adult are present in the yolk and the egg white. The cleidoic condition necessitated the development of additional extraembryonic membranes, the amnion, chorion and allantois, to provide for an aqueous environment within the egg shell, the exchange of gases with the environment through the shell and the uptake of nutrients from the large store of yolk. This resulted in a great many changes in the development of such animals, as described in Chapter 5. The viviparity of the mammals is possible because of the evolution of the extraembryonic membranes that arose intially as adaptations for cleidoic eggs.

In many animals the organism hatching out of the egg membrane is not a larva but a miniature adult. This phenomenon, called *direct development,* or larval suppression (Fig. 20.7), has occurred in many marine and freshwater animals and most terrestrial ones. The adaptive reasons for larval suppression in different forms are quite varied. The larvae of terrestrial insects, however, are a secondary specialization, for larvae are not present among primitive insects.

20.7   PROTECTION AND NUTRITION OF THE EMBRYO

The egg and developing embryo are generally very susceptible to unfavorable environmental conditions, and a variety of adaptations have evolved in invertebrates and vertebrates to tide the embryo over this critical period. The eggs of many parasitic worms are covered with shells that enable them to survive exposure to heat, cold, desiccation or digestive juices. The skate egg is covered by a tough leathery case that protects the developing embryo within. The eggs of most fish and amphibians are surrounded by a jelly coat that is of some value in protecting against mechanical shock. The cleidoic eggs of reptiles and birds are protected by tough leathery or calcareous shells. The developing embryos of birds take in oxygen and give off carbon dioxide through the hard shell. The eggs of fishes and amphibians are fairly large and contain yolk that supplies the nutrients for the developing embryo. These eggs are laid and typically develop in water, from which the oxygen, salts and water required for development are obtained. The embryos develop a pouchlike outgrowth of the digestive tract, the **yolk sac,** that grows around the yolk, elaborates enzymes to digest it and transports the products in its blood vessels to the rest of the embryo.

The eggs of reptiles and birds are usually laid on land and are enclosed within a shell that protects them from excessive drying. They have additional membranes, the **amnion, chorion** and **allantois,** for the protection and nutrition of the embryo. These membranes are sheets of tissue that grow out of the embryo itself. The amnion and chorion develop as folds of the body wall and surround the embryo; the allantois grows out of the digestive tract and functions along with the yolk sac in nutrition, excretion and respiration. Each of these membranes is composed of two germ layers in close apposition. The familiar white of the hen's egg is an extra store of protein and water to support the embryo until the time of hatching. Both the shells and the whites of eggs of reptiles and birds are secreted by glands located in the wall of the oviducts while the egg is passing down these tubes.

After being implanted in the lining of the uterus, the mammalian embryo continues to develop, obtaining its nourishment at first by enzymically breaking down the cells of the uterine lining immediately around it and later by extracting nutrients from the blood stream of the mother via the blood vessels of the placenta.

The mammalian embryo develops only from the **inner cell mass** of the blastocyst, that is, the cells lying on one side of the hollow ball originally implanted in the uterus. The other cells form the membranes that nourish and protect the developing mammal. Thus the problem of supplying the embryo with food during development has been solved in different ways by the various groups of vertebrates and invertebrates. The embryos of reptiles, birds and mammals develop in the liquid-filled amniotic cavity, in their own private pond within the shell or uterus. This arrangement permits the embryo to move around to some extent but protects it from bumps and shocks.

The chorion of reptiles and birds comes to lie next to the shell, and that of mammals is in contact with the maternal tissues of the uterus. The allantois, an outgrowth of the digestive tract between the amnion and chorion, largely fills the chorionic cavity. The allantois of the bird and reptile typically fuses with the chorion to form a compound membrane equipped with many blood vessels by means of which the embryo takes in oxygen, gives off carbon dioxide and excretes certain wastes.

The mammalian allantois is usually small and has no function as a membrane but supplies blood vessels to the **placenta,** the organ formed from chorion, allantois and maternal tissue. Finger-like projections, or villi, of the chorion grow into and become imbedded within the lining of the uterus. These villi, their blood vessels and the uterine tissues with which they are in contact constitute the placenta. This organ, in which fetal blood vessels come in close contact with maternal blood vessels, provides the developing mammalian fetus with nutrients and oxygen from the maternal blood and eliminates carbon dioxide and waste products into the mother's blood stream. The two blood streams do not mix at all; they are always separated by one or more tissues. However, substances can diffuse or be transported by some active process from mother to fetus or the reverse. The form of the placenta and the intimacy of the connection between maternal and fetal tissues varies from one species of mammal to another.

The growth of the embryo and the amnion brings the edges of the amniotic folds together to form a tube that encloses the yolk sac, the allantois, the two umbilical arteries and the umbilical vein that pass to the placenta. This tube, the **umbilical cord,** is composed of a peculiar jelly-like material unique to the cord.

The amnion, chorion and allantois, together with the egg shell or placenta, are adaptations that permit the embryos of the higher vertebrates to develop on land. They are a substitute for the pond or seawater in which the embryos of the lower vertebrates develop.

The property of reproduction, one of the outstanding characteristics of living things, involves a great many complex and interdependent processes. These include the elaboration of hormones that regulate the development of gonads, secondary sex structures and the production of gametes by the parents; behavior patterns that bring the parents together and have them release their gametes at such a time and in such a place as to make their fusion probable; the union of male and female pronuclei, followed by cleavage, gastrulation and morphogenesis; and devices for the care and protection of the developing young. Our descriptive knowledge of these phenomena is extensive, but our understanding of the fundamental mechanisms involved in each of these processes is still rudimentary. This is a fertile field for further investigation.

20.8 DEVELOPMENT AND IMPLANTATION OF MAMMALIAN BLASTOCYSTS

The first cleavage of the fertilized egg occurs about 30 hours after insemination, and the succeeding mitoses occur every 10 hours or so. By the time the developing ovum reaches the uterus, perhaps three or seven days after fertilization, it is a tight ball of some 32 cells called a **morula.** If the fertilized egg passes through the fallopian tube too rapidly and reaches the uterus prematurely, it cannot be implanted in the wall. One effect of the intrauterine device (IUD) may be to stimulate muscular contractions of the fallopian tube and uterus with the result that the fertilized ovum reaches the uterine cavity prematurely and dies before it can undergo implantation.

When the developing ovum reaches the uterine cavity, it begins to differentiate into a

blastocyst composed of an outer envelope of cells, the **trophoblast,** and an **inner cell mass,** a ball of cells at one pole within the trophoblast that is the precursor of the embryo. The cells of the inner cell mass have a more basophilic cytoplasm, reflecting the rapid synthesis of RNA occurring within them. The cavity of the blastocyst may be compared to the blastocoele, but the embryo is not a blastula, for its cells are differentiated into two types.

The cells of the inner cell mass differentiate further into a thin layer of flat cells, the **hypoblast,** located on the interior surface of the mass, adjacent to the blastocoele and representing presumptive endoderm (Fig. 20.8). The remaining cells of the inner cell mass become the **epiblast.** The cells of the hypoblast spread along the inner surface of the trophoblast and eventually surround the cavity of the blastocyst, forming a **yolk sac,** although the cavity is filled with fluid, not yolk.

As the hypoblast spreads out, the inner cell mass does also and becomes a disc-shaped

plate of cells similar to the blastodisc of avian and reptilian eggs. The blastodisc, composed of an epiblast of thick columnar cells and a hypoblast of more irregular flat cells, becomes delimited from the rest of the embryo. Gastrulation begins with the formation of a primitive streak and Hensen's node, in which cells migrate downward, laterally and anteriorly between the epiblast and hypoblast. Some of these migrating cells join the hypoblast and become presumptive endoderm; others, migrating laterally between the epiblast and hypoblast, become presumptive mesoderm. The amniotic cavity appears as a crevice between the cells of the inner cell mass, which then enlarges. The cavity of the blastocyst becomes filled with mesenchymal cells that are presumptive extraembryonic mesoderm. A secondary small yolk sac cavity appears just beneath the hypoblast. The bilaminar embryonic disc comes to lie as a plate between the two cavities, connected to the trophoblast at the posterior end by cells of the extraembryonic

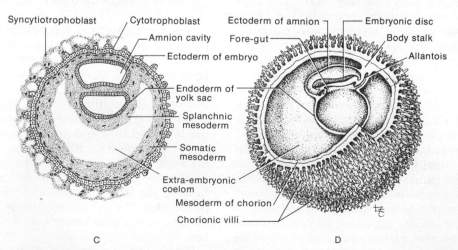

Figure 20.8 Diagrams of human embryos 10 (*A*) to 20 (*D*) days old, showing the formation of the amniotic and yolk sac cavities and the origin of the embryonic disc. (From Villee, C. A.: Biology. 7th ed. Philadelphia, W. B. Saunders Co., 1977.)

mesoderm, the body stalk or **allantoic stalk.** The nonfunctional endodermal part of the allantois, which develops as a tube from the yolk sac, is rudimentary and never reaches the trophoblast. Thus after two weeks of development, the human embryo is a flat two-layered disc of cells, about 250 micrometers across and connected by a stalk to the trophoblast.

The blastocyst implants in the endometrial lining of the uterus by secreting enzymes that erode the cells of the endometrium, permitting the imbedded blastocyst to establish close contact with the maternal blood stream. The cells of the trophoblast grow and divide rapidly. They and the adjacent cells of the uterine lining, the **decidua,** form the **placenta** and fetal membranes. The cells of the endometrium heal over the site of entry of the blastocyst so that it lies wholly within the endometrium and out of the uterine lumen.

The imbedding reaction of the endometrium can be elicited by pricking with a glass needle the uterine lining of a female suitably primed with estradiol and progesterone. This stimulus leads to "pseudopregnancy" and the uterus develops for a short period, just as though an embryo were present.

The trophoblast initially consists of two layers of cells: an inner **cytotrophoblast** composed of individual cells and an outer **syncytiotrophoblast** composed of a multinucleate syncytium. The trophoblastic cells digest and phagocytize materials in the endometrium that were stored prior to implantation. The trophoblast soon is bathed in and nourished by the maternal blood. Menstruation normally occurs about 14 days after ovulation in the nonpregnant woman; to prevent this, the embedded embryo must signal the maternal organism in some way. Because of the time required for traversing the oviduct and because the fertilized egg remains in the lumen of the uterus for some days, about 11 days elapse between ovulation and implantation. Thus the embryo has only a very few days to provide the signal to prevent menstruation. Fairly frequently, the signal does not arrive in time and menstruation sweeps out the fertilized egg. The woman has been pregnant in the sense that she had a fertilized egg in her reproductive tract, but she was never aware of it and menstruated at her usual time. One of the major contributions of the trophoblast is its secretion of **chorionic gonadotropin** by the cells of the syncytiotrophoblast. Chorionic gonadotropin has properties similar to those of luteinizing hormone and luteotrophic hormone of the pituitary. It prevents the corpus luteum from undergoing involution. The secretion of chorionic gonadotropin begins the day the trophoblast is imbedded in the endometrial lining.

This process of implantation has implicit in it two questions of general biological interest. Why does the trophoblast generally cease invading the endometrial lining once it has formed a connection with the maternal blood? Why doesn't it continue to invade as a group of cancer cells would? And why, since the cells of the trophoblast have the genotype of the developing fetus, a different one from that of the mother, do the maternal cells not react as though the trophoblast were a transplant and reject it, as an animal rejects a skin graft from another, genetically different member of the same species?

20.9 DEVELOPMENT OF THE GONADS AND REPRODUCTIVE PASSAGES

In many vertebrates a number of accessory structures have developed that facilitate the transfer of sperm from the male to the female reproductive tract and provide a place for the fertilized egg to develop. These structures have evolved either from or in close association with the urinary system, and the two are frequently referred to as the **urogenital system.**

The sex of an embryo is determined at the time of fertilization by the XX-XY mechanism, but early in development the embryo has the potential of differentiating into either a male or female, for the primordia of both duct systems are present (Fig. 20.9).

The paired paramesonephric or **mullerian ducts** develop just lateral to the wolffian ducts (Fig. 20.10) and are the precursors of the fallopian tubes, uterus and part of the vagina. The gonads develop as thickenings, **germinal ridges,** on the mesodermal epithelium lining the coelom. The ridge becomes composed of several layers of cells and protrudes into the coelomic cavity, suspended from the peritoneal wall by a double layer of peritoneum, the **mesorchium** or **mesovarium.** The cells in the germinal ridge are of two types: one is very similar to other cells of the peritoneal epithelium, and the second is a very different type, the primordial germ cells (Fig. 20.11). The latter are larger cells with large vesicular nuclei and clear cytoplasm that stains deeply with an alkaline phosphatase reaction. These primordial germ cells appear in the endodermal epithelium of the yolk sac in the vicinity of

Figure 20.9 A ventral view of the urogenital organs of the sexually indifferent stage of an embryo. (From Villee, C. A.: Biology. 7th ed. Philadelphia, W. B. Saunders Co., 1977.)

Figure 20.10 Diagrams showing stages in the development of the urinary and reproductive systems in humans. *A*, Early in the fifth week of development; *B*, early in the sixth week; *C*, Seventh week; *D*, eighth week. *E* ♂, Three months male; *E* ♀, three months female. (From Villee, C. A.: Biology. 7th ed. Philadelphia, W. B. Saunders Co., 1977.)

Figure 20.11 Diagram showing development of gonads in higher vertebrates. *A*, Germinal ridge stage; primordial germ cells partly embedded in epithelium of the ridge and located partly in the adjacent mesenchyme. *B*, Indifferent gonad; germ cells in the cortex and in primary sex cords. *C*, Gonad differentiating as testis; cortex reduced; germ cells in sex cords (future seminiferous tubules). *D*, Gonad differentiating as ovary; primary sex cords reduced; proliferating cortex contains the germ cells. (From Balinsky, B. I.: An Introduction to Embryology. 4th ed. Philadelphia, W. B. Saunders Company, 1975.)

the allantoic stalk. They migrate by ameboid motion into the mesenchyme and eventually reach the germinal ridge. There are fewer than 100 primordial germ cells when they are first evident in the yolk sac, but they multiply during their migration and some 5000 or more reach the germinal ridge. The mesenchyme of the germinal ridge is gradually replaced by finger-like accumulations of cells, the **sex cords,** which migrate in from the mesonephric cords or the epithelium of the germinal ridge. The epithelium of the germinal ridge forms the cortex of the gonad and the sex cords compose its medulla. Development to this stage, that of the indifferent gonad, is similar in ovary and testis.

In male embryos, the primordial germ cells migrate from the cortex into the primitive sex cords of the medulla, which become organized into the seminiferous tubules. The primordial germ cells become the spermatogonia, and the Sertoli cells develop from the sex cords. The medulla of the testis becomes its functional part, and the cortex is reduced and converted into a thin epithelial layer covering the coelomic surface of the testis.

In female embryos, the medulla is reduced, the primitive sex cords are resorbed and the central part of the gonad becomes filled with loose mesenchyme permeated by blood vessels. The primordial germ cells remain in the cortex, which increases in thickness. Masses of cortical cells form primary follicles that surround one or several primordial germ cells;

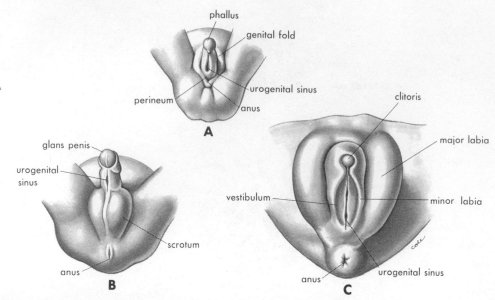

Figure 20.12 Development of external genitalia in human embryos. *A,* Indifferent stage in an embryo of 29 mm. (beginning third month). *B,* Male embryo of 145 mm. (16 weeks). *C,* Female embryo of 150 mm. (16 weeks). (After Balinsky, B. I.: An Introduction to Embryology. 3rd ed. Philadelphia, W. B. Saunders Company, 1970.)

these swell and become oögonia. The oögonia undergo rapid mitotic divisions, and after 20 weeks of development in the human fetus some 7,000,000 oögonia are present. Thereafter, mitosis ceases and no further ova are produced. The number is reduced by **atresia** (cell degeneration) to about 1,000,000 at the time of birth and to about 400,000 at the time of puberty. Only about 450 of these are destined to undergo ovulation during a woman's reproductive period; all the rest undergo atresia.

The seminiferous tubules in male embryos become connected to the **rete testis,** a system of thin tubules that develop from the dorsal part of the gonad and form connections with the adjoining mesonephric tubules (which become the **epididymis**) and, through them, with the wolffian duct, which becomes the **vas deferens.** This development proceeds under the stimulation of testosterone secreted by the interstitial cells of the developing testis. The testis seems to secrete a second hormone, possibly a peptide, that inhibits the development of the mullerian ducts so that no oviduct or uterus appears.

The wolffian ducts in female embryos degenerate when the mesonephros stops functioning as an excretory organ. The mullerian ducts open at the anterior end through a funnel-shaped **ostium** into the coelomic cavity, grow posteriorly and connect with the cloaca. The cloaca is subsequently divided by a **cloacal**

septum into a ventral **urogenital sinus** and a dorsal rectum. The latter opens to the exterior by the anus. The mullerian ducts are lateral to the wolffian ducts initially, but the posterior ends of the mullerian ducts later pass over the wolffian ducts. They fuse to form the uterus and cervix and combine with part of the urogenital sinus to form the vagina. The ventral part of the urogenital sinus into which the ureters open, together with the adjoining part of the allantoic stalk, expands to form the **urinary bladder.** The caudal part of the ventral urogenital sinus becomes a narrow tube, the **urethra.**

The external opening of the urogenital sinus is flanked on both sides by elongated thickenings, the **genital folds,** that meet anteriorly in front of the sinus and form a median outgrowth, the **genital tubercle,** the rudiment of the **phallus** (Fig. 20.12). From this sexually indifferent condition, under the stimulation of the sex hormones, development occurs toward the male or female condition. In the male, the rudiment of the phallus grows and becomes the penis. The genital folds fuse on the posterior surface of the penis and form a tube, the penile urethra, continuous with the urethra leading from the urinary bladder. In the female, the phallus grows slightly and becomes the clitoris, and the genital folds do not fuse but remain as the **labia minora.** The outer folds become the **labia majora,** comparable to the skin of the scrotum.

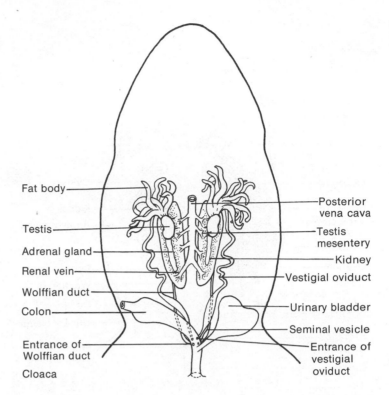

Figure 20.13 Ventral view of the urogenital system of a male frog. The vestigial oviduct shown in this figure tends to be absent in many male frogs.

Fat body

Testis

Adrenal gland

Renal vein

Wolffian duct

Colon

Entrance of
Wolffian duct

Cloaca

Posterior
vena cava

Testis
mesentery

Kidney

Vestigial oviduct

Urinary bladder

Seminal vesicle

Entrance of
vestigial
oviduct

20.10 VERTEBRATE GONADS

From a biological point of view, all the structures and processes that permit an animal to survive are of no avail unless the species can reproduce its kind. The general aspects of reproduction, including the production of gametes in the gonads, fertilization and the early development of the embryo, were considered in Chapter 5. At this time we shall deal more specifically with the reproductive organs of vertebrates and their role in reproduction.

Reproduction is sexual in vertebrates, and the sexes are separate in all but a few bony fishes. The **testes** (Figs. 20.13 and 20.14) are paired organs of modest size, each consisting of numerous highly coiled **seminiferous tubules** whose total length in man has been estimated at 250 meters! This provides an area large enough for the production of billions of sperm. As the sperm mature, they enter the lumen of the tubule and move toward the genital ducts. Finger-shaped **fat bodies** are associated with the gonads of frogs. They contain

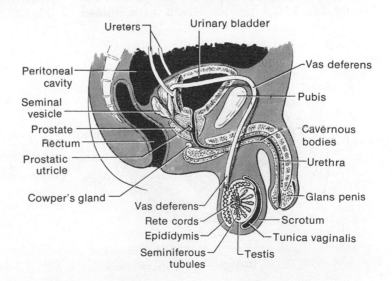

Figure 20.14 A diagrammatic sagittal section through the pelvic region of a man to show the genital organs. The prostatic utricle is a vestige of the oviduct that is present in the sexually indifferent stage of the embryo. (Modified after Turner.)

Ureters

Urinary bladder

Peritoneal
cavity

Seminal
vesicle

Prostate

Rectum

Prostatic
utricle

Cowper's gland

Vas deferens

Rete cords

Epididymis

Seminiferous
tubules

Vas deferens

Pubis

Cavernous
bodies

Urethra

Glans penis

Scrotum

Tunica vaginalis

Testis

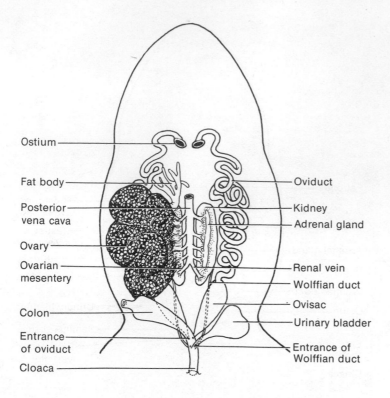

Ostium

Fat body

Posterior
vena cava

Ovary

Ovarian
mesentery

Colon

Entrance
of oviduct

Cloaca

Oviduct

Kidney

Adrenal gland

Renal vein

Wolffian duct

Ovisac

Urinary bladder

Entrance of
Wolffian duct

Figure 20.15 Ventral view of the uro-genital system of a female frog. The left ovary has been removed.

food reserves utilized during hibernation and in the spring breeding season when the animals become very active.

The ovaries of fishes or amphibians, which produce thousands or hundreds of eggs, fill much of the body cavity (Fig. 20.15). Higher vertebrates produce fewer eggs, for fertilization is internal and the eggs are deposited in situations where there is a greater chance for survival of the developing embryos. The ovaries of reptiles and birds are still large, and the eggs contain much yolk. Mammalian eggs contain very little yolk, and the ovaries are quite small, 2.5 cm. in length in a human

being (Fig. 20.16). The eggs are not free within the ovary, for each one is surrounded by a **follicle** of epithelial and connective tissue cells. When the egg is ripe, the follicle bursts and the egg is discharged into the coelom, a process known as **ovulation** (Fig. 20.18).

In nearly all vertebrates, the gonads are suspended by mesenteries in the abdominal cavity, and they remain there throughout life. But in the males of most mammals, the testes undergo a posterior migration, or descent, and move out of the main part of the abdominal cavity into a sac of skin known as the **scrotum** (Fig. 20.14). As they move into the scrotum,

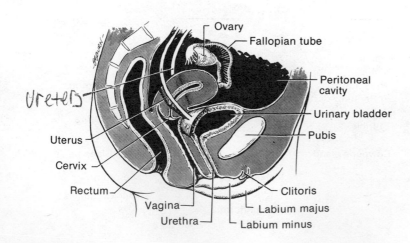

Ovary
Fallopian tube

Peritoneal
cavity

Urinary bladder

Pubis

Ureter

Uterus

Cervix

Rectum

Clitoris

Vagina   Labium majus

Urethra   Labium minus

Figure 20.16 A diagrammatic sagittal section through the pelvic region of a woman to show the genital organs. (Modified after Turner.)

they carry a coelomic sac, the **tunica vaginalis,** down with them so that, despite their superficial position, they lie within a portion of the coelom.

In most mammals, spermatogenesis does not go to completion unless the testes are descended. They remain descended in the majority of species, but in rabbits and rodents they are migratory—descending into the scrotum during the breeding season, withdrawing into the abdominal cavity at other times. Spermatogenesis, like other vital processes, can only occur within a limited temperature range. This range is exceeded by the temperature in the abdominal cavity but not by the temperature in the scrotum, which is approximately 4°C. lower. In order to test this hypothesis, Dr. Carl R. Moore of the University of Chicago confined the testes of rats to the abdominal cavity and found that spermatogenesis did not occur. Indeed, the seminiferous tubules underwent regression. He also insulated the scrotum of a ram in which the testes were descended. This raised the temperature, and again spermatogenesis did not occur. Apparently, during the evolution of endothermism in mammals, spermatogenesis did not become adapted to the higher body temperatures.

20.11 REPRODUCTIVE PASSAGES

Once the sperm and eggs have been produced, they must be removed from the body and brought together to form a zygote. This is a simple procedure in primitive vertebrates, such as cyclostomes. No reproductive ducts are present, and both eggs and sperm simply break out of the gonad into the coelom. Ciliary currents carry them to the posterior end of the coelom, where they are discharged through a pore into the cloaca. Fertilization and development occur outside the body.

Male Vertebrates. In the males of most fishes and amphibians (frog, Fig. 20.13), the rete cords become the **vasa efferentia,** which extend through the mesentery, carrying sperm from the seminiferous tubules in the testis to the anterior part of the kidney. The frog's kidney is an opisthonephros, but its anterior portion develops from the embryonic mesonephros. Sperm pass through kidney tubules into the wolffian duct, which carries both sperm and urine to the cloaca, although not at the same time.

Reptiles, birds and mammals (man, Fig. 20.14) have metanephric kidneys, and sperm pass from each testis to an **epididymis,** thence

out a **vas deferens** to the urethra. This, seemingly, is a different pattern, but it is not so different as it first appears. Rete cords connect the seminiferous tubules with the epididymis, and the epididymis represents those mesonephric tubules that are associated embryonically with the testis, together with a highly convoluted portion of the wolffian duct. The vas deferens represents the rest of the wolffian duct, and most of the urethra represents the ventral part of a divided cloaca. Man thus utilizes passages homologous to those of a frog.

Other differences between the male reproductive organs of lower and higher vertebrates are correlated with differences in mode of reproduction. Frogs mate in the water. The male grasps the female in an embrace termed amplexus (Fig. 20.17) and sprays sperm over the eggs as they are discharged. Fertilization is external. This mating procedure is perfectly satisfactory for species that mate in water, but the gametes are too delicate for external fertilization in the terrestrial environment. To accomplish internal fertilization, male mammals have a **penis** with which sperm are deposited in the female reproductive tract and a series of **accessory sex glands** that secrete a fluid in which the sperm are carried. The penis develops around the urethra and contains three **cavernous bodies** composed of spongy **erectile tissue.** Arterial dilatation coupled with a restriction of venous return causes the vascular spaces within the erectile tissue to become filled with blood during sexual excitement, making the penis turgid and effective as a copulatory organ. The accessory sex glands are a pair of **seminal vesicles** that connect with the distal end of the vasa deferentia, a **prostate gland** surrounding the urethra at the point of entrance of the vasa deferentia and a pair of **Cowper's glands** located more distally along the urethra.

Female Vertebrates. Eggs are removed from the coelom in most female vertebrates by a pair of oviducts, but the oviducts are modified for various modes of reproduction. Fishes and nearly all amphibians reproduce in the water. Most are oviparous, fertilization is external, and the eggs develop into larvae that can care for themselves. In the frog (Fig. 20.15), each oviduct is a simple coiled tube that extends from the anterior end of the coelom to the cloaca. The oviducts contain glandular cells that secrete layers of jelly about the eggs, and their lower ends are expanded for temporary storage of the eggs, but they are not otherwise specialized.

Figure 20.17   Leopard frogs in amplexus.

Fertilization is internal in reptiles, birds and mammals. They reproduce on the land, and the free larval stage has been replaced by the evolution of a cleidoic egg. Most reptiles and all birds are oviparous, and the eggs develop externally. The oviducal glands, which secrete the albumin and a shell around the egg, are more complex in the oviducts of reptiles than in those of amphibians and fishes, but in other respects the oviducts of reptiles have not changed greatly. Birds have lost the right oviduct along with the right ovary, but the remaining left oviduct is essentially similar to a reptilian one.

Most mammals and a few fishes and reptiles have become viviparous; they retain the fertilized egg within the reproductive tract until embryonic development is complete. The embryo receives its nutrients from the mother. The oviducts are modified accordingly. In the human female (Figs. 20.16 and 20.18), the **ostium** lies adjacent to the ovary and may even partially surround it. When ovulation occurs, the discharged eggs are close enough to the ostium to be easily carried into it by ciliary currents. The anterior portion of each oviduct is a narrow tube known as the **fallopian tube,** and eggs are carried down it by ciliary action and muscular contractions. The remainder of the primitive oviducts have fused with each other to form a thick-walled muscular **uterus** and part of the **vagina.** The terminal portions of the vagina and urethra develop from a further subdivision of the ventral part of the cloaca. The vagina is a tube specialized for the recep-

tion of the penis, and it is lined with stratified squamous epithelium. It is separated from the main body of the uterus, in which the embryo develops, by the sphincter-like neck of the uterus known as the **cervix.** The orifices of the vagina and urethra are flanked by paired folds of skin, the **labia minora** and the **labia majora.** A small bundle of sensitive erectile tissue, the **clitoris,** lies just in front of the labia minora. Structures comparable to these are present in the sexually indifferent stage of the embryo and develop into more conspicuous organs in the male. The labia majora are comparable to the scrotum; the labia minora and clitoris, to the penis. A pair of glands, homologous to Cowper's glands in the male, discharge a mucous secretion near the orifice of the vagina. A fold of skin, the **hymen,** partially occludes the opening of the vagina but is ruptured during the first intercourse.

20.12 MAMMALIAN REPRODUCTION

Fertilization.   During copulation, the sperm that have been stored primarily in the epididymis are ejaculated by the sudden contraction of muscles in and around the male ducts, and the accessory sex glands concurrently discharge their secretions. The seminal fluid that is deposited may contain as many as 400,000,000 sperm. Mucus in the seminal fluid serves as a conveyance for the sperm;

proteolytic enzymes break it down into a more watery fluid after the semen has been deposited in the vagina and permit the sperm to become highly motile. Fructose provides a source of energy, alkaline materials prevent the sperm from being killed by acids normally in the vagina, and certain fatty acids (prostaglandins) promote the contraction of the smooth muscle in the walls of the uterus and fallopian tubes.

Sperm move from the vagina through the uterus and up the fallopian tube in a little over one hour. How they do this is not entirely understood. They can swim, tadpole fashion, by the beating of the tail, but muscular contraction of the uterus and fallopian tubes and ciliary currents in the tubes must help considerably. Fertilization occurs in the upper part of the fallopian tube (Fig. 20.18), but the arrival of an egg and the sperm in this region need not coincide exactly. Sperm retain their fertilizing powers for a day or two, and the egg moves slowly down the fallopian tube, retaining its ability to be fertilized for about a day. The chance of fertilization is further increased in many species of mammals (but not in human beings) by the female coming into "heat" and receiving the male only near the time of ovulation. Ovulation, "heat," and changes in the uterine lining in preparation for the reception of a fertilized egg are controlled by an intricate endocrine mechanism, considered in Chapter 17.

Only one sperm fertilizes each egg, yet unless millions are discharged fertilization does not occur. One reason for this is that only a fraction of the sperm deposited in the vagina reach the upper part of the fallopian tube. The others are lost or destroyed along the way. When the egg enters the fallopian tube, it is still surrounded by a few of the follicle cells that encased the egg within the ovary (Fig. 3.27), and a sperm cannot penetrate the egg until these are dispersed. This requires the enzyme **hyaluronidase,** which can break down **hyaluronic acid,** a component of the intercellular cement. Hyaluronidase is believed to be produced by the sperm themselves, and large numbers are apparently necessary to produce enough of it. After the follicle cells are dispersed and one sperm has fertilized the egg, a **fertilization membrane** is raised from the surface of the egg, and additional sperm penetration is not possible.

Establishment of the Embryo in the Uterine Lining. The fertilized egg passes down the fallopian tube into the uterus, undergoing cleavage along the way. Energy for early development is supplied by the small amount of food within the egg and by secretions from glands in the uterine lining. About a week after fertilization, the embryo of most mammals penetrates the uterine lining, apparently by secreting digestive enzymes, and the lining folds over it. The placenta is formed by the union of the chorioallantoic membrane of the embryo with the uterine lining. The degree of union of fetal and maternal tissues differs among the various groups of mammals. In some, including the pig, the fetal chorion simply rests against the uterine lining. There is no breakdown of fetal or maternal tissue. In human beings, the union is more intimate, for microscopic **chorionic villi,** which contain the fetal capillaries, penetrate the uterine lining, break down maternal tissue and become bathed in maternal blood (Fig. 20.19). Since maternal blood and fetal chorion are in contact, this type of union is called a **hemochorial placenta.** It should be emphasized that, except for occasional breaks in the placental membrane, no blood is exchanged between fetus and mother. Most gases and waste products simply diffuse across the membrane. However, active transport may play a role in the transfer of certain substances; the fetal blood contains a higher concentration of amino acids, calcium and ascorbic acid than the maternal blood.

Birth. As the embryo develops, the uterus enlarges considerably to accommodate it. At the time of conception, the human uterus does not protrude far above the pubic symphysis (Fig. 20.16), but nine months later, when em-

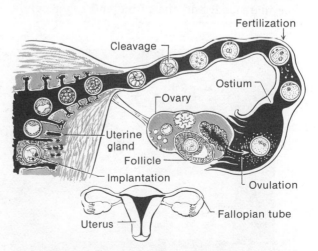

Figure 20.18 A diagram to show the path of an egg from the ovary to the uterus and the changes that occur en route. The last stage is about a week and one-half old. (Modified after Dickinson.)

Figure 20.19 A young human embryo surrounded by its extraembryonic membranes and lying within the uterus. Notice that the whole complex of embryo and membranes is embedded in the uterine lining. Villi are present all over the surface of the chorion at this stage, but only those on the side toward the uterine wall enlarge and contribute to the definitive placenta. (Modified after Patten.)

bryonic development has been completed, it extends up in the abdominal cavity nearly to the level of the breasts. During this enlargement, the individual muscle fibers in its wall increase in size, and additional muscle develops from undifferentiated cells in the uterine wall. The uterus becomes a powerful muscular organ ready to assume its role in childbirth, or **parturition.**

The factors that initiate birth are not entirely clear, but hormones secreted by the ovary and pituitary and perhaps by the fetal adrenal gland may play a role. Hormones produced by the pituitary, ovary and also the placenta have prepared the mother's body for the birth. The mammary glands have enlarged and are ready for milk production, the uterine musculature has increased, and the pubic and other pelvic

ligaments have relaxed so that the pelvic canal can enlarge slightly. Birth begins by a series of involuntary uterine contractions, "labor," that gradually increase in intensity and push the fetus, generally head first, against the cervix (Fig. 20.20). The cervix gradually dilates, but in human beings as much as 18 hours or more may be required to open the cervical canal completely at the first birth. The sac of amniotic fluid that surrounds the fetus acts as a wedge and also helps to open the cervix. The amnion normally ruptures during this process, and the amniotic fluid is discharged. When the head begins to move down the vagina, particularly strong uterine contractions set in, and the baby is born within a few minutes. A few more contractions of the uterus force most of the fetal blood from the placenta to the baby,

Figure 20.20 Photographs of two models from the Dickinson-Belski series on human birth. (From the *Birth Atlas*, published by Maternity Center Association, New York.)

and the umbilical cord can be cut and tied, although tying is unnecessary since contraction of the umbilical arteries would prevent excessive bleeding of the infant. Other mammals simply bite through the cord. Within a week the stump of the cord shrivels, drops off and leaves a scar known as the **navel.**

Uterine contractions continue for a while after birth, and the placenta and remaining extraembryonic membranes are expelled as the "afterbirth." Much of the uterine lining is lost at birth, for the human placenta is an intimate union of fetal membranes and maternal tissue. Uterine contractions prevent excessive bleeding at this time. Following the birth, the uterine lining is gradually reconstituted and the uterus decreases in size, though it does not become as small as it was originally.

ANNOTATED REFERENCES

Attention is again called to the general references on vertebrate organ systems cited at the end of Chapter 11.

Asdell, S. A.: Patterns of Mammalian Reproduction. 2nd ed. Ithaca, N.Y., Comstock Publishing Co., 1964. An important source book on differences in reproduction and reproductive cycles that occur in the various kinds of mammals, from the aardvark to the zebu.

Austin, C. R.: Fertilization. Englewood Cliffs, New Jersey, Prentice-Hall, Inc., 1965. A brief text summarizing this interesting aspect of zoology.

Balinsky, B. I.: An Introduction to Embryology. 4th ed. Philadelphia, W. B. Saunders Co., 1976. Contains excellent accounts of gametogenesis, fertilization and early development.

Diamond, M. (ed.): Perspectives in Reproduction and Sexual Behavior. Bloomington, Indiana, Indiana University Press, 1968. A collection of papers analyzing a wide variety of problems dealing with the endocrine control of reproduction and sexual behavior.

Masters, W. H. and V. F. Johnson: Human Sexual Response, Boston, Little, Brown and Co., 1966. A careful analysis of the physiological aspects of human mating.

Page, E. W., C. A. Villee, and D. B. Villee: Human Reproduction. 2nd ed. Philadelphia, W. B. Saunders Co., 1976. A general account of human reproduction ranging from the behavioral to the endocrine, genetic and biochemical aspects of the subject.

Pincus, G.: Control of conception by hormonal steroids. Science 153:493, 1966. A review of the benefits and potential dangers of oral contraceptives.

Rugh, R., and L. B. Shettles: From Conception to Birth: The Drama of Life's Beginnings. New York, Harper and Row Publishers, Inc., 1971. A detailed description with many illustrations, including some color photographs of the developing fetus, of human development from the moment of conception until birth. Changes in the mother and the processes of labor and delivery are discussed.

Turner, C. D. and J. T. Bagnara: General Endocrinology, 5th ed. Philadelphia, W. B. Saunders Co., 1971. Contains an excellent chapter on the biology of sex and reproduction.

Villee, C. A. (Ed.): The Control of Ovulation. London, Pergamon Press, 1961. Summarizes studies on the process of ovulation and its control in a series of papers prepared by leading students of the subject.

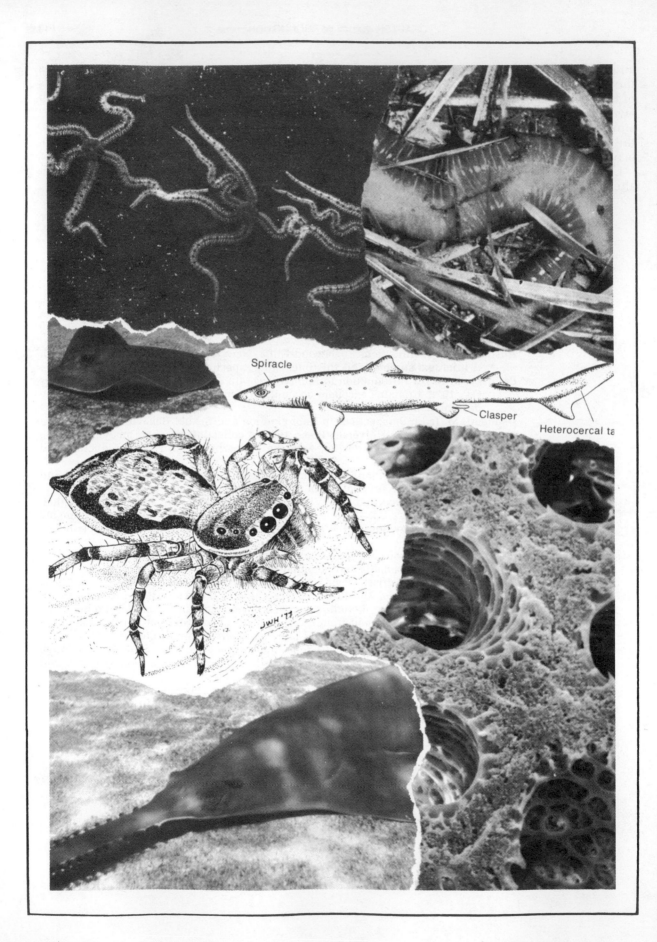

Spiracle

Clasper

Heterocercal ta

Part Four

ANIMAL GROUPS

THE CLASSIFICATION OF ANIMALS

Over 1,500,000 different kinds of organisms have been described, about two-thirds of which are multicellular, heterotrophic and motile and are termed animals, or **metazoans.** Some of these, such as sponges, corals and barnacles, cannot move about; such attached animals, however, have clearly evolved from motile ancestors and still retain the ability to move parts of the body. The various members of the Animal Kingdom are all related through some degree of common ancestry and have varying degrees of similarity in structure and function.

The great number and diversity of animals require some framework of order within which they can be viewed. That framework is a system of classification, and the branch of zoology concerned with classification is called **taxonomy** or **systematic zoology.**

21.1 THE SPECIES CONCEPT

The basic unit in the classification of animals and other organisms is the **species.** A species is a population of organisms that is distributed over certain geographical areas, its **range,** and occurs in certain kinds of **habitats** within that geographical area. For example, the giraffe, *Giraffa camelopardalis,* inhabits the open bush and savannah country of East Africa from the Sudan to South Africa and up to Angola to the west. All of the giraffes constitute the species population.

The science of genetics provides the theoretical basis of the species concept. All members of the same species population have a common gene pool and are capable of breeding with one another. The genetic constitution determines the ways in which one species differs from another. There will be structural, biochemical, physiological and behavioral differences. Those differences that can be used for easy recognition are said to be **diagnostic.** For example, it is easier to recognize a giraffe by its long neck than by its blood proteins. However, it is important to remember that species differ from one another in all the ways expressed by their genotypes and not just by a few conspicuous features.

Although the members of a species can breed with one another, they normally cannot breed with members of other species. Structural differences eliminate the possibility of mating between most species of animals. Closely related species with similar structure may be separated by different ranges, or if they occur within the same geographical area they may be isolated by different habitats, different breeding times, different behavior or by genetic incompatibility, which may result in lethal or sterile offspring. Species that do not interbreed in the wild because of such isolating mechanisms occasionally interbreed in captivity. The lion-tiger hybrids that occasionally appear in zoos are an example.

21.2 THE GROUPING OF SPECIES

A species population may be thought to have not only a dimension in space (range) but also a dimension in time. The population extends backward in time, merging with other species populations much like the branches of a tree (Fig. 21.1). Thus, species have varying degrees of evolutionary relationships with one another depending upon the distance to the point of population mergence or common ancestry. The evolutionary relationship of organisms is known as **phylogeny,** and this relationship provides a theoretical basis for the grouping, or classification, of organisms. Closely related species are grouped together within a **genus.** Related genera are grouped together within a **family;** related families within **orders;** orders within **classes;** and classes within **phyla** (Fig. 21.2). Theoretically at least, any category in this hierarchy of classification should constitute a **monophyletic** grouping; i.e., all the species or subgroups within any group should have a common ancestry.

The evolutionary history of only a very few species is known directly from fossil records, and species relationships must therefore be determined indirectly. Related species should have similar gene pools and thus similar phenotypes. The degree of similarity between species, reflected by their similar gene pools, can be used as a measure of the closeness of their relationship. Although the degree of similarity can be studied at the biochemical level, morphological similarities are more commonly used to determine relationships because morphological characteristics are generally simpler to work with.

As a category, the species is relatively easy to recognize and delimit. If the individuals sampled from a population of organisms are fairly similar and their diagnostic characteristics do not intergrade with those of other related species, the population constitutes a species. Higher categories, however, are much more subjective and difficult to define. For example, in Figure 21.1, should species A–F be placed within a single genus or do they represent three genera, A–B, C–E and F? (All of the groupings would be monophyletic.) The same type of problem exists in determining the limits of family, order, class and phylum. In general, the greater the gap (i.e., the extent of dissimilarity) between species, the more likely a separate grouping is justified.

The approach to classification just described assumes that species are actual biological entities that can be determined and studied. This is known as a *phyletic* viewpoint and is the approach taken by many zoologists. Other taxonomists, however, hold a *phenetic* view of classification, meaning that they believe that species are only artificial abstractions devised by man. They claim that a classification based on phylogeny will never be attained and that artificiality is unavoidable and should be simply accepted. It is important to note that phyleticists and pheneticists differ only in their interpretation of what classification means, not in the method of classification. Both would agree

Figure 21.1 Diagrammatic representation of the evolutionary relationships of six (hypothetical) monophyletic species. Circular cross sections of the branches represent the species at the present time; junctions of branches represent points of common ancestry. Species G is extinct.

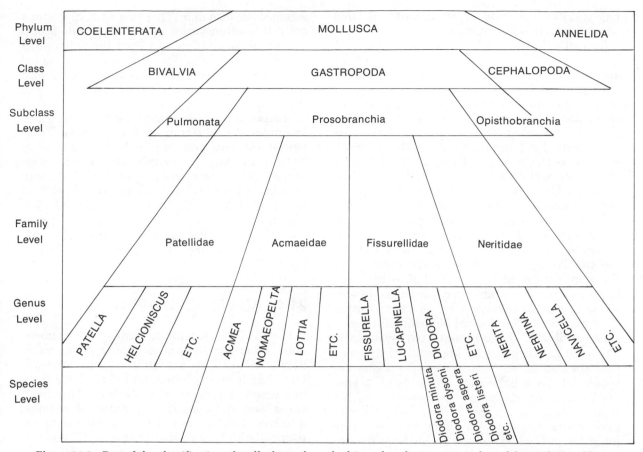

Figure 21.2   Part of the classification of mollusks to show the hierarchy of categories. (Adapted from Blackwelder.)

that the best classification is one based upon all discernible characteristics.

Over the past decade there has developed an approach to taxonomy, called *numerical taxonomy*, that emphasizes the use of quantitative methods. Systems of classification or relationships are based upon similarity between taxonomic units as reflected by the statistical analysis of a large number of characters, all weighted equally. The computer is an important tool in numerical taxonomy.

21.3 TAXONOMIC NOMENCLATURE

The naming of species and higher taxonomic categories is another aspect of taxonomy. The scientific name of a species is composed of words usually derived from Latin or Greek, of which the first is the **generic** name and the second is the **species** name. Together the generic and specific names constitute the scientific name, which is therefore a binomial; i.e., composed of two words. All species in

the same genus have the same generic name. As an illustration, the scientific names of the categories to which the familiar box turtles belong are listed below. Note that the family name ends in *idae* and that the species name is always italicized.

Phylum Chordata
　Subphylum Vertebrata
　　Class Reptilia
　　　Order Chelonia—Turtles and tortoises
　　　　Family Emydidae—Freshwater and marsh turtles
　　　　　Genus *Terrapene*—Box turtles
　　　　　Terrapene carolina—Common box turtle
　　　　　Terrapene ornata—Ornate box turtle

Scientific names are derived from many sources—descriptive adjectives, mythological figures, names of persons and geographical place names. However, there is no necessity to know the derivation of a name. The name has significance only as an identifying label or tag, similar to a person's name.

Law of Priority. All taxonomic names must be unique; i.e., the same name cannot be applied to more than one species or higher taxonomic category. When duplication occurs, the first species or category to which the name was applied takes priority in the claim to the name. The application of the **Law of Priority** for animals begins with the tenth edition of the *Systema Naturae*, a catalog of the plants and animals known to the Swedish biologist Carolus Linnaeus in 1758. The *Systema Naturae* represents the first uniform application of the present system of binomial nomenclature, although the system was not invented by Linnaeus. Detailed rules governing the construction of scientific names and the application of the Law of Priority have been formulated in an international code and are continually reviewed by an International Commission on Zoological Nomenclature.

Role of the Taxonomist. No systematic zoologist could possibly know all of the million species of animals, so most workers are specialists in a particular group. Such a specialist can recognize a new species in his group when one is encountered. A description of the new species along with its name must be published before it can be officially numbered among the known species. A number of specimens of the new species are designated as **type specimens** and are deposited in the collections of a museum for future reference.

An important task of many taxonomists is the identification of animal species. Identification, mistakenly termed classification by many people, is the determination of the species to which specimens belong. Ecological and other types of research may require the collection of large numbers of animals. If the collector is unfamiliar with the groups involved, he may ask a specialist in that group for help with identification.

The largest collections of animals for taxonomic purposes are located in museums, which may have educational as well as research facilities. Research collections are rarely on public display, but they are available to zoologists who visit the museum or who wish to borrow specimens. Museum specimens are usually obtained from scientific collecting expeditions, or they are given to the museum by zoologists following the completion of a research project. The collections are catalogued and maintained by a staff of museum taxonomists, called curators, who normally are also engaged in research. The size of a museum's collection often varies greatly from one group of animals to another and from the collections of other museums. The differences in holdings simply reflect the sources of material and interests of the curators in the past history of the museum.

Museums may be a part of a university, for example, the Museum of Comparative Zoology at Harvard and the Peabody Museum at Yale; they may be private institutions supported by endowments, like the American Museum of Natural History in New York and the Philadelphia Academy of Natural Sciences; or they may be supported by state or federal government, as is true of the National Museum of Natural History of the Smithsonian Institution.

21.4 ADAPTIVE DIVERSITY

The Animal Kingdom displays a tremendous adaptive diversity. Every conceivable habitat within the ancestral oceanic environment has been exploited, and there have been numerous invasions of organisms into fresh water and onto land. The entry of animals into new habitats posed new problems of existence, which were met with modification of old structures and new systems for new functions. Since different animal groups invaded the same habitats, the evolutionary history of animals is filled with convergent adaptations. The varied success of the invading groups is reflected in part by the number of individuals and the number of species.

Adaptation to new conditions rarely has resulted in the loss of all of a species' ancestral characteristics. Some are modified to the demands of the new environment; some remain unchanged and are said to be **primitive.** Thus every species possesses some primitive and some specialized features. A primitive species is one that has retained many ancestral characteristics. But the term primitive is useful only in a comparative sense. One can speak of a primitive animal, a primitive vertebrate, a primitive mammal or a primitive man. But a primitive man is not a primitive mammal, nor is a primitive mammal a primitive vertebrate. Animals may be primitive because they lack certain structures, but they do not lack the structure because they are primitive. Hydras do not lack an anus, gills and blood vessels because they are primitive but because the hydra's small size and mode of existence makes them unnecessary. The ab-

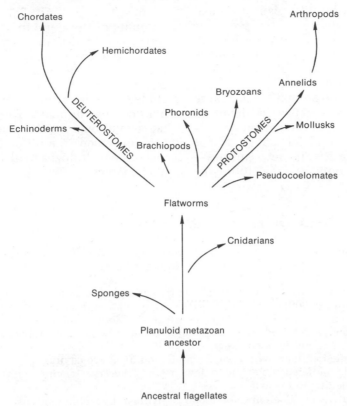

Figure 21.3 A phylogenetic tree of the Animal Kingdom. This is but one of many that might be constructed, to reflect different ideas about the evolutionary relationships of different groups of animals.

sence of such structures probably reflects a primitive condition, since it is unlikely that hydras evolved from animals that possessed them. However, structures may be lost as well as gained through evolution.

The evolutionary history of the Animal Kingdom can be depicted as having a treelike form (a **phylogenetic tree**), the diverging lines dividing trunks and branches representing diverging evolutionary lines (Fig. 21.3). Those groups of animals that diverged from the main line early in the history of the Animal Kingdom (i.e., from the lower part of the phylogenetic tree) are often described as being "lower animals." Others that arose later at a higher level of the tree are often said to be "higher animals." However, these terms are often carelessly used and seem to imply that all animals and animal structures fit into a hierarchy of simple to complex. Even the "lowest" animals, such as sponges, may display certain specialized and unique features in addition to their primitive characteristics.

21.5 HOW TO STUDY ANIMAL GROUPS

The study of animal groups, or any group of organisms for that matter, can be a tedious and painful task if it amounts to nothing more than committing to memory a list of characteristics of one group after another. A much more useful and interesting approach is one that emphasizes relationships and adaptations. This is the approach that will be utilized in the following chapters. Two questions should be continually asked as these chapters are studied: (1) How is the structure and physiology of the group correlated with its mode of existence or the environmental demands of the habitat in which it lives? (2) How does the structure and physiology of the group reflect the group's evolutionary origin and its relationship to other groups? These questions can provide a meaningful framework into which facts about animal groups can be fitted. Moreover, with a little knowledge of the structural ground plan of animal phyla and an understanding of the problems posed by different modes of existence and different environments, it is possible to make intelligent guesses about possible adaptations that might be encountered.

There are about thirty animal phyla, more than we can possibly discuss in detail in this book. Our discussion of animal groups will be limited to the major phyla. Minor phyla—those with relatively few species— will be mentioned only in the following synopsis of the Animal Kingdom. Detailed accounts of these groups can be found in the references cited at the end of the chapter.

Protozoa are often treated as animals, for many are able to move and are heterotrophic. But protozoa are unicellular organisms, excluded by definition from the Metazoa, or multicellular animals. They have had a separate evolutionary history, and their complexity has developed in a different way from that of metazoans. Protozoa will be discussed in the next chapter.

SYNOPSIS OF THE PHYLA OF METAZOAN ANIMALS

The following diagnoses are limited to distinguishing characteristics. The approximate number of species described to date is indicated in parentheses.

PARAZOA. Many-celled animals with poorly defined tissues and no organs.

Phylum Porifera (10,000). Sponges. Sessile; no anterior end; some primitively radially symmetrical, but most are irregular. Mouth and digestive cavity absent; body organized about a system of water canals and chambers. Marine, but a few found in fresh water.

EUMETAZOA. Many-celled animals with organs, mouth and digestive cavity.

RADIATA. Tentaculate radiate animals with few organs. Digestive cavity, with mouth the principal opening to the exterior.

Phylum Cnidaria, or Coelenterata (9000). Hydras, hydroids, jellyfish, sea anemones and corals. Free-swimming, or sessile, with tentacles surrounding mouth. Specialized cells bearing stinging organoids called nematocysts. Solitary or colonial. Marine, with a few in fresh water.

Phylum Ctenophora (90). Comb jellies. Free-swimming; biradiate, with two tentacles and eight longitudinal rows of ciliary combs (membranelles). Marine.

BILATERIA. Bilateral animals.

Protostomes. Cleavage spiral and determinate; mouth arising from or near blastopore.

ACOELOMATES. Area between body wall and internal organs filled with mesenchyme.

Phylum Platyhelminthes (12,700). Flatworms. Body dorsoventrally flattened; digestive cavity (when present) with a single opening, the mouth. The turbellarians are free-living, a few terrestrial. Trematodes and cestodes are parasitic.

Phylum Mesozoa (50). An enigmatic group of minute parasites of marine invertebrates. No organs, and body composed of few cells.

Phylum Rhynchocoela, or Nemertina (650). Nemerteans. Long, dorsoventrally flattened body. Digestive cavity with mouth and anus. Marine, with a few terrestrial and in fresh water.

Phylum Gnathostomulida (43). Gnathostomulids. Minute wormlike animals. Body covered by a single layer of epithelial cells, each of which bears a single cilium. Anterior end with bristle-like sensory cilia. Mouth cavity with a pair of cuticular jaws. Marine.

PSEUDOCOELOMATES. Animals in which the blastocoel persists, forming a body cavity. Digestive tract with mouth and anus. Body usually covered with a cuticle.

Phylum Rotifera (1500). Rotifers. Anterior end bearing a ciliated crown; posterior end tapering to a foot. Pharynx containing movable cuticular pieces. Microscopic. Largely in fresh water, some marine, some inhabitants of mosses.

Phylum Gastrotricha (175). Gastrotrichs. Slightly elongated body with flattened ciliated ventral surface. Few to many adhesive tubes present; cuticle commonly ornamented. Microscopic. Marine and in fresh water.

Phylum Kinorhyncha (64). Kinorhynchs. Slightly elongated body. Cuticle segmented and bearing recurved spines. Spiny, retractile anterior end. Less than 1 mm. in length. Marine.

Phylum Nematoda (10,000). Roundworms. Slender cylindrical worms with tapered anterior and posterior ends. Cuticle often ornamented. Free-living species usually only a few millimeters or less in length. Marine, fresh water, terrestrial and parasitic forms.

Phylum Nematomorpha (230). Hairworms. Extremely long threadlike bodies. Adults free-living in damp soil, in fresh water and a few marine. Juveniles parasitic.

Phylum Acanthocephala (500). Acanthocephalans. Small wormlike endo-parasites of arthropods. Anterior retractile proboscis bearing recurved spines.

Phylum Entoprocta (60). Entoprocts. Body attached by a stalk. Mouth and anus surrounded by a tentacular crown. Mostly marine.

SCHIZOCOELOUS COELOMATES. Body cavity a coelom; or, if a body cavity is absent, the coelom has been lost. Digestive tract with mouth and anus.

Phylum Priapulida (8). Priapulids. Cucumber-shaped marine animals, with a large anterior proboscis. Body surface covered with spines and tubercles. Peritoneum of coelom greatly reduced.

Phylum Sipunculida (250). Sipunculids. Cylindrical marine worms. Retractable anterior end, bearing lobes or tentacles around mouth.

Phylum Mollusca (65,000). Snails, chitons, clams, squids and octopods. Ventral surface modified in the form of a muscular foot, having various shapes; dorsal and lateral surfaces of body modified as a shell-secreting mantle, although shell may be reduced or absent. Marine, in fresh water, and terrestrial.

Phylum Echiurida (60). Echiurids. Cylindrical marine worms, with a flattened nonretractile proboscis. Trunk with a large pair of ventral setae.

Phylum Annelida (8700). Segmented worms—polychaetes, earthworms and leeches. Body wormlike and metameric. A large longitudinal ventral nerve cord. Marine, in fresh water and terrestrial.

Phylum Pogonophora (80). Marine, deepwater animals, with a long body housed within a chitinous tube. Anterior end of body bearing from one to many long tentacles. Digestive tract absent.

Phylum Tardigrada (350). Water bears. Microscopic segmented animals. Short cylindrical body bearing four pairs of stubby legs terminating in claws. Fresh water and terrestrial in lichens and mosses; few marine.

Phylum Onychophora (65). Onychophorans. Terrestrial, segmented, worm-like animals, with an anterior pair of antennae and many pairs of short conical legs terminating in claws. Body covered by a thin cuticle.

Phylum Arthropoda (923,000). Crabs, shrimp, mites, ticks, scorpions, spiders and insects. Body metameric with jointed appendages and encased within a chitinous exoskeleton. Vestigial coelom. Marine, in fresh water, terrestrial or parasitic.

Phylum Pentastomida (70). Wormlike endoparasites of vertebrates. Anterior end of body with two pairs of leglike projections terminating in claws and a median snoutlike projection bearing the mouth.

LOPHOPHORATE COELOMATES. Mouth surrounded or partially surrounded by a crown of hollow tentacles (a lophophore).

Phylum Phoronida (70). Phoronids. Marine, wormlike animals with the body housed within a chitinous tube.

Phylum Bryozoa (4000). Bryozoans. Colonial, sessile; the body usually housed within a chitinous or chitinous-calcareous exoskeleton. Mostly marine, a few in fresh water.

Phylum Brachiopoda (260). Lamp shells. Body attached by a stalk and enclosed within two unequal dorsoventrally oriented calcareous shells. Marine.

Deuterostomes, or Enterocoelous Coelomates. Cleavage radial and indeterminate; mouth arising some distance (anteriorly) from blastopore. Mesoderm and coelom develop primitively by enterocoelic pouching of the primitive gut.

Phylum Chaetognatha (50). Arrow worms. Marine planktonic animals with dart-shaped bodies bearing fins. Anterior end with grasping spines flanking a ventral preoral chamber.

Phylum Echinodermata (5300).   Starfish, sea urchins, sand dollars and sea cucumbers. Secondarily pentamerous radial symmetry. Most existing forms free-moving. Body wall containing calcareous ossicles usually bearing projecting spines. A part of the coelom modified into a system of water canals; external projections used in feeding and locomotion. Marine.

Phylum Hemichordata (80).   Acorn worms. Body divided into proboscis, collar and trunk. Anterior part of trunk perforated with varying number of pairs of pharyngeal clefts. Marine.

Phylum Chordata (39,000).   Pharyngeal clefts, notochord and dorsal hollow nerve cord present at some time in life history. Marine, in fresh water and terrestrial.

 Subphylum Urochordata **(1600).**   Sea squirts, or tunicates. Sessile, non-metameric invertebrate chordates enclosed within a cellulose tunic. Notochord and nerve cord present only in larva. Solitary and colonial. Marine.

 Subphylum Cephalochordata **(1).**   Amphioxus. Fishlike metameric invertebrate chordates.

 Subphylum Vertebrata **(37,000).**   The vertebrates—fish, amphibians, reptiles, birds and mammals. Metameric. Trunk supported by a series of cartilaginous or bony skeletal pieces (vertebrae) surrounding or replacing notochord in the adult.

ANNOTATED REFERENCES

Blackwelder, R. E.: Taxonomy. New York, John Wiley and Sons, Inc., 1967. A general account of the theories and methods of zoological taxonomy.

Mayr, E.: Principles of Systematic Zoology. New York, McGraw-Hill Book Co., 1969. A general account of zoological taxonomy including the text of the International Code of Zoological Nomenclature.

Ross, H. H.: Biological Systematics. Reading, Mass., Addison-Wesley Publ. Co., 1974. A short text on systematics with emphasis on evolution.

Sneath, P. H. A. and R. R. Sokal: Numerical Taxonomy. San Francisco, W. H. Freeman and Co., 1973. An elaboration of the methods of numerical taxonomy.

The following invertebrate zoology texts and reference works contain general discussions of all of the invertebrate phyla, including the minor phyla omitted from this text. References that deal solely with particular groups will be listed at the end of each chapter of the following survey.

Barnes, R. D.: Invertebrate Zoology. 3rd ed. Philadelphia, W. B. Saunders Co., 1974.

Bayer, F. M. and H. B. Owre: The Free-Living Lower Invertebrates. New York, Macmillan, 1968.

Hickman, C. P.: Biology of the Invertebrates. St. Louis, The C. V. Mosby Co., 1967.

Hyman, L. H.: The Invertebrates. Six volumes. New York, McGraw-Hill Book Co., 1940–1967. Vol. I (1940): Protozoa through Ctenophora; Vol. II (1951): Platyhelminthes and Rhynchocoela; Vol. III (1951): Acanthocephala, Aschelminthes, and Entoprocta; Vol IV (1955): Echinodermata; Vol. V (1959): Smaller Coelomate Groups; Vol VI (1967): Mollusca, Part 1.

Kaestner, A.: Invertebrate Zoology. Three volumes. New York, Wiley—Interscience, 1967–69.

Meglitsch, P. A.: Invertebrate Zoology, 2nd ed. Oxford, Oxford University Press, 1972.

THE PROTOZOANS

Protozoans are unicellular, **eukaryotic** (nucleated) organisms. The animals that are the subject of the greater part of this book are **metazoans**-multicellular, motile **heterotrophs.** But motility and heterotrophic nutrition are also characteristic of many protozoans. For these reasons they are commonly considered to be animals and traditionally are the starting point for surveys of the Animal Kingdom.

Whether one considers protozoans to be "animals" is simply a matter of definition, and there are many biologists who would prefer to restrict the concept of animals to metazoans. However, there are other reasons that make it useful to examine protozoans before surveying the more typical animals that comprise the Metazoa. Metazoans undoubtedly evolved from some group of protozoans, most probably the flagellated protozoans, and the metazoan characteristics of motility and heterotrophic nutrition are probably a protozoan inheritance. Despite these similarities, however, protozoans and metazoans are very different from each other, especially in the way they have become complex.

The evolution of complexity in multicellular plants and animals has occurred through a division of labor among the cells, certain cells becoming specialized for certain functions. Complexity in unicellular organisms has evolved through the specialization of different parts of the cell, for although protozoans are single cells, they are also complete organisms. The cell must perform not just certain functions but also must retain the ability to perform all of the functions demanded of an organism.

Specialized parts of the cytoplasm are called **organelles.** Organelles are found in the cells of all eukaryotic organisms, unicellular or mul-

ticellular, but are developed most elaborately in unicellular organisms. In many groups the cells are fantastically complex. *Paramecium,* for example, far exceeds in complexity any cell found within the human body.

22.1 CLASSIFICATION

The classification of protozoans poses a number of problems. The principal subgroups are distinctive and easily defined, but they share few characteristics with each other. They are similar chiefly in being **unicellular.** Traditionally the protozoans have been considered a phylum, but many biologists today believe that their unicellular condition alone is too general a feature with which to define a phylum. They advocate raising the various subphyla to the level of phyla so that what was a single phylum becomes a group of phyla. The name Protozoa refers to the group and can be used like Metazoa to designate a subkingdom of the Animal Kingdom, if one wishes to consider protozoans to be animals.

Many modern biologists favor classifications that divide living organisms into a number of kingdoms rather than the traditional two kingdoms embracing plants and animals Of the many schemes proposed, the popular protistan concept raises the old protozoan classes to the rank of phyla and places them within the kingdom Protista. The members of this kingdom, which includes the protozoans and algae, are characterized by unicellular or multicellular bodies composes of nucleated cells. Gametes are produced by single cells, not by multicellular gonads, and development never involves an embryo. The **Metazoa** (animals),

Embryophyta (plants), **Fungi** (slime molds, water molds, mildews, yeasts, mushrooms and other fungi) and **Monera** (bacteria and blue-green algae) constitute separate kingdoms.

We will consider the Protozoa to be a group of phyla but will include them in our survey because of their animal-like nature and traditional inclusion within the Animal Kingdom.

The flagellated protozoans are generally considered to be primitive and ancestral to the other protozoans, but let's break with tradition and start our examination of protozoan organisms with the ciliates. The ciliates are the most homogeneous, the most specialized and the most animal-like protozoans and will serve especially well to contrast with the metazoans we will be studying.

22.2 PHYLUM CILIATA (Ciliophora)

Members of the phylum Ciliata are characterized by possessing at some time in their life history ciliary organelles for locomotion and feeding. Most ciliates are the size of the familiar *Paramecium*, although a few species are large enough to be barely visible to the naked eye. Ciliates are widespread in the sea and in fresh water, and a few species are commensals within the gut of vertebrates.

There is also always a distinct anterior end, and primitive species are radially symmetrical (Fig. 22.1). However, most ciliates are asymmetrical. The constant shape is maintained by a **pellicle**, a living outer layer of denser cytoplasm containing the peripheral and surface organelles. Studies with electron microscopy have revealed that this region is extremely complex (Fig. 22.3). The ciliate peripheral organelles, like many other cellular organelles, are largely derived from folding of the plasma membrane.

The cilia, the most conspicuous surface organelles, arise from subsurface basal granules, or **kinetosomes** (Fig. 22.3). The kinetosomes are connected together in longitudinal rows by fibrils, and all the fibrils and kinetosomes of a row make up a **kinety.** The kineties constitute the subsurface, or infraciliature.

The layer of cilia covering the general body surface is called the **somatic ciliature.** Primitively longitudinal rows of somatic cilia cover the entire surface of the body, but in many species the somatic ciliature is reduced to girdles, tufts and bristles or is lacking altogether (Fig. 22.1). However, even those species with no somatic cilia possess an infraciliature persisting from cilia of earlier developmental stages.

The pellicle of many ciliates, including *Paramecium*, contains large vesicles (alveoli) around the cilia (Fig. 22.3). Although lying beneath the cell membrane, they contribute to the surface configuration characteristic of the species. Other common pellicular organelles of ciliates are bottle- or rod-shaped bodies, termed **trichocysts** (Fig. 22.3). In some species these can be discharged and transformed into fine threads. Their functional significance varies according to species. They may serve in anchoring the organism during feeding, in defense or in prey capture. Other pellicular organelles found in many ciliates include vesicles that secrete mucoid substances, perhaps used in the formation of protective encasements (cysts) during adverse environmental conditions.

Locomotion.   The majority of ciliates swim by ciliary propulsion. Ciliary beat is planar (i.e., more or less in one plane), with an effective stroke followed by a recovery stroke. When the effective stroke is directed backward, the ciliate moves forward, but the beat can be reversed, permitting backward movement (Fig. 22.4). In normal forward swimming, the entire body of the organism spirals because the cilia beat somewhat obliquely to the long axis of the body. Beating occurs in synchronized waves down the length of the body. It is presumed that the subsurface kinety system is in some way responsible for the synchronization and direction of the beat.

Some of the most specialized ciliates, members of the order **Hypotricha,** have the somatic ciliature restricted to isolated tufts of fused cilia located in rows or groups on one side of the body. Fused cilia have more limited movements than do separate cilia, and commonly the undulations are restricted to one plane, producing a simple lashing motion forward and backward.

Some ciliates are sessile. *Stentor*, a trumpet-shaped form, is attached by the tapered end and shortens by the contraction of pellicular fibrils similar to muscle myofibrils. *Stentor* can also release itself and swim about. *Vorticella* has a bell-shaped body connected to the substratum by a long stalk that contains a contractile fibril (Fig. 22.1). The ciliate retracts, coiling the stalk like a spring, and extends by a sudden release and popping movement. Some related forms are colonial and the individuals of the colony are connected together by a common stalk. Some other sessile ciliates are "tubicolous," living in shells composed of foreign material cemented together (Fig. 22.1).

Figure 22.1   Ciliates. *A, Prorodon,* a primitive ciliate; *B, Vorticella; C, Pleuronema; D, Stentor; E, Tintinnopsis,* a marine ciliate with a test composed of foreign particles. (*A* and *E* after Faure-Fremiet from Corliss; *B* after Pierson from Kudo; *C* after Noland from Corliss; *D* after Tartar from Manwell.)

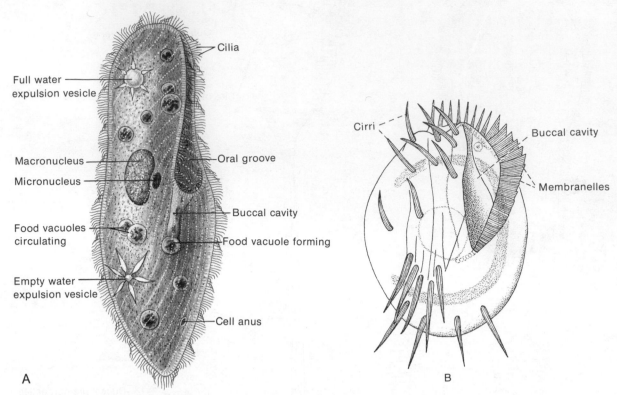

Figure 22.2   *A, Paramecium* (generalized drawing combining features of several species); *B, Euplotes.* (*B* after Pierson from Kudo.)

Figure 22.3   Pellicular system in *Paramecium.* (After Ehret and Powers from Corliss.)

Nutrition. Ciliates possess a mouth and it is this structure, along with motility, that gives them such an animal character. However, the ciliate mouth, or **cytostome,** is an organelle, an aperture within a single cell. It opens into a short canal, the **cytopharynx,** which leads into the interior, more fluid cytoplasm, or endoplasm. Primitively, the mouth is located at the anterior end (Fig. 22.1A), but in most ciliates it has been displaced posteriorly to varying degrees (Fig. 22.2). Species in which the oral apparatus is located at the anterior end are usually carnivores, and this is probably the primitive mode of feeding in ciliates. They prey upon rotifers, gastrotrichs and other protozoans, including other ciliates, and the mouth can be opened to a great diameter to ingest prey (Fig. 22.5). The prey or contents of the prey pass into a food vacuole, which forms within the endoplasm at the end of the cytopharynx.

The higher ciliates are suspension feeders and possess a more complex buccal apparatus. Typically, the mouth lies at the bottom of a preoral chamber, the buccal cavity, which contains compound ciliary organelles (Figs. 22.2 and 22.5). These organelles, which constitute the buccal as opposed to the somatic ciliature, consist of two types: undulating membranes

Figure 22.4 Two waves of movement passing down a series of cilia. The cilia at either end are in the effective stroke; those in the middle are in the recovery stroke. Dashed arrow indicates direction of effective stroke; solid arrow direction of propulsion.

and membranelles. An undulating membrane is a long row of fused cilia that forms a movable membrane (Fig. 22.1C). A membranelle is a short row of fused cilia, which forms a plate (Fig. 22.2B). Membranelles are typically arranged in fairly large numbers one behind another. The function of the buccal ciliature is to produce a feeding current and drive suspended food particles—detritus, bacteria and so forth—into the cytopharynx.

In the sessile *Vorticella* and the trumpet-shaped *Stentor*, the ciliary organelles wind around the distal end of the animal and spiral down into a pit on one side (Fig. 22.1B and D). In the familiar *Paramecium*, the buccal cavity, cytostome and cytopharynx are located at the

Figure 22.5 *A,* Buccal organelles of *Paramecium; B, Didinium,* a raptorial ciliate, attacking a *Paramecium.* (*A* after Yusa from Manwell; *B* after Mast from Dogiel.)

posterior end of a lateral oral groove. The somatic cilia within the oral groove produce a feeding current that sweeps from front to back, and the compound ciliary organelles of the funnel drive food particles down into the mouth (Figs. 22.2 and 22.5).

In these suspension feeders, the food particles collect at the end of the cytopharynx within a food vacuole, which gradually increases in size like a soap bubble at the end of a pipe (Figs. 22.2 and 22.5). When the vacuole reaches a certain size it breaks free from the cytopharynx and circulates within the fluid endoplasm. Like other organelles food vacuoles are composed of lipid bilayer membranes.

Studies on digestion in ciliates have demonstrated that the contents of the vacuole first become increasingly acid. The acid phase is followed by an alkaline phase, which results from the secretion of enzymes within an alkaline medium into the vacuole. This phase coincides with an increase in the size of the vacuole. Can you advance a theory as to why the food vacuole increases in size as digestion proceeds? The digestive enzymes include all of the major classes—**amylases, lipases, nucleases** and **proteases.** Products of digestion are absorbed from the vacuole into the surrounding cytoplasm. Electron microscopy has revealed that during digestion and absorption there is extensive pinocytosis of the vacuole membrane into the exterior surrounding cytoplasm. The small food vacuole containing residual undigestible waste eventually connects with the **cytoproct,** a fixed cell anus in the pellicle (Fig. 22.2A). The contents of the vacuole are discharged, and the vacuole itself disappears.

Although only a few ciliates are parasitic, many are commensals. Among the most interesting commensals are the highly specialized species living within the stomach of some hoofed mammals. These are capable of digesting cellulose to acetate.

Water Balance. Because of their microscopic size, ciliates have no special structures for gas exchange or excretion. They do have one or more organelles for water balance, called **contractile vacuoles** or **water expulsion vesicles,** which have fixed positions within the body. *Paramecium,* for example, has a water expulsion vesicle at each end (Fig. 22.2). In most ciliates the organelle is composed of a ring of radiating tubules that deliver droplets of water centrally (Fig. 22.2). These droplets coalesce into a single vesicle, which gradually increases in size. On reaching a definitive size, the vacuole rapidly discharges through a canal in the pellicle. Then a new vacuole forms.

One might expect that only freshwater ciliates would have water expulsion vesicles, but they are also found in marine species. They apparently serve in marine forms to rid the body of water taken in during feeding. That this is true is indicated by the fact that the water expulsion vesicles of marine ciliates pulsate at a slower rate than do those in fresh water and that in forms such as *Paramecium,* which have one anterior and one posterior vesicle, the posterior organelle, which is near the site of food intake, pulsates at a more rapid rate than does the anterior vesicle.

Reproduction. Ciliates are distinguished from other protozoans in possessing two types of nuclei. One type, the **macronucleus,** is large and governs the nonreproductive functions of the cell. It is highly polyploid and its shape varies greatly: in *Paramecium* it is bean-shaped (Fig. 22.2); in *Vorticella* it is U-shaped; and in *Stentor* it has the form of a string of beads (Fig. 22.1).

The **micronuclei,** which range in number from one to 80, are small, round and typically located in the vicinity of the macronucleus. They are diploid and function in reproduction.

Asexual. Ciliates reproduce asexually by means of transverse fission, the fission plane cutting across the longitudinal rows of kinetosomes (Fig. 22.6). Mitotic spindle fibers form for each micronucleus, but the macronucleus usually divides by constriction. Regeneration

Figure 22.6   Transverse fission in *Paramecium.*

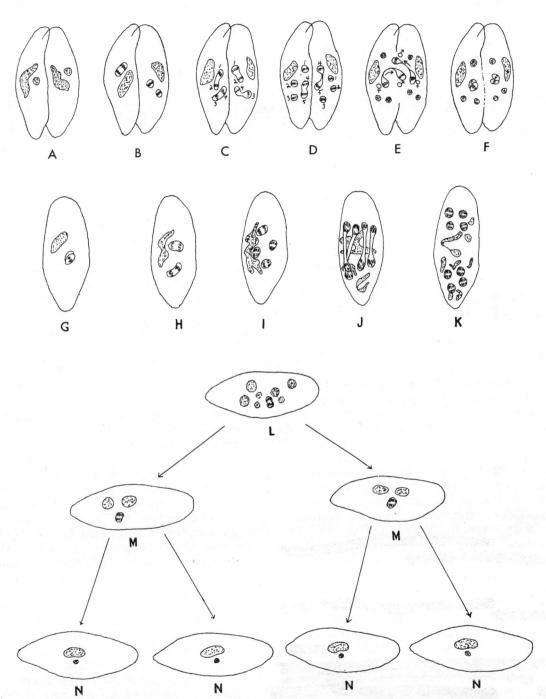

Figure 22.7 Sexual reproduction in *Paramecium caudatum*. *A* to *F*, Conjugation. *B* to *D*, Micronuclei undergo three divisions, the first two of which are meiotic. *E*, "Male" micronuclei are exchanged. *F*, They fuse with the stationary micronucleus of the opposite conjugant. *G*, Exconjugant with macronucleus and zygote micronucleus; other micronuclei have been resorbed. *H* to *K*, Three divisions of zygote, forming eight micronuclei; old macronucleus is resorbed. *L*, Four micronuclei form macronuclei; three are resorbed. *L* and *M*, Remaining micronucleus divides twice in course of two cytoplasmic divisions. Resulting daughter cells each receive one of the four macronuclei in *L. N*, Normal nuclear condition restored. (Modified after Calkins from Wichterman.)

TABLE 22.1   BREEDING RELATIONS IN *PARAMECIUM BURSARIA*

Variety →		I				II								III			
Mating	**Type**	I	II	III	IV	I	II	III	IV	V	VI	VII	VIII	I	II	III	IV
I	I	—	+	+	+	—	—	—	—	—	—	—	—	—	—	—	—
	II	+	—	+	+	—	—	—	—	—	—	—	—	—	—	—	—
	III	+	+	—	+	—	—	—	—	—	—	—	—	—	—	—	—
	IV	+	+	+	—	—	—	—	—	—	—	—	—	—	—	—	—
II	I	—	—	—	—	—	+	+	+	+	+	+	+	—	—	—	—
	II	—	—	—	—	+	—	+	+	+	+	+	+	—	—	—	—
	III	—	—	—	—	+	+	—	+	+	+	+	+	—	—	—	—
	IV	—	—	—	—	+	+	+	—	+	+	+	+	—	—	—	—
	V	—	—	—	—	+	+	+	+	—	+	+	+	—	—	—	—
	VI	—	—	—	—	+	+	+	+	+	—	+	+	—	—	—	—
	VII	—	—	—	—	+	+	+	+	+	+	—	+	—	—	—	—
	VIII	—	—	—	—	+	+	+	+	+	+	+	—	—	—	—	—
III	I	—	—	—	—	—	—	—	—	—	—	—	—	—	+	+	+
	II	—	—	—	—	—	—	—	—	—	—	—	—	+	—	+	+
	III	—	—	—	—	—	—	—	—	—	—	—	—	+	+	—	+
	IV	—	—	—	—	—	—	—	—	—	—	—	—	+	+	+	—

A plus sign indicates the possibility of mating, and a minus sign indicates that mating does not occur. No mating type will mate with another individual of its own type, but it will mate with any other mating type of its variety. Three varieties are found in the United States, and no mating type of one variety will mate with any mating type of any other variety. (From Sonneborn, 1947.)

of organelles lost in fission is a complex process and depends largely upon replication of existing structures. Of great importance in the reformation process are the kinetosomes. In many species, these bodies replicate early in the division process, and each half receives an equal number. In some highly specialized ciliates, such as the hypotrichs, all of the organelles are resorbed during division and new organelles are formed from a small number of persisting "germinal" kinetosomes.

Sexual. Sexual reproduction involves a process of **conjugation** and an exchange of nuclear material. Two individuals, called **conjugants,** meet, probably by random contact, adhere and their cytoplasm fuses in the region of adhesion (Fig. 22.7). The macronucleus is not involved in conjugation and is resorbed during the course of the process. All of the micronuclei undergo two meiotic divisions. Then all but one of these haploid nuclei disappear. The remaining nucleus divides mitotically to form two haploid nuclei: a stationary and a wandering nucleus (Fig. 22.7). The wandering nucleus of each conjugant migrates to the opposite conjugant and fuses with the stationary nucleus to form a **synkaryon,** or **"zygote,"** nucleus. The conjugants separate, and the zygote nucleus undergoes a number of mitotic divisions to restore the number of micronuclei characteristic of the species. The

macronucleus develops from one of these micronuclei.

Restoration of nuclear number is commonly associated with cytosomal divisions. The process is highly variable and is best illustrated with an example. The adult *Paramecium caudatum* has one macronucleus and one micronucleus. Following conjugation the "zygote" nucleus divides three times to produce eight nuclei (Fig. 22.7). Four become macronuclei and four micronuclei. However, three of the micronuclei degenerate. The remaining micronucleus undergoes two mitotic divisions accompanying two cytosomal divisions (Fig. 22.7). Each of the resulting four daughter individuals obtains one micronucleus from these divisions and inherits one of the four macronuclei formed prior to fission.

In some species of ciliates, including *Paramecium,* the individuals of the population belong to genetically determined mating types. In *P. bursaria,* for example, there are a number of mating types within each variety, and conjugation can only occur between individuals of different mating types (Table 22.1). Apparently adhesion of the cytoplasm will not take place between individuals of the same mating type.

Note that conjugation itself does not result in any increase in the number of individuals but does provide for an exchange of genetic material within the population. All of the individuals that result from the asexual reproduction of an original parent without intervening conjugation are genetically identical and constitute a **clone.** There have been numerous studies to determine how frequently conjugation must occur. Some species can undergo several hundred asexual divisions and perhaps can continue this indefinitely. In other species, there does seem to be an absolute requirement for conjugation after a given period of time or the clone will perish.

Most ciliates are capable of forming cysts under adverse environmental conditions. The body becomes encased within a protective secreted covering; there is some loss of water; and the metabolic rate is sharply reduced. Encystment is very important for the survival of ciliates when pools and ditches dry up and for dispersion by wind and on the muddy feet of aquatic birds and mammals.

22.3 PHYLUM MASTIGOPHORA

The phylum Mastigophora, which literally means *whip-bearer*, includes those protozoan organisms that possess one to many flagella as locomotor organelles. However, other than possession of flagella, the members of the phylum are very heterogeneous, which probably indicates that the phylum is an assemblage of only distantly related groups.

The flagellates are generally believed to be the oldest of the eukaryotic organisms and the ancestors, directly or indirectly, of the other major groups of organisms—ciliates, algae, fungi, plants, metazoan animals and others. The ciliates, for example, probably derive from some group of multinucleated mastigophorans that possessed many flagella.

A flagellum has an ultrastructure similar to that of a cilium but is longer (Fig. 22.8). In contrast to the lashing planar beat of a cilium,

Figure 22.8 *Euglena. A*, A lateral view of *Euglena viridis. B*, A diagram showing successive positions in the spiral swimming pattern of *Euglena*. The position of the pigment spot shows the rotation that occurs.

the beating of a flagellum describes spiral or planar undulations. When the undulations pass from the base to the tip of the flagellum, a pushing force is created like that of a boat propellor. When the undulations pass from the tip to the base, a pulling force is produced like that of an airplane propellor. Whether the flagellate body is pulled or pushed also depends upon where the flagellum is attached or how it is held in relation to the body. In *Euglena*, for example, the flagellum is attached anteriorly but trails to the side; it pushes the body forward. Many flagellates tend to move in a spiral path, as do ciliates (Fig. 22.8).

Phytoflagellates. The phylum is composed of two groups: the phytoflagellates and the zooflagellates. The phytoflagellates are plantlike marine and freshwater forms. Most possess chlorophyll and exhibit autotrophic nutrition; the ten orders are classified with different groups of algae by algologists. The body is commonly asymmetrical, and the structure varies greatly. They possess one or two flagella, which are generally carried at the anterior end. A nonliving cell wall or envelope is often present. Descriptions of a few of the more common groups will illustrate the diversity of the phytoflagellates.

The mostly freshwater **euglenids** have rather spindle-shaped bodies covered by a living pellicle (Fig. 22.8). The pellicle is flexible, and many euglenids are capable of bending or ameboid movement. One or two flagella arise from a deep recess of the anterior end, and in the same vicinity there is a light sensitive "eye" spot containing a red carotene pigment. One or two water expulsion vesicles are also present.

Euglenids are typically green, but there are some colorless species, such as *Peranema*. This common little flagellate swims against the bottom, and its large, conspicuous, forward-projecting flagellum undergoes lateral undulations in propelling the organism forward. Another smaller flagellum trails. *Peranema* is predaceous and feeds on other protozoans, including *Euglena*. The prey is ingested through an anterior cytostome, which can be greatly distended in swallowing (Fig. 22.9).

The members of the Volvocida are small green phytoflagellates possessing two flagella and a cellulose wall. Although some species, such as *Chlamydomonas*, are solitary (Fig. 22.10), the group as a whole tends to be colonial, forming platelike or spherical aggregations. The most highly developed colonies are the large hollow spheres of *Volvox* (Fig. 22.10). The entire colony moves, rotating through the water, by the synchronized beating of the flagella.

The brownish or yellowish dinoflagellates are common members of marine plankton, although there are also freshwater species. The various species of dinoflagellates have a variety of shapes, but most are more or less oval or

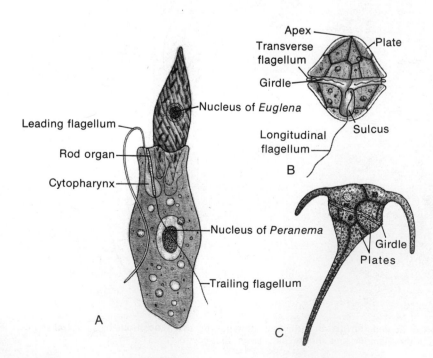

Figure 22.9 *A*, A *Peranema* swallowing a *Euglena*; *B*, a freshwater dinoflagellate, *Glenodinium cinctum*; *C*, *Ceratium*, a genus of common marine dinoflagellates. (*A* after Chen; *B* after Pennak; *C* after Jorgenson from Hyman.)

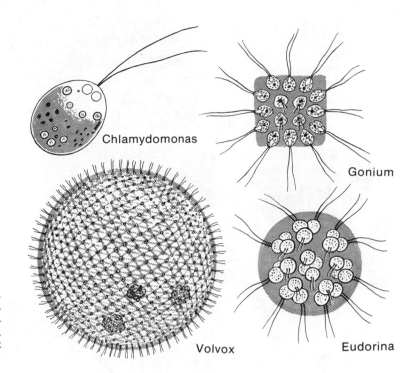

Figure 22.10 Examples of the order Volvocida showing solitary form (*Chlamydomonas*), simple colonies (*Gonium, Eudorina*) and a complex colony (*Volvox*). Colonial forms are embedded in a matrix of transparent jelly.

Chlamydomonas

Gonium

Volvox

Eudorina

shaped like a top (Fig. 22.9*B*). Many are encased within a nonliving cellulose wall, which is often divided into plates and sculptured (Fig. 22.9*A* and *B*). One flagellum is located within a transverse groove that rings the body, and the other is located posteriorly in a longitudinal groove (Fig. 22.9*B*).

Many species of dinoflagellates, as well as some other phytoflagellates, lose their flagella and pass into a nonmotile vegetative, or palmella, phase. The symbiotic algae of many corals, scyphomedusae and other invertebrates are dinoflagellates in a palmella state.

The red tides that periodically plague the Gulf coast and other parts of the world are commonly caused by species of dinoflagellates. Optimum environmental conditions result in tremendous population growths, or blooms, of certain species. They occur in such great densities that their metabolic wastes poison fish and other marine animals.

Zooflagellates. The zooflagellates lack chlorophyll and are heterotrophic. Most species are either commensal or parasitic, and they constitute a much smaller part of the flagellate fauna than the phytoflagellates. They are undoubtedly a polyphyletic assemblage that evolved from different groups of phytoflagellates through loss of chlorophyll and autotrophic nutrition. The following examples will serve to illustrate their diversity.

The choanoflagellates are a strange little group of free-living marine and freshwater forms with bodies resembling the collar cell of sponges (Fig. 22.11). Most are colonial and attached, but one species has the collar cells embedded in a gelatinous mass (Fig. 22.11). Some zoologists have suggested that the sponges may have evolved from the choanoflagellates.

Trypanosoma is the most notable parasitic zooflagellate. It lives in the blood stream of vertebrate hosts, particularly mammals, and is transmitted by an insect host, usually tsetse flies. In the blood stream the organism has an elongate, flattened body. A single flagellum arises posteriorly and extends anteriorly and laterally along the side of the body as an undulating membrane (Fig. 22.12). When ingested by the bloodsucking insect host, the trypanosome losses its flagellum, curls up and becomes an intracellular parasite of the insect gut cells. Some get into the buccal and salivary glands and are passed back into the vertebrate host during another feeding.

Although most species of *Trypanosoma* are probably not pathogenic, a few African species cause serious diseases in domestic mammals and humans. In humans, the flagellate invades the fluid of the cerebrospinal canal, causing the fatal disease known as **sleeping sickness.** The development of drugs that attack the flag-

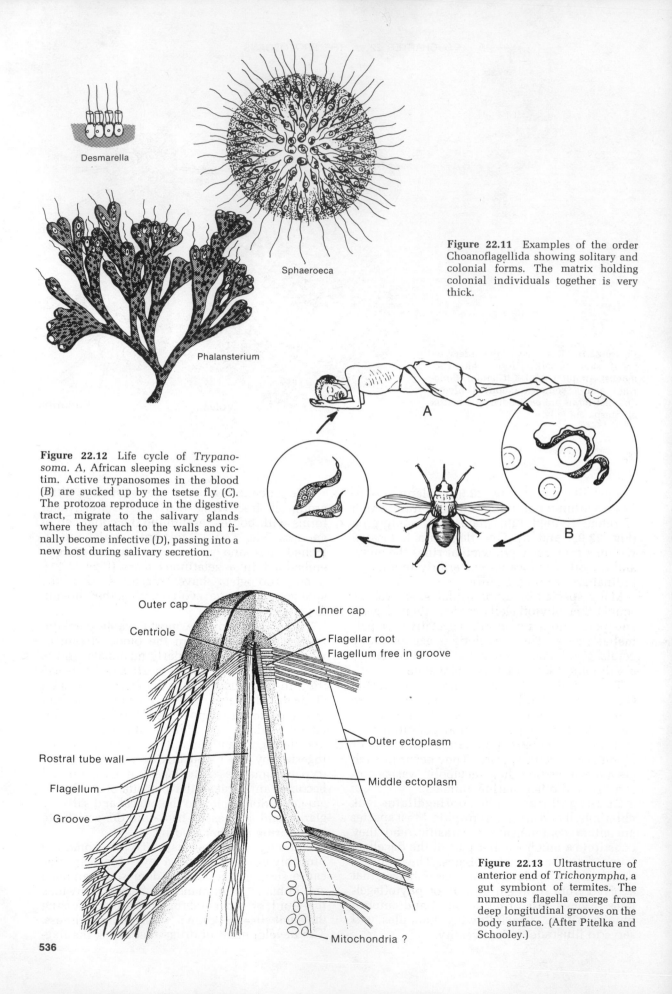

Desmarella

Sphaeroeca

Phalansterium

Figure 22.11 Examples of the order Choanoflagellida showing solitary and colonial forms. The matrix holding colonial individuals together is very thick.

Figure 22.12 Life cycle of *Trypanosoma*. *A*, African sleeping sickness victim. Active trypanosomes in the blood (*B*) are sucked up by the tsetse fly (*C*). The protozoa reproduce in the digestive tract, migrate to the salivary glands where they attach to the walls and finally become infective (*D*), passing into a new host during salivary secretion.

Outer cap

Inner cap

Centriole

Flagellar root

Flagellum free in groove

Outer ectoplasm

Middle ectoplasm

Rostral tube wall

Flagellum

Groove

Mitochondria ?

Figure 22.13 Ultrastructure of anterior end of *Trichonympha*, a gut symbiont of termites. The numerous flagella emerge from deep longitudinal grooves on the body surface. (After Pitelka and Schooley.)

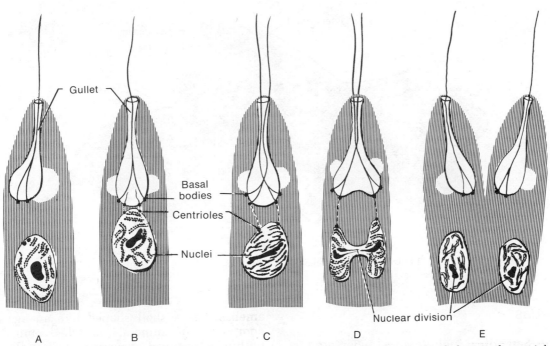

Figure 22.14 Details of asexual reproduction in *Euglena*. *A*, The centriole has already divided. *B*, Each centriole produces a new basal body and flagellum. The nucleus is in prophase, and the contractile vacuole is double. *C*, The old pair of flagellar roots separate and fuse with the new roots. *D*, Mitosis proceeds, and the gullet begins to divide. *E*, Anterior end dividing following duplication of organelles. (Redrawn from Ratcliffe, 1927.)

ellate and of pesticides to control the tsetse fly has reduced but not eradicated the disease.

The most complex flagellates, indeed among the most complex protozoans, are the gut symbionts of termites and wood roaches. The body is commonly saclike, and the anterior end bears a cap and a kinetosome complex from which spring the large number of flagella (Fig. 22.13). These flagellates engulf at the posterior end bits of wood ingested by the host. The flagellates have enzymes, β-glucosidases, to transform the cellulose to glucose within their food vacuoles. The product, glucose, is shared with the host, which is incapable of digesting cellulose. The insect obtains a new gut fauna following each molt by licking other individuals, by rectal feeding or by eating fecal cysts.

Reproduction. Asexual reproduction is typically by binary fission, but in contrast to ciliates, flagellates divide longitudinally to produce two more or less equal halves (Fig. 22.14). The kinetosomes and other organelles duplicate or re-form before or after actual division.

Not much is known about sexual reproduction in most flagellates. The volvocids are an exception. This group beautifully illustrates the probable origin of **gametes,** an evolution

that may have occurred more than once among flagellates. In *Chlamydomonas* the motile **isogametes** are identical and only a little smaller than the adults. In *Platydorina* the gametes are flagellate and motile, but one is much larger than the other. *Volvox* possesses typical **heterogametes.** The sperm are small, flagellate and motile, and the egg is large and nonmotile., As in most algae and plants, meiosis in the volvocids is post-zygotic; i.e., it occurs *after* fertilization, not before, as in metazoans and ciliates.

22.4 PHYLUM SARCODINA

The phylum Sarcodina contains the amebas and other protozoans that possess flowing extensions of the body known as **pseudopodia.** Pseudopodia are used in feeding in all Sarcodina; in some, the pseudopodia have a locomotor function as well. Sarcodina possess fewer organelles than ciliates and flagellates and are therefore relatively simple in structure. However, skeletal structures have reached a degree of development that is equalled by few other protozoans.

Figure 22.15   An ameba. The animal is flowing to the right.

Amebas

The Sarcodina include four major groups. The most familiar are the freshwater and marine amebas, some of which are naked and some of which are enclosed within a shell (Fig. 22.16). Amebas have straplike or large blunt pseudopodia, used in locomotion and in feeding (Fig. 22.15).

In the shelled species, the shell is secreted or composed of mineral particles cemented together. For example, the freshwater *Arcella* possesses a brownish dome-shaped shell composed of a secreted keratinoid material (Fig. 22.16). *Difflugia*, another genus of freshwater

amebas, has a shell shaped like an egg and composed of mineral particles embedded within a secreted cementing material (Fig. 22.16). A large opening in the shell permits extension of the body and the pseudopodia.

Locomotion.   The pseudopodia and other parts of the body are bounded by a thick layer of gelatinous ectoplasm. Ectoplasm and the more fluid interior endoplasm are different molecular states of cytoplasm, and ameboid flow, which is described on page 252, involves a rapid change of the molecular organization at the pseudopodial tip.

The factors initiating flow are still uncertain. Several pseudopodia may form simulta-

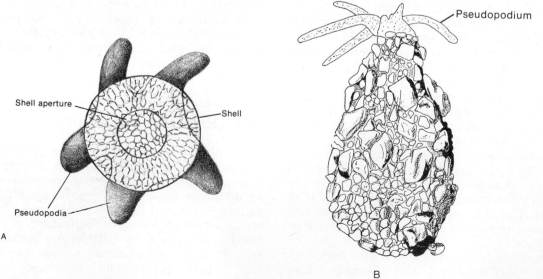

Figure 22.16   Shelled amebas. *A, Arcella vulgaris*, dorsal view; *B, Difflugia oblonga* with test of mineral particles. (*B* after Deflandre.)

neously, but one eventually dominates. Also, pseudopodial movement occurs in more than the single plane usually depicted.

Nutrition. Amebas feed on diatoms, algae, rotifers and other protozoans. The food is surrounded by pseudopodia and eventually enclosed within a food vacuole along with some water (Fig. 22.17). Digestion proceeds as in ciliates, passing through an acid and then an alkaline phase. The vacuole containing undigestible residue ruptures at the posterior end.

A number of commensal and parasitic amebas inhabit the gut of different animals, including humans. The commensal species, such as *Entamoeba coli* of humans, feed on bacteria and intestinal debris. The parasitic species invade the intestinal tissue. *Entamoeba histolytica* parasitizes the large intestine of man and is the cause of amebic dysentery, or amebiasis (Fig. 22.18). Both commensal and parasitic species leave the host as cysts in the feces, and reinfestation occurs through the mouth.

A water expulsion vesicle is present in freshwater amebas but is absent from marine forms. This is true of other Sarcodina besides the amebas. Unlike that of ciliates, the vesicle does not form from radiating tubules nor does it have a fixed position within the body.

Foraminiferans

Foraminiferans are marine Sarcodina that secrete a chambered calcareous shell (Fig.

Figure 22.17 An ameba capturing a large flagellate. The flagellate hits the side of the ameba and slips into the crevasse at the base of a pseudopod. The ectoplasm of this region erupts and rapidly engulfs the prey. Stages shown are at about one second intervals.

22.19). The shell is perforated by many small openings through which the pseudopodia extend. Foram pseudopodia are delicate anastomosing strands that form a foodtrapping net over the shell surface. The pseudopodial cytoplasm is adhesive, and any small organism swimming into the net is held and slowly surrounded by cytoplasm. Digestion occurs in vacuoles outside of the shell, and the products of digestion diffuse to the interior through the shell perforations.

Although there are some planktonic forams, such as *Globigerina* (Fig. 22.19), the majority are benthic and move slowly over the bottom by means of pseudopodia.

In all forams the shell is composed at first of a single chamber with an opening at one end. When the foram outgrows the first chamber, it secretes another one. This process continues, producing a multichambered shell that is occupied entirely by one individual. The shape

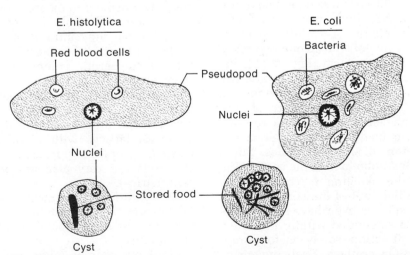

Figure 22.18 A parasite, *Entamoeba histolytica* (left), and a commensal, *Entamoeba coli* (right), of the human large intestine. Active amebas above, cysts below, that are passed in the feces and can infect new individuals.

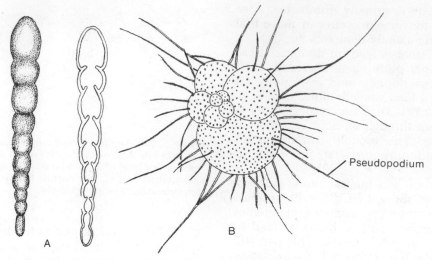

Figure 22.19   Foraminiferans. *A*, Shell of *Rheophax nodulosa* (entire, and in section); *B*, living *Globigerina*. (After Cushman from Calvez.)

and arrangement of the chambers vary according to species (Fig. 22.19).

Accumulations of foram shells are an important constituent of ocean bottom sediments and are sometimes termed "globigerina ooze" because of the large number of individuals of the genus *Globigerina*. Forams have an extensive fossil record that begins in the Cambrian, and there are great limestone deposits composed largely of foram shells. The paleontology of these organisms is especially important in the search for oil deposits.

Heliozoans

Heliozoans are mostly freshwater Sarcodina. Their spherical bodies are free or attached by a stalk (Fig. 22.20). The body is composed of a central cytoplasmic core, or medulla, that contains one to many nuclei, and an outer cortex of highly vacuolated cytoplasm (Fig. 22.20). Radiating from the cortex are many needle-like pseudopodia, called axopods, from which the name Heliozoa, sun animals, is derived. The axopods contain a central cytoplasmic rod that extends from the medulla. Many heliozoans possess a skeleton of siliceous scales, tubes, spheres or needles embedded in the cortex (Fig. 22.20). Where the skeleton is composed of needles, they radiate out of the cortex like the axopods. Some species even have sand grains or living diatoms embedded within the cortex.

The axopods function solely as food-trapping organelles. On contact, small organisms adhere to the axopods, which then liquefy and withdraw. The prey is covered by cytoplasm and is gradually withdrawn into the cortex to be digested within a food vacuole.

Radiolarians

The radiolarians are another group of marine Sarcodina in which skeletal structures are highly developed. They are planktonic and reach a fairly large size, several millimeters in some species. The body is somewhat like that of heliozoans in being more or less spherical with a central nucleated core of cytoplasm and a broad outer vacuolated cortex (Fig. 22.21). The cortical cytoplasm of many species contains symbiotic dinoflagellates in a palmella state. The pseudopodia are axopods, or netlike as in forams. Radiolarians differ from heliozoans in that the central cytoplasm is encased within a chitin-like capsule, which is perforated to permit communication with the outer cortex (Fig. 22.21).

Some radiolarians have a skeleton of radiating strontium sulfate needles, but most have a skeleton of silicate arranged as concentric spherical lattices within and outside of the cortical cytoplasm (Fig. 22.21). The beautiful skeletons are very elaborate and not always symmetrical.

The pseudopodia project through the skeletal openings and function as foodtrapping structures in the same way as in heliozoans and forams.

Plankton samples taken at different depths reveal a distinct vertical stratification of radio-

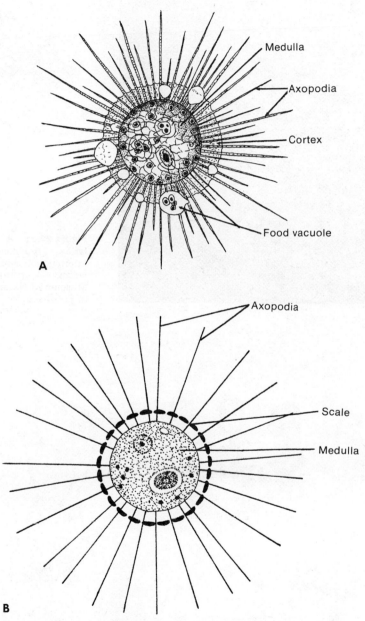

Figure 22.20 Heliozoans. *A, Actinosphaerium; B, Pinaciophora*, with skeleton of scales. (*A* after Doflein from Tregouboff; *B* after Penard from Hall.)

Figure 22.21 *A*, Glass model of a radiolarian, *Trypanosphaera regina*; *B*, the internal, siliceous skeleton of a radiolarian. (*A*, courtesy of the American Museum of Natural History; *B*, courtesy of E. Giltsch, Jena.)

Figure 22.22 *A*, Fission in a naked ameba. *B* and *C*, Two stages in the division of *Euglypha*, a shelled ameba: *B*, formation of skeletal plates on cytoplasmic mass protruding from aperture; *C*, division of nucleus. (*B* and *C* after Sevajakov from Dogiel.)

larians, which are capable of some depth regulation through the retraction or extension of the radiating skeletal rods or the cortical cytoplasm. Some are found only at great depths (4600 m.); some species undergo seasonal movements, dropping from the surface to lower depths in the summer. This could be passive movement following a density gradient, since the density of sea water increases with decreasing temperature. At great depths, where calcium carbonate shells of forams dissolve, radiolarian skeletons often accumulate and predominate in the ooze.

The fossil record of radiolarians extends back into the pre-Cambrian, and they have contributed to great sedimentary deposits.

Reproduction in Sarcodina. Asexual reproduction is generally by binary fission (Fig. 22.22). In shelled amebas, heliozoans and radiolarians, the skeleton is divided, or else one daughter cell gets the skeleton and the other secretes a new one (Fig. 22.22). Sexual reproduction usually involves the fusion of isogametes, which in radiolarians and foraminiferans are flagellated. Forams have a complex life cycle in which a sexual generation alternates with an asexual one. Both stages spawn numerous individuals, each of which produces a shell. Old shells never divide.

Considerable evidence suggests that the Sarcodina evolved from flagellate ancestors. Radiolarians and forams have flagellated developmental stages, and some flagellates lose their flagella and undergo ameboid stages. An interesting little group of mastigophorans, called the Rhizomastigida, look like naked amebas but have a long flagellum (Fig. 22.23).

22.4 SPOROZOANS

Two phyla of parasitic organisms, the Sporozoa and the Cnidospora, were once placed together within the old class Sporozoa because of the presence of a sporelike infective stage in the life cycle of many species. The similarities, however, appear to represent convergence associated with parasitism, and the members of the phylum Sporozoa are probably more closely related to the Sarcodina and the phylum Cnidospora to the flagellates than they are to each other.

Most sporozoan species are intracellular parasites, infecting both invertebrate and vertebrate hosts. The life cycles typically involve (1) a stage in which spores are formed, (2) a stage of asexual multiplication by fission and (3) a sexual stage leading to gamete formation and a zygote. Some sporozoans require only

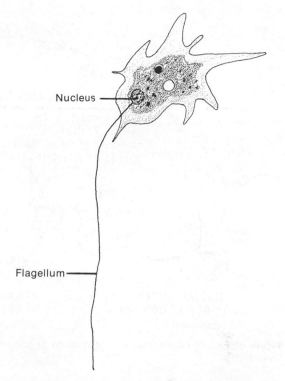

Nucleus

Flagellum

Figure 22.23 *Mastigameba.* (After Calkins, modified from Hyman.)

one host to complete the life cycle; others require two.

The nature and life cycle of sporozoans can be illustrated by the **coccidians,** which include the parasites causing malaria in humans and coccidiosis in domesticated animals.

Although in decline today, malaria was once widespread throughout the world and was one of the worst scourges of mankind. The untreated disease can be long-lasting and terribly debilitating. As late as 1957 it was estimated that 250 million persons were suffering from malaria, of which 2.5 million cases would be fatal. Malaria has been a continual and important factor in the shaping of human history.

Malaria is caused by species of the genus *Plasmodium.* There are more than 50 species of *Plasmodium,* all of which require a mosquito as one host and various vertebrates, mostly birds, for the other. Humans are the vertebrate host for four species.

The introduction of the parasite into a human host is brought about by the bite of certain species of mosquitoes, which inject the spore stages (sporozoites) along with their salivary secretions into the capillaries of the skin (Fig. 22.24). The parasite is carried by the blood stream to the liver, where it invades a liver cell. Here further development results in asexual reproduction through multiple fission. These

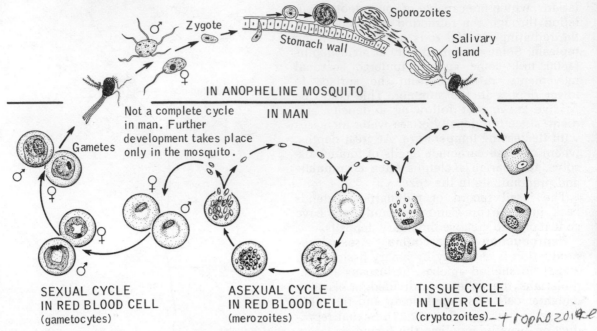

Sporozoites

Salivary gland

Stomach wall

Zygote

IN ANOPHELINE MOSQUITO

IN MAN

Not a complete cycle in man. Further development takes place only in the mosquito.

Gametes

SEXUAL CYCLE
IN RED BLOOD CELL
(gametocytes)

ASEXUAL CYCLE
IN RED BLOOD CELL
(merozoites)

TISSUE CYCLE
IN LIVER CELL
(cryptozoites) —trophozoite

Figure 22.24 The life cycles of *Plasmodium* in a mosquito and in man. (Redrawn and modified from Blacklock and Southwell.)

daughter cells (cryptozoites) invade other liver cells and continue to reproduce. After a week or so there is an invasion of red blood cells by parasites produced in the liver. Within the red cell the parasite increases in size and undergoes multiple fission. These individuals (merozoites), produced by fission within the red cell, escape and invade other red cells. The liberation and reinvasion does not occur continually but occurs simultaneously from all infected red blood cells. The timing of the event depends upon the period of time required to complete the developmental cycle within the host's cells. The release causes chills and fever, the typical symptoms of malaria. A specific periodicity is characteristic of different species. For example, in *Plasmodium vivax*, 48 hours are required to complete the development of erythrocytic stages, i.e., the period from invasion of the red cell to release of daughter cells. The chills and fever produced by this species thus recur at approximately 48-hour intervals.

Eventually some of the parasites invading red cells do not undergo fission but become transformed into **gametocytes.** The gametocyte remains within the red blood cell. If such a cell is ingested by a mosquito, the gametocyte is liberated within the new host's gut. After some further development, a male gametocyte (microgametocyte) fuses with a female gametocyte (macrogametocyte) to form a zygote. The zygote enters the stomach wall and gives rise to a large number of spore stages (sporozoites). It is these stages that are introduced into the human host by mosquitoes.

Species of *Eimeria* cause coccidiosis in domesticated animals, especially chickens. This parasite has a life cycle that involves stages similar to those of malaria, but all of which take place in the cells of the gut wall of a single host. The intestinal tissue may be so badly damaged that sloughing and hemorrhaging occur and may lead to death. Spore stages are passed in the feces and the next host is invaded through the mouth.

CLASSIFICATION OF PROTOZOA

Protozoa

Unicellular eukaryotic organisms. Most are heterotrophic and move by means of flagella, cilia or pseudopodia.

PHYLUM MASTIGOPHORA, OR FLAGELLATA. Unicellular organisms possessing one or more flagella.

    Subphylum Phytomastigina. Plantlike flagellates. Autotrophic. (These organisms might also be placed within the algal phyla.)

    Subphylum Zoomastigina. Animal-like flagellates. Heterotrophic.

PHYLUM SARCODINA. Unicellular organisms possessing pseudopodia.

    CLASS AMEBIDA. Naked amebas. Large pseudopodia; no shell.

    CLASS TESTACIDA. Shelled amebas. Straplike pseudopodia; shell present.

    CLASS FORAMINIFERA. Forams. Strandlike interconnecting pseudopodia; calcareous perforated shell that is usually chambered.

    CLASS HELIOZOA. Heliozoans. Needle-like pseudopodia, or axopods. Sessile or free spherical bodies. Siliceous skeletons of separate pieces.

    CLASS RADIOLARIA. Radiolarians. Axopods or strandlike pseudopodia. Cell with central capsule surrounded by a vacuolated cortex. Skeleton usually siliceous; composed of strontium sulfate in some.

PHYLUM SPOROZOA. Intracellular parasitic organisms, with nonflagellated sporelike stages. *Plasmodium, Eimeria.*

PHYLUM CNIDOSPORA. Mostly intracellular parasitic organisms with flagellated sporelike stages.

PHYLUM CILIATA. Ciliates. Ciliated unicellular organisms possessing two types of nuclei.

    Subphylum Holotrichia. With simple and uniform somatic cilia. Buccal cilia absent or inconspicuous. *Didinium, Paramecium.*

    Subphylum Peritrichia. Adults usually lack body cilia, but the apical end of the body typically bears a conspicuous buccal ciliature. Mostly attached ciliates. *Vorticella.*

    Subphylum Spirotrichia. With reduced body cilia and well developed conspicuous buccal ciliature. *Stentor, Euplotes* (hypotrichs).

    Subphylum Suctoria. Sessile, stalked tentaculate ciliates lacking cilia as adults.

ANNOTATED REFERENCES

Accounts of the protozoans can be found in the references cited at the end of Chapter 21. The parasitology texts listed at the end of Chapter 24 cover the parasitic protozoans. The references cited below are devoted exclusively to protozoans.

Dodson, E. O.: The kingdoms of organisms. Systematic Zoology 20:265–281, 1971. This paper and Whittaker's review present various schemes for the classification of living organisms.

Jones, A. R.: The Ciliates. New York, St. Martin's Press, 1974. A general biology of the ciliated protozoans.

Poindexter, J. S.: Microbiology: An Introduction to Protists. New York, Macmillan, 1971.

Sleigh, M. A.: The Biology of the Protozoa. New York, American Elsevier Publishing Co., 1973. A good, well balanced general account of the protozoans.

Whittaker, R. H.: New Concepts of the kingdoms of organisms. Science 163:150–159, 1971.

CHAPTER 23

SPONGES

1. Most sponges are marine animals that live attached to the bottom. A few species live in fresh water.

2. Body structure is organized around a system of canals and chambers through which water flows. Water enters through small surface pores and exits through a larger opening, the osculum. Internal mineral spicules or organic fibers, or both, contribute to support.

3. The water current provides for filter feeding, gas exchange and waste removal. Flagellated collar cells not only create the water current but also filter out fine food particles and digest them intracellularly.

4. Sponges are either hermaphroditic or the sexes are separate, and development includes a free-swimming larva.

5. Eggs are fertilized internally by sperm brought in with the water current.

6. The primitive nature of sponges is reflected in their lack of organs, head, mouth and gut cavity.

7. Their system of water canals and chambers is a specialization correlated with their sessile life style.

Sponges, which compose the phylum **Porifera,** lack most of the typical animal features. There is no head or anterior end; there is no mouth or gut cavity; and the body is immobile. In fact, some zoologists believe that sponges had an evolutionary origin separate from other animals. Certainly they departed early from the main line of animal evolution, and their structure reflects a combination of primitive and specialized characteristics.

With the exception of two freshwater families, the 10,000 species of sponges are marine. They live attached to the bottom, most commonly to rock, shell, coral, pilings and other hard surfaces. Bottom-dwelling animals, both marine and freshwater, are called **benthic,** and all organisms that are attached to the substratum are said to be **sessile.** Many of the peculiarities of sponges are correlated with their sessile mode of existence.

23.1 STRUCTURE OF SPONGES

The simplest sponges are shaped like little vases (Fig. 23.1). The interior cavity, called the

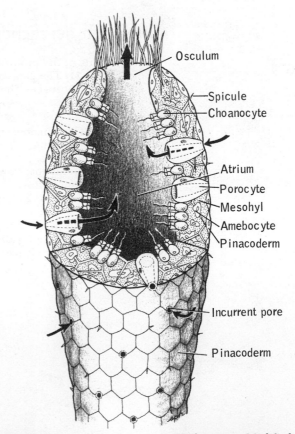

Figure 23.1 Structure of an asconoid sponge. (Modified from Buchsbaum, from Barnes, R. D.: Invertebrate Zoology. 3rd ed. Philadelphia, W. B. Saunders Company, 1974.)

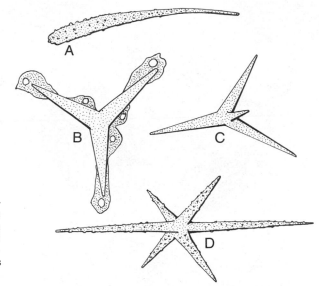

Figure 23.2   Sponge spicules. *A* and *C*, Two types of calcium carbonate spicules found in members of the class Calcispongiae. Similar types, made of silicate, occur in members of the class Demospongiae. *B*, A three-pronged calcareous spicule being secreted by amebocytes. *D*, A six-pronged spicule characteristic of the members of the class Hexactinellida. (*B* modified from Minchin.)

atrium or **spongocoel,** opens to the outside through a large opening at the top, the **osculum.** The body of the sponge surrounding the spongocoel is perforated by pores; the phylum name, Porifera, means pore-bearer. The outer surface of the sponge body is covered by flattened epidermal cells, **pinacocytes,** and the pores are formed by **porocytes,** cells that are perforated like a ring. The spongocoel is lined with flagellated **collar cells,** so called because of a collar-like extension of the cell membrane around the base of the flagellum. Between the epidermis and the flagellated cells lies a layer called the mesohyl or mesenchyme, containing

ameboid cells of different types and skeletal pieces embedded within a gelatinous protein matrix. The skeleton of most sponges consists of **spicules** of calcium carbonate or silicon dioxide secreted by the amebocytes (Fig. 23.2). Each spicule may be a single needle or several needles fused at certain angles to each other. They are generally microscopic and unconnected, but in some sponges the spicules are fused together to form a complex skeleton. The shape of the spicule and the number of rays present are characteristic for each species of sponge, and spicule structure is an important characteristic in the classification of sponges.

Figure 23.3   Spongin fibers (they appear translucent in photograph). (Courtesy of the General Biological Supply House, Inc.)

A

Figure 23.4 *A*, Sponge cluster showing an osculum at the end of each oval mass.
Illustration continued on opposite page.

Some sponges possess an organic skeleton of spongin fibers (Fig. 23.3) instead of or in addition to spicules, and it is from species with only a spongin skeleton that commercial sponges are obtained. In preparing commercial sponges, the living tissue is allowed to decompose and is then washed away from the spongin skeleton (Fig. 23.4). The natural sponge industry has greatly declined as a result of the development of synthetic substitutes.

Many sponges are brightly colored red, orange, yellow or purple. The color is derived from pigment located within the amebocytes. Many of the freshwater species are green from symbiotic algae living within the amebocytes.

23.2 PHYSIOLOGY OF SPONGES

The flagellum of the collar cells describes a spiral in its beat, the motion characteristic of most flagellar movement. As a result, water is driven from a collar cell in the same manner as air is blown from a fan (Fig. 23.5). The beating of flagella lining the atrium (spongocoel) causes water to be sucked in through the pores and driven up and out of the spongocoel through the osculum. The rate of water flow can be regulated to some extent by changes in the diameter of the osculum. Cells with contractile fibrils are located around the osculum, and their contraction brings about a reduction in the size of the opening.

Many aspects of sponge physiology depend upon the water current passing through the body. Sponges are filter feeders, removing bacteria and other very fine suspended organic matter from the water. An initial screening is provided by the pores, which permit only very small particles to enter. When such particles pass over the collar cells, they may become trapped on the collar surface. Electron microscopy has revealed that the collar is actually composed of fine parallel fibrils between which water passes (Fig. 23.5). The trapped particle passes down to the base of the collar where it is engulfed by the cell. Digestion occurs within the collar cell (Fig. 23.5), or if the collar cell is small, the food particle is transferred to an amebocyte for intracellular digestion. In either case the products of digestion are passed by diffusion to the other cells of the body.

The current of water passing through the sponge body not only provides a source of food material but also serves for gas exchange, for the removal of wastes and for the transfer of gametes.

23.3 HYDRODYNAMICS AND BODY FORM

Sponges with the simple vaselike structure described earlier are called **asconoid** sponges and are small, not more than a few centimeters high. The asconoid structure im-

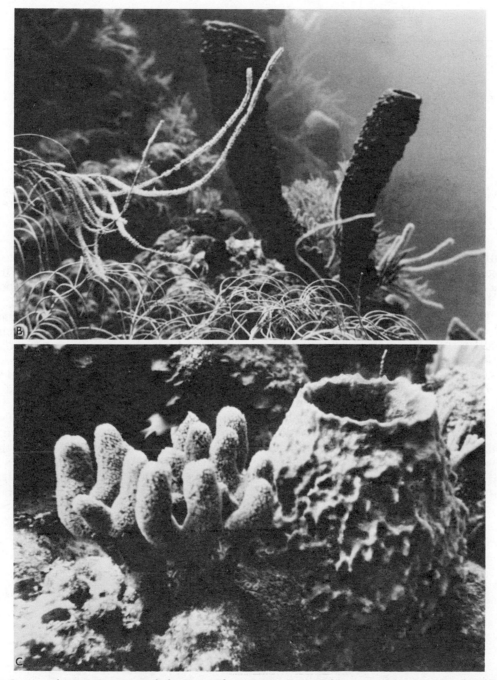

Figure 23.4 *Continued.* *B, Verongia,* a tubular West Indian sponge. *C,* A vaselike species of *Ircinia* from the West Indies. A scleractinian coral is located to the left of the sponge. (*B* and *C* by David Barnes.)

Flagellum

Collar

Trapped food particle

Food vacuole

Mesenchymal
ameboid cell

Figure 23.5 Section of flagellated layer and underlying mesenchyme, showing three choanocytes, or collar cells. Arrows indicate direction of water current.

poses limitations in size, for as the volume of the spongocoel increases, the flagellated surface area does not increase proportionally (volume increases by the cube; surface area only by the square). Consequently, a large asconoid sponge would contain more water than its collar cells could efficiently move. In the evolution of sponges this problem was solved by repeated folding of the flagellated layer to increase its surface area. The first stage of folding is exhibited by **syconoid** sponges, in which the flagellated layer is evaginated outward into finger-like projections (Fig. 23.6). The evaginations are called **flagellated canals** and the corresponding invaginations of the external surface are called **incur-**

rent canals. Pores are located between the incurrent and flagellated canals. The flagellated canals of syconoid sponges open into a central atrium devoid of collar cells.

In the great majority of sponges the surface area of the flagellated layer has been further increased by the formation of many small chambers within which the collar cells are located. In these so-called **leuconoid** sponges, water enters dermal pores on the body surface and passes through a system of incurrent canals, eventually reaching the flagellated chambers (Fig. 23.6). Many leuconoid sponges have no spongocoel; water leaves the body through converging excurrent canals opening to the exterior through an osculum. The

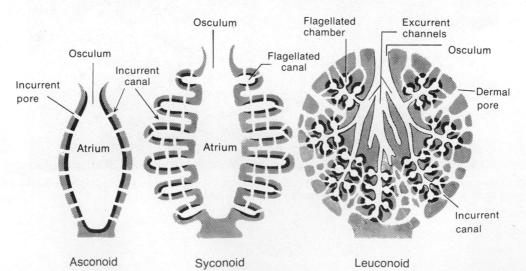

Osculum

Flagellated
chamber

Excurrent
channels

Osculum

Flagellated
canal

Osculum

Incurrent
canal

Dermal
pore

Incurrent
pore

Atrium

Atrium

Incurrent
canal

Asconoid   Syconoid   Leuconoid

Figure 23.6 The three structural types of sponges. In each, the choanocytes are shown in black.

development of flagellated chambers greatly increases the water-moving ability of a sponge. The number of flagellated chambers can range from 10,000 to 18,000 per mm.³, and one leuconoid sponge 10 cm. × 3 cm. × 3 cm. was found to possess some 2,250,000 flagellated chambers and could pass 22.5 liters of water through its body in 24 hours.

Most leuconoid sponges are irregular in shape, with many oscula located over the body surface (Fig. 23.4). A far greater size is attained than in asconoid and in syconoid forms. The largest leuconoid sponge would more than fill a bushel basket. Most of the sponges commonly encountered in shallow water are leuconoid.

The large, radially symmetrical glass sponges, members of the class **Hexactinellida,** are more or less syconoid in structure with an atrium. The glass spicules of these sponges are often highly organized and fused together, forming a beautiful lattice skeleton (Fig. 23.7). Most glass sponges occur in water deeper than 50 meters, and some are adapted for living on soft bottoms by means of a basal tuft of spicules that anchors the animal into the substratum.

23.4 REGENERATION AND REPRODUCTION

The truly remarkable regenerative powers of sponges are well illustrated by the classical experiment of forcing a small bit of sponge through bolting silk and dissociating the cells. Within a short period of time, the dissociated cells reaggregate in the proper relationship. Ability to regenerate is closely correlated with asexual reproduction, which is also highly developed in sponges. A bud or small fragment of sponge breaks free of the parent and gives rise to a new sponge. Some sponges produce special asexual reproductive bodies consisting of an aggregate of essential cells, especially amebocytes. Certain amebocytes are **totipotent,** i.e., capable of giving rise to any other type of cell, and these play an important role in sponge regeneration and growth. Commercial sponges are sometimes propagated by attaching a piece of sponge to a bit of rock or concrete. A marketable sponge is obtained after several years' growth.

Figure 23.7 Photograph of the skeleton of the glass sponge, *Euplectella*, Venus's-flower-basket. The spicules are fused to form intersecting girders. (Courtesy of the American Museum of Natural History.)

Figure 23.8 Sponge development. *A*, An amphiblastula larva. *B* and *C*, Parenchymula larvae. *D*, Gastrulation of an amphiblastula larva following settling. (*C* and *D* redrawn from Hyman.)

The majority of sponges are hermaphroditic. Hermaphroditism is another common adaptation to a stationary existence in animals. Any individual that settles near another can provide sperm or eggs. Cross fertilization can occur—this is typical of most hermaphroditic animals—but there is no requirement to be near an individual of the opposite sex.

Sperm and eggs originate from amebocytes and develop within the mesenchyme; they are not located within a gonad. Mature sperm are shed into the water canals and exit through the osculum. Some are carried into the water canals of neighboring sponges. These sperm are trapped and engulfed by collar cells, or amebocytes, which then move to an adjacent ripe egg. The carrier cell plasters itself against the egg and transfers the sperm. Fertilized eggs may be released into the water canals, carried out with the water stream and undergo development in the sea water; or the eggs may be brooded and develop within the parent mesenchyme.

In either case, embryonic development leads to a free-swimming larva, a stage that is important for species dispersion in sessile animals. In most species, the larval condition is reached at the blastula stage. In some species, the larva (an **amphiblastula** larva) is hollow and one hemisphere is flagellated (Fig. 23.8). In other sponges, the larva (a **parenchymula** larva) is solid and covered with flagellated cells (Fig. 23.8). After a free-swimming existence among the plankton, the larva settles to the bottom, undergoes gastrulation (Fig. 23.8) and develops into an adult sponge. During development of the adult form after settling, some leuconoid sponges pass through asconoid and syconoid stages, reflecting perhaps the evolutionary sequence of sponge structure.

Although sponges are primitive in that they lack organs, including a gut, and have only a small number of different kinds of cells, they are highly specialized animals in other respects. The specializations are largely adaptations to a stationary mode of existence—the absence of a head or anterior end, the circulation of water through the body for filter feeding, gas exchange and water removal, and the condition of hermaphroditism. Certainly sponges evolved early in the evolution of the Animal Kingdom, and it seems highly unlikely that they gave rise to any other groups of animals.

CLASSIFICATION OF PHYLUM PORIFERA

CLASS CALCISPONGIAE. Calcareous sponges. Spicules are usually separate and composed of calcium carbonate. Body form is asconoid, syconoid or leuconoid. Calcareous sponges are mostly drab in color and small, not exceeding 10 cm. in height.

CLASS HEXACTINELLIDA. Glass sponges. Spicules are siliceous, six-pointed and often fused together to form a highly organized skeleton. Although leuconoid, many species are cup-, vase- or urnlike in shape, reaching a height of 10 to 30 cm. Most glass sponges occur in deeper water (450 to 900 m.) than do other sponges and are thus less frequently encountered.

CLASS DEMOSPONGIAE. This class contains the greatest number of sponge species and includes most of the commonly encountered North American sponges. The skeleton of Demospongiae is composed of separate siliceous spicules. But some species, the commercial sponges, possess a skeleton of spongin fibers and some possess both spongin fibers and siliceous spicules. The body structure is always leuconoid with many oscula, and an irregular symmetry prevails. Many Demospongiae are brightly colored. The small number of freshwater sponges belong to this class.

ANNOTATED REFERENCES

Detailed accounts of sponges may be found in the references listed at the end of Chapter 21.

CHAPTER 24

CNIDARIANS

1. Most members of the phylum **Cnidaria** or **Coelenterata** are marine animals and include jellyfish, sea anemones and corals. The familiar hydras are among the few freshwater species.

2. The body is radially symmetrical, with the mouth and surrounding tentacles located at one end of the radial axis.

3. Some cnidarians are columnar in shape (polypoid form) and live attached to the bottom, with the oral end directed upward. Others are bowl-shaped (medusoid, or jellyfish, form) and are free-swimming, with the oral end directed downward.

4. There are few organs, and the body wall is composed of two principal layers: an outer epidermis and an inner gastrodermis, separated by the mesoglea. The mesoglea may be an acellular membrane or a thick gelatinous layer with or without cells.

5. The epidermis and gastrodermis are composed of a number of different cell types; the explosive stinging cells called cnidocytes are unique to the phylum.

6. Most cnidarians are carnivorous, and cnidocytes are important in prey capture. The mouth is the only opening into the digestive cavity, where digestion is both extracellular and intracellular.

7. The nervous system is commonly in the form of a net with receptor cells dispersed over the body surface. The statocysts and ocelli of jellyfish are the only cnidarian sense organs.

8. Cnidarians are hermaphroditic or the sexes are separate. The gonads are only aggregations of developing gametes, and there are no gonoducts.

9. Fertilization is usually external, and development leads to a free-swimming planula larva.

24.1 CNIDARIAN STRUCTURE

Cnidarians are radially symmetrical, and the oral end of the axis terminates in the mouth and a circle of tentacles (Fig. 24.1). The mouth opens into a blind **gastrovascular cavity.** A layer of cells, the **epidermis,** covers the outer surface of the body; another layer, the **gastrodermis,** lines the gastrovascular cavity (Fig. 24.1). The **mesoglea,** located between these two layers, varies from a thin noncellular membrane, as in hydras, to a thick jelly layer with or without cells, as in jellyfish. The epidermis and gastrodermis contain several kinds of cells, and each would be functionally equivalent to several different cell layers in higher animals. Cnidarians have no organs.

Two types of cnidarian body form can be distinguished. **Polypoid** cnidarians, such as hydras and sea anemones, have a cylindrical body with the oral end (bearing the mouth and tentacles) directed upward and the aboral end attached to the substratum (Figs. 24.1 and 24.20). **Medusoid** cnidarians, such as jellyfish (Figs. 24.1 and 24.15), have bell- or saucer-shaped bodies with the aboral end convex and directed upward and the oral end concave and directed downward. Medusae are usually unattached and free-swimming.

24.2 MOVEMENT

The body and tentacles of cnidarians can be extended or contracted and bent to one side or the other. Movement is brought about by the contraction of longitudinal and circular muscle fibers. However, the fibers are located not in true muscle cells but in basal extensions of

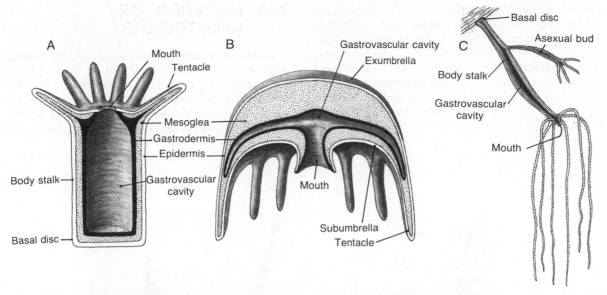

Figure 24.1 *A*, Polypoid body form; *B*, medusoid body form; *C*, an attached hydra with tentacles hanging in water. (*C* after Hyman.)

Figure 24.2 *A*, Body wall of hydra (longitudinal section); *B*, an epitheliomuscle cell (After Gelei from Hyman.)

epidermal and gastrodermal cells (Fig. 24.2). These **epitheliomuscle cells** and **nutritive muscle cells** have characteristics of the epithelial and muscle tissues of most other animals. The contractile layers are variously developed in different cnidarians. The gastrodermal fibrils of hydras are poorly developed and movement is largely the result of contraction of the epidermal cells.

24.3 NUTRITION AND NEMATOCYSTS

Cnidarians are carnivorous. Small forms, such as hydras and corals, feed upon **planktonic** animals, especially crustaceans, but the larger jellyfish and sea anemones can consume small fish. Gland cells lining the gut secrete proteolytic enzymes that rapidly digest the

Figure 24.3 *A*, Undischarged penetrant nematocyst within cnidocyte of hydra; *B*, a discharged penetrant nematocyst; *C*, a volvent nematocyst of hydra; *D*, discharged penetrant and volvent nematocysts on a crustacean bristle. (*C* and *D* after Hyman.)

tissues of the prey. Mixing of the gut contents is aided by the beating of the flagella of the gastrodermal cells. Small fragments of tissue are then engulfed by another type of gastrodermal cell, and digestion is completed intracellularly (Fig. 24.2). Undigestible waste materials are ejected through the mouth when the animal contracts.

Cnidarians capture their prey with the aid of special stinging cells, called **cnidocytes,** located largely in the epidermal layer (Fig. 24.2). A cnidocyte contains a surface projection, the cnidocil, and the **nematocyst,** the actual stinging element. The undischarged nematocyst is composed of a bulb and a long thread coiled within the bulb (Fig. 24.3). When discharged, the nematocyst is expelled from the cnidocyte, and the thread is everted out of the bulb in the process. The mechanism of discharge is not completely understood, but it is believed that stimulation by contact or chemical substances produced by the prey changes the permeability of the bulb wall and fluid rushes into the interior. The elevated fluid pressure both everts the thread and hurls the nematocyst from the cnidocyte. Some types of nematocysts function by entanglement, the thread wrapping around the prey **(volvent nematocysts).** Other nematocysts **(penetrant nematocysts)** are driven into the body of the prey and may inject a toxin (Fig. 24.3). It is these toxic penetrants that produce the sting of jellyfish and other cnidarians. Hydras produce nematocysts with sticky threads **(glutinant nematocysts)** that function in adhesion to the bottom when the animal is moving upon its tentacles. Nematocysts vary in detail in different groups of cnidarians and sometimes are of importance in identification.

Cnidocytes are typically embedded within the surface epidermal cells (Fig. 24.2). Although they may be located throughout the epidermis, they are especially prevalent on the tentacles, where they may be concentrated in ringlike or wartlike batteries (Fig. 24.10). In hydras, contact with prey may result in the discharge of 25 per cent of the tentacular nematocysts. Following discharge, the old cnidocytes are resorbed and new cnidocytes differentiate from **interstitial cells** (Fig. 24.2), totipotent cells similar to certain amebocytes of sponges.

After the prey has been quelled by the nematocysts, it is pulled by the tentacles toward the mouth, which often dilates enormously in swallowing. Mucus secreted by glands in the mouth region facilitates the process of swallowing.

There are no special systems for internal transport, gas exchange or secretion. The size, surface area and structure of cnidarians permit all of these processes to take place by diffusion.

24.4 NERVOUS SYSTEM

The cnidarian nervous system displays a number of primitive features. The neurons, located at the base of the epidermis and gastrodermis, are usually arranged as **nerve nets** rather than as nerve bundles (Fig. 24.4). The association of nerve cells to form conducting chains between receptors and effectors shows varying degrees of complexity. Sensory neurons may connect directly with motor neurons, or sensory and motor neurons may be separated by interneurons. There are even neurons with both sensory and motor processes. The neurons may possess from two to many processes; the axons and dendrites of interneurons are not differentiated. Conduction occurs in a radiating manner, with interneurons capable of transmitting impulses in both directions. Neuronal junctions are synaptic in all cnidarians, with transmission in some occurring in more than one direction.

24.5 GROWTH AND REPRODUCTION

Polypoid cnidarians exhibit a high level of regenerative ability. In fact, some polypoid species appear to undergo continuous cell replacement. In hydras, for example, old cells at the basal disc and ends of the tentacles are continually being replaced by new cells arising by mitosis within the body stalk. When a major part of the body, such as the oral end, is lost, the remaining part undergoes reorganization to form a new mouth region and tentacles. The

Figure 24.4 Diagram of the cnidarian epidermis, showing epitheliomuscle cells, sensory cell, and nerve net. (From Mackie, G. O. and L. M. Passano: Epithelial conduction in hydromedusae. J. Gen. Physiol. *52:*600, 1968.)

Figure 24.5 *Gonionemus:* planula larva that develops from the egg. (After Hyman.)

original oral-aboral polarity, however, is retained, and tentacles and mouth will form only at the reorganized oral end.

Asexual reproduction is very common in cnidarians, especially in polypoid species. New individuals are usually formed by **budding.** A bud arises as an outpocketing of the body wall and thus contains an extension of the gastrovascular cavity and all of the body wall layers (Fig. 24.1). The distal, or outer, end acquires the oral end polarity and forms a mouth and tentacles. The body separates from the parent, or in colonial species may remain attached as a new individual of the colony.

The sexes of most cnidarians are separate, but some are hermaphroditic. The gametes develop from interstitial cells and form aggregations in specific locations in the epidermis or gastrodermis. Since there is no surrounding wall of sterile cells, the cnidarian gonads are not comparable to the gonads of higher animals.

Fertilization is commonly external with development occurring in the plankton. At the completion of gastrulation, a characteristic larval stage, termed a **planula,** is attained. The planula is slightly elongate and radially symmetrical (Fig. 24.5), composed of a solid interior mass of cells surrounded by an outer layer of flagellated cells. The course of development from the planula stages varies with different groups of cnidarians.

24.6 CLASS HYDROZOA

The class Hydrozoa includes the hydras and many colonial species called **hydroids.** While hydrozoan cnidarians are very abundant, they are small and not as conspicuous as the larger jellyfish, sea anemones and corals. Most hydrozoans are marine, but the few species of freshwater cnidarians—the hydras and a few jellyfish—are members of this class.

The members of the class Hydrozoa may exhibit a medusoid body form or a polypoid body form, or both, during the life history. The mesoglea is never cellular, nematocysts are limited to the epidermis, and the gametes develop within the epidermis.

The hydrozoans are characterized by colony formation, skeleton formation, polymorphism and reduction of the medusoid stage. All degrees of each condition occur, and apparently each evolved independently a number of times within the class. An understanding of these conditions will be helpful in gaining an appreciation of hydrozoan diversity.

Colony Formation. Some hydrozoans—the hydras and the small hydrozoan jellyfish—are solitary, but many species are colonial. The colonial forms are mostly polypoid and are called **hydroids** (Fig. 24.6). If one can imagine hydras budding but the bud remaining attached to the parent, some notion of a hydroid colony can be attained. The individuals, or polyps, of a hydroid colony are usually attached to a main stalk, which is in turn anchored to the substratum (algae, rock, shell or wharf piling) by a rootlike stolon. The arrangement of polyps on the stalk and the branching of the stalk vary with the species (Fig. 24.6). In some hydroids, separate polyps spring directly from the stolon (Fig. 24.6). The tissue layers of the stalks and stolons are continuous with the tissue layers of the polyps. Thus all of the polyps are interconnected, and there is a common gastrovascular cavity for the entire colony.

Skeleton Formation. The solitary hydrozoans have no skeletons and some hydroid colonies in which the polyps arise directly from the substratum have only anchoring skeletons. Most hydroid colonies, however, are 3 to 10 cm. high, and support is provided by an external chitinous skeleton secreted by the epidermis. In some species, the skeletal envelope is limited to the stalk (Fig. 24.11); in others, the skeleton extends around the body of the polyps but is open at the oral end for the emergence of the tentacles and mouth (Fig. 24.7). In a few species the opening at the oral end is closed by a lid.

Polymorphism. The existence of two or more structurally and functionally different kinds of individuals within the same species is termed polymorphism. The unspecialized,

Figure 24.6 Hydroid colonies. *A, Coryne tubulosa,* a dimorphic form in which medusoids are formed directly on gastro-zooids. *B, Hydractinia.* Polyps arise separately from mat of stolons, and skeleton is limited to stolons. *C,* Skeleton of a branching type of hydroid colony. (*A* modified after Naumov; *B* after Hyman; *C* after Kuhn from Naumov.)

commonest type of individual in hydroid colonies is the feeding polyp **(gastrozooid)**, resembling a hydra (Figs. 24.6 and 24.7). These polyps capture food and carry out the initial extracellular phase of digestion. Intracellular digestion and absorption occur throughout the colony since the gastrovascular cavities are continuous.

Some hydroid colonies include defensive individuals, highly modified polyps with club-like bodies bearing great numbers of cnidocytes (Fig. 24.6). Tentacles and mouth have disappeared.

Gametes are produced by reproductive individuals. These are not polyps but medusae, or jellyfish. The medusae are produced from buds formed by polypoid individuals. In some hydroids, medusae are budded from the body of feeding polyps (Fig. 24.6). In many hydroids, however, there has been a further division of labor and medusae are budded from special reproductive polyps (gonozooids) no longer involved in feeding (Figs. 24.6 and 24.7).

Hydroid colonies with feeding polyps that produce medusoids are said to be dimorphic, i.e., having two kinds of individuals. Those with feeding polyps plus special polyps for the budding of medusoids are termed trimorphic. And those which in addition have defensive polyps are tetramorphic.

Figure 24.7 Life cycle of *Obelia*, showing structure of the hydroid colony. (Adapted from various sources.)

Figure 24.8 *A, Physalia,* the Portuguese man-of-war; *B,* a cluster of polyps from *Physalia,* showing the various modifications of the individual polyps. (*A,* courtesy of the New York Zoological Society; *B* after Hyman.)

Most colonial hydrozoans are hydroids, i.e., polypoid. However, one group of colonial species, the **Siphonophora**, includes large pelagic colonies composed of both medusoid and polypoid individuals. There are individuals adapted as floats and pulsating swimming bells, from which hang feeding polypoid individuals with long tentacles. The siphonophoran **Portuguese man-of-war** has a single gas-filled float from which the tentacles of the polyps may hang down several meters (Fig. 24.8). The nematocysts can produce painful stings, and entanglement with the tentacles can be a dangerous encounter for a swimmer in deep water.

Medusa Reduction. Sexual individuals are medusae. Hydrozoan jellyfish **(hydromedusae)** are small, usually not more than a few centimeters in diameter (Figs. 24.9 and 24.10). The large jellyfish often encountered at beaches

Figure 24.9 Medusae of five genera of hydrozoans. All are only a few centimeters or less in diameter. (*A* after Mayer, from Naumov; *B* after Hyman; *C* and *D* after Mayer; *E* after Mayer from Hyman.)

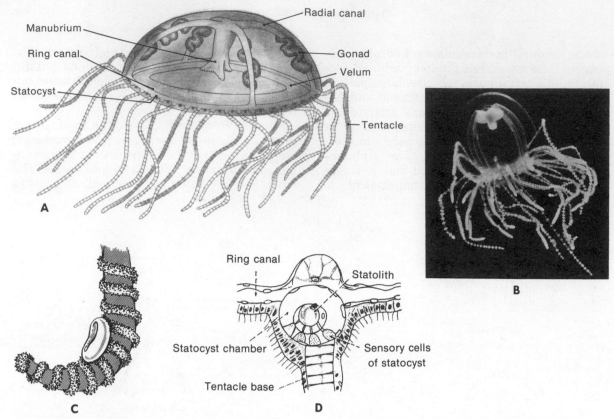

Figure 24.10 *A*, Lateral view of the hydromedusa *Gonionemus*. *B*, *Gonionemus vertens* actively swimming. *C*, A portion of a tentacle from *Gonionemus*, showing rings of nematocysts. *D*, Statocyst of hydromedusa. (*B*, courtesy of D. P. Wilson; *C* after Hyman; *D* after the Hertwigs from Hyman.)

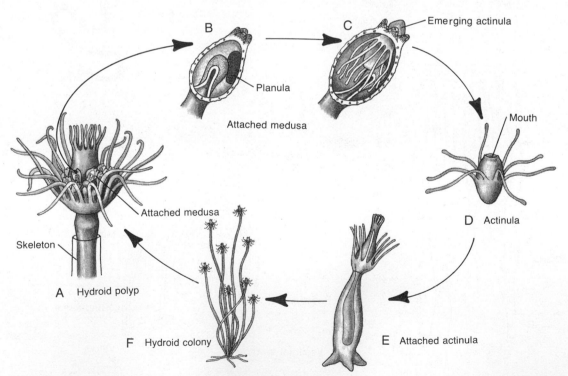

Figure 24.11 Life cycle of *Tubularia*. *A*, Skeleton is restricted to stalk of polyp; medusae are formed on gastrozooid and remain attached. *B*, Egg develops into planula within attached parent medusa. *C*, An actinula larva is released from medusa (*D*) and eventually settles to the bottom and develops into a new hydroid colony (*E* and *F*). (After Allman from Bayer and Owre.)

belong to another class of coelenterates. A
varying number of tentacles hang from the bell
margin of the hydromedusa, and the **manubrium,** a fold of body wall surrounding the
mouth, hangs from the center of the convex
undersurface. The contraction of a ring of
muscle fibers within the bell margin reduces
the diameter of the bell and drives water from
beneath the animal. A shelf-like inward projecting fold of the bell margin, the velum, increases the force of the water jet. These pulsations propel the animal mostly in an upward
direction and maintain the animal at a specific
depth.

Hydromedusae feed on other small planktonic organisms. The gastrovascular cavity is
not a single sac, as in polyps. The mouth opens

into a central stomach from which extend four
radial canals (Figs. 24.9 and 24.10). The radial
canals connect in turn with a ring canal that
extends around the bell margin.

The gametes of hydromedusae develop
within the epidermis beneath the radial canals
(Figs. 24.9 and 24.10). The eggs are fertilized
when shed into the sea water and development
to the planula larva occurs in the plankton. In
some species, fertilization and early development occur within the epidermis.

Hydrozoa exhibit great variations in their life
cycles. Some species have no polypoid individuals; the planulae from the free-swimming
medusoid adults develop directly into hydromedusae (Fig. 24.13). In the course of that
development, there is a stage called the **actin-**

A <u>Craspedacusta</u>

B <u>Hydra</u>

Figure 24.12 Life cycles of the freshwater *Craspedacusta* (*A*) and *Hydra* (*B*). Both have solitary polypoid stages; the
medusoid stage is absent in *Hydra*.

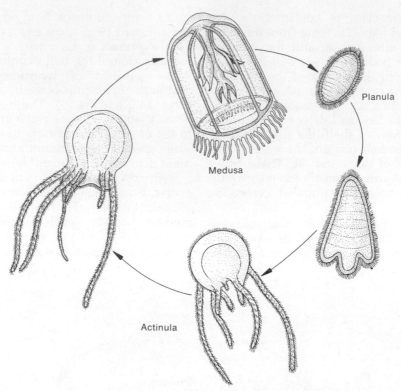

Figure 24.13 Life cycle of *Aglaura*, a hydrozoan that has no polypoid stage. Planula larva develops into an actinula, which develops directly into a medusa. (From Bayer and Owre.)

ula that looks like either a polyp or a medusa, depending upon whether the mouth is turned upward or downward (Fig. 24.13).

Some hydroids produce free-swimming medusae by budding from polypoid individuals (Figs. 24.7). Even more common in hydroids is the formation of medusae that remain attached to the parent polyp (Fig. 24.11).

Finally, there are polypoid species, such as hydra, in which there is no medusa at all. Eggs and sperm are produced in the epidermis of the body column (Fig. 24.12).

We can account for the diversity of hydrozoan life cycles by postulating that the cnidarian polyp represents a persistent larval stage and that in the evolution of the Hydrozoa there has been a tendency for the medusa to become reduced. Living hydrozoans reflect varying degrees of polypoid development and medusoid reduction.

According to this view, the medusa is the more primitive cnidarian body form. The hydrozoan life cycle initially lacked a polypoid stage, and the planula developed first into an actinula and then into an adult medusa (Fig. 24.13). Such a life cycle is exhibited by some living hydrozoans. In other hydrozoans, the actinula stage perhaps became attached to the bottom by the aboral end and produced more

actinulae by budding. The latter developed directly into medusae (Fig. 24.13). This attached actinula became the first polyp, and secondary budding would represent the formation of medusa buds. Such a life cycle has the advantage of permitting a long life for an actinula, which can give rise to not one but many medusae. This type of life cycle occurs in such hydrozoans as the marine *Gonionemus* and the little freshwater jellyfish *Craspedacuspa* (Fig. 24.12).

With the evolution of the polyp, there now occurred a progressive reduction of the medusa. The medusa developed by budding but was not free or detached. There was a gradual reduction of medusoid tissues until the polyp epidermis itself produced gametes.

Present evidence indicates that colonies evolved independently in different lines of hydrozoans, and this occurred independent of the evolutionary trend toward medusa reduction. A solitary polyp would produce more polyps by budding during certain times of the year and then produce medusa buds at another season. If the first type of buds remained attached to each other, this could have led to the evolution of a hydroid colony. Colony formation did not evolve in polyps of all hydrozoans. The hydras, for example, probably evolved from some line of solitary polyps that underwent reduction of the medusoid form.

24.7 CLASS SCYPHOZOA

The medusae of the class Scyphozoa are the cnidarians to which the name jellyfish is usually applied; they are considerably larger and more conspicuous than hydrozoan medusae (Fig. 24.14). The medusoid body form is dominant in the Scyphozoa, the polypoid form being strictly a larval stage. In contrast to hydromedusae, the mesoglea of scyphomedusae may be **cellular,** cnidocytes are present in the gastrodermis, and the gametes develop within the gastrodermis rather than within the epidermis.

Scyphomedusae are about the size of a saucer, and the translucent body is often tinted with orange, pink, purple and other colors. Tentacles of varying numbers and length hang from the margin of the bell. Four long divisions of the manubrium, called oral arms, hang from the center of the underside of the bell and are often more conspicuous than the tentacles (Figs. 24.14 and 24.15). The nematocysts, especially those in the tentacles and oral arms, can produce painful stings. The sea nettles, such as the Chesapeake *Chrysaora*, can be a painful nuisance to swimmers at certain times of the year when they occur in large numbers (Fig. 24.14). The sea wasps of the Indo-Pacific have such virulent nematocysts that their sting causes severe lesions and even death.

The epidermis of the scyphozoan bell margin contains two types of sensory structures. Clusters of photoreceptor cells form simple eyes, or **ocelli,** that detect general light intensity. Balancing organs, called **statocysts,** are composed of a vesicle with an interior body **(statolith)** in contact with hairlike endings of receptor cells located in the vesicle wall (Fig. 24.10). The gravitational responses of the statolith stimulate different receptor cells depending upon the orientation of the animal.

Ocelli and statocysts occur in hydromedusae, but in the scyphomedusae they are associated at specific sites between the scallop-like folds of the bell margin to form a **rhopalium** (Figs. 24.15 and 24.16). The scyphomedusae are larger than hydromedusae. Their swimming movements are similar and are primarily vertical. Horizontal movement depends largely upon water currents. If the bell becomes tilted, stimuli from the statocysts bring about asymmetrical contractions and the bell is righted. There is no velum in scyphomedusae.

Scyphozoans feed on animals of various sizes, including fish, that come in contact with the tentacles and oral arms. Some species, such as *Aurelia*, feed on plankton that adheres to the undersurface of the bell. The plankton is then swept by flagellated surface cells to the bell margin, which is wiped by the oral arms. There is a central stomach, as in the hydromedusae, but the stomach floor bears filaments containing cnidocytes (Fig. 24.15). These gastrodermal nematocysts perhaps function to quell any prey still alive when it enters the stomach. In most of the commonly encountered north temperate jellyfish, numerous canals extend radially to the bell margin. A ring canal may or may not be present.

Gametes develop in the gastrodermis of the stomach floor, and the "gonads" are often conspicuous within the transparent body (Fig. 24.15). The eggs and sperm are shed through the mouth. Fertilization and early development may occur in the sea water, or the eggs may be brooded on the oral arms. In either case, a free-swimming planula larva is attained. The planula develops into a polypoid larval stage,

Figure 24.14 The sea nettle, *Chrysaora quinquecirrha*, a common scyphozoan along the Atlantic coast. (Courtesy of William H. Amos.)

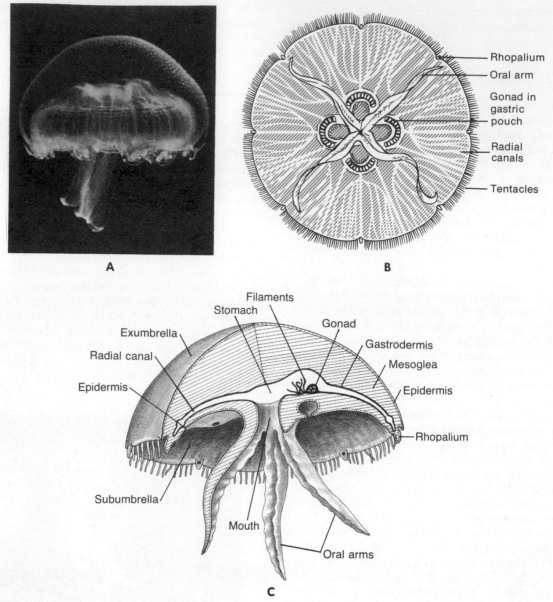

Figure 24.15 *Aurelia,* a scyphozoan medusa. *A,* Young specimen with bell contraction; *B,* ventral view; *C,* side view in section. (*A,* courtesy of D. P. Wilson.)

called a **scyphistoma,** that is about the size of a hydra and lives attached to the bottom (Fig. 24.17). The scyphistoma has a life span of one to several years during which it feeds and produces more scyphistomae by budding. At certain seasons it ceases feeding and undergoes a special form of budding to produce young jellyfish. The buds are produced at the oral end of the body and in the course of formation are stacked up like plates **(strobila).** When detached, the young jellyfish **(ephyra)** is tiny and displays only a rudimentary medusoid form

(Fig. 24.17). There are some species in which the entire scyphistoma transforms into a little medusa.

In describing the evolution of hydrozoans, we theorized that the polyps evolved from a larval stage. Note that the scyphozoan life cycle displays just such a polypoid larva, which by budding produces both more polyps and medusae. Thus this theory of the origin of the polypoid stage is supported by evidence from comparative development of scyphozoans as well as hydrozoans.

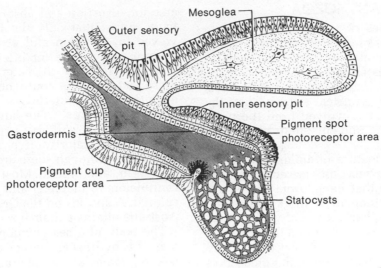

Figure 24.16 Section through a rhopalium showing the hood and the various sensory areas. (Modified from Hyman after Schewiakoff.)

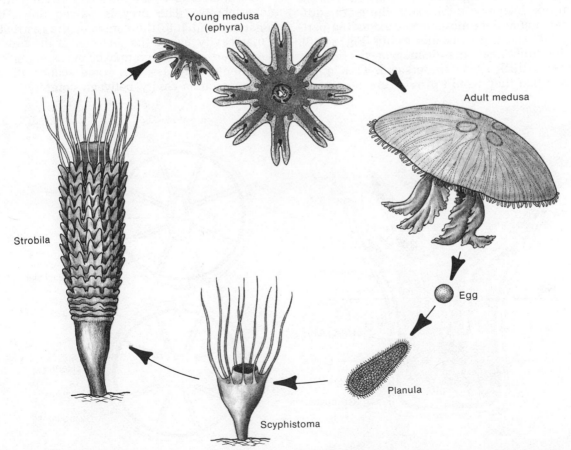

Figure 24.17 Life cycle of the scyphozoan *Aurelia*. The polypoid stage is a larva and produces medusae by transverse budding.

24.8 CLASS ANTHOZOA

The 6000 species of anthozoans constitute the largest class of cnidarians and include the familiar sea anemones and various types of corals. The class is entirely polypoid, and the cellular mesoglea, gastrodermal nematocysts and gastrodermal gametes suggest that the anthozoans may have evolved from the polypoid larva of some ancient group of scyphozoans.

The distinctive feature of anthozoans is the presence of a **pharynx** and **mesenteries**. The pharynx is a tube that hangs from the mouth into the gastrovascular cavity like a sleeve (Fig. 24.18). Since the pharynx is derived from an infolding of the body wall around the mouth, it possesses the same layers as the outer body wall but reversed; the lumen of the pharynx is lined with epidermis. The mesenteries are sheetlike partitions that extend into the gastrovascular cavity from the outer body wall toward the pharynx (Fig. 24.18). Each is composed of two layers of gastrodermis separated by a layer of mesoglea. At least eight of the mesenteries, called complete mesenteries, connect with the mesoglea and gastrodermis of the pharynx. In many anthozoans, there are additional incomplete mesenteries extending partway into the gastrovascular cavity (Fig. 24.20).

The functional significance of the mesenteries is difficult to understand. One might expect that they would provide greater surface area for digestion and absorption, but studies have shown that only the free margin is involved in digestion and absorption. The mesenteries may serve for internal gas exchange and may limit the diameter of the body, but these factors alone could not account for the evolution of mesenteries.

Sea Anemones. The largest of the anthozoans are the solitary sea anemones (Fig. 24.19). The majority live attached to hard substrates, although there are some forms that burrow in sand or mud. Most are one to several centimeters in diameter and are often brightly colored. A species on the Great Barrier Reef of Australia attains a diameter of one meter.

The body of a sea anemone is heavier than that of a hydrozoan polyp, and the oral end, bearing tentacles peripherally and a central slit-shaped mouth, is disclike. When sea anemones contract, the upper rim of the body is pulled over the oral disclike collar (Fig. 24.19). The muscle fibers are entirely gastrodermal. Circular fibers are located in the gastrodermis of the body wall, and longitudinal retractor fibers are located in the mesenteries.

Sea anemones feed upon small fish and other invertebrates. The prey is passed down the pharynx and into the center of the gastrovascular cavity below the pharynx. If the prey is large, partial digestion may be necessary before fragments can pass into some section of the gastrovascular cavity between complete mes-

Figure 24.18 Structure of an anthozoan polyp. *A*, Longitudinal section; *B*, cross section taken at level of pharynx; *C*, cross section taken below level of pharynx.

A

B

Figure 24.19 Sea anemones. *A,* View of expanded oral end. *B,* A contracted sea anemone in which the oral surface is partially covered by the upper surface of column. This species is typically clothed with shell fragments. (By Betty M. Barnes.)

enteries. Sea anemones possess at least 12 mesenteries arranged as six couples that extend across the gastrovascular cavity between the body wall and the pharynx (complete mesenteries) and usually six or more couples that extend from the body wall only part way into the gastrovascular cavity (incomplete mesenteries) (Fig. 24.20). The free edge of the mesenteries bears a glandular ciliated trilobed band called the mesenterial filament. This is the location of the gastrodermal cnidocytes, of enzyme production for extracellular digestion and of intracellular digestion and absorption. A perforation in the upper part of each complete mesentery permits circulation of the gastrovascular contents.

Some sea anemones reproduce asexually by splitting of the body or by the regeneration of fragments of tissues separated from the base of the body. The sexes are usually separate, and the gametes develop in gastrodermal bands just behind the free edge of the mesenteries (Fig. 24.20). Fertilization and early development may occur externally in the sea water or within the gastrovascular cavity. The planula larva develops into a ciliated planktonic polypoid larva in which mesenteries and pharynx make their appearance. The polyp soon settles and becomes attached as a young sea anemone.

Scleractinian Corals. Most other anthozoans are various types of corals. The species with which the name coral is most generally associated are the scleractinian, or stony, corals. The polyps of scleractinian corals are

Figure 24.20 Structure of a sea anemone. *A*, Longitudinal section; *B*, cross section at the level of the pharynx. (*A* after Hyman.)

similar to those of sea anemones but are usually smaller and almost always are connected together in colonies (Fig. 24.21). The connection is by means of a lateral fold of the body wall, and the entire colony has the form of a sheet. A calcium carbonate skeleton is secreted by the epidermis of the undersurface of the connecting sheet and the lower part of the polyp. The living colony is thus sitting on top of a skeleton that is actually external. The lower part of the polyp is situated within a skeletal cup, the bottom of which contains radiating septa that project up into folds in the base of the polyp (Figs. 24.21 and 24.22).

Species of corals display various growth forms. Some are low and encrusting; others are upright and branching (Fig. 24.22). The surface configuration of the skeleton depends upon the relative positions of the polyps. The coral appears pockmarked when the polyps are well separated. In brain coral, which has a surface configuration of troughs and ridges, the polyps are joined together in rows (Fig. 24.22). No skeletal material is deposited between the polyps within a row, but skeleton is deposited between the rows.

The gastrodermal cells of most scleractinian corals contain symbiotic algae (zooxanthellae). The algae utilize the nitrogenous and carbon dioxide wastes of the coral and facilitate the deposition of calcium carbonate ($CaCO_3$). The coral obtains glycerol from the algae. Phosphate is cycled between the two. When either light or algae are absent, $CaCO_3$ deposition is greatly reduced. Corals with symbiotic algae do not occur at depths below the penetration of light.

Octocorallian Corals. Most other an-

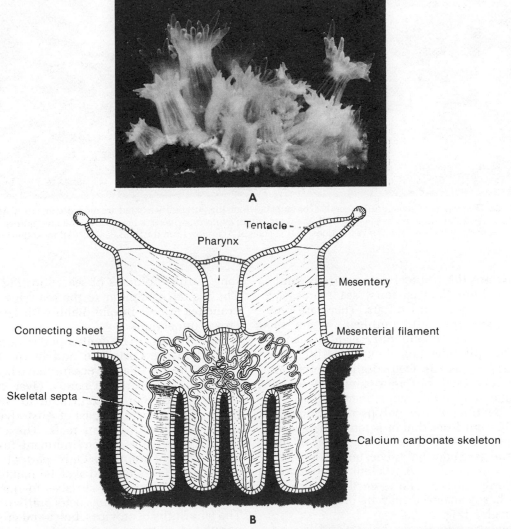

Figure 24.21 *A*, Polyps of the star coral; *B*, a coral polyp within its skeletal cup. (*A*, courtesy of the American Museum of Natural History; *B* after Hyman.)

Figure 24.22   Skeletons of scleractinian corals. The cups in which the polyps are located are most distinct in *A, Montastrea cavernosa*, a West Indian platelike species. In *B*, the West Indian *Diploria strigosa*, a brain coral, the polyps are arranged in rows. *C*, A species of *Porites* from the West Indies. *D*, A species of *Acropora* from the Indo-Pacific. (All by Katherine Barnes.)

thozoans are the tropical octocorallian corals, which include the sea fans, sea whips, sea pens, sea pansies and others. There are only eight mesenteries, all complete, within the body of the polyp and only eight tentacles, which are **pinnate**; i.e., they have little side branches (Fig. 24.23). Octocorallian polyps are usually very small but are organized into colonies that may attain considerable size. The interconnection of the polyps of a colony is quite different from that of scleractinian coral. A common mass of tissue, composed of mesoglea and gastrodermal tubes, unites the lower half of the colony. A skeleton of separate microscopic pieces and sometimes also an organic horny material may be present *within* the mesoglea (Fig. 24.23).

A skeleton supports the colony and is especially important in large upright forms. A central horny skeletal core extends through the long rodlike colonies of sea whip (Fig. 24.24). The sea fan is similar to the sea whip, but its branches are all in one plane with lattice-like cross connections.

Coral Reefs.   The contemporary coral reefs, located in the Caribbean and in the tropical parts of the Indo-Pacific oceans, have been built chiefly by scleractinian corals. Over 200 species of living corals contribute to the Great Barrier Reef along the coast of Australia and 35 species to the Caribbean reefs. The coralline foundation creates an environment for a vast assemblage of animals. Only part of the exposed coral is living. Old skeletal remains form attachment and boring surfaces for many sessile animals, including octocorallian corals. The labyrinths of crevices, holes and grottos are populated with a great number of different invertebrates and vertebrates.

Coral reefs can be classified into three major

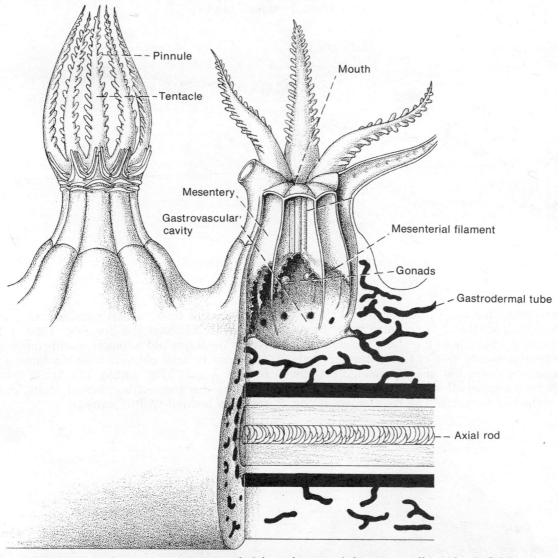

Figure 24.23 Structure of a gorgonian coral of the anthozoan subclass Octocorallia. (After R. C. Moore.)

Figure 24.24 An octocorallian coral. This highly branched gorgonian looks very much like a plant. Just to the left of its base is a scleractinian coral. (By David Barnes.)

Figure 24.25   Zonation of a coral reef.

types, depending upon their location. **Fringing reefs,** which are the most common type of reef, are located adjacent to the shore of islands or continental coasts (Fig. 24.25). **Barrier reefs** parallel the coast but are separated from the shore by a lagoon of varying width. There are not many barrier reefs. The best known is the

Great Barrier Reef, which parallels the northeast coast of Australia for over 1000 miles. **Atolls** lie above old submerged volcanoes and are more or less circular reefs containing an interior lagoon (Fig. 24.25). The South Pacific contains the greatest number of atolls. **Patch reefs** are located within lagoons.

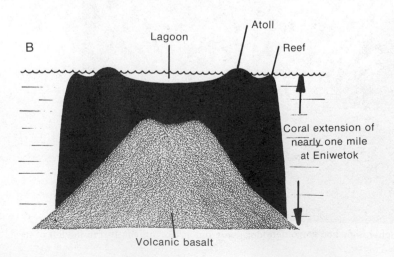

Figure 24.26 Formation of an atoll. A, Fringing reef around an emergent volcano. B, Continuous deposition of coral as volcanic cone subsides leads to the formation of a great coralline cap; emergent part of cap is atoll.

Most reefs have one side, the **reef front,** facing the open ocean (Fig. 24.25). Behind the reef front, which rises close to the surface, is a **reef flat** covered by only a meter or less of water. The greatest growth of large massive corals is on the reef front. The shallow reef flat may contain pools, sand beds and rocky surfaces and supports a very different fauna from that on the reef front.

As the reef increases in thickness, there is a gradual compaction of the lower deposits. Core drillings of coral reefs have disclosed coralline deposits of great depths. On the Pacific atoll of Eniwetok, coral limestone extends downward for almost a mile before reaching basaltic rock. Since the deposition of coral by living colonies occurs only in the upper lighted zone, thick deposits of coral can only be explained by fluctuations in sea level and by subsidence of the bedrock upon which the reef is resting. Both have occurred at Eniwetok. The geologic history of Eniwetok began during the early Cenozoic Period, when a fringing reef developed around an emergent volcanic cone (Fig. 24.26). This oceanic peak gradually subsided to below sea level. Coral deposition occurred at a rate equalling subsidence, and the original fringing reef was transformed into an atoll resting on vertical walls almost one mile in height (Fig. 24.26). Fluctuating sea level periodically caused emergence and submergence. The rise in sea level from the last Pleistocene ice age low 120 meters below the present level gave Eniwetok its present emergent form.

Subsidence of the Gulf Basin and fluctuating sea level have also contributed to reef formation in the Caribbean. Submarine caverns with stalactites in the Bahamas indicate a period of past emergence and subsequent submergence.

Although existing coral reefs have been formed by scleractinian coral, reefs in the past have been produced by other organisms. Two extinct groups of anthozoans, the tabulate and rugose corals, formed great reefs during the Paleozoic. Reefs have also been formed in the past by algae that secrete calcium carbonate. Coralline algae are still prevalent today and are common members of reef communities, but there are no existing reefs composed primarily of living coralline algae.

CLASSIFICATION OF CNIDARIANS

PHYLUM CNIDARIA

CLASS HYDROZOA. Common but inconspicuous forms, usually with both polypoid and medusa stages. There are many colonial polypoid forms, called hydroids, that are surrounded by a chitinous skeleton.

Order Trachylina. Medusoid hydrozoans lacking or having a poorly developed polypoid generation. This order contains perhaps the most primitive members of the class.

Order Hydroida. Hydrozoans with a well developed polypoid generation. Medusoid stage present or absent. The majority of hydrozoans belong to this order.

Order Siphonophora. Pelagic hydrozoan colonies of polypoid and medusoid individuals. Colonies with float or large swimming bells. Largely in warm seas. *Physalia* (Portuguese man-of-war). *Stephalia, Nectalia.*

Order Milleporina. Colonial polypoid forms having a heavy calcareous skeleton. Free, minute medusa. Single genus *Millepora,* stinging coral or fire coral, common in tropical waters.

Order Stylasterina. Colonial polypoid hydrozoans provided with a calcareous skeleton. Usually with an upright, branching growth form. Medusa absent. Mostly tropical. *Stylaster.*

CLASS SCYPHOZOA. Medusa stage is most conspicuous, with bell diameters commonly from 2 to 40 cm. Widespread marine forms, mostly free-swimming.

Order Stauromedusae, or *Lucernariida.* Sessile scyphozoans attached by a stalk on the aboral side of the trumpet-shaped body. Chiefly in cold littoral waters.

Order Cubomedusae, or *Carybdeida.* Scyphomedusae with bells having four flattened sides and simple margins. Tropical and subtropical oceans.

Order Coronatae. Bell of medusa with a deep groove or constriction, the coronal groove, extending around the exumbrella. Many deep sea species.

Order Semaeostomae. Scyphomedusae with bowl-shaped or saucer-shaped bells having scalloped margins. Gastrovascular cavity with radial canals or channels extending from central stomach to bell margin. Occur throughout the oceans of the world.

Order Rhizostomae. Bell of medusa lacking tentacles. Oral arms of manubrium branched and bearing secondary mouth openings that lead into arm canals. Arm canals of manubrium pass into stomach. Original mouth lost through fusion of oral arms. Tropical and subtropical littoral scyphozoans.

CLASS ANTHOZOA. Either solitary or colonial, with medusoid stage completely absent. Includes sea anemones and corals, plus many other forms.

Subclass Octocorallia. Polyps with eight mesenteries and eight pinnate tentacles. All colonial.

Order Telestacea. Lateral polyps on simple or branched stems. Skeleton of calcareous spicules.

Order Alcyonacea. Soft corals. Coenenchyme forming a rubbery mass and colony having a massive, mushroom or a variously lobate growth form. Skeleton of separate calcareous spicules. Largely tropical.

Order Coenothecalia. Contains only the Indo-Pacific blue coral, *Heliopora*, having a massive calcareous skeleton.

Order Gorgonacea. Horny corals or gorgonian corals. Common tropical and subtropical octocorallian coelenterates having a largely upright plant-like growth form and a skeleton of a hornlike organic material. Separate or fused calcareous spicules may be present.

Order Pennatulacea. Sea pens. Colony having a fleshy, flattened or elongate body, or rachis. Skeleton of calcareous spicules.

Subclass Zoantharia. Polyps with more than eight tentacles, and tentacles rarely pinnate. Solitary or colonial.

Order Zoanthidea. Small colonial anemone-like anthozoans having one siphonoglyph and no skeleton.

Order Actiniaria. Sea anemones. Solitary anthozoans with no skeleton, with mesenteries in hexamerous cycles, and usually with two siphonoglyphs.

Order Scleractinia or *Madreporaria.* Stony corals. Mostly colonial anthozoans secreting a heavy external calcareous skeleton. Sclerosepta arranged in hexamerous cycles. Many fossil species.

Order Rugosa, or *Tetracoralla.* An extinct order of mostly solitary corals possessing a system of major and minor radiating sclerosepta. Cambrian to Permian.

Order Corallimorpharia. Tentacles radially arranged. Resemble true corals but lack skeletons.

Order Ceriantharia. Anemone-like anthozoans with greatly elongate bodies adapted for living in sand burrows. One siphonoglyph; mesenteries all complete.

Order Antipatharia. Black or thorny corals. Gorgonian-like species with upright, plantlike colonies. Polyps arranged around an axial skeleton composed of a black thorny material and bearing thorns. Largely in deep water in tropics.

Subclass Tabulata. Extinct colonial anthozoans with heavy calcareous skeletal tubes containing horizontal platforms, or tabulae, on which the polyps rested. Sclerosepta absent or poorly developed.

ANNOTATED REFERENCES

Detailed accounts of cnidarians may be found in the references listed at the end of Chapter 21.

Muscatine, L. and H. M. Lenhoff: Coelenterate Biology. Academic Press, N.Y., 1974. Review papers on many aspects of the biology of cnidarians.

THE FLATWORMS

1. *Members of the phylum Platyhelminthes, called flatworms, are free-living or parasitic animals whose bodies are greatly flattened dorso-ventrally.*

2. *Free-living species are usually less than 3 cm. in length and are found in the sea and in fresh water.*

3. *Parasitic species, called flukes and tapeworms, are external and internal parasites of many kinds of animals, including vertebrates. Many species require more than one host to complete the life cycle.*

4. *The body of a free-living flatworm is covered with cilia, which are used in locomotion. Adult parasitic forms lack external cilia, and the body is covered by a cuticle.*

5. *Free-living flatworms are carnivores or scavengers. The mouth is the only opening into the digestive tract and digestion is both extracellular and intracellular. A digestive tract is absent in tapeworms.*

6. *There is no internal transport system; protonephridia are present; and the nervous system is composed of a varying number of longitudinal cords.*

7. *Flatworms are hermaphroditic, with relatively complex reproductive systems. Sperm transfer is reciprocal and fertilization is internal.*

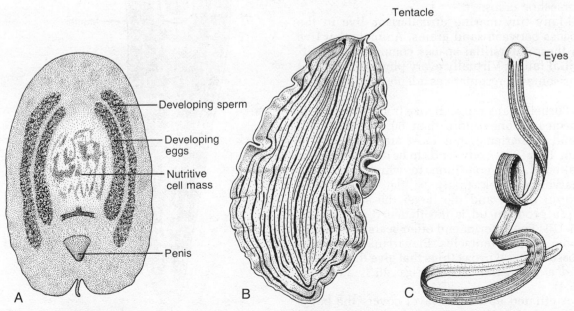

- Tentacle
- Eyes
- Developing sperm
- Developing eggs
- Nutritive cell mass
- Penis

A   B   C

Figure 25.1 Turbellarian flatworms. *A, Polychoerus,* an acoel; *B, Prostheceraeus,* a polyclad; *C, Bipalium kewense,* a cosmopolitan land planarian found in Florida, Louisiana, California and in greenhouses throughout the United States. (*A* from Barnes: Invertebrate Zoology. 3rd ed. Philadelphia, W. B. Saunders Company, 1974; *B* redrawn from a photograph by D. P. Wilson; *C* after Hyman.)

The flatworms and all the remaining members of the Animal Kingdom are either bilaterally symmetrical or, if radially symmetrical, have clearly evolved from bilateral ancestors. The body can be cut along only one plane—anterior to posterior, dorsal to ventral—to produce two mirror image halves. Bilaterality and cephalization (head development) are widespread animal features correlated with motility.

Most zoologists believe that the flatworms, members of the phylum **Platyhelminthes,** are probably the most primitive of all the bilateral animals. The phylum is composed of three classes. The class **Turbellaria** contains the **free-living flatworms.** The class **Trematoda,** the **flukes,** and the class **Cestoda,** the **tapeworms,** are both entirely parasitic and probably evolved independently from turbellarians.

25.1 CLASS TURBELLARIA

Members of the class Turbellaria, the free-living flatworms, occur both in the sea and in fresh water, and a few species are terrestrial in humid forests. The aquatic turbellarians are bottom dwellers, living within algal masses and beneath stones and other objects. The common and familiar freshwater **planarians** inhabit the undersurface of stones in springs, streams and lakes. There are also marine flatworms that live with other animals. *Bdelloura,* for example, lives on the gills of horseshoe crabs.

Many tiny marine turbellarians live in the spaces between sand grains. Animals that live in these interstitial spaces compose the **interstitial fauna.** Virtually every phylum of animal has some representatives adapted for this habitat.

Turbellarians range in size from microscopic species to ones more than 60 cm. in length (land planarians), but most are less than 10 mm. long. The body tends to be dorso-ventrally flattened, the condition to which the name Platyhelminthes (*platy* = flat; *helminth* = worm) refers, and the larger the species, the more pronounced is the flattened shape (Fig. 25.1*B*). The anterior end often bears eyes and in some species tentacles. Planarians commonly possess lateral projections that give the anterior end a triangular shape (Figs. 25.1, 25.2 and 25.4).

A ciliated epidermal layer covers the body, although in the planarians only the ventral surface is ciliated. Beneath the epidermis is a muscle layer of circular, diagonal and longitudinal fibers (Fig. 25.3). A network of loosely connected cells, called **mesenchyme,** surrounds the gut and other organs and fills the interior of the body (Fig. 25.3). Flatworms are therefore said to possess a solid, or **acoelomate,** body structure, there being no cavity between the body wall and the internal organs.

Figure 25.2 *Dugesia.* Dorsal view, showing the digestive and nervous systems. The mouth opens ventrally. (After Hyman).

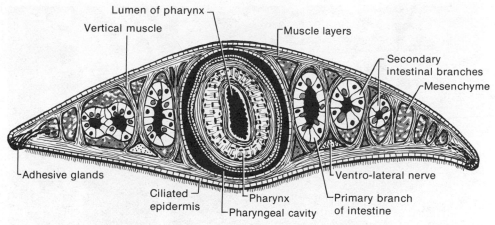

Figure 25.3 Diagrammatic cross section of the planarian *Dugesia* at the level of the pharynx.

25.2 LOCOMOTION

Very small flatworms are not greatly flattened and swim or crawl about bottom debris by ciliary propulsion. Contractions of the muscle layer permit turning, twisting and folding of the body (Fig. 25.4). The movement of turbellarians over 3 mm. in length involves delicate undulatory waves of muscle contraction in addition to cilia. The dorso-ventral flattening of the body is probably in part an adaptation for locomotion. With increased size, a flattened shape preserves a large surface area upon which the body can be carried. Glands in the epidermis and underlying tissue secrete a mucous film over which the animal glides.

25.3 NUTRITION

Turbellarians are **carnivorous,** feeding upon other small invertebrates. The mouth is located along the mid-ventral line. In most species, the gut is differentiated into a pharynx adapted for ingestion, and an intestinal sac that functions in digestion and absorption. No anus is present

and wastes are ejected through the mouth, as in cnidarians.

In small turbellarians the intestine is a simple unbranched sac and the pharynx is a short nonmuscular ciliated tube or a muscular bulb that can be projected from the mouth (Fig. 25.5). Most of the larger flatworms, including the freshwater planarians, have a tubular pharynx and a highly branched intestinal sac (Fig. 25.2). The tubular pharynx is muscular and, except for the end attached to the intestinal sac, lies free within a pharyngeal cavity. The cavity opens to the exterior through the mouth. When the animal feeds, the pharyngeal tube is projected out of the mouth (Fig. 25.7). In some flatworms, the pharynx is directed anteriorly (Fig. 25.6); in others, including the planarians, the pharynx is directed posteriorly. In flatworms with the latter condition, the mouth is located behind the middle of the body (Fig. 25.2).

During feeding, flatworms crawl upon their prey, pinning them down and enveloping them with mucus (Fig. 25.4). The prey is often swallowed whole. However, planarians, as well as many other species with a tubular

Figure 25.4 Hunting and feeding in *Dugesia*. A small crustacean (*Daphnia*) is captured and eaten, its tough exoskeleton remaining as an empty shell.

1 CM

A

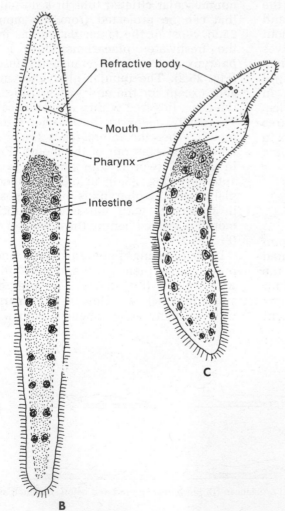

Refractive body

Mouth

Pharynx

Intestine

C

B

Figure 25.5 *A*, A large West Indian polyclad crawling on a sea fan. The light marginal band is orange. (By Betty M. Barnes.) The catenulid *Stenostomum*, a common microscopic freshwater turbellarian. *B*, Dorsal view. *C*, Contracted specimen with anterior end turned in lateral view. A nonmuscular ciliated pharynx connects the anterior ventral mouth and the intestine, which is a simple elongated sac.

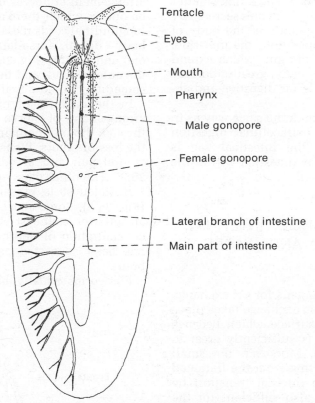

Figure 25.6 Digestive system of a polyclad. (After Hyman.)

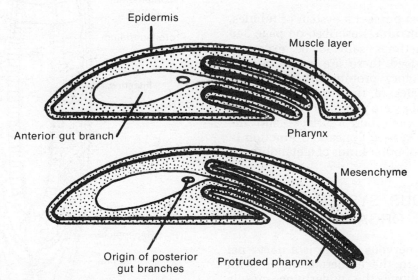

Figure 25.7 Longitudinal section of the freshwater triclad, *Polycelis*, showing retracted and protruded tubular pharynx. (After Jennings.)

pharynx, do not swallow the prey entirely. Rather, the extended pharynx is inserted into the body of the prey or into dead animal matter with the aid of proteolytic enzymes secreted by pharyngeal glands. Fragments of the body of the prey are then pumped into the intestine. The many divisions of the gut, which extend throughout much of the body, greatly increase the surface area available for digestion and absorption.

Digestion occurs in much the same way as in cnidarians. An initial extracellular digestion within the lumen of the intestinal sac is followed by intracellular digestion within the cells of the intestinal wall.

25.4 GAS EXCHANGE, INTERNAL TRANSPORT AND WATER BALANCE

There are no special organs for gas exchange or internal transport. Gas exchange takes place across the general body surface, which, because of its flattened shape, is sufficiently great to meet oxygen demands. Moreover, the small vertical distances resulting from the flattened shape greatly facilitate internal transport by diffusion. Diffusion is also sufficient for the movement of food materials, for in the large flatworms the gut branches are so extensive that no tissue is very far from some intestinal cells.

Most flatworms possess a system of tubules, called **protonephridia,** described on page 346. These are sometimes said to be excretory organs, and indeed there are some animals possessing excretory protonephridia, but the nitrogenous wastes of flatworms, which are chiefly ammonia, are removed by general diffusion across the body surface, and the protonephridial system appears to function in the elimination of other kinds of metabolites.

25.5 NERVOUS SYSTEM AND SENSE ORGANS

The simplest nervous system is a **nerve net** system located in the epidermal layer, with processes supplying the underlying muscle fibers.

In the majority of flatworms, however, the neurons are organized in bundles, or nerves, which lie just below the epidermis. The neuronal processes are not well differentiated into dendrites or axons, and there are few

ganglia; that is, the cell bodies of the neurons are not clustered together. The most primitive arrangement of nerves in flatworms appears to be one in which there are four or five pairs of longitudinal cords radially arranged around the body—dorsal, dorso-lateral, lateral, ventro-lateral and ventral (Fig. 25.8). The cords merge at the anterior end, and the slight enlargement is sometimes called a brain.

In the majority of turbellarians, the radial arrangement of nerve cords has been lost through the predominance of certain pairs and the loss of others. Planarians have a very large ventral pair and a reduced lateral pair (Fig. 25.3). The other pairs are absent.

A statocyst is present in many flatworms (Fig. 25.8). Some acoels with a nerve net have an anterior statocyst surrounded by a heavy concentration of neurons. This concentration has been suggested as the forerunner of a brain.

Flatworms usually have two or more eyes

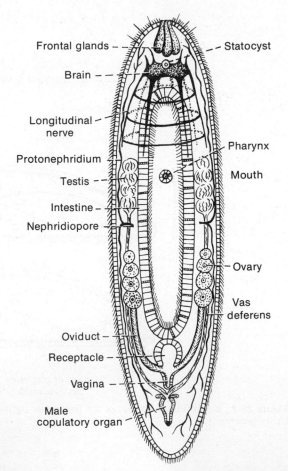

Figure 25.8 Dorsal view of a generalized rhabdocoel turbellarian. (Modified from Ax.)

- Pigment cup
- Light sensitive portions of photoreceptor cells
- Nuclei of photoreceptor cells
- Nerve to brain

Figure 25.9 Diagrammatic section through the eye of a planarian. Light reaches the sensitive elements from the right.

(Fig. 25.2), but they are never highly developed. They are generally located at the anterior end, sometimes on or around tentacles, but in some flatworms, the eyes are located along the sides of the body. The simplest eyes, found in acoels, are merely patches of photoreceptors and pigment cells in the epidermis. Usually, however, the eye has a cuplike form resulting from the invagination of the pigment cells (Fig. 25.9). In some animals the opening of the retinal cup becomes filled in and covered with a lens-cornea, but lenses are not present in most flatworm eyes.

Flatworms avoid bright light and the response is probably an adaptation to remain under cover.

25.6 REGENERATION AND REPRODUCTION

Many flatworms are capable of asexual reproduction, and this ability is closely correlated with the ability to regenerate. The most common method of asexual reproduction is by **transverse fission.** In many freshwater planarians, for example, a fission plane forms behind the pharynx, and during movement the animal breaks in two. Following fission, the two halves regenerate the missing parts. In a number of small flatworms, such as *Stenostomum* and *Microstomum*, fission planes and regeneration proceed more rapidly than the separation of the parts, and chains of individuals are formed as a result (Fig. 25.10).

Mesenchyme is the principal source of new cells for regeneration, which involves both reorganization of remaining tissues and addition of new cells. As in cnidarians, the regenerative parts retain their original polarity (Fig. 25.11). A piece taken from the middle of the body always regenerates a new anterior end at the severed anterior surface and a new posterior end at the posterior surface. There is

also a lateral polarity; thus, a two-headed planarian can be produced by cutting the anterior end longitudinally along the midline (Fig. 25.11). The rate of regeneration of pieces taken at different levels reflects a distinct metabolic gradient along the anterior-posterior axis. Anterior pieces regenerate more rapidly than do posterior pieces of equal size. Or, reflecting the same phenomenon, posterior pieces must be larger than anterior pieces for regeneration to be completed in the same length of time.

Flatworms are hermaphroditic, but most exhibit cross-fertilization (a condition true of hermaphroditic animals in general; self-fertilization is rare). There is usually simultaneous and mutual copulation (Fig. 25.12), the penis of each animal being received by the female system of the other animal, and the sperm are stored for a period of time. The eggs are fertilized internally and deposited on the bottom in jelly masses or in cocoons.

The reproductive systems of most turbellarians are complex and highly variable but reflect adaptations for the reproductive conditions just described. The male system (Figs. 25.6, 25.8 and 25.13) may contain one testis or many pairs of testes. A small duct connects each testis with a main sperm duct that extends along each side of the body. The sperm ducts join together to form an ejaculation duct, which exits through a penis. The penis is muscular

Figure 25.10 Chain of attached daughter individuals resulting from incomplete transverse fission in *Stenostomum*. (After Child from Hyman.)

Figure 25.11 Polarity and regeneration in *Dugesia. Left*, each of five pieces regenerates, but the rapidity with which the head develops depends upon the level of the piece. *Right*, a two-headed form produced by repeated splitting of the anterior end.

and often armed with a stylet (but not in planarians). Parts of the sperm ducts are frequently modified as seminal vesicles. Glands are present, most commonly associated with the ejaculation duct. Commonly, the penis is located within a chamber, the genital atrium, which may also contain the terminal part of the female system. The atrium opens to the outside through a gonopore.

The female system (Figs. 25.6, 25.8 and 25.13) contains either a single ovary or one to many pairs of ovaries, but only a single pair of oviducts is present. In many turbellarians, including the planarians, there are yolk glands located along the length of the oviduct. Yolk cells are released as the eggs travel down the oviduct, and the eggs, when deposited, are surrounded by yolk. The deposition of yolk outside an egg cell (**ectolecithal**) is an unusual condition. In most animals the eggs are **entolecithal;** that is, the yolk is contained within the egg cytoplasm. A seminal receptacle, a vagina or atrium, and glands for the production of egg envelopes are typically present. There may be a separate female gonopore, or there may be a common genital atrium and gonopore for both systems.

The eggs of some turbellarians are deposited in gelatinous masses. Many species, especially freshwater forms, lay their eggs within capsules, or cocoons, which in the planarians are attached to stones or other objects by little stalks (Fig. 25.12).

Some freshwater turbellarians produce two kinds of eggs. Summer eggs have a thin shell and hatch within a short period of time; winter or dormant eggs have a thick resistant shell and are capable of withstanding cold and desiccation. Having two types of eggs is an adaptation shared with many other freshwater animals, especially those living in pools and small streams.

Spiral cleavage is present in turbellarians with entolecithal eggs but has been lost in the many flatworms with ectolecithal eggs. Some marine flatworms have larvae.

egg

Figure 25.12 *A*, The planarian *Dugesia* copulating; *B*, an egg cocoon of a freshwater planarian. (Both after Hyman.)

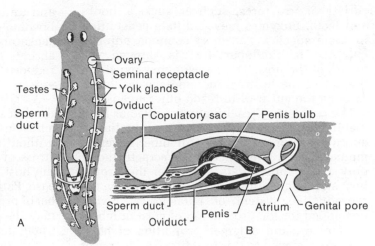

Figure 25.13 Reproductive system of the freshwater planarian *Dugesia*. *A*, A dorsal view showing the male organs on the left and the female organs on the right. *B*, A side view of the copulatory organs. (After Hyman.)

25.7 SYMBIOSIS AND PARASITISM

Before describing the two parasitic classes of flatworms, we need to consider briefly the general nature of parasitism. Many organisms depend for their existence upon an intimate physical relationship with other organisms. Such a relationship is known as **symbiosis.** The larger organism is termed the host and the smaller, the **symbiote.** The symbiote always derives some benefit from the relationship, but the consequence to the host varies. When both the symbiote and the host benefit, the relationship is called **mutualism.** Mutualism is the least common type of symbiosis, but we have already encountered examples among the sponges and cnidarians. Algae live within the amebocytes of freshwater sponges and within the gastrodermal cells of scleractinian corals. The metabolism of each species is enhanced by the presence of the symbiotic partner.

Commensalism, a more common symbiotic relationship, has evolved many times within the Animal Kingdom. In commensalism, the host is neither benefited nor harmed. The symbiote, or commensal, is protected by the host or utilizes excess food of the host. The shrimp living within the spongocoel of glass sponges and the clown fish living among the tentacles of sea anemones are commensals. The sponge and sea anemones can exist without their symbiotes, but the commensal shrimp and fish cannot exist alone.

The most common symbiotic relationship is **parasitism.** Here the host is harmed by the presence of the symbiote, or parasite. The relationship may not necessarily be seriously detrimental to the host species. Indeed, if the host species suffers too greatly from its parasites and becomes extinct, the parasite species is adversely affected.

Parasites that live on the outside of their host are said to be **ectoparasitic;** those on the inside, **endoparasitic.** Most parasites can utilize several closely related or ecologically similar species as hosts. Some parasites require two hosts to complete their life cycles, in which case the host for the larval or developmental stage is termed the **intermediate host;** the host for the adult stage is termed the **primary host.**

The principal benefit derived by a parasite from his relationship with the host is nutritive. The parasite feeds upon the host's tissues or body fluids or utilizes the food ingested by the host. When the parasite is enclosed by some part of the host's body, there may be secondary advantages, such as protection.

Although the symbiote gains an easy source of food material from a parasitic existence, the relationship poses problems and is not without cost. The primary problem of parasites is that of reaching and penetrating new hosts.

From the few generalizations made thus far about the nature of a parasitic existence, many of the adaptations that are encountered in parasites can be predicted.

The anterior end of the body, the mouth region and the anterior parts of the gut of parasites are commonly adapted for attaching, piercing and sucking. Ectoparasites which feed infrequently may have a part of the gut modified for storage. On the other hand, some endoparasites utilize digested food of the host and have lost the gut entirely.

The problems of penetration and attachment to the host have resulted in the evolution of a variety of structures, such as suckers, hooks and teeth. Enzymes may facilitate penetration in some species. The most common point of entrance for endoparasites is through the mouth of the host. The parasite may remain in the gut of the host or, in some species, break through the gut wall to reach other organs.

The problem of reaching new hosts has been met in most parasites through the production of enormous numbers of eggs or other developmental stages. The species is perpetuated if only a few individuals encounter the proper host and reach adulthood. Needless to say, the reproductive system of parasites is highly developed for the production of great numbers of gametes, and a larger proportion of body tissue is assigned to this function than is true of free-living animals. In fact, there are some parasites in which most of the body is concerned with reproduction.

The high reproductive output of parasites demands a high expenditure of energy. The energy costs are met by the host that is providing food for the parasite.

The structure of endoparasites usually reflects the less rigorous demands of the limited and uniform environment in which they live. Locomotor processes are reduced, and the sense organs found in free-living species are absent. The nervous system in turn is greatly reduced. Gas exchange organs are never present.

A "good" parasite doesn't kill its host, for a dead host results in a dead parasite. Moreover, a host may carry a small population of parasites without any serious consequences. Where the stress of a parasitic infection is manifested in the host, the condition is recognized as a disease. Parasitic disease may be caused by a number of processes of the parasite. Cells and tissues may be destroyed when the parasite feeds upon, penetrates into or moves within the host. Blood vessels, ducts or the gut may be clogged by parasites. The host may be robbed of food. The parasite may produce a substance that has a toxic or allergic effect on the host.

Some parasites cause the host's tissues to undergo abnormal mitotic activity and consequent growth. Plant galls, for example, represent the response of the surrounding plant tissues to the deposition of an egg by certain insects.

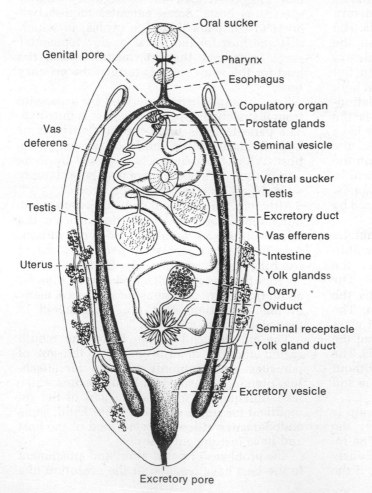

Figure 25.14 Structure of a generalized trematode. (After Chandler and Read.)

Oral sucker
Genital pore
Pharynx
Esophagus
Copulatory organ
Prostate glands
Vas deferens
Seminal vesicle
Ventral sucker
Testis
Excretory duct
Testis
Vas efferens
Intestine
Uterus
Yolk glandss
Ovary
Oviduct
Seminal receptacle
Yolk gland duct
Excretory vesicle
Excretory pore

Not only has symbiosis evolved many times in the Animal Kingdom, but it probably originated in different ways. Some symbiotes probably evolved from ancestral species that used the host simply as a substratum on which they attached or moved about; others evolved from species that used the host as one of a number of protective retreats, or habitations. Many parasites undoubtedly evolved from predatory ancestors, but some may have originated from forms that lived in decaying organic matter. For example, the bot flies, which lay their eggs in the wounds of cattle, probably evolved from flies that deposited their eggs in carrion.

Once having shifted over to a dependent symbiosis, many species, especially parasites, exploited the relationship and underwent adaptive radiation resulting in thousands of species.

The parasitic flatworms comprise two classes: the Trematoda, or flukes, and the Cestoda, or tapeworms. These two groups of flatworm parasites, along with the parasitic roundworm, include the great majority of parasitic worms of great economic and medical significance. The Trematoda and Cestoda evolved independently from the free-living turbellarians.

25.8 CLASS TREMATODA

The class Trematoda, to which the flukes belong, contains over 6000 species of ectoparasities and endoparasites. The majority are

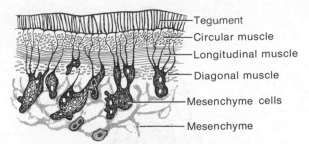

Figure 25.15 Part of the body wall of a trematode. Note that an epidermis is missing and that the covering tegument lies directly on tissue of mesodermal origin.

parasites of vertebrates, especially fish, but immature stages are harbored by invertebrates.

Trematodes are flattened and oval in shape (Fig. 25.14), the majority being not more than a few centimeters long. The body is covered by a nonciliated cytoplasmic syncytium, the tegument, the nuclei of which are sunken into the mesenchyme (Fig. 25.15). The mouth is located at the anterior end. Adhesive suckers are usually present around the mouth and may also be present mid-ventrally and at the posterior end. The digestive tract is composed of a muscular pumping pharynx behind the mouth, a short esophagus and usually two long blind intestinal sacs (Fig. 25.14). Trematodes feed on cell fragments, mucus, tissue fluids and blood of the host. Protonephridia are present, and the nervous system is similar to that of turbellarians, except that sensory structures are absent in endoparasitic forms.

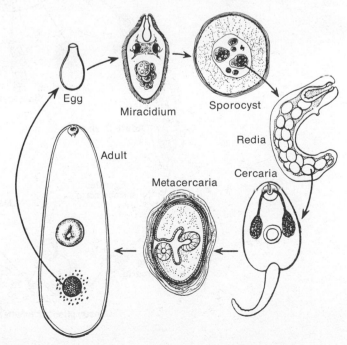

Figure 25.16 Life cycle of a digenetic fluke. The stages shown here belong to various species. The arrows indicate whether one stage becomes the next or whether it produces the next by reproduction. (After Hyman.)

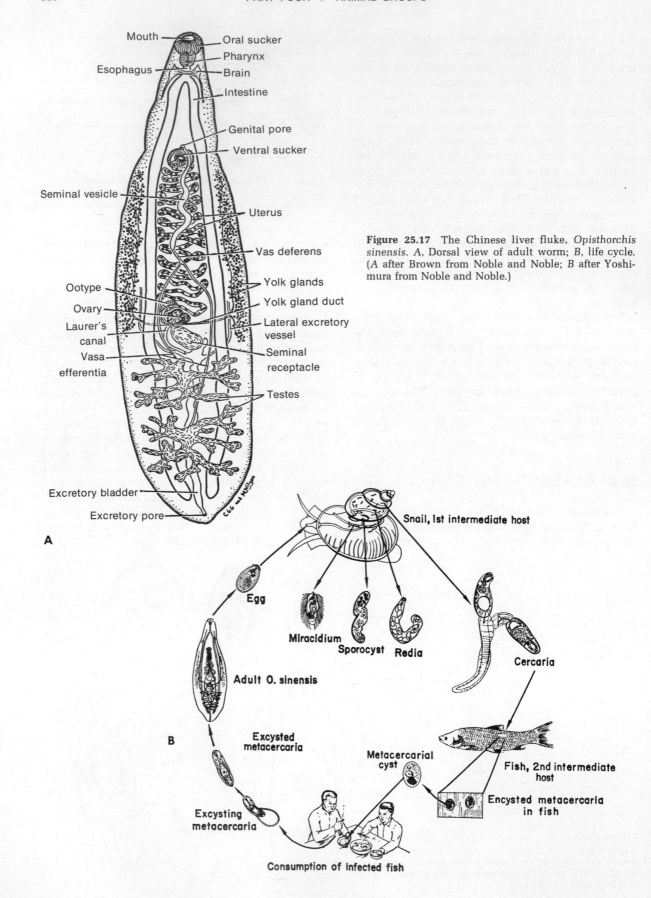

Figure 25.17 The Chinese liver fluke, *Opisthorchis sinensis*. *A*, Dorsal view of adult worm; *B*, life cycle. (*A* after Brown from Noble and Noble; *B* after Yoshimura from Noble and Noble.)

Mouth
Oral sucker
Pharynx
Esophagus
Brain
Intestine
Genital pore
Ventral sucker
Seminal vesicle
Uterus
Vas deferens
Yolk glands
Yolk gland duct
Ootype
Ovary
Laurer's canal
Lateral excretory vessel
Vasa efferentia
Seminal receptacle
Testes
Excretory bladder
Excretory pore

A

Snail, 1st intermediate host
Egg
Miracidium
Sporocyst
Redia
Cercaria
Adult O. sinensis
Fish, 2nd intermediate host
Encysted metacercaria in fish
Excysted metacercaria
Metacercarial cyst
Excysting metacercaria
Consumption of infected fish

B

Mutual copulation and cross fertilization is the rule, and the reproductive system is adapted for the production of a tremendous number of eggs, which are ectolecithal and surrounded by separate shells. Egg production has been estimated to be 10,000 to 100,000 times greater in trematodes than in the free-living turbellarians.

Life Cycles. Eggs are passed out of the host and hatch as mobile aquatic ciliated **miracidium** larvae (Fig. 25.16), which provide the means of reaching a new host. In many trematodes, development requires an intermediate host. The miracidium enters an invertebrate, commonly a snail, and develops further into a tailed **cercaria** larva (Fig. 25.16). The cercaria passes from the first intermediate to a second intermediate host, where it develops into a developmental stage, called a **metacercaria.** If the second intermediate host is eaten by the primary host, the metacercaria is freed from the tissues of the intermediate host and develops into an adult. There are many variations of this generalized outline of the life cycles and space only permits a few examples.

Bendenia melleni. This ectoparasite lives on the skin and eyes of a number of different marine fish. The worm can cause blindness and serious damage to the host's integument. The eggs, when shed, fall to the bottom. A larva, similar to a miracidium and having a preco-cious adhesive organ, escapes from the egg, and if a host fish is encountered, attaches to the skin. Following attachment and while feeding upon the host, the miracidium develops into an adult parasite.

Echinostoma revolutum. A great many trematodes infect the gut or gut derivatives of their primary host. Lungs, bile ducts, pancreatic ducts and intestines are common sites. The life cycle of *Echinostoma revolutum*, which infects the intestine of a variety of aquatic birds and mammals, is a good illustration of the generalized life cycle described earlier. The eggs pass out with the host's feces and hatch in water. The free-swimming miracidium can penetrate and develop within species of three different genera of snails. A cercaria larva leaves this snail and penetrates another snail, where it undergoes further development. If the second intermediate host is ingested by the primary host, the immature worm is liberated and develops into an adult.

Opisthorchis sinensis. This species, commonly called the Chinese liver fluke, lives in the bile ducts of humans, dogs, cats, foxes and other fish-eating mammals (Fig. 25.17). The intermediate hosts for the miracidium and cercaria are a snail and a fish respectively. Human infestation has been common in the Orient because human feces are used to fertilize ponds and raw fish from these ponds are eaten. A few worms cause no disease symp-

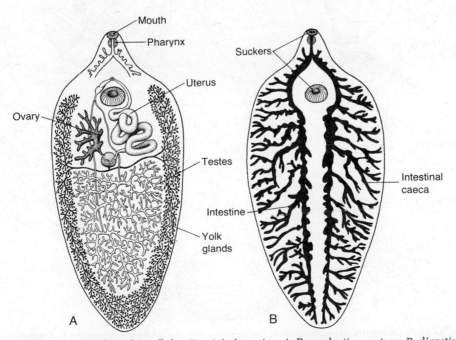

Figure 25.18 Structure of the sheep liver fluke, *Fasciola hepatica*. A, Reproductive system; B, digestive system.

toms, but several hundred can cause destruction of liver tissue, clogging of ducts, formation of bile stones and hypertrophy of the liver.

Fasciola hepatica. Known as the sheep liver fluke, this species is also an inhabitant of its host's bile ducts, and humans, cattle, pigs, rodents and other mammals may be hosts in addition to sheep (Fig. 25.18). The life cycle is similar to that of the Chinese liver fluke, but a snail is the only intermediate host. The cercariae leave the snail and encyst as metacercariae on vegetation along the edges of ponds and streams.

25.9 CLASS CESTODA

About 3400 species of cestodes, or tapeworms, have been described. All are endoparasites, and the majority are adapted for living in the gut of vertebrates.

The body form of adult tapeworms is quite different from that of turbellarians and trematodes. The anterior end consists of a knoblike **scolex** containing suckers and hooks that anchors the worm to the host (Fig. 25.19).

A narrow neck region connects the scolex to the **strobila,** which makes up the greater part of the tapeworm body. The strobila is composed of flattened sections, called **proglottids,** arranged in a linear series. New proglottids are continually being formed in the neck region and old ones detached at the end of the strobila. Tapeworms are generally long, and some species, with hundreds of proglottids, reach lengths of 40 feet.

As in trematodes, the body is covered by a tegument. However, a digestive system is completely absent and digested food of the host is absorbed through the tegument. Longitudinal nerve cords and excretory ducts run the length of the strobila. Much of the body tissue is given over to the reproductive system, which is complete within each proglottid (Fig. 25.19). A common gonopore is present on one edge of each proglottid. Mutual copulation between the proglottids of two different worms generally occurs, but copulation between proglottids on the same strobila and self-fertilization within one proglottid are known.

The eggs, which are surrounded by a shell as in trematodes, may be continually shed

Figure 25.19 The pork tapeworm, *Taenia solium.* Insets show the head and an immature and a mature section of the body. (Villee, C. A.: Biology. 7th ed. Philadelphia, W. B. Saunders Company, 1977.)

through the gonopore into the host's intestine, or they may be stored in a blind sac, called the uterus. In the latter case, terminal proglottids packed with eggs break away from the strobila and may rupture within the host's intestine, or they may leave with the feces and rupture later.

Life Cycles. One or more arthropod or vertebrate intermediate hosts are required to complete the life cycle, which involves an **oncosphere** and a **cystocercus** larva. The following examples illustrate the basic pattern of the life cycle, but there are many variations.

Species of the genus *Taenia* are among the best known tapeworms. *Taenia solium*, the pork tapeworm, lives in the intestine of humans and reaches a length of up to 10 feet. Humans are the only primary host. The proglottids are passed into the host's feces, and the eggs may lie dormant in moist soil for long periods of time. The eggs do not hatch unless swallowed by a pig, dog or certain other mammals, including humans. On hatching, a spherical oncosphere larva bearing a number of hooks bores into the intestinal wall, where it is picked up by the circulatory system and transported to striated muscle. Here the larva leaves the blood stream and develops into a cystocercus stage (Fig. 25.19). The cystocercus, sometimes called a bladder worm, is an oval stage about 10 mm. in length, with the scolex invaginated into the interior. If raw or insufficiently cooked pork is ingested by humans, the cystocercus is freed, the scolex evaginates and the larva develops into an adult worm within the gut.

A severe infection of adult tapeworms may cause diarrhea, weight loss and occasionally perforation of the intestinal wall. The worms may be eliminated with drugs. Much more serious is cystocercus infection, and humans may be the intermediate host for the cystocercus of other tapeworms besides that of *Taenia solium*. Sometimes cystocerci lodge in the brain or in tissues other than striated muscle. Cystocerci can be removed only by surgery.

Soil mites are the intermediate hosts of sheep tapeworms, and copepods—tiny aquatic crustaceans—are the intermediate hosts for the tapeworms of many fish. Insects are intermediate hosts of many tapeworms.

CLASSIFICATION OF FLATWORMS

PHYLUM PLATYHELMINTHES (FLATWORMS)

Of the nine orders of turbellarians, the following are commonly encountered and most easily described.

CLASS TURBELLARIA. Free-living ciliated flatworms.

Order Acoela. Small marine flatworms, usually measuring less than 2 mm. in length. Mouth and sometimes a simple pharynx present, but no digestive cavity. Protonephridia absent. No distinct gonads. Oviducts and yolk glands absent.

Order Macrostomida and Catenulida. Small marine and freshwater species having a nonmuscular pharynx and a simple saclike intestine.

Order Seriata, Suborder Tricladida. Relatively large marine, freshwater and terrestrial flatworms, known as planarians, ranging from 2 to more than 60 cm. in length in the terrestrial forms. Digestive system with a tubular muscular pharynx and a three-branched intestinal sac. Both mouth and pharynx located in the middle of the body. Eyes present in most species.

Order Polycladida. Marine flatworms of moderate to large size, averaging from 2 to 20 mm. in length with a greatly flattened and more or less oval shape. Intestinal sac elongate and centrally located with many highly branched diverticula. Eyes numerous. Yolk glands absent.

CLASS TREMATODA. Parasitic flatworms known as flukes. Body oval. Ciliated epidermis absent; a living cuticle overlying muscle layer of body wall. Digestive tract present; mouth at anterior end.

Order Monogenea. Mostly ectoparasites of fish. Anterior end with adhesive glands; posterior end with adhesive organ-bearing suckers. Life cycle does not require an intermediate host.

Order Digenea. This order of endoparasites contains the majority of flukes. Mouth within an anterior sucker; ventral or posterior sucker also present. Life cycle requires one or more intermediate hosts.

CLASS CESTODA. Tapeworms. Endoparasitic flatworms lacking a digestive system. Ciliated epidermis lacking; body covered by a cuticle.

Subclass Eucestoda. Body composed of a scolex and strobila. Larva possesses six hooks. Includes almost all tapeworms.

Subclass Cestodaria. Fifteen species of fish parasites. Body leaf-shaped without a scolex and strobila.

ANNOTATED REFERENCES

Detailed accounts of flatworms may be found in the references listed at the end of Chapter 21. The works listed below deal specifically with the biology of animal parasites and commensals and the general biology of symbiosis.

Cheng, T. C.: The Biology of Animal Parasites. New York, Academic Press, 1973.

Gotto, R. V.: Marine Animals: Partnerships and Other Associations. New York, American Elsevier Publishing Co., 1969.

Noble, E. R. and G. A. Noble: Parasitology. 4th ed. Philadelphia, Lea and Febiger, 1976.

Read, C. P.: Parasitism and Symbiology. New York, Ronald Press, 1970.

PSEUDOCOELOMATES

Flatworms are **acoelomates.** Almost all other bilateral animals possess a body cavity or at least are derived from forms that had a body cavity. The body cavity of most animals is a **coelom,** but in others the body cavity, termed a **pseudocoel,** has a different embryonic origin. When gastrulation occurs, the blastocoel is not obliterated but remains to become the adult body cavity (Fig. 26.1). A pseudocoel therefore does not have a peritoneal lining as does a coelom.

Pseudocoelomates are a diverse assemblage of free-living and parasitic animals, but they share a number of features in addition to the pseudocoel. They are all poorly cephalized; i.e., they lack a distinct head with well developed sensory structures. Free-living species are all essentially aquatic. Even the nematodes that live in soil are dependent upon water films around soil particles. Most pseudocoelomates are tiny, usually microscopic, and they possess neither a blood vascular system nor organs for gas exchange.

The organs, such as the intestine and gonads, of many pseudocoelomates have fixed numbers of cells, which are constant for a particular species. For example, the intestine of the roundworm *Rhabditis longicauda* is always

Figure 26.1 Embryonic origin of the pseudocoel. *A,* Frontal section of the gastrula, showing the remains of the blastocoel and the developing mesoderm; *B,* cross section of *A* taken at level B–B; *C,* diagrammatic cross section of adult pseudocoelomate.

composed of 18 cells. Nuclear divisions are completed at hatching, and subsequent growth involves an increase in the size of the cells but no cell division occurs. This strange condition of cell constancy, called **utely,** is possibly correlated with the small size of pseudocoelomates, for it is also found in certain other minute but unrelated animals. The cells of these tiny animals are perhaps near the lower limit of effective size for metazoans, and cell constancy may be an adaptation to preserve an adequate cell size.

The body of pseudocoelomates is covered by a flexible nonliving **collagen cuticle,** which is often sculptured (Fig. 26.9). A varying number of adhesive glands is usually present and these open through the cuticle individually as small tubes (Fig. 26.6). The adhesive tubes are used for temporary anchorage to the substratum.

Water balance is achieved by a pair of **protonephridia** or by structures probably derived from protonephridia. The nervous system is relatively simple, composed of a varying number of longitudinal nerves.

The digestive tract of pseudocoelomates, like that of higher animals, is a tube with two openings: a **mouth** and an **anus.** A muscular pharynx and a stomach or intestine are the principal specializations of the gut tube.

Reproduction in most pseudocoelomates is entirely sexual, and with few exceptions, the sexes are separate. Cleavage is determinate, but the spiral condition is often lost. Development is typically direct in free-living species.

There are seven groups of pseudocoelomate animals. Some zoologists believe that many of these groups are closely enough related to constitute classes of a single phylum, called the **Aschelminthes.** Other zoologists object to the phylum Aschelminthes, pointing out that although pseudocoelomates share a number of features, there are no distinctive unifying characteristics. These objections seem justified to us and we will treat each of the pseudocoelomate groups as a separate phylum. Limitations of space permit us to describe only the three largest phyla of pseudocoelomates.

26.1 PHYLUM ROTIFERA

The phylum Rotifera consists of some 1500 species of microscopic animals, which are very common in fresh water. There are some marine rotifers and some that live in the water films of mosses.

The body of a rotifer is composed of a somewhat cylindrical trunk and foot (Figs. 26.2

Figure 26.2 Ventral view of a rotifer, *Philodina roseola,* showing many of the internal structures. (Redrawn from Hyman.)

and 26.3). The anterior end of the trunk bears a crown of **cilia,** a distinguishing characteristic of the phylum. The distribution of crown cilia varies in different rotifers (Fig. 26.2 and 26.3). A common type of crown is one in which the cilia are arranged in the form of two discs (Fig. 26.2). The two discs of cilia beat in a circular manner, one clockwise, and one counterclockwise, and look like two wheels spinning, from which the name *Rotifera*—wheel bearer—is derived.

The foot is a narrow posterior extension of the trunk and commonly terminates in a pair of adhesive glands, which open to the exterior through tubes, called toes (Fig. 26.3A). In some rotifers, the foot is divided into ringlike sections, which can telescope into one another (Fig. 26.2).

Figure 26.3 Three rotifers. *A, Notommata,* lateral view; *B, Floscularia,* a sessile rotifer with funnel-shaped anterior end; trunk and stalk surrounded by gelatinous mass; *C, Testudinella,* a turtle-shaped swimming rotifer. (*A* and *C* after Hyman, *B* after Hudson.)

B ----- Muscles

A

Figue 26.4 Two types of mastax teeth. *A,* Mastax with teeth adapted for grinding. Note the two large ridged plates that provide grinding surfaces. *B,* Mastax with grasping, forceps-like teeth. (*A* after Beauchamp from Hyman; *B* after Hyman.)

Rotifers swim by means of the ciliated crown and crawl in a leechlike fashion, using the terminal adhesive glands as a means of attachment. Most rotifers are found on algae and bottom debris and swim and crawl intermittently. Some rotifers are planktonic and are never found on the bottom (Fig. 26.3C), and a few are sessile and live in tubes (Fig. 26.3B).

A second distinguishing feature of rotifers is the pharyngeal apparatus, called the **mastax.** The mastax differs from the pharynx of the other pseudocoelomates in possessing seven cuticular teethlike pieces projecting into the lumen (Fig. 26.2). The teeth vary in size and shape depending upon the feeding habit. Predatory rotifers have a mastax that can project from the mouth and grasp and seize prey, mostly protozoans and other rotifers (Fig. 26.4B). Rotifers with disclike ciliated crowns are filter feeders. The beating cilia drive small particles into the mouth. The mastax of these rotifers contains broad flattened pieces adapted for grinding (Fig. 26.4A).

Digestion is probably extracellular within the large stomach. A short intestine passes from the stomach to the anus, which is located on the ventral side at the end of the trunk.

Reproduction is entirely sexual. The male, which is usually smaller than the female, stabs the female on any part of her body with his penis. This type of copulation, called **hypodermic impregnation,** occurs in other animals, including some flatworms. The sperm then migrate to the eggs for fertilization. During the life of a female, 10 to 20 shelled eggs, one for each ovarian nucleus, are deposited singly on the bottom or are attached to the body of the female. When the females hatch, they have all of the adult features and attain sexual maturity after a growth period of a few days. The smaller

males are sexually mature on hatching. Sessile rotifers are free-swimming for a brief period after hatching.

Many freshwater species produce both rapidly hatching thin shelled eggs and dormant thick shelled eggs. **Parthenogenesis** is common. In some species, both parthenogenesis and development from fertilized eggs occur. But in other species, there are no known males, and all individuals are females produced parthenogenetically. Parthenogenesis is perhaps an adaptation for a rapid expansion in number of a population. Populations of few individuals would be common in freshwater pools and streams that are subject to desiccation and other extreme environmental conditions.

As might be expected, the rotifers living on

Figure 26.5 A desiccated rotifer from a dried-up puddle. It can absorb water and become active again in a few minutes.

mosses are capable of withstanding extreme environmental conditions. During the short periods of time when these plants are filled with water, the animals swim about in the water films on the leaves and stems. They are capable of undergoing desiccation, usually without the formation of protective covering, and can remain in a dormant state for as long as three to four years (Fig. 26.5). The resistance to both low temperatures and lack of moisture in such a dormant state is remarkable. Some species have been placed in liquid helium (−272°C) and in vacuum desiccators without suffering damage. The few other kinds of animals that live on mosses have similar abilities.

26.2 PHYLUM GASTROTRICHA

Gastrotricha constitutes one of the smaller phyla of pseudocoelomates. Its members are found in the sea and in fresh water but are less common than rotifers. Gastrotrichs are about the same size as rotifers. The body is elongate, and the ventral surface is flattened and ciliated, from which the name *Gastrotricha*—stomach hairs—is derived. The anterior end may bear bristles or tufts of cilia. The posterior end is sometimes forked (Fig. 26.6). The **cuticle** of gastrotrichs is commonly in the form of scales, which are sometimes ornamented with spines (Fig. 26.6*A*). Adhesive tubes are present, often in rows along the sides of the body.

A

Pharynx

Salivary gland

Nerve

Protonephridium

Pseudocoel

Intestine

Muscle

Egg

Ovary

Rectum

Anus

Adhesive gland

B

Figure 26.6 Gastrotrichs. *A*, Gastrotrich in which the cuticle is ornamented with spines; *B*, Internal structure of a gastrotrich, ventral view. (*B* modified after Zelinka from Pennak.)

Gastrotrichs glide over the bottom propelled by the ventral cilia and may temporarily attach with the adhesive tubes. They feed on bacteria, small protozoa, algae and detritus. Food is swept into the anterior mouth by cilia or is pumped in by the muscular pharynx.

In contrast to most other pseudocoelomates, gastrotrichs are **hermaphroditic.** However, in most freshwater species, the male system is degenerate, and all individuals are thus parthenogenetic females. As in freshwater rotifers, both dormant and nondormant eggs are produced.

26.3 PHYLUM NEMATODA

The more than 10,000 species of nematodes, or roundworms, constitute the largest phylum of pseudocoelomates and one of the most widespread and abundant groups of animals. They occur in enormous numbers in the interstitial spaces of marine and freshwater bottom sediments and in the water films around soil particles. One square meter of bottom mud off the Dutch coast has been reported to contain as many as 4,420,000 nematodes. Several hundred billion nematodes may be present in an acre of good farmland. Decomposing plant and animal bodies often contain large populations of nematodes. For example, examination of one decomposing apple revealed some 90,000 nematodes belonging to several different species.

The phylum includes many parasitic species, exhibiting all degrees of parasitism on both plant and animal hosts. Food crops, domestic animals and humans are parasitized by many species, and consequently roundworms are of tremendous economic and medical importance.

Most free-living nematodes are less than a millimeter in length. While parasitic nematodes may also be minute, species parasitic in the gut of vertebrates may attain much larger sizes. Many are about the length of a pencil, and the horse nematode, *Parascaris equorum*, reaches a length of 35 cm.

The body of nematodes is long, cylindrical and tapered at both ends (Fig. 26.7). Lips, small sensory bristles, and papillae encircle the mouth at the anterior end (Fig. 26.8A). The anus is located a short distance in front of the posterior end, which in free-living species usually terminates in an adhesive gland. The complex nematode cuticle is composed of

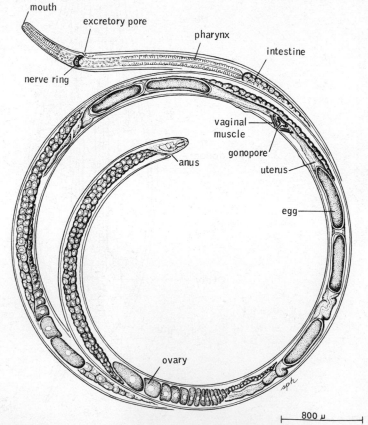

Figure 26.7 Female of the marine nematode *Pseudocella*. (After Hope.)

Figure 26.8 *A*, Anterior end of a marine nematode; *B*, anterior end of a predaceous nematode, with a retracted, spearlike stylet; *C*, stylet projected. (*A* after de Coninck, from Hyman; *B* and *C* after Thorne, from Hyman.)

several layers, often with a sculptured outer layer. The cuticle is periodically shed, but unlike the exoskeleton of insects and crustaceans, the nematode cuticle grows between molts. Beneath the cuticle is an epidermis composed of cells whose nuclei are restricted to the mid-dorsal, mid-ventral and mid-lateral lines (Fig. 26.9). The subepidermal muscle layer is composed only of longitudinal fibers. Contractions of the muscle fibers produce undulatory or thrashing locomotor movements that drive nematodes through the spaces between sand or soil particles.

The mouth opens into a **buccal cavity,** which may be provided with teeth or a **stylet** (Fig. 26.8). The buccal cavity is connected to a tubular muscular **pharynx,** which functions as a pump. The remainder of the gut, a long straight intestine, is the site of digestion and absorption (Fig. 26.7).

Many free-living nematodes are predaceous; others feed on the contents of plant cells. Some consume dead organic matter. Carnivorous nematodes use the teeth or stylet to puncture the prey. For example, *Mononchus papillatus,* a soil nematode, consumes as many as 1000 other nematodes during its life span of approximately 18 weeks by attaching its lips to its prey, making an incision with one of the buccal teeth and then pumping out the contents with the pharynx. Many herbivorous nematodes use a stylet to puncture plant cells, the contents of which are then pumped out with the pharynx (Fig. 26.8B).

The female reproductive system is paired and tubular and includes two ovaries (Fig. 26.7). Each oviduct empties into a uterus and then into a common vagina that opens to the exterior in the midregion of the body. Male nematodes are usually smaller than females, and the posterior end of the male is curled like a hook (Fig. 26.10B). The male reproductive system is a long coiled tube composed of a testis, sperm duct, seminal vesicle and muscular ejaculatory duct. The latter opens into the rectum so that the anus functions as a gonopore as well as for egestion of wastes. The rectum contains two short curved spicules, which can be projected from the anus (Fig. 26.10A). During copulation, the posterior end of the male is curled around the body of the female in the region of the gonopore. The spicules are used to hold open the gonopore of the female during sperm transmission. The sperm of nematodes are peculiar in lacking a flagellum. The eggs possess a thick shell, which may be sculptured, and the sculpturing is often important in identifying the particular kind of infection present in a fecal smear (Fig. 26.11). Free-living nematodes deposit their eggs in the bottom debris and soil in which they live. The young have most of the adult features on hatching and undergo several molts before attaining maturity. Adults do not molt but continue to grow.

Parasitic Nematodes. Parasitic nematodes attack both plants and animals and exhibit all degrees of complexity in their relationship with the host and in their life cycle. Some species are parasitic only during the juvenile stages; others are parasitic only during adulthood; still others are parasitic throughout their

Figure 26.9 Cross section through the middle region of a male *Ascaris lumbricoides*. The testes are sectioned several times because they lie folded in the body.

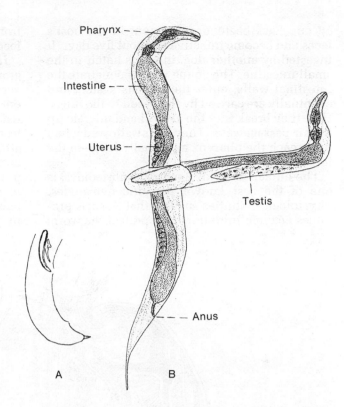

Figure 26.10 *A,* Posterior of a male *Aphelenchoides,* showing right penile spicule. *B,* Termite parasite, *Pristionchus aerivora,* in copulation. (*A* after Cobb, from Chitwood and Allen; *B* after Merrill and Ford, from Hyman.)

life cycle. The life cycle may be completed within a single host, or one or two intermediate hosts may be required.

Ascarids. The **ascaroid** nematodes, which are intestinal parasites of humans, dogs, cats, pigs, cattle, horses, chickens and other vertebrates, are entirely parasitic within a single host. The life cycle typically involves transmission by the ingestion of eggs passed in the feces of another host. The juvenile stages, usually called larvae, are not confined to the intestinal lumen and frequently involve temporary invasion of other tissues of the host.

A typical ascarid life cycle is that of *Toxicara canis,* a parasite of dogs. The adult male of this species reaches 10 cm. in length and the female

Figure 26.11 Human fecal smear containing eggs of the hookworm *Necator americanus.* (From Davidsohn I. and J. B. Henry: Todd-Sanford Clinical Diagnosis by Laboratory Methods. 15th ed. W. B. Saunders Company, 1974.)

18 cm. Eggs begin development in the host's feces and become infective in about five days. If ingested by another dog, the eggs hatch in the small intestine. The young worms penetrate the intestinal walls, enter the venous system and eventually are carried by the blood to the lungs. Here they break into the alveoli and migrate up the air passageways. They are swallowed when they reach the pharynx and finally lodge in the intestine.

The human ascarid, *Ascaris lumbricoides*, is one of the best known parasitic nematodes. Physiological studies suggest that *Ascaris* produces enzyme inhibitors that protect the worm from the host's digestive enzymes. The ascarids feed on the host's intestinal contents.

Hookworms. The **hookworms** are another group of parasites of the digestive tract of vertebrates. Most members of this group feed on the host's blood. The mouth region is usually provided with cutting plates, hooks, teeth or combinations of these structures for attaching and lacerating the gut wall (Fig. 26.12). A heavy infestation of hookworms can produce serious danger to the host through loss of blood and tissue damage. Hookworms are one of the parasites causing serious infections in humans. It is estimated that millions of

Figure 26.12   Hookworm. *A*, Longitudinal section through head of adult showing mouthful of intestinal wall being sucked. Eggs (*B*) pass out in the host feces, hatch in the soil (*C*) and grow to the infective stage (*D*). These penetrate the host skin and migrate by way of the blood, lungs and throat to the small intestine. *E*, Anterior end of a dog hookworm, showing buccal region and teeth. (*A* after Ash and Spitz; *B*, *C* and *D* after Chandler; *E* from Barnes, R. D.: Invertebrate Zoology. 3rd ed. Philadelphia, W. B. Saunders Co., 1974.)

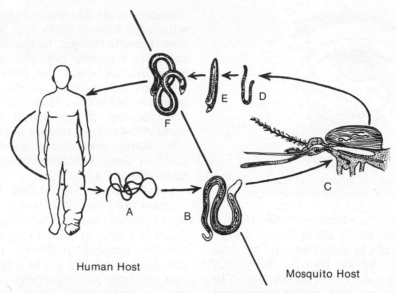

Figure 26.13 *Wuchereria bancrofti.* Adult worms in human lymphatic tissue (A) release microscopic larvae into the blood (B). If these are taken up by a mosquito (C), they migrate to the thoracic muscle, where they metamorphose and grow (D, E and F). The infective stage (F) migrates to the proboscis where it can penetrate into man while the insect is feeding.

people in India, Pakistan and Indonesia are infected with the Old World species, *Ancylostoma duodenale.*

The life cycle of hookworms involves an indirect migratory pathway by the juveniles, as in ascaroids. The fertilized eggs leave the host in its feces and hatch outside the host's body on the ground. The larva gains re-entry by penetrating the host's skin (feet in humans) and is carried in the blood to the lungs. From the lungs the larva migrates to the pharynx where it is swallowed and passes to the intestine. Not all species of this group gain re-entry by skin penetration. Some, such as dog hookworms, have a pattern similar to the dog ascarid.

Filarioids. The **filarioid** nematodes have life cycles requiring an intermediate host. The filarioids are threadlike worms inhabiting the lymphatic glands, coelom and some other sites in the vertebrate host, especially birds and mammals. The female is **ovoviviparous,** and the larvae are called **microfilariae.** Bloodsucking insects, such as fleas, certain flies and especially mosquitoes, are the intermediate hosts. A number of species parasitize humans, producing **filariasis.**

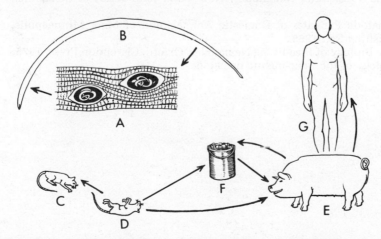

Figure 26.14 *Trichinella spiralis.* Larvae encysted in muscle (A) mature into intestinal worms when eaten (B). These give birth to larvae that burrow into the host, encysting in muscle. The natural reservoir is rodents (C) and similar animals that eat their dead (D). Pigs (E) will also eat dead rodents. Furthermore, killed rodents and pig scraps are fed to them in garbage (F). Man can become infected by eating insufficiently cooked meat containing larvae.

Figure 26.15 Adult of the African eye worm, *Loa loa*, visible in the white of the eye. (After Fülleborn.)

The chiefly Asian *Wuchereria bancrofti* illustrates the life cycle. The male is 40 mm. by 0.1 mm., and the female is about 90 mm. by 0.24 mm. (Fig. 26.13). Adults live in the lymph glands of man. The microfilariae are found in the blood and are present in the peripheral blood stream. When certain species of mosquitoes bite the host, the microfilariae enter the mosquito with the host's blood. Development within the intermediate host involves a migration through the gut to the **thoracic** muscles and after a certain period into the **proboscis.** From the proboscis the microfilariae are introduced back into the primary host when the mosquito feeds (Fig. 26.13). In severe filariasis, the blocking of the lymph vessels by large numbers of worms results in extreme edema of host tissues, particularly in the legs, breast and scrotum. Such enlargement is called **elephantiasis** (Fig. 26.13).

Some other important nematode parasites of humans are *Enterobius vermicularis*, intestinal pinworms of children; *Trichinella spiralis*, the juvenile stages of which encyst in the striated muscles, causing trichinosis (Fig. 26.14); *Loa loa*, the African eye worm (Fig. 26.15); and the 120 cm. long guinea worm, *Dracunculus medinesis*, which in the final part of its life cycle migrates to the subcutaneous tissue and produces an ulcerated opening to the exterior.

CLASSIFICATION OF PSEUDOCOELOMATES

(Classification of the pseudocoelomate phyla is technically difficult; therefore, we will supply no explicit listing of classification here. The interested student can consult the references below and at the end of Chapter 21 for further information.)

ANNOTATED REFERENCES

Detailed accounts of the pseudocoelomate phyla may be found in the references listed at the end of Chapter 21. Parasitic forms are described in the parasitology texts at the end of Chapter 25. The following are a few works devoted to nematodes.

Goodey, T.: Soil and Freshwater Nematodes. New York, John Wiley & Sons, Inc., 1951.
Levine, N. D.: Nematode Parasites of Domestic Animals and of Man. Minneapolis, Burgess Publishing Co., 1968.
Nicholas, W. L.: The Biology of Free-living Nematodes. Oxford, Clarendon Press, 1975.
    A general biology of the nonparasitic nematodes.

MOLLUSKS

1. The phylum Mollusca is the second largest phylum in the Animal Kingdom and contains such familiar forms as snails, slugs, clams, oysters, scallops, squids and octopods. Although the greatest number of mollusks are marine, there are snails and clams that inhabit fresh water and many species of snails and slugs that are terrestrial.

2. The body of a mollusk is usually covered by a shell of one piece (snails), two pieces (clams) or eight pieces (chitons) secreted by the underlying integument, called the mantle.

3. A ventral muscular foot is the locomotor organ of most mollusks. Primitively, the foot is a flat creeping sole, as in snails and chitons, but it is laterally compressed in clams and divided into arms or tentacles in squids and octopods.

4. The shell and mantle overhang the body, creating a mantle cavity that houses the gills of aquatic species and forms a lung in land snails

5. A great variety of diets and many modes of feeding are found among mollusks. Except for clams, most species employ as a feeding organ a unique beltlike structure bearing chitinous teeth called a **radula.** The radula can be protruded to scrape or tear and pull food into the mouth.

6. The circulatory system is open, and the heart, which lies within a pericardial coelom, consists of a pumping ventricle and paired auricles that receive blood from the gills.

7. The excretory organs are metanephridia, usually two, that drain the coelom and empty into the mantle cavity.

8. The nervous system consists of a pair of pedal cords to the foot and visceral cords supplying the organs of the mantle and visceral mass. The pedal and visceral cords unite anteriorly in a cerebral ganglion, which supplies nerves to various head structures. Sense organs include one or two statocysts in the foot, one or two pairs of head tentacles, a pair of eyes and a pair of osphradia, the last being sense organs that monitor the ventilating current passing through the mantle cavity.

9. Mollusks may have separate sexes or may be hermaphroditic, and paired gonads are adjacent to the pericardial coelom. Primitively, the gametes exit through the nephridia, but separate and often complex gonoducts are usually present. In many groups, copulation occurs and the eggs are fertilized internally or within the mantle cavity. The earliest larval stage is a trochophore, but in many snails and clams there is only a later larval stage, called a veliger.

27.1 COELOM AND COELOMATES

All the remaining members of the Animal Kingdom either have a coelom or have evolved from coelomate ancestors. The coelom appears during development as a cavity within the mesoderm (Fig. 27.1) and has a mesodermal lining, the **peritoneum.** Internal organs may bulge into the coelom, but they are still covered by peritoneum and not literally "in" the coelomic cavity.

Neither the manner in which the coelom evolved nor its original adaptive significance is known. Given the varieties of development discussed in Chapter 5, it can be seen that the coeloms of living animals arise in a number of different ways. They also serve a variety of functions: (1) as a medium for transport of substances; (2) as a site for the deposition of nitrogenous wastes before the removal of these wastes by excretory organs; (3) as a space in which gametes may undergo development; or (4) as a space for growth and development of organs during embryogeny, and (5) for the functional movements of organs in adults.

Figure 27.1   Origin of the coelom in protostomes. *A*, Frontal section of a late gastrula. *B*, Cross section of gastrula at level indicated in *A*. *C*, Coelom developing as a split (schizocoel) in the mesodermal mass. *D*, Further development of the mesoderm and coelom. Note that the old blastocoel is gradually obliterated. *E*, Cross section through an adult coelomate protostome. Muscle layer develops from mesoderm. Internal organs lie behind peritoneum (retroperitoneal), which lines coelom.

Comparative studies of the development of coelomates led to the recognition of two major groups of metazoan phyla: the **protostomes** and the **deuterostomes.** These groups probably represent two principal lines of evolution within the Animal Kingdom. The protostome phyla—mollusks, annelids, arthropods and certain lesser phyla—are characterized by **determinate spiral cleavage;** the **mesoderm** can be traced back to a single cell of the **blastula** (p. 100, Fig. 5.13). The **coelom** is a schizocoele, formed by an internal splitting of the mesodermal masses, a process called schizocoely (Fig. 27.1C). The **mouth** originates from the blastopore, hence the name protostome—first mouth. Protostomes commonly possess a **trochophore** larva (Fig. 27.6), described on page 610. Although flatworms lack a coelom, their development by spiral cleavage justifies classifying them as acoelomate protostomes.

The features just described do not characterize the embryogeny of all protostomes; rather, they represent the primitive or basic plan of protostome development and are most commonly encountered among marine species. They have been partly or completely lost in the many species whose reproductive circumstances have led to marked changes in the amount and distribution of yolk material (in

turn, the amount and distribution of yolk greatly influence the cleavage pattern) and to the suppression of larval stages, a characteristic of most freshwater and terrestrial animals and even of some marine species.

In contrast to protostomes, the deuterostome coelomates are characterized by **indeterminate radial cleavage.** The mouth appears during development as a new opening at the opposite end of the embryo from the blastopore, hence the name deuterostome—second mouth. The formation of the mesoderm and coelom is quite different from that of protostomes (p. 733).

27.2   GENERAL FEATURES OF THE PHYLUM MOLLUSCA

The phylum **Mollusca** includes some 65,000 living and 35,000 fossil species. Mollusks are about as well known taxonomically as mammals and birds. This reflects, at least in part, the long popularity of shell collecting.

Two distinguishing characteristics of the mollusks are the adaptation of the ventral surface of the body as a muscular foot for locomotion and the modification of the integument of the dorsal surface, the **mantle,** for the secretion of a protective calcareous shell. The shell has been lost in some mollusks, but the mantle almost always remains.

27.3 THE ANCESTRAL MOLLUSKS

We have no direct evidence of the first mollusks. The phylum evolved in the Archeozoic seas and by the time of the Paleozoic, when the fossil record first becomes clear, the different molluscan classes were already defined. Nevertheless, we can make some inferences about the nature of the ancestral mollusks on the basis of what can be observed in living forms. The characteristics shared by some or all of the seven classes of mollusks probably were present in the ancestral form.

Although it is not particularly important that we know the nature of the first mollusk, such a hypothetical ancestor can serve as an introduction to the general features of the phylum.

Biologists postulate that the ancestral mollusk was small, perhaps about two centimeters in length, and adapted for living on hard rocky bottoms in the Archeozoic seas (Fig. 27.2). By piecing together bits of evidence from living mollusks, we can infer that the animal's ventral surface formed a broad, flat, muscular creeping **foot**. The **head** was probably poorly developed, bearing a pair of tentacles with a simple **eye** at the base of each. The dorsal surface of the body was covered by a low shield-shaped **shell** secreted by the underlying mantle and composed of one to several layers of calcium carbonate covered on the outside by

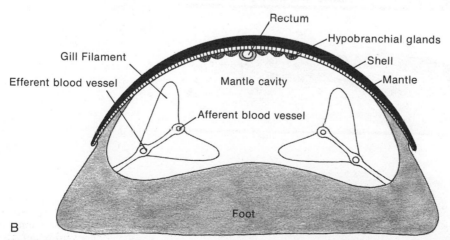

Figure 27.2 *A*, Hypothetical ancestral mollusk (lateral view). Arrows indicate path of water current through mantle cavity. (Adapted from various authors.) *B*, Transverse section through body of ancestral mollusk at level of mantle cavity.

an organic layer, the **periostracum.** Much of the shell of living mollusks is secreted by the mantle margin so that the shell grows in diameter as well as in thickness. The shell of the ancestral mollusks provided protection as long as the animal was attached to the rocky substratum; pairs of retractor muscles extending from the foot to the shell permitted the shell to be pulled down over the body.

The overhanging shell and mantle at the posterior end created a large mantle cavity containing two or several pairs of gills and two excretory openings (Fig. 27.2). The **anus** was located at the dorsal side of the cavity opening. The bipectinate **gill** structure postulated for the ancestral mollusk is similar to that found in many living species. Each gill was composed of a longitudinal axis to which was attached on either side flattened gill filaments (Fig. 27.2). Blood circulated through the filaments, flowing from a supply to a drainage vessel in the gill axis. The filaments were ciliated and produced a ventilating current of water that entered the lower part of the mantle cavity, made a U-turn across the gills and then flowed out of the cavity dorsally and posteriorly (Fig. 27.2). Wastes discharged by the **nephridia** and anus were removed in the exhalant stream of water.

As in most living mollusks, the mouth cavity of the ancestral mollusk probably contained a **radula,** a unique rasping organ consisting of a beltlike membrane bearing a large number of teeth (Fig. 27.3). The radula rested upon a cartilaginous supporting skeleton, the **odontophore,** supplied with protractor and retractor muscles. The ancestral mollusks were probably **microphagous**—they fed upon small particles of algae scraped with the radula from rocks.

In living mollusks, the end of the radula and odontophore can project from the mouth and lick the adjacent surface, like the tongue of a mammal. New teeth are secreted at the posterior end of the radula as old teeth are worn away anteriorly. Particles of algae brought into the mouth are enveloped by mucus secreted by the **salivary glands.** A rope of mucus containing the algal particles then passes posteriorly through an **esophagus** to the **stomach.** The posterior end of the stomach, the **style sac,** is ciliated and rotates the mucous mass, winding in the string of mucus like a windlass winds in a rope (Figs. 27.2 and 27.4). Particles dislodged from the mucus are sorted by a ridged ciliated area at the anterior end of the stomach (Fig. 27.4). Large undigestible particles are conducted posteriorly to the **intestine,** and fine digestible particles are conveyed

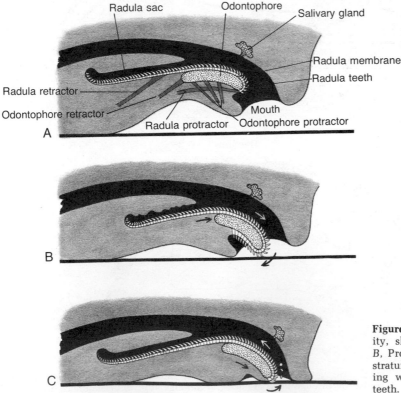

Figure 27.3 Molluscan radula. *A,* Mouth cavity, showing radula apparatus (lateral view). *B,* Protraction of the radula against the substratum. *C,* Forward retracting movement during which substratum is scraped by radula teeth.

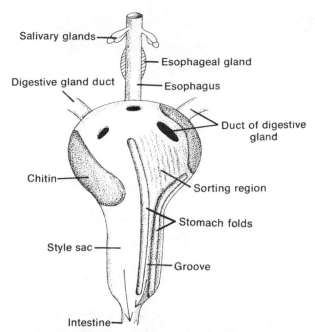

Figure 27.4 Diagram of a primitive molluscan stomach. (Modified from Owen.)

to the ducts of a pair of **digestive glands** located to either side of the stomach. Digestion occurs intracellularly within the digestive gland. The long intestine functions only in the formation of feces.

An open blood vascular system provides for internal transport in all mollusks and was undoubtedly characteristic of the ancestral form. The heart, located within the coelomic cavity, is composed of a **ventricle** and two lateral **auricles,** which receive blood from each gill (Fig. 27.2). The contractile ventricle forces blood via an anterior aorta to the tissue spaces of the body—head, foot, visceral mass and mantle. Blood then collects within larger sinuses from which it is returned to the heart by way of the gills (Fig. 27.5A).

The pair of excretory organs, often called **kidneys,** are **metanephridia.** In contrast to the blind protonephridia of flatworms and pseudocoelomates, these tubes open at the inner end into the coelom by way of a ciliated **nephrostome** (Fig. 27.5B). The outer end opens into the mantle cavity. Like most excre-

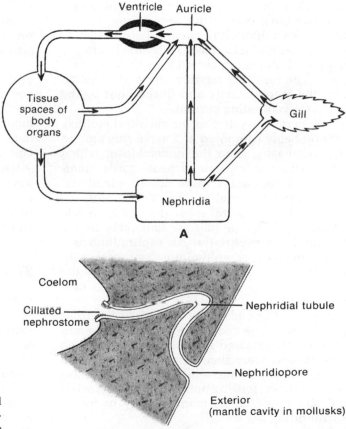

Figure 27.5 *A,* Diagram showing the principal features of the molluscan circulatory system. Auricles, gills and nephridia are usually paired. *B,* Diagram of a metanephridium.

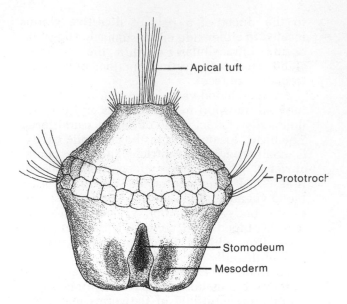

Figure 27.6 The trochophore larva of *Patella*, a marine snail. (After Patten.)

tory organs, the molluscan nephridia function to concentrate metabolic wastes in urine. Special glandular tissue lining the coelom secretes waste into the coelomic fluid. Fluid containing wastes collected by the blood from body tissues filters from the heart into the coelomic fluid. Coelomic fluid containing wastes passes into the inner end of the nephridium. As the fluid passes through the nephridial tube, some substances are selectively reabsorbed into the body and others are secreted by the cells into fluid. The resulting mixture, urine, is expelled into the mantle cavity and flushed out by the exhalant ventilating current.

The nervous system of the ancestral mollusk was probably composed of a nerve ring around the esophagus, from the underside of which extended a ventral pair of pedal cords innervating the foot and a more dorsal pair of visceral cords innervating the mouth and the organs of the visceral mass (Fig. 27.2). Sense organs included a pair of **statocysts** in the foot and the **osphradia**. An osphradium is a patch of sensory epithelium near each gill, which monitors chemical substances in the water current passing through the mantle cavity (Fig. 27.2).

A pair of **gonads** were located to the front and sides of the coelom (Fig. 27.2). Eggs and sperm were released from the gonads into the coelom and were then carried to the mantle cavity by way of the nephridia. In the ancestral mollusks, fertilization was probably external either within the mantle cavity or in the surrounding sea water.

In living mollusks, cleavage is typically spiral, and the resulting gastrula develops into a free-swimming trochophore larva (Fig. 27.6). The larval body is ringed about the middle with a girdle of **cilia**, the **prototroch**, and the anterior pole bears an apical tuft of cilia. A digestive tract is present, and the mouth is located below the prototroch. The anus opens at the posterior pole. The beating cilia of the prototroch provide for locomotion and may also serve to collect fine plankton for food.

In the ancestral mollusks, the trochophore probably developed directly into the adult body; the body lengthened behind the prototroch, the foot formed behind and below the mouth and mantle and shell developed on the dorsal side. The young animal settled to the bottom to take up the adult mode of existence.

Using this hypothetical ancestral form as a basis for comparison, we will examine five of the seven classes of mollusks, omitting two small classes, the **Scaphopoda**, or tooth shells, and the **Aplacophora**.

27.4 CLASS GASTROPODA

Among the 35,000 living species of gastropods, the largest and most diverse class of mollusks, are snails, whelks, conchs, limpets and sea slugs. Most gastropods are marine, but some live in fresh water and others have become adapted for life on land.

Torsion and Shell Spiraling. Gastropods possess a well developed **head** bearing two tentacles with an eye at the base of each and, like the ancestral mollusk, a broad, flat, creeping **foot** and a **shell** composed of a single piece. However, gastropods are distinguished from all other mollusks by the curious twisting of the visceral mass that occurs during development (Fig. 27.7). This condition, called **torsion,** involves a 180° counterclockwise (viewed from above) twist that results in the **mantle cavity** and **anus** being located at the anterior end of the body. The gut, nervous system and blood vascular system are correspondingly twisted. During embryonic development the mantle cavity forms first at the posterior end of the body but shifts to an anterior position (Fig. 27.7, C and D). The evolutionary significance of torsion is uncertain, although several theories have been put forward to account for it.

The spiral shell of gastropods is *not* a consequence of torsion; indeed, the spiral shell is formed during development before torsion occurs. Paleontological evidence also indicates that shell spiraling is not a direct consequence of torsion. The shieldlike shell of the ancestral mollusk protected only its dorsal surface, and the animal would have been vulnerable when dislodged from its substratum. In the course of evolution the gastropod shell became higher and more conelike, with a reduced aperture (Fig. 27.8). Such a shell provided a protective retreat into which the entire body could be withdrawn. The spiral condition appears to have been an adaptation for making the shell more compact and less awkward to carry than if it were a long cone.

The gastropod shell is balanced obliquely upon the body, with the spire directed posteriorly and upward on the right side (Figs. 27.9 and 27.12). The large first whorl of the shell bulges into the right side of the mantle

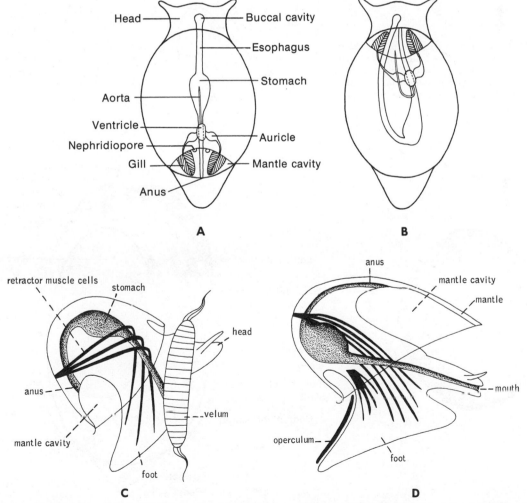

Figure 27.7 Dorsal view of hypothetical ancestral gastropod. *A,* Prior to torsion; *B,* after torsion; *C* and *D,* embryonic torsion in the gastropod *Patella* (a limpet). (*C* and *D* from Barnes, R. D.: Invertebrate Zoology. 3rd ed. Philadelphia, W. B. Saunders Co., 1974.)

Figure 27.8 *A,* Evolution of a plano-spiral shell. Height of the shieldlike shell of hypothetical ancestral mollusk increases and a peak forms. Peak is pulled forward and coiled under. Aperture is reduced, and animal can withdraw into spiral shell, which is more compact and less awkward to carry than would be a straight conical shell. Note that shell is bilaterally symmetrical. *B,* Diagram of a gastropod larva before torsion. Velum is the ciliated larval locomotor organ. *C,* Hypothetical pretorsion "gastropod" with a plano-spiral shell. *D,* Post-torsion gastropod. Torsion does not affect plano-spiral shell except to place coils of shell posteriorly. Mantle cavity is now anterior.

Figure 27.9 Position in which shell is carried in gastropods. Slot (*A*) that is found in some primitive gastropods indicates the original mid-dorsal line of the mantle cavity. Note that the mantle cavity is restricted largely to the left side. (*B* after Yonge.)

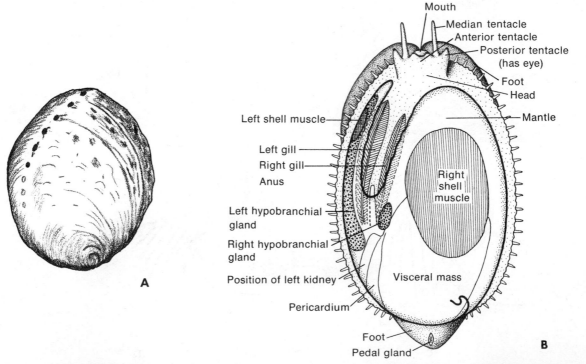

Figure 27.10 The abalone, *Haliotis.* *A,* Dorsal view of shell, which consists largely of a single expanded whorl. *B,* Dorsal view, showing two bipectinate gills. Note that the right gill, which is lost in most gastropods, is a little smaller than the left. (After Bullough.)

cavity (Fig. 27.9). The partial occlusion of the mantle cavity has resulted in the reduction and loss of the right gill and in turn the reduction and loss of the right auricle and right kidney. Some primitive gastropods still have two gills (Fig. 27.10), two auricles and two kidneys, but most have only those on the left side.

Evolution of Water Circulation and Gas Exchange. Whatever the evolutionary significance of torsion may have been, the anterior position of the anus imposed a sanitation problem on the early gastropods, for waste would have been dumped on top of the head. The problem has been solved in most aquatic gastropods by the evolution of structures to generate an oblique water current. Except in certain primitive species, which have the ancestral bipectinate gills (Fig. 27.10), the single left gill lacks the filaments on one side and is anchored by the axis to the mantle wall. Water enters the mantle cavity on the left side of the head, makes a turn across the gills and exits on the right side of the head (Fig. 27.11). The anus is located at the right edge of the

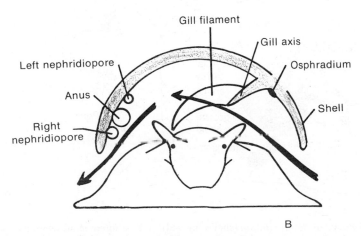

Figure 27.11 Typical prosobranch gastropod, with a single gill and mantle cavity on left side of animal. Arrows indicate path of water current through mantle cavity. *A*, Dorsal view; *B*, diagrammatic transverse section through head and mantle cavity. (After Graham.)

mantle cavity, and wastes are removed by the exhalant water current. Commonly, the edge of the mantle on the left side is drawn out to form a siphon into which the inhalant water stream passes (Fig. 27.12).

Most marine snails, including the common whelks, conchs, slipper shells and limpets, compose the subclass **Prosobranchia.** The name Prosobranchia—"front gills"—refers to the position of the gills in the anterior mantle cavity. From the prosobranchs two other large subclasses of gastropods evolved. Members of the subclass **Opisthobranchia**—the bubble shells, sea butterflies, sea hares and sea slugs—are characterized by a degree of detorsion, or untwisting of the visceral mass, from which the name Opisthobranchia (back gill) is derived. The significance of detorsion is as uncertain as torsion, but it has been accompanied by reduction of the shell and mantle cavity. Primitive opisthobranchs, such as the bubble shells, possess a shell and look like proso-

Figure 27.12 Anatomy of *Busycon canaliculatum* (shell removed). *A,* Position of shell when foot is extended. *B,* Left side, showing external organs and internal organs visible through the integument, *C,* Same view with digestive, respiratory, circulatory and nervous systems indicated. *D,* Female, showing portion of the right side. *E,* Male, portion of the right side with mantle and retractor muscle cut short. In *D* and *E* the proboscis is withdrawn.

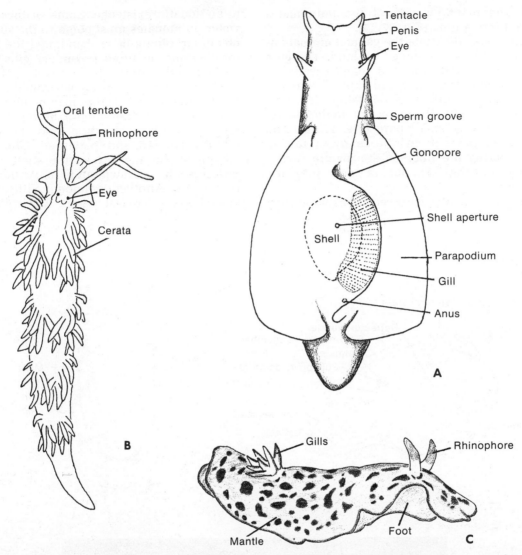

Figure 27.13 Opisthobranch gastropods. *A*, The sea hare, *Aplysia* (dorsal view). *B*, *Eolidia papillosa*, a sea slug (nudibranch). *C*, *Glossodoris*, a sea slug with secondary anal gills. Color in life is yellow with red spots. (*A* after Bullough; *B* after Pierce; *C* after a photograph by Zahl.)

branchs, but the mantle cavity and gill are located on the right side as a result of detorsion. Other opisthobranchs exhibit such a degree of shell reduction or loss that they have secondarily become bilaterally symmetrical. Bilateral opisthobranchs include sea hares (Fig. 27.13) and the beautiful sea slugs. The shell and mantle cavity of sea slugs have disappeared, and gas exchange occurs either through secondary gills located around the now posterior anus or across the external mantle surface. In the latter group, the mantle surface is greatly increased by a large number of club-shaped projections.

The term *secondary* refers to the reattainment of a habit, condition or structure that is similar to some ancestral condition. Most mollusks exhibit a primary bilateral symmetry inherited from the ancestral form. As a result of torsion, gastropods became asymmetrical and lost the primary bilateral symmetry. With detorsion, opisthobranchs, such as sea slugs, reacquired bilateral symmetry, but since the bilateral symmetry is a new feature evolved from asymmetrical ancestors, it is said to be secondary.

The Prosobranchia gave rise to a third subclass, the **Pulmonata**, adapted for living on land. The mantle cavity of the pulmonates evolved into a lung (Fig. 27.14).

Pulmonates illustrate the general animal adaptations for gas exchange on land. The gills have disappeared, and the mantle wall has become highly vascular and functions as the gas exchange surface. Desiccation is minimized by reducing the opening into the mantle cavity to a very small pore (Fig. 27.14). The animal ventilates its mantle cavity by raising and lowering the floor of the mantle cavity, thereby moving air out of and into the chamber.

Many pulmonates have secondarily returned to an aquatic existence. Some of these freshwater pulmonates must come to the surface to obtain air; others have abandoned the lung altogether and acquired secondary gills outside of the mantle cavity.

Shell. The shells of gastropods exhibit great variations. The spire may be high or low; the surface may be smooth or highly sculptured; and the coloration may be dull or brightly colored and patterned. The ventral margin of the aperture of the shell of many prosobranchs is drawn out to house the siphon (Fig. 27.12). Another common modification is a greatly expanded last whorl. The spiral apex

Figure 27.14 Pulmonates. *A,* The land snail, *Helix,* partially dissected to show porelike opening (pneumostome) into lung, or mantle cavity. *B,* The freshwater pulmonate *Ferrissia,* a limpet, with secondary gill (ventral view). *C,* A terrestrial slug. Opening into lung seen at the lower edge of the saddle-like mantle. (*A* after Rowett: Histology and Embryology; *B* after Pennak; *C* by Betty M. Barnes.)

A

B

C

D

Figure 27.15 Gastropod shells. A, Longitudinal section through shell. B, *Acmaea*, a common intertidal prosobranch limpet on rocky shores. C, Vermetid worm shells. Specimen on right shows typical gastropod spire of youngest whorls. Specimen on left is a mass of fused individuals. D, Slipper shell, *Crepidula*, a semi-sessile prosobranch in which the shell represents a single expanded whorl. These commonly occur in clusters, attaching to each other. (*B* and *C* by Betty M. Barnes; *D* after Coe.)

develops during juvenile stages. Then the last whorl becomes extremely large and covers most of the body of the animal as a low convex plate. Abalones, slipper shells and limpets have shells of this type. The shell of limpets has become secondarily bilateral.

The central axis of the asymmetrical spiral shell is the **columella,** to which the single large retractor muscle of the body is anchored (Fig. 27.15A). The gastropod withdraws into the shell by folding the middle of the foot. The anterior half of the foot and the head are withdrawn first, followed by the posterior half of the foot. Many prosobranchs have a round plate, the **operculum,** on the back of the foot, which plugs the aperture when the animal is withdrawn (Fig. 27.12).

Locomotion. Gastropods creep upon the broad, flat, ventral foot by a process of body deformation. Small transverse waves of contrac-

tion sweep along the foot from back to front, one wave closely following another wave (Fig. 27.16). The area of the foot in the contracted region is lifted; at relaxation it is replaced on the substratum a little in front of the point from which it was raised (Fig. 27.16B). During each wave of contraction a small section of the foot performs a little step; the summation of all of these little steps gives the appearance of a gliding motion. The substratum is lubricated by large amounts of mucus produced by a large gland in the foot. In some gastropods, the waves sweep from back to front (direct waves), and the region of the foot lifted from the substratum is in a state of contraction. In many other species, the waves move from front to back (retrograde waves), opposite to the direction in which the animal is moving. In retrograde waves, the area lifted is in a state of relaxation, with contraction immediately fol-

Figure 27.16 Pedal creeping in gastropods. *A*, Ventral view of foot, showing two monotactic waves moving from back to front of foot (direct waves). *B*, Ventral view of foot, showing ditactic waves moving from front to back of foot (retrograde waves). *C*, Diagrammatic sections of a region of the foot during the course of movement of a direct wave (arrow). In the region of the wave, the foot is lifted from substratum. In the passage of the wave, a region of the foot is placed back upon the substratum in advance of the point from which it was lifted. The foot surface anterior to the wave crest is contracted (crinkled line); posterior to crest surface is elongated (straight lines). Letters indicate fixed positions of substratum. Diagonal dashed line shows forward movement of foot. *D*, Diagrammatic sections illustrating a retrograde wave. Contracting surface lies behind crest of wave.

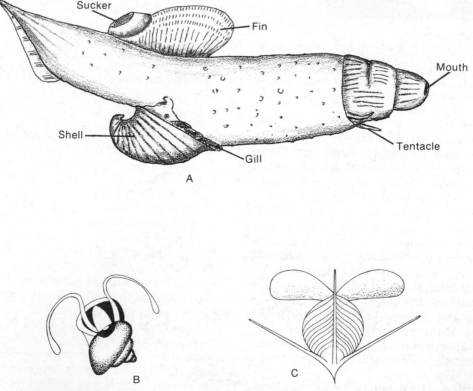

Figure 27.17 Swimming gastropods. *A*, *Carinaria*, a pelagic prosobranch. *B* and *C*, Two opisthobranch sea butterflies in which the feet are modified as fins. (*A* after Abbott; *B* and *C* after Morton.)

lowing. Waves may extend all the way across the foot (monotaxic), or those on one side of the foot may alternate with those on the other side (ditaxic). The most common mode of muscular pedal creeping in gastropods is by ditaxic retrograde waves. Gastropods living on soft substrata, such as sand and mud, move by means of the cilia that cover the foot surface. Many species living on soft bottoms can burrow into the substratum.

The foot of the abalone, limpet, slipper shell and other gastropods living on hard substrata functions as an adhesive organ (Fig. 27.15). The sucker-like foot and the low platelike shell enable the animal to live attached to rocks that are swept by currents. Some gastropods are semi-sessile or sessile. Slipper shells, for example, move about very little (Fig. 27.15). Worm shells have lost the foot, and the shell, which has uncoiled and become tubelike, is attached to rocks and other objects (Fig. 27.15).

The ability to swim has evolved in a number of different gastropod groups, all of which have the foot modified in some way as a fin. The best swimmers are the prosobranch heteropods and the opisthobranch sea butterflies (Fig. 27.17). The sea hares swim intermittently, moving by beautiful undulations of the lateral fin folds of the foot.

Nutrition. Primitive gastropods inhabiting rocks are **microphagous,** feeding on fine algae scraped up with the radula. Most gastropods are **macrophagous,** feeding on large food masses, with the radula retained as the feeding organ. Some species, especially pulmonates, feed on algae or terrestrial vegetation; other gastropods are carnivorous. The radula of a carnivore usually has large heavy teeth and may be located at the end of an extensible proboscis (Figs. 27.12 and 27.18). The prey —other invertebrates, especially other mollusks—is held with the foot, and the tissues of the victim are torn and devoured with

A

B

Figure 27.18 Scanning electron photomicrographs of radula teeth. A, Radula of a carnivorous drilling gastropod. There are only three teeth to a transverse row, but each tooth bears several cusps. B, Harpoon radula tooth of the poisonous gastropod *Conus imperialis.* (A from Carriker, M. R.: Excavation of boreholes by the gastropod, *Urosalpinx.* Am. Zool. 9:917–933, 1969; B from Kuhn, A. J., et al.: Radula tooth structure of the gastropod *Conus imperialis.* Science 176:49–51. Copyright 1972 by the American Association for the Advancement of Science.)

A   B

Figure 27.19 *A*, A bivalve shell that has been drilled by a gastropod. Note the beveled edge of the hole. *B, Nassarius obsoletus* feeding at low tide. This intertidal prosobranch scavenger and deposit feeder occurs in enormous numbers on protected beaches on the east coast of the United States. (By Betty M. Barnes.)

the radula. Some carnivores use the radula as a drill to penetrate the shells of bivalve prey that have been previously softened with secretions from a gland on the foot or proboscis (Fig. 27.19A). Another group of carnivores, the tropical cone shells, kill prey by hurling a long, detached, harpoon-like radula tooth carrying a poison into the victim (Fig. 27.18B).

Many aquatic gastropods are scavengers or detritus feeders, and some, such as the Atlantic coast mud snail *Nassarius obsoletus*, may occur in enormous numbers on intertidal beaches (Fig. 27.19). A few filter-feeding gastropods, like the semi-sessile slipper shells, use the gills to filter plankton from the ventilating current.

Some gastropods are ectoparasites of bi-

valves (Fig. 27.20); others are endoparasites in the body wall of echinoderms, such as starfish, or in the coelom of sea cucumbers. The foot of endoparasites is greatly reduced, and the shell is lacking in coelomic parasites. In all, the buccal region is modified for piercing and sucking.

The stomach of macrophagous gastropods has undergone profound changes. The style sac and sorting region have disappeared, and the stomach has the form of a simple sac in which extracellular digestion occurs. Enzymes are provided by the salivary glands, glands along the esophagus, or by the digestive glands. Absorption occurs within the digestive glands, and the intestine functions only in the formation of feces.

A   B

Figure 27.20 *A, Brachystomia*, an ectoparasitic snail, feeding on the body fluids of a clam. *B, Stylifer*, an endoparasitic snail in the body wall of a starfish. (Both after Abbott.)

Circulation, Gas Exchange, Excretion and Water Balance. Except for the asymmetry produced by torsion and the loss of the right gill, the circulatory system of gastropods is like that described for the ancestral mollusks. The blood of most gastropods contains **hemocyanin,** a respiratory pigment.

Ammonia is the excretory waste of aquatic gastropods, but terrestrial pulmonates excrete uric acid. By adding a **ureter** formed from the mantle wall, the excretory opening of land pulmonates is located outside of the mantle cavity. The anus also opens outside the mantle cavity; thus, the lung is not fouled with waste. Pulmonates are not especially adapted for avoiding water loss through desiccation, and large amounts of water are lost in the mucus secreted when crawling. Most pulmonates are therefore restricted to humid environments or must be nocturnal. During the winter or during very dry periods the animals hide in leaf mold or beneath wood or stone or attach the shell to vegetation by mucous cords. The aperture of the shell is then covered with a mucous film, which dries to form a protective diaphragm. The animal is inactive during this time and the metabolic rate drops to a very low level. Such a period of dormancy (termed **estivation**) is a common adaptation of semitropical and tropical animals living in regions with long dry seasons. Nocturnal activity and estivation permit some pulmonates to inhabit deserts and other arid regions.

Sense Organs. The head of a gastropod

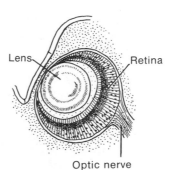

Figure 27.21 Section through the eye of a whelk. (After Helger.)

bears one or two pairs of sensory tentacles. In limpets and abalones, the mantle edge may be fringed with tentacles.

The two eyes located at the base of the tentacles (Figs. 27.12 and 27.21) are not highly developed; indeed, in most species they are not capable of object discrimination. The **osphradium** is an important sense organ, especially in carnivorous gastropods. The mobile siphon moves about, drawing in water from different areas, and the chemoreceptors of the osphradium enable the animal to determine the location of prey or other food.

Reproduction. Most species of prosobranchs have separate sexes. Pulmonates and opisthobranchs are hermaphroditic. In primitive prosobranchs, the gametes are discharged through the nephridia. In most gastropods,

Figure 27.22 The egg case of *Busycon.* (Photo by Hugh Spencer.)

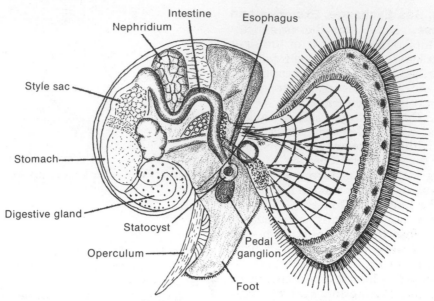

Nephridium   Intestine   Esophagus
Style sac
Stomach
Digestive gland
Statocyst
Operculum
Pedal ganglion
Foot

Figure 27.23   Lateral view of the veliger larva of gastropod (slipper shell). (After Werner from Raven.)

however, a complex reproductive tract has developed to which the original right nephridium contributes only a small part. Copulation followed by internal fertilization is the usual mode of reproduction. The eggs of aquatic species are deposited in strings or masses or within special cases molded by the foot (Fig. 27.22). The large yolky eggs of pulmonates are laid singly and deposited in small clusters in the soil and leaf mold or beneath bark, logs or stones.

Cleavage is typically spiral. In primitive marine prosobranchs the initial trochophore larva develops further into a **veliger** larva. Most marine prosobranchs pass through the trochophore stage within the egg membranes and hatch as veligers. The veliger larva displays many gastropod features, such as a spiral shell and foot (Fig. 27.23). Torsion occurs during the veliger stage (Fig. 27.27). Like the trochophore, the veliger is planktonic and swims by means of the **velum,** a large ciliated organ derived from the prototroch of the trochophore. The velum also collects fine suspended particles upon which the larva feeds. In many marine and in all freshwater and terrestrial species, development is direct; i.e., no larval stage is present. Little snails emerge from the eggs or cases. In all of these forms with direct development, the eggs contain more yolk than the eggs of species with a feeding larval stage.

27.5   POLYPLACOPHORA AND MONOPLACOPHORA

The **Polyplacophora** and the **Monoplacophora** have a broad, flat, creeping foot. The class Polyplacophora (or Amphineura) includes about 600 species of **chitons** ranging from a few centimeters to over 35 cm. in length and adapted for living on rocks and other hard substrata. In many ways chitons parallel the prosobranch limpets. The head is reduced, and much of the ventral surface is occupied by the broad foot (Fig. 27.24 and 27.25). Chitons, however, are distinguished from all other mollusks in possessing a shell composed of eight plates, hence the name Polyplacophora—bearer of many plates. The plates are arranged linearly from anterior to posterior and overlap one another (Fig. 27.24). The lateral margins of the plates are overgrown by the mantle to varying degrees; in some chitons the plates are completely buried in the mantle (Fig. 27.25).

Beyond the plates the mantle forms a heavy girdle surrounding the body (Fig. 27.24). The girdle may be smooth or covered with scales or calcareous spines; it is an important area of sensory reception.

Chitons crawl about much like snails but may be immobile for long periods of time. When the chiton is clinging to rocks and other objects, the mantle margin is held tightly

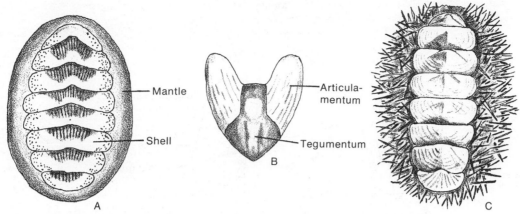

Figure 27.24 A, Common Atlantic coast chiton, *Chaetopleura apiculata. B,* Single shell plate of *Katharina. C, Chiton.* (*A* after Pierce; *C* after Borradaile and others.)

against the substratum. The inner edge of the mantle is then lifted to create a partial vacuum. This vacuum, combined with the adhesion of the foot, enables the chiton to hold tenaciously to the substratum. A knife is often required to remove the animals from rocks or shells.

The mantle cavity of chitons is limited to two lateral troughs, or grooves, one on each side of the body between the foot and mantle edge (Fig. 27.26). Within each groove are many gills.

The ventilating current, propelled by cilia on the gills, flows through the mantle grooves from anterior to posterior (Fig. 27.26). The cur-

rent enters each groove through an opening created by the raised mantle margin. After passing the length of the groove and over the gills, the two currents exit as a single stream at the posterior end of the body, where the mantle edge is raised.

Chitons feed largely on algae scraped from rock surfaces with a radula. The stomach is unlike that described for the ancestral mollusks or gastropods, but a large digestive gland is present.

A single pair of auricles collects blood from all of the gills on each side. A single pair of nephridia opens into the posterior half of the mantle groove (Fig. 27.26).

Figure 27.25 Photograph of two species of chitons from the northwest Pacific. The shell plates of the larger species are completely covered by the mantle. (By Betty M. Barnes.)

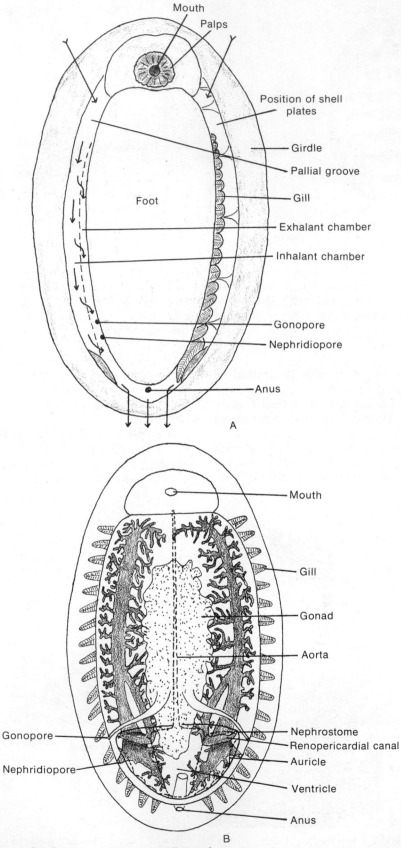

Figure 27.26   Structure of a chiton, ventral views. *A,* External structure, showing direction of water currents; *B,* internal structure. (*A* after Yonge; *B* afer Lang and Haller.)

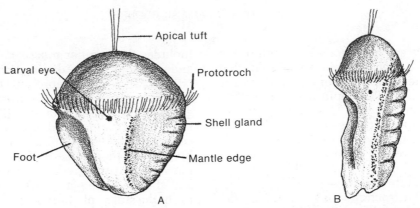

Figure 27.27 Development of chitons. *A*, Trochophore; *B*, metamorphosing trochophore. (Both after Heath from Dawydoff.)

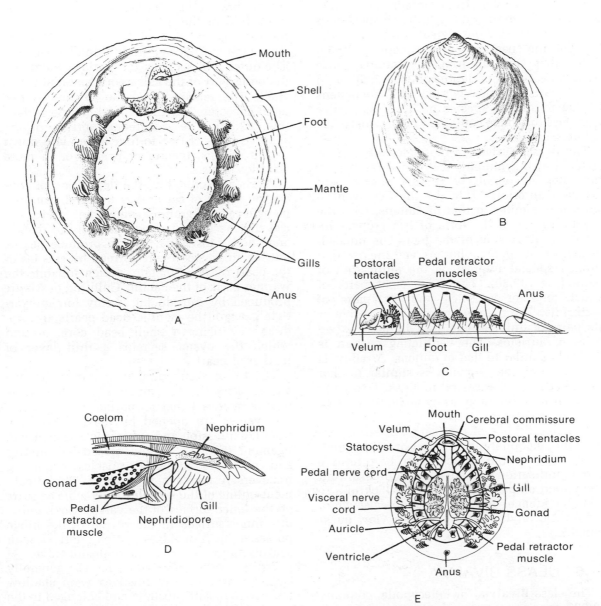

Figure 27.28 The monoplacophoran, *Neopilina*. *A*, Ventral view; *B*, dorsal view of shell; *C*, side view; *D*, transverse section through one half of body; *E*, internal anatomy. (All adapted from Lemche and Wingstrand.)

Chitons have separate sexes and each of the two gonads is provided with a gonoduct that opens near the nephridiopore in the mantle trough (Fig. 27.26). The nephridia do not serve as gonoducts in chitons. The animals do not copulate, but sperm released by a male fertilize the eggs in the sea, or fertilization occurs in the mantle trough of the female. Eggs are shed singly or in strings, or they may be brooded within the mantle cavity. Spiral cleavage results in a trochophore larva, which develops directly into the adult without an intervening veliger (Fig. 27.27).

The members of the class Monoplacophora were first known from fossils and were thought to be extinct until 10 living specimens were dredged up from a great depth in the Pacific, off the coast of Central America in 1952. Since that time additional specimens have been collected from depths of 2000-7000 m. in other parts of the world. The specimens belong to several species of *Neopilina*, relics of a class that once contained more widely distributed species.

The dorsal surface of monoplacophorans is covered by a symmetrical shield-shaped shell, the apex of which is a little peaked and directed anteriorly (Fig. 27.28). The ventral surface is similar to that of chitons, with the mantle cavity in the form of two grooves located to either side of the foot. The unusual feature of monoplacophorans, and the one that evoked special interest, is the replication of parts (Fig. 27.28). There are eight pairs of pedal retractor muscles and six pairs of nephridia. The mantle groove contains five or six pairs of unipectinate gills drained by two pairs of auricles. The ventilating current is probably similar to that of chitons. A radula is present, and the stomach is similar to that described for the ancestral mollusk. Two pairs of nephridia function as gonoducts for two pairs of gonads.

The significance of replication of structures in monoplacophorans is not clear. Some zoologists consider the replication to be primitive and an indication that ancestral mollusks were segmented animals. However, the lack of any other evidence of segmentation elsewhere within the phylum makes such a view questionable.

27.6   CLASS BIVALVIA

The class **Bivalvia,** or Pelecypoda, contains 20,000 species of marine and freshwater mollusks commonly called clams or bivalves; it includes the familiar mussels, cockles, oysters and scallops. The distinguishing characteristics of the class represent adaptations for burrowing in a soft substratum. The body is greatly compressed laterally (Fig. 27.29). The head is very much reduced, and the shell is composed of two lateral pieces, or valves, hinged together dorsally. Lateral compression has resulted in a great overhang of the mantle and shell, and the large mantle cavity extends to both sides of the body. The anteriorly directed foot is laterally compressed and somewhat bladelike, hence the name Pelecypoda—hatchet foot.

Mantle and Shell. The mantle margin possesses three folds (Fig. 27.30). The muscular inner fold of the mantle edge of one side is pressed against that of the other when the valves are closed. The middle fold is sensory. The outer fold secretes the **periostracum** and the outer part of the calcareous material of the shell. The remainder of the calcareous material is secreted by the general mantle surface (Fig. 27.30). The calcareous portion of the shell consists of two to four layers laid down as thin sheets, (nacre) or prisms or in more complex ways.

Pearls are formed by the deposition of concentric layers of calcareous material around a parasite, a sand grain or some other foreign object that becomes lodged between the mantle and shell. Most bivalves can produce pearls, but the finest pearls are formed by species of the pearl oyster, *Pinctata*, which inhabits the warmer parts of the Pacific. Cultured pearls are produced by introducing a tiny foreign particle beneath the mantle. Bead pearls are produced by using a shell bead core, around which the oyster secretes a thin layer of mother-of-pearl.

The hinge ligament that connects the two valves dorsally is composed of the same organic material that forms the periostracum. The valves are opened by tension resulting from compression and stretching of the elastic ligament (Fig. 27.30) and closed by anterior and posterior adductor muscles extending transversely between the valves. The attachment points of the muscles are visible as scars on the inner surface of the valves. Commonly, this inner surface in the region of the hinge possesses teeth or ridges, which interlock with sockets and grooves on the opposite valve.

Protobranch Bivalves. It is generally agreed that the early bivalves were shallow burrowers in soft substrata and belonged to the subclass Protobranchia, some species of which are still living today. The protobranchs

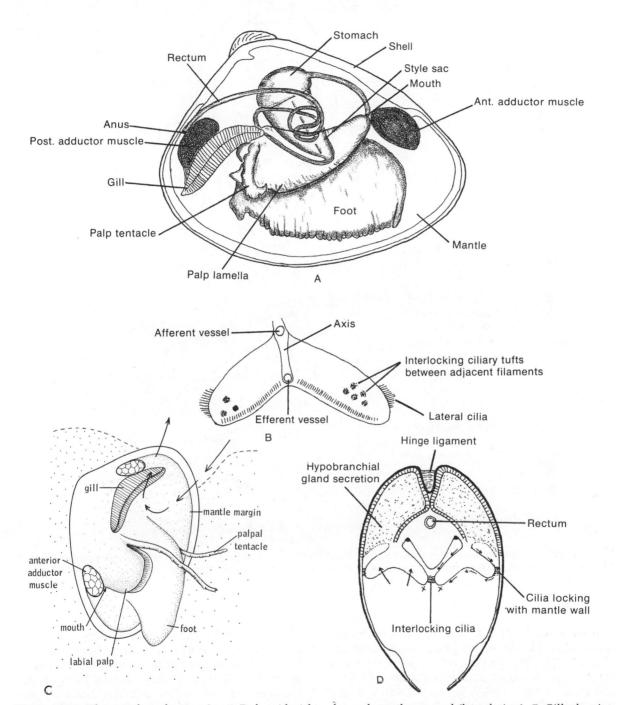

Figure 27.29 The protobranch, *Nucula*. *A*, Body with right valve and mantle removed (lateral view). *B*, Gill, showing lateral filaments (transverse section). Note similarity to gill of the ancestral mollusk shown in Figure 27.2. *C*, Feeding position. Arrows indicate direction of water current through mantle cavity. *D*, Position of gills in mantle cavity (transverse section). (All after Yonge.)

Figure 27.30   *A,* Diagrammatic transverse section through a bivalve to show hinge ligament and antagonistic adductor muscle. *B,* Transverse section through margin of shell and mantle of a bivalve, showing mantle lobes. (After Kennedy.)

possess a single pair of posterior-lateral bipectinate gills like those of primitive gastropods (Fig. 27.29). The ventilating current enters the mantle cavity between the posterior and ventral gape of the valves, passes up through the gills and exits posteriorly and dorsally.

The early protobranch bivalves, like some living species, were probably selective deposit feeders. Deposit feeding is the consumption of dead organic particles—fragments of plant and animal bodies—that have settled (deposited) to the bottom and become mixed with sand particles (p. 655). Protobranchs utilize a pair

of long tentacles for deposit feeding. Each tentacle is associated with two large flaplike folds, called **labial palps,** located to either side of the mouth (Fig. 27.29). During feeding, the tentacles are extended into the bottom sediments. Deposit material adheres to the mucous-covered surface of the tentacle, and then is transported by cilia back to the palps. Each pair of palps functions as a sorting device. The inner opposing surfaces are ridged and ciliated (similar to Fig. 27.4). Light particles are carried by certain cilia to the mouth; heavy particles are carried by other cilia to the palp

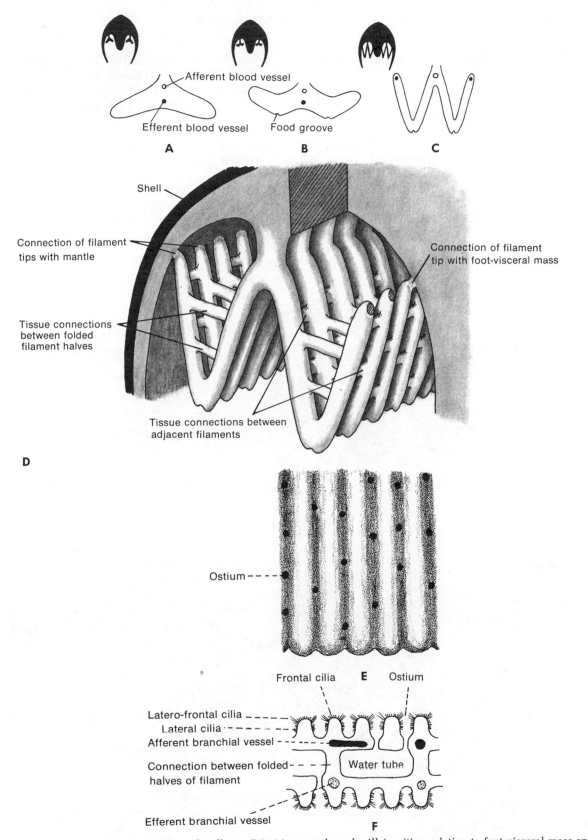

Figure 27.31 Evolution of lamellibranch gills. *A*, Primitive protobranch gill (position relative to foot-visceral mass and mantle indicated in cross section; see Figure 27.29). *B*, Development of food groove in hypothetical intermediate condition. *C*, Folding of filaments at food groove to produce the lamellibranch condition. *D*, Tissue connections that provide support for the folded lamellibranch filaments. *E*, Surface view of a lamellibranch gill showing five fused adjacent filaments. The extensive tissue connections between filaments leave only small openings (ostia) for the entrance of water. *F*, Frontal section of *E*.

margins, where they are ejected to the mantle cavity.

The evolution of selective deposit feeding in protobranch bivalves was correlated with the invasion of a soft bottom environment and the shift of the mouth away from the substratum as the body became laterally compressed for burrowing. The radula disappeared.

Evolution of Lamellibranchs. In some group of early protobranch bivalves, filter feeding evolved. An explosive evolution followed, and the filter feeders, called **lamellibranchs,** came to dominate the bivalve fauna.

The gills and ventilating current of protobranchs preadapted them for filter feeding. The term preadaptation implies that the basic features of a new condition existed in the ancestral stock but served some other function. As the lamellibranchs evolved, plankton in the ventilating current came to be utilized as a source of food, the gills became the filters, and the gill cilia (which originally served for the transport of sediment) became adapted for the transport of the trapped plankton from the filter to the mouth.

The principal modification of the gills for

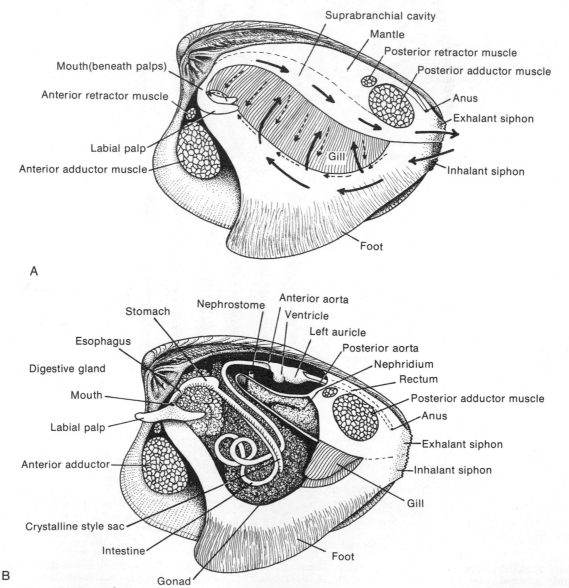

Figure 27.32 Anatomy of *Mercenaria mercenaria*. *A,* Interior of the right valve. *B,* Partial dissection, showing some of the internal organs.

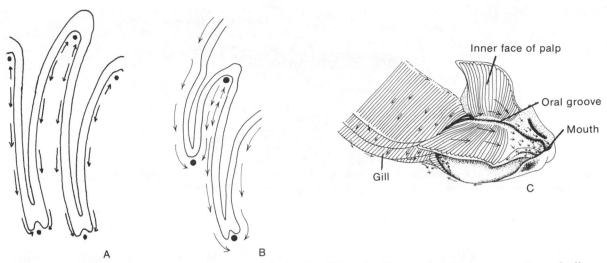

Figure 27.33 Transport and sorting of filtered particles by lamellibranch gills. A and B, Transverse sections of gills on one side (inner gill on right) showing direction of frontal cilia beat and position of anteriorly moving food tracts (black dots). A shows primitive condition with five tracts; B, condition in many lamellibranchs. C, One pair of palps spread apart and anterior section of gills of oyster. Arrows indicate ciliary tracts. (A and B after Atkins; C after Yonge.)

filtering was the lengthening and folding of the gill filaments, which greatly increased their surface area (Fig. 27.31). The long folded filaments are supported by the development of cross connections between the two halves, by connections between adjacent filaments and by connection of the tips of the filaments to the foot or mantle wall (Fig. 27.31D). The folding converted the original single gill into two gills, each new gill formed by one series of folded filaments. The lengthened filaments and their attachment to one another give the gills a sheetlike form, hence the name Lamellibranchia—sheet gills. Four large broad filtering surfaces are present, two on each new gill (Figs. 27.31 and 27.32).

Although adjacent filaments are connected, the points of attachment are behind the original ciliated margin, which remains distinct, and openings (**ostia**) remain for the passage of water between the filaments (Fig. 27.31E and F). The interior space between the two folded halves of the filaments forms water tubes, which connect with the **suprabranchial cavity,** the portion of the mantle cavity above the gills (Figs. 27.31D and 27.32). In lamellibranchs, the ventilating current, now also the feeding current, enters posteriorly and ventrally as in protobranchs. On reaching the gills, the water, propelled by the lateral gill cilia, enters openings between the filaments on all four of the gill surfaces. Within the interior water tubes, the water stream flows upward to the suprabranchial cavity, where it turns posteriorly and flows outward through the shell gape (Fig. 27.32).

Suspended particles are filtered out by special latero-frontal cilia as the ventilating current passes between the filaments (Fig. 27.31F). The filtered particles are passed to short frontal cilia, become covered with mucus and are transported downward or upward to food grooves. The food grooves are located ventrally along the gill margins and are dorsally adjacent to the points where the gill is attached (Fig. 27.33). Few lamellibranchs possess the five food grooves that could be present, but the number and distribution varies from group to group (Fig. 27.33B).

The food grooves carry the collected particles to the labial palps, which retain their original sorting function (Fig. 27.33C). Fine particles, mostly phytoplankton, are conveyed to the mouth.

A cord of mucus filled with phytoplankton is carried down the esophagus and wound into the stomach, which is much like that of protobranchs and the ancestral mollusk (they are all microphagous). However, the mass of mucus within the style sac has become compacted into a stiff rod, the **crystalline style** (Fig. 27.34). The rod is secreted by the style sac and rotated by the style sac cilia. In addition to mucus, the style contains carbohydrate-splitting enzymes. The rotating end of the style is abraded against a chitinous piece, the **gastric shield,** located at the anterior end of the stomach. The rotation of the crystalline style winds in the mucous rope from the esophagus and stirs the contents of the stomach; the abrasion of the end liberates enzymes and initiates the digestion of carbohydrates within the stomach.

Figure 27.34 Diagram of the stomach and digestive gland duct of a lamellibranch, showing rotation of crystalline style and winding in of mucous food string. Arrows indicate ciliary pathways. (After Morton.)

The churning of the stomach mass throws particles against the sorting region, which separates and conveys fine particles to the digestive glands. Here digestion is completed intracellularly. The rejected coarse particles are carried along a groove to the intestine, where they are compacted into fecal pellets and eventually ejected.

Adaptive Groups of Bivalves. The evolution of filter feeding enabled lamellibranchs to exploit habitats other than shallow depths with soft bottoms. Although bivalves are not nearly as diverse as gastropods, several groups have evolved, each adapted for survival in a specific environment.

Soft-Bottom Burrowers. Most lamellibranchs have continued to inhabit sand and mud bottoms, but the ability to utilize the ventilating current in feeding has enabled many species to survive while burrowing deeper into the substratum.

Burrowing is accomplished with the foot, which is extended anteriorly between the gape of the valves into the substratum (Fig. 27.35). Initial extension of the foot is provided by a pair of protractor muscles, which extends from the foot to a point on the shell below the anterior adductor muscle. The substratum is simultaneously softened by water ejected from the mantle cavity as the valves are closed. The closing valves exert pressure on the water remaining within the mantle cavity, which in turn exerts pressure on the visceral mass, driving blood into the foot (Fig. 27.35). The elevated blood pressure further extends the foot and dilates the distal end, anchoring it in the substratum. Two pairs of pedal retractors, one anchored on the shell near the posterior adductor muscle and one near the anterior adductor muscle, pull the valves down upon the anchored foot. The pedal retractors may contract somewhat alternately, causing the valves

to rock and facilitating their movement through the substratum.

Some bivalves can burrow much more rapidly than others. Species of *Donax* that live on surf beaches may be washed up by one wave and burrow back into the sand as the wave recedes. Some bivalves burrow quite deeply; the common razor clam *Tagelus* burrows to depths of one meter and can anchor tenaciously with the engorged dilated foot (Fig. 27.36).

Burrowers in soft bottoms must cope with the tendency of the surrounding sediment to enter with the ventilating and feeding water stream. The problem becomes greater as the animal burrows deeper. Various adaptations have evolved to reduce the entrance of sediment into the mantle cavity. Relatively permanent burrows with mucus-compacted walls are formed by many species, especially those that burrow deeply. There has also been an evolutionary tendency for the opposing mantle edges to fuse together. The most common fusion points are around the apertures for inhalant and exhalant water currents, but fusion may also occur ventrally, leaving only an opening for the protrusion of the foot (Figs. 27.35 and 27.36). Ventral mantle fusion undoubtedly aids in the elevation of hydrostatic pressure within the mantle cavity during burrowing. **Siphons,** fused tubelike extensions of the mantle around the inhalant and exhalant openings, are yet another adaptation to reduce the intake of sediment (Fig. 27.36). The siphons are extended upward to the surface and permit the animal to obtain water relatively free of sediment.

Attached Surface Dwellers. Many bivalves have abandoned burrowing and live on the surface of the sea bottom, especially on hard surfaces, such as rock and coral. The animal attaches itself in one of two ways. It

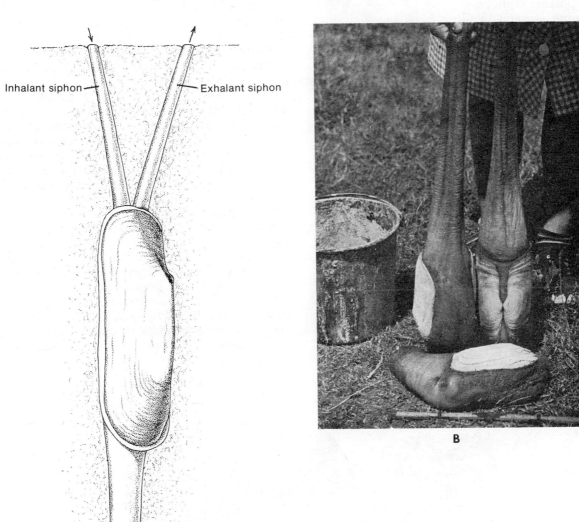

Figure 27.35 *A,* Diagrammatic cross section of a bivalve, showing hydrostatic forces that produce dilation of foot. Central vertical arrow indicates flow of blood into foot. *B,* Areas of mantle fusion in bivalves: (1) between inhalant and exhalant apertures or siphons, the most common point of fusion; (2) fusion below inhalant aperture or siphon; and (3) fusion between inhalant aperture and foot aperture. (*A* modified after Trueman.)

Figure 27.36 *A,* The razor clam *Tagelus,* a common deep-burrowing bivalve in shallow water along the east coast of the United States. *B,* The geoduck, *Panope generosa,* a giant California bivalve in which body and siphons cannot be enclosed within valves; mantle fused. (From Milne and Milne: Animal Life. Prentice-Hall, Inc., 1959.)

may, like the oyster, lie on its side with one valve fused to the substratum (Fig. 27.37). The foot is absent and the anterior adductor muscle is reduced or absent. Other bivalves, such as the common mussels, are attached by means of horny **byssal threads,** secreted by a gland in the foot (Fig. 27.37). The foot is small and finger-like and contains a groove for the placement and molding of the thread. A large number of byssal threads may be secreted and radiate out from the ventral gape of the valves like guy wires.

Sediment is much less a problem for bivalves living attached to the surface of hard substrata. It is not surprising therefore that these species exhibit no mantle fusion and do not possess siphons.

Some surface-dwelling bivalves, including both mussels and oysters, inhabit the intertidal zone and may be found attached to rocks, pilings and sea walls that are exposed at low tide. During this period the animals are inactive and keep the valves closed to reduce desiccation.

Unattached Surface Dwellers. A few bivalves, such as scallops and file shells, rest unattached on the bottom or are attached only temporarily (Fig. 27.38). These bivalves can swim in a jerky manner for short distances by rapidly clapping the valves and driving a jet of water from the mantle cavity. Such unattached surface dwellers exhibit many adaptations similar to those of attached species. The foot is reduced. Only the posterior adductor muscle is present; it is located in the middle of the valves. (This is the part of the scallop that is eaten.) Correlated with the mobility of unattached surface dwellers is the fact that the sensory lobe of the mantle margin is highly developed and may bear eyes and tentacles (Fig. 27.38).

Hard-Bottom Burrowers. Several groups of bivalves have evolved the ability to burrow into peat, clay, sandstone, coral and even limestone rock. The invasion of this type of habitat occurred independently within a number of different lamellibranch groups. The ancestors of some were probably soft-bottom burrowers and acquired the ability to burrow in rock through adaptations to successively harder substrates. The ancestors of others were probably surface dwellers attached to the substrate by byssus threads.

Boring bivalves use the anterior margins of the valves to drill. Cutting occurs when the two valves are opened, pulled down or rocked against the head of the burrow. The drilling force is provided by pulling the body against the attachment of the suckerlike foot or against the attachment of byssal threads.

Shipworms are highly modified to drill into wood. The valves, which function as the drill, are very tiny and no longer cover the greatly elongated body and siphons (Fig. 27.39). The animal occupies the entire burrow, which is lined with calcium carbonate secreted by the mantle, and the ends of the siphons are located at the burrow opening. Shipworms are no longer filter feeders but ingest the sawdust that they excavate in drilling and digest it with cellulase. Shipworms can do extensive damage to marine timbers.

Reproduction and Development in Bivalves. Most bivalves have separate sexes. The small number of hermaphroditic forms includes some oysters and scallops. The two nephridia serve as gonoducts in the protobranchs, but separate gonoducts are present in most lamellibranchs (Fig. 27.32). The gametes are shed in the exhalant water current, and fertilization occurs in the surrounding sea

Byssal thread

Left valve

Figure 27.37 Sessile bivalves: mussels attached by byssal threads; oysters attached by left valve.

Figure 27.38 *A*, File shell, *Lima*, a swimming bivalve. *B*, Burrows of rock-boring bivalves excavated in limestone. (Both courtesy of D. P. Wilson.)

water. Some bivalves, including most fresh-water species, brood their eggs within the water tubes of the gills.

Successive trochophore and veliger larvae are typical of the development of marine bivalves. A number of characteristics of the class appear in the veliger (Fig. 27.40). There is some lateral compression of the body and a bivalve shell. The finger-like foot provides the larva with considerable mobility when it first settles.

Many freshwater bivalves have a highly specialized developmental history. There is no veliger, but a modified larval stage, a **glochidium,** is released from the brood chambers of the female. The glochidium settles to the bottom of the stream or lake. When certain species of fish swim over the bottom, the glochidia become attached to the fins or gills by means of an attachment thread or a hook on the ventral margin of each valve. The tissues of the host then overgrow the glochidium, which now becomes a parasite for the remainder of larval development. The fish serves for the dispersion of the bivalves. When development is complete, the young clam breaks free of the host, drops to the bottom and becomes a free-living adult.

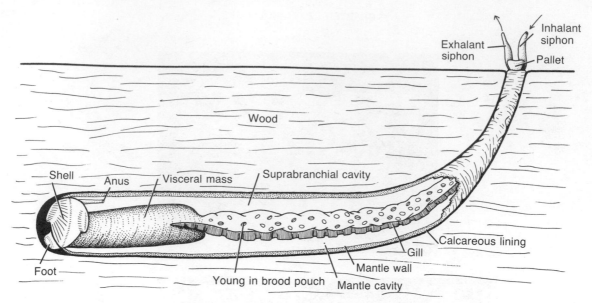

Figure 27.39 A shipworm, a wood-boring bivalve.

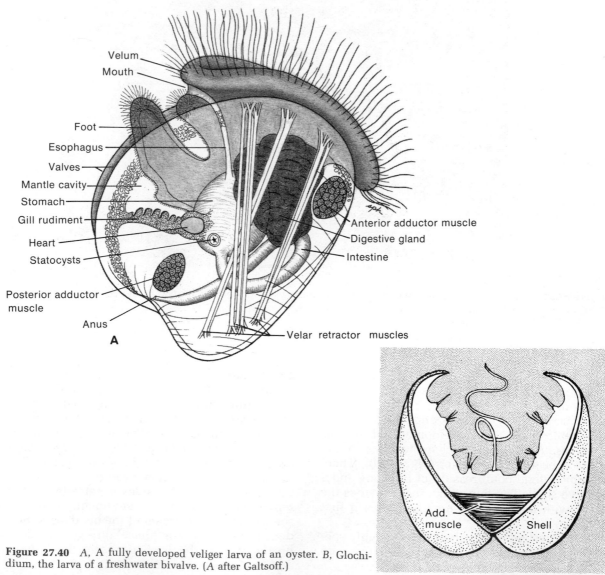

Figure 27.40 *A*, A fully developed veliger larva of an oyster. *B*, Glochidium, the larva of a freshwater bivalve. (*A* after Galtsoff.)

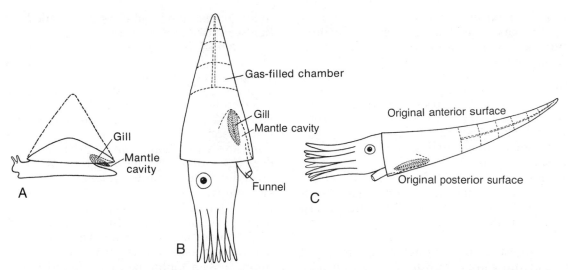

Figure 27.41 Evolution of a cephalopod. *A*, Lateral view of hypothetical ancestral mollusk. Dashed line indicates increase in length of dorso-ventral axis and a change in the shape of the shell from a shield to a cone. *B*, An early cephalopod oriented in a position that is comparable to the ancestral mollusk. *C*, Actual swimming position of an early cephalopod.

27.7 CLASS CEPHALOPODA

The class **Cephalopoda** includes *Nautilus*, **squids, cuttlefish** and **octopods.** There are only some 200 living species, but the rich fossil record indicates that in past geologic periods the class contained thousands of members.

The cephalopod characteristics represent adaptations for a swimming, predatory mode of existence, although there are many species that have secondarily adopted other life styles. The dorsoventral axis has become greatly lengthened (Fig. 27.41). The foot has become divided into **tentacles,** or arms, and has shifted somewhat anteriorly around the mouth, hence the name cephalopod—"head foot." Cephalopods swim by means of jet propulsion. The force is generated by the contraction of the **mantle,** which becomes locked about the **head,** and water is expelled from the mantle cavity through a short funnel derived from a part of the foot.

The great increase in the dorsoventral axis and the assumption of a swimming habit have led to a shift in the orientation of the body. The tentacles (which represent the original ventral surface) are directed forward, and the **visceral mass** (which represents the original dorsal surface) is now directed posteriorly.

Cephalopod Shells. Only *Nautilus* among living cephalopods possesses a well developed **shell,** but a shell was characteristic of thousands of fossil species (Figs. 27.41 and 27.42). The cephalopod shell, unlike that of other mollusks, is divided by transverse septa into interior chambers. The living animal occupies only the outer chamber, which opens to the exterior. The posterior chambers are filled

Figure 27.42 Lateral view of *Nautilus*. A diagrammatic section of the chambered shell is shown. The mantle of the left side is cut away to show the mantle cavity and two of the gills. When the animal retracts, the leathery hood protects the shell opening. (Combined from several sources.)

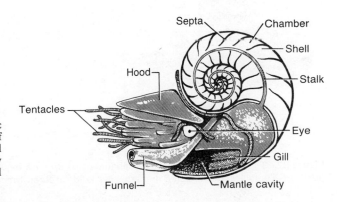

with gas, which provides buoyancy for swimming. The gas is secreted by a cord of mantle tissue that extends back through the septa. *Nautilus* possesses a plano-spiral shell; that is, the shell is coiled but still bilaterally symmetrical. Many fossil species had plano-spiral shells, but others had straight tusklike conical shells, which is probably the more primitive condition. These early cephalopods swam with the straight conical shell inclined to varying degrees (Fig. 27.41C).

In squids and cuttlefish, the shell has become greatly reduced and completely overgrown by the mantle (Figs. 27.43 and 27.44). In octopods, the shell has disappeared completely.

The largest living cephalopods are giant squids of the genus *Architeuthis,* which may reach 16 meters in length and are bottom dwellers at 200 to 400 meters. There are no giant octopods. The largest shelled fossil species were ones with straight conical shells up to five meters long and ones with coiled shells that were two meters in diameter.

Locomotion.   The squids are the most powerful cephalopod swimmers. Flying squids of the genus *Onchyteuthis,* which can shoot out of water and glide for short distances, have been collected from the decks of ships. The body of a squid is torpedo-shaped; its posterior lateral fins are used as stabilizers. The arms are held together and function as a rudder. The siphon can be directed anteriorly or posteriorly to permit backward or forward swimming.

In very rapid swimming, contractions of the mantle muscles are synchronized by a system of **giant neurons.** The greater the diameter of a neuronal process, the faster it will conduct an impulse. The motor neurons radiating out of the mantle ganglia are of different diameters depending upon the distance that the impulse must be transmitted. Long neurons have larger diameters than do short neurons. As a result of

Figure 27.43   *A,* The cuttlefish *Sepia* seizing a shrimp with its tentacles. *B,* Dorsal view of the squid *Loligo* in swimming position. Tentacles are retracted and arms are held together, functioning as a rudder.

A

B

Figure 27.44 *A*, A "small" relative of the giant squid, the oceanic squid, *Ommastrephes caroli*. This remarkably intact specimen was stranded. A meter ruler gives the scale. *B*, Octopus pursuing a crab. (*A*, courtesy of Douglas P. Wilson; *B* by Fritz Goro—courtesy of LIFE Magazine, © 1955, Time Inc.)

these structural differences, an impulse originating in the mantle ganglion reaches all of the muscle fibers at the same time, ensuring simultaneous contraction.

Cuttlefish have shorter bodies than squids and are agile but not powerful swimmers. Their fins undulate and contribute to propulsion. Many species live in shallow water, hovering and darting over the bottom. Some lie on the bottom covered with sand during the day or while waiting for prey.

The body of octopods is rather globular (Fig. 27.44B). They are largely bottom dwellers and crawl about with the arms, swimming only to escape. They are most commonly found on rocky and coralline bottoms. Some deep-sea octopods have become pelagic. The arms are connected together by webbing and the animals swim like jellyfish (Fig. 27.45).

Nutrition. Cephalopods are highly adapted for raptorial feeding and a carnivorous diet. The prey is seized by the many tentacles. Squids and cuttlefish have eight arms and two long prehensile tentacles; octopods possess only the eight arms. Except in *Nautilus*, the arms and tentacles are provided with suckers.

The principal ingestive organ is a pair of powerful, highly mobile, beaklike, horny **jaws,** which can cut and tear prey (Fig. 27.46B). The radula functions as a tongue for pulling in pieces of flesh bitten off by the jaws. As a further aid to dispatching the victim, a pair of salivary glands have been modified as poison glands (Fig. 27.46).

Fish, shrimp, crabs and other mollusks serve as food. The two prehensile tentacles of squids and cuttlefish can be shot out to seize prey. Some squids can dart backward into a school of fish, seize one and quickly bite out a chunk of flesh behind the head. The fish is rapidly eaten until the only remains are the intestines, which are dropped to the bottom. Octopods feed on crabs and snails and may remove the occupant by injecting poison through a hole drilled in the shell with the radula.

Internal Structure and Physiology. Many features of their internal structure and physiology reflect the active, raptorial mode of life of cephalopods. The stomach is quite different from that of other mollusks. Digestion is extracellular, and the digestive glands produce large amounts of powerful proteolytic en-

Figure 27.45 *Histioteuthis,* a pelagic squid with webbed arms. (After Chun from Lane, F. W.: Kingdom of the Octopus, Sheridan House, New York, 1954.)

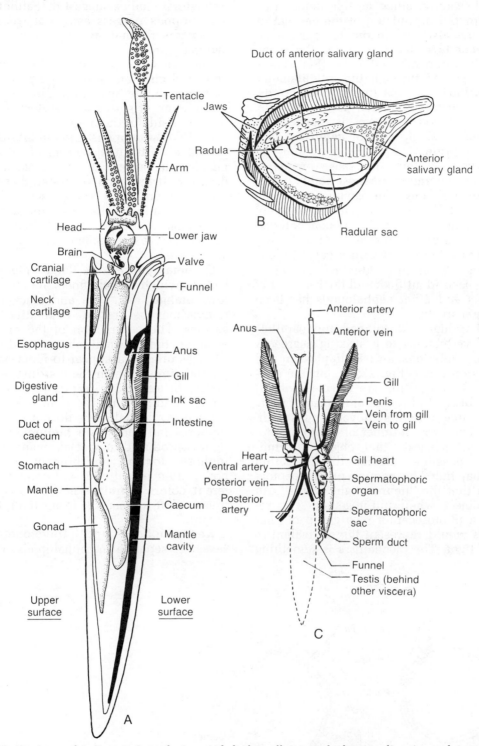

Figure 27.46 Anatomy of *Loligo*. *A*, Lateral view with body wall removed, showing digestive and nervous systems. *B*, Sagittal section through buccal mass of *Loligo*. *C*, Ventral view of part of the circulatory system and of the male reproductive system (drawn to the same scale as *A*, so that it can be turned on edge and fitted into *A*). (*A* and *C* after Williams; *B* after Owen.)

zymes. Cilia are no longer needed in the gills and mantle cavity, since muscle contraction provides for the propulsion of the ventilating current, now also the locomotor current. A pair of **secondary hearts** elevates the blood pressure on entrance into the gills (Fig. 27.45), and blood passes through the gill filaments within **capillaries,** increasing the rate of transport and reflecting a higher metabolic rate. Within the blood, oxygen is carried by **hemocyanin.** A single pair of **nephridia** provides for excretion.

The nervous system and sense organs of cephalopods are among the most highly developed of invertebrates. The ganglia, which in many other mollusks are located at different points along the nerve cords, are concentrated in cephalopods at the anterior end to form a complex **brain** housed within a cartilaginous capsule (Fig. 27.46). Functional centers of the brain have been identified, and the behavior of the octopus and other cephalopods has been studied extensively.

The **eye** is highly developed and parallels the eye of vertebrates to a striking degree. A **retina,** a movable **lens,** an **iris diaphragm** and a **cornea** are present (Fig. 27.47), but the photoreceptors in the retina are directed *toward* the source of light (*direct eye*) instead of *away* from the source of light as in vertebrates (*indirect eye*). The large number of photoreceptor cells present suggests that object discrimination is possible. Object discrimination requires that there be enough photoreceptors present so that the "neuron light points" can form an object when put together. The response of a photoreceptor or a group of photoreceptors would represent the equivalent of a point of light. The mechanism is something like image formation on a television screen. Behavioral studies suggest that although many cephalopods possess eyes that would permit object discrimination, they are very nearsighted animals.

The arms are provided with many tactile cells and chemoreceptors and play an important role in the structural and chemical discrimination of surfaces. A **statocyst** is located on either side of the brain.

An **ink gland** opens into the intestine and is provided with an ejector mechanism, which enables the animal to discharge a cloud of dark ink. It has been suggested that the cloud functions not as a smoke screen but as a dummy, confusing the predator and enabling the animal to escape. The alkaloids in the ink may narcotize the chemoreceptors of predators such as fish.

Chromatophores are large, pigment-laden cells to which muscle fibers are attached. Yellow, orange, red, blue and black chromatophores may be present, depending upon the species. The contraction of the muscle cells expands the chromatophores, and the color changes produced appear to function as background simulation in the courting behavior of males, in aggressive responses between individuals and in alarm reactions to intruders.

Deep-sea cephalopods do not have chromatophores but many are **bioluminescent.** The light is produced in special organs called **photophores** located on different parts of the body, often in patterns and composed of different colors. Luminescent ink may also be present, but in deep-sea forms the ink gland is absent.

Reproduction and Development. The sexes are separate in cephalopods. The gonad

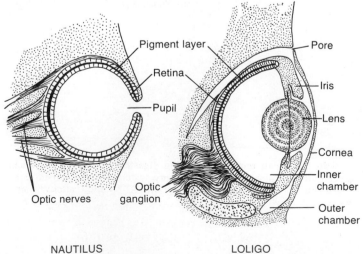

Pigment layer

Retina

Pupil

Optic nerves

Optic ganglion

Pore

Iris

Lens

Cornea

Inner chamber

Outer chamber

NAUTILUS   LOLIGO

Figure 27.47 Cephalopod eyes. *Left,* The pinhole camera type in *Nautilus.* (After Borradaile et al.) *Right,* The lens type complete with shutter (iris) in *Loligo.* (After Williams.)

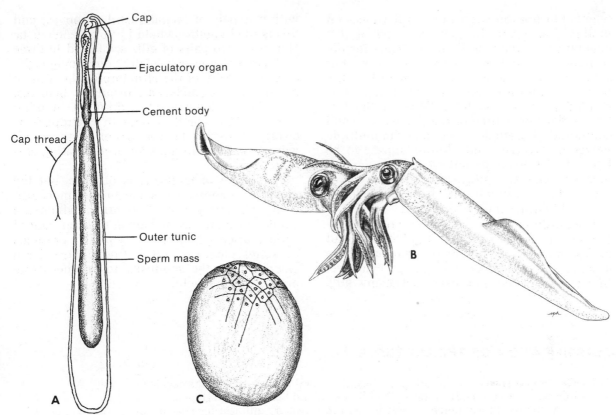

Figure 27.48 *A*, Spermatophore of the squid *Loligo*; *B*, copulating squids; *C*, incomplete cleavage in the yolky egg of *Loligo*. (*B* based on a photograph by Robert F. Sisson; *C* after Watase from Dawydoff.)

in each sex is provided with a relatively complex gonoduct (Fig. 27.46). Copulation occurs during swimming and involves the transfer of sperm in packets termed **spermatophores.** Masses of sperm are cemented to one another or encased within special secretions in certain regions of the sperm duct. The shape of the spermatophore is often determined by the region of the sperm duct that molds it. The spermatophores of many cephalopods look like baseball bats (Fig. 27.48).

In copulation, the male seizes the female head on and interlocks his arms with hers. One especially modified arm of the male reaches into his mantle cavity and plucks a mass of spermatophores from the gonoduct. The copulatory arm of the male then deposits the spermatophores in the mantle cavity of the female or (in some species) within a special pouch on the membrane around the mouth. The spermatophores discharge their sperm shortly after being deposited within the mantle cavity, and the eggs are fertilized on leaving the female gonoduct. Where the sperm are

deposited on the buccal membrane, the eggs are fertilized as they leave the female siphon for deposition.

In some pelagic (deep-sea) cephalopods, the eggs are planktonic. In most, however, the eggs are deposited in strings on the bottom or are attached to stones or other objects. Octopods remain with their eggs, guarding and cleaning them of sediment.

Cephalopod eggs are telolecithal, with large amounts of yolk material. Cleavage is therefore meroblastic and discoidal as in the eggs of reptiles and birds. Development is direct, and the young have the adult form on hatching.

27.8 THE ANCESTRAL MOLLUSK: A REVIEW

Having studied members of five of the seven molluscan classes, you are now in a better position to appreciate the way a hypothetical ancestor can be derived and to determine whether or not the description of the ancestral mollusks as given in the beginning of this chapter is justified. Is there evidence for describing the first mollusks differently? For some features, there would seem to be little basis for question. Four of the five classes studied possess a **radula.** Three of the five have a flat **foot.** Three of the five have a **shell** composed of a single piece, and the multiple **valves** of bivalves and chitons appear to be later adaptations to permit articulation. Or chitons, as some malacologists believe, may have diverged early in molluscan evolution before a single shieldlike shell had formed.

The most debatable derivations are the position of the **mantle cavity** and the number and nature of the **gills.** One might argue, using the chitons and monoplacophorans as evidence, that the ancestral mollusks had many gills within a pair of lateral mantle grooves, and this is the hypothesis held by some zoologists. However, two pairs of gills are found in three of the five classes. Especially convincing is the close similarity of the structure of the gills of primitive gastropods and primitive bivalves, two classes that are quite different in other respects. The mantle cavity in gastropods is clearly posterior before torsion, and in bivalves there is still a posterior ventilating current.

The origin of mollusks is obscure, but the most likely ancestor is some ancient metazoan group stemming from flatworms, which are similar to mollusks in having a broad ventral surface upon which they creep. The presence of spiral cleavage and a trochophore larva clearly links the mollusks to another large phylum, the Annelida.

CLASSIFICATION OF THE MOLLUSCA

CLASS GASTROPODA. Snails. Mantle and visceral mass exhibit some degree of torsion. Shell typically spiraled. Foot flattened and head well developed.

Subclass Prosobranchia. Marine and freshwater, gill-bearing species in which the mantle cavity and contained organs are located anteriorly. Shell present and operculum is common. Most marine snails with well developed shells belong to this subclass.

Subclass Opisthobranchia. Visceral mass has undergone detorsion and some degree of reduction of shell and mantle cavity is common. Hermaphroditic. Entirely marine. Bubble shells, sea hares and sea slugs.

Subclass Pulmonata. Gills absent and mantle cavity converted into a lung. Shell usually present but never an operculum. Hermaphroditic. Terrestrial and freshwater. Most land snails of temperate regions are pulmonates.

CLASS POLYPLACOPHORA, OR AMPHINEURA. Chitons. Body greatly flattened dorso-ventrally. Head reduced. Shell composed of eight linearly arranged, overlapping plates. Marine.

CLASS APLACOPHORA. Small group of aberrant marine mollusks with worm-like bodies and no shells.

CLASS MONOPLACOPHORA. Small group of marine deep-water species. Body flattened dorso-ventrally. Shieldlike shell. Various organs replicated — eight pairs of retractor muscles, five or six pairs of gills, six pairs of nephridia and two pairs of auricles.

CLASS BIVALVIA, OR PELECYPODA. Clams, or bivalves. Body greatly flattened laterally. Shell composed of two lateral valves hinged dorsally. Head reduced.

Subclass Protobranchia. Primitive marine bivalves with one pair of unfolded gills.

Subclass Lamellibranchia. Marine and freshwater filter feeding bivalves with folded gills. Most bivalves belong to this subclass.

Subclass Septibranchia. Small group of marine species having the gills modified as a pumping septum.

CLASS SCAPHOPODA. Tusk or tooth shells. Burrowing marine mollusks having a tusklike shell open at each end.

CLASS CEPHALOPODA. *Nautilus*, cuttlefish, squids and octopods. Marine, mostly swimming mollusks, having the foot divided into tentacles, or arms. Shells, when present, divided into chambers, of which only the most recent is occupied by the living animal.

Subclass Nautiloidea. Mostly fossil cephalopods with straight or coiled chambered shells. *Nautilus* is only living species.

Subclass Ammonoidea. Fossil cephalopods with coiled shells having wrinkled septa.

Subclass Coleoidea. Shell internal or absent. Cuttlefish, squids and octopods.

ANNOTATED REFERENCES

The references listed below are devoted solely to mollusks. Accounts of mollusks may also be found in the general references listed at the end of Chapter 21.

Abbott, R. T.: American Seashells. 2nd Ed. Princeton, New Jersey, D. Van Nostrand Co., 1974. A guide to the marine mollusks of the Atlantic and Pacific coasts of North America.

Morton, J. E.: Molluscs. 4th ed. London, Hutchinson University Library, 1967. A general account of the mollusks.

Purchon, R. D.: The Biology of the Mollusca. Oxford, Pergamon Press, 1968. A general account of the mollusks.

Solem, G. A.: The Shell Makers. New York, John Wiley and Sons, 1974. An introduction to the mollusks.

Wilbur, K. M. and C. M. Yonge (Eds.): Physiology of Mollusca. New York, Academic Press, Vol. I, 1964 (Vol. II, 1966); A detailed account of all aspects of molluscan physiology.

Yonge, C. M. and J. E. Thompson: Living Marine Molluscs. London, Wm. Collins and Sons, 1976. A general biology of the mollusks.

CHAPTER 28

THE ANNELIDS

1. Members of the phylum Annelida, called segmented worms, inhabit the sea, fresh water and land (earthworms and some leeches).

2. The long, more or less cylindrical body is metameric, and externally, segments are demarcated by integumental grooves and serial repetition of appendages, setae and pores.

3. Internally, metamerism is reflected in the segmental organization of the body wall musculature and in the compartmentation of the coelom by a septum between each segment.

4. Annelids that live in burrows and tubes move by peristaltic locomotion, which is believed to be the primitive mode of movement for the phylum and the one for which metamerism is an adaptation. Other annelids crawl by means of lateral segmental appendages (parapodia) or, in the case of leeches, by lengthening and shortening the entire body while alternately adhering to the substratum with anterior and posterior suckers.

5. Annelids are characterized by a great variety of diets and employ many different modes of feeding. A pharynx and intestine are usually the most conspicuous parts of the generally straight gut tube.

6. The excretory organs are paired segmental nephridia. The distal end of each nephridium lies in the segment anterior to the one that houses most of the tubule and the nephridiopore.

7. The circulatory system is at least partially closed in most annelids and is composed of a dorsal vessel that pumps blood anteriorly, a ventral vessel that transports blood posteriorly and segmental vessels that supply various organs and deliver blood back to the dorsal vessel.

8. The nervous system consists of a pair of dorsal anterior cerebral ganglia (brain), a pair of connectives around the gut and a pair of longitudinal ventral nerve cords that extend the length of the body. The ventral cords are commonly fused together. Paired lateral nerves arise from a ganglionic swelling of the ventral nerve cord in each segment. Those annelids with well developed heads may possess eyes and antennae as sense organs.

9. Marine annelids (polychaetes) have the sexes separated, and the gametes usually arise from the peritoneum of most segments, mature in the coelom

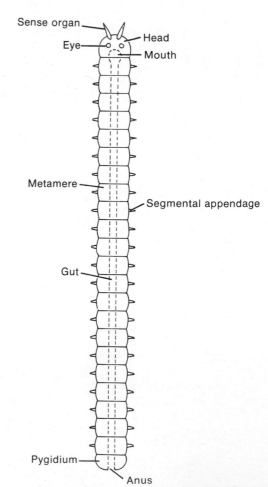

Figure 28.1 Diagram of a metameric animal, such as an annelid worm.

and exit by nephridia or by rupture. Fertilization is external, and a top-shaped trochophore is the larval stage. Freshwater and terrestrial annelids (oligochaetes and leeches) are hermaphroditic, with gonads and gonoducts in a few segments. There is copulation with reciprocal sperm transfer, and the eggs are deposited within a cocoon secreted by certain glandular segments called the clitellum.

The phylum **Annelida,** the segmented worms, includes the familiar earthworms and leeches as well as many marine and freshwater worms. The most striking annelid characteristic is the division of the body into similar parts, or **segments,** arranged in a linear series along the anterior-posterior axis. Each segment is termed a **metamere** (Fig. 28.1), and this condition is called **metamerism.** Only the trunk is segmented. Neither the **head,** or **prostomium,** containing the brain and bearing the mouth, nor the terminal part of the body, the **pygidium,** which carries the **anus,** is a segment. New segments arise in front of the pygidium; the oldest segments lie just behind the head. Some longitudinal structures, such as the gut and the principal blood vessels and nerves, extend the length of the body, passing through successive segments; other structures are repeated in each segment, reflecting the metameric organization of the body.

28.1 METAMERISM AND LOCOMOTION

Metamerism appears to have evolved twice in the evolutionary history of the Animal Kingdom, once in the evolution of annelids and arthropods and once in the evolution of chordates. In both instances, the condition appears to have evolved as an adaptation for locomotion, but for different kinds of locomotion. In chordates, metamerism probably represented an adaptation for undulatory swimming (p. 234). The ancestors of the annelids were probably elongate coelomate animals that inhabited marine sand and mud; metamerism probably arose as an adaptation for burrowing (p. 245).

The **coelom** is compartmented by transverse **septa,** and the longitudinal and circular muscles are organized into segmental blocks corresponding to the coelomic compartmentation. A coelomic compartment and its surrounding muscle blocks constitute a segment, and the body is composed of a series of such segments. Each segment is a functional, as well as a structural, unit, and the elongation and shortening

resulting from the interaction of the fluid skeleton and the body wall muscles is restricted to a small part of the body (Fig. 28.2). In locomotion, the segments function in groups, and waves of muscle contraction sweep down the length of the body, affecting a number of segments simultaneously. The wave of *longitudinal* contraction is followed by a wave of *circular* contraction. Each section of the body is thus alternately elongated and shortened.

It is important to note that annelidan metamerism is basically a modification of the body wall muscles and the coelom. The **nervous, circulatory,** and **excretory** systems are also metameric, i.e., parts are repeated in each segment. But the metamerism of these structures probably reflects an adjustment of these supply systems to the primary metamerism of the muscles and coelom (Fig. 28.3).

The typical annelidan has a **nervous system** with a dorsal **brain** and a ventral longitudinal **nerve cord,** a straight tubular **gut** and a closed circulatory system (Fig. 28.3). The excretory organs are usually **metanephridia** (see p. 346), typically one pair per segment (Fig. 28.3).

28.2 CLASSIFICATION

The phylum is divided into three classes. The **Polychaeta,** the largest of the three classes, contains the marine annelids. The **Oligochaeta** includes the freshwater annelids and earthworms, and the **Hirudinea** comprises the leeches. The latter clearly evolved from the oligochaetes, and the oligochaetes and polychaetes are believed to have evolved separately from some commom marine burrowing ancestor.

28.3 CLASS POLYCHAETA

The members of the class Polychaeta are distinguished from other annelids by the presence of paired segmented appendages, **parapodia,** that carry projecting bundles of chitinous rods, the **setae** (Fig. 28.4). The setae grip the substratum when the animal is crawling.

The parapodia vary greatly in shape and size, but when well developed, each is composed of an upper **notopodium** and a lower **neuropodium.** Each contains setae, a small, tentacle-like process at the base called a **cirrus** and an internal heavy chitinous rod (**aciculum**) to which the parapodial muscles are attached (Fig. 28.5).

Septum

Coelomic fluid

Body wall muscles

A

B

Figure 28.2 *A* and *B*, Diagrammatic frontal section through two annelid segments. External arrows indicate direction of force exerted by body wall muscles. Internal arrows indicate direction of force exerted by coelomic fluid pressure: (*A*) during circular muscle contraction and (*B*) during longitudinal muscle contraction. *C*, Diagram showing mode of locomotion of an earthworm. Segments undergoing longitudinal muscle contraction are marked with larger dot and drawn twice as wide as those undergoing circular muscle contraction. The forward progression of a segment during the course of several waves of circular muscle contraction is indicated by the horizontal lines connecting the same segments. (After Gray and Lissmann.)

10
c.m.

0 1 2 3 4 5 6 7 8 9 10 11 12 13 14 15 16

sec.

C

Longitudinal muscles
Epidermis
Circular muscles
Septum
Intestine
Nephrostome
Nerve cord
Peritoneum
Ventral mesentery

Dorsal blood vessel
Dorsal mesentery

Epidermis
Circular muscles
Longitudinal muscles

Junction of two segments

Dorsal blood vessel
Septum
Intestine

Ventral blood vessel
Nephridium
Nephridiopore

Figure 28.3 Diagrammatic lateral view of a series of annelid segments. (After Kaestner.)

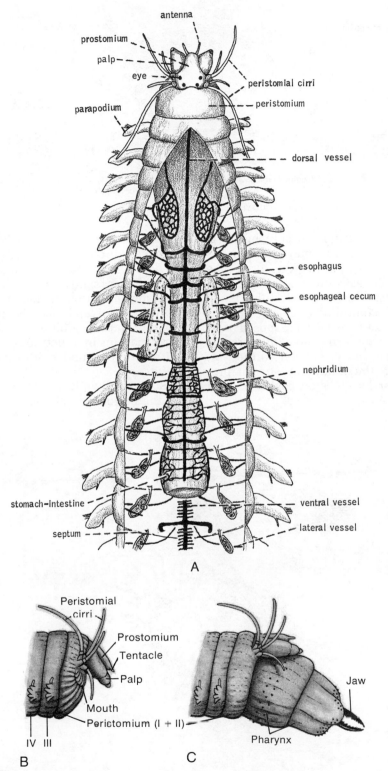

Figure 28.4 The polychaete *Nereis*. *A*, Dorsal view of anterior end. *B*, Lateral view of head. Roman numerals designate segments. *C*, Lateral view of head with pharynx everted.

Figure 28.5 Transverse section of *Nereis* at level of intestine.

The head, or prostomium, of polychaetes is often highly developed and may bear various sensory and feeding structures (Fig. 28.4). The mouth is located beneath the prostomium, just in front of the first one or two modified trunk segments (the **peristomium**).

The digestive tract is usually composed of an eversible muscular **pharynx** and a long straight **intestine.** The pharynx is everted (turned inside out) through the mouth by special protractor muscles or by elevated coelomic fluid pressure (Fig. 28.4).

The more than 5000 species of marine polychaetes exhibit great diversity both in form and in habit. All types of marine habitats contain polychaetes; sandy and muddy bottoms harbor an especially rich fauna. An appreciation of polychaete diversity can be gained by examining the adaptive groups that compose the class, for variations in structure are closely correlated with different modes of existence and feeding habits.

Surface Dwellers. Many polychaetes live on the bottom of the sea but never on exposed

Figure 28.6 Two surface-dwelling polychaetes. Note the well-developed head and parapodia. *A, Phyllodoce maculata; B, Trypanosyllis zebra.*

Figure 28.7 Parapodial movement in polychaete crawling. Waves of movement on one side of body alternate with the opposite side. (After Mettam.)

surfaces. They are found beneath stones and shells, in rock and coral crevices and in algae and sessile animals. The heads of surface-dwelling polychaetes are well developed and carry several kinds of sensory structures—one or two pairs of **eyes,** up to five **antennae** and a pair of **palps** (Fig. 28.6).

The parapodia are large and function like legs in crawling. At the beginning of the power stroke, the parapodium is moved forward and placed against the substratum. The parapodial muscles then pull the parapodium backward. The parapodium does not lose contact with the substratum during the back sweep; consequently the body is driven forward. Each stroke constitutes a small step, and the movements of the parapodium occur in sequence as a result of **waves of contraction** passing along the length of the body (Fig. 28.7). The waves of contraction on one side of the body alternate with those on the opposite side. In rapid movement, alternating contractions of the longitudinal muscles of the body wall throw the body into undulatory waves that generate additional thrust against the substratum.

The crawling mode of locomotion of surface-dwelling polychaetes is reflected in the structure of the body wall and coelom. The longitudinal muscles are more highly developed than the circular muscles (Fig. 28.5), and the septa are somewhat reduced, since the coelom is not as important as a localized hydrostatic skeleton as it is in peristaltic movement.

Many surface dwellers are carnivores and feed on other small invertebrates, including polychaetes. Others are algae eaters, scavengers or detritus feeders. Almost all have a muscular pharynx provided with one or more chitinous jaws used for seizing prey or tearing off pieces of algae.

Pelagic Polychaetes. Several families of polychaetes are adapted for living in the open ocean. These worms never come to the bottom but resemble the surface-dwelling polychaetes in some ways. They have well developed heads, and the large parapodia are used as paddles in swimming (Fig. 28.8). They are commonly carnivorous. Like many other small pelagic or planktonic animals, these polychaetes tend to be pale or transparent.

Figure 28.8 *Tomopteris renata,* a pelagic polychaete.

Figure 28.9 Burrow system of *Glycera alba*, showing worm lying in wait for prey. (After Oekelmann and Vahl.)

Gallery Dwellers. Many polychaetes are adapted for burrowing in sand or mud. Some excavate extensive burrow systems, or galleries, that open to the surface at numerous points (Fig. 28.9). The wall of the burrow is kept from caving in by its lining of mucus. The body of a gallery dweller is usually long and the prostomium is commonly conical or a simple lobe devoid of eyes and other sensory structures (Fig. 28.10).

Gallery dwellers may use the parapodia to crawl through the burrow system. However, many crawl by peristaltic contractions; the parapodia of these forms are somewhat reduced and function primarily to anchor the segments against the burrow wall. Septa and circular muscles are well developed.

Some gallery dwellers are carnivores. The bloodworm *Glycera*, for example, feeds on small surface-dwelling invertebrates. The worm lies in wait below the surface and can detect slight changes in water pressure created by the movement of an animal above (Fig. 28.9). When prey is in the vicinity, *Glycera* slowly creeps toward the opening and then everts the long proboscis-like pharynx with great force. The victim is seized with four teeth at the proboscis tip (Fig. 28.10).

Other gallery dwellers are direct deposit feeders and consume the substratum through which the gallery system penetrates. The organic material is then digested and the mineral material is defecated.

Sedentary Burrowers. Some burrowing polychaetes construct simple vertical burrows with only one or two openings to the surface (Fig. 28.11). These sedentary burrowers, in contrast to gallery dwellers, move about relatively little. Peristaltic crawling is the rule, and the parapodia are reduced to ridges that bear special hooklike setae that aid in gripping the burrow wall. The prostomium lacks sensory structures, although special feeding appendages may be present.

Some sedentary burrowers are nonselective deposit feeders. *Arenicola*, the lugworm, for

Figure 28.10 Gallery dwellers. *A*, Anterior end of *Glycera* with proboscis everted. *B*, Anterior end of *Drilonereis*.

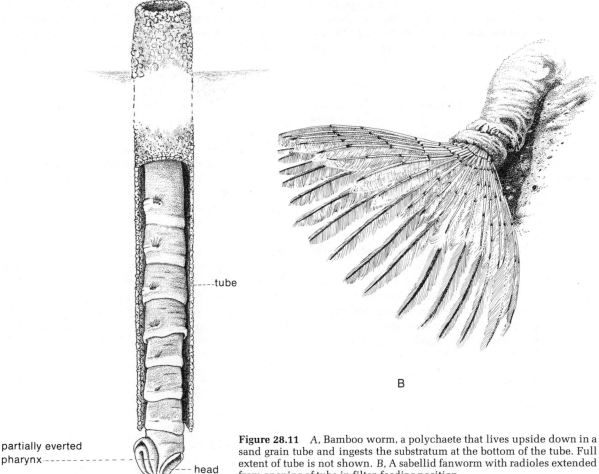

----tube

partially everted
pharynx---------

--- head

A

Figure 28.11 *A,* Bamboo worm, a polychaete that lives upside down in a sand grain tube and ingests the substratum at the bottom of the tube. Full extent of tube is not shown. *B,* A sabellid fanworm with radioles extended from opening of tube in filter-feeding position.

B

example, lives head down within its **L**-shaped burrow and eats the sand at the end (Fig. 12.4). Its soft eversible pharynx lacks jaws. With rhythmic periodicity, the animal backs up to the surface and defecates castings of sand. An intertidal sand flat may display thousands of castings, reflecting the large population of worms living below the surface (Fig. 28.12).

Selective deposit feeding is a common adaptation of sedentary burrowers, and such worms have typically evolved tentaclelike head appendages. The head of *Amphitrite,* for example, is provided with a great mass of long tentacles. These spread over the surface from the opening of the burrow (Fig. 12.5). Detritus material adheres to the mucus on the tentacles and then is conveyed to the mouth in a ciliated tentacular gutter and by tentacular contraction. There is no eversible pharynx.

Tubiculous Polychaetes. Many polychaetes live in tubes. A tube-dwelling, or tubiculous, habit is found in other animals be-

sides polychaetes, but the polychaetes have exploited this particular mode of existence to a greater degree than any other group of animals.

The tube may function as a protective retreat

Figure 28.12 Lugworm (*Arenicola*) and castings on a mud flat exposed at low tide in Puget Sound. (By Betty M. Barnes.)

or as a lair from which the worm may pounce upon passing prey. In soft bottoms, the opening of the tube projects above the sand surface and thus provides the worm with access to water free of sediment. Tubes may also permit worms to live on hard exposed substrates, such as rock, coral or shell.

The tube may be composed entirely of hardened material secreted by the worm, or it may be composed of foreign material cemented together. Secreted tubes are commonly horny or parchment-like, but one family of fanworms secretes tubes of calcium carbonate (Fig. 28.14). Sand grains are the usual building materials of cemented tubes (Fig. 28.14), although some worms may use pieces of shell. The tubes and the cementing material are commonly secreted by glands on the ventral surface of a number of segments. The construction methods vary greatly in different species.

Tube-dwelling polychaetes may be divided into **carnivorous** tube dwellers and **sedentary** tube dwellers. The carnivorous group contains predatory species that feed on other poly-

chaetes and small invertebrates. They lie in wait at the mouth of the tube and seize the victim as it passes by (Fig. 28.13). As might be expected, these worms are very agile and crawl within the tube using the parapodia. The body may be projected some distance out of the tube opening. Their adaptations — sensory structures on the head and well-developed parapodia — are essentially like those of surface dwellers.

The sedentary tubiculous polychaetes move about within the tube less actively than do the carnivorous tube-dwellers, and only the extreme ends of the body project from the tube opening. This group is similar in many respects to sedentary burrowers. The head usually lacks special sensory structures, although feeding appendages may be present. The animal moves within the tube by peristaltic contractions; hence the parapodia tend to be reduced to ridges with hooked setae.

The structural diversity of tube-dwelling polychaetes is in large part correlated with their different modes of feeding. A few, such as the

A   **B**

Figure 28.13   *A*, Head and first two gill-bearing segments of *Diopatra*, a carnivorous tube dweller (dorsal view). *B*, Funnel-shaped parchment tube of *Diopatra*, to which pieces of shell and algae are attached.

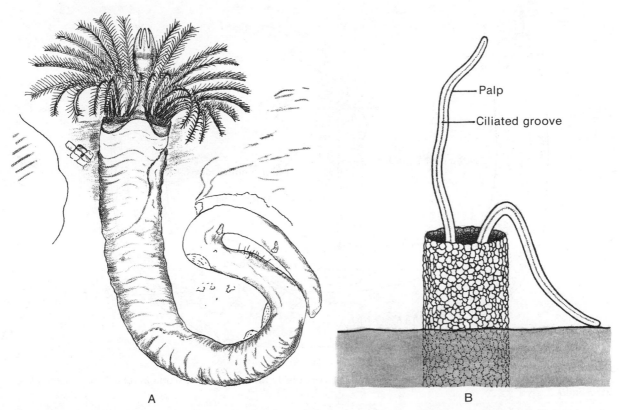

A   B

Figure 28.14 *A*, A fanworm with radioles extending from end of calcareous tube, which is attached to a shell. *B*, Palps of a spionid polychaete projecting from its sand grain tube. Ciliated groove of palp conveys detritus material picked from substratum back to mouth.

bamboo worms, are nonselective **deposit feeders.** Like *Arenicola*, the bamboo worm lives head down in sand grain tubes and eats the substratum at the bottom of the tube (Fig. 28.11). Periodically, the worm backs to the surface and defecates the mineral material. The eversible pharynx lacks jaws.

Some sedentary tube-dwellers are selective deposit feeders. There are species that feed in the same manner as described for *Amphitrite*, a sedentary burrower (Fig. 12.5). Members of the family Spionidae are common tubiculous polychaetes with two long, tentacle-like palps (Fig. 28.14). The palps of these tiny worms are projected from the opening of the tube and pick up deposit material from the surrounding substratum. The collected particles are then conveyed back to the mouth along the palp within a ciliated groove.

Filter feeding has evolved in several families of sedentary, tube-dwelling polychaetes. The beautiful fanworms are a notable example. The members of the two fanworm families (one of which secretes a calcareous tube and the other a parchment tube or a tube of cemented materials), have feather-like head structures called **radioles.** When feeding, the radioles project from the opening of the tube, often in the form of a funnel (Figs. 12.3 and 28.11). The cilia on the radioles create a water current, and plankton and other suspended particles passing through the radioles are filtered out. The trapped particles are then transported by cilia down the radiole into the mouth.

The members of the family **Chaetopteridae** use a mucous bag to filter the water pumped through the U-shaped or straight tube (Fig. 28.15). This remarkable filter is secreted by certain parapodia held out from the body like a basketball hoop (Fig. 28.15). Water passes through the mucous film leaving suspended particles behind. Mucus is continuously spun out as the end of the bag is gathered and rolled up within a ciliated **cupule.** Periodically, the animal halts the secretion, and the remainder of the mucous bag is rolled up within the cupule. The resulting ball containing trapped food particles is transported forward along a ciliated groove to the mouth (Fig. 28.15).

Internal Transport, Gas Exchange, and Excretion in Polychaetes. The closed blood vascular system of polychaetes has a relatively simple basic plan (Fig. 28.16). A contractile, longitudinal vessel located above the gut func-

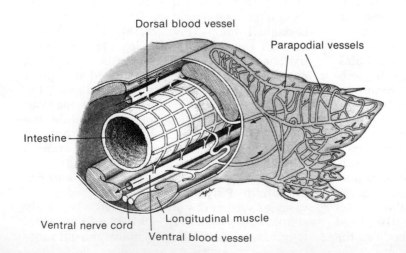

Figure 28.15 *Chaetopterus* during feeding. *A*, Anterior part of body (dorsal view). *B*, Worm in tube (lateral view). (After MacGinitie.)

tions as the heart and conveys blood anteriorly. At the anterior end of the body, paired vessels carry the blood around the gut to a ventral longitudinal vessel in which blood passively flows posteriorly. In each segment the ventral vessel gives rise to a vessel supplying the gut and to paired vessels supplying the body wall, parapodia and nephridia. These segmental vessels eventually return blood to the dorsal vessel.

Very small polychaetes and those with long, threadlike bodies have no gills. The varying structure and the distribution of the gills present in the larger species indicate that they have evolved independently in different groups. Most commonly, the gills are modifications of the parapodia (Fig. 28.13) or outgrowths of some part of the parapodium. However, gills are found in other parts of the body in some polychaetes. The gills of the sedentary burrower, *Amphitrite,* for example, are located dorsally at the anterior end of the body (Fig.

Figure 28.16 Vascular system within a segment of *Nereis*. Arrows indicate direction of blood flow; anterior is to the right. (Modified from Nicoll.)

28.11). The radioles of fanworms serve as gills in addition to functioning as filters.

The ventilating current may be produced by cilia or by contractions of the gills. Gallery dwellers, sedentary burrowers and tube dwellers drive a ventilating current through the burrow or tube, usually by undulations of the body or by peristaltic contractions (Fig. 28.11B).

Polychaetes with gills always possess a respiratory pigment, usually **hemoglobin.** Polychaete hemoglobin occurs either dissolved in the blood plasma or within corpuscles in the coelomic fluid. When present in solution in the plasma, it is always a very large molecule. The hemoglobin of *Arenicola,* for example, is composed of 96 heme units, whereas vertebrate hemoglobin contains only four heme units. In contrast, the hemoglobin in the coelomic corpuscles of some polychaetes, such as the carnivorous gallery-dweller *Glycera,* is a small molecule.

A second respiratory pigment found in some polychaetes, especially fanworms, is **chlorocruorin.** Chlorocruorin is also an iron porphyrin but differs from heme in one of the side chains. This difference is sufficient to make this pigment green rather than red in color. It

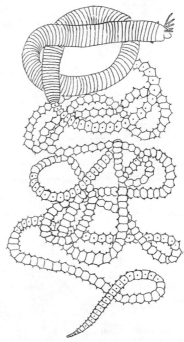

Figure 28.18 *Eunice viridis,* the Samoan palolo worm, with posterior region of gamete bearing segments. (After Woodworth from Fauvel.)

is always a large molecule dissolved in the blood plasma.

It seems likely that the gas-transporting function of hemoglobin and chlorocruorin evolved a number of times within annelids.

A third kind of respiratory pigment, **hemerythrin,** is found in one family of polychaetes, Hemerythrin is not a porphyrin. Rather, it is similar to hemocyanin in that the oxygen molecule is bound between two iron atoms. This pigment gives the blood a pink color.

Polychaete excretory organs are metanephridia (p. 346). A pair of nephridia is present in each segment, but the nephrostome of each nephridium opens through the septum into the segment anterior to the one containing the tubule (Fig. 28.3). The nephridiopore opens near the neuropodium.

Reproduction and Development. The sexes of polychaetes are separate. The **gametes** are produced by the coelomic **peritoneum** and mature within the coelomic fluid. Primitively, most segments produce gametes, but in many polychaetes, gamete production is restricted to the segments in certain regions of the body. The **nephridia** commonly function as the **gonoducts,** but some species have separate gonoducts.

The animals do not copulate. Instead, the gametes are shed into the sea water where fertilization occurs. The likelihood of fertiliza-

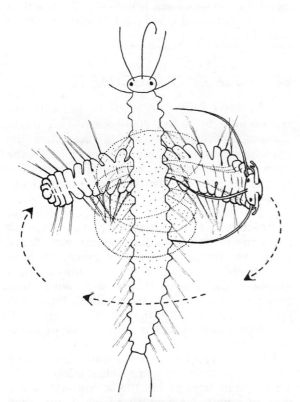

Figure 28.17 Syllid polychaetes during swarming. Male is swimming around female and releasing sperm. (After Gidholm.)

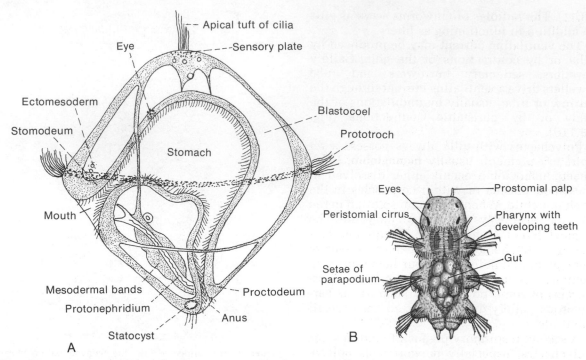

Figure 28.19 *A,* An annelid trochophore; *B,* a metamorphosing polychaete. (*A* after Shearer from Hyman; *B* from life.)

tion in some polychaetes is increased by swarming behavior. Males and females with ripe gametes swim to the surface at the same time in large numbers, shedding their eggs and sperm simultaneously (Fig. 28.17). Swarming usually occurs at specific times of the year, determined by annual, lunar and tidal periodicities. The times at which swarming will occur in certain species can be predicted. The most notable example is the Samoan palolo worm. At the beginning of the last lunar quarter of October or November the worms release the posterior gamete-bearing regions of the body (Fig. 28.18), which come to the surface in enormous numbers and rupture.

Some polychaetes undergo structural modification for the swarming reproductive period, and the head and parapodia take on a different appearance from the nonreproductive individuals. This condition is called **epitoky.**

The eggs, especially those of swarming polychaetes, are often planktonic. However, there are species that deposit their eggs within jelly masses attached to the tube or burrow.

Cleavage is typically spiral and leads to a trochophore larva, which is essentially like that of mollusks (Fig. 28.19). During the course of larval life, the trochophore lengthens, and segments with parapodia and setae appear behind the prototroch and mouth (Fig. 28.19). As

development proceeds, the young worm settles to the bottom and takes up the adult mode of existence.

28.4 CLASS OLIGOCHAETA

In fresh water and on land, the annelids are represented by some 3000 species of oligochaetes. These are far less diverse than the polychaetes and in many respects parallel the burrowing polychaete gallery dwellers. The simple conical or lobelike **prostomium** lacks sensory structures (Fig. 28.20). Oligochaetes crawl by peristaltic contractions. Parapodia are completely absent, but **setae** serve to anchor segments in crawling (Figs. 28.20 and 28.21). There are fewer setae per segment than in polychaetes, hence the name Oligochaeta—"few setae." Correlated with their peristaltic locomotion, the **coelom, septa** and circular and longitudinal **muscles** are well developed and reflect a high order of metamerism.

Aquatic oligochaetes are usually less than 3 cm. in length and a few are almost microscopic. They live in algae and in the bottom debris and mud of ponds, lakes and streams. Some, such as the members of the family Tubificidae, are tubiculous (Fig. 28.22), but this habit is much less widespread than in polychaetes. Setae are generally short, but species living in algae and loose debris may have long setae (Fig. 28.22). Most aquatic oligochaetes live in shallow water, but some Tubificidae occur in enormous numbers in the lower anaerobic regions of deep lakes and have been reported from deep ocean bottoms.

Many oligochaetes live in wet boggy soil and intergrade with the more terrestrial earthworms, of which there are five major families. The largest annelids are Australian earthworms of the genus *Megascolides*, which may attain a length of three meters.

Earthworms are gallery dwellers with short heavy setae. They display few structural modifications for a terrestrial existence. Earthworms, like most other oligochaetes, lack special organs for gas exchange. The high concentration of oxygen in air and the highly vascularized body wall permit adequate gas exchange across the general body surface despite the large size that earthworms may attain. The surface is kept moist for gas exchange by epidermal mucous glands and by coelomic fluid, which is released through a dorsal pore between each segment.

Desiccation is minimized by behavioral adaptations. Earthworms come to the surface only at night or during rains. During dry periods or freezing winter weather, earthworms burrow deeper in the soil. Some tropical Asian earthworms migrate down to depths of more than three meters. Estivation and hibernation are characteristic of many species.

Well-drained soils with a large amount of organic matter provide the best environment for earthworms. Within such soils the population is vertically stratified. Small species and small individuals live in or near the upper humus layer; larger species, such as *Lumbricus*, tunnel down to depths of one to two meters.

Wet soils may lack adequate oxygen, and soils that are too acidic lack the free calcium ions necessary for the removal of excess carbon dioxide.

Nutrition. Decomposing plant tissue is the principal food of most oligochaetes, although some freshwater species feed upon algae, and a few are raptorial on other small invertebrates. Earthworms feed on organic matter, especially dead vegetation, in the humus layer of the soil. They are nonselective deposit feeders, cycling soil through the **gut** and digesting the organic matter. The castings may be deposited at the surface.

Food is ingested by means of a muscular **pharynx.** In many aquatic species the pharynx is everted as a mucus-laden pad; in earthworms it functions as a pump (Fig. 28.21). From the pharynx food passes into the **esophagus,** which is often modified at different levels as a **crop** and one or more **gizzards** (Fig. 28.21). The thin-walled crop functions in storage and the muscular gizzard grinds the food to smaller particles. *Lumbricus* has one crop and one gizzard at the beginning of the intestine (Fig. 28.21).

The remainder of the gut is a long straight **intestine,** the site of extracellular digestion and absorption. The surface area of the earthworm's intestine is nearly doubled by a dorsal longitudinal fold, the **typhlosole** (Fig. 28.21).

Surrounding the intestine is a layer of chloragogen cells. These, like the vertebrate liver, are an important site of intermediary metabolism. The synthesis and storage of glycogen and fat, the deamination of amino acids and synthesis of urea and the removal of silicates (in terrestrial species) occur in chloragogen cells.

The circulatory system, nephridia and nervous system of oligochaetes are essentially like those of polychaetes. The contractile dorsal

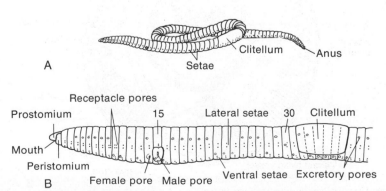

Figure 28.20   Earthworm. *A*, Entire worm. *B*, Lateral view of the anterior 40 segments of *Lumbricus*. Reproductive openings are found on segments 9, 10, 14 and 15. On each segment the excretory pore is either ventral, near the ventral setae, or lateral, above the lateral setae, with much variability between worms.

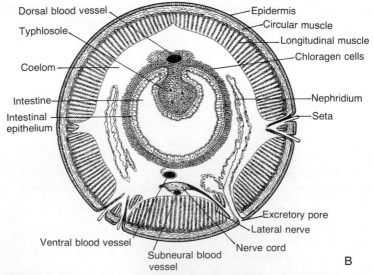

Figure 28.21 Structure of the earthworm *Lumbricus. A,* Dorsal view of anterior section of body; *B,* cross section taken at the level of the intestine.

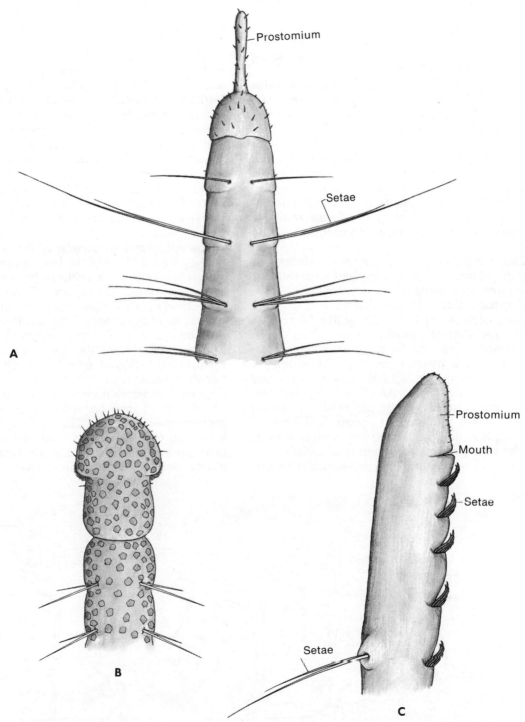

Figure 28.22 Freshwater oligochaetes. All three species are very small, no more than a centimeter in length. They live within algae and loose debris. *A, Stylaria,* dorsal view. *B, Aeolosoma,* dorsal view. *C, Dero,* lateral view.

blood vessel provides the principal force for blood propulsion. In some oligochaetes, certain anterior commissural vessels (five pairs around the esophagus in *Lumbricus*) are conspicuously contractile (Fig. 28.21). These "hearts" function as accessory pumps.

Hemoglobin is present in solution in the plasma of the larger oligochaetes, including most of the earthworms.

The **nephridia** play a role in salt and water balance, for in freshwater species and in earthworms with a plentiful water supply the urine is hypotonic to the coelomic fluid. Like other freshwater animals, oligochaetes must eliminate the water entering by osmosis.

Giant fibers are present in the ventral nerve cord, and earthworms have three especially large giant fibers grouped together in the dorsal part of the cord. The giant fibers transmit impulses controlling rapid contraction of the body.

Sense organs are lacking in most species, but the integument is richly supplied with sensory receptors, including photoreceptors.

Reproduction. In contrast to polychaetes, all oligochaetes are **hermaphroditic.** Moreover, there are distinct **gonads,** restricted to a few segments in the anterior third of the body.

The paired **ovaries** are located within a single segment (Fig. 28.23). The eggs mature within the coelomic fluid of that segment and exit through a pair of simple **oviducts.** These have ciliated funnels that penetrate the posterior septum and open onto the ventral surface of the next posterior segment (Fig. 28.23).

Seminal receptacles are present, but in oligochaetes these structures are inpocketings of the ventral body wall and are not connected to the other parts of the female system (Fig. 28.23).

There are one or two testicular segments, each containing a pair of **testes** (Fig. 28.23). In some earthworms, a transverse partition separates the lower part of the coelomic cavity containing the testes, and the sperm mature within this special chamber. The **septum** of the testicular segment is usually greatly evaginated to form a pouchlike **seminal vesicle.** The paired sperm ducts penetrate the posterior septum and may pass through several segments before opening onto the ventral surface (Figs. 28.20 and 28.23). In animals with two pairs of testes, the sperm ducts from the two testes of one side unite. The exact number of reproductive structures and their location in specific segments varies greatly.

The epidermis of certain adajcent segments in the anterior third of the body contains many glands that produce secretions for various parts of the reproductive process. These segments, collectively called the **clitellum,** are characteristic of oligochaetes. The position, number of segments included and degree of glandular development of the clitellum vary considerably in different oligochaetes. The clitellum of aquatic species is usually composed of only two segments, those that contain the genital pores. The clitellum may be visible only during the reproductive period. The earthworm *Lumbricus,* in contrast, has a conspicuous, permanent, girdle-like clitellum of six to seven segments located behind the genital pores (Fig.

Figure 28.23 Reproductive segments of the earthworm *Lumbricus* (lateral view). (After Hesse from Avel.)

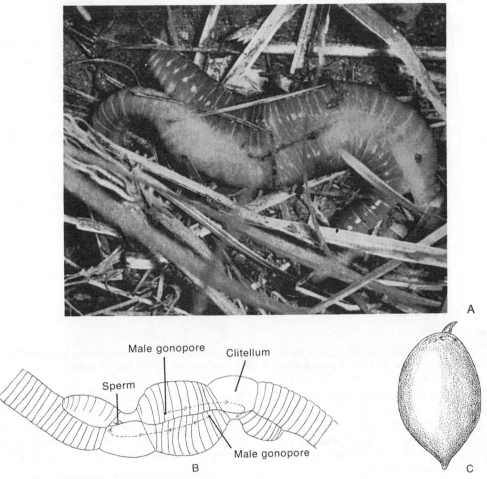

Figure 28.24 *A*, Two earthworms copulating. (Photograph of living animals made at night, courtesy of General Biological Supply, Chicago, Ill.) *B*, Movement of sperm during copulation in lumbricid earthworms. *C*, Earthworm cocoon. (*B* after Grove and Cowley from Avel; *C* after Avel.)

28.20). The clitellum of some earthworms includes as many as 60 segments.

Copulation with mutual transfer of sperm is characteristic of oligochaetes. Two worms come together with ventral surfaces opposed and with their anterior ends facing in opposite directions. The worms are held together by mucus secreted by the clitellum and sometimes by special genital setae. In most oligochaetes, sperm are passed directly from the male gonopores into the seminal receptacles. However, in *Lumbricus* and other members of the same family, the seminal receptacles of one worm are located some distance anterior to the male gonopores of the copulating partner (Fig. 28.24). The sperm, on being released, pass down a pair of grooves on the ventral surface and then cross over at the level of the seminal receptacles. The grooves are arched over by mucus and thus separated from the grooves of the opposite worm. Copulation in *Lumbricus* is a process requiring two to three hours.

A few days after copulation the clitellum secretes a dense material that will form the **cocoon.** Albumin is secreted inside the cocoon. The secretions form an encircling band that slips forward over the body. Eggs from the female gonopores and sperm from the seminal receptacles are collected en route. The cocoon eventually slips off the anterior end of the worm, the two ends sealing in the process (Fig. 28.24). Fertilization takes place within the cocoon that is left in the bottom mud and debris or in the soil. Development is direct and young worms emerge from the end of the cocoon. Only one egg completes development within the cocoons of *Lumbricus terrestris*.

28.5 CLASS HIRUDINEA

The members of the class Hirudinea, the **leeches,** constitute the smallest and the most

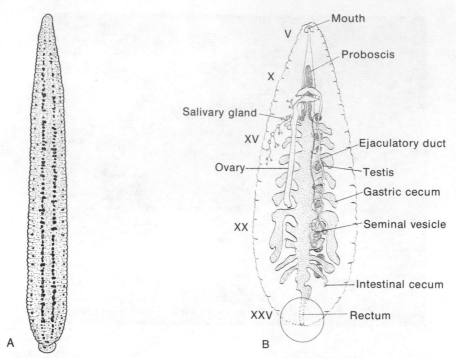

Figure 28.25 *A, Erpoodella punctata,* a common North American scavenger and predatory leech; *B,* internal structures of *Glossiphonia complanata,* ventral view. *A* after Sawyer; *B* after Harding and Moore from Pennak.)

aberrant of the annelid classes. Only some 500 species have been described. The body is dorso-ventrally flattened, with the anterior segments modified as a small **sucker** surrounding the mouth and the posterior segments forming a larger sucker behind the anus (Fig. 28.25). The head is greatly reduced, setae are absent and there is little external evidence of metamerism. The body is ringed with many annuli. but they do not correspond to the 34

Figure 28.26 *A,* Sagittal section through anterior end of *Glossiphonia,* showing protruded tubular pharynx. (Modified after Scribin.) *B,* Ventral view of oral region of a terrestrial blood sucking haemadipsid leech. Jaws are not exposed. (From Keegan, H. L., et al., 1968: 406th Medical Laboratory Special Report. U.S. Army Med. Command, Japan.) *C,* Anterior end of *Hirudo,* the medicinal leech (ventral dissection). *D* and *E,* Ingestion by *Hirudo.* Outward movement of jaws (*D*) followed by medial movement of teeth and dilation of pharynx (*E*). (*C* modified after Pfurtscheller from Pennak; *D* and *E* after Herter from Harant and Grasse.)

segments that compose the leech body (Fig. 28.25). Only the nephridia and nervous system indicate the original segmentation.

Most leeches are two to five cm. in length and are never as small as many polychaetes and other annelids. *Hirudo medicinalis*, the medicinal leech (Fig. 28.26), may reach a length of 20 cm. Leeches are commonly olive green and brown in color.

Locomotion. Leeches crawl in a looping manner. The body is extended and the anterior sucker attached. Then the posterior sucker is released, pulled forward, and reattached. Many leeches can swim by dorso-ventral undulations of the flattened body. Correlated with these modes of movement, the longitudinal muscles of the body wall are powerfully developed. The septa have been lost, and the coelom is reduced by the invasion of connective tissue to a system of interconnecting spaces, or sinuses (Fig. 28.27).

Nutrition. Contrary to popular notion, not all leeches are bloodsuckers. Many are predaceous and feed upon other small invertebrates, especially annelids, snails and insect larvae. The bloodsucking leeches are ectoparasitic on invertebrates, such as snails and crustaceans, and on vertebrates. All classes of vertebrates may be hosts, but fish and other aquatic forms, including water fowl, are the most common victims. In humid areas of the tropics, members of the family Haemadipsidae attack terrestrial birds and mammals, attaching to the moist thin membranes of the nose and mouth. They climb upon the host when it brushes through vegetation on which the leeches are located. Some ectoparasitic leeches, such as the fish leeches of the family Piscicolidae, may remain on the host throughout most of their lives; others are located on the host only when feeding.

Many leeches, both predatory and parasitic, ingest food through a tubular jawless proboscis, which can be extended through the mouth from a proboscis cavity (Fig. 28.25). The method by which the proboscis penetrates the prey is not understood, although it is thought that enzymes may be utilized. Predatory species usually swallow the prey whole.

The **pharynx** of other leeches serves as a pump but cannot be protruded. The mouth cavity in many of these forms contains three bladelike **teeth** (Fig. 28.26). When the anterior sucker is attached, the teeth slice through the skin of the host. A salivary secretion, called **hirudin,** prevents the coagulation of blood, which is ingested by the pumping action of the pharynx.

An **esophagus** connects the pharynx with a large stomach, or **crop,** that commonly possesses lateral pouches, or **caeca** (Fig. 28.25). An **intestine** composes the posterior third of the digestive tract and opens through the **anus** just above the posterior sucker. The stomach functions both in digestion and as a storage center. Many leeches may consume two to 10 times their weight in blood at a single feeding. Water is removed from the ingested blood, and the remaining material may require as long as six months to be digested. Symbiotic bacteria provide the enzymes for a part of the digestive process and produce some substance that prevents decomposition of the blood cells. Thus a meal may be stored for a long time between infrequent feedings. Leeches have survived without feeding for as long as a year and a half.

Some leeches have a blood vascular system like that of their oligochaete ancestors. Others have lost the original blood vessels and the interconnecting coelomic channels have been converted into a blood vascular system. Hemoglobin provides for gas transport, but gills are

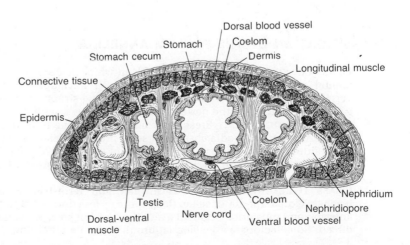

Figure 28.27 Cross section of a leech showing the filling of the coelom by connective tissue.

Stomach cecum

Stomach

Dorsal blood vessel

Coelom

Dermis

Longitudinal muscle

Connective tissue

Epidermis

Nephridium

Nephridiopore

Ventral blood vessel

Coelom

Testis

Nerve cord

Dorsal-ventral muscle

A

—Sperms

B

Figure 28.28 *A*, Two copulating *Hirudinaria. B*, Spermatophore of the rhynchobdellid, Haementeria. (*A* after a photograph by Keegan et al.; *B* after Pavlovsky from Harant and Grassé.)

present in only a few leeches. Although the head is reduced, some leeches possess eyes. Special sensory papillae are arranged in rings on the annuli.

Reproduction. The presence of a **clitellum** and a **hermaphroditic** reproductive system is taken as evidence that leeches evolved from oligochaetes. Copulation involves a mutual exchange of **spermatophores** (Fig. 28.28), which in some leeches are transferred by means of hypodermic impregnation. The spermato-

phores are injected into the ventral surface of the opposite leech (Fig. 28.28), and the injected sperm make their way to the **ovaries** of the female system, where fertilization occurs.

The clitellum produces a **cocoon** as in oligochaetes. Many fish leeches attach their cocoon to the host. Other leeches deposit the cocoon in water, beneath stones and other objects and in soil. The members of one family of leeches brood their eggs within a modified cocoon attached to the ventral surface of the body.

CLASSIFICATION OF THE PHYLUM ANNELIDA

CLASS POLYCHAETA. Marine annelids in which setae are carried on lateral segmental parapodia. Metamerism usually well developed. Prostomium (head) variable, but commonly bears sensory or feeding structures. Sexes separate and gametes produced by the peritoneum of numerous segments.

CLASS OLIGOCHAETA. Freshwater annelids and the terrestrial species known as earthworms. Parapodia absent but setae present. Metamerism well developed. Prostomium a simple lobe without sensory or feeding structures. Hermaphroditic, with gonads present in a few specific segments. Epidermis of certain segments modified as a clitellum for the secretion of a cocoon.

CLASS HIRUDINEA. Marine, freshwater and terrestrial annelids known as leeches. No parapodia or setae. Body more or less dorso-ventrally flattened with anterior and posterior segments modified as suckers. Metamerism greatly reduced. Some ectoparasitic. Hermaphroditic, with a clitellum.

ANNOTATED REFERENCES

Detailed accounts of the annelids may be found in many of the references listed at the end of Chapter 21. The following works are devoted solely to annelids.

Dales, R. P.: Annelids. London, Hutchinson University Library, 1963. A brief general account of the annelids.

Edwards, C. A. and J. R. Lofty: Biology of Earthworms. London, Chapman and Hall, Ltd., 1972. A general account of the earthworms.

Mann, K. H.: Leeches (Hirudinea). New York, Pergamon Press, 1962. A general biology of leeches.

THE ARTHROPODS

1. The phylum Arthropoda is the largest phylum of the Animal Kingdom and includes such familiar forms as spiders, mites, scorpions, shrimps, crabs, insects, centipedes and millipedes.

2. The body is covered by an exoskeleton of chitin and protein. This is divided into plates and cylinders, permitting movement, and is periodically shed, permitting growth.

3. Muscles are attached to the inside of the skeleton and the skeletomuscular system functions as a lever system.

4. Arthropods are metameric, reflecting their annelidan ancestry, but most species exhibit some degree of reduction in metamerism as a consequence of loss, fusion and differentiation of segments.

5. The segments carry paired jointed appendages, which serve not only in locomotion but also have been adapted for many other kinds of functions.

6. All internal structures derived from invaginations of the body wall have a chitinous lining. The anterior and posterior parts of the gut, called the foregut and hindgut, possess such a lining. The midgut, derived from endoderm, is more restricted than in most animals.

7. The blood vascular system is open, and the dorsal heart is primitively tubular.

8. The plan of the nervous system is like that of annelids.

9. The eggs are generally centrolecithal, and cleavage is commonly superficial.

The phylum **Arthropoda** is a vast assemblage of animals. At least three quarters of a million species have been described; this is more than three times the number of all other animal species combined. The tremendous adaptive diversity of arthropods has enabled them to survive in virtually every habitat; they are perhaps the most successful of all the invaders of the terrestrial environment.

A number of features indicate that arthropods evolved from annelids, perhaps from some group of surface-dwelling polychaetes. Arthropods are **metameric** and the segments bear lateral appendages, which may be homologous to the parapodia of polychaetes. The nervous system, with its large ventral **nerve cord,** is essentially like that of annelids, and the dorsal tubular **heart** may be the homologue of the dorsal contractile blood vessel of annelids.

29.1 SUPPORT AND MOVEMENT

The distinguishing feature of arthropods, one that has been an important factor in the evolutionary success of the group, is the **chitinous exoskeleton** (p. 241) that covers the entire body. The exoskeleton is divided into plates on the trunk and head and forms cylinders on the appendages (Fig. 29.1). At the junction points between skeletal pieces, the chitin is thin and folded, permitting movement.

Distinct **muscle bundles** are anchored to the inner side of the skeleton and extend or flex parts at a joint, depending upon the point of attachment (Fig. 29.1). Thus the skeletal muscle system of arthropods operates by using a system of levers in much the same way as the skeletal muscle system of vertebrates.

Despite the locomotor and supporting advantages of an external skeleton, it poses problems for a growing animal. The solution to this problem evolved by the arthropods has been the periodic shedding of the skeleton, a process called **molting** or **ecdysis.** The process of molt-

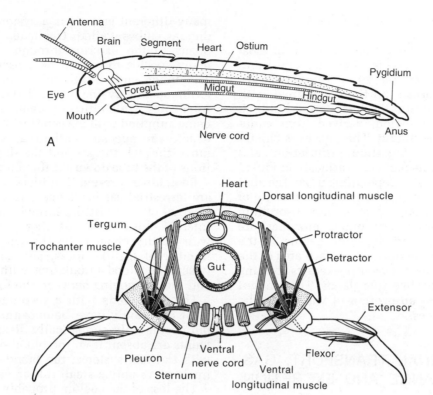

Figure 29.1 Structure of a generalized arthropod. *A*, Sagittal section; *B*, cross section.

ing was described earlier (see Section 11.1.).

The stages between molts are known as **instars,** and the length of the instars becomes longer as the animal becomes older. Some arthropods, such as lobsters, continue to molt throughout their life, although the event becomes less and less frequent. Other arthropods, such as insects and spiders, have fixed numbers of instars, the last being attained with sexual maturity.

Only vestiges of the coelom remain in adult arthropods, although it appears during the course of embryonic development. The loss is probably related to the shift from a fluid internal skeleton to a solid external skeleton.

The primitive arthropod trunk was composed of a large number of segments, each bearing a pair of similar appendages (Fig. 29.5). The evolution of different modes of existence with different types of locomotion — swimming, crawling, running, pushing and climbing — and with different types of feeding behavior has led to a reduction in the number of segments. Segments dropped out, fused together or became specialized. Appendages became specialized for many functions other than locomotion — prey capture, filter feeding, food handling, gas exchange, ventilation, copula-

tion, egg brooding and so forth. In the crayfish, for example, the oral appendages are adapted for feeding, the large claws are adapted for grasping, the legs for crawling, the abdominal appendages for egg brooding and the last appendages form a flipper with the terminal section of the trunk (Fig. 29.24*B*). Although these appendages are structurally quite different, they are all derived from originally similar segmental appendages. They are therefore said to be **serially homologous.** The fusion, loss and differentiation of segments and appendages has been repeated many times in the history of different arthropod groups.

The legs of arthropods are much longer and more slender than the parapodia of polychaetes, and the body sags down between them. However, there are some similarities to the crawling gait of polychaetes. Each leg performs an effective and a recovery stroke in which the leg is lifted, swung forward and placed down upon the substratum. The legs on one side of the body carry out these movements in sequence, composing a locomotor wave; the locomotor waves on the two sides of the body alternate with each other, as in crawling polychaetes. The polyneuronal innervation of arthropod muscles is described on page 260.

29.2 DIGESTIVE SYSTEM

The arthropod gut differs from that of most other animals in having large stomodeal and proctodeal regions (Fig. 29.1). The derivatives of these ectodermal portions are lined with chitin and constitute the **foregut** and **hindgut**. The intervening region, derived from endoderm, forms the **midgut**. The foregut is chiefly concerned with ingestion, trituration and storage of food; its parts are variously modified for these functions depending upon the diet and mode of feeding. The midgut is the site of enzyme production, digestion and absorption; however, in some arthropods enzymes are passed forward, and digestion begins in the foregut. Very commonly the surface area of the midgut is increased by outpocketings forming pouches or large digestive glands. The hindgut functions in the absorption of water and the formation of feces.

29.3 INTERNAL TRANSPORT, GAS EXCHANGE AND EXCRETION

The circulatory system of arthropods is open; a large **pericardial sinus** surrounds the dorsal **heart** (Fig. 29.1). Blood flows from the pericardial sinus into the heart through small lateral openings called **ostia**. In primitive tubular hearts there is one pair of ostia per body segment. When the heart contracts, the ostia close and blood is propelled anteriorly or posteriorly out of the heart through arteries. The arterial system varies greatly in the various kinds of arthropods but generally is more extensive in large species. The arteries deliver the blood to tissue spaces where various exchanges take place. From the tissue spaces (collectively termed the **hemocoel**) blood drains into a system of larger sinuses and eventually returns to the pericardial sinus around the heart.

Most arthropods have organs for gas exchange. Their **gills** are usually modifications of appendages or outgrowths of the integument associated with an appendage (Fig. 29.2). The surface area of the gill is increased by flattening, folding or branching in various ways. The chitin over the exchange surface is thin and offers no obstacle to the passage of gases.

The gas exchange organs of terrestrial arthropods are typically internal. However, they are derived from invaginations of the integument and thus are lined with chitin. The commonest gas exchange organ in terrestrial arthropods, and one that has evolved independently in many different groups, is a system of air-conducting tubes called **tracheae.** The system opens to the exterior through **spiracles;** in many species these can be closed by valves. The inner end of the tubes are blind and terminate in very fine fluid-filled tubules called **tracheoles,** which directly penetrate the tissue being supplied with oxygen (Fig. 29.2). There is usually no special ventilating system. Gases move through the air and the fluid-filled sections of the tube down a diffusion gradient.

Book lungs, present in spiders and scorpions, are terrestrial gas exchange organs resembling internal gills. A slitlike spiracle opens into an internal chitin-lined chamber (Fig. 29.2). One wall of the chamber is thrown into a large number of leaflike folds, and gases pass between the blood circulating *within* the folds and the air moving *between* the folds.

The arthropods with a respiratory pigment most commonly have **hemocyanin.** Hemoglobin occurs only sporadically. Respiratory pigments are absent from the blood of arthropods with tracheal systems; the blood in these animals plays only a small role in gas transport.

The loss of the coelom probably accounts for the disappearance of nephridia and the evolution of the diverse excretory organs of arthropods. Most of the excretory organs are blood-bathed sacs opening by a duct to the body surface. The excretory organs of insects and some arachnids, such as spiders, consist of a few to many blind **tubules** lying free in the **hemocoel** and opening into the posterior section of the **gut** (Fig. 29.52). Waste materials are removed from the blood and secreted into the lumen of the excretory organ. Aquatic arthropods excrete ammonia. Some terrestrial species, such as insects, excrete uric acid; others, such as spiders, excrete guanine.

29.4 NERVOUS SYSTEM AND SENSE ORGANS

The nervous system of arthropods exhibits the same basic design as that of annelids—a dorsal anterior **brain** and a ventral **nerve cord** (Fig. 29.1 and 29.52). However, the fusion and loss of segments is reflected in a corresponding forward migration and fusion of the ventral **ganglia** of many arthropods. For example, in a crab all of the ganglia of the ventral nerve cord have fused together anteriorly. This ventral mass of ganglia plus the circumesophageal connectives and the brain form a large collar that encircles the esophagus.

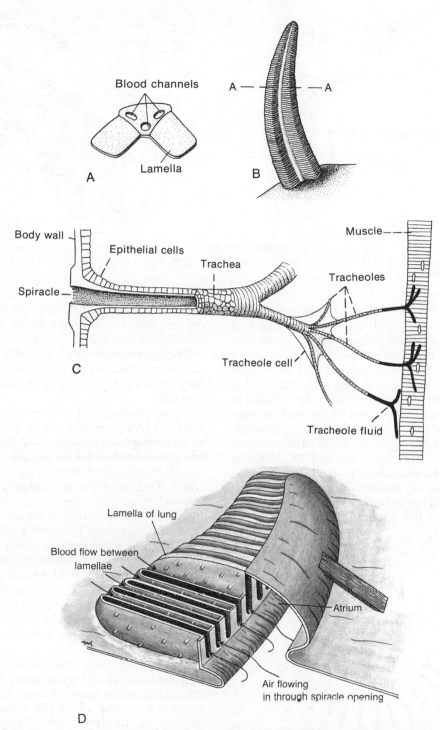

Figure 29.2 Arthropod gas exchange organs. *A* and *B*, Crab gill. Gill consists of a series of flat plates (lamellae) stacked upon one another and attached to a central axis (*B*). Cross section shown in *A* taken at the level indicated in *B*. *C*, Diagrammatic section of an insect tracheal system showing relationship of spiracle and tracheoles to tracheae. *D*, Section through a book lung. (*A* and *B* modified from Calman; *C* modified from Ross.)

Figure 29.3   *A*, Sensory hair of a mite. *B*, Section through part of a lyriform organ of a spider. The distal end of the sensory neuron is shown attached to the incurved cuticular membrane of the slit. (*A* after Gossel from Millot; *B* modified from Kaston and others.)

The **sensory receptors** of arthropods are usually associated with some modification of the chitinous exoskeleton, which otherwise would act as a barrier to the detection of external stimuli. An important and very common type of receptor is one connected with **hairs, bristles** or **setae.** When the bristle is moved, the receptor ending in the shaft or base is stimulated (Fig. 29.3). Some arthropods, such as spiders, have hairs so delicate and sensitive that they can respond to airborne vibrations. Other common modifications for receptors are slits, pits or other openings in the exoskeleton. The opening is covered by a thin membrane, to the underside of which is at-tached a nerve ending (Fig. 29.3). Such sense organs detect vibrations or other forces that change the tension of the skeleton.

Chemoreceptors are especially concentrated on the appendages or mouthparts, where they may be lodged in setae or the general skeletal surface.

The appendages of arthropods contain many kinds of receptors and are of great importance in receiving information about the surrounding environment. A leg, for example, may serve not only for movement but also for touch, taste and smell. One or more appendages (e.g., the **antennae**) may be given over entirely to the function of monitoring environmental signals.

Most arthropods have **eyes,** which can vary greatly in complexity. Some are simple and have only a few photoreceptors. Others are large, with thousands of retinal cells, and can form a crude image. A transparent **lens-cornea** is the usual skeletal contribution to the eye, and the focus is always fixed since the immovable lens is continuous with the surrounding skeleton (Fig. 29.4).

Insects and many crustaceans, such as crabs and shrimp, have a **compound eye,** so called because it is composed of many long cylindrical units possessing all of the elements for light reception. The structure and functioning of this type of eye are described on page 384.

29.5   REPRODUCTION AND DEVELOPMENT

Arthropods have separate sexes, each with a

Figure 29.4   Section through the eye of a scorpion.

single pair of **gonads,** but the nature of the reproductive system and the location of reproductive openings are too variable to permit generalization. Arthropod eggs are usually centrolecithal, and cleavage is superficial (p. 96).

29.6 ARTHROPOD CLASSIFICATION

Living arthropods have traditionally been divided into two subphyla. Those that lack antennae have been commonly placed in the subphylum **Chelicerata,** which takes its name from the fact that the first post-oral appendages are feeding appendages called chelicerae. To this group belong the horseshoe crabs, scorpions, spiders, mites and ticks. Arthropods with antennae have been placed within the subphylum **Mandibulata;** the first post-oral appendages in the members of this group are feeding appendages called mandibles. This group includes all of the insects as well as shrimp, crabs, millipedes and centipedes.

Most zoologists today agree, however, that the Mandibulata is an artificial assemblage of unrelated groups and that there are probably four main lines of arthropod evolution rather than two, as the Mandibulata and Chelicerata suggest. These lines are believed to be represented by the extinct **Trilobita,** the **Chelicerata,** the **Crustacea** and the **Uniramiata.** The uniramiates include the centipedes, millipedes and insects. In contrast to the marine origin of the other three arthropod lines, the uniramiates appear to have evolved on land. They are mandibulate and antennate, but the name uniramiate refers to the fact that the limbs, or legs, are basically unbranched appendages and there is no evidence that they were derived from a branched condition.

There are a number of zoologists who believe that each of these four arthropod groups had a separate origin from different annelidan ancestors and that arthropodization (i.e., the arthropod features of a chitinous exoskeleton and jointed appendages) evolved independently four times. If this is true, the Arthropoda should be considered a superphylum, and the Trilobita, Chelicerata, Crustacea and Uniramiata should each be raised to phylum rank.

29.7 TRILOBITES

The most primitive known arthropods are the extinct trilobites, which inhabited the Paleozoic oceans from the Cambrian through the Permian. They are usually placed within a separate subphylum, the **Trilobita.**

The flattened oval body of most trilobites was about 3 to 10 cm. in length and composed of an anterior **cephalon,** a middle **thorax** and a posterior **pygidium** (Fig. 29.5). The cephalon, originating from the fusion of the head with four trunk segments, formed a solid shield-shaped **carapace,** which bore a pair of eyes on the dorsal surface. The middle section of the body, the thorax, was composed of many unfused segments, and the posterior pygidium contained a number of fused segments. Two dorsal longitudinal furrows extended the length of the body and divided it into three transverse sections, from which the name trilobite is derived.

A pair of antennae was located on the undersurface of the body, just in front of the mouth. Behind the mouth, each segment, whether fused with other segments or separate, carried a pair of identical appendages (Fig. 29.5). In all other arthropods, at least some of the segmental appendages are different from others. Each appendage contained two branches—one leglike, on which the animal crawled, and one that contained parallel filaments and was believed to have functioned as a gill (Fig. 29.5).

Most trilobites were bottom dwellers; however, variations in their body form and size suggest that some species burrowed through sand, some were swimmers and some were planktonic. Surprisingly, we know a great deal about the development of trilobites, for complete series of larval stages of numerous species have been preserved in the fossil record.

29.8 THE CHELICERATES

Members of the subphylum Chelicerata have no antennae and the first post-oral appendages are feeding appendages called **chelicerae.** Most chelicerates are further distinguished from other arthropods in having the body composed of a **cephalothorax** and an **abdomen** (Fig. 29.6). The cephalothorax contains a number of trunk segments that have fused with the head, and the fused region is covered dorsally by a single skeletal piece, the **carapace.** Behind the chelicerae the cephalothorax carries a pair of appendages, called **pedipalps,** and four pairs of legs. There are no distinct abdominal appendages.

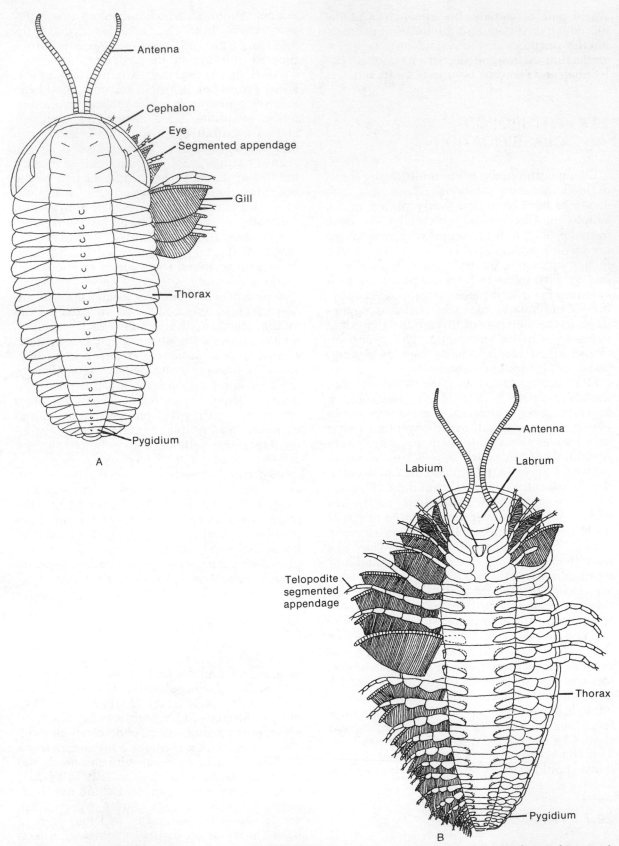

Figure 29.5 The Ordovician trilobite, *Triarthrus eatoni*. *A*, Dorsal view; *B*, ventral view. (Both after Walcott and Raymond from Störmer.)

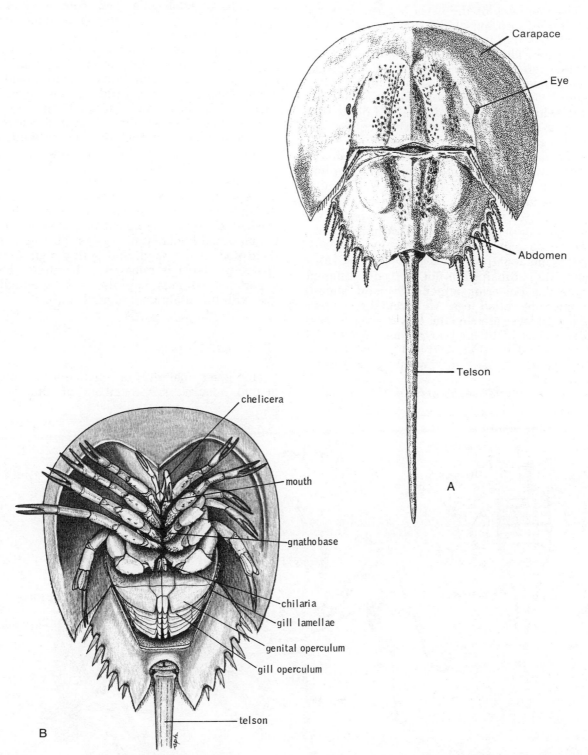

Figure 29.6 A horseshoe crab, *Limulus polyphemus. A,* Dorsal view; *B,* ventral view showing appendages. (*A* after Van der Hoeven from Fage.)

Class Merostomata

The members of the class Merostomata are marine gill-bearing chelicerates called horseshoe crabs that inhabit sandy bottoms in shallow water. The class has been known from the Ordovician, but there are only five species living today. Four are Asian and one, *Limulus polyphemus,* is found along the Atlantic coast of the United States and in the Gulf of Mexico. They are among the largest living chelicerates, reaching a length of 60 cm.

The name horseshoe crab comes from the large, convex, horseshoe-shaped carapace that covers the cephalothorax (Fig. 29.6). Dorsally, the carapace carries a pair of eyes, and the underside of the cephalothorax bears the six pairs of appendages typical of most chelicerates—a pair of chelicerae, a pair of pedipalps and four pairs of legs (Fig. 29.6). However, the pedipalps are not markedly different from the posterior appendages and should properly be called legs. All except the last pair of appendages are **chelate;** that is, they possess pincers formed by the two terminal segments of the appendage (Fig. 29.6). The last pair of appendages is used for sweeping away sand and mud.

The abdominal segments are fused together and terminate in a long spikelike **telson** (Fig. 29.6). The underside of the abdomen carries five pairs of **book gills,** each composed of a large number of thin plates **(lamellae)** protected by a flaplike cover (Fig. 29.6).

Horseshoe crabs are harmless animals. They crawl over or push through sand in shallow water, and small individuals swim upside-down using the gills as paddles. The telson is used to right the animal if flipped over.

Horseshoe crabs are omnivores and scavengers, and their diet includes the softbodied invertebrates and algae they encounter as they plow through the bottom.

Adults congregate in shallow water during the reproductive period. Fertilization occurs at the time of egg deposition while the male is clasped to the back of the female abdomen with his modified hooklike pedipalps. The eggs are deposited in the sand, and a tiny swimming "trilobite" larva is released at hatching. The superficial trilobite appearance is gradually lost with the attainment of adult form.

Class Arachnida

The great majority of chelicerates, some 60,000 species, are members of the class

Figure 29.7   *A,* Dorsal view of a eurypterid, an extinct group of Paleozoic merostomes, which may have been ancestral to the scorpions. *B,* The North African scorpion, *Androctonus australis,* capturing a grasshopper. (*A* after Nieszkowski; *B* after Vachon from Kaestner.)

Arachnida. In contrast to merostomes, arachnids are terrestrial and lack gills. They are widely distributed, living in vegetation, in leaf mold and beneath bark, logs and stones.

The class is divided into ten orders, five of which contain the most common and familiar species. Scorpions (order **Scorpionida**) are large arachnids with big chelate pedipalps and a long segmented abdomen terminating in a **sting** (Fig. 29.7). Paired eyes are mounted on tubercles in the middle of the carapace, and two to five additional pairs of eyes may be present along the anterior lateral margins. Scorpions are secretive, largely nocturnal animals of the tropics and semitropics. In the United States they are common only in the Gulf and southwestern states.

Scorpions are the most ancient arachnids, known from the Silurian, and probably were the first terrestrial arthropods. Their ancestors may have been the extinct giant eurypterids, a group of Paleozoic merostomes (Fig. 29.7). Although most eurypterids were marine, some invaded brackish and fresh water. Fossil evidence suggests that the first scorpions were also aquatic.

Pseudoscorpions (order **Pseudoscorpionida**) are only a few millimeters in length and are common inhabitants of leaf mold in both tropical and temperate regions. These tiny arachnids have large chelipeds like scorpions, but the segmented abdomen is short and lacks a terminal sting (Fig. 29.8).

Spiders (order **Araneae**) compose the largest

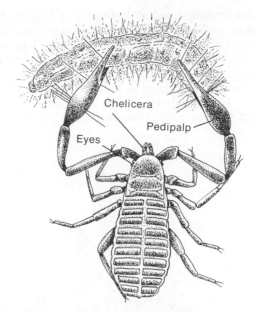

Figure 29.8 Dorsal view of a pseudoscorpion catching an insect larva with its pedipalps.

of the arachnid orders. Over 30,000 species have been described, and they frequently occur in far greater numbers than most people are aware. An ungrazed meadow, for example, may support as many as 2,250,000 spiders per acre. The abdomen is unsegmented and connected to the cephalothorax by a narrow waist (Figs. 29.11 and 29.13). The pedipalps are small and leglike. Eight eyes are arranged in two rows of four each across the front of the carapace.

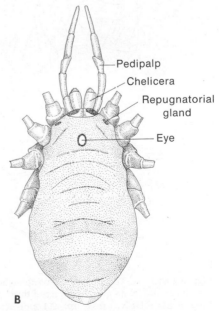

A   B

Figure 29.9 *A*, Dorsal view of a harvestman. (By Betty M. Barnes.) *B*, Dorsal view of a harvestman showing only bases of legs. Repugnatorial gland emits a noxious substance, which repels some possible predators.

Daddy-long-legs, or harvestmen (order **Opi-liones,** or **Phalangida**), are distinguished from other arachnids by their very long legs and segmented abdomen broadly joined to the cephalothorax (Fig. 29.9). A tubercle on the center of the carapace bears a single pair of eyes. The members of this order are common arachnids in both temperate and tropical regions.

The order containing mites and ticks (order **Acarina**) is the second largest and most diverse group of arachnids. Over 20,000 species have been described to date, and the order is still poorly known taxonomically.

Mites are usually less than a millimeter in length, and their adaptive diversity may in part be attributed to their small size, which has enabled them to exploit many types of mi-

Figure 29.10 *A, Tuckerella,* a phytophagous mite with symmetrically arranged clublike setae. *B,* Ventral view of *Analges,* a feather mite. Note greatly enlarged third pair of legs and claws. (*A* and *B* from Krantz, G. W., 1971: Manual of Acarology. Oregon State Univ. Press, pp. 211 and 275.) *C,* An oribatid mite, *Belba jacoti,* carrying five shed nymphal skins. (After Wilson from Baker and Wharton.) *D,* Water mite, *Mideopsis orbicularis.* (After Soar and Williamson from Pennak.)

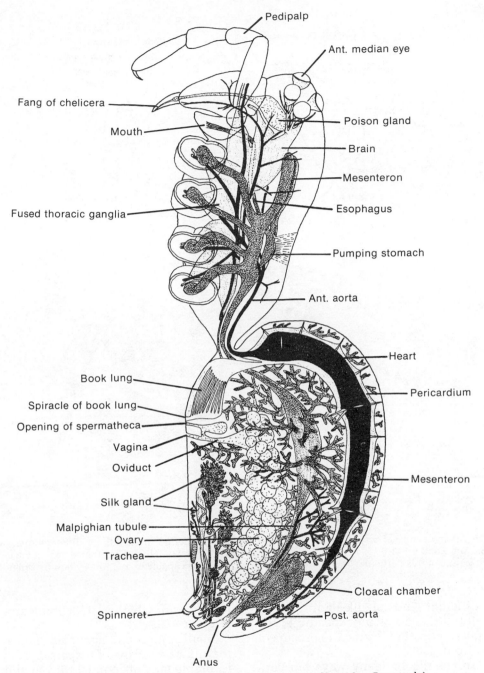

Figure 29.11 Internal anatomy of a two-lunged spider. (After Comstock.)

crohabitats. However, such habitats must contain adequate water vapor, for the small size of mites also makes them vulnerable to desiccation. Where conditions are favorable, as in leaf mold, mites occur in enormous numbers.

The abdomen of mites is unsegmented and broadly fused with the cephalothorax. The entire body is thus covered dorsally by a single skeletal piece (Fig. 29.10). The pedipalps are small and usually leglike.

Silk. Two groups of arachnids evolved the means of producing and utilizing silk. The tiny pseudoscorpions have silk glands in the chelicerae and use the silk to build retreats or nests, especially during the time of reproduction.

The **silk glands** of spiders are located in the abdomen and open through conical **spinnerets** at the end of the abdomen (Figs. 29.11 and 29.12A), each spinneret bearing numerous spigots. A particular species may possess from

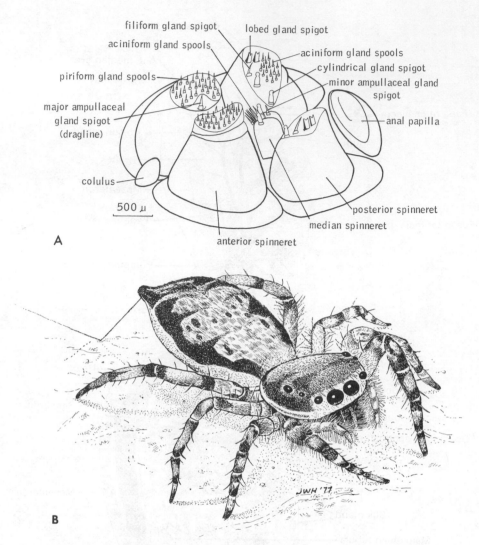

Figure 29.12 *A,* Spinnerets of an orb-weaving spider showing distribution of spigots for different silk glands. (From Wilson, R. S.: Am. Zool. 9:103–111, 1969.) *B,* Jumping spider laying down dragline. The arrangement of the eight eyes is a characteristic feature of this particular family of hunting spider, which is capable of jumping by rapidly extending the legs.

two to six different kinds of silk glands. Silk is a protein and hardens not on exposure to the air but by polymerization during the process of being drawn out.

Spiders utilize silk in many ways but contrary to popular notion, only some spiders use silk to construct webs for trapping prey. Silk is used to build nests, which are used as retreats, for reproduction or for overwintering; in all spiders the eggs are encased within a silken **cocoon.**

Most spiders lay down a **dragline** behind them, anchoring it at intervals to the substratum. The dragline not only functions as a safety line for the spider, but also is an important means of communication between members of the species. A spider may determine from another dragline whether its owner is male or

female and whether it is immature or adult (Fig. 29.12*B*).

Small spiders and newly hatched spiders use the silk as a means of dispersal. They climb to favorable take-off points, tilt the abdomen upward and release a strand of silk. When air currents produce sufficient pull, the spider lets go and sails out to whatever new habitat and fate the wind will take it. The wide distribution of many species of spiders is undoubtedly correlated with this ballooning phenomenon. The larger, more primitive tarantulas, which do not balloon, have a relatively limited distribution.

Feeding. Most arachnids are predatory animals, and other arthropods are their usual prey. As an aid to dispatching prey, three orders of arachnids have independently evolved **poison glands.** The poison glands of scorpions are located in the terminal sting at the end of the abdomen (Fig. 29.7). The prey is caught with the large pedipalps and then stabbed by the poison barb with a forward thrust of the abdomen, which is curled back over the body. Although the sting of scorpions may be painful, few species have a poison dangerous to man. Members of the genus *Centruroides*, which occurs in Mexico, New Mexico and Arizona, are dangerous. The poison is neurotoxic but would be fatal only to children.

Pseudoscorpions have poison glands in their large scorpion-like pedipalps. The glands open at the tips of the fingers, and the poison is injected as the prey is held. The pedipalps are too tiny to be of any danger to man.

Spiders have a poison gland associated with each chelicera, which consists of a terminal **fang** that folds down against a larger basal piece (Fig. 29.13). The gland, which may extend back into the head, opens by a duct through the end of the fang, and the poison is injected through a bite.

The poison of a very small number of spiders is dangerous to man. Few tarantulas have a toxic bite, despite popular mythology. The members of the cosmopolitan genus *Latrodectus*, which includes the black widows, are

perhaps the most notorious of the poisonous species. *Latrodectus mactans* is widely distributed in the United States. The poison is neurotoxic, and the symptoms, which include pain in the legs and abdomen, are somewhat like those for appendicitis. However, the consequences of a bite, although severe and painful, are not likely to be fatal except in children.

The poison of the brown recluse spider, *Loxosceles reclusa*, which occurs in midwestern and southeastern United States, causes necrosis of the tissues around the wound, and healing is difficult (Fig. 29.14B).

The method of catching prey can be a basis for dividing spiders into two adaptive groups: **hunters** and **web builders.** Hunting spiders include the tarantulas, wolf spiders, crab spiders, jumping spiders and others. They spin silk draglines, nests and cocoons, but they do not use silk to capture prey. Rather, the prey is stalked, pounced upon and bitten. Hunting spiders generally have heavier legs and more highly developed eyes than do web builders (Fig. 29.12B). Wolf spiders and jumping spiders have the largest eyes of any arachnids, and some species appear to be capable of object discrimination.

Web-building spiders construct webs to trap prey. Sheet webs, tangle webs, funnel webs and

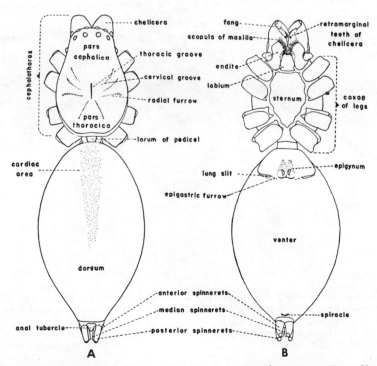

Figure 29.13 Dorsal (*A*) and ventral (*B*) views of a spider, showing external structure. (From Kaston, B. J.: Spiders of Connecticut. Bull. 70, State Geol. Nat. Hist. Survey, 13, 1948.)

Figure 29.14 *A*, Copulating pair of black widow spiders, *Lactrodectus mactans*. The bite of the larger female is neurotoxic but in most instances would only be fatal to a small child. These spiders build tangle webs in protected places, a habit they share with the other members (nonpoisonous) of the large family to which they belong. A common house spider is a member of this family. (Adapted from figures by Kaston.) *B*, The brown recluse spider, *Loxosceles reclusa*, whose bite produces an ulcerated wound that is difficult to heal.   *Illustration continued on opposite page.*

C

Figure 29.14 *Continued. C,* Female wolf spider carrying her egg case. When the eggs hatch, she will carry the spiderlings on her back. Most other spiders leave the egg case in the webbing or attached to some object, and on hatching, the young fend for themselves. (Based on a photograph by William H. Amos.)

orb webs are common types, and each is characteristic of a different family. Note that only one major family builds the orb web that is so commonly associated with spiders. Web-building spiders are aerialists and have long slender legs for climbing about the silken lines (Fig. 29.14). The thread is hooked beneath a small middle claw that lies between the two main terminal claws characteristic of the legs of all spiders. Eyesight is poor, but web builders are able to detect and interpret the various vibrations of the web with great facility. Web vibrations inform an orb weaver, for example, about the size of the struggling prey and whether it is securely caught. The spider approaches the prey, throws out additional silk and then gives it a fatal bite.

The behavior of web-building spiders, especially orb-weavers, has been studied by a number of ethologists in recent years. Each species exhibits a distinct prey-catching program involving a precise sequence of responses, as shown in Figure 29.15.

The web-building habit of spiders may have evolved from random draglines laid down in close quarters by some ancestral species. An occasional insect caught in such a primitive snare could have provided the selective advantage that led to the evolution of the more complex traps of modern spiders.

Arachnids are unusual in that they begin digestion of their prey outside their bodies. While the tissues are torn by the chelicerae, enzymes are secreted by the midgut, passed forward through the foregut and poured out of the mouth into the prey. The partly digested tissues are then sucked in by the pumping action of a part of the foregut (Fig. 29.11). Digestion is completed and absorption occurs in the midgut, which may be greatly evaginated and ramify into various parts of the body.

There are some exceptions to the predatory feeding habit of most arachnids. Harvestmen are omnivores and feed upon vegetable material and dead animal remains in addition to live invertebrates. The greatest diversity in feeding is displayed by mites. Some are herbivorous and have mouthparts adapted for piercing the cells of plants and sucking out the contents. Such plant-feeding mites are often considered to be parasites.

Members of several groups of mites are scavengers. The oribatids, called beetle mites because of their convex and highly sclerotized dorsal surface (Fig. 29.10C), occur in large numbers in leaf mold and eat decaying vegetation and animal remains. Hair and feather mites spend their lives on the skin of mammals and birds, where they feed upon sloughed skin cells, gland secretions and fragments of hair and feathers.

Parasitic mites are largely ectoparasites, although some species attack the respiratory membranes of vertebrates. The animal parasites feed upon the blood of the host or upon skin or lymph. Ticks are bloodsucking parasites of mammals and birds, and the chelicerae are adapted for penetrating and anchoring into the skin of the host (Fig. 29.20). A tick feeds until it is engorged. Then it drops off the host and does not feed again until the next instar.

Chiggers, or redbugs, are ectoparasites on the skin of mammals during their posthatching instar. In feeding, the minute mite secretes

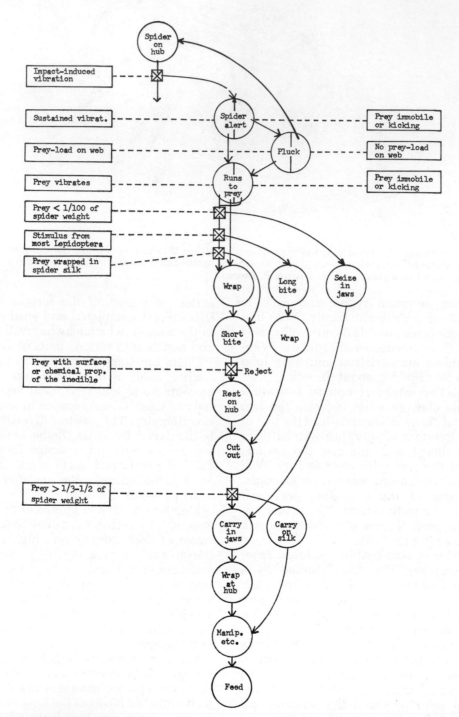

Figure 29.15 Predatory behavior of the orb-weaver *Argiope argentata*, showing sequence and relationship to prey size. Circle depicts action of spider; square indicates prey. Sequence begins with spider waiting in center of web (on hub). When insect strikes web, spider becomes alert, following which it may run to prey or pluck web. Vibrations from web-plucking provide information on nature of prey. Large prey are given a long bite, injecting a larger amount of poison than in a short bite. They are then wrapped with silk. Small prey are first wrapped and then given a short bite. (From Robinson, M. H.: Am. Zool. 9:161–173, 1969.)

Mange mite burrowing in skin

Figure 29.16 The mange mite, *Sarcoptes scabiei*. These pass their entire life cycle on the host. Eggs laid in the burrows hatch into young mites that begin burrows of their own. Note the suckers on the anterior legs. (After Craig and Faust.)

Adult female

enzymes that produce a deep well from which it sucks out the digested contents. The mite secretions produce an irritating reaction resembling a mosquito bite but lasting much longer. Following feeding the chigger falls off the host and is predaceous as an adult.

Mange and itch mites (Fig. 29.16) spend their entire lives tunneling through the skin of their mammalian host. Eggs are deposited in the burrows and the young, upon hatching, follow the same existence as the adults.

Gas Exchange. Four pairs of book lungs provide for gas exchange in scorpions. The slitlike openings are located on the ventral surface of the anterior abdominal segments. Primitive spiders, such as tarantulas, have two pairs of book lungs, but most spiders have one pair of book lungs and one pair of tracheae. The openings are on the ventral surface of the abdomen (Fig. 29.11).

All other arachnids utilize tracheae for gas exchange, but the different locations of the spiracles in different groups suggest that tracheae evolved independently a number of times within the class.

Reproduction and Development. The paired gonads are located in the abdomen, and in both sexes a median **gonopore** opens onto the anterior ventral surface.

Indirect transfer of **spermatophores** appears to be a primitive condition in arachnids and perhaps represents the early arthropod solution to the problem of sperm transmission on land. The male deposits a stalked spermatophore on the ground (Fig. 29.17). The female then moves over the spermatophore and takes it up within her gonopore. Chemical, tactile and sometimes visual signals are important in attracting and orienting the female to the spermatophore.

Among the orders of arachnids we are examining, scorpions, pseudoscorpions and some mites produce spermatophores, and the peculiar mode of sperm transfer in spiders probably evolved from indirect sperm transfer by spermatophores. Spermatophore transfer in scorpions is preceded by a "courtship" dance, during which the large pedipalps of the male are locked with those of the female. In the course of the dance, the male deposits a spermatophore and then maneuvers the female so that it is taken up into her gonopore (Fig. 29.17).

The primitive pattern in pseudoscorpions

A

B

C

Figure 29.17 Sperm transfer in scorpions. *A*, While holding female's pedipalps with his own, male on left deposits spermatophore on ground. *B*, Female is pulled over spermatophore. *C*, Spermatophore taken up into female gonopore. (After Angermann.)

appears to be one in which the male deposits a spermatophore in the absence of a female. Later a passing female is attracted chemotactically to the spermatophore. Most pseudoscorpions have evolved various mechanisms to increase the likelihood of the female encountering the spermatophore. The males of some species, on encountering a female, lay down a long silk signal line from the spermatophore. The female follows the chemical signals of the line to the spermatophore. Other species have a courtship pattern in which the spermatophore is not deposited until the male encounters the female. He then entices her over the spermatophore with a courtship dance.

Our knowledge of reproductive behavior in mites is still very poor. Some species transfer sperm indirectly by spermatophores; others transfer sperm directly, utilizing a penis.

The process of sperm transfer in spiders is remarkable and is paralleled in few other animals. The copulatory organs of the male are the ends of the pedipalps, which resemble a pair of boxing gloves (Fig. 29.18). Although

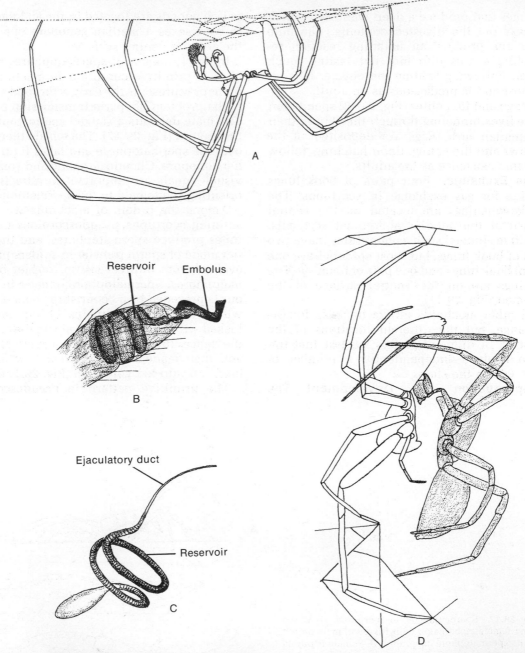

Figure 29.18 *A,* Male tetragnath spider in sperm web, filling palps from globule of semen. *B,* Simple palp of *Filistata hibernalis. C,* Receptaculum seminis, the semen-containing part of the male palp. *D,* Mating position of *Chiracanthium* (male shaded). (*A* after Gerhardt from Millot; *B* and *C* after Comstock; *D* after Gerhardt from Kaston.)

Figure 29.19 Female scorpion carrying young. (Courtesy of H. L. Stahnke.)

complex in structure, the basic element is a blind coiled tube, the inner end of which functions as a storage chamber; the distal end serves as an ejaculatory duct that opens to the exterior through a long projecting process (Fig. 29.18).

Prior to mating, the male spins a tiny sperm web on which a droplet of semen is secreted. The two pedipalps are then rapidly dipped into the droplet until the semen is taken up within the reservoir of the palpal organ. The male now seeks a female.

The male is frequently smaller than the female, and the predatory nature of spiders makes it important for the male to ensure that the female does not mistake him for potential prey. The precopulatory behavior of any given species thus displays features that are corre-

lated with its feeding adaptations. For example, hunting spiders depend upon visual cues in addition to chemical and tactile stimuli in their sexual encounters. Elaborate courtship dances involving posturing by the male occur in some wolf spiders and jumping spiders that have highly developed eyesight. Male web-building spiders use the vibrations of the web to determine the presence of the female and to advertise his presence and desire to enter.

Chemical and tactile signals are important in locating the female and in the precopulatory behavior of all spiders. An amputated leg of a female will evoke courting postures in the male. The same leg washed in ether and dried initiates no responses, but the male will posture before the apparently empty watchglass containing the evaporated washings. Such

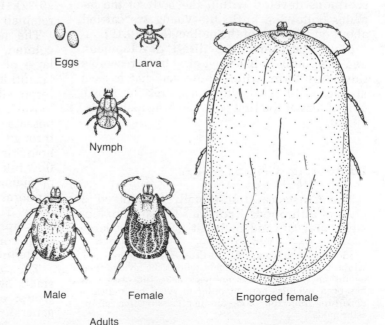

Eggs   Larva

Nymph

Male   Female   Engorged female

Adults

Figure 29.20 The common tick, *Dermacentor andersoni.* Eggs laid on the ground hatch into six-legged larvae that feed on small mammals. These drop off, molt into eight-legged nymphs that return to small mammals. After molting on the ground again the adults attack large mammals. The females become enormous after mating and eventually fall to the ground to lay a thousand or more eggs. (After Chandler.)

chemical substances produced by one individual that trigger behavioral responses in another, called **pheromones** (p. 457), are important in evoking and directing behavioral patterns in many arthropods.

Following final precopulatory stroking and palpating by the male, the palpal organ is locked onto the chitinous plate containing the female reproductive openings by means of special palpal projections, and the ejaculatory process is inserted into the seminal receptacles (Fig. 29.18). Sperm is transferred from one palp at a time. This unusual mode of sperm transfer in spiders probably had its origins in transfer by spermatophore. The male of some arachnid ancestral to the spiders may have used the palp to place a spermatophore into the female gonopore. The claw of the palp (the palpal organ of the male is homologous to the claw) might then have become adapted for the transfer process. The deposition of sperm within a spermatophore disappeared, and the palpal claw became further modified to handle semen.

Harvestmen and many mites transfer sperm directly without spermatophores. These arachnids deposit their fertilized eggs in soil, in leaf mold or beneath bark, but spiders place their eggs in silk cocoons, which are then usually left beneath stones, bark or leaf mold or are attached to vegetation.

Brooding is common. Wolf spiders and fisher spiders carry their cocoons about with them. After hatching, wolf spiderlings are carried on the back of their mother. Pseudoscorpion eggs develop within a **brood sac** attached to the female reproductive openings. The eggs of scorpions develop within the body of the female. Following birth, the young are carried about on the back of the mother (Fig. 29.19).

Most arachnids have direct development, and the young at hatching or at birth resemble the adult form. Mites, however, hatch as a six-legged "larva," which acquires the fourth pair of legs at the first post-hatching instar (Fig. 29.20).

29.9 THE CRUSTACEA

The Crustacea, traditionally considered a class of arthropods,* contains the shrimp, lobsters, crayfish and crabs; it is the only large group of aquatic arthropods. Most are marine, but there are many species found in fresh water, and some, notably the sow bugs, are adapted for a terrestrial existence.

Crustaceans differ from all other mandibulates in possessing two pairs of antennae. The first pair is probably homologous to the antennae of insects, centipedes, millipedes and other mandibulates, but the second pair is unique to crustaceans. Behind the mandibles are two pairs of maxillae. Thus the head appendages, which are constant for all members of the class, are two pairs of antennae, one pair of mandibles and two pairs of maxillae (Fig. 29.21).

The trunk varies too greatly to permit many generalizations. A **carapace** formed from a posterior fold of the head is commonly present. It may cover only a small part of the dorsal surface of the trunk, or it may greatly overhang the sides of the body (Fig. 29.21). In some crustaceans, the entire body is enclosed within a carapace. The terminology of crustacean appendages is based on the assumption that, primitively, each appendage was a simple two-branched **(biramous)** structure. The outer branch **(exopodite)** and inner branch **(endopodite)** were attached to a basal piece **(protopodite)** (Fig. 29.21). The trunk appendages have been adapted for a wide range of functions, and there is little uniformity in the way differentiation has occurred.

Crustaceans possess either compound eyes or a simple median eye, but not both. The **median,** or **nauplius,** eye is a little cluster of three or four pigment cups containing photoreceptors. It is characteristic of the nauplius larva (Fig. 29.22), but in some groups the larval eye is retained in the adult.

The larger crustaceans are often brightly colored, the pigment being located in the outer layer of the exoskeleton. A blue pigment, astaxanthin, is denatured when heated and is responsible for the red color of boiled shrimp, lobsters and crabs. The larger crustaceans also possess **chromatophores,** which are visible through the exoskeleton. These are cells with branching processes, and the pigment granules flow into and out of the processes. Many colors of pigments may be apparent; some chromatophores contain pigments of more than one color. The chromatophores enable the animal to adapt to the color of its background; the most common change is simple darkening or lightening.

The ancestral crustaceans were probably small swimming animals, and many living groups are planktonic. The larger species are generally benthic.

---

*Some current works treat the Crustacea as a subphylum, which probably more accurately reflects the level of their evolution, as mentioned on page 673. The class rank has been retained here in order to avoid too radical a departure from the more familiar classification of the arthropods.

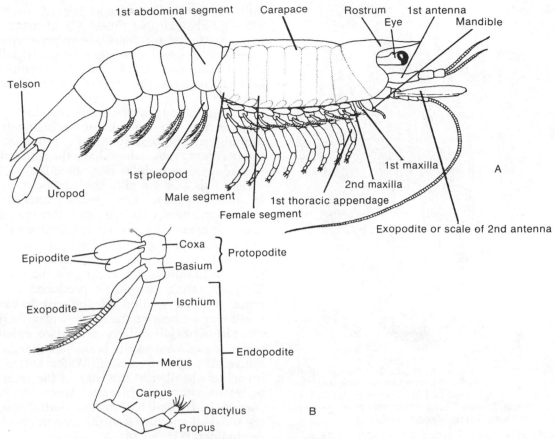

Figure 29.21 A generalized malacostracan crustacean. *A*, Lateral view of body; *B*, a thoracic appendage. (Both after Calman.)

Although crustaceans have a wide range of diets and feeding habits, filter feeding is very common, especially among small crustaceans. The **filter** is formed by closely spaced **setae** on certain appendages, usually head or anterior trunk appendages (Fig. 29.23).

A number of generalizations can be made about crustacean reproduction. Copulation is the rule, and certain appendages are adapted for clasping the female and transferring sperm. Fertilization may be internal or it may occur externally at the time of egg deposition, and the eggs are usually brooded on certain parts of the body (Fig. 29.30). The earliest hatching stage, the **nauplius larva,** possesses three pairs of appendages—first antennae, second antennae, and mandibles (Fig. 29.22). The acquisition and differentiation of additional segments and appendages occur in subsequent developmental instars, and these later larval stages have been given various names. Bottom-dwelling species settle upon attaining the adult form.

The structural diversity of crustaceans makes them difficult to describe without considering some of the different subgroups which compose the class.

Decapods. The **Decapoda,** the largest of the crustacean orders, contains over 8500 species. It also contains the biggest and most familiar crustaceans—shrimp, lobsters, crayfish and crabs. The name Decapoda refers to the five pairs of legs, including the first pair, which are frequently modified as large claws. In front of the legs are three pairs of small appendages

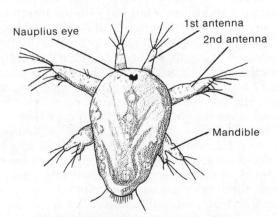

Figure 29.22 Nauplius larva of *Cyclops fuscus,* a copepod. (After Green.)

Figure 29.23 Filtering setae on the left maxilla of the copepod, *Calanus.* Three diatoms, representing species of different sizes, are drawn to scale to indicate the filtering ability of setae. Diatom *B* is approximately 25 micrometers long. (After Dennell from Marshall and Orr.)

called **maxillipeds,** which, like the maxillae, function in food handling (Figs. 29.24 and 29.25). The three pairs of maxillipeds and the five pairs of legs are appendages of the anterior trunk region, called the **thorax.** The thorax is covered by a carapace, and the head-thorax region together is usually called the **cephalothorax** (Fig. 29.24).

Primitively, the **abdomen** is large, as in shrimp, lobsters and crayfish, and carries six pairs of biramous appendages. The last pair is flattened and forms a **tail fan** with the terminal **telson.** The first five pairs are called **swimmerets** or **pleopods** (Fig. 29.24). In crabs the abdomen is greatly reduced and flexed beneath the thorax (Fig. 29.25). This change to a short body form shifts the center of gravity forward beneath the cephalothorax and represents an adaptation for greater mobility in crawling.

The abdomen of hermit crabs is housed within an empty gastropod shell; the soft twisted abdomen bears reduced appendages (Fig. 29.26). The hermit crab carries the gastropod shell by clasping the columella with a modified uropod and jamming the fourth and fifth pairs of legs against the inside of the shell. When the hermit crab becomes too large, it finds another shell and leaves the old one. Most hermit crabs can use the shells of many different gastropod species. Only empty shells are used; the original occupant is rarely killed. Hermit crabs probably evolved from forms that backed the body into crevices and other retreats.

Most decapods are bottom dwellers, but many shrimp can swim, using the pleopods for propulsion. The best swimmers, however, are the portunid crabs, which have the fifth pair of legs adapted as paddles (Fig. 29.25D).

Decapods possess **gills** that project upward from near the base of the thoracic appendages and are enclosed within a protective branchial chamber created by the great lateral overhang of the carapace (Fig. 29.27). Each gill consists of a central axis to which are attached either **flat plates** or **filaments,** depending on the species. The ventilating current is produced by the rapid sculling motion of the **gill bailer,** a semilunar process of the second maxilla (Fig. 29.24B). The gill bailers drive two exhalant water streams out of the branchial chamber to either side of the mouth. Water enters the branchial chamber at the back of the carapace or between the legs. In crabs, where the carapace is tightly sealed along the ventral margin, the inhalant aperture is at the base of the **claw,** or **cheliped** (Fig. 29.28).

Several groups of crabs have invaded the land with varying degrees of success. Amphibious fiddler crabs (*Uca*) burrow in the intertidal zone (Fig. 29.29). At high tide the crab remains within its burrow, but at low tide it emerges from the burrow to feed, utilizing water contained within the branchial chambers for gas exchange.

The related ghost crabs (*Ocypode*) are more terrestrial. These crabs live above the high tide mark and in many areas are common inhabitants of dunes. In the tropics, land crabs live in forests and thickets well back from the sea, as much as hundreds of miles inland. Almost all terrestrial species continue to use the gills as organs of gas exchange, which may explain the restriction of land invasions to crabs, for only in these decapods is the branchial chamber sufficiently closed to make possible the retention of the moisture needed for gas exchange. Most land crabs are nocturnal and remain in burrows during the day.

Although the majority of decapods are predators or scavengers, there are many exceptions. Recently fallen fruits and leaves are an important food source for many land crabs. Some decapods are filter feeders. Certain crabs living within the U-shaped tubes of *Chaetopterus*

A

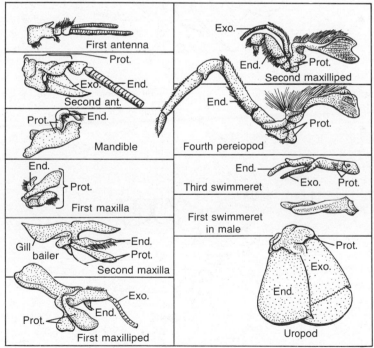

B

Figure 20.24 A, Lateral view of a crayfish. B, Appendages of the crayfish: prot. = protopodite; exo. = exopodite. Those on the left are drawn to a larger scale than those on the right. First antenna is not considered to be a segmental appendage. (Both after Howes.)

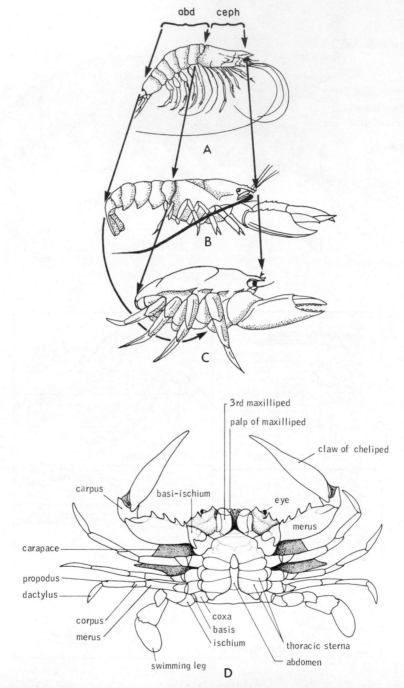

Figure 29.25 *A–C*, Evolution of the short body form in decapod crustaceans. In forms with long bodies, such as shrimp (*A*) and lobsters (*B*), a large abdomen extends posterior to the cephalothorax. In forms with short bodies, such as crabs (*C*), the abdomen is reduced and folded beneath the cephalothorax. (From Glaessner, M. F.: Treatise on Invertebrate Paleontology, Part R4(2), p. R401.) *D*, Ventral view of a brachyuran crab of the family Portunidae. The fifth pair of legs is adapted for swimming.

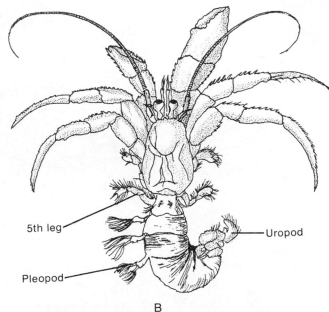

A

B

Figure 29.26   Hermit crab. *A*, In shell; *B*, out of shell. (*A* courtesy of the American Museum of Natural History; *B* after Calman from Kaestner.)

filter the worm's water current with out-stretched maxillipeds. Mole crabs are also filter feeders. These animals burrow backward in sand and project their plumose antennae as filters. Some species, including the Atlantic coast mole crab, live on wave-swept beaches and filter the current of the receding waves with their second antennae (Fig. 29.29).

The amphibious fiddler crabs feed on fine organic matter deposited onto the surface of the intertidal zone of protected beaches, estuaries and marshes. A mass of sand or mud is placed in the mouthparts, which are then flooded with water pumped forward from the gill chambers. Organic material is floated off and separated by fine setae on the feeding appendages. The remaining mineral material is ejected as a small spitball (Fig. 29.29).

The decapod digestive tract is complex. In lobsters, crayfish and crabs, the foregut includes a very large chitin-lined cardiac **stomach** containing one dorsal and two lateral **teeth** that form a **gastric mill** (Fig. 29.27). A narrow constriction separates the **cardiac** stomach from the **pyloric** stomach, part of which is derived from the foregut and part from the midgut. Associated with the midgut section and derived from it is a pair of large **digestive glands** (Fig. 29.27).

The cardiac stomach functions both as a **crop** and a **gizzard.** The grinding action of the gastric mill plus the action of **enzymes** passed forward from the digestive glands reduces the food

masses to fluid and fine particulate matter, which is then conducted along channels through the pyloric stomach to the digestive glands. Here absorption takes place.

The excretory organs of decapods are **antennal glands,** also called green glands. The organ consists of a blood-bathed sac in the head and a duct that opens to the outside by a pore at the base of the second antennae.

Most decapods are marine and are **osmoconformers,** but some shrimp and crabs and the widely distributed crayfish inhabit estuaries and freshwater. The osmoregulatory ability of these species is described on page 355.

Decapods have **compound eyes,** and some species, such as crabs, have a high degree of visual acuity. The eyes are located on **stalks** and have a wide visual arc (Fig. 29.25).

In the base of each first antenna is a **statocyst** that opens by a **pore** to the exterior. The animal uses sand grains as **statoliths** and must replace them following each molt. A newly molted crayfish given iron filings instead of sand grains will place them within the statocyst chamber and then will orient itself to a magnetic field rather than to gravity.

The **hormones** and **neurosecretions** of crustaceans are perhaps better known than those of any other invertebrates. There is experimental evidence of hormones that regulate chromatophores, molting, growth, sexual development and reproduction. A number of hormones, such as those controlling chromatophores, are re-

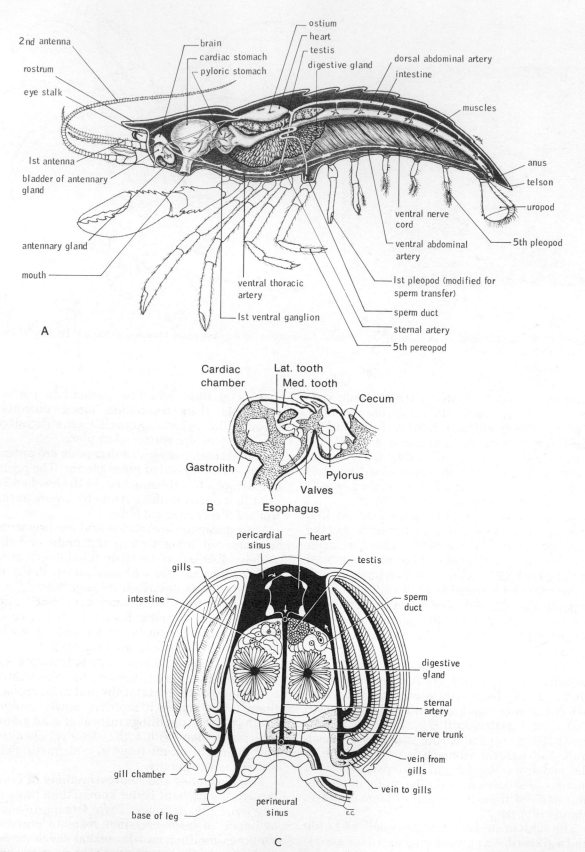

Figure 29.27 Internal anatomy of the crayfish. *A*, Lateral view; *B*, stomach enlarged; *C*, cross section just behind the third pair of legs. (All after Howes.)

Figure 29.28 Path of ventilating current through gill chamber of (A) a crayfish and (B) a crab.

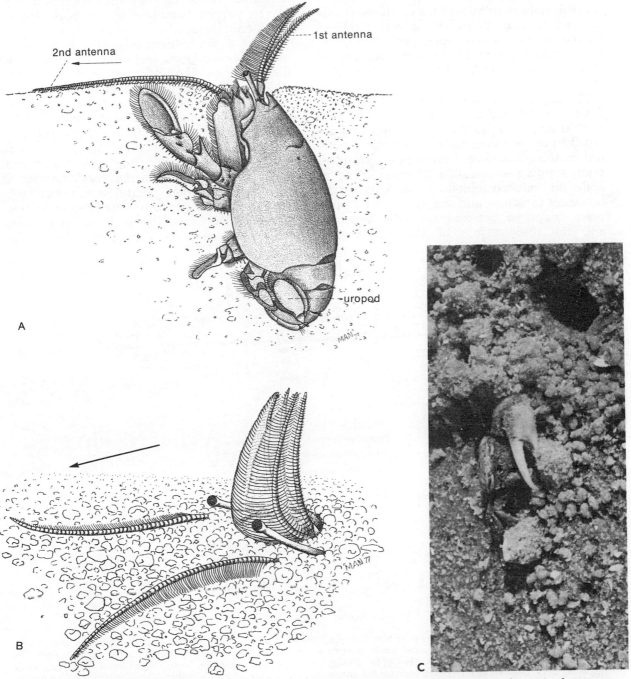

Figure 29.29 A and B, Mole crab buried in sand, lateral view (A) and surface view (B). Biramous first pair of antennae form a screening siphon for ventilating current. Setose second antennae are extended into water current (arrow) and function in filter feeding. Eyes are on long stalks. C, Fiddler crab by burrow. The large balls of sand were excavated from the burrow; the small pellets are composed of sand that has been filtered of organic material and ejected by the mouth parts.

leased from a gland in the eye stalk. These are easily studied, since the eye stalk can be removed and extracts of the eye stalk can be prepared and tested.

The single pair of **testes** or **ovaries** is located in the dorsal part of the thorax. The male **sperm duct** descends to the ventral side and opens at the base of the fifth pair of legs (Fig. 29.30). The **oviduct** opens to the exterior at the base of the third pair of legs (Fig. 29.30). During copulation the male assumes various positions astride the female and, using the greatly modified anterior two pairs of pleopods, transfers sperm to the female gonopores or to a median seminal receptacle between the fourth or fifth pair of legs.

The eggs are fertilized internally or on release from the oviduct. The fertilized eggs pass to the ventral surface of the abdomen, where they are attached to the pleopods by an adhesive material on the egg surface. Here they are brooded, often forming a conspicuous mass. Even in crabs the reduced folded abdomen retains its brooding function, and the eggs form a large mass, or sponge, between the folded abdomen and the undersurface of the thorax (Fig. 29.30).

Some shrimp hatch as a nauplius larva; but in most decapods the hatching stage is a later planktonic larval stage, called a **zoea,** in which all of the thoracic appendages are present (Fig. 29.30). The adult form is gradually attained through a series of instars, and the young shrimp, lobster or crab settles to the bottom and takes up the adult mode of existence.

Copulation and brooding occur on land in terrestrial crabs, but most must go back to the sea to permit hatching of the eggs and a planktonic larval life.

Direct development takes place in most freshwater decapods, including the crayfish. In these species, the eggs are brooded on the abdomen throughout development.

Many species of shrimp and crabs are eaten by humans throughout the world. In the United States, the blue crab, *Calinectes sapidus,* a swimming portunid (Fig. 29.30), is the principal commercial species on the Atlantic and Gulf coasts. On the Pacific coast, *Cancer,* a member of another family, is commercially important. The Alaskan King crab is a large decapod that is crablike in form but is more closely related to the hermit crabs than to the true crabs.

Figure 29.30 Life cycle of the commercial blue crab, *Callinectes sapidus.* Adults are common off shore as well as in very brackish water. Females living in brackish water must return to water of high salinity to permit hatching of the eggs. The hatching stage is a zoea larva, which lives in the plankton for a period of time. Successive molts transform the zoea to a megalops larva, possessing many crab features but with a still unfolded abdomen. Further development leads to the adult body form. Prior to the female's last molt she is carried about beneath a male, and copulation occurs after molting. Eggs are laid some time later and are attached to pleopod setae beneath the folded abdomen. (Courtesy of C. Piling; adapted from various sources.)

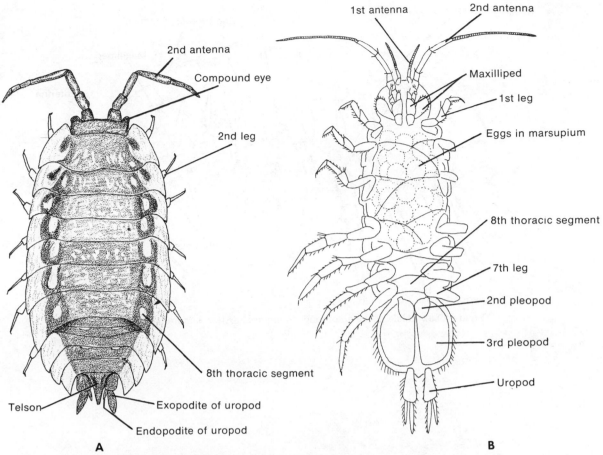

Figure 29.31 Isopods. *A*, Dorsal view of *Oniscus asellus*, a terrestrial isopod. *B*, Ventral view of the freshwater isopod *Asellus*. (*A* after Paulmier from Van Name; *B* after Van Name.)

Species of *Penaeus* are important commercial shrimp throughout the world, including the United States. The Japanese have developed the art of shrimp farming, raising the animals from egg to a marketable size of 20 to 25 cm. The shrimp are packed live and shipped in seaweed. The cost of shrimp farming is very high and is profitable in Japan only because of the high prices the Japanese will pay for live shrimp. Declining catches and rising demand will eventually make shrimp farming in the United States economically feasible.

The class Crustacea is divided into subclasses, and the decapods belong to the largest subclass, the **Malacostraca.** In these crustaceans, the trunk is composed of a thorax of eight segments, which bear the legs, and an abdomen of six segments, which usually carry five pairs of biramous **pleopods** and a terminal pair of flattened **uropods** (Fig. 29.21).

Amphipods and Isopods. The two largest orders of malacostracans other than decapods are the **Amphipoda** and **Isopoda.** These mala-

costracans, which are only about a centimeter in length, share a number of features. The **compound eyes** are on the sides of the head and not on stalks, as in decapods. No carapace is present, and the thoracic and abdominal regions are not sharply demarcated on the dorsal side (Fig. 29.31). As in decapods, there are eight pairs of **thoracic appendages,** but only the first pair are food-handling maxillipeds; the other seven are legs. In both orders development is direct, and the young are brooded beneath the thorax in a chamber, the **marsupium,** formed by inward shelflike projections from the bases of the legs (Fig. 29.31).

Isopods tend to be dorso-ventrally flattened, and the pleopods are modified for gas exchange (Fig. 29.31). Amphipods, in contrast, are laterally flattened and look somewhat like shrimp (Fig. 29.32). The gills of amphipods are simple processes from the bases of the legs.

The majority of isopods and amphipods are marine, and most are bottom dwellers, swimming only intermittently with the pleopods. They live in bottom debris, in algae and among

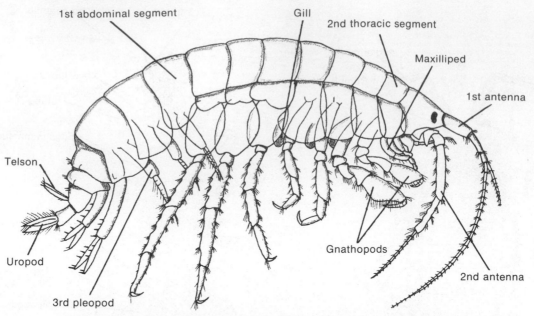

Figure 29.32 Male of the amphipod, *Gammarus*. (After Sars from Calman.)

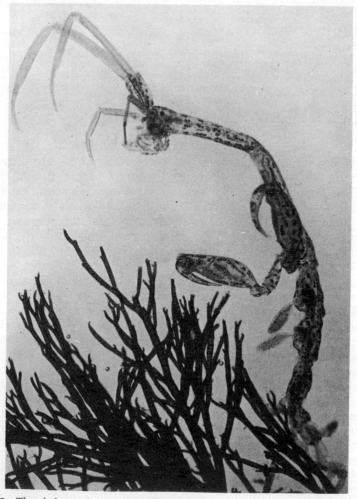

Figure 29.33 The skeleton shrimp, *Caprella equilibra*, clinging to seaweed. (By D. P. Wilson.)

sessile animals. The common skeleton shrimp are very aberrant amphipods, found climbing about hydroids and bryozoans (Fig. 29.33). Many amphipods burrow in sand or mud and some are tube dwellers. The tubes are composed of mud, sand grains, shell or plant fragments bound together with cement from glands on certain appendages.

Both orders include freshwater species; in favorable habitats, amphipods may occur in great numbers. Both groups have also invaded land, but the beach fleas, which occur beneath drift at the high tide mark, represent the limits of the amphipod incursion. Beach fleas get their name from their ability to jump by a rapid backward extension of the abdomen.

The pill bugs, sow bugs and woodlice, all isopods, represent the only truly successful crustacean invasion of land. These isopods are widely distributed in both temperate and tropical regions, where they are commonly found beneath stones and wood and in leaf mold. They reduce desiccation by living in protected habitats, by avoiding light and by their ability to roll up into a ball, covering the less highly sclerotized ventral surface. The gills remain as gas exchange organs, and the marsupium is filled with water when developing eggs are carried.

Most isopods and amphipods are omnivores and scavengers, but among members of both orders there are predaceous species, such as some skeleton shrimp. Filter feeding has evolved in some amphipods, with the various anterior appendages used as filters. The terrestrial sow bugs feed chiefly on dead plant and animal remains.

Some isopods that burrow at the high tide mark and certain species of amphipod beach fleas are able to return to their habitat when they move down to the intertidal zone to feed or are for other reasons displaced. Orientation is achieved by sensing the slope of the beach and the angle of the sun. The latter ability, however, requires a "map sense," i.e., knowledge of the orientation of the beach with regard to compass points, and an internal clock to compensate for the changing sun angles as the day passes. Animals can be misdirected experimentally by altering the sun's apparent position with a mirror. A comparable orientation ability, using the angle of the sun, has been demonstrated in certain wolf spiders living on the banks of lakes and streams and in bees that return to particular flowers to obtain pollen and nectar (p. 475).

Branchiopods. Most of the nonmalacostracan crustaceans are very small, averaging only a few millimeters in length. They are found in all sorts of habitats, but many species are planktonic. Most are filter feeders, and the filter feeding planktonic species constitute an important link in the food chain between the photosynthetic level and the larger predators.

The majority of nonmalacostracan crustaceans belong to one of four subclasses. The subclass **Branchiopoda** contains the water fleas, fairy shrimp and brine shrimp (Figs. 29.34 and 29.35). All possess flattened **leaflike trunk appendages** bordered by fine setae, but the body form is quite diverse. Fairy shrimp have an elongate body without a carapace, and there are many segments and appendages. Water fleas look like plump little birds. The trunk, but not the head, is enclosed within a bivalved carapace that greatly overhangs the body. The head bears large biramous second antennae and a single median compound eye.

The flattened foliaceous appendages of branchiopods function in gas exchange, from which the name Branchiopoda—**gill foot**—is derived. The appendages also serve in filter feeding, the fine setae bordering the appendages functioning as filters. Fairy shrimp also use the appendages for swimming, but water fleas swim with the antennae.

In contrast to most of the other subclasses, branchiopods are chiefly inhabitants of fresh water. Only the water fleas are found in the sea, where they may occur in large numbers in the plankton. Water fleas are the only branchiopods inhabiting lakes. Most other branchiopods are restricted to ponds and temporary pools and are often distributed sporadically. The brine shrimp *Artemia*, which belong to the same group as fairy shrimp, are an exception. They are adapted for living in salt lakes and can tolerate salinities near the saturation point of salt.

The internal structure of water fleas can be easily observed through the transparent carapace when these animals are placed on a depression slide and studied under a microscope. Most of the structures illustrated in Figure 29.35 can be seen in a specimen of *Daphnia*. The ease with which water fleas can be collected, maintained and observed has made them the subjects of many experimental studies of their orientation to light, their heart beat and their reactions to various environmental factors.

Water fleas brood their eggs in the space beneath the carapace above the back of the trunk (Fig. 29.35). Fairy shrimp brood their eggs with **ovisacs** attached to the female gonopore and secreted at the time of egg deposition.

Water fleas have direct development, but in

Figue 29.34 Branchiopods. *A*, A fairy shrimp (they swim upside down). *B*, A clam shrimp. One of the valves of the carapace has been removed. *C*, A tadpole shrimp. (All after Borradaile et al.)

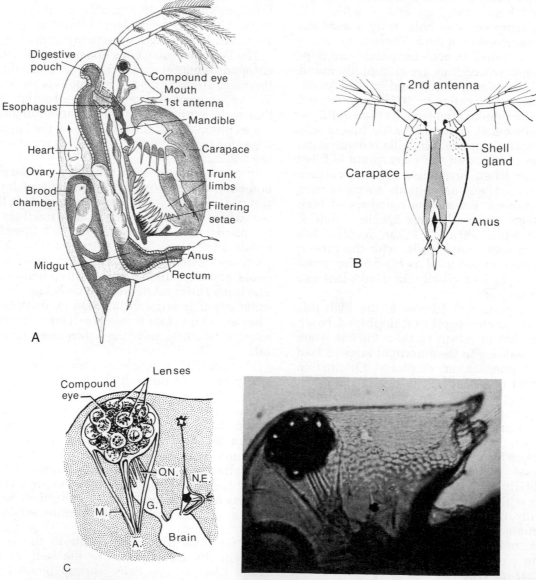

Figure 29.35 *Daphnia*, the water flea. *A*, Side view, with one side of the carapace removed to show enclosed body and organs. *B*, Ventral view, with trunk appendages omitted. *C*, Part of the head of *Daphnia* showing compound eye with protruding lenses and muscles (M.) of the right side attached to the side of the head (at A.). Also shown are the optic nerves (O.N.), optic ganglion (G.), brain and nauplius eyes (N.E.). Figure at right is photograph of area shown in drawing.

other branchiopods the egg hatches as a nauplius larva. The retention of larval stages in freshwater branchiopods is an exception to the general rule of suppression of larval stages in fresh water.

Branchiopods exhibit a number of reproductive and developmental adaptations to the environmental stresses common in freshwater lakes and ponds. Parthenogenesis is common. Many species produce thin-shelled eggs that hatch rapidly and thick-shelled eggs that remain dormant during periods of drought or freezing. Note that these adaptations are the same as those exhibited by freshwater flatworms and rotifers.

Copepods. The subclass **Copepoda** is the largest of the nonmalacostracan subclasses. The 4500 species are mostly marine, but there are also many freshwater forms. The copepod body, usually less than 3 mm. in length, is cylindrical and tapered (Fig. 29.36). The head bears a single **median nauplius eye** but no compound eyes. The first pair of antennae are very large. The trunk is composed of an anterior **thorax** bearing biramous appendages and a narrower posterior **abdomen,** which lacks appendages but usually carries a pair of terminal processes.

Copepods may be planktonic; they may live near the bottom, where they swim about over debris; they may be interstitial (live in the spaces between sand particles); and some are parasitic. Each swims or crawls with his thoracic appendages. The antennae of swimming copepods serve to decrease the tendency to sink when the thoracic appendages are not

beating. Planktonic species have many setae on especially long antennae. Interstitial copepods are tiny with elongate bodies and short antennae.

Although some copepods are grazers or are predatory, most planktonic forms are filter feeders. Certain of the post-oral head appendages produce the water current and function as the filter. Diatoms are the principal source of food, and copepods play an important role in the food chain. Studies on the marine planktonic copepod *Calanus finmerchious,* which is 5 mm. long, have shown that as many as 373,000 diatoms may be filtered out and digested every 24 hours.

About one quarter of the copepod species are parasitic, attacking the skin, fins and gills of marine and freshwater fish. These are commonly referred to as fish lice (Fig. 29.37). All degrees of modification from the typical copepod body form exist, and some species are so aberrant that they no longer look like crustaceans.

Copepods have no gills, and many species also lack a heart and blood vessels.

Some copepods shed their eggs singly into the water, but many brood their eggs within secreted ovisacs attached to the female gonopore. A nauplius larva is the hatching stage in both marine and freshwater forms.

Ostracods. Members of the subclass Ostracoda are tiny marine and freshwater crustaceans with the body entirely enclosed within a bivalved carapace (Fig. 29.38). Ostracods are sometimes called mussel shrimp and indeed parallel the bivalve mollusks in many ways.

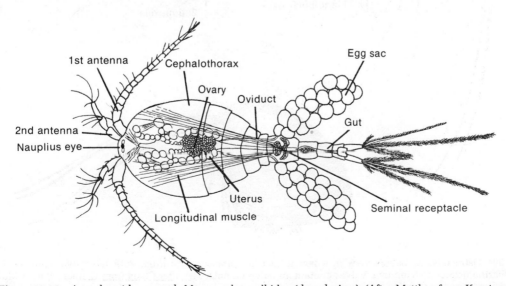

Figure 29.36   A cyclopoid copepod, *Macrocyclops albidus* (dorsal view). (After Matthes from Kaestner.)

Figure 29.37 Parasitic copepods, *Penella exocoeti*, on a flying fish. The copepods are in turn carrying the barnacle, *Conchoderma virgatum* (striped body). (Modified after Schmitt.)

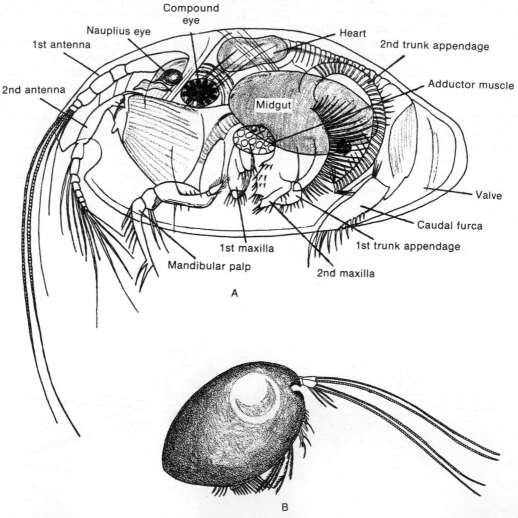

Figure 29.38 Ostracods. *A*, Lateral view of a female marine ostracod, *Cypridina*, with left valve removed. *B*, Lateral view of a marine ostracod swimming. Valves contain an antennal notch. (*A* after Claus from Calman; *B* after Müller from Schmitt.)

The two **valves,** usually 2 mm. or less in length, may have interlocking teeth and are held together dorsally by an elastic hinge. The valves are closed by a bundle of **adductor muscle fibers** extending transversely between the two valves near the middle of the body (Fig. 29.38). The chitinous skeleton is impregnated with calcium carbonate. As a consequence, ostracods have an extensive fossil record dating from the Cambrian. Sculpturing of the valves and the location of the adductor muscle scars are the principal taxonomic characteristics of fossil species.

Within the two valves, the ostracod body is mostly head, for the trunk is greatly reduced (Fig. 29.38). The two antennae are large, and the other head appendages are well developed. However, the number of trunk appendages is reduced to two, one or zero.

A nauplius eye is characteristic of all ostracods; sessile compound eyes are present in the members of one order.

There are some planktonic ostracods, including the 23 mm. giant *Gigantocypris,* but most species are benthic and scurry over the bottom, crawling and swimming with the large antennae. Along with copepods and water fleas, ostracods are very common crustaceans in freshwater ponds and small pools.

Many ostracods are filter feeders utilizing their different post-oral appendages as filters, but some species are predators or scavengers.

There are no gills. Only in one order is there a heart or blood vessels, but blood is propelled through tissue spaces, including the valves of the carapace, by the movements of the body.

Eggs are released singly or brooded within the carapace, and a nauplius with a bivalved carapace is the hatching stage.

Barnacles. The subclass **Cirripedia** contains the barnacles. These marine animals differ from other crustaceans and most other arthropods in being attached to the substratum. Correlated with the sessile habit, the body is enclosed within a bivalved carapace covered with protective **calcareous plates** (Figs. 29.39 and 29.40). Their attached condition and calcareous covering makes them so unlike other members of the class that it was not until 1830, when the larvae were discovered, that barnacles were recognized as crustaceans.

If one can imagine an ostracod, which is also enclosed within a bivalved carapace, attached to the substratum by the antennae, and the valves covered by calcareous plates, one has some idea of the structure of barnacles. The late larval stage of barnacles is enclosed within a bivalved carapace and looks much like an ostracod. It settles to the bottom and attaches by means of a **cement gland** in the base of the first antennae.

All barnacles are attached, but the group may be divided into **stalked** and **sessile** forms. The body of stalked barnacles is composed of a fleshy **peduncle,** which represents the pre-oral part of the body, and a **capitulum,** which represents the post-oral part of the body (Fig. 29.40). The capitulum carries the bivalve cara-

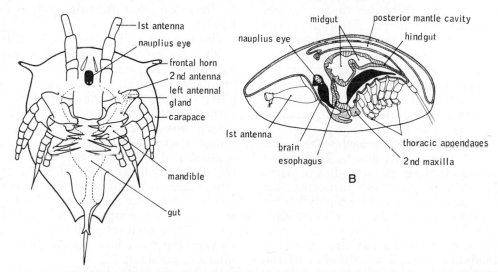

Figure 29.39 *A,* Nauplius larva of *Balanus,* setae omitted. *B,* Cypris larva of *Balanus,* side view with left half of carapace removed. (From Walley, L. J.: Philos. Trans. R. Soc. Lond. (Biol) *256,807*:237–280, 1970.)

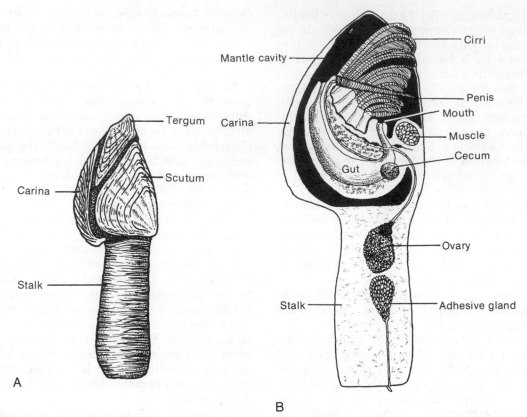

Figure 29.40 *Lepas,* a stalked barnacle. *A,* Entire animal showing calcareous plates. *B,* Internal structure. (After Broch from Kaestner.)

pace, often called the **mantle,** and the **gape** is directed upward. The plates covering the carapace vary in number, but there are always two pairs of plates, **scuta** and **terga,** that flank either side of the gape.

The body of sessile barnacles consists mostly of the capitulum, for the pre-oral part is reduced to a platform on which the capitulum rests and is attached to the substratum (Fig. 29.41). The platform of sessile barnacles and the peduncle of stalked species both contain the antennal cement gland, indicating that the two structures are homologous. In sessile barnacles the basal plates of the capitulum form a rigid circular wall surrounding the upper movable lidlike terga and scuta.

Although some stalked barnacles are found on rocks, many species live attached to floating objects, such as timbers, or the bodies of larger swimming animals. Whales, porpoises, sea turtles, sea snakes, crabs and even jellyfish are utilized as substrata by different species. Correlated with the utilization of other animals as attachment surfaces is the fact that there has been a tendency toward a reduction of the protective calcareous covering so that some barnacles have become almost naked.

Most sessile barnacles have become adapted for a life on rock or other hard substrata, and the heavy, somewhat fused, ring of wall plates probably represents an adaptation for protection against currents, pounding waves and browsing fish. However, like stalked barnacles, some sessile barnacles have become adapted for living on other animals, such as crabs and whales.

The tendency of barnacles to utilize other animal bodies as substrata probably led to parasitism. Approximately one-third of the members of the class are parasitic and are so highly modified that they are recognizable as barnacles only by their larval stages. Other crustaceans are the principal host.

Both stalked and sessile barnacles are filter feeders, and the trunk bears six pairs of long, coiled, biramous appendages called **cirri,** from which the name of the class is derived. In feeding, the cirri unroll and project through the gape of the carapace as a large basket (Fig. 29.42). They perform a rhythmic scooping motion and the many fine setae of the appendages filter out plankton.

Neither eyes, gills, heart nor blood vessels are

present. Correlated with their attached existence is the fact that barnacles, unlike most other arthropods, are **hermaphroditic.** At copulation, a long, extensible, tubular penis is projected from the gape of one individual into the mantle cavity of a neighboring barnacle. Fertilization and brooding occur within the mantle cavity, and a nauplius is the usual

hatching stage. The nauplius develops into a bivalved **cypris larva** that settles and attaches to the bottom. The many changes involved in metamorphosis include the loss of antennae and the development of calcareous plates. In the molts that accompany growth the plates are not shed; only the inner chitinous lining of the mantle is molted.

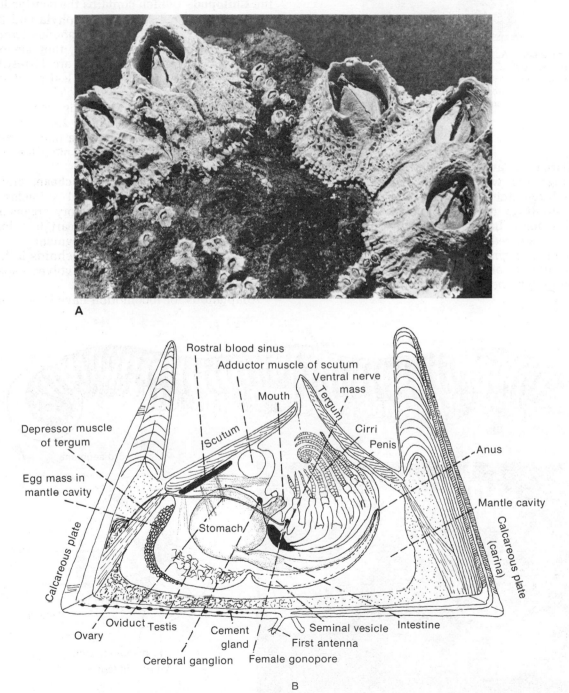

Figure 29.41 *Balanus,* a sessile barnacle. *A,* A cluster of animals attached to a mussel shell. They are viewed from above and show the movable terga and scuta within the surrounding immovable wall of plates. *B,* Lateral view of animal in section. (*A* by Betty M. Barnes; *B* after Gruval from Calman.)

Figure 29.42 A feeding *Chelonebia*, showing the outstretched cirri. *Chelonebia* is a sessile barnacle, which lives attached to the carapaces of swimming crabs. The conspicuous plates at the base of the cirri are the two terga. One scutum can be seen on the right side. (By E. Ryan.)

Barnacles are major fouling organisms on pilings, sea walls, buoys and ship bottoms. Much research has been expended to develop anti-fouling measures, such as anti-fouling paint, for a badly fouled ship may have its speed reduced by as much as 35 per cent. Recently, barnacle cement has been the subject of a number of investigations in the hope that there might be practical applications for dentistry.

29.10 UNIRAMIATES: MYRIAPODOUS ARTHROPODS

The myriapodous arthropods comprise four classes of uniramiates that were once placed within a single class, the **Myriapoda.** They are the **Diplopoda** (which contains the millipedes), the **Chilopoda** (which contains the centipedes) and two small classes, the **Symphyla** and the **Pauropoda.** Although the myriapodous arthropods belong to separate classes, they share a number of characteristics. All are terrestrial and secretive, living in soil and leaf mold and beneath stones, logs and bark. The body is composed of a **head** and a long **trunk** with many segments and legs (Fig. 29.43A). The head bears a pair of **antennae,** a pair of **mandibles** and one or two pairs of **maxillae.** The eyes are not compound.

Gas exchange organs are **tracheae,** and a separate system of tubules and spiracles is present in each segment. Excretory organs are **malpighian tubules,** and the heart is a long dorsal tube with ostia in each segment.

Reproduction parallels the arachnids in that sperm transfer is indirect and involves a spermatophore.

Adaptations for locomotion have been a pri-

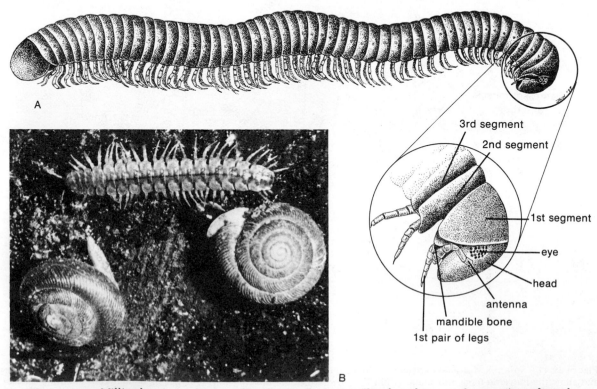

3rd segment

2nd segment

1st segment

eye

head

antenna

mandible bone

1st pair of legs

Figure 29.43 Millipedes. *A,* A juliform millipede; *B,* a flat-back millipede and two snails on a piece of wood.

Spiracles

Figure 29.44 Ventral view of a diplosegment of *Aphel-oria,* a flat-back millipede. (After Snodgrass.)

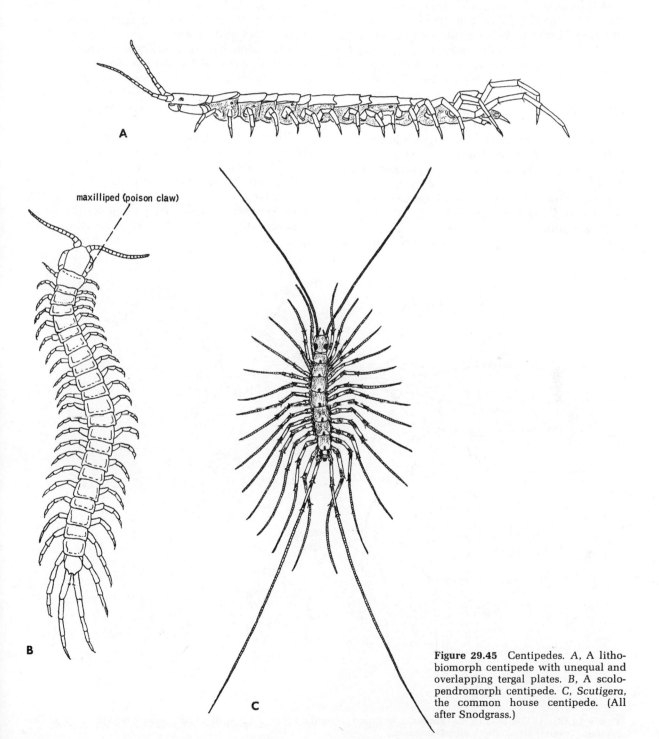

A

maxilliped (poison claw)

B

C

Figure 29.45 Centipedes. *A*, A litho-biomorph centipede with unequal and overlapping tergal plates. *B*, A scolo-pendromorph centipede. *C*, *Scutigera*, the common house centipede. (All after Snodgrass.)

mary theme in the evolutionary history of centipedes and millipedes. These animals have become adapted for running, climbing, pushing and wedging beneath objects. Yet, in contrast to most other arthropods where increased motility is correlated with a reduction in number of legs and compaction of the body, myriapods have retained a long trunk with many appendages.

Millipedes. Millipedes, members of the class Diplopoda, have the body adapted for pushing, the force being generated by the large number of legs. As many as 12 to 52 legs may be involved in one wave of movement sweeping down the length of the body. The large number of legs requires a large number of segments, but a large number of segments results in a weakened trunk column. This problem was solved by the formation of double segments, or **diplosegments;** i.e., each trunk section of a millipede represents two fused segments. Each double segment contains two pairs of legs, two sets of spiracles, two pairs of heart ostia and two ventral nerve ganglia (Fig. 29.44).

The common cylindrical (juliform) millipedes are best adapted for pushing through leaf mold and other loose debris (Fig. 29.43). The head is rounded and the many legs are attached near the mid-ventral line of the cylindrical smooth body. Some species may become as large as pencils in size.

Flatbacked millipedes are adapted for wedging into confining places, such as beneath stones or bark. The body is dorso-ventrally flattened, and the laterally projecting terga create a protective working space for the legs (Figs. 29.43 and 29.44).

Both flatbacked and juliform millipedes possess **repugnatorial (stink) glands** on the diplosegments. These glands secrete compounds that contain iodine, quinone or hydrocyanic acid and are believed to serve a defensive function.

Millipedes are chiefly scavengers, feeding on living and decaying vegetation and on dead animal remains.

Eggs are deposited freely in the soil, in protective mud capsules or in a nest composed of excrement or soil. The young, on hatching, possess only a small number of segments and legs.

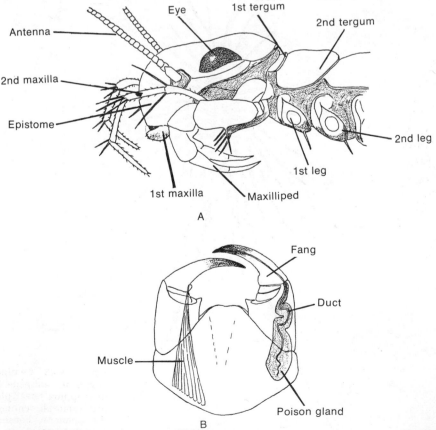

Figure 29.46   *A,* The head of the centipede *Scutigera coleoptrata* (lateral view). *B,* Maxillipeds, or poison claws, of *Otocryptops* (ventral view). (Both after Snodgrass.)

Centipedes. Centipedes, members of the class Chilopoda, are predaceous and most are adapted for running. Behind the mandibles and the two pairs of maxillae are a pair of large poison claws used for seizing and killing prey (Fig. 29.46). Centipedes feed mostly on other arthropods, and although large specimens can inflict a painful bite, only a few tropical species are actually dangerous.

Only one pair of legs is present per trunk segment, for the segments are not doubled as in millipedes. To reduce its tendency to undulate or wobble when running, the trunk of many species is strengthened by having overlapping terga of different lengths (Fig. 29.45).

The centipedes most highly adapted for running are members of the order **Scutigeromorpha** (Fig. 29.45). They inhabit rocky open areas and can catch moving insects. Some species occur in human habitations, commonly in bathtubs and sinks. The scutigeromorphs have overlapping trunk terga; the legs are very long, which provides for a greater step; the leg length increases along the trunk from anterior to posterior, preventing interference; and the distal part of the leg is composed of many small segments, providing greater traction through increased contact with the ground.

Although some centipedes deposit their eggs in the soil, most brood their eggs, wrapping the body around the developing young.

29.11 THE UNIRAMIATES: CLASS INSECTA

The insects are not only the largest class of arthropods but also the largest group of animals, including more than three quarters of a million species. The great adaptive radiation that the class has undergone has led to the occupation of virtually every type of terrestrial habitat, and some groups have even invaded fresh water. The great success of insects over other terrestrial arthropods can be attributed in part to the evolution of flight, which provided advantages for dispersal, escape, access to food or more favorable environmental conditions. The ability to fly evolved in reptiles, birds and mammals, but the first flying animals were insects.

Insects are of great ecological significance in the terrestrial environment. Two-thirds of all flowering plants are dependent upon insects for pollination. The principal pollinators are bees, wasps, butterflies, moths and flies, and the three orders represented by these insects have an evolutionary history that is closely tied

to that of the flowering plants, which underwent an explosive evolution in the Cretaceous.

Insects are of enormous importance for humans. Mosquitoes, lice, fleas, bedbugs and a host of flies can contribute directly to human misery. Some contribute indirectly as vectors of human diseases or diseases of their domesticated animals: mosquitoes (malaria, elephantiasis and yellow fever); tsetse fly (sleeping sickness); lice (typhus and relapsing fever); fleas (bubonic plague); and the housefly (typhoid fever and dysentery). Our domesticated plants are dependent upon some insects for pollination but are destroyed by others. Vast sums are expended to control insect pests, which can greatly reduce the high agricultural yields necessary to support large human populations. But the overzealous use of pesticides can in turn be hazardous to the environment.

External Structure. Despite the great diversity of insects, the general structure is relatively uniform. The body is composed of **head, thorax** and **abdomen** (Fig. 29.47). A large pair of **compound eyes** occupies the lateral surface of the head. Between the compound eyes are three small **simple eyes,** or **ocelli,** and a pair of **antennae** (Fig. 29.48). The feeding appendages consist of a pair of **mandibles** and two pairs of **maxillae.** The second pair of maxillae are fused together and called the lower lip, or **labium.** An upper lip, or **labrum,** formed by a shelflike projection of the head, covers the mandibles anteriorly. Near the base of the labium a median process, the **hypopharynx,** projects from the floor of the oral cavity.

The thorax is composed of three segments: a **prothorax, mesothorax** and **metathorax.** Each segment bears a pair of legs, and the last two segments may each carry a pair of wings (Fig. 29.47).

The abdomen is composed of nine to 11 segments, which lack appendages; however, the terminal reproductive structures are believed by some entomologists to be derived from segmental appendages.

Flight. Most insects have wings and compose the subclass **Pterygota.** Primitive insects, such as proturans, thysanurans and collembolans (members of the subclass **Apterygota**), are wingless and apparently diverged from the main stream of insect evolution before the origin of flight (Fig. 29.49). Some insects — ants, termites, fleas and lice — are secondarily wingless.

The wings develop as folds of the body wall and are thus composed of two layers of skeletal material applied together. Support is provided

Figure 29.47 *A*, A parasitic female ichneumon fly (order Hymenoptera). *B*, A gall gnat (Order Diptera). (*A* after Lutz; *B* after Usda from Borror and Delong.)

Figure 29.48 External structure of an insect. *A*, Anterior surface of the head of a grasshopper. *B*, Lateral view of the head of a grasshopper. *C*, Lateral view of a wingless thoracic segment. *D*, Leg of a grasshopper. *E*, Lateral view of the abdomen of a male cricket. (*C–E* after Anodgrass.)

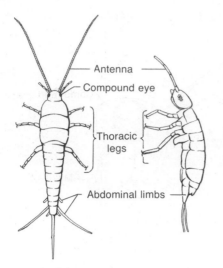

Figure 29.49 Primitive wingless insects (Apterygota), showing a silverfish (*left*) and a springtail (*right*). (After Lubbock [*left*] and Carpenter and Folsom [*right*].)

by a netlike or strutlike arrangement of tubular thickenings called **veins.**

Each wing articulates with the edge of the **tergum,** but its inner end rests on a dorsal pleural process, which acts as a fulcrum (Fig. 29.50). The wing is thus somewhat analogous to a seesaw off center. Upward movement of the wings results *indirectly* from the contraction of vertical muscles within the thorax, depressing the tergum. Downward movement of the wings is produced either *directly*, by contraction of muscles attached to the wing base (dragonflies and roaches), or *indirectly*, by the contraction of transverse horizontal muscles raising the tergum (bees, wasps and flies).

Up and down movement alone is not sufficient for flight. The wings must at the same time be moved forward and backward. A complete cycle of a single wing beat describes an ellipse (grasshoppers) or a figure eight (bees and flies), during which the wings are held at different angles to provide both lift and forward thrust. In many insects the two pairs of wings are locked together in flight.

The raising or lowering of the wings resulting from the contraction of one set of flight muscles stretches the antagonistic muscles, which then also contract. Insect wing beat thus involves the alternate contraction of the antagonistic elastic systems. The beat frequency varies greatly—4 to 20 beats per second in butterflies and grasshoppers; 190 beats per second in the honeybee and housefly; and 1000 beats per second in certain gnats. At low frequencies (30 beats per second or less) there is usually one

nerve impulse to one muscle contraction. At higher frequencies, however, the contraction is **myogenic,** originating from the stretching caused by the contraction of the antagonistic muscles, and there are a number of beats, or oscillations, between each nerve impulse.

Rapid contraction is facilitated by the nature of the muscle insertion. A very slight decrease in muscle length during contraction can bring about a large movement of the wing (as a seesaw with the fulcrum near one end). A large part of insect flight muscles is occupied by giant mitochondria, which provide for the high rate of respiration in these cells.

Flying ability varies greatly. Many butterflies and damselflies have a relatively slow wing beat and limited maneuverability. At the other extreme, some flies, bees and moths can hover and dart. The fastest flying insects are hummingbird moths and horseflies, which have been clocked at over 33 miles per hour. Gliding, an important form of flight in birds, occurs in only a few large insects.

There is no flight control center in the insect nervous system, but the eyes and the sensory receptors on the antennae, on the wings and in the wing muscles themselves provide continual feedback information for flight control. Horizontal stability is maintained in part by a dorsal light reaction: the insect keeps the dorsal ommatidia of the eyes under maximum illumination from above. Deviation because of rolling is corrected by slight changes in wing position to bring the dorsal part of the eyes back to maximum illumination.

Members of the order **Diptera** (flies, gnats and mosquitoes) have the second pair of wings reduced to knobs, called **halteres.** The halteres beat with the same frequency as the forewings and function as gyroscopes for the control of flight instability (pitching, rolling and yawing).

Flight speed is probably determined by air flow over receptors on the antennae and movement of objects from front to back across the eyes.

Flight is inhibited by contact of the **tarsi** with the ground. At rest the wings are either held outstretched, directed back over the abdomen or folded up. The anterior pair of wings of beetles (order **Coleoptera**) are heavy shieldlike covers for the membranous second pair of wings, which are folded over the abdomen at rest.

Nutrition. Primitive insects were herbivores that chewed plants with heavy jawlike **mandibles.** In the evolution of various groups, the mouthparts have become adapted for many

Figure 29.50 Diagrams showing relationship of wings to tergum and pleura and the mechanism of the basic wing strokes in an insect. *A*, Upstroke resulting from the depression of the tergum through the contraction of vertical muscles. *B*, Downstroke resulting from the arching of the tergum through the contraction of longitudinal muscles. *C*, An insect in flight, showing the figure 8 described by the wing during an upstroke and a downstroke. *D*, Changes in the position of the forewing on a grasshopper during the course of a single beat. Short arrows indicate direction of wind flowing over wing and numbers indicate consecutive wing positions. (*A* and *B* after Ross; *C* after Magnan from Fox and Fox; *D* after Jensen from Chapman: The Insects: Structure and Function. London, University of London Press Ltd., 1971. Courtesy of the American Elsevier Publishing Co., Inc.)

different modes of feeding, and now a wide range of diets is utilized. Sucking, piercing and sucking, and chewing and sucking insects have the mouthparts elongated to form a heavy **beak**, but since this feeding habit evolved independently a number of times, the beak is formed in different ways.

The primitive herbivorous chewing habit has been retained in many members of a number of major orders, such as grasshoppers, crickets and beetles. The raptorial dragonflies and some beetles, on the other hand, use their chewing mouthparts for seizing and tearing prey. A sucking proboscis that rolls up when not in use is characteristic of butterflies and moths (Fig. 29.51); bees have both chewing and sucking mouthparts. Bugs, aphids, cicadas and many

mosquitoes, gnats and flies have piercing and sucking mouthparts to pierce plant tissue or to penetrate the bodies of other animals.

Salivary secretions play an important role in feeding. Lubrication of food, digestion of sugar and pectins, anticoagulation and production of venom are provided by the salivary glands of different groups.

The **foregut** of insects is variously modified to suit the diet and mode of feeding, but a **pharynx, crop** and **proventriculus** are most commonly encountered (Fig. 29.52). The proventriculus may function as a gizzard or as a valve into the midgut.

The insect **midgut** (the **ventriculus,** or **stomach**) is the site of enzyme production, digestion and absorption, as in other arthropods (Fig.

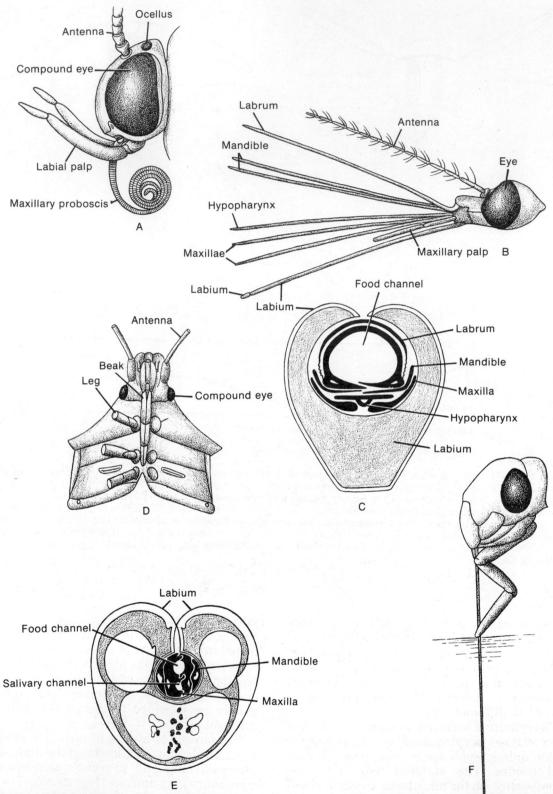

Figure 29.51 Mouth parts of sucking and piercing insects. *A*, Lateral view of the head of a moth, a sucking insect. *B*, Lateral view of the head of a mosquito, showing separated mouth parts. *C*, Cross section of mouth parts of a mosquito in their normal functional position. *D*, Ventral view of anterior half of a hemipteran, showing beak. *E*, Cross section through a hemipteran beak, showing the food and the salivary channels enclosed within the stylet-like maxillae and the mandibles. *F*, A hemipteran penetrating plant tissue with its stylets. (*A* after Snodgrass; *B* and *C* after Waldbauer from Ross; *D* after Hickmann; *E* after Poisson; *F* after Kullenberg.)

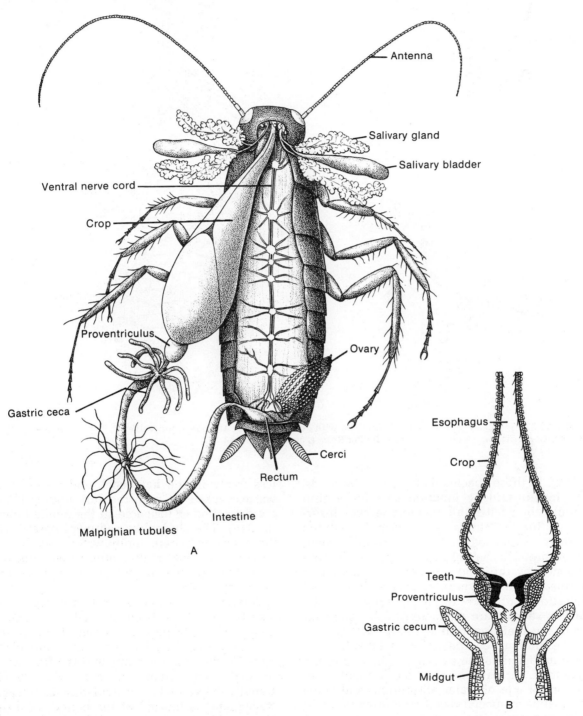

Figure 29.52 *A*, Internal structure of a cockroach. *B*, Longitudinal section through the foregut and anterior part of the midgut of a cockroach. (*A* after Rolleston; *B* after Snodgrass.)

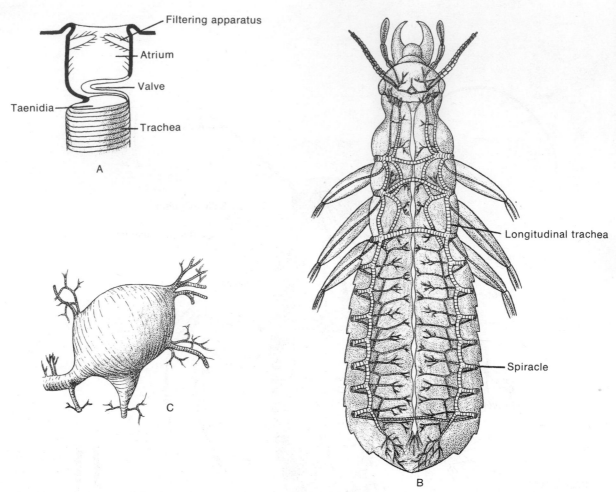

Figure 29.53   Insect tracheal system. *A,* Spiracle with atrium, filtering apparatus and valve. *B,* Typical plan of tracheal system. *C,* An air sac. (*A* after Snodgrass; *B* after Ross; *C* after Snodgrass.)

29.52). In those species that ingest solid foods, the foregut-midgut junction secretes a thin cuticle, the **peritrophic membrane,** which surrounds the food mass as it passes through the midgut. Supposedly, the peritrophic membrane protects the delicate midgut walls from abrasion by the food mass. The membrane is permeable to enzymes and the products of digestion. Outpocketings of the midgut, called **gastric caecae,** are characteristic of many insects, but their function is still uncertain.

The **hindgut,** composed of an **intestine** and **rectum,** opens to the exterior at the end of the abdomen. In many insects, the hindgut is an important site of water absorption, and some species have special rectal structures to facilitate the process.

Excretion, Gas Exchange and Internal Transport. The excretory organs are **malphigian tubules.** Two to several hundred tubes arise from the hindgut-midgut junction. Sometimes they are long and coiled. They lie more or

less freely in the **hemocoel,** and nitrogenous wastes picked up from the surrounding blood are deposited as uric acid in the tubule lumen (p. 346). The sludgelike sediment passes into the intestine where more water may be removed. The fecal matter eliminated from the anus thus consists of undigested residues and excretory wastes.

Tracheae are the gas exchange organs of insects; they have been studied more extensively in this group of animals than in any others. A **spiracle** is located on each side of the last two thoracic segments and the first seven or eight abdominal segments (Fig. 29.53). Usually the spiracle is provided with a closing device and is located at the bottom of a pit. Various filtering devices reduce clogging from dust. Internally, the tracheal system exhibits a more or less ladder-like plan, with longitudinal and transverse trunks. The larger tubes may have the walls strengthened by ringlike thickenings. The fluid-filled **tracheoles** are minute, usually less than a micrometer in diameter, and a number of tracheoles may arise from a single tracheal cell. The tracheoles are not shed at molting, but the new cuticle fuses with the old cuticle of the persisting tracheole. The distribution of tracheoles reflects the oxygen demands of the tissues served. In the wing muscles, for instance, the tracheoles actually push into the muscle cells.

Diffusion along concentration and pressure gradients is the principal means by which oxygen reaches the tracheoles and carbon dioxide passes to the outside. Insects weighing more than a gram and extremely active insects possess additional ventilating mechanisms. The movement of body muscles may compress the larger trachea or special dilated regions called **air sacs.** Compression coupled with a pattern of opening and closing of spiracles moves air from one part of the system to another.

Adult aquatic insects continue to use tracheae and obtain oxygen by contact with the surface or from bubbles or films of air trapped on the surface of the body. Aquatic larvae may have gills, but even here, tracheae provide for internal gas transport. Oxygen dissolved in the water diffuses across the gill cuticle into underlying tracheae. Within the tracheal system it follows a diffusion gradient to the body tissues.

The **heart** is usually a long abdominal tube with nine pairs of ostia, and the only vessel is an anterior **aorta** leading into the thorax and head. Blood flows from posterior to anterior, although reversal of flow is known to occur in some forms.

Insect blood is unusual in containing large amounts of organic compounds, especially amino acids. High concentrations of uric acid, organic phosphates and a nonreducing sugar, trehalose, are also present.

Reproduction and Development. The single pair of gonads is located in the abdomen, and the gonoducts from each side unite posteriorly before opening at the end of the abdomen through a short median **vagina** in the female or an **ejaculatory duct** in the male. In addition, the female system usually includes a **seminal receptacle** and **accessory glands** associated with the vagina, and the male system includes paired **seminal vesicles** and **accessory glands** (Figs. 29.54 and 29.55).

The sperm of many insects are transferred

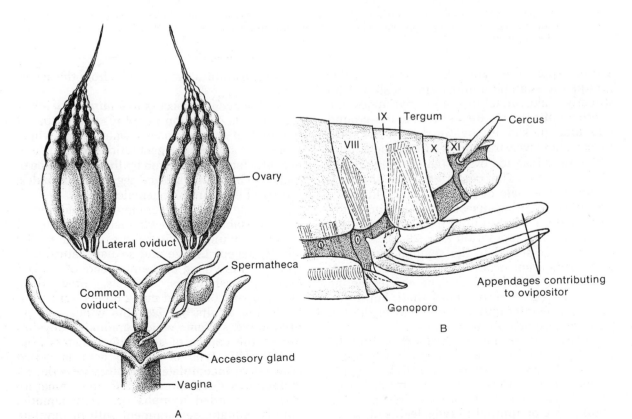

A

B

Figure 29.54 *A,* Reproductive system in a female insect. *B,* Lateral view of the posterior end of the abdomen, showing reproductive opening and appendages, forming ovipositor. (Both after Snodgrass.)

Figure 29.55   Reproductive system in a male insect. *A*, General plan of system. *B*, Lateral view of the posterior end of the abdomen, showing reproductive opening and other structures. *C*, Posterior view of the abdomen, showing penis and claspers. (All after Snodgrass.)

within spermatophores. In a few primitive groups the spermatophores are transferred indirectly. Like arachnids and myriapods the spermatophores are deposited on the ground and then picked up by the female. In most insects, however, transfer is direct, and a tubular **penis** is inserted by the male into the vagina of the female. The posterior abdominal segments of the male of many moths, butterflies, true flies and other insects bear clasping structures that are used to hold the abdomen of the female during copulation (Fig. 29.55).

Sperm are stored in the seminal receptacle, and fertilization occurs internally as the eggs pass through the oviduct. The egg leaves the ovary encased within a hard shell, but a tiny opening, or **micropyle,** at one end of the egg permits entrance of the sperm.

Certain parts of the terminal segments of the female are modified as an **ovipositor,** through which the eggs pass upon leaving the gonopore. By means of the ovipositor, eggs can be buried in soil, excrement or carrion, injected into plant tissue or applied to twig, leaf, soil, water or other surfaces. The eggs are generally deposited in batches, cemented to each other and

to the substratum by secretions from the accessory glands.

Parthenogenesis occurs in a number of insect groups. The condition in aphids closely parallels that of the crustacean water fleas, where there are successive generations of parthenogenetic females followed by the appearance of males. In bees, unfertilized eggs produce males; fertilized eggs produce females.

The degree of development at hatching is quite variable. The newly hatched young of primitive wingless insects are similar to the adults, except in size and sexual maturity. In contrast, the young of the members of winged orders—grasshoppers, crickets, dragonflies, leaf hoppers, bugs and many others—are similar to the adults but lack the wings and reproductive system, which gradually develop during the course of subsequent instars (Fig. 29.56). This type of development is called **gradual,** or **incomplete, metamorphosis (hemimetabolous development),** and the immature stages are called **nymphs** or, when aquatic, **naiads.** Aquatic development with incomplete metamorphosis is characteristic of dragonflies, stoneflies and mayflies.

The higher orders of insects, such as beetles, butterflies, moths, bees, wasps and flies, undergo **complete metamorphosis (holometabolous development)** involving more radical changes between the immature and adult forms. The hatching stage is an active feeding wormlike **larva** called a caterpillar (butterflies and moths), grub (beetles) or maggot (flies). The larva increases in size through successive instars, and at the end of the larval period, passes into a quiescent **pupal** stage, in which a radical transformation takes place. In many groups, most of the larval structures break down and are resorbed, and residual germinal areas **(imaginal discs)** give rise to adult structures. A completely formed adult emerges at the end of pupation, which typically occurs within a protective location (soil) or within some sort of encasement (cocoon of some moths). The transformation of immature insects into reproducing adults is known to be under endocrine control (p. 454).

INCOMPLETE METAMORPHOSIS   COMPLETE METAMORPHOSIS

Figure 29.56 Comparison of gradual, or incomplete, metamorphosis of a grasshopper and complete metamorphosis of a cecropia moth. (Villee, C. A.: Biology. 7th ed., Philadelphia, W. B. Saunders Co., 1977.)

Insect larvae are a specialized developmental condition and, as we have seen, are absent in primitive groups. Larvae are able to utilize food sources and habitats that would be unavailable to the adults and vice versa. For example, the caterpillars of butterflies are chewing herbivores, in contrast to the nectar-feeding habit of the adults. In some insects, certain moths for instance, the short-lived adults have lost the ability to feed at all.

An unusual form of "twinning," called **polyembryony,** occurs in some small wasps. The eggs develop parasitically within the body of caterpillars. At an early stage of development the embryonic mass fractures into a number of smaller masses, each of which develops into a complete individual. In *Litomastix,* as many as 3000 larvae may develop from one egg!

Parasitism. There are many parasitic insects, and the condition has undoubtedly evolved several times within the class.

Insect parasitism represents an adaptation to meet the habitat-nutrition needs of different stages in the life cycle. For some insects, parasitism provides a new food source and habitat for the adults; for others, a new food source for the larvae.

Adult fleas and lice (Fig. 29.57) are blood-sucking ectoparasites on the skin of birds and mammals (there is one group of chewing lice). The eggs and immature stages of fleas may develop on the host or in the host's nest or habitation.

Many species of wasps and flies illustrate larval parasitism. Certain small wasps insert an egg into the leaves and stems of plants. The surrounding plant tissue reacts to form a large mass, or gall. The egg develops within the gall and the larval stage feeds upon the surrounding plant tissue. Other wasps deposit their eggs in the bodies of insects, especially insect larvae.

The screw-worm fly, a species of blowfly and a pest of domestic animals, lays its eggs in the wounds and nostrils of mammals, and the larvae feed on living tissue. The parasitic condition was probably preceded by the deposition of eggs in carrion, for this is the habit of many nonparasitic species of blowflies.

Communication. Both social and nonsocial insects utilize chemical, tactile, visual and auditory signals as methods of communication. Many examples of chemical communication by pheromones (p. 457) are now known, and the following are just a few of the many striking examples in insects. The males of some moths can locate females by means of airborne substances detected from a distance as great as 4.5 km. Substances deposited on the ground by ants returning from a foraging trip serve as a trail marker for other ants. This type of communication is especially important in the complex movements of tropical army ants. Substances produced by the death and decomposition of the body of an ant within the colony stimulate other workers to remove the body. If a live ant is painted with an extract from a decomposing body, the painted ant will be carried live and struggling from the nest.

Among the more unusual visual signals are the luminescent flashings of fireflies, which play a role in sexual attraction. In species of *Photinus,* for example, flying males flash at definite intervals. Females located on vegetation will flash in response if the male is sufficiently close. The male will then redirect his flight toward her and further flashing will occur.

Sound production is especially notable in grasshoppers, crickets and cicadas. The chirping sounds of the first two are produced by rasping. The front margin of the forewing or the hind leg acts as a scraper and is rubbed over a file formed by veins of the forewing. Where scraper and file are both located on the forewing, as in crickets and katydids, the wings cross over and one forewing functions as a scraper and the other as a file. Each species of cricket produces a number of songs that differ from the songs of the other species. Cricket songs function in sexual attraction and aggres-

Head louse   Pubic louse   Body louse

Figure 29.57 Anoplura. The three varieties of human lice. The head louse, *Pediculus humanus* var. *capitis,* and body louse, *P. h.* var. *corporis* are interfertile varieties of one species that rarely interbreed because one lives on the head, laying eggs on the hairs, while the other lives on the clothed portion of the body, laying eggs in the clothing. The pubic louse, *Phthirus pubis,* lives in the pubic region and occasionally in the armpits. (After Patton and Evans.)

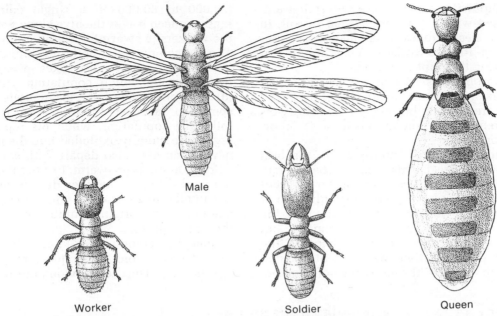

Male

Worker   Soldier   Queen

Figure 29.58   Castes of the common North American termite, *Termes flavipes*. (After Lutz.)

sion. The staticlike sounds of cicadas, which serve to aggregate individuals, are produced by vibrations of special chitinous abdominal membranes.

The remarkable mechanisms of communication in bees are described in the next section.

Social Insects.   Colonial organization has evolved in a number of animal phyla, but only among some insects and vertebrates are individuals functionally interdependent yet morphologically separate. The condition is therefore usually described as being a social organization.

Social organizations have evolved in two orders of insects; the **Isoptera,** which contains the termites, and the **Hymenoptera,** which includes the ants, bees and wasps. In all social insects, no individual can exist outside of the colony nor can it be a member of any colony but the one in which it developed. There is cooperative brood care and an overlap of generations. All social insects exhibit some degree of **polymorphism**, and the different types of individuals of a colony are termed **castes.** The principal castes are **male, female (or queen)** and **worker.** Males function largely for the insemination of the queen, which produces new individuals for the colony. The workers provide for the support and maintenance of the colony. Worker castes are sterile individuals, and caste determination is a developmental phenomenon.

Termites have been called social cockroaches, for the two groups are related and in both the gut harbors symbiotic flagellates, which are important in the digestion of cellulose. The contact necessary between individuals for the transfer of the symbionts may have been a factor in the evolution of termite social behavior.

Termites live in galleries constructed in wood or soil, and in some species, the colonies may be huge and structurally very complex. Although many species feed upon wood, others have exploited a wide range of cellulose sources, such as seeds, grasses and humus. The colony is built and maintained by workers and soldiers (Fig. 29.58). The soldiers possess large heads and mandibles and serve for the defense of the colony. Workers and soldiers are sterile and wingless and include both males and females. Wings are present in the fertile males and queens only during a brief nuptial flight. The male, or king, remains with the queen, copulating intermittently and aiding her in the construction of the first nest.

The colonies of ants, bees and wasps differ from those of termites in being essentially female societies, for all the workers are sterile females. Caste determination of the female is regulated by the presence or absence of certain substances provided in the immature stages by other members of the colony. However, only in bees is much known about the actual determination process. Males have a brief existence, functioning only in the copulatory nuptial flight. Unlike termite males, they neither contribute to the construction of the first nest nor remain with the queen.

Ant colonies resemble those of termites and are housed within a gallery system in soil, in wood or beneath stones. There may be a soldier caste in addition to workers. Wings are present only during the nuptial period.

Polymorphism is less highly developed in wasps and bees. There is no soldier caste and workers are winged, but many of these insects exhibit remarkable adaptations for a social organization.

The honeybee, *Apis mellifera*, is the best known social insect. This species is believed to have originated in Africa and to be a recent invader of temperate regions, for unlike other social bees and wasps of temperate regions, the honeybee colony survives the winter, and multiplication occurs by the division of the colony, a process called swarming. Stimulated at least in part by the crowding of workers

(20,000 to 80,000 in a single colony), the mother queen leaves the hive along with part of the workers (a **swarm**) to found a new colony. The old colony is left with developing queens. On hatching, a new queen takes several nuptial flights during which copulation with males **(drones)** occurs, and she accumulates enough sperm to last her lifetime. The male dies following copulation, when his reproductive organs are literally exploded into the female. A new queen may also depart with some of the workers as an after-swarm, leaving the remaining workers to yet another developing queen. Eventually the old colony will consist of about one-third of the original number of workers and their new queen.

Honeybee colonies are large. The workers' life span is not long, and a queen may lay 1000 eggs per day. The diet provided these larvae by

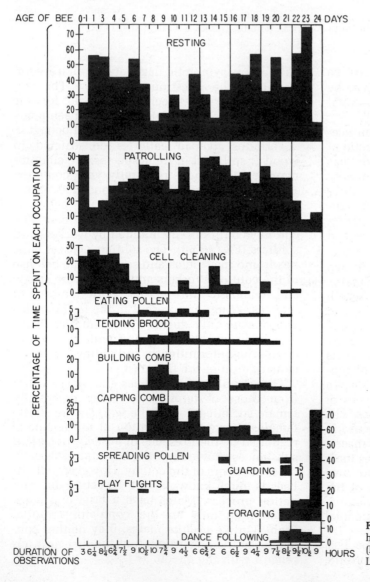

Figure 29.59 The activities of a single worker honeybee during the first 24 days of her adult life (Redrawn from Ribbands, 1953; based on data of Lindauer, 1952.)

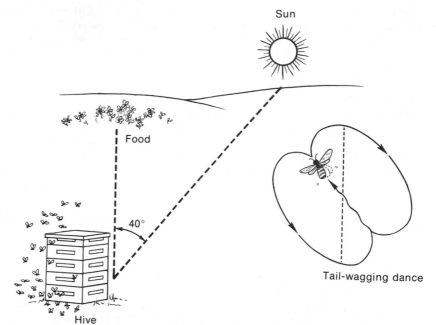

Figure 29.60 Diagram illustrating the inclination of the straight tail-wagging run by a scout bee to indicate the location of a food source by reference to the sun. The food source is located at an angle 40 degrees to the left of the sun. The tail-wagging run of the scout bee is therefore upward (indicating that food is toward sun) and inclined 40 degrees to the left (indicating the angle of the food source to the sun). (After Von Frisch.)

the nursing workers results in their developing into sterile females, i.e., additional workers. The nursing behavior of the workers is a response to a pheromone ("queen substance") produced by the queen's mandibular glands. At the advent of the swarming or when the vitality of the queen diminishes, the production of this pheromone declines. In the absence of the inhibiting effect of the pheromone, the nursing workers construct royal cells into which eggs and royal jelly are placed. The exact composition of this complex food is still unknown, but those larvae fed upon it develop into queens in about 16 days. At the same time that queens are being produced, unfertilized eggs are deposited in cells similar to those for workers. These haploid eggs develop into drones.

A remarkable feature of honeybee social organization is the temporal division of the workers. As can be seen from the graph on Figure 29.59, the first activities of the worker are maintenance tasks within the hive. During this period there is secretion by wax, mandibular and other glands involved in comb construction, food storage and larval care. After about three weeks, such glandular activity declines, and the bee begins a period of foraging outside of the hive, its final service to the colony. The many functions performed in the lifetime of a worker are not strictly sequential, but a worker shifts from one task to another (Fig. 29.59). A large amount of time is spent by the older worker bees in resting and patrolling. Patrolling, or "determination" of hive needs, plus the ability of workers to change tasks,

enables a colony to adjust to changing environmental conditions and is believed to have contributed to the success of the species.

Communication between members of a honeybee colony is highly evolved; some aspects, such as the tail-wagging dance, set the honeybees apart from all other social insects. A successful foraging scout returns to the hive and communicates to other workers the nature, direction and distance of a food source. The nectar and pollen on the scout's body provide the information about the kind of food that has been found. The scout bee also executes an excited dance that is a ritualization of the flight path. The dancing bee circles to the right and to the left, with a straight-line run between the two semi-circles (Fig. 29.60). During the straight-line run, the bee wags her abdomen and emits audible pulsations. Von Frisch, a pioneer in the study of communication in bees, discovered that the orientation of the circular movements shows the direction of the food, and that the frequency of the tail-wagging runs indicates the distance. The closer the food source is to the hive, the greater the frequency of tail-wagging runs. Bees use the angle of the sun and light polarization as a means of orientation, and the dance of the scout bee indicates the location of the food in reference to the sun's position. If the tail-wagging run is directed upward, the food is located toward the sun; if the tail-wagging run is directed downward, the food is located away from the sun. The inclination of the run to the right or to the left of vertical indicates the angle of the sun to the right or to the left of the food source. An internal "clock" compensates for the passage of time between discovery of the food and the start of the dance so that the information is correct even though the sun has moved during the interval. In cloudy days, the polarization of the light rays and ultra-violet light act as indirect references in the absence of the sun. If the food source is closer than 80 meters, the clues provided by chemoreception are sufficient for finding the food, and the tail-wagging dance is not performed by the scout bee.

Although the tail-wagging dance has been decoded, the sensory modality by which it is transmitted to other bees is still uncertain. The hive is dark so that the dance cannot be easily detected visually. The surrounding bees must receive the dancer's vibrations through their antennae or legs. The sound pulsations of the dancer apparently also indicate the distance of the food source from the hive. The average number of vibrations is proportional to the distance of the food from the hive.

CLASSIFICATION OF THE PHYLUM ARTHROPODA*

Subphylum Trilobita
The fossil trilobites. A single pair of antennae present; the many post-oral trunk appendages are all similar.

Subphylum Chelicerata
Antennae absent; the first post-oral appendages are a pair of chelicerae.

CLASS MEROSTOMATA. Body composed of a cephalothorax and abdomen. Marine, with abdominal gills.

Subclass Xiphosura. Horseshoe crabs. Abdominal segments fused.

Subclass Eurypterida. Extinct merostomes. Abdominal segments not fused.

CLASS ARACHNIDA.* Body composed of a cephalothorax and abdomen with four pairs of legs. Terrestrial; gills absent.

Order Scorpionida. Scorpions. Large chelate pedipalps and long segmented abdomen with terminal sting.

Order Pseudoscorpionida. Tiny arachnids with large pedipalps and short segmented abdomens.

Order Araneae. Spiders. Abdomen unsegmented and attached to cephalothorax by a narrow waist.

Order Opiliones. Harvestmen, or daddy-long-legs. Legs usually very long; the short, segmented abdomen is broadly joined to the cephalothorax.

Order Acarina. Mites and ticks. Very small arachnids with unsegmented abdomen broadly fused with cephalothorax.

CLASS PYCNOGONIDA. Sea spiders. Marine chelicerates(?) having a very narrow trunk and four pairs of long legs.

Subphylum Mandibulata
At least one pair of antennae is present, and first post-oral appendages are a pair of mandibles.

CLASS CRUSTACEA.† Head with two pairs of antennae.

Subclass Branchiopoda. Fairy shrimp, tadpole shrimp, clam shrimp and water fleas. Freshwater and marine crustaceans with leaflike setose appendages.

Subclass Ostracoda. Mussel, or seed shrimp. Tiny marine and freshwater crustaceans in which the entire body is enclosed within a bivalved carapace.

Subclass Copepoda. Copepods. Very small marine and freshwater crustaceans having a cylindrical tapered body with long first antennae.

Subclass Cirripedia. Barnacles. Marine, sessile crustaceans in which the body is enclosed within a bivalved carapace that is typically covered with calcareous plates.

Subclass Malacostraca.‡ Trunk composed of an eight-segmented thorax, on which the legs are located, and a six-segmented abdomen.

Order Amphipoda. Amphipods. Small marine and freshwater crusta-

---

*Abbreviated.

†In the absence of any yet widely accepted new scheme for the classification of higher categories of arthropods, this traditional system has been retained.

‡Greatly abbreviated.

ceans in which the body is laterally compressed and the eggs are brooded within a ventral thoracic marsupium.

Order Isopoda. Isopods. Marine, freshwater and terrestrial (wood lice and pill bugs) crustaceans in which the body is dorso-ventrally flattened and the eggs are brooded within a ventral thoracic marsupium.

Order Decapoda. Shrimp, crayfish, lobsters and crabs. Thorax covered by a carapace; thoracic appendages consist of three pairs of maxillipeds and five pairs of legs.

CLASS DIPLOPODA. Millipedes. Elongate trunk composed of many similar doubled segments, each bearing two pairs of legs.

CLASS PAUROPODA. Minute grublike animals that inhabit leaf mold. Elongate trunk of eleven segments, nine of which bear legs.

CLASS SYMPHYLA. Small centipede-like mandibulates that inhabit leaf mold. Trunk contains 12 leg-bearing segments.

CLASS CHILOPODA. Centipedes. Elongate trunk of many leg-bearing segments. First trunk segment carries a pair of large poison claws.

CLASS INSECTA. Insects. Body composed of head, thorax and abdomen. Thorax bears three pairs of legs and usually two pairs of wings.

Subclass Apterygota. Primary wingless insects. The five orders include the silverfish and springtails.

Subclass Pterygota.* Winged insects, or if lacking wings, the wingless condition is secondary.

Order Ephemeroptera. Mayflies.

Order Odonata. Dragonflies and damselflies.

Order Orthoptera. Grasshoppers and crickets.

Order Isoptera. Termites.

Order Plecoptera. Stoneflies.

Order Dermaptera. Earwigs.

Order Mallophaga. Chewing lice and bird lice.

Order Anoplura. Sucking lice.

Order Thysanoptera. Thrips.

Order Hemiptera. True bugs.

Order Homoptera. Cicadas, leaf hoppers and aphids.

Order Neuroptera. Lacewings, ant lions, snakeflies and dobsonflies.

Order Coleoptera. Beetles and weevils.

Order Trichoptera. Caddisflies.

Order Lepidoptera. Butterflies and moths.

Order Diptera. True flies, midges, gnats and mosquitoes.

Order Hymenoptera. Ants, bees, wasps and sawflies.

Order Siphonaptera. Fleas.

ANNOTATED REFERENCES

Detailed accounts of arthropods may be found in the references listed at the end of Chapter 21. The works listed below deal exclusively with specific arthropod groups.

Borror, D. J. and D. M. De Long: An Introduction to the Study of Insects. 4th ed. New York, Holt, Rinehart and Winston, 1976. A general entomology text with emphasis on the taxonomy of insects.

Borror, D. J. and R. E. White: A Field Guide to the Insects. Boston, Houghton Mifflin Co., 1970. A good guide for the identification of common North American insects.

Chapman, R. F.: The Insects: Structure and Function. New York, American Elsevier Publishing Co., 1969. A detailed account of the anatomy and physiology of insects.

Englemann, F.: Physiology of Insect Reproduction. New York, Pergamon Press, 1970.

Gertsch, W. J.: American Spiders. New York, Van Nostrand, 1949. An old but still valuable account of the natural history of spiders.

Kaestner, A.: Invertebrate Zoology. Vols. II (Chelicerates and Myriapods) and III

---

*Abbreviated.

(Crustaceans). New York, John Wiley and Sons, Inc., 1968 and 1970. The best general accounts in English of the arthropods other than insects.

Lockwood, A. P. M.: Aspects of the Physiology of Crustacea. San Francisco, W. H. Freeman, 1967.

Romoser, W. S.: The Science of Entomology. New York, Macmillan, 1973. A good balanced general biology of insects.

Schaller, F.: Soil Animals. Ann Arbor, Mich., University of Michigan Press, 1962. An account of the various invertebrates found in soil and leaf mold.

Schmitt, W. L.: Crustaceans. Ann Arbor, Mich., University of Michigan Press, 1965. A natural history of the crustaceans.

Weygoldt, P.: The Biology of Pseudoscorpions. Cambridge, Mass., Harvard University Press, 1969.

Wilson, E. O.: The Insect Societies. New York, Academic Press, 1971. A superb general biology of the social insects.

CHAPTER 30

BRYOZOANS

The members of the phylum Bryozoa, sometimes called moss animals, are common marine organisms, but their small size and atypical form make them virtually unknown to the layman. They live attached to rocks, shells, pilings, jetties and ship bottoms, and a few species occur in fresh water.

Bryozoans can be used to illustrate the predictive value of some of the information we have developed about animals thus far. The members of this phylum are **sessile, colonial** and **coelomate,** and individuals are usually less than 0.5 mm. in length. On the basis of these four facts, what other features — structural or functional — would you expect to find in bryozoans? Think a few minutes and make some educated guesses before you continue reading.

The following bryozoan features are correlated with their sessile life, colonial organization and minute size and are therefore not unexpected:
Correlated with their colonial organization:
*1. Bryozoans are **polymorphic.***
Correlated with their sessile condition:
*2. Bryozoans are encased within an **exoskeleton.** A protective and supporting skeleton is characteristic of many sessile animals, such as sponges, hydrozoans, corals and barnacles.*
*3. Bryozoans are **filter feeders.** Filter feeding, a common adaptation for a slow-moving or immobile life, is illustrated by such animals as sponges, sedentary tubiculous polychaetes, worm shells, slipper shells, bivalves and barnacles.*
*4. Bryozoans are **hermaphroditic.** Hermaphroditism is a common adaptation of animals that cannot move about, such as many sponges, some sea anemones, some corals, some bivalves, slipper shells and barnacles.*
*5. A **larval stage** is present in the development of*

bryozoans. Larvae are the principal means of dispersion for most sessile animals.
Correlated with their very small size:
6. Bryozoans have no special internal transport system, for internal distances are short enough to permit transport solely by diffusion.
7. Bryozoans have no gas exchange organs. The ratio of surface area to volume is sufficiently favorable to permit gas exchange across the general body surface.
8. Bryozoans have no excretory system. The excretion of highly soluble ammonia and the large ratio of surface area to volume permit elimination by diffusion across the general body surface.

30.1 STRUCTURE OF A BRYOZOAN INDIVIDUAL

Individuals (**zooids**) of a bryozoan colony are shaped something like a little rectangular box or coffin, in which each of the sides represents one of the usual morphological surfaces — anterior, posterior, ventral and so on (Fig. 30.1). The body is covered with an exoskeleton commonly composed of **chitin.** A layer of calcium carbonate is typically found just beneath the chitin; both are secreted by a single layer of epidermal cells. Immediately beneath the epidermis is a layer of **peritoneum;** the rigidity of the exoskeleton would make ordinary body wall muscles useless. The interior of the body is occupied by a spacious **coelom.**

The principal organ in the coelom is the **gut,** which is U-shaped to avoid the obstruction resulting from the attachment of the posterior end to other members of the colony. At the anterior end, the skeletal housing is perforated

727

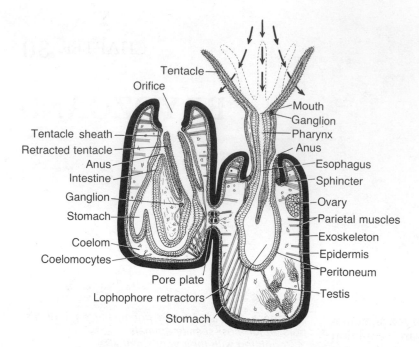

Figure 30.1 Structure of two zooids of a marine bryozoan. (After Marcus from Hyman.)

by a circular orifice through which the two ends of the gut project.

The feeding organ of bryozoans is a crown of ciliated tentacles, the **lophophore.** When the lophophore is protruded, the outstretched tentacles usually form a funnel with the **mouth** in the center of the base and the **anus** projecting outside of the base (Figs. 30.1, 30.2 and 30.4). The hollow tentacles contain an extension of the coelom. When the lophophore is protruded, a sheath of body wall extends from the rim of the orifice up to the base of the lophophore. On retraction of the lophophore into the orifice, this tentacular sheath is reversed

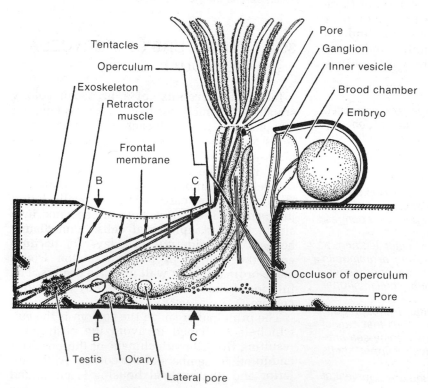

Figure 30.2 Lateral section through an encrusting bryozoan. This individual possesses a brood chamber containing an embryo. The orifice of this species is closed by a lid, or operculum. (After Ryland.)

and surrounds the bunched tentacles within the skeletal housing (Fig. 30.1).

The lophophore is protruded by increased pressure in the coelomic fluid. The pressure is elevated by the inbowing of the body wall on contraction of certain transverse muscle bands. In many bryozoans the muscle bands attach to the inner side (**frontal membrane**) of the ventral surface, which has become thin and easily depressed (Fig. 30.2). Withdrawal of the lophophore back into the orifice is brought about by a special retractor muscle.

During feeding, the **cilia** on the outstretched tentacles drive water down into and out the sides of the lophophore funnel, and food particles suspended in the incoming water stream are driven by cilia downward into the mouth (Fig. 30.1).

The nervous system consists of a **ganglion** and **nerve ring** around the anterior end of the gut. Fibers extend from the nerve ring to the tentacles and other parts of the body. No special systems for internal transport, gas exchange or excretion are present.

30.2 ORGANIZATION OF COLONIES

In most bryozoans the individuals are so attached to each other that the resulting colonies

Figure 30.3 *Zoobotryon*, a stoloniferous bryozoan. The zooids are very tiny and arranged around the cylindrical stolons, which compose the conspicuous part of the colony.

form erect plantlike growths or encrusting sheets over rocks and shells (Fig. 30.3). In either case, the ventral, or frontal, surface is exposed and the dorsal surface is attached. To visualize an encrusting colony, imagine a large number of rectangular boxlike individuals lying on their backs on a rock. The dorsal surface is attached to the substratum, the lateral, anterior and posterior surfaces are attached to surrounding individuals, and the ventral surface is exposed (Fig. 30.4). Since the anterior end is generally blocked by adjacent individu-

Figure 30.4 Lateral view of a portion of a colony of an encrusting bryozoan, *Electra*. (Modified from Marcus.)

Figure 30.5 *A, Bowerbankia,* a marine bryozoan in which the feeding individuals arise from a creeping stolon, formed by highly modified members of the colony. *B, Bugula,* a genus of common marine bryozoans that form erect branching colonies. (*A* after Mature; *B* after Hyman.)

als, the orifice is commonly shifted over onto the ventral (exposed) surface. Such an encrusting colony follows the contours of the shell or rock on which it is growing and may reach a diameter of many centimeters. The colonies are commonly white or pale hues of red or other colors.

Erect colonies may be foliaceous, dendritic or straplike, but the ventral surface is still the one that is exposed; individuals are attached to one another by the dorsal surface. *Bugula,* one of the most common and widely distributed bryozoans and an important marine fouling organism, illustrates the erect growth form. The brownish colonies are branching and dendritic, and the individuals spiral around an axis formed by their attaching dorsal surfaces (Fig. 30.5). An overlapping arrangement keeps the anterior end bearing the orifice clear of other individuals.

Members of a colony are connected to each other more intimately than by the fusion of their exoskeletons. Pores in the walls permit diffusion of substances from one individual to another, and food material can be distributed to nonfeeding members of the colony (Figs. 30.1 and 30.4).

Bryozoans are **polymorphic,** but the degree

and nature of polymorphism vary. The most common members of the colony are the feeding individuals already described. The colonies of some species include highly modified defensive individuals that protect the colony against other settling organisms (Fig. 30.5*B*). Some species have vegetative stoloniferous individuals that creep over and anchor to the substratum (Fig. 30.5*A*). Reproductive individuals with special brooding structures are still another type found in many bryozoan colonies (Fig. 30.2).

30.3 REPRODUCTION

Most bryozoans are **hermaphroditic.** The gonads bulge into the coelom, with one or two ovaries at the anterior end and testes at the posterior end. There are no gonoducts; **gametes** are shed into the coelom and exit through a pore in one or two of the tentacles or through a special opening in the region of the lophophore. In many bryozoans, the latter opening is mounted on an elongate process called the **intertentacular organ,** and it is here that the eggs are released (Fig. 30.4)

In the very few species in which fertilization

has been observed, sperm swept in with the feeding current fertilized the eggs as they were released, or else the sperm entered and fertilized the eggs within the intertentacular organ.

Most bryozoans brood their few eggs during the early stages of development, commonly in special brooding chambers. For example, *Bugula* and other related forms brood their eggs within a large hoodlike chamber that develops at the anterior end of reproductive individuals (Fig. 30.2).

The developing eggs are released from the brood chambers as larvae that are somewhat like the trochophore of annelids and mollusks. Following a planktonic existence of varying length, the larva settles to the bottom to become the parent member of a new colony. The body wall evaginates and separates as a new individual, the particular way in which subsequent budding occurs accounting for the growth form of the colony.

During their life span, the individuals of a colony undergo phases of degeneration and regeneration, and a large colony typically exhibits zones of individuals in various stages of development. At the outer perimeter of an encrusting colony, for example, budding and development of new individuals occur. Further inward is a zone of fully developed individuals, which are feeding and reproducing.

Still further inward is a zone of degenerating members. The gut and lophophore are phagocytized, and the remains of these organs are lodged in the coelom as dark balls. A new gut and lophophore form from the persistent body wall in regenerating zooids, which form still another zone to the inside of the degenerating one.

30.4 OTHER LOPHOPHORATES

Bryozoans are not the only animals having a lophophore. Two other small phyla, the **Phoronida** and the **Brachiopoda,** are lophophorates. The phoronids are tubiculous, wormlike, marine animals (Fig. 30.6A and B). The brachiopods, although represented today by only a small number of species, were very abundant in the past, and their rich fossil record dates back to the Cambrian. Brachiopods superficially resemble bivalve mollusks in having the body enclosed within two calcareous valves (Fig. 30.6C). However, in brachiopods the valves are dorso-ventrally oriented and the ventral valve is larger than the dorsal one. A fleshy stalk, which emerges from a hole in the back of the ventral shell, attaches these animals upside down to the bottom.

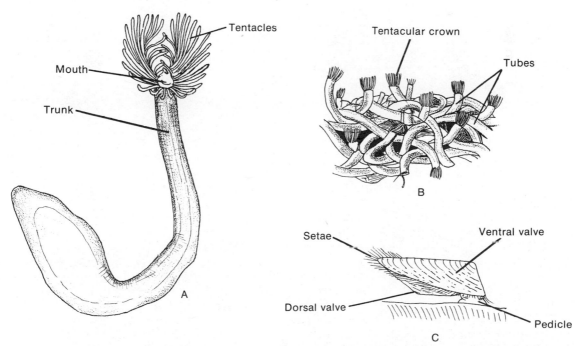

Figure 30.6 *A,* The phoronid, *Phoronis architecta,* removed from tube. *B,* Part of a cluster of *Phoronis hippocrepia.* *C,* Lateral view of the brachiopod *Discinisca* attached to substratum. (*A* after Wilson from Hyman; *B* after Shipley from Hyman; *C* after Morse from Hyman.)

ANNOTATED REFERENCES

Detailed accounts of bryozoans, phoronids and brachiopods can be found in many of the references listed at the end of Chapter 21. The works listed below deal exclusively with the biology of two of these lophophorate phyla.

Rudwick, M. J. S.: Living and Fossil Brachiopods. London, Hutchinson University Library, 1970.

Ryland, J. S.: Bryozoans. London, Hutchinson University Library, 1970.

ECHINODERMS

1. The Echinodermata is a phylum of marine animals including such familiar forms as starfish, brittle stars, sea urchins, sand dollars and sea cucumbers.

2. The symmetry is usually radial, typically pentamerous, or five-parted, but is secondarily derived. The larva is bilateral, and in the course of later development the symmetry shifts to a radial plan.

3. The body wall contains a skeleton of calcareous ossicles, which are usually movable against one another but may be fused together (sea urchins and sand dollars) or reduced to microscopic size (sea cucumbers). External spines are commonly present and are a part of the skeleton.

4. Echinoderms possess a unique water vascular system of internal canals and external appendages called tube feet. The system functions in locomotion, gas exchange, feeding and sensory reception in different groups of echinoderms.

5. Movement is by means of tube feet or spines or arm movement. Sea lilies are attached, and many fossil echinoderms also lived attached.

6. The mouth is in the center of one side of the radial body, and in most echinoderms this oral surface is directed toward the substratum. Diet, feeding mechanism and digestive tract vary greatly within the phylum.

7. Coelomic fluid is the principal means of internal transport, and exchange of gases and wastes between sea water and coelomic fluid takes place across various surface structures.

8. The nervous system is pentamerous and closely associated with the integument on the oral side of the animals.

9. The sexes are separate, and the reproductive system is simple. Fertilization is external, and early development is usually planktonic and leads to a bilateral larva bearing bands of cilia used in locomotion and feeding.

Echinoderms, hemichordates and chordates constitute the **deuterostomes,** the second great evolutionary line of the Animal Kingdom. In contrast to the protostomes, the mouth arises as a new opening located opposite the blastopore, which forms the anus. Cleavage is radial, not spiral (Fig. 5.7), and the fate of the blastomeres is fixed much later in development than in protostomes. Primitively, the mesoderm and coelom arise as paired outpocketings of the embryonic gut; the coelom is an **enterocoel** (Fig.

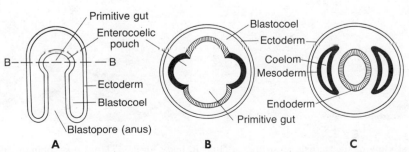

Figure 31.1 Mesoderm and coelom formation in deuterostomes. A, Lateral view of a gastrula. B, Section of gastrula (A) taken at level B–B and showing developing enterocoelic pouches, outpocketings of the primitive gut. C, Cross section of a later gastrula in which pouches have separated as mesodermal vesicles each containing a coelomic cavity.

Figure 31.2 Aboral view of six starfish, showing variation in body form. *A, Linckia*, a common genus with long arms; *B, Culcita*, a genus with such short arms that body is pentagonal; *C, Oreaster*, a large starfish with an elevated aboral surfaces; *D*, the coral-eating *Acanthaster; E, Astropecten*, a soft-bottom starfish; *F*, the sun star, *Crossaster. ,A, B* and *D* courtesy of the National Museum of Natural History; *C, E* and *F* by D. P. Wilson.)

31.1), in contrast to the schizocoelous mode of coelom formation in many protostomes (p. 604).

The more than 5000 species of the phylum Echinodermata represent the largest group of invertebrate deuterostomes, for most of the other deuterostomes are vertebrates. Echinoderms are entirely marine and include the familiar starfish, which are perhaps the invertebrate symbols of the sea, as well as brittle stars, sea urchins, sand dollars, sea cucumbers and sea lilies.

Echinoderms are almost entirely bottom dwellers, and hard substrata of rock, shell and coral were the habitats of many extinct and contemporary forms. But within each class of echinoderms are species that invaded soft bottoms and became adapted for life in sand. The adaptations for soft substrata are quite diverse and reflect the different prior adaptations of their ancestors.

31.1 CLASS STELLEROIDEA: ASTEROIDS

The class **Stelleroidea** contains those free-moving echinoderms in which the body is drawn out into arms. The most familiar stelleroids are members of the subclass Asteroidea, which include the starfish, or sea stars. In the members of this group the body is drawn out into arms that are not sharply set off from a central disc. There typically are five arms, which vary in length from species to species, but the sun stars have many arms (Fig. 31.2). The body surface may appear smooth or granular or may bear conspicuous spines, as in the coral-eating crown-of-thorns, *Acanthaster*. Starfish are usually brightly colored, with hues of red, orange and blue predominating. Most are 12 to 24 cm. in diameter.

The mouth is located in the center of the oral surface, which is directed downward. An **ambulacral groove** radiates from the mouth to the tip of each arm and contains the tube feet, the locomotor structures of sea stars (Fig. 31.3).

Body Wall. The body surface of asteroids is covered by a ciliated epithelium. Beneath the epithelium is a thick layer of connective tissue, which secretes the skeleton of small calcareous ossicles. The ossicles of all echinoderms are somewhat similar to vertebrate bone, for they are perforated by irregular canals filled by cells. In starfish, the ossicles form a movable skeletal lattice (Fig. 31.4).

A muscle layer below the dermis enables the arms to bend. Ciliated peritoneum forms the innermost layer of the body wall and lines the very large coelom that fills the interior of the arms and disc (Fig. 31.6).

A number of special structures are associated with the body wall. All asteroids bear calcareous spines that may be projections of the deeper dermal skeleton or special ossicles resting on top of the deeper skeleton (Figs. 31.2 and 31.5). The spines vary greatly in size and prominence. Even those species that seem to be smooth and spineless in fact have the body surface covered with many tiny, closely fitting spines.

Located between the spines are small finger-like projections of the body wall called **papulae,** which function in gas exchange and excretion (Figs. 31.5 and 31.6). The thin wall of each papula is composed of an inner peritoneum and outer epidermis, and coelomic fluid circulates within the interior.

Other structures associated with the body

Figure 31.3 Aboral view of two arms of the tropical starfish *Coscinasterias*. Tube feet project from the oral surface. (By Betty M. Barnes, from Barnes, R. D.: Invertebrate Zoology. 3rd ed. Philadelphia, W. B. Saunders Company, 1974.)

Figure 31.4 Lattice-like arrangement of the skeletal ossicles (enlarged) of a starfish. Surface view after overlying tissue had been removed. (After Fisher from Hyman.)

wall of many starfish are minute jawlike appendages called **pedicellariae.** A pedicellaria contains two ossicles that form the jaws; these are embedded in epithelium and provided with muscles for opening and closing the gape. The pedicellariae are believed to function in the killing of small organisms that might settle in the bodies of starfish. Such organisms, as well as sediment, are swept clear by the ciliated surface epithelium.

The Water Vascular System. The water vascular system, unique to echinoderms, is composed of tubular outpocketings of the body wall—the tube feet, or podia—and an internal system of canals derived from the coelom. The system opens to the exterior through the button-like aboral **madreporite,** which is perforated by tiny canals. The canals from the madreporite converge into a vertical **stone canal** that extends orally to a **ring canal** embedded in the ossicles around the mouth

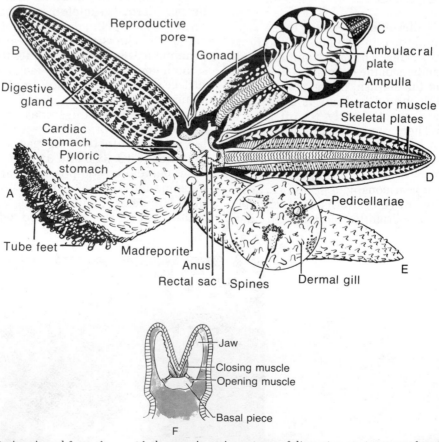

Figure 31.5 *Asterias* viewed from above with the arms in various stages of dissection. *A,* Arm turned to show lower side. *B,* Upper body wall removed. *C,* Upper body wall and digestive glands removed, with a magnified detail of the ampullae and ambulacral plates. *D,* All internal organs removed except the retractor muscles, showing the inner surface of the lower body wall. *E,* Upper surface, with a magnified detail showing surface features. *F,* Structure of a pedicellaria of the starfish *Asterias.*

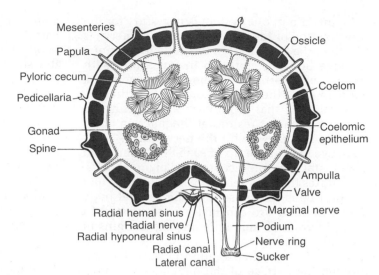

Mesenteries
Papula
Pyloric cecum
Pedicellaria
Gonad
Spine
Ossicle
Coelom
Coelomic epithelium
Ampulla
Valve
Marginal nerve
Radial hemal sinus
Radial nerve
Radial hyponeural sinus
Podium
Nerve ring
Sucker
Radial canal
Lateral canal

Figure 31.6 Diagrammatic cross section through the arm of a sea star.

(Fig. 31.7). From the ring canal a **radial canal** extends into each arm, passing between the ossicles at the top of the ambulacral groove. At frequent intervals, **lateral canals** leave the radial canals, those on one side alternating with those on the other (Figs. 31.6 and 31.7). Each lateral canal terminates in an aborally directed bulblike **ampulla,** which bulges into the coelom, and an orally directed tube foot, which projects into the ambulacral groove (Fig. 31.6).

Fluid pressure plays an important role in the functioning of the tube feet. Muscular contraction of the ampulla forces fluid into the podium and at the same time closes a valve in the lateral canal, preventing backflow. Hydraulic pressure extends the podium, and the sucker at its tip is brought into contact with the substratum. Following adhesion, longitudinal muscles of the podium contract, shortening the tube foot, forcing water back into the ampulla and

pulling the body forward. When a podium is extended, it is also swung forward. Thus each podium performs a little step and the combined activity of all the podia enables a starfish to grip objects tenaciously and to crawl about. The podia do not move synchronously, but they are coordinated to the extent that they all step in the same direction. One arm acts as the leading arm and exerts a temporary dominance over the other arms.

Nervous System, Gas Exchange and Excretion. The nervous system can be regarded as primitive in that it is intimately associated with the epidermal layer. The epidermis surrounding the mouth contains a pentagonal **nerve ring,** which at each angle supplies a **radial nerve** to an arm (Fig. 31.6). The radial nerve is located in the epidermis in the middle of the ambulacral groove. Fibers in the nerve ring and the radial nerve make connection with neurons of a general epidermal nerve plexus.

Madreporite
Stone canal
Ampulla
Podium
Sucker
Ring canal
Lateral canal
Radial canal

Figure 31.7 Diagram of the asteroid water vascular system.

An intact nervous system is necessary for the coordination of the tube feet. If a radial nerve is cut, the tube feet distal to the cut will continue to move but may not step in synchrony with those in other arms. If the nerve ring is cut, the podia in all of the arms become uncoordinated. It is believed that temporary arm dominance during movement resides at the junction point of the radial nerve with the nerve ring.

An **eye spot** at the tip of each arm, composed of a cluster of photoreceptor and pigment cells, constitutes the only sense organ, but individual receptor cells are present in the general body epidermis and are especially concentrated in the epidermis of the podia and the margins of the ambulacral groove.

Gas exchange and excretion in starfish occur through the surface of the papulae and podia, and internal transport is provided by the coelomic fluid. In almost all echinoderms, gas exchange organs are infoldings or outpocketings of the body wall, and coelomic fluid is the medium for internal transport. These gas exchange organs must have evolved independently in different echinoderm groups, for there is little consistency in their structure or location.

Soft-Bottom Starfish. Podia with suckers represent an adaptation for life on hard substrata such as rock, shell and coral — habitats in which starfish often abound. Many starfish are adapted for living on sandy bottoms and can even burrow to some degree. Soft-bottom starfish have doubled ampullae that provide greater pressure for thrusting the suckerless pointed tube feet into the sand.

The papulae of soft-bottom starfish are protected from the surrounding sand and sediment by special table-like spines that create a protective cover under which the papulae are lodged and a ventilating current flows (Fig. 31.8). It is these covering spines that give a smooth appearance to the surface of many soft-bottom starfish.

Nutrition. The digestive system of starfish reflects the radial plan of body structure. The mouth opens into a large thick-walled **cardiac**

Figure 31.8 Diagrammatic section through the surface structures of a soft-bottom asteroid. The large table-like spines bear smaller spines above and protect the papulae below. (From Barnes, R. D.: Invertebrate Zoology. 3rd ed. Philadelphia, W. B. Saunders Company, 1974.)

stomach that fills most of the central disc. The cardiac stomach opens in turn into a smaller aboral **pyloric stomach** (Fig. 31.5). A pair of digestive glands located in each arm discharges separately or by a common duct into the periphery of the pyloric stomach, giving it a star-shaped appearance. A short intestine extends from the top of the pyloric stomach to the inconspicuous anus at the center of the aboral surface. Associated with the intestine are rectal ceca, outpocketings of unknown function (Fig. 31.5).

Most starfish are carnivores and scavengers. Crustaceans, mollusks and other echinoderms are common prey. Species with short arms swallow their prey whole; those with longer arms wrap them about the body of the prey and by increased pressure of the coelomic fluid evert the cardiac stomach over the victim.

Asterias and other species of the same family feed largely on bivalve mollusks, penetrating their prey in a remarkable way. The starfish humps over the clam, the mouth directed over some part of the gape. The pull exerted by the arms produces a very slight opening between the valves, and through this opening the stomach of the starfish is slithered. These animals can pass the stomach through a gape of no more than 0.1 mm. It is this ability, rather than the ability to pull, that enables a starfish to prey upon clams. Many bivalves cannot close the valves tightly enough to prevent entrance of the starfish's stomach. The Pacific starfish *Pisaster* can penetrate the gapes of bivalves that have been wired closed!

Asteroids can be a serious pest of oyster beds and are sometimes removed by dragging a large mop across the bottom. The pedicellaria attach to the threads, and the starfish can be brought to the surface and destroyed.

Enzymes produced by the digestive glands are passed to the pyloric and cardiac stomachs. In those species that evert the stomach, the enzymes flow out onto the everted surface and initiate digestion outside the body. Thus, when a starfish attacks a bivalve, the digestion of the clam's adductor muscle causes the valves to gape widely. The stomach is later retracted, bringing with it the partially digested prey. The digestive glands appear to be the principal site of absorption.

Reproduction and Development. Like most other echinoderms, starfish have separate sexes; there are usually two gonads to an arm, with a simple gonoduct from each leading to an inconspicuous gonopore at the base of the arm (Fig. 31.5). Correlated with the simple nature of the reproductive system is the fact that there is

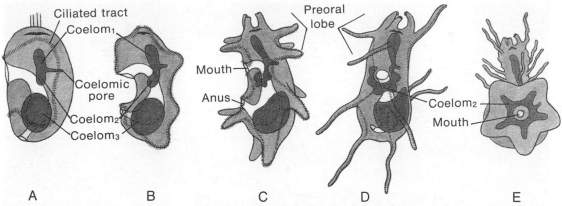

Figure 31.9 Diagrammatic lateral views of larval development and metamorphosis of a starfish, showing development of the coelom and water vascular system. *A* and *B*, Early bipinnaria larva. *C*, Brachiolaria larva. *D*, Attached metamorphosing larva. *E*, Young starfish developing from posterior part of old larva. (Adapted from various sources.)

no copulation, and the eggs are shed freely into the sea water where fertilization takes place. Development occurs in the plankton. The eggs contain little yolk material, and radial equal cleavage results in a hollow blastula. Gastrulation occurs by the invagination of cells of the vegetal hemisphere, resulting in a narrow finger-like archenteron. From the tip of the archenteron arise two lateral outpocketings, the enterocoelic pouches, which eventually separate as coelomic vesicles (Fig. 31.1). These primary vesicles later divide to form three pairs of vesicles. Their walls give rise to all of the mesodermal derivatives of the body. The remainder of the archenteron contributes to the gut proper and connects anteriorly with a stomodeal invagination from the ectoderm.

The appearance of ciliated bands on the body surface indicates that development has reached a larval stage, called a **bipinnaria** (Fig. 31.9). The bipinnaria larvae and the larvae of all other echinoderms are distinctly bilateral and are believed to represent recapitulations of the symmetry of ancestral echinoderms. The course of further development is marked by the appearance of **larval arms,** extensions of the body over which the ciliated locomotor bands run. The larval arms bear no relationship to the arms of the adult.

With the formation of an attachment device—three small anterior arms and a sucker—the bipinnaria larva becomes a **brachiolaria** larva. On encountering a suitable substratum, the brachiolaria anchors by the anterior adhesive structures, and a radical and complex metamorphosis ensues. The larval arms and much of the larval gut degenerate, and the adult structures, including the arms, arise from the posterior part of the larval body. After its metamorphosis the young starfish

becomes detached from the substratum and takes up the adult benthic existence.

Some arctic and antarctic starfish brood their eggs beneath or on various parts of the body. There are only a few large yolky eggs and development is direct. Such brooding behavior is found in many cold-water species of other echinoderm classes.

The regenerative ability of starfish is well known. Any fragment of the body that contains a portion of the central disc is capable of regeneration. But the process is slow and may take as long as a year. Only in some starfish is regeneration coupled with asexual reproduction, when the central disc is cleaved along certain predetermined lines and each part grows into a new starfish.

31.2 CLASS STELLEROIDEA: OPHIUROIDS

Closely related to the Asteroidea is the largest group of echinoderms, the **Ophiuroidea,** which contains the basket stars and brittle stars or serpent stars. The ophiuroid body, like that of the asteroids, is composed of arms and a central disc, but the arms are long, slender and sharply set off from the disc (Fig. 31.10). The arms are highly mobile and easily broken, characteristics that led to the names serpent star and brittle star. In basket stars, the arms are branched.

Although the central disc of basket stars may reach a diameter of 12 cm., most brittle stars are small and have a central disc no larger than a dime. They are commonly mottled or banded but are usually not as brilliantly colored as starfish.

A

B

Figure 31.10 *A,* A Caribbean brittle star, shown in repetitive flash photographs, pulls itself along with its two anterior arms and shoves with the other three. Ophiuroids are far more agile and flexible than starfish. *B,* The basket star, *Gorgonocephalus,* showing the branched arms. (*A* by Fritz Goro, courtesy of LIFE Magazine. © 1955, Time, Inc.; *B,* courtesy of the American Museum of Natural History.)

The arms of brittle stars are largely composed of a string of large ossicles, called **vertebrae** because of their superficial similarity to the bones in the vertebrate backbone (Fig. 31.11). The articulation of the arm ossicles and their musculature give the arms great mobility: brittle stars and basket stars move about by flexing their arms. The body is propelled forward by the backward movements of which-

ever arms are located laterally. Spines along the sides of the arms increase traction. An arm can be broken at any point if seized by a predator. This ability to undergo self-amputation, or **autotomy,** is an escape adaptation also found in some arthropods, such as crabs.

Ophiuroids are especially common in rock and coral habitats with rough uneven surfaces. Brittle stars and basket stars can climb, and

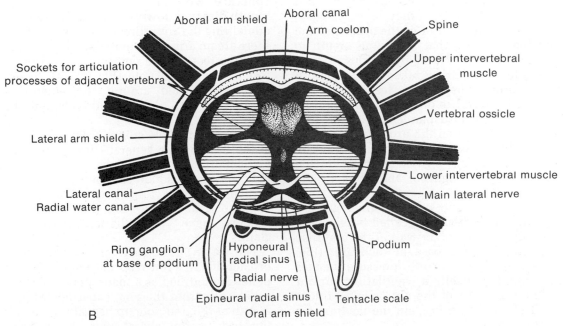

Figure 31.11 *A,* Oral view of disc of a brittle star. *B,* Cross section of a brittle star arm. (*A* after Streklov.)

Figure 31.12 Pluteus larva of the brittle star, *Ophiothrix fragilis.* (Courtesy of Douglas P. Wilson.)

some species clamber about sponges and branched gorgonian corals.

Some echinologists consider the ophiuroids the most successful of modern echinoderms and attribute their success in part to their motility, small size and ability to utilize the protective cover of crevices, holes and other natural retreats.

The **madreporite** of ophiuroids is in an ossicle located on the oral rather than the aboral surface. Correlated with the evolution of a different mode of locomotion and the reduction of the arm coelom is the fact that ampullae are absent, and the suckerless podia are extended by water pressure generated by the contraction of a section of the podial canal (Fig. 31.11).

Ophiuroids employ a variety of feeding methods; in fact, several methods may be used by a single species. They may be scavengers, using the looping motion of the arms and the five jawlike ossicles that frame the mouth to ingest the bodies of dead animals (Fig. 31.11). They may feed on detritus raked into the mouth with the arms. Suspension feeding is also common. The arms are often lifted up into the water current and suspended plankton and detritus are trapped on the arm surfaces or on strands of mucus draped between spines. The collected material is pushed as balls by the podia along the length of the arms to the mouth.

Ophiuroids have no papulae; their gas exchange organs are pouches, called **bursae**, that are infoldings of the body wall of the oral disc to either side of the base of each arm (Fig. 31.11). By means of cilia or by pressure changes within the disc itself, a ventilating current passes into and out of the slit opening into a bursa. Coelomic fluid is again the medium for internal transport.

Most brittle stars have separate sexes, but it is within this class that some hermaphroditic echinoderms are encountered. There are from one to many gonads attached to the inner wall of each of the 10 bursal sacs. Gametes reach the exterior by rupturing into the sac and passing out with the ventilating current. Brooding species use the bursae as brood chambers.

Fertilization occurs in the sea water and a bilateral larva develops as in asteroids. However, the number and position of the arms are different, and the larva can be recognized as an ophiuroid (Fig. 31.12). There is no attachment stage and the entire course of development occurs in the plankton. A tiny brittle star settles to the bottom on completion of metamorphosis.

31.3 CLASS ECHINOIDEA

The class **Echinoidea** contains the sea urchins, heart urchins, sand dollars and sea biscuits. The members of this class have a body that is not drawn out into arms but is spherical or discoidal (Fig. 31.13). The skeletal ossicles are flattened plates fused together to form a rigid internal shell, or **test,** and the body surface is covered with movable **spines** mounted upon tubercles on the test (Fig. 31.15).

The class contains two adaptive groups: the sea urchins, most of which are adapted for life on hard bottoms, and the sand dollars and heart urchins, which are adapted for burrowing in sand.

Sea Urchins. The body of a sea urchin is spherical with the oral pole containing the mouth directed downward, and the aboral pole containing the anus directed upward. Although there are no arms, five ambulacral areas bearing the tube feet radiate out and upward over the test to the aboral pole (Fig. 31.14). Thus the body is divided around the equator into alternating ambulacral and interambulacral meridians.

The long movable spines are covered at least at the base by the general surface epithelium. Pedicellariae are located between the spines, but, unlike those of asteroids, the echinoid pedicellariae have three jaws and are mounted at the end of a long stalk (Fig. 31.15). Some species have pedicellariae with poison glands.

The water vascular system differs from that of asteroids in a few respects. One of the aboral plates is modified as a madreporite. The radial canals run from the ring canal encircling the mouth at the oral pole up beneath the inside of the test to the aboral pole. Each ampulla connects with the podium by a pair of little canals, which actually perforate the ossicle (Fig. 31.15). Thus an ambulacral area can be recognized on a dried test by the two lines of paired holes.

Sea urchins use the podia to move about in the same manner as asteroids. The podia are long and can reach beyond the spines. Locomotion may also be aided by the pushing movement of the spines. Some sea urchins tend to settle themselves in rock depressions, and some actually excavate burrows in rock. Some kinds of sea urchins cover themselves with shells, pebbles and other debris.

When a sea urchin is alarmed, the spines become erect and immobile. Bending and erection are operated by two different sets of

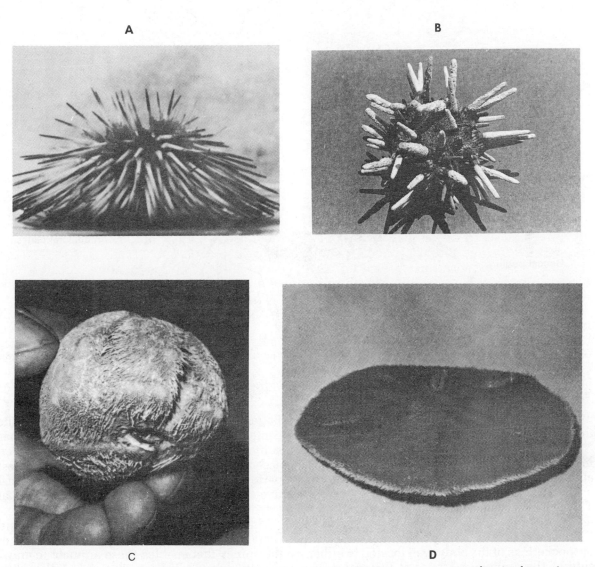

Figure 31.13 Echinoids. *A*, Side view of the common Atlantic sea urchin, *Arbacia punctulata*, showing long spines and podia. *B*, *Heterocentrotus*, a sea urchin with large spines and many small short spines. *C*, A heart urchin viewed from the anterior, or oral end. Mouth can be seen on the lower side. *D*, The Atlantic five-slotted sand dollar, *Mellita quinquies-perforata*. (By Betty M. Barnes.)

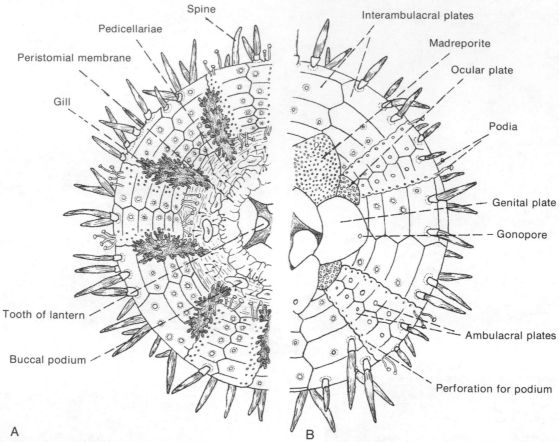

Spine
Pedicellariae
Peristomial membrane
Gill
Tooth of lantern
Buccal podium
A

Interambulacral plates
Madreporite
Ocular plate
Podia
Genital plate
Gonopore
Ambulacral plates
Perforation for podium
B

Figure 31.14   The common Atlantic coast sea urchin, *Arbacia punctulata. A,* Oral view; *B,* aboral view. (Modified from Petrunkevitch.)

muscle fibers and motor innervations around the socket of the spine.

Sea urchins feed on algae and encrusting animals, that are scraped up with a complex movable apparatus (known as **Aristotle's lantern**) containing five projecting teeth (Fig. 31.16). The gut is tubular and loops about inside the test.

Five pairs of gills, which are highly dissected outpocketings of the body wall located to either side of the ambulacral areas at the oral pole, provide for gas exchange (Fig. 31.14). Coelomic fluid is pumped into and out of the gills.

Sand Dollars and Heart Urchins. Sand dollars and heart urchins, the soft-bottom echinoids, burrow in sand. These animals, when moving, always keep the same meridian forward and thus have a definitive anterior end. Shifts in the position of the oral center or anus, or both, have led to a degree of bilateral symmetry, so that sand dollars and heart urchins are sometimes called **irregular echinoids.** In sand dollars, for example, the oral-aboral axis is so depressed that the body is

flattened (Fig. 31.13). The mouth is still in the center of the oral surface, but the anus has shifted out of the aboral center and is located eccentrically. Many sand dollars have the test perforated by slots, and in the Atlantic Coast five-slotted sand dollar, the anus is located at the inner margin of the posterior slot.

The spines of irregular echinoids are tiny and literally clothe the body; it is the movement of the spines that enables these animals to crawl slowly through the sand. Sand dollars live below the intertidal zone and burrow into the surface layer of sand. Heart urchins construct semipermanent burrows with a small opening to the surface (Fig. 31.18).

On the aboral surface, the podia of irregular echinoids are very wide and flattened as a modification for gas exchange. They form five petal-like areas, which are especially conspicuous in sand dollars. The dense clothing of spines extends just beyond the podial gills and prevents the gills from being smothered by sand.

Soft-bottom echinoids are **selective deposit**

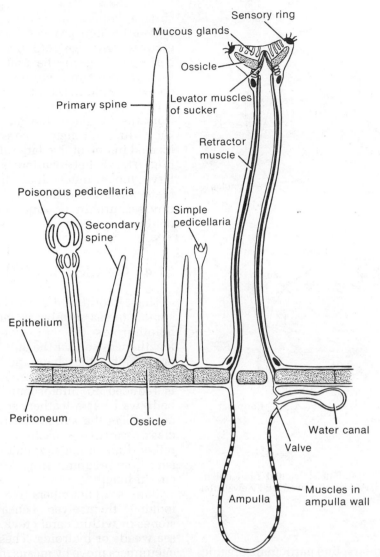

Figure 31.15 Diagrammatic section through the body wall of a sea urchin, showing one ambulacral and one interambulacral ossicle and associated structures. (After Nichols, in part.)

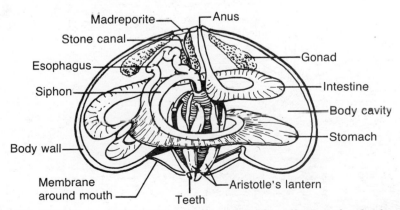

Figure 31.16 A sea urchin, *Arbacia punctulata*, with one side of the body wall removed and only some of the structures shown. The teeth protrude from Aristotle's lantern, of which only the outer structures are indicated. The digestive tract circles twice around the body, once in each direction. A second tube, the siphon, by-passes the esophagus and stomach. Each of the five gonads opens above, near the anus.

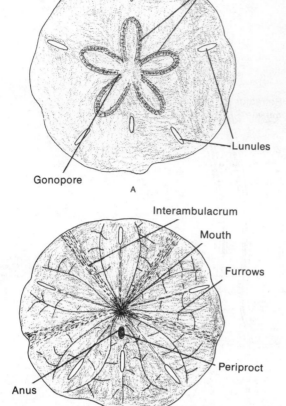

Figure 31.17 The West Indian six-slotted sand dollar, *Mellita sexiesperforata. A,* Aboral view; *B,* oral view. (After Hyman.)

feeders. In sand dollars, fine particulate organic matter, but not sand grains, can drop down between the dense clothing of spines. The particles are then driven by cilia to the oral surface, where they pass into a branching system of food grooves (Fig. 31.17). The food grooves converge into five principal grooves corresponding to the ambulacral areas. The food grooves are lined by podia that push the collected food masses to the mouth.

Four or five gonads are suspended radially from the inside of the test; each opens by a short duct through a conspicuous gonopore located in one of five large aboral genital plates (Fig. 31.14). Fertilization is external, and the larva looks much like that of ophiuroids. Metamorphosis occurs in the plankton, and a tiny sea urchin or sand dollar settles to the bottom.

31.4   CLASS HOLOTHUROIDEA

The Holothuroids, or **sea cucumbers**, are similar to sea urchins in lacking arms, but in holothuroids the oral-aboral axis is greatly lengthened, so that the animal has a wormlike or cucumber-like shape and lies on its side (Fig. 31.19). Unlike other echinoderms the skeleton is reduced to microscopic ossicles, and the body wall has a leathery texture. In many sea cucumbers the surface bears numerous tubercles or wartlike projections. Sea cucumbers are rather drab animals, with greenish and brownish colors predominating. Most are from 6 to 30 cm. in length.

Many sea cucumbers live on hard substrata, lodging themselves beneath and between stones or within coral crevices. Some attach to sea weeds or hydroids. These hard-bottom sea cucumbers move by means of tube feet; in some species, three ambulacra are kept against the substratum as a sole and the two upper ambulacra have reduced podia (Fig. 31.19).

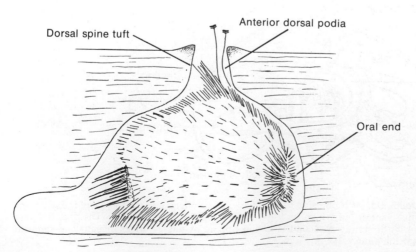

Figure 31.18 Heart urchin, *Echinocardium flavescens,* in its sand burrow (lateral view). (Modified after Gandolfi-Hornyold.)

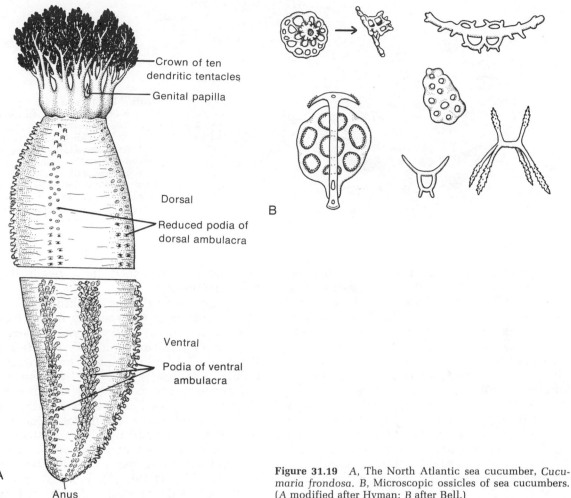

Crown of ten
dendritic tentacles

Genital papilla

Dorsal

Reduced podia of
dorsal ambulacra

Ventral

Podia of ventral
ambulacra

A

Anus

B

Figure 31.19 *A,* The North Atlantic sea cucumber, *Cucumaria frondosa. B,* Microscopic ossicles of sea cucumbers. (*A* modified after Hyman; *B* after Bell.)

Soft-bottom sea cucumbers bury themselves in sand. Many construct temporary burrows from which the oral end of the animal projects to the surface. The very wormlike synaptid sea cucumbers burrow beneath the surface by peristaltic contractions (Fig. 31.20). The podia have completely disappeared.

At the oral end of the body, a circle of **tentacles** representing modified podia surrounds the mouth (Fig. 31.19 and 31.21). The tentacles are outstretched and collect plankton or deposit material from the surrounding sea bottom. They are then retracted and stuffed one at a time into the mouth. The wormlike synaptid sea cucumbers are nonselective feeders; a column of sand is often visible through the transparent body wall.

The gut of sea cucumbers is tubular and terminates before the anus in a muscular **cloaca** that is involved in gas exchange. The gas exchange organs are unusual tubular branching structures called **respiratory trees** that arise as evaginations of the cloacal wall and extend up

into the coelom (Fig. 31.21). The pumping action of the cloaca moves a ventilating current of sea water into and out of the respiratory trees.

In the tropics, a little commensal pearlfish

Figure 31.20 A burrowing wormlike synaptid sea cucumber.

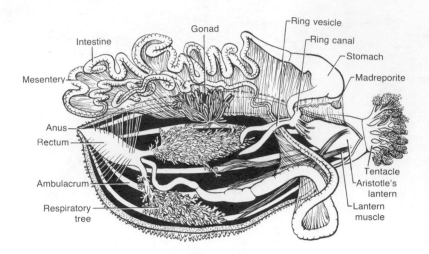

(*Carapus*) uses the base of the respiratory tree as a home. To enter its home, the fish nudges the anus of the sea cucumber with its snout and then backs in, tail first, sometimes twisting against the pressure of the closing anus. The fish has reduced scales and no pelvic fins, and the anus has shifted far forward.

Some sea cucumbers possess a cluster of tubular evaginations from the base of the respiratory trees. These tubules, which look like thin spaghetti, are called **tubules of Cuvier** and can be shot out of the anus. They elongate in the process and are very adhesive. An intruder or predator that disturbs a sea cucumber enough to evoke the discharge of the tubules can become enmeshed in a death trap of adhesive threads.

Another strange process exhibited by many sea cucumbers is the spewing out of internal organs, a process called **evisceration.** The respiratory trees and gut may be discharged through a ruptured cloaca or through the mouth. The loss is followed by regeneration. Evisceration has been most commonly observed in crowded laboratory conditions; its significance and the extent to which it is a normal event are still poorly understood.

There is only one gonad in sea cucumbers, and the gonopore opens between two of the tentacles (Fig. 31.19). The bilateral larva possesses ciliated locomotor bands but never develops the long larval arms characteristic of other echinoderm larvae (Fig. 31.22). Transformation to a young sea cucumber occurs before settling.

31.5 CLASS CRINOIDEA

The echinoderm classes we have examined thus far have the mouth directed downward and the oral surface placed against the substratum. However, in one living class of echinoderms, the **Crinoidea,** the reverse is true: the oral surface is directed upward. Moreover, many members of this class are attached to the substratum by a **stalk.** The attached crinoids are believed to be the most primitive of the living echinoderms and probably illustrate the ancestral mode of existence of the phylum and the original function of the water vascular system.

Crinoids consist of two groups: the sessile **sea lilies,** in which the aboral surface is connected to the substratum by a stalk (Fig. 31.23),

Figure 31.22 Auricularia larva of the sea cucumber *Labidoplax*. (Courtesy of Douglas P. Wilson.)

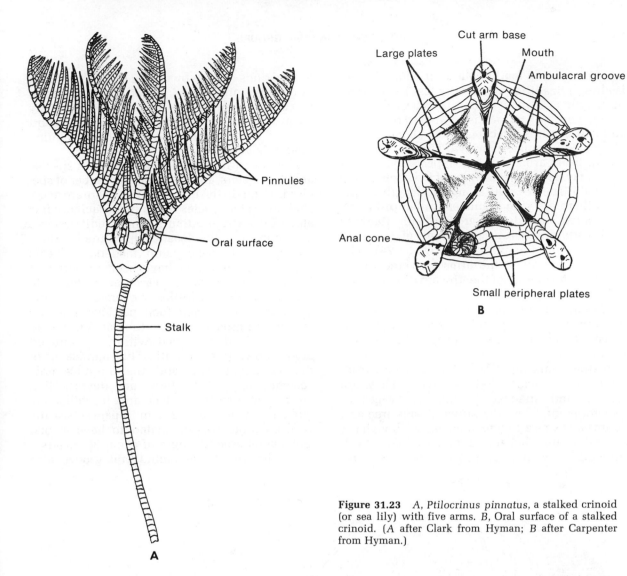

Pinnules

Oral surface

Stalk

A

Large plates

Cut arm base

Mouth

Ambulacral groove

Anal cone

Small peripheral plates

B

Figure 31.23 *A, Ptilocrinus pinnatus,* a stalked crinoid (or sea lily) with five arms. *B,* Oral surface of a stalked crinoid. (*A* after Clark from Hyman; *B* after Carpenter from Hyman.)

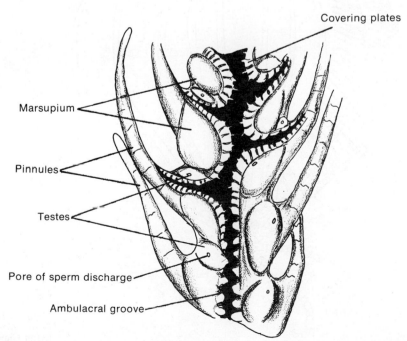

Covering plates

Marsupium

Pinnules

Testes

Pore of sperm discharge

Ambulacral groove

Section of arm and pin-
showing ambulacral

the free-moving **feather stars,** in which a stalk is absent. Sea lilies are usually less than 70 cm. in length and most occur in relatively deep water. The stalk, which is composed of ossicles and can bend, is attached to the aboral surface of the body proper, or **crown.** The pentamerous crown bears arms that fork repeatedly, so that there may be many multiples of five, depending upon the species. All along the length of the arms there are side branches, called **pinnules.** A ciliated ambulacral groove with flanking suckerless podia runs the length of the arms and up onto the pinnules (Fig. 31.24). The grooves from all of the arms converge to the mouth in the center of the upward directed oral surface.

The arms are composed mostly of ossicles, but the articulation and musculature permit considerable movement, especially vertical movement.

Feather stars are similar to sea lilies, from which they clearly have evolved. They are stalked and attached in the last stages of development. Then the crown breaks free and swims away as a tiny feather star. Although the animal is unattached, its oral surface is still directed upward (Fig. 31.25). They perch for

long periods of time by means of an aboral ring of clawlike projections, called **cirri,** on rocks, on coral or even on soft substrata. Feather stars swim intermittently. The arms are raised and lowered in cycles of five, and the swimming motion is very graceful.

The largest number of living crinoid species are feather stars, and the greatest number of species is found in the Indo-Pacific. There are many shallow-water species in the Indo-Pacific, a few along the eastern Atlantic and Mediterranean but none along the coast of the United States.

Crinoids are suspension feeders, and this mode of feeding may have been the original function of the water vascular system. On contact, rapid whiplike movements of the podia toss suspended food particles into the ambulacral grooves, where they are entrapped in mucus and conveyed within the ciliated groove down to the mouth. The significance of the branching arms and the pinnules now becomes apparent. These are devices that greatly increase the surface area for collecting food. For example, a 24 cm. Japanese sea lily with 56 arms was estimated to have a total ambulacral groove length of about 80 meters.

The length of the ambulacral groove also

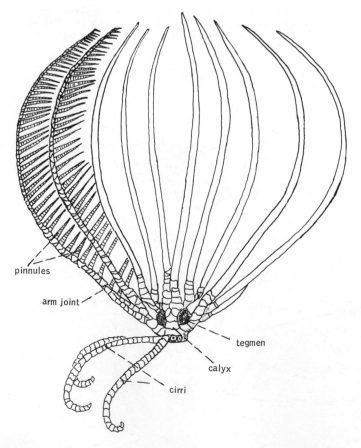

pinnules

arm joint

tegmen

calyx

cirri

Figure 31.25 A Philippine thirty-armed comatulid (or feather star), *Neometra acanthaster.* (After Clark from Hyman.)

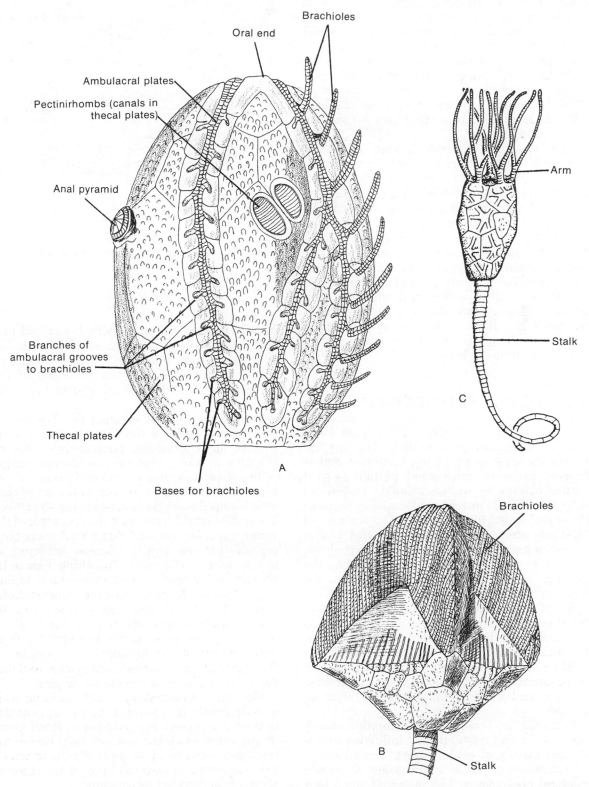

Figure 31.26   Fossil echinoderms. *A*, Lateral view of a cystoid; *B*, lateral view of a blastoid; *C*, an eocrinoid. (*B* after Jackel from Hyman; *C* after Nichols.)

reflects the differences in the biological productivity of tropical and cold-water oceans. In general, biological productivity as reflected by plankton density is greater in coastal waters than in the open ocean and is greater in cold than in warm oceans. These differences are reflected in crinoid structure. In general, cold-water crinoids have fewer and shorter arms than do tropical species; the greater plankton content demands less surface area for suspension feeding.

The anus of crinoids is located on a short elevation to one side of the central oral surface (Fig. 31.23). Feces can thus be more readily swept away without fouling the ambulacral grooves. Podia are the gas exchange organs.

Gametes develop from coelomic epithelium in the arms and pinnules, and spawning takes place by rupture of the body wall. Brooding species retain their eggs on the arms. The larval stage has ciliated bands but no larval arms. The larva eventually settles and attaches, and metamorphosis occurs during attachment, as in asteroids. Sea lilies, of course, remain permanently attached, whereas the crown of feather stars is eventually freed.

31.6 FOSSIL ECHINODERMS

Echinoderms have one of the richest fossil records of any group of animals. The phylum makes its appearance in the Cambrian and is present in every subsequent period. Several extinct classes, as well as crinoids, flourished in great numbers during the Paleozoic. Some of these earlier echinoderms, such as the extinct blastoids and cystoids, had spherical bodies with the ambulacral grooves extending down over the body surface. Significantly, the grooves were bordered with **brachioles,** the equivalent of crinoid pinnules (Fig. 31.26).

All of this seems to suggest that these early fossil echinoderms fed in the same manner as do modern crinoids and that the original function of the water vascular system was feeding. It could scarcely have functioned for locomotion, for the wrong side was directed upward and most of these animals were attached.

From the fossil record and embryonic development we can postulate the following evolutionary history of echinoderms. The ancestors of echinoderms were some group of motile bilateral coelomates. The ancestral stock took

Figure 31.27 One of several projected hypothetical ancestors of echinoderms. (After Bury from Hyman.)

up an attached existence, which resulted in a shift from a bilateral to a more adaptive radial symmetry. Also correlated with an attached mode of existence was the evolution of the calcareous skeleton and of suspension feeding.

We are not certain whether the first echinoderms had arms for the oldest known echinoderms, the eocrinoids, possessed brachioles. We also do not know the evolutionary origin of the water vascular system and podia. Structures seldom arise in evolution *de novo;* they arise as modifications of pre-existing structures. What then could have been the precursor of the water vascular system? Some zoologists have argued that the motile bilateral ancestors of echinoderms possessed something like a lophophore for feeding and that such a lophophore-like structure became the water vascular system (Fig. 31.27). They suggest that the cavity of each tentacle became a radial water canal. As evidence, they point to the fact that the coelomic cavity of the lophophore is similar to the cavity of the water vascular system and that the two have similar embryonic origins.

Modern free-moving echinoderms — asteroids, brittle stars and echinoids — appear late in the fossil record, and we have no fossil forms or any other evidence that can help bridge the great gap between those echinoderms in which the oral surface is directed upward and those in which it is directed downward.

CLASSIFICATION OF THE PHYLUM ECHINODERMATA

Subphylum Echinozoa. Radially symmetrical, commonly globoid echinoderms without arms or armlike appendages.

CLASS HELICOPLACOIDEA. Extinct Paleozoic echinoderms. Spindle-shaped, with a body wall composed of pleated plates.

CLASS EDRIOASTEROIDEA. Extinct Paleozoic echinoderms. Attached or free; oral surface directed upward.

CLASS OPHIOCYSTIOIDEA. Extinct Paleozoic echinoderms. Helmet-shaped test; podia covered with scales.

CLASS HOLOTHUROIDEA. Sea cucumbers. Mouth and anus at opposite ends of cucumber-shaped body. Oral end bearing tentacles derived from podia. Ossicles microscopic.

CLASS ECHINOIDEA. Sea urchins and sand dollars. Body globose or greatly flattened dorso-ventrally. Ossicles fused to form a rigid test.

Subphylum Homalozoa. Extinct irregular Paleozoic echinoderms. In some the stalked body was bent over so that one side was directed toward the substratum.

Subphylum Crinozoa. Radially symmetrical globoid echinoderms with arms or brachioles. Oral surface directed upward.

CLASS EOCRINOIDEA. Extinct Paleozoic echinoderms. Body attached directly or by a stalk to the substratum. The upper or oral end carried five to many brachioles.

CLASS CYSTOIDEA. Extinct Paleozoic echinoderms. Body attached directly or by stalk to substratum. Three or five ambulacra extended down over body surface. Ambulacra commonly bordered by brachioles. Body wall perforated by pores, which may have constituted gas exchange system.

CLASS BLASTOIDEA. Extinct Paleozoic echinoderms. Body attached directly or by stalk to substratum. Closely placed brachioles bordered the five ambulacra, which were typically elevated on a ridge, giving the body a pentagonal shape.

CLASS CRINOIDEA. Sea lilies and feather stars. Living and fossil echinoderms dating from the early Paleozoic. Body attached by a stalk (sea lilies) or free (feather stars). Ambulacra located on arms, which are typically branched.

Subphylum Asterozoa. Radially symmetrical, unattached star-shaped echinoderms. Oral surface directed downward.

CLASS STELLEROIDEA

Subclass Somasteroidea. Extinct starfish.

Subclass Asteroidea. Living starfish. Arms containing a large coelomic cavity not sharply set off from central disc. Tube feet located within an ambulacral groove.

Subclass Ophiuroidea. Brittle stars. Arms containing a canal-like coelomic cavity, sharply set off from central disc. Ambulacra not constructed as a groove.

ANNOTATED REFERENCES

Detailed accounts of the echinoderms can be found in many of the references listed at the end of Chapter 21. The following works deal with echinoderms alone.

Boolootian, R. A. (Ed.): Physiology of Echinodermata. New York, John Wiley, 1966. An excellent review of echinoderm ecology, behavior and physiology.

Clark, A. M.: Starfishes and Their Relations. London, British Museum, 1962.

Nichols, D.: Echinoderms. London, Hutchinson University Library, 1969. A good short biology of the phylum.

CHAPTER 32

CHORDATES AND HEMICHORDATES

The Chordata, the largest of the deuterostome phyla, includes animals with three distinguishing characteristics (see Fig. 32.8): (1) a dorsal hollow **nerve cord,** which arises as an infolding of the surface ectoderm; (2) a dorsal longitudinal skeletal rod, the **notochord,** located beneath the nerve cord; and (3) paired lateral openings, commonly referred to as **gill clefts** or **slits,** through the pharyngeal wall of the gut. Although these three structures are found in the early developmental stages of all chordates and were probably characteristic of the ancestral chordates, not all persist in the adults. The notochord disappears during development in most vertebrates, and only the hollow nerve cord and the pharyngeal clefts, or their derivatives, remain in the adult.

Most of the chordates are vertebrates, but there are two interesting small subphyla of invertebrate chordates, i.e., chordates lacking a vertebral column or backbone.

32.1 SUBPHYLUM UROCHORDATA

The members of the subphylum **Urochordata,** called **tunicates** or **ascidians,** are the larger group of invertebrate chordates. Most tunicates are sessile, attached to rocks, shells, pilings and ship bottoms.

The planktonic larva of urochordates is about 0.7 mm. in length and looks like a tiny fish or tadpole. In fact, it is sometimes called a "tadpole" larva. The finned tail is the locomotor organ, and it contains longitudinal muscle fibers, the nerve cord and the notochord (Fig. 32.1). The notochord prevents a shortening, or telescoping, of the tail when the muscle fibers contract and converts muscle contraction into lateral undulations of the tail. The anterior

mouth, which may not be open during the larval stage, is connected to a large **pharynx** that is perforated by two paired lateral clefts. The clefts lead into another chamber, the **atrium.**

At the end of its planktonic existence, the larva settles to the bottom and becomes attached by anterior adhesive papillae. A radical metamorphosis ensues and the larva develops into the very different adult body form. In the course of the transformation, the tail, including the notochord and neural tube, is resorbed and disappears; the pharynx develops into a large food-collecting chamber; and growth on the anterior dorsal sides of the body is so much greater than elsewhere that the mouth ends up almost 180° from its original position (Fig. 32.1).

Adult tunicates have more or less spherical bodies, which range in size from a pinhead to a small potato; in fact, some species actually bear a strong resemblance to potatoes. The body is covered by a thick envelope, or **tunic,** containing **tunicine,** a form of cellulose. Cellulose is rarely encountered in animals. The consistency of the tunic may be gelatinous, but quite often it is heavy and leathery, even rigid. Many tunicates are drab, but some are bright orange, lavender, green or other colors.

One end of the urochordate body is attached to the substratum; the opposite end contains two openings: the **buccal** and **atrial siphons** (Fig. 32.2). The buccal siphon opens into a large pharyngeal chamber, which fills the upper half or two-thirds of the body of the animal. The wall of the pharynx is perforated by such a vast number of openings that it resembles a grid. The perforations open into an outer surrounding chamber, the **atrium,** connected to the exterior through the atrial siphon. Through the pharyngeal and atrial

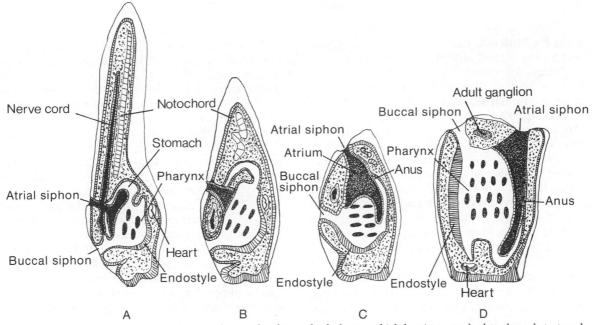

Figure 32.1 *A,* Diagrammatic lateral view of a urochordate tadpole larva, which has just attached to the substratum by the anterior end. *B* and *C,* Metamorphosis. *D,* A young individual just after metamorphosis. (*B* modified after Seeliger from Brien.)

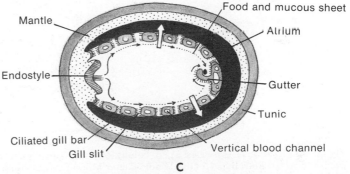

Figure 32.2 *A,* External view of a solitary tunicate. Diagrammatic lateral view (*B*) and cross section (*C*) of a tunicate showing major internal organs. Large arrows represent the course of the current of water; small arrows that of the food and mucous sheet. The stomach, intestine and other visceral organs are embedded in the mantle.

chambers flows a current of water, entering through the buccal siphon and leaving through the atrial siphon. The body below the pharynx, called the **abdomen**, contains the **stomach, intestine, heart** and other internal organs (Fig. 32.2)

Nutrition. Tunicates are **filter feeders,** removing plankton from the water stream passing through the pharyngeal chamber. The current is produced by the beating of cilia that border the margins of the pharyngeal perforations. A deep groove, the **endostyle,** extends down the length of the pharyngeal chamber on the side into which the buccal siphon opens (Fig. 32.2). Mucus secreted by the endostyle is driven by cilia out of the groove and over the inner pharyngeal surface as a film. Plankton is trapped in the film and carried across the pharynx by frontal cilia on the gridlike bars

that form its walls. The plankton-laden film eventually collects within a deep gutter on the opposite side of the pharynx. Cilia within the gutter carry the collected material downward to the opening of the esophagus at the bottom of the pharynx (Fig. 32.2).

An enormous amount of water passes through the pharyngeal basket of a tunicate and serves not only for feeding but also for gas exchange. A specimen of *Phalusia* only several centimeters long was estimated to circulate 173 liters of water through the body in 24 hours. The current can be regulated or halted by opening or closing the siphons. Moreover, some small contraction of the body is possible in species that do not have an especially heavy and rigid tunic. Exposed intertidal species may suddenly eject a spurt of water from the siphons, hence one of their names, sea squirts.

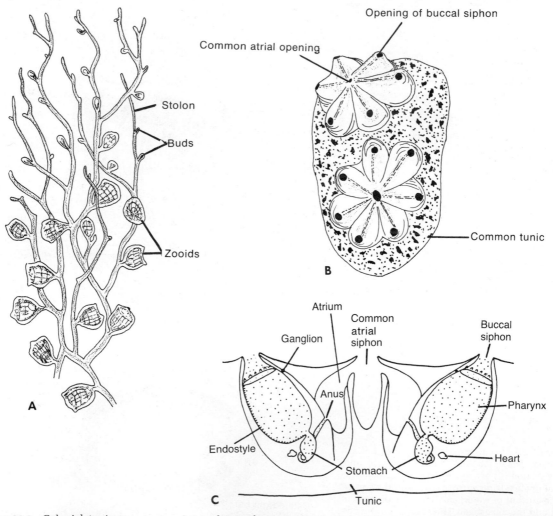

Figure 32.3 Colonial tunicates. *A,* A colony of *Perophora viridis; B,* surface view of *Botryllus schlosseri; C,* vertical section through *Botryllus.* (*A* after Miner; *B* after Milne-Edwards from Yonge; *C* after Delage and Herouard.)

The gut is **U**-shaped and a large stomach occupies the bend of the U (Fig. 32.2). An intestine leads back upward and opens into the atrium. Fecal matter is swept away by the water current in the atrial siphon.

Excretion and Internal Transport. There are no special organs for excretion of nitrogenous waste. Most escapes by diffusion; some accumulates as inert uric acid crystals in various parts of the body. Tunicates have an open internal transport system, but the blood follows distinct channels, which course through the pharyngeal basket, abdominal organs and, in some species, even the tunic. The heart, located near the stomach, periodically reverses its beat and the flow of blood, a most unusual phenomenon. Apparently, excitation centers at each end of the heart alternately dominate each other, but the significance of the reversal of flow is not understood.

The blood of some tunicates possesses green cells, which contain large amounts of vanadium. These animals have the ability to remove and concentrate an element that is present in sea water in only trace amounts. The vanadocytes are involved in the production of components of the tunic.

Colonial Tunicates. Many common tunicates are solitary, but a considerable number are colonial. Some colonial tunicates are connected by **stolons** and resemble a vine creeping over the substratum (Fig. 32.3). Others have bodies arranged in clusters; still others are so intimately connected that they are embedded within a common tunic. Such colonial tunicates have the individuals symmetrically arranged, and the atrial siphons open into a common cloacal chamber. Some colonial tunicates are unattached and swim using the collective feeding current expelled through a common opening as a water jet for locomotion.

Reproduction. Most tunicates are capable of asexual reproduction by an often extremely complex process of **budding.** All species are **hermaphroditic** with a single abdominal ovary and testis; the oviduct and sperm duct open into the atrium (Fig. 32.2). Fertilization may be external with planktonic development, especially in solitary species. Alternatively, fertilization may take place within the atrium and the eggs brooded there through early embryonic stages. Development eventually leads to the tadpole larva already described.

32.2 CHORDATE METAMERISM

Although members of another subphylum, the **Cephalochordata,** lack a backbone, they

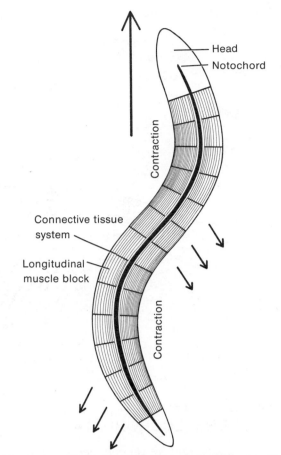

Figure 32.4 Diagrammatic frontal section of a primitive segmented chordate showing pattern of muscle contraction in undulatory swimming. Large arrow indicates direction of movement; small arrows indicate direction of propulsive force generated by wave of muscle contraction sweeping from anterior to posterior.

do share with vertebrates the condition of **metamerism.** Metamerism appears to have evolved early in chordate history in the line leading to the cephalochordates and vertebrates but after the divergence of the urochordates. As in annelids, segmentation developed as an adaptation for locomotion—but for undulatory swimming rather than burrowing (Fig. 32.4).

As in annelids, it is the musculature that exhibits the primary segmentation. The segmentation of the nervous system and blood vascular systems represents an adjustment of these systems to supply the muscle blocks. Segmentation in chordates does not involve the coelom. The coelom in chordates is not utilized as a localized hydrostatic skeleton; rather it is the notochord against which the muscle blocks are indirectly pulling.

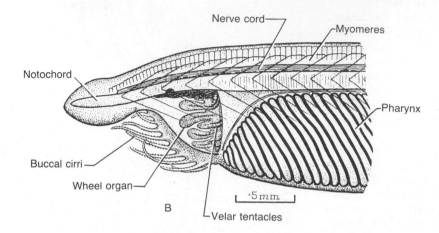

Figure 32.5 *Amphioxus. A,* A diagrammatic lateral view. White arrows represent the course of the current of water; black arrows that of the food. Further details of the course of food particles in the gut are shown in Figure 32.7. *B,* An enlarged view of the anterior end of *Amphioxus,* cleared and stained to show internal structures. (*B* modified after Yonge.)

32.3 SUBPHYLUM CEPHALOCHORDATA

Amphioxus and a related genus of small superficially fish-shaped chordates constitute the subphylum **Cephalochordata.** Species occur in United States coastal waters south from the Chesapeake (*Amphioxus virginiae*) and Monterey (*A. californiense*) Bays. They usu-

ally lie buried in sand with only their anterior end protruding, but they can also swim fairly well.

The body of *Amphioxus* (Figs. 32.5 and 32.6) is elongated, tapering at each end, and compressed from side to side. Small median fins and a pair of lateral finlike **metapleural folds** are present, but apparently they are neither large nor strong enough to keep the

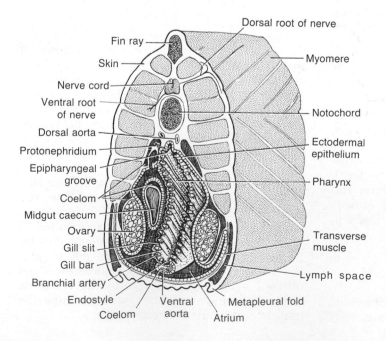

Figure 32.6 A diagrammatic cross section through the posterior part of the pharynx of *Amphioxus.* Branchial arteries extend from the ventral aorta through the gill bars to the dorsal aortas. The portion of the coelom ventral to the endostyle is connected through alternate gill bars with the pair of coelomic canals lying dorsal to the atrium. Other parts of the coelom are associated with the midgut caecum and gonads.

animal on an even keel, for *Amphioxus* spirals as it swims. Swimming is accomplished by the contraction of the muscle blocks, or **myomeres.** Shortening of the body is prevented by an unusually long notochord that extends farther anteriorly than in any other chordate, an attribute after which the subphylum is named.

Nutrition. *Amphioxus* is a **filter feeder.** Water and minute food particles are taken in through the **oral hood,** whose edges bear a series of delicate projections, the **cirri,** that act as a strainer to exclude larger particles (Fig. 32.5). The inside of the oral hood is lined with bands of cilia, the **wheel organ.** These, together with cilia in the pharynx, produce a current of water that enters the mouth. The mouth proper lies deep within the oral hood surrounded by **velar tentacles.**

Food is entrapped within the pharynx in mucus secreted by an endostyle just as it is in urochordates. Water in the pharynx escapes into an atrium through numerous gill slits. Some gas exchange occurs in the pharynx, but the skin is the major respiratory surface. The pharynx is primarily a food-gathering device. Food particles carried back into the intestinal region are sorted out by complex ciliary currents. Large particles continue posteriorly, but small ones are deflected into the **midgut caecum,** a ventral outgrowth from the floor of the gut, and drift ventrally along the lateral walls of the caecum (Fig. 32.7). Many of these particles are ingested by caecal cells and are digested intracellularly. Undigested material, together with enzymes secreted by certain caecal cells, is carried posteriorly to join the mass of larger food particles. As this mass rotates and is degraded by enzymic action, small particles are carried forward to the caecum to be engulfed and digested. Material not broken down in the midgut is carried posteriorly. Much of this is undigestible detritus, but some additional digestion and absorption may occur in the posterior part of the intestine. Fecal material is discharged through an **anus,** which, as in vertebrates, lies slightly anterior to the posterior tip of the body.

Internal Transport. Absorbed nutrients are distributed by a circulatory system. A series of veins returns blood from the various parts of the body to a sinus that is located ventral to the posterior part of the pharynx and may be comparable to the posterior part of the vertebrate heart. No muscular heart is present; blood is propelled by the contraction of the arteries. A ventral aorta extends from the sinus forward beneath the pharynx and connects with branchial arteries that extend dorsally through the gill bars to a pair of dorsal aortas (Fig. 32.6). The dorsal aortas in turn carry the blood posteriorly to spaces within the tissues. True capillaries are absent, but the general direction of blood flow, i.e., anteriorly in the ventral part of the body and posteriorly in the dorsal part, is similar to that of a vertebrate and different from that of other animals.

The excretory organs are segmentally arranged ciliated **protonephridia** that lie dorsal to certain gill bars and open into the atrium (Fig. 32.6).

Nervous System. The nervous system of *Amphioxus* consists of a tubular nerve cord located dorsal to the notochord (Figs. 32.5 and 32.6). Its anterior end is differentiated slightly but is not expanded to form a brain. Paired segmental nerves, consisting of dorsal and ventral roots as in vertebrates, extend into the tissues. The ventral roots go directly into the myomeres, and the dorsal roots pass between myomeres to supply the skin, gut wall and ventral parts of the body. *Amphioxus* is sensitive to light and to chemical and tactile stimuli, but no elaborate sense organs are present. The cirri on the oral hood and a flagellated pit in the skin near the front of the nerve cord appear to be chemoreceptors. Photoreceptive cells, which are partly masked with pigment, lie in the nerve cord. The prominent pigment spot at the anterior end of the cord apparently does not function in light reception.

Reproduction. The sexes are separate in *Amphioxus,* and numerous testes or ovaries bulge into the atrial cavity. Actually, they lie within a portion of a highly modified coelom

Figure 32.7 Diagram of food particle movements in the intestine and midgut caecum. (From Barrington.)

(Fig. 32.6). The gametes are discharged into the atrium upon the rupture of the gonad walls. Fertilization and development are external.

32.4   SUBPHYLUM VERTEBRATA

The Vertebrata is by far the largest and most important of the chordate subphyla, for all but about 2000 of the approximately 41,000 living species of chordates are vertebrates. The eight classes of vertebrates (p. 764) share with the lower chordates the three diagnostic characteristics of the phylum. These are clearly represented at some stage in the life history of the various groups. The dorsal tubular nerve cord has differentiated into a brain and spinal cord, present in the embryos and adults of all species (Fig. 32.8). Embryonic vertebrates have a notochord lying ventral to the nerve cord and extending from the middle of the brain nearly to the posterior end of the body, but a vertebral column replaces the notochord in most adults. All embryonic vertebrates have a series of **pharyngeal pouches** that grow laterally from the walls of the pharynx, but these pouches break through the body surface to form gill slits only in fishes and larval amphibians.

Vertebrates evolved as a group of chordates that became more active and developed more aggressive ways of feeding. Most of the distinctive characteristics of vertebrates are related to these changes in mode of life. Their greater activity is reflected in the replacement of the notochord by a vertebral column, in the continued elaboration of the segmented muscular system seen beginning in *Amphioxus*, in an aggregation of nervous tissue (the brain) and elaborate sense organs at the anterior end of the body and in the protection of these organs by a brain case, or **cranium.** An alternate name for the subphylum, the Craniata, emphasizes this last feature.

Early vertebrates probably continued a filter-feeding mode of life but used muscular movements of their pharynx, and not simply ciliary currents, to draw in water and food. This was certainly more efficient and made possible an increase in size. Larger size and the eventual evolution of jaws permitted a yet more active and aggressive mode of life. Respiratory organs (gills or lungs) replaced the general body surface as sites of gas exchange. A muscular **heart** developed that pumps blood effectively through a closed circulatory system. Wastes are excreted by a pair of **kidneys** (Fig. 32.8) composed of numerous tubules quite unlike any invertebrate excretory organ. Vertebrates have become the most successful and dominant group of chordates.

32.5   PHYLUM HEMICHORDATA

We must now examine one last group of deuterostomes, the phylum **Hemichordata**, for the members of this phylum provide some additional evidence linking echinoderms and chordates and must be considered in speculating about chordate origins.

Figure 32.8   A diagrammatic sagittal section through a generalized vertebrate to show the characteristics of vertebrates and the arrangement of the major organs.

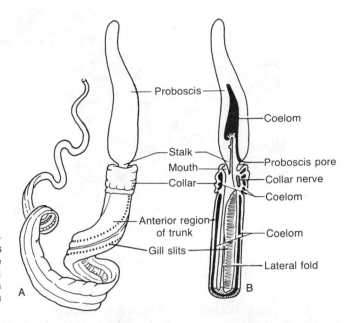

Figure 32.9 Phylum Hemichordata (genus *Saccoglossus*). *A,* External view showing external features (after Bateson). *B,* A diagrammatic section through the anterior part of the body showing some of the internal organs. A lateral fold subdivides the pharynx into a ventral channel along which the sand passes and a dorsal channel containing the gill slits.

Most hemichordates are marine wormlike animals that are found beneath stones and shells, in burrows in sand and mud and in tubes in a few species. The body is composed of three regions: an anterior **proboscis,** a middle **collar** and a long posterior **trunk** (Fig. 32.9). The rounded or pointed proboscis is simply the anterior end of the body and in this respect is equivalent to the prostomium of annelids. The shape of the proboscis of some species has resulted in their being called **acorn worms.** At

the back of the proboscis on the ventral side is the mouth. It lies just in front of the short bandlike collar region. The trunk follows the collar and makes up the greater part of the body.

Hemichordates possess **gill clefts.** These are present as a line of many perforations on either side of the anterior end of the trunk (Fig. 32.9). A stream of water enters the mouth, passes into the pharynx, and then exits through the gill pores, driven by cilia. Al-

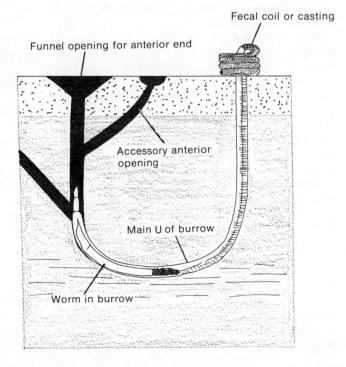

Figure 32.10 Burrow system of the Mediterranean *Balanoglossus clavigerus.* (After Stiasny from Hyman.)

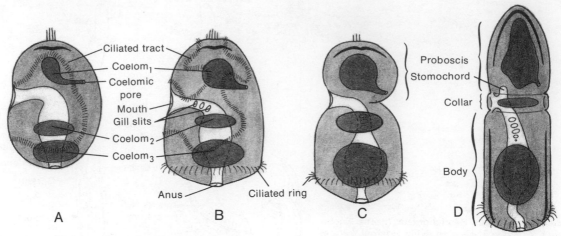

Figure 32.11 Diagrammatic side views of the larval development of a hemichordate. Compare with Figure 31.9. *A*, Early larva; *B*, later larva; *C* and *D*, metamorphosis.

though hemichordates do not contain gills, gas exchange occurs as water passes through the pharyngeal clefts; this appears to be their only function.

Nutrition. Hemichordates feed largely on suspended and deposit material that adheres to mucus on the surface of the proboscis and is driven by cilia into the mouth along with the ventilating current. Within the pharynx, the water is separated from the ingested food, which passes into the posterior part of the gut. Some hemichordates (*Balanoglossus*, for example) that live in burrows ingest large quantities of sand and extract the organic detritus from it. The castings of these animals are often a conspicuous feature of a sand flat exposed at low tide (Fig. 32.10).

Nervous System. Despite the presence of pharyngeal clefts, hemichordates are not chordates, as they were once thought to be, for they have no notochord, nor is there a true dorsal nerve cord. However, within the collar region of some hemichordates, a dense concentration of dorsal epidermal nerve fibers sinks inward and forms a tubelike arrangement (Fig. 32.9); some zoologists have thought this might be the homolog of the chordate nerve cord.

Reproduction. Hemichordates have separate sexes, and the numerous paired lateral gonads, each with a separate gonopore, are located within either side of the trunk. The eggs are released in mucous masses and fertilization is external. Development follows the typical deuterostome pattern and leads in many species to a **tornaria larva** (Fig. 32.11), which is strikingly like that of echinoderms. After a planktonic existence, the tornaria larva lengthens and settles to the bottom as a young worm.

The preceding discussion has been devoted to the most common hemichordates, members of the class **Enteropneusta.** A small number of species comprises another class, the **Pterobranchia.** These are mostly deep-water hemichordates that live attached within secreted tubes. The collar region bears tentacles, and some species lack pharyngeal clefts. Many zoologists consider the pterobranchs to be the most primitive members of the phylum.

32.6 DEUTEROSTOME RELATIONSHIPS AND CHORDATE ORIGINS

The common pattern in the early embryogeny of echinoderms, hemichordates and chordates is an important basis for recognizing the deuterostome line of evolution, but there is other evidence of the close evolutionary relationship between these three principal deuterostome phyla. The strikingly similar larval stages clearly indicate a connection between hemichordates and echinoderms. On the other hand, the presence of gill clefts and perhaps the dorsal collar nerve cord relates hemichordates to chordates. Although the precise relationship of the three phyla is unknown, there seems little doubt that they share a common evolutionary history.

The nature of the common ancestral form, as well as the origins of each phylum, is obscure. We speculated about echinoderm origins in the previous chapter. No more is known about chordate origins. An idea elaborated by N. J. Berrill and others, and held by many zoologists, postulates that the ancestral chordates were sessile and that urochordates reflect the

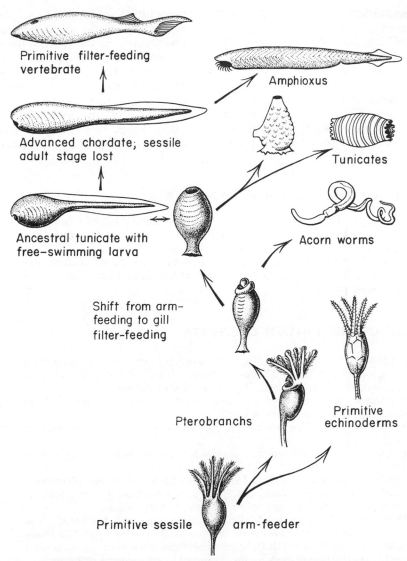

Primitive filter-feeding
vertebrate

Amphioxus

Advanced chordate; sessile
adult stage lost

Tunicates

Ancestral tunicate with
free-swimming larva

Acorn worms

Shift from arm-
feeding to gill
filter-feeding

Pterobranchs

Primitive
echinoderms

Primitive sessile   arm-feeder

Figure 32.12   A diagram showing Berrill's hypothesis of chordate evolution. (From Romer, A. S.: The Vertebrate Body. 4th ed. Philadelphia, W. B. Saunders Company, 1971.)

primitive condition of the phylum (Fig. 32.12). Pharyngeal clefts were originally an adaptation for filter feeding in the adult, and the notochord and finned tail were adaptations for swimming during larval life. According to this theory, the cephalochordates and vertebrates evolved from the urochordate tadpole larva through neoteny and paedogenesis, i.e., prolongation of larval life, precocious sexual development and suppression of adult features. Metamerism developed as an adaptation for swimming in the neotenous cephalochordate-vertebrate ancestor. The sessile existence of primitive chordates, pterobranch hemichordates and primitive echinoderms is viewed as a feature resulting from common ancestry.

However, like most theories that attempt to

explain evolutionary relationships between classes and phyla, this theory of chordate origins is not perfect. A principal difficulty stems from making the urochordates the ancestral members of the phylum, for they are certainly highly specialized. Indeed, sessility is a specialized condition wherever it occurs in the Animal Kingdom. Larval neoteny is used as a way of getting around this problem and in effect "starting fresh" on the cephalochordate and vertebrate line.

There is also no certainty that the common ancestor of echinoderms and chordates was sessile. Echinoderms appear to have evolved from motile ancestors, and their sessile existence and subsequent radiation may well have been an independent evolutionary event unrelated to the evolution of a sessile existence in urochordates.

Other views of chordate origins assume that the common ancestral deuterostome was a swimming bilateral coelomate. Pharyngeal clefts evolved as a mechanism for gas exchange or filter feeding in the line leading to hemichordates and chordates. The ancestral chordates may have been small swimming bilateral animals having a dorsal nerve tube, a notochord skeleton and pharyngeal clefts. Whether for gas exchange or filter feeding, pharyngeal clefts would have enabled these ancestral chordates to take up a sessile existence, and some became attached to give rise to the urochordates. In another stock, locomotor adaptations were developed, leading to metamerism and an undulatory mode of swimming. From this group the cephalochordates and vertebrates could have arisen.

CLASSIFICATION OF THE PHYLUM HEMICHORDATA

CLASS ENTEROPNEUSTA. Acorn worms. Wormlike hemichordates having a cylindrical proboscis and collar and a long trunk. Many species burrow in sand or mud; others live beneath stones or within algae.

CLASS PTEROBRANCHIA. Minute hemichordates having a shield-shaped proboscis and a collar bearing tentaculate arms. Most are colonial and live in secreted tubes in deep water.

CLASSIFICATION OF THE PHYLUM CHORDATA

Subphylum Urochordata. Mostly sessile, nonmetameric invertebrate chordates enclosed within a tunic containing cellulose. Pharynx highly developed and used in filter feeding; notochord and nerve cord present only in the larva.

CLASS ASCIDIACEA. Sea squirts or tunicates. Sessile; solitary or colonial. This class contains most species of urochordates.

CLASS THALIACEA. Salps. Free-swimming planktonic urochordates; solitary or colonial.

CLASS LARVACEA. Small, neotenous, planktonic urochordates living within a delicate gelatinous "house."

Subphylum Cephalochordata. Small fishlike invertebrate chordates. Metameric but trunk supported by well-developed notochord; no vertebrae present. Jawless suspension feeders; mouth surrounded by an oral hood. *Amphioxus.*

Subphylum Vertebrata. Vertebrates. Metameric chordates with the trunk supported by a linear series of cartilaginous or bony skeletal pieces (vertebrae) surrounding or replacing notochord in adult. Most vertebrates have mouth surrounded by jaws. (Detailed classification of vertebrates provided in subsequent chapters.)

CLASS AGNATHA. Lamprey eels and hagfishes.

CLASS PLACODERMI. Extinct primitive jawed fishes.

CLASS CHONDRICHTHYES. Sharks and rays.

CLASS OSTEICHTHYES. Bony fishes.

CLASS AMPHIBIA. Frogs, toads and salamanders.

CLASS REPTILIA. Alligators, snakes, lizards, turtles.

CLASS AVES. Birds.

CLASS MAMMALIA. Mammals.

[handwritten annotation: Superclass Tetrapoda]

ANNOTATED REFERENCES

Accounts of the hemichordates and chordates can be found in many of the references listed at the end of Chapter 21. Those below deal solely with these groups.

Barrington, E.: The Biology of Hemichordates and Protochordates. San Francisco, W H. Freeman, 1965. A general account emphasizing the behavior, physiology and reproduction of hemichordates, urochordates and cephalochordates.

Berrill, N. J.: The Origin of the Vertebrates. London, Oxford University Press, 1955. This work contains Berrill's theory of chordate and vertebrate origins.

Clark, R. B.: Dynamics in Metazoan Evolution. Oxford, Clarendon Press, 1964. A detailed synthesis of the ideas regarding the origins of the coelom and metamerism and their phylogenetic implications.

VERTEBRATES: FISHES

Fishes are aquatic vertebrates whose ancestors were also aquatic. This is in contrast to whales, sea turtles and certain other vertebrates who had terrestrial ancestors but underwent a secondary adaptation to aquatic life. Each of the three living classes and one extinct class of fishes has unique characteristics, but all share certain primitive vertebrate features and others that adapt them to life in the water.

1. Fishes are ectothermic vertebrates.

2. Their integument contains many simple mucous glands and, usually, bony scales.

3. Their skeleton is less massive than that in terrestrial vertebrates. There is little regional differentiation of the vertebral column other than trunk and tail; the visceral skeleton is well developed; and the appendicular skeleton consists only of fin supports.

4. The eye lacks movable eyelids and tear glands. Fishes have an inner ear but no cochlea; middle and external ears are absent. Most groups lack an internal nostril. The lateral line system is well developed.

5. Their cerebrum is small and primarily an olfactory center; the optic lobes are the primary integrating center.

6. Fishes do not have a fleshy muscular tongue, and the intestine is not divided into small and large segments. A cloaca is present.

7. Adults exchange gases with the environment through gills located in gill pouches; lungs or other accessory respiratory organs are present in a few groups.

8. The heart of most fishes receives only venous blood, and the atrium and ventricle are not divided. Pressures in peripheral vessels are low.

9. Nitrogenous wastes, primarily ammonia and urea, are eliminated through the gills in most groups and also by an opisthonephric kidney.

10. Most males lack a copulatory organ, and fertilization is external.

11. Great numbers of yolk-laden eggs are produced by large ovaries in most fishes.

12. Most species are oviparous and pass through an aquatic larval stage during development.

33.1 AQUATIC ADAPTATIONS

The integument of fishes has the basic form characteristic of vertebrates (see Section 10.02). However, there is little deposition of horny keratin in the epidermal cells. **Bony scales** develop in the dermis of most groups. **Mucous glands** are very abundant. Most are modified single epidermal cells; others are groups of cells that form saclike bulges into the dermis. Discharged mucus spreads over the body surface, where it helps to protect the body from ectoparasites and undue exchange of water with the environment (loss or gain, depending on osmotic circumstances). It probably also serves as a lubricant in fast swimming fishes.

Fishes have a streamlined, or fusiform, body shape that enables them to move through a dense medium, water, with a minimum of effort. The body is supported by a skeleton, but the skeleton is not so strong as in terrestrial vertebrates, for the buoyancy of the water also provides considerable support. Trunk muscles are segmented myomeres whose successive activation causes the lateral undulations of the trunk and tail by which the animal swims (see Section 11.11). Primitive fishes have rather heavy bodies and tend to sink, but the forward movement also generates lift forces. Lift is generated at the anterior end of the body by the planing of a head that is somewhat flattened ventrally and by large pectoral fins; it is developed at the posterior end by a **heterocercal tail** (Fig. 33.1*A*). The vertebral axis turns dorsally into the upper lobe of such a tail and stiffens this part of the tail in relation to the lower lobe. As the tail moves from side to side, the lower lobe has a sculling action and generates an upward thrust. The more advanced fishes have evolved a swim bladder, a sac of air dorsal to the body cavity, that gives them a density equal to that of the surrounding water. They do not sink so easily, the pectoral fins are smaller and the tail is symmetrical. The medial fins, as well as the paired pectoral and pelvic fins, provide stability against rolling, pitching and yawing (Fig. 33.1 *B*). The paired fins are also used in turning and in changing elevation.

The sense organs of fishes are adapted for receiving stimuli from the water. Movable eyelids and tear glands are absent; the eyes are bathed and cleansed by the surrounding water. Most light refraction is accomplished by a spherical lens; little occurs in the cornea, for its refractive index is similar to that of water. An inner ear is present, but fishes lack the external and middle ear found in terrestrial vertebrates. The pressure waves in the water pass easily through the tissues of the fish to the inner ear. Low frequency vibrations and certain water movements are detected by the lateral line system, which has been defined as a system of distant touch (p. 368). The nose is simply an olfactory organ, and water does not pass through it on the way to the pharynx in most species. A few fishes with lungs do have a connection between nose and mouth cavity.

Gas exchange between the water and blood occurs by diffusion across the gills. Water typically enters the pharynx through the mouth and is discharged through the gill pouches (Fig. 13.7, p. 299). The heart, located just posterior and ventral to the gills, receives blood low in oxygen content, and its contraction drives the blood through the gill capillaries. After leaving the gills, oxygenated blood is distributed to the body under relatively low pressure.

The movable muscular tongue characteristic of terrestrial vertebrates is absent in fishes, but the floor of the mouth and pharynx is specialized in some groups to aid in holding and swallowing prey. In many species teeth are present on the roof of the mouth and gill arches, as well as on the margins of the jaws. The respiratory current of water also helps carry food posteriorly into the pharynx. Digestion and absorption are similar in all vertebrates, but the ectothermic fishes require less absorptive surface than more active endothermic vertebrates.

Nitrogenous metabolic wastes are eliminated in most fishes by diffusion through the gills, and also by kidneys, which are drained by archinephric ducts. The kidneys also play an important role in water and salt balance. Excess salt swallowed by marine fishes is eliminated by special salt-excreting cells or glands.

Most fishes are oviparous. As the female lays eggs, the male discharges sperm. Fertilization is external, and there is often an aquatic larval stage during the development.

Figure 33.1 Role of the tail and fins of a dogfish in generating lift and providing stability. *A*, Lateral view and transverse section of the tail; *B*, composite transverse section through dorsal and pectoral fins. (Modified after Marshall and Hughes.)

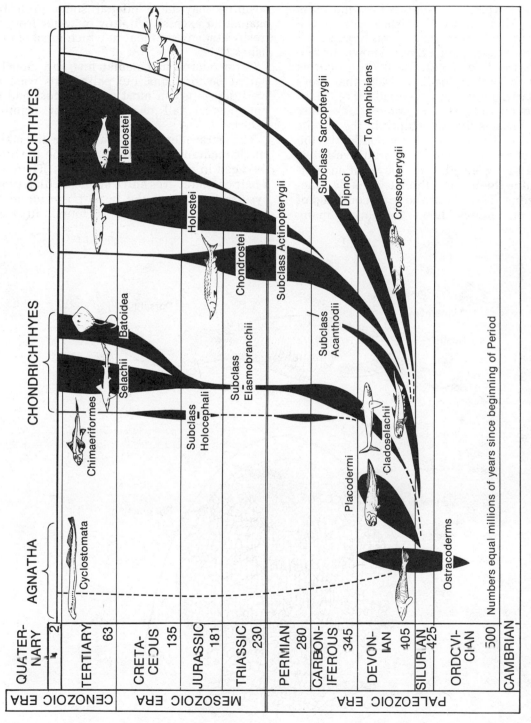

Figure 33.2 An evolutionary tree of fishes. The relationships of the various groups, their relative abundance and their distribution in time are shown.

33.2 VERTEBRATE BEGINNINGS

Although vertebrate origins are obscure, we have a reasonably complete fossil record of their subsequent evolution. The most primitive fishes are jawless types placed in the class **Agnatha.** This group flourished during the middle Paleozoic era, when it was represented by several orders collectively known as the **ostracoderms** (Fig. 33.2). Most ostracoderms were small, bottom-feeding fishes that were somewhat flattened dorsoventrally (Fig. 33.3). They had an extensive armor of thick bony plates and scales that developed in the dermis of the skin. These were very similar to the **cosmoid scales** of some other primitive fishes (Fig. 33.4), for beneath a thin layer of enamel-like **ganoine** there was a thick layer of dentine-like **cosmine.** The rest of the scale consisted of a layer of **spongy bone** containing many

vascular spaces and a layer of more compact **lamellar bone.** These heavy scales certainly offered some mechanical protection, possibly against aquatic scorpion-like eurypterids of the period; in addition, the layers of ganoine and cosmine may have offered some protection against an excessive inflow of water from the freshwater environment in which most of these fishes lived.

Ostracoderms had median fins, and some had paired pectoral fins, but paired fins were not well developed in most groups. Most had the heterocercal tail characteristic of primitive fishes.

The ostracoderm head was rather unusual. A single **median nostril** was present on the top of the head in the best known species. There was a pair of lateral eyes and a single median **pineal eye** on the top of the head posterior to the nostril. A median eye is found in many

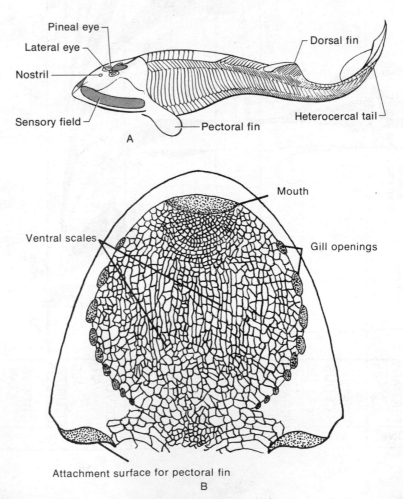

Figure 33.3 *Hemicyclaspis*, a representative ostracoderm of the early Devonian period. This fish was about 20 cm. long. *A*, Laterodorsal view; *B*, underside of head. (Modified after Stensiö.)

A Cosmoid scale

Ganoine
Cosmine layer
Spongy layer
Lamellar bone

B Ganoid scale

Layers of
ganoine
Cosmine
Vascular
spaces
Lamellar
bone

C Placoid scale

Dentine
Epidermis
Dermis
Pulp cavity

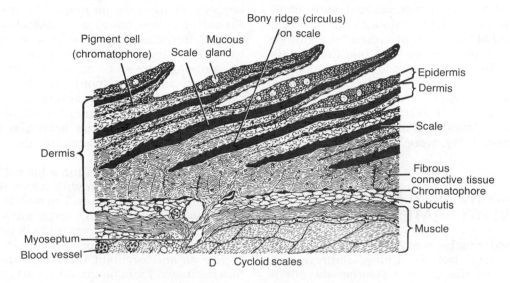

Pigment cell
(chromatophore)

Scale

Mucous
gland

Bony ridge (circulus)
on scale

Epidermis
Dermis
Scale

Fibrous
connective tissue
Chromatophore
Subcutis
Muscle

Dermis

Myoseptum
Blood vessel

D Cycloid scales

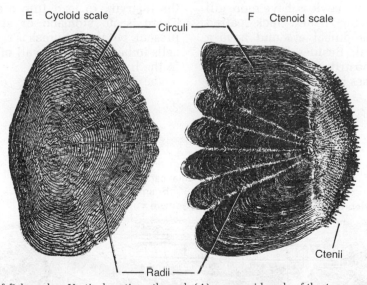

E Cycloid scale

Circuli

F Ctenoid scale

Ctenii

Radii

Figure 33.4 Types of fish scales. Vertical sections through (A) a cosmoid scale of the type seen in certain ostracoderms and sarcopterygians, (B) a ganoid scale of the type seen in acanthodians and primitive actinopterygians, (C) a placoid scale of a shark, (D) the skin and cycloid scales of a teleost. Surface view of (E) the cycloid scale of a primitive teleost and (F) the ctenoid scale of an advanced teleost. (A–C from Romer; D, courtesy of General Biological Supply House; E and F from Hubbs and Lagler.)

primitive fishes and terrestrial vertebrates, and it is retained in a few living fishes and reptiles. It is a photoreceptor rather than an image-forming eye and receives stimuli that enable the organism to adjust its physiological activity to the diurnal cycle. Three regions of the head, one dorsomedial and a pair of dorsolateral areas, contained small plates beneath which were enlarged cranial nerves. It is believed that these areas were sensory fields, possibly a part of the lateral line system. Much of the ventral surface of the head was covered with small plates forming a flexible floor to the pharynx. Movement of this floor presumably drew water and minute particles of food into the jawless mouth. The water then left the pharynx through as many as 9 pairs of small **gill slits**, but the food particles were somehow trapped in the pharynx. It seems probable that the ancestral vertebrates, like the lower chordates of the present day, were filter-feeders.

33.3 LIVING JAWLESS VERTEBRATES

Ostracoderms became extinct by the end of the Devonian, but the living lampreys and hagfishes of the order **Cyclostomata** are a specialized remnant of the class Agnatha (Figs. 33.2 and 33.5). They are jawless, have more gill slits than other living fishes, lack paired appendages, retain a pineal eye and have a single median nostril. Besides leading to an olfactory sac, this nostril opens into a **hypophyseal sac** that passes beneath the front of the

brain. Much of the pituitary gland of higher vertebrates is derived from an embryonic hypophysis. Cyclostomes differ from ostracoderms in several respects: they have an eel-like shape and a slimy scaleless skin, and they are predators or scavengers. Many lampreys, like the ostracoderms, live in fresh water, but some spend their adult life in the ocean and return to fresh water only to reproduce. The hagfishes are exclusively marine.

A familiar example of the group is the sea lamprey, *Petromyzon marinus*. The chief axial support for the body is a **notochord** that persists throughout life and is never replaced by vertebrae (Fig. 33.5). Rudimentary vertebrae are present, however, on each side of the notochord and spinal cord. The brain is encased by a cartilaginous **cranium**, and the gills are supported by a complex cartilaginous lattice-work known as the **branchial basket**, which may be homologous to the visceral skeleton of other fishes.

The **mouth** lies deep within a **buccal funnel**, a suction-cup mechanism with which the lamprey attaches to other fishes. The mobile **tongue** armed with horny "teeth" rasps away at the prey's flesh, and the lamprey sucks in the blood and bits of tissue. It has special **oral glands** that secrete an anticoagulant that keeps the blood flowing freely. From the mouth cavity, the food enters a specialized **esophagus** that bypasses the pharynx to lead into a straight **intestine**. There is no stomach or spleen. A **liver** is present, but the pancreas is represented only by cells imbedded in the wall of the intestine and in the liver.

The respiratory system consists of seven

Figure 33.5 A diagrammatic representation of the more important organs found in the anterior part of the lamprey.

Figure 33.6 The life cycle of the sea lamprey in the Great Lakes. The lamprey spends all but a year or two of its six and one-half to seven and one-half years of life as a larva. (From The Sea Lamprey by Applegate and Moffett. Copyright © by Scientific American, Inc., April, 1955. All rights reserved.)

pairs of gill pouches that connect to a modified pharynx known as the respiratory tube. The respiratory tube is isolated from the food passage so the animal can pump water in and out of the gill pouches through the external gill slits while it is feeding. The pumping of the pharyngeal region also changes the pressure in the hypophyseal sac in a way that resembles the compressing and relaxing of the bulb of an eye dropper, and water is thereby circulated across the olfactory sac.

The **kidneys** are drained by **archinephric ducts**. These ducts carry only urine, for sperm or eggs pass from the large **testis** or **ovary** into the coelom. A pair of **genital pores** leads from the coelom into a **urogenital sinus** formed by the fused posterior ends of the archinephric ducts and thence to the cloaca and outside. The absence of genital ducts may be a very primitive feature.

The eggs are laid on the bottom of streams in a shallow nest, which the lampreys make by removing the large stones (Fig. 33.6). Fertilization is external and the adults die after spawning.

Developing sea lampreys pass through a larval stage that lasts five to six years. The larva is so different in appearance from adult lampreys that originally it was believed to be a different kind of animal and was named **Ammocoetes**. The ammocoetes larva is eel-shaped but lacks the specialized feeding mechanism of the adult. It lies within burrows in the mud at the bottom of streams and sifts minute food particles from water passing through the pharnyx. Like the lower chordates, it has a mucus-producing **endostyle** that aids in trapping food.

Adult lampreys injure and kill many other fishes. The sea lamprey was able to pass the Niagara barrier, presumably through the Welland Canal, and extended its range from Lake Ontario into the other Great Lakes. The lake fishing industry was severely harmed during the late 1940's, and lake trout, which had been an important economic resource, were nearly eliminated. Development of a chemical larvicide that selectively kills larval lampreys gives some hope for a restoration of the fishing industry.

The hagfishes resemble lampreys in major respects, though differing, of course, in certain details. Hags are primarily scavengers feeding upon dead fish along the ocean bottom, but they also attack disabled fish of any sort, including those hooked or netted. They burrow into the fish and eat out the inside, leaving little but a bag of skin and bone. They are a commercial nuisance, but their over-all damage is not great, since they are abundant in only a few localities.

33.4 PRIMITIVE JAWED FISHES

During the Silurian and Devonian periods, other descendants of the ostracoderms became more active and predaceous. A more streamlined body shape, some reduction of the heavy trunk armor and the presence of paired pelvic as well as pectoral fins suggest greater activity. The presence of jaws enabled these fishes to feed upon a wider variety of food than the jawless ostracoderms.

Jaws evolved as the mouth was displaced posteriorly and became associated with the anterior part of the gill or visceral skeleton (p. 246). One or two visceral arches may have been lost in the process, for jawed fishes have fewer gill slits and arches than jawless ones. However, the most anterior of the remaining visceral arches, known as the **mandibular arch**, became enlarged and, together with dermal bones developed in the skin adjacent to it, formed the jaws. Among the earliest jawed fishes were the "spiny sharks," or **acantho-**

dians (Figs. 33.2 and 33.7). When bony plates are removed from the cheek and gill region, it is evident that the mandibular arch is a modified gill arch. The second visceral arch, known as the **hyoid arch**, lay close behind the mandibular arch and in some species, extended as a prop from the brain case to the posterior end of the upper jaw as it does in many contemporary species (Fig. 11.8, p. 246). The remaining visceral arches are **branchial arches** that supported the gills.

The **placoderms** were a second group of early jawed fishes (Figs. 33.2 and 33.8). Many were freshwater species of modest size, but the most spectacular genus, *Dunkleosteus*, was a 2- to 3-meter-long monster of upper Devonian Ohio seas. The head and anterior part of the trunk were encased in heavy bony plates, but the rest of the body was nearly naked. The most conspicuous parts of the jaws were sharp edged plates of dermal bone, but these were attached to a deeper lying mandibular arch.

The development of jaws in these early fishes was an important step in the evolution of vertebrates, for jaws enabled them to adapt to many modes of life. The success of jawed fishes probably contributed to the extinction of ostracoderms and to the limitation of cyclostomes to rather specialized ecological niches.

Other jawed fishes can be sorted out into two distinct evolutionary lines (Fig. 33.2). Members of one class, the sharks and skates, have a cartilaginous skeleton and hence are known as the cartilaginous fishes, or **Chondrichthyes.** Species in the other class have at least a partly ossified internal skeleton and hence are known as bony fishes, or **Osteichthyes.** These are our

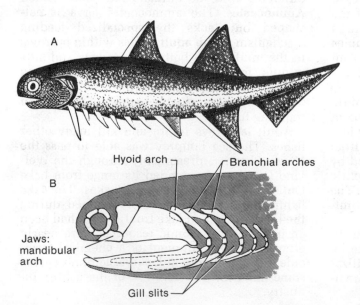

Figure 33.7 *A*, The "spiny shark," *Climatius*, was among the first jawed vertebrates. This fish was about 8 cm. long. After removal of the gill covering and superficial bony scales and plates on the head of a related genus (*Acanthodes*, *B*), it can be seen that the jaws are modified gill arches. (*A*, after Watson; *B* modified after Watson.)

Figure 33.8 A lateral view of the anterior end of the placoderm *Dunkleosteus*. The dermal plates composing the upper and lower jaws were attached to the mandibular arch. This fish attained a length of 3 meters.

most familiar fishes, including salmon, minnows, perch, lungfishes and the like. Most paleontologists believe that the acanthodians were an early subclass of the Osteichthyes. Although the placoderms had a partly ossified internal skeleton, other features suggest that they were more closely related to the cartilaginous fishes. They are usually placed in a distinct class of their own, the **Placodermi.**

33.5 CHARACTERISTICS OF CARTILAGINOUS FISHES

The early cartilaginous fishes appeared in the Devonian period (Fig. 33.2). Unlike most of their contemporaries, they were a marine group and, with few exceptions, sharks, skates, and their allies have remained a marine group. The cartilaginous internal skeleton is sometimes strengthened by the deposition of calcium salts, but there is never any internal ossification. A skeleton composed of cartilage is believed to represent the retention by the adult of the embryonic skeletal material. It is not regarded as the primitive adult condition, for at least parts of the skeleton were ossified in most early fishes. The only bonelike materials present in cartilaginous fishes are teeth and the small spiny **placoid scales** imbedded in the skin (Fig. 33.4), giving it a sandpaper-like quality. Placoid scales are structurally comparable to the enamel-like and dentine-like outer parts of the heavy bony scales of more primitive fishes. The triangular teeth of sharks closely resemble enlarged placoid scales and have doubtless evolved from them (Fig. 33.9).

Cartilaginous fishes resemble more primitive fishes in being heavy-bodied and denser than water, for they lack air-filled lungs or a swim bladder—features found in bony fishes.

The visceral organs of the dogfish, *Squalus*

Figure 33.9 A vertical section through the lower jaw of a shark. Teeth, which are structurally very similar to placoid scales, are continually developing, moving to the surface, and sloughing off. (From Rand.)

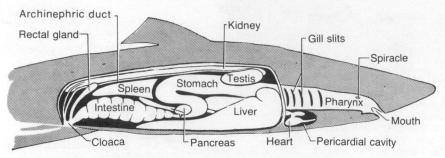

Figure 33.10   The visceral organs of the dogfish.

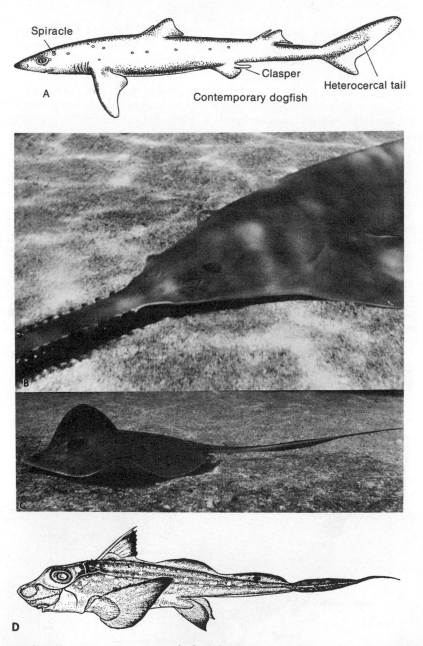

Figure 33.11   A group of cartilaginous fishes. *A*, A male dogfish, *Squalus acanthias*; *B*, the sawfish, *Pristis*; *C*, the stingray, *Dasyatis*; *D*, the ratfish, *Chimaera*. (*A* modified after Bigelow and Schroeder; *C*, courtesy of Marine Studios; *D* from Romer and Parsons after Dean.)

acanthias (Fig. 33.10), are in many ways more characteristic of primitive fishes than those of the specialized cyclostomes. The **mouth cavity** is continuous posteriorly with a long **pharynx**. A spiracle, containing a vestigial gill, and the gill slits, containing functional gills, open from the pharynx to the body surface. A wide **esophagus** leads from the back of the pharynx to a J-shaped **stomach**. A short, straight **valvular intestine** continues back to the **cloaca**. The valvular intestine receives secretions from the **liver** and **pancreas**. It contains an elaborate spiral fold known as the **spiral valve**; this helical fold serves both to slow the passage of food and to increase the digestive and absorptive surface of the intestine. Excess salt taken in with the food is eliminated by a salt-excreting **rectal gland** that empties into the end of the intestine. Intestine and urogenital ducts discharge into a common **cloaca** that opens on the underside of the body.

Cartilaginous fishes are unique in their method of reproduction. Part of the male's pelvic fin is modified as a **clasper** with which to transfer sperm to the female reproductive tract (Fig. 33.11*A*). The eggs are fertilized in the upper part of the oviduct, and a horny protective capsule is secreted around them by certain oviducal cells. Skates are **oviparous,** but there is no free larval stage as there is in frogs and many other oviparous animals. The eggs are very heavily laden with yolk, and the embryos develop within the protective capsule. Some sharks are also oviparous, but most have departed from this primitive egg-laying habit. A few are truly **viviparous**, for the fertilized eggs develop in a modified portion of the oviduct known as the **uterus**; an intimate association is established between each embryo's yolk sac and the uterine lining, forming a **yolk sac placenta**; and the embryos have a greater dependence for their nutrient requirements upon the mother than upon food stored in the yolk. However, most sharks, including our common dogfish, are **ovoviviparous**; the eggs also develop within a uterus, but there is a greater dependence upon food stored in the yolk. In some cases, a portion of the nutritional requirements is derived by the absorption of materials secreted by the mother into the uterine fluid, but an intimate placental relationship is not established. It is frequently difficult to make a sharp distinction between viviparous and ovoviviparous reproduction, for the degree of dependence of the embryo for its nutrients upon the mother ranges along a continuum, and in both, the young are born at an advanced stage of development as miniature adults.

33.6 EVOLUTION OF CARTILAGINOUS FISHES

The ancestral Chondrichthyes were essentially sharklike in body shape, but the structure of their fins and jaws suggests that they were not as efficient in swimming and feeding as are contemporary species. In their subsequent evolution (Fig. 33.2), the cartilaginous fishes have diverged widely and have become adapted to many modes of life within the aquatic environment. One line of evolution (subclass **Holo-**

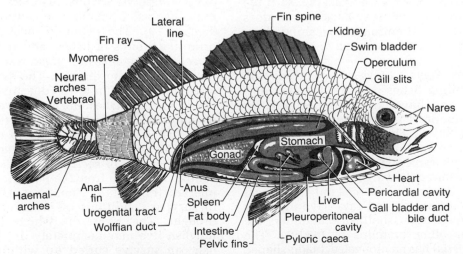

Figure 33.12 The visceral organs of the perch.

cephali) has led to our present-day, rather rare, deep-water ratfish (*Chimaera*) (Fig. 33.11*D*). In these fishes, the gill slits are covered by an operculum, so there is a common external orifice, and the tail is long and ratlike. The other line of evolution (subclass **Elasmobranchii**) is distinguished by having separate external openings for each gill slit. Elasmobranchs have been far more successful and have diverged into two contemporary orders—**Selachii** (sharks and dogfish) and **Batoidea** (skates and rays).

Sharks. Most selachians are active fishes with protrusible jaws that enable them to feed voraciously with their sharp triangular teeth upon other fishes, crustaceans and certain mollusks. The largest sharks, such as the whale shark (*Rhineodon*), which may reach a length of about 12 meters, have minute teeth and feed entirely upon small crustaceans and other organisms that form the drifting plankton. They gulp mouthfuls of water and as the water passes out of the gill slits, the food is kept in their pharynx by a branchial sieve. Whale sharks are the largest living fishes.

Although rare, attacks on humans by sharks receive considerable notoriety. In order to obtain reliable data as to their frequency and the circumstances of their occurrence, the Shark Research Panel of the American Institute of Biological Sciences has been maintaining a world-wide "Shark Attack File." There seem to be some 30 unprovoked attacks per year, occurring mostly between latitudes 35° N and 35° S, where the waters are warm and the more dangerous species live. It is well established that sharks are attracted to fresh blood, even in small quantities, coming from injuries to a swimmer or from speared fish a skin diver may be towing. Most sharks circle their intended victim several times before attacking, so if the victim is aware of their presence and he does not attract undue attention by splashing, there is a fair chance of his avoiding or warding off any attack.

Skates and Rays. Most skates and rays are bottom-dwelling fishes that are flattened dorsoventrally and have enormous pectoral fins whose undulations propel the fish along the bottom (Fig. 33.11). Their mouth is often buried in the sand or mud, and water for respiration enters the pharynx via the pair of enlarged spiracles. A spiracular valve in each one is then closed, and the water is forced out the typical gill slits. Most skates and rays have crushing teeth and feed upon shellfish, but others are adapted for other methods of feeding. The sawfish (*Pristis*) has an elongated, blade-shaped

snout armed with toothlike scales. By thrashing about in a shoal of small fishes, it can disable many and eat them at leisure. As in the sharks, the largest members of the group (the devilfish, *Manta*) have reduced teeth and are plankton feeders. Some devilfish have a "wing spread" of 6 meters and can easily upset small boats. Harpooning these is an exciting sport!

33.7 CHARACTERISTICS OF BONY FISHES

While early sharks were becoming dominant in the ocean, the bony fishes, class **Osteichthyes**, became dominant in fresh water. Near the middle of the Mesozoic Era they began to enter the ocean and became the most successful group there as well.

Bony fishes differ from the cartilaginous fishes in having an ossified internal skeleton and in retaining more of the primitive dermal bony scales and plates. All bony fish also have an **operculum**, a lateral flap of the body wall that extends posteriorly from the head and covers the gill region. The gill pouches lead to an opercular chamber that opens on the body surface at the posterior margin of the operculum (Fig. 13.7, p. 299). An operculum is found only in the holocephalans among the cartilaginous fishes. Male bony fishes never have the pelvic clasper found in all male cartilaginous fishes.

The soft parts of most bony fishes, the perch for example (Fig. 33.12), show a peculiar mixture of primitive and highly specialized characters. Most need not concern us, but one of great interest is the **swim bladder**. This is primarily a hydrostatic organ (see Section 11.11), but in some cases it also serves as an oxygen reserve. Swim bladders and **lungs** probably had a common evolutionary origin (Fig. 33.13). The organ is most lunglike in primitive groups of fishes, and it is generally agreed that the ancestral organ was more like a lung than a swim bladder. The swim bladder loses its connection to the digestive tract only in adult teleost fishes, the most advanced group of bony fishes. We believe that the ancestral bony fishes had lungs similar to those of the living African lungfish (*Protopterus*). In the lungfish, a pair of saclike lungs develops as a ventral outgrowth from the posterior part of the pharynx (Fig. 33.13). The lungs enable the fish to survive conditions of stagnant water and drought. The rivers in which the African lungfish live may completely dry up, but the fish can survive curled up within a mucous

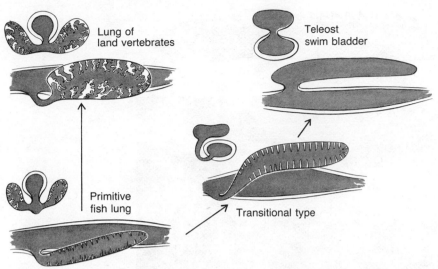

Lung of
land vertebrates

Teleost
swim bladder

Primitive
fish lung

Transitional type

Figure 33.13 A diagram to illustrate the evolution of lungs and the swim bladder. The connection to the digestive tract is lost in most adult teleosts. (After Dean.)

cocoon that it secretes around itself in the dried mud. A small opening from the cocoon to the surface of the mud enables the fish to breathe air during this period. The African lungfish has become so dependent upon its lungs that it will die if it cannot occasionally reach the surface to gulp air.

Air breathing probably evolved in fishes as a supplement to gill respiration. Presumably, early bony fishes evolved lungs as an adaptation to the unreliable freshwater conditions of the Devonian period. Geologic evidence indicates that the Devonian was a period of frequent seasonal drought. Bodies of fresh water undoubtedly either became stagnant swamps with a low oxygen content or dried up completely. Only fishes with such an adaptation could survive these conditions. The others became extinct or migrated to the sea, as did many later placoderms and the cartilaginous fishes. Groups of bony fishes that have remained in fresh water throughout their history tended to retain lunglike organs, but those that went to sea no longer needed lungs, for ocean waters are rich in oxygen. Their useless lungs evolved into useful hydrostatic organs. What are presumed to be intermediate stages in this shift can still be seen in certain species. Later, when conditions were more favorable, many salt-water bony fishes re-entered fresh water but retained their swim bladders. The fresh-water perch has such a history. Its ancestors first evolved lungs in a fresh-water environment, then went into the ocean, where the lungs changed into swim bladders; later the fish re-entered fresh water and retained the swim bladders.

33.8 EVOLUTION OF BONY FISHES

Bony fishes can be traced back to the early Devonian and late Silurian (Fig. 33.2). At that time several lines of evolution within the class were already established. Since all share such features as a partly ossified internal skeleton, well developed bony scales and an operculum (a combination of characters not present in other classes of fish), it is assumed that they had a common evolutionary origin, possibly from some ostracoderm group. Yet each evolutionary line was already well established. The acanthodians (p. 772) were characterized by having a prominent supporting spine at the anterior edge of the dorsal, anal and paired fins. Often there were more than two pairs of paired fins. Acanthodians were the earliest of the bony fishes, and it is possible that other lines of bony fish evolution were derived from them. They flourished during the Devonian period but were replaced by more progressive types by the end of the Paleozoic era. Other bony fishes are usually grouped into the subclasses **Actinopterygii** and **Sarcopterygii.**

Actinopterygians. The actinopterygians are the familiar ray-finned fishes, such as the perch. Their paired fins are fan-shaped (Fig. 33.14). Skeletal elements enter their base, but most of the fin is supported by numerous dermal rays that evolved from rows of bony scales. Their paired olfactory sacs connect only with the outside and not with the mouth cavity.

Actinopterygian evolution presents a good

Figure 33.14 A group of primitive ray-finned fishes that have survived to the present day. *A*, The Nile bichir, *Polypterus*; *B*, the shovel-nosed sturgeon, *Scaphirhynchus platorhynchus*; *C*, the longnose gar, *Lepisosteus oseus*; *D*, the bowfin, *Amia calva*. (A after Dean; B, C and D, courtesy of the American Museum of Natural History.)

example of a succession in which early dominant groups became replaced by more successful types. Most authorities recognize three infraclasses, and each, in turn, had its day (Fig. 33.2). Currently, the infraclass **Chondrostei** has dwindled to a few species, of which the Nile bichir (*Polypterus*) and the sturgeon (*Scaphirhynchus*) are examples (Fig. 33.14). The infraclass **Holostei** has also dwindled and is represented today by such relict species as the gar (*Lepisosteus*) and bowfin (*Amia*). The infraclass **Teleostei**, in contrast, has been continuously expanding since its origin in the Mesozoic era. It is to this group that the minnows, perch and most familiar fishes belong.

Various evolutionary tendencies can be traced through this succession, most of which are related to improvements in locomotion and feeding mechanisms. The vertebral column became more thoroughly ossified and stronger, thereby providing a better attachment for powerful axial muscles. The functional lungs of early actinopterygians (still retained in *Polypterus*) became transformed into swim bladders with little respiratory function. In connection with increased buoyancy and better streamlining, we find that the primitive heterocercal tail of most chondrosteans (*Polypterus* is an exception) became superficially symmetrical in teleosts, but the caudal skeleton still shows indications of the upward tilt of the vertebral column. The caudal fin rays attach to **hypural bones**, which are modifications of the haemal spines that attach to the ventral surface of the tail vertebrae (Fig. 33.15). Such a tail is said to be **homocercal**. Holosteans have an intermediate **abbreviated heterocercal** tail. Increased buoyancy relieves the fins of their primitive hydroplaning function. Their base becomes narrower, and the point of attachment of the pectoral fins shifts dorsally so that they are closer to a horizontal plane passing through the center of gravity. These changes improve the braking and turning functions of the fins. Early actinopterygians were clothed with thick bony scales of the **ganoid type** (Fig. 33.4B). During subsequent evolution, the scales became thinner and lighter. Superficial layers were lost, and the bone was reduced to a thin disc that develops in the dermis of the skin (Fig. 33.4D). Such a scale is termed **cycloid** if its surface is smooth, **ctenoid** if the posterior portion bears minute spiny processes, or ctenii. As the fish grows, increments of bone are added to the scale, and these appear as rings, or **circuli**. In some fish, in which the rate of growth slows down in the winter, the circuli are much closer together and form a check mark. By counting such marks, and not the circuli themselves, estimates of a fish's age can be made. Improvements in feeding mechanisms are reflected in a shortening of the mouth

and forward shifting of the jaw joint. The jaws become shorter and develop a more powerful bite.

Teleosts are an exceedingly large and diverse infraclass, for there are 30,000-odd species. The evolutionary relationships between the various groups are far from certain, but certain evolutionary trends are quite apparent. Many of these are correlated with changes in locomotion and feeding patterns. The more primitive teleosts, the tarpons and herrings (Fig. 33.16), are characterized by having elongate streamlined bodies, a single dorsal fin, pelvic fins located near the posterior part of the trunk, fins supported by flexible and branching bony rays rather than by spines, cycloid scales covering the tail and trunk but not extending onto the head, and an air, or pneumatic, duct connecting the swim bladder and digestive tract. The jaws are short but do not extend forward when the mouth is opened. The most advanced are the spiny-finned teleosts, such as the sunfishes and perch. These are often rather short and deep-bodied (from dorsal to ventral) fishes. There is a tendency for the dorsal fin to split into two parts, the anterior being supported by spines, the posterior by flexible bony rays. The pelvic fins have shifted forward to a point beneath the pectoral fins, and spines are present in the anterior border of these fins. The scales have become ctenoid and have extended onto the head and gill covering. The pneumatic

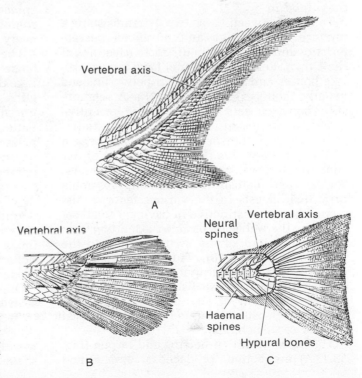

Figure 33.15 Caudal fin types of bony fishes *A*, The primitive heterocercal tail seen in the sturgeon, *Acipenser; B*, the abbreviated heterocercal tail of the garpike, *Lepisosteus; C*, the homocercal tail of a teleost. (*A* and *B* from Jordan; *C* from Romer.)

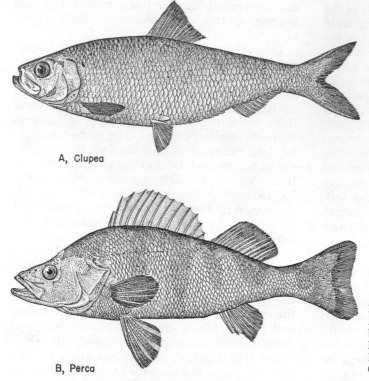

A, Clupea

B, Perca

Figure 33.16 *A*, A primitive teleost, the herring; *B*, an advanced teleost, the perch. Differences in body proportions and fins are apparent. (From Romer, A. S., and T. S. Parsons: The Vertebrate Body. 5th ed. Philadelphia, W. B. Saunders Company, 1977.)

duct is lost. The jaws are highly specialized and protrude rapidly as the mouth opens. This and the concomitant expansion of the buccal and pharyngeal cavities create a suction that helps draw food in.

Teleosts between these two extremes show a mixture of primitive and advanced characteristics and often certain distinctive features of their own. Eels, for example, have long snake-like bodies and have lost their pelvic fins and usually their scales (Fig. 33.17*F*). A relationship between eels and the more primitive tarpon is suggested, for both have a similar ribbon-shaped, transparent larval stage known as the **leptocephalus** (Fig. 33.18). Most of our common freshwater teleosts, such as minnows, suckers and catfish (Fig. 33.19*C*), resemble primitive teleosts in the various features discussed but are distinctive in having a small set of bones, the **weberian ossicles**, extending from the anterior end of the swim bladder to the inner ear (Fig. 33.20). This mechanism apparently acts as a hydrophone, for the acoustical sensitivity of these fishes is considerably greater than that of any other group. Other groups of teleosts are defined in the synopsis at the end of this chapter.

The more primitive herring and tarpon (Fig. 33.17*A*) are active, predaceous, streamlined

fishes of the open waters. But there have been many interesting departures from these generalized types, for teleosts have undergone an extensive adaptive radiation (p. 195), have spread out into all parts of the aquatic environment and have become adapted to nearly every conceivable ecologic niche.

The halibut, soles and flounders have become specialized for a bottom-dwelling life. Like the skates and rays, they are flattened and glide along the bottom with up-and-down undulatory movements. But instead of being flattened dorsoventrally, they are greatly compressed from side to side and swim turned over on one side (Fig. 33.17*B*). During larval development, the eye that would be on the "ventral" side migrates to the top surface, but the mouth does not change position. The skates and flounders present another good example of convergent evolution (p. 198), by which animals that are widely separated in the evolutionary scale independently adapt to similar modes of life.

Other teleosts have adapted to a life among seaweeds and in coral reefs. The sea horses with their monkey-like prehensile tails, the sargassum fish with its camouflaging color and weedlike protuberances, and the elongate, snakelike moray eels are examples (Fig. 33.17).

Figure 33.17 Diversity among the teleosts. *A*, A tarpon, *Tarpon*, one of the more primitive types of teleosts; *B*, a halibut, *Hippoglossus*, one of the flatfish that feed along the bottom; *C*, a male sea horse, *Hippocampus*, with the brood pouch in which the female deposits her eggs, young shown at right; *D*, the sargassum fish, *Histrio*, appears very bizarre out of its natural environment but is well concealed among seaweed; *E*, the sharksucker, *Remora*; *F*, the moray eel, *Gymnothorax*, normally lurks within interstices of coral reefs. (*A, C, D, E,* and *F*, courtesy of Marine Studios; *B*, courtesy of the American Museum of Natural History.)

Figure 33.18 The transparent leptocephalus larva of the eel. A similar larval stage occurs during the development of the tarpon. (From Jordan after Eigenmann.)

Figure 33.19 Further diversity among the teleosts. *A*, The deep-sea gulper eel, *Saccopharynx*; *B*, *Mormyrus*, one of the primitive fresh-water osteoglossomorphs; *C*, the catfish, *Ictalurus*, one of the ostariophysians; *D*, the flying fish, *Exocoetus*, a specialized top minnow; *E*, the clingfish, *Caularchus*, with pelvic fins modified as a sucking disc; *F*, the angler, *Lophius*, with its tassel-like lure; *G*, the John Dory, *Zeus*; *H*, the oarfish, *Regalecus*. (*C* and *E* from Jordan; others from Romer.)

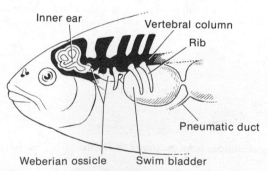

Figure 33.20. A lateral view of the head of a carp to show the connection between inner ear and swim bladder made by the weberian ossicles. (From Portman.)

A few teleosts have adapted to life in the ocean depths. Such fish often have light-producing luminescent organs, presumably for species recognition, and large mouths and greatly distensible stomachs that enable them to take full advantage of the occasional meal that comes their way (Fig. 33.19A).

Some teleosts live in intimate association with other fishes. The remora has an anterior dorsal fin that is modified as a suction cup and is used to attach to sharks. It feeds upon crumbs of the larger fish's meals or obtains free rides to favorable feeding grounds. Relationships of this type, in which one organism benefits and the other receives neither benefit nor harm, are known as **commensalism** (Fig. 33.17E).

A few teleosts have become amphibious. The Australian mudskipper frequently hops about on the mud flats of mangrove swamps at low tide in search of food and may even bask in the sun. It has unusually muscular pectoral fins with which it pulls itself along the land, and it can close its opercular chamber and extract oxygen from the air with its gills.

A number of fishes, both cartilaginous and bony, have evolved (from muscular tissue) electric organs capable of discharging pulses of electrical energy. In such fishes as the electric ray *(Torpedo)* of the North Atlantic and the electric eel *(Electrophorus)* of the Amazon basin, the pulses are strong enough to stun prey or predator. *Electrophorus* can emit pulses in excess of 500 volts! In other fishes, the pulses are very weak and their biologic significance was long unknown. Researchers have found that they are part of an electric guidance system. Dr. H. W. Lissmann of the University of Cambridge has found that *Gymnarchus*, a nocturnal fresh-water fish of West Africa, gives off a continuous stream of weak electric pulses that set up an electric field in the surrounding water resembling a dipole. Objects having an electrical conductivity different from that of the water distort the field, and these changes can be detected by the fish by way of modified portions of its lateral line system. Dr. Lissmann has postulated that electric organs of this type represent an intermediate evolutionary stage between typical muscle with its action potential measured in millivolts and the stunning electric organs. Darwin could not explain how natural selection could favor the evolution of stunning electric organs for, at the lower levels of efficiency, they would be useless. Probably they had a different function when they were first evolving. After a certain threshold of power was reached, the function could change.

Many other fascinating adaptations are found among teleosts, but we must not dwell upon them, for these fishes are only a side issue in the total picture of vertebrate evolution. The branch toward the higher vertebrates passed through the less spectacular Sarcopterygians of ancient Devonian swamps.

Sarcopterygians. The sarcopterygians include the lungfishes and crossopterygians (Figs. 33.21 and 33.22). Their paired appendages are typically elongate and lobe-shaped and are supported internally by an axis of flesh and bone. In many species, each of the olfactory sacs connects to the body surface through an **external nostril** and to the front part of the roof of the mouth cavity through an **internal nostril**. Sarcopterygian evolution diverged at an early time into two lines—the lung-fishes (order **Dipnoi**) and the crossopterygians (order **Crossopterygii**). The primitive crossopterygians had a well-ossified internal skeleton, a unique jointed chondrocranium and small conical teeth suited for seizing prey. It is from this group that the amphibians arose. Throughout their evolutionary history, lung-fishes have had a weak skeleton with little ossification of the vertebral column. They developed specialized crushing tooth plates, which enabled them to feed effectively upon shellfish, and showed tendencies toward reduction of the paired appendages. In certain other features, lungfishes and crossopterygians have paralleled actinopterygian evolution. They evolved symmetrical tails, though of a type that is symmetrical internally as well as externally (**diphycercal**), and their primitive,

Figure 33.21 The Australian lungfish, *Epiceratodus*. (After Norman.)

Figure 33.22 *Latimeria*, a living coelacanth found off the coast of the Comoro Islands. (From Millot.)

thick bony scales, which were characterized by having a thick layer of dentine-like **cosmine** (Fig. 33.4), have tended to thin to the cycloid type.

Both crossopterygians and lungfishes were successful in the fresh waters of the Devonian but have dwindled to a few relict species today. Lungfishes retained their lungs and have survived in the unstable freshwater environments of tropical South America, Africa and Australia. It was long believed that all crossopterygians had become extinct, as indeed the primitive freshwater ones have. However, a few specimens of a somewhat specialized side branch (the coelacanths) have been found in recent years near the Comoro Islands between Africa and Madagascar (Fig. 33.22).

*CLASSIFICATION OF FISHES

CLASS AGNATHA. Jawless fishes. The oldest and most primitive class of vertebrates; all members lack jaws and pelvic fins; many lack pectoral fins.

†*Ostracoderms.* Several orders of small Paleozoic fishes are collectively called the ostracoderms; most were heavily armored. *Hemicyclaspis.*

Order Cyclostomata. Lampreys and hagfishes. Body eel-shaped; round mouth without jaws; no paired appendages; scaleless slimy skin. The sea lamprey, *Petromyzon.*

†**CLASS PLACODERMI.** Several orders of primitive jawed fish of the Paleozoic; bony head shield movably articulated with thoracic shield; most of trunk without scales; probably related to cartilaginous fishes. *Dunkleosteus.*

CLASS CHONDRICHTHYES. Cartilaginous fishes. Skeleton entirely of cartilage; placoid scales usually present; lungs or swim bladder absent; pelvic clasper in male; nearly all are marine.

Subclass Elasmobranchii. Gill slits open independently on the body surface; placoid scales present; lateral line system imbedded in the skin.

†*Order Cladoselachii.* Ancestral sharks with broad-based paired fins; vertebral centra poorly developed; long jaws. *Cladoselache.*

Order Selachii. Modern sharks with narrow-based paired fins; well formed vertebral centra; streamlined bodies; heterocercal tail; short protrusible jaws usually provided with sharp triangular teeth. The dogfish, *Squalus;* whale shark, *Rhineodon.*

Order Batoidea. Sawfish, skates and rays. Body dorsoventrally flattened; pectoral fins enlarged; tail reduced; teeth usually in the form of crushing plates. Common skate, *Raja;* devilfish, *Manta;* sawfish, *Pristis.*

---

*Based on the classification in A. S. Romer: Vertebrate Paleontology. 3rd ed. Chicago, University of Chicago Press, 1966. The classification is taken down to orders and, in a few cases, suborders. All living groups are included, but only the more important extinct groups are mentioned.

†Extinct groups.

Subclass Holocephali. Gill slits covered by an operculum; scales absent; lateral line system in the form of open grooves.

    Order Chimaeriformes. The ratfish, *Chimaera.*

CLASS OSTEICHTHYES. Bony fishes. Skeleton includes considerable bone; various types of bony scales, other than placoid, usually present; lungs or swim bladder usually present; abundant in fresh and salt water.

    *Subclass Acanthodii. Several orders of Paleozoic fishes. Oldest jawed fishes; a prominent spine in front of dorsal, anal and paired fins; usually more than two pairs of paired fins. The "spiny shark," *Climatius.*

    Subclass Actinopterygii. Ray-finned fishes. Paired fins fan-shaped, supported by radiating bony rays or spines; each nasal cavity has an entrance and exit on snout surface.

    INFRACLASS CHONDROSTEI. Two extinct and two living orders of primitive ray-finned fishes. Scales usually ganoid; mouth opening large; tail usually heterocercal.

        Order Polypteriformes. The bichir. Body slender, numerous dorsal fins; lungs present; tropical Africa. *Polypterus.*

        Order Acipenseriformes. Sturgeon and allies. Long sharklike snout; subterminal mouth; several rows of heavy bony scales. Common sturgeon, *Acipenser;* paddlefish, *Polydon.*

    INFRACLASS HOLOSTEI. Three extinct and two living orders of intermediate ray-finned fishes. Scales ganoid or cycloid; mouth opening smaller than in chondrosteans; abbreviated heterocercal tail.

        Order Semionotiformes. Garpikes. Elongated trunk and snout; scales ganoid. *Lepisosteus.*

        Order Amiiformes. The bowfin. Snout rounded; scales cycloid; long dorsal fin. *Amia.*

    INFRACLASS TELEOSTEI. Advanced ray-finned fishes. Scales cycloid or ctenoid; tail homocercal; a hydrostatic swim bladder usually present; jaws short and usually protrusible.

    *Superorder Leptolepimorpha. One order of primitive herring-like teleosts still retaining some holostean characters, including traces of ganoine on scales.

    Superorder Elpomorpha. Primitive teleosts with a "leptocephalous" larva.

        Order Elpoiformes. Tarpons. Often large, rapidly swimming oceanic fishes; much prized for sport. *Tarpon.*

        Order Anguilliformes. Eels. Body elongate; scales rudimentary or absent; dorsal and anal fins long and frequently fused with reduced caudal fin; pelvic fins absent. American eel, *Anguilla.*

        Order Notacanthiformes. Gulper eels. Deep-sea, somewhat eel-shaped fishes; large mouths. *Saccopharynx.*

    Superorder Clupeomorpha.

        Order Clupeiformes. Herring, sardines, shad and their allies. Primitive teleosts; scales cycloid; head and operculum not scaled; fins without spines; trunk not greatly shortened; single dorsal fin; pelvic fins abdominal; a duct connects swim bladder and digestive tract; most marine. Common herring, *Clupea.*

    Superorder Osteoglossomorpha.

        Order Osteoglossiformes. Primitive freshwater teleosts of the Old and New World tropics with a unique bony tongue that pushes against palatal teeth. The North American mooneye, *Hiodon;* and the African elephant-snouted fish, *Mormyrus.*

    Superorder Protacanthopterygii. Primitive teleosts, some of which are beginning to acquire certain of the advanced features of acanthopterygians.

        Order Salmoniformes. Salmon, trout and pike. A fleshy adipose fin is often present on the back. The salmon, *Salmo.*

        Order Cetomimiformes. A few unusual deep-sea fishes.

        Order Ctenothrissiformes. Possibly ancestral spiny-finned teleosts; most are extinct.

---

*Extinct groups.

Order Gonorhynchiformes. Milkfishes.

Superorder Ostariophysi. A large group of primarily freshwater teleosts; primitive in most respects, but have webberian ossicles.

Order Cypriniformes. Carp, minnows, suckers and their allies. Primitive ostariophysians; usually retain cycloid scales; most lack spines in fins. Carp, *Cyprinus*; shiner, *Notropis*; sucker, *Catostomus*.

Order Siluriformes. Catfish. Specialized ostariophysians; scales usually lost; a large erectile spine in dorsal and pectoral fins; long sensory barbels on head. Channel catfish, *Ictalurus*.

Superorder Atherinomorpha.

Order Atheriniformes. Top minnows, flying fishes and their allies. Teleosts with a mixture of primitive and advanced characters; scales usually cycloid; scales beginning to spread onto head and operculum; some spines present in fins; pelvic fins abdominal in position. Killifish, *Fundulus*; silverside, *Atherina*; flying fish, *Exocoetus*.

Superorder Paracanthopterygii. Several small orders of specialized teleost fishes with advanced characteristics. Scales usually ctenoid; spines usually present in fins; pelvic fins located far forward.

Order Amblyopsiformes. Pirate perch. Anus located far forward under gill region. Pirate perch, *Aphredoderus*.

Order Batrachoidiformes. Toadfishes. Bizarre teleosts with large somewhat flattened heads and long tapering trunks. *Batrachoides*.

Order Gobiesociformes. Clingfishes. Pelvic fins located far forward and modified as a sucking disc; often found in tidepools. *Caularchus*.

Order Lophiiformes. Anglers. First dorsal spine carries a tassel that dangles as a lure over large mouth; most deep sea. Common angler, *Lophius*.

Order Gadiformes. Cod and haddock. Elongate and subdivided dorsal and anal fins. Cod, *Gadus*; burbot, *Lota*.

Superorder Acanthopterygii. Spiny-finned teleosts. Advanced teleosts; scales ctenoid; head and operculum usually scaled; spines present in fins; trunk often short; often two dorsal fins; pelvic fin thoracic in position; swim bladder without a duct.

Order Beryciformes. Squirrelfish. Moderately deep-bodied fishes; retain 18 or 19 rays in the tail fin in contrast to 17 in more advanced acanthopterygians.

Order Zeiformes. John Dory, *Zeus*.

Order Lampidiformes. Ribbonfish and oarfish. Elongate oceanic fish; some species 6 meters long. The oarfish, *Regalecus*.

Order Gasterosteiformes. Sticklebacks, pipefish and sea horses. Trunk encased in bony armor; small mouth at end of tubular snout. The sea horse, *Hippocampus*.

Order Channiformes. Snakeheads. Small Asiatic coastal fish with lunglike extensions from gill chambers; can live out of water. *Channa*.

Order Synbranchiformes. Eel-like, tropical coastal fish. Swamp eel.

Order Scorpaeniformes. Scorpion fish and sculpins. Enlarged heads and pectoral fins; projecting spines from gill covering. The sculpin, *Acanthocottus*.

Order Dactylopteriformes. Enlarged pectoral fins; glide in a manner similar to the more familiar flying fishes.

Order Pegasiformes. Dragonfishes. Small marine fishes of tropical seas; trunk and tail encased in a bony box or rings. *Pegasus*.

Order Perciformes. The largest order of spiny-finned teleosts; includes perch, bass, mackerel, gobies, barracuda and their allies. A relatively unspecialized group of spiny-finned teleosts. The perch, *Perca*.

Order Pleuronectiformes. Flatfishes. Body highly compressed; fish lie on ocean bottom on one side of body; both eyes on upper side of body. Halibut, *Hippoglossus*; flounder, *Paralichthys*.

Order Tetraodontiformes. Triggerfish, trunkfish and puffers. Strong jaws with a sharp beak; scales often spiny; some inflate by swallowing water. Porcupine fish, *Diodon*.

Subclass Sarcopterygii. Fleshy-finned fishes. Paired fins lobe-shaped with a central axis of flesh and bone; internal nostrils usually present; primitive members had cosmoid scales.

Order Crossopterygii. Crossopterygians. Primitive fleshy-finned fishes, conical teeth present; median eye usually present.

    SUBORDER RHIPIDISTIA. Primitive freshwater crossopterygians; includes the ancestor of amphibians. *Osteolepis.*

    SUBORDER COELACANTHINI. More specialized freshwater and marine crossopterygians; one living genus. *Latimeria.*

Order Dipnoi. Lungfishes. Specialized fleshy-finned fishes; teeth forming crushing tooth plates; median eye usually absent, three living genera confined to Australia (*Neoceratodus*), Africa (*Protopterus*) and South America (*Lepidosiren*).

---

*Extinct groups.

ANNOTATED REFERENCES

Vertebrates

The following references contain useful information on many aspects of the biology of fishes and other groups of vertebrates.

Alexander, R. McN.: The Chordates. Cambridge, Cambridge University Press, 1975. An outstanding discussion of the major vertebrate groups in which biomechanical, physiological and ecological factors are skillfully integrated with structure.

Blair, W. F., A. P. Blair, P. Brodkorb, F. R. Cagle and G. A. Moore: Vertebrates of the United States. 2nd ed. New York, McGraw-Hill Book Company, 1968. A very useful taxonomic treatment of the vertebrates of the United States, apart from Alaska and Hawaii. Keys for all groups are included, and each group and species is discussed briefly.

Fishbein, S. L. (Ed.): Our Continent, A Natural History of North America. Washington, National Geographic Society, 1976. Five chapters in this superbly illustrated and authoritative book deal with the evolution of vertebrates.

Gans, C.: Biomechanics, An Approach to Vertebrate Biology. Philadelphia, J. B. Lippincott Company, 1974. The application of physical principles to the solution of biological problems is developed by an analysis of several specific examples of feeding and locomotion in vertebrates.

Gregory, W. K.: Evolution Emerging, A Survey of Changing Patterns from Primeval Life to Man. New York, Macmillan, 1951. A valuable and superbly illustrated source book on vertebrate evolution; includes much original material.

Olson, E. E.: Vertebrate Paleozoology. New York, Wiley-Interscience, 1971. A useful reference on vertebrate paleontology; discusses the adaptive nature of many evolutionary changes.

Orr, R. T.: Vertebrate Biology. 4th ed. Philadelphia, W. B. Saunders Co., 1976. A valuable text and reference on many aspects of vertebrates; a chapter is devoted to each major group of vertebrates and to such general topics as territory, dormancy, and population dynamics.

Romer, A. S.: Notes and Comments on Vertebrate Paleontology. Chicago, University of Chicago Press, 1968. An informal discussion of interesting points and speculations deemed inappropriate for the more formal book on vertebrate paleontology.

Romer, A. S.: Vertebrate Paleontology. 3rd ed. Chicago, University of Chicago Press, 1900. The most recent edition of a standard text and source book.

Romer, A. S. and T. S. Parsons: The Vertebrate Story. 5th ed. Chicago, University of Chicago Press, 1977. A very well written and nontechnical account of the evolution of vertebrates.

Stahl, B. J.: Vertebrate History: Problems in Evolution. New York, McGraw-Hill Book Company, 1974. A fascinating account of vertebrate history with some emphasis upon alternative interpretations of the data and unresolved problems.

Young, J. Z.: The Life of Vertebrates. 2nd ed. Oxford, Clarendon Press, 1963. A fascinating account of the evolution and adaptations of vertebrates. One or more chapters are devoted to each class; morphological and physiological perspectives are skillfully interwoven.

Fishes

Breder, C. M. Jr.: Field Book of Marine Fishes of the Atlantic Coast from Labrador to Texas. 2nd ed. New York, S. P. Putnam's Sons, Inc., 1948. A useful guide for the identification of the more common fishes of our Atlantic Coast.

Herald, E. S.: Living Fishes of the World. Garden City, N.Y., Doubleday & Co., 1961. A very well written, nontechnical account of the biology of the various groups of fish. As with other books in this series, the photographs are superb; many are in color.

Hoar, W. S. and D. J. Randall (Eds.): Fish Physiology. New York, Academic Press, 1969–1970. A valuable source book; chapters on various aspects of fish physiology have been written by leading investigators.

Hubbs, C. L. and K. F. Lagler: Fishes of the Great Lakes Region. 2nd ed. Bloomfield Hills, Mich., Cranbrook Institute of Science, 1958. The definitive guide to fishes of the Great Lakes Basin, this book would also be useful, though not infallible, for adjacent parts of the United States.

Lagler, K. F., J. E. Bardach and R. R. Miller: Ichthyology. New York, John Wiley & Sons, Inc., 1962. A very useful text covering the basic anatomy, physiology, systematics and ecology of fishes.

Marshall, N. B.: Explorations in the Life of Fishes. Cambridge, Harvard University Press, 1971. Deep sea fishes and convergent evolution among fishes are emphasized.

Moy-Thomas, J. A.: Paleozoic Fishes. 2nd ed. by R. S. Miles. Philadelphia, W. B. Saunders Company, 1971. An extensive revision of the classic book on primitive fishes.

Norman, J. R.: A History of Fishes. 2nd ed. Revised by P. H. Greenwood. New York, Hill and Wang, Inc., 1963. A recent edition of a classic on the biology of fishes.

Tee-van, J., H. B. Bigelow and G. W. Mead (Eds.): Fishes of the Western North Atlantic. New Haven, Sears Foundation for Marine Research, Yale University Press, 1948, 1953, 1963, 1964, 1966. Five parts of this definitive treatise on the taxonomy and biology of Atlantic fishes have been published.

VERTEBRATES: AMPHIBIANS AND REPTILES

34.1 THE TRANSITION FROM WATER TO LAND

The transition from fresh water to land was a momentous step in vertebrate evolution that opened up vast new areas for exploitation. It was an extremely difficult step because the physical conditions on land are so very different from those in water. Air neither affords so much support nor offers so much resistance as water. The terrestrial environment provides little of the essential body water and salts. Oxygen is more abundant in the air than in water, but it must be extracted from a different medium. The ambient temperature fluctuates much more on the land than in the water. Air and water have different refractive indices.

Successful adaptation to the terrestrial environment necessitated changes throughout the body. Stronger skeletal support and different methods of locomotion evolved. Changes occurred in the equipment for sensory perception, and changes in the nervous system were a natural corollary of the more complex muscular system and altered sense organs. An efficient method of obtaining oxygen from the air evolved, as did adaptations to prevent desiccation. The delicate, free-swimming, aquatic larval stage was suppressed, and reproduction upon land became possible. Finally, the ability to maintain a fairly constant and high body temperature was achieved, and terrestrial vertebrates could then be active under a wide range of external temperatures.

In view of the magnitude of these changes, it is not surprising that the transition from water to land was not abrupt but took millions of years and involved the participation of many groups. Indeed, a main theme in the evolution of the terrestrial vertebrates, or **tetrapods,** has been a continual improvement in their adjustment to terrestrial conditions.

The crossopterygians unwittingly made the first steps in this transition. Their lungs, as we have seen, were probably an adaptation to survive conditions of stagnant water or temporary drought. Their relatively strong, lobate paired fins may have enabled them to squirm from one drying and overcrowded swamp to another, more favorable one. Crossopterygians were not trying to get onto the land, but, in adapting to their own environment, they evolved features that made them viable in a new and different environment. That is, they became **preadapted** to certain terrestrial conditions (see Section 8.11). Given this preadaptation, an abundance of food (various invertebrates, stranded fishes and plants) upon the land or the shores of swamps, little competition upon the land and a constriction of favorable freshwater habitats, it is not hard to imagine some of the crossopterygians making the adaptive shift from water to land and becoming the amphibians (Fig. 34.1). No one knows how long the transition from crossopterygians to amphibians took, but the first amphibian fossils are found in strata that were formed nearly 50 million years later than those containing the first crossopterygians.

Amphibians in turn acquired additional terrestrial features, and reptiles still more. But the pinnacle of terrestrial adaptation is achieved only by the reptiles' descendants—the birds and mammals.

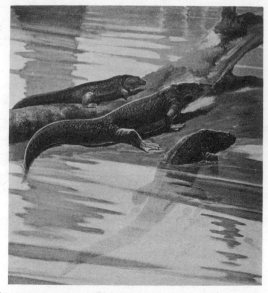

Figure 34.1 A restoration of life in a Carboniferous swamp 345 million years ago. The labyrinthodont amphibians were the first terrestrial vertebrates. (Courtesy of the American Museum of Natural History.)

34.2 GENERAL CHARACTERISTICS OF AMPHIBIANS

Since they are a transitional group between an aquatic and a terrestrial mode of life, frogs, salamanders and other amphibians have a mixture of characteristics. Some of them are adaptations to terrestrial life and are found in all terrestrial vertebrates; other have been retained from the amphibians' freshwater piscine ancestors.

1. Amphibians continue to be ectothermic.
2. Their epidermis is slightly cornified but remains permeable to water. Large mucous glands are abundant. Bony scales are absent in all but a few primitive groups.
3. Successive vertebrae interlock to form a strong yet flexible vertebral column. Ribs are very short and frequently fused to the vertebrae in contemporary species. One pair of sacral ribs articulates the pelvic girdle to the vertebral column. In species living today, the carpus and tarsus are not ossified, and there are never more than four toes in the front foot. The skull tends to be short, broad and incompletely ossified.
4. Movable eyelids and tear glands protect and cleanse the eye. A tympanic membrane and a middle ear cavity are present in frogs but not in salamanders. Vibrations are transmitted to the inner ear by a stapes. Internal nostrils are present. The

lateral line occurs in aquatic larvae but is lost at metamorphosis.
5. Amphibians have a muscular and protrusible tongue. Their intestine is divided into small and large segments.
6. Larval external gills are lost at metamorphosis and gas exchange with the environment is by means of moist membranes in the lungs, skin and buccopharyngeal cavity.
7. Separate right and left atria are present in the heart and receive primarily venous and arterial blood, respectively. These blood streams remain separated to a large extent in their passage through the single ventricle.
8. Most nitrogen is eliminated as urea by opisthonephric kidneys. Amphibians have a urinary bladder.
9. Males lack copulatory organs. Fertilization is external or takes place with the cloaca.
10. Many yolk-laden eggs are produced in large ovaries. Jelly layers are secreted around the eggs as they pass down the oviducts.
11. Most amphibians are oviparous. Aquatic larvae usually undergo metamorphosis into terrestrial adults.

34.3 THE EVOLUTION AND ADAPTATIONS OF AMPHIBIANS

Amphibians (Gr. *amphi* = double + *bios* = life) are well adapted to their double mode of life. The larvae are aquatic and the adults usually are terrestrial. The adults are well suited for life on the land in many respects, but the retention of many freshwater fishlike features limits most species to habitats that are moist and have a high relative humidity. Salamanders, frogs and toads are found near the shores of freshwater ponds and streams, in damp meadows and woods and beneath rocks and logs. One group of rare amphibians (the caecilians, see below) burrows in soft ground.

Labyrinthodonts and the Evolution of Terrestrial Adaptations. The best known of the early amphibians were the **labyrinthodonts,** a subclass that diverged from the crossopterygian fishes during the Devonian period (Fig. 34.2) and became extinct early in the Mesozoic Era. An interesting detail they shared with crossopterygians was a peculiar labyrinthine infolding of the enamel of their teeth. Their name is derived from this feature.

Although labyrinthodonts had certain distinctive characteristics, as a group they evolved most of the features that characterize all amphibians. Most were rather large (as much as 1 meter long) salamander-shaped creatures with rudimentary necks and heavy muscular

Figure 34.2 An evolutionary tree of terrestrial vertebrates. The relationship of the various groups, their relative abundance, and their distribution in time are shown.

tails inherited from their piscine ancestors. Their vertebral column was much stronger than that in fishes. Vertebral centra were more thoroughly ossified, and successive vertebrae interlocked securely by means of overlapping **articular processes** on the vertebral arches (Fig. 11.11, p. 249). The paired fins of fishes were transformed into legs that at first were not powerful. The humerus and femur moved in the horizontal plane rather than being nearly vertical and more or less under the body as they are in mammals. The muscles of the ribs and trunk were powerful, and lateral undulations of the trunk and tail were still important in locomotion. Undulations helped to advance the limbs to a position favorable for placing them on the ground and also helped to develop the thrust of the limb (Fig. 34.3). The footfall pattern seen in labyrinthodonts and living amphibians, and even in the slower moving higher vertebrates that do not undulate their trunk, follows naturally from the positions assumed by the fins of a fish as a result of the undulations of the trunk. Advancement of the left front foot is followed in turn by the right rear foot, right front foot and left rear foot. During most stages of a stride, just one foot is advanced, leaving a tripod of three feet to support the body.

Figure 34.3 A comparison of locomotion in (A) a fish and (B) a labyrinthodont or salamander. Fishlike undulations of trunk and tail continue in the salamander and help to determine the placement of the limbs upon the ground. Numbers indicate the sequence of limb movements. (A after Marshall and Hughes; B after Romer.)

Many changes occurred in the sensory apparatus during the evolution of the amphibians because problems associated with sensory perception on land are quite different from those in water. Early amphibians retained the lateral line system, and it is present in the aquatic larvae of amphibians living today, but it is lost in terrestrial adults. We know little of the details of other sense organs in labyrinthodonts, but in contemporary amphibians, the eye is transformed from an aquatic to an aerial organ at metamorphosis. The lens flattens as the cornea assumes more importance in light refraction, and eyelids and tear glands that protect and cleanse the eye develop. An ear sensitive to air- or ground-borne vibrations appears.

The brain remains essentially fishlike, although subtle changes occur in relation to changing patterns of sensory perception and locomotion. Lateral line centers are absent in adults.

Early labyrinthodonts must have spent more time in the water than most contemporary salamanders and frogs. Their legs were relatively small, and one very early species had fishlike rays supporting a tail fin.

Small bony scales were present in the skin of early labyrinthodonts. This would have limited cutaneous respiration so that air breathing must have depended primarily on lungs, which probably were ventilated by movements of the well developed ribs along with a pumping action of the floor of the mouth and pharynx. The loss of scales and the decrease in body size, which increases body surface area relative to mass, made cutaneous respiration more efficient. The broadening of the head of later labyrinthodonts, with the concomitant widening of the sheets of muscle in the floor of the mouth and pharynx, suggests the increasing importance of the buccopharyngeal pump in lung ventilation (p. 304). In modern amphibians, the ribs are short and frequently fused to the vertebrae so that they have no function in respiratory movements.

The substitution of lungs for gills as the major site of gas exchange has important consequences in the circulatory system. Two types of blood, oxygen-depleted blood from the body and oxygen-rich blood from the lungs, return to the heart of contemporary amphibians. These two blood streams are kept separate to some extent as they flow through the heart (p. 327). Depleted blood for the most part is sent to the lungs and skin, and oxygen-rich blood is distributed to the body. Since blood leaving the heart on its way to the body does not first go through the gills, as it would in a fish, there is little loss of blood pressure. Blood pressure in the dorsal aorta of a frog is nearly double that in the dorsal aorta of a dogfish. In this respect, terrestrial vertebrates have a more efficient circulatory system than do fishes. The presence of a lymphatic system is a correlate of increased blood pressure.

Catching and swallowing food on land is somewhat more complex than in the water, and amphibians have evolved a muscular **tongue** and **oral glands** that aid in these processes.

Limitations of Amphibians' Adaptations. Inability to conserve body water and to control fluctuations in body temperature and the necessity of reproducing in water, or in a very wet environment, are features that restrict all amphibians to moist habitats and prevent them from exploiting fully the entire spectrum of terrestrial environments. Most amphibians are unable to prevent a large loss of body water when on the land. Although there has been a slight cornification of the epidermis, cutaneous respiration necessitates a thin, moist and vascular skin. Considerable evaporation takes place through it. There has been some shift in the nitrogen metabolism of amphibians so that less of the waste products are removed as highly toxic ammonia, and more are removed as **urea** (Table 34.1). Amphibians do not need as rapid a turnover of water as the freshwater fishes do, yet a fairly large volume is still needed to flush these products from the tissues and body. The structure of their kidney tubules is not yet modified for the reabsorption of a large volume of water from the urine during the process of urine formation. Some water is reabsorbed from a urinary bladder.

Amphibians are ectotherms; hence they cannot be active at low environmental and body temperatures. This is not too serious for a fish, for in most bodies of water, temperatures, even beneath the ice, remain above freezing. But terrestrial ectotherms living in temperate regions must move during the winter to areas that do not freeze and enter a dormant state known as **hibernation.** Amphibians bury themselves in the mud at the bottom of ponds or burrow into soft ground below the frost line. During hibernation metabolic activities are at a minimum. The only food utilized is that stored within the body; respiration and circulation are very slow. Some tropical amphibians during the hottest and driest parts of the year go into a comparable dormant state known as **aestivation.**

Amphibians are unable to reproduce under truly terrestrial conditions. Most species must

TABLE 34.1 TYPES OF NITROGEN EXCRETION IN REPRESENTATIVE VERTEBRATES*

Animal	Ammonia	Urea	Uric Acid	Other
Freshwater minnow, Cyprinus	60.0	6.2	0.2	33.6
Bullfrog, Rana	3.2	84.0	0.4	12.4
European tortoise, Testudo	4.1	22.0	51.9	22.0
Hen	3.0	10.0	87.0	0.0
Adult human	3.5	85.0	2.2	9.3

*Data from Prosser and Brown. Figures in per cent of total nitrogen excretion.

return to the water to lay their eggs. Even the terrestrial toad returns to this medium, for it has no means of internal fertilization and sperm cannot be sprayed over eggs upon the land. In most species, an egg develops into an aquatic larva that has a powerful muscular tail, external gills, well developed mouth parts, kidneys and other organs needed to lead an active and independent life in the water. Larval features are lost and adult terrestrial ones are acquired during a period of rapid developmental change known as **metamorphosis.**

Evolution of Amphibians. Labyrinthodonts were the most important group of extinct amphibians, for they included not only the ancestors of other amphibians but also those of reptiles, and hence of all higher tetrapods. Another group that became extinct, the subclass **Lepospondylii** (Fig. 34.2), probably diverged very early from labyrinthodonts. Lepo-

Figure 34.4 A burrowing caecilian, Ichthyophis, with her egg mass. (From Romer after the Sarasins.)

Figure 34.5 Three stages in the leap of a bullfrog. Observe that the pelvic girdle is flexed when the frog is at rest and this causes a characteristic bump in the back (A); the hind legs and pelvic girdle extend at take-off (B); and the frog lands on its front legs (C). (Courtesy of Mr. Earle R. Edmiston.)

spondyls differ from labyrinthodonts in having spool-shaped vertebral centra rather than centra composed of chunks of bone surrounding the notochord. They became extinct at the end of the Carboniferous period without leaving descendants.

Paleontologists believe that the three orders of contemporary amphibians descended from labyrinthodont ancestors, although the structure of their vertebral centra is somewhat similar to that of lepospondyls. They are grouped into a common subclass, the **Lissamphibia,** for all share such features as reduced ribs and, when legs are present, an unossified carpus and tarsus and never more than four toes in the front foot.

Salamanders (order **Urodela**) retain the primitive body form with short legs, long trunk and well developed tail. Caecilians (order **Apoda**) are legless, burrowing, wormlike forms confined to the tropics (Fig. 34.4). Frogs and toads (order **Anura**), with their short trunk and powerful enlarged hind legs, are specialized for a hopping and jumping mode of life (Fig. 34.5). The fossil history of all these groups is fragmentary, but we do have one frog fossil from the early Triassic (Fig. 34.6) that is intermediate between specialized contemporary frogs and more primitive amphibians. Its

Figure 34.6 Restoration of the skeleton of *Triadobatrachus*, a Triassic froglike amphibian possibly ancestral to modern frogs. (After Estes in Vial (Ed.): Evol. Biol. Anurans, University of Missouri Press, 1973.)

skull shape, shortened trunk, elongated ilia in the pelvic girdle and slightly enlarged hind legs are suggestive of a frog. But the trunk is not as short as in modern frogs, there is a bit of a tail and there has been no fusion of the bones of the lower arm and lower leg.

34.4   SALAMANDERS

Salamanders are not such familiar amphibians as frogs and toads, for most have secretive habits. They may be found beneath stones and logs in damp woods or beneath stones along the side of streams, and some are entirely aquatic. A rather generalized type is Jefferson's salamander, *Ambystoma jeffersonianum* (Fig. 34.7), of the eastern United States. This species is terrestrial as an adult but returns to the water in early spring to reproduce. The males deposit sperm in clumps called **spermatophores** on sticks and leaves in the water. Later, the females pick these up with their cloacal lips. Fertilization occurs inside the cloaca, and the fertilized eggs are deposited in masses attached to sticks submerged beneath the surface of the water. This habit is adaptive, for these salamanders reproduce very early in the spring, at a time when the surface of the water may freeze over. The larvae of salamanders differ from those of frogs and toads in retaining external gills throughout their larval life and in having true rather than horny teeth.

There have been many special adaptations among salamanders. The most abundant of our American species are woodland types like the red-backed salamander (*Plethodon cinereus*) of the family Plethodontidae. A particularly interesting feature of plethodonts is their complete loss of lungs; gas exchange occurs entirely across the moist membranes lining the mouth and pharynx and the skin. The skin is a more effective respiratory organ than in frogs because the epidermis is very thin and capillaries come close to the surface. Loss of lungs may seem to be a curious adaptation for a terrestrial vertebrate, but it has been postulated that early in their evolution plethodonts became adapted for life in rapid mountain streams. As we have seen (p. 294), the oxygen content of cold water is rather high, and the low temperature reduces the general rate of metabolism of the animals. Air in the lungs would be disadvantageous under these conditions, for the animals would float and be washed away. Lungs may have been lost in adapting to this habitat. Many plethodonts still live in streams; others have moved onto the land but never regained the lost lungs.

Several groups of salamanders, including the mudpuppy (*Necturus maculosus*, Fig. 34.8A), have become entirely aquatic. This is accomplished by the evolution of neoteny, that is, acquiring sexual maturity relatively early in development and suppressing metamorphosis so that aquatic larval features are retained throughout life. The hormone of the thyroid gland, thyroxine, is necessary for metamorphosis; however, failure of *Necturus* to metamor-

Figure 34.7 Jefferson's salamander, *Ambystoma jeffersonianum*, reproduces in the water. The four white structures attached to sticks are spermatophores. A clump of eggs can be seen in the upper right hand corner.

Figure 34.8 Neotenic salamanders. *A*, The mudpuppy, *Necturus maculosus*, is a permanent larva. *B*, The tiger salamander. *Ambystoma tigrinum*, metamorphoses in most environments, but fails to do so in certain mountain lakes. *C*, The axolotl, or neotenic form of *Ambystoma tigrinum*. (*A* courtesy of Shedd Aquarium, Chicago; *B* and *C* courtesy of the Philadelphia Zoological Society.)

phose appears to result from the inability of the tissues to respond to thyroxine rather than from an absence of this hormone. Thyroxine is produced, for the thyroid of *Necturus* hastens metamorphosis when transplanted to frog tadpoles. In some other neotenic salamanders, the failure to metamorphose may result from an inhibition of the mechanism that releases thyroxine. The tiger salamander, *Ambystoma tigrinum* (Fig. 34.8 *B*), metamorphoses under most conditions, but those living at high altitudes in the Rocky Mountains fail to do so and remain permanent larvae known as **axolotls.** Apparently, cold inhibits the release of thyroxine, for when axolotls are fed thyroxine or when they are brought to warmer climates, metamorphosis occurs normally.

34.5 FROGS AND TOADS

Most anurans are amphibious as adults, living near water to which they frequently go to feed or escape danger, but some are more terrestrial in habits, and others have become adapted to an arboreal life. The terms frog, toad

and tree frog or tree toad ordinarily imply amphibious, terrestrial and arboreal modes of life, not natural evolutionary groups. Members of several distinct families of anurans, for example, have become adapted independently to an arboreal life.

Toads are better adapted to terrestrial life than are frogs, but we do not know all of the modifications that have brought this about. Toad skin is somewhat more cornified than frog skin, but toads lose as much water by the skin as frogs do under comparable experimental conditions The behavior pattern of the toads tend to minimize water loss through the skin. Most toads are crepuscular in habit; they burrow or take shelter by day and come out to feed in the evening when the relative humidity is greater. If need be, toads can tolerate a greater loss of body water than frogs can.

The chief adaptation to arboreal life has been the evolution of **digital pads** upon the tips of the toes (Fig. 34.9). The surface epithelium of the pads is rough and grips the substratum by friction.

A particularly fascinating aspect of anuran biology is the evolution of methods by which development can proceed elsewhere than in the open water. In some cases, development takes place within a part of the body of one parent, and the young emerge as miniature adults. There has been a convergence with the viviparity of some reptiles and mammals, but the site of development is never the uterus and a placenta is not present. Modifications in development have occurred primarily among tropical frogs and probably as a protection against seasonal drought and numerous aquatic enemies, such as predaceous insect larvae. A Brazilian tree frog (*Hyla faber*) protects its young by laying its eggs in mud craters that the male builds by circling in shallow water and pushing up mud (Fig. 34.10A). A more striking means of protection is seen in a small Chilean frog, *Rhinoderma darwinii*. The male of this species stuffs the fertilized eggs into his vocal sacs where they remain until metamorphosis is complete. In the completely aquatic Surinam toad, *Pipa pipa*, of tropical South America, the male pushes the fertilized eggs into the puffy skin on the back of the female. The eggs gradually sink deeper into the skin, and the outer membrane of each one forms a protective cap over its temporary skin pocket. The young emerge as little froglets. The eggs and larvae of all these frogs are equipped with a large supply of yolk, but in other respects the larvae are fairly typical and simply develop in a sheltered environment.

In certain species, the vulnerable larval stage is omitted, and the embryo develops directly into a miniature adult. Anurans with **direct development** include the marsupial frog, *Gastrotheca*, and *Eleutherodactylus*—both of the New World tropics. The former carries her eggs in a dorsal brood pouch (Fig. 34.10C); the latter lays eggs in protected damp places, such as beneath stones or in the axil of leaves. The jelly layers about the egg of *Eleutherodactylus* help to prevent desiccation; sufficient yolk is stored within the egg for the nutritive requirements of the embryo; such larval features as horny teeth, gills and opercular fold are vestigial or absent; the fins of the larval tail are expanded, become highly vascular and form an organ for gas exchange; and the period of development is shortened.

Something similar to what has taken place among these frogs today may have occurred among the amphibians that were ancestral to reptiles. If the aquatic larvae of the ancestral amphibians were subjected to a very high predation, any variation that tended toward the suppression of the defenseless larval stage and toward the direct development of their embryos in a less vulnerable environment would have a selective advantage. It is possible that the labyrinthodont amphibians that gave rise to the reptiles developed means of terrestrial reproduction before the adults completely left the water to live on the land.

Figure 34.9 The tree frog *Hyla faber* clings to trees by means of its expanded digital pads. (Courtesy of G. A. Lutz, Natural History, Vol. LXX, No. 1.)

Figure 34.10 Adaptations of frogs that protect the larvae from aquatic predators. *A*, The circular mud crater nests of the Brazilian tree frog, *Hyla faber*; *B*, *Pipa pipa*, the female's back is covered with eggs that by the third day have begun the process of sinking into the skin; *C*, the marsupial frog, *Gastrotheca*, carries eggs in a brood pouch on her back. (*A* from G. A. Lutz, Natural History, Vol. LXX, No. 1; *B* from G. Rabb, Natural History, Vol. LXX, No. 5; *C* from E. S. Ross in Orr, R. T.: Vertebrate Biology. 4th ed. Philadelphia, W. B. Saunders Co., 1976.)

34.6 GENERAL CHARACTERISTICS OF REPTILES

Turtles, lizards, snakes, crocodiles and other reptiles are adapted more completely for terrestrial life than are amphibians. They have successfully occupied most of the terrestrial habitats, except for those where temperatures fall very low. No reptiles live beyond the line of permafrost, for example. Some have readapted very well to an aquatic life.

1. Some reptiles are ectothermic; others are heliothermic.

2. Their heavily cornified epidermis forms horny scales or plates over the body surface. Cutaneous glands are absent except for a few scent glands.

3. Most of the trunk ribs are long, and the anterior ones articulate with a sternum. Typically, two sacral vertebrae and rib pairs articulate with the pelvic girdle. The carpus and tarsus are ossified, and five toes are present on the front foot. The toes are clawed. The skull is deeper and narrower than that in amphibians.

4. A nictitating membrane helps protect the eye. The tympanic membrane frequently is located at the bottom of an external auditory meatus. A lateral line system is never present.

5. Enlargement of the cerebrum results primarily from the expansion of deep-seated masses of gray matter (the corpus striatum). No cerebral cortex is present.

6. The internal surface of the lungs is quite large, and the lungs are ventilated in most groups by rib movements.

7. The heart ventricle is partly divided. A complex tripartite division of the conus arteriosus shunts venous blood to the lungs, arterial blood to the head and both arterial and mixed blood to the rest of the body.

8. Most nitrogenous wastes are eliminated as urea and uric acid through metanephric kidneys. A large amount of water is reabsorbed in the urinary bladder.

9. Males have a copulatory organ, and fertilization is internal.

10. The large ovaries produce fewer eggs than those in lower vertebrates, but the eggs contain much more yolk. Albuminous materials and a shell are secreted around the fertilized eggs as they pass down the oviduct.

11. Primitive reptiles are oviparous and lay their eggs on land. The cleidoic egg has sufficient yolk, and the embryo soon develops important extraembryonic membranes so that development can proceed directly to a miniature adult before hatching. A few reptiles have become ovoviviparous and viviparous.

34.7 MAJOR ADAPTATIONS OF REPTILES

Reptiles are more active and agile vertebrates than are amphibians. Improvement in adaptations that permit a more complete exploitation of the land can be seen in their locomotion, motor coordination and most other systems, as summarized in the list of general characteristics. Particularly significant advances have occurred in their ability to conserve body water, regulate their body temperature to some extent and reproduce upon the land.

The surface of the skin is covered with dry **horny scales** that minimize water loss by this route. These scales develop through the deposition of **keratin** (a very insoluble and hence waterproofing protein) in the superficial layers of the epidermis. Some lizards and crocodilians also have small plates of dermal bone imbedded in the dermis of the skin beneath many of the horny scales.

The kidney tubules of reptiles are modified in such a way that less water is initially removed from the blood than in amphibians, and much of the water that is removed is later reabsorbed by other parts of the kidney tubule and by the urinary bladder. In reptiles, a large proportion of the nitrogenous waste products is excreted as **uric acid** (Table 34.1). Uric acid is much less toxic and water soluble than ammonia or urea, hence it does not have to be flushed out of the tissues so rapidly nor with such a large volume of water. The urine of animals excreting uric acid typically has a pastelike consistency. The reptilian kidney also differs from that of lower vertebrates for it is a **metanephros.** Kidney functions are limited to the caudal portion of the more primitive opisthonephros, which is now drained by the **ureter.** The anterior parts of the opisthonephros and the archinephric duct become exclusively reproductive passages in the male (epididymis and ductus deferens, respectively) and are lost in females.

The dry horny skin of reptiles reduces cutaneous respiration to a negligible amount, but an increase in the respiratory surface of the lungs not only compensates for this but also provides for the increased volume of gas exchange necessitated by a general increase in activity. Mechanisms for moving air into and out of the lungs are also more efficient. Instead of pumping air into the lungs by froglike throat movements, reptiles decrease the pressure within their body cavity, and atmospheric pressure drives in air. A subatmospheric pressure is created around the lungs during inspiration by the forward movement of the ribs and the concomitant increase in size of the body cavity. The contraction of abdominal muscles and the elastic recoil of the lungs force out air. Circulatory changes, previously discussed, further separate the oxygenated and unoxygenated blood leaving the heart and make the oxygen supply to the tissues more effective.

Although reptiles are still ectotherms, many can control their body temperature within narrow limits during the daytime by changes in behavior. Studies made on spiny lizards of the southwestern deserts of the United States show that they can maintain their body temperature at about 34° C (93° F) during much of the day. If body temperatures fall below the threshold for normal activity, the lizards lie at right angles to the sun's rays, thereby exposing a maximum body surface to the sun. If body temperatures rise too far, they seek shelter or lie parallel to the sun's rays. The chuckwalla (*Sauromalus*) can also eliminate body heat by panting (Fig. 34.11). Color changes are an additional means of temperature regulation in some lizards. In cold weather, the skin is dark and absorbs a maximum amount of heat; as temperatures rise, pigment migrates to the center of the chromatophores and the skin becomes lighter in color. All these methods are useful only during the daytime. At night, body temperature falls to ambient levels, and in prolonged cold weather reptiles become dormant.

Major changes have come about in the method of reproduction. Male reptiles have evolved **copulatory organs** that introduce the sperm directly into the female reproductive tract. Fertilization is internal, and the delicate sperm are not exposed to the external environment. A large quantity of nutritive **yolk** is stored within the egg while it is still in the ovary. As the eggs pass down the oviduct after ovulation, they are fertilized, and additional substances and a **shell** are secreted around each one by certain oviducal cells. **Albumin** and similar materials around the egg provide additional food, ions and water. The leathery or calcareous shell serves for protection against mechanical injury and desiccation, yet it is porous enough to permit gas exchange. Such an egg, which contains, or has the means of providing, all substances necessary for the complete development of the embryo to a miniature adult, is called a **cleidoic egg.** Reptiles lay fewer eggs than lower vertebrates but the eggs are larger, better equipped and laid in sheltered situations so that the mortality is low. A collared lizard lays only four to 24 eggs in contrast to the 2000 or 3000 of the leopard frog.

As the embryo develops, it separates from the yolk, which becomes suspended in a **yolk sac** (Fig. 34.12). A protective **chorion** and **amnion** develop as folds over the embryo (see Section 20.7). These two membranes appear to be derived phylogenetically from something similar to the superficial layers covering the yolk sac of certain large yolked fish embryos. Another membrane, the **allantois,** is a saclike outgrowth from the embryo's hindgut. It is homologous to the urinary bladder but extends beyond the body wall, passing between the amnion and the chorion. It functions in gas exchange and excre-

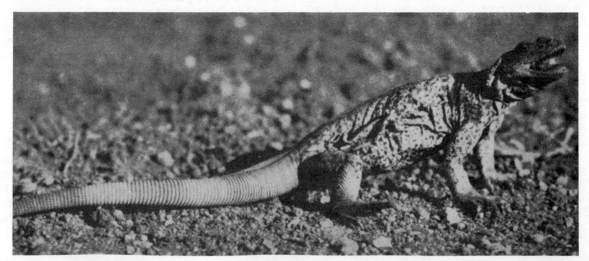

Figure 34.11 The chuckwalla (*Sauromalus obesus*) lives in the deserts of the southwestern United States. When unable to shelter, it prevents overheating by panting and placing its body parallel to the sun's rays. (From How Reptiles Regulate Their Body Temperature by C. M. Bogert. Copyright © by Scientific American, Inc. All rights reserved.)

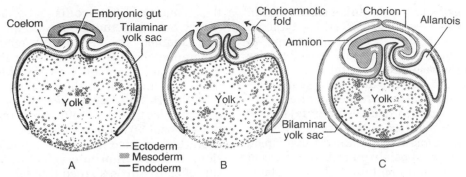

Figure 34.12 Sections of vertebrate embryos to show the extraembryonic membranes. *A*, The trilaminar yolk sac of a large yolk fish embryo consists of all three germ layers. *B*, The chorioamniotic folds of an early embryo of a reptile appear to have evolved from the ectoderm and part of the mesoderm of a trilaminar yolk sac. *C*, A later reptile embryo in which the extra-embryonic membranes are complete. Notice that the yolk sac is bilaminar. The albumin and shell, which surround the reptile embryo and extraembryonic membranes, have not been shown.

tion. It has been postulated that the evolution of the cleidoic egg was correlated with a shift in nitrogen metabolism. Little water is available to carry off ammonia and urea in the terrestrial environment in which the eggs are laid, and nitrogenous wastes cannot be eliminated in gaseous form; consequently much of it must be converted to inert uric acid, which can be stored. These adaptations for terrestrial reproduction are found in the embryos of all reptiles, birds and mammals. These groups of vertebrates are often called **amniotes.** In contrast, the various fish groups and amphibians lacking an amnion are called **anamniotes.**

34.8 EVOLUTION OF REPTILES

Stem Reptiles. Having "solved" the essential problems of terrestrial life at a time when there were few competitors upon the land, the

reptiles multiplied rapidly, spread into all of the ecological niches available to them, and became specialized accordingly. Much of their adaptive radiation has involved adaptation to different methods of locomotion and feeding. Different feeding patterns have entailed, among other things, modification of the jaw muscles, and this in turn has affected the structure of the temporal region of the skull. Skull morphology, therefore, provides a convenient way to sort out the various lines of reptile evolution. Only a small part of the skull of a primitive tetrapod actually surrounds the brain (Fig. 34.13A). A large space lies lateral to the brain case and posterior to the eyes, and it is largely filled with jaw muscles that extend down to the lower jaw. In ancestral reptiles belonging to the order **Cotylosauria,** as in earlier vertebrates, a solid roof of dermal bone covered these muscles dorsally and laterally (Fig. 34.13A and B). This is referred to as the **anapsid** condition. As a

Figure 34.13 Diagrams to show the types of temporal roofs found in reptile skulls. *A*, Posterior view of the skull of a cotylosaur to show the relationship between temporal roof and brain case; *B*, the anapsid condition; *C*, the synapsid condition; *D*, The euryapsid condition; *E*, the diapsid condition. (*A* from Romer after Price; *B–E* from Romer.)

consequence of some brain enlargement and an enlargement of jaw muscles, various types of openings have evolved in the temporal roof of most later reptilian groups. Jaw muscles arise from the brain case and from the periphery of the temporal fenestrae, and they can bulge through the openings when they contract.

Turtles. Turtles (order **Testudinata**) are believed to be direct descendants of cotylosaurs (Fig. 34.2), for most have retained the anapsid skull, but they are specialized by being encased in a protective shell composed of bony plates overlaid by horny scales. The bony plates have ossified in the dermis of the skin, but they have also fused with the ribs and some other deeper parts of the skeleton. The portion of the shell covering the back is known as the **carapace;** the ventral portion, the **plastron.**

Ancestral turtles were stiff-necked creatures, unable to retract their heads, but modern species can withdraw theirs into the shell. This is accomplished by bending the neck in an S-shaped loop in either the vertical plane (North American species such as the red-eared turtle, *Pseudemys scripta elegans*) or in the horizontal plane (Australian side-necked tur-

tle, *Chelodina longicollis*, Fig. 34.14). Sea turtles belong to the former group. They have also adapted to an aquatic mode of life; swimming about by means of oarlike flippers. They come ashore only to lay their eggs in holes that they dig on the beaches.

Marine Blind Alleys. Sea turtles are not the only reptiles that have returned to the ocean. In the Mesozoic, two important lines of reptilian evolution adapted to marine conditions. Plesiosaurs (order **Sauropterygia,** Fig. 34.15) were superficially turtle-shaped (though they lacked the shell), with squat heavy bodies and long necks. Some species reached a length of 12 meters. They propelled themselves by means of large paddle-shaped appendages. Members of the other line, the ichthyosaurs (order **Ichthyosauria,** Fig. 34.15), were porpoise-like in size and shape, and probably in habits. They moved with fishlike undulations of the trunk. Both groups are characterized by having a temporal opening high up on the side of the skull (Fig. 34.13D).

Plesiosaurs probably came onto the beaches to lay their eggs, but the extreme aquatic adaptation of the ichthyosaurs would have precluded

Figure 34.14 *A,* Copulating loggerhead sea turtles, *Caretta caretta; B,* sea turtle laying eggs; *C,* diagram showing retraction of head into the shell by bending the neck in the vertical plane; *D,* the Australian side-necked turtle, *Chelodina longicollis.* (*A,* photograph by Frank Essapian, courtesy of Marine Studios; *B,* LIFE photo by Fritz Goro. © Time, Inc.; *C* from Gregory, William King: Evolution Emerging, Volume II, The Macmillan Co., 1951; *D* courtesy of the New York Zoological Society.)

Figure 34.15 Aquatic reptiles of the Mesozoic era. Plesiosaurs on the left; ichthyosaurs on the right. Plesiosaurs reached a length of 12 meters; ichthyosaurs, a length of about 3 meters. (Courtesy of the Chicago Museum of Natural History.)

their doing so. How then did they reproduce, for cleidoic eggs cannot develop submerged in water? In an unusual fossil, several small ichthyosaurs are lodged in the posterior part of the mother's abdominal cavity, and one individual is part way out the cloaca. These must have been offspring about to be born, for the skeletons of individuals that had been eaten would not remain intact during a passage through the digestive tract. Apparently, these reptiles, like some modern lizards and snakes, were viviparous or ovoviviparous, the eggs being retained in the oviduct until embryonic development was complete.

These marine reptiles flourished during the Mesozoic, competing with the more primitive kinds of fishes. Just why they became extinct near the close of this era is uncertain, but their extinction coincides with the evolution and increase of the teleosts. Possibly they could not compete successfully with these fishes.

Lizard-like Reptiles. The most abundant of our present-day reptiles are the lizard-like ones, of which lizards and snakes are the most familiar examples. The most primitive living member of this group is the tuatara (*Sphenodon*, Fig. 34.16)—the only surviving representative of the order **Rhynchocephalia.** Rhyn-

Figure 34.16 The tuatara, *Sphenodon*, is one of the most primitive of living reptiles. (Courtesy of the New York Zoological Society.)

chocephalians are lizard-like in general appearance and are characterized by having a skull with two temporal openings (**diapsid** condition), one above and one below the squamosal-postorbital bar (Fig. 34.13E). At one time, the group was very widespread, but now it is limited to a few small islands off the coast of New Zealand. *Sphenodon* is a surviving "fossil," for it has not changed greatly from species that were living 150 million years ago.

Lizards and snakes, though superficially different from each other, are similar enough in basic structure to be placed in the single order **Squamata.** Lizards (suborder **Lacertilia**) are the older and more primitive. They probably evolved from some rhynchocephalian-like ancestor early in the Mesozoic era but have lost more of the temporal roof including the bar of bone just beneath the lower temporal opening. For the most part, lizards are diurnal terrestrial quadrupeds, but, like other successful groups, they have undergone an extensive adaptive radiation (Fig. 34.17).

Several groups have become arboreal and evolved interesting adaptations for climbing. The true chameleon of Africa (not to be confused with the circus chameleon of our Southeast) has a prehensile tail and an odd foot structure in which the toes of each foot are fused together into two groups that oppose each other like the jaws of a pair of pliers. Geckos, in contrast, cling to trees by means of expanded digital pads. Numerous fine ridges on the undersurface of the pads increase the friction.

Many lizards, including the horned toads of our Southwest (*Phrynosoma*), burrow to some extent for protection, and some have taken to a burrowing mode of life. Appendages are lost in many burrowing lizards, though vestiges of girdles are present. The eyes may be reduced, and the body form becomes wormlike. The glass snake (*Ophisaurus*), although it burrows only part of the time, is a lizard of this type. The glass snake derives its name from its ability to break off its tail when seized. The tail, which constitutes about two-thirds of the animal's length, fragments into many pieces that writhe about, attracting attention while the lizard moves quietly away. Other lizards also have this ability, though developed to a less spectacular degree. Lost tails are regenerated, but

Figure 34.17 Adaptive radiation among lizards. *A*, The Old World chameleon has grasping feet and a prehensile tail with which to climb about the trees. *B*, The gecko climbs by means of digital pads. *C*, The horned-toad, *Phrynosoma*, is a ground-dwelling species that often burrows. *D*, The glass snake, *Ophisaurus*, also burrows. *E*, The Gila monster, *Heloderma*, and a related Mexican species are the only poisonous lizards in the world. (Courtesy of the New York Zoological Society.)

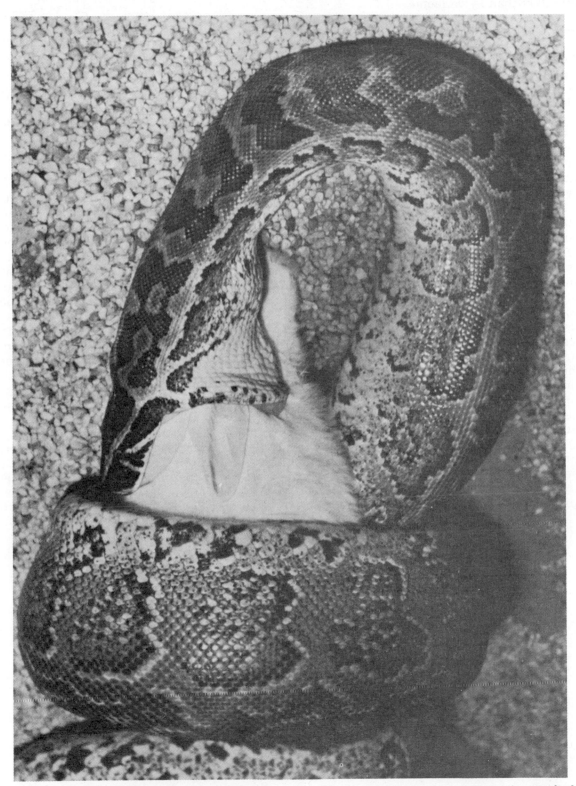

Figure 34.18 An Indian python *(Python molurus)* swallowing a rabbit. (From Bellairs: The Life of Reptiles, Weidenfeld and Nicolson.)

the new tails are supported by a cartilaginous rod rather than by vertebrae.

The only poisonous lizards are the beaded lizards, such as the Gila monster (*Heloderma*) of the Southwestern United States. Modified glands in the floor of the mouth discharge a neurotoxic poison, which is injected into the victim by means of grooved teeth. This is a relatively inefficient method, so the bite is not so dangerous as the bite of most poisonous snakes. Charles Bogert of the American Museum of Natural History reports that eight of 34 bites that have come to his attention were fatal, and he believes that the majority of minor bites are never reported. It is probable that the poison is used for defense rather than for killing prey, for the Gila monster crushes its food with its powerful jaws.

Snakes (suborder **Ophidia**) differ from lizards most notably in being able to swallow animals several times their own diameters (Fig. 34.18). This is made possible by an unusually flexible jaw mechanism, which results in part from the further "erosion" of the temporal roof. Snakes have also lost the squamosal-postorbital bar of bone, and this permits the quadrate bone (the skull bone that in all reptiles and lower vertebrates articulates with the lower jaw) to participate in jaw movements. In a sense, snakes have five jaw joints (Fig. 34.19): (1) the usual one between quadrate and lower jaw, (2) one between quadrate and squamosal, (3) one between squamosal and brain case, (4) one about half-way along the lower jaw and (5) one at the chin, for the two lower jaws are not united at this point. Other features that charac-

Figure 34.19   Lateral view of the skull of a python to show the five major points at which motion can occur when the jaws are opened. (From Romer after M. Smith.)

terize snakes are the absence of movable eyelids, of a tympanic membrane and middle ear cavity and of legs and girdles. The absence of the pectoral girdle is a necessary correlate of swallowing animals larger than the diameter of the body. There are exceptions to these generalizations, for geckos do not have movable eyelids, glass "snakes" lack legs and some of the more primitive snakes, such as the python, have vestigial hind legs.

Snakes doubtless evolved from some primitive lizard group, and very probably from burrowing members of that group. The most primitive living snakes are burrowing species, and many details of ophidean anatomy suggest a fossorial ancestry. The absence of a muscular ciliary body and other features of their eyes, for example, indicate that the eyes redeveloped from eyes that had undergone marked retrogressive changes. Their forked tongue, which is often seen darting from the mouth (Fig. 34.20A), is an organ concerned with touch and smelling. Odorous particles adhere to it, the

Figure 34.20   *A,* A coachwhip snake protruding its tongue; *B,* "milking" a rattlesnake to get poison for the production of antivenom. The tongue is a tactile and olfactory organ that is perfectly harmless; it should not be confused with fangs, which are specialized teeth. (*A* courtesy of the New York Zoological Society; *B* courtesy of Ross Allen's Reptile Institute.)

tongue is withdrawn into the mouth, and the tip is projected into a specialized part of the nasal cavity (Jacobson's organ). The great elaboration of such a device would seem to be an adaptation to a burrowing mode of life in which other senses would be less useful.

In their subsequent evolution some snakes gave up the burrowing habit, adjusted to epigean life and underwent an extensive adaptive radiation. A pattern of locomotion evolved that is dependent primarily upon undulatory movements of the long trunk. When a snake moves through grass, for example, loops of the trunk form behind the head and move posteriorly. When these loops meet protuberances from the ground, they push upon them and the resultants of these forces move the snake forward.

Of particular interest has been the evolution in several distinct groups of snakes of methods of prey immobilization. If a snake is to swallow an animal larger in diameter than itself, there is a distinct advantage in having the animal quiet, particularly if it is an animal, such as a rodent, capable of inflicting a severe bite! Boa constrictors, pythons and many harmless rodent-eating snakes, such as king snakes, quickly entwine their prey in loops of the trunk and gradually suffocate the prey by stopping their respiratory movements. A number of other groups have evolved a poison mechanism consisting of specialized oral glands associated with grooved or hollow hypodermic-like teeth—the **fangs.** Old World vipers and New World pit vipers (rattlesnakes, copperheads, cottonmouths and water moccasins) have a pair of large hollow fangs at the front of the mouth that are articulated to bones of the upper jaw and palate in such a way that they are folded against the roof of the mouth when the mouth is closed and automatically brought forward when the mouth is opened (Fig. 34.20B). The poison of many of the snakes is hemolytic and causes a breakdown of the red blood cells in the animal bitten. Coral snakes belong to a group related to the Old World cobras. Their poison is neurotoxic, and their fangs are a pair of hollow, short, stationary teeth attached to the front of the upper jaw.

Pit vipers differ from other poisonous snakes in having a prominent **sensory pit** on each side of their heads between the nostril and eye. Experiments in which other sense organs have been rendered functionless show that the pit enables these snakes to seek out and strike accurately at objects warmer than their surroundings. The pits probably help the snakes to feed upon small nocturnal mammals.

Dinosaurs and Their Allies. Lizards and snakes are the successful reptiles today, but during the Mesozoic era the land was dominated by another offshoot of primitive rhynchocephalian-like reptiles. These "ruling reptiles" were the **archosaurs**—an assemblage of several orders that shared many features, including a diapsid skull and a tendency to evolve a two-legged gait. Reduced pectoral appendages, enlarged pelvic appendages and a heavy tail that could act as a counterbalance for the trunk were correlated with this method of locomotion.

Saurischian dinosaurs (order **Saurischia**) evolved from ancestors that were approximately 1 meter long, but later saurischians became giants of the land and swamps. *Tyrannosaurus* (Fig. 34.21A) was the largest terrestrial carnivore that the world has ever seen. It stood about 6 meters high and had large jaws armed with dagger-like teeth 15 cm. long—a truly formidable creature! Other saurischian dinosaurs were herbivores that reverted to a quadruped gait, but their bipedal ancestry is reflected in their retention of a heavy tail and in having hind legs larger than the front ones. *Brontosaurus* (Fig. 34.21B and C) was an enormous herbivore that attained a length of 25 meters and an estimated weight of 45 metric tons. Its huge size led to the hypothesis that it lived in swamps where it would be supported partly by the buoyancy of the water. A more recent view is that it fed in a giraffe-like fashion upon high vegetation. Its limb structure resembled that of elephants, and its tail was not flattened, as is usually the case in aquatic vertebrates.

Dinosaurs belonging to the order **Ornithischia** were all herbivores. Some lived in swamps, others in the uplands. Many reverted to a quadruped gait and increased in size, though none was so large as the saurischians. These animals undoubtedly formed much of the diet of carnivores, such as *Tyrannosaurus*, and many evolved protective devices, such as spiked tails, bony plates on the body and horned skulls. *Stegosaurus* and *Triceratops* (Fig. 34.21D and E) are examples of this group.

The reasons for the evolution of large size are not entirely clear. Within limits, large size has a protective value, but it may also have been a way of achieving a more nearly constant body temperature. Ectothermic and heliothermic reptiles derive a great deal of their body heat in a warm climate from the external environment. As mass increases, the relative amount of body surface available for the absorption of heat

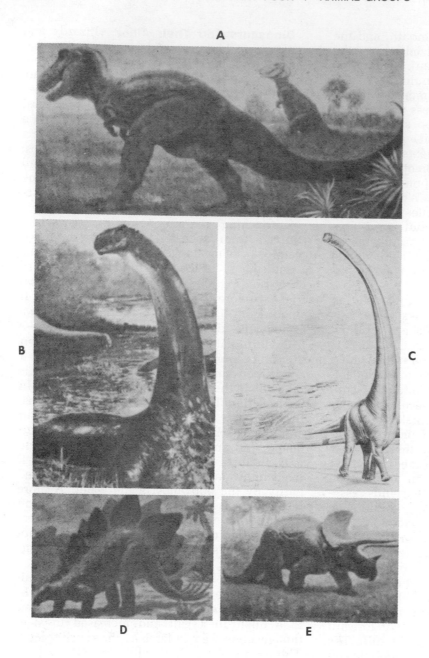

Figure 34.21 Representatives of the main groups of dinosaurs that flourished during the late Mesozoic era. *A, Tyrannosaurus,* a carnivorous saurischian; *B* and *C, Brontosaurus* reconstructed as a swamp dweller or browser of high vegetation; *D, Stegosaurus,* an ornithischian; *E, Triceratops,* another ornithischian. (*C* courtesy of Fauna; others, courtesy of the Chicago Museum of Natural History.)

decreases. An adaptation of this type may have been particularly important for animals that lived in a warm climate and were too big to shelter by burrowing or hiding beneath debris, for it would help to prevent body temperature from reaching a lethal point. As explained earlier, prolonged high temperature destroys most enzyme systems.

Dinosaurs of both orders carried their limbs in mammal-like fashion, more or less under their bodies. John Ostrom of Yale University believes that this implies that they were active animals capable of rapid locomotion. The greater energy expenditure in turn suggests

some degree of endothermism. Robert Bakker of Johns Hopkins University believes that dinosaurs were indeed active endotherms. He points out that their bone histology was closer to mammalian than to reptilian bone; a few may have lived in rather cool climates, and analysis of fossil communities suggests that the ratio of carnivorous dinosaurs to their herbivorous prey was closer to the ratio we see for a lion to its prey than that for a carnivorous lizard to its prey. A warm-blooded predator requires about 10 times as much food supply as a cold-blooded one does. These lines of evidence should be explored further, but not all paleontologists agree that the underlying data are sufficiently clear at this time to be conclusive.

A bipedal gait naturally freed the front legs from use in terrestrial locomotion. The front legs became reduced in many dinosaurs, but in one group of archosaurs, they evolved into wings. The wings of the flying reptiles (order **Pterosauria**) consisted of a membrane of skin supported by a greatly elongated fourth finger (Fig. 34.22). The fifth finger was lost, and the others probably were used for clinging to cliffs. The hind legs were very feeble, and the animal must have been nearly helpless on the ground. Certain pterosaurs became very large, one having a wing spread of over 6 meters. Again the energy needed for flight imples endothermism, and the pterosaurs may have been warm-blooded. Support for this notion has come from A. G. Sharvov's discovery in 1970 in Russia of a specimen in which the fossil shows clear impressions of fur!

Figure 34.22 *Pteranodon,* the largest of the flying reptiles, or pterosaurs, that lived during the late Mesozoic era. (Courtesy of the American Museum of Natural History.)

Most of the archosaurs became extinct toward the end of the Mesozoic, but the reason for this is not entirely clear. Perhaps the pterosaurs succumbed in competition with birds, which also evolved from primitive bipedal archosaurs. The extinction of the dinosaurs may have resulted from climatic changes. At the end of the Cretaceous period, environmental temperatures became somewhat cooler, and large inland seas that would have buffered temperature fluctuations disappeared. Seasonal or even diurnal temperature changes were probably more pronounced than earlier in the Mesozoic Era. The inability of large dinosaurs to retreat to shelter or to hibernate would make them very vulnerable, especially if their ability to conserve body heat was rudimentary. Changes in vegetation that accompanied climatic changes may also have been a factor. An inability of the specialized herbivores to adapt to changes in vegetation would have led to their death. Their disappearance, in turn, would deprive the huge carnivores of most of their food supply. The factors mentioned may not have wiped out a population but simply reduced it to a size at which it became very difficult for the species to maintain itself.

Only one group of archosaurs survived this wholesale extinction—the alligators and crocodiles (order **Crocodilia**). Crocodiles have reverted to a quadruped gait (though their hind legs are much longer than the front) and an amphibious mode of life. Only two species occur in the United States—the American alligator, which can be distinguished by its rounded snout, and the American crocodile, which has a much more pointed snout (Fig. 34.23).

Mammal-like Reptiles. Another line of evolution, which was destined to lead to mammals, diverged from the cotylosaurs millions of years before the advent of archosaurs. Early mammal-like reptiles (order **Pelycosauria**) were very similar to cotylosaurs, differing primarily in having a skull with a temporal opening ventral to the squamosal and postorbital bones (**synapsid** condition, Fig. 34.13C). They were medium-sized, somewhat clumsy, terrestrial quadrupeds with limbs sprawled out at right angles to the body. Some pelycosaurs (*Dimetrodon,* Fig. 34.24A) had bizarre sails upon their backs supported by elongated neural spines. There has been much speculation as to the sail's function, but it did help to maintain a constant ratio of body surface to mass as members of this evolutionary line became larger. It has been suggested that the sail was a primitive thermoregulator, acting as a heat receptor or radiator.

Figure 34.23 The broad-nosed caiman (*Caiman latirostris*) from South America. The nose, eye, and ear (arrow) of a crocodilian can remain above the surface when the rest of the body is submerged. (From Bellairs: The Life of Reptiles, Weidenfeld and Nicolson.)

Figure 34.24 Two mammal-like reptiles. *A*, *Dimetrodon*, an early member of the group; *B*, *Lycaenops*, a later mammal-like reptile similar to those that gave rise to mammals. (Courtesy of the American Museum of Natural History.)

Later mammal-like reptiles (*Lycaenops*, order **Therapsida,** Fig. 34.24*B*) came to resemble mammals more closely. Their limbs were more nearly beneath the body, where they could provide better support and move more rapidly back and forth. Their jaws were strong, and their teeth were specialized, like those of a mammal, into ones suited for cropping, stabbing, cutting and sometimes grinding. A secondary palate separated the mouth and nasal passages so that breathing could continue as the animal chewed its food. Long ribs were limited to the thorax, as in mammals, and lumbar ribs were very short. These features are indicative of a diaphragm. The existence of numerous pits for blood vessels and nerves in the snout bones suggests a bare and moist nasal area (rhinarium), seen in mammals, and possibly the presence of modified facial hairs, the whiskers or vibrissae. The late mammal-like reptiles were very active creatures that could eat and breath efficiently enough to sustain a high level of metabolism. Quite possibly they were haired endotherms. Increased activity would generate heat internally, and hairs, if they were present, would have reduced its loss. Mechanisms for dissipating excess heat, such as sweat glands and panting, may have been rudimentary, as they are in contemporary primitive mammals—the spiny anteater and duckbilled platypus. During the early Mesozoic Era, climates moderated, and we see a decrease in body size in the later mammal-like reptiles. This may have been an adaptation to lose body heat because surface area would increase relative to mass.

Why then were late therapsids not mammals? The matter is largely one of definition. We know nothing about the presence or absence of milk-producing mammary glands, one character used to define contemporary mammals. The major osteological character that separated them from mammals was the nature of the jaw joint and the sound-transmitting apparatus. In all nonmammalian vertebrates, the jaw joint lies between the posterior ends of the mandibular arch, which are usually ossified as a **quadrate bone** in the upper jaw and an **articular bone** in the lower jaw. In most terrestrial vertebrates, a **stapes** (a derivative of the fish hyoid arch) transmits airborne vibrations from a tympanic membrane to the inner ear located within the skull (cf. frog, Fig. 16.8, p. 369). This may have been the case in mammal-like reptiles, although we cannot be sure, for a soft part, such as a tympanic membrane, is not preserved in fossils. In mammals, the jaw joint is between two dermal bones (the **dentary** of the lower jaw and **squamosal** of the upper jaw) that lie just anterior to the quadrate and articular. The mammalian homologues of the quadrate and articular (the **incus** and **malleus,** respectively) are covered by a tympanic membrane and form, with the stapes, a chain of three delicate auditory ossicles that transmit airborne vibrations from the tympanic membrane to the inner ear (Fig. 16.9). This character had not been achieved by the late therapsids, but changes in feeding mechanisms that resulted in a stronger bite had led to an enlargement of the dentary, which grew posteriorly close to the squamosal, and the quadrate and articular became very small. In a few very late therapsids a dentary-squamosal joint was present beside the quadrate-articular joint. The final change to the mammalian condition was made late in the Triassic period; shortly afterward, the mammal-like reptiles became extinct. Their mammalian descendants remained a rather inconspicuous part of the fauna until the disappearance of the dinosaurs.

CLASSIFICATION OF AMPHIBIANS AND REPTILES

CLASS AMPHIBIA. The amphibians. Ectothermic vertebrates usually having an aquatic larval stage; adults typically terrestrial; skin moist and slimy; scales absent in most groups.

    * **Subclass Labyrinthodontia.** Several orders of extinct and primitive amphibians collectively called labyrinthodonts; vertebral centra of the "arch" type, each usually consisting of two or three arches of bone encasing the notochord.

    * **Subclass Lepospondyli.** Several orders of extinct amphibia with vertebral centra of the "spool" type; each centrum a single structure, spool-shaped, pierced by a longitudinal canal for persistent notochord.

---

*Extinct.

Subclass Lissamphibia. Modern amphibians, probably evolved from labyrinthodonts. Skull usually flat and broad; pineal opening lost; brain case poorly ossified; pubis never ossified; carpals and tarsals cartilaginous; never more than four toes on front foot.

Superorder Salientia. Frogs and their allies. Trunk short; tail reduced or absent; ilia elongated; hind legs elongated.

* Order Proanura. Ancestral frogs of the Triassic. Tail still present; limbs not highly specialized. *Triadobatrachus.*

Order Anura. Modern frogs and toads. Short trunk; tail absent or very small; candal vertebrae form a urostyle; legs specialized for jumping. The leopard frog, *Rana*; tree frog, *Hyla*; American toad, *Bufo*.

Superorder Caudata. Tailed amphibians. Trunk long; tail present; legs, if present, not elongated.

Order Urodela. Salamanders. Tail long; legs usually present. The spotted salamander, *Ambystoma*; redbacked salamander, *Plethodon*; mudpuppy, *Necturus*.

Order Apoda. Caecilians. Wormlike trunk; limbs absent; tail very short; vestiges of dermal scales in the skin. Confined to the tropics.

CLASS REPTILIA. The reptiles. Ectothermic or heliothermic terrestrial vertebrates that reproduce upon the land; body covered with horny scales or plates.

Subclass Anapsida. Primitive reptiles characterized by a solid roof in the temporal region of the skull.

* Order Cotylosauria. The cotylosaurs. Ancestral reptiles retaining a large number of primitive features.

Order Testudinata (Chelonia). The turtles. Anapsid reptiles with a short trunk encased in a bony shell that in turn is covered by horny plates; teeth absent. Red-eared turtle, *Pseudemys*; green sea turtle, *Chelonia*; side-necked turtle, *Chelodina*.

* **Subclass Euryapsida.** Ancient marine reptiles that propelled themselves with long paddle-shaped limbs; single temporal opening high on the skull.

* Order Sauropterygia. The plesiosaurs.

* **Subclass Ichthyopterygia.** Ancient, marine, fishlike reptiles; single temporal opening high on the skull.

* Order Ichthyosauria. The ichthyosaurs.

Subclass Lepidosauria. Primitive reptiles with two temporal openings in the skull (diapsid); pineal foramen usually retained; teeth on palate as well as jaw margins, usually not set in sockets; usually quadrupeds.

* Order Eosuchia. Ancestral lepidosaurians.

Order Rhynchocephalia. Primitive lizard-like reptiles. Temporal openings completely bounded by arches of bone; distinctive overhanging beak on upper jaw; now confined to certain islands off New Zealand. The tuatara, *Sphenodon*.

Order Squamata. Lizards and snakes. Advanced lepidosaurians in which there has been a loss of one or two arches of bone bordering the temporal openings.

SUBORDER LACERTILIA. Lizards. Only the lower arch of bone has been lost in the temporal region; most groups retain legs, eyelids and an auditory opening. The chuckwalla, *Sauromalus*; horned toad, *Phrynosoma*; Gila monster, *Heloderma*.

SUBORDER OPHIDIA. Snakes. Both arches of bone bounding temporal openings have been lost; jaw mechanism extremely flexible so that food larger than head can be swallowed; pineal foramen, legs and auditory opening are lost; eyes covered by a transparent scale, the spectacle. The garter snake, *Thamnophis*; water snake, *Natrix*; rattlesnake, *Crotalus*.

Subclass Archosauria. Advanced diapsid reptiles. Pineal foramen usually absent; teeth usually restricted to jaw margin, set in sockets; an extra fenestra usually present in front of the eye or on the lower jaw, or both; most are bipedal reptiles, or show signs of bipedal ancestry.

---

*Extinct.

Order Thecodontia. Ancestral archosaurs. Show the beginning of archosaur tendencies but not the specializations of later archosaurs.

Order Saurischia. Saurischian dinosaurs. Pelvis three-pronged with pubis located entirely anterior to ischium; some carnivorous, others herbivorous. *Tyrannosaurus* and *Brontosaurus.*

Order Ornithischia. Ornithischian dinosaurs. Pelvis birdlike with part of pubis growing posteriorly parallel to ischium; all herbivorous. *Stegasaurus* and *Triceratops.*

Order Pterosauria. The flying reptiles. Wing a skin membrane supported by an elongated fourth finger. *Pteranodon.*

Order Crocodilia. Alligators and crocodiles. Quadrupeds but with hind legs noticeably larger than forelegs; adapted for an aquatic mode of life. The American alligator, *Alligator*; the American crocodile, *Crocodilus.*

Subclass Synapsida. Mammal-like reptiles characterized by a single temporal opening on the lateral surface of the skull.

Order Pelycosauria. Early mammal-like reptiles retaining many primitive features; limbs in primitive tetrapod position with humerus and femur moving in the horizontal plane. *Dimetrodon.*

Order Therapsida. Advanced mammal-like reptiles beginning to show many mammalian characteristics; limbs rotated beneath the body with humerus and femur moving close to the vertical plane. *Lycaenops.*

ANNOTATED REFERENCES

Many of the general references on vertebrates cited at the end of Chapter 33 contain considerable information on the biology of amphibians and reptiles.

Bellairs, A.: The Life of Reptiles. London, Weidenfeld and Nicolson, 1969. Deals with the anatomy, physiology, development, ecology and evolution of reptiles. The modern successor to Gadow's famous book (see below).

Bishop, S. C.: Handbook of Salamanders of the United States, of Canada and of Lower California. Ithaca, N.Y., Comstock Publishing Co., 1943. A standard reference work on the taxonomy and natural history of salamanders.

Bogert, C. M.: How reptiles regulate their body temperature. Scientific American 200:105 (April) 1959. Many interesting experiments on temperature regulation are discussed.

Cochran, D. M.: Living Amphibians of the World. New York, Doubleday & Co., 1961. A fascinating and superbly illustrated account of the groups of living amphibians.

Conant, R.: A Field Guide to Reptiles and Amphibians of Eastern North America. Boston, Houghton Mifflin Co., 1958. A very useful guide for the field identification of amphibians and reptiles; similar to the Peterson bird guides.

Desmond, A. J.: The Hot-Blooded Dinosaurs. New York, The Dial Press, 1976. A popular account of the history of archosaur discovery and of changing interpretations of their mode of life.

Ernst, C. H., and R. W. Barbour: Turtles of the United States. Lexington, University Press of Kentucky, 1972. The most recent comprehensive reference work on the taxonomy and natural history of turtles.

Gadow, H.: Amphibia and Reptilia. Weinheim, Germany, Engelmann, 1958. A recent reprint of a book originally published as a part of the "Cambridge Natural History" in 1901; it is still a valuable account of the anatomy and evolution of amphibians and reptiles.

Gans, C., A. d'A. Bellairs and T. S. Parsons (Eds.): Biology of the Reptilia. New York, Academic Press, 1969. The skeleton, sense organs, blood and circulation, endocrine glands and muscles are considered in the first four volumes. Other volumes are in preparation.

Moore, J. A., and B. Lofts (Eds.): Physiology of the Amphibia. New York, Academic Press, 1964–1976. Most aspects of amphibian physiology are covered in this very valuable source book.

---

*Extinct.

Noble, G. K.: The Biology of the Amphibia. New York, Dover Publications, Inc., 1956.
    This is a recent reprint of an older but still very valuable reference work on the
    classification, anatomy, physiology and ecology of amphibians.
Porter, K. R.: Herpetology. Philadelphia, W. B. Saunders Co., 1972. An excellent ac-
    count of most aspects of the biology of amphibians and reptiles.
Schmidt, K. P. and R. F. Inger: Living Reptiles of the World. New York, Doubleday &
    Co., 1957. A beautifully illustrated account of the natural history of the families
    of living reptiles.
Smith, H. M.: Handbook of Lizards of the United States and Canada, Ithaca, N.Y., Com-
    stock Publishing Co., 1946. A standard reference work on the taxonomy and nat-
    ural history of lizards.
Wright, A. H., and A. A. Wright: Handbook of Frogs of the United States and Canada.
    3rd ed. Ithaca, N.Y., Comstock Publishing Co., 1949. A standard reference work
    on the taxonomy and natural history of frogs.
Wright, A. H., and A. A. Wright: Handbook of Snakes of the United States and Canada.
    Ithaca, N.Y., Comstock Publishing Co., 1957. A standard reference work on the
    taxonomy and natural history of snakes.

VERTEBRATES: BIRDS

Reptiles are able to occupy the terrestrial environment more successfully than amphibians largely because they evolved ways of conserving body water, reproducing upon the land and regulating to some degree their body temperature by behavioral patterns. Birds (class **Aves**) and mammals (class **Mammalia**) are descendants of reptiles, and both groups underwent further adaptations to the terrestrial environment. Birds are very active flying vertebrates. Most of their distinctive characteristics are adaptations for flight, which in turn requires the maintenance of a very high level of metabolism in a body of minimum weight.

35.1 GENERAL CHARACTERISTICS OF BIRDS

1. Birds are endothermic vertebrates.

2. Horny scales are retained on the feet, but feathers cover most of the body. Cutaneous glands are absent except for a uropygeal oil gland.

3. Skeletal bones are exceptionally light and frequently pneumatic. The pectoral appendages are wings, the sternum is broad and usually keeled, and the reduced number of caudal vertebrae form a pygostyle.

4. The sense of smell is poorly developed; the eyes and visual centers in the brain are very important and large. The inner ear contains a cochlea, but it is not long and coiled.

5. Expansion of the cerebrum results primarily from an enlargement of the corpus striatum. The cerebellum is large.

6. The narrow jaws form a horn-covered beak in contemporary species. Teeth are absent. Villi are present in the small intestine.

7. The lungs are relatively small, but an unusual pattern of air passages within them, together with interconnections with air sacs, produces an exceptionally efficient one-way passage of air across the respiratory surfaces.

8. The flows of venous and arterial blood are completely separated; the atrium and ventricle of the heart are divided; and the reptilian left systemic arch has been lost.

9. Nitrogenous wastes are eliminated primarily as uric acid by metanephric kidneys. The urinary bladder has been lost.

10. Although the testes lie within the abdomen, sperm mature at lower temperatures because the testes lie beside certain air sacs and parts of the sperm ducts are coiled close to the cloaca. Males do not have a copulatory organ.

11. One ovary and oviduct have been lost. Eggs are heavily laden with yolk and are very large. Albuminous materials and a calcareous shell are secreted around the eggs as they pass down the oviduct.

12. Birds are oviparous, and the parents incubate the eggs.

35.2 PRINCIPLES OF FLIGHT

A group of extinct reptiles, the pterosaurs, and a group of mammals, the bats, have evolved true flight, but neither group has been as successful fliers as have birds. Bird wings are modified pectoral appendages, and the flying surfaces are composed of feathers. A wing is shaped like an airfoil, thick in front and thin and tapering behind, and sometimes cambered so that it is slightly concave on the undersurface and convex on the upper surface. As the air stream flows across the wing, the stream moves faster along the longer upper surface than the shorter lower surface. In accordance with Bernoulli's law in physics, which states that in a fluid stream the pressure

Figure 35.1 The effect of a wing on the airstream. *A*, Air moving more rapidly along the longer upper surface of the wing reduces the pressure on the upper surface and generates a force that can be resolved into lift and drag forces; *B*, if the angle of attack of the wing becomes too great, air swirls into the low pressure area causing turbulence and a reduction in lift; *C*, turbulence can be reduced by raising the alula and forming a slot through which the air moves rapidly and smoothly.

is least where the velocity is greatest, this differential in air speed decreases the pressure above the wing in relation to the underside. This produces a force acting perpendicular to the wing surface that can be resolved into a lift component perpendicular to the air stream and a drag component parallel to the air stream (Fig. 35.1*A*). For the bird to fly, the lift force must equal the force of gravity on the bird, and there must be a propulsive force (discussed below) that can overcome the drag force. The somewhat teardrop shape of the wing allows a smooth flow of air across the surface and keeps to a minimum lift-reducing eddies and drag.

Various factors can increase lift. Lift increases in direct proportion to the surface area of the wing. Wing areas differ among different species of bird and can be varied in individual

birds by the degree to which the wing is stretched out or unfolded. Lift increases greatly as the speed of airflow across the wing increases, for lift is proportional to the square of the speed. Fast flying birds, such as a swift, can and do have relatively smaller wings than slower flying species. When a bird is flying at low speeds, or during taking off and landing, other mechanisms are used to increase lift. The wing can be tilted so that its anterior edge is considerably higher, a procedure known as increasing the angle of attack (Fig. 35.1*B*). This increases lift but also tends to create lift-reducing turbulence above the wing. Separating certain feathers, which produces slots through which the air moves very rapidly, can reduce this turbulence and make it possible for the wing to generate a very high lift. A small group of feathers supported by the first digit and known as the **alula** can produce a slot at the front of the wing (Figs. 35.1C and 35.2). Additional slots are often formed along the trailing margin of the wing and at the wing tip. The latter slots reduce the turbulence known as tip vortex. Some birds obtain additional lift on landing by fanning out the tail feathers and bending them down. The tail, then, acts both as a brake and as a high-lift, low-speed airfoil.

There are many kinds of flight. The simplest is **gliding,** in which the wings provide some lift and the forward motion comes from falling through the air. Altitude is lost in a glide of this type, but altitude can be maintained or even increased if the bird also soars. Land birds, such as the turkey vulture or the osprey (Fig. 35.3A), circle within a rising current of warm air or above a bluff where air is deflected upward. Birds that engage in **static soaring** of this type have relatively short, broad wings that enable them to maneuver easily in the capricious air currents yet have enough surface area to provide lift. Flight is slow, and additional lift comes from considerable slotting of the wing, particularly near the tip. Oceanic birds engage in **dynamic soaring,** which makes use of the increase in air speed with increasing elevation above the ocean surface. Friction with the ocean causes air speed to be slowest at the ocean surface (Fig. 35.3C). Starting at a high elevation, these birds glide rapidly downward with the wind. Just above the ocean, they wheel into the wind and use the momentum gained in the glide to start to gain altitude. As they gain altitude, they encounter increasingly fast air speeds that in turn generate additional lift. In this way, the birds regain their original altitude. Dynamic soarers have long narrow wings.

In the familiar **flapping flight** used by such birds as the pigeon (Fig. 35.3B), the wings not only provide the lift but also serve as propellers. The up and down movement of the wings in relation to the body of the bird is responsible for the forward movement, but the wings do not simply push back against the air as a swimmer would push back against the water. On the downstroke, they move down and forward; on the upstroke, up and back. As a wing moves down, the air pushes up against it and the more flexible posterior margin of the distal part of the wing is twisted up. The distal portion of the wing twists the opposite way on the upstroke. The twisting of the distal portion of the wing gives it a pitch comparable to that of a propeller, and this, together with the movements of this part of the wing, is responsible for the forward motion. In all types of flight, the tail helps to support and balance the body and is used as a rudder.

Most of the features of bird wings also apply

Figure 35.2 Photograph of a red-shouldered hawk landing. Air speed is low and lift is increased by several slots: 1, alulae; 2, separation of primary wing feathers; 3, separation of feathers on leading edge of wing; 4, auxiliary feathers on upper surface of wing. (From Hertel.)

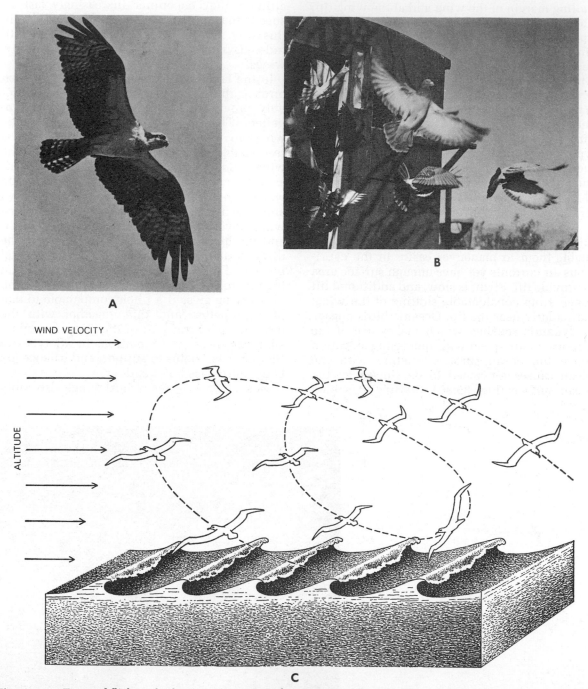

Figure 35.3 Types of flight in birds. *A*, Static soaring of an osprey; *B*, flapping flight of pigeons; *C*, dynamic soaring of oceanic birds. (*A*, photograph by Allan D. Cruickshank, National Audubon Society; *B*, U.S. Army photograph; *C*, from The Soaring Flight of Birds, by C. D. Cone, Jr. Copyright © by Scientific American, Inc. All rights reserved.)

to the wings of airplanes. But a bird's wings and tail have one great advantage over those of an airplane in that they can be varied considerably to adjust to different speeds and types of flight, for different angles of attack and for many other variables. Even though recently designed planes have wings that can be varied to some extent, any bird is more versatile than any single type of airplane.

35.3 THE ADAPTIVE FEATURES OF BIRDS

There are few features of the anatomy of birds that are not directly or indirectly related to endothermy and flight. They are adapted structurally and functionally to provide a high energy output in a body of low weight.

Thermoregulation. Birds have retained the horny scales of reptiles on parts of their legs, on their feet and, in modified form, as a covering for their beaks, but the scales that cover the rest of the reptilian body have been transformed into feathers. Feathers overlap and entrap air between them. This reduces loss of body heat because it prevents air convection currents across the body surface that would carry heat away, and the entrapped air is itself a poor conductor of heat. Water, which is a very good conductor of heat, is prevented from penetrating the feathers by an oily secretion produced by the **uropygeal gland** located on the back near the tail base (Fig. 35.9). When a bird preens, it spreads this secretion over the feathers. As would be expected, water fowl have very large uropygeal glands. When temperatures fall, the feathers are fluffed out, and this increases the thickness of the insulating layer of air. If temperatures fall very low, the bird must produce more heat by raising its metabolic level, as mammals do (Fig. 36.1, p. 837). When heat is to be lost, the feathers are held closer to the body, more blood is directed through the skin (and especially to uninsulated areas, such as the legs) and panting starts. Birds have no sweat glands. The interaction of these mechanisms enables birds to maintain their body temperature at constant and relatively high levels, 40° to 43°C.

Feathers. Feathers, like horny scales, are epidermal outgrowths whose cells have accumulated large amounts of keratin and are no longer living. Pigments deposited in these cells during the development of the feather, together with surface modifications that reflect certain light rays, are responsible for the brilliant colors of birds. Although feathers cover a bird, they fan out in most species from localized **feather tracts** rather than growing out uniformly from all of the body surface. Feathers, more than any other single feature, characterize birds. They form an insulating layer, and those on the tail and wings form the flying surfaces.

The **contour feathers** that cover the body and provide the flying surface consist of a stiff central **shaft** bearing numerous parallel side branches, the **barbs**, that collectively form the **vane** (Fig. 35.4). Each barb bears minute hooked branches, **barbules,** along its side that interlock with the barbules of adjacent barbs to hold the barbs together. If the barbs separate, the bird can preen the feather with its bill until they hook together again; thus, the vane is a strong, light and easily repaired surface ideal for flight. In birds that have lost the power of flight, such as the ostrich, hooklets are not present upon the barbules, and the feather is very fluffy. The proximal end of the shaft, known as the **quill**, is lodged within an epidermal follicle in the skin. Blood vessels enter it during the development of the feather. A small **aftershaft** bearing a few barbs may arise from the distal end of the quill.

Other types of feathers include the hair-like **filoplumes,** sometimes visible on plucked fowl, and **down feathers.** Down covers young birds and is found under the contour feathers in the adults of certain species, particularly aquatic ones. It is unusually good insulation, for it has a reduced shaft and long fluffy barbs arising directly from the distal end of the quill.

Most birds **molt** once a year, usually after the breeding season. Feathers are lost and replaced in a characteristic sequence for each species. The process is gradual in most species, and the birds can move about normally during molting. Certain male ducks, however, shed the large flight feathers on their wings so rapidly that they are unable to fly for a while.

Skeleton. Many adaptations for flight are apparent in the skeleton of birds. The bones are very light in weight, for they are hollow and remarkably thin. Extensions from the lungs enter the limb bones in many species. Robert Cushman Murphy of the American Museum of Natural History has reported that the skeleton of a frigate bird having a wingspread of over 2 meters weighed only 115 gm., which was less than the weight of its feathers! This is an extreme example, but the skeletons of all birds weigh less in relation to their body weight than do the skeletons of mammals. The bones are very strong because most of the bone substance is located at the periphery of the bone, where it gives better structural support. A bird bone may be compared to a metal tube, which is

Figure 35.4 Contour feather of a bird. *A*, Entire feather; *B*, photomicrograph of barbs and interlocking barbules; *C*, further enlargement to show hooks and spines on barbules. (*A* modified after Young; *B* and *C* from Welty.)

more resistant to certain types of stress than a metal rod of equal weight. The rod would be much narrower and could be bent more easily than the tube. Many bird bones are further strengthened by internal struts of bone arranged in a manner similar to the trusses inside the wing of an airplane (Fig. 35.5).

The **skull** is notable for the large size of the cranial region, the large orbits and the toothless beak (Fig. 35.6). The neck region is very long, and the **cervical vertebrae** are articulated in such a way that the head and neck are very mobile. Since the bird's bill is used for feeding, preening, nest building, defense and the like, freedom of movement of the head is very important. The trunk region, in contrast, is shortened, and the **trunk vertebrae** are firmly united to form a strong fulcrum for the action of the wings and a strong point of attachment for the pelvic girdle and hind legs. The hind legs

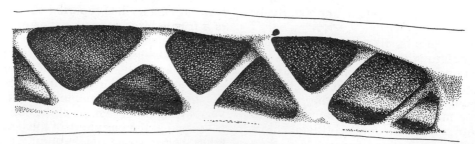

Figure 35.5 Longitudinal section of the metacarpal bone from a vulture's wing. Notice the internal trussing similar to that in an airplane's wing. (From D'Arcy Thompson.)

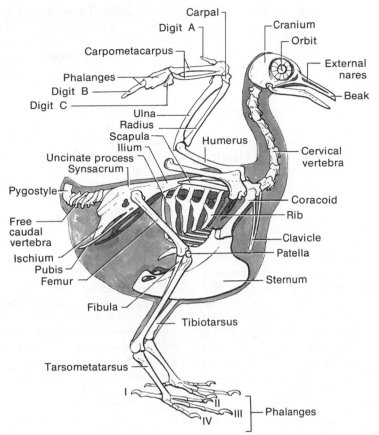

Figure 35.6   Skeleton of a pigeon. The distal part of the right wing has been omitted. (Modified after Heilmann.)

bear the entire weight of the body when the bird is on the ground. In the pigeon, 13 of the more posterior vertebrae (some of the trunk, all of the sacral and some of the caudal vertebrae) are fused together to form a **synsacrum** with which the pelvic girdle is fused. Several free **caudal vertebrae,** which permit movement of the tail, follow the synsacrum. The terminal caudal vertebrae are fused together as a **pygostyle** and support the large tail feathers.

The last two cervical vertebrae of the pigeon and the thoracic vertebrae bear distinct **ribs.** The thoracic basket is firm yet flexible. Extra firmness is provided by the ossification of the ventral portions of the thoracic ribs (the parts articulating with the sternum and comparable to our costal cartilages) and by posteriorly projecting **uncinate processes** on the dorsal portions of the ribs, which overlap the next posterior ribs. Flexibility, needed in respiratory movements, is made possible by the joints between the dorsal and ventral portions of the ribs. The **sternum,** or breastbone, is greatly expanded and, in all but the flightless birds,

has a large midventral **keel** that increases the area available for the attachment of the flight muscles.

The bones of the wing are homologous to those of the pectoral appendage of other tetrapods. The **humerus, radius** and **ulna** can be recognized easily, but the capacity of the radius to rotate upon the ulna has been lost. The bones of the hand have been greatly modified. Two free **carpals** are present and a **carpometacarpus** (a safety-pin–shaped complex of bone representing the fused carpals and metacarpals of three fingers) lies distal to them. The end of the most anterior finger is represented by a spur-shaped **phalanx** articulated to the proximal end of the carpometacarpus. This supports the alula. The main axis of the hand passes through the next finger, and it has two distinct phalanges articulated to the distal end of the carpometacarpus. Another small, spur-shaped phalanx at the distal end of the carpometacarpus represents the end of the last finger. Paleontological evidence suggests that the fingers are homologous to the first three of reptiles. The pectoral girdle, which supports the wing, consists of a narrow dorsal **scapula,** a stout **coracoid** extending as a prop from the shoulder joint to the sternum, and a delicate **clavicle,** which unites distally with its mate of the opposite side to form the wishbone.

The legs of birds resemble the hind legs of bipedal archosaurs. The **femur** articulates distally with a reduced **fibula** and a large **tibiotarsus** (fusion of the tibia with certain tarsals). The remaining tarsals and the elongated metatarsals have fused to form a **tarsometatarsus.** The fifth toe has been lost in all birds and the fourth in some species. The first toe is turned posteriorly in the pigeon and many other birds. It serves as a prop and increases the grasping action of the foot when the bird perches. The efficiency of the leg in running on the ground and jumping at take-off is increased by the elongation of the metatarsals and by the elevation of the heel off the ground. The various fusions of the limb bones reduce the chance of dislocation and injury, for birds' legs must act as shock absorbers when they land. The pelvic girdle is equally sturdy; the **ilium, ischium** and **pubis** of each side are firmly united with each other and with the vertebral column. The pubes and ischia of the two sides do not unite to form a midventral pelvic symphysis as they do in other tetrapods. This permits a more posterior displacement of the viscera, which, together with the shortened trunk, shifts the center of gravity of the body nearer to the hind legs. The absence of a symphysis also makes possible the laying of large eggs with calcareous shells.

The feet of birds have undergone a variety of modifications as birds have adapted to particular modes of life (Fig. 35.7). The foot and toes become particularly sturdy in ground-dwelling species, and the power of grasping is especially well developed in such perching specialists as our songbirds. In perching birds, the tendons of the foot are so arranged that the weight of the body automatically causes the toes to flex and grasp the perch when the bird alights upon a branch. The woodpeckers have sharp claws, and the fourth toe is turned backward with the first to form a foot ideally suited for clinging onto the sides of trees. Swimming birds have a web stretching between certain of their toes—the three anterior toes in loons, albatrosses, ducks, gulls and many others; all four toes in pelicans, cormorants and their relatives. The marsh-dwelling jacana of the tropics has a foot with exceedingly long toes and claws that enable it to scamper across lily pads and other floating vegetation. Swifts and hummingbirds have very small feet barely strong enough to grasp a perch. These birds spend most of their time on the wing and almost never alight on the ground.

Muscles. The intricate movements of the neck and the support of the body by a single pair of legs entail numerous modifications of

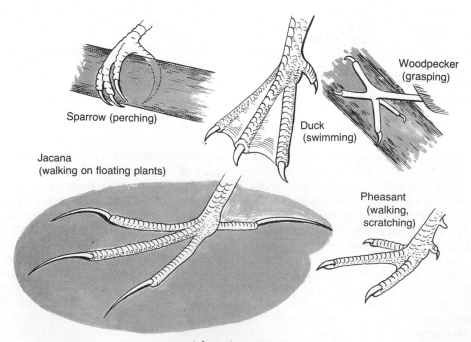

Figure 35.7 Adaptations of the feet of birds.

the muscular system, but the flight muscles are of particular interest. A **pectoralis,** which originates on the sternum and inserts on the ventral surface of the humerus, is responsible for the powerful downstroke of the wings (Fig. 35.8). One might expect that dorsally placed muscles would be responsible for the recovery stroke, but, instead, another ventral muscle, the **supracoracoideus** is responsible for the upstroke by virtue of a peculiar pulley-like arrangement of its tendon of insertion. The origin of the supracoracoideus is on the sternum dorsal to the pectoralis. Its tendon passes through a canal in the pectoral girdle near the shoulder joint and inserts on the dorsal surface of the humerus. These two muscles are exceptionally large and together may make up as much as 25 per cent of body weight in birds that are powerful flyers. In ducks and other birds that fly a great deal, most of the fibers in these muscles are of the tonic type (p. 259). They are dark, for they contain a great deal of myoglobin. The wing muscle fibers are phasic and lighter in color in chickens and other birds that use their wings intermittently.

Muscles within the wing are responsible for its folding and unfolding and the regulation of its shape and angles during flight. Other muscles attach to the follicles of the large flight feathers of the wings and tail and control their positions.

Major Features of the Visceral Organs. Less obvious but no less important adaptations for increased activity and flight are

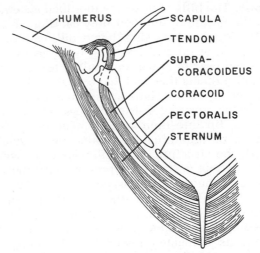

Figure 35.8 A diagrammatic cross section through the shoulder region and sternum showing the arrangement of major flight muscles. (From Welty after Storer.)

present in many of the internal organs (Fig. 35.9). Increased activity and a high metabolic rate necessitate a large intake of food. The digestive system is compact, but it is so effective that, in some of the smaller birds, an amount of food equivalent to 30 per cent of the body weight can be processsed each day! Moreover, most of the food that is selected has a high caloric value. Birds eat a variety of insects and other animals and such plant food as fruit and seeds. They do not attempt to eat such bulky, low caloric foods as leaves and

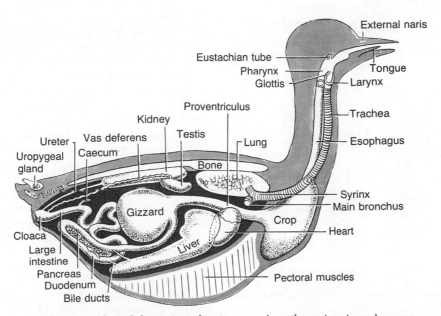

Figure 35.9 A lateral dissection of a pigeon to show the major visceral organs.

grass. The bills of birds are modified according to the nature of the food they eat (Fig. 38.1, p. 875). Finches have short heavy bills well suited for picking up and breaking open seeds. The hooked beak of hawks is ideal for tearing apart small animals that they have seized with their powerful talons. Herons use their long sharp bills for spearing fish and frogs, which they deftly flip into their mouths. The length and shape of hummingbirds' bills are correlated with the structure of the flowers from which they extract nectar. Whippoorwills fly about in the evening catching insects with their gaping mouths. Bristle-like feathers at the base of the bill help them to catch their prey. When feeding, the skimmer flies just above the ocean with its elongated lower jaw skimming the surface. Any fish or other organisms that are hit are flicked into its open mouth. The wood-cock's long and sensitive bill is adapted for probing for worms in the soft ground. The woodcock can open the tip of its bill slightly to grasp a worm without opening the rest of the mouth!

Food taken into the **mouth** is mixed with a lubricating saliva and passes through the

pharynx and down the **esophagus** without further treatment, for birds have no teeth. In grain-eating species, such as the pigeon, the lower end of the esophagus is modified to form a **crop** in which the seeds are temporarily stored and softened by the uptake of water. Food is mixed with peptic enzymes in the **proventriculus,** or first part of the stomach, and then passes into the **gizzard,** the highly modified posterior part of the stomach characterized by thick muscular walls and modified glands that secrete a horny lining. Small stones that have been swallowed are usually found in the gizzard and aid in grinding the food to a pulp and mixing it with the gastric juices. The **intestinal region** is relatively short compared to the intestine of mammals. As in mammals, it is lined with microscopic, finger-like projections, the **villi,** that greatly increase the absorptive surface area.

The **lungs** of birds are unique. They are relatively small and compact organs but lead to an extensive system of **air sacs** that extend into many parts of the body, some even entering the bones (Fig. 35.10). The air sacs connect with a large **mesobronchus** that extends through each lung and also with **recurrent bronchi** that lead back to the lungs. The larger bronchi connect with secondary bronchi (**dorsobronchi** and **ventrobronchi**) that, in turn, are interconnected by smaller **parabronchi.** Minute branching and anastomosing **air capillaries** radiate from the parabronchi, and it is through their vascular walls that gas exchange with the blood occurs (Fig. 35.11). There are no blind sacs, such as the alveoli of mammal lungs, in bird lungs.

By tracing the flow of certain gases through the lungs and air sacs and detecting the direction of flow by small probes, Dr. Knut Schmidt-Nielsen and his students have determined the course of air through this complex system. During inspiration, the sternum is lowered, the air sacs expand and the lungs, surprisingly, are compressed slightly by the contraction of muscles that pull a membranous "diaphragm" against their ventral surface. Air high in oxygen content moves through the trachea and mesobronchi into the posterior air sacs, and stale air, already in the lungs, moves from the lungs into the anterior air sacs. During expiration, the sternum is raised, the air sacs are compressed and the lungs expand. Air high in oxygen content now moves from the posterior air sacs into the lungs; stale air in the anterior sacs moves into the mesobronchi, the trachea, then out of the body.

This unusual mechanism of ventilation provides a one-way flow of air through the

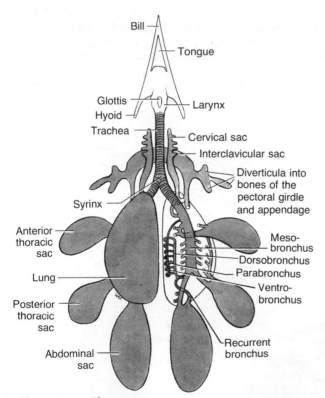

Figure 35.10 The respiratory organs of a bird as seen in a dorsal view. Only one ventrobronchus and one dorsobronchus and their interconnecting parabronchi are shown. (Partly after Welty.)

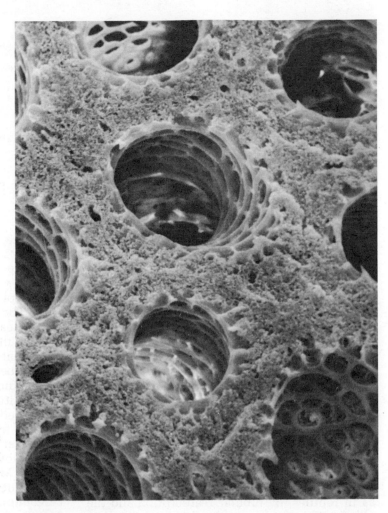

Figure 35.11 Scanning electron micrograph of a section through the lung of a domestic fowl showing parabronchi in cross section and associated air capillaries. (Photograph by H. R. Duncker from How Birds Breathe, by K. Schmidt-Nielsen. Copyright © by Scientific American, Inc. All rights reserved.)

parabronchi and air capillaries, but it takes two cycles of inspiration and expiration for a given unit of air to move through the system: inspiration 1, fresh air mixed with some stale air remaining in the trachea enters the posterior air sacs; expiration 1, this air enters the lungs where it gives up oxygen and takes on more carbon dioxide; inspiration 2, stale air moves from the lungs to the anterior air sacs; expiration 2, the stale air leaves the anterior sacs and most of it is expelled from the body. The residual carbon dioxide remaining in the trachea mixes with the incoming air, and it prevents too much carbon dioxide being removed from the blood. The amount of carbon dioxide in the blood must be kept above a certain threshold value to maintain the activity of the respiratory center (p. 307) and the proper acid-base balance of the body.

Birds and mammals have evolved quite different ways of providing sufficient gas exchange to sustain a high level of metabolism. Bird lungs are the more efficient, for air moves through the lung tissue in only one direction, and the direction is counter to that of the blood flow. The system is analogous to the countercurrent exchange mechanism found in the gills of many fishes (p. 298). The great efficiency of bird lungs permits some species to fly freely at altitudes in excess of 4000 meters.

A mechanism for the production of sounds is associated with the air passages. Membranes are set vibrating by the movement of air; however, the vibratory membranes are not in the larynx at the anterior end of the trachea but in a **syrinx** at its posterior end (Fig. 35.10). Muscles associated with the syrinx vary the pitch of the notes.

Birds resemble mammals in having evolved a double circulation through the heart. The complete separation of venous and arterial blood within the heart, the rapid heart beat (400 to 500 times per minute in a small bird such as a sparrow when it is at rest) and an increase in blood pressure make for a very

Figure 35.12 The salt-excreting glands of the herring gull. (Modified after Fänge, Schmidt-Nielsen and Osaki.)

rapid and efficient circulation. This is of the utmost importance in an endotherm, for the tissues need a large supply of food and oxygen, and waste products of metabolism must be removed quickly.

Nitrogenous wastes are removed from the blood by a pair of **metanephric kidneys,** which are basically similar to those of reptiles (Fig. 35.9). The high rate of metabolism of birds, however, requires a much greater number of kidney tubules. Indeed, most birds have relatively more tubules than do mammals; a cubic millimeter of tissue from the cortex of a bird's kidney contains 100 to 500 renal corpuscles in contrast to 15 or less in a comparable amount of mammalian kidney. Body water is conserved, as in reptiles, by tubular reabsorption and the elimination of most of the nitrogenous waste as uric acid. Birds have lost the urinary bladder present in ground-dwelling vertebrates, possibly as one means of reducing body weight.

Body salts are generally conserved by terrestrial vertebrates, although any excess can be eliminated by the kidneys. Sea birds, however, have a large salt intake with their food and water and must eliminate more salts than can be disposed of by the kidneys. This is done by special **salt-excreting glands,** which in the herring gull, for example, are located above the eyes and discharge a concentrated salt solution into the nasal cavities. This solution leaves these cavities through the external nares and drips from the end of the beak (Fig. 35.12). Each gland consists of many lobes, each of which is composed of many vascularized secretory tubules radiating from a central canal.

The sense of smell is less important in vertebrates that spend a considerable part of their life off the ground than it is in terrestrial species, so it is not surprising to find that the olfactory organ and olfactory portions of the

brain are reduced in birds. Sight, on the other hand, is very important, and the eyes and optic regions of the brain are unusually well developed. The eyes of birds occupy a large portion of the head, and both eyes together are often heavier than the brain. The visual acuity of birds, that is, their ability to distinguish objects as they become smaller and closer together, is several times greater than that of human beings. The ability to accommodate rapidly is also well developed in birds' eyes, for birds must change quickly from distant to near vision as they maneuver among the branches of a tree or swoop down to the ground from a considerable height. Muscular coordination is also very important in the bird way of life, and the **cerebellum** is correspondingly well developed. The **cerebral hemispheres** are large; however, their size is not a result of the expansion of the surface cortex, as it is in mammals, but rather the enlargement of a more deeply situated mass of gray matter, the **corpus striatum.**

35.4 THE ORIGIN AND EVOLUTION OF BIRDS

One might infer simply from the structure of modern birds that they have evolved from archosaurian reptiles, but we need not stretch our inferences, for five specimens of a Jurassic bird are known that are clearly intermediate between archosaurs and modern birds. Some of the fossils are preserved with remarkable detail in a fine-grained, lithographic limestone.

Archaeopteryx lithographica (Fig. 35.13A) was about the size of a crow. Its skeleton is reptilian in having toothed jaws, no fusion of trunk or sacral vertebrae, a long tail and a poorly developed sternum. Birdlike tendencies are evident in the enlarged orbits, some expansion of the brain case and particularly in the winglike structure of the hand. As in modern birds, the "hand" is elongated and only three "fingers" are present; however, there is little fusion of bones and each finger bears a claw. The first fossils to be discovered showed impressions of feathers (Fig. 9.1, p. 207), and this points clearly to a relationship with birds. If the skeleton alone had been discovered, this creature probably would have been regarded as a peculiar archosaur. Indeed, some of the later specimens of *Archaeopteryx* to be described were at first misidentified as small dinosaurs. The ratio of its wing surface to its body size, together with the poorly developed sternum, indicates that it was at best a weak flier. These most primitive birds are placed in the subclass **Archaeornithes.**

For many years it was believed that birds and dinosaurs had a common ancestor in the thecodonts, the stem group of archosaurs, but in 1975, John H. Ostrum of Yale University presented compelling evidence that birds evolved from coelurosaurs, a group of early saurischian dinosaurs (Fig. 34.2). Coelurosaurs were small bipedal creatures with a long tail, a long flexible neck and arms bearing only three clawed fingers that may have been used in some specialized method of food gathering. Structurally, some species resembled **Archaeopteryx** very closely. We do not know just how flight evolved in the ancestors of *Archaeopteryx*, but the early stages in the evolution of flight must have been adaptive in some

way, even though the "proavis" would not at first have been able to fly. It is possible that the ancestors of birds were becoming more active and possibly warm-blooded, and feathers may have first been of value in helping to conserve body heat. Professor Ostrum believes that their enlargement, along the posterior edge of the forelimb, enabled proavis to use these limbs as nets to seize insects. Further enlargement of wing and tail feathers may have conferred stability in running rapidly along the ground and in rudimentary gliding from low branches. Feathers and wings may have evolved gradually in this way and may have been adaptive at all stages of their evolution. On reaching a certain threshold of size, they could be used for true flight.

The next fossil birds have been found in Cretaceous deposits. These birds had lost the long reptilian tail, had evolved a well-developed sternum and were modern in many other ways. A true pygostyle had not yet evolved, and teeth were present. *Hesperornis* (Fig. 35.13B) was a large diving species with powerful hind legs and vestigial wings. *Ichthyornis* was a tern-sized flying species. Although they are placed in the subclass **Neornithes** along with modern birds, the more primitive nature of these Cretaceous species is recognized by placing them in a distinct superorder—the **Odontognathae.**

All later birds have lost the reptilian teeth, but a few (superorder **Palaeognathae**) re-

Figure 35.13 Extinct birds. *A*, A restoration of *Archaeopteryx*, the earliest known bird; *B*, a restoration of *Hesperornis*, a large diving bird of the Cretaceous. (*A* from Heilmann; *B* courtesy of the American Museum of Natural History.)

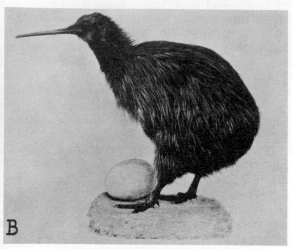

Figure 35.14   Representative paleognathous birds. *A*, Ostrich; *B*, a kiwi with its relatively huge egg. (*A* from Grzimek, B., in *Natural History*, Vol. LXX, No. 1; *B* courtesy of the American Museum of Natural History.)

tain a somewhat reptilian palate, whereas others (superorder **Neognathae**) have a more specialized palatal structure. Living paleognathous birds are for the most part ground-dwelling flightless species, such as the ostriches of Africa, the rheas of South America, the cassowaries of Australia and the peculiar kiwi of New Zealand (Fig. 35.14). The legs are well developed and powerful, the wings are vestigial and the feathers are fluffy. Presumably, these birds evolved from flying ancestors but readapted to a terrestrial mode of life in areas where there were an abundant food supply upon the ground and few competitors or enemies. The ancestry of certain of them can be traced back to the early Cenozoic era. A number of large ground-dwelling neognathous birds also lived then, which suggests that there might have been a competition at this time between birds and early mammals for the conquest of the land surface, which had recently, geologically speaking, been vacated by the large reptiles. Mammals won, and only a few ground-dwelling birds survived.

All other birds, including the vast majority of living species, are neognathous types (Fig. 35.15). Twenty-three orders of these birds are recognized (cf. Classification at the end of this chapter). Most can be distinguished by specializations of their bills, wings, tails and feet that reflect their adaptive radiation with respect to feeding and method of locomotion. Some, including the loons, ducks and gulls are aquatic as well as good fliers. Albatrosses and

petrels are oceanic birds that spend much of their life at sea. It is probable that the penguins evolved from certain diving members of this group. Penguins have lost their ability to fly, and their wings are modified as paddles for swimming under water. The herons, cranes and coots have become specialized for a wading, marsh-dwelling mode of life. Hawks, eagles and owls are birds of prey. The grouse, pheasants and fowl are predominantly terrestrial forms, though they can fly short distances, and the perching and songbirds are well adapted for life in the trees. The songbirds are members of the order **Passeriformes**.

35.5 MIGRATION AND NAVIGATION

An aspect of bird biology of particular interest is the seasonal migration of many species and their uncanny ability to navigate. Many vertebrates migrate and find their way home if displaced, but birds' power of flight has endowed them with a more spectacular range of movement than in other vertebrates.

The extensive movement of populations back and forth between different areas has certain advantages. The migration of birds from winter quarters in temperate or tropical regions to breeding areas in the north permits them to spread into an area where the days are long in the summer and a large food supply develops

Figure 35.15 A group of neognathous birds. *A*, Penguins use their modified wings as flippers; *B*, courtship of albatrosses; *C*, the young cormorant has to reach into the throat of its parents to get its food; *D*, an American egret, or heron, a wading bird; *E*, barn owl strikes; *F*, noddy terns on nest. (*A* courtesy of Smithsonian Institute; *B* courtesy of Lt. Col. N. Rankin; *C*, photo by L. Walker from National Audubon Society; *D* and *F* courtesy of the American Museum of Natural History; *E* from Payne, R. S., and Drury, W. H., Jr., in Natural History, Vol. LXVII, No. 6.)

for a few months. Birds can establish territories with a minimum of effort, and they have long hours of daylight to obtain food at a time when their population is increasing greatly. As conditions become inclement, the birds return to winter quarters. Migration also prevents predator populations from increasing greatly, for the predators of a particular region do not have a sustained food supply if the birds move out at intervals. In addition, migration helps birds avoid harsh climatic extremes and is, in a sense, a homeostatic mechanism.

Although there are advantages to migration, there are also hazards to a long trip. Many migrants are caught in storms and perish. It is not surprising then that not all species of birds migrate. Even within a single species there may be populations that do and others that do not. Barn owls (*Tyto alba*) from the northern part of the United States, for example, tend to migrate, whereas more southern populations are sedentary. Presumably, each group of birds has adapted to the particular environmental stresses it encounters. If the advantages of movement outweigh the hazards, the population migrates; if not, it remains more sedentary. Those that migrate are simply using their great capacity of movement to spread out in space biological functions that in another species occur in different parts of a more restricted range. It is common for many animals to feed and breed in somewhat different areas, even though the areas may not be very far apart.

The stimulus for migration has been a topic of intensive research. It is generally agreed that neuroendocrine changes closely tied to reproduction predispose a bird to migrate. In the spring of the year in the northern hemisphere, the hypothalamus stimulates the pituitary gland to secrete more gonadotropic hormones, and the gonads increase in size. There is also a rapid increase in fat deposits and an increased activity, or restlessness, of the birds. In the fall, the gonad decreases in size, but the birds again become restless and accumulate food reserves for the return trip. The interrelation of these factors to each other and to changes in day length is not entirely clear. Once birds are in a migratory condition, favorable weather and other external factors trigger the onset of migration.

Most passerine migrants travel at night, stopping to feed and rest during the day. Some may fly several hundred miles during a single night but then may rest for several days. Many larger birds, including hawks and herons, migrate by day, and ducks and geese may migrate either at night or in the day. Flight speeds during migration vary greatly among species, ranging from about 60 km. per day for the great tit to nearly 1000 km. per day for the white-crowned sparrow. Many species tend to follow the advance of certain temperature lines, or isotherms (Fig. 35.16). The length of migration and the route taken are very consistent for each type of bird but vary with the species. The Canada goose winters in the United States from the Great Lakes south, breeds in Canada as far north as the arctic coast and migrates along a broad front between the two areas. The scarlet tanager winters in parts of South America and breeds in the area from Nova Scotia, southern Quebec and southern Manitoba south to South Carolina, northern Georgia, northern Alabama and Kansas. In contrast to the Canada goose, it has a narrow migration route that extends through southern Central America and then across the center of the Gulf of Mexico, passing between Yucatan and Cuba. The longest migration is that of the arctic tern; some of these birds travel 40,000 km. in a year. This species breeds in the Arctic, then migrates to its winter quarters in the South Atlantic (Fig. 35.17).

The season, speed and routes of migration have been carefully described for most migratory species, but how birds navigate and find their way remains one of the intriguing, incompletely solved problems of animal behavior. Obviously, the birds must know where they are going; there must be some feature of the environment that is related to the goal of the bird; and the bird must have some way of perceiving this feature. It is very likely that some combination of methods is used in navigating and that the specific choices vary between bird species or even within one species according to the bird's experience and

Figure 35.16 The northward migration of the Canada goose keeps pace with spring, following the isotherm of 2°C (35°F) (Modified after Lincoln.)

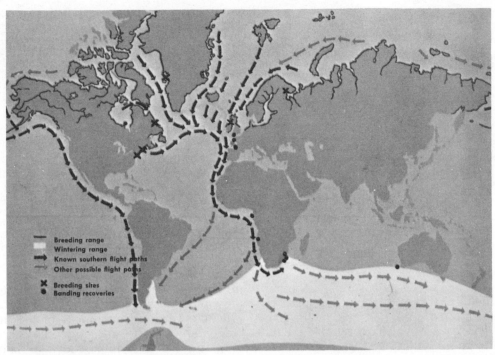

Figure 35.17 Migration routes of different populations of the arctic tern. (From The Natural Science Encyclopedia, Vol. 2. Published by Golden Press. Copyright © 1962 by Western Publishing Company, Inc.)

environmental conditions. Some experiments have shown that certain species have an innate sense of direction and migrate along a predetermined compass course. During their fall migration, certain populations of European starlings fly from Northwestern Europe to the southeast. In 1958, Perdeck caught and banded thousands of specimens in Holland and released them in Switzerland about 600 km. southeast. Young birds continued in their "innate" direction and ended up far to the east of their normal winter quarters, whereas most of the older birds made an appropriate correction and ended in their normal winter range.

Birds have an exceptionally good sense of sight, so it is not surprising that many species use visual landmarks to some extent in migrating, or in finding their way home if they have been displaced. There are many visual clues that birds could follow on long trips: topographic features, such as coast lines and mountain ranges, or such ecological features as deserts, prairies and forests. In 1949, Griffin displaced gannets (a sea bird) in Maine, 340 km. southwest from their nests on an island off the coast. He followed their return in an airplane. They flew in widening circles until they came in sight of the coast and then flew more directly home.

Some birds use a sun compass to find their

way. Professor Matthews released lesser black-backed gulls some distance south of home and observed their direction as they vanished from view. In a "vanishing point" diagram (Fig. 35.18), the number of birds that took a particular compass course at the vanishing point is shown by the length and direction of lines. Most birds took off in the direction of home on sunny days, but became disoriented on cloudy days. Additional support for a sun compass theory has come from experiments by others in which migrating birds are caught and placed outside in large circular cages. On a sunny day, they tend to orient on their perches, or aggregate on the side of the cage, appropriate to their migratory direction. Orientation can be changed by altering the apparent position of the sun with mirrors. It can also be changed by resetting the internal clocks of the birds by prior dark-light exposures so that the birds think it is many hours earlier or later than the real time. Birds orient themselves to the sun on the basis of the reset time.

Studies by Sauer suggest that night migrants might use the star pattern to navigate. Spring migrants heading north were caught and placed in a cage in a planetarium where they oriented themselves to the north of the artificial sky regardless of true north.

Other investigators have shown that some

Sun

Cloud

Figure 35.18 Vanishing point diagrams of the compass course taken by Lesser Black-backed Gulls on sunny and cloudy days. North is toward the top of the diagram and the shortest line represents one bird. The direction of home is shown by the arrow. (From Matthews.)

species can perceive the earth's or other magnetic fields and that this might be used in navigation. In 1965, Merkel and Wiltschko found that caged European robins caught during migration will orient appropriate to their migratory direction in the absence of any celestial clues, but their choice of direction could be altered by subjecting them to artificial magnetic fields. More recent work has con-

firmed a magnetic sensitivity. We do not know how birds sense magnetic fields, but it appears that the vertical component of the field (the angle between the magnetic force and the pull of gravity) is more important than its compass direction. We have learned a great deal about migration, but more observations and experiments will be needed before the full story is known.

CLASSIFICATION OF BIRDS

CLASS AVES.   The birds. Endothermic, typically flying vertebrates covered with feathers.

*Subclass Archaeornithes. Ancestral birds retaining many reptilian features including jaws with teeth, long tail and three unfused fingers, each bearing a claw. *Archaeopteryx.*

Subclass Neornithes. Birds with a reduced number of caudal vertebrae; wing composed of three highly modified fingers partly fused together.

*Superorder Odontognathae. Cretaceous birds; some at least, retaining teeth. *Hesperornis, Ichthyornis.*

Superorder Palaeognathae. Modern toothless birds with a primitive palate. Most are flightless.

Order Tinamiformes. Tinamous. Largely a ground-dwelling group, but with weak powers of flight; sternum retains a keel. Mexico to South America, *Tinamus.*

Order Struthioniformes. Ostriches. Huge flightless birds with small wings; unkeeled sternum; head and neck largely devoid of feathers; large powerful legs with only two toes. Africa and western Asia, *Struthio.*

Order Rheiformes. Rheas. Large flightless birds with unkeeled sternum; head and neck feathered; heavy legs with three toes. South America, *Rhea.*

Order Casuariformes. Cassowaries and emus. Large flightless birds with small wings and unkeeled sternum; long hairlike feathers with long aftershaft; heavy legs with three toes. New Guinea and Australia, *Casuarius.*

Order Aepyornithiformes. Elephant birds. Turkey to ostrich-sized flightless birds of Africa and Madagascar; became extinct in historic times; laid largest eggs known, 33 × 23 cm. *Aepyornis.*

---

*Extinct.

Order Dinornithiformes. Moas. Largest of the flightless birds, attained a height of 3 meters. Became extinct about 700 years ago. New Zealand, *Dinornis.*

Order Apterygiformes. Kiwis. Hen-sized flightless birds with unkeeled sternum and vestigial wings; four toes on feet; long bill with nostrils near the tip used in probing soft ground for food. New Zealand, *Apteryx.*

Superorder Neognathae. Modern birds with a less reptilian palate.

Order Gaviformes. Loons. Legs located far back on the body; webbed feet; reduced tail; long, compressed and sharply pointed bill; very good divers. The common loon, *Gavia.*

Order Podicipediformes. Grebes. Legs located far back on the body; lobate toes; reduced tail; very good divers. Eared grebe, *Colymbus.*

Order Procellariiformes. Albatrosses, shearwaters, fulmars, petrels and tropic birds. Webbed feet; fourth toe vestigial; long narrow wings; hooked beak; tubular nostrils. The petrel, *Oceanodroma.*

Order Sphenisciformes. Penguins. Flightless oceanic birds with four anteriorly directed toes with a web between three of them; wings modified as paddles; excellent divers. Confined to the southern hemisphere, chiefly Antarctica. The emperor penguin, *Aptenodytes.*

Order Pelecaniformes. Pelicans, gannets, cormorants, water-turkey and man-o-war bird. Totipalmate swimmers with four toes included in the webbed foot; tendency for the development of a gular sac. The pelican, *Pelecanus.*

Order Ciconiiformes. Herons, bitterns, storks, ibises and flamingos. Long-legged and long-necked wading birds; feet broad, but usually not webbed; the portion of the head between the eye and nostril (the lores) usually devoid of feathers. Great blue heron, *Ardea.*

Order Anseriformes. Ducks, geese and swans. Short-legged, web-footed swimming and diving birds; bill usually broad and flat with transverse horny ridges adapted for filtering mud. The mallard, *Anas;* white-fronted goose, *Anser.*

Order Falconiformes. Vultures, kites, hawks, falcons and eagles. Diurnal birds of prey with strong hooked bill; sharp curved talons. Cooper's hawk, *Accipiter;* duck hawk, *Falco.*

Order Galliformes. Grouse, quails, partridges, pheasants, turkeys and chickens. Seed- and plant-eating, largely ground-dwelling birds; short stout bill; heavy feet with short strong claws, adapted for running and scratching in the ground; wings relatively short. The chicken, *Gallus.*

Order Gruiformes. Cranes, rails, gallinules and coots. Marsh birds; feet not webbed, but toes sometimes lobed; legs elongate in some groups; lores feathered. European crane, *Grus.*

Order Diatrymiformes. Large flightless birds of the early Cenozoic. *Diatryma.*

Order Charadriiformes. Plovers, woodcock, snipe, sandpipers, stilts, phalaropes, gulls, terns, skimmers, auks and puffins. A diverse group of shore birds. The killdeer, *Charadrius.*

Order Columbiformes. Pigeons and doves. Short slender bill with a fleshy cere at its base overhanging the slitlike nostrils; short legs. The domestic pigeon, *Columba.*

Order Psittaciformes. Parrots. Feet adapted for grasping, with fourth toe capable of being turned back beside the first toe; bill heavy and hooked; often brilliantly colored plumage. Carolina parakeet, *Conurus.*

Order Cuculiformes. Cuckoos and road-runners. Foot zygodactylous with fourth toe permanently reversed beside tho first, tail long. The cuckoo, *Cuculus.*

Order Strigiformes. Owls. Nocturnal birds of prey with strong hooked bills; sharp curved talons; feathers arranged as a facial disc around the large, forwardly turned eyes. The barred owl, *Strix.*

Order Caprimulgiformes. Nighthawks and whippoorwills. Twilight flying birds with small bills, but large mouths surrounded by bristle-like feathers that help net insects; legs and feet small. The whippoorwill, *Caprimulgus.*

Order Apodiformes. Swifts and hummingbirds. Fast flying birds with long narrow wings; legs and feet very small. The chimney swift, *Chaetura.*

---

*Extinct.

Order Coliiformes. The colies of Africa. Small birds with long tails; first and fourth toes can be turned posteriorly.

Order Trogoniformes. Trogons. Short stout bill; small feet with first and second toes directed backward; often green iridescent plumage. The coppery-tailed trogon, *Trogon.*

Order Coraciiformes. Kingfishers. Strong sharp bill; foot syndactylous with third and fourth toes fused at their bases; feathers often forming a crest on the head. The belted kingfisher, *Megaceryle.*

Order Piciformes. Woodpeckers and toucans. Bill chisel-like (woodpeckers) or very large (toucans); zygodactylous foot with fourth toe permanently turned posteriorly. The flicker, *Colaptes.*

Order Passeriformes. The perching birds and songbirds. The largest order of birds, it includes the flycatchers, larks, swallows, crows, jays, chickadees, nuthatches, creepers, wrens, dippers, thrashers, thrushes, robins, bluebirds, kinglets, pipets, waxwings, shrikes, starlings, vireos, wood warblers, weaver finches, blackbirds, orioles, tanagers, finches, sparrows and so forth. Foot adapted for perching; three toes in front opposed by one well-developed toe behind. The English sparrow, *Passer.*

ANNOTATED REFERENCES

Many of the general references cited at the end of Chapter 33 contain considerable information on the biology of birds.

Bent, A. C.: Life Histories of North American Birds. New York, Dover Publications, 1961–1968. A reprinting of Bent's famous multivolume study of the natural history of birds. Originally published between 1919 and 1958 as Bulletins of the U.S. National Museum.

Emlen, S. T.: The stellar-orientation system of migratory birds. Scientific American *233*:2 (Aug.) 1975. The interaction of stellar and other clues used by the indigo bunting.

Farner, D. S. and J. R. King (Eds.): Avian Biology. New York, Academic Press, 1971–1974. A four volume treatise covering most aspects of the biology of birds.

Gilliard, E. T.: Living Birds of the World. New York, Doubleday & Co., 1958. The major groups of birds are summarized and superbly illustrated.

Hertel, H.: Structure, Form, Movement. New York, Reinhold Publishing Corp., 1966. One section of this important book on animal locomotion is devoted to various aspects of bird flight.

Hinde, R. A. (Ed.): Bird Vocalizations. Cambridge, Cambridge University Press, 1969. Many authors discuss the nature, production and biological uses of songs.

Howard, E.: Territory in Bird Life. New York, Atheneum, 1964. A reprint of a classic book on bird behavior.

Marshall, A. J. (Ed.): Biology and Comparative Physiology of Birds. New York, Academic Press, 1960–1961. An important two volume source book on many aspects of the anatomy, physiology, reproduction, migration and behavior of birds.

Matthews, G. V.: Bird Navigation, 2nd ed. Cambridge, Cambridge University Press, 1968. A thorough account is given of the different theories of navigation and homing.

Peterson, R. T.: A Field Guide to the Birds. 2nd ed. Boston, Houghton Mifflin Co., 1962. The standard and widely used guide for the field identification of birds from the Great Plains to the East Coast.

Peterson, R. T.: A Field Guide to Western Birds. Revised ed. Boston, Houghton Mifflin Co., 1961. A companion to the preceding volume, it covers the birds from the Pacific Coast to the western parts of the Great Plains.

Pettingill, O. S., Jr.: A Laboratory and Field Manual of Ornithology. 4th ed. Minneapolis, Burgess Publishing Co., 1970. A manual on the structure, habits and ecology of birds for the serious student of ornithology.

Schmidt-Nielsen, K.: How birds breathe. Scientific American *225*:73 (Dec.) 1971. A summary is given of important new findings on the unique lung ventilating mechanism of birds.

Sturkie, P. D.: Avian Physiology. 2nd ed. Ithaca, N.Y., Comstock Publishing Co., 1965. A very important source book; covers most aspects of avian physiology.

Thomson, Sir A. L. (Ed.): A New Dictionary of Birds. New York, McGraw-Hill Book Company, 1964. A very useful one volume encyclopedia on all aspects of the biology of birds prepared under the auspices of the British Ornithologist Union. Some entries are only a few lines; others run to several pages.

Van Tyne, J., and A. J. Berger: Fundamentals of Ornithology. New York, John Wiley & Sons, Inc., 1959. An important text for the serious student; emphasizes the groups of birds.

Welty, J. C.: The Life of Birds, 2nd ed. Philadelphia, W. B. Saunders Co., 1975. A comprehensive one-volume work on all aspects of the biology of birds.

CHAPTER 36

VERTEBRATES: MAMMALS

36.1 GENERAL CHARACTERISTICS OF MAMMALS

Mammals **(class Mammalia)** are the familiar hair-covered creatures, including cats, mice and human beings. Most of their distinctive features are related to the evolution of increased activity and greater care of the young. In this, they resemble birds, but mammals evolved from reptiles independently and achieved endothermy and greater care of the young in somewhat different ways.

1. Mammals are endothermic vertebrates.
2. Hair forms an insulating layer in most groups. Cutaneous glands are abundant.
3. The trunk vertebrae are differentiated into thoracic and lumbar vertebrae. Typically, there are three sacral vertebrae. The limbs of most mammals are situated more or less under the body. The skull is of the synapsid type and has a relatively large brain case. The jaw joint lies between the dentary and the squamosal (temporal) bones.
4. There are three auditory ossicles in the middle ear and a spiral cochlea in the inner ear. Enlarged nasal cavities are separated from the mouth cavity by a hard palate and contain folded turbinate bones.
5. The cerebrum is large and has a gray cortex. Large cerebellar hemispheres are present.
6. Teeth are heterodont and their replacement is limited. The small intestine has numerous multicellular intestinal glands and microscopic villi. Most species have no cloaca.
7. Respiratory and digestive passages are nearly completely separated in the oral and pharyngeal regions. Numerous lung alveoli greatly increase surface area. A muscular diaphragm plays a major role in lung ventilation.

8. The heart is completely divided so that pulmonary and systemic circulations are separate.
9. Nitrogenous wastes are eliminated primarily as urea by metanephric kidneys. Long loops of Henle in the renal tubules make possible the production of a urine hyperosmotic to the blood.
10. The testes of most mammals either lie permanently within a scrotum or descend into the scrotum during the reproductive season. Males have a copulatory organ, and fertilization is internal.
11. Except for primitive egg-laying mammals, the ovaries are small and produce fewer eggs than in lower vertebrates; and little yolk is deposited in the eggs. The oviducts have differentiated into vaginal, uterine and uterine tube regions.
12. Monotremes are oviparous; other mammals viviparous. The uterine lining and extraembryonic membranes unite to form a placenta. Mammary glands are always present in females.

36.2 MAJOR ADAPTATIONS OF MAMMALS

Temperature Regulation. Their relatively high and constant body temperature permits mammals to be very active animals. As we explained earlier (p. 259), heat loss, or gain, between the organism and its environment is reduced by the hairs that entrap an insulating layer of still air next to the skin. Heat is produced internally by a high level of oxidative metabolism. Body temperature is maintained by physiological controls of the mechanisms for heat loss and production. Important aspects of the control mechanisms can be seen in Figure 36.1, which shows the amount of metabolic work, measured by oxygen consumption, needed to maintain body temperature over a wide range of environmental temperatures. Mammals have a minimum and constant oxygen consumption over a certain range of ambient temperatures that extends downwards from their normal body temperature (37°C. in human beings). This range is called the **thermal neutral zone** and is bounded by **upper** and **lower critical temperatures.** Little extra metabolic work is required to maintain body temperature within the thermal neutral zone. As ambient temperatures fall slightly below body temperature, heat conservation mechanisms that do not require much energy output are utilized. The hairs are elevated by the contraction of the arrector pili muscles (p. 233), thus increasing the thickness of the insulating layer; blood flow through the skin is reduced by the constriction of cutaneous capillaries; and heat loss by water evaporation is reduced by a decreased activity of the sweat glands. Opposite changes occur so that less heat is conserved as the ambient temperature rises toward body temperature.

Beyond the critical temperatures, however, body temperature can be maintained only by the expenditure of significantly more metabolic work. The rise in oxygen consumption below the lower critical temperature reflects the extra heat production now needed to maintain body temperature. Shivering, which is an involuntary activity of superficial muscles, is one mechanism of increased heat production, but there is also a general increase in metabolic activity in many parts of the body. The rise in oxygen consumption above the upper critical temperature reflects the added metabolic work needed to dissipate heat when ambient temperature exceeds body temperature. There is now an increase in rate of heart beat and rate of circulation through the skin. Profuse sweating occurs in some mammals and panting in others. Panting, which involves the evaporation of water from the respiratory passages, is an important cooling mechanism in heavily furred mammals that have few sweat glands and in some rodents that have none. In general, mammals have more difficulty in dissipating excess heat when ambient temperatures rise far above the upper critical temperature than they do in generating more heat as ambient temperatures fall below the lower critical temperature.

Neural mechanisms by which body temperature is regulated are very complex. There are thermal receptors in the skin whereby mammals become aware of environmental temperature changes, and they may modify behavior accordingly. The major thermal control center is in the hypothalamus. It responds to slight changes in blood temperature and initiates the changes needed to adjust heat loss and production to the environmental context.

Mammals that live in areas where the environment is rigorous have evolved adaptations that supplement thermal regulation. Arctic mammals, for example, have features that minimize the metabolic cost of living in a cold climate. Typically, a thick fur lowers both the

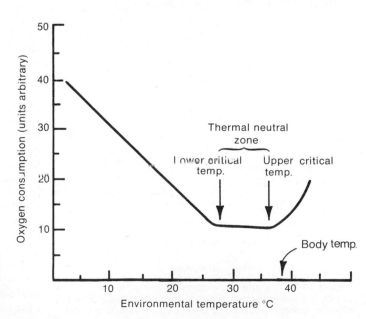

Figure 36.1 A graph to show the relationship between oxygen consumption and environmental temperature in a typical mammal with a body temperature of 37° C. (Modified after Gordon: Animal Function: Principles and Adaptations, Macmillan.)

lower critical temperature and the slope of the environmental temperature-oxygen consumption curve. The temperature of the appendages, which cannot be insulated as well as the rest of the body, is permitted to fall below that of the body core. Arteries carrying blood to the limbs are sometimes closely intermeshed with the veins returning blood so that a countercurrent exchange mechanism is set up whereby much body heat moves from the arteries to the veins and is not lost. Enough heat must be permitted to enter the appendages, however, to keep them from freezing.

Some arctic and temperate mammals, notably many insectivores, bats and rodents, adjust to winter weather by going into a period of dormancy known as **hibernation.** During this period, they lose considerable control over the mechanisms regulating body temperature, and their body temperature approaches the ambient temperature. Metabolism is very low during hibernation, yet it is sufficient to sustain life and to keep the body from freezing. There are certain advantages to hibernation for a small endotherm. Small mammals have a relatively higher rate of metabolism than large ones because they have more surface area in proportion to their mass. They lose a great deal of heat through their surface areas and must consume much food just to maintain body temperature. In many regions, insects and certain types of plant food are not available in quantity during the winter. If an animal can permit its body temperature to drop, it can get by on less food, or even on the food reserves within its body. Hibernation in mammals should not be confused with hibernation in certain ectotherms (p. 793). Body temperature always fluctuates with ambient temperatures in ectotherms, and when it is low enough the animals naturally are inactive. In a warm-blooded animal, a normally high and constant body temperature is permitted to drop; in a sense, the "thermostat" in the hypothalamus is turned down. Physiologic adjustments are more complex in this situation.

Although mammals in general are endothermic, there are differences among them in the level at which they maintain body temperature. As might be expected, the primitive egg-laying monotremes maintain body temperatures at a lower level ($31° \pm 2°$C.) than the marsupials ($36° \pm 2°$C.) or true placental mammals ($38° \pm 2°$C.). There are also differences among mammals in their capacity to maintain their body temperature as ambient temperatures increase beyond the normal range. The Central American opossum (a marsupial), for example, can nearly stabilize its body temperature over an external temperature range of $10°$ to $30°$C. but loses control of body temperature and in effect becomes an ectotherm when external temperatures exceed $30°$C. (Fig. 36.2). It maintains a higher and more constant body temperature during its period of nocturnal activity than when it is sleeping during the day. Mammals of this type are sometimes described as **heterothermic.**

Metabolic Systems. In order to sustain their high level of metabolism, mammals must be very efficient in obtaining large supplies of food and oxygen, eliminating waste products and transporting materials throughout the body. The dentition of mammals enables them to obtain and utilize a wide variety of foods. Their teeth are not all the same shape, as they are in most reptiles, but are differentiated into various types (Fig. 36.3). Chisel-shaped **incisors** are present at the front of each jaw and are used for nipping and cropping. Next is a single **canine** tooth, which is primitively a long, sharp tooth, useful in attacking and stabbing prey or in defense. A series of **premolars** and **molars** follow the canine. These teeth tear, crush and grind up the food. In primitive mammals, the premolars are sharper than the molars and have more of a tearing function. A primitive placental mammal, such as an insectivore, has three incisors, one canine, four premolars and three molars in each side of the upper and lower jaw. This can be expressed as a dental formula: I_3^3, C_1^1, Pm_4^4, M_3^3. No placental mammal has more teeth than this, but the number of teeth is reduced in many groups. We, for example, have a dental formula of I_2^2, C_1^1, Pm_2^2, M_3^3. Considerable variation also occurs in the structure of particular teeth in the different groups. The differen-

Figure 36.2 A graph showing relationship of body temperature to external temperature in a nocturnally active Central American opossum, *Metachirus.* (From Morrison.)

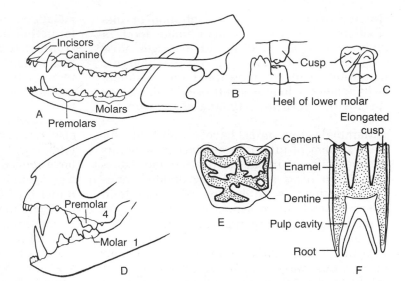

Figure 36.3 Teeth of mammals. *A*, The relatively unspecialized teeth of a primitive insectivore; *B* and *C*, lateral and crown views of the left upper and lower molars of an insectivore to show their occlusion; *D*, the stabbing and carnassial teeth of a cat; *E* and *F*, a crown view and a vertical section through the left upper molar of a horse to show its adaptation for crushing and grinding.

tiated, or **heterodont,** teeth of mammals need to occlude more precisely than the undifferentiated homodont teeth of lower vertebrates. Limited tooth replacement permits this. Milk, or deciduous, incisors, canines and premolars are replaced once by permanent ones. The molars, which appear sequentially during the growth of the animal, are not replaced. This pattern of replacement allows for larger and more teeth as the jaws grow yet maintains the needed occlusion.

Most mammals do not swallow their food whole but break it up mechanically with their teeth and mix it with saliva that, in addition to lubricating the food, usually contains an enzyme that begins the digestion of carbohydrates. As we have seen (Sections 12.1 and 12.4), digestion is completed in the stomach and intestinal region. Numerous microscopic

villi line the small intestine, as they do in birds, and increase the surface area available for absorption.

As discussed in Section 13.4, a greater exchange of oxygen and carbon dioxide is made possible by the evolution of pulmonary alveoli that greatly increase the respiratory surface of the lungs and by the evolution of a diaphragm that increases the efficiency of ventilation. Of particular interest from an evolutionary viewpoint has been the development of a **secondary palate,** a horizontal partition of bone and flesh that separates the air and food passages in the mouth cavity and pharynx (Fig. 36.4). In lower tetrapods, the nasal cavities lead directly into the front of the mouth, but in mammals they open more posteriorly. The secondary palate permits nearly continuous breathing, which is a neces-

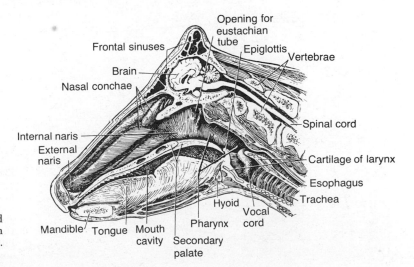

Figure 36.4 A sagittal section of the head of a cow showing the relationship between the digestive and respiratory systems. (Modified after Sisson and Grossman.)

sity for organisms with a high rate of metabolism. Mammals can manipulate food in their mouths, for food and air passages cross only in the laryngeal part of the pharynx. Breathing need be interrupted only momentarily when the food is swallowed, and in some species not even then (young of opossum, p. 844).

Mammals, like birds, have evolved an efficient system of internal transport of materials between sites of intake, utilization and excretion. Their heart is completely divided internally so there is no mixing of venous and arterial blood. Increased blood pressure also contributes to a more rapid and efficient circulation.

Although mammals eliminate most of their nitrogenous wastes as **urea,** which is more soluble and requires more water for its removal than does the uric acid excreted by some reptiles and birds, approximately 99 per cent of the water that starts down the kidney tubules is later reabsorbed, so that the net loss of water in these animals is minimal. The generally high

metabolic rate of mammals results in the formation of a large amount of wastes to be eliminated. An increase in blood pressure, and hence in blood flow through the kidney, and an increase in the number of kidney tubules have enabled mammals to increase the rate of excretion.

Locomotion and Coordination. Mammals also move about with greater agility than lower tetrapods. In most species, the elbow and knee have moved in close to the trunk so that the legs extend down to the ground more or less under the body. The humerus and femur move back and forth closer to the vertical plane than in amphibians and contemporary reptiles in which the elbow and knee lie far lateral to the trunk, and humerus and femur move close to the horizontal plane. The changed posture provides better mechanical support and the potential for a longer swing of the appendage, increased stride length and greater speed. Most mammals have three sacral vertebrae, in contrast to the two found in most reptiles. This strengthens the articulation between the pelvic

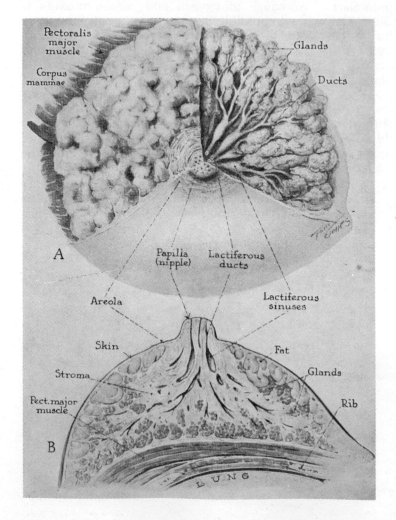

Figure 36.5 *A*, Successive stages of the dissection of the mammary gland of a lactating human being are shown clockwise. *B*, Vertical section through the nipple of a mammary gland. (From King, B. G. and M. J. Showers: Human Anatomy and Physiology. 5th ed. Philadelphia, W. B. Saunders Co., 1963; courtesy of S. H. Camp Co.)

girdle and vertebral column. Arboreal species use the tail for balancing, and it plays a major role in the propulsion of aquatic mammals, such as the whales, but in most mammals it has lost its primitive locomotor function and is frequently reduced in size or absent.

Changes in locomotion also entailed changes in the neuromuscular apparatus. Shifts in the position of many of the muscles concerned with support and locomotion are correlated with the new limb posture. Moreover, the muscular system of mammals is considerably more elaborate than that of reptiles. This, together with more highly developed sensory and nervous systems, permits more varied responses and adjustments to environmental conditions.

Care of the Young. The evolution by reptiles of the cleidoic egg was a successful adjustment to terrestrial reproduction so long as vertebrates were ectothermic. However, embryos that are to become endothermic adults must develop under warm and constant temperatures; birds and mammals cannot lay eggs and then ignore them. Birds incubate their eggs and one group of primitive mammals, which includes the duckbilled platypus of Australia, does the same. All other mammals are viviparous. The eggs are retained within a specialized region of the female's oviduct, the **uterus,** and the young are born as miniature adults.

All of the extraembryonic membranes characteristic of reptiles are present in viviparous mammals, but albuminous materials are not ordinarily secreted about the egg. The allantois, or in a few species the yolk sac, unites with the chorion, thereby carrying the fetal blood vessels over to this outermost membrane. The vascularized chorion unites in varying degrees with the uterine lining to form a **placenta,** in which fetal and maternal blood streams come close together, though they remain separated by some layers of tissue (Fig. 20.19). The embryo derives its food and oxygen and eliminates its carbon dioxide and nitrogenous wastes across these membranes.

Different species of mammals are born at different stages of maturity. Certain mice, for example, are extremely **altricial,** being born naked and with closed eyes and plugged ears. Newborn deer and other large herbivores are quite **precocial** and can run about and largely care for themselves. But regardless of maturity at birth, all newborn mammals feed upon milk secreted by specialized **mammary glands** of the female (Fig. 36.5). When the young finally leave their mother, they are at a relatively advanced stage of development and are equipped to care for themselves.

36.3 PRIMITIVE MAMMALS

Early Mammalian Evolution. Mammals evolved from therapsid reptiles early in the Triassic period and replaced them in the fauna. From a geological perspective, mammallike reptiles became extinct soon after the origin of mammals (Fig. 34.2, p. 791). Ancestral mammals, known as **triconodonts,** were very similar to the late therapsids but did have the characteristic mammalian jaw joint between the dentary and squamosal bones. They differed from later mammals in having molar teeth with three conical cusps arranged in a linear series (Fig. 36.6A). From reconstructions of their skeleton and a comparative study of primitive mammals living today, F. A. Jenkins, Jr., of Harvard University and F. R. Parington of the University of Cambridge have been able to infer some aspects of the mode of life of our remote Triassic ancestors (Fig. 36.7). They were small mouse-sized creatures whose tooth structure indicated that they fed upon small invertebrates that they must have sought in the ground litter of plant debris and in low vegetation. Limb joint surfaces indicate that they had flexible limbs and a thumb and great toe that, although carried nearly parallel to the other digits, could be abducted and adducted to grasp an object. Sharp claws also helped in climbing. These are adaptations that would have enabled the triconodonts to clamber over objects on the ground and to climb into vegetation. The differences between a terrestrial and arboreal habitat are not great for a small animal, for they must be able to climb over the obstacles and irregularities that they encounter on the ground.

Monotremes. Paleontologists know of many lines of mammalian evolution during the Mesozoic Era. It is possible that all diverged from triconodonts, but the fossil record is not complete enough to be certain. One line led to contemporary **monotremes,** the duckbilled platypus (*Ornithorhychus*) and spiny anteater (*Tachyglossus*) of the Australian region (Fig. 36.8A and B). These are certainly the most primitive mammals living today, but their relationships are somewhat uncertain. Few fossil monotremes are known, and living species lack teeth, which have been used extensively in sorting out lines of mammalian evolution. They lay eggs and retain many other reptilian characteristics, including a cloaca. The ordinal name for the group, **Monotremata,** refers to the presence of a single opening for the discharge of feces, excretory and genital products. In other mammals, the cloaca has become

Figure 36.6 Lower molar teeth of primitive mammals as viewed from the lingual side and crown. *A*, linear cusps of a triconodont; *B*, the triangular molar of a symmetrodont (infraclass Trituberculata); *C*, the addition of a low posterior heel (talanid) to the trigonid of a pantothere (infraclass Trituberculata); *D*, the tribosphenic molar of a primitive insectivore (infraclass Eutheria). *C* and *D* are from opposite sides of the jaw; in each case the talonid is on the posterior surface of the tooth. (From Stahl, B. J.: Vertebrate History: Problems in Evolution. New York, McGraw-Hill Book Company, 1974. After Simpson.)

divided, and the opening of the intestine, the **anus,** is separate from that of the urogenital ducts.

Monotremes are curious animals that have survived to the present only because they have been isolated from serious competition. The platypus is a semiaquatic species with webbed feet, short hairs and a bill like a duck's used in grubbing in the mud for food. Claws are retained, and they dig long burrows in muddy banks, Spiny anteaters have large claws and a long beak adapted for feeding upon ants and

1 cm

Figure 36.7 Skeletal reconstruction of a primitive Triassic triconodont based primarily on *Megazostrodon*. Head–body length was about 10 cm. (From Jenkins, F. A., Jr., and F. R. Parrington: The postcranial skeletons of the Triassic mammals *Eozostrodon, Megazostrodon* and *Erythrotherium*. Philos. Trans. R. Soc. Lond., B. *273*:387–431.)

termites. The animal can burrow very effectively with its claws, completely burying itself even in fairly hard ground in a few minutes. Many of its hairs are modified as quills. Facial muscles are poorly developed in both species, and they do not have fleshy lips. Young monotremes cannot suckle in the usual mammalian way; rather they lap up milk discharged from teatless mammary glands onto tufts of hair.

When the first skins of the platypus were shipped to Europe in the late eighteenth century, many zoologists viewed them as skillful fakes, such as the then current Chinese mermaids (the forepart of a monkey sewn onto the tail of a fish). After the authenticity of the platypus was established, a long controversy ensued as to whether to consider it a reptile or a mammal. Monotremes were finally regarded as mammals, but as such primitive and unusual ones that they are placed in a separate subclass—the **Prototheria.**

Figure 36.8 Monotremes and marsupials. *A*, The duckbilled platypus; *B*, the spiny anteater; *C*, opossum and young; *D*, koala bear; *E*, kangaroo. The platypus and anteater are monotremes; the others are marsupials. (*A* and *B* courtesy of the New York Zoological Society; *C* and *D* courtesy of American Museum of Natural History; *E* from Australian News and Information Bureau.)

Marsupials. In another line of evolution from triconodonts, the three linearly arranged molar cusps shifted to a triangular pattern, and a low heel was added to the posterior edge of each lower molar (Fig. 36.6*B* and *C*). These changes, first seen in the extinct **trituberculates,** improved the efficiency of the molar teeth in cutting and crushing insects and other small invertebrates. Primitive marsupials and true placental mammals living today have similar molar teeth (Fig. 36.6*D*) and doubtless evolved from trituberculates. Because of similarities in their tooth structure, trituberculates, marsupials and placental mammals are grouped together in the subclass **Theria.**

We know nothing about the reproduction of trituberculates, and they are set apart in their own infraclass. The opossum, kangaroo and other pouched mammals of the order Marsupialia (Fig. 36.8) are placed in the infraclass **Metatheria** and the true placental mammals in the infraclass **Eutheria.** Members of both groups are viviparous, though the placental arrangements of marsupials are less effective than those of eutherians. In most marsupials, the extraembryonic membranes, and chiefly the yolk sac, simply absorb a "uterine milk" secreted by the mother. There is no intimate union between the extraembryonic membranes and the uterine lining as there is in most eutherians.

Marsupials are born in what we would regard as a very premature stage. Their front legs, however, are well developed at birth, and the young pull themselves into a **marsupium,** or pouch, on the belly of the mother, attach to a nipple and there complete their development. An opossum, for example, is born after only 13 days' gestation, but it continues its development in the pouch until it is about 70 days old. A forward extension of the tubular epiglottis of the infant opossum dorsal to its secondary palate completely separates the digestive and respiratory tracts, and breathing and feeding can take place concurrently.

Marsupials were distributed world-wide during the late Mesozoic and early Cenozoic, but as eutherians began to spread out, marsupials became restricted. They have been most successful in those parts of the world where they have been isolated from competition with eutherians. They are the dominant type of mammal in Australia, have undergone an adaptive radiation and have become specialized for many modes of life. There are carnivorous marsupials, such as the Tasmanian wolf, ant-eating types, molelike types, semiarboreal phalangers and koala bears (the original "Teddy-bear"), plains-dwelling kangaroos and rabbit-like bandicoots. In contrast, the only marsupial present in North America is the opossum.

36.4 ADAPTIVE RADIATION OF EUTHERIANS

The eutherians, or placental mammals as they are frequently called, are the most successful mammals in all the parts of the world that they have reached. They have radiated widely and adapted to nearly every conceivable ecological niche upon the land. Others have readapted successfully to an aquatic mode of life, and some have evolved true flight.

Insectivores. The most primitive eutherians, that is, the stem group from which the other lines of descent evolved, were rather generalized, insect-eating types of the order **Insectivora.** Among modern types are the shrews, moles, the European hedgehog and the tree shrews of India and Southeastern Asia (Fig. 36.9). All are small mammals that retain many primitive mammalian features. The limbs of moles are specialized for digging, but limb structure is quite generalized in the other species. Five clawed toes are retained, and the entire sole of the foot is placed upon the ground, a foot posture termed **plantigrade** (Fig. 36.10). Insectivores have a primitive dentition in which the three primary cusps of each molar tooth are arranged in a triangle known as the **trigon** (upper molar) or **trigonid** (lower molar). Trigon and trigonid are mirror images of each other, so there is a cutting action as they slide past each other during occlusion. Some crushing occurs when the primary cone of the trigon falls upon a low heel, or **talonid,** on the posterior surface of the trigonid. We call these **tribosphenic** molars. They are seen first in some of the trituberculate mammals of the Mesozoic and are retained in primitive mammal groups living today. They are well suited for an insectivorous diet but become modified in groups specialized for carnivorous or herbivorous diets.

Flying Mammals. Bats, order **Chiroptera,** are closely related to this stem group and are sometimes characterized as flying insectivores. As in other flying vertebrates, the pectoral appendages have been transformed into wings (Fig. 36.11). Bat wings are structurally closer to those of pterosaurs than to birds' wings, for the flying surface is a leathery membrane, but the wing of a bat is supported by four elongated

Figure 36.9 Insectivores. *A*, A shrew eats more than its own weight every day; *B*, a mole in its burrow; *C*, a tree shrew upon a log. (*A* from Conoway, C. H., in Natural History, Vol. LXVIII, No. 10; *B*, courtesy of the American Museum of Natural History; *C* from Vaughan, T. A.: Mammalogy. Philadelphia, W. B. Saunders Co., 1972, photograph by M. W. Sorenson.)

Figure 36.10 Lateral and anterior views of the skeleton of the left hind foot of representative mammals. *A*, The primitive plantigrade foot of a lemur; *B*, the digitigrade foot of a cat; *C* and *D*, the unguligrade foot of a pig, an even-toed ungulate; *E*, the unguligrade foot of a horse, an odd-toed ungulate. The digits are indicated by Roman numerals, the metatarsals are black and the tarsals are stippled.

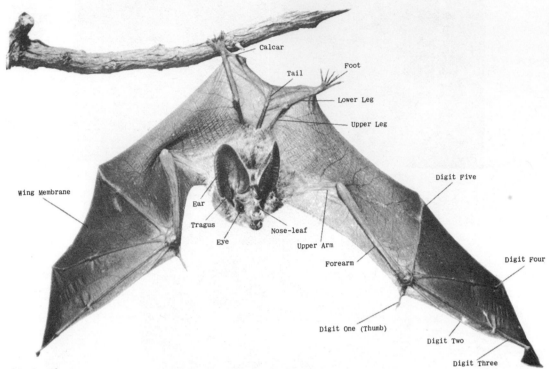

Figure 36.11   The Mexican big-eared bat (*Macrotus mexicanus*), about to take off. (From Walker, E. P.: Mammals of the World. Baltimore, The Johns Hopkins Press, 1964. Photograph by E. P. Walker.)

fingers (the second to fifth) rather than by a single one as in the pterosaur. The wing membrane attaches onto the hind leg and, in some bats, the tail is included in the membrane. The first finger is free of the wing, bears a small claw and is used for grasping and clinging. The hind legs are small and are of little use upon the ground, but they, too, are effective grasping organs and are used for clinging to a perch from which the bats hang upside down when at rest.

Our familiar bats are insect eaters that fly about at dusk in search of their prey. They retain primitive tribosphenic molars, but bats have diverged considerably in feeding habits and details of molar structure. Among the foods that various groups are specialized to take are fruit, pollen and nectar, blood (vampire bats), small mammals, birds and fish. Fish-eating bats catch their prey near the surface of the water by means of hooked claws on their rather powerful feet.

Insectivorous, nocturnally flying bats have evolved a system of echolocation that enables them to avoid obstacles in the dark and also enables many species to find their prey. As early as 1793, Spallanzani observed that a blinded bat could find its way about, but that one in which the ears had been plugged was helpless. However, it was not until the avail-

ability of sophisticated electronic apparatus about the time of World War II that Dr. Donald Griffin and others were able to show that bats emitted ultrasonic sounds that bounced off an object and returned as an echo. By analyzing the echoes, many bats can determine the distance, direction, size and possibly the texture of the object.

There are many differences in method of echolocation among the various groups of bats, which suggests that there has been some independent evolution of mechanisms. As the familiar bats of the family Vespertilionidae fly about at dusk searching for insects, they emit ultrasonic pulses through their mouths that range from 25 to 100 kHz, well above our threshold of hearing. Sounds at these frequencies have short wave lengths, and hence can produce sharp echoes from small objects. The pulses are frequency modulated and drop about an octave (i.e., from 60 or 80 kHz to 30 or 40 kHz) during their 1 to 4 millisecond duration. While the bat is searching, pulses are emitted at the rate of about 10 per second, but the frequency of emission increases tenfold, and the duration of the pulses shortens when a bat detects an insect and homes in on it. The emitted sounds are at very high energy levels, 60 dynes per square cm., which is over twice that in a boiler factory. Tiny muscles in the middle ear contract during sound emission, thereby damping the movement of the auditory ossicles and protecting the inner ear. These muscles relax as the echo returns. Distance appears to be perceived by the time interval between the emitted pulse and the echo. The potential problem of an overlap between the pulse and echo from a close object is avoided in some bats by decreasing the duration of the pulses as they get nearer to the object. In some other cases, there is an overlap, and the resolution of the problem is not clear. Directionality appears to be determined by a comparison of the differences in intensity of the echo between the two ears. Bats can perceive meaningful signals in the presence of considerable extraneous noise, but the mechanisms for this are not entirely clear.

Bats are the only mammals to have evolved true flight, but some other mammals can stretch a loose skin fold between their front and hind legs and glide from tree to tree. The colugo (order **Dermoptera**) of the East Indies and Philippines has been observed to glide 136 meters while losing no more than 12 meters in elevation (Fig. 36.12A). The flying squirrel is another example of a gliding mammal (Fig. 36.18A).

Toothless Mammals. Since primitive mammals were insectivorous, it is not surprising that certain ones became specialized to feed upon ants and termites, which are very abundant in certain regions. The South American anteater, order **Edentata** (Fig. 36.12B), is representative of this mode of life. Its large claws enable it to open ant hills, and then it laps up the insects with its long tongue, which is covered by a very sticky saliva secreted by enlarged salivary glands. In contrast to a primitive insectivore, which crushes its insect food with its teeth, an anteater swallows whole the insects that it eats. Ants are crushed and ground up by the very muscular pyloric region of the stomach. Its teeth were not needed for survival and have been lost. The tree sloth and armadillo belong to this same order, though they retain vestiges of teeth.

The pangolins of Africa and Asia (order **Pholidota**) and the aardvark of Africa (order **Tubilidentata**) are similar to edentates superficially, but this results from an independent adaptation to an anteating and termite-eating mode of life (Fig. 36.13). The scales of the pangolin are composed of hairs cemented together. When disturbed, the animal rolls into a ball.

Carnivores. As mammals increased in number and diversity, the opportunity arose for them to feed upon one another. Certain ones became specialized for a flesh-eating mode of life, and these constitute the order **Carnivora**. Familiar living carnivores are the weasels, dogs, raccoons, bears and cats (Fig. 36.14). The shift from an insectivorous to a flesh-eating diet was not difficult. Primarily, it required specializations of the teeth for killing and cutting prey and of limb structure for the speed needed to run down prey. Canines are large, and in contemporary species the last premolar tooth and first molar are specialized to form a set of shearing **carnassial** teeth (Fig. 36.3). The trigonid of the lower molar has flattened, two

A   **B**

Figure 36.12   *A*, A calugo; *B*, a giant anteater. (*A* from Walker, E. P.: Mammals of the World. Baltimore, The Johns Hopkins Press, 1964. Photograph by J. N. Hamlet. *B*, courtesy of the American Museum of Natural History.)

Figure 36.13 (A) The African aardvark and (B) the African tree pangolin resemble the South American anteater in their adaptations for insect eating, but these common features have evolved in these mammals independently. (B from Walker, E. P.: Mammals of the World. Baltimore, The Johns Hopkins Press, 1964. Photograph by Jean-Luc Perret.)

cusps are enlarged and the third is very small or absent. Teeth posterior to the carnassials are reduced or lost. One of the most conspicuous adaptations for speed has been a lengthening of the distal part of the limb, especially the foot, which increases stride length. Most carnivores have shifted from the primitive plantigrade to a **digitigrade** foot posture (Fig. 36.10B). They stand upon their toes (though not on their toe tips) with the rest of the foot raised off of the ground in the manner of a sprinter.

Most carnivores are semiarboreal or terrestrial, but one branch of the order, which includes the seals, sea lions and walruses, early became specialized for exploiting the resources of the sea. In addition to their adaptations as carnivores, which include the large canine tusks of the walrus used in gathering shellfish, these species evolved flippers and other aquatic modifications. When they swim, the large pelvic flippers are turned posteriorly and are moved from side to side in the manner of a fish's tail.

Ungulates. Horses, cows and similar mammals have become highly specialized for a plant diet. This has entailed a considerable change in their dentition, for plant food must be thoroughly ground by the teeth before it can be acted upon by the digestive enzymes. The molars of plant-eating mammals (and those of omnivorous species, such as human beings) have become square, as seen in a surface view. A fourth cusp is added to the upper trigon, the heel of the lower molar and its two cusps are elevated and the anteriormost cusp of the trigonid is lost. Four primary cusps are present on each molar, and sometimes these fuse to form complex ridges (Fig. 36.3E). Premolar teeth frequently acquire the form of molars. Upper and lower molars no longer slide partly past each other to give some cutting action, as do the tribosphenic molars of primitive mammals, but meet and crush food between them. A simple squaring of the molars is sufficient for herbivorous mammals that browse upon soft vegetation. But those that feed upon grass and other hard and gritty fare, as do the grazing species, are confronted with the additional problem of the wearing away of the teeth. Two adaptations have occurred; the height of the cusps of the teeth has increased, and cement (a hard material previously found only on the roots of the teeth) has grown up over the surface of the tooth and into the "valleys" between the elongated cusps (Fig. 36.3F). More tooth is provided to wear away, and the tooth is more resistant to wear. Teeth of this type are referred to as high-crowned, in contrast to the more primitive low-crowned type.

Herbivores constitute the primary food supply of carnivores and protect themselves primarily by the simple expedient of running away. Adaptations for speed have entailed a lengthening of the legs, especially their distal portions, and a relative shortening of the proximal parts of the limbs. The feet are very long, and the animals walk upon their toe tips, a gait termed **unguligrade** (Fig. 36.10 C, D and E). Those toes that no longer reached the ground became vestigial, or disappeared, and the primitive claws on the remaining ones were transformed into hoof—a characteristic that

Figure 36.14 Representative carnivores and cetaceans. A, Raccoon; B, walrus; C, the birth of a porpoise; D, the whalebone plates of a toothless whale hang down from the roof of the mouth; E, weasels in summer pelage. The porpose and whale are cetaceans; the others are carnivores. (A, B, D and E, courtesy of the American Museum of Natural History; C, courtesy of Marine Studios.)

gives the name ungulate to these mammals. The relatively shorter proximal segment of a limb places the retractor muscles closer to the fulcrum (Fig. 11.27); therefore a slight contraction of the muscle can induce a rather extensive movement of the distal end of a limb. Capacity for limb rotation is lost. The limbs are in effect jointed pendulums that can swing rapidly back and forth in the vertical plane.

The numerous and varied contemporary ungulates (Fig. 36.15) are grouped into two orders that can be separated on the basis of the type of toe reduction. In the order **Perissodactyla,** the axis of the foot passes through the third toe, and this is always the largest. Ancestral perissodactyls, including the primitive forest-dwelling horses of the early Tertiary, had three well developed toes (the second,

Figure 36.15   *A*, Tapir in Rangoon Zoo, Burma; *B*, expedition camel, Kalgan, China; *C*, cattle egret warns the weak-sighted rhinoceros of approaching danger; *D*, hippopotamus. (*A, B* and *D*, courtesy of the American Museum of Natural History; *C* from Natural History, Vol. LXVIII, No. 10, 1959.)

third and fourth) and sometimes a trace of a fourth toe (the fifth). The tapir and rhinoceros, which still walk upon soft ground, retain the middle three toes as functional toes, but only the third is left in modern, plains-dwelling horses (Fig. 36.10E). Perissodactyls are characterized by having an odd number of toes.

In the order **Artiodactyla,** the axis of the foot passes between the third and fourth toes, which are equal in size and importance. Ancestral artiodactyls had four toes (the second, third, fourth and fifth). Pigs and their allies, which move across soft ground, retain these four toes, though the second and fifth are

reduced in size (Fig. 36.10C and D). Vestiges of the second and fifth toes, the dew claws, are present in some deer, but camels, giraffes, antelope, sheep and cattle retain only the third and fourth toes. Artiodactyls, then, are even-toed ungulates. It is probable that these two orders have had a separate evolutionary origin and owe their points of similarity to parallel evolution.

In addition to running away, ungulates protect themselves by kicking with their powerful legs and hooves. Many artiodactyls also have evolved weapons for defense or for combat among males. Wild boars have canine

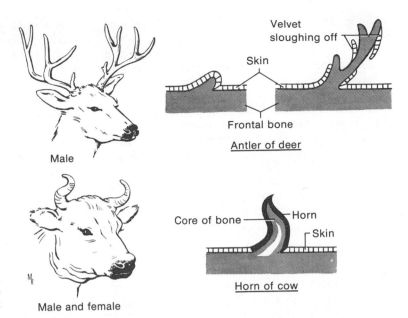

Figure 36.16 A diagram to show the differences between antlers (deer) and horns (cow). Antlers are annual growths that are shed in the winter; horns are permanent outgrowths.

tusks; male deer, **antlers;** and sheep and cattle of both sexes, **horns** (Fig. 36.16). Both antlers and horns are bony outgrowths from the skull. Antlers branch, are covered by skin (the velvet) only during their growth, and are shed annually; horns are permanent nonbranching structures covered by heavily cornified skin.

Subungulates. Subungulates are a group of plant eaters that have certain incipient ungulate tendencies. Elephants (order **Proboscidea,** Fig. 36.17A), for example, have five toes, each ending in a hooflike nail. They also walk to some extent upon their toe tips, but a pad of elastic tissue posterior to the digits supports most of the body weight. Elephants are noted for their enormous size, which must approach

the maximum for a completely terrestrial animal. Though large mammals have a relatively lower metabolic rate than small mammals, the huge mass of elephants necessitates their obtaining large quantities of food. The trunk, which represents the drawn out upper lip and nose, is an effective food-gathering organ. Elephants have a unique dentition in which all of the front teeth are lost except for one pair of incisors, which are modified as tusks. Their premolars, which have come to resemble molars, and their molars are very effective organs for grinding up large quantities of rather coarse plant food. They are high-crowned and so large that there is room for only one in each side of the upper and lower jaws at

Figure 36.17 Subungulates. The elephant (A) and the manatee (B) are believed to have had a common ancestry. (Courtesy of the American Museum of Natural History.)

a time. Premolars and molars are replaced sequentially. As one is worn down, a new one moves in. By using up their premolars and molars one at a time, elephants have evolved an interesting way of prolonging total tooth life.

Living elephants are restricted to Africa and tropical Asia and are only a small remnant of a once worldwide and varied proboscidean population. During the Pleistocene Epoch, mastodons, mammoths and other proboscideans were abundant in North America.

The conies of the Middle East (order **Hyracoidea**), though superficially rabbit-like animals, show an affinity to the elephants in their foot structure and in certain features of their dentition.

A final group of contemporary subungulates are the sea cows or manatees (order **Sirenia** Fig. 36.17*B*). These animals live in warm coastal waters and feed upon seaweed, grinding it up with molars that are replaced from behind in elephant-like fashion. Sea cows have a powerful, horizontally flattened tail and well developed pectoral flippers. These features, together with a very mobile and expressive snout and a single pair of pectoral mammary glands, led mariners of long ago to regard them as mermaids.

Rodents and Lagomorphs. Other herbivorous mammals gnaw and, in addition to high-crowned grinding molars, have an upper and lower pair of enlarged, chisel-like incisor teeth that grow out from the base as fast as they wear away at the tip. Gnawing has been a very successful mode of life; in fact, there are more species, and possibly more individuals, of gnawing mammals, or rodents (order **Rodentia**, Fig. 36.18), than of all other mammals combined. Rodents have undergone their own adaptive radiation and have evolved specializations for a variety of ecological niches. Rats, mice and chipmunks live on the ground, gophers and woodchucks burrow, squirrels and porcupines are adept at climbing trees and muskrats and beavers are semiaquatic.

Rabbits and the related pika of our Western mountains are superficially similar to rodents and were at one time placed in this order. True rodents, however, have only one pair of incisors in each jaw, whereas rabbits have a reduced second pair hidden behind the large pair of upper incisors. It is now believed that rabbits and the pika belong to a separate order, the **Lagomorpha,** and that their resemblance to rodents is a result of parallel evolution.

Whales. Whales, dolphins, and porpoises,

Figure 36.18   Rodents and lagomorphs. *A*, A flying squirrel; *B*, the pika; *C*, a chipmunk shelling a nut; *D*, a group of beavers. (Courtesy of the American Museum of Natural History.)

of the order **Cetacea,** are highly specialized marine mammals that may have evolved from primitive terrestrial carnivores. They have a fish-shaped body, pectoral flippers for steering and balancing, no pelvic flippers and horizontal flukes on a powerful tail that is moved up and down to propel the animal through the water. Some species have even re-evolved a dorsal fin. Despite these fishlike attributes, cetaceans are airbreathing, viviparous and suckle their young (Fig. 36.14C). Some hair is present in the fetus, but it is vestigial or lost in the adult stage in which its insulating function is performed by a thick layer of blubber.

Certain species have evolved sonar-like systems that help them to avoid obstacles and find their prey even in muddy river waters that a few enter.

Most cetaceans have a good complement of conical teeth well-suited for feeding upon fish, but the largest whales have lost their teeth and feed upon plankton. With fringed horny plates (the whalebone) that hang down from the palate (Fig. 36.14D), a toothless whale strains these small organisms from water passing through its mouth. The richness of the plankton together with the buoyancy of the water has enabled these whales to attain enormous size. The blue whale, which reaches a length of 30 meters and a weight of 135 metric tons, is the largest animal that has ever existed.

CLASSIFICATION OF MAMMALS

CLASS MAMMALIA. The mammals. Endothermic tetrapods, generally covered with hair; mammary glands present; jaw joint between dentary and squamosal bones; three auditory ossicles.

Subclass Uncertain.

Order Triconodonta. Ancestral mammals of the early Mesozoic having molar teeth with three linearly arranged cusps. Triconodonts; *Eozostrodon, Megazostrodon.*

Subclass Prototheria. One order of primitive mammals retaining many reptilian features including the egg-laying habit and cloaca; now confined to the Australian and New Zealand regions.

Order Monotremata. The monotremes. The platypus, *Ornithorhynchus;* spiny anteater, *Tachyglossus.*

*Subclass Allotheria. One order of gnawing mammals from the Mesozoic and early Tertiary. The multituberculates.

Subclass Theria. Typical mammals. All living ones are viviparous.

*INFRACLASS TRITUBERCULATA. Two orders of Mesozoic mammals with molar teeth having three cusps arranged in a triangle; a talonid sometimes present on lower molars. Ancestral to higher therians. Symmetrodonts and pantotheres.

INFRACLASS METATHERIA. Pouched mammals. Young are born at an early stage of development and complete their development attached to teats that are located in a skin pouch; usually three premolar teeth and four molars in each jaw.

Order Marsupialia. Marsupials. The opossum, *Didelphis.*

INFRACLASS EUTHERIA. Placental mammals. Young develop to a relatively mature stage in the uterus; primitive dental formula is $I_3^3, C_1^1, Pm_4^4, M_3^3$.

Order Insectivora. Insectivores including shrews, moles and hedgehog. Small mammals, usually with long pointed snouts; sharp cusps on molar teeth adapted for insect eating; feet retain five toes and claws. The common shrew, *Sorex.*

Order Dermoptera. The colugo, or flying lemur, of the East Indies and the Philippines. A gliding animal with a lateral fold of skin.

Order Chiroptera. The bats. Pectoral appendages modified as wings; hind legs small and included in wing membranes. The little brown bat, *Myotis.*

Order Primates. The primates. Rather generalized mammals retaining five digits on hands and feet; first digit usually opposable; claws usually replaced by finger- and toenails; eyes typically large and turned forward; often considerable reduction in length of snout.

---

*Extinct.

SUBORODER PROSIMII. Five families of primitive primates that show the beginnings of primate specializations: nails, enlarged brain and cranium, short snout and square molars. Lemurs (*Lemur*) of Madagascar; indrids of Madagascar; the aye-aye of Madagascar; lorises (*Galago*) of Africa, India and Southeast Asia; tarsier (*Tarsius*) of East Indies and Philippine Islands.

SUBORDER ANTHROPOIDEA. Five families of higher primates in which primate characters are well developed, snout much reduced, face at least partly free of fur, brain quite large and locomotor patterns variable (quadrupeds, brachiators and terrestrial bipeds). Marmosets and monkeys of the New World, Old World monkeys, apes and human beings. The marmoset (*Saguinus*), howler monkey (*Alouatta*), macaque (*Macaca*), chimpanzee (*Pan*), human being, *Homo*.

Order Carnivora. The carnivores. Flesh-eating mammals; large canines; certain premolars and molars modified as shearing teeth; claws well developed.

*SUBORDER CREODONTIA. Ancestral carnivores.

SUBORDER FISSIPEDIA. Modern terrestrial carnivores, including the dogs, wolves, foxes, raccoons, pandas, bears, weasels, martens, wolverines, badgers, skunks, minks, otters, cats, lions, tigers, mongooses, and hyenas. The domestic cat, *Felis*.

SUBORDER PINNIPEDIA. Marine carnivores, including the seals, sea lions and walruses. Members of this group have many aquatic specializations, such as paddle-like limbs and reduced tail. The harbor seal, *Phoca*.

**Order Condylarthra.* Ancestral ungulates. Five toes were retained, but each bore a small hoof; except for loss of clavicle, limb skeleton little modified; dentition complete; molars slightly modified for plant eating.

Order Proboscidea. Elephants and related extinct mammoths and mastodons. Massive ungulates retaining five toes, each with a small hoof; two upper incisors elongated as tusks; nose and upper lip modified as a proboscis. African elephant, *Loxodonta*; Indian elephant, *Elephas*.

Order Sirenia. Sea cows. Marine herbivores; pectoral limbs paddle-like; pelvic limbs lost; large horizontally flattened tail used in propulsion. Florida manatee, *Trichechus*.

Order Hyracoidea. Coneys. Small, guinea pig-like herbivores of the Middle East; four toes on front foot, three on hind foot, each with a hoof. *Procavia*.

Order Perissodactyla. Odd-toed ungulates. Axis of support passes through third digit; lateral digits reduced or lost.

SUBORDER CERATOMORPHA. Tapirs and rhinoceroses.

SUBORDER HIPPOMORPHA. Horses and their allies. *Equus*.

Order Artiodactyla. Even-toed ungulates. Axis of support passes between third and fourth toes; first toe lost; second and fifth toes reduced or lost.

SUBORDER SUINA. Pigs, peccaries and hippopotamuses. The pig, *Sus*.

SUBORDER RUMINANTIA. The cud-chewing artiodactyls, including camels, llamas, chevrotains, deer, giraffes, pronghorns, antelopes, cattle, sheep and goats. American buffalo, *Bison*.

Order Edentata. New World edentates, including sloths, anteaters and armadillos. Teeth reduced or lost; large claws on toes. The armadillo, *Dasypus*.

Order Pholidota. The pangolin, *Manis*, of Africa and southeastern Asia. Teeth lost: long tongue used to feed on insects; body covered with overlapping horny plates.

Order Tubulidentata. The aardvark, *Orycteropus*, of South Africa. Teeth reduced; long tongue used to feed on insects.

Order Cetacea. The whales and their allies. Large marine mammals; pectoral limbs reduced to flippers; pelvic limbs lost; large tail bears horizontal flukes, which are used in propulsion.

---

*Extinct.

SUBORDER ODONTOCETI. Toothed whales. The bottlenosed dolphin. *Tursiops.*

SUBORDER MYSTICETI. Whalebone whales. Teeth lost; strain small animals from water with horny whalebone plates that hang down from upper jaws. The blue whale, *Balaenoptera.*

Order Rodentia. The rodents. Gnawing mammals with two pairs of chisel-like incisor teeth. The largest order of mammals, it includes the squirrels, chipmunks, marmots, gophers, beavers, rats, mice, muskrats, lemmings, voles, porcupines, guinea pigs, capybaras and chinchillas. The woodchuck, *Marmota.*

Order Lagomorpha. Hares, rabbits, and pikas. Gnawing mammals with two pairs of chisel-like incisors and an extra pair of small upper incisors that lie behind the enlarged first pair. The rabbit, *Lepus.*

ANNOTATED REFERENCES

Many of the general references on vertebrates cited at the end of Chapter 33 contain considerable information on the biology of mammals.

Andersen, H. T. (Ed.): The Biology of Marine Mammals. New York, Academic Press, 1969. Chapters deal with the swimming, diving, echolocation and other aspects of the biology of cetaceans and other marine mammals.

Bourlière, F.: The Natural History of Mammals. 2nd ed. New York, Alfred A. Knopf, Inc., 1956. A fascinating account of the natural history of mammals; originally published in French as Vie et Moeurs des Mammifères.

Burt, W. H. and R. P. Grossenheider: A Field Guide to the Mammals. Boston, Houghton Mifflin Co., 1952. A useful guide, in the style of the Peterson bird guides, for the field identification of mammals.

Hall, E. R. and K. R. Kelson: The Mammals of North America. New York, The Ronald Press Co., 1959. A two-volume monograph on mammals for the serious student.

Sanderson, I. T.: Living Mammals of the World. Garden City, N.Y., Doubleday & Co., 1955. A superbly illustrated account of the major groups of mammals.

Schmidt-Nielsen, K.: Desert Animals. Oxford, Clarendon Press, 1964. The adaptations of camels, the kangaroo rat, man and other animals to desert life are thoroughly analyzed.

Scott, W. B.: A History of Land Mammals in the Western Hemisphere. 2nd ed. New York, Macmillan, 1937. An old but still very valuable source book on the evolution of horses, camels, mastodons and other groups that roamed the New World in ages past.

Slijper, E. J.: Whales. London, Hutchinson & Co., 1962. A valuable source book on the natural history and the specialized anatomy and physiology of cetaceans.

Vaughan, T. A.: Mammalogy. Philadelphia, W. B. Saunders Co., 1972. An excellent textbook with chapters on the origins of mammals, the various groups, ecology, zoogeography, behavior and various aspects of mammalian physiology.

Walker, E. P., et al.: Mammals of the World. 3rd ed. Revised by J. L. Paradiso. Baltimore. The Johns Hopkins Press, 1975. Each known genus of mammals is discussed and illustrated in the first two volumes of this treatise. A third volume is devoted to a classified bibliography of the literature regarding mammalian groups and their anatomy, physiology, ecology and so forth.

Wimsatt, W. A. (Ed.): Biology of Bats. New York, Academic Press, 1970. Includes chapters on the evolution, anatomy and physiology of the organ systems, thermoregulation and hibernation, development, echolocation and ecology of bats.

Young, J. Z. and M. J. Hobbs: The Life of Mammals, their Anatomy and Physiology. 2nd ed. Oxford, Clarendon Press, 1975. A very valuable source book emphasizing the anatomy and physiology of mammals.

---

*Extinct.

CHAPTER 37

PRIMATES, NONHUMAN AND HUMAN

37.1 PRIMATE ADAPTATIONS

Of all of the mammals, the order **Primates** is of particular interest to us, for we are members of this group and have designated it as of foremost importance (Latin *primus* = first). Primates, which also include the lemurs, monkeys and apes, diverged from insectivorous ancestors probably as early as the late Cretaceous. Certain primitive insectivores and primates are difficult to distinguish. The tree shrews, which we have described with insectivores, are considered to be primates by some investigators. As a group, primates have become more adapted to an arboreal life. Baboons, human beings and a few others have reverted to a terrestrial life, but they too bear the stamp of a prior arboreal adaptation. Our flexible limbs and grasping hands are fundamentally adaptations for life in the trees. Claws were transformed into finger- and toenails when grasping hands and feet evolved. Molar teeth became square, each with four low crowned cusps. This is a configuration well adapted for crushing the variety of soft food encountered in the trees: insects, leaves and shoots and fruit. The reduction of the olfactory organ and olfactory portion of the brain and the development of stereoscopic, or binocular, vision represent other adaptations of primates to arboreal life. Keen vision and the ability to appreciate depth are very important for animals moving through trees, whereas smell is less important for organisms living some distance from the ground than it is for terrestrial species. Muscular coordination is also very important, and the cerebellum of primates is unusually well developed. The evolution of stereoscopic vision, increased agility and particularly the influx of a new sort of sensory information gained by the handling of objects with a grasping hand was accompanied by an extraordinary development of the cerebral hemispheres. The cerebrum is the chief center for the integration of sensory information and the initiation of appropriate motor responses in all mammals, but it is particularly prominent in primates. It is believed that higher mental functions, such as conceptual thought, could only have evolved in organisms with a grasping hand. Life for young primates in the trees is more precarious than for young mammals living upon the ground. Primates have fewer young to care for, sometimes only a single offspring per birth, and the young stay with the parents for a longer period of time. This close association has led to the evolution of rather elaborate social structures in many species. In a very real sense, we and other primates are products of the trees.

37.2 THE GROUPS OF PRIMATES

Prosimians. Primates are usually divided into two suborders. The first, suborder **Prosimii**, includes five families of small, mostly arboreal, primates of the Old World tropics, extending from Africa to the Philippine Islands. Madagascar has a particularly diverse assemblage. (See classification at end of Chapter 36). Lemurs, indrids, lorises and the aye-aye of Madagascar and the tarsier (*Tarsius*) of the East Indies and Philippines (Fig. 37.1) all show the beginning of primate adaptations, but in general these features are not developed to the

Figure 37.1 Prosimians. *A,* Lemur; *B,* tarsier. (*A,* Courtesy of the San Diego Zoo; *B,* courtesy of the American Museum of Natural History.)

extent that they are in the higher primates: monkeys, apes and human beings of the suborder **Anthropoidea.** Some lemurs retain rather long snouts, for olfaction is still important in their social structure. The tribosphenic molars are beginning to become square, grasping feet have evolved and most of the claws have been transformed into nails. Some prosimians are highly specialized for unique modes of life. The tarsier has large eyes adapted for nocturnal life. Its elongated hind feet enable it to hop through the trees. Elongation has been accomplished through the lengthening of certain tarsal bones rather than by the toes as we see in carnivores and ungulates. The needed leverage is provided and the grasping digits are retained. Digital pads are borne upon the ends of the toes.

Anthropoids. Most monkeys, apes and human beings have a relatively flat face that is at least partly devoid of fur, well developed stereoscopic vision, the capacity to sit on their haunches and examine objects with their hands and an unusually large brain and globular brain case. Anthropoids appear in the fossil record early in the Tertiary period. It is believed that they evolved from certain prosimians, but the record is not clear enough to know which group.

New World monkeys and marmosets retain three premolar teeth. Most are arboreal quadrupeds and climbers. They do not have an opposable thumb, but the hind foot is grasping, and some species of monkey have evolved a prehensile tail that serves as a fifth limb (Fig. 37.2*A*). Marmosets retain claws on most of their toes.

Old World monkeys, apes and human beings have only two premolar teeth. Contemporary Old World monkeys (family **Cercopithecidae**) constitute a large and diverse group, including the forest-dwelling langurs and green monkeys, the macaques (who are only partially arboreal) and the terrestrial baboons and mandrills. All are quadrupeds, but they tend to sit upright upon **ischial callosities**—hardened skin pads upon their buttocks (Fig. 37.2*B* and *C*). Sometimes the skin around the callosities is brilliantly colored. The thumb is opposable. Cusps of each molar tooth have fused to form two transverse ridges that help in grinding the plant food on which this group largely feeds. Many species have cheek pouches in which food can be stored temporarily. Social structure is highly developed in many species; baboons travel in troops and cooperate in obtaining food and protecting the females and young.

Contemporary apes (family **Pongidae**) are the gibbon (*Hylobates*) of Malaysia, the orangutan (*Pongo*) of Borneo and Sumatra and the chimpanzee (*Pan*) and gorilla (*Gorilla*) of tropical Africa (Fig. 37.3). All apes lack a tail, have broad chests and limb specializations that enable them to **brachiate,** that is, to swing from branch to branch using their arms alternately. Their arms are longer than their legs. They retain the grasping hind foot, but the thumb is short in relation to the elongated palm (Fig. 37.4*A*). The hand is not as effective in grasping as in most monkeys and human beings; rather it is used as a hook when brachiating. Gibbons are relatively small apes, weighing between 5 and 13 kg. They are the best brachiators and can clear 3 meters or more with each swing. When walking along a branch or on the ground, they stand nearly erect with arms outstretched

Figure 37.2 A group of monkeys. A, Woolly monkey from South America; B, baboon from Africa; C, langur from Indo-china. (A, from Walker, E. P. and J. L. Paradiso: Mammals of the World. Baltimore, The Johns Hopkins Press, 1975; B, from Campbell, B. G.: Human Evolution. Chicago, Aldine Publishing Company, 1974; C, courtesy of the Chicago Museum of Natural History.)

Figure 37.3 Apes. A, Gibbon; B, gorilla. (From Campbell, B. G.: Human Evolution. Chicago, Aldine Publishing Company, 1974.)

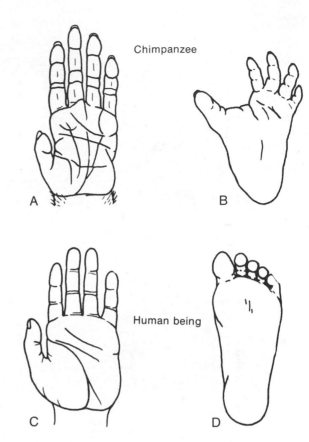

Chimpanzee

A   B

C   D

Human being

Figure 37.4 The hand of a chimpanzee (*A*) is adapted for brachiating and knuckle walking, and its foot (*B*) for grasping; the human hand (*C*) is adapted for grasping, and the foot (*D*) for bipedal locomotion. (*A* and *C* modified from Biegert; *B* and *D* from Morton.)

Sagittal crest

Nuchal crest

Brow ridge

A

Mastoid process

B

Figure 37.5 Lateral view of the skull and lower jaw of *A*, a gorilla; *B*, *Australopithecus*. The arrow indicates the inclination of the foramen magnum. (From LeGros Clark: The Antecedents of Man, Quadrangle Books, Inc.)

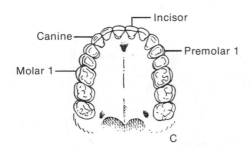

Figure 37.6 Palate and upper teeth of *A*, gorilla; *B*, *Australopithecus*; *C*, modern human. (From LeGros Clark: The Antecedents of Man, Quadrangle Books, Inc.)

as balancers. The remaining apes, known as the great apes, are larger. Some gorillas attain weights of 270 kg. Orangutans are primarily arboreal climbers, but they can brachiate. The pigmy chimpanzee and young gorillas spend considerable time in the trees and are good brachiators, but larger chimps and gorillas spend more time on the ground. They can walk bipedally over short distances but prefer a modified quadrupedal gait in which they support the front of the body upon the knuckles of their elongated hands (Fig. 37.3*B*).

Apes are herbivores, feeding upon fruit, young leaves and other plant material. The teeth are robust and the jaws are powerful. Prominent bony brow ridges above the orbits help resist the stresses set up in the skull by the powerful jaw mechanisms (Fig. 37.5). The gorilla's skull also has large sagittal and nuchal crests that increase the area available for the attachment of jaw muscles. The cusps of ape molars remain distinct and do not fuse to form transverse crests as they do in Old World monkeys. A fifth small cusp is present on the posterior edge of the lower molars. The tooth row has a somewhat squarish or U-shaped appearance, for the molars are in parallel rows and the canine, which is used in defense, is large (Fig. 37.6).

Apes are very intelligent creatures. They live in groups, have a social structure, communicate with each other by primitive sounds and facial expressions, and they can use sticks and other objects as tools to reach food.

37.3 HUMAN CHARACTERISTICS

Human beings differ enough from apes to be placed in a separate anthropoid family, the **Hominidae.** Yet apes and hominids have much in common, and this is recognized by placing both families in the superfamily **Hominoidea.** We differ from contemporary apes in being well adapted to a bipedal gait, using the hands not for locomotion but for tool manipulation and manufacture and having an omnivorous diet that includes both plant and animal material. Many features are correlated with our upright stance. We have a lumbar curve in our back that places our center of gravity over the pelvis and hind legs. Our ilium is broad and flaring, providing a large surface for the attachment of gluteal and other muscles that hold us erect (Fig. 37.7). Our legs are longer and stronger than our arms. The distal ends of our femora are brought close to the midline, giving us a knock-kneed appearance but placing the

Figure 37.7 Lateral views of the pelvis of a chimpanzee, *Australopithecus*, and modern human. (From C. Coon: The Origin of Races. Copyright © 1962 by Carleton S. Coon. Reprinted by permission of Alfred A. Knopf, Inc.)

foot under the projection of the body's center of gravity, thereby enabling us to balance easily on one foot when the other is off the ground (Fig. 37.8). Our foot has lost its primitive grasping ability, for the toes are short and parallel each other (Fig. 37.4). The heel bone is large, the tarsals and metatarsals form strong supporting arches and the great toe, with which we push off when walking, is enlarged. Our head is balanced upon the top of the vertebral column. The foramen magnum is far under the skull; the nuchal area on the back of the skull

Figure 37.8 Femora of *A*, an extinct ape (*Dryopithecus*); *B*, a modern human being. The axis of the shaft of the human femur inclines medially and is no longer perpendicular to the distal articular surface. (Modified from Campbell, B. G.: Human Evolution. Chicago, Aldine Publishing Co., 1974.)

for the attachment of neck muscles is much reduced; and the mastoid process is enlarged (Fig. 37.5). Body hair is very much reduced in human beings, and a great increase in the number of sweat glands is correlated with this. It has been estimated that each of us has between two and five million sweat glands that play a vital role in cooling the body. Reduction of hair and increased numbers of sweat glands appear to be adaptations associated with increased muscular activity on the part of our ancestors, who may have run down game in hot open country. Hominids must exert more muscular effort to attain a given speed than carnivores or ungulates do because our locomotor apparatus is not as well adapted for running.

Since the hands were no longer needed in locomotion, they could be used in other ways. Our ancestors retained and improved upon the grasping ability of the primitive primate hand as hominids first began to use large bones and stones as clubs and then to make stone tools. Our thumb is longer and the metacarpal portion of our hand shorter than in apes (Fig. 37.4C).

Early hominids probably began to eat a more varied diet than apes that included insects, lizards, rodents and other small animals as well as plant material. Later, fire was used to cook and soften food. Our teeth and jaws are less massive than in apes, and our face does not protrude as much. Our tooth row is more rounded and the canines are small (Fig. 37.6). Hominids defend themselves with tools and weapons rather than with teeth.

Our brain differs from an ape's not only in size but also in organization. The parietal, frontal and temporal areas of the cortex are enlarged in relation to the ape's. These are regions that are important in sensory and motor integration and in association, memory and speech. Although not much larger than an ape's in absolute size, the brain of early hominids was already quite large in relation to body size. Endocranial casts indicate that many of the characteristic human features were beginning to appear. Brain size and complexity evolved rapidly, probably during the period when hominids began to hunt larger animals that required more sophisticated weapons, a cooperative social structure and effective means of communicating and sharing ideas.

37.4 EARLY EVOLUTION OF APES AND HOMINIDS

Although contemporary apes and human beings can be distinguished easily, differences become blurred as we follow lines of descent back in time. At some point, pongids and hominids had a common ancestry, but when was this? Fossil anthropoids are found in early Oligocene deposits of Egypt. *Parapithecus*, a monkey-sized creature known from a number of mandibles and upper jaw fragments, had the two premolars characteristic of Old World anthropoids. It probably was on the line of ancestry to Old World monkeys, but it may not have been far from the line to pongids because it lacked certain dental specializations of later monkeys (Fig. 37.9). The separation between monkeys and apes occurred very early, for the same beds yield fossils (*Propliopithecus*, *Aegyptopithecus*) that show the beginnings of ape specializations. Gibbons appear to have diverged from other apes soon thereafter, for *Pliopithecus* of Miocene age already resembled gibbons in skull characteristics.

Many ape remains have been recovered from Miocene deposits of Europe, Africa and South Asia. Their ages range from about 10 to 22 million years ago. Originally assigned to different genera, most are now considered to represent different species of *Dryopithecus*. Remains of the limb skeleton indicate that the dryopithecines, although clearly apes, were not as specialized for brachiating as modern species. Their foot structure showed some adaptations toward bipedalism.

Until about 10 million years ago, East Africa was heavily forested. Shortly thereafter climatic changes accompanied such geologic disturbances as rifting and volcanic activity. Forests began to break up, open grasslands and savannas appeared and the habitat became much more diverse. Many apes remained in forest country, but some must have lived on the fringes of the forests and entered the grass lands. Opportunities existed for adaptive di-

vergence. In China and India, where similar climatic changes were taking place, *Gigantopithecus* specialized for a diet of grass seeds, stems and other coarse material. If other parts of its body were proportional to its huge jaws and teeth, which have been preserved, the creature must have weighed 300 kg. and stood 2.5 meters tall! It is likely that hominids also diverged from forest apes in adapting to life in a more open habitat and a diet that included small animals along with plant material. Changes in pattern of locomotion toward a more erect posture and bipedal gait, for which apes were already preadapted, would have a selective advantage in the open country. Dental changes would accompany a change in diet.

Ramapithecus known from jaw fragments found in Asian, African and European deposits that are 10 to 15 million years old, shows some hominid characters in its dentition. It appears to have been distinct from dryopithecines living at the same time. Early interpretations of the jaw fragments suggested that *Ramapithecus* had the somewhat rounded dental arcade seen in hominids (Fig. 37.6), but recently discovered material casts doubt on this interpretation. Still, the size of its incisor teeth, wear patterns on its molars and other dental features indicate some changes in diet. *Ramapithecus* may represent a stage transitional between apes and hominids, but the evidence is not clear enough to be certain.

37.5 THE APE MEN

In 1924, before many of the ape fossils discussed above had been found, an endocranial cast and part of the skull of a young hominoid were discovered in cave deposits in South Africa. Raymond Dart, Professor of Anatomy at the University of Witwatersrand, described it as *Australopithecus* (Latin *australis* = south + *pithecus* = ape) *africanus*. Subsequently, other specimens have been discovered in South and East Africa. Particularly fruitful have been Olduvai Gorge in Kenya, which was explored by Louis and Mary Leakey; Lake Turkana (formerly known as Lake Rudolf) in Kenya, being investigated by a team led by Richard E. F. Leakey; the Omo River basin in Ethiopia being explored by F. C. Howell; and the Afars Territory being studied by a group led by Donald Johanson. Several hundred fossil fragments of australopithecines have been recovered that range in age from the early Miocene (four or five million years ago) into the Pleistocene (one million years ago). Many of

Figure 37.9 A phylogeny of the major groups of Old World anthropoids. Time, shown on the vertical axis, has been greatly fore-shortened after the Pliocene.

the fossils were assigned originally to different genera, but most physical anthropologists now agree that they represent varieties of two species, or possibly only of one (Fig. 37.9). *Australopithecus africanus* was a lightly built, gracile creature that stood about 1.3 meters tall; the other, *Australopithecus robustus*, was a larger creature perhaps 1.6 meters in height.

The structural features of the australopithecines were closer to those of modern human beings than to those of apes. The foramen magnum was far under the skull; the mastoid process was well developed (Fig. 37.5B); the ilium, like ours, was broad, but the acetabulum to which the femur articulates faced somewhat more anteriorly than in modern human beings (Fig. 37.7). Even parts of a characteristic human foot have been found. These and other features indicate that they were terrestrial bipeds, although their gait may have been more shambling than ours. The masticatory apparatus was powerful, so the face protruded somewhat (Fig. 37.5). Jaws were large, teeth were massive and brow ridges were present. The teeth, although large, were very close to ours in details of their configuration and in the parabolic shape of the dental arcade (Fig. 37.6). Their cranial capacity ranged from about 430 to 750 ml. This overlaps the cranial capacity of the larger apes (340 to 650 ml. in the gorilla) and is far below that of modern human beings (1200 to 2000 ml.).

It seems clear that human limb, dental and some facial features evolved before there was a great increase in brain size. Darwin and other late 19th century scholars made the opposite assumption, which for a time appeared to be confirmed by the discovery of *Eoanthropus* in an Early Pleistocene gravel pit in Piltdown, England. The fragments suggested a person with a modern cranial capacity and the face, jaws and teeth of an ape. As the australopithecines were discovered, *Eoanthropus* appeared more and more as an anachronism. Reinvestigation of the fragments with modern methods of dating revealed a hoax; someone had "salted" the gravel pit with cranial fragments of a Late Pleistocene human being and part of the mandible of a contemporary orangutan stained to make it look old!

Australopithecus africanus and *robustus* differed somewhat in their dentition and probably in their mode of life. The teeth of *robustus* were larger and the jaws more massive. A sagittal crest upon the skull indicates a very powerful chewing mechanism. *Robustus* may have been more of a herbivore, feeding upon tough roots and young shoots along with berries and fruit; *africanus* was a more agile omnivore eating insects and small game along with softer plant

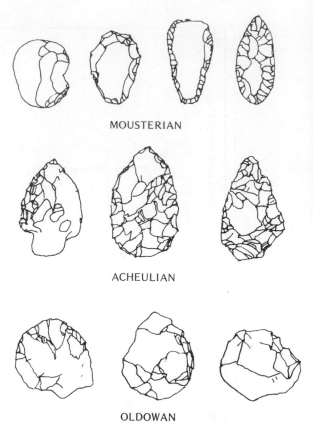

MOUSTERIAN

ACHEULIAN

OLDOWAN

Figure 37.10 Paleolithic tool kits. From bottom to top, these three sets of tools illustrate major stages in the early refinement of stoneworking. (From Keesing and Keesing: New Perspectives in Cultural Anthropology. Copyright © 1971 by Holt, Rinehart, and Winston, Inc. Reprinted by permission of Holt, Rinehart and Winston, Inc.)

material. There is no evidence that *robustus* was a toolmaker, but primitive stone tools of the Oldowan culture are found in the campsites of *africanus*. These choppers were little more than rounded stones sharpened by chipping one end (Fig. 37.10), but they, and the stone flakes broken from them, could have been used for killing and skinning small game, for sharpening branches as weapons and for cracking open bones for the marrow. Many cracked bones and bashed-in skulls have been found at the campsites.

When the australopithecines were first discovered, it was uncertain whether to interpret them as apes or early hominids. The question is largely semantic. Although *robustus* appears to have been a side line that died out by the Middle Pleistocene, many physical anthropologists believe that the line of evolution to more advanced hominids passed through *africanus*. Hunting and toolmaking probably set up selective forces that favored an increase in complexity of brain organization and size in *africanus.* Middle Pleistocene specimens (originally described as *Homo habilis*) are those with the largest cranial capacity.

37.6 HOMO ERECTUS

Long before the australopithecines were discovered in Africa, Haeckel in Germany had postulated the existence of a missing link between apes and modern human beings. He was presumptuous enough to give a name to his hypothetical creature—*Pithecanthropus alalus*—which literally means ape man without speech. This stirred the imagination of the Dutchman Eugene Dubois, who searched diligently for the missing link in Java, finally discovering in 1894 fossils in a river bank of a primitive hominid he called *Pithecanthropus erectus.* Shortly afterwards, Davidson Black discovered the remains of similar creatures in caves near Peking, which he named *Sinanthropus pekinensis.* Subsequently, other fossils of similar hominids have been found in the East Indies, China, Africa and Europe. A host of names are associated with them (Algerian man, Heidelberg man), but physical anthropologists regard all of them as representing a single, widespread species now called *Homo erectus.* This hominid is considered to be sufficiently close to modern human beings (*Homo sapiens*) to be included in the same genus.

The femur discovered by Dubois indicates that *Homo erectus* stood upright and was probably about 1.5 meters tall. Brain size ranged from 725 to 1225 ml., which overlaps that of the australopithecines on the one extreme and that of modern people on the other. Frontal areas of the brain, however, were poorly developed, for *Homo erectus* had a low sloping forehead (Fig. 37.11). His face was

somewhat brutish, protruding slightly, with rather heavy brow ridges and no chin. Teeth, though large, were essentially modern in their configuration.

Discoveries from Lake Turkana in Kenya, reported by Richard Leakey and Alan Walker in 1976, indicate that *Homo erectus* is at least 1.5 million years old. Possibly the genus *Homo* is even older than this. In 1973, Richard Leakey described a skull which is simply known by its museum number KNM-ER 1470, from deposits at Lake Turkana that are nearly three million years old (Fig. 37.12). Leakey has provisionally attributed this skull, which has a very large cranial capacity, to the genus *Homo*, but not all anthropologists agree with his interpretation because many of the features of the skull are within the range of variation of the australopithecines. In any case, the genus to which we belong is very old, and *Homo erectus* overlapped in space and time at least the later australopithecines. *Homo erectus* may have evolved from them, or both may have come from an older and as yet undiscovered common ancestor.

Some of the oldest specimens of *Homo erectus* are associated with the Oldowan stone culture, but this hominid soon developed a much more sophisticated Acheulian culture. *Homo erectus* was chipping flint and other fine-grained stones on all surfaces to fashion hand axes (Fig. 37.10) and was producing large stone flakes for a variety of cutting tools. Charred bones indicated the use of fire. Fire, and perhaps crude hide clothing, would have been essential for this hominid to penetrate central Europe and Asia in the Pleistocene, when continental glaciers were advancing. It is clear that *Homo erectus* had become a hunter of large game, for campsites contain the bones of bears, horses and even elephants. Such animals must have been stampeded over cliffs or into

A   **B**   **C**

Figure 37.11 Lateral views of the skull and lower jaw of *A, Homo erectus* (Java); *B, Homo sapiens* (Neanderthal); *C, Homo sapiens* (Cro-Magnon). (From T. H. Eaton, Jr.: Evolution. Copyright © 1970 by W. W. Norton & Co., Inc.)

Figure 37.12 Lateral view of skull KNM-ER 1470 from deposits 2.9 million years old at Lake Turkana, Kenya, discovered by Richard E. F. Leakey. This may be the oldest representative of *Homo*. (From Leakey, R. E. F.: Evidence for an advanced Plio-Pleistocene hominid from East Rudolf, Kenya. Nature *242*:447–450, 1973.)

bogs to have been killed by primitive man. From a study of early hunting sites in Spain, F. Clark Howell of the University of Chicago has postulated that fire was sometimes used in driving game. All this implies rather intelligent persons, living in groups, with an ability to communicate and to teach the young to make tools and to hunt, and with a knowledge of the seasons and habits of game (Fig. 37.13).

37.7 HOMO SAPIENS

Our species, *Homo sapiens*, evolved from and replaced *Homo erectus* about 300,000 years ago. The earliest specimens from Swanscombe, England, and Steinheim, Germany,

indicated that the first *sapiens* were a rather robust people with heavily built faces and jaws. Traces of brow ridges were present, and the forehead was somewhat sloping, but all features are within the range of variation of modern people. Acheulian tools are associated with Swanscombe man.

During most of the last glacial stage, western Europe was occupied by a stocky brutish-looking fellow called Neanderthal man because his first remains were discovered in the Neander valley (German, *thal* = valley). Originally considered to be a distinct species, Neanderthals are now regarded as an extinct race of *Homo sapiens*. These "classic" Neanderthals had large brains, and they were powerful individuals with strong jaws, heavy brow ridges, receding chins and short, bandy legs

Figure 37.13 Reconstruction of the life of *Homo erectus* living in the Pleistocene at Peking. (By permission of the trustees of the British Museum [Natural History].)

(Fig. 37.11B). They often built their fires in hearths in cave floors, and their Mousterian culture included a large tool kit of stone axes, scrapers, borers, knives, spear points and saw-edged and notched tools probably used in making spear handles and other simple wooden implements. The stones show considerable secondary chipping to refine the shapes and sharpen the edges (Fig. 37.10). The Neanderthals probably had developed a belief in the supernatural and in an afterlife, for they buried their dead with food and tools.

Some physical anthropologists limit the term Neanderthal to these classic fellows of western Europe, but others use the term in a far broader sense for all men living in Europe, Africa and Asia during the last part of the Pleistocene. All used variants of the Mousterian culture, and some, including Solo man of Java and Rhodesian man of southern Africa, were nearly as brutish in appearance as classic Neanderthal. However, many people during this period had more delicate features; they had good chins, high foreheads and no or weak brow ridges.

It is doubtless from such as these that Cro-Magnon man evolved (Fig. 37.11C). Cro-Magnons abruptly replaced the classic Neanderthals in western Europe about 35,000 years ago. Physically, Cro-Magnons are indistinguishable from many present-day Europeans. This hominid developed a very sophisticated Aurignacian culture, using delicate stone, bone and wooden tools, and he has left fine examples of painting in caves in France and Spain.

Mankind today can be divided into dozens of geographic races, that is, populations found in particular parts of the world and sharing certain genes and traits. Examples of races are the Ainu of northern Japan, the Nordics of northern Europe, the Eskimos of arctic America and the American Indians. Races differ from one another not in single features but in having different frequencies of many alleles and characteristics affecting body proportions, skull shape, degree of skin pigmentation, texture of head hair, abundance of body hair, form of eyelids, thickness of lips, frequency of various blood groups, ability to taste phenylthiocarbamide and many other anatomical and physiological traits. While certain of these differences, such as skin pigmentation, probably are adaptive, the significance of many is unknown.

Anthropologists believe that the numerous races represent localized differences in larger geographic groups. On the basis of skeletal remains and the present-day differences in the characteristics and distribution of races, Carleton Coon recognizes five major racial groups: (1) the **Caucasoids,** which include the Nordic, Alpine and Mediterranean races of Europe; the Armenoids and Dinarics of Eastern Europe, the

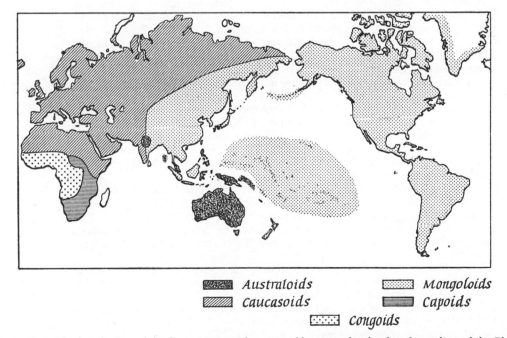

Figure 37.14 Probable distribution of the five major racial groups of humans shortly after the ending of the Pleistocene. (From C. Coon: The Origin of Races. Copyright © 1962 by Carleton S. Coon. Reprinted by permission of Alfred A. Knopf, Inc.)

Near East and North Africa; and the Hindus of India; (2) the **Mongoloids,** which include the Chinese, Japanese, Ainu, Eskimos and American Indians; (3) the **Congoids,** or Negroes and Pygmies; (4) the **Capoids,** or Bushmen and Hottentots of Africa; and (5) the **Australoids,** which include the Australian aborigines, Negritos, Tasmanians and Papuo-melanesians.

Fossil evidence indicates that early hominid evolution occurred in Africa, but by mid-Pleistocene, hominids had spread widely through the Old World. It is uncertain at which time the racial groups began to differentiate. Many anthropologists believe that it was in the latter part of the Pleistocene, after the *Homo sapiens* level had been attained, but Professor Coon and some others believe that they can find skeletal evidence for the beginning of racial differentiation as far back as *Homo erectus.* If this was the case, then many human populations in different parts of the world must have crossed the *sapiens* line independently. *Homo erectus* and *sapiens* would then be viewed as a large interbreeding complex changing in space and time.

Exactly when Mongoloids first crossed the Bering Strait from Asia to the New World is uncertain. Presumably it was during one or more of the glacial periods of the Pleistocene, for at that time much water would have been utilized in forming continental glaciers, the sea level would have been lower and Siberia and Alaska would have been connected by at least a series of close islands. Geologic evidence indicates that parts of Siberia and Alaska were unglaciated during the ice ages, so ice would not have blocked the route. The oldest sign of people in the New World is a group of primitive tools found in Venezuela and associated with the remains of extinct mammals, such as mastodons and glyptodons (giant anteaters). These have been dated at 16,000 B.C. The earliest Indian sites in North America are in Colorado and Arizona, and are dated at 8800 and 9300 B.C., respectively. The presumed distribution of racial groups shortly after the ending of the Pleistocene is shown in Figure 37.14. Within historic times there has been an increasing movement of peoples and a resultant increase in racial mixture.

ANNOTATED REFERENCES

Binford, S. R. and L. R. Binford: Stone tools and human behavior. Scientific American *220*:4 (April) 1969. An analysis of the social structure of Neanderthal man from a study of the groups of Mousterian tools and their use.

Campbell, B. C.: Human Evolution. 2nd ed. Chicago, Aldine Publishing Co., 1974. An excellent account of human evolution with an emphasis on man's unique anatomical adaptations.

Clark, W. E. LeGros: The Antecedents of Man. 3rd ed. Chicago, Quadrangle Books, 1971. An excellent presentation of primates and their evolution.

Coon, C. S.: The Origin of Races. New York, Alfred A. Knopf, 1962. Describes contemporary human races and traces their origin back to the Middle Pleistocene; all human fossils known to date of publication are described.

Day, M.: Guide to Fossil Man. Cleveland, World Publishing Co., 1965. A tabulation and description of all known human fossils; an excellent guide to the primary literature on human evolution.

Editors of Time-Life Books: The Emergence of Man. New York, Time-Life Books, 1972–1973. A series of authoritative and superbly illustrated books on human evolution written by the editors of Time-Life Books in consultation with leading anthropologists. Volumes on Life Before Man, The Missing Link (Australopithecines), The First Men (Homo erectus), and The Neanderthals are particularly interesting.

Holloway, R. L.: The casts of fossil hominid brains Scientific American *231*:106, (July) 1974. The major features of human brain organization and relative size were attained very early in our evolution.

Howell, F. C. and the editors of *Life*: Early Man. New York, Time Inc., 1965. A Life Nature Library book; a superbly illustrated account of man's evolution prepared by a leading authority on the subject.

Napier, J.: The antiquity of human walking. Scientific American *216*:4 (April) 1967. An analysis of man's striding gait, its anatomical prerequisites and its probable origin.

Pilbeam, D. L.: Ascent of Man. New York, Macmillan Publishing Company, 1972. An excellent textbook on physical anthropology.

Simons, E. L.: The earliest apes. Scientific American *217*:6(Dec.) 1967. A discussion of fossil apes and their interrelationships.

Simons, E. L., and P. C. Ettel: Gigantopithecus. Scientific American *222*:1 (Jan.) 1970. A study of this largest of all apes and its interrelationships and probable mode of life.

Tobias, P. V.: The Brain in Hominid Evolution. New York, Columbia University Press, 1971. A thorough review of brain evolution and its relation to cultural development.

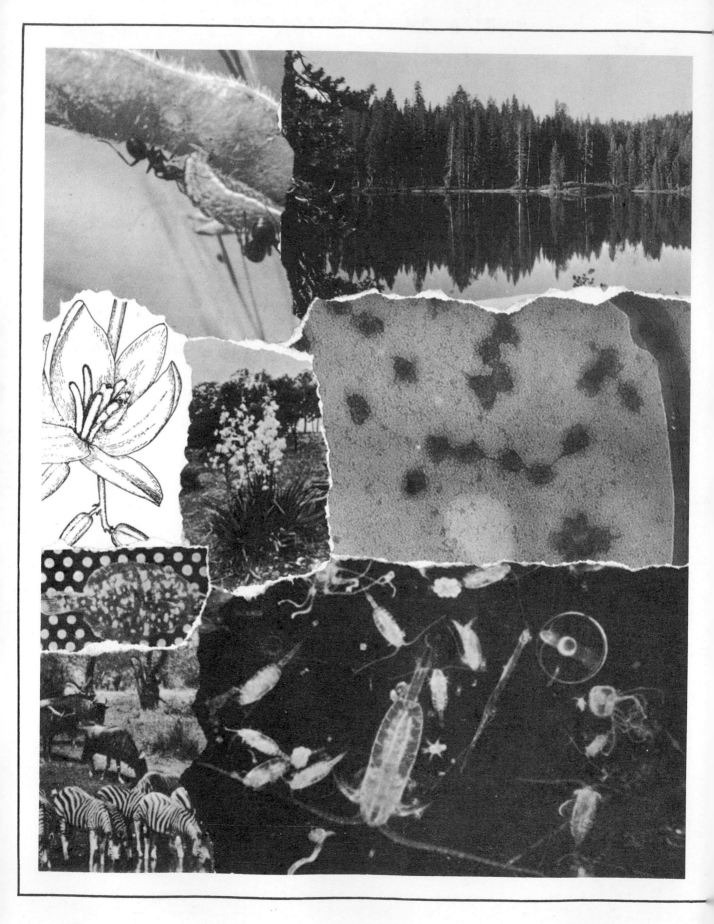

Part Five

ANIMALS AND THEIR ENVIRONMENT

CHAPTER 38

ECOLOGY

As more is learned about any kind of plant or animal, it becomes increasingly clear that each species has undergone adaptations to survive under a particular set of environmental circumstances. Each may show adaptations to wind, sun, moisture, temperature, salinity and other aspects of the physical environment as well as adaptations to specific plants and animals that live in the same region. Studies of the interrelationships of organisms with their physical and biotic environments are termed **"ecology."** This word is very much in the public consciousness today as we become aware of some of our past and current ecological malpractices. It is important for everyone to know and appreciate the principles of this aspect of biology so that he can form an intelligent opinion regarding topics such as insecticides, detergents, mercury pollution, sewage disposal, power dams and their effects on mankind, on our civilization and on the world we live in.

The Greek *oikos* means "house" or "place to live," and ecology (*oikos logos*) is literally the study of organisms "at home," in their native environment. The term was proposed by the German biologist Ernst Haeckel in 1869, but many of the concepts of ecology antedated the term by a century or more. Ecology is concerned with the biology of groups of organisms and their relations to the environment. The term **autecology** refers to studies of individual organisms, or of populations of single species, and their relations to their environment. The contrasting term, **synecology,** refers to studies of groups of organisms associated to form a functional unit of the environment. Groups of organisms may be associated in three different levels of organization—populations, communities and ecosystems. In ecological usage, a **population** is a group of individuals of any one kind of organism, a group of individuals of a single species. A community in the ecological sense, a **biotic community**, includes all of the populations occupying a given defined physical area. The community together with the physical, nonliving environment compose an **ecosystem.** Thus synecology is concerned with the many relationships within communities and ecosystems. The ecologist deals with such questions as who lives in the shade of whom, who eats whom, who plays a role in the propagation and dispersion of whom and how energy flows from one individual to the next in a food chain. The ecologist attempts to define and analyze those characteristics of populations that are distinct from the characteristics of individuals and the factors responsible for the aggregation of populations into communities.

In earlier chapters, we examined the reactions by which energy and materials flow through cells and through individual organisms. We can now consider the thesis that the flow of energy and materials through all the organisms living in a given area and composing a specific ecosystem is regulated by control systems analogous to those operating in the cells and tissues of a single organism.

38.1 THE CONCEPTS OF RANGES AND LIMITS

Probably no species of plant or animal is found everywhere in the world; some parts of the earth are too hot, too cold, too wet, too dry or too something else for the organism to survive there. The environment may not kill the adult directly, but effectively keeps the species from becoming established if it prevents its reproducing or kills off the egg, embryo or some other stage in the life cycle.

Most species of organisms are not even found in all the regions of the world where they could survive. The existence of barriers prevents their further dispersal and enables us to distinguish the major biogeographic realms (p. 221) characterized by certain assemblages of plants and animals.

Biologists early in the nineteenth century were aware that each species requires certain materials for growth and reproduction and can be restricted if the environment does not provide a certain minimal amount of each one of these materials. Justus Liebig stated in 1840 what is now known as his **Law of the Minimum:** the rate of growth of each organism is limited by whatever essential nutrient is present in a minimal amount. Liebig, who studied the factors affecting the growth of plants, found that the yield of crops was often limited not by a nutrient required in large amounts, such as water or carbon dioxide, but by something needed only in trace amounts, such as boron or manganese. Liebig's law is strictly applicable only under steady-state conditions, when the inflow of energy and materials equals the outflow. In addition, there may be interactions between factors such that a very high concentration of one nutrient may alter the rate of utilization of another (the rate-limiting one) and hence alter the effective minimal amount required. Certain plants, for example, require less zinc when growing in the shade than when growing in the sunlight.

V. E. Shelford pointed out in 1913 that *too much* of a certain factor would act as a limiting factor just as well as too little of it and that the distribution of each species is determined by its **range of tolerance** to variations in each environmental factor. Much work has been done to define the limits of tolerance, the limits within which species can exist, and this concept, sometimes called Shelford's **Law of Tolerance,** has been helpful in understanding the distribution of organisms.

It has usually been found that certain stages in the life cycle are critical in limiting organisms — seedlings and larvae are usually more sensitive than adult plants and animals. Adult blue crabs, for example, can survive in water with a low salt content and can migrate for some distance up river from the sea, but their larvae cannot, and the species cannot become permanently established there.

Some organisms have very narrow ranges of tolerance to environmental factors; others can survive within much broader limits. Any given organism may have narrow limits of tolerance for one factor and wide limits for another. Ecologists use the prefixes **steno-** and **eury-** to refer to organisms with narrow and wide, respectively, ranges of tolerance to a given factor. A stenothermic organism is one that will tolerate only narrow variations in temperature. The housefly is a eurythermic organism, for it can tolerate temperatures ranging from 5° to 45° C. The adaptation to cold of the antarctic fish *Trematomus bernacchi* is remarkable. It is extremely stenothermic and will tolerate temperatures only between −2° C. and +2° C. At 1.9° C. this fish is immobile from heat prostration!

Temperature. Temperature is an important limiting factor, as the relative sparseness of life in the desert and arctic demonstrates. Most of the animals that do live in the desert have adapted to the rigors of the environment by living in burrows during the day and coming out to forage only at night. The temperature in the burrow of a kangaroo rat 60 cm. underground may be only 16° C. when the temperature at the surface is over 38° C. Many animals escape the bitter cold of the northern winter not by migrating south but by burrowing beneath the snow. Measurements made in Alaska show that when the surface temperatue is −55°C., the temperature 60 cm. under the snow, at the surface of the soil, is a "warm" −7° C.

Although the ringnecked pheasant has been introduced into the southern states a number of times and the adults survive well, the developing eggs are apparently killed by the high daily temperatures and arc unable to complete development.

Light. The amount of light is an important factor in determining the distribution and behavior of both plants and animals. Light is, of course, the ultimate source of energy for life on this planet, yet prolonged direct exposure of cells to light of high intensity may be fatal. Both plants and animals have evolved mechanisms and responses to protect them against too much (or too little) light.

The amount of daylight per day, known as the **photoperiod,** has a marked influence on the time of flowering of plants, the time of migration of birds, the time of spawning of fish and the seasonal change of color of certain birds and mammals. The effects of the photoperiod on vertebrates are mediated by neurohormonal mechanisms involving the hypothalamus, the pituitary and the pineal. Knowledge of photoperiod phenomena has proved to be of considerable economic importance. Chicken farmers have found that artificial illumination in the hen house, by extending the photoperiod, stimulates the hens to lay more eggs.

Water. Water is a physiologic necessity for all living things but is a limiting factor primarily for land organisms. The total amount of rainfall, its seasonal distribution, the humidity and the supply of ground water are some of the factors limiting distribution of animals and plants. Some lakes and streams, especially in the western and southwestern United States, periodically become dry or almost dry, and the fish and other aquatic animals are killed. During periods of low water, the water temperature may rise sufficiently to kill off the aquatic forms. Many of the protozoa survive the drying of the puddles in which they normally live by forming thick-walled cysts. Some animals have adapted to desert conditions by digging and living in burrows where the temperature is lower and the humidity is higher than at the surface.

An excess of water is fatal to certain animals; earthworms, for example, may be driven from their burrows by heavy rainfall because oxygen is only sparingly soluble in water and they are unable to get enough oxygen when immersed. Knowledge of the limits of water tolerance is helpful in attacking insect and other pests. Wire worms have rather narrow limits of tolerance to water and are most sensitive as larvae and pupae. They can be killed by flooding the infested fields or by planting alfalfa or wheat to dry out the soil below the limit of tolerance of the wire worm larvae.

Other Evironmental Factors. The supply of oxygen and carbon dioxide is usually not limiting for land organisms, except for animals living deep in the soil, on the tops of mountains or within the bodies of other animals. Animals living in aquatic environments may be limited by the amount of dissolved oxygen present; the oxygen tension in stagnant ponds or in streams fouled by industrial wastes may become so low as to be incompatible with many forms of life. Some parasites have adapted to the low oxygen tension within the host's body by evolving special metabolic pathways by which energy can be released from foodstuffs without the utilization of free oxygen.

The trace elements necessary for plant and animal life are limiting factors in certain parts of the world. The soil in certain parts of Australia, for example, is deficient in copper and cobalt and is unsuitable for raising cattle or sheep. Other trace elements that may be a limiting factor are manganese, zinc, iron, sulfur, selenium and boron.

The amount of carbon dioxide in the air is remarkably constant, but the amount dissolved in water varies widely. An excess of carbon dioxide may be a limiting factor for fish and insect larvae. The hydrogen ion concentration, pH, of water is related physicochemically to the carbon dioxide concentration, and it too may be an important limiting factor in aquatic environments.

Water currents are limiting for a number of kinds of animals and plants; the fauna and flora of a still pond and of a rapidly flowing stream are quite different. Winds may have a comparable limiting effect upon land organisms.

The type of soil, the amount of topsoil, its pH, porosity, slope, water-retaining properties and so on are limiting factors for a variety of plants, and hence indirectly for animals. The ability of many animals to survive in a given region depends upon the presence of certain plants to provide shelter and cover, as well as food. Grasses, shrubs and trees on land each provide shelter for certain kinds of animals, and seaweeds and freshwater aquatic plants have a similar role for aquatic animals. Some animals require special shelter for breeding places and the care of the young. In many different kinds of birds, mammals, crustaceans and other animals, each animal or pair establishes a **territory,** a region that supplies food and shelter for it and its offspring and that it defends vigorously against invasion by other members of the same species (p. 472).

In summary, whether an animal can become established in a given region is the result of a complex interplay of such physical factors as temperature, light, water, winds and salts and biotic factors, such as the plants and other animals in that region that serve as food, compete for food or space or act as predators or parasites.

38.2 STRUCTURAL ADAPTATIONS

In the course of evolution, organisms have undergone successive structural adaptations and readaptations as the environment changed or as they migrated to a new environment. As a result, many organisms today have structures or physiologic mechanisms that are useless or even deleterious but were advantageous at an earlier time when the organism was adapted to a different environment.

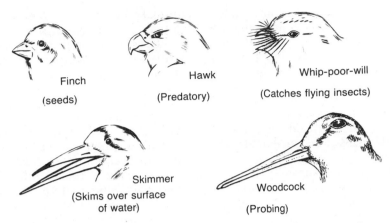

Figure 38.1 Diagrams of the bills of a variety of birds, illustrating their adaptation to the type of food eaten.

The adaptations of the mouthparts of certain animals to the kind of food eaten are among the most striking that can be cited. The mouthparts of certain insects are adapted for sucking nectar from certain species of plants; others are specialized for sucking blood, for biting or for chewing vegetation. The bills of various kinds of birds (Fig. 38.1) and the teeth of various kinds of mammals may be highly adapted for particular kinds of food.

In many animals, the specialized adaptation to a certain way of life is simply the latest stage in a series of adaptations. For example, both the human and the baboon, whose immediate ancestors were tree-dwellers, have returned to the ground and have become readapted to walking rather than to climbing trees.

Readaptation may be a very complicated process. The present-day Australian tree-climbing kangaroos are the descendants of an original ground-dwelling marsupial. From the ground-dwellers evolved forms that, in adaptive radiation, took to the trees and developed limbs adapted to tree climbing (or perhaps the sequence of events was the reverse—first the evolution of specialized limbs, then the adoption of a tree-dwelling mode of life).

Some of these tree-dwellers eventually left the trees and became readapted to ground life, and the hind legs became lengthened, strengthened and adapted for leaping. But finally some of these kangaroos went back to the trees, although now their legs were so highly specialized for leaping that they could not be used for grasping a tree trunk. Consequently, the present-day tree kangaroos must climb like bears, by bracing their feet against the tree trunk. A comparison of the feet of existing Australian marsupials reveals all the stages in this complicated, shifting process of adaptation.

38.3 PHYSIOLOGICAL ADAPTATIONS

One of the major struggles among organisms stems from the competition for food; hence a mutation enabling an animal to use a new type of food is extremely advantageous. This may be accomplished in a number of ways—by the evolution of a new digestive or energy-liberating enzyme system, for example. A mutation resulting in a new energy-liberating enzyme enabled the sulfur bacteria to obtain energy from hydrogen sulfide, a substance that is poisonous to almost all other organisms. The evolution of a special enzyme for reducing disulfide bridges ($-S-S-$) gave the clothes moth its unique ability to digest wool. The protein molecules of wool are cross-linked with many disulfide bridges.

Another type of favorable mutation is one that decreases the growing season of a plant or the total length of time required for an insect to develop. Such mutations enable the organism to survive farther from the equator and open up new areas of living space and new sources of food.

Any mutation that increases a species' limits of temperature tolerance—i.e., makes it more eurythermic—may enable it to live in a new region, at a higher latitude or a higher altitude. Other organisms have solved the problem of living in arctic regions by becoming dormant during the cold season or by developing behavior patterns of migration. Many birds, but only a few mammals, migrate south to avoid the cold northern winter.

Marine fish are usually adapted to survive

within a certain range of pressures and hence at a particular depth. Surface animals are crushed by the terrific pressures of the deep, and deep-sea animals usually burst when brought to the surface. The whale, however, is able to withstand great changes in pressure, and can dive to depths of 800 meters without injury. Presumably, its lung alveoli collapse when the pressure on the body reaches a certain point and gases are no longer absorbed into the blood.

Humans can survive pressures as high as six atmospheres if the pressure is increased and then decreased slowly. The increase in pressure increases the amount of gases dissolved in the blood and body fluids. If the pressure is decreased suddenly, these gases come out of solution and form bubbles throughout the body. Those in the blood impede circulation and bring about the symptoms of diver's disease, or "the bends." The pilot of a jet plane may climb so rapidly that the atmospheric pressure is reduced fast enough to bring bubbles of gas out of solution in his blood and cause a type of the bends.

38.4 COLOR ADAPTATIONS

Adaptations for survival are evident in the color and pattern of plants and animals as well as in their structure and physiologic processes. We can distinguish three types of color adaptation: concealing or **protective coloration,** which enables the organism to blend with its background and be less visible to predators; **warning coloration,** which consists of bright, conspicuous colors and is assumed by poisonous or unpalatable animals to warn potential predators not to eat them; and **mimicry,** in which the organism resembles some other living or nonliving object—a twig, leaf, stone or perhaps some other animal which, being poisonous, has warning coloration.

Concealing coloration may serve to hide an animal that wants to escape the notice of potential predators or it may be assumed by a predator in order to be unnoticed by his potential prey. Examples of concealing coloration are legion—the white coats of arctic mammals and the stripes and spots of tigers, leopards, zebras and giraffes which, though conspicuous in a zoo, blend imperceptibly with the moving pattern of light and dark typical of their native savanna. Some animals—frogs, flounders, chameleons, crabs and others—have a remarkable ability to change their color and pattern as they move from a dark to a light background or from one that is uniform to one that is mottled (Fig. 38.2).

To demonstrate experimentally that concealing coloration does have survival value, F. B. Isely fastened grasshoppers with different body colors to plots of different colored soils—light, dark, grassy, sandy and so on. After these plots had been exposed to the predatory activities of wild birds, or chickens for a given length of time, the survivors were tabulated. It was found that there was a significantly higher percentage of survivors among those grasshoppers that matched their background.

H. B. D. Kettlewell of Oxford University made careful field observations of the frequency of capture of light-colored and dark-colored mutants of the moth *Biston betularia* by birds. In the woods of Dorsetshire, where there is little pollution and the tree trunks are light-colored, the light-colored moths were significantly less likely to be caught and eaten by birds. In woods in the industrial Midlands, where leaves and tree trunks are blackened with soot, the dark-colored moths were significantly less likely to be caught and eaten.

When an animal is equipped with poison fangs, a stinging mechanism or some chemical that gives it a noxious taste, it is probably to its advantage to have the fact widely advertised, and, in fact, many animals so equipped do have warning colors.

A species of European toad with a bright

Figure 38.2 An experiment to show the remarkable ability of the flounder to change its color and pattern to conform with its background. Left, a flounder on a uniform, light background; right, the same fish after being placed on a spotted, darker background. (From Villee, C. A.: Biology. 7th ed. Philadelphia, W. B. Saunders Company, 1977.)

scarlet belly has certain chemicals secreted by its skin glands that make it extremely unpalatable. Whenever a potential predator, such as a stork, swoops over a congregation of toads, they flop on their backs, exposing their scarlet bellies as a warning. The storks and other birds apparently become conditioned by the association of the red color and the bad taste and avoid the toads assiduously.

Other animals survive by mimicking one of these protectively colored animals; for instance, some harmless, defenseless and palatable animals are identical in shape and color with a poisonous or noxious animal of quite a different family or order and, being mistaken for it by predators, are left alone. Many tropical insects have evolved this type of protection termed **Batesian mimicry.** If the animal is noxious enough, there will be significant protection afforded to its mimics even if the mimics outnumber their noxious models. The evolution of markedly similar shapes and color patterns by two species, both of which are noxious, is termed **Müllerian mimicry.** It is believed that the similar appearance of the two species increases the probability that predators will learn to avoid that particular shape and color pattern.

The reality of the selective advantage of color adaptations has been debated. It has been argued that animal vision may be so different from human vision (certain animals may be colorblind or may be able to see ultraviolet or infrared light) that an animal that appears to be protectively colored to a human may be quite evident to its natural predators. However, many experimental studies, such as the experiments with grasshoppers and moths, have shown that protective coloration does have survival value.

Color and patterns may serve to attract other organisms when it is necessary for survival. The red and blue sexual skin around the ischial callosities of certain monkeys and the gay extravagant plumage of various birds apparently have an attraction for members of the opposite sex that plays an important role as a prelude to mating. And vividly colored flowers seem to attract the birds and insects whose activities are needed to insure the pollination of the plant or the dispersal of its seed.

38.5 ADAPTATIONS OF SPECIES TO SPECIES

The evolution and adaptation of each species have not occurred in a biological vac-

uum, independent of other forms. On the contrary, the adaptation of each species has been influenced markedly by the concurrent adaptations of other species. As a result of this, many types of cross-dependency between species have arisen. Some of the clearest and best understood of these involve insects.

Insects are necessary for the pollination of a great many plants; some plants are so dependent on certain insects that they are unable to survive in a given region unless those particular insects are present. For example, the Smyrna fig could not be grown in California, even though all climatic conditions were favorable, until the fig insect that pollinates the plant was introduced.

Birds, bats and even snails serve as pollen transporters for some plants, but insects are the prime pollinators. Flowering plants have evolved bright colors and fragrances, presumably to attract insects and birds and ensure pollination. There had been some doubt as to whether insects can detect different odors and colors. The experiments of Karl von Frisch show that honeybees, at least, are able to differentiate colors and scents and that they are guided in their visits to flowers by these stimuli.

Some of the species-to-species adaptations are so exact that neither form can exist in a region without the other. The yucca plant and the yucca moth, like the fig and fig insect, have evolved to a point of complete interdependence (Fig. 38.3A). The yucca moth, by a series of unlearned acts, goes to a yucca flower, collects some pollen and takes it to a second flower. There it pushes its ovipositor (egg-laying organ) through the wall of the ovary of the flower and lays an egg. It then carefully places some pollen on the stigma. The yucca plant in this way is sure to be fertilized and produce seeds; the larva of the yucca moth feeds on these yucca seeds. The yucca produces a large number of seeds and is not injured by the loss of the few eaten by the moth larva.

38.6 HABITAT AND ECOLOGIC NICHE

Two basic concepts useful in describing the ecological relations of organisms are the habitat and the ecologic niche. The **habitat** of an organism is the place where it lives, a physical area, some specific part of the earth's surface, air, soil or water. It may be as large as the ocean or a prairie or as small as the

Figure 38.3 The yucca plant (A) is pollinated only by the yucca moth, one of which is shown in the open flower at the right. B, A larva of the butterfly family Lycaenidae eating a lupine seed pod while being tended by ants. The species to which the larva belongs has adapted to overcome toxic substances in the food plant, and the ants earn the sticky substance given off by the larva when they fend off parasites or predators. It is interesting to note that the ants eat the plant in the absence of caterpillars and aphids. (A from Weatherwax, P.: Botany. 3rd ed. Philadelphia, W. B. Saunders Co., 1956; B from Villee, C. A.: Biology. 7th ed. Philadelphia, W. B. Saunders Co., 1977.)

underside of a rotten log or the intestine of a termite, but it is always a tangible, physically demarcated region. More than one animal or plant may live in a particular habitat.

The **ecologic niche** is a more inclusive term that includes not only the physical space occupied by an organism, but its functional role as a member of the community—that is, its trophic position and its position in the gradients of temperature, moisture, pH and other conditions of the environment. The ecologic niche of an organism depends not only on where it lives but also on what it does—that is, how it transforms energy, how it behaves in response to and modifies its physical and biotic environment and how it is acted upon by other species. A common analogy is that the habitat is the organism's "address" and the ecologic niche is the organism's "profession," biologically speaking. To define completely the ecologic niche of any species would require detailed knowledge of a large number of biological characteristics and physical properties of the organism and its environment. Since this is very difficult to obtain, the concept of ecologic niche is used most often to describe differences between species with regard to one or a few of their major features.

Charles Elton, in 1927, was one of the first to use the term in the sense of the functional status of an organism in its community. The term "ecologic niche" has somewhat different meanings to different ecologists. To some it is the ultimate distributional unit within which each species is held by its structural and behavioral limitations. To others it is the functional status of an organism in its community, and these ecologists emphasize the energy relations of the species. To others, such as G. E. Hutchinson, the niche is a multidimensional space within which the environment permits an individual or species to survive indefinitely. The ecologic niche is an abstraction that includes all of the physical, chemical, physiologic and biotic factors that an organism needs in order to survive. To describe an animal's ecologic niche, we must know what it eats and what eats it, what its activities and range of movements are and what effects it has on other organisms and on the nonliving parts of the surroundings.

In the shallow waters at the edge of a lake, you might find many different kinds of water bugs. They all live in the same place and hence have the same habitat. Some of these water bugs, such as the "back-swimmer" *Notonecta*, are predators, catching and eating other animals of about their size (Fig. 38.4). Other water bugs, such as *Corixa*, the water boatman, feed on dead and decaying organisms. Each has quite a different role in the biological economy of the lake, and each occupies an entirely different ecologic niche.

A single species may occupy somewhat different niches in different regions, depending on such things as the available food supply and the number and kinds of competitors. Animals with distinctly different stages in their life history may occupy different niches in succession. The frog tadpole is a primary consumer, feeding on plants, but an adult frog is a secondary consumer, feeding on insects and other animals. In contrast, young river turtles are secondary consumers, eating snails, worms and insects, whereas the adult turtles are primary consumers and eat green plants such as tape grass.

Two species of organisms that occupy the same or similar ecologic niches in different geographical locations are termed **ecological equivalents.** The array of species present in a given type of community in different biogeographic regions may differ widely. However, similar ecosystems tend to develop wherever there are similar physical habitats; the equivalent functional niches are occupied by whatever biological groups happen to be present in the region. Thus, a savanna biome tends to develop wherever the climate permits the development of extensive grasslands, but the species of grass and the species of animals eating the grass may be quite different in different parts of the world. On each of the four continents there are grasslands with large grazing herbivores present. These herbivores are all ecological equivalents. However, in North America the grazing herbivores were bison and prong-horn antelope; in Eurasia, the saga antelope and wild horses; in Africa, other species of antelope and zebra; and in Australia, the large kangaroos. In all four regions, these native herbivores have been replaced to a greater or lesser extent by domesticated sheep and cattle. As examples of ecological equivalents, the species occupying three marine ecologic niches in four different regions of the coast are listed in Table 38.1. The same kinds of ecologic niches are usually present in similar habitats in different parts of the world. Comparisons of such habitats and analyses of the similarities and differences in the species that are ecological equivalents in these different habitats have been helpful in clarifying the interrelations of these different ecologic niches in any given habitat.

Figure 38.4 *Notonecta,* the "back-swimmer" (*left*), and *Corixa,* the "water boatman" (*right*), are two aquatic bugs occupying the same habitat—the shallow, vegetation-choked edges of ponds and lakes—but having different ecologic niches.

TABLE 38.1 ECOLOGICAL EQUIVALENTS: SPECIES OCCUPYING
COMPARABLE ECOLOGIC NICHES ON NORTH AMERICAN COASTS*

	Grazers on Intertidal Rocks	Fish Feeding on Plankton	Bottom-Dwelling Carnivores
Northeast coast	*Littorina littorea* (periwinkle)	Alewife, Atlantic herring	*Homarus* (lobster)
Gulf coast	*Littorina irrorata*	Menhaden, threadfin	*Menippe* (Stone crab)
Northwest coast	*Littorina danaxis*, *Littorina scutelata*	Sardine, Pacific herring	*Paralithodes* (King crab)
Tropical coast	*Littorina ziczac*	Anchovy	*Panulirus* (spiny lobster)

*Adapted from Odum, E. P.: Fundamentals of Ecology. 3rd ed. Philadelphia, W. B. Saunders Co., 1971.

38.7 THE PHYSICAL ENVIRONMENT

The numbers and distribution of plants are influenced by climate **(climatic factors)** and by soil **(edaphic factors).** The kinds of plants together with climatic and edaphic factors, influence the number and distribution of the various kinds of animals, and this, in turn, may influence plants. The soil on the surface of the earth is quite nonuniform. The original crust of the earth, the igneous rock, was relatively uniform, but weathering, erosion and sedimentation have produced marked geochemical differentiation of the earth's surface. The distribution of organisms is greatly affected by the kind of soil present. The converse relationship is also true—namely, that the organisms present make a major contribution to the kind of soil. Organisms are the sources of the great deposits of fossil fuels, such as peat, coal and oil.

The weathering of the earth's crust has led to the deposition on the surface of the earth of a coat, the characteristics of which depend on the kind of parent rock that was weathered, the kind of weathering processes to which it has been subjected, and its overall age. The term **soil** is applied to this mixture of weathered rock plus organic debris. Most soils remain where they were formed from the parent rock, but some have been carried from their place of origin to another location by the wind (sand dunes or loess), by water (alluvial deposits at the deltas of rivers) or by glaciers. Most of the soils of Canada and the northeastern United States were deposited by the action of glaciers.

After its formation, a soil undergoes development, controlled both directly and indirectly by the climate. The temperature and rainfall determine the rate at which materials in solution and suspension are transported by percolating water out of the soil to places where they accumulate. The nature of the climate determines the kind of vegetation present, and this, in turn, determines what kind of organic materials will be available to be incorporated into the soil. In cold humid regions where rainfall is greater than evaporation, the vegetation produces an acid humus, giving an ash-grey soil termed **podzol.** In tropical regions with high temperatures and heavy rainfall, there is little acidity from the decay of tropical vegetation and this results in red **lateritic** soil with a high iron content. In regions with low rainfall, unevenly distributed over the year, and with high rates of evaporation, the soil tends to be calcified, rich in calcium carbonate. The different characteristics of the various types of soil determine the kinds of plants that can grow in the region. Soils supply anchorage for the plants, water, mineral nutrients and aeration of the roots. Some of the important characteristics of a soil are its texture—whether it is gravel, sand, silt or clay, its organic content, the amount of soil water present, the amount of air entrapped in the soil and its acidity and salinity.

As soils develop, they tend to become stratified. The uppermost layer is one from which nutrients have been removed by water percolating through it. Below this is a layer of accumulated materials derived from the layer above. The bottom layer is composed of unweathered parent material.

38.8 THE CYCLIC USE OF MATTER

The total mass of all the organisms that have lived on the earth in the past 1.5 billion years is much greater than the mass of carbon and nitrogen atoms present. According to the Law of Conservation of Matter, matter is neither created nor destroyed; obviously the carbon and nitrogen atoms must have been used over and over again in the course of time. The earth neither receives any great amount of matter from other parts of the universe nor does it lose significant amounts of matter to outer space. The atoms of each element—carbon, hydrogen, oxygen, nitrogen and the rest—are taken from the environment, made a part of some cellular component and finally, perhaps by a quite circuitous route involving several other organisms, are returned to the environment to be used over again. An appreciation of the roles of green plants, animals, fungi and bacteria in this cyclic use of the elements can be gained from considering the details of the more important cycles.

The Carbon Cycle. There are about six tons of carbon (as carbon dioxide) in the atmosphere over each acre of the earth's surface. Yet each year an acre of plants, such as sugar cane, will extract as much as twenty tons of carbon from the atmosphere and incorporate it into the plant bodies. If there were no way to renew the supply, the green plants would eventually, perhaps in a few centuries, use up the entire atmospheric supply of carbon. Carbon dioxide fixation by bacteria and animals is another, but quantitatively minor, drain on the supply of carbon dioxide. Carbon dioxide is returned to the air by the decarboxylations that occur in cellular respiration. Plants carry on respiration continuously. Plant tissues are eaten by animals which, by respiration, return more carbon dioxide to the air. But respiration alone would be unable to return enough carbon dioxide to the air to balance that withdrawn by photosynthesis. Vast amounts of carbon would accumulate in the compounds making up the dead bodies of plants and animals. The carbon cycle is balanced by the decay bacteria and fungi that cleave the carbon compounds of dead plants and animals and convert the carbon to carbon dioxide (Fig. 38.5).

When the bodies of plants are compressed under water for long periods of time, they are not decayed by bacteria but undergo a series of chemical changes to form **peat,** later brown coal or **lignite** and finally **coal.** The bodies of certain marine plants and animals may undergo somewhat similar changes to form **petroleum.** These processes remove some carbon from the cycle but eventually geologic changes or man's mining and drilling bring the coal and oil to the surface to be burned to carbon dioxide and restored to the cycle.

Most of the earth's carbon atoms are present in limestone and marble as carbonates. The rocks are very gradually worn down and the carbonates in time are added to the carbon cycle. But other rocks are forming at the bottom of the sea from the sediments of dead animals and plants, so that the amount of

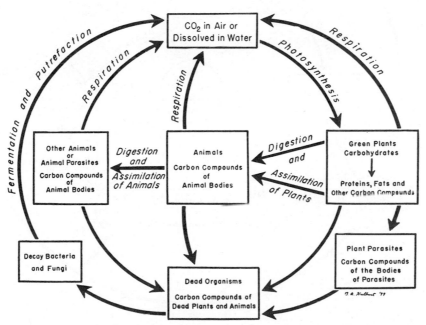

Figure 38.5 The carbon cycle in nature. See text for discussion.

carbon in the carbon cycle remains nearly constant.

The Nitrogen Cycle. The nitrogen for the synthesis of amino acids and proteins is taken up from the soil and water by plants as nitrates (Fig. 38.6). The nitrates are converted to amino groups and used by the plant cells in the synthesis of amino acids and proteins. Animals may eat the plants and utilize the amino acids from the plant proteins in the synthesis of their own proteins and other cellular constituents. When animals and plants die, the decay bacteria convert the nitrogen of their proteins and other compounds into ammonia. Animals excrete several kinds of nitrogen-containing wastes—urea, uric acid, creatinine and ammonia—and the decay bacteria convert these wastes to ammonia. Most of the ammonia is converted by nitrite bacteria to nitrites, and this in turn is converted by nitrate bacteria to nitrates, thus completing the cycle. Denitrifying bacteria convert some of the ammonia to atmospheric nitrogen. Atmospheric nitrogen can be "fixed," converted to organic nitrogen compounds, such as amino acids, by some blue-green algae (*Nostoc* and *Anabaena*) and by the soil bacteria *Azotobacter* and *Clostridium*.

Other bacteria of the genus *Rhizobium*, although unable to fix atmospheric nitrogen themselves, can do this when in combination with cells from the roots of legumes, such as peas and beans. The bacteria invade the roots and stimulate the formation of root nodules, a sort of harmless tumor. The combination of legume cell and bacteria is able to fix atmospheric nitrogen (something neither one can do alone), and for this reason legumes are often planted to restore soil fertility by increasing the content of fixed nitrogen. Nodule bacteria may fix as much as 50 to 100 kilograms of nitrogen per acre per year, and free

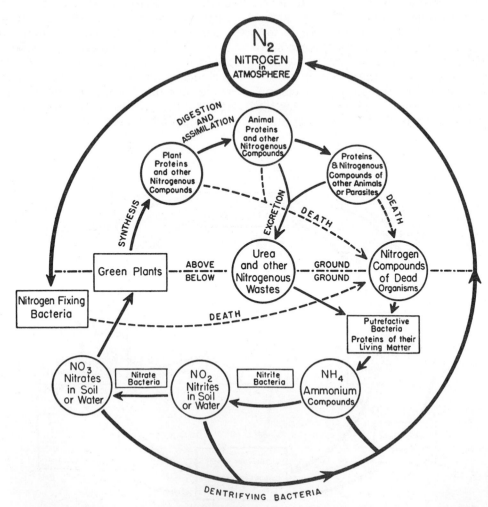

Figure 38.6   The nitrogen cycle in nature. See text for discussion.

soil bacteria as much as 12 kilograms per acre per year.

Atmospheric nitrogen can also be fixed by electrical energy, supplied either by lightning or by the electric company. Although 80 per cent of the atmosphere is nitrogen, no animals and only these few plants can utilize it in this form. When the bodies of the nitrogen-fixing bacteria decay, the amino acids are metabolized to ammonia, and this in turn is coverted by the nitrite and nitrate bacteria to nitrates to complete the cycle.

The Water Cycle. The great reservoir of water is the ocean. The sun's heat vaporizes water and forms clouds. These, moved by winds, may pass over land, where they are cooled enough to precipitate the water as rain or snow. Some of the precipitated water soaks into the ground, some runs off the surface into streams and goes directly back to the sea. The ground water is returned to the surface by springs, by pumps and by transpiration—the movement of water in plants from roots to leaves. Water inevitably ends up back in the sea, but it may become incorporated into the bodies of several different organisms, one after another, en route. The energy to run the cycle—the heat needed to evaporate water—comes from sunlight.

The Phosphorus Cycle. As water runs over rocks, it gradually wears away the surface and carries off a variety of minerals, some in solution and some in suspension. Some of these minerals, such as phosphates, sulfates, calcium, magnesium and others, are necessary for the growth of plants and animals. Phosphorus, an extremely important constituent of all cells, is taken in by plants as inorganic phosphate and converted to a variety of organic phosphates (which are intermediates in the metabolism of carbohydrates, nucleic acids and fats). Animals get their phosphorus as inorganic phosphate in the water they drink or as inorganic plus organic phosphates in the food they eat.

The phosphorus cycle is not completely balanced, for phosphates are being carried into the sediments at the bottom of the sea faster than they are being returned by the actions of marine birds and fish. Sea birds play an important role in returning phosphorus to the cycle by depositing phosphate-rich **guano** on land. Man and other animals, by catching and eating fish, also recover some phosphorus from the sea. In time, geologic upheavals bring some of the sea bottom back to the surface as new mountains are raised, and in this way minerals are recovered from the sea bottom and made available for use once more.

The Energy Cycle. The cycles of all these types of matter are closed: the atoms are used over and over again. To keep the cycles going does not require new matter, but it does require energy, for the energy cycle is not a closed one. The Law of the Conservation of Energy, or the First Law of Thermodynamics, states that energy is neither created nor destroyed but only transformed from one kind to another (p. 64). However, the Second Law of Thermodynamics states that whenever energy is transformed from one kind to another, there is an increase in entropy and a decrease in the amount of useful energy. Some energy is degraded into heat and dissipated.

Only a small fraction of the light energy reaching the earth is trapped; considerable areas of the earth have no plants, and plants can utilize in photosynthesis only about 3 per cent of the incident energy. This radiant energy is converted into the potential energy of the chemical bonds of the organic substances made by the plant. When an animal eats a plant (or when bacteria decompose it) and these organic substances are oxidized, the energy liberated is just equal to the amount of energy used in synthesizing the substances (First Law of Thermodynamics), but some of the energy is heat and not useful energy (Second Law of Thermodynamics). If this animal in turn is eaten by another one, a further decrease in useful energy occurs as the second animal oxidizes the organic substances of the first to liberate energy to synthesize its own cellular constituents.

In the successive steps of such a **food chain** (p. 887)—photosynthetic autotroph, herbivorous heterotroph, carnivorous heterotroph, decay bacteria—the number and mass of the organisms in each step is limited by the amount of energy available. Since some energy is lost as heat in each transformation, the steps become progressively smaller near the top. This relationship is sometimes called a "food pyramid" or pyramid of numbers, to emphasize that in each successively higher section of the food chain the number (or more precisely the total mass) of the predators decreases.

Eventually, all the energy originally trapped by plants in photosynthesis is converted to heat and dissipated to outer space, and all the carbon of the organic compounds ends up as carbon dioxide. The only important source of energy on earth is sunlight—energy derived from nuclear reactions, largely the conversion of hydrogen to helium, occurring at extremely high temperatures (about 10,000,000° C.) in the interior of the sun. When this energy is exhausted and the radiant energy of the sun can no longer support photosynthesis, the carbon cycle will stop, all plants and animals will die and organic carbon will be converted to carbon dioxide. There is no immediate cause for alarm, however; the sun will continue to shine for several billions of years!

38.9 SOLAR RADIATION

Perhaps the most outstanding feature of the earth is the nonuniformity of its physical conditions, which range from Arctic tundra to tropical rain forests. Even the oceans are very patchy, nonuniform places. The earth derives nearly all its energy from the sun, but even the sun's energy is not uniformly distributed over the face of the globe. The solar radiation reaching the surface of the earth varies with the length of the path that the sun's rays take through the atmosphere (whether it is vertical or at an angle), the area of horizontal surface over which is spread a "bundle" of the sun's rays of a given cross-sectional area, the distance of the earth from the sun (which changes seasonally because of the elliptical orbit of the earth around the sun), the amount of water vapor, dust and pollutants in the atmosphere, and the total length of the day (the photoperiod). At higher latitudes the

angle of incidence of the sun rays is less than at middle latitudes, and the energy is spread more thinly. The rays must pass through a thicker layer of atmosphere (Fig. 38.7); consequently, the polar regions receive less radiant energy in the course of a year than do equatorial regions.

The major variations in the amount of incoming solar energy are related to the movements the earth makes with respect to the sun (Fig. 38.8). One complete annual orbit of the earth around the sun requires 365¼ days. The axis of the earth is tilted 23½ degrees in relation to its plane of orbit, and therefore the distribution of energy varies throughout the year. The northern hemisphere receives more radiant energy during the period between March 21 and September 21 than in the other half of the year, not only because there are more hours of daylight but also because the angle of incidence of sunlight is more nearly vertical during that period.

The rotation of the earth on its own axis every 24 hours produces day and night and the energy changes associated with these periods. The changes in temperature lag behind the changes in the amount of light energy received. The maximum temperature during the day is usually in midafternoon, and the lowest temperature is just before sunrise. The temperature of the soil tends to lag even more than this. The atmosphere changes the distribution of the energy of different wave lengths so that the nature of sun-

Figure 38.7 Circle of illumination, areas of daylight and darkness, angles of sun's rays at different latitudes and differences in areas affected and thickness of atmosphere penetrated at time of summer solstice. (After Ward. H. B., and W. E. Powers: Introduction to Weather and Climate. Evanston, Illinois, Northwestern University Press, 1942.)

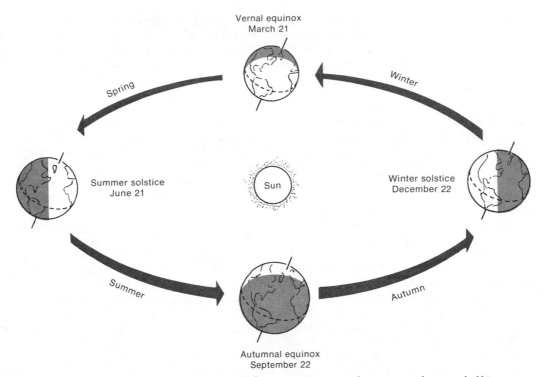

Vernal equinox
March 21

Spring

Winter

Summer solstice
June 21

Sun

Winter solstice
December 22

Summer

Autumn

Autumnal equinox
September 22

Figure 38.8 The sunlit portions of the Northern Hemisphere are seen to vary from greater than one-half in summer to less than one-half in winter. The proportion of any latitude that is sunlit is also the proportion of the 24-hour day between sunrise and sunset. (From MacArthur, R. H., and J. H. Connell: The Biology of Populations. New York, John Wiley & Sons, Inc., 1966.)

light actually reaching the earth is different from the sunlight some 100 miles above the atmosphere. Radiant energy of short wave lengths is not absorbed by water or water vapor; it thus passes through the atmosphere with little diminution. Some of the energy of sunlight is absorbed by the earth; some is reflected back as longer wave lengths or heat. Snow, water and light-colored soils reflect heat, whereas bare ground with dark soils absorbs it. This situation has been altered further by man's activities in paving large areas of the earth and in building, plowing, removing trees and causing air pollution. It is estimated that as much as 40 per cent of the heat of the atmosphere is derived from the condensation of water vapor derived from the evaporation of water from the surface of the ocean. The moisture-laden air rises, moves to higher latitudes where it is cooled and gives up its moisture as clouds or rain. The heat is then absorbed by the wet atmosphere. The atmosphere, heated from below and radiating heat back to the surface of the earth, serves as a heat trap as does the roof of a greenhouse whose glass substitutes for clouds and water vapor.

38.10 ENERGY FLOW

Most of the sun's energy that reaches the earth is eventually lost as heat. A small proportion of the energy of sunlight is absorbed by plants, and a small portion of this is transformed into the potential energy of stored food products. The rest of the energy leaves the plant and becomes part of the earth's general heat loss. All living things, except green plants, obtain their energy by taking in the products of photosynthesis, carried out by green plants, or the products of chemosynthesis, carried out by microorganisms. Each organism is in a dynamic state and its constituents are constantly being degraded and rebuilt. Thus each organism can be considered as a sort of durable framework through which energy and matter flow in this dynamic state. When carbon or nitrogen atoms enter an organism, they are synthesized into the compounds characteristic for that organism, then are subsequently returned to the environment. The entire earth has a finite amount of carbon, nitrogen and other atoms, and these must constantly be recycled. The degradation of organic molecules by decomposer organisms is of great importance in preventing the catastrophe that would occur if all of the carbon or nitrogen atoms became bound in some life form and could no longer be used.

The various kinds of organisms in nature are balanced with their environment, but for many organisms this balance is a precarious one. Members of the human race have upset the initial balance of nature in an alarming number of instances, and many of these, once undone, are very difficult, if not impossible, to reestablish. The reasons for this will become more apparent when we discuss community successions in the next chapter. In tropical Africa and South America, the natives clear portions of the rain forest for fields to grow crops. Without fertilization the cleared area may produce a crop for only a few years. When it becomes unproductive, the area is abandoned and more forest is felled to provide new crop land. The abandoned area can probably never again be covered with a mature rain forest. The thin tropical soils have very meager supplies of mineral nutrients; these are leached out of the soil by the heavy rains. The mature rain forest is in precarious balance with the soil and can maintain itself only as long as the balance is not disturbed. Once the balance has been upset, the forest is irretrievably lost. Civilized man in the more technologically advanced countries has done similar things much more efficiently and with equally devastating results. Many tragic examples can be given of our blunders in upsetting critical ecological balances. The organisms living in nature are parts of complex interacting communities of many species and are not single isolated species. The evolution of organisms is determined not simply by the limitations and peculiarities of the temperature, soil, pH, salinity and other factors of the abiotic environment but also by their relationships with other organisms living in that region. It is true that organisms make the community, but it is equally true that the community makes the organisms.

The carbon, nitrogen and various other cycles in nature operate to conserve the limited amount of usable matter on the earth. In contrast, the amount of energy available is very great and is constantly being renewed in sunlight. The flow of energy is not cyclic but is one-way. By measuring the amount of energy taken up and given off by each kind of organism, the ecologist can determine the functional structure of the organisms living together in a community. From this can be calculated how much life can be supported in a given area and how many individuals of each species the area can support. As the potential energy of sunlight is transferred from plants and other primary producers through herbivores and their carnivorous predators and parasites and ultimately, following their deaths, through the decomposer microorganisms, a large portion of the energy is lost at each step as heat. Because of this progressive loss of energy as heat, the total energy flow at each succeeding level is less and less. When an animal eats food, less than 20 per cent of the foodstuff is eventually converted into the flesh of the animal that is doing the eating. The domestic pig is one of the more efficient converters; under the best feeding practices, a pig will convert about 20 per cent of the mass of the food that it eats into pork chops and bacon.

The transfer of energy through a biological community begins when the energy of sunlight is fixed in a green plant by photosynthesis. It is estimated that only 8 per cent of the energy of the sun reaching the planet strikes green plants and that only 3 per cent of this is utilized in photosynthesis. Part of this energy is used by the plant itself to drive the many processes required for maintenance. The amount left over that is stored and expressed as growth represents the **net primary production.**

The net primary production of a field of sugar cane in Hawaii was 190 Kcal. per square meter per day. The average insolation was about 4000 Kcal. per square meter per day. From this we can calculate that the net efficiency of the sugar cane is about 4.8 per cent. Such values can be achieved only by crops under intensive cultivation during a favorable growing season. On an over-all annual basis, sugar cane fields have an efficiency of about 1.9 per cent and tropical forests have an efficiency of about 2 per cent. The stored energy accumulates as living material or biomass. Part is recycled each season by death and decomposition of the organisms; the part that is alive at any particular moment is called the **standing crop biomass.** This, of course, can vary greatly with the season. In grasslands, there is an annual turnover of the biomass, but in forests, much of the energy is tied up in wood. The most productive ecosystems on an energy basis are coral reefs and the estuaries of rivers (Fig. 38.9). The least productive are deserts and the open ocean. By and large the production of

Figure 38.9 The world distribution of primary production, in grams of dry matter per square meter day, as indicated by average daily rates of gross production in major ecosystems. (From Odum, E. P.: Fundamentals of Ecology. 3rd ed. Philadelphia, W. B. Saunders Co., 1971.)

plant material in each area of the earth has reached an optimum level that is limited only by soil and by the climate. The various kinds of consumers depend upon the production of green plants. Of the net production that is available to herbivores, not all is assimilated. For example, a grasshopper assimilates only about 30 per cent of its food, but some mice assimilate nearly 90 per cent. Most of this goes into the maintenance of the organism and is eventually lost as heat in the process of respiration. A small residue is stored in the form of new tissue and new individuals; it is this stored energy in the herbivore that is available to the next trophic layer, the carnivores. This enormous decrease in biomass at each stage is the basis for the concept of food chains and the pyramidal nature of the successive levels in the food chain.

38.11 FOOD CHAINS AND PYRAMIDS

The number of organisms of each species—or more precisely their total mass—is determined by the rate of flow of energy through the biological part of the ecosystem that includes them.

The transfer of energy from its ultimate source in a plant through a series of organisms, each of which eats the preceding and is eaten by the following, is known as a **food chain.** The number of steps in the series is limited to perhaps four or five because of the great decrease in available energy at each step. The percentage of food energy consumed that is converted to new cellular material, and thus is available as food energy for the next animal in the food chain, is known as the percentage efficiency of energy transfer.

The flow of energy in ecosystems, from sunlight through photosynthesis in autotrophic producers, through the tissues of herbivorous primary consumers and the tissues of carnivorous secondary consumers, determines the number and total weight **(biomass)** of organisms at each level in the ecosystem. The flow of energy is greatly reduced at each successive level of nutrition because of the heat losses at each transformation of energy. This decreases the total biomass produced in each level.

Some animals eat but one kind of food and, therefore, are members of a single food chain. Other animals eat many different kinds of food and not only are members of different food chains but also may occupy different positions in different food chains. An animal may be a primary consumer in one chain, eating green plants, but a secondary or tertiary consumer in other chains, eating herbivorous animals or other carnivores.

Humans are at the end of a number of food chains; for example, we eat a fish, such as a black bass, that ate little fish that ate small invertebrates that ate algae. The ultimate size of the human population (or the population of any animal) is limited by the length of the food chain, the per cent efficiency of energy transfer at each step in the chain and the amount of light energy falling on the earth.

Since humans can do nothing about increasing the amount of incident light energy and very little about the per cent efficiency of energy transfer, they can increase their supply of food energy only by shortening the food chain, i.e., by eating the primary producers, plants, rather than animals. In overcrowded countries such as India and China people are largely vegetarians because this food chain is shortest and a given area of land can in this way support the greatest number of people. Steak is a luxury in both ecological and economic terms, but hamburger is just as much an ecological luxury as steak is.

In addition to predator food chains, such as the man–black bass–minnow–crustacean one, there are parasite food chains. For example, mammals and birds are parasitized by fleas; in the fleas live protozoa that are, in turn, the hosts of bacteria. Since the bacteria might be parasitized by viruses, there could be a five-step parasite food chain.

A third type of food chain is one in which plant material is converted into dead organic matter, **detritus,** before being eaten by animals, such as millipedes and earthworms on land, by marine worms and mollusks, or by bacteria and fungi. In a community of organisms in the shallow sea, about 30 per cent of the total energy flows via detritus chains, but in a forest community, with a large biomass of plants and a relatively small biomass of animals, as much as 90 per cent of energy flow may be via detritus pathways. In an intertidal salt marsh, where most of the animals—shellfish, snails and crabs—are detritus eaters, 90 per cent or more of the energy flow is via detritus chains.

Since, in any food chain, there is a loss of energy at each step, it follows that there is a smaller biomass in each successive step. H. T. Odum has calculated that 8100 kg. of alfalfa plants are required to provide the food for 1000 kilograms of calves, which provide enough food to keep one 12-year-old, 48-kg. boy alive for one year. Although boys eat many things other than veal, and calves other things besides alfalfa, these numbers illustrate the principle of a food chain. A food chain may be visualized as a **pyramid;** each step in the pyramid is much smaller than the one on which it feeds. Since the predators are usually larger than the ones on which they prey, the pyramid of numbers of individuals in each step of the chain is even more striking than the pyramid of the mass of individuals in successive steps: one boy requires 4.5 calves, which require 20,000,000 alfalfa plants. However, the trophic relationships of populations are not necessarily pyramidal if we consider the standing crop biomass. The food webs in the pelagic region of the sea may exhibit a pyramid of numbers but an inverse pyramid of biomass. At any moment there may be more shark biomass than prey and more menhaden biomass than copepod biomass. These are among the relatively rare exceptions to the general rule of a pyramid of both numbers and biomass.

38.12 INTERACTIONS BETWEEN SPECIES

Two species may compete for the same space, food or light, or in avoiding predators or disease. They are, in a sense, competing for the same ecologic niche. Competition may result in one species dying off or being forced to change its ecologic niche—to move away or to utilize a different source of food. Careful ecologic studies usually confirm Gause's Rule: that there is only one species in an ecologic niche. The cormorant and the shag are two fish-eating, cliff-nesting sea birds that seemed at first glance to have survived despite the fact that they occupy the same ecologic niche. However, the cormorant feeds on bottom-dwelling fish and shrimp, whereas the shag hunts fish and eels in the upper levels of the sea. Further study showed that these birds typically choose slightly different nesting sites on the cliffs.

Studies of hundreds of different examples of parasite-host and predator-prey interrelations show that, in general, where the associations are of long standing, the long-term effect on the host or prey is not very detrimental and may even be beneficial. Conversely, newly acquired predators or parasites are usually quite damaging. The plant parasites and insect pests that are most troublesome to humans and their crops are usually those that have recently been carried into some new area and thus have a new group of organisms to attack.

A striking example of the result of upsetting a long-standing predator-prey relationship occurred on the Kaibab plateau, on the north side of the Grand Canyon of the Colorado River. In this area in 1907 there were some 4000 deer and a considerable population of their predators, mountain lions and wolves. When a concerted effort was made to "protect" the deer by killing off the predators, the deer population increased tremendously. By 1925 there were some 100,000 deer on the plateau, far too many for the supply of vegetation. The deer ate everything in reach, grass, tree seedlings and shrubs, and there was marked damage to the vegetation. There was no longer enough vegetation to support the deer population over the winter, and in the next two winters vast numbers of deer starved to death. Finally the deer population fell to about 10,000. The original predator-prey interaction had been maintaining a fairly stable equilibrium, with the number of deer being kept at a level within the available food supply.

The size of the predator population in the wild varies with the size of the population preyed upon. The swings in the size of the predator population lag a bit behind those of the prey.

38.13 INTRASPECIFIC RELATIONS

In addition to the association between the members of two different species just described, aggregations of animals or plants of a *single* species frequently occur. Some of these aggregations are temporary, for breeding; others are more permanent. Despite the fact that the crowding that accompanies dense aggregations of animals is ecologically undesirable and deleterious, both laboratory experiments and field observations show that such aggregations of individuals may be able to survive when a single individual of the same species placed in the same environment would die. A herd of deer, with many noses and pairs of eyes, is less likely to be surprised by a predator than is a single deer. Wolves hunting in a pack are more likely to make a kill than is a lone wolf. The survival value of intraspecific aggregations is less obvious, but nonetheless real, in some of the lower animals. A group of insects is less likely to dry up and die in a dry environment than is a single insect, and a group of planaria is less likely to be killed by a given dose of ultraviolet light than is a single flatworm. When a dozen goldfish are placed in one bowl and a single one in a second bowl, and the same amount of a toxic agent, such as colloidal silver, is added to each bowl, the single fish will die but the group will survive. The explanation for this has proved to be that the slime secreted by the group of fish is enough to precipitate much of the colloidal silver and render it nontoxic, whereas the amount secreted by a single fish is not.

When genes governing a tendency toward aggregation arise in a species and prove to have survival value, natural selection will tend to preserve this inherited behavior pattern. The occurrence of many fish in schools, of birds in flocks and so on are examples of this "unconscious cooperation," which occurs very widely in the Animal Kingdom.

From such simple animal aggregations there may evolve complex animal societies, composed of specialized types of individuals, such as the colonies of bees, ants and termites. Human beings provide another example of social animals.

ANNOTATED REFERENCES

Brady, N. C. and H. O. Buckman: The Nature and Properties of Soil. 6th ed. New York, The Macmillan Co., 1960. A classic and complete text of soil science.

Brock, T. D.: The Principles of Microbial Ecology. Englewood Cliffs, N.J., Prentice-Hall, Inc., 1966. Emphasizes the roles of microorganisms in ecological relationships, a subject frequently passed over lightly in general ecologic texts.

Browning, T. O.: Animal Populations. New York, Harper & Row, Publishers, 1963. A discussion of factors regulating the size of animal populations.

Cott, H. B.: Adaptive Coloration in Animals. Oxford, Oxford University Press, 1940. A wonderfully illustrated account of animal camouflage.

Kendeigh, S. C.: Animal Ecology. Englewood Cliffs, N.J., Prentice-Hall, Inc., 1961. A concise treatment of the principles of animal ecology.

Krebs, C. J.: Ecology: The Environmental Analysis of Distribution and Abundance. New York, Harper and Row, 1972. A fine modern text emphasizing the population aspects of ecology.

Lack, D.: The Natural Regulation of Animal Numbers. New York, Oxford University Press, 1954. A concise presentation of the principles of ecologic regulation.

Leopold, Aldo: A Sand County Almanac. New York, Oxford University Press, 1949. An eloquent essay on the relevance of ecology to human affairs.

Moran, J. M., M. D. Morgan and J. H. Wiersma: An Introduction to Environmental Sciences. Boston, Little, Brown and Co., 1973. A general discussion of the biosphere, emphasizing present concerns about pollution of the environment.

Odum, E. P.: Fundamentals of Ecology. 3rd ed. Philadelphia, W. B. Saunders Co., 1971. A broad coverage of the field that emphasizes especially the energy relationships in ecology.

Smith, Robert L.: Ecology and Field Biology. New York, Harper & Row, Publishers, 1966. An excellent general text emphasizing the natural history aspects of the science.

CHAPTER 39

SYNECOLOGY: COMMUNITIES, BIOMES AND LIFE ZONES

Each region of the earth—sea, lake, forest, prairie, tundra, desert—is inhabited by a characteristic assemblage of animals and plants that are interrelated in many and diverse ways as competitors, commensals, predators and so on. The members of each assemblage are not determined by chance but by the total effect of the many interacting physical and biotic factors of the environment. Ecologists refer to the organisms living in a given area as a **biotic community.** The intermeshings of the food chains in any biotic community are very complex and are sometimes called a food web, or "web of life." Some of the interrelated food chains of a deciduous forest in eastern North America are indicated in Figure 39.1. The basic principles of the ecological relations within biotic communities have been elucidated by

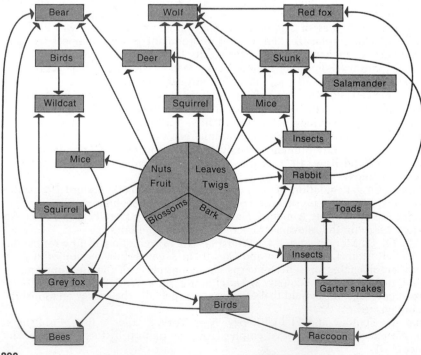

Figure 39.1 Diagram of the interrelationships in a food chain in a deciduous forest in Illinois. (After Shelford, V. E.: Ecological Monographs 21:183–214, 1951.)

the study of somewhat simpler communities, such as the arctic tundra or desert. The producer organisms of the tundra are lichens, mosses and grasses. Reindeer and caribou feed on the lichens and are preyed upon by wolves and humans. Grasses are eaten by the arctic hare and the lemming, which are eaten by the snowy owl and the arctic fox, which is preyed upon by humans for its fur. During the brief arctic summer, the food web is enlarged by many insects and by migratory birds which feed upon them.

39.1 POPULATIONS AND THEIR CHARACTERISTICS

A **population** may be defined as a group of organisms of the same species that occupy a given area. It has characteristics that are a function of the whole group and not of the individual members; these are **population density, birth rate, death rate, age distribution, biotic potential, rate of dispersion** and **growth form.** Although individuals are born and die, individuals do not have birth rates or death rates; these are characteristics of the population as a whole. Modern ecology deals especially with communities and populations; the study of community organization is a particularly active field at present. Population and community relationships are often more important in determining the occurrence and survival of organisms in nature than are the direct effects of physical factors in the environment.

One important attribute of a population is its **density**—the number of individuals per unit area or volume, e.g., human inhabitants per square mile, trees per acre in a forest, millions of diatoms per cubic meter of sea water. This is a measure of the population's success in a given region. Frequently, in ecological studies it is important to know not only the population density but also whether it is changing and, if so, what the rate of change is.

Population density is often difficult to measure in terms of individuals, but measures, such as the number of insects caught per hour in a standard trap, the number of sea urchins caught in a standard "sea mop" or the number of birds seen or heard per hour, are usable substitutes. A method that will give good results when used with the proper precautions is that of capturing, let us say, 100 animals, tagging them in some way and then releasing them. On some subsequent day, another 100 animals are trapped and the proportion of tagged animals is determined. This assumes that animals caught once are neither more likely nor less likely to be caught again and that both sets of trapped animals are random samples of the population. If the ratio of untagged to tagged animals in the second sample is 100/20, then the ratio of total population size in the area (x) to the first sample is x/100, and x = 500, for x/100 = 100/20.

For many types of ecological investigations, an estimate of the number of individuals per total area or volume, known as the "crude density," is not sufficiently precise. Only a fraction of that total area may be a suitable habitat for the population, and the size of the individual members of a population may vary tremendously. Ecologists, therefore, calculate an **ecologic density,** defined as the number, or more exactly as the mass, of individuals per area or volume of habitable space. Trapping and tagging experiments might give an estimate of 500 rabbits per square mile, but if only half of that square mile actually consists of areas suitable for rabbits to live in, then the ecologic density will be 1000 rabbits per square mile of rabbit habitat. With species whose members vary greatly in size, such as fish, live weight or some other estimate of the total mass of living fish is a much more satisfactory estimate of density than simply the number of individuals present.

A graph in which the number of organisms or the logarithm of that number is plotted against time is a **population growth curve** (Fig. 39.2). Such curves are characteristic of populations,

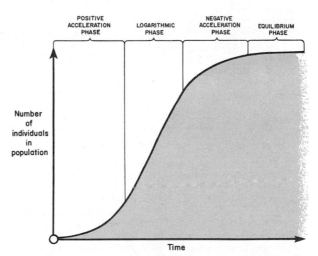

Figure 39.2 A typical sigmoid (S-shaped) growth curve of a population, one in which the total number of individuals is plotted against time. The absolute units of time and the total number in the population would vary from one species to another, but the shape of the growth curve would be similar for all populations.

rather than of a single species and are amazingly similar for populations of almost all organisms from bacteria to human beings.

Population growth curves have a characteristic shape. When a few individuals enter a previously unoccupied area, growth is slow at first (the positive acceleration phase), then becomes rapid and increases exponentially (the logarithmic phase). The growth rate eventually slows down as environmental resistance gradually increases (the negative acceleration phase) and finally reaches an equilibrium or saturation level. The upper asymptote of the sigmoid curve is termed the **carrying capacity** of the environment.

The birth rate, or natality, of a population is simply the number of new individuals produced per unit time. The **maximum birth rate** is the largest number of organisms that could be produced per unit time under ideal conditions, when there are no limiting factors. This is a constant for a species, determined by physiological factors such as the number of eggs produced per female per unit time, the proportion of females in the species and so on. The

actual birth rate is usually considerably less than this, for not all the eggs laid are able to hatch, not all the larvae or young survive and so on. The size and composition of the population and a variety of environmental conditions affect the actual birth rate. It is difficult to determine the maximum natality, for it is difficult to be sure that all limiting factors have been removed. However, under experimental conditions, one can get an estimate of this value that is useful in predicting the rate of increase of the population and in providing a yardstick for comparison with the actual birth rate.

The mortality rate of a population refers to the number of individuals dying per unit time. There is a theoretical **minimum mortality**, somewhat analogous to the maximum birth rate, that is the number of deaths that would occur under ideal conditions, deaths due simply to the physiological changes of old age. This minimum mortality rate is also a constant for a given population. The actual mortality rate will vary depending upon physical factors and on the size, composition and density of the population.

By plotting the number of survivors in a population against time, one gets a **survival curve** (Fig. 39.3). If the units of the time axis are expressed as the percentage of total life span, the survival curves for organisms with very different total life spans can be compared. Modern medical and public health practices have greatly increased the average life expectancy in developed countries, and the curve for human survival approaches the curve for minimal mortality. From such curves one can determine at what stage in the life cycle a particular species is most vulnerable. Reducing or increasing mortality in this vulnerable period will have the greatest effect on the future size of the population. Since the death rate is more variable and affected to a greater extent by environmental factors than the birth rate, it has a primary role in controlling the size of a population.

It is obvious that populations that differ in the relative numbers of young and old will have different characteristics, different birth and death rates and different prospects. Death rates typically vary with age, and birth rates are usually proportional to the number of individuals able to reproduce. Three ages can be distinguished in a population in this respect: pre-reproductive, reproductive and post-reproductive. A. J. Lotka has shown from theoretical considerations that a population will tend to become stable and have a constant proportion of individuals of these three ages.

Figure 39.3 Survival curves of four different animals, plotted as number of survivors left at each fraction of the total life span of the species. The total life span for man is about 100 years; the solid curve indicates that about 10 per cent of the babies born die during the first few years of life. Only a small fraction of the human population dies between ages 5 and 45, but after 45 the number of survivors decreases rapidly. Starved fruit flies live only about five days, but almost the entire population lives the same length of time and dies at once. The vast majority of oyster larvae die, but the few that become attached to the proper sort of rock or to an old oyster shell survive. The survival curve of hydras is one typical of most animals and plants, in which a relatively constant fraction of the population dies off in each successive time period.

Censuses of the ages of plant or animal populations are of value in predicting population trends. Rapidly growing populations have a high proportion of young forms. The age of fishes can be determined from the growth rings on their scales, and studies of the age ratios of commercial fish catches are of great use in predicting future catches and in preventing overfishing of a region.

The term **biotic potential,** or reproductive potential, refers to the inherent power of a population to increase in numbers when the age ratio is stable and all environmental conditions are optimal. The biotic potential is defined mathematically as the slope of the population growth curve during the logarithmic phase of growth. When environmental conditions are less than optimal, the rate of population growth is less. The difference between the potential ability of a population to increase and the actual change in the size of the population is a measure of environmental resistance.

Even when a population is growing rapidly in number, each *individual* organism of the reproductive age carries on reproduction at the same rate as at any other time; the increase in numbers is due to increased survival. At a conservative estimate, one man and one woman, with the cooperation of their children and grandchildren, could give rise to 200,000 progeny within a century, and a pair of fruit flies could multiply to give 3368×10^{52} offspring in a year. Since optimal conditions are not maintained, such biological catastrophes do not occur, but the situations in India, Africa and elsewhere indicate the tragedy implicit in the tendency toward human overpopulation.

The sum of the physical and biological factors that prevent a species from reproducing at its maximum rate is termed the **environmental resistance.** Environmental resistance is often low when a species is first introduced into a new territory so that the species increases in number at a fantastic rate, as when the rabbit was introduced into Australia and the English sparrow and Japanese beetle were brought into the United States. But as a species increases in number the environmental resistance to it also increases, in the form of organisms that prey upon it or parasitize it, and the competition between the members of the species for food and living space.

In an essay in 1798 the Englishman Robert Malthus pointed out this tendency for populations to increase in size until checked by the environment. He realized that these same principles apply to human populations and suggested that wars, famines and pestilences are inevitable and necessary as brakes on population growth. Since Malthus' time the earth's productive capacity has increased tremendously as has the total human population. But Malthus' basic principle, that there are physical limits to the amount of food that can be produced for any species, remains true. The earth has a finite carrying capacity for human beings just as it does for any other animal. As environmental resistance increases, the rate of increase of the human population will eventually have to decrease. An equilibrium will be reached either by decreasing the birth rate or by increasing the mortality rate.

39.2 POPULATION CYCLES

Once a population becomes established in a certain region and has reached the equilibrium level, the numbers will vary up and down from year to year, depending on variations in environmental resistance or on factors intrinsic to the population. Some of these population variations are completely irregular, but others are regular and cyclic.

One of the best known of these is the regular nine to ten year cycle of abundance and scarcity of the snowshoe hare and the lynx in Canada that can be traced from the records of the number of pelts received by the Hudson's Bay Company. The peak of the hare population comes about a year before the peak of the lynx population (Fig. 39.4). Since the lynx feeds on the hare, it is obvious that the lynx cycle is related to the hare cycle.

Lemmings and voles are small mouselike animals living in the northern tundra region. Every three or four years there is a great increase in the number of lemmings; they eat all the available food in the tundra and then migrate in vast numbers looking for food. They may invade villages in hordes, and finally many reach the sea and drown. The numbers of arctic foxes and snowy owls, which feed on lemmings, increase similarly. When the lemming population decreases, the foxes starve and the owls migrate south—thus there is an invasion of snowy owls in the United States every three or four years.

Although some cycles recur with great regularity, others do not. For example, in the carefully managed forests of Germany the numbers of four species of moths whose caterpillars feed on pine needles were estimated from censuses made each year from

Figure 39.4   Changes in the abundance of the lynx and the snowshoe hare, as indicated by the number of pelts received by the Hudson's Bay Company. This is a classic case of cyclic oscillation in population density. (Redrawn from MacLulich, D. A.: Fluctuations in the numbers of the varying hare (*Lepus americanus*). Univ. Toronto Studies, Biol. series, no. 43, 1937.)

1880 to 1940. The numbers varied from less than one to more than 10,000 per thousand square meters. The cycles of maxima and minima of the four species were quite independent and were irregular in their frequency and duration. Ecologically speaking, each species was marching to its own tune.

Attempts to explain these vast oscillations in numbers on the basis of climatic changes have been unsuccessful. At one time it was believed that these were caused by sunspots, and the sunspot and lynx cycles do appear to correspond during the early part of the nineteenth century. However, the cycles are of slightly different lengths and by 1920 were completely out of phase, sunspot maxima corresponding to lynx minima. Attempts to correlate these cycles with other periodic weather changes and with cycles of disease organisms have been unsuccessful.

The snowshoe hares die off cyclically even in the absence of predators and in the absence of known disease organisms. The animals apparently die of "shock," characterized by low blood sugar, exhaustion, convulsions and death, symptoms that resemble the "alarm response" induced in laboratory animals subjected to physiologic stress.

This similarity led J. J. Christian (1950) to propose that their death, like the alarm response, is the result of some upset in the adrenal-pituitary system. As population density increases, there is increasing physiologic stress on individual hares, owing to crowding and competition for food. Some individuals are forced into poorer habitats where food is less abundant and predators more abundant. The physiologic stresses stimulate the adrenal medulla to secrete epinephrine, which stimulates the pituitary via the hypothalamus to secrete

more ACTH. This, in turn, stimulates the adrenal cortex to produce corticosteroids, an excess or imbalance of which produces the alarm response or physiologic shock.

In the latter part of the winter of a year of peak abundance, with the stress of cold weather, lack of food and the onset of the new reproductive season putting additional demands on the pituitary to secrete gonadotropins, the adrenal-pituitary system fails, becomes unable to maintain its normal control of carbohydrate metabolism, and low blood sugar, convulsions and death ensue. This is a reasonable hypothesis, but the appropriate experiments and observations in the wild needed to test it have not yet been made.

39.3 POPULATION DISPERSION AND TERRITORIALITY

Populations have a tendency to disperse, or spread out in all directions, until some barrier is reached. Within the area, the members of the population may occur at random (this is rarely found), they may be distributed more or less uniformly throughout the area (this occurs when there is competition or antagonism to keep them apart) or, most commonly, they may occur in small groups or clumps.

Aggregation in clumps may increase the competition between the members of the group for food or space, but this is more than counterbalanced by the greater survival power of the group during unfavorable periods. A group of animals has much greater resistance than a single individual to adverse conditions, such as desiccation, heat, cold or poisons. The combined effect of the protective mechanisms of the group is effective in countering the adverse environment, whereas that of a single individual is not.

Aggregation may be caused by local habitat differences, weather changes, reproductive urges or social attractions. Such aggregations of individuals may have a definite organization involving social hierarchies of dominant and subordinate individuals arranged in a "peck-order."

Other species of animals are regularly found spaced apart; each member tends to occupy a certain area or **territory**, which it defends against intrusion by other members of the same species and sex. Usually a male establishes a territory of his own (perhaps by fighting with other males) and then, by making himself conspicuous, tries to entice a female to share the territory with him.

It has been suggested that territoriality may have survival value for a species in ensuring an adequate amount of food, nesting materials and cover for the young, in protecting the female and young against other males and in limiting the population to a density that can be supported by the environment. Many species of birds and some mammals, fish, crabs and insects establish such territories, either as regions for gathering food or as nesting areas.

39.4 BIOTIC COMMUNITIES

A biotic community is an assemblage of populations living in a defined area or habitat; it can be either large or small. The interactions of the various kinds of organisms maintain the structure and function of the community and provide the basis for the ecological regulation of community succession. The concept that animals and plants live together in an orderly manner, not strewn haphazardly over the surface of the earth, is an important principle of ecology.

Sometimes adjacent communities are sharply defined and separated from each other; more frequently they blend imperceptibly together. Why certain plants and animals compose a given community, how they affect each other and how humans can control them to their advantage are some of the major problems of ecological research.

In trying to control some particular species, it has frequently been found more effective to modify the community rather than to attempt direct control of the species itself. For example, the most effective way to increase the quail population is not to raise and release birds, nor even to kill off predators, but to maintain the particular biotic community in which quail are most successful.

Although each community may contain hundreds or thousands of species of plants and animals, most of these are relatively unimportant, and only a few exert a major control of the community owing to their size, numbers or activities. In land communities these major species are usually plants, for they both produce food and provide shelter for many other species. Many land communities are named for their dominant plants — sagebrush, oak-hickory, pine and so on. Aquatic communities, containing no conspicuous large plants, are usually named for some physical characteristic — stream rapids community, mud flat community or sandy beach community.

In ecological investigations, it is unnecessary (and indeed usually impossible) to consider all of the species present in a community. Usually a study of the major plants that control the community, the larger populations of animals and the fundamental energy relations (the food chains) of the system will define the ecological relations within the community. For example, in studying a lake one would first investigate the kinds, distribution and abundance of the important producer plants and the physical and chemical factors of the environment that might be limiting. Then the reproductive rates, mortality rates, age distributions and other population characteristics of the important game fish would be determined. A study of the kinds, distribution and abundance of the primary and secondary consumers of the lake, which constitute the food of the game fish, and the nature of other organisms that compete with these fish for food would elucidate the basic food chains in the lake. Quantitative studies of these would reveal the basic energy relationships of the system and show how efficiently the incident light energy is being converted into the desired end product, the flesh of game fish. On the basis of this knowledge the lake could intelligently be managed to increase the production of game fish.

Detailed studies of simpler biotic communities, such as those of the arctic or desert, where there are fewer organisms and their interrelations are more evident, have provided a basis for studying and understanding the much more varied and complex forest communities.

A thorough ecological investigation of a particular region requires that the region be studied at regular intervals throughout the year for a period of several years. The physical, chemical, climatic and other factors of the region are carefully evaluated, and an intensive study is made of a number of carefully delimited areas that are large enough to be representative of the region but small enough to be studied quantitatively. The number and kinds of plants and animals in these study areas are estimated by suitable sampling techniques. Estimates are made periodically throughout the year to determine not only the components of the community at any one time but also their seasonal and annual variations. The biological and physical data are correlated, the major and minor communities of the region are identified and the food chains and other important ecological relationships of the members of the community are analyzed. The particular adaptations of the animals and plants for their respective roles in the community can then be appreciated.

39.5　COMMUNITY SUCCESSION

Any given area tends to have an orderly sequence of communities that change together with the physical conditions and lead finally to a stable mature community or **climax community.** The entire sequence of communities characteristic of a given region is termed a **sere,** and the individual transitional communities are called **seral stages** or seral communities. In successive stages there is not only a change in the species of organisms present but also an increase in the number of species and in the total biomass.

These series are so regular in many parts of the world that an ecologist, recognizing the particular seral community present in a given area, can predict the sequence of future changes. The ultimate causes of these successions are not clear. Climate and other physical factors play some role, but the succession is directed in part by the nature of the community itself, for the action of each seral community is to make the area less favorable for itself and more favorable for other species until the stable climax community is reached. Physical factors, such as the nature of the soil, the topography and the amount of water, may cause the succession of communities to stop short of the expected climax community in what is called an **edaphic climax.**

Occasionally, the organisms that we want to encourage for our own ends—timber, game birds, freshwater game fish—are members of a seral stage in community succession rather than of the climax community. Then the ecologist has the difficult problem of trying to manipulate the community to halt the succession and maintain the desired seral community.

One of the classic studies of ecological succession was made on the shores of Lake Michigan (Fig. 39.5). As the lake has become smaller it has left successively younger sand dunes, and one can study the stages in ecological succession as one goes away from the lake. The youngest dunes, nearest the lake, have only grasses and insects; the next older ones have shrubs, such as cottonwoods, then evergreens and finally there is a beech-maple climax community, with deep rich soil full of earthworms and snails.

As the lake retreated it also left a series of ponds. The youngest of these contain little rooted vegetation and lots of bass and bluegills. Later the ponds become choked with vegetation and smaller in size as the basins fill. Finally the ponds become marshes and then dry ground, invaded by shrubs and ending in the beech-maple climax forest. Man-made ponds, such as those behind dams, similarly tend to become filled up.

Another dramatic example of community succession began August 7, 1883, when a volcanic explosion occurred on the Indonesian island Krakatoa, causing part of the island to disappear. The remainder was covered with hot volcanic debris to a depth of 60 meters, and all life was obliterated. A year later some grass and a single spider were found. By 1908, 202 species of animals had taken up residence on the island. This increased to 621 species by 1919, and to 880 species by 1934, when there was a young forest on one part of the island.

Ecological succession can be demonstrated in the laboratory. If a few pieces of dry grass are placed in a beaker of pond water, a population of bacteria will appear in a few days. Next, flagellates appear and eat the bacteria, then ciliated protozoa, such as paramecia, appear and eat the flagellates. Finally, predator protozoa, such as *Didinium*, will appear and eat the paramecia. The protozoa, present as spores or cysts attached to the grass, emerge in a definite succession of protozoan communities.

Biotic communities show marked **vertical stratification,** determined in large part by vertical differences in physical factors, such as temperature, light and oxygen. The operation of such physical factors in determining vertical stratification in lakes and in the ocean is quite evident. In a forest, there is a vertical stratification of plant life, from mosses and herbs on the ground to shrubs, low trees and tall trees.

Figure 39.5 Diagram of the succession of communities with time along the shores of Lake Michigan in northern Indiana. (From Allee, W. C., et al.: Principles of Animal Ecology. Philadelphia, W. B. Saunders Co., 1949. After Buchsbaum.)

Each of these strata has a distinctive animal population. Even such highly motile animals as birds are restricted, more or less, to certain layers. Some species of birds are found only in shrubs, others only in the tops of tall trees. There are daily and seasonal changes in the populations found in each stratum, and some animals are found first in one, then in another layer as they pass through their life histories. These strata are interrelated in many diverse ways, and most ecologists consider them to be subdivisions of one large community rather than separate communities. Vertical stratification, by increasing the number of ecologic niches in a given surface area, reduces competition between species and enables more species to coexist in a given area.

39.6 THE CONCEPT OF THE ECOSYSTEM

All the living organisms that inhabit a certain area compose the biotic community. A larger unit, termed the **ecosystem,** includes the organisms in a given area and the encompassing physical environment. In the ecosystem, a flow of energy, derived from organism-environment interactions, leads to a clearly defined trophic structure with biotic diversity and to the cyclic exchange of materials between the living and nonliving parts of the system. From the trophic (nourishment) standpoint, an ecosystem has two components: an **autotrophic** part, in which light energy is captured or "fixed" and used to synthesize complex organic compounds from simple inorganic ones, and a **heterotrophic** part, in which the complex molecules undergo rearrangement, utilization and decomposition. In describing an ecosystem, it is convenient to recognize and tabulate the following components: (1) the inorganic substances, such as carbon dioxide, water, nitrogen and phosphate, that are involved in material cycles; (2) the organic compounds, such as proteins, carbohydrates and lipids, that are synthesized in the biotic phase; (3) the climate, temperature and other physical factors; (4) the producers, autotrophic organisms (mostly green plants) that can manufacture complex organic materials from simple inorganic substances; (5) the macroconsumers, or **phagotrophs,** heterotrophic organisms (mostly animals) that ingest other organisms or chunks of organic matter; and (6) the microconsumers, or **saprotrophs,** heterotrophic organisms (mostly fungi and bacteria) that break down the complex compounds of dead organisms, absorb some of the decomposition products and release inorganic nutrients that are made available to the producers to complete the various cycles of elements.

The producers, phagotrophs and saprotrophs make up the biomass of the ecosystem—the living weight. In analyzing an ecosystem, the investigator studies the energy circuits present, the food chains, the patterns of biological

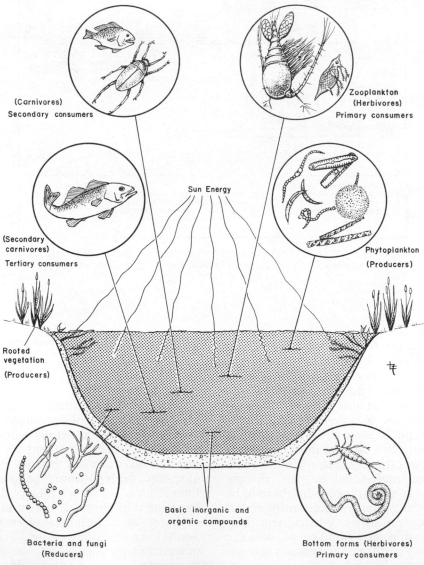

Figure 39.6 A small freshwater pond as an example of an ecosystem. The component parts, producer, consumer and decomposer or reducer organisms, plus the nonliving parts are indicated.

diversity in time and space, the nutrient cycles, the development and evolution of the ecosystem and the factors that control the composition of the ecosystem. It is important to appreciate that the ecosystem is the basic functional unit in ecology and includes both the biotic communities and the abiotic environment in a given region, each of which influences the properties of the other and both of which are needed to maintain life on the earth.

A classic example of an ecosystem compact enough to be investigated in quantitative detail is a small lake or pond (Fig. 39.6). The nonliving parts of the lake include the water, dissolved oxygen, carbon dioxide, inorganic salts such as phosphates, nitrates and chlorides of sodium, potassium and calcium, and a multitude of organic compounds. The living part of the lake can be subdivided according to the functions of the organisms, i.e., what they contribute toward keeping the ecosystem operating as a stable, interacting whole. In a lake there are two types of producers: the larger plants growing along the shore or floating in shallow water, and the microscopic floating plants, most of which are algae, that are distributed throughout the water as deep as light will penetrate. These tiny plants, collectively known as **phytoplankton,** are usually not visible unless they are present in great abundance and give the water a greenish tinge. They are usually much more important as food producers for the lake than are the more readily visible plants.

The macroconsumers or phagotrophs include insects and insect larvae, crustacea, fish and perhaps some freshwater clams. Primary consumers are the plant eaters and secondary consumers are the carnivores that eat the primary consumers. There might be some tertiary consumers that eat the carnivorous secondary consumers.

The ecosystem is completed by saprotrophs or decomposer organisms, bacteria and fungi, that break down the organic compounds of cells from dead producer and consumer organisms either into small organic molecules, which they utilize themselves, or into inorganic substances that can be used as raw materials by green plants.

The term "ecosystem" was proposed by the British ecologist A. G. Tansley in 1935, but the concept of the unity of organisms and environment can be traced back to very early biological literature. The concept of the ecosystem is a very broad one and emphasizes the obligatory relationships, the causal relationships and the interdependence of biotic and abiotic components. The ecosystem is the level of biological organization particularly suitable for the application of the techniques of systems analysis. An entity may be considered an ecosystem if its major components are present and operate together to achieve some sort of functional stability, even though this may persist for only a short time. Even a temporary pond is a definite ecosystem. It contains characteristic organisms and undergoes characteristic processes even though its existence may be limited to a relatively brief period of time.

The interaction of autotrophic and heterotrophic components is a universal feature of all ecosystems whether they are located on land, in fresh water or in the ocean. Frequently, the autotrophic and heterotrophic components are partially separated spatially. The greatest amount of autotrophic metabolism occurs in a "green belt" stratum in which light energy is available. Below this lies a "brown belt" in which the most intense heterotrophic metabolism takes place. In the brown belt, organic matter tends to accumulate both in soils and in sediments. The two functions may also be partially separated in time, for there may be a considerable delay in the heterotrophic utilization of the products of autotrophic organisms. In a forest, for example, the products of photosynthesis tend to accumulate in the form of leaves, wood and the food stored in seeds and roots. A relatively long time may elapse before these materials become litter and soil and available to the heterotrophic system.

The two major energy circuits in any ecosystem are the **grazing circuit,** in which animals eat living plants or parts of plants, and the contrasting **organic detritus circuit,** in which dead materials accumulate and are decomposed by bacteria and fungi. From an operational standpoint, the living and nonliving parts of ecosystems are tightly interwoven and difficult to separate. Both inorganic compounds and organic compounds are not only found within and without living organisms, but also are in a constant state of flux between living and nonliving conditions. A few substances, such as ATP, are found uniquely inside living cells. In contrast, humic substances, the resistant end products of decomposition, are never found inside living cells, yet are a major and characteristic component of all ecosystems. DNA and chlorophyll may occur both inside and outside organisms but are nonfunctional when outside the cell. Ecologists can measure the amount of ATP, humus and chlorophyll in a given area or volume to provide an index of the biomass, the decomposition and the production, respectively, in that ecosystem.

The three living components of an ecosystem, the producers, phagotrophs and saprotrophs, are roughly equivalent to plants, animals and bacteria plus fungi, respectively. However, it should be realized that these are functional classifications and that some species of organisms occupy intermediate positions, while others shift their mode of nutrition according to the circumstances of the environment. The heterotrophs can be separated into large and small consumers in an arbitrary fashion.

An excellent way to study an ecosystem is to investigate a small pond, a meadow or an old field. Any area exposed to light—a lawn, a flower-box in the window or a laboratory aquarium—can be used as long as the physical dimensions and biotic diversity of the area are

not so great that observation of the whole is difficult.

39.7 THE HABITAT APPROACH

The subject of ecology can be approached through discussions of the principles and concepts of the science as they apply to different levels of organization, the individual, population, community and ecosystem. Another general approach, the **habitat approach,** describes the distinctive features of the major habitats and their subdivisions, how they are organized, the organisms present in each and the ecological role of these organisms in that region (i.e., the identity of the major producers, consumers and decomposers).

Four major habitats can be distinguished: **marine, estuarine, fresh water** and **terrestrial.** No plant or animal is found in all four major habitats, and indeed, no animal or plant is

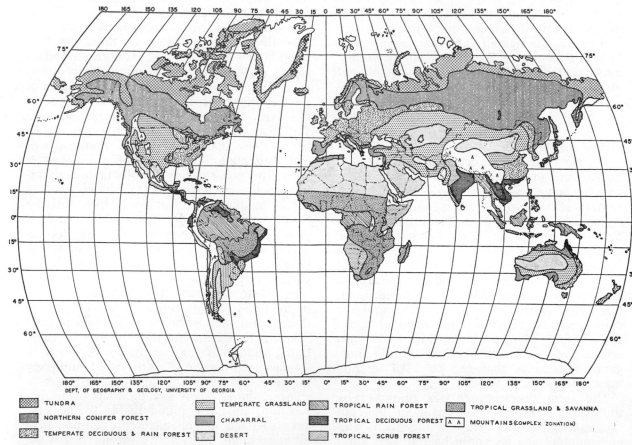

Figure 39.7 A map of the biomes of the world. Note that only the tundra and northern coniferous forest are more or less continuous bands around the world. Other biomes are generally isolated in different biogeographic realms and may be expected to have ecologically equivalent but taxonomically unrelated species. (From Odum, E. P.: Fundamentals of Ecology. 3rd ed. Philadelphia, W. B. Saunders Co., 1971.)

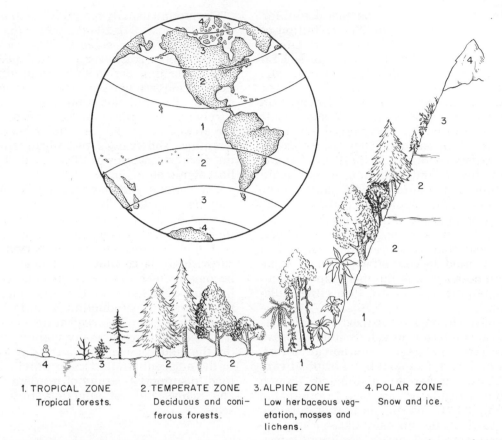

1. TROPICAL ZONE   2. TEMPERATE ZONE   3. ALPINE ZONE   4. POLAR ZONE
Tropical forests.   Deciduous and coni-   Low herbaceous veg-   Snow and ice.
                     ferous forests.   etation, mosses and
                                           lichens.

Figure 39.8 Diagram showing correspondence of life zones at successively higher altitudes at the same latitude (1 to 4, right) and at successively higher latitudes at the same altitude (1 to 4, left, and inset).

found everywhere within any one of these. Every species of animal and plant tends to produce more offspring than can survive within the normal range of the organism. There is strong **population pressure** tending to force the individuals of each species to spread out and become established in new territories. Competing species, predators, lack of food, adverse climate and the unsuitability of the adjacent regions, perhaps owing to the lack of some requisite physical or chemical factor, all act to counterbalance the population pressure and to prevent the spread of the species. Since all of these factors are subject to change, the range of a species tends to be dynamic rather than static and may change quite suddenly. The spread of a species is prevented by geographic **barriers,** such as oceans, mountains, deserts and large rivers, and is facilitated by **highways,** such as land connections between continents. The present distribution of plants and animals is determined by the barriers and highways that exist now and those that have existed in the geologic past.

The biogeographic realms, discussed on page 221, are regions made up of whole continents, or of large parts of a continent, separated by major geographic barriers and characterized by the presence of certain unique animals and plants. Within these biogeographic realms and established by a complex interaction of climate, other physical factors and biotic factors, are large, distinct, easily differentiated community units called **biomes.** A biome is a large community unit characterized by the kinds of plants and animals present. This may be contrasted to the ecosystem, which is a natural unit of living and nonliving components that interact to form a stable system in which the exchange of materials follows a circular path. Thus an ecosystem might be a small pond or a large area coextensive with a biome, but it includes the *physical* environment as well as the populations of animals, plants and microorganisms.

In each biome the *kind* of climax vegetation is uniform—grasses, conifers, deciduous trees—but the particular *species* of plant may vary in different parts of the biome. The kind of climax vegetation depends upon the physical environment, and the two together determine the kind of animals present. The definition of biome includes not only the actual climax community of a region but also the several intermediate communities that precede the climax community.

There is usually no sharp line of demarcation between adjacent biomes; instead each blends with the next through a fairly broad transition region termed an **ecotone.** There is, for example, an extensive region in northern Canada where the tundra and coniferous forests blend in the tundra coniferous forest ecotone. The ecotonal community typically consists of some organisms from each of the adjacent biomes plus some that are characteristic of, and perhaps restricted to, the ecotone. There is a tendency (called the **edge effect)** for the ecotone to contain both a greater number of species and a higher population density than either adjacent biome.

Some of the biomes recognized by ecologists are **tundra, coniferous forest, deciduous forest, broad-leaved evergreen subtropical forest, grassland, desert, chaparral** and **tropical rain forest.** These biomes are distributed, though somewhat irregularly, as belts around the world (Fig. 39.7), and as one travels from the equator to the pole one may traverse tropical rain forests, grassland, desert, deciduous forest, coniferous forest and finally reach the tundra in northern Canada, Alaska or Siberia.

Since climatic conditions at higher altitudes are in many ways similar to those at higher latitudes, there is a similar succession of biomes on the slopes of high mountains (Fig. 39.8). For example, as one goes from the San Joaquin Valley of California into the Sierras, one passes from desert through grassland and chaparral to deciduous forest and coniferous forest, then, above timberline, to a region resembling the tundra of the Arctic.

39.8 THE TUNDRA BIOME

Between the Arctic Ocean and polar icecaps and the forests to the south lies a band of treeless, wet, arctic grassland called the **tundra** (Fig. 39.9).

Some five million acres of tundra stretch across northern North America, northern Europe and Siberia. The primary characteristics of this region are the low temperatures and a short growing season. The amount of precipitation is rather small but water is usually not a limiting factor because the rate of evaporation is also very low.

The ground usually remains frozen except for the uppermost 10 or 20 cm., which thaw during the brief summer season. The permanently frozen deeper soil layer is called **permafrost.** The rather thin carpet of vegetation includes lichens, mosses, grasses, sedges and a few low shrubs. The animals that have adapted to survive in the tundra are caribou or reindeer, the arctic hare, arctic fox, polar bear, wolves, lemmings, snowy owls, ptarmigans and, during the summer, swarms of flies, mosquitoes and a host of migratory birds.

The caribou and reindeer are highly migratory because there is not enough vegetation produced in any one local area to support them. Although casual inspection might suggest that tundras are rather barren areas, a surprisingly large number of organisms have become adapted to survive the cold. During the long daylight hours of the brief summer, the rate of primary production is quite high. The production from the vegetation on the land, from the plants in the many shallow ponds that dot the landscape and from the phytoplankton in the adjacent Arctic Ocean provides enough food to support a variety of resident mammals and many kinds of migratory birds and insects.

39.9 THE FOREST BIOMES

Several different types of forest biomes can be distinguished. These are generally arranged on a gradient from north to south or from high altitude to lower altitude. Adjacent to the tundra region either at high latitude or high altitude is the **northern coniferous forest** (Fig. 39.10), which stretches across both North America and Eurasia just south of the tundra. This is characterized by spruce, fir and pine trees and by animals such as the snowshoe hare, the lynx and the wolf.

The evergreen conifers provide dense shade throughout the year; this tends to inhibit development of shrubs and a herbaceous undergrowth. The continuous presence of green leaves permits photosynthesis to occur throughout the year despite the low temperature during the winter and results in a fairly high annual rate of primary production.

These coniferous forests are the major source of commercial lumber around the world. After they have fallen, the needles decay very slowly and the soil develops a characteristic condition with relatively little humus. The northern coniferous forest, like the tundra, shows a

Figure 39.9 The tundra biome. *Above*: View of the low tundra near Churchill, Manitoba, in July. Note the numerous ponds. *Below*: View of tundra vegetation showing "lumpy" nature of low tundra and a characteristic tundra bird, the willow ptarmigan. (Lower photo by C. Lynn Haywood.)

Figure 39.10 The coniferous forest biome covers parts of Canada, northern Europe and Siberia and extends southward at higher altitudes on the larger mountain ranges. (From Orr, R. T.: Vertebrate Biology. 3rd ed. Philadelphia, W. B. Saunders Co., 1971.)

marked seasonal periodicity, and the populations of animals undergo striking peaks and depressions in numbers.

A distinctive subdivision of the northern coniferous forest biome, perhaps distinctive enough to be considered a separate biome, is the pigmy conifer or **piñon-juniper biome** found in west central California and in the Great Basin and Colorado River regions of Nevada, Utah, Colorado, New Mexico and Arizona. This occupies a belt between the desert or grasslands at lower altitudes and the true northern coniferous forest found at higher altitudes where there is more rainfall. In this region, the annual rainfall of 25 to 50 cm. is irregularly distributed throughout the year. The small piñon pines and cedars tend to be widely spaced, and the biome has an open, parklike appearance (Fig. 39.11). The pine nuts and cedar berries are eaten by the piñon jay, the gray titmouse, and the bush tit.

Along the west coast of North America from Alaska south to central California is a region termed the **moist coniferous forest biome,** characterized by a much greater humidity, somewhat higher temperatures and smaller seasonal ranges than the classic coniferous forest farther north. There is high rainfall, from 75 to 375 cm. per year, and, in addition, a great deal of moisture is contributed by the frequent fogs. There are forests of Sitka spruce in the northern section, western hemlock, arbor vitae and Douglas fir in the Puget Sound area and the coastal redwood, *Sequoia sempervirens*, in California. These forests typically have a luxuriant ground cover of ferns and other herbaceous plants. The potential production of this region is very great, and with careful foresting and replanting, the annual crop of lumber is very high.

The **temperate deciduous forest biome** (Fig. 39.12) is found in areas with abundant evenly distributed rainfall (75 to 150 cm. annually) and moderate temperatures with distinct summers and winters. Temperate deciduous forest biomes originally covered eastern North America, all of Europe, parts of Japan and Australia and the southern portion of South America.

The trees present—beech, maple, oak, hickory and chestnut—lose their leaves during half the year; thus, the contrast between winter and summer is very marked. The undergrowth of shrubs and herbs is generally well developed. The animals originally present in the forest were deer, bears, squirrels, gray foxes, bobcats, wild turkeys and woodpeckers. Much of this forest region has now been replaced by cultivated fields and cities.

In regions of fairly high rainfall but where temperature differences between winter and summer are less marked, as in Florida, the **broad-leaved evergreen subtropical forest biome** is found. The vegetation includes live oaks, magnolias, tamarinds and palm trees, with many vines and epiphytes, such as orchids and Spanish moss.

The variety of life reaches its maximum in the **tropical rain forests** (Fig. 39.13), which occupy low lying areas near the equator with annual rainfalls of 200 cm. or more. The thick rain forests, with a tremendous variety of plants and animals, are found in the valleys of the Amazon, Orinoco, Congo and Zambesi

Figure 39.11 The piñon-juniper biome in Arizona. The small piñon pines and cedars each grow some distance from the neighboring trees, giving an open, parklike appearance to the woodland. (U.S. Forest Service Photo.)

Figure 39.12 An example of a temperate deciduous forest, Noble County, Ohio. The dominant trees are white and red oaks with an understory of hickory. (Photograph by U.S. Forest Service.)

rivers and in parts of Central America, Malaya, Borneo and New Guinea.

The extremely dense vegetation makes it difficult to study or even photograph the rain forest biome. The vegetation is vertically stratified with tall trees often covered with vines, creepers, lianas and epiphytes. Under the tall trees is a continuous evergreen carpet, the canopy layer, some 25 to 35 meters tall. The lowest layer is an understory that becomes dense where there is a break in the canopy.

No single species of animal or plant is present in large enough numbers to be dominant. The diversity of species is remarkable; there may be more species of plants and insects in a few acres of tropical rain forest than in all of Europe. The trees of the tropical rain forest are usually evergreen and rather tall. Their

Figure 39.13 The rain forest biome: border of a clearing in the Ituri Forest of Nala, The Congo. (Photograph by Herbert Lang. Courtesy of the American Museum of Natural History, New York.)

roots are often shallow and have swollen bases or flying buttresses.

The tropical rain forest is the ultimate of jungles, although the low light intensity at the ground level may result in sparse herbaceous vegetation and actual bare spots in certain areas. Many of the animals live in the upper layers of the vegetation. Among the characteristic animals are monkeys, sloths, termites, ants, anteaters, many reptiles and many brilliantly colored birds—parakeets, toucans and birds of paradise.

39.10 THE GRASSLAND BIOME

The **grassland biome** (Fig. 39.14) is found where rainfall is about 25 to 75 cm. per year, not enough to support a forest, yet more than that of a true desert. Grasslands typically occur in the interiors of continents—the prairies of the western United States, and those of Argentina, Australia, southern Russia and Siberia. Grasslands provide natural pasture for grazing animals, and our principal agricultural food plants have been developed by artificial selection from the grasses.

The mammals of the grassland biome are either grazing or burrowing forms—bison, antelope, zebras, wild horses and asses, rabbits, ground squirrels, prairie dogs and gophers. These characteristically aggregate into herds or colonies; this aggregation probably provides some protection against predators. The birds characteristic of grasslands are prairie chickens, meadow larks and rodent hawks.

The species of grasses present in any given grassland may range from tall species, 150 to 250 cm. in height, to short species of grass that do not exceed 15 cm. in height. Some species of grass grow in clumps or bunches and others spread out and form sods with underground rhizomes. The roots of the several species of grass found in grasslands all penetrate deeply into the soil and the weight of the roots of a healthy plant will be several times the weight of the shoot.

Trees and shrubs may occur in grasslands either as scattered individuals or in belts along the streams and rivers. The soil of grasslands is very rich in humus because of the rapid growth and decay of the individual plants. The grassland soils are well suited for growing cultivated food plants, such as corn and wheat, that are species of cultivated grasses. The grasslands are also well adapted to serve as natural pastures for cattle, sheep and goats. However, when grasslands are subjected to consistent overgrazing and overplowing, they can be turned into man-made deserts.

There is a broad belt of tropical grassland, or **savanna,** in Africa lying between the Sahara desert and the tropical rain forest of the Congo basin. Other savannas are found in South America and Australia. Although the annual rainfall is high, as much as 125 cm., a distinct, prolonged dry season prevents the development of a forest. During the dry season there may be extensive fires, which play an important role in the ecology of the region. In this region (Fig. 39.15) are great numbers and great varieties of grazing animals and predators, such as lions.

How to make best use of these African

Figure 39.14 A region of short-grass grassland with a herd of bison, originally one of the major grazing animals in the grassland biome of western United States and Canada. The bison in the center is wallowing. (From Odum, E. P.: Fundamentals of Ecology. 3rd ed. Philadelphia, W. B. Saunders Co., 1971.)

Figure 39.15   The savanna biome; characteristic animals of the African grasslands, zebra and wildebeest, Kruger National Park, Transvaal, Republic of South Africa. (From Odum, E. P.: Fundamentals of Ecology. 2nd ed. Philadelphia, W. B. Saunders Co., 1959. Photograph by Herbert Lang.)

grasslands is a problem now facing the new nations of Africa as they work to raise the level of nutrition in their human populations. Many ecologists are of the opinion that it would be better to harvest the native herbivores—antelope, hippopotamuses and wildebeests—on a sustained yield basis rather than try to exterminate them completely and substitute cattle. The diversity of the natural population would mean broader use of all the resources of primary production, and the native species are immune to the many tropical parasites and diseases that plague the cattle that have been introduced.

39.11   THE CHAPARRAL BIOME

In mild temperate regions of the world with relatively abundant rain in the winter but with very dry summers the climax community includes trees and shrubs with hard, thick evergreen leaves (Fig. 39.16). This type of vegetation is called "chaparral" in California and Mexico, "macchie" around the Mediterranean and "mellee scrub" on Australia's south coast.

The trees and shrubs common in California's chaparral are chamiso and manzanita. Eucalyptus trees introduced from Australia's south coast into California's chaparral region have prospered mightily and have replaced to a considerable extent the native woody vegetation in areas near cities.

Mule deer and many kinds of birds live in the chaparral during the rainy season but move north or to higher altitudes to escape the hot, dry summer. Brush rabbits, wood rats, chipmunks, lizards, wren-tits and brown towhees are characteristic animals of the chaparral biome. During the hot dry season, there is an ever present danger of fire that may sweep rapidly over the chaparral slopes. Following a fire, the shrubs sprout vigorously after the first rains and may reach maximum size within twenty years.

Figure 39.16   An example of the chaparral biome in California. The shrubs in the picture are *Eriodictyon tomentosum*. (Photograph by U.S. Forest Service.)

39.12 THE DESERT BIOME

In regions with less than 25 cm. of rain per year or in certain hot regions where there may be more rainfall but with an uneven distribution in the annual cycle, vegetation is sparse and consists of greasewood, sagebrush or cactus. The individual plants in the desert are typically widely spaced with large bare areas separating them. In the brief rainy season, the California desert becomes carpeted with an amazing variety of wild flowers and grasses, most of which complete their life cycle from seed to seed in a few weeks. The animals present in the desert are reptiles, insects and burrowing rodents, such as the kangaroo rat and pocket mouse, both of which are able to live without drinking water by extracting the moisture from the seeds and succulent cactus they eat.

The small amount of rainfall may be due to continued high barometric pressure, as in the Sahara and Australian deserts; a geographical position in the rain shadow of a mountain, as in the western North American deserts; or high altitude, as in the deserts in Tibet and Bolivia. The only absolute deserts, where little or no rain ever falls, are those of northern Chile and of the central Sahara.

Careful measurements of the amount of dry matter produced for a given area in the course of a year show a clear linear relationship with the amount of rainfall, at least up to 60 cm. per year. This illustrates the primary role of moisture as a limiting factor in the productivity of the desert. Where the soil is favorable, an irrigated desert can be extremely productive because of the large amount of sunlight.

Two types of deserts can be distinguished on the basis of their average temperatures: "hot" deserts, such as that found in Arizona, characterized by the giant saguaro cactus, palo verde trees and the creosote bush; and "cool" deserts, such as that present in Idaho, dominated by sagebrush (Fig. 39.17).

Certain reptiles and insects are well adapted for survival in deserts because of their thick, impervious integuments and the fact that they excrete dry waste matter. A few species of mammals have become secondarily adapted to the desert by excreting very concentrated urine. They avoid the sun by remaining in their burrows during the day. The camel and the desert birds must have an occasional drink of water but can go for long periods of time using the water stored in the body.

When deserts are irrigated, the large volume of water passing through the irrigation system may lead to the accumulation of salts in the soil

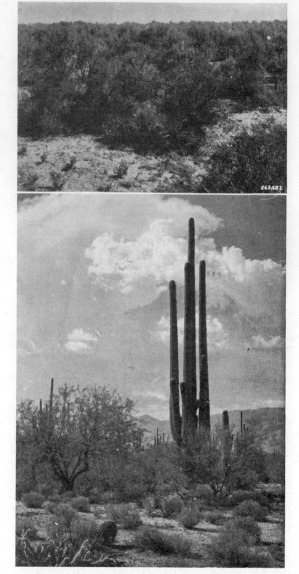

Figure 39.17 Two types of desert in western America: a "cool" desert in Idaho, dominated by sagebrush (*above*), and (*below*) a rather luxuriant "hot" desert in Arizona, with giant cactus (Saguaro) and palo verde trees, in addition to creosote bushes and other desert shrubs. In extensive areas of desert country the desert shrubs alone dot the landscape. (Upper photograph by U.S. Forest Service; lower photograph by U.S. Soil Conservation Service.)

as some of the water is evaporated, and this will eventually limit the area's productivity. The water supply itself can fail if the watershed from which it is obtained is not cared for appropriately. The ruins of old irrigation systems and of the civilizations they supported in the deserts of North Africa and the Near East remind us that the irrigated desert will retain its productivity only when the entire system is kept in appropriate balance.

39.13 THE EDGE OF THE SEA: MARSHES AND ESTUARIES

Where the sea meets the land there may be one of several kinds of ecosystems with distinctive characteristics: a rocky shore, a sandy beach, an intertidal mud flat or a tidal estuary containing salt marshes. The word estuary refers to the mouth of a river or a coastal bay where the salinity is less than in the open ocean, intermediate between the sea and fresh water. Most estuaries, particularly those in temperate and arctic regions, undergo marked variations in temperature, salinity and other physical properties in the course of a year. To survive there, estuarine organisms must have a wide range of tolerance to these changes (they must be euryhaline, eurythermic and so on).

The waters of estuaries are among the most naturally fertile in the world, frequently having a much greater productivity than the adjacent sea or the fresh water up the river. This high productivity is brought about by the action of the tides, which promote a rapid circulation of nutrients and aid in the removal of waste products, and by the presence of many kinds of plants, which provide for an extensive photosynthetic carpet. These include the phytoplankton, the algae living in and on the mud, sand, rocks or other hard surfaces, and the large attached plants, "seaweeds," eel grasses and marsh grasses.

Some of the marsh grass is eaten by insects and other terrestrial herbivores, but most of it is converted to detritus and is consumed by the clams, crabs and other marine detritus eaters. Estuaries may have a high productivity of fish, oysters, shrimp and other seafood, which can be tapped by "mariculture," such as the oyster farms of Japan where oysters are grown suspended on rafts hanging from floats. This is an excellent way of obtaining protein foods as a harvest of the natural productivity of the estuaries. The farms must be spaced well apart and protected from pollution.

Estuaries and marshes are high on the list of ecological regions of the world that are seriously threatened by man's activities. They were long considered to be worthless regions in which waste materials could be dumped. Many have been irretrievably lost by being drained, filled and converted to housing developments or industrial sites. Man is just beginning to appreciate that the best interests of all are served by maintaining estuaries in their natural state and protecting them from waste material and thermal and oil pollution.

39.14 MARINE LIFE ZONES

There has recently been a great upsurge of interest in oceanography in general and in marine ecology in particular as we have begun to appreciate that there is much to learn about the mysterious sea. The oceans, which cover 70 per cent of the earth's surface, constitute one of the great reservoirs of living things and of the essential nutrients needed by both land and marine organisms. It is clear that the total weight of living things (the "biomass") in the ocean far exceeds that of all living things on land and in fresh water.

The seas are continuous one with another, and marine organisms are restrained from spreading to all parts of the ocean only by factors such as temperature, salinity and depth. The salinity of the open ocean is about 35 parts per thousand. In the western Baltic it is 12 parts per thousand and 0.6 parts per thousand in the Gulf of Finland owing to the inflow of fresh water. In the Red Sea, with no source of fresh water and a high rate of evaporation, it reaches 46.5 parts per thousand. The temperature of the oceans ranges from about $-2°$ C. in the polar seas to $32°$ C. or more in the tropics, but the annual range of variation in any given region is usually no more than $6°$ C.

The waters of the seas are continually moving in vast currents, such as the Gulf Stream, the North Pacific Current and the Humboldt Current, which circle in a clockwise fashion in the northern hemisphere and counterclockwise in the southern hemisphere. These currents not only influence the distribution of marine forms but also have marked effects on the climates of the adjacent land masses. In addition, there are very slow currents of cold dense water flowing at great depths from the polar regions toward the equator.

Where the wind consistently moves surface water away from steep coastal slopes, water from the deep is brought to the surface by a process termed **upwelling**. This water is cold and rich in nutrients that have accumulated in the depths. Regions of upwelling typically occur on the western coasts of continents, as in California, Peru and Portugal, and are the most productive of all marine areas. The upwelling produced by the Peru Current has created one of the richest fisheries in the world and supports large populations of seabirds that deposit nitrate- and phosphate-rich **guano** on the headlands and adjacent coastal islands. These upwellings are very important in returning elements, such as phosphorus, to the surface for recycling.

Although the saltiness of the open ocean is relatively uniform, the concentrations of phosphates, nitrates and other nutrients vary widely in different parts of the sea and at different times of the year and are usually the major factors limiting the biological productivity of the seas in a given region.

Like the land, the ocean consists of regions characterized by different physical conditions and consequently inhabited by specific kinds of plants and animals. All the phyla except Onychophora and all the classes except amphibians, centipedes, millipedes and insects are well represented in the oceans. Ctenophores, brachiopods, echinoderms, chaetognaths and a few lesser phyla are found only in the oceans.

A gently sloping continental shelf usually extends some distance offshore; beyond this, the ocean floor (the **continental slope**) drops steeply to the abyssal region. The region of shallow water over the continental shelf is called the **neritic zone;** it can be subdivided into **supratidal** (above the high tide mark), **intertidal** (between the high and low tide lines, a region also known as the "littoral") and **subtidal** regions (Fig. 39.18).

The open sea beyond the edge of the continental shelf is the **oceanic zone.** The upper part of the ocean, into which enough light can penetrate to be effective in photosynthesis, is known as the **euphotic zone.** The average lower limit of this is at about 100 meters, but in a few regions of clear tropical water this may extend to twice that depth. The regions of the ocean

beneath the euphotic zone are called the **bathyal zone** over the continental slope to a depth of perhaps 2000 meters; the depths of the ocean beyond that compose the **abyssal zone.**

The floor of the ocean is not flat but is thrown into gigantic ridges and valleys. Some of the ridges rise nearly to the surface (or above it where there are oceanic islands), and some of the valleys lie 10,000 meters below the surface of the sea. Huge underwater avalanches occur from time to time as parts of the ridges tumble into the valleys.

Some organisms are bottom dwellers, called **benthos,** and creep or crawl over the bottom or are **sessile** (attached to it). Others are **pelagic,** living in the open water, and are either active swimmers, **nekton,** or organisms that float with the current, **plankton.** The plankton includes algae, protozoa, small larval forms of a variety of animals and a few worms. The nekton includes jellyfish, squid, fishes, turtles, seals and whales. Some of the benthic animals—crabs, snails, starfish, certain worms—crawl over the substrate. Clams and worms burrow into the sand, mud or rock of the sea bottom. Sponges, sea anemones, corals, bryozoa, crinoids, oysters, barnacles and tunicates are attached to the substratum.

The intertidal zone is one of the most favorable of all the habitats in the world and many biologists believe that life may have originated here. The abundance of water, light, oxygen, carbon dioxide and minerals makes it extremely salutary for plants. The dense growth of plants, providing food and shelter,

Figure 39.18 Zonation in the sea. (Redrawn from Hedgpeth, J. W.: The Classification of Estuarine and Brackish Waters and the Hydrographic Climate. Rpt. 11. National Research Council Committee on a Treatise on Marine Ecology and Paleoecology, 1951.)

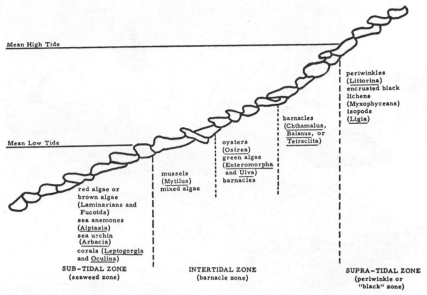

Figure 39.19 Distribution of plants and animals on a rocky shore at Beaufort, North Carolina. (From Odum, E. P.: Fundamentals of Ecology. 3rd ed. Philadelphia, W. B. Saunders Co., 1971. Data from Stephenson and Stephenson, 1952.)

makes it an excellent habitat for animals. The plants of the region are primarily a wide variety of algae plus a few grasses. Many of the animals are sessile and are more or less permanently fixed to the sea bottom, though they may be pelagic at some stage of their life cycle. These sessile animals are usually restricted to certain depths of the intertidal zone (Fig. 39.19). There is keen competition among the plants for space and among the animals for space and food, so the forms living here have had to evolve special adaptations to survive.

The gravitational pulls of sun and moon each cause two bulges of the water on opposite sides of the earth, the high tides. The tides advance westward, since the earth is rotating eastward on its axis. Since the earth rotates once a day on its axis, there are two high tides and two low tides per day. The periodicity of the tides is about 12.5 hours; hence the tides are about 50 minutes later on each succeeding day. Twice a month, at full and new moon, the earth, sun and moon are in line; the pulls of the sun and moon on the waters of the earth are additive; and the difference between high and low tide is greater than normal (the spring tides). At the quarter moons, when sun and moon are pulling at right angles, the difference between high and low tide is less than usual (neap tides). The difference in tidal range between high and low tides ranges from about 30 cm. in the open ocean to 15 meters or more in the Bay of Fundy and the English Channel.

Since the intertidal zone is exposed to air twice daily, its inhabitants have had to develop some sort of protection against desiccation. Some animals avoid this by burrowing into the damp sand or rocks until the tide returns; others have developed shells that can be closed, retaining a supply of water inside. Many plants contain jelly-like substances, such as agar, that absorb large quantities of water and retain it while the tide is out.

One of the outstanding characteristics of this region is the ever present action of the waves, and the organisms living on a sandy or rocky beach have had to evolve ways of resisting wave action. The many seaweeds have tough pliable bodies, able to bend with the waves without breaking, while the animals either are encased in hard calcareous shells, such as those of mollusks, bryozoa, starfish, barnacles and crabs, or are covered by a strong leathery skin that can bend without breaking, like that of the sea anemone and octopus.

The successive zones of the intertidal region can be seen clearly on most rocky shores (Fig. 39.20). At the uppermost end is a zone of bare rock marking the transition between land and sea. Next is a spray zone with dark patches of algae on which the periwinkles (*Littorina*) graze. Below this is the zone regularly covered by the high tide; rocks in this zone are encrusted with barnacles, limpets and mussels. On the layer of rocks below this, there is a cover of rock kelp and other algae containing small

Figure 39.20 Zonation along a rocky shore—mid-Atlantic coast line. (I) Bare rock with some black algae and swamp periwinkle; (II) barnacle zone; (III) oyster zone: oysters, Enteromorpha, sea lettuce and purple laver; (IV) muddy zone: mussel beds; (V) infralittoral fringe: starfish and so on. Note absence of kelps. (Zonation drawing based on Stephenson, 1952; sketches done from life or specimens. From Smith, R. L.: Ecology and Field Biology. New York, Harper & Row, Publishers, 1966.)

crabs and snails. Rocky coasts typically have tide pools that contain characteristic assemblages of animals and plants.

The sandy shore may be even more harsh than the rocky shore. It is subject to all the extremes of the latter plus the inconvenience of a constantly shifting substratum. The last makes life on the surface almost impossible; life has retreated below the surface. Zonation on a sandy beach is illustrated in Figure 39.21. It does not, however, conform to a universal pattern as does that of rocky shores. In the example depicted, the supralittoral zone is inhabited by ghost crabs and beach hoppers. These animals spend most of the daytime hidden in damp burrows. They forage at night. Ghost crabs nightly go to the water to dampen their gill chambers. The intertidal zone is not as rich here as on the rocky shore, but it is the home of ghost shrimps, clams and bristle worms. Lower down the beach are lugworms, trumpet worms and other species of clams. On surf beaches, the mole crab and the coquina clam live in this zone. As waves roll up the beach these two small creatures emerge from the sand, ride the waves up the beach and, as the velocity of the water decreases, burrow quickly into the sand as the waves retreat. Once

settled, the crab extends its antennae and the clam its siphon to extract particulate food from the receding waves.

The subtidal zone is also thickly populated, for it has plenty of light and the nutrients required by plants. The absence of the periodic exposure to air and the diminished wave action permit many plants and animals to live here that could not survive in the tidal zone. Here live many species of fish and many single-celled algae; the larger seaweeds, which require a substratum for attachment, are found only in the shallower parts of the region.

The subtidal zone is the home of cockles, razor clams, moon shells and other mollusks. Still farther out are starfish, sand dollars, sea cucumbers, killifish, silversides and flounders.

In this area of the marine environment, as in all others, the **primary producer organisms,** equivalent to the flowering plants on land, are the **phytoplankton** (Fig. 39.22), consisting principally of diatoms and dinoflagellates. It is difficult to appreciate their importance because they are so small. An absolutely minimal estimate would place their density at 375,000,000 individuals per cubic meter. In temperate regions, the phytoplankton undergoes two seasonal population explosions or

Figure 39.21 Life on a sandy ocean beach along the Atlantic coast. Although strong zonation is absent, organisms still change on a gradient from land to sea (I) Supratidal zone: ghost crabs and sand fleas; (II) flat beach zone: ghost shrimp, bristle worms and clams; (III) infratidal zone: clams, lugworms and mole crabs; (IV) subtidal zone: the dotted line indicates high tide. (From Smith, R. L.: Ecology and Field Biology. New York, Harper & Row, Publishers, 1966.)

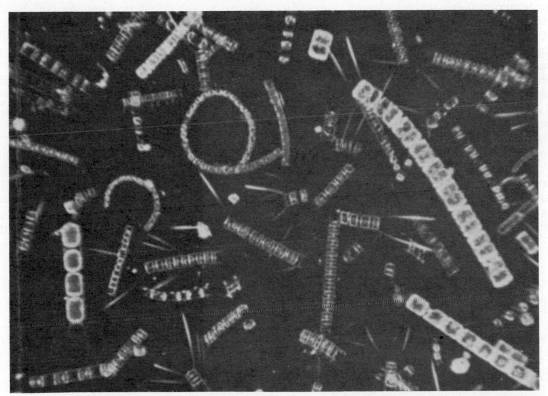

Figure 39.22 Living plants of the plankton (phytoplankton). ×110. Chains of cells of several species of *Chaetoceros* (those with spines), a chain of *Thalassiosira condensata* (at and pointing to bottom right corner) and a chain of *Lauderia borealis* (above the last named). By electronic flash. (From Hardy, A.: The Open Sea. Vol. 1. London, William Collins Sons & Co., Ltd., 1966.)

"blooms," one in the spring, the other in late summer or fall. The mechanism is similar to that responsible for "blooms" in lakes. In the wintertime, low temperatures and reduced light restrict photosynthesis to a low level; however, these factors do not prevent bacteria and other microorganisms from generating high concentrations of nitrogen, phosphorus and other nutrient elements. When spring brings higher temperatures and more light, photosynthesis accelerates. The nutrient supply is ample because the winter mixing of surface and deep water brings up nutrients that have fallen to and accumulated at the bottom. Within a fortnight, the diatoms multiply ten thousand-fold. This prodigious growth accounts for the spring bloom. Soon, however, the nutrients are exhausted. Replacement from lower layers no longer occurs because warming of the surface water keeps it on top and prevents mixing. Nutrients are now locked in the bodies of animals that have eaten the phytoplankton or are slowly falling to the bottom in dead bodies. Whereas temperature and light were the limiting factors during the winter, nutrient level is the limiting factor during the summer, especially since existing phytoplankton is now being consumed by animals. Now nutrients begin to accumulate again in lower layers. As fall approaches, the upper layers of water begin to cool again. The accompanying density change, together with the autumn equinoctial gales, begins mixing the water again. Water rich in phosphates and nitrates is brought up from below. Other forms of phytoplankton, especially nitrogen-fixing blue-green algae, now bloom until reduced nutrients or temperature again intercedes (Fig. 39.23).

Tropical oceans, except for coral reefs and coastal waters, are less productive than temperate and cold seas. The permanently warm upper layer of water (warm water is lighter than cold water) receives only a slow replacement of nutrients from the colder, heavier layers below.

Productive seas are green or grey in color because of the high content of plankton. The blue color of clear water is sometimes called the color of ocean deserts.

The other important population in the ocean is the **zooplankton** (Fig. 39.24). The zooplankton comprises all the animals carried passively by moving water. Every major phylum of the Animal Kingdom is represented, if not as adults, at least as eggs or larval stages. All the organisms are small, but not so small as not to be visible through a 6× hand lens. The beauty and almost limitless variety of forms of these animals beggar description. A sample haul near the Isle of Man gave an average of 4500 animals per cubic meter. The same hauls gave about 727,000 planktonic plants per cubic meter.

The active surface dwellers of the **neritic** zone are many bony fish, large crustacea, turtles, seals, whales and a host of sea birds. All these **consumers** are restricted in their distribution by temperature, salinity and nutrients. There is horizontal zonation and vertical stratification. The greatest numbers of fish (individuals — not species) are found in northern waters and where there are cold upwellings. Only a few species make up the bulk of commercial fisheries. Three-fourths of the world's catch consists of herring, cod, haddock, pollock, salmon, flounder, sole, plaice, halibut, mackerel, tuna and bonito groups. The cod and flounder groups are bottom fish. Upon these and other marine organisms feed a great number of birds. Sandpipers, plovers, herons, curlews and others search the supra- and intertidal zones; cormorants, sea ducks, and pelicans, the subtidal zones; petrels and shearwaters, the lower neritic zone farther out to sea.

The **oceanic region** is less rich in species and numbers than the coastal areas, but it has its characteristic species. Many of these are transparent or bluish, and since the sediment-free water of the open sea is marvelously transparent, these animals are nearly invisible. Animals

Figure 39.23 The probable mechanisms for phytoplankton "blooms." See text for explanation. (From Odum, E. P.: Fundamentals of Ecology. 3rd ed. Philadelphia, W. B. Saunders Company, 1971.)

Figure 39.24 Living animals of the plankton (zooplankton). × 16. The copepods *Calanus finmarchicus* (the largest animal) and *Pseudocalanus elongatus* (similar in shape, but much smaller than *Calanus*, and the one with the cluster of eggs); two small anthomedusae with long tentacles; a fish egg (the circular object); a young arrow-worm *Sagitta* (to the right of the fish egg); small nauplius (larval stage) of copepod (close to left side of *Calanus*) and the planktonic tunicate *Oikopleura* (curly objects top right and middle bottom). By electronic flash but partially narcotized. (From Hardy, A.: The Open Sea. Vol. 1. London, Williams Collins Sons & Co., Ltd., 1966.)

that are too thick to be transparent frequently have smooth shiny and silvery bodies that make them invisible by mirroring the water in which they swim. Among the most characteristic animals of the open ocean are the baleen and toothed whales, the only mammals that are truly and absolutely marine. The baleen or whalebone whales live on plankton that they strain from the water. The toothed whales live on nekton. Giant squids compose part of their diet. The distribution of these whales is correlated directly or indirectly with the distribution of plankton. The flora and fauna of arctic seas differ from those of tropical seas. The boundaries between the two are not sharp and shift with the seasons. The major tropical oceans, the Atlantic and the Indo-Pacific, are separated from each other by regions of cold water. Each contains a large number of unique species of animals and plants.

Above the surface of the open sea are the oceanic birds—the petrels, albatrosses and frigate birds. They are not truly marine because they must come to land to breed. Apart from

that they may spend all their lives at sea out of sight of land. Like the whales, their distribution may be worldwide but not uniform. They are restricted by the distribution of fish, which in turn are dependent on the phytoplankton.

Eighty-eight per cent of the ocean is more than 1 mile deep. It is continuous throughout the world except for the deep water of the Arctic Ocean, which is cut off from the rest by a narrow submerged mountain range connecting Greenland, Iceland and Europe. This is the area of great pressure and of perpetual night. Since no photosynthesis is possible, the only source of energy is the constant rain of organic debris, the bodies and waste products of organisms in the surface layers, that falls toward the bottom. The other prerequisite for life, oxygen, gets to the bottom by means of the oceanic circulation discussed earlier.

Life in the deep is poorly known because of the enormous difficulty of observation. The **pelagic** animals are strong swimmers, not easily caught in nets. The scant knowledge available has been gleaned from studies of net hauls and by observation from special undersea

Figure 39.25 Sexual parasitism in the deep-sea angler fish, *Photocorynus spiniceps*. The small male is permanently attached to the female; he has no independent existence but is nourished by the blood of the female. (From Allee, W. C., et al.: Principles of Animal Ecology. Philadelphia, W. B. Saunders Co., 1949. After Norman.)

Figure 39.26 Photographs of the ocean bottom at three different depths off southern California (San Diego Trough). *A,* At 95 meters. Note abundant sea urchins (probably *Lytechinus*) appearing as globular, light-colored bodies and the long curved sea whips (probably *Acanthoptilum*). Burrowing worms have built the conical piles of sediment at the mouth of their burrows. (Emery, 1952.) *B,* At 1200 meters. Vertical photograph of about 4 square meters of bottom composed of green silty mud having a high organic content. Note the numerous brittle stars (Ophiuroidea) and several large sea cucumbers (Holothuroidea). (Official Navy photo, courtesy U.S. Navy Electronics Laboratory, San Diego.) *C,* At 1400 meters. Note in the right foreground the ten arms of what is probably a crinoid (a relative of the starfish that is attached to the bottom by a stalklike part). Small worm tubes and brittle stars litter the surface, and two sea cucumbers may be seen in the left foreground. Continual activity of burrowing animals keeps the sea bottom "bumpy." The bottom edge of the picture represents a distance of about 2 meters. (From Emery, K. O.: Submarine photography with the benthograph. Scient. Monthly *75*:3–11, 1952.)

craft or via underwater television. In general, the animals of this world are either filter-feeders that sieve out particles before they reach the bottom, grubbers that ingest sediment or predators on the filter-feeders and grubbers.

Most of the fish of the abyssal region are rather small and peculiarly shaped; many are equipped with luminescent organs, which may serve as lures for the forms preyed upon. The majority of the deep-sea creatures are related to shallow-sea forms, and they must have migrated to their present habitat recently (geologically speaking), for none is older than the Mesozoic.

Since the number of members of any one species in these vast depths is small, reproduction is more of a problem than in any other region, and certain fish have evolved a curious adaptation to ensure that reproduction will occur. At an early age the male becomes attached to and fuses with the head of the female, where he continues to live as a small (2.5 cm.) parasite (Fig. 39.25). In due course, he becomes mature and when the female lays her eggs, he releases his sperm into the water to fertilize them.

The bottom of the sea is a soft ooze, made of the organic remains and shells of foraminifera, radiolaria and other animals and plants. Organic detritus is an important food source for many of the invertebrates that live on the ocean floor (Fig. 39.26). Apparently even the greatest "deeps" are inhabited, for tube-dwelling worms have been dredged up from depths of 8000 meters, and sea urchins, starfish, bryozoa and brachiopods have been found at depths of 6000 meters. Arthropods living on these bottom oozes typically have long thin appendages.

39.15 FRESHWATER LIFE ZONES

Freshwater habitats can be divided into **standing water**—lakes, ponds and swamps—and **running water**—rivers, creeks and springs—each of which can be further subdivided. The biological communities of freshwater habitats are in general more familiar than the salt water ones, and many of the animals used as specimens in zoology classes are from fresh water—amebas, hydras, planarias, crayfish and frogs.

Standing water, such as a lake, can be divided (much as the zones of the ocean were distinguished) into the shallow water near the shore (the **littoral zone**), the surface waters away from the shore (the **limnetic zone**) and the

deep waters under the limnetic zone (the **profundal zone**).

Aquatic life is probably most prolific in the littoral zone. Within this zone the plant communities form concentric rings around the pond or lake as the depth increases (Fig. 39.27). At the shore proper are the cattails, bulrushes, arrowheads and pickerelweeds—the emergent, firmly rooted vegetation linking water and land environments. Out slightly deeper are the rooted plants with floating leaves, such as the water lilies. Still deeper are the fragile thin-stemmed water weeds, rooted but totally submerged. Here also are found diatoms, blue-green algae and green algae. Common green pond scum is one of the latter.

The littoral zone is also the scene of the greatest concentration of animals (Fig. 39.28), distributed in recognizable communities. In or on the bottom are various dragonfly nymphs, crayfish, isopods, worms, snails and clams. Other animals live in or on plants and other objects projecting up from the bottom. These include the climbing dragonfly and damsel fly nymphs, rotifers, flatworms, bryozoa, hydra, snails and others. The zooplankton consists of water fleas, such as *Daphnia*, rotifers and ostracods. The larger freely swimming fauna (**nekton**) includes diving beetles and bugs, dipterous larvae (e.g., mosquitoes) and large numbers of many other insects. Among the vertebrates are frogs, salamanders, snakes and turtles. Floating members of the community (**neuston**) include whirligig beetles, water striders and numerous protozoa. Many pond fish (sunfish, top minnows, bass, pike and gar) spend much of their time in the littoral zone.

The limnetic or open-water zone is occupied by many microscopic plants (dinoflagellates, *Euglena* and *Volvox*), many small crustaceans (copepods, cladocera and so on) and many fish.

Deep (profundal) life consists of bacteria, fungi, clams, blood worms (larvae of midges), annelids and other small animals capable of surviving in a region of little light and low oxygen.

As compared to ponds where the littoral zone is large, the water usually shallow and temperature stratification usually absent, lakes have large limnetic and profundal zones, a marked **thermal stratification** and a seasonal cycle of heat and oxygen distribution. In the summertime, the surface water (**epilimnion**) of lakes becomes heated while that below (**hypolimnion**) remains cold. There is no circulatory exchange between upper and lower layers, with the result that the lower layers frequently

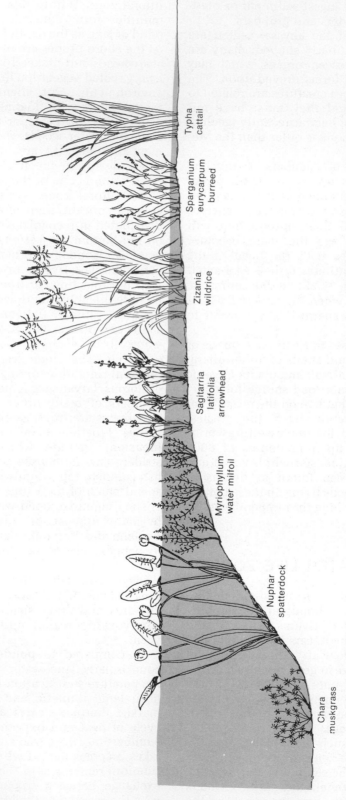

Figure 39.27   Zonation of vegetation about ponds and along river banks. Note the changes in vegetation with water depth. (After Dansereau, P., Biogeography: An Ecological Perspective. New York, The Ronald Press Company, 1959; from Smith, R. L.: Ecology and Field Biology. New York, Harper & Row, Publishers, 1966.)

Typha
cattail

Sparganium
eurycarpum
burreed

Zizania
wildrice

Sagittaria
latifolia
arrowhead

Myriophyllum
water milfoil

Nuphar
spatterdock

Chara
muskgrass

Figure 39.28   Some representative animals of the littoral zone of ponds and lakes. Series 1 to 4 are primarily herbivorous forms (primary consumers); series 5 to 8 are predators (secondary consumers). *1,* Pond snails (*left to right*): *Lymnaea* (*pseudosuccinea*) *columella; Physa gyrina; Helisoma trivolvis; Campeloma decisum. 2,* Small arthropods living on or near the bottom or associated with plants or detritus (*left to right*): a water mite, or Hydracarina (*Mideopsis*); an amphipod (*Gammarus*); an isopod (*Asellus*). *3,* A pond caddis fly larva (*Triaenodes*), with its thin, light portable case. *4,* (*Left to right*) A mosquito larva (*Culex pipiens*); a clinging or periphytic mayfly nymph (*Cloeon*); a benthic mayfly nymph (*Caenis*)—note gill covers that protect gills from silt. *5,* A predatory diving beetle, *Dytiscus,* adult and (*right*) larva. *6,* Two predaceous Hemipterans, a water scorpion, *Ranatra* (Nepidae) and (*right*) a back-swimmer, *Notonecta. 7,* A damsel fly nymph, *Lestes* (Odonata-Zygoptera); note three caudal gills. *8,* Two dragonfly nymphs (Odonata-Anisoptera), *Helocordulia,* a long-legged sprawling type (benthos) and (*right*) *Aeschina,* a slender climbing type (periphyton). (After Pennak, R. W.: Freshwater Invertebrates of the United States. New York, The Ronald Press Company, 1953.)

become deprived of oxygen. Between the two is a region of steep temperature decline (**thermocline**). As the cooler weather of fall approaches, the surface water cools, the temperature is equal at all levels, the water of the whole lake begins to circulate and the deep is again oxygenated. This is the **"fall overturn."** In winter, as the surface temperature drops to 4° C., the water at the surface becomes less dense, remains at the surface and impedes circulation. The bottom is now warmer than the top. Because bacterial decomposition and respiration are less at low temperatures and cold water holds more oxygen, there is usually no great winter stagnation. The formation of ice may, however, cause oxygen depletion and result in a heavy winterkill of fish. The **"spring overturn"** occurs when the ice melts and the heavier surface water sinks to the bottom (Fig. 39.29).

Moving waters differ in three major aspects

from lakes and ponds: current is a controlling and limiting factor; land-water interchange is great because of the small size and depth of moving water systems, as compared with lakes; oxygen is almost always in abundant supply except when there is pollution. Temperature extremes tend to be greater than in standing water. Plants and animals living in streams are usually attached to surfaces or, in the case of animals, are exceptionally strong swimmers. Characteristic stream organisms are: caddis fly larvae, blackfly larvae, attached green algae, encrusting diatoms and aquatic mosses.

Freshwater habitats change much more rapidly than other life zones; ponds become swamps, swamps become filled in and converted to dry land and streams erode their banks and change their course. The kinds of plants and animals present may change markedly and show ecological successions similar to those on land. The large lakes, such as the

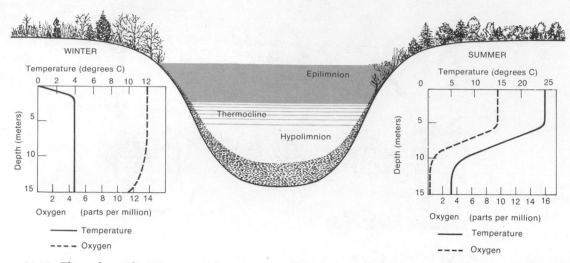

Figure 39.29 Thermal stratification in a north temperate lake (Linsley Pond, Conn.). Summer conditions are shown on the right, winter conditions on the left. Note that in summer the oxygen-rich circulating layer of water, the epilimnion, is separated from the cold oxygen-poor hypolimnion waters by a broad zone, called the thermocline, that is characterized by a rapid change in temperature and oxygen with increasing depth. (From Life in the Depths of a Pond by E. S. Deevey, Jr. Copyright © by Scientific American, Inc. All rights reserved.)

Great Lakes, are relatively stable habitats and have more stable populations of plants and animals. Lake Baikal in the Soviet Union is the oldest and deepest lake in the world, formed during the Mesozoic Era and containing many species of fish and other animals found nowhere else.

39.16 THE DYNAMIC BALANCE OF NATURE

The concept of the dynamic state of the cellular constituents was discussed in Chapter 4, and we learned that the protein, fat, carbohydrate and other constituents of the cells are constantly being broken down and resynthesized. A biotic community undergoes an analogous constant reshuffling of its constitu-

ent parts; the concept of the dynamic state of biotic communities is an important ecological principle. Plant and animal populations are constantly subject to changes in their physical and biotic environment and must adapt or die. In addition, communities undergo a number of rhythmic changes—daily, tidal, lunar, seasonal and annual—in the activities or movements of their constituent organisms. These result in periodic changes in the composition of the community as a whole. A population may vary in size, but if it outruns its food supply, like the Kaibab deer or the lemmings, equilibrium is quickly restored. Communities of organisms are comparable in many ways to a many-celled organism and exhibit growth, specialization and interdependence of parts, characteristic form and even development from immaturity to maturity, old age and death.

ANNOTATED REFERENCES

Aubert de la Rue, E. F. Boulière and J. P. Harroy: The Tropics. New York, Alfred A. Knopf, 1957. A well illustrated account of the ecological relations of animals and plants in the tropic rain forest.

Bates, M.: The Forest and the Sea. New York, Random House, 1960. Presents a more detailed discussion of these specific biological communities.

Braun, E. L.: Deciduous Forests of Eastern North America. Philadelphia, The Blakiston Company, 1950. A fine summary of the vegetation present in our deciduous forests.

Carson, R. L.: The Sea Around Us. New York, Oxford University Press, 1951. A beautifully written, nontechnical presentation of certain aspects of marine life.

Carson, R. L.: The Edge of the Sea. Boston, Houghton Mifflin Co., 1955. A companion book describing animal and plant interrelations at the seashore.

Coker, R. E.: Lakes, Streams and Ponds. Chapel Hill, N.C., University of North Carolina Press, 1964. A classic presentation of limnology.

Farb, P.: The Forest. New York, Time-Life Nature Library, 1961. Beautifully illustrated, nontechnical accounts of life in these biomes.

Gross, M. G.: Oceanography. 2nd ed. New York, Prentice-Hall, 1977. An excellent text of oceanography and marine biology.

Hardy, A. C.: The Open Sea: The World of Plankton. New York, Houghton Mifflin Co., 1957. An interesting presentation of some of the microscopic forms living in the sea.

Jaeger, E. C.: The North American Deserts. Palo Alto, Stanford University Press, 1957. An interesting presentation of our deserts and the animals that live there.

Leopold, A. C.: The Desert. New York, New York, Time-Life Nature Library, 1961.

Ley, W.: The Poles. New York, Time-Life Nature Library, 1962.

MacArthur, R. H. and J. H. Connell: The Biology of Populations. New York, John Wiley & Sons, 1966. A text that emphasizes the population aspects of ecology.

Malin, J.: The Grasslands of North America. Lawrence, Kan., James Malin, 1956. Presents an interesting discussion of the ecology of the grassland biome that covers much of the midwest.

Odum, E. P.: Fundamentals of Ecology. 3rd ed. Philadelphia, W. B. Saunders Co., 1971. This popular text has an excellent account of each of the major biomes and life zones.

Reid, G. and R. Wood: Ecology of Inland Waters and Estuaries. 2nd ed. New York, Reinhold Publishing Corp., 1976. Describes the life zones in lakes, rivers and estuaries.

Richard, P. W.: The Tropical Rain Forest. New York, Cambridge University Press, 1952. A classic account of life in the jungle.

Tait, R. V. and R. S. DeSanto: Elements of Marine Ecology. 2nd ed. New York, Springer-Verlag, 1972. A good general textbook of marine ecology.

Van Dyne, G. (Ed.): The Ecosystem Concept in Natural Resource Management. New York, Academic Press, 1969. An excellent source of material relating to a variety of ecosystems.

Wetzel, R. G.: Limnology. Philadelphia, W. B. Saunders Co., 1975. A detailed treatment of the ecology of lakes, rivers and estuaries.

Wilson, D. P.: Life of the Shore and Shallow Sea. London, Nicholson and Watson, 1951. A beautifully illustrated description of animals and plants that live in the sea and on the shore.

CHAPTER 40

HUMAN ECOLOGY

The dramatic upsurge of popular interest in ecology in recent years has brought to everyone's attention the cycles of water and nutrients, the factors controlling biological productivity, the nature of food chains and the problems of global pollution. The explosive growth of the human population has forced us to consider the ultimate limitations of the earth, rather than simply local limitations. In his famous *Essay on Population* (1798), Thomas Malthus predicted that human populations would increase faster than their food supply. We are beginning to realize that the earth is simply a big spaceship with finite limits to its productivity and its ability to cope with pollutants. A key question in ecology is whether humans are in greater danger of running out of resources or of being overcome by polluted air and water.

The technology to solve many aspects of population and pollution problems is at hand; what is urgently needed is more general appreciation of the nature of the problems by informed laymen and a willingness on everyone's part to apply the principles of ecology for the ultimate benefit of mankind. In the immediate future, this probably depends not so much on advances in the environmental sciences but rather on the application of economics, law, politics, planning and other areas of humanistic science to ecological problems.

The threat to the future of mankind posed by overpopulation and its resulting ecological and economic problems continues to grow, but at long last governments in some portions of the world are beginning to take drastic steps to meet this problem. One of the most remarkable success stories of the family planning program occurred in mainland China where the crude birth rate dropped from 32 in 1970 to 19 in 1975. Over this same period, there was a 33 per cent decline in the birth rate in the United

States, coupled with a drop in the marriage rate, the increased employment of women and the increased enrollment of women in graduate and professional schools in this country. In many parts of the world the decreased rate of population growth is due not simply to falling birth rates but to increasing death rates from starvation. The world-wide demand for fish has led to overfishing, decreased catches and a greatly increased price of fish. Overgrazing, deforestation and overplowing have led to soil erosion, the encroachment of deserts and the abandonment of crop land. Under intense population pressure, both land-based and marine-based food resources give way. This has been seen in the most spectacular fashion in Ethiopia and in Haiti. In the once fertile island of Haiti, the land produces less and less each year because of deforestation and overplowing while the population soars. Haiti's remaining forests are being depleted and flash floods following the rain pour down scarred arroyos, emptying their silted waters into the sea. The eroded and leached mountain soil can no longer support tree growth. The danger of famine in Haiti in 1975 was avoided by massive help from the United States, which provided supplies to feed 120,000 people.

The rapid decrease in birth rates in many parts of the world reflects the widening availability of family planning services, such as contraception and abortion, and the growing desire of people to use these services. The use of the contraceptive pill has increased on every continent, along with greatly increased numbers of vasectomies for the male. The liberalization of abortion laws has made abortion available to women in two-thirds of the world. At present the regions with the highest birth rates are certain countries of Latin America, Africa and Southeast Asia. In contrast, the populations of East and West Germany, Austria, Luxembourg, the United Kingdom and Belgium have achieved stability. Certain other countries in Scandinavia, France and Italy are approaching a balance of births and deaths.

The severity of the food problem is reflected in the sharp decline in the reserves of grain. For two years, in 1966 and 1967, the United States gave one-fifth of its wheat crop to India to stave off major famines when there were two consecutive failures of the monsoon in India. Unilateral rescue efforts such as that are no longer possible because the food reserves are gone. In famine areas, such as Bangladesh in 1971–72 and again in 1974–75, large numbers of deaths—some 300,000 to 400,000 per year—have resulted from hunger. Truly draconian solutions to the overpopulation problem are now being advanced. For example, one of the Indian states had a law for the compulsory sterilization of any male with three or more living children and the compulsory abortion of any pregnancy that would result in a fourth child.

Earlier predictions that the world's population will double in the next generation may prove to be quite unrealistic. There are very few countries that could meet even the basic needs, such as food, water and energy, for the doubling of its population. In most countries, an increase would result in potentially unmanageable ecological, economic and political stresses.

There are many ways in which a knowledge of the principles of ecology can be used to further human society, one of the most important of which is the rational conservation of our natural resources. Conservation does not mean simply hoarding—not using the resources at all—nor does it imply a simple rationing of our supplies so that some will be left for the future. True conservation implies taking full advantage of our knowledge of ecology and managing our ecosystems so as to establish a balance of harvest and renewal, thus ensuring a continuous yield of useful plants, animals and other material. Further, this should at the same time guarantee the preservation of an environment with high quality that offers aesthetic and recreational uses as well as products.

The record of human squandering of natural resources is indeed a dark one. The slaughter of the bison that once roamed the western plains, the decimation of the whales, the depletion of our supplies of many kinds of fresh water and marine fishes, the extinction of the passenger pigeon and other birds, the razing of thousands of square miles of forest and the burning of even more by careless use of fire, the pollution of streams with sewage and industrial wastes, the careless cultivation of land that has resulted in the complete ruin of many square miles of land and the silting of streams—these are some of the more flagrant examples of natural resources wasted beyond hope of rapid reclamation. Although countermeasures have been taken by state and federal departments of conservation, the chief task at present is to make the population at large realize the urgency and magnitude of the job to be done and to obtain general support for the measures that must be taken. For many aspects of the conservation problem, additional basic ecological research is needed to determine the possible effects of a particular conservation measure on the ecology of the entire region.

40.1 AGRICULTURE

At the end of the Pleistocene, some 10,000 years ago, our human ancestors gradually changed from nomadic hunters and fishers into farmers. The earliest farmers raised wheat and barley and domesticated sheep and goats. Agricultural societies first appeared in the Near East in what is now Turkey, Iran and Iraq; they then spread across Europe and by about 6000 years ago had reached Britain. Agricultural communities apparently originated separately in Central and South America and in China and India. The consequences of the change from a hunter society to an agricultural society were truly profound. The populations were no longer wanderers; they could store food, accumulate possessions and own land. The change to an agricultural economy led to an increase in the population. Estimates place the number of human beings living some 10,000 years ago at about five million. In the next 4000 years the population increased to about 85 million, and by the time of Christ there were some 133 million people scattered over the globe. Thus the human populaion increased more than twenty-five fold in an 8000 year period.

The intensive cultivation of soil needed to feed this ever increasing population has taken its toll of the land available for farming. Wind and water have caused soil erosion through all geologic ages, but unwise farming and forestry practices have greatly increased the rate of erosion in certain parts of the globe. After decades of destructive exploitation of farmlands by planting one crop such as corn or cotton year after year, the soil conservation program sponsored jointly by federal and local agencies and based on sound ecological principles is beginning to be effective in countering erosion. The rotation of crops, contour farming, the establishment of windbreaks to prevent soil erosion by winds and the use of proper fertilizers to renew the soil are all measures that have proved effective in maintaining a balanced ecosystem. Successful farming must follow the principles of good land use. It is not conservation to reclaim marginal land for agricultural purposes or to build expensive dams and canals to irrigate land unless the land can produce crops that make the irrigation worthwhile.

If the grasslands of regions with slight rainfall are plowed and planted with wheat, a "dust bowl" will inevitably develop. If the land is kept as grassland and grazed in moderation, the soil will be kept in place, no dust bowl will develop and the land can be used economically year after year. By destroying the grass covering the soil, overgrazing can lead to destructive erosion just as improper plowing does. Overgrazing also leads to the invasion of the grassland by undesirable weeds and desert shrubs that are subsequently difficult to eradicate so that grass may grow again. Poor land use affects not only the unwise farmer but the whole population, which is eventually taxed to pay for rehabilitation of the land.

The ecologists specializing in the management of land have classified land into eight categories on the basis of its slope, kind of soil and natural biotic communities. These categories range from type I, which is excellent for farming and can be cultivated continuously, through three classes that can be used for farming only with special care and another three classes that are suitable only for perma-

nent pasture or forest, to type VIII, suitable only to be left as it is for game, for recreational and scenic purposes or for watershed protection (Fig. 40.1).

The control of insect pests by chemicals, such as DDT, must be carried out cautiously, with possible ecological upsets in mind. Spraying orchards, forests and marshes may destroy not only the pests but also useful insects such as honeybees that pollinate many kinds of fruit trees and crops and useful insect parasites.

DDT and related chemicals kill other animals in addition to insects; amphibians and reptiles are the most vulnerable vertebrates. The vertebrates are less sensitive than insects, and DDT applied at a level of about one pound per acre is effective in insect control without endangering the vertebrates. However, when applied at a level of 5 to 10 pounds per acre, some of the useful larger animals are killed along with the insects. Insect pests may actually increase after the use of DDT; the chemical may kill greater numbers of insect enemies of the pest than of the pests themselves. A number of strains of insects resistant to DDT have developed. However, the World Health Organization warns that banning DDT before cheap, safe and effective substitutes are found would be a disaster to world health. The agricultural production of the United States might decrease about 30 per cent if pesticides were banned, and the poor nations of the world would suffer even more from decreased food supplies and increased incidences of insect-borne diseases.

The great increase in agricultural productivity, the "green revolution," has been accomplished with the use of fertilizers, irrigation, pest control and genetic selection of specific

Figure 40.1 Classification of land according to its usefulness. Types I and II may be cultivated continuously; types III and IV are subject to erosion and must be cultivated with great care; types V, VI and VII are suitable for pasture or forests but not for cultivation; type VIII is productive only as a habitat for game. (U.S. Soil Conservation Service photograph.)

strains of crop plants. This industrialization of agriculture has required the input of considerable amounts of fuel energy. Doubling the crop yield, for example, is achieved only by a tenfold increase in the use of fertilizers, pesticides and fuel energy. As a result, industrialized agriculture is a major cause of both air and water pollution.

40.2 FORESTRY

The management of our forests is an important aspect of applied ecology. Until very recently most of the timber cut came from the accumulated growth of trees over the centuries. The lumber industry must adjust itself to bring the amount harvested into balance with the annual growth of the forests. Careful forest management has been carried on in Europe for many decades but is only beginning in this country. Proper timber management in our national and state forests has been important in demonstrating to the owners of private forests the results that can be obtained in this way. Since, in some regions, the desirable timber trees are members of the climax community, the ecological problem is simply to find the best way to speed the return of the climax community after the trees have been cut. In other regions, the desirable trees are earlier seral stages of the ecological succession, and forest management involves establishing means of preventing the succession to the climax community. This is also true of many kinds of animals; most game birds and many of the most valuable game fish are members of, and thrive best in, an early seral stage of their community.

40.3 WILDLIFE MANAGEMENT

The management of our fish and wildlife resources is a field of applied ecology supported by wide public interest, especially by sportsmen's clubs and associations. "Wildlife" used in this connection usually means game and fur-bearing animals. Since the various types of wildlife are adapted to different stages of ecological succession, their management requires a knowledge of and the proper use of these stages. As the Middle West became more and more intensively farmed, and the original forests and prairies were reduced to small patches, the prairie chickens and ruffed grouse that were adapted to these habitats were greatly decreased in numbers. However, this region has been partially restocked with game birds by introducing ring-necked pheasants and Hungarian partridges, which had become adapted to living in the intensively farmed regions of Europe.

Of the three general methods used to increase the population of game animals — laws restricting the number killed, artificial stocking and the improvement of the habitat — the last is the most effective. If the game habitats are destroyed or drastically altered, protective laws and artificial stocking of the region are useless. Protective laws must operate to prevent a population from getting too large as well as too small. Deer populations, in the absence of natural predators but subject to a constant, moderate amount of hunting, may increase to a point where they actually ruin the vegetation of the forest. Hunting should be restricted, of course, when populations are small and increased when they are larger. This requires accurate annual estimates of the population density of the game species.

Stocking a region artificially with game animals is effective only if they are being introduced into a new region or into one from which they have been killed off. Beavers, for example, were trapped to extermination in Pennsylvania, but restocking with Canadian beavers has been very successful, and it is estimated that there are some 15,000 to 20,000 beavers busy building dams in Pennsylvania. These are now an important factor in flood control in that region. The principles of population growth make it clear that if game animals of a certain species are already present, artificially stocking that region with additional members of the species will be futile. Stocking a region with a completely new species must be done cautiously, or the species may succeed so well as to become a pest and upset the biotic community, as has happened with rabbits in Australia and the English sparrow in the United States.

The management of the fish in a pond may be directed toward providing sport for hook and line fishermen or toward raising a crop of food fish and draining the pond at regular intervals to harvest the crop. To provide the best sport fishing, a lake or pond should be stocked with a combination of the sport fish and its natural prey; stocking a pond with large-mouth bass plus bluegills gives seven to 10 times more bass in three years than does stocking with bass alone. Stocking with fish must be done with care, for if a lake that already has about as many fish occupying a certain ecologic niche as possible is stocked with more of the same kind, there will be a decrease in the rate of growth and the average size of the fish. It has been found that sport fishing with hook and line is not likely to overfish a lake; the lake is more likely to be underfished, and the resulting crowding leads to a decrease in the average size of the fish population.

40.4 AQUACULTURE

The building of dams raises intricate ecologic problems, for dams may be intended for power, for flood control, for the prevention of soil erosion, for irrigation or for the creation of recreational areas. Since no one dam can satisfactorily accomplish all of these objectives, the primary objective must be clearly delineated and the secondary results must be understood. A contrast of two proposals for dealing with the same watershed (Table 40.1) shows that the multiple dam plan costs less, destroys a smaller area of productive farmland, impounds more water and is more effective in controlling floods and soil erosion. Since the productivity of a lake is inversely proportional to its depth, the series of small reservoirs would have greater productivity of fish.

To be most effective, the building of a dam must be accompanied by measures to decrease soil erosion upstream or the storage reservoir will fill with silt in a few years.

The management of the fish population in the lakes created by large dams is more difficult than the management of a pond. Sport fishing is usually very good when a dam has first been built, but gradually the silting up of the reservoir and the decrease in productivity change the nature of the fish community from game fish to less desirable catfish and shiners.

The term "ecologic backlash" has come into use to describe unforeseen, detrimental consequences of a project that either cancel out the anticipated gains of a particular modification of the environment or create more problems than were solved. A dam built primarily for hydroelectric power on the Zambesi River in Africa has had many undesirable side effects. The large lake shore created by the dam has greatly increased the habitat for the tsetse fly and has caused a severe outbreak of disease in cattle. Similar dams in other parts of Africa have increased the incidence of sleeping sickness among humans. Because the temperature of the water in the lake is high, the lake does not undergo spring and fall overturns, which mix the waters of lakes in the temperate zone. This greatly decreases the lake's productivity, and the fish catch has not compensated for the lost productivity of the grazing and agricultural land covered by the waters of the lake. The displacement of people from the rich river bank lands covered by the dammed waters to less suitable land has led to increased soil erosion. Some of the displaced people have moved to cities and caused further social upheaval. The regulated flow of water downstream of the dam has proved to be more damaging than the natural flooding that previously inundated the bottom lands annually. Maintenance of the fertility of these lands will require the input of expensive chemical fertilizers. There is evidence that, similarly, the Aswan dam on the Nile is not an unmixed blessing.

Lake Baikal in Siberia is a very deep and ancient lake that was formed by earth movement during the Mesozoic Era, some 150,000,000 years ago. It contains some 20 to 25 per cent of the world's fresh water supply. It is often called the "Australia" of fresh water lakes because of its long isolation and its unique collection of plants and animals. Three hundred and seventy-six of its 384 species of arthropods and 29 of its 36 species of fish are

TABLE 40.1 A COMPARISON OF A SINGLE MAIN RIVER RESERVOIR PLAN WITH A PLAN FOR MULTIPLE SMALLER HEADWATERS RESERVOIRS*

	Main Stream Reservoir	Multiple Headwaters Reservoirs
Number of reservoirs	1	34
Drainage area, square miles	195	190
Flood storage, acre feet	52,000	59,100
Surface water area for recreation, acres	1,950	2,100
Flood pool, acres	3,650	5,100
Bottom farmland inundated, acres	1,850	1,600
Bottom farmland protected, acres	3,371	8,080
Total cost	$6,000,000	$1,983,000

*From Odum, E. P.: Fundamentals of Ecology. 3rd ed. Philadelphia, W. B. Saunders Co., 1971. Data from Peterson, 1952; the costs in 1978 dollars would be greatly inflated!

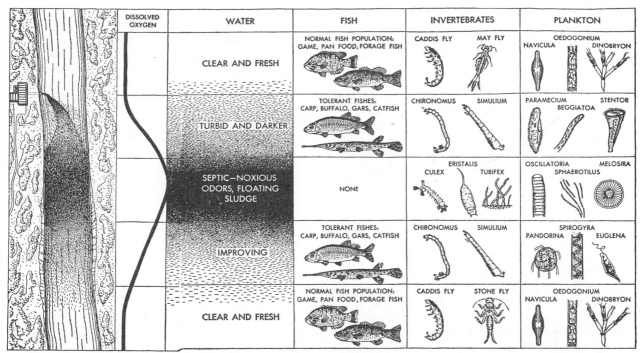

DISSOLVED OXYGEN	WATER	FISH	INVERTEBRATES		PLANKTON	
	CLEAR AND FRESH	NORMAL FISH POPULATION: GAME, PAN FOOD, FORAGE FISH	CADDIS FLY	MAY FLY	OEDOGONIUM NAVICULA DINOBRYON	
	TURBID AND DARKER	TOLERANT FISHES: CARP, BUFFALO, GARS, CATFISH	CHIRONOMUS	SIMULIUM	PARAMECIUM STENTOR BEGGIATOA	
	SEPTIC—NOXIOUS ODORS, FLOATING SLUDGE	NONE	ERISTALIS CULEX TUBIFEX		OSCILLATORIA MELOSIRA SPHAEROTILUS	
	IMPROVING	TOLERANT FISHES: CARP, BUFFALO, GARS, CATFISH	CHIRONOMUS	SIMULIUM	SPIROGYRA PANDORINA EUGLENA	
	CLEAR AND FRESH	NORMAL FISH POPULATION: GAME, PAN FOOD, FORAGE FISH	CADDIS FLY	STONE FLY	OEDOGONIUM NAVICULA DINOBRYON	

Figure 40.2 Pollution of a stream with untreated sewage and its subsequent recovery. As the amount of oxygen dissolved in the water decreases (*left*), fishes disappear and only organisms able to obtain oxygen from the surface, or those that tolerate low oxygen tensions, are able to survive. When the sewage has been reduced by bacteria, the population of animals and plants in the stream returns to normal. (From Odum, E. P.: Fundamentals of Ecology. 3rd ed. Philadelphia, W. B. Saunders Co., 1971. After Eliassen.)

found nowhere else in the world. The ecology of Lake Baikal is now threatened by Soviet industry. Its water is polluted by wastes discharged from an enormous paper factory and the trees of its watershed are being felled for conversion to paper. The soil in the watershed is very thin and soil erosion occurs rapidly after the trees are felled. The soil is washed into the lake, making its originally crystal clear waters turbid and further altering the ecosystem. The region south and east of the lake is desert and with this change in the watershed the desert may spread to the shores of the lake.

The three chief sources of stream pollution are industrial materials that either are directly toxic themselves or reduce the oxygen supply in the water; sewage and other materials that decrease the oxygen content of the water and introduce bacteria and other septic organisms (Fig. 40.2); and turbidity caused by soil erosion in the watershed. As the silt settles out downstream, it may cover up the spawning grounds of fish and have other direct deleterious effects. Erosion can be prevented by proper soil management, industrial wastes can be prevented by suitable design of the manufacturing process and properly treated sewage can be emptied into a stream without deranging

its ecological relations. The proper design and operation of "waste management parks" can utilize heated water from power plants (thermal pollution) and certain types of treated and diluted domestic and industrial wastes as energy subsidies for adapted species of fish and other animals. These can ultimately be harvested to provide food and other useful products.

40.5 MARINE FISHERIES AND ESTUARINE MARICULTURE

The primary productivity of the sea, as measured by the pounds of organic carbon produced per year per acre of surface, is very high. The productivity of the western Atlantic off the coast of North America is 2.5 to 3.5 tons of organic carbon per acre and that of Long Island Sound is 2.5 to 4.5 tons per acre. The productivity of the average forest is about 1 ton per acre; most cultivated land fixes only about $3/4$ ton of organic carbon per acre; and only the rich, intensively cultivated cornfields of Ohio produce as much as 4 tons per acre. Despite this high productivity, man's actual harvest from the ocean, in terms of pounds of fish caught per acre of surface, is very low. Only the rich fishing grounds of the North Sea produce as much as 15 pounds of fish per acre. The ecological reasons for this are clear; the fish are secondary or tertiary consumers and are on top of a vast pyramid of producers. There are many organisms competing for the food energy fixed by the algae in addition to the edible fish and crustacea harvested from the sea.

The annual harvest of marine fish over the entire globe was about 18 million tons in 1938, reached 70 million tons in 1970 but has been declining since then. Almost 80 per cent of the total comes from the North Atlantic, the northern and western Pacific and just off the west coasts of Peru and Ecuador in South America. About half of the total harvest is used as human food; the remainder is used as food for pets, poultry and livestock.

Humans could undoubtedly recover for their use much more of the biological productivity of the sea. Although we might be reluctant to eat marine algae ourselves, the algae might be filtered from sea water and processed so as to be suitable feed for cattle or some other gastronomically acceptable animal.

Careful studies by the U.S. Fish and Wildlife Service of the fish population of George's Bank and other commercially fished areas have led to recommendations about the rate of fishing and the size of nets used that ensure that the fish are harvested at an optimal size for greatest yield at present and in the future. These areas, which had been fished so extensively that some of the most desirable species were reduced greatly in numbers, are now beginning to revive under careful management.

The shellfish—oysters, clams, shrimps and lobsters—present somewhat different and more difficult problems, for their habitat is more limited than that of commercial fish, and they are more affected by adverse environmental changes. Oysters, whose food consists of algae or detritus of a certain size filtered from the sea water by their gills, are unable to use algae of a different size. Oysters were unable to survive in certain bays of Long Island Sound when ducks were raised in large numbers commercially on the adjacent shore. The wastes from the duck farms were washed into the bays, and the addition of this organic matter changed the community ecology in such a way that the normal food of the oyster, diatoms, was replaced by other algae that could not be used by the oysters. Once an oyster bed has been seriously depleted it may fail to recover even if seeded with oyster larvae, because the larvae require a favorable surface for attachment and the most favorable is the shell of an old oyster.

In commercial oyster farms, the larvae are provided with artificial sites for attachment. Once they have become attached they may be moved to other waters, even from one ocean to another, to complete their growth in waters that are favorable for feeding although not favorable for the reproduction of the species. The intensive cultivation (mariculture) of oysters, clams and other shellfish in estuaries and other marine environments has provided an important source of protein-rich food in Japan, Indonesia and Australia. It does require a considerable input of energy, at present largely in the form of human hand labor. By using the raft method of oyster culture, rather than simply harvesting the wild population, the production of oyster meat in a given area can be increased fivefold to tenfold.

40.6 MINERAL RESOURCES

Because it was assumed that the supply of mineral resources would last for centuries to come and that there was no way of conserving them, little attention has been paid to this aspect of conservation. However, both of these assumptions are quite wrong. The relationship between the resources available and their utilization is expressed by the demographic quotient, Q:

$$Q = \frac{\text{total resources available}}{\text{population density} \times \text{per capita consumption}}$$

The quality of modern life goes down as the quotient goes down, and it is decreasing rapidly at present. Even if the total resources available could be kept constant by recycling, the value of Q decreases as the population density increases and as the per capita consumption increases. The utilization of natural resources in the United States is increasing at a rate of about 10 per cent per year. The demands for iron, copper, lead, tin and even the relatively abundant metal aluminum are such that the country is no longer self-sufficient either in mineral resources or in the fossil fuels needed to extract the new mineral resources. At the moment biotic and mineral resources are more critical and will be exhausted sooner than energy sources. Natural gas may soon be depleted and oil will follow a few decades later; but coal will last for a considerably longer time, hopefully until completely safe methods for producing large amounts of electricity in nuclear reactors have been devised. The limiting factor in the utilization of industrial energy will be pollution rather than supply. The report of the Committee on Pollution of the National Academy of Sciences, entitled "Resources and Man (1969)," made a total of 26 recommendations. These can be summarized in the statement that control of the size of human populations and improved management and recycling of our resources are urgently needed, beginning immediately.

40.7 PUBLIC HEALTH

Many aspects of the field of public health require the application of ecological principles; the prevention of the spread of diseases carried by animals is an ecological as well as a medical problem. The most effective way of eliminating malaria, for example, is to eliminate the particular species of mosquito that is the vector of the malaria parasite, yet this must be done without destroying the useful insects of the region. The mosquitoes that transmit malaria in different parts of the world have quite different ecologic niches, and, therefore, measures that may be effective in mosquito control in one region may be quite ineffective in another. The malaria of the southeastern United States is transmitted by mosquitoes living in marshes; Italian malarial mosquitoes live in cool running water in the uplands; and Puerto Rican malarial mosquitoes live in brackish (slightly salty) water. Careful ecological surveys of each region are necessary to formulate the proper measures to control the insects.

The size of the populations of rats, mice and many insect pests increases with the size of cities and the correlated tendency toward the development of slums in the older parts of the town. A survey in England in 1953 revealed that only 0.1 per cent of the houses in towns with less than 25,000 houses were infested with bedbugs, but over 1.0 per cent were infested in towns with more than 100,000 houses! Careful ecological studies in Baltimore showed that although professional crews of rat trappers might catch as much as half the rat population, it quickly returned to its former level. Cats proved to be much overrated as rat predators and were not effective in controlling the rat population. However, by changing the essential elements of the rats' habitat, by improving sanitation, thus decreasing the garbage on which the rats fed and the wastes in which they hid, the rat population was reduced to about 10 per cent of its former size. It remained at this lower level because that was the total number of rats that could survive in the altered environment.

40.8 POLLUTION

The ultimate cause of pollution is people, and as the number of people increases there is a corresponding increase in the amount of pollution. As the number of people has increased and the amount of energy used by each person has also increased, the total energy demand has increased at a very high rate. This is reflected in the air pollution caused by the exhaust stacks of industrial plants and electric generating plants burning fossil fuels, and by the exhaust of automobiles transporting people to and from their work and play.

The National Academy of Sciences, in the report by the Committee on Pollution, defined pollution as follows:

Pollution is an undesirable change in the physical, chemical or biological characteristics of our air, land and water that may or will harmfully affect human life or that of desirable species; industrial processes, living conditions and cultural assets; or that may or will waste or deteriorate our raw material resources. Pollutants are residues of the things we make, use and throw away. Pollution increases not only because as people multiply the space available to each person becomes smaller, but also because the demands per person are continually increasing so that each throws away more year by year. As the earth becomes more crowded there is no longer an "away." One person's trash basket is another's living space.*

Pollution is undoubtedly the most important limiting factor for humans. It is difficult to put a price tag on the cost of pollution, but it is clear that pollution places a burden on human society (1) from the loss of resources owing to unnecessarily wasteful exploitation, (2) from the cost of abating and controlling pollution and (3) from the cost in human health.

From an ecological viewpoint two types of pollution can be recognized—pollution involving biodegradable pollutants and pollution involving nondegradable pollutants. Biodegradable pollutants, such as domestic sewage, can be decomposed rapidly by natural processes or by carefully engineered systems, such as a community's sewage treatment plant. Problems arise when the input of degradable pollutants into the environment exceeds the environment's capacity to decompose or disperse them. The problems with disposable sewage result in general from the fact that urban populations have grown much faster

---

*From Resources and Man (1969), the National Research Council, National Academy of Sciences, Washington, D.C.

than their sewage treatment facilities. Degradable pollutants can be dealt with by a combination of mechanical and biological treatments; but there are limits to the total amount of organic matter that can be decomposed in a given area, and there is an overall limit to the amount of carbon dioxide (one of the decomposition products) that can be released into the air.

The nondegradable pollutants include metals, such as mercury, trace metals, steel and aluminum cans, and organic chemicals, such as DDT, that are degraded only very slowly. Some of these nondegradable pollutants accumulate as they proceed along the food chains. Dealing with these pollutants is a much more difficult and expensive problem. Some of these pollutants, such as aluminum cans, are actually a resource if they are recycled instead of thrown away. Others, such as DDT, can be replaced by more degradable substitutes when they are found.

The design and operation of an efficient waste treatment plant is a very complex subject, beyond the scope of this book, but it is obvious that such operations must be carried out with the best possible application of ecological principles. It is perfectly feasible to produce drinkable water from sewage waste, but this is a very expensive procedure. (However, it is less expensive than producing drinking water from sea water.)

The pollution of the air in industrialized countries such as the United States is a problem of first magnitude. Although the pollution of the air over the entire country is serious enough, the most striking aspect of air pollution is the local concentration of pollutants that may occur over cities, such as Los Angeles and New York, during a temperature inversion. This occurs when air is trapped under a warm upper layer that prevents a vertical rise of the pollutants. Certain combinations of pollutants may react in the environment to produce additional pollution. Certain components in automobile exhaust can combine in the presence of sunlight to produce even more toxic substances called **photochemical smog.** One of the components of photochemical smog blocks a key reaction in photosynthesis and will kill plants by inhibiting their production of food.

The subject of pesticides and herbicides has been an extremely controversial one in the past decade. Early in this century, when farms were small and diversified and there was plenty of farm labor, the increase in the numbers of insect pests was blocked by simple inorganic salts. The industrialization of agriculture that has occurred within the last few decades required, and was made possible by, broad spectrum poisons, such as the organochlorides and organophosphates. Unfortunately, these poisons were used much too enthusiastically, and since many are persistent and are degraded very slowly, these broad spectrum poisons have accumulated to such an extent in many parts of the environment that they can no longer be used.

Two highly undesirable effects of the saturation of the environment with DDT and other organochlorides are the development of strains of insects that are resistant to the DDT and the accumulation of these toxic materials in the food chain. Most of the problems can be traced to the fact that new insecticides have been tested, usually superficially, at the level of the organism against which they are directed and then are used at the ecosystem level without controlled tests of their effects on the entire ecosystem.

It is clear that if humans are to raise enough food crops to supply our burgeoning population we must have some way of dealing with the insect pests that are competing for this same supply of food. To protect ourselves against the diseases transmitted by insects, such as malaria, yellow fever and dengue (mosquitoes), sleeping sickness (tsetse flies), plague (fleas), typhus (fleas, ticks, lice or mites) and typhoid (frequently transmitted by flies), we must continue to use effective insecticides. Chemical insecticides must be used only in appropriate places and amounts and with adequate controls. Certain kinds of insects can be removed by introducing appropriate predators or parasites, by supplying hormonal factors, such as juvenile hormone, that will prevent the insect from completing its life cycle, by using sex lures and other pheromones that will confuse the reproductive efforts of the insects, or by sterilizing large numbers of male insect pests by chemicals or radiation and then releasing the sterile males into the natural environment. For other insects, however, the only effective counter measures are chemical insecticides.

If all chemical pesticides were banned, the agricultural production of the United States would decrease by 30 per cent or more, according to the United States Department of Agriculture. Some partisans have claimed that 50 million Americans would be in danger of starving to death if chemical pesticides were withheld from agriculture. The poor and undeveloped nations of the world would probably suffer even more from disease and hunger if DDT were banned. In pest control programs on many kinds of crops, the organochlorine pesticides are being replaced by organophosphorus and carbamate compounds, which are less resistant and more easily biodegradable. However, larger amounts of these pesticides are usually required to control pests in a given area, and they may be more toxic to humans than DDT. Maintaining agricultural production with the aid of chemical pesticides and herbicides without damaging the ecosystem irretrievably will be a difficult task.

Recently much interest has centered on the discovery of abnormally large quantities of mercury in both fresh water and marine fish. The ocean contains a great many metals, including gold, silver and platinum as well as mercury. These metals have been washed into the ocean over the course of billions of years and are not primarily the result of industrial pollution. In contrast, the mercury in certain freshwater ecosystems such as the Great Lakes is primarily the result of industrial pollution, and measures can and should be taken to eliminate further input. Mercury from sea water undoubtedly accumulates in the flesh of all kinds of fish. It is also present in flesh of lobsters, shrimp, clams, oysters and other shellfish. It should be noted that one commonly prescribed medicine, calomel, contains mercury.

Other pollutants in the environment are the various radioactive materials, such as strontium-90. Strontium-90 is a beta emitter, having a half-life of 28 years. Strontium is very similar to calcium in its chemical and physical properties and moves with it in its natural cycle. Calcium and strontium are washed out of rocks and move down rivers to the ocean. In the far north, radioactive fall-out of strontium-90 has been absorbed by lichens, and the caribou and reindeer that eat the lichens have concentrated the strontium in their flesh. Humans who eat the flesh of the animals or drink milk from them may then accumulate strontium in their bones. In other parts of the world strontium-90 has accumulated in the bones of both adults and children who obtain it in cow's milk; the cows obtained it by eating vegetation polluted with strontium-90.

Little strontium-90 has been added to the environment since 1963. If the ban on nuclear testing can be continued, strontium-90, and the related cesium-137, will gradually disappear from the environment. Both are byproducts of fission reactions and enter the environment through fall-out or through faulty waste disposal; they are especially hazardous because they accumulate in successive living organisms along the food chain. Studies in Great Britain showed that strontium-90 was 21 times more concentrated in grass than in soil and some 700 times more concentrated in sheep than in the grass.

Most human beings now live in urban areas, and urban areas not only pollute the air but modify the climate. Because of the absorption of solar radiation by vertical surfaces and the production of heat by the machines in a city, the air temperature in a city will be 1 or 2° C. higher and the humidity will be about 6 per cent lower than in the surrounding countryside. Because of particulate air pollution, cloudiness will be 10 per cent greater and there will be 30 to 100 per cent more fog, the higher value occurring in the winter. Precipitation will be 10 per cent greater, there will be 15 per cent less sunshine and 10 to 30 per cent less ultraviolet radiation in the city than in the countryside.

In 1970, the average population density in the United States was one person to every 10 acres of ice-free land area. The average population of the world was at the same level. Even if the anticipated reduction in birth rate occurs, the population of the United States will double in the next 30 years. By the year 2000 there will be only five acres of land for every man, woman and child. Although as little as $1/3$ acre can produce enough calories to sustain a person, the kind of optimal quality diet that includes meats, fruits and vegetables requires at least $1\frac{1}{2}$ acres per person. Another acre per person is needed to produce the paper, wood and cotton needed by each person and an additional $1/2$ acre is needed for the roads, airports, buildings and other areas not involved in food production. Only about 2 acres per person remain for all the many other uses that humans have for land. An affluent developed nation actually requires more space and resources per person than an undeveloped nation. From this it follows that the key in determining the optimum human population density should be the adequate and usable *pollution-free living space* and not simply the food needed. E. P. Odum, in his book *Fundamentals of Ecology* (1971), has said, in essence, that this earth can feed more warm bodies sustained as so many domestic animals in a polluted feed lot than it can support quality human beings who have a right to a pollution-free environment, a reasonable chance for personal liberty and a variety of options for the pursuit of happiness. To obtain this latter goal, at least one-third of all land should remain protected for use as national, state or municipal parks, greenbelts, refuges, wilderness areas and so on. If the land is under private ownership, it should be protected by zoning or other legal means.

40.9   HUMAN ECOLOGY

No great amount of thought is required to realize that the ecological principles apply to human populations as well as to populations of animals and plants. Human ecology deals not only with the dynamics of human populations but also with the relationship of human beings to the many physical and biotic factors that impinge upon him. By appreciating that human populations are a part of larger units—of biotic communities and ecosystems—we can deal with our own special problems more intelligently. Human beings have wittingly and unwittingly exercised a great deal of control over their environment and have modified the communities and ecosystems of which they are a part. However, this control is far from complete, and humans must, like other animals and plants, adapt to those situations that we cannot change. By understanding and cooperating with the various cycles of nature, we will have a better chance of surviving in the future than if we blindly attempt to change and control them.

The human population is clearly in danger of multiplying beyond the ability of the earth to support it. In the past several centuries, the population of the world has increased tremendously as new territories have been opened for exploitation and as methods of producing food have become more efficient. Most biologists and social scientists believe that the danger of human overpopulation is both great and imminent. It has been amply shown that the Malthusian Principle (that populations have an inherent ability to grow exponentially) is true for organisms generally, and the growth of the human population in the past 300 years follows an exponential curve. The productivity and carrying capacity of the earth for human beings can be maintained and perhaps increased somewhat, but eventually the human biomass must be brought into equilibrium with the space and food available. The human population is probably in greater danger of running out of drinkable water and breathable air than it is of exhausting its food supply! Some limitation of human reproduction is clearly inevitable. It remains to be seen whether the human species will do this voluntarily or involuntarily.

ANNOTATED REFERENCES

Adams, A.: Eleventh Hour: A Hard Look at Conservation. New York, G. P. Putnam's Sons, 1970. An overview of the conservation movement.

Benarde, M. A.: Our Precarious Habitat, An Integrated Approach to Understanding Man's Effect on His Environment. New York, W. W. Norton & Co., 1970. This presents another viewpoint of the problems arising from man's misuse of the environment.

Carson, R. L.: Silent Spring. Boston, Houghton Mifflin Co., 1962. The beautifully written book that did much to touch off popular concern about preserving our environment.

Cox, G. W. (Ed.): Readings in Conservation Ecology. New York, Appleton-Century-Crofts, 1969. An anthology; an excellent source book for topics related to conservation of natural resources.

Doutt, R. L. and W. W. Kilgore: Pest Control. New York, Academic Press, 1967.

Draybill, H. F. (Ed.): Biological Effects of Pesticides in Mammalian Systems. New York, N.Y. Acad. Sci., 1969. Two excellent books on pests and their control by pesticides.

Ehrlich, P. and A. Ehrlich: Population Resources and Environment: Issues in Human Ecology. San Francisco, W. H. Freeman & Co., 1970. A thorough ecological analysis of man's impact on the productivity of the biosphere, presented by two of the outstanding figures in the field.

Graham, F., Jr.: Since Silent Spring; Boston, Houghton Mifflin Co., 1970. A follow-up, eight years later, of what had happened in popular ecology since the publication of Rachel Carson's book.

Hardin, G. (Ed.): Population, Evolution and Birth Control. San Francisco, W. H. Freeman & Co., 1969. A thoughtful analysis of the present human predicament.

Marine, G.: America The Raped. New York, Simon and Schuster, 1969. Some examples of the outrageous devastation resulting from poor ecological planning.

McHarg, I.: Design With Nature. Garden City, N.Y., Natural History Press, 1969. An outline of the author's proposals for improving the design of human communities.

Miller, M. W. and G. G. Berg (Eds.): Chemical Fallout. Springfield, Ill., Charles C Thomas, 1969. A collection of two dozen papers discussing pesticides — how they work and their effects on animal and human populations.

Roueche, B.: What's Left. Boston, Little, Brown & Co., 1968. A particularly well written account of the damage to our environment by lack of consideration of ecologic issues.

GLOSSARY

Important and widely used terms and concepts whose meanings or pronunciations may not be well known are included in this glossary. Relatively common and better known terms have not been included, nor have many of those not so widely used; pages on which these are defined can be found in the index. Names of the major taxa, generally down to class, are included; others can be found through the index.

A simplified, phonetic respelling of terms appears in parentheses. Unless indicated to the contrary, an unmarked vowel ending a phonetic syllable is long; an unmarked vowel in a syllable ending with a consonant is short. In cases where these rules do not apply, the vowel is marked long (ē) or short (ĕ). Primary (') and secondary (") accents are indicated.

Derivations of terms are given in brackets. The student's understanding of the terms will be facilitated if he studies these and learns to recognize the common roots used in etymology.

abductor (ab-duk'tōr) [L. *ab* away + *ducere* to lead]. A muscle that draws a structure away from a point of reference, such as the midventral line of the body.

absorption (ab-sorp'shun) [L. *ab* away + *sorbere* to suck in]. The taking up of a substance, as by the skin, mucous surfaces or lining of the digestive tract.

absorption spectrum A measure of the amount of energy at specific wave lengths that has been absorbed as light passes through a substance. Each type of molecule has a characteristic absorption spectrum.

accommodation (ah-kom"o-da'shun) [L. *accommodere* to fit to]. Adjustment of the lens of the eye for various distances; accomplished in fishes and amphibians by lens movements and in higher vertebrates by a change in the shape of the lens.

acetylcholine (as"ĕ-til-ko'lēn). The acetic acid ester of the organic base choline, normally secreted at the ends of many neurons; responsible for the transmission of a nerve impulse across a synapse.

achondroplasia (ah-kon"dro-pla'ze-ah) [*a* neg. + Gr. *chondros* cartilage + *plassein* to form + *-ia*]. A hereditary disturbance of growth and maturation of the bones which results in inadequate bone formation and a characteristic type of dwarfism.

acid (as'id) [L. *acidus*, from *acere* to be sour]. A substance whose molecules or ions release hydrogen ions (protons) in water. Acids have a sour taste, turn blue litmus paper red and unite with bases to form salts.

acidosis (as"i-do'sis). A pathologic condition resulting from the accumulation of acid or the loss of base in the body; characterized by an increased hydrogen ion concentration (decreased pH).

acoelomate (a-se'lo-mate) [*a* neg. + *koilia* cavity]. The condition of having no body cavity, as in flatworms.

acromegaly (ak"ro-meg'ah-le) [Gr. *akron* extremity + Gr. *megalē* great]. A condition characterized by overgrowth of the extremities of the skeleton, the nose, jaws, fingers and toes. This may be produced by excessive secretion of growth hormone from the pituitary.

acron (ak'ron) [Gr. *akron* extremity). The anterior nonsegmental part of the body of a metameric animal.

acrosome (ak'ro-sōm) [*acro-* + Gr. *soma* body]. A caplike structure covering the head of the spermatozoon.

actin (ak'tin) [Gr. *aktis* a ray]. Minute protein filaments found in most contractile cells. In striated muscle cells, thin filaments of actin overlap and alternate with thicker filaments of myosin to form myofilaments. The interactions between thick and thin filaments lead to muscle contraction and relaxation.

actinopterygian (ak″tĭ-nop′tēr-ij-ĭ-an) [Gr. *actinos* ray + *pterygos* wing]. A member of a subclass of the Osteichthyes in which the fins are supported by bony rays; includes sturgeons, gars and teleosts.

action potential A slight current which can be detected with appropriate sensitive devices when any tissue becomes active—as when a muscle contracts, a gland secretes or a nerve conducts an impulse.

active transport The transfer of a substance into or out of a cell across the cell membrane against a concentration gradient by a process which requires the expenditure of energy.

adaptation The fitness of an organism for its environment; the process by which it becomes fit; a characteristic which enables the organism to survive in its environment.

adaptive radiation The evolution from a single ancestral species of a variety of species which occupy different habitats.

adaptive value A measure of the success of one genotype in a population relative to the most successful one, which is given the adaptive value of 1.0. The complement of selection coefficient (q.v.).

adductor (ad-duk′ter) [L. that which draws toward]. A muscle that draws a structure toward the midventral line of the body, as in bringing the legs together or closing a shell.

adenine (ad′ĕ-nīn) [Gr. *aden* a gland]. A purine (nitrogenous base) which is a component of nucleic acids and of nucleotides important in energy transfer—adenosine triphosphate (ATP), adenosine diphosphate (ADP) and adenylic acid (AMP).

adenosine triphosphate (ah-den′o-sin). An organic compound containing adenine, ribose and three phosphate groups; of prime importance for energy transfers in biological systems.

adipose (ad′ĭ-pōs) [L. *adiposus* fatty]. Referring to the tissue in which fat is stored or to the fat itself.

aerobic (a-er-o′bik) [Gr. *aero* air]. Growing or metabolizing only in the presence of molecular oxygen.

aestivation (es″tĭ-va′shun) [L. *aestivus* summer]. The dormant state of decreased metabolism in which certain animals pass hot, dry seasons.

afferent (af′er-ent) [L. *ad* to + *ferre* to carry]. Conveying toward a center; designating vessels or neurons that transmit blood or impulses toward a point of reference; afferent neurons are sensory neurons conducting impulses toward the central nervous system.

agglutination (ah-gloo″tĭ-na′shun) [L. *agglutinare* to glue to a thing]. The collection into clumps of cells or particles distributed in a fluid.

agnatha (ag′na-thah) [*a-* without + Gr. *gna-*

thos jaw]. The jawless fishes. A class of vertebrates including lampreys, hagfishes and many extinct forms.

allantois (ah-lan′to-is) [Gr. *allas* sausage + *eidos* form]. One of the extraembryonic membranes of reptiles, birds and mammals; a pouch growing out of the posterior part of the digestive system and serving as an embryonic urinary bladder or as a source of blood vessels to and from the chorion or placenta.

allele (ah-lēl′) [Gr. *allēlōn* of one another]. One of a group of alternative forms of a gene that may occur at a given site (locus) on a chromosome.

allergy A hypersensitivity to some substance in the environment, manifested as hay fever, skin rash or asthma.

allopolyploidy (al′o-pol″e-ploi-de) [Gr. *allos* other + *polys* many + *ploos* folds]. A type of speciation involving the hybridization of two species and the continuation of the hybrids as a third species.

alveolus (al-ve′o-lus) [L. dim. of *alveus* hollow]. A small saclike dilatation or cavity; the terminal chamber of air passages in the mammalian lung.

amebocyte (ah-me′bo-sĭt″) [Gr. *amoibe* change + *kytos* hollow vessel]. In a metazoan, a cell that wanders through the body in ameboid fashion.

ameboid motion (ah-me′boid) [Gr. *amoibé* change + *eidos* form]. The movement of a cell by means of the slow oozing of the cellular contents.

amensalism (a-men′sal-izm). A relationship between two species whereby one is adversely affected by the second, but the second species is unaffected by the presence of the first.

amino acid (am′ĭ-no). An organic compound containing an amino group ($-NH_2$) and a carboxyl group ($-COOH$); amino acids may be linked together to form the peptide chains of protein molecules.

amnion (am′ne-on) [Gr. *amnion* lamb]. One of the extraembryonic membranes of reptiles, birds and mammals; a fluid-filled sac around the embryo.

amniote (am′ne-ōt) [Gr. *amnion* lamb]. A vertebrate characterized by having an amnion during its development; a reptile, bird or mammal.

amphibian (am-fib′e-an) [Gr. *amphi* on both sides + *bios* life]. A member of a class of vertebrates in which the larvae are usually aquatic and the adults terrestrial; includes frogs, toads, salamanders and their allies.

amphiblastula (am″fĭ-blas′tu-lah) [Gr. *amphi* on both sides + *blastos* germ]. Larval sponge blastula composed of two kinds of cells.

amylase (am′ĭ-lās) [L *amylum* starch]. An enzyme that catalyzes the hydrolysis of starches; it cleaves α-1 → 4 glucosidic bonds of polysaccharides.

anabolism (ah-nab′o-lizm) [Gr. *anabolē* a

throwing up]. Chemical reactions in which simpler substances are combined to form more complex substances, resulting in the storage of energy, the production of new cellular materials and growth.

anaerobic (an″a-er-o′bik) [*an* neg. + Gr. *aero* air + *bios* life]. Growing or metabolizing only in the absence of molecular oxygen.

analogous (ah-nal′o-gus) [Gr. *analogos* according to a due ratio, conformable, proportionate]. Similar in function or appearance but not in origin or development.

anamniote (an-am′ne-ōt) [Gr. *an* without + *amnion* lamb]. A vertebrate characterized by the absence of an amnion during its development; a fish or amphibian.

anaphase (an′ah-fāz) [Gr. *ana* up, back, again + *phasis* phase]. Stage in mitosis or meiosis, following the metaphase, in which the chromosomes move apart toward the poles of the spindle.

androgen (an′dro-jen) [Gr. *andros* man + *gennan* to produce]. Any substance which possesses masculinizing activities, such as testosterone or one of the other male sex hormones.

anion (an′i-on) [Gr. *ana* up + *iōn* going]. An ion carrying a negative charge.

annelida (a-nel′ĭ-dah) [F. *anneler* to arrange in rings]. The phylum of segmented worms, including earthworms, clamworms and leeches.

antenna (an-ten′ah) [L. a spar projecting from the mast of a boat, supporting a sail; a sail yard]. A projecting, usually filamentous organ equipped with sensory receptors.

anthozoa (an″tho-zo′ah) [Gr. *anthos* flower + *zoia* animal]. A class of coelenterates with large polyps and no medusa stage; includes corals.

anthropoid (an′thro-poid″) [Gr. *anthropos* man + *oeides* like]. A member of the more advanced of two primate suborders; a monkey, ape or human being.

antibiotics (an″ti-bi-ot′ik) [Gr. *anti* against + *bios* life]. Substances produced by microorganisms which have the capacity, in dilute solutions, to inhibit the growth of or to destroy bacteria and other microorganisms; used largely in the treatment of infectious diseases of man, animals and plants.

antibody (an′ti-bod″e). A protein produced in response to the presence of some foreign substance in the blood or tissues.

anticodon (ăn′ti-kōdŏn). A sequence of three nucleotides in transfer RNA that is complementary to, and combines with, the three nucleotide codon on messenger RNA, thereby binding the amino acid-transfer RNA combination to the mRNA.

antidiuretic hormone (an″tĭ-di″u-ret′ik hor′-mōn) [Gr. *anti* against + *diouretikos* promoting urine; *hormaein* to set in motion, spur on]. A hormone produced in the hypothalamus and stored and released from the posterior lobe of the pituitary which controls the rate at which water is reabsorbed by the kidney tubules.

antigen (an′tĭ-jen) [Gr. *anti* against + *gennan* to produce]. A foreign substance, usually protein or protein-polysaccharide complex in nature, which elicits the formation of specific antibodies within an organism.

antimetabolites (an″tĭ-mĕ-tab′o-līt) [Gr. *anti* against + *metaballein* to turn about, change, alter]. Substances bearing a close structural resemblance to ones required for normal physiological functioning; exert their effect by replacing or interfering with the utilization of the essential metabolite.

antitoxin (an″ti-tok′sin) [Gr. *anti* against + *toxicon* poison]. An antibody produced in response to the presence of a toxin (usually protein) released by a bacterium.

aorta (a-or′tah) [Gr. *aortē* the aorta]. One of the primary arteries of the body, e.g., the ventral aorta of fishes which distributes blood to the gills or the dorsal aorta which distributes blood to the body.

aphasia (ah-fa′ze-ah) [*a* neg. + Gr. *phasis* speech]. Inability to recognize certain kinds of symbols (such as writing or speech) due to injury or disease of the brain centers.

aplacophora (a plah kof′o-rah) [*a* neg + Gr. *plax* plate + *pherein* to bear]. A class of wormlike mollusks having no shell.

apodeme (ap′o-dēm) [Gr. *apo* from + *dermas* body]. A tendon-like extension of cuticle that crosses arthropod joints and onto which muscle fibers attach.

apoenzyme (ap″o-en′zīm) [Gr. *apo* from + *en* in + *zymé* leaven]. Protein portion of an enzyme; requires the presence of a specific coenzyme to become a complete functional enzyme.

arachnida (ah-rak′nĭ-dah) [Gr. *arachnē* spider]. Class of arthropods containing the spiders, mites and scorpions. Anterior half of body bears four pairs of legs and no antennae.

archenteron (ar-ken′ter-on) [Gr. *archē* beginning + *enteron* intestine]. The central cavity of the gastrula, lined with endoderm, which forms the rudiment of the digestive system.

archinephric duct (ar″ke-nef′ric) [Gr. *archaios* ancient + *nephros* kidney]. The primitive kidney duct; drains the pronephros, mesonephros and opisthonephros.

arteriole (ar-te′re-ōl) [Gr. *arteria* artery]. A minute arterial branch, especially one just proximal to a capillary.

artery A vessel through which the blood passes away from the heart to the various parts of the body; typically has thick, elastic walls.

arthropoda (ar-throp′o-dah) [Gr. *arthron* joint + *podos* foot]. The phylum of invertebrates with jointed legs, including insects and crustaceans.

asconoid (as′kon-oid) [Gr. *askos* bladder +

eidos form]. In sponges, a simple vaselike body plan with no folding of the body wall.

asteroidea (as"ter-oi'de-ah) [Gr. *asteroeides* like a star]. The starfishes a subclass of common and familiar echinoderms.

atom The smallest quantity of an element which can retain the chemical properties of the element, composed of an atomic nucleus containing protons and neutrons together with electrons that circle the nucleus in specific orbits.

atomic orbital Distribution of an electron around the atomic nucleus.

atresia (ah-tre'ze-ah) [Gr. *a* neg. + *trēsis* a hole + -*ia*]. Absence or closure of a normal body orifice, passage or cavity.

atrium (a'tre-um) [Gr. *atrion* hall]. A chamber affording entrance to another structure or organ; a chamber of the heart receiving blood from a vein and pumping it into a ventricle.

auditory ossicle (os'sĭ-k'l) [L. *ossiculum* a small bone]. One of three small bones (malleus, incus and stapes) which transmit vibrations across the middle ear cavity from tympanic membrane to the inner ear.

auricle (aw'ri-k'l) [L. *auricula* small ear]. The ear-shaped portion of the mammalian atrium (q.v.); sometimes used as a synonym for atrium. The external ear flap of mammals.

autosome (aw'to-sōm) [Gr. *autos* self + *soma* body]. Any ordinary paired chromosome, as distinguished from a sex chromosome.

autotrophs (aw-to-trofs) [Gr. *autos* self + *trophis* to nourish]. Organisms that manufacture organic nutrients from inorganic raw materials.

avian (a've-an) [L. *avis* bird]. Pertaining to the Aves, the class of flying vertebrates consisting of birds.

avicularia (a-vik"u-la'ri-a) [L. *avicula* small bird]. Specialized members of a colony of bryozoa which resemble the head of a bird.

axon (ak'son) [Gr. *axōn* axle]. Nerve fiber which conducts nerve impulses away from the cell body.

bacteriophage (bak-te're-o-fāj") [L. *bactērion* little rod + Gr. *phagein* to eat]. Virus which infects and may kill bacteria.

bacterium (bak-te're-um) [L. *bactērion* little rod]. Small, typically one-celled microorganisms characterized by the absence of a formed nucleus.

balanced polymorphism (pol"e-mor'fizm) [Gr. *poly* many + *morphē* form]. An equilibrium mixture of homozygotes and heterozygotes maintained by separate and opposing forces of natural selection.

basal body See *kinetosome*.

base A compound which releases hydroxyl ions (OH⁻) when dissolved in water; turns red litmus paper blue.

benthos (ben'thos) [Gr. *benthos* bottom of the sea]. The flora and fauna of the bottom of oceans or lakes.

bicuspid (bi-kus'pid) [L. *bi* two + *cuspis* point]. Having two cusps, flaps or points.

binary fission (bi'na-re) [L. *binarius* two at a time]. Simple cell division into two equal daughter cells.

binomial system System of naming organisms by the combination of the names of genus and species.

bioassay (bi"o-as-sa') [Gr. *bios* life]. Determination of the effectiveness of a biologically active substance by noting its effect on a living organism.

biogenesis (bi"-o-jen'e-sĭs). The generalization that all living things come only from pre-existing living things.

biologic clock Means by which activities of plants or animals are adapted to the regularly recurring changes in the external physical conditions, and perhaps to changes in internal milieu as well.

biological oxidation Process in which electrons removed from an atom or molecule are transferred through the electron transmitter system of the mitochondrion.

bioluminescence (bi"-o-loo"mĭ-nes'ens) [Gr. *bios* life + L. *lumen* light]. Emission of light by living cells or by enzyme systems prepared from living cells.

biomass The total weight of all the organisms in a particular habitat.

biome (bi'ōm) [Gr. *bios* life + *ome* mass]. Large, easily differentiated community unit arising as a result of complex interactions of climate, other physical factors and biotic factors.

biosphere The entire zone of air, land and water at the surface of the earth that is occupied by living things.

biotic potential Inherent power of a population to increase in numbers when the age ratio is stable and all environmental conditions are optimal.

birefringence (bi"re-frin'jens) [L. *bi* two + *refringere* to break up]. Property of a substance in solution to refract light differently in different planes.

bivalvia (bi-val've-ah) [*bi* two + L. *valva* valve]. Class of mollusks containing clams and oysters, in which the body is enclosed within two calcareous valves, or shells.

blastocoele (blas'to-sēl) [Gr. *blastos* germ + *koilos* hollow]. The fluid-filled cavity of the blastula, the mass of cells produced by cleavage of a fertilized ovum.

blastocyst (blas'to-sist) [Gr. *blastos* germ + *kystis* bladder]. The modified blastula stage of embryonic mammals; it consists of an inner cell mass, which develops into the embryo, and a peripheral layer of cells, the trophoblast, which contributes to the placenta.

blastopore (blas'to-pōr) [Gr. *blastos* germ +

poro opening]. The opening, in the gastrula stage of development, from the archenteron to the surface.

blastula (blas'tu-lah) [Gr. *blastos* germ]. Usually spherical structure produced by cleavage of a fertilized ovum, consisting of a single layer of cells surrounding a fluid-filled cavity.

blood plasma (plaz'mah) [Gr. *plasma* to mold]. The liquid portion of the blood in which the corpuscles are suspended; differs from blood serum in that it contains fibrinogen.

book lung A lunglike invagination on the underside of the abdomen of scorpions and larger spiders. Opens to the surface by a spiracle and contains many leaflike lamellae.

Bowman's capsule Double-walled, hollow sac of cells which surrounds the glomerulus at the end of each kidney tubule.

brachiopoda (bra″ke-op'o-da) [L. *brachium* arm]. A phylum of marine organisms which possess a pair of shells and, internally, a lophophore.

brachydactyly (brak″e-dak'tĭ-le) [Gr. *brachys* short + *daktylos* finger]. Abnormal shortness of the fingers and toes.

brachyphalangy (brak″e-fah-lan'je) [Gr. *brachy* short + *phalanx* log]. Abnormal shortness of one or more of the phalanges of a finger or toe.

branchial (brang'ke-al) [Gr. *branchion* a gill]. Pertaining to the gills or gill region.

brownian movement Motion of small particles in solution or suspension resulting from their being bumped by water molecules.

Bruce effect The blocking of pregnancy in a newly impregnated mouse by a pheromone, the odor of a foreign male.

brush border The many fine hairlike processes extending from the free surface of certain epithelial cells, e.g., the cells of the proximal convoluted tubules of the mammalian kidney.

bryozoa (bri″o-zo'a) [Gr. *bryo* moss + *zōon* animal]. Moss animals; the colonies of some species, delicately branched and beautiful, are sometimes mistaken for seaweed; other species form colonies which appear as thin, lacy encrustations on rocks.

budding Asexual reproduction in which a small part of the parent's body separates from the rest and develops into a new individual, eventually either taking up an independent existence or becoming a more or less independent member of the colony.

buffers Substances in a solution which tend to lessen the change in hydrogen ion concentration (*p*H), which otherwise would be produced by adding acids or bases.

byssus (bis'us) [Gr. *byssos* flax]. Organic secretion, usually in the form of a thread, that anchors certain bivalve mollusks to the substratum.

calcarea (kal-ka're-ah). Class of sponges having calcareous spicules.

calcitonin (kal″sĭ-to'nin). A polypeptide hormone composed of 32 amino acids in a single chain, secreted by parafollicular cells in the thyroid; it counters the effect of parathyroid hormone and causes the deposition of calcium and phosphate in bones.

calorie The amount of heat required to raise one gram of water one degree centigrade (strictly, from 14.5° to 15.5° C.) A kilocalorie or Calorie is a unit 1000 times larger, the amount of heat required to raise one kilogram of water one degree centrigrade.

capillaries (kap'ĭ-lar″e) [L. *capillaris* hairlike]. Microscopic thin-walled vessels located in the tissues, connecting arteries and veins and through the walls of which substances pass to the interstitial fluid.

carapace (kar'ă-pās) [Sp. *carapacho*]. A bony or chitinous shield covering the back of an animal.

carbohydrate (kar″bo-hi'drāt). Compounds containing carbon, hydrogen and oxygen, in the ratio of 1C:2H:1O; e.g., sugars, starches and cellulose.

carbonic anhydrase (kar'bon-ik an-hi'drās) [Gr. *an* not + *hydor* water]. An enzyme which catalyzes the reaction carbon dioxide + water ⇌ carbonic acid; abundant in erythrocytes.

carnivore (kar'nĭ-vōr) [L. *carno* flesh + *vorare* to devour]. An animal that eats flesh.

carotene (kar'o-ten) [L. *carota* carrot]. Yellow to orange-red pigments found in carrots, sweet potatoes, leafy vegetables, etc., which can be converted in the animal body to vitamin A.

cartilage replacement bone (kar'tĭ-lij) [L. *cartilago* cartilage]. Bone which develops in and around a cartilaginous rudiment, which it gradually replaces.

catabolism (kah-tab'o-lizm) [Gr. *katabolé* a throwing down]. Chemical reactions by which complex substances are converted, within living cells, into simpler compounds with the release of energy.

catalyst (kat'ah-list) [Gr. *katalysis* dissolution]. A substance which regulates the speed at which a chemical reaction occurs without affecting the end point of the reaction and without being used up as a result of the reaction.

cation (kat'i-on) [Gr. *kata* down + *ion* going]. An ion bearing a positive charge.

ceboids New World prehensile-tailed monkeys.

cecum (se'kum) [L. *caecum* blind]. A blind pouch into which open the ileum, the colon and the vermiform appendix.

cell constancy An extreme example of mosaic development which results in all individuals in a species having the same number of cells in comparable tissues performing similar functions.

cell lineage   In embryos with mosaic development, the tracing of cell histories through successive cleavages.

cell theory   The generalization that all living things are composed of cells and cell products, that new cells are formed by the division of preexisting cells, that there are fundamental similarities in the chemical constituents and metabolic activities of all cells, and that the activity of an organism as a whole is the sum of the activities and interactions of its independent cell units.

cells   The microscopic units of structure and function that comprise the bodies of plants and animals.

centriole   (sen″trĭ-ōl) [L. *centrum* center]. Small, dark-staining organelle lying near the nucleus in the cytoplasm of animal cells, and forming the spindle during mitosis and meiosis.

centrolecithal   (sen″tro-les′ĭ-thal) [*centro* + Gr. *lekithos* yolk]. Type of arthropod egg having the yolk arranged as a large sphere around the central nucleus.

centromere   (sen′tro-mēr) [Gr. *kentro* center + *meros* part]. The point on a chromosome to which the spindle fiber is attached; during mitosis or meiosis it is the first part of the chromosome to pass toward the pole.

cephalization   (sef″al-i-za′shun) [Gr. *kephalé* head]. Head formation, concentration of nervous tissue and sense organs at the anterior end of the body.

cephalochordata   (sef″ah-lo-kor-dā′-tah) [Gr. *kephalē* head + L. *chorda* string]. A subphylum of chordates characterized by the extension of the notochord to the anterior end of the body; includes *Amphioxus.*

cephalopoda   (sef″ah-lop′o-dah) [Gr. *kephalē* head + *podos* foot]. The squids and octopuses, a class of mollusks with arms or tentacles surrounding the head.

cercaria   (ser-ka′re-ah) [Gr. *kerkos* tail]. The final free-swimming larval stage of a trematode parasite.

cercopithecoid   An Old World monkey; has a tail but does not use it as a limb.

cerebellum   (ser″e-bel′um) [L. dim of *cerebrum* brain]. The part of the vertebrate brain which controls muscular coordination.

cerebrum   (ser′e-brum) [L. *cerebrum* brain]. A major portion of the vertebrate brain, occupying the upper part of the cranium; the two cerebral hemispheres, united by the corpus callosum, form the largest part of the central nervous system in humans.

cestoda   (ses-to′dah) [Gr. *kestos* girdle]. The tapeworms, a class of parasitic flatworms.

character displacement   The adaptive characters that evolve and enable one species to exclude another from its ecological niche. See *competitive exclusion.*

chelate   (ke′lāt) [Gr. *chélé* claw]. Having the terminal parts of certain appendages in the form of pincers, as in crabs and scorpions.

chelicera   (ke-lis′er-a) [Gr. *chēlē* claw + *keras* horn]. A pair of pincer-like head appendages found in spiders, scorpions and other arachnids.

chelicerae   (ke-lis′-er-ee). Anterior feeding appendages of certain arthropods, such as spiders, mites and scorpions.

chelicerata   (ke-lis′-uh-rah-tah). Subphylum of arthropods containing spiders, mites, and scorpions in which there are no antennae and the anterior feeding appendages are chelicerae.

chemoreceptor   (ke″mo-re-sep′tor). A sense organ or sensory cell that responds to chemical stimuli.

chemotropism   (ke-mot′ro-pizm) [Gr. *chemeia* chemistry + *tropos* a turning]. A growth response to a chemical stimulus.

chilopoda   (ki-lop′o-dah) [Gr. *cheilos* lip + *podos* foot]. Centipedes, a class of arthropods whose first pair of legs have evolved as poison fangs.

chimaera   (ki-me′rah) [Gr. *chimaira* a mythological fire-spouting monster with a lion's head, goat's body and serpent's tail]. An individual organism whose body contains cell populations derived from different zygotes of the same or of different species; occurring spontaneously, as in twins, or produced artificially, as an organism which develops from combined portions of different embryos, or one in which tissues or cells of another organism have been introduced. Also, a cartilaginous fish belonging to the subclass Holocephali.

chitin   (ki′tin) [Gr. *chiton* tunic]. An insoluble horny polysaccharide bound to protein to form a glycoprotein. Abundant in the exoskeleton of arthropods and found in the cell walls of many fungi.

choanocyte   (ko′ă-no″sīt) [Gr. *choane* funnel + *kytos* hollow vessel]. A unique cell type with a flagellum surrounded by a thin cytoplasmic collar; characteristic of sponges and one group of protozoa.

chondrichthyes   (kon″drik′thi-ēz) [Gr. *chondros* cartilage + *ichthys* fish]. A class of fishes with cartilaginous skeletons; includes sharks, skates and their allies.

chordate   (kor′dāt) [L. *chorda* string]. A member of the phylum of animals characterized by the presence of a notochord at some stage of development; includes urochordates, *Amphioxus* and vertebrates.

chorion   (ko′re-on). An extraembryonic membrane in reptiles, birds and mammals that forms an outer cover around the embryo and in mammals contributes to the formation of the placenta.

choroid coat   (ko′roid) [Gr. *chorion* the outermost extraembryonic membrane of an amniote embryo + *eidos* form]. The middle, pigmented and vascular layer of the eyeball; its anterior portion contributes to the iris and ciliary body.

chromatin (kro'mah-tin) [Gr. *chroma* color]. The readily stainable portion of the cell nucleus, forming a network of fibrils within the nucleus; composed of DNA and proteins.

chromatin spot An aggregation of chromatin at the periphery of the nucleus, evident in cells of human skin or from the mucosal lining of the mouth; makes possible the determination of the "nuclear sex" of an individual. Most of the cells of a female and none of the cells of a male have a chromatin spot.

chromatophore (kro'mah-to-for) [Gr. *chroma* color + *pherein* to bear]. Any pigmentary cell or color-producing plastid, such as those of the deep layers of the epidermis.

chromomere (kro'mo-mēr) [Gr. *chroma* color + *meros* part]. One of a linear series of bead-like structures composing a chromosome.

chromosomes (kro'mo-sōm) [Gr. *chroma* color + *soma* body]. Filamentous or rod-shaped bodies in the cell nucleus which contain the hereditary units. the genes.

cilium (sil'e-um) [L. *cilium* eyelash]. One of numerous hairlike cytoplasmic processes of the free surface of many cells. Composed of a sheaf of nine microtubules just beneath the plasma membrane and a core of two central microtubules. Cilia typically are shorter and more numerous than flagella (q.v.) and have a more oarlike action.

circadian rhythms (ser"kah-de'an) [L. *circa* about + *dies* a day]. Repeated sequences of events which occur at about 24 hour intervals.

cistron (sis'tron). The genetic unit of biochemical function; the sequence of nucleotide pairs in DNA that specify the amino acid sequence of a single peptide chain.

class In taxonomy, a major subdivision of a phylum. Each class is composed of one or more related orders.

cleidoic egg (kli-do'ik) [Gr. *kleidouchos* holding the keys]. The eggs of reptiles, birds and primitive mammals which are self-sufficient, and in which the embryo develops directly to the miniature adult without passing through a larval stage.

cline (klin) [Gr. *klin-*, stem of *klinein* to slope, and of *klinē* bed]. Continuous series of differences in structure or function exhibited by the members of a species along a line extending from one part of their range to another.

clitellum (kli tel'um) [L. *clitellae* pack-saddle]. Glandular segments in earthworms and leeches that secrete the cocoon.

clitoris (kli'to-ris) [Gr. *kleitoris*]. A small, erectile body at the anterior part of the vulva which is homologous to the male penis.

cloaca (klo-a'kah) [L. a sewer]. A common chamber receiving the discharge of the digestive, excretory and reproductive systems in most of the lower vertebrates.

clone A population of cells descended by mitotic division from a single ancestral cell.

cnidaria (ni-dah're-ah) [Gr. *knidé* a nettle]. Phylum of animals containing hydras, jellyfish, sea anemones and corals, in which the radially symmetrical body bears tentacles around the mouth.

cnidocyte (ni'dŏ site). Cells of cnidarians containing explosive stinging structures, called nematocysts.

cnidospora (ni dos'po-rah) [Gr. *knidē* nettle + L. *spora* seed]. Class of parasitic protozoans having spores with filaments.

cobalamin (ko-bal'ah-min). Vitamin B_{12}; substance essential to the manufacture of red cells.

cocci (kok'si) [L.; Gr. *kokkos* berry]. Spherical bacterial cells, usually less than 1 μm in diameter.

cochlea (kok'le-ah) [Gr. *kochlias* snail]. The part of the inner ear consisting of the cochlear duct, which contains the receptive organ of Corti.

codon A sequence of three adjacent nucleotides that code for a single amino acid.

coelom (se'lom) [Gr. *koilia* cavity]. Body cavity of triploblastic animals lying within the mesoderm and lined by it.

coenzyme (ko-en'zīm) [L. *cum* with + Gr. *en* in + *zymē* leaven]. A substance which is required for some particular enzymatic reaction to occur; participates in the reaction by donating or accepting some reactant; loosely bound to enzyme.

collagen (kol'ah-jen) [Gr. *kolla* glue + *gennan* to produce]. Protein in connective tissue fibers which is converted to gelatin by boiling.

colony (kol'o-ne) [L. *colonus* farmer]. An association of unicellular or multicellular organisms of the same species; each individual is separate or essentially so, but sometimes there are connections among the members of the colony and some division of labor among them, e.g., feeding and reproductive polyps of certain hydrozoa.

commensalism (kŏ-men'sal-izm") [L. *cum* together + *mensa* table]. A relationship between two species in which one is benefited and the second is neither harmed nor benefited by existing together.

community An assemblage of populations that live in a defined area or habitat, which can either be very large or quite small. The organisms constituting the community interact in various ways with one another.

competitive exclusion The tendency for the better adapted species to exclude another related species from its particular ecological niche (q.v.).

compound eye Type of arthropod eye composed of units, each of which contains all of the structural and functional elements of the eye—lens, photoreceptor cells, and so forth.

cone (kōn) [L. *conus*]. In zoology, the cone-shaped photoreceptive cells of the retina which are particularly sensitive to bright light and, by distinguishing light of various wave lengths, mediate color vision.

conjugation (kon″ju-ga′shun) [L. *conjugatio* a blending]. The act of joining together; form of sexual reproduction in which nuclear material is exchanged during the temporary union of two cells; occurs in many ciliate protozoa and in bacteria.

conservation of energy, law of A fundamental law of physics which states that in any given system the amount of energy is constant; energy is neither created nor destroyed, but only transformed from one form to another.

conservation of matter, law of A fundamental law of physics which states that in any chemical reaction atoms are neither created nor destroyed but simply change partners.

"consumer" organisms Those elements of an ecosystem, plants or animals, that eat other plants or animals.

contractile vacuole Osmoregulatory organelle of protozoans and sponges.

contraception (kon-trah-sep′shun) [L. *contra* against + *conceptus* conceiving]. Method of birth control which involves the use of mechanical or chemical agents to prevent the sperm from reaching and fertilizing the egg.

conus arteriosus (ko′nus ar-te″re-o′sus) [L. *conus* cone + *arteriosus* arterial]. The terminal chamber of the heart of many fishes and amphibians; receives blood from the ventricle and delivers it to the ventral aorta.

convergent evolution (kon-ver′jent) [L. *cum* together + *vergere* to incline]. The independent evolution of similar structures, which carry on similar functions, in two or more organisms of widely different, unrelated ancestry.

copulation (kop″u-la′shun) [L. *copulatio*]. Sexual union; act of physical joining of two animals during which sperm cells are transferred from one to the other.

cornea (kor″ne-ah) [L. *corneus* horny]. The transparent structure forming the anterior part of the fibrous layer of the eyeball; it is continuous posteriorly with the sclera.

corpus allatum (kor′pus al-la′tum) [L. *corpus* body + *adlatus* added]. An endocrine gland located in the head of insects just behind the brain; it secretes juvenile hormone.

corpus callosum (kōr′pus kah-lo′sum) [L. *corpus* body + *callosus* hard]. A large commissure of fibers interconnecting the two cerebral hemispheres in mammals.

corpus luteum (kor′pus lu′tĭ-um) [L. *corpus* body + *luteus* yellow]. A yellow glandular mass in the ovary formed by the cells of an ovarian follicle that has matured and discharged its ovum.

corpus striatum (stri-a′tum) [L. *corpus* body + *striatum* striped]. A large subcortical mass of neuron cell bodies and fibers in the base of each cerebral hemisphere.

cortex (kor′teks) [L. *cortex* bark]. The outer layer of an organ.

covalent bond Chemical bond involving one or more shared pairs of electrons.

cranium (kra′ne-um) [Gr. *kranion* head]. The part of a skull which surrounds the brain; the brain case.

cretinism (kre′tin-izm). A chronic condition in the young due to congenital lack of thyroid secretion; retarded physical and mental development.

crinoidea (kri-noi′de-ah) [Gr. *krinoeides* like a lily]. The sea lilies, a class of ancient echinoderms, some of which survive today.

crop A sac in the digestive tract in which food is stored before digestion begins.

crossing over Process during meiosis in which the homologous chromosomes undergo synapsis and exchange segments.

crossopterigian (krŏ-sop′tĕ-rijĭ-an) [Gr. *krossoi* a fringe + *pterygos* wing]. A member of an order of sarcopterygian fishes ancestral to terrestrial vertebrates; now nearly extinct.

crustacea (krus-ta′she-ah) [L. *crusta* a hard surface]. A large subphylum (or class) of arthropods having typically biramous limbs and two pairs of antennae.

ctenophores (tēn′o-fors) [Gr. *ktenos* comb]. Marine animals ("comb jellies") whose bodies consist of two layers of cells enclosing a mass of jelly; the outer surface is covered with eight rows of cilia, resembling combs, by which the animal moves through the water.

cutaneous (ku-ta′ne-us) [L. *cutis* skin]. Pertaining to the skin.

cytochromes (si′to-krom) [Gr. *kytos* hollow vessel + *chroma* color]. The iron-containing heme proteins of the electron transmitter system that are alternately oxidized and reduced in biological oxidation.

cytokinesis (si″to-ki-ne′sis) [Gr. *kytōs* hollow vessel + *kinesis* motion]. The division of the cytoplasm during mitosis or meiosis.

cytopharynx (si″to-far′inks). Gullet-like organelle of ciliate and certain other protozoans.

cytostome (si′to-stōm) [Gr. *kytos* hollow vesicle + *stoma* mouth]. The mouth opening of ciliate and certain other protozoans.

dalton [John Dalton, eighteenth century British physicist]. The unit of molecular weight; the weight of one hydrogen atom.

deamination (de-am′ĭ-na′shun). Removal of an amino group ($-NH_2$) from an amino acid or other organic compound.

decarboxylation (de″kar-bok″sĭ-la′shun). Removal of a carboxyl group ($-COOH$) from an organic compound.

defecation (def″e-ka′shun) [L. *defaecare* to deprive of dregs]. The elimination of excrement, much of which is undigested refuse that has not taken part in metabolism.

dehydrogenation (de-hi″dro-jen-a′shun) [L. *de* apart + Gr. *hydōr* water]. A form of oxidation in which hydrogen atoms are removed from a molecule.

delamination (de″lam-ĭ-na′shun) [L. *de* apart + *lamina* plate]. Separation of the blastoderm into an upper ectoderm and a lower endoderm during embryonic development.

deme (dēm) [Gr. *demos* people]. A population of very similar organisms interbreeding in nature and occupying a circumscribed area.

denaturation (de-na-tūr-a′shun). Alteration of physical properties and three dimensional structure of a protein, nucleic acid or other macromolecule by mild treatment that does not break the primary structure.

dendrite (den′drīt) [Gr. *dendron* tree]. Nerve fiber, typically branched; the part of the neuron specialized for receiving excitation either from environmental stimuli or another cell.

deoxyribose (de-ok″se-ri′bōs). A five-carbon sugar with one less oxygen atom than the parent sugar, ribose; constituent of DNA.

deposit feeding Utilization as a food source of organic detritus that has become mixed with mineral particles in aquatic and terrestrial habitats.

dermal bone (der′mal) [Gr. *derma* skin]. Bone that develops directly in connective tissue and is not preceded by a cartilaginous rudiment; found in the dermis of the skin, or just beneath the skin.

dermis (der′mis) [Gr. *derma* skin]. The deeper layer of the skin of vertebrates and some invertebrates.

desmosomes Discontinuous button-like plaques present on the two opposing cell surfaces and separated by the intercellular space; they apparently serve to hold the cells together.

determinate cleavage Condition in which the embryonic fate of the blastomeres is fixed very early in cleavage.

detoxification (de-tok″sĭ-fi-ka′shun). Enzymatic processes which reduce the toxicity of a substance.

deuterostome (du″ter-o-stōm″) [Gr. *deuteros* second + *stoma* mouth]. A major branch of the animal kingdom containing animals in which the site of the blastopore is posterior— far from the mouth, which forms anew at the anterior end.

diapause (di′a-pawz) [Gr. *dia* through + *pausis* a stopping]. Inactive state of an insect during the pupal stage.

diaphragm (di′ah-fram) [Gr. *diaphragnynai* to fence by a partition]. The fibromuscular partition separating the thoracic and abdominal cavities of mammals; its contraction plays the major role in inspiration.

diastole (di-as′to-le) [Gr. *diastole* a drawing asunder; expansion]. Relaxation of the heart muscle, especially that of the ventricle, during which the lumen becomes filled with blood.

diencephalon (di″en-sef′ah-lon) [Gr. *dia* through + *enkephalos* brain]. A major subdivision of the brain lying between the telencephalon and mesencephalon; includes the epithalamus, thalamus and hypothalamus.

differentiation Development toward a more mature state; a process changing a relatively unspecialized cell to a more specialized cell.

diffusion The movement of molecules from a region of high concentration to one of lower concentration, brought about by their kinetic energy.

dinoflagellates (di″no-flaj″ĕ-la′tah) [Gr. *dinos* whirl + *flagellum* whip]. Flagellated unicellular protozoans (or algae) found in the sea and in fresh water.

diploid (dip′loid) [Gr. *diploos* twofold]. A chromosome number twice that found in gametes; containing two sets of chromosomes.

diplopoda (dĭ-plop′o-dah) [Gr. *diploos* double + *podos* foot]. Millipedes, a class of arthropods in which the trunk segments each bear two pairs of legs.

disaccharides (di-sak′ah-rid). Sugars which yield two monosaccharides on hydrolysis; e.g., sucrose, lactose and maltose.

distal (dis′tal) [L. *distans* distant]. Remote; farther from the point of reference.

DNA Deoxyribose nucleic acid; is present in chromosomes and contains genetic information coded in specific sequences of its constituent nucleotides.

ductus arteriosus (duk′tus ar-te″re-o′sus) [L. *ductus* conduit + *arteriosus* arterial]. A short vessel, derived from the dorsal part of the sixth aortic arch, which connects the pulmonary artery and dorsal aorta; normally present only in larval or embryonic stages and serving to divert blood from the developing lungs.

ecdysone (ek-di′son) [Gr. *ekdysis* a getting out]. The hormone that induces molting (ecdysis) in arthropods.

ecdysis (ek′dĭ-sis) [Gr. *ekdysis* a getting out]. The shedding, or molting, of the arthropod exoskeleton.

echinoderm (e-kin′o-derm) [L. *echinus* hedgehog, sea urchin + Gr. *derma* skin]. Phylum of spiny-skinned marine animals; e.g., starfish, sea urchins, sea cucumbers.

echinoidea (ek″in-oi′de-ah). The sea urchins, a class of echinoderms with rigid body walls and long moveable spines.

ecologic niche The status of an organism within a community or ecosystem; depends on the organism's structural adaptations, physiologic responses and behavior.

ecology (e-kol′o-je) [Gr. *oikos* house + *logos* word, discourse]. The study of the interrelations between living things and their environment, both physical and biotic.

ecosystem (ek″o-sis′tem) [Gr. *oikos* house]. A natural unit of living and nonliving parts that interact to produce a stable system; all of the organisms of a given area and the encompassing physical environment.

ecotone A fairly broad transition region between adjacent biomes; contains some organisms from each of the adjacent biomes plus some that are characteristic of, and perhaps restricted to, the ecotone.

ectoderm (ek′to-derm) [Gr. *ektos* without + *derma* skin]. The outer of the two germ layers of the gastrula; gives rise to the skin and nervous system.

ectothermic Condition in which the internal body temperature of an animal is dependent upon external heat.

effector (ef-fek′tor). Structures of the body by which an organism acts; the means by which it reacts to stimuli; e.g., muscles and glands.

efferent (ef′er-ent) [L. *ex* out + *ferre* to carry]. Conveying away; denoting certain vessels or neurons that transmit blood or impulses away from a point of reference; efferent neurons are motor neurons conducting impulses away from the central nervous system.

elasmobranch (e-las′mo-brank) [Gr. *elasmos* plate + L. *branchia* gill]. A member of a subclass of cartilaginous fishes in which the gills are platelike and each gill slit opens independently on the body surface; sharks, skates and their allies.

electrolyte (e-lek′tro-līt) [Gr. *elektron* amber + *lytos* soluble]. Substance which dissociates in solution into charged particles, ions and thus permits the conduction of an electric current through the solution.

electron transmitter system System of enzymes localized within the mitochondria which transfer electrons from foodstuff molecules to oxygen.

element (el′ĕ-ment). One of the hundred or so types of matter, natural or man-made, composed of atoms all of which have the same number of protons in the atomic nucleus and the same number of electrons circling in the orbits.

embolus (em′bo-lus) [Gr. *embolos* plug]. A thrombus or any other particle carried by the bloodstream which blocks a blood vessel.

embryo (em′bre-o) [Gre. *en* in + *bryein* to swell]. The early stage of development of an organism; the developing product of fertilization of an egg.

endergonic (end″er-gon′ik) [Gr. *endon* within + *ergon* work]. A reaction characterized by the absorption of energy; requires energy to occur.

endocrine (en′do-krin) [Gr. *endon* within + *krinein* to separate]. Secreting internally; applied to organs whose function is to secrete into the blood or lymph a substance that has a specific effect on another organ or part.

endoderm (en′do-derm) [Gr. *endon* within + *derma* skin]. The inner germ layer of the gastrula, lining the archenteron; forms the epithelial cells of the digestive tract and its outgrowths—the liver, lungs and pancreas.

endolymph (en′do-limf) [Gr. *endon* within + L. *lympha* lymph]. The lymphlike fluid contained within the semicircular canals, cochlear duct and other parts of the membranous labyrinth of the inner ear.

endopeptidase (en″do-pep′tĭ-dās) [Gr. *endon* within + *peptos* digested]. A proteolytic enzyme which cleaves peptide bonds within a peptide chain.

endopodite (en-dop′o-dīt) [Gr. *endon* within + *podos* foot]. The inner branch of a biramous crustacean limb.

endoskeleton (en″do-skel′ĕton) [Gr. *endon* within + *skeleton* a dried body]. Bony and cartilaginous supporting structures within the body; provide support from within.

endostyle (en′do-stīl) [Gr. *endon* within + *stylos* a pillar]. A longitudinal groove in the floor of the pharynx of certain lower chordates and larval cyclostomes; its glandular and ciliated cells secrete mucus which is moved anteriorly and entraps minute food particles.

endothermic Condition in which internal body temperature is derived from the metabolic heat produced by the animal.

engram (en′gram) [Gr. *en* in + *gramma* mark]. The term is applied to the presumed change that occurs in the brain as a consequence of learning; a memory trace.

enterocoel (en″ter-o-sēl″) [Gr. *enteron* intestine + *koilia* cavity]. A body cavity formed by outpouches from the primitive gut.

enteropneusta (en″ter-op-noos′tah) [Gr. *enteron* intestine + *phesta* to breathe]. Class of the phylum Hemichordata containing the acorn worms in which the body is wormlike and divided into a proposcis, collar and trunk.

enzyme (en′zīm) [Gr. *en* in + *zymē* leaven]. A protein catalyst produced within a living organism which accelerates specific chemical reactions.

epiboly (e-pib′o-le) [Gr. *epibolē* cover]. A method of gastrulation by which the smaller blastomeres at the animal pole of the embryo grow over and enclose the cells of the vegetal hemisphere.

epidermis (ep′ĭ-der′mis) [Gr. *epi* on + *derma* skin]. The outermost layer of cells of an organism.

epididymis (ep′ĭ-did′ĭ-mis) [Gr. *epi* on + *didymos* testis]. Complexly coiled tube adjacent to the testis where sperm are stored.

epigenesis (ep′ĭ-jen′e-sis) [Gr. *epi* on + *genesis* to be born]. The theory that development proceeds from a structureless cell by the successive formation and addition of new parts which do not preexist in the fertilized egg.

epiglottis (ep′ĭ-glot′is) [Gr. *epi* on + *glottis*

the tongue]. The lidlike structure which covers the glottis, the entrance to the larynx.

epilimnion (ĕp-i-lĭm'nĭ-ŏn). The uppermost layer, or surface water, of a lake or pond.

epithelium (ep'ĭ-the'le-um) [Gr. *epi* on + *thēlē* nipple]. The layer of tissue covering the internal and external surfaces of the body, including the lining of vessels and other small cavities; consists of cells joined by small amounts of cementing substances.

equilibrium (e"kwĭ.lib're-um) [L. *aequus* equal + *libra* balance]. A state of balance; a condition in which opposing forces exactly counteract each other.

erythrocyte (e-rith'ro-sīt) [Gr. *erythros* red + *kytos* hollow vessel]. Red blood cells; they contain hemoglobin and transport gases.

estrogen (es'tro-jen). One of the female sex hormones, produced by the ovarian follicle, which promotes the development of the secondary sex characteristics.

estrus (es'trus) [L. *oestrus* gadfly; Gr. *oistros* anything that drives mad, any vehement desire]. The recurrent, restricted period of sexual receptivity in many female mammals, marked by intense sexual urge.

ethology [Gr. *ethos* custom + *logus* study of]. The study of the whole range of animal behavior under natural conditions.

eukaryotic [Gr. *eu* true + *karyon* kernel]. Applied to organisms that have nuclei surrounded by membranes, Golgi apparatus and mitochondria.

euryhaline (u-re-ha'-līn) [Gr. *eurys* wide + *halinos* salt]. Ability of an animal to tolerate a relatively wide range of environmental salt concentrations (salinities).

eustachian tube (u-sta'ke-an) [*Eustachio* an Italian anatomist of the sixteenth century]. The auditory tube passing between the middle ear cavity and pharynx of most terrestrial vertebrates; it permits the equalization of pressure on the tympanic membrane.

eutherian (u-thēr'ĭ-an) [Gr. *eu* true + *therion* beast]. One of the placental mammals in which a well formed placenta is present and the young are born at a relatively advanced stage of development; includes all living mammals except monotremes and marsupials.

excretion (eks-kre'shun) [L. *ex* out + *cernere* to sift, separate]. Removal of metabolic wastes by an organism.

exergonic (ek"ser-gon'ik) [L. *ex* out + Gr. *ergon* work]. A reaction characterized by the release of onorgy.

exopeptidase (ek"so-pep'tĭ-dās) [Gr. *exo* outside and *peptos* digested]. A proteolytic enzyme which cleaves only the bond joining a terminal amino acid to a peptide chain.

exopodite (eks-op'o-dīt) [Gr. *exo* outside + *podos* foot]. The outer branch of a biramous crustacean limb.

exoskeleton (ek"so-skel'ĕ-ton) [Gr. *exo* outside + *skeleton* a dried body]. Calcareous, chitinous or other hard material covering the body surface and providing protection or support.

expressivity (eks"pres-siv'ĭ-te [L. *expressus*]. The extent to which a heritable trait is manifested by an individual carrying the principal gene conditioning it.

extensor (eks-ten'ser) [L. one that stretches]. A muscle that serves to extend or straighten a limb.

facilitation (fah-sil"i-ta'shun) [L., *facilis* easy]. The promotion or hastening of any natural process; the reverse of inhibition.

fallopian tube (fal-lo'pe-an) [*Fallopius* an Italian anatomist of the sixteenth century]. The uterine tube of mammals which extends from a point near the ovary to the uterus; evolved from a part of the oviduct of lower vertebrates.

family In taxonomy, a major subdivision of an order. Each family is composed of one or more related genera.

feedback control System in which the accumulation of the product of a reaction leads to a decrease in its rate of production or a deficiency of the product leads to an increase in its rate of production.

fenestra (fĕ-nes'trah) [L. *fenestra* a window]. A moderate-sized opening in a structure, e.g., the fenestra ovalis (oval window) in the middle ear cavity.

fermentation (fer"men-ta'shun) [L. *fermentum* leaven]. Anaerobic decomposition of an organic compound by an enzyme system; energy is made available to the cell for other processes.

fertilization (fer"tĭ-lĭ-za'shun) [L. *fertilis* to bear, produce]. The fusion of a spermatozoon with an ovum to initiate development of the resulting zygote.

fetus (fe'tus) [L. *fetus* fruitful]. The unborn offspring after it has largely completed its embryonic devclopmcnt.

fibrin (fĭ'brin) [L. *fibra* fiber]. Delicate protein threads derived from soluble prothrombin in the presence of the enzyme thrombin (q.v.) during the formation of a blood clot.

filtration (fil-tra'shun) [Fr. *filtre* a filter]. The passage of a liquid through a filter following a pressure gradient; occurs in capillary beds, including the glomeruli of the kidney.

fission (fish'un) [L. *fissio* to cleave]. Process of asexual reproduction in which an organism divides into two approximately equal parts.

flagellates (flaj'ĕ-lāt) [L. *flagellum* whip]. Microorganisms furnished with one or more slender, whiplike processes termed flagella.

flagellum (flă-jel'um) [L. *flagellum* whip]. A whiplike cytoplasmic process on the free surface of many cells. Structurally similar to a cilium (q.v.), but typically flagella are longer, fewer in number and have a more sinuous movement.

flexor (flek'sor). A muscle that serves to bend a limb.

fluke Common name of the parasitic flatworms belonging to the class Trematoda.

fluorescence (floo″o-res′ens). The emission of light by a substance which has absorbed radiation of a different wave length; results when an excited singlet state decays to the ground state, an extremely rapid process which is independent of temperature.

follicle (fol′lĭ-k′l) [L. *folliculus* small bag]. A small sac of cells in the mammalian ovary which contains a maturing egg.

food chain A sequence of organisms through which energy is transferred from its ultimate source in a plant; each organism eats the preceding and is eaten by the following member of the sequence.

foramen (fo-ra′men) [L. *forare* to bore]. A small opening or perforation in a body structure.

foramen ovale The oval window between the right and left atria, present in the fetus and by means of which blood entering the right atrium may enter the aorta without passing through the lung.

foraminifera (fo-ram″ĭ-nif′er-ah) [L. *forare* to bore + *ferre* to bear]. Ameboid protozoa which secrete chalky, many-chambered shells with pores through which the animal extends its pseudopods.

fossa (fos′ah) [L., *fossa* a ditch]. A depression in a structure, e.g., the fossa ovalis or depression in the interatrial septum of the adult mammal; located at the site of the embryonic foramen ovale.

fossils (fos′ils) [L. *fossilis* to dig]. Any remains of an organism that have been preserved in the earth's crust.

fovea (fo′ve-ah) [L. a small pit]. A small pit in the surface of a structure or organ; specifically, a pit in the center of the retina which contains only cones and provides for keenest vision.

frustule (frus′tūl) [L. a small piece]. In cnidarians, a cylinder of tissue formed asexually that creeps away and forms a new individual.

fundus (fun′dus) [L. *fundus* bottom]. The bottom or base of an organ; the part of a hollow organ farthest from its opening.

gamete (gam′ēt) [Gr. *gametē* wife]. A reproductive cell; an egg or sperm whose union, in sexual reproduction, initiates the development of a new individual.

ganglion (gang′gle-on) [Gr. *gangli* knot]. A knotlike mass of the cell bodies of neurons located outside the central nervous system; in invertebrates, includes the swellings of the central nervous system.

gastrodermis (gas″tro-der′mis) [Gr. *gastēr* stomach + *derma* skin]. The tissue lining the gut cavity that is responsible for digestion and absorption.

gastropoda (gas-trop′o-dah) [Gr. *gaster* stomach + *podos* foot]. The snails, a class of mollusks that creep on a broad ventral foot.

gastrotricha (gas″tro-trĭ′kah) [Gr. *gastēr* stomach + *trichos* hair]. A phylum of minute pseudocoelomates with ciliated regions on the ventral surface.

gastrula (gas′troo-lah) [Gr. *gastēr* stomach]. Early embryonic stage which follows the blastula; consists initially of two layers, the ectoderm and the endoderm, and of two cavities, the blastocoele between ectoderm and endoderm and the archenteron, formed by invagination, lying within the endoderm, and opening to the exterior through the blastopore.

gastrulation (gas″troo-la′shun) [Gr. *gastēr* stomach]. The process by which the young embryo becomes a gastrula and acquires first two and then three layers of cells.

gel (jel) [L. from *gelare* to congeal]. A colloidal system in which the solid phase is continuous and the liquid phase is dispersed.

gemmule (jem′ūl) [L. *gemmula*, small bud]. In sponges, an asexually produced reproductive body surrounded by a protective cover.

gene (jēn) [Gr. *gennan* to produce]. The biologic unit of genetic information, self-reproducing and located in a definite position (locus) on a particular chromosome.

gene pool The aggregation of all the genes and their alleles of all of the individuals of a species.

genetic drift The tendency, within small interbreeding populations, for heterozygous gene pairs to become homozygous for one allele or the other by *chance* rather than by selection.

genetic equilibrium The situation in which the distribution of alleles in a population is constant in successive generations (unless altered by selection or mutation).

genome (je′nom) [Gr. *gennan* to produce + *ōma* mass, abstract entity]. A complete set of hereditary factors, contained in the haploid assortment of chromosomes.

genotype (jen′o-tip) [Gr. *geno-* from *gennan* to produce + *typos* type]. The fundamental hereditary constitution, assortment of genes, of any given organism.

genus (je′nus) [L. birth, race, kind, sort]. A rank in taxonomic classification in which closely related species are grouped together.

gill Gas exchange organs whose surface area is increased by filaments, lamellae or other folds evaginated from a surface.

gill slit An opening to the outside from the pharynx that arises during development. Water taken in at the mouth passes out through the gill slits, aiding respiration and the filtering of food.

gizzard (giz'erd) [L. *gigeria* cooked entrails of poultry]. A portion of the digestive tract specialized for mechanical digestion.

globulin (glob'u-lin) [L. *globulus* globule]. One of a class of proteins in blood plasma, some of which (gamma globulins) function as antibodies.

glomerulus (glo-mer'u-lŏs) [L. *glomus* ball]. A tuft of minute blood vessels or nerve fibers; specifically, the knot of capillaries at the proximal end of a kidney tubule.

glottis (glot'is) [Gr. *glossa* tongue]. The opening between the pharynx and larynx; it is bounded by the vocal cords in higher vertebrates.

glycolysis (gli-kol'ĭ-sis) [Gr. *glykys* sweet + *lysis* solution]. The metabolic conversion of sugars into simpler compounds.

goiter (goi'ter). An enlargement of the thyroid gland, causing a swelling in the front part of the neck; may result from overactivity of the thyroid or from deficiency of iodine.

Golgi bodies A type of cell organelle found in the cytoplasm of all cells except mature sperm and red blood cells; believed to play a role in the secretion of cell products.

gonad (gon'ad) [Gr. *gonē* seed]. A gamete-producing gland; an ovary or testis.

gonoduct Principal duct providing exit for sperm and eggs of any reproductive system.

gonopore External opening of any reproductive system.

green gland Excretory organ of certain crustaceans, such as shrimp and crabs; also called antennal gland.

habitat (hab'ĭ-tat) [L. *habitus*, from *habere* to hold]. The natural abode of an animal or plant species; the physical area in which it may be found.

habituation A gradual decrease in response to successive stimulation due to changes in the central nervous system.

haploid (hap'loid) [Gr. *haploos* simple, single]. Having a single set of chromosomes, as normally present in a mature gamete.

Hardy-Weinberg Law The relative frequencies of the members of a pair of allelic genes in a population are described by the expansion of the binomial equation, $a^2 + 2ab + b^2$.

haversian canals (ha-ver'shan). Channels extending through the matrix of bone and containing blood vessels and nerves.

hemerythrin (hēm"ē-rith'ren) [Gr. *haima* blood + *erythros* red]. Protein respiratory pigment of a small number of invertebrates in which the oxygen molecule is carried between two atoms of iron.

hemichordata (hem"ĭ-kor-dah'tah) [Gr. *hēmi* half + L. *chorda* cord]. A marine phylum related to the chordates; includes the acorn worms.

hemocoel (he'mo-sēl) [Gr. *haima* blood + *koilia* cavity]. The spaces between the cells and tissues of many invertebrates. Derived from the blastocoel. Differs from the coelom (q.v.) in the absence of a definitive epithelial lining of mesodermally derived cells.

hemocyanin (he"mo-si'an-in) [Gr. *haima* blood + *kyanos* dark blue]. A copper-containing respiratory pigment found in the blood of various invertebrates.

hemoglobin (he"mo-glo'bin) [Gr. *haima, haimatos* blood]. The red, iron-containing, protein pigment of the erythrocytes that transports oxygen and carbon dioxide and aids in regulation of pH.

hemolymph (he'mo-limf) [Gr. *haima* blood + L. *lympha* water]. The bloodlike fluid of animals with open circulatory systems; combines the properties of blood and lymphlike interstitial fluid.

hemophilia (he"mo-fil'e-ah) [Gr. *haima, haimatos* blood + *philein* to love]. Hereditary disease in which the formation of thromboplastin is impaired due to a deficiency of the so-called antihemophilic globulin; blood does not clot properly; "bleeder's disease."

hemostatis (hē"mo-stā'sis) [Gr. *haima* blood + *stasis* standstill]. The stopping of blood flow from an injured blood vessel; includes the contraction of the vessel and the formation of a blood clot.

hepatic (he-pat'ik) [Gr. *hepatikos*]. Pertaining to the liver.

herbivore (her'bĭ-vōr) [L. *herba* herb + *vorare* to devour]. A plant-eating animal.

hermaphroditism (her-maf'ro-dit-izm) (Gr. god and goddess, Hermes and Aphrodite, whence *hermaphroditos* a person having the attributes of both sexes]. A state characterized by the presence of both male and female sex organs in the same organism.

heterogamy (het"er-og'ah-me) [Gr. *heteros* other + *gamōs* marriage]. Reproduction involving the union of two gametes which differ in size and structure; e.g., egg and sperm.

heterografts (het'er-o-grafts) [Gr. *heteros* other]. Grafts of tissue obtained from the body of an animal of a species other than that of the recipient.

heterosis (het"e-rō'sis) [Gr. *heteros* other]. Hybrid vigor; the offspring from the mating of individuals of totally unrelated strains frequently are much better adapted for survival than either parent.

heterotrophs (het'er-o-trofs) [Gr. *heteros* other + *trophos* feeder]. Organisms which cannot synthesize their own food from inorganic materials and therefore must live either at the expense of autotrophs or upon decaying matter.

heterozygous (het"er-o-zi'gus) [Gr. *heteros* other + *zygos* yoke]. Possessing two different alleles for a given character at the corresponding loci of homologous chromosomes.

hexapoda (heks-ah'po-dah) [Gr. *hex* six + *podos* foot]. The insects, a class of arthropods in which adults have three pairs of legs.

hibernation (hi″ber-na′shun) [L. *hiberna* winter]. The dormant state of decreased metabolism in which certain animals pass the winter.

hirudin (hi-ru′din) [L. *hirudo* leech]. A substance secreted by leeches that prevents the clotting of blood; used in medicine.

hirudinea (hi″ru-din′e-ah) [L *hirudo* leech]. The leeches, a class of annelids many of which suck blood.

holothuroidea (ho″lo-thu-roi′de-ah) [Gr. *holothurion* a sea polyp]. The sea cucumbers, a class of soft-bodied echinoderms.

homeostasis (ho″me-o-sta′sis) [Gr. *homois* unchanging + *stasis* standing]. The tendency to maintain uniformity or stability in the internal environment of the organism.

hominid (hom′ĭ-nid) [L. *homo* man + Gr. *eidos* resemblance]. Pertaining to the family of human beings; a living or extinct member of the group.

hominoid (hom′ĭ-noid) [L. *homo* man + Gr. *oeidos* likeness]. Pertaining to the group (superfamily) that includes both apes and human beings.

homograft reaction The rejection by the host organism of a graft of tissue from an organism of the same species but a different genotype.

homoiothermic (ho-moi′o-ther″mik) [Gr. *homois* unchanging + *thermē* heat]. Constant-temperature animals; e.g., birds and mammals which maintain a constant body temperature despite variations in environmental temperature.

homologous structures (ho-mol′o-gus) [Gr. *homologos* agreeing, corresponding]. Those structures of various animals which arise from common rudiments and are similar in basic plan and development.

homoplastic (ho″mo-plas′tik) [Gr. *homos* same + *plasma* a thing molded]. Pertaining to a type of analogy in which there is a superficial structural similarity between analogous organs, but there is no community of evolutionary origin.

homozygous (ho″mo-zi′gus) [Gr. *homos* same + *zygos* yoke]. Possessing an identical pair of alleles at the corresponding loci of homologous chromosomes for a given character or for all characters.

hormones (hor′mōns) [Gr. *hormaein* to set in motion, spur on]. Substances produced in cells in one part of the body which diffuse or are transported by the blood stream to cells in other parts of the body where they regulate and coordinate their activities.

humus (hu′mus) [L.]. Organic matter in the soil; a dark mold of decayed vegetable tissue which gives soil a brown or black color.

hybrid vigor (hi′brid) [L. *hybrida* mongrel]. The mating of genetically dissimilar individuals of totally unrelated strains; may yield offspring better adapted to survive than either parental strain.

hydrogen bond A weak bond between two molecules formed when a hydrogen atom is shared between two atoms, one of which is usually oxygen; of primary importance in the structure of nucleic acids and proteins.

hydroids (hī′droids) [Gr. *hydra* water serpent + *odes* like]. Colonial polypoid cnidarians belonging to the class Hydrozoa.

hydrolysis (hi-drol′ĭ-sis) [Gr. *hydōr* water + *lysis* dissolution]. The splitting of a compound into parts by the addition of water between certain of its bonds, the hydroxyl group being incorporated in one fragment and the hydrogen atom in the other.

hydroskeleton (hy′drō-skel″e-tun) [Gr. *hydōr* water + *skeletos* dried up]. A turgid column of liquid within one of the body spaces that provides support or rigidity to an organism or a part, e.g., the coelom and coelomic fluid of annelid worms.

hydrozoa (hi″dro-zo′äh) [Gr. *Hydra* a mythical nine-headed monster + *zōon* animal]. A class of cnidarians usually with colonial branching polyps and small medusae.

hyoid (hi′oid) [Gr. *hyo* the letter *v* + *eidos* form]. One or more bones of visceral arch origin lodged in and supporting the base of the tongue.

hypersensitivity (hi″per-sen′sĭ-tiv′ĭ-te [Gr. *hyper* above]. A state of altered reactivity; abnormally increased sensitivity; ability to react with characteristic symptoms to the presence of certain substances (allergens) in amounts innocuous to normal individuals.

hypertonic (hi″per-ton′ik) [Gr. *hyper* above + *tonos* tone]. Having a greater concentration of solute molecules and a lower concentration of solvent (water) molecules and hence an osmotic pressure greater than that of the solution with which it is compared.

hypothalamus (hi″po-thal′ah-mus) [Gr. *hypo* under + *thalamos* inner chamber]. A region of the forebrain, the floor of the third ventricle, which contains various centers controlling visceral activities, water balance, temperature, sleep and so on.

hypothesis (hi-poth′ĕ-sis) [Gr. *hypo* under + *thesis* setting down]. A supposition assumed as a basis of reasoning which can then be tested by further controlled experiments.

hypotonic (hi″po-ton′ik) [Gr. *hypo* under + *tonos* tone]. Having a lower concentration of solute molecules and a higher concentration of solvent (water) molecules and hence an osmotic pressure lower than that of the solution with which it is compared.

ileum (il′e-um) [L. groin]. The terminal portion of the small intestine of higher vertebrates lying between the jejunum and colon.

immune reaction (ĭ-mūn′). [L. *immunis* safe]. The production of antibodies in response to antigens.

immunologic tolerance (im-mu″no-loj′ik). The ability of an organism to accept cells transplanted from a genetically distinct organism; results from exposure of the organism to an antigen before it has developed the capacity to react to it, thereafter development of capacity to react may be delayed or postponed indefinitely.

implantation (im″plan-ta′shun) [L. *in* into + *plantare* to set]. The insertion of a part of tissue in a new site in the body; the attachment of the developing embryo to the epithelial lining (endometrium) of the uterus.

imprinting A form of rapid learning by which a young bird or mammal forms a strong social attachment to an object within a few hours after hatching or birth.

indeterminate cleavage Condition in which the embryonic fate of blastomeres is fixed late in cleavage.

induction (in-duk′shun) [L. *inductio* from *inducere* to lead in]. The production of a specific morphogenetic effect in one tissue of a developing embryo through the influence of an organizer or another tissue.

inflammation (in″flah-ma′shun) [L. *inflammare* to set on fire]. The reactions of tissues to injury: pain, increased temperature, redness and accumulation of leukocytes.

ingestion (in-jes′chun) [L. *in* into + *gerere* to carry]. The act of taking food into the body by mouth.

instar The stage of an insect or arthropod between successive molts.

integument (in-teg′u-ment) [L. *integumentum* from *in* on + *tegere* to cover]. Skin, the covering of the body.

intergradation (in″ter-grā-dā′-shun) [L. *inter* between + *gradus* step]. The interbreeding of two subspecies of the same species in the area where they meet (primary intergradation) or of closely related species whose ranges overlap before isolating mechanisms (q.v.) are fully developed (secondary intergradation).

interferon (in″ter-fēr′on). A protein formed during the interaction of animal cells with viruses, which is capable of conferring on fresh animal cells of the same species resistance to infection with a wide range of viruses.

intermedin (in″ter-me′din) [L. *inter* between + *medius* middle]. A hormone secreted by the intermediate lobe of the pituitary which regulates the degree of extension of pigment cells in the skin of certain vertebrates such as frogs.

interstitial fauna The minute animals inhabiting the spaces between sand grains in the sea, fresh water or soil.

interstitial fluid (in″ter-stish′-al) [L. *inter* between + *sistere* to stand]. The liquid that bathes the cells and tissues of the body. In vertebrates, it is a filtrate of the blood plasma and differs from it in lacking most of the large protein molecules of plasma.

invagination (in-vaj″ĭ-na′shun) [L. *invaginatio* from *in* within + *vagina* sheath]. The infolding of one part within another, specifically a process of gastrulation in which one region infolds to form a double-layered cup.

inversion, **chromosomal** Turning a segment of a chromosome end for end and attaching it to the same chromosome.

ion (i′on) [Gr. *iōn* going]. An atom or a group of atoms bearing an electric charge, either positive (cation) or negative (anion).

isogamy (i-sog′ah-me) [Gr. *isos* equal + *gamos* marriage]. Reproduction resulting from the union of two gametes that are identical in size and structure.

isolating mechanism The means by which two closely related species living together in the same geographic area are prevented from mating or producing successful offspring.

isomer (i′so-mer) [Gr. *isos* equal + *meros* part]. Molecule with the same molecular formula as another but a different structural formula; e.g., glucose and fructose.

isotonic or **isosmotic** (i-so-ton′ik, i-sos-mot′-ik). Having identical concentrations of solute and solvent molecules and hence the same osmotic pressure as the solution with which it is compared.

isotopes (i′so-tōps) [Gr. *isos* equal + *topos* place]. Alternate forms of a chemical element having the same atomic number (that is, the same number of nuclear protons and orbital electrons) but possessing different atomic masses (that is, different numbers of neutrons).

isozymes (i′so-zīms) [Gr. *isos* equal + *zymē* leaven]. Different molecular forms of proteins with the same enzymatic activity.

jejunum (je-joo′num) [L. *jejunus* empty]. The middle portion of the small intestine of higher vertebrates lying between the duodenum and ileum.

juvenile hormone An arthropod hormone that preserves juvenile morphology during a molt. Without it, metamorphosis toward the adult form takes place.

karyokinesis (kar″e-o-ki-ne′sis) [Gr. *karyon* nucleus or nut + *kinesis* motion]. The phenomena involved in division of the nucleus in mitosis.

karyotype [Gr. *karyon* nucleus or nut + *typos* type]. A characterization of a set of chromosomes of an individual with regard to their number, size and shape.

keratin (ker′ah-tin) [Gr. *keratos* horn]. A horny, water-insoluble protein found in the epidermis of vertebrates and in nails, feathers, hair, horn and the like.

ketone bodies Incompletely oxidized fatty acids which are toxic in high concentrations; excreted in the urine, causing an acidosis.

kinesis (ki-ne″sis) [Gr. *kinēsis* movement]. The activity of an organism in response to a stimulus; the direction of the response is not controlled by the direction of the stimulus (in contrast to a taxis).

kinesthesis (kin″es-the′sis) [Gr. *kinésis* movement + *aisthésis* perception]. Sense which gives us our awareness of the position and movement of the various parts of the body.

kinetosome (kĭ-net′o-sōm) [Gr. *kinétos* moving + *soma* body]. A minute basal body, similar to a centriole (q.v.) associated with the bases of cilia and flagella.

kinins (ki′nins). Group of polypeptides produced in blood and tissues and acting on blood vessels, smooth muscles and certain nerve endings; e.g., bradykinin or kallidin; one of a group of compounds containing adenine that stimulates cell division and growth of plant cells in tissue culture.

labyrinthodont (lab″ĭ-rin′tho-dont) [Gr. *labyrinthos* labyrinth + *odontos* tooth]. A member of a subclass of extinct amphibians in which the enamel of the tooth was complexly invaginated into the dentin; included the first terrestrial vertebrates and the ancestors of modern amphibians and reptiles.

lamella (lah-mel′ah) [L. dim. of *lamina* plate, leaf]. A thin leaf or plate, as of bone.

larva (lar′vah) [L.]. A motile and sometimes feeding stage in the early development of certain animals.

larynx (lar′inks) [Gr. "the upper part of the windpipe"]. The cartilaginous structure located at the entrance of the trachea which functions secondarily as the organ of voice.

latent period An interval, lasting about 0.01 second, between the application of a stimulus and the beginning of the visible shortening of a muscle.

leuconoid (lu′kon-oid) [Gr. *leukos* white + *eidos* form]. In sponges, a body plan similar to that of the sponge leucon, in which the flagellated cells are located within chambers.

leukemia (lu-ke′me-ah) [Gr. *leukos* white + *haima* blood]. A type of cancer, characterized by the abnormally rapid growth of white blood cells.

leukocytes (lu′ko-sīts) [Gr. *leukos* white + *kytos* cell]. White blood cells; colorless cells exhibiting phagocytosis and ameboid movement.

linkage The tendency for a group of genes located in the same chromosome to be inherited together in successive generations.

lipase (lip′ăss [Gr. *lipos* fat]. An enzyme that catalyzes the hydrolysis of fats; it cleaves the ester bonds joining fatty acids to glycerol.

liter (lē′ter) [Gr. *litra* a unit of weight]. The unit of volume in the metric system. Defined for most practical purposes as the volume of 1000 grams of pure water under specified conditions. Subdivisions of the liter use the same prefixes, factors and abbreviations as used for subdivisions of the meter (q.v.).

littoral (lit′o-ral) [L. *litoralis* the seashore]. The region of shallow water near the shore between the high and low tide marks.

locus (lo′kuss) [L. place]. The particular point on the chromosome at which the gene for a given trait occurs.

loop of Henle (hen′lē) [*Henle* German anatomist of the nineteenth century]. The U-shaped loop of a mammalian kidney tubule which dips down into the medulla; lies between the proximal and distal convoluted tubules.

lophophore (lof′o-for) [Gr. *lophos* crest, tuft + *phore*, from *phorein*, to bear]. The circular or horseshoe-shaped ridge with a set of ciliated tentacles around the mouth of bryozoans, phoronids and brachiopods.

luciferin (lu-sif′er-in) [L. *lux* light + *ferre* to bear]. Substrate present in certain organisms capable of bioluminescence, producing light, when acted upon by the enzyme luciferase.

lumen (lu′men) [L. *lumen* light or an opening through which light passes]. The cavity of a tubular organ such as the intestine.

lymph (limf) [L. *lympha* lymph]. The colorless fluid which is derived from blood plasma and resembles it closely in composition; contains white cells, some of which enter the lymph capillaries from the interstitial fluid, others of which are manufactured in the lymph nodes.

lymph node (limf nōd) One of the nodules of lymphatic tissue that occur in groups along the course of the lymphatic vessels; produces and contains lymphocytes and phagocytic cells.

lysis (li′sis) [Gr. *lysis* loosening]. The process of disintegration or solution of a cell or some other structure.

lysosome (lī′so-sōm). Intracellular organelle present in many animal cells; contains a variety of hydrolytic enzymes that are released when the lysosome ruptures.

macromere (mak′ro-mēr) [Gr. *makros* long in extent + *meros* a portion]. The larger cell resulting from unequal cell division.

macronucleus (mak″ro-nu′kle-us) [Gr. *makros* + L. *nucleus* kernel]. The large nucleus in ciliates that governs activities not associated with reproduction.

macrophage (mak″rō-făge) [Gr. *makros* large + *phagein* to eat]. Large phagocytic cells that differentiate from certain leukocytes at the site of an infection and phagocytize cellular debris.

make-ready reaction A term for metabolic reactions that result in the formation of a configuration of the substrate molecule suitable for a subsequent dehydrogenation reaction.

malpighian tubule The excretory organ of many arthropods, named for the seventeenth century Italian anatomist Marcello Malpighi.

mammal (mam′al) [L. *mamma* breast]. A member of a class of vertebrates characterized by having hair and mammary glands; includes such diverse types as shrews, bats, cats, whales, cattle and humans.

mandibulata (man-dĭb′-u-lah-tah). Subphylum of arthropods containing shrimp, crabs and insects in which there are antennae and the anterior feeding appendages are mandibles.

mantle Integument of mollusks that is covered by and secretes the shell.

marsupials (mar-su′pe-als) [L. *marsupium* a pouch]. A group of mammals characterized by the possession of an abdominal pouch in which the young are carried for some time after being born in a very immature condition.

mastigophora (mas″ti-gaf′o-rah) [Gr. *mastix* whip + *pherein* to bear]. Phylum of protozoans containing species that possess one or more flagella.

mating type In Protozoa, a sex. As many as eight sexes are known in some species.

matrix (ma′triks) [L. *mater* mother]. Nonliving material secreted by and surrounding the connective tissue cells; frequently contains a thick, interlacing matted network of microscopic fibers.

medulla (me-dul′lah) [L. from *medius* middle]. The inner part of an organ, e.g., the medulla of the kidney; the most posterior part of the brain, lying next to the spinal cord.

medusa (me-du′sah) [Gr. *medousa*]. A jellyfish; a free-swimming, umbrella-shaped form in the life cycle of certain cnidarians.

meiosis (mi-o′sis) [Gr. *meiōsis* diminution]. Kind of nuclear division, usually two successive cell divisions, which results in daughter cells with the haploid number of chromosomes, one half the number of chromosomes in the original cell.

melanin (mel′ah-nin) [Gr. *melas* black]. A dark-brown to black pigment common in the integument of many animals and sometimes found in other organs; usually occurs within special pigment cells.

membranelle An organelle formed by the fusion of a row of cilia to produce a small membrane.

menopause (men′o-pawz) [Gr. *men* month; *meniaia* the menses + *pausis* cessation]. The period (from 40 to 50 years of age) when the recurring menstrual cycle ceases.

menstruation (men″stroo-a′shun) [L. *menstrualis* monthly]. The cyclic, physiologic uterine bleeding which normally recurs, usually at approximately four-week intervals, in the absence of pregnancy during the reproductive period of the female.

merozoite (mer″o-zo′ĭt) [Gr. *meros* part + *zōon* animal]. One of the young forms derived from the splitting up of the schizont in the human cycle of the malarial parasite, plasmodium; it is released into the circulating blood and attacks new erythrocytes.

mesencephalon (mes″en-sef′ah-lon) [Gr. *mesos* middle + *enkephalos* brain]. The middle subdivision of the brain lying between the diencephalon and metencephalon; includes the superior and inferior colliculi.

mesenchyme (mes′eng-kīm) [Gr. *mesos* middle + *enchyme* an infusion]. A meshwork of loosely associated, often stellate cells; found in the embryos of vertebrates and the adults of some invertebrates.

mesentery (mes′en-ter″e) [Gr. *mesos* middle + *enteron* intestine]. One of the membranes in vertebrates that extend from the body wall to the visceral organs or from one organ to another; consists of two layers of coelomic epithelium and enclosed connective tissue, vessels and nerves.

mesoderm (mes′o-derm) [Gr. *mesos* middle + *derma* skin]. The middle layer of the three primary germ layers of the embryo, lying between the ectoderm and the endoderm.

mesoglea (mes″o-gle′ah) [Gr. *mesos* middle + *gloia* glue]. A gelatinous matrix located between the epidermis and gastrodermis of cnidarians.

mesonephros (mes″o-nef′ros) [Gr. *mesos* middle + *nephros* kidney]. An embryonic vertebrate kidney which succeeds the pronephros; its tubules develop adjacent to the middle portion of the coelom and drain into the archinephric duct.

messenger RNA A particular kind of ribonucleic acid which is synthesized in the nucleus and passes to the ribosomes in the cytoplasm; combines with RNA in the ribosomes and provides a template for the synthesis of an enzyme or some other specific protein.

metabolism (mĕ-tab′o-lizm) [Gr. *metaballein* to turn about, change, alter]. The sum of all the physical and chemical processes by which living organized substance is produced and maintained; the transformations by which energy and matter are made available for the uses of the organism.

metamerism (met-am′er izm) [Gr. *meta* with + *meros* part]. The division of the body into a linear series of similar parts or segments, as in annelids and chordates.

metamorphosis (met″ah-mor′fo-sis) [Gr. *meta* after, beyond, over + *morphosis* a shaping, bringing into shape]. An abrupt transition from one developmental stage to another, e.g., from a larva to an adult.

metanephridium (met″ah-nĕ-frid′e-um) [Gr. *meta* after + *nephros* kidney]. An excretory tubule of invertebrates in which the inner end opens into the coelom by way of a ciliated funnel.

metanephros (met″ah-nef′ros) [Gr. *meta* after, beyond, over + *nephros* kidney]. The adult kidney of reptiles, birds and mammals.

metaphase (met′ah-fāz) [Gr. *meta* after, beyond, over + *phasis* to make to appear]. The middle stage of mitosis during which the chromosomes line up in the equatorial plate and separate lengthwise.

metazoa (met″ah-zo′ah) [Gr. *meta* after, beyond, over + *zōon* animal]. All multicellular animals whose cells become differentiated to form tissues; all animals except the protozoa.

metencephalon (met″en-sef′ah-lon) [Gr. *meta*

after + *enkephalos* brain]. A major subdivision of the brain lying between the mesencephalon and myelencephalon; includes the cerebellum and pons.

meter (mē′ter) [Gr. *metron* measure]. The unit of length in the metric scale; abbreviated m. Defined as 1,650,763.73 wave lengths of krypton under specified conditions; equal to 39.3701 inches. Common multiples and divisions of the meter and their abbreviations are:

> kilometer, 10^3m, km
> centimeter, 10^{-2}m, cm
> millimeter, 10^{-3}m, mm
> micrometer (micron), 10^{-6}m, μm
> nanometer, 10^{-9}m, nm

micromere (mi′kro-mēer) [Gr. *mikros* small + *meros* part]. The smaller cell, following unequal cell division.

micronucleus (mi″kro-nu′kle-us) [Gr. *mikros* + L. *nucleus* kernel]. A small nucleus that governs reproduction in ciliates.

microtubule A cytoplasmic organelle, an elongate slender tube; contains a specific protein, tubulin.

mimicry (mim′ik-re″) [Gr. *mimos* to imitate]. An adaptation for survival in which an organism resembles some other living or nonliving object.

miracidium (mir″ah-sid′ĭ-um) [Gr. *meirakidion* youthful person]. The first larval stage of parasitic flukes.

mitochondria (mīt″-o-kon′dre-ah) [Gr. *mitos* thread + *chondrion* granule]. Spherical or elongate intracellular organelles which contain the electron transmitter system and certain other enzymes; site of oxidative phosphorylation.

mitosis (mi-to′sis) [Gr. *mitos* thread + *osis* state or condition]. A form of cell or nuclear division by means of which each of the two daughter nuclei receives exactly the same complement of chromosomes as the parent nucleus had.

mixture A solution made up of two or more kinds of atoms or molecules which may be combined in varying proportions.

mole (mōl) [L. *moles* a shapeless mass]. The amount of a chemical compound whose mass in grams is equivalent to its molecular weight, the sum of the atomic weights of its constituent atoms.

molecule (mol′ĕ-kul) [L. *molecula* little mass]. The smallest particle of a covalently bonded element or compound having the composition and properties of a larger part of the substance.

mollusca (mol-lus′kah) [L. *molluscus*, soft]. A phylum of soft-bodied animals usually enclosed in shells, including snails and clams.

molting [L. *mutare* to change]. The shedding and replacement of an outer covering such as hair, feathers and exoskeleton.

monera (mo-ne′rah) [Gr. *monērēs* single]. A category of organisms that includes the simplest microorganisms, the bacteria and blue-green algae, forms lacking true nuclei or plastids and in which sexual reproduction is very rare or absent.

mongolism (mon′go-lizm) A congenital malformation in which individuals have abnormalities of the face, eyelids, tongue and other parts of the body and are greatly retarded in both their physical and mental development; results from a trisomy of chromosome 21 or 18.

monomer (mon′o-mer) [Gr. *monos* single + *meros* part]. A simple molecule of a compound of relatively low molecular weight which can be linked with others to form a polymer.

monophyletic (mon″o-fi-let′ik) [Gr. *monos* single + *phyle* tribe]. Referring to a group of organisms evolved from the same ancestor.

monoplacophora (mon″-o-pla-kof′-o-rah) [Gr. *monos* one + *plax* plate + *pherein* bearer]. Class of the phylum Mollusca containing a small number of deep-sea species possessing a single shield-shaped shell.

monotreme (mon′o-trēm) [Gr. *monos* one + *trema* hole]. A member of the most primitive order of living mammals characterized by the retention of the cloaca, into which the digestive and urogenital tracts discharge, and an egg-laying habit; includes the duckbilled platypus and spiny anteater.

morphogenesis (mor″fo-jen′ĕ-sis) [Gr. *morphe* form + *gennan* to produce]. The development of form, size and other features of a particular organ or part of the body.

mosaic development Embryonic development in which the capacities of the cells are restricted to the structures they normally form.

motor unit (mo′tor u′nit). All the skeletal muscle fibers that are stimulated by a single motor neuron.

mucosa (mu-ko′sah). Mucous membrane; e.g., the lining of the digestive tract.

multiple alleles Three or more alternate conditions of a single locus which produce different phenotypes.

mutation A stable, inherited change in a gene.

mutualism (mu′tu-al-izm). An association whereby two organisms of different species each gain from being together and are unable to survive separately.

myelencephalon (mi″ĕ-len-sef′ah-lon) [Gr. *myelos* marrow + *enkephalos* brain]. The most posterior of the five major subdivisions of the brain; comprises the medulla oblongata.

myelin (mi′ĕ-lin) [Gr. *myelos* marrow]. The fatty material which forms a sheath around the axons of nerve cells in the central nervous system and in certain peripheral nerves.

myofibrils (mi″o-fi′brils) [Gr. *mys* muscle + L. *fibrilla* small fiber]. The contractile fibrils visible in muscle tissue with light microscopy. Composed of groups of myofilaments of actin (q.v.) and myosin.

myomere (mi′o-mēr) [Gr. *mys* muscle + *meros* part]. The muscle segment of an animal; sometimes restricted to the adult segment.

myopia (mi-o'pe-ah) [Gr. *myein* to shut + *ōps* eye]. Nearsightedness; the eyeball is too long and the retina too far from the lens; light rays converge at a point in front of the retina, and are again diverging when they reach it, resulting in a blurred image.

myosin (mi'o-sin) [Gr. *mys* muscle]. Minute protein filaments found in most contractile cells. Usually associated with actin (q.v.).

myotome (mi'o-tōm) [Gr. *mys* muscle + *tomos* section]. The muscle segment of animal; sometimes restricted to the embryonic segment.

myxedema (mik"sĕ-de'mah) [Gr. *myxa* mucus + *oidema* swelling]. A condition which results from a deficiency of thyroxin secretion in an adult; characterized by a low metabolic rate and decreased heat production.

nares (na'rēz) [L. *nares* nostrils]. The openings of the nasal cavities. External nares open to the body surface; internal nares, to the pharynx.

nauplius (no'plĭ-us) [L. a kind of shellfish]. A larva with three pairs of appendages—future head limbs—characteristic of the crustaceans.

nekton (nek'ton) [Gr. *nēktos* swimming]. Collective term for the organisms which are active swimmers.

nematocyst (nem'ah-to-sist) [Gr. *nematos* thread + *kystis* bladder]. A minute stinging structure found on cnidarians and used for anchorage, for defense and for the capture of prey.

nematoda (nem"ah-to'dah) [Gr. *nematos* thread]. The roundworms, a large phylum of pseudocoelomates, many of which are parasitic.

nemertea (nem-ur'te-ah) [After Nemertes, one of the Nereids of Greek mythology]. The ribbon worms, a phylum with many marine and a few freshwater species.

nephridium (nĕ-frid'e-um) [Gr. *nephros* kidney]. The excretory organ of the earthworm and other annelids which consists of a ciliated funnel, opening into the next anterior coelomic cavity and connected by a tube to the outside of the body.

nephron (nef'ron) [Gr. *nephros* kidney]. The anatomical and functional unit of the vertebrate kidney.

nerve (nerv) [L. *nervus* nerve]. A cordlike collection of neurons and associated connective tissue that extends between the central nervous system and other parts of the body; most nerves contain both afferent and efferent neurons.

nerve net A relatively unorganized, diffuse net of nerve cells with no obvious directionality in the transmission of impulses.

neuroglia (nū-rŏg'lĭ-ă). Connecting and supporting cells in the central nervous system surrounding the neurons.

neurohumor (nu"ro-hu'mor) [Gr. *neuron* nerve + L. *humor* a liquid]. A substance secreted by the tip of a neuron which is able to activate a neighboring neuron or muscle.

neuron (nu'ron) [Gr. *neuron* nerve]. A nerve cell with its processes, collaterals and terminations; the structural unit of the nervous system.

neurosecretion (nu"ro-se-kre'shun) [Gr. *neuon* nerve + L. *secretio*, from *secrenere* to secrete]. The production of hormones by nerve cells.

neuroses (nu-ro'sēs) [Gr. *neuron* nerve + *osis* state or condition]. Comparatively mild and common psychic disorders with a great variety of symptoms: anxiety, fear, shyness and oversensitiveness.

neurula (nu'roo-lah) [Gr. *neuron* nerve]. The early embryonic stage during which the primitive nervous system forms.

neutrons (nu'trons). Electrically uncharged particles of matter existing along with protons in the atomic nucleus of all elements except the mass 1 isotope of hydrogen.

nondisjunction (non"dis-junk'shun). The failure of a pair of homologous chromosomes to separate normally during the reduction division at meiosis; both members of the pair are carried to the same daughter nucleus and the other daughter cell is lacking in that particular chromosome.

notochord (no'to-kord) [Gr. *noton* back + *chorde* cord]. The rod-shaped body in the anteroposterior axis which serves as an internal skeleton in the embryos of all chordates and in the adults of some; replaced by a vertebral column in most adult vertebrates.

notum (no'tum) [Gr. *noton* back]. The dorsal part of the body. In arthropods, the dorsal element of each segment.

nuclease (nu'kle ās) [L. *nucleus* a kernel]. An enzyme that facilitates the hydrolysis of nucleic acids.

nucleolus (nu-kle'o-lus) [L. dim. of *nucleus*, dim. of *nux* nut]. A spherical body found within the cell nucleus; rich in ribonucleic acid and believed to be the site of synthesis of ribosomes.

nucleotide (nu'kle-o-tīd). A molecule composed of a phosphate group, a 5-carbon sugar—ribose or deoxyribose—and a nitrogenous base—a purine or a pyrimidine; one of the subunits into which nucleic acids are split by the action of nucleases.

nucleus (nu'kle-us) [L. *nucleus* a kernel]. The organelle of a cell containing the hereditary material; a group of nerve cell bodies in the central nervous system.

nutrient (nu'tre-ent) [L. *nutriens*]. A general term for any substance which can be used in the metabolic processes of the body.

nymph (nimf) [L. *nympha* young woman]. A juvenile insect that often resembles the adult and that will become an adult without an intervening pupal stage.

ocellus (o-sel′us) [L. diminutive of *oculus* eye]. A simple light receptor found in many different types of invertebrate animals.

olfaction (o-fak′shun) [L. *olfacere* to smell]. The act of smelling.

oligochaeta (ol″ig-o-ke′tah) [Gr. *oligos* scant + *chaite* hair]. A class of annelids bearing few setae; includes earthworms.

ommatidium (om″ah-tid′ĭ-um) [Gr. diminutive of *omma* eye]. One of the elements of a compound eye, itself complete with lens and retina.

ontogeny (on-toj′ĕ-ne) [Gr. *ōn* existing + *gennan* to produce]. The complete developmental history of the individual organism.

onychophora (on-ĭ-kof′o-rah) [Gr. *onyx* nail + *phoros* bearing]. Rare, tropical, caterpillar-like animals structurally intermediate between annelids and arthropods, possessing an annelid-like excretory system, an insect-like respiratory system and claw-tipped short legs.

oögenesis (o″o-jen′e-sis) [Gr. *ōon* egg + *genesis* production] The origin and development of the ovum.

oögonium (o″o-go′ne-um) [Gr. *ōon* egg + *gonē* generation]. The primordial cell from which the ovarian egg arises; undergoes growth to become a primary oöcyte.

operon The genes whose codes are transcribed on a single mRNA molecule and are under the control of a single repressor.

ophiuroidea (o″fi-u-roi′de-ah) [Gr. *ophis* snake + *oura* tail + *eidos* form]. The brittle stars, a subclass of echinoderms with snakelike arms.

opisthonephros (o″pis-tho-nef′ros) [Gr. *opisthen* behind + *nephros* kidney]. The adult kidney of most fishes and amphibians; its tubules extend from the mesonephric region to the posterior end of the coelom; drained by the archinephric duct and sometimes also by accessory urinary ducts.

orbital (or′bĭ-tal) [L. *orbitalis* mark of a wheel]. The distribution of an electron around the atomic nucleus.

order In taxonomy, a major subdivision of a class. Each order is composed of one or more related families.

organ of Corti (cōr′tĭ) [*Corti* Italian anatomist of the nineteenth century]. The organ attached to the basilar membrane in the cochlear duct which contains the cells receptive to sound.

organelle (or″gan-el′) [Gr. *organon* bodily organ]. One of the specialized structures within a cell, e.g., the mitochondria, Golgi complex, ribosomes, contractile vacuole and so on.

organizer A part of an embryo which influences some other part and directs its histological and morphological differentiation.

orthogenesis (or″tho-jen′ĕ-sis) [Gr. *orthos* straight + *genesis* production]. Evolution progressing in a given direction; straight-line evolution.

osculum (os′ku-lum) [L. little mouth]. In sponges, the large excurrent openings.

osmoconformers (oz″mo-con-for′mers). Marine animals in which the salt concentration of the extracellular body fluids conforms to, or is the same as, that of the environment.

osmosis (os-mo′sis) [Gr. *ōsmos* impulsion]. The passage of solvent molecules from the lesser to the greater concentration of solute when two solutions are separated by a membrane which selectively prevents the passage of solute molecules but is permeable to the solvent.

osphradium (oz-fra′di-um) [Gr. *oshradion* strong scent]. Sense organ in the mantle cavity of many mollusks.

osteichthyes (os″te-ik′thĭ-ēz) [Gr. *osteon* bone + *ichthys* fish]. A class of fishes in which the skeleton is composed, at least in part, of bone; includes sturgeons, teleosts, lungfish.

ostracoderm (os′trä-ko-derm″) [Gr. *ostrakon* shell + *derma* skin]. A member of one of several orders of primitive jawless fishes which were abundant during the Devonian; the ancestral vertebrates.

outbreeding The mating of individuals of unrelated strains.

ovary (o′vah-re) [L. *ovaria*]. The female gonad which produces eggs.

ovulation (ōv″u-la′shun) [L. *ovulum* little egg + *atus* process or product]. The discharge of a mature ovum from the graafian follicle of the ovary.

ovum (o′vum) [L. *ovum* egg]. The female reproductive cell, which after fertilization by a sperm, develops into a new member of the same species.

oxidation The process in which electrons are removed from an atom or molecule.

oxidative phosphorylation (ok″sĭ-da′tiv fos″fōr-ĭ-la′shun). The conversion of inorganic phosphate to the energy-rich phosphate of ATP by reactions coupled to the transfer of electrons in the electron transmitter system of the mitochondria.

oxygen debt The amount of oxygen required to oxidize completely the lactic acid that accumulates in the muscle tissue during vigorous exercise.

pacemaker The part whose rate of reaction sets the pace for a series of interrelated reactions; e.g., the sinoatrial node initiates the heart beat and regulates the rate of contraction of the heart.

palps Appendages located in the mouth region involved in feeding.

parallel evolution (par′ă-lel) [Gr. *para* beside + *allos* other]. The independent evolution of similar structures in two or more rather closely related organisms, e.g., the independent evolution of quills from hair by American and African porcupines.

paramylum Carbohydrate storage compound present in the euglenoids, chemically distinct from both starch and glycogen.

paramyosin (par-ă-mi'o-sin) [Gr. *para* beside + *mys* muscle]. A type of protein found in the shell adductor muscle of many bivalved mollusks; characterized by the ability to sustain a contraction for long periods with minimal energy expenditure.

parapodia (par"ah-po'de-ah) [L. *para* beyond + Gr. *podion* little foot]. Paired appendages extending laterally from each segment of polychaete worms and bearing setae.

parasitism (par'ah-sīt"izm) [Gr. *parasitos* one who eats at the table of another + *ismos* condition]. A type of heterotrophic nutrition found among both plants and animals; a parasite lives in or on the living body of a plant or animal (host) and obtains its nourishment from it.

parasympathetic (par'ah-sim"pah-thet'ik) [Gr. *para* beyond + *sym* with + *pathos* feeling]. A segment of the autonomic nervous system; fibers originate in the brain and the pelvic region of the spinal cord and innervate primarily the internal organs.

parathyroids (par"ah-thi'roids) [Gr. *para* beyond + *thyreoeides* shieldlike]. Small, pea-sized glands situated in the substance of the thyroid gland; their secretion is concerned chiefly with regulating the metabolism of calcium and phosphorus by the body.

parthenogenesis (par"thĕ-no-jen'ĕ-sis) [Gr. *parthenos* virgin + *genesis* production]. The development of an unfertilized egg into an adult organism; common among honeybees, wasps and certain other arthropods.

parturition (par"tu-rish'un) [L. *parturito* childbirth]. The process of giving birth to a child.

pelagic (pe-laj'ik) [Gr. *pelagios* living in the sea]. An organism which inhabits open water, as in midocean.

pelecypoda (pel"e-sip'o-dah) [Gr. *pelekys* hatchet + *podos* foot]. The bivalves, a class of mollusks with a spadelike foot for digging.

penetrance (pen'ĕ-trans) [L. *penetrare* to enter into]. The expression of the frequency with which a heritable trait is shown in individuals carrying the principal gene or genes conditioning it.

penis (pe'nis) [L.]. The copulatory organ of the male; found in most of those species of animals in which fertilization is internal.

pepsin (pep'sin) [L. *pepsinum* from Gr. *pepsis* digestion]. A proteolytic enzyme secreted by the cells lining the stomach: functions only in a very acid medium and works optimally at pH 2.

pericardial sinus An enlarged portion of the hemocoel (q.v.) surrounding the heart in many invertebrates with open circulatory systems.

pericardium (per"ĭ-kar'de-um) [Gr. *perikardios*, near the heart]. The lining of that part of the coelom that forms a separate chamber containing the heart.

perilymph (per'ĭ-limf) [Gr. *peri* around + L. *lympha* lymph]. The lymphlike fluid which lies between the membranous labyrinth and the bone or cartilage encapsulating the inner ear; the scala tympani and scala vestibuli of the mammalian choclea are perilymphatic channels.

peripheral resistance The state of constriction or relaxation of the blood vessels; plays an important role in determining blood pressure.

peristalsis (per"ĭ-stal'sis) [Gr. *peri* around + *stalsis* contraction]. Powerful, rhythmic waves of muscular contraction and relaxation in the walls of hollow tubular organs such as the ureter or the parts of the digestive tract; serve to move the contents through the tube.

peritoneum (per"ĭ-to-ne'um) [Gr. *peritonos* stretched over]. Coelomic epithelium and supporting connective tissue which lines a coelomic cavity.

peritrophic membrane (per"ĭ-trof'ic) [Gr. *peri* around + *trophe* food]. In insects, a cylindrical sheath of chitin continuously secreted from the posterior edge of the foregut (stomadeum); encloses the gut contents.

permeability (per"me-ah-bil'ĭ-te) [L. *per* through + *meare* to pass]. The property of a membrane that permits the passage of a given substance.

pH The negative logarithm of the hydrogen ion concentration, by which the degree of acidity or alkalinity of a fluid may be expressed.

phagocytosis (fag"o-si-to'sis) [Gr. *phagein* to eat + *kytos* hollow vessel + *osis* state or condition]. The engulfing of microorganisms, other cells and foreign particles by a cell such as a white blood cell.

pharynx (far'inks) [Gr.]. Anterior region of the gut, generally muscular and adapted for ingestion. In vertebrates, that part of the digestive tract from which the gill pouches or slits develop; in higher vertebrates it is bounded anteriorly by the mouth and nasal cavities and posteriorly by the esophagus and larynx.

phenocopy (fe'no-kop"e) [Gr. *phainein* to show + L. *copia* abundance, number]. The simulation by an individual of traits characteristic of another genotype; results from physical or chemical influences in the environment which change the course of development and produce a trait which mimics that of an individual with a different genotype.

phenotype (fe'no-tīp) [Gr. *phainein* to show + *typos* type]. The outward, visible expression of the hereditary constitution of an organism.

pheromone (fēr'o-mōn) [Gr. *phorein* to carry]. A substance secreted by one organism to the external environment which influences the development or behavior of other members of the same species.

phoronida (fo-ro'nid-ah) [From Phoronis, the surname of Io in Greek mythology]. A small phylum of lophophorate worms related to brachiopods.

phosphorescence (fos″fo-res′ens) [Gr. *phōs* light + *phorein* to carry]. The emission of light without appreciable heat, caused by the decay of a molecule in the triplet state to the ground state.

phosphorylation (fos″fōr-ĭ-la′shun) [Gr. *phōs* light + *phorein* to carry]. The introduction of a phosphate group into an organic molecule.

photon (fo′ton) [Gr. *phōs* light + *ton* slice]. A particle of electromagnetic radiation, one quantum of radiant energy.

photoperiodism (fo″to-pe′re-od-izm) [Gr. *phōs* light + *peri* around + *hodos* way + *ismos* state]. The physiologic response of animals and plants to variations of light and darkness.

photosynthesis (fo″to-sin′the-sis) [Gr. *phōs* light + *synthesis* putting together]. The process of synthesizing carbohydrates from carbon dioxide and water, utilizing the radiant energy of light captured by the chlorophyll in plant cells.

phylogeny (fi-loj′e-ne) [Gr. *phylon* tribe + *genesis* generation]. The evolutionary history of a group of organisms.

phylum (fi′lum) [Gr. *phylon* race]. A primary, large, main division of the animal or plant kingdom, including organisms which are assumed to have a common ancestry.

phytoplankton (fi″to-plank′ton) [Gr. *phyton* plant + *planktos* wandering]. Microscopic floating plants, most of which are algae, which are distributed throughout the ocean or a lake.

pi electrons Mobile electrons located in a system of conjugated single and double bonds which are associated not with a single atom or bond but with the conjugated system as a whole.

pinocytosis (pi″no-si-to′sis) [Gr. *pinein* to drink + *kytos* cell + *osis* state or condition]. "Cell drinking"; the engulfing and absorption of droplets of liquids by cells.

pituitary (pĭ-tu′ĭ-tār″e) [L. *pituitarius* secreting phlegm]. A small gland which lies just below the hypothalamus of the brain, to which it is attached by a narrow stalk; the anterior lobe forms in the embryo as an outgrowth of the roof of the mouth and the posterior lobe grows down from the floor of the brain.

placenta (plah-sen′tah) [L. a flat cake]. A structure formed in part from tissues derived from the embryo and in part from maternal tissues—the lining of the uterus—by means of which the embryo receives nutrients and oxygen and eliminates wastes.

placodermi (plak′o-der″mi) [Gr. *plakos* a flat plate + *derma* skin]. Several orders of Paleozoic jawed fishes with a bony head shield movably articulated with a thoracic shield; probably related to the cartilaginous fishes.

plankton (plank′ton) [Gr. *planktos* wandering]. Minute plants and animals suspended in fresh or salt water.

planula (plan′u-lah) [L. *planus* flat]. Larval stage of cnidarians (hydras, jellyfish, sea anemones and corals).

plasma membrane (plaz′mah mem′brān) [Gr. *plasma* anything formed or molded + L. *membrana* skin covering]. A living, functional part of the cell through which all nutrients entering the cell and all waste products or secretions leaving it must pass.

plasmodium (plaz-mo′de-um) [Gr. *plasma* anything formed + *odēs* like]. Multinucleate, ameboid mass of living matter that comprises the diploid phase of slime molds; single-celled animals that reproduce by spore formation and cause malaria.

plasmolysis (plaz-mol′ĭ-sis) [Gr. *plasma* anything formed + *lysis* dissolution]. Contraction of the cytoplasm of a cell due to the loss of water by osmotic action.

platelet (plāt′let) [Gr. *platē* a flattened surface]. A small colorless blood corpuscle of mammals that plays an important role in blood coagulation and in the contraction of the clot.

platyhelminthes (plat″ĭ-hel-min′thēz) [Gr. *platys* flat + *helminthos* worms]. The phylum of flatworms, flukes and tapeworms.

pleiotropic gene (plī″ō-trō′pik) [Gr. *pleiōn* more + *tropos* turn]. A gene that affects a number of different characteristics in a given individual.

pleuron (ploor′on). [L. one of the side of an animal]. In arthropods, a lateral skeletal piece of any segment.

plexus (plek′sus) [L. *plexus* a braid]. A network of interconnecting structures such as nerves, e.g., the brachial plexus of nerves supplying the arm.

ploidy (ploi′de) [Gr. *ploos* fold + *odēs* like, resembling]. Relating to the number of sets of chromosomes in a cell.

poikilothermic (poi″kī-lo-ther′mik) [Gr. *poikilos* varied + *thermē* heat]. Having a body temperature that flucutates with that of the environment; "cold-blooded."

polar body Small cell which consists of practically nothing but a nucleus; formed during oögenesis, maturation of the egg, and appears as a speck at the animal pole of the egg.

polarity In biology, the tendency of a piece of an organism to retain its original body orientation, regenerating a head at the original anterior end, and so on.

polychaeta (pol″e-ke′tah) [Gr. *polys* many + *chaite* hair]. A class of annelids with parapodia bearing numerous bristles.

polygenes (pol″e-jēns′) [Gr. *polys* many + *gennan* to produce]. Two or more pairs of genes that affect the same trait in an additive fashion.

polymorphism (pol″e-mor′fizm) [Gr. *poly* + *morphē* form]. Differences in form among the members of a species; occurrence of several distinct phenotypes in a population.

polyphyletic (pol″e-fi-let′ik) [Gr. *polys* many +

phylē tube]. Referring to a group of organisms that are classified together but have evolved from different ancestors.

polyplacophora (pol"e-pla-kof'-o-rah [Gr. *polys* many + *plax* plate + *pherein* bearer]. Class of the phylum Mollusca containing species known as chitons.

polyploids (pol"e-ploids') [Gr. *polys* many + *ploos* folds]. Organisms which have more than two full sets of homologous chromosomes.

polyps (pol'ips) [Gr. *polypous* morbid excrescences]. Hydra-like animals; the sessile form of many cnidarians; protruding growths from a mucous membrane.

pons (ponz) [L. bridge]. The ventral portion of the metencephalon; it relays certain impulses from the cerebrum to the cerebellum and interconnects the two sides of the cerebellum.

population The group of individuals of a given species inhabiting a specified geographic area.

porifera (po-rif'ĕ-rah) [L. *porus* pore + *ferre* to bear]. The phylum of sponges; the body is perforated with many pores to admit water, from which food is strained.

portal system (por'tal) [L. *porta* a gate]. A group of veins which drain one region and lead to a capillary bed in another organ rather than directly to the heart, e.g., the renal portal system and hepatic portal system.

preadaptation The acquisition in an ancestral group of certain characteristics that usually are adaptive to the ancestral mode of life yet at the same time enable a shift in mode of life, e.g., lungs in fish ancestral to tetrapods.

precursor (pre-kur'sor) [L. *praecursor* to run before]. A substance which precedes another substance in a metabolic pathway; a substance from which another substance is synthesized.

predation (pre-da'shun) [K. *praedatio* to plunder]. Relationship in which one species adversely affects the second but cannot live without it; the first species kills and devours the second.

primitive streak (prim'ĭ-tiv) [L. *primitivus* first in point of time]. A longitudinal groove which develops on the embryonic disc of the eggs of fishes, reptiles, birds and mammals as a consequence of the movement of cells and formation of mesoderm; it is homologous to the lips of the blastopore and marks the future longitudinal axis of the embryo.

primordium (pri-mor'de-um) [L. "the beginning"]. The earliest discernible indication during embryonic development of an organ or part.

proboscis (pro-bos'is) [Gr. *pro* before + *boskein* to feed, graze]. Any tubular process of the head or snout of an animal, usually used in feeding.

progeny selection (proj'e-ne se-lek'shun) [L. *progignere* to bring forth]. A breeding program in which the genotype is determined by making test matings and observing the offspring.

progesterone (pro-jes'ter-ōn) [L. *pro* before + *gestus* to bear, carry, conduct]. The hormone produced in the corpus luteum of the ovary and in the placenta; acts with estradiol to regulate estrous and menstrual cycles and to maintain pregnancy.

proglottid (pro-glot'id) [L. *pro* before, in front of + *glottis* the tongue]. The body sections of a tapeworm.

prokaryotic [L. *pro* before + Gr. *karyon* nucleus or nut]. Applied to organisms that lack membrane-bound nuclei, plastids and Golgi apparatus; the bacteria and blue-green algae.

pronephros (pro-nef'ros) [Gr. *pro* before + *nephros* kidney]. The first formed kidney of embryonic or larval vertebrates; its tubules develop adjacent to the cranial end of the coelom and form the archinephric duct.

prophase (pro'fāz) [L. *pro* before + Gr. *phasis* an appearance]. The first stage in mitosis, during which the chromatin threads condense, distinct chromosomes become evident and a spindle forms.

prosimian (pro"sim'ĭ-an) [L. *pro* before, in front of + *simia* an ape]. A member of the more primitive of two primate suborders; an ancestral primate or a contemporary lemur, loris or tarsier or one of their allies.

prostate (pros'tāt) [Gr. *prostates* one who stands before]. The largest accessory sex gland of male mammals; it surrounds the urethra at the point where the vasa deferentia join it and secretes a large portion of the seminal fluid.

prosthetic group (pros-thet'ik) [Gr. "a putting to, addition"]. A cofactor tightly bound to an enzyme.

protean behavior (pro'te-un) [From Proteus, the sea god of Greek mythology who changed shape unpredictably when seized]. An irregular, unpredictable sequence of movements by prey when pursued by predators.

protease (pro'te-ās). An enzyme that catalyzes the digestion of proteins.

proteins (pro'te-ins) [Gr. *prōtos* first]. Macromolecules containing carbon, hydrogen, oxygen, nitrogen and usually sulfur and phosphorus; composed of chains of amino acids bound in peptide bonds; one of the principal types of compounds present in all cells.

protista (pro-tis'tah([Gr. *protista* the very first, from *prōtos* first]. Kingdom of living organisms, including the protozoa, flagellates, slime molds, certain algae and fungi.

protocooperation (pro"to-co-op"er-a'shun) [Gr. *prōtos* first + L. *cooperatio* to work]. Relationship in which each of two populations benefits by the presence of the other but can survive in its absence.

proton (pro'ton) [Gr. *prōtos* first]. A basic physical particle present in the nuclei of all atoms which has a positive electric charge and a mass similar to that of a neutron; a hydrogen ion.

protonephridium (pro"to-nef-rid'e-um) [Gr.

prōtos first + *nephridios* kidneys]. The flame-cell excretory organs of certain invertebrates.

protopodite (pro-top′o-dīt) [Gr. *prōtos* + *podos* foot]. The basal portion of a biramous crustacean limb.

protostome (pro′to-stōm) [Gr. *prōtos* + *stoma* mouth]. An animal in which the blastopore is in a ventral site and contributes to the formation of the mouth as well as the anus.

protozoa (pro″to-zo′ah) [Gr. *prōtos* first + *zōon* animal]. Single-celled animals, including amebas, ciliates, flagellates and sporozoa.

pseudocoelom (su″do-se′lom) [Gr. *pseudēs* false + *koilia* cavity]. A body cavity between the mesoderm and endoderm; a persistent blastocoel.

pseudopod (su′do-pod) [Gr. *pseudes* false + *podos* foot]. A temporary cytoplasmic protrusion of an ameba or ameboid cell, functions in locomotion and feeding.

pterobranchia (ter″-o-brang″ke-ah [Gr. *pteron* wing + *branchia* gills]. Class of the phylum Hemichordata containing minute tube-dwelling species.

pterygota (ter″ĭ-go′tah) [Gr. *pteryos* wing]. The group of insects having wings, including species that have secondarily lost wings.

pupa (pu′pah) [L. "a doll"]. A stage in the development of an insect, between the larva and the imago (adult); a form which neither moves nor feeds.

purines (pu′rēns) [blend of *pure* and *urine*]. Organic bases with carbon and nitrogen atoms in two interlocking rings; components of nucleic acids, ATP, NAD and other biologically active substances.

putrefaction (pu″trĕ-fak′shun) [L. *putrefactio* decaying]. The enzymatic anaerobic degradation of proteins and amino acids.

pygidium (pi-jid′i-um) [Gr. *pyge* rump]. The nonsegmental posterior part of a metameric animal that usually bears the anus.

pyrenoid (pi′rĕ-noid) [Gr. *pyrēn* fruit stone + *eidos* form]. Starch-containing granular bodies seen in the chromatophores of certain protozoa.

pyrimidines (pi-rim′ĭ-dins). Nitrogenous bases composed of a single ring of carbon and nitrogen atoms; components of nucleic acids.

quantum (kwon′tum) [L. "as much as"]. A unit of radiant energy; has no electric charge and very little mass; the energy of a quantum is an inverse function of the wavelength of the radiation.

race A division of a species; a population which differs from other populations with respect to the frequency of one or more genes; a subgroup of a species distinguished by a certain combination of morphologic and physiologic traits.

radial cleavage Type of cleavage pattern characteristic of echinoderms and vertebrates in which the spindle axes are parallel or at right angles to the polar axis.

radula (raj′oo-la) [L. a scraper]. A rasplike structure in the alimentary tract of chitons, snails, squids and certain other mollusks.

range The portion of the earth in which a given species is found.

reabsorption Term applied to the selective removal of certain substances from the glomular filtrate by the cells of the convoluted tubules of the kidney and their secretion into the blood stream.

recapitulation The tendency for embryos in the course of development to repeat, perhaps in an abbreviated fashion, the sequence of stages in the embryonic development of their evolutionary ancestors.

receptor A sensory cell, or sometimes a free nerve ending, which responds to a given type of stimulus.

recessive genes Genes which do not express their phenotype unless carried by both members of a set of homologous chromosomes, i.e., genes which produce their effect only when homozygous, when present in "double dose."

recon (re′kon) [From *recombination* + Gr. *on* neuter ending]. The ultimate unit of genetic recombination; the smallest element in a one-dimensional array of genetic material that is interchangeable, but not divisible, by recombination.

redia (re′dĭ-ah) [After Francesco Redi, seventeenth century Italian naturalist]. The second stage of flukes. It reproduces asexually in snails.

reduction The addition of electrons to an atom or molecule; opposite of oxidation.

reflex (re′fleks) [L. *reflexus* bent back]. An inborn, automatic, involuntary response to a given stimulus which is determined by the anatomic relations of the involved neurons; the functional units of the nervous system.

reflex arc A sequence of sensory, internuncial and motor neurons which conduct the nerve impulses for a given reflex.

refractory period The period of time which elapses after the response of a neuron or muscle fiber to one impulse before it can respond again.

regeneration Regrowth of a lost or injured tissue or part of an organism.

releasing hormones Short peptides synthesized in the hypothalamus, secreted into the hypothalamo-hypophyseal portal system and carried to the pituitary, where they initiate the synthesis and release of specific pituitary hormones.

renal (re′nal) [L. *renalis* the kidney]. Pertaining to the kidney.

renal corpuscle (re′nal) [L. *renalis* kidney]. The complex formed by a glomerulus and the surrounding Bowman's capsule of a kidney tubule; filtration, the first step in urine formation, occurs here.

rennin (ren'in). Enzyme secreted by the gastric mucosa which converts the milk protein, casein, from a soluble to an insoluble substance, thereby curdling the milk.

repressor The protein substance produced by a regulator gene that represses protein synthesis in a specific gene.

reptile (rep'tĭl) [L. *repere* to creep]. A member of a class of terrestrial vertebrates which are covered with horny scales or plates; living representatives include turtles, lizards, snakes and crocodiles.

resonating system A system of atoms bonded together which includes many different ways of arranging the external electrons without moving any of the constituent atoms.

respiration (res"pĭ-ra'shun) [L. *respirare* to breathe]. Process by which animal and plant cells utilize oxygen, produce carbon dioxide and conserve the energy of foodstuff molecules in biologically useful forms such as ATP; the act or function of breathing.

respiratory pigment Metal-containing proteins that bind reversibly with oxygen and sometimes with carbon dioxide. They transport gases in the blood: hemoglobin (q.v.), hemocyanin (q.v.), hemerythrin, chlorocruorin.

reticulum (re-tik'u-lum) [L. dim. of *rete* net]. A network of fibrils or filaments, either within a cell or in the intercellular matrix.

retroperitoneal Located behind peritoneum, referring to organs that may bulge into the coelom but are still covered by the coelomic lining, or peritoneum.

retina (ret'ĭ-nah) [L. *rete* net]. The innermost of the three tunics of the eyeball, surrounding the vitreous body and continuous posteriorly with the optic nerve; contains the light-sensitive receptor cells, rods and cones.

Rh factor An agglutinogen, originally discovered in the rhesus monkey, which is found in the erythrocytes of about 85 per cent of the white population.

rhodopsin (ro-dop'sin) [Gr. *rhodon* rose + *opsis* sight]. A substance in the retina of the eye (visual purple) made up of retinene, a derivative of vitamin A, and a protein, opsin; undergoes a chemical reaction triggered by light which stimulates the receptor cell to send an impulse to the brain, resulting in the sensation of sight.

ribonucleic acid (RNA) (ri"bo-nu'kle-ik as'id). Nucleic acid containing the sugar ribose; present in both nucleus and cytoplasm and of prime importance in the synthesis of proteins.

ribosomes (ri'bo-sōms). Minute granules composed of protein and ribonucleic acid either free in the cytoplasm or attached to the membranes of the endoplasmic reticulum of a cell; the site of protein synthesis.

rickettsia (rik-et'se-ah). A type of disease organism, discovered by Howard Ricketts, intermediate in size and complexity between a virus and a bacterium; parasitic within cells of insects and ticks, transmitted to man by the bite of the infected insect or tick.

rod (rod) [AS. *rodd*]. In zoology, the rod-shaped photoreceptive cells of the retina which are particularly sensitive to dim light and mediate black and white vision.

rotifera (ro-tif'er-ah) [L. *rota* wheel + *ferre* to bear]. A phylum of abundant small pseudocoelomates, the rotifers.

saccule (sak'ūl) [L. *sacculus* a little sac]. Small, hollow sac in the inner ear lined with sensitive hair cells and containing small stones made of calcium carbonate; contains receptors for the sense of static balance.

saprophytic nutrition (sap"ro-fit'ik nutrish'un) [Gr. *sapros* rotten + *phyton* plant + L. *nutritio* nourishing]. A type of heterotrophic nutrition in which organisms absorb their required nutrients through the cell membrane following the extracellular digestion of nonliving organic material.

sarcodina (sar"ko-di'nah) [Gr. *sarkos* flesh + *eidos* form]. A phylum of protozoa that form pseudopods.

sarcopterygian (sar'kop-tĕ-rij'ĭ-an) [Gr. *sarkos* flesh + *pterygos* wing]. A member of a subclass of Osteichthyes in which the fins are supported by a central axis of flesh and bone; includes lungfish and crossopterygians.

saturation deficit The difference between the amount of water vapor a given volume of air could contain at a specific temperature and the amount it actually contains.

scaphopoda (skaf-op'o-dah) [Gr. *skaphe* boat + *podos* foot]. The tooth-shells, a small class of mollusks.

schizocoel (skiz'o-cēl) [Gr. *schizein* to divide + *koilia* cavity]. A body cavity formed by the splitting of embryonic mesoderm into two layers.

sclera (skle'rah) [Gr. *skleros* hard]. The tough, fibrous supporting wall of the eyeball forming approximately the posterior five-sixths of the wall; it is continuous anteriorly with the cornea.

scrotum (skro'tum) [L. *scrotum* bag]. The pouch which contains the testes in most mammals; its wall is composed of integument and muscular and connective tissue layers of the body wall which are everted during the descent of the testes.

scyphozoa (si"fo-zo'ah) [Gr. *skyphos* cup + *zōon* animal]. A class of cnidarians with small larval polyps and large medusae; jellyfish.

secondary response A rapid production of antibodies induced by a second injection of antigen several days, weeks or even months after the primary injection.

secretion (se-kre'shun) [L. *secretio*, from

secernere to secrete]. The production and release by a cell of some substance that is used elsewhere in the body in some process.

segmentation (seg″men-ta′shun) [L. *segmentum* a piece cut off]. Division of a body or structure into more or less similar parts.

selection coefficient A measure of the elimination of one genotype in a population in relation to the most successful one, which is given a selection coefficient of zero.

semicircular canals The three canals or ducts of the membranous labyrinth which lie at right angles to each other in the vertical and horizontal planes of the body; they detect changes in the angular acceleration of the body.

seminal receptacle A portion of the female reproductive tract in which sperm are stored after mating.

seminal vesicle A portion of the male reproductive tract in which sperm are stored before mating; in mammals produces part of the seminal fluid.

seminiferous tubule (se″mĭ-nif′er-us) [L. *semen* seed + *ferre* to bear]. One of the tubules in the testis; cells in its walls multiply and differentiate into spermatozoa.

senescence (se-nes′ens) [L. *senescere* to grow old]. The gradual loss of vigor and physiological capacity through the aging process.

sere (sēr). A sequence of communities that replace one another in succession in a given area; the transitory communities are called seral stages. The series ends with a climax community typical of the climate in that part of the world.

serum (se′rum) [L. "whey"]. The clear portion of a biological fluid separated from its particulate elements; light yellow liquid left after clotting of blood has occurred.

sessile (ses′il) [L. *sessilis* attached]. Living attached to the substratum like sea anemones and sponges, without a stalk as in barnacles.

seta (se′tah) [L. *seta* bristle]. Bristle-like projection of the body surface composed of some skeletal material, most commonly chitin.

sinoatrial node (si″no-a′tre-al nōd) [L. *sino* "a hollow" + *atrium* hall + *nodus* knot]. A small mass of nodal tissue located at the point where the superior vena cava empties into the right atrium; initiates the heart beat and regulates the rate of contraction.

sinus venosus (si′nus ve-no′sus) [L. *sinus* a bent surface + *venosus* venous]. The first chamber of the heart of lower vertebrates; it receives the systemic veins and opens into the atrium.

solute (so′lūt) [L., from *solvere* to dissolve]. A substance dissolved in a true solution; a solution consists of a solute and a solvent.

solvent (sol′vent) [L., from *solvere* to dissolve]. The fluid medium in which the solute molecules are dissolved in a true solution; a liquid that dissolves or that is capable of dissolving.

somatic (so-mat′ik) [Gr. *sōma* the body . Pertaining to the body wall of an organism, e.g., the somatic muscles, or muscles in the body wall.

somites (so′mīts) [Gr. *sōma* body]. Paired, blocklike masses of mesoderm, arranged in a longitudinal series alongside the neural tube of the embryo, forming the vertebral column and dorsal muscles.

species (spe′shēz) [L. *species* sort, kind]. The unit of taxonomic classification for both plants and animals; a population of similar individuals, alike in their structural and functional characteristics, which in nature breed only with each other and have a common ancestry.

spermatophore (sper-mat′o-fōr) [Gr. *spermatos* seed + *phorein* to bear]. A package secreted by part of the male reproductive tract, enclosing a number of sperm.

sphincter (sfingk′ter) [Gr. *sphinkter* to bind tight]. A group of circularly arranged muscle fibers whose contractions close an opening, e.g., the pyloric sphincter at the end of the stomach.

spicules (spik′ŭls). Microscopic mineral needle-like or rodlike skeletal pieces found in sponges and certain other animals.

spiracle (spir′ah-k′l) [L. *spirare* to breathe]. A breathing opening such as the opening on the body surface of a trachea in insects or a modified gill opening in cartilaginous fishes through which some water enters the pharynx.

spiral cleavage A cleavage pattern characteristic of a number of invertebrate phyla, such as annelids and mollusks, in which the cleavage planes are oriented obliquely to the polar axis of the egg.

spongin (spun′jin) [Gr. *spongia* sponge]. A flexible skeletal material or protein fiber found in some sponges, especially those of commerce.

spore (spōr) [Gr. *sporos* seed]. An asexual reproductive element, usually unicellular, of an organism, such as a protozoan or a cryptogamic plant, which can develop directly into an adult.

sporozoa (spo′ro-zo′ah) [Gr. *sporos* seed + *zōon* animal]. Phylum of protozoa (spore formers); have no special method of locomotion and are parasitic; one kind is the human parasite causing malaria.

standard metabolism The amount of energy expended by an animal at rest.

statocyst (stat′o-sist) [Gr. *statos* standing + *kystis* sac]. A cellular cyst containing one or more granules that is used in a variety of animals to sense the direction of gravity.

stenohaline (sten′o-ha′lĭn) [Gr. *stenos* narrow + *halinos* salt]. Descriptive of animals restricted to a narrow range of environmental salt concentrations.

sternum (ster′num) [L. the chest]. The chest region, especially the breastbone of a vertebrate, or the ventral skeletal piece of any arthropod segment.

steroids (ste'roids) [Gr. *stereos* solid + *eides* like]. Complex molecules containing carbon atoms arranged in four interlocking rings, three of which contain six carbon atoms each and the fourth of which contains five; the male and female sex hormones and the adrenal cortical hormones.

stimulus (stim'u-lus) [L. "goad"]. Any agent, act or influence that produces functional or trophic reaction in a receptor or in an irritable tissue.

stomodeum (sto″mo-de'um)[Gr. *stoma* mouth + *hadaios* on the way]. An ectodermal pouch invaginated at the front of an embryo; contributes to the mouth cavity.

subgerminal space (sub-jer'mĭ-nal spās) [L. *sub* under + *germinalis* germ]. The shallow cavity under the dividing cells of a hen's egg, not homologus to the blastocoele of the frog egg.

summation The adding of one response to another, as in types of muscle contraction in which stimuli follow one another before the effects of previous ones have dissipated.

swim bladder A gas-filled chamber, homologous to a lung, situated in the dorsal part of the abdominal cavity of teleost fishes. By controlling its gas content, a fish can equalize its density to that of the surrounding water.

syconoid (si'kon-oid) [Gr. *sykon* fig + *eidos* form]. In sponges, a body plan in which the wall is indented from both sides, giving the sponge greater bulk.

symbiosis (sim″bi-o'sis) [Gr. *symbiōsis* to live together]. The living together of two dissimilar organisms; association may form mutualism, commensalism, parasitism or amensalism.

sympathetic system (sim″pah-thet'ik) [Gr. *sym* with + *pathos* feeling]. A division of the autonomic nervous system in which fibers leave the central nervous system with certain thoracic and lumbar nerves and go to the sweat glands and visceral organs; has an effect on most organs that is antagonistic to parasympathetic stimulation.

synapse (sin'aps) [Gr. *synapsis* conjunction]. The junction between the axon of one neuron and the dendrite of the next.

synapsis (sĭ-nap'sis) [Gr. *synapsis* conjunction]. The pairing and union side by side of homologous chromosomes from the male and female pronuclei early in meiosis.

syncytium (sin-sit'e-um) [Gr. *syn* together + *kytos* a hollow vessel]. A multinucleate mass of cytoplasm produced by the merging of cells.

synergistic (sin″er-jis'tik) [Gr. *syn* with + *ergon* work]. Acting together; enhancing the effect of another force or agent.

syngamy (sin'gah-me) [Gr. *syn* with + *gamos* marriage]. Sexual reproduction; the union of the gametes in fertilization.

systole (sĭs'to-le) [Gr. *systolē* a drawing together]. The contraction of the heart; the interval between the first and second heart sounds during which blood is forced into the aorta and pulmonary arteries.

taiga (ti'ga) [Russian]. Northern coniferous forest biome found particularly in Canada, northern Europe and Siberia.

taxis (tak'sis) [Gr. a drawing up in rank and file]. An orientation movement in response to a stimulus in a direction determined by the direction of the stimulus; found in animals, some lower plants and the male sex cells of mosses or ferns.

taxonomy (taks-on'o-me) [Gr. *taxis* a drawing up in rank and file + *nomos* law]. The science of naming, describing and classifying organisms.

tectorial membrane (tek-to're-al mem'brān) [L. *tectum* roof + *membrana* skin covering]. The roof membrane of the organ of Corti in the cochlea of the ear.

telencephalon (tel″en-sef'ah-lon) [Gr. *tele* end + *enkephalos* brain]. The most anterior of the five major subdivisions of the brain; includes the olfactory bulbs and cerebral hemispheres.

teleost (tel'e-ost) [Gr. *tele* end + *osteon* bone]. A member of the most advanced groups of actinopterygian fishes; includes most of the familiar species, such as herring, salmon, eels, minnows, suckers, catfish, bass and perch.

telolecithal (tel″o-les'ĭ-thal) [Gr. *telos* end + *lekithos* yolk]. Type of egg in which the yolk material is concentrated on one side.

telophase (tel'o-fāz) [Gr. *tele* end + *phasis* phase]. The last of the four stages of mitosis, during which the two daughter nuclei appear and the cytoplasm usually divides.

template (tem'plāt) [L. *templum* a small timber]. A pattern or mold which guides the formation of a duplicate.

territoriality (ter″i-tor'ĭ-al'ĭ-te) [L. *territorium* the earth]. Behavior pattern in which one organism (usually a male) delineates a territory of his own and defends it against intrusion by other members of the same species and sex.

testis (tes'tis) [L.]. The male gonad which produces spermatozoa; in man and certain other mammals the testes are situated in the scrotal sac.

tetanus (tet'ah-nus) [Gr. *tetanos*, from *teinein* to stretch]. Sustained, steady maximal contraction of a muscle, without distinct twitching, resulting from a rapid succession of nerve impulses.

tetany (tet'ah-ne) [Gr. *tetanos* stretched]. A syndrome manifested by sharp flexion of the wrist and ankle joints, muscle twitchings, cramps and convulsions; occurs in parathyroid hypofunction.

tetrad (tet'rad) [Gr. *tetra* four]. A bundle of

four homologous chromatids produced at the end of the first meiotic prophase.

tetraploid (tet″rah-ploid′) [Gr. *tetra* four + *ploos* fold]. An individual or cell having four sets of chromosomes.

tetrapod (tet′ra-pod) [Gr. *tetra* four + *podos* foot]. Four-limbed vertebrates, or ones with a secondary reduction in limb number that evolved from quadrupeds; amphibians, reptiles, birds and mammals.

thalamus (thal′ah-mus) [Gr. *thalamos* inner chamber]. The lateral walls of the diencephalon; it is the main relay center for sensory impulses going to the cerebrum, and it also interacts with the cerebrum in complex ways.

theory (the′o-re) [Gr. *theoria* speculation as opposed to practice]. A formulated hypothesis supported by a large body of observations and experiments.

therapsids (ther-ap′sids). An order of mammal-like reptiles of the Permian period from which mammals evolved.

thermodynamics, first law of (ther″mo-di-nam′iks) [Gr. *thermē* heat + *dynamis* power]. Law which states that energy is neither created nor destroyed but only transformed from one kind to another.

thigmotropism (thig-mot′ro-pizm) [Gr. *thigma* touch + *tropē* turn]. The orientation of an organism in response to the stimulus of contact or touch.

threshold (thresh′old). The value at which a stimulus just produces a sensation, is just appreciable or comes just within the limits of perception.

thrombin (throm′bin) [Gr. *thrombos* lump, curd, clot]. The enzyme derived from prothrombin which converts fibrinogen to fibrin; participates in blood clotting.

thrombus (throm′bus) [Gr. *thrombos* lump, curd, clot]. A clot in a blood vessel or in one of the cavities of the heart that remains at the point of its formation.

thymus (thī′mus) [Gr. *thymos* a warty lump]. An organ of pharyngeal pouch origin located in the base of the neck. Site of production of T-lymphocytes in young mammals; atrophies at maturity.

tissue (tish′u) [Fr. *tissue* to weave]. Group of similarly specialized cells which together perform certain special functions; e.g., muscle tissue, bone tissue, nerve tissue.

tonus (to′nus) [Gr. *tonos* strain, tone]. The continuous partial contraction of muscle.

tornaria (tor-na′re-ah) [L. *tornare* to turn]. The free-swimming hemichordate larva that shows many similarities to echinoderm larvae.

toxin (tok′sin) [L. *toxicum* poison]. Poisonous substance produced by one organism which usually affects one particular organ or organ system, rather than the body as a whole, of another organism.

trachea (tra′ke-ah) [Gr. *trachelos* the throat]. An air conducting tube: in terrestrial vertebrates, the main trunk of the system of tubes through which air passes to and from the lungs; in terrestrial arthropods, one of a system of minute tubes that permeate the body and deliver air to the tissues.

transducers (trans-du′sers) [L. *transducere* to lead across]. Devices receiving energy from one system in one form and supplying it to a second system in a different form; e.g., converting radiant energy to chemical energy.

transduction (trans-duk′shun) [L. *transducere* to lead across]. The transfer of a genetic fragment from one cell to another; e.g., from one bacterium to another by a virus.

transfer RNA A form of RNA composed of about 70 nucleotides which serve as adaptor molecules in the synthesis of proteins. An amino acid is bound to a specific kind of transfer RNA and then arranged in order by the complementary nature of the nucleotide triplet (codon) in template or messenger RNA and the triplet anticodon of transfer RNA.

transforming agents Substances isolated from pneumococci and certain other bacteria which bring about a permanent, inherited change when applied to another strain of that type of bacteria.

transverse plane A section in a bilaterally symmetrical animal which includes a dorsoventral axis and a left-right axis but is at right angles to the anteroposterior axis.

trematoda (trem″ah-to′dah) [Gr. *trematodes* having holes]. The flukes, a class of parasitic flatworms.

trichocyst (trik′o-sist) [Gr. *trichos* hair + *kryptos* concealed]. A cellular organelle in the cytoplasm of ciliated protozoa such as *Paramecium* which can discharge a filament that may aid in trapping and holding prey.

trilobite (tri′lo-bīt) [L. *tres* three + *lobus* lobe]. Marine arthropods of the Paleozoic era characterized by two dorsal longitudinal furrows that separated the body into three lobes.

triplet code The sequences of three nucleotides which comprise the codons, the units of genetic information in DNA which specify the order of amino acids in a peptide chain.

triplet state The state resulting when an electron is activated by absorbing a photon, moves to an outer orbital of higher energy and pairs with an electron of like spin.

triploid (trip′loid) [Gr. *triploos* triple + *eidēs* like]. An individual or cell having three sets of chromosomes.

trochophore (tro′ko-fōr) [Gr. *trochos* wheel + *phoros* bearing]. The top-shaped larva of marine mollusks and polychaete annelids that bears a girdle of cilia.

troop The social unit of many primate species, consisting of several males, three to many females, and their offspring.

trophallaxis (tro″fah-lak′sis) [Gr. *trephein* to nourish + *allaxis* exchange]. The mutual exchange of food and secretions among the members of an insect colony.

tropism (tro′pizm) [Gr. *tropē* a turning]. A

growth response in a nonmotile organism, elicited by an external stimulus.

tubicolous (tu-bīk′o-lūs). Living within a tube.

tubulin (tub′il-in) [L. *tubus* tube]. The protein filaments that make up the walls of the microtubules of cilia and flagella.

tundra [Russian]. A treeless plain between the taiga in the south and the polar ice cap in the north; characterized by low temperatures, a short growing season and ground that is frozen most of the year.

turbellaria (tur″bel-a′rĭ-ah)[L. *turbella* a stir]. The class of free-living flatworms.

turnover number The number of molecules of substrate acted upon by one molecule of enzyme per minute.

ubiquinone Coenzyme Q, a component of the electron transmitter system; consists of a head, a six-membered carbon ring, which can take up and release electrons, and a long tail composed of a chain of carbon atoms.

umbilical cord (um-bil′i-k′l) [L. *umbilicus* navel]. The stalk, or cord, attached to the navel and connecting the embryo with the placenta; it contains the umbilical arteries and veins and the remnants of the allantois and yolk stalk.

umbilicus (um-bil′i-cus) [L.] The navel; the scar marking the site of attachment of the umbilical cord in the fetus.

ungulates (ung″gu-lātes) [L. *ungula* hoof]. Four-legged mammals which walk on the tips of their digits and which have lost one or more toes. The end of each digit is protected by a hoof.

urea (u-re′ah) [Gr. *ouron* urine]. One of the end products of protein metabolism; the diamide of carbonic acid, NH_2CONH_2; soluble in water.

ureter (u-re′ter) [Gr.]. The fibromuscular tube which conveys urine from a metanephric kidney to the cloaca or exterior of the body.

urethra (u-re′thrah) [Gr.]. The membranous canal conveying urine from the bladder to the exterior of the body.

uric acid (u′rik) [Gr. *ouron* urine]. An end product of nucleic acid and protein metabolism with a low solubility in water; particularly abundant in certain terrestrial animals; $C_5H_4N_4O_3$.

urochordate (u-ro-kor′dāt) [Gr. *oura* tail + L. *chorda* string]. A member of a subphylum of chordates characterized by the restriction of the notochord to the larval tail; includes sea squirts and their allies; adults are usually sessile.

uterus (u′ter-us) [L.]. The womb; the hollow, muscular organ of the female reproductive tract in which the fetus undergoes development.

utricle (u′tre-k′l) [L. *utriculus* a bag]. The larger of the two divisions of the membranous labyrinth; contains the receptors for dynamic body balance.

vaccine (vak′sēn) [L. *vaccinus* a cow]. The commercially produced antigen of a particular disease, strong enough to stimulate the body to make antibodies but not sufficiently strong to cause the disease's harmful effects.

vacuole (vak′u-ōl) [L. *vacuus* empty + *-ole* dim. ending]. Small space within a cell, filled with watery liquid and separated by a vacuolar membrane from the rest of the cytoplasm.

vagina (vah-ji′nah) [L. a scabbard]. In many kinds of animals, the terminal portion of the female reproductive tract; receives the male copulatory organ.

valence (va′lens) [L. *valentia* strength]. An expression of the number of atoms of hydrogen (or its equivalent) which one atom of a chemical element can hold in combination, if negative, or displace in a reaction, if positive; the number of electrons gained, lost or shared by the atom in forming bonds with one or more other atoms.

vas deferens (vas def′er-enz) [L. *vas* duct + *deferens* carrying away]. The duct which carries sperm from the testis and epididymis to the cloaca or penis; it develops from the archinephric or wolffian duct.

vasa efferentia (va′sa ef′er-ent″ĭ-ah) [L. *vasa* ducts + *exferre* to carry out]. In lower vertebrates, the cords of the urogenital union which extend between the testis and kidney; in mammals, the tubules in the head of the epididymis which have developed from certain kidney tubules.

vein (vān) [L. *vena*]. A vessel through which the blood passes from the tissues toward the heart; typically has thin walls and contains valves that prevent a reverse flow of blood.

veliger (vel′-e-jer) [L. *velum* covering + *gerere* to carry]. Larval stage of many marine clams and snails.

velum (ve′lum) [L. an awning or veil]. In hydrozoan medusae, a shelflike membrane partially closing the subumbrellar opening.

ventricle (ven′trĭ-k′l) [L. *ventriculus*, dim. of *venter* the stomach]. A cavity in an organ, such as one of the several cavities of the brain or one of the chambers of the heart that receive blood from the atria.

vertebrate (ver′tĕ-brāt) [L. *vertebratus* having a backbone]. A member of a subphylum of chordates characterized by the presence of a vertebral column; includes fishes, amphibians, reptiles, birds and mammals.

vestigial (ves-tĭj′e-al) [L. *vestigium* footprint, trace, sign]. Useless, incomplete or undersized; said of an organ present in one organism which is a remnant of a homologous organ that functioned in an ancestral organism.

villus (vil′lus) [L. "tuft of hair"]. A small, fingerlike vascular process or protrusion, especially a protrusion from the free surface of a membrane such as the lining of the intestine.

virus (vi′rus) [L. slimy liquid, poison]. Minute infectious agent, composed of a nucleic acid core and a protein shell; may reproduce and mutate within a host cell.

visceral (vis′er-al) [L. *viscera* internal organs]. Pertaining to the internal organs; e.g., the visceral muscles of the gut wall.

visceral arch (vis′er-al) [L. *viscera* internal organs]. An arch of cartilage or bone which develops in the wall of the pharynx between the gill slits; they supported the gills in ancestral vertebrates, but certain ones became incorporated into the skull of higher vertebrates.

vital capacity (vi′tal kah-pas′ĭ-te) [L. *vita* life + *capacitas,* from *capere* to take]. The total amount of air displaced when one breathes in as deeply as possible and then breathes out as completely as possible.

vitamin (vi-tah-min) [L. *vita* life]. An organic substance necessary in small amounts for the normal metabolic functioning of a given organism; must be present in the diet because the organism cannot synthesize an adequate amount of it.

vitreous (vit′re-us) [L. *vitreus* glassy]. Glass-like or hyaline; designates the vitreous body of the eye which contains clear transparent jelly which fills the posterior part of the eyeball.

viviparous (vi-vip′ah-rus) [L. *vivus* alive + *parere* to bring forth, produce]. Bearing living young which develop from eggs within the body of the mother, deriving nutrition from the maternal organism through a special organ, the placenta, which is a union of certain fetal membranes and the uterine lining.

warning coloration Adaptation for survival which consists of bright, conspicuous colors and is assumed by poisonous or unpalatable animals, or their mimics, to warn potential predators not to eat them.

water expulsion vesicle Osmoregulatory organelle of protozoans and sponges.

wolffian duct See *archinephric duct.*

X organ Organ present in crustacea which produces hormones that regulate molting, metabolism, reproduction, the distribution of pigment in the compound eyes and the control of pigmentation of the body.

yolk sac A pouchlike outgrowth of the digestive tract of certain vertebrate embryos which grows around the yolk, digests it and makes it available to the rest of the organism.

zooflagellates Heterotrophic, or animal-like, flagellated protozoans.

zooxanthellae (zo″an-thel′-e) [Gr. *zoon* animal + *xanthos* yellow]. Dinoflagellates living symbiotically with certain marine animals, especially corals.

zygote (zi′got) [Gr. *zygotos* yoked together]. The cell formed by the union of two gametes; a fertilized egg.

zymogen (zi′mo-jen) [Gr. *zyme* leaven + *gennan* to produce]. The inactive state of an enzyme.

INDEX

Prolactin, 443, 450
Pronephros, 106, 349
Prophase, 48
Propliopithecus, 862
Proprioception, 367
Prosimians, 856
Prosimii, 854
Prosobranchia, 614
Prostaglandins, 451
Prostate gland, 509
Prosthetic groups, 72
Prostomium, 647
of oligochaetes, 658
Protacanthopterygii, 785
Protandry, 499
Protean behavior, 471
Proteases, 268
Protective coloration, 876
Protein kinases, 427
Protein molecules, 203
Proteins, 23, 26
primary structure of, 26
Proterozoic era, 211
Prothoracic gland, 455
Prothoracicotropic hormone, 454
Protobranchia, 626
Proton acceptors, 68
Proton donors, 68
Protonephridia, 346
of flatworms, 582
Protons, 18
Protoplasm, 31
Protopterus, 776
Protostomes, 522, 606
Prototheria, 843, 853
Protozoa, 31, 525
classification of, 525, 544
Proventriculus, 274, 713, 824
Proximal convoluted tubule, 350
Pseudemys, 802
Pseudocoel, 593
Pseudocoelomates, 522, 593
Pseudopodia, 252, 537
Pseudopregnancy, 503
Pseudoscorpionida, 677
Psittaciformes, 833
Pterosauria, 809, 813
Pterygota, 709
Ptilocrinus, 749
Pulmonata, 618
Pulp cavity, 278
Pulse, 338
Pulse rate, 337
Punishing centers, 416
Punnett square, 115
Pupa, 453, 719
Pycnogonida, 724
Pygidium, 234, 647
Pygostyle, 821
Pyloric sphincter, 281
Pyramid of numbers, 883
Pyramidal pathway, 418
Pyridoxine, 288
Pyruvic acid, 344
Python, 805

Quaternary period, 213
Quill, 819

Rabbits, coat color of, 132
Radial symmetry, 230
Radiata, 522
Radiolaria, 545
Radiolarians, 243, 540
Radioles, 272
Radula, 608
Railroad worm, 83
Ramapithecus, 862
Range, 517
Range of tolerance, 873
Raptorial feeding, 271
Rathke's pouch, 439
Ray, John, 13
Rays, 776
Razor clam, 632
Recapitulation, 218
Receptor cells, 362
Receptor mechanisms, 362
Receptor potential, 364, 366
Receptors, hormonal, 425
Recombination, 181
Rectal gland, 775
Rectum, 270
Red blood cells, 60
Red tides, 535
Redbugs, 683
Redi, Francesco, 17, 202
Reduction, 69
Reefs, coral, 572
Reflex arc, monosynaptic, 405
Reflex ovulators, 448
Reflexes, 405, 461
Refractory period, absolute, 398
Regulator genes, 162
Releaser pheromones, 458
Releasing factors, 416
Renal corpuscle, 350
Renal pelvis, 350
Renal portal system, 327
Renal threshold, 352
Rennin, 281
Replication, semi-conservative, 149
Repressors, 162
Reproduction, 17
mammalian, 510
sexual, 493
types of, 491
Reproductive fighting, 481
Reproductive isolation, 194
Reproductive passages, 509
Reproductive tissue, 62
Reptiles, 789
adaptations of, 799
characteristics of, 799
evolution of, 801
mammal-like, 809
Residual bodies, 92
Respiratory assemblies, 79
Respiratory center, 307
Respiratory membrane, 294
Respiratory organ, 294
Respiratory pigments, 308
Respiratory trees, 296, 747
Resting membrane potential, 396
Rete testis, 506
Reticular fibers, 58
Reticular formation, 414, 467
Reticulocytes, rabbit, 157
Reticulum, 271

Retina, 378, 379
development of, 102
organization of, 382
Retinal, 381
Retinol, 288
Retinula, 384
Retrograde amnesia, 487
Reverse transcription, 152
Rh factor, 133
Rhabdome, 385
Rheas, 828
Rheiformes, 832
Rhinoceros, 850
Rhinoderma, 797
Rhizobium, 882
Rhizostomae, 576
Rhodopsin, 381
Rhopalium, 387, 565
Rhynchocephalia, 803, 812
Rhynchocoela, 522
Rib, 248
Riboflavin, 288
Ribonuclease, structure of, 26
Ribonucleic acid (RNA), 24, 28
synthesis of, 151
Ribose, 24
Ribosomal functions, 153
Ribosomal RNA, 151
Ribosome(s), 37
structure of, 153
subunits of, 39
Ring canal, 736
RNA. See *Ribonucleic acid.*
Roach, 715
Rock deposits, radioactive dating of, 207
Rocky Mountain revolution, 212
Rodentia, 852, 855
Rods, 379
Root, of tooth, 278
Rose combs, 130
Rotifera, 522, 594
Round dance, 475
Roundworms, 598
Ruffini's endings, 367
Rugose corals, 575
Rumen, 271
Ruminants, 271
Running water, 917

S phase of cell cycle, 43, 89
Sabella, 272
Saccoglossus, 761
Saccopharynx, 782
Saccule, 388
Salamanders, 795
Salientia, 812
Saliva, 278
Salivary amylase, 279
Salivary glands, vertebrate, 278
Salmon, homing in of, 489
Salps, 764
Salt, 22
Salt-excreting glands, 826
Salt glands, 356
Saltatory conduction, 399
Sand dollars, 744
Saprophytic nutrition, 267